ECOLOGY
INDIVIDUALS, POPULATIONS
AND COMMUNITIES

ECOLOGY
INDIVIDUALS, POPULATIONS AND COMMUNITIES

MICHAEL BEGON
Department of Environmental and Evolutionary Biology,
The University of Liverpool, Liverpool, United Kingdom

JOHN L. HARPER
Cae Groes, Glan-y-coed Park, Dwygyfylchi, Penmaenmawr, North Wales

COLIN R. TOWNSEND
Department of Zoology, University of Otago, Dunedin, New Zealand

THIRD EDITION

Blackwell
Science

© 1986, 1990, 1996 by
Blackwell Science Ltd
Editorial Offices:
Osney Mead, Oxford OX2 0EL
25 John Street, London WC1N 2BL
23 Ainslie Place, Edinburgh EH3 6AJ
350 Main Street, Malden
  MA 02148 5018, USA
54 University Street, Carlton
  Victoria 3053, Australia
10, rue Casimir Delavigne
  75006 Paris, France

Other Editorial Offices:
Blackwell Wissenschafts-Verlag GmbH
Kurfürstendamm 57
10707 Berlin, Germany

Blackwell Science KK
MG Kodenmacho Building
7–10 Kodenmacho Nihombashi
Chuo-ku, Tokyo 104, Japan

First published 1986
Reprinted with corrections 1987
Second edition 1990
Third edition 1996
Reprinted 1996, 1997, 1999

Set by Semantic Graphics, Singapore
Printed and bound in Italy
by Rotolito Lombarda S.p.A., Milan

The Blackwell Science logo is a
trade mark of Blackwell Science Ltd,
registered at the United Kingdom
Trade Marks Registry

For further information on
Blackwell Science, visit our website:
www.blackwell-science.com

DISTRIBUTORS

Marston Book Services Ltd
PO Box 269
Abingdon, Oxon OX14 4YN
(*Orders*:   Tel: 01235 465500
               Fax: 01235 465555)

USA
Blackwell Science, Inc.
Commerce Place
350 Main Street
Malden, MA 02148 5018
(*Orders*:   Tel: 800 759 6102
                        781 388 8250
               Fax: 781 388 8255)

Canada
Login Brothers Book Company
324 Saulteaux Crescent
Winnipeg, Manitoba R3J 3T2
(*Orders*:   Tel: 204 837-2987)

Australia
Blackwell Science Pty Ltd
54 University Street
Carlton, Victoria 3053
(*Orders*:   Tel: 3 9347 0300
               Fax: 3 9347 5001)

A catalogue record for this title
is available from the British Library

ISBN 0-632-03801-2 (pbk)
ISBN 0-86542-845-X (hbk)
ISBN 0-632-04393-8 (pbk with CD)

Library of Congress
Cataloging-in-publication Data

Begon, Michael.
   Ecology: individuals, populations and
communities/Michael Begon, John L. Harper,
   Colin R. Townsend. — 3rd ed.
         p.   cm.
   Includes bibliographical references
   (p.   ) and indexes.
       ISBN 0-632-03801-2 (pbk)
       ISBN 0-86542-845-X (hbk)
       ISBN 0-632-04393-8 (pbk with CD)
   1. Ecology.   I. Harper, John L.
   II. Townsend, Colin R.   III. Title
   QH541.B415   1996
   574.5–dc20   95-24627
                                        CIP

Contents

Preface, vii

Introduction: Ecology and its Domain, x

Part 1: Organisms

Introduction, 3

1   The Match between Organisms and their Environments, 4

2   Conditions, 48

3   Resources, 90

4   Life and Death in Unitary and Modular Organisms, 135

5   Dispersal, Dispersion and Migration in Space and Time, 173

Part 2: Interactions

Introduction, 211

6   Intraspecific Competition, 214

7   Interspecific Competition, 265

8   The Nature of Predation, 313

9   The Behaviour of Predators, 334

10   The Population Dynamics of Predation, 369

11   Decomposers and Detritivores, 402

12   Parasitism and Disease, 429

13   Symbiosis and Mutualism, 482

Part 3: Three Overviews

Introduction, 525

14   Life-History Variation, 526

15 Abundance, 567

16 Manipulating Abundance: Killing and Culling, 621

Part 4: Communities

Introduction, 677

17 The Nature of the Community, 679

18 The Flux of Energy through Communities, 711

19 The Flux of Matter through Communities, 744

20 The Influence of Competition on Community Structure, 775

21 The Influence of Predation and Disturbance on Community Structure, 801

22 Food Webs, 828

23 Islands, Areas and Colonization, 861

24 Patterns in Species Richness, 884

25 Conservation and Biodiversity, 913

Glossary, 953

References, 971

Organism Index, 1031

Subject Index, 1049

Preface

This book is about the distribution and abundance of different types of organism over the face of the earth, and about the physical, chemical but especially the biological features and interactions that determine these distributions and abundances.

Unlike some other sciences, the subject matter of ecology is apparent to everybody; to the extent that most people have observed and pondered nature, most people are ecologists of sorts. But ecology is not an easy science, and it has particular subtlety and complexity. It must deal explicitly with three levels of the biological hierarchy—the organisms, the populations of organisms, and the communities of populations—and, as we shall see, it ignores at its peril the details of the biology of individuals, or the pervading influences of historical, evolutionary and geological events. It feeds, in a peripheral way, on advances in our knowledge of biochemistry, behaviour, climatology, plate tectonics and so on, but it feeds back to our understanding of vast areas of biology too. If, as T.H. Dobzhansky said, 'Nothing in biology makes sense, except in the light of evolution', then, equally, very little in evolution makes sense, except in the light of ecology.

Ecology has the distinction of being peculiarly confronted with uniqueness: millions of different species, countless billions of genetically distinct individuals, all living and interacting in a varied and ever-changing world. The beauty of ecology is that it challenges us to develop an understanding of very basic and apparent problems, in a way that recognizes the uniqueness and complexity of all aspects of nature but seeks patterns and predictions within this complexity rather than being swamped by it. As L.C. Birch has pointed out, the physicist Whitehead's recipe for science is never more apposite than when applied to ecology: seek simplicity, but distrust it.

Ecology is not a science with a simple linear structure: everything affects everything else. Different chapters contain different proportions of descriptive natural history, physiology, behaviour, rigorous laboratory and field experimentation, careful field monitoring and censusing, and mathematical modelling (a form of simplicity that it is essential to seek but equally essential to distrust). These varying proportions to some extent reflect the progress made in different areas. They also reflect intrinsic differences in various aspects of ecology. Whatever progress is made, ecology will remain a meeting-ground for the naturalist, the experimentalist, the field biologist and the mathematical modeller. We believe that all ecologists should to some extent try to combine all these facets.

The book is aimed at all those whose degree programme includes ecology. Certain aspects of the subject, particularly the mathematical ones, will prove difficult for some, but our coverage is designed to ensure that wherever our readers' strengths lie—in field or laboratory, in theory or in practice—a balanced and up-to-date view should emerge. One technical feature of the book is the incorporation of marginal notes as signposts throughout the text. These, we hope, will serve a number of purposes. In the first place, they constitute a series of subheadings highlighting the detailed structure of the text. However, because they are numerous and often informative in their own right, they can also be read in sequence along with the conventional subheadings, as an outline of each chapter. They should act too as a revision aid for students—indeed, they are similar to the annotations that students themselves often add to their textbooks. Finally, because the marginal notes generally summarize the take-home message of the paragraph or paragraphs that they accompany, they can act as a continuous assessment of comprehension: if you can see that the signpost is the take-home message of what you have just read, then you have understood.

In this third edition we have tried to take account of the comments and criticisms made by reviewers. We have made a number of changes, particularly in emphasis. Some of these are implied in the modern 'cave painting' that we have used for the cover of this edition, chosen to reflect some of the priorities of the science of ecology at the end of the 20th and the start of the 21st centuries. A shoal of fish represents concerns about overfishing and its consequences in unsustainable yield. The domestic beast, the ears of wheat and the human figure with tool in hand are there to remind us of the demand for food from growing populations and the loss of natural communities that is forced by the growth of agriculture to feed this demand. A whale is in the picture as a representative of all those species that are at risk of extinction as a result of human activities and the dodo stands for those species for which the risk has become reality.

Homo sapiens is just one of the species with a unique ecology but we now give it more special concern. We have given more emphasis to the ways in which our species affects the densities of others and how the communities that inhabit areas of land and water perform in nature when managed to cater for our ambitions. Idealists may long to study nature in an undisturbed wilderness. But we have deliberately accepted that the activities of *H. sapiens* are ecologically at least as interesting as those of other species and we have tried to set these activities (especially cultivation, harvesting and conservation) in the social and economic contexts that are peculiar to the human species. We have deliberately made the text more anthropocentric.

In this edition we have paid much more attention to the spatial patchiness of communities and populations—Peter Kareiva has called the study of space 'the final frontier in ecological theory'. Community patchiness appears as a crucial element in many chapters, including those concerned with dispersal, inter- and intraspecific competition and most particularly the role of predators and pathogens in the distribution and abundance of their hosts.

Since the publication of the first and second editions of the book there have been major advances in modelling and predicting the effects of the rising levels of carbon dioxide and other 'greenhouse' gases that result from industrial and domestic activities, especially global warming. We have therefore given more attention to the predicted ecological consequences of these changes. They may be expected to affect the net primary production of both terrestrial and aquatic communities and produce changes in the distribution of species as great or greater than those that occurred during the Quaternary ice-ages.

In the face of the dramatic perturbations in the physical environment produced by human activity there has been great interest in the stability of natural communities and the rate at which species are becoming extinct. Chapter 22 attempts a new synthesis of our knowledge of food webs and the intricate interactions within them that appear to contribute to the stability of communities. Chapter 25 puts current rates of species extinction into a historic perspective and discusses the causes of extinctions and the scientific underpinning of methods of nature conservation.

In previous editions we thanked the great many friends and colleagues who helped us by commenting on various drafts of the text. The effects of their contributions are still strongly evident in the present text. This third edition was also read by reviewers who remained anonymous but contributed so much that we wish we knew and could record their names. We are grateful to Claire Harper for the artwork used on the cover and part and chapter titles. At Blackwell Science we were again helped and encouraged by Bob Campbell, Simon Rallison and Susan Sternberg and, especially in the later stages, by Katrina McCallum.

Introduction: Ecology and its Domain

The word 'ecology' was first used by Ernest Haeckel in 1869. Paraphrasing Haeckel we can define ecology as the scientific study of the interactions between organisms and their environment. The word is derived from the Greek *oikos*, meaning 'home'. Ecology might therefore be thought of as the study of the 'home life' of living organisms. A more informative, much less vague definition has been suggested by Krebs (1972): 'Ecology is the scientific study of the interactions that determine the distribution and abundance of organisms.' This definition has the merit of pinpointing the ultimate subject matter of ecology: the distribution and abundance of organisms—where organisms occur, how many there are and what they do. However, Krebs's definition does not use the word 'environment', and to see why, it is necessary to define the word. The environment of an organism consists of all those factors and phenomena outside the organism that influence it, whether those factors be physical and chemical (abiotic) or other organisms (biotic). But the 'interactions' in Krebs's definition are, of course, interactions with these very factors. The environment had therefore retained the central position that Haeckel gave it in the definition of ecology.

As far as the subject matter of ecology is concerned, 'the distribution and abundance of organisms' is pleasantly succinct. But we need to expand it. Ecology deals with three levels of concern: the individual *organism*, the *population* (consisting of individuals of the same species) and the *community* (consisting of a greater or lesser number of populations).

At the level of the organism, ecology deals with how individuals are affected by (and how they affect) their biotic and abiotic environment. At the level of the population, ecology deals with the presence or absence of particular species, with their abundance or rarity, and with the trends and fluctuations in their numbers. There are, however, two approaches at this level. One deals first with the attributes of individual organisms, and then considers the way in which these combine to determine the characteristics of the population. The alternative deals directly with the characteristics of populations, and tries to relate these to aspects of the environment. Both approaches have their uses, and both will be used. Indeed, these two approaches are also both useful at the level of the community. Community ecology deals with the composition or *structure* of communities, and with the pathways followed by energy, nutrients and other chemicals as they pass through them. We can pursue an understanding of these patterns and processes from a consideration of the component populations; but we can also look directly at properties of the communities themselves, like species diversity, the rate of biomass production and so on. Again, both approaches will be used.

Ecology is a central discipline in biology, and therefore, not surprisingly, it overlaps with many other disciplines, especially genetics, evolution, behaviour and physiology. The particular concern of ecology, however, is with those characteristics that most affect distribution and abundance—the processes of birth, death and migration.

Ecologists are concerned not only with communities, populations and organisms *in nature*, but also with man-made or man-influenced environments (orchards, wheat fields, grain stores, nature reserves and so on), and with the consequences of man's influence *on* nature (pollution, global warming and so on). Perhaps less obviously, ecologists also often become interested in laboratory systems and mathematical models. These have played a crucial role in the development of ecology, and they are bound to continue to do so. They will be examined frequently in this book. A major aim of science is to simplify, and thereby make it easier to understand the complexity of the real world. Ultimately, therefore, it is the real world that we are interested in, and the worth of models and simple laboratory experiments must always be judged in terms of the light they throw on the working of more natural systems. They are a means to an end—never an end in themselves.

Explanation, description, prediction and control

At all levels of the ecological hierarchy we can try to do a number of different things. In the first place we can try to *explain* or *understand*. This is a search for knowledge in the pure scientific tradition. In order to do this, however, it is necessary first to describe. This, too, adds to our knowledge of the living world. Obviously, in order to understand something, we must first have a description of whatever it is that we wish to understand. Equally, but less obviously, the most valuable descriptions are those carried out with a particular problem or 'need for understanding' in mind. All descriptions are selective; but undirected description, carried out for its own sake, is often found afterwards to have selected the wrong things.

Ecologists also often try to *predict* what will happen to an organism, a population or a community under a particular set of circumstances; and on the basis of these predictions we try to *control* or *exploit* them. We try to minimize the effects of locust plagues by predicting when they are likely to occur and taking appropriate action. We try to protect crops by predicting when conditions will be favourable to the crop and unfavourable to its enemies. We try to preserve rare species by predicting the conservation policy that will enable us to do so. Some prediction and control can be carried out without explanation or understanding. But confident predictions, precise predictions and predictions of what will happen in unusual circumstances can be made only when we can also explain what is going on.

It is important to realize that there are two different classes of explanation in biology: proximal and ultimate explanations. For example, the present distribution and abundance of a particular species of bird may be 'explained' in terms of the physical environment that the bird tolerates, the food that it eats and the parasites and predators that attack it. This is a *proximal* explanation. However, we may also ask how this species of bird has come to have these properties that now appear to govern its life. This question has to be answered by an explanation in evolutionary terms. The *ultimate* explanation of the present distribution and abundance of this bird lies in the ecological experiences of its ancestors.

There are many problems in ecology that demand evolutionary, ultimate explanations. 'How does it come about that coexisting species are often similar but rarely the same?' (Chapter 7), 'What causes predators to adopt particular patterns of foraging behaviour?' (Chapter 9), and 'How have organisms come to possess particular combinations of size, developmental rate, reproductive output and so on?' (Chapter 14). These problems are as much a part of modern ecology as are the prevention of plagues, the protection of crops and the preservation of rarities. Our ability to control and exploit cannot fail to be improved by an ability to explain and understand. In the search for understanding, we must combine both proximal and ultimate explanations.

PART 1
ORGANISMS

Introduction

We have chosen to start this book with chapters about organisms, then to consider the ways in which they interact with each other, and lastly to consider the properties of the communities that they form. We could, quite sensibly, have treated the subject the other way round—starting with a discussion of the complex communities of natural and man-made habitats, proceeding to analyse them at ever finer scales, and ending with chapters on the characteristics of the individual organisms. The problem is a little like writing a book about clocks and watches. We could start with a survey and classification of the different types and end with chapters on the springs, cogwheels, batteries and crystals that make them work. We chose to start with organisms because it is these that are acted upon by evolutionary forces (they live or die, leave descendants or fail to do so); and the nature of communities must ultimately be explained (rather than simply described) in terms of their parts—individual populations composed of individual organisms.

We consider first the sorts of correspondences that we can detect between organisms and the environments in which they live. It would be trite and superficial to start with the view that every organism is in some way ideally fitted to live where it does. We emphasize that organisms are as they are, and live where they do, because of the specializations and constraints imposed by their evolutionary history (Chapter 1). All species are absent from almost everywhere and we consider the ways in which environmental conditions vary from place to place and from time to time, and how these put limits on the distribution of particular species (Chapter 2). Next we look at the resources that different types of organisms consume, and the nature of their interaction with these resources (Chapter 3).

The particular species present in a community, and their abundance, give that community much of its ecological interest. Abundance and distribution (variation in abundance from place to place) are determined by the balance between birth rates and death rates, and between immigration and emigration. In Chapter 4 we consider some of the variety in the patterns of birth and death that we will later see having profound effects on the behaviour of populations.

Every species of plant and animal has a characteristic ability to disperse. This determines the rate at which individuals escape from environments that are or become unfavourable, and the rate at which they discover sites that are ripe for colonization and exploitation. The abundance or rarity of a species may be determined by its ability to disperse (or migrate) to unoccupied patches, islands or continents. The dispersal and migration of organisms is considered in Chapter 5.

Chapter 1
The Match between Organisms and their Environments

1.1 Introduction

Ecology deals with organisms and their environments, and it is important that we understand the relationship between them. Probably the most important statement that we can make about this relationship is that different kinds of organism are not distributed at random amongst different kinds of environment: there is a correspondence between the two. This correspondence is part of our sense of the order of things. But, what exactly is the nature of the match between organisms and their environment? It is quite impossible to think of an organism without an environment, but it is easily possible to think of environments without organisms. It is convenient therefore to consider first the variation that exists in environments.

1.1.1 The nature of environments

Even a sand grain on the surface of the moon has an environment. The climate is one component of this environment and is determined by the radiation that drives the variations in its temperature and moves its atmosphere, for example its winds. The earth is exposed to solar radiation that varies across its surface depending on its distance from, movement around and changing inclination to the sun. Therefore, the surface of our planet, like that of any other, has a range of environments that vary from place to place. A very special character of our planet is that the temperatures experienced at its surface are in the range in which water can change between solid, liquid and gaseous phases. Moreover, in its gaseous and liquid phases, water is continually moving under the influence of the radiation, which is unequally distributed at the earth's surface. The least radiation is received near the poles (less than $0.04\ \mathrm{J\ cm^{-2}\ min^{-1}}$), and most of the water there is solid ice. The most radiation is received in the tropics (0.10–$0.11\ \mathrm{J\ cm^{-2}\ min^{-1}}$), where much of it provides the latent heat of evaporation that converts liquid water into gas. Radiation also provides the advective energy that moves clouds of water in gaseous form about the atmosphere and redistributes it on the surface in liquid (rain) or solid form (snow). The surface of the earth is uneven and liquid water drains under gravity from the highlands to the hollows—hence, the environment is partly exposed 'terrestrial' surface and partly 'aquatic' (the oceans, lakes, rivers and puddles).

The planet earth has a history of climatic change that has interacted with its geological history. Various geological formations contribute to the character of the earth's surface. They have different mineral compositions, and they weather and decompose at different rates. The rock formations extend to different heights and, because temperature falls with altitude, the transitions in the state of water may occur with height (i.e. up a mountain), as well as from the equator towards the poles. Water at the top of a mountain may be predominantly solid, yet at the bottom evaporation (loss of water as gas) may exceed precipitation. Near the poles, in parts of the tropics and at both the top and bottom of mountains in the tropics, liquid water may therefore be absent from the environment for most of the time. The variations in the environment of the planet earth are mainly defined by variations in the radiation received in its different parts, and by the topology and nature of its geological formations.

1.1.2 Environmental conditions necessary for life

It is conceivable that all the biological activities on a planet such as earth could be powered by energy obtained by inorganic transformations—such as the oxidation of methane, sulphur compounds or ammonia. There are archaebacteria (*Archaea*) that can indeed do just this, although most of them live in extremely localized and peculiar environments, such as those that are extremely hot or extremely acid (see Chapter 2, Sections 2.4 and 2.6). However, the mainstream of biological activity on our planet has depended on solar energy fixed by photosynthesis as its source of power. The result has been that, overwhelmingly, the biological activities on the planet are limited by the efficiency of the photosynthetic process, and depend on incident radiation. But, the intensity of radiation also determines the physical state of water and can therefore act in opposing ways: the distribution of solar radiation determines where and when photosynthesis might occur, but it also determines whether the liquid water is present that photosynthetic organisms require abso- lutely. Where radiation is abundant (e.g. in the tropics) liquid water may be scarce because it is volatilized, and where there is much less radiation, liquid water is scarce because it is solidified! The ecology of our planet is caught 'between the frying pan and the freezer'!

1.1.3 The diversity of organisms and the patchiness of their distribution

It is not too difficult to imagine a planet like earth that is inhabited by just one sort of organism. It might be limited in its distribution to just a very tiny subset of the multiplicity of environments. It might range widely over many physical environ- ments if it possessed broad temperature tolerance. It could range even more widely if it could tolerate periods of desiccation, and even further if it could also function when immersed in water. But, it could live only where there was liquid water, a source of energy and access to inorganic resources for growth. (When, or if, it died it would need to be self decomposing (autolyse) and release its components back to the environment, otherwise it would exhaust its resources and become extinct leaving undecomposed bodies littering the habitat!) Even one such 'ideal' organism would give the planet an ecology and a biogeography. A visiting ecologist would find plenty to keep himself busy, studying the relationships between this unique organism and its environment! In fact, of course, the earth is inhabited by a multiplicity of types of organism that are neither distributed randomly nor as a homogeneous mixture over the surface of the globe.

Any sampled area, even on the scale of a whole continent, contains only a subset of the variety of species present on earth. These restricted patterns of species distribution occur despite the fact that individuals (or the progeny) of all species are capable of some dispersal, and, for many, dispersal may be on an intercontinental scale (e.g. birds and the seeds of orchids). A great part of the science of ecology tries to explain why every type of organism does not live everywhere. One of the greatest of all ecological generalizations is that: *all species are always absent from almost everywhere.*

It is natural to look for correlations between the biology of different species and the nature of the environment in the special places in which each is found. This approach emphasizes the ways in which the peculiar biologies of different species 'match' the features of the environments in which they live. We try to explain not only how the properties of different sorts of species make their life possible in particular environments, but also to explain their failure in others. A further task of the ecologist is to try to explain how the diversity of organisms has evolved. One of the greatest of ecologists, Evelyn Hutchinson, pointed out that ecology provides the theatre in which the evolutionary play is performed and asked the tantalizing question 'Why are there so many sorts of animal?' (Hutchinson, 1959, 1965).

1.1.4 Natural selection: adaptation or abaptation?

The phrase that, in everyday speech, is most commonly used to describe the match between organisms and environment is: 'organism X is adapted to' followed by a description of the conditions where the organism is found. Thus, we often hear that 'fish are adapted to live in water', or 'cacti are adapted to live in conditions of drought'.

Unfortunately, biologists use the words 'adapted' or 'adaptation' to mean various, quite different things.

The phrase 'X is adapted to live in Y' may not mean much more than one of the following.

1   X lives in environment Y. For example, in this most simple form it says that because most fishes are found living in water, they are adapted to live in water (they possess the properties necessary for life in water). The word 'adapted' here says nothing about how the properties were acquired, only about the positive consequences of having them.

2   The above statement may mean that properties of X constrain it to live only in Y. The statement that plants are 'adapted' to live in calcareous soils may often be taken to imply that they have properties that exclude them from other soil types.

3   For an evolutionary ecologist, X is adapted to live in Y will usually mean that environment Y has provided forces of natural selection that have affected the life of X's ancestors and so have moulded and specialized the evolution of X. 'Adaptation' here means that genetic change has occurred.

4   In complete contrast, for an ecologist with physiological leanings the statement can mean that individuals of X have themselves had some prior experience (e.g. been cold-hardened) that makes life in Y (e.g. an extremely cold winter) possible. Adaptation here is definitely not genetic, but is phenotypic change.

5   Alternatively, the statement 'X is adapted to live in Y' can mean that a property (an adaptation) has been identified that X possesses and that has been shown to (or at a reasonable guess might) explain why X can live in Y (e.g. a detoxifying enzyme for an organism that can tolerate a toxin). Such a property could be either genotypic or phenotypic.

A word that has so many contrasting meanings does not contribute much precision to science, but all the meanings imply something about the way in which organisms match their environments. Very unfortunately, the word '*ad*aptation' both inside and outside science implies that the way that organisms react to present circumstances, prepares them for the future, is predictive and implies some sort of forward planning, or at the very least, of design. But, organisms are not designed for, or matched to or fitted to, the present or the future—their character or properties are entirely consequences of the past—they reflect the successes and failures of ancestors. *Ab*aptation (or exaptation) would be a better word than adaptation because its etymology brings the implication that the aptness (match) of organisms for their environment is a product of their past rather than a programme for the future. The properties of organisms appear to be apt for the environments that they live in at present only because present environments tend to be similar to those of the past.

evolution by natural selection

Darwin's (1859) theory of evolution by natural selection is an ecological theory and rests on a series of propositions.

1 The individuals that make up a population of a species are *not identical*: they vary, although sometimes only slightly, in size, rate of development, response to temperature, and so on.

2 Some, at least, of this variation is *heritable*. In other words, the characteristics of an individual are determined to some extent by its genetic make-up. Individuals receive their genes from their ancestors and therefore tend to share their characteristics.

3 All populations have the *potential* to populate the whole earth, and they would do so if each individual survived and each individual produced its maximum number of descendants. But, they do not: many individuals die prior to reproduction, and most (if not all) reproduce at a less than maximal rate and some of these differences are heritable.

4 Different ancestors leave *different numbers of descendants*. This means much more than saying that different individuals produce different numbers of offspring. It includes also the chances of survival of individuals to reproductive age, the survival and reproduction of their offspring, the survival and reproduction of *their* offspring in turn, and so on.

5 Finally, the number of descendants that an individual leaves depends, not entirely but crucially, on *the interaction between the characteristics (or properties) of the individual and its environment*.

An individual will survive, reproduce and leave descendants in some environments but not in others. This is the sense in which nature may be loosely (and anthropomorphically) thought of as *selecting*. It is in this sense that some environments may be described as favourable or unfavourable, and it is in this same sense that some organisms can be considered to be fit or not. If, because some individuals leave more descendants than others the heritable characteristics of a population change from generation to generation, then evolution by natural selection is said to have occurred.

Of course, the present properties of an organism have not all been selected in the type of environment in which it now lives. Over the course of its evolutionary history (its phylogeny) an organism's remote ancestors may have been single-celled aquatic eukaryotes, and as its evolutionary pathway has led to the present specialized organism—peacock, orchid, shrimp or elephant—it has accumulated some, and shed

fitness is the proportionate
contribution of individuals to
future generations

other properties that it acquired in its evolutionary progress. This baggage of evolutionary history places limits on, and constrains, future evolution and much of what we now see as precise matches between an organism and its environment are either these inherited constraints (e.g. that koala bears can live only on *Eucalyptus* foliage), or detailed fine-tuning. Jacob's phrase 'Evolution is tinkering' captures the essence of the process (Jacob, 1977). Like 'tinkering' with a broken bicycle, the process of evolution involves imperfect machines and a restricted supply of gadgets that might be used to make them work better or even do something different (tinkering a toy trolley out of the parts of the bicycle).

gobioid fishes illustrate the
match of related species to
highly diverse environments

The ecological radiation of gobioid fishes (Figure 1.1) illustrates the ways in which a basic plan—the generalized benthic goby—has radiated into (been tinkered into) a wide diversity of forms (more than 2500 nominal species), that 'match' highly specialized environments and lifestyles. Each of these specialized forms carries with it the essential qualities that make it a goby (a basic teleost structure modified by evolutionary loss of certain skull bones, complicated reproductive organs and curious leaf-like sperm duct glands). There is strong evidence that the gobies form a monophyletic lineage, within which there have been many lines of specialized divergence. But with the gobies, as in many other cases of ecological radiation, we lack the details of phylogeny that could put the ecological radiation fully into its evolutionary context. Such examples sound a loud and clear warning that each species is not an independently evolved match to its particular present environment. Its evolution has been constrained by its ancestry.

1.1.5 Fitness

The fittest individuals in a population are by definition those that leave the greatest number of descendants. In practice, the term is often applied not to a single individual, but to a typical individual or a type, for example we may say that in sand dunes, yellow-shelled snails are fitter than brown-shelled snails (i.e. they are more likely to survive and to leave more descendants).

Fitness is a relative not an absolute term. The numbers of seeds produced by a plant, or eggs produced by an insect, are not direct measures of their fitness; nor indeed are the *numbers* of descendants that they leave. Rather, it is the proportionate contribution that an individual makes to future generations that determines its fitness: the fittest individuals in a population are those that leave the greatest number of descendants *relative to* the number of descendants left by other individuals in the population. Those individuals that leave the greatest proportion of descendants in a population have the greatest influence on the heritable characteristics of that population.

When we marvel at the diversity of complex specializations that ensure the match between organisms and their environment, there is a temptation to regard each case as an example of evolved perfection. But, there is nothing in the process of evolution by natural selection that implies perfection. For example, no population of
organisms can contain all the possible genetic variants that might exist and might influence fitness. The evolutionary process works on the genetic variation that is available. It follows from this that natural selection is unlikely to lead to the evolution of perfect, 'maximally fit' individuals. It can favour only those that are fittest from *amongst the range of variety available*, and this may be a very restricted choice. The very essence of natural selection is that organisms come to match their environments

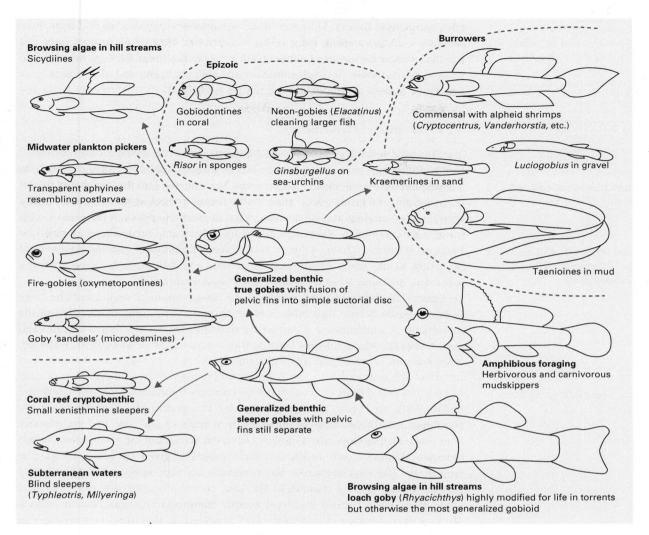

Browsing algae in hill streams
Sicydiines

Epizoic

Gobiodontines
in coral

Neon-gobies (*Elacatinus*)
cleaning larger fish

Burrowers

Commensal with alpheid shrimps
(*Cryptocentrus, Vanderhorstia,* etc.)

Midwater plankton pickers

Risor in sponges

Ginsburgellus on
sea-urchins

Kraemeriines in sand

Luciogobius in gravel

Transparent aphyines
resembling postlarvae

Fire-gobies (oxymetopontines)

**Generalized benthic
true gobies** with fusion of
pelvic fins into simple suctorial disc

Taenioines in mud

Goby 'sandeels' (microdesmines)

Amphibious foraging
Herbivorous and carnivorous
mudskippers

Coral reef cryptobenthic
Small xenisthmine sleepers

**Generalized benthic
sleeper gobies** with pelvic
fins still separate

Subterranean waters
Blind sleepers
(*Typhleotris, Milyeringa*)

**Browsing algae in hill streams
loach goby** (*Rhyacichthys*) highly modified for life in torrents
but otherwise the most generalized gobioid

Figure 1.1 Ecological radiation of the gobioid fishes into a variety of highly specialized forms and habitats. The figure describes ecological radiation that has occurred within systematically closely related forms, but it does not necessarily imply the detailed phylogenetic sequences (e.g. gains and losses of structures) by which this radiation occurred; these are still not completely understood. (After Miller, 1993; see also Nelson, 1994.)

by being 'the fittest available' or 'the fittest yet': they are not 'the best imaginable'. The reasons why we should not expect to find the evolution of perfection in nature are developed further in papers by Jacob (1977) and Gould and Lewontin (1979).

1.2 Historical factors

It is particularly important to realize that past events on the earth can have profound effects on the nature and distribution of organisms. Our world has not been constructed by someone taking each organism in turn, testing it against each environment and moulding it so that every organism finds its perfect place. It is a world in which organisms live where they do for reasons that are often, at least in

9   ORGANISMS AND THEIR ENVIRONMENTS

part, accidents of history. Moreover, they have lived in environments that were quite different from the present: some of the features that they then acquired hang like millstones around their necks, limiting and constraining where they can now live and what they can now do. Environments are still changing and the matching of organisms to their environments will always lag behind.

1.2.1 Movements of land masses

land masses have moved ...

The curious distributions of organisms between continents, seemingly inexplicable in terms of dispersal over vast distances, led biologists, especially Wegener (1915), to suggest that the continents themselves must have moved. This was vigorously denied by geologists until geomagnetic measurements required the same, apparently wildly improbable, explanation. The discovery that the tectonic plates of the earth's crust move and carry with them the migrating continents, reconciles geologist and biologist (Figure 1.2b–e). Thus, whilst major evolutionary developments were occurring in the plant and animal kingdoms, populations were being split and separated, and land areas were moving across climatic zones.

and divided populations which have then evolved independently

Figure 1.3 shows just one example of a major group of organisms (the large flightless birds), whose distributions begin to make sense only in the light of the movement of land masses. It would be unwarranted to say that the emus and cassowaries are where they are because they represent the best match to Australian environments, whereas the rheas and tinamous are where they are because they represent the best match to South American environments. The evolution of each has been moulded in its own continent and past environments. Their very disparate distributions are essentially determined by the prehistoric movements of the continents, and the subsequent impossibility of geographically isolated evolutionary lines reaching into each others' environment. An account of the evolutionary trends amongst mammals over much the same period is given by Janis (1993). She emphasizes the dependence of the mammals on the types and distribution of vegetation, e.g. climate changes in the Late Eocene disrupted the global tropical forest of the Early Tertiary and most archaic mammalian lineages became extinct. Most modern families of mammals first appeared at that time and changes in sea-level (indicated in Figure 1.2a) allowed major episodes of dispersal to take place.

for example, large flightless birds

1.2.2 Climatic changes

climates have changed ...

Changes in climate have occurred on shorter time scales than the movements of land masses, and a great variety of technologies has been drafted into describing and analysing these changes (IGBP, 1990a; Boden *et al.*, 1990). Much of what we see in the present distribution of species represents phases in a recovery from past climatic shifts. Changes in climate during the Pleistocene ice-ages in particular, bear a lot of the responsibility for present patterns of distribution of plants and animals.

The extent of climatic and biotic change in the past 2 million years (the Pleistocene) is only beginning to be unravelled as the technology for discovering, analysing and dating biological remains becomes more sophisticated (particularly by the analysis of buried pollen samples). These methods increasingly allow us to determine just how much of the present distribution of organisms represents a precise local match to present environments, and how much is a fingerprint left by the hand of history.

Techniques for the measurement of oxygen isotopes in ocean cores indicate that

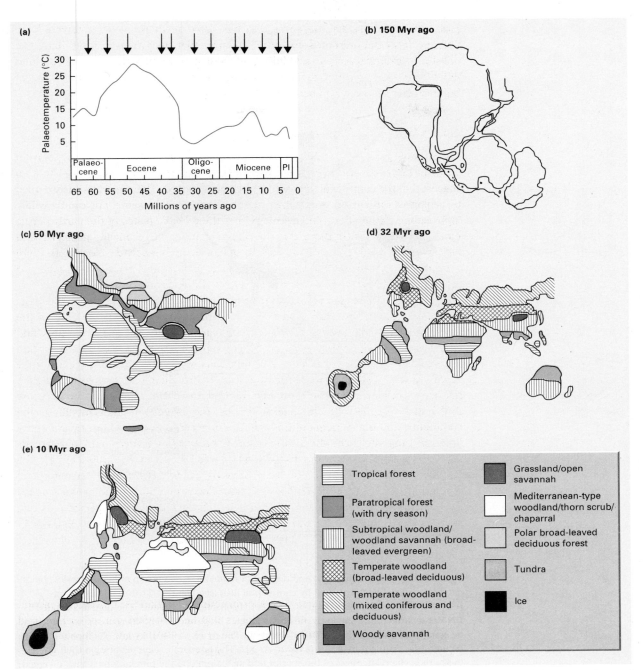

Figure 1.2 (a) Changes in temperature in the North Sea over the past 60 million years (Myr). The arrows indicate times when there were large falls in sea-level that allowed major dispersal events to occur between land masses. (b) The beginning of the break-up of the ancient supercontinent of Gondwanaland at about 150 Myr ago. (c) Palaeogeography and global vegetation in the Early Middle Eocene, at about 50 Myr ago. (d) Palaeogeography and global vegetation in the Early Oligocene, at about 32 Myr ago. (e) Palaeogeography and global vegetation in the Early Miocene, at about 10 Myr ago. (The position of the Antarctic ice-cap is highly schematic.) (After Norton & Sclater, 1979; Janis, 1993; and a variety of other sources.)

11   ORGANISMS AND THEIR ENVIRONMENTS

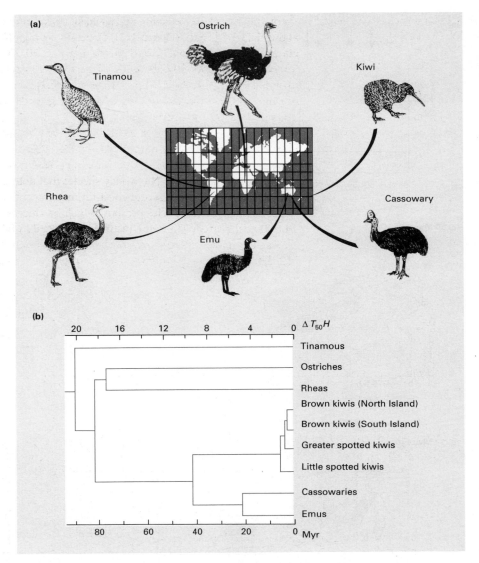

Figure 1.3 (a) The distributions and degrees of relatedness of a group of large flightless birds can be at least partly explained by continental drift (see Figure 1.1). (b) The degrees of relatedness have been measured by a DNA hybridization technique. The double-stranded DNA is separated into single strands by heating. The strands from different species can then be combined, and again separated by heating. The more similar they are, the higher is the temperature required to separate them, $\Delta T_{50}H$. The temperature for separation then gives a measure of the relatedness of the species and an estimate of the time at which they diverged (Myr = millions of years). The earliest divergence was that of tinamous from the remainder (the ratites). The subsequent divergences agree well with the timing of the break-up of Gondwanaland and the subsequent continental drift (see Figure 1.2): (i) the rift between Australia and the other southern continents; (ii) the opening of the Atlantic between Africa and South America; (iii) the opening of the Tasman Sea about 80 Myr ago, probably followed by island hopping by ancestors of the kiwi across this divide to New Zealand 40 Myr ago. The divergence of the various species of kiwi appears to be very recent. (After Diamond, 1983; from data of Sibley & Ahlquist.)

there may have been as many as 16 glacial cycles in the Pleistocene, each lasting for about 125 000 years (Figure 1.3). Contrary to popular belief, it seems that each glacial phase may have lasted for as long as 50 000–100 000 years, with brief intervals of 10 000–20 000 years when the temperatures rose close to those we experience today. If these time scales are correct, it is present floras and faunas that are unusual, because they have developed towards the end of one of a series of unusual catastrophic warm events!

During the 20 000 years since the peak of the last glaciation, global temperatures have risen by about 8°C, and the rate at which vegetation has changed over much of this period has been detected by examining pollen records (e.g. as at Rogers Lake in Connecticut) (Figure 1.4). The woody species that dominate such pollen profiles have arrived in turn, spruce first and chestnut most recently. Each new arrival has added to the number of the species present, which has increased continually over the past 14 000-year period. The same picture is repeated in European profiles.

As the number of pollen records has increased, it has become possible not only to plot the changes in vegetation at a point in space, but to begin to map the movements

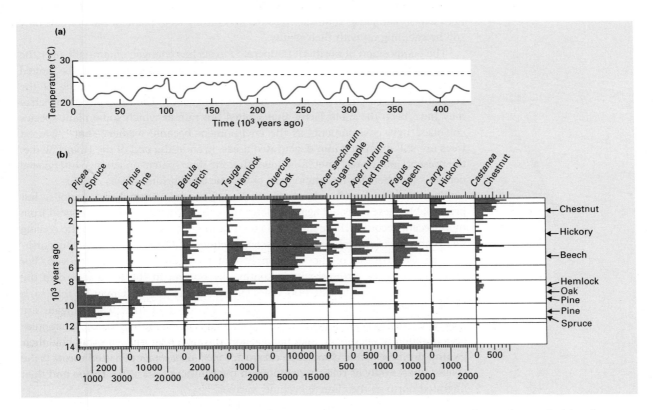

Figure 1.4 (a) An estimate of the temperature variations with time during glacial cycles over the last half million years. The estimates were obtained by comparing oxygen isotope ratios in fossils taken from ocean cores. The dashed line corresponds to the ratio at the surface. This suggests that interglacial periods as warm as the present have been rare events, and most of the past half million years has experienced glacial climates. (After Emiliani, 1966; Davis, 1976.) (b) The profiles of pollen accumulation from late glacial times to the present in the sediments of Rogers Lake, Connecticut. Approximate time of arrival of each species in Connecticut is shown by arrows at the right of the figure. (After Davis, 1976.)

13   ORGANISMS AND THEIR ENVIRONMENTS

the distributions of trees have
changed gradually since the
last glaciation

of the various species as they have spread across the continents. It appears that
different species have spread at different speeds, and not always in the same
direction. The composition of the vegetation types has, at the same time, been
shifting in balance, and is almost certainly continuing to do so.

In the invasions that followed the retreat of the ice in eastern North America,
spruce invaded first, followed by jack pine or red pine, which spread northwards at
a rate of 350–500 m year^{-1} for several thousands of years. White pine started its
migration about 1000 years later, at the same time as oak. Hemlock was also one of
the rapid invaders (200–300 m year^{-1}), and arrived at most sites about 1000 years
after white pine. Chestnut moved slowly (100 m year^{-1}), but became a dominant
species once it had arrived; it appeared on the Allegheny Mountains some 3000
years before it reached Connecticut. (Even 100 m year^{-1} seems a remarkably rapid
rate of advance for a plant species that bears heavy seeds and has a long period of
juvenile growth before it starts to produce them. See Bennet (1986) for a discussion
of postglacial spread and increase of forest trees in the postglacial period.)

We do not have such good records for the postglacial spread of the animals
associated with the changing forests, but it is at least certain that many species could
not have spread faster than the trees on which they feed. Some of the animals may
still be catching up with their plants.

The composition of northern temperate forests has changed continually over the
past 10 000 years, and change continues. The rate of change appears to be limited
by the rates of spread of the main species, and the northwards march of the
vegetation lags behind the northwards retreat of the ice. The rate of seed dispersal
may have been the main factor that limited the rate at which some plant species
colonized new environments as the environment became warmer. That '... forest
trees are still migrating into deglaciated areas, even at the end of the Holocene (i.e.
the present) interglacial interval, implies that the time span of an average interglacial
period is too short for the attainment of floristic equilibrium' (Davis, 1976).

Very little is known about the floras of previous interglacial periods, but
presumably these repeated again and again some of what has been discovered from
more recent records. There may be a continuum of types of response to the cooling
and warming of glacial episodes. Reconstructions of the vegetation in the north-
western Great Basin of the USA imply that some species scarcely moved. For
example, *Juniperus osteospermum* is continually present in the fossil record of the
Great Basin over the last 30 000 years (Nowak *et al.*, 1994). It may be that such
species that remain in place as the climate freezes and then warms again are
genetically highly diverse populations, or even species containing many subpopulations of different tolerances, and that it is only the most tolerant forms that hold their
ground during climatic change. In contrast, *Pinus monophylla* persisted through the
coolest phases only as relict populations in the southern part of the region and then
migrated north as the climate became warmer.

predicted global warming by
the 'greenhouse effect' is
nearly 100 times faster than
postglacial warming

Evidence of changes in vegetation that followed the last retreat of the ice hint at
the consequence of the global warming (3 °C in the next 100 years), which is
predicted to result from continuing increases in atmospheric carbon dioxide (see
references to the 'greenhouse effect' in Chapter 2, Section 2.11.1). But, the scales are
quite different. Postglacial warming of about 8 °C occurred over 20 000 years.
Vegetational change failed to keep pace with the speed of environmental warming.
The predicted pace of global warming in the 'greenhouse effect' is nearly 100 times
faster and must presumably result in widespread waves of species extinction.

A classic example of species' distributions that may need to be explained in terms of history is the presence of isolated pockets of often very specialized cold-tolerant species of flowering plant in the Arctic—alpine floras of North America and northern Europe. Frequently, such species are found in only one locality. Other species occur in two or more curiously isolated regions (i.e. they have bicentric or polycentric disjunct distributions). The bicentric distribution of *Campanula uniflora* in Norway (Figure 1.5) is a typical example.

isolated distributions and nunataks

One view of these separate centres of distribution is that they represent locally suitable habitats, providing the unique specialized match to their environment required for the growth of the species. On this theory, propagules are supposed to have dispersed to these isolated sites across intervening areas where the habitats are unsuitable. There may be other areas in which the needs of the species are matched, but they have not yet been reached. Certainly, many of these oddly distributed plant species are found in localized patches of calcareous soils or associated with particular temperature conditions (Figure 1.5), but this is not always the case.

A different interpretation is that the present distributions are relicts of populations once distributed more widely. As the ice sheet moved down from the north, some

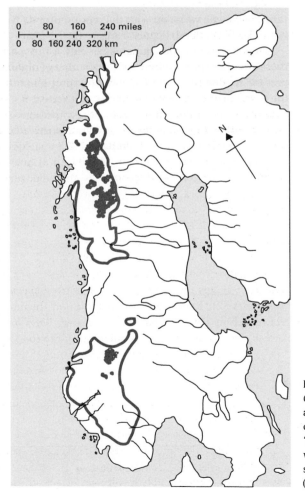

Figure 1.5 The distribution of *Campanula uniflora* in Norway: a characteristic bicentric distribution. The thick lines are 'isotherms': lines joining places where the mean maximum summer temperature is 22°C. (After Ives, 1974.)

high-altitude areas remained free of ice (although still intensely cold). Populations of some plant species persisted in these unglaciated areas or *nunataks*, and when the ice retreated they remained in their isolated fastnesses. Geological evidence can be found that supports this idea (although it is not conclusive). There is also some intriguing evidence from the distribution of flightless beetles in supposed nunataks that adds support to the 'relict' argument. (These ideas are reviewed by Ives, 1974 and Birks (1993) questions the need for the nunatak hypothesis to explain Arctic floras.)

Molecular genetic analysis might be expected to give clean answers to such questions. Two plant species, a campion (*Silene acaulis*) and a saxifrage (*Saxifraga oppositifolia*), are found in the high Arctic and also on more southern mountains. If the high-latitude populations are derived largely from stocks that survived the last glaciation in ice-free refugia they would be expected to be genetically differentiated from the more southern populations. Genetic variation in both species was analysed by a survey of chloroplast DNA and that in *Silene* was also analysed by surveying enzyme polymorphism (Abbott *et al.*, 1995). The results suggest that the saxifrage populations in the high Arctic were indeed derived, at least in part, from ancient relicts in nunataks, because they contain a greater range of DNA variation than the more southern forms. In marked contrast the campion was probably a postglacial immigrant because the most northern Spitsbergen populations share most of their range of genetic variation with the more southerly Norwegian populations (although not those of Scotland) (Figure 1.6).

The argument about nunataks is only a tiny part of a broader debate between those who interpret the present distribution of organisms as showing their match to present conditions, and those who interpret much of the present as fingerprints of the past. Species populations that did not retreat fast enough when the ice advanced (and it advanced faster than it withdrew) presumably became extinct, with a few perhaps surviving in nunataks. Those that survived (those that we now know), were simply the ones that could keep pace with (or survive through) the successive cycles of the ice's invasion and retreat. There is, of course, no reason to suppose that the genetic composition of those species that did survive the migrations remained

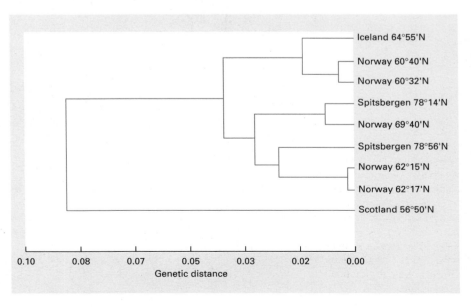

Figure 1.6 A dendrogram illustrating genetic relationships between populations of *Silene acaulis*, based on the unbiased genetic distance of Nei (1978). (After Abbott *et al.*, 1995.)

unchanged. We know that quite rapid genetic changes can be produced in laboratory populations of both animals and plants by artificial selection for cold tolerance and that some species contain much ecotypic variation for frost tolerance (see Chapter 2, Section 2.3.4). In addition to being selected for changing temperature tolerances during glacial advances and retreats, we might also expect that after an episode of migration, each species would have been selected for increased rates of colonization.

climatic change in the tropics The records of climatic change in the tropics are far less complete than those for temperate regions. There is therefore the temptation to imagine that whilst dramatic climatic shifts and ice invasions were dominating temperate regions, the tropics persisted in the state we know today. This is almost certainly wrong. Data from a variety of sources indicate that there were abrupt fluctuations in postglacial climates in Asia and Africa. In continental monsoon areas (e.g. Tibet, Ethiopia, western Sahara and sub-equatorial Africa) the postglacial period started with an extensive period of high humidity followed by a series of phases of intense aridity (Zahn, 1994). In South America a picture is emerging of vegetational changes that parallel those occurring in temperate regions, as the extent of tropical forest increased in warmer, wetter periods, and contracted as savannah dominated the vegetation in the cooler and drier periods (Prance, 1981, 1985, 1987). The present distributions of both plant and animal species provide evidence of the positions once occupied by these 'tropical islands in a sea of savannah' (Figure 1.7).

Even on the limited time scale of the recurrent ice-ages, the plant communities of the planet must have been ephemeral, risk ridden and perpetually readjusting to change. They never had time to settle down. This is even more obviously the case over shorter time scales of catastrophes and recurrent disasters, such as hurricanes, storms and volcanic eruptions, and the massive scales of human intervention through deforestation, slash and burn agriculture and severe overfishing. Communities spend much of their time recovering from the last crisis and what returns may differ subtly or even profoundly from what existed before (see Chapter 2, Section 2.9.1).

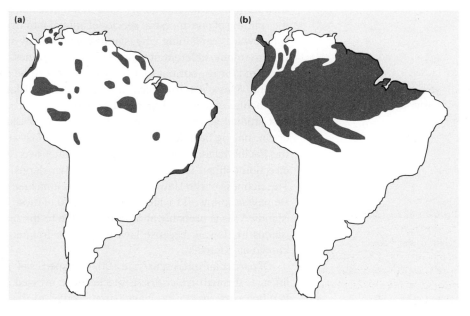

Figure 1.7 (a) The estimated distribution of tropical forest in South America at the time when the last glaciation was at its highest. Gallery forest along riversides may have prevented the patches from being quite as isolated as the map suggests. (b) The present distribution. (After Prance, 1981.)

1.2.3 Island patterns

The fauna and flora of islands (whether surrounded by ocean or by a 'sea' of different vegetation) have several features that distinguish them from the fauna and flora of continents. These are discussed lucidly and in some detail in Williamson's (1981) book, and are touched on in several chapters, and in greater depth in Chapter 22. In the present context though, island biotas illustrate three important, related points: (i) that there is a historical element in the match between organisms and environments; (ii) that there is not just one perfect organism for each type of environment; and (iii) that natural selection acts on disparate types of organism and may match them to the same type of environment.

Put very simply, the fauna and flora of islands have two distinguishing features. There are fewer species on islands than in comparable areas of mainland of the same size; and many of the species on islands are either subtly or profoundly different from those on the nearest comparable area of mainland. Again put simply, there are two main reasons for this.

1   The fauna and flora of an island are limited to those types having ancestors that managed to disperse to the island, although the extent of this limitation depends on the distance of the island from the mainland (or other islands), and varies from group to group of organism depending on their intrinsic dispersal ability.

2   Because of this isolation, the rate of evolutionary change on an island may often **the isolation of islands favours** be fast enough to outweigh the effects of the exchange of genetic material between **the formation of new species** the island population and the parent population on the mainland.

Let us explain this second point more fully. Biologists recognize organisms as members of the same species if the individuals are capable of breeding with each other, freely exchanging genetic material and producing fertile offspring. Such genetic exchange tends to result in homogeneity in the genetic character of a population (as does the reassortment of genes resulting from genetic recombination). Reproductive isolation on the other hand (e.g. between island and mainland populations), allows the more localized evolution of matches between organisms and their environments. Indeed, it seems that reproductive isolation is an essential step in the splitting of one ancestral species of animal into two. This undoubtedly goes a long way towards explaining why islands contain many species unique to themselves, as well as many differentiated 'races' or 'subspecies' that are distinguishable from mainland forms, but not to a sufficient extent to be called separate species.

the picture-winged *Drosophila* The *Drosophila* fruit-flies of the Hawaiian islands provide one of the most **of Hawaii** spectacular examples of species formation on islands, and it is certainly the example best studied genetically. The Hawaiian chain of islands (Figure 1.8) is volcanic in origin, having been formed gradually over the last 40 million years, as the centre of the Pacific tectonic plate has moved steadily over a 'hot-spot' in a south-easterly direction (Niihau is the most ancient of the islands, Hawaii itself the most recent). The richness of the Hawaiian *Drosophila* is amazing: there are probably about 1500 *Drosophila* spp. world-wide, but at least 500 of these are found only in the Hawaiian islands. This is probably at least partly due to the fact that there are islands within islands in Hawaii, because lava flows have frequently isolated areas of vegetation known as 'kipukas'.

Of particular interest are the 100 or so species of 'picture-winged' *Drosophila*. The lineages through which these species have evolved can be traced by analysing the banding patterns on the giant chromosomes in the salivary glands of their larvae.

The evolutionary tree that emerges is shown in Figure 1.8, with each species lined up above the island on which it is found (there are only two species found on more than one island). The historical element in 'what lives where' is plainly apparent: the more ancient species live on the more ancient islands and, as new islands have been

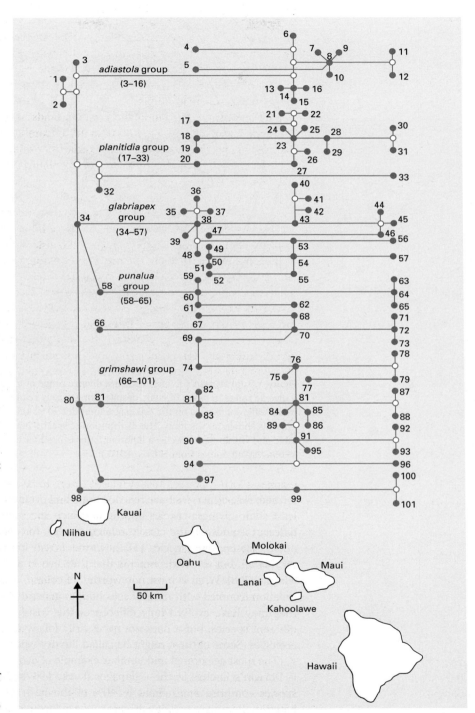

Figure 1.8 An evolutionary tree linking the picture-winged *Drosophila* of Hawaii, traced by the analysis of chromosomal banding patterns. The most ancient species are *D. primaeva* (species 1) and *D. attigua* (species 2), found only on the island of Kauai. Other species are represented by solid circles; hypothetical species, needed to link the present-day ones, are represented by open circles. Each species has been placed above the island or islands on which it is found (although Molokai, Lanai and Maui are grouped together). Niihau and Kahoolawe support no *Drosophila*. (After Carson & Kaneshiro, 1976; Williamson, 1981.) The first *Drosophila* colonist probably reached the Hawaiian archipelago 40 million years ago, before any of the present islands existed. (After Beverly & Wilson, 1985.)

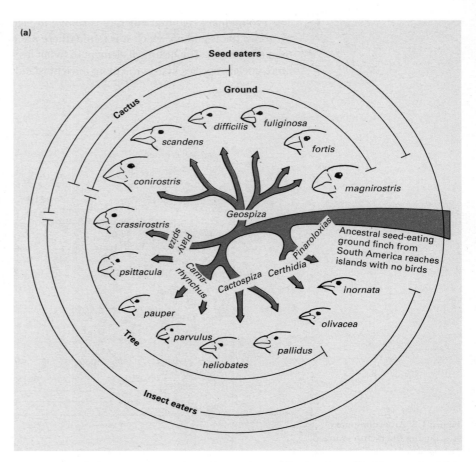

(a)

Figure 1.9 (a) Darwin's finches exploit a diverse range of food types and habitats, and exhibit a diverse range of forms of beak, despite being closely related. (b) (*page 21*) The distribution of the different species on the Galapagos (and Cocos) islands. The number of species on each island is shown on the map. The distribution of species between the islands is shown in the table and within each species a different letter is used for each subspecies. (After Lack, 1969a. Island names from Harris, 1973.)

formed, rare dispersers have reached them and eventually evolved to new species. At least some of these species appear to match the same environment as others on different islands. Of the closely related species, for example *D. adiastola* (species 8) and *D. setosimentum* (species 11), the former is only found on Maui and the latter only on Hawaii, but the environments that they live in are apparently indistinguishable (Heed, 1968). What is most noteworthy, of course, is the power and importance of isolation (coupled with natural selection) in generating new species. In some cases, they may have evolved only differences that justify a taxonomist in calling them different species, but it does not necessarily follow that they have evolved different ecologies. Some of them might be called 'luxury' species.

The most celebrated and familiar example of evolution and speciation on islands is Darwin's finches in the Galapagos (Lack, 1947a; Williamson, 1981). These 13 species comprise approximately 40% of the bird species on the Galapagos, and between them they exploit a diverse range of food types and habitats (Figure 1.9); but

Darwin's finches on the Galapagos

Plate 1 Arctic tundra, Greenland.
(Courtesy of J.A. Vickery.)

(a)

(b)

Plate 2 Coniferous forest. (a) Aerial view of conifer forest, Alberta, Canada. (Courtesy of Planet Earth Pictures/Martin King.) (b) Pine forest in autumn, Sweden. (Courtesy of Planet Earth Pictures/Jan Tove Johansson.)

Plate 3 Temperate forest. (a) Mixed woodland in autumn, North Carolina, USA. (Courtesy of The Image Bank/Arthur Mayerson.) (b) Late summer in Beechwood, Harburn, Scotland. (Courtesy of Ecoscene/Wilkinson.)

(a)

(b)

Plate 4 (*right*) Savannah.
(a) Huge herds of wildebeest
and common zebra seen from
Naabi Hill, Serengeti, Tanzania.
(b) Grassland savannah with
scattered trees. Common zebra
and wildebeest in the western
corridor of Serengeti, Tanzania.
(Courtesy of David Keith Jones/
Images of Africa.)

Plate 5 (*below*) Desert.
(a) Summertime in
Namaqualand. (b) Spring
flowers in Namaqualand,
western South Africa.
(Courtesy of Planet Earth
Pictures/J. MacKinnon.)

(a)

(b)

(a)

(b)

Plate 6 Rainforest. (a) and (b) Impenetrable forest in south west Uganda. (Courtesy of David Keith Jones/Images of Africa.)

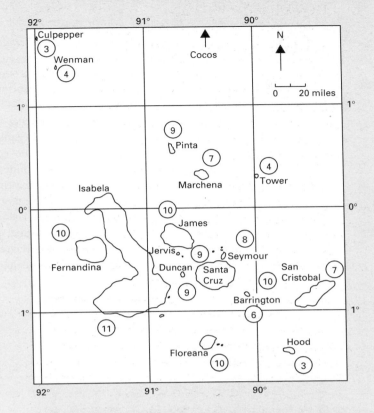

	Large central				Small central				Medium outlying				Small outlying			
	James	Santa Cruz	Isabela	Fernandina	Jervis	Seymour	Duncan	Barrington	Pinta	Marchena	San Cristobal	Floreana	Culpepper	Wenman	Tower	Hood
Geospiza																
magnirostris	A	A	A	A	A	A	A	A	A	A	–	B	–	A	A	–
fortis	A	A	A	A	A	A	A	A	A	A	A	A	–	–	–	–
fuliginosa	A	A	A	A	A	A	A	A	A	A	A	A	–	–	–	A
difficilis	A	A	A	A	–	–	–	–	B	–	–	C	D	D	B	–
scandens	A	B	B	–	A	B	B	B	C	D	E	B	–	–	–	–
conirostris	–	–	–	–	–	–	–	–	–	–	–	–	A	B?	B	C
Platyspiza																
crassirostris	A	A	A	A	A	–	A	–	A	A	A	A	–	–	–	–
Camarhynchus																
psittacula	A	A	B	B	A	A	A–B	A	C	C	–	A	–	–	–	–
pauper	–	–	–	–	–	–	–	–	–	–	–	A	–	–	–	–
parvulus	A	A	A	A	A	A	A	A	A	–	B	A	–	–	–	–
Cactospiza																
pallidus	A	A	B	B	A	A	A	–	–	–	C	–	–	–	–	–
heliobates	–	–	A	A	–	–	–	–	–	–	–	–	–	–	–	–
Certhidea																
olivacea	A	A	A	A	A	A	A	B	C	C	D	E	F	F	G	H
Number of resident species	10	10	11	10	9	8	9	6	9	7	7	10	3	3-4	4	3

they are all closely related, and probably evolved from something similar to the present-day Cocos finch.

1.3 Convergents and parallels

A match between the nature of organisms and their environment can often be seen as a similarity in form and behaviour between organisms living in a similar environment, but belonging to different phyletic lines (i.e. different branches of the evolutionary tree). Such similarities also undermine further the idea that for every environment there is one, and only one, perfect organism. The evidence is particularly persuasive when the phyletic lines are far removed from each other, and when similar roles are played by structures that have quite different evolutionary origins, i.e. when the structures are *analogous* (similar in superficial form or function), but not *homologous* (derived from an equivalent structure in a common ancestry). When this is seen to occur, we speak of *convergent evolution*. Large swimming carnivores have evolved into four quite distinct groups: amongst fish, reptiles, birds and mammals. The convergence of form (Figure 1.10) is remarkable because it conceals profound differences in internal structures and metabolism,

large swimming carnivores ...

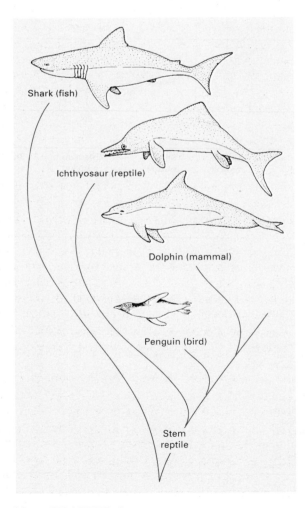

Figure 1.10 An example of convergent evolution of body form amongst large marine carnivores from different evolutionary lines. (After Hildebrand, 1974; and others.)

indicating that the organisms are far removed from each other in their evolutionary history.

… and the climbing organs of plants

Many flowering plants (and some ferns) use the support of others to climb high in canopies of vegetation, and so gain access to more light than if they depended on their own supporting tissues. The ability to climb has evolved in many different families, and quite different organs have become modified into climbing structures (Figure 1.11a): they are analogous structures but not homologous. In other plant species the same organ has been modified into quite different structures with quite different roles: they are therefore homologous, although they may not be analogous (Figure 1.11b).

… show convergent evolution

In these examples (there are very many more) we can argue that similar selective forces have acted so that the same property has been acquired from quite different evolutionary starting points. A comparable series of examples can be used to show the parallels in the evolutionary pathways of phylogenetically related groups that have radiated after they were isolated from each other.

marsupial and placental mammals show parallel evolution

The classic example of such parallel evolution is the radiation amongst the placental and marsupial mammals. Marsupials arrived on the Australian continent in the Cretaceous period (around 90 million years ago), when the only other mammals present were the curious egg-laying monotremes (now represented only by the spiny anteaters and the duck-billed platypus, *Ornithorhynchus anatinus*). An evolutionary process of radiation then occurred that in many ways accurately paralleled what occurred in the placental mammals on other continents (Figures 1.12a, b). The subtlety of the parallels in both the form of the organisms and their lifestyle is so striking that it is hard to escape the view that the environments of placentals and marsupials contained 'ecological niches' into which the evolutionary process has neatly 'fitted' ecological equivalents. (It is important to remember that, in contrast with convergent evolution, the marsupials and placentals started their radiative evolution with an essentially common set of constraints and potentials because they sprang from a common ancestral line.)

do the megaphytes show parallel evolution?

It is a fascinating exercise to collect examples to illustrate convergent or parallel evolution in plant and animal groups, but there are serious risks. One such risk is illustrated by the 'rosette treelets' or megaphytes. These are plants that usually bear only one main stem (usually unbranched), which continues growth for a long period and is characteristically topped with a mop of very large leaves. When megaphytes flower, each plant produces gigantic inflorescences and then either dies or there is a long delay before the next flowering episode. They often carry a mass of dead leaves or leaf bases, so that their trunks often falsely appear to be enormously thick (this is particularly apparent in many palms, *Espeletia* spp. and giant species of *Senecio*). The megaphyte habit has clearly arisen repeatedly in widely different phylogenetic lines, for example it occurs in both monocotyledons (e.g. palms) and dicotyledons (e.g. *Lobelias*).

We might expect that such striking similarity in form amongst quite unrelated plants implies an equally striking similarity in their environments. In fact, they 'fit' into a very odd diversity of vegetation types (Table 1.1). The exposed high-mountain and often island habitats of the dicotyledonous megaphytes seem to have little in common with the shaded protected environments of the understorey palms and tree ferns, or the arid habitats of many aloes and agaves.

A partial explanation for the resemblance amongst megaphytes may be that a correlated set of characters comes together as a package and selection cannot favour

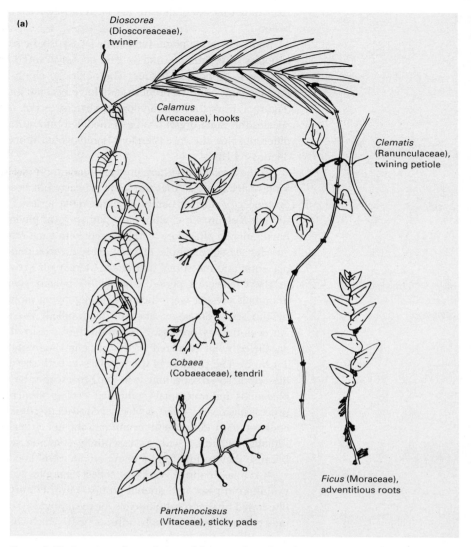

(a)

Dioscorea (Dioscoreaceae), twiner

Calamus (Arecaceae), hooks

Clematis (Ranunculaceae), twining petiole

Cobaea (Cobaeaceae), tendril

Ficus (Moraceae), adventitious roots

Parthenocissus (Vitaceae), sticky pads

Figure 1.11 A variety of morphological features that allow flowering plants to climb. (a) Structural features that are analogous, i.e. derived from the modification of quite different organs, e.g. leaves, petioles, stems, roots and tendrils, and (b) (*facing page*) structural features that are homologous, i.e. derived by modifications of a single organ, the leaf, shown by reference to an idealized leaf in the centre of the figure. (Drawn by Alan Bryant.)

one character without carrying other changes with it (see also the discussion of allometry in Chapter 14). It may be that in some habitats natural selection has operated to increase the size of inflorescences and that this requires large shoot meristems which as a necessary consequence produce bigger leaves. In other habitats it may be that selection has favoured large leaves and morphological constraints lead necessarily to larger inflorescences. There are examples from plant breeding that tend to support this view. The ancestral form of the sunflower is branched, bears small leaves, many small flowering capitula and small seeds. The cultivated form has been selected to produce large seeds which are easily harvested.

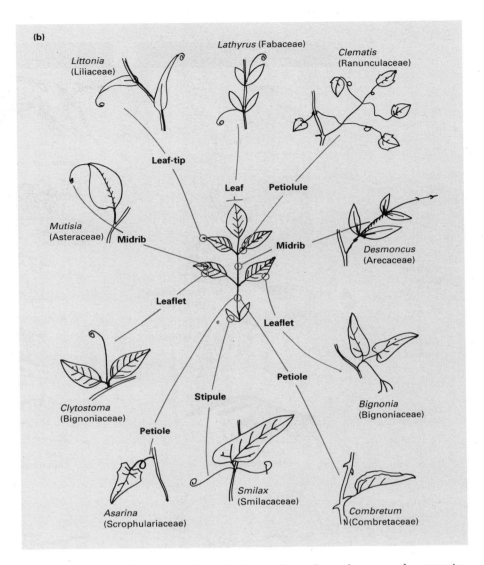

The cultivated sunflower is unbranched and bears large leaves and a massive terminal flowering capitulum very like a typical megaphyte.

Plant breeders have also produced a variety of megaphyte forms within the genus *Brassica*. From wild branched forms with relatively small leaves the plant breeder has produced forms with a huge inflorescence (cauliflower and broccoli), and these same forms have a stout, unbranched stem and very large leaves. In another cultivated *Brassica*, a mop of large leaves is at the top of an unbranched stem that is so strongly developed (2–2.5 m high) that it is used commercially to make walking sticks!

Both convergent and parallel evolution are often recognized by striking visual resemblances. The marsupial bandicoot looks rather like a placental rabbit (the long ears make the resemblance particularly striking), and both species excavate burrows, but the bandicoot is largely carnivorous, feeding on insect larvae, whilst the rabbit is a herbivore. By contrast, kangaroos and sheep look remarkably different although both are large grazing herbivores which, when they live side-by-side, have very similar diets (Griffiths & Barker, 1966). Root (1967) introduced the word 'guild'

different species may look alike but have very different ecologies

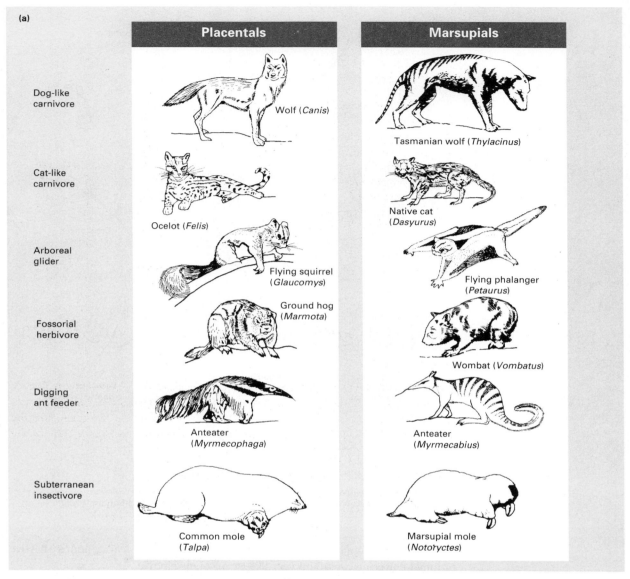

Figure 1.12 (a) Parallel evolution of marsupial and placental mammals. The pairs of species are similar in both appearance, habit and usually (but not always) in lifestyle. (b) (*facing page*) Phylogeny of marsupial mammals suggested by analysis of DNA sequences in protamine P1 genes from 21 mammalian species representing all extant orders of marsupials. Comparable analyses for two placental (eutherian) mammals, mouse and human, and the monotreme platypus (*Ornithorhynchus anatinus*) are also included. None of the marsupial protamines contains cysteine residues in contrast with those of placentals which contain six to nine cysteines. The branch length scale is substitutions per site and the asterisks indicate internal lengths that are significantly positive (*P* < 0.01). (After Retief *et al.*, 1995, who also give examples of alternative phylogenies suggested by different methods.)

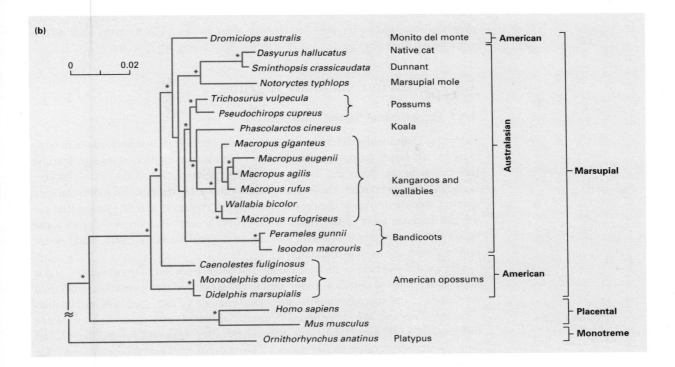

different species may look very different but have very similar ecologies (belong to the same guild)

to describe a group of species that exploit the same class of environmental resources in a similar way. The kangaroo and the sheep are members of the same guild, but look very different. The rabbit and the bandicoot look very similar but belong to quite different guilds. There is an important sense in which an ecologist's view of the

Table 1.1 Megaphytes or rosette treelets and their habitats.

Tree ferns	Tropical and subtropical forest— usually understorey
Cycads	Tropical and subtropical forest— usually understorey
Angiosperms	
Monocotyledons	
Palms	Tropical and subtropical forest— often understorey
Liliaceae and Amaryllidaceae, e.g. *Yucca, Agave, Aloe, Cordyline, Dracaena*	Tropical and subtropical—many in arid zones and deserts but others forming understorey in tropical and subtropical forest
Xanthorrhoeaceae (grass trees)	Arid regions—exposed or understorey
Dicotyledons	
Boraginaceae, e.g. *Echium*	Canary Islands
Campanulaceae, e.g. tree *Lobelias*	Hawaii and African mountains
Compositae, e.g. *Espeletia, Senecio*	High Andes and African mountains

match of organisms to their environment may be reflected in their lifestyle (e.g. what they feed on and what feeds on them), than by an evolutionist's view which emphasizes their phylogeny.

1.4 Patterns in community structure

1.4.1 The terrestrial biomes of the earth

biomes

Before we examine the differences and similarities between communities we need to consider the larger groupings, 'biomes', in which biogeographers recognize marked differences in the flora and fauna of different parts of the world. These biomes occupy the areas between the frozen wastes of the polar regions and mountain tops and the drought deserts of the continental tropics. They extend from regions that are too dry because most of the water is frozen, to regions that are too dry at some seasons because water evaporates too fast. In tropical rainforest, both liquid water and radiation are usually abundant.

eight terrestrial biomes ...

The number of biomes that are recognized differs between biogeographers, and it is rather a matter of taste how many should be recognized. They certainly grade into one another, and sharp boundaries are a convenience for cartographers, rather than a reality of nature. We describe eight terrestrial biomes and illustrate their global distribution in Figure 1.13, and show how they may be related to annual temperature and precipitation (Figures 1.14 and 1.15). See also Woodward (1987) for a more detailed account.

Tundra (Plate 1, facing p. 20) occurs around the Arctic Circle, beyond the tree line. Small areas also occur on sub-Antarctic islands in the Southern hemisphere. Alpine tundra is found under similar conditions, but at high altitude (including some tropical mountains) rather than high latitude. The environment is characterized by the presence of permafrost, water permanently frozen in the soil. Liquid water is present for only short periods of the year. The typical flora includes lichens, mosses, sedges and dwarf trees. Insects are extremely seasonal inhabitants and the native bird and mammal fauna is enriched by species that migrate from warmer latitudes in the summer.

Taiga or northern coniferous forest (Plate 2, facing p. 20) occupies a broad belt across North America and Eurasia. Liquid water is unavailable for much of the winter, and plants and many of the animals have a conspicuous winter dormancy in which metabolism is very slow. The trees are mostly evergreen conifers and vast tracts of land are commonly occupied by just one or two species of tree.

Temperate forests (Plate 3, facing p. 20) range in type from the mixed conifer and broad-leaved forests of much of North America and northern central Europe (where most of the habitat has been modified by humans), to the moist dripping forests of broad-leaved evergreen trees found, for example, in Florida and the southern tip of the South Island, New Zealand. In most such temperate forests there are periods of the year when liquid water is in short supply, because potential evaporation exceeds precipitation plus the water available from the soil.

Grassland occupies somewhat drier parts of temperate and tropical regions. Temperate grassland has many local names: prairie, steppe, pampas, veld. Tropical grassland or savannah (Plate 4, facing p. 20) is the name applied to tropical vegetation ranging from pure grassland to some trees with much grass. Almost all of these temperate and tropical grasslands experience seasonal drought, but the role of

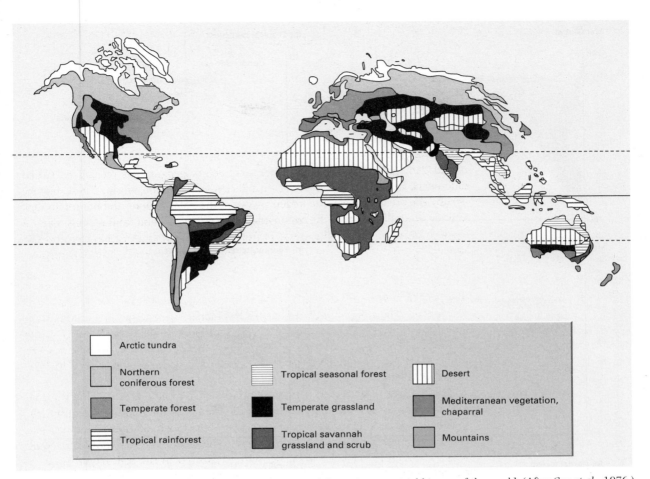

Figure 1.13 Distribution of the major terrestrial biomes of the world. (After Cox *et al.*, 1976.)

climate in determining their vegetation is almost completely overridden by the effects of grazing animals which limit the species present to those that can recover from frequent defoliation.

Chaparral occurs in Mediterranean-type climates (mild, wet winters and summer drought) in Europe, California and north-west Mexico, and in a few small areas in the Southern hemisphere in Australia, Chile and South Africa. Chaparral develops in regions with less rainfall than temperate grasslands and is dominated mainly by drought-resistant, hard-leaved scrub of low-growing woody plants.

Deserts (Plate 5, facing p. 20) are found in areas that experience extreme water shortage: rainfall is usually less than about 25 cm year^{-1}, is usually very unpredictable and is considerably less than potential evaporation. The desert biome spans a very wide range of temperatures, from hot deserts, such as the Sahara, to very cold deserts, such as the Gobi in Mongolia.

Tropical rainforests (Plate 6, facing p. 20) are very species-rich communities that rarely suffer from a shortage of liquid water, and certainly not because it is ever frozen! The annual rainfall is greater than 200 cm, and characteristically at least 12 cm falls in even the driest month. This biome occurs between the tropics of Cancer and Capricorn.

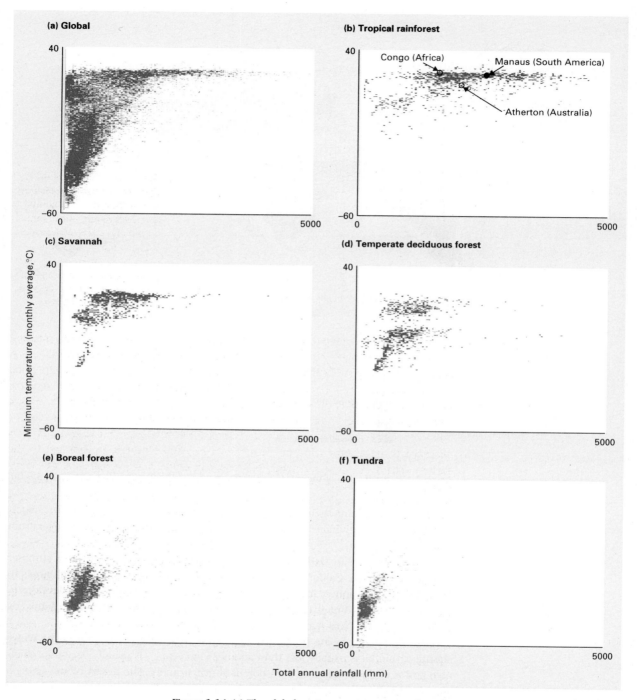

Figure 1.14 (a) The global environment represented on the two dimensions of monthly average minimum temperature and total annual rainfall, and (b)–(f) the global environment partitioned according to five major geographical biomes. (After Heal *et al.*, 1993, © UNESCO 1995.)

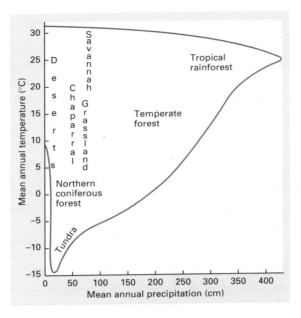

Figure 1.15 The distribution of eight major terrestrial biomes with respect to mean annual temperature and mean annual precipitation. It is not possible to draw sharp boundaries between biome types—local climatic effects, soil types and fire can each shift the balance. (After Whittaker, 1975.)

... and two aquatic biomes

All of these biomes are terrestrial. We can also recognize two aquatic biomes, *marine* and *freshwater*. The oceans cover about 71% of the earth's surface and reach depths of more than 10 000 m. They extend from regions where precipitation exceeds evaporation to regions where the opposite is true. There are massive movements within this body of water that prevent major differences in salt concentrations developing (the average concentration is about 3%). Two main factors influence the biological activity of the oceans. Photosynthetically active radiation is absorbed in its passage through water and so photosynthesis is confined to the surface region. Mineral nutrients, especially nitrogen and phosphorus, are commonly so dilute that they limit the biomass that can develop. Shallow waters (e.g. coastal regions and estuaries) tend to have high biological activity, because they receive mineral input from the land and less incident radiation is lost than in passage through deep water. Intense biological activity also occurs where nutrient-rich waters from the ocean depths come to the surface; this accounts for the concentration of many of the world's fisheries in Arctic and Antarctic waters.

Freshwater biomes occur mainly on the route from land drainage to the sea. The chemical composition of the water varies enormously, depending on its source, its rate of flow and the inputs of organic matter from vegetation that is rooted in or around the aquatic environment. In water catchments where the rate of evaporation is high, salts leached from the land may accumulate and the concentrations may far exceed those present in the oceans and form brine lakes or even salt pans in which little life is possible. Even in aquatic situations liquid water may be unavailable, as is the case in polar regions.

The differences between biomes allow only a very crude recognition of the sorts of differences and similarities that occur between communities of organisms. Within biomes there are both small- and large-scale patterns of variation in the structure of communities and in the organisms that inhabit them. It has proved quite difficult to devise ways in which even very obvious similarities and differences between communities can be described and measured.

31   ORGANISMS AND THEIR ENVIRONMENTS

1.4.2 The 'life-form' spectra of communities

Only in the rarest situations do we find natural communities occupied by just one unique species. (This is when there is one overwhelmingly important physical factor in the environment, so that, for example, a curious species of square bacterium is often the sole occupant of concentrated brine (Walsby, 1980).) In the vast majority of cases, natural communities contain a spectrum of life forms and biologies. A consequence of evolutionary divergence is that taxonomically closely related species may live in different environments. A consequence of evolutionary convergence is that taxonomically unrelated species may live in the same environment. For these two reasons a list of the names of species, genera or even families present in two areas may tell us virtually nothing about how similar the areas are ecologically.

A classic vegetation type such as the 'Mediterranean' chapparal can be recognized around the Mediterranean itself, in California, Chile, South Africa and south Australia. These are all areas with very similar climates. The similarity in vegetation is apparent in aerial photographs or during a fast car ride through these areas. Yet, the taxonomist's list of the species present (even the families they represent) gives no indication of the similarity. Often, it is the diversities in the 'architectures' of the various plants present that give the vegetation its character and it is difficult to devise ways to quantify this diversity. Our vocabularies are quite inadequate and we usually content ourselves with loose, qualitative terms to describe plant form, such as shrub, hummock, tussock, sward and scrub.

classification of organisms by their ecologies

However, serious attempts have been made to develop and refine the ways in which the life and form of higher plants can be described, irrespective of their taxonomy and systematics. Still the simplest, and in many ways the most satisfying, classification of plant forms that disregards their systematics is that of the Danish botanist Raunkiaer (1934). He emphasized that the growth of higher plants depends on the initiation of tissues at apices (meristems), and he classified plants according to their 'life forms' defined by the way in which these meristems were held and protected (Table 1.2). The relative frequency of different life forms is then used to construct spectra from the flora of different regions. He compared these with a 'global spectrum' obtained by sampling from a compendium of all species known and described in his time (the *Index Kewensis*), biased by the fact that the tropics were, and still are, relatively unexplored (Figure 1.16). Striking differences between vegetation types are exposed in this way which demonstrate not only matches of organisms to environments but of whole community complexes.

the life-form spectra of vegetation . . .

Raunkiaer's methods are clearly a step on the way to a definitive analysis of the ecological diversity of communities, and attempts have been made to improve on Raunkiaer's methods (e.g. see Box, 1981; Barkman, 1988). Most of these share a number of important problems: the difficulties of pigeon-holing species, and the subjective nature of the criteria by which they and their environments are classified. Objective statistical methods are, however, being developed for classifying plants into ecologically relevant groups. A multivariate analysis was made of the distribution of 43 traits amongst 300 species from semi-arid woodlands in western New South Wales, Australia (Leishman & Westoby, 1992). These included vegetative characters and features of life history, phenology and seed biology. Just five main groups emerged from the analysis: (i) perennial forbs and C_3 grasses; (ii) subshrubs of the Chenopodiaceae; (iii) perennial C_4 grasses; (iv) trees and shrubs; and (v) annual forbs and grasses, and there was no evidence of any other significant natural grouping. We

Table 1.2 The life-form classification of Raunkiaer (1934).

Phanerophytes	The surviving buds or shoot apices are borne on shoots which project into the air (a) Evergreens without bud covering ⎫ (b) Evergreens with bud covering ⎬ More than 2 m high (c) Deciduous with bud covering ⎭ (d) Less than 2 m high
Chamaephytes	The surviving buds or shoot apices are borne on shoots very close to the ground (a) Suffruticose chamaephytes, i.e. those bearing erect shoots which die back to the portion that bears the surviving buds (b) Passive chamaephytes with persistent weak shoots that trail on or near the ground (c) Active chamaephytes that trail on or near the ground because they are persistent and have horizontally directed growth (d) Cushion plants
Hemicryptophytes	The surviving buds or shoot apices are situated in the soil surface (a) Protohemicryptophytes with aerial shoots that bear normal foliage leaves, but of which the lower ones are less perfectly developed (b) Partial rosette plants bearing most of their leaves (and the largest) on short internodes near ground level (c) Rosette plants bearing all their foliage leaves in a basal rosette
Cryptophytes	The surviving buds or shoot apices are buried in the ground (or under water) (a) Geocryptophytes or geophytes which include forms with: (i) rhizomes; (ii) bulbs; (iii) stem tubers; and (iv) root tubers (b) Marsh plants (helophytes) (c) Aquatic plants
Therophytes	Plants that complete their life cycle from seed to seed and die within a season (this group also includes species that germinate in autumn and flower and die in the spring of the following year)

now need comparable analyses for a range of different communities so that we can begin to classify and compare them objectively and so make quantitative measures of the match between communities and environment.

... and of mammals

Studies of the match between animal communities and their environment are rare, but a fascinating attempt has been made to quantify the ecological diversity (life-form spectra) of mammalian faunas (see Andrews *et al.*, 1979, who used the information to make some predictions about the environments of fossil faunas). They classified the mammals of forest communities in Malaya, Panama, Australia and Zaire according to their means of locomotion: (i) aerial (bats and flying squirrels); (ii) arboreal (species found on small branches in the upper canopy, e.g. primates); (iii) scansorial (clawed species of large branches such as squirrels); and (iv) small ground mammals (species living mainly on the ground, although coming a little way into the lower canopy). They also classified the mammals according to their feeding habits: (i) plant eaters (herbivores and fruit eaters (fructivores) (HF)); (ii) insectivores (I); (iii) carnivores (C); and (iv) mixed feeders (M). The percentages of species falling into the various categories were then presented in graphical form as 'life-form' spectra (Figure 1.17).

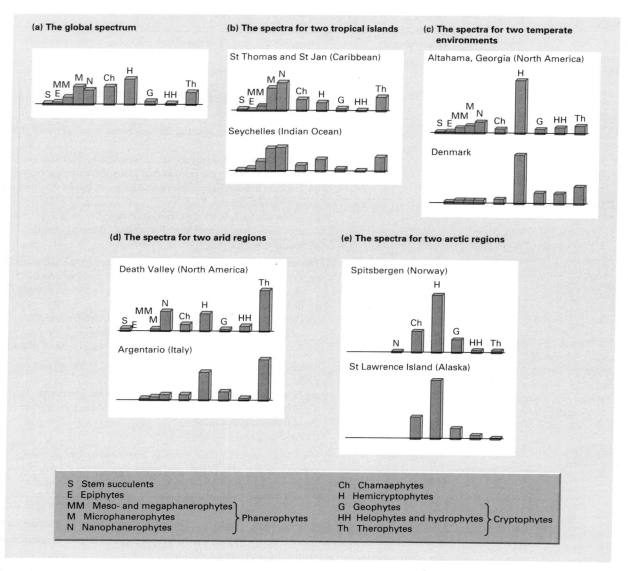

Figure 1.16 Whole plant communities can be compared by the method of Raunkiaer (1934). (a) The global life-form spectrum obtained by sampling from *Index Kewensis*. (b) Life-form spectra for the vegetation of two tropical islands. (c) Life-form spectra for two temperate environments. (d) Life-form spectra for two arid environments. (e) Life-form spectra for two arctic regions. The various categories of 'life form' are described in Table 1.2.

The spectra from the four continents are strikingly similar. Particularly remarkable is the very close similarity of the ecological diversity spectra for the Australian and Malayan forests, bearing in mind that the fauna of Australia is marsupial and that of Malaya is placental. This is a spectacular confirmation that the parallel evolutionary divergence within marsupial and placental phylogenies has resulted in ecologically matched faunas (see Figures 1.12a, b). Ecological diversity spectra could readily be constructed for other animal groups and prove very rewarding—although it must be remembered that constructing 'life-form' spectra depends on essentially

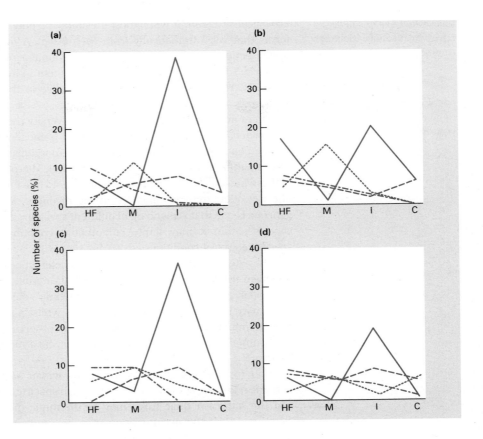

Figure 1.17 The percentages of forest mammals in various locomotory and feeding habit categories in communities in: (a) Malaya, all forested areas, 161 species; (b) Panama dry forest, 70 species; (c) Australia, Cape York forest, 50 species; (d) Zaire, Irangi forest, 96 species. (After Andrews *et al.*, 1979.) C, carnivores; HF, herbivores and fructivores; I, insectivores; M, mixed feeders; (——), aerial; (– – –), arboreal; (- - -), scansorial; (– – -), small ground mammals.

subjective decisions about which properties best define a 'life form'. A further example of the comparison of patterns of life form can be found in a study of animal communities in Mediterranean environments by Cody and Mooney (1978).

Perhaps the most serious problem in defining and classifying communities and the groupings of species that compose them is that species are not truly independent points to be used in such an analysis. 'That two related species share a character state does not provide two pieces of independent evidence that a particular ecological attribute is responsible: the species share many other attributes as well' (Harvey & Pagel, 1991). For this reason it is argued that the phylogenetic history of species needs to be taken into account when they are used to describe and compare communities. There is a vigorous argument about this issue between Westoby *et al.* (1995) and Harvey (1995). A wider consideration of the uses of phylogenetic information is given in two symposia, one edited by Harvey *et al.* (1995) and another by Franco *et al.* (1996). The importance of phylogeny in studies of comparative ecology is obviously particularly important if evolutionary convergence, parallel evolution and phylogenetic constraint have all contributed, together with present-day attributes, to determine the nature of floras and faunas.

1.4.3 The diversity of matches within communities

Although a particular type of organism is often characteristic of a particular

species that live in similar
environments have much in
common ...

but, species that live in the
same environment are usually
very different

ecological situation, it will almost inevitably be only part of a diverse community of
forms that vary dramatically from each other. A fully satisfactory account of the
match between organisms and their environment must do more than identify the
similarities between organisms that allow them to live in the same environment—it
must also try to explain why species that live in the same environment are often
profoundly different.

There are many explanations for the diversity that exists within communities.

1 *Environments are patchy.* There are no homogeneous environments in nature.
Even a continuously stirred culture of microorganisms is heterogeneous because it
has a boundary: the walls of the culture vessel. Microbiologists are often made aware
of this when cultured organisms subdivide into two forms: one that sticks to the walls
and the other that remains free in the medium. An environment that is heterogeneous is one that is made up of different specialized environments. The diversity of
species within it may imply only that it contains organisms that match each
specialized set of conditions.

The extent to which an environment is heterogeneous depends on the scale of the
organism that senses it. To a mustard seed, a grain of soil is a mountain; and to a
caterpillar, a single leaf may represent a lifetime's diet. A seed lying in the shadow of
a leaf may be inhibited in its germination, whilst a seed lying outside that shadow
may germinate freely. What appears to the observing ecologist as a homogeneous
environment may, to an organism within it, be a mosaic of the intolerable and the
adequate.

2 *Most (perhaps all) environments contain within them gradients of conditions or of
available resources.* These may be gradients in space (e.g. altitude) or gradients in time,
and the latter, in their turn, may be rhythmic (like daily and seasonal cycles),
directional (like the accumulation of a pollutant in a lake) or erratic (like fires,
hailstorms and typhoons). Some of the diversity of species in a community may
represent specialization to different conditions along such gradients.

3 *The existence of one type of organism in an area immediately diversifies it for others.*
During its life an organism increases the diversity of its environment by contributing
dung, urine, dead leaves and ultimately its dead body, and so makes the
environment more patchy in resources and conditions. During its life its body may
serve as a place in which other species find homes or it may serve as a resource for
predators, pathogens and parasites that had no place in the community until it was
there.

The diversity within natural communities may be due to all three of these types
of heterogeneity. To some extent, this 'explanation' of diversity is a trivial exercise. It
comes as no surprise to anyone that a plant utilizing sunlight, a fungus living on the
plant, a herbivore eating the plant and a parasitic worm living in the herbivore
should all coexist in the same community. On the other hand, most communities also
contain a variety of different species that are all constructed in a fairly similar way
and all living (at least superficially) a fairly similar life. Understanding or explaining
this type of diversity is a far from trivial exercise.

the Antarctic seals and the
coexistence of similar species

The Antarctic seals illustrate some of the differences that exist between species
that occur together in the same community. It is thought that the ancestral seals
evolved in the Northern hemisphere, where they are present as Miocene fossils, but
one group of seals moved south into warmer waters and probably colonized the
Antarctic in the Late Miocene or Early Pliocene (about 5 million years ago). When
they entered the Antarctic, the Southern Ocean was probably rich in food and free

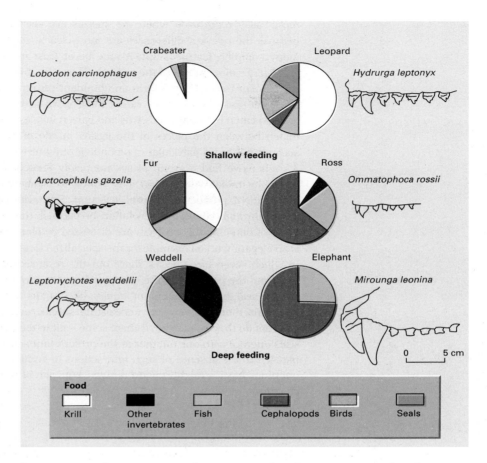

Figure 1.18 The food, feeding depth and jaw structure of a group of Antarctic seals. (After Laws, 1984.)

from major predators, as it is today. It was within this environment that the group appears to have undergone radiative evolution (Figure 1.18). For example, the Weddell seal feeds primarily on fish and has unspecialized dentition; the crabeater seal feeds almost exclusively on krill and its teeth are suited to filtering these from the seawater; the Ross seal has small, sharp teeth and feeds mainly on pelagic squid; and the leopard seal has large, cusped, grasping teeth and feeds on a wide variety of foods, including other seals and, in some seasons, penguins. Indeed, the differences between these seal species could possibly determine the extent to which they coexist in mixed communities. Where two or more species feed on similar diets at similar depths (e.g. the elephant and Ross seals), they are separated geographically with only slight overlap. However, when mixed communities of several seal species occur, they comprise species that exploit different food organisms (crabeater, Ross and leopard, or leopard and elephant, or Weddell and leopard—Figure 1.18).

The interpretation of patterns such as this (or at least as full an interpretation as we can currently achieve) must be postponed until coexistence and competition are discussed in later chapters. Even at this early stage, though, a number of important points can be made.

1 Coexisting species commonly differ in subtle ways, but each matches its environment and finds a style of life within it because the environment itself is heterogeneous.

2 The differences in the ecology of species that coexist within communities are most easily recognized when the species are closely related (as with the seals), because the relevant differences are most clearly exposed against a background of many common features. But, it may be at least as important to understand the coexistence of members of the same guild (see p. 25) that are not necessarily closely related. For instance, krill form an abundant resource in Antarctic waters, and the crabeater seal and to a lesser extent the leopard seal specialize on this diet. But krill are also eaten by squid, fish, birds and baleen whales, and a complete analysis of the match between the pieces in the jigsaw puzzle of such a community must take account not just a particular taxonomic group but whole guilds.

3 When we find a group of systematically related species that have apparently arisen by radiative evolution it is wrong to treat each as if it has arisen as an independent product of natural selection. Similarities between species may have arisen through convergent evolution or because they have ancestors in common. Ways of handling this problem are discussed by Harvey and Pagel (1991) and there is an elegant worked example for the specialized foraging of species of tits (*Parus* spp.) in which seven bird species share out the resources of conifers and broad-leaved trees and the positions in the canopy where they feed (Suhonen *et al.*, 1994).

The seal example and many others, like the tits referred to above, provide more questions than answers. Are coexisting species necessarily different? If so, how different do they need to be: is there some limit to their similarity? Do species like the seals interact with one another at the present time, or has radiative evolution in the past led to the absence of such interactions in contemporary guilds? None of these questions has a straightforward answer, but each of them will be addressed in later chapters.

1.5 Specialization within species

Up to this point, we have been considering species (or larger taxa) as the units that match the environment, but it is within species that some of the most specialized fits to the environment can be traced. When islands were discussed, it was stressed that the homogenizing effects of genetic exchange (and recombination) tend to mean that most of the more obvious heterogeneities in a population arise only when parts become isolated geographically and cease to interbreed. However, if the local forces of natural selection are very powerful, they can override the homogenizing forces of sexual reproduction and recombination. Then, even where there is some interbreeding, locally favoured genotypes may be at such an advantage that ill-favoured combinations are continually eliminated. Gene flow continues to occur, populations remain part of the same species, but local specialized races appear within it.

The exchange of genetic material through a population depends on the mobility of whole organisms or, in the case of sessile organisms, on the mobility of their gametes, pollen or seeds. If the members of a population wander freely in search of mates (or their gametes, spores or seeds are spread widely), then local, special forces of natural selection need to be stronger to produce local, special matches between the organisms and their local environments.

Local, specialized populations become differentiated most conspicuously amongst organisms that are sessile for most of their lives. Motile organisms have a large measure of control over the environment in which they live; they can recoil or retreat from a lethal or unfavourable environment and actively seek another. However, the sessile higher plants, for example moss, seaweed and coral, have no

sessile organisms must match their environments ...

such freedom. After dispersal they must live, or die, in the conditions where they settle. The most that a higher plant can do is to search out resources or escape from an unfavourable site by growing from one place to another; it can never uproot itself and choose to transplant itself elsewhere. Its descendants (seeds, pollen or gametes) are hazarded to the vagaries of passive dispersal on the wind or water, or in or on the bodies of animals. Populations of non-mobile organisms are therefore exposed to forces of natural selection in a peculiarly intense form.

... but, mobile organisms can match their environment to themselves

The contrast between the ways in which mobile and fixed organisms 'match' their environment is seen at its most dramatic on the seashore, where the intertidal environment continually oscillates between being terrestrial and being aquatic. The fixed algae, hydroids, sponges, bryozoans, mussels and barnacles all meet and tolerate life in the two extremes. But, the mobile members of the community, for example the shrimps, crabs and fish, travel with and track their aquatic habitat as it moves, whilst the shore-feeding birds move back and forth, following the advance and retreat of their terrestrial habitat. The fixed organisms have to tolerate the whole daily cycle of change in their environment, but those that are mobile have no need for such tolerance—they move with the tides. There is a sense in which the match of such mobile organisms to their environment enables them to escape many of the forces of natural selection. Mobility enables the organism to match its environment to itself. The immobile organism must match itself to its environment.

1.5.1 Ecotypes

ecotypes

The word 'ecotype' was first coined for plant populations (Turesson, 1922) to describe genetically determined differences between populations within a species that reflect local matches between the organisms and their environments. The existence of different ecotypes within a species was demonstrated by transplanting plants from a variety of natural habitats into a common habitat, and allowing them to develop through one or more growing seasons. It is essential in such studies that the plants from the different places of origin should be grown and *compared in the same environment*, since some of the differences between field populations may be environmentally determined phenotypic responses that are not due to differences in genotype. The kinds of differences between races of a species that have been recognized as ecotypes are illustrated in Table 1.3.

ecotypic specialization can be on a very fine spatial scale ...

The extremely fine scale over which local specialization can occur amongst plants has been shown in studies of tolerance to toxic heavy metals (lead, zinc, copper, and so on) (Antonovics & Bradshaw, 1970). At the edge of areas that have been contaminated after mining activities, the intensity of selection against intolerant genotypes changes abruptly, and populations on contaminated areas may differ sharply in their tolerance of heavy metals over distances of less than 100 m (in sweet vernal grass (*Anthoxanthum odoratum*), populations are differentiated over less than 1.5 m). This occurs in spite of the fact that pollen is freely borne across such boundaries by wind or insects.

heavy metal tolerant ecotypes

and may evolve very fast

In some cases, it has been possible to date the time over which such selection has occurred. Heavy metal tolerance has apparently evolved in Swansea (South Wales) since mining started 300 years ago; zinc tolerance has evolved beneath a galvanized iron fence within 25 years; in an experiment a single generation of selection has been sufficient to establish a metal-tolerant population when a high density of seed was applied to a heavily contaminated soil. Both the speed of selection and the sharpness

Table 1.3 Examples of differences between ecotypes in higher plants.

Aspects of form or life cycle	Difference between ecotypes	Investigator(s) associated with the studies
Growth form	Contrast between upright and prostrate *Plantago maritima*	Gregor (1930)
Water requirements } Earliness of growth }	Contrasts between alpine, subalpine and lowland forms of *Poa alpina*	Turesson (1922)
Annual growth cycle	Contrasts between forms along an altitudinal gradient with different flowering and fruiting times, e.g. *Hemizonia*, *Achillea*	Clausen *et al.* (1941)
Longevity and vegetative vigour	Contrasts between grasses of various species in managed grasslands	Stapledon (1928)
Flowering time	Geographic differences in photoperiodic requirement for flowering	McMillan (1957)
Nutritional responses	Variation within species of grass and white clover in response to nitrogen, phosphorus, calcium, etc.	Snaydon and Bradshaw (1969)
Tolerance of toxic metals	Contrasts between forms on heavy metal mine spoils and elsewhere	Antonovics and Bradshaw (1970)
Chill and freeze tolerance Acclimation	Contrasts between populations of *Opuntia fragilis* in northern USA and Canada	Loik and Nobel (1993)

of the boundaries show how powerless is the mixing power of gene flow against very strong forces of natural selection.

1.5.2 Genetic polymorphism

On a finer scale even than ecotypes, biologists are increasingly able to detect levels of selectively relevant variation within small local populations. Such variation is known as polymorphism. Specifically, genetic polymorphism is 'the occurrence together in the same habitat of two or more discontinuous forms of a species in such proportions that the rarest of them cannot merely be maintained by recurrent mutation or immigration' (Ford, 1940). Not all such variation represents a match between organism and environment. Indeed, some of it may represent a mismatch. Mismatches may arise through the dispersal of propagules from one specialized form into the habitat of another. They may also arise if conditions in a habitat have changed so that one form is in the process of being replaced by another. Such polymorphisms are called transient. As all communities are always changing, a lot of the polymorphism that we observe in nature may be of this transient character, representing the extent to which the genetic response of populations to environmental change will always lag and be out of step with the environment and unable to anticipate change.

transient polymorphisms and mismatches

Many polymorphisms, however, are actively maintained in a population by natural selection, and there are a number of ways in which this may occur.

the active maintenance of genetic polymorphism

1 *Heterozygotes may be of superior fitness*, but because of the mechanics of Mendelian genetics they continually generate less fit homozygotes within the population. Such 'heterosis' is seen in human sickle-cell anaemia where malaria is prevalent. Sickle-cell heterozygotes suffer only slightly from anaemia and are little affected by malaria; but they continually generate homozygotes that are either dangerously anaemic or susceptible to malaria.

2 *There may be gradients of selective forces* favouring one form (morph) at one end of the gradient, and another form at the other. This can produce polymorphic populations at intermediate positions in the gradient.

3 *There may be frequency dependent selection* in which each of the morphs of a species is fittest when it is rarest (Clarke & Partridge, 1988). This is believed to be the case when rare colour forms of prey are fit because they go unrecognized and are therefore ignored by their predators.

4 *Selective forces may operate in different directions within different patches* of a fine mosaic in the population. In this case the maintenance of a fit between organisms and environment has to depend on large numbers of propagules being dispersed, so that each type has a reasonable chance of landing in the part of the patchwork in which it is fittest. Alternatively, the organism may be long lived and able to wander between and exploit the appropriate parts of the patchwork.

polymorphism in a field of clover

A striking example of polymorphism in a natural population is provided by a series of studies of white clover (*Trifolium repens*) in a 1 ha field of permanent grassland (at least 80 years old) in North Wales. In one study, 50 plants of this species were sampled from the field. The plants were grown in a glasshouse, and each plant was 'fingerprinted' with respect to the genes it carried for a number of characters, almost all of which were of known selective importance (Figure 1.19). Amongst the 50 clones, almost all differed from each other in the combination of qualities that might be expected to affect the fitness of plants in the field.

Turkington and Harper (1979) attempted to test how far the different variants in the clover population represent local matches with local features of the field. They removed plants from the field, multiplied them in the glasshouse and then transplanted them back into the field. They returned a sample of each clone to the place from which it had been taken, and also to the places from which the others had come. Samples had been taken from patches dominated by different species of pasture grass. Each clover plant was transplanted back into patches dominated by its own grass species and into patches of the other grass species. Each type of clover grew best when it was returned to the species of grass neighbour from which it had been taken. This was strong and direct evidence that clover genotypes in the pasture were distributed in such a way that they matched their local biotic environment. It was shown later that this very local match could also be detected amongst the grasses at the site (Gliddon & Trathan, 1985).

In the same field, Dirzo and Harper (1982) studied the distribution of a specific polymorphism in white clover that controls whether the plant releases hydrogen cyanide when it is damaged (e.g. nibbled). Molluscs are known to avoid eating cyanogenic clover, and a highly significant association was found within the field between the distribution of cyanogenic forms and areas of high mollusc density (Figure 1.20). Clearly, some at least of the variation in white clover within this small field is genetic, and represents a match between organisms and environment.

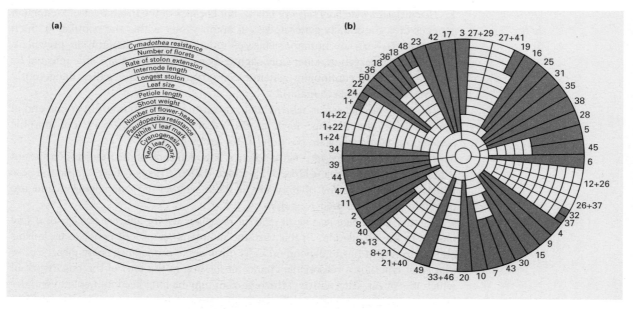

Figure 1.19 The diversity of genotypes of 50 plants (1–50) of white clover (*Trifolium repens*) sampled from a 1-ha (1×10^4 m^2) field of permanent pasture. The variety of factors examined and under known genetic control is shown in (a). In (b) we start at the centre with 50 plants and each genetic character is used in turn to subdivide this population. Thus, the initial population is divided into approximately equal numbers with and without a red-leaf mark. Each of these populations is then further subdivided into four classes according to the genotype controlling hydrogen cyanide production. The process continues to the outermost circle, which divides resistance from and susceptibility to the disease, *Cymadothea*. Each unique genotype identified in this way is represented by a shaded sector. (After Burdon, 1980.)

1.6 The match of organisms to varying environments

No environments are constant over time, but some are more constant than others. No form or behaviour of an organism can match a changing environment unless it too changes. Three major categories of environmental change can be recognized.

1 *Cyclic changes*—rhythmicly repetitive, like the cycles of the seasons, the movements of the tides and the light and dark periods within a day.

2 *Directional changes*—in which the direction of a change is maintained over a period that may be long in relation to the life span of the organisms that experience it. Examples are the progressive erosion of a coastline, the progressive deposition of silt in an estuary and the cycles of glaciation.

3 *Erratic change*—this includes all those environmental changes that have no rhythm and no consistent direction, for example the erratic course and timing of hurricanes and cyclones, flash storms and fires caused by lightning.

The optimal fit of organisms to varying environments must involve some compromise between matching the variation and tolerating it. The repeated experience of cyclic change by successive generations of an organism has selected many patterns of behaviour that are in themselves cyclic: diapause in insects, the annual shedding of leaves from deciduous trees, the diurnal movements of leaves, the rhythm of tidal movement in crabs, the annual cycle of breeding systems and the

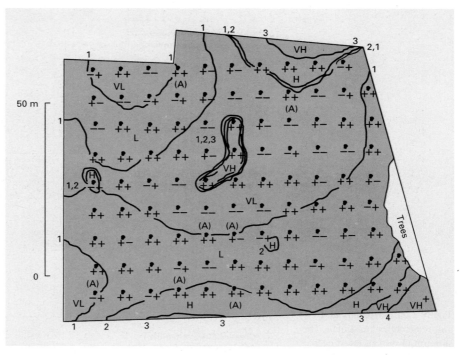

Figure 1.20 A map of the distribution of the slug *Dendroceras* in a small field of permanent grassland. The contours (1, 2, 3) represent the density of slugs and distinguish zones with very high (VH), high (H), low (L) and very low (VL) densities. On the same map are shown the genotypes of white clover sampled at intersections of a grid across the field. Plants are distinguished as + + (AcLi), + – (Acli), – + (acLi) and – – (acli); (A = ambiguous result). Only + + plants release hydrogen cyanide when they are damaged; they carry the alleles Ac and Li, which determine the production of the cyanogenic glycoside and the enzyme that releases hydrogen cyanide from it. χ^2 test of the null hypothesis that clover genotypes and the density of slugs are unrelated firmly rejects the hypothesis. (After Dirzo & Harper, 1982.) The three outlined islands of high slug density are associated with colonies of *Urtica dioica*, which provided refuge for the slugs.

seasonal cycle of fur colour and/or thickness in mammals (see Figure 2.16b) and of plumage in birds.

organisms may respond directly to change or use a cue

There are two main ways in which organisms time their responses to cyclic changes in their environment: (i) by changing in response to the environmental change; or (ii) by using a cue that anticipates the change. If the cycle in conditions is weak and contains much variation, the organisms may best match the changing conditions by responding to them directly. Figure 1.21 illustrates this unpredictability of rainfall, even at locations where there is a marked seasonal rhythm. This variation can be seen most strongly and dramatically in the bottom diagram, which shows rainfall figures for Cairo in January. Under such circumstances a plant that is cued to germinate at the start of the 'rainy season' would rarely survive. Instead, most desert plants germinate in direct response to the arrival of rain, rather than responding to some cue that indicates that the rainy season is coming.

There is, however, a disadvantage, a price to be paid by organisms that respond directly to environmental change. A mammal that changes the thickness of its coat as a reaction to the weather becoming cold (see Chapter 2, Figure 2.16b) will have to

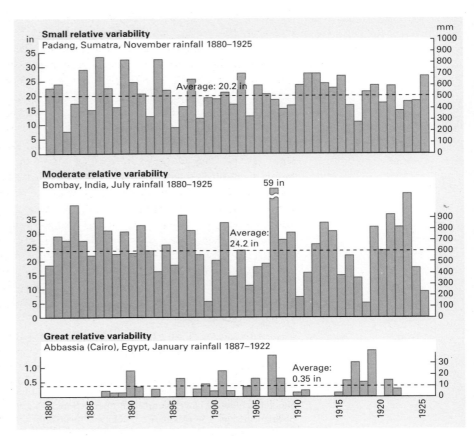

Figure 1.21 Variability and unpredictability in rainfall. The amount of rain received each year at three locations, during the month which, on average, is the rainiest at that location. (After Stahler, 1960; data from H.H. Clayton, Smithsonian Institute.)

shiver until the process of replacing its coat has taken place. But, if it reacts not to the onset of cold, but to an environmental cue that is correlated with, and therefore predicts, the onset of cold, such as the shortening of day length, it may start to develop the thicker coat in advance of the event. The use of such cues is common amongst both plants and animals that live in environments with a strong and repeated cycle of environmental change, and where the variation in the cycle is relatively weak.

There is a temptation to write about organisms as if they make predictions (or about environments having predictability), and about cues that enable the organism to plan ahead. Patterns of cyclic behaviour give the appearance of being predictive—but, of course they are not. They are properties selected by the repeated experiences of the past. The form and behaviour of an organism is programmed by its ancestors' experience of life and death in the past, rather than by its own anticipation of the future.

Events like the darkness of night and the cold of winter are tied to the passage of time, and they can be measured and timed by using the sun itself as a clock (the height of the sun during the day, or the length of the photoperiod during a year). Some organisms have added to this external clock their own internal physiological clock (an 'endogenous rhythm'). Such clocks require to be set right repeatedly by reference to the solar clock, but they may continue for long intervals without such a reference.

the way change is experienced
depends on the length of the
life cycle

The way in which any organism experiences the rhythmic changes of the environment depends on the length of its life cycle. One cell in a multiplying population of a unicellular algae like *Chlorella* may live over only one cycle of day and night before dividing to form two new cells. An oak tree, an elephant, a whale or a human includes within one life cycle the repeated experiences of spring, summer, autumn and winter, and the associated famines and gluts in the seasonal rarity and abundance of food. If the life cycle is short, the organism may spend the whole of its active life concentrated in a small segment of the year, and may be constrained within a sharply defined programme of development and behaviour restricted to one short season. The gall wasps have just this type of life, emerging at a time that is tightly synchronized with the availability of the leaf or flower of their particular host species. If, on the other hand, the life cycle is long relative to the annual cycle of the seasons, the organism cannot so easily become specialized. It may endure the periods of drought or cold by aestivating or hibernating, sleeping if it is a dormouse or dropping its leaves if it is a deciduous tree; or it may have an all-purpose physiology that merely continues faster or slower through the seasons (as does the polar bear or the evergreen tree). There must be a conflict in the evolution of most organisms between being active for a very short part of the year and tightly matching the environment of that part, or being a generalist, a 'man for all seasons' but perhaps hero of none.

The life cycles of most birds and mammals are longer than a calendar year, and many of the smaller mammals in seasonal climates pass through winter by hibernating: they slow down their rate of metabolism and rest in protected places (see Chapter 2, Section 2.5.1). However, there are many others that meet the variation in conditions by changing the extent to which their bodies are insulated. The most dramatic changes of fur thickness are found in the larger mammals (Hart, 1956). The winter fur of the black bear has an insulating power 92% greater than that of its summer coat. This sort of cyclic change in thickness (and colour) of insulation is characteristic of the seasonal responses of animals that cannot escape from the adverse seasons; but by far the most conspicuous seasonal rhythms in the behaviour of mobile animals involve movement from place to place. This may be movement to a place of shelter, or migration to another climate (the annual migrations of reindeer, bison and many birds). But, of course, such movement is denied to the rooted higher plant or sessile animal, and it is in these organisms that some of the most profound seasonal rhythms can be seen, affecting most particularly their form.

For many aquatic plants, changes in the depth of the water represent a major seasonal cycle in the environment; but these seasonal rhythms may be affected by erratic change during serious drought or in flash floods. In the zoned habitat of the water's edge there is a variety of characteristic plant forms. At the margins, most plants bear aerial foliage that is photosynthetically inefficient or even damaged when it is submerged. In somewhat deeper water, species commonly bear floating foliage. In deeper water still, species have entirely submerged foliage with long, flexible, band-like leaves that are not damaged in a fast water-flow. In the deep water zone, there are species that have submerged, finely dissected foliage. There are many species that bear one or other of these specialized leaf types. In addition, there are many aquatic plants capable of bearing more than one sort of leaf on a single plant. Such somatic polymorphism allows these species a potentially wider range of water levels at which they can function effectively (Figure 1.22).

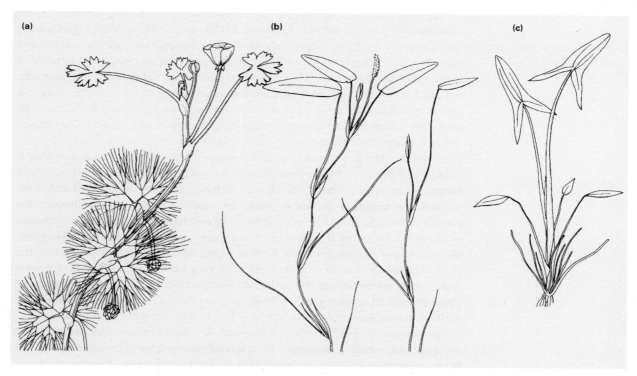

Figure 1.22 Leaf polymorphism in aquatic plants. (a) *Ranunculus aquatilis*. (Courtesy of S. Ross-Craig.) (b) *Potamogeton natans*. (Courtesy of W. Keble-Martin.) (c) *Sagittaria sagittifolia*. (Courtesy of S. Ross-Craig.)

Even within the single genus *Ranunculus*, subgenus *Batrachium* (the water crowfoots), various types of match to a varying environment can be found. There are species that are monomorphic, and match the water level that they most commonly experience. Two species, *Ranunculus hederaceus* and *R. omiophyllus*, bear only floating foliage and are usually found growing in mud or shallow water. Three species, *R. fluitans*, *R. circinatus* and *R. trichophyllus*, produce only submerged, finely dissected leaves. These species are usually found in deeper or fast-flowing water.

Within a further group of *Ranunculus* spp. there are differences in the ways by which the transition from one leaf form to another is initiated. In *R. aquatilis*, the type of leaf is determined by the photoperiod and the temperature at the time of leaf initiation, not by the level of the water but by a seasonal cue (e.g. see Bradshaw, 1965). The plant begins growing in spring, and grows for a time producing submerged leaves. It reaches the surface, but may continue extending its submerged leaves for some time. Then, in response to the stimulus of temperature and photoperiod, it suddenly produces floating leaves on the surface. Yet, if such plants are kept under a regime of short days, they continue to produce finely divided leaves, irrespective of whether they are submerged or exposed. In other species, for instance *R. flabellaris*, it is the prevailing conditions themselves that influence the type of leaf that is developed. The form of leaf (either floating or submerged and band-like) is determined by the conditions affecting the plant when the leaves are initiated, and if the environment suddenly changes, the plant finds itself with the 'wrong' leaf form when the water level falls.

For an organism that cannot run away from adverse conditions, seasonal changes in its form may be the most effective solution to problems of survival in a changing environment. In arid environments, somatic polymorphism may be even more extreme than in the case of aquatic plants. Some species produce three crops of leaves within 1 year, each of different morphology. In *Teucrium pollium*, relatively large leaves are formed in the wetter season. These fall off and are replaced by small leaves or scales during the drier season. This second foliage may itself be lost, and the plant then spends the driest period with only green stems and thorns (Orshan, 1963).

... and in desert plants

1.7 Pairs of species

Some of the most strongly developed matches between organisms and their environment are those in which one species has developed a dependence upon another. This is the case in many relationships between consumers and their foods, such as the dependence of koala bears on *Eucalyptus* foliage or giant pandas on bamboo shoots. Whole syndromes of form, behaviour and metabolism constrain the animal within its narrow food niche, and deny it access to what might otherwise appear suitable alternative foods. Similar tight matches are characteristic of the relationships between some parasites and their hosts (see Chapter 12). For example, the host-specific rust fungi fit narrow and precisely defined environments: their unique hosts.

Where two species have evolved a mutual dependence, the fit may be even tighter. The mutualistic association of nitrogen-fixing bacteria with the roots of leguminous plants, and the often extremely precise relationship between insect pollinators and their flowers are two good examples (see Chapter 13). The closest matches between organisms and their environments have evolved where the most critical factor in the life of one species is the presence of another: the whole environment of one organism may then be another organism.

When a population has been exposed to variations in the physical factors of the environment, for example a short growing season, a high risk of frost or drought or the repeated application of a herbicide, a once-and-for-all tolerance may ultimately evolve. The physical factor cannot itself change or evolve as a result of the evolution of the organisms. By contrast, when members of two species interact, the change in each produces alterations in the life of the other, and each may generate selective forces that direct the evolution of the other. In such a coevolutionary process the interaction between two species may continually escalate. What we then see in nature may be pairs of species that have driven each other into ever-narrowing ruts of specialization—an ever-closer match.

two species may 'coevolve' and sharpen the match of each to the presence of the other

1.8 The interpretation of nature

When we look at the diversity of nature it is difficult not to be overwhelmed by feelings of wonder, admiration and enchantment at what can so easily be interpreted as perfection. We need to remember that the abilities to wonder and admire are special features of our own biology! As a scientist the ecologist searches for causes and effects and must not be satisfied by 'explanations' that seek only to show how, at this moment, the match between organism and environment is 'the best in the best of all possible worlds'.

Chapter 2
Conditions

2.1 Introduction

In order to understand the distribution and abundance of a species we need to know many things: (i) its history (see Chapter 1); (ii) the resources that it requires (see Chapter 3); (iii) the individuals' rates of birth, death and migration (see Chapters 4 and 5); (iv) their interactions with their own and other species (see Chapters 6–13); and (v) the effects of environmental conditions. This chapter deals with the limits placed on organisms by environmental conditions.

We define a condition as an abiotic environmental factor which varies in space and time. Examples include temperature, relative humidity, pH, salinity and the concentration of pollutants. The 'conditions' of an environment also include a variety of hazards such as hurricanes and volcanoes and, especially in aquatic communities, destructive storms and turbulence. A condition may be modified by the presence of other organisms, for example temperature, humidity and soil pH may be altered under a forest canopy. But, unlike resources (see Chapter 3), conditions are not consumed or used up by organisms.

conditions (unlike resources) are not consumed

For some conditions we can recognize an optimum concentration or level at which an organism performs best, with its activity tailing off at both lower and higher levels (Figure 2.1a). But, we need to define what we mean by 'performs best' and this poses semantic and operational problems. For an evolutionary ecologist 'optimal' conditions are most likely to be those under which the individuals of the species leave most descendants (are fittest), but these are often quite impossible to determine in practice because measures of fitness should be made over several (ideally many) generations. Instead, we more often measure the effect of conditions on some chosen properties like the activity of an enzyme, the respiration rate of a tissue, the growth rate of individuals, their rate of reproduction or survivorship or on physiological states (Figure 2.1b). However, the effect of variation in conditions on these various properties will often not be the same; organisms can usually survive over a wider range of conditions than permit them to grow or reproduce (Figure 2.1a).

The precise shape of the curve of response to a condition—whether it is symmetrical or skewed, broad or narrow—will vary from condition to condition. The generalized form of response, shown in Figure 2.1a, is appropriate for conditions like temperature and pH in which there is a continuum from an adverse or lethal level (e.g. freezing or very acid conditions), through favourable levels of the condition to a further adverse or lethal level (heat damage or very alkaline conditions). A particular problem arises at very high temperatures which may increase the activity of enzymes

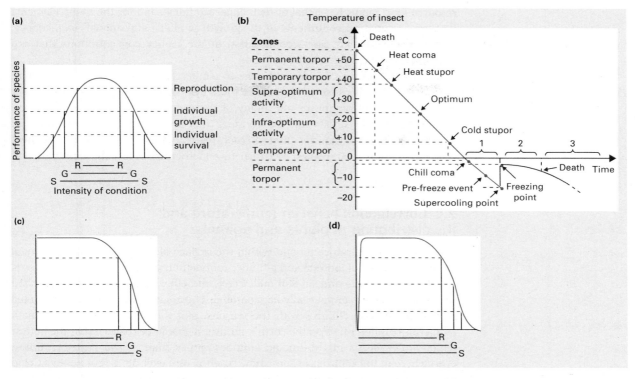

Figure 2.1 (a) A generalized graphical representation of the manner in which the activities of an organism may relate to the intensity of an environmental condition, such as temperature or pH. The narrow range over which reproduction can occur (R–R) usually dictates where continued existence of the species is possible (although the patterns of distribution of some species may be maintained by the repeated dispersal and recruitment of individuals into areas where they could not otherwise maintain themselves). (b) The thermobiology of an arthropod which does not tolerate freezing. Note that a symmetrical sequence of physiological states is recognized passing through stupor into temporary torpor and coma to permanent torpor and death as conditions become much hotter or cooler than the optimum. (After Vannier, 1987; Block, 1990.) (c) A generalized response curve of the form that relates the activities of an organism to the level, concentration or intensity of a condition (e.g. toxin, radioactive emission or pollutant which is harmful only at high levels). (d) As (c) but as the level, concentration or intensity of a condition is increased it changes from being an essential resource for growth at low levels to becoming damaging or lethal at higher concentrations (e.g. micronutrients such as copper and zinc).

but also the rate at which they become inactivated. A very short exposure to high temperature may then increase an activity but a longer exposure may be lethal.

There are many environmental conditions for which the response curve is better generalized by Figure 2.1c. This is appropriate for the response to most toxins, radioactive emissions and chemical pollutants, for which a low-level intensity or concentration of the condition has no detectable effect, but an increase begins to cause damage and a further increase may be lethal. There is also a quite different form of response curve to conditions that are toxic at high levels but essential for growth at low levels (Figure 2.1d). This is the case for sodium chloride—an essential

49   CONDITIONS

resource for animals but lethal at high concentrations—and for the many elements that are essential micronutrients in the growth of plants and animals (e.g. copper, zinc, manganese), but can become lethal at the higher concentrations that are sometimes caused by industrial pollution.

In this chapter we consider responses to temperature in much more detail than other conditions because it is perhaps the single most important condition that affects the lives of organisms and many of the generalizations that we make are relevant to other conditions. We consider first the effects of temperature on ectotherms—those organisms that do not regulate their temperatures by heat produced within their bodies. We turn to endotherms, especially mammals and birds, in Section 2.5.

2.2 Correlations between temperature and the distribution of plants and animals

Variations in temperature on and within the surface of the earth have a variety of causes: the effects of latitude and altitude, continental, seasonal and diurnal effects, microclimatic effects and, in soil and water, the effects of depth. Latitudinal and seasonal variations cannot really be separated. The angle at which the earth is tilted relative to the sun changes with the seasons, and this drives some of the main temperature differentials on the earth's surface. It should be realized that the hottest temperatures occur in the middle latitudes rather than at the Equator: almost everywhere in the USA has at some time reached well over 38°C, but neither Colon in Panama nor Belem on the Equator in Brazil has ever exceeded 35°C (MacArthur, 1972).

Superimposed on these broad geographical trends are the influences of altitude and 'continentality'. There is a drop of 1°C for every 100 m increase in altitude in dry air, and a drop of 0.6°C in moist air. This is the result of the 'adiabatic' expansion of air as atmospheric pressure falls with increasing altitude. The effects of continentality are largely attributable to different rates of heating and cooling of the land and the sea. The land surface reflects less heat than the water, so the surface warms more quickly, but it also loses heat more quickly. The sea therefore has a moderating, 'maritime' effect on the temperatures of coastal regions and especially islands; both daily and seasonal variations in temperature are far less marked than at more inland, continental locations at the same latitude. Moreover, there are comparable effects within land masses: dry, bare areas like deserts suffer greater daily and seasonal extremes of temperature than do wetter areas like forests. Thus, global maps of temperature zones hide a great deal of local variation.

It is much less widely appreciated, however, that on a smaller scale still there can be a great deal of microclimatic variation. For example, the sinking of dense, cold air into the bottom of a valley at night can make it as much as 30°C colder than the side of the valley only 100 m higher; the winter sun, shining on a cold day, can heat the south-facing side of a tree (and the habitable cracks and crevices within it) to as high as 30°C; and the air temperature in a patch of vegetation can vary by 10°C over a vertical distance of 2.6 m from the soil surface to the top of the canopy (Geiger, 1955). Hence, we need not confine our attention to global or geographical patterns when seeking evidence for the influence of temperature on the distribution and abundance of organisms.

There are very many examples of plant and animal distributions that are

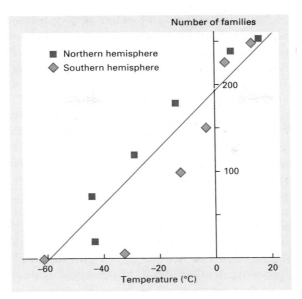

Figure 2.2 The relationship between absolute minimum temperature and the number of families of flowering plants in the Northern and Southern hemispheres. (After Woodward, 1987, who also discusses the limitations to this sort of analysis and how the history of continental isolation may account for the odd difference between Northern and Southern hemispheres.)

evidence for the effects of temperature on distribution

strikingly correlated with some aspect of environmental temperature even at gross taxonomic and systematic levels (Figure 2.2). The distribution of many species closely matches the map of some aspect of temperature (Figure 2.3). For example, the northern limit of the distribution of wild madder plants *(Rubia peregrina)* is closely correlated with the position of the January 4.5 °C isotherm (Figure 2.3a).

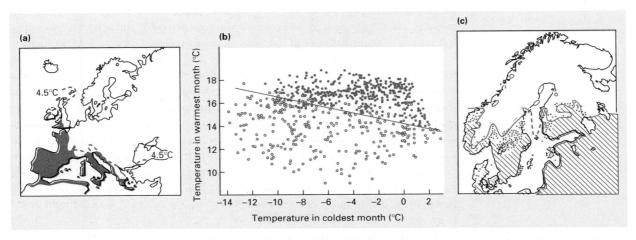

Figure 2.3 (a) The northern limit of the distribution of the wild madder *(Rubia peregrina)* is closely correlated with the position of the January 4.5 °C isotherm. (After Cox *et al.,* 1976.) (b) A plot of places within the range of *Tilia cordata* (●), and outside its range (○) in the graphical space defined by the minimum temperature of the coldest month and the maximum temperature of the warmest month. (c) The margin of the geographical range of *T. cordata* in northern Europe defined by the straight line in Figure 2.3b. ((b, c) after Hintikka, 1963; from Hengeveld, 1990.)

51   CONDITIONS

Similarly, the limit to the eastern range of holly (*Ilex aquifolium*) in Europe is quite accurately defined by the 0°C January isotherm and the isotherm of maximum temperatures exceeding 0°C for more than 345 days in 1 year. However, limits defined by at least 21 other climatic measures have been shown to fit equally well. We need to be very careful how we interpret such relationships: they can be extremely valuable in predicting where we might and might not find a particular species; they may suggest that some feature related to temperature is important in the life of the organisms; but, they do not prove that temperature is responsible for (causes) the limits to a species distribution. The literature relevant to this and many other correlations between temperature and distribution patterns is given in Hengeveld (1990), who also describes a more subtle graphical procedure devised by biogeographers in Fennoscandia. They estimate the minimum temperature of the coldest month and the maximum temperature of the hottest month for many places within and outside the range of a species, and then plot these graphically. A line is then drawn that optimally discriminates between the presence and absence records (Figure 2.3b), and this line is then used to define the geographical margin of the species distributions (Figure 2.3c). There is much less arbitrariness about this procedure, which may have powerful predictive value, but it still tells us nothing about the underlying forces that cause the distribution patterns.

One reason why we need to be cautious about reading too much into correlations of species distributions with maps of temperature is that the temperatures measured for constructing isotherms for a map are only rarely those that the organisms experience. In nature an organism may choose to lie in the sun, or hide in the shade and, even in a single day, may experience a baking midday sun and a freezing night. Moreover temperature varies from place to place on a far finer scale than will usually concern a geographer, but it is the conditions in these 'microclimates' that will be crucial in determining what is habitable for a particular species. For example, the surface of soil directly exposed to sunlight experiences much sharper diurnal cycles of temperature than the air above. The local temperature cycles on exposed soil surfaces are exaggerated by soil colour (pale soils are cold) and aspect (Figure 2.4), but largely disappear deeper in the soil and under a closed canopy of vegetation.

As temperature varies geographically, its effects may often be to change the frequency of microhabitats that are particularly protected from (or exposed to) the more extreme variations in temperature. For example, the rufous grasshopper (*Gomphocerripus rufus*) is distributed widely in Europe but in Great Britain reaches its northern limit only 150 km from the south coast where it is restricted to steep, south-facing and therefore relatively sun-drenched and warm, grassy slopes. Yet, in the more southerly and generally warmer parts of Europe there is no such restriction. Similarly, the prostrate shrub (*Dryas octopetala*) is restricted to altitudes exceeding 650 m in North Wales, where it is close to its southern limit. But, to the north, in Sutherland in Scotland, where it is generally colder, it is found right down to sea-level.

The environments that matter to an organism are those that individuals experience: these occur as spatial mosaics of favourable and unfavourable conditions that themselves shift and change with the seasons and even within a single day. The temperatures that organisms experience may bear little relationship to the temperatures in a biogeographer's maps, which are best regarded as showing the boundaries to the areas in which suitable inhabited microenvironments occur. (Of course, not all suitable microenvironments may have been colonized and inhabited.)

the limits to species distributions can often be defined by isotherms ...

... but, these do not prove that temperature has a causal role

the temperatures that an organism experiences are seldom those that a geographer or meteorologist maps

microhabitats may differ in temperature ...

... and form mosaics of favourable and unfavourable conditions

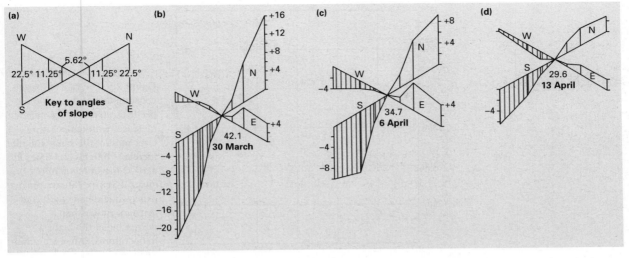

Figure 2.4 The influence of aspect on the speed of germination and emergence of corn (maize) seedlings. Pyramids of soil were built near Oxford, England, with the sides of slope 5.62°, 11.25° and 22.5° facing north, south, east and west. The base of each pyramid was approximately 7 m × 7 m. Corn (*Zea mays*) was sown in plots on the sides of the pyramids at weekly intervals through the spring from 30 March. Seedlings were recorded as they emerged and the time taken for 50% of the seedlings to emerge on the slopes and on horizontal ground was determined. (a) Shows the coordinates of slope and aspect as a plane on which the experimental results are plotted in (b–d) for seed sown on 30 March, 6 April and 13 April, respectively. In each figure seedling emergence that is slower than on horizontal ground is shown unshaded above the coordinate plane, and accelerated emergence is shown shaded as if below the plane. Thus, 50% of the seedlings from seed sown on horizontal ground on 30 March had emerged after 42.1 days, but on the steepest south-facing slope emergence occurred 22 days earlier and on the steepest north-facing slope was 17 days later. The effects of aspect become less marked later in the season. (After Ludwig *et al.*, 1957.)

Cartographers usually use arithmetic mean temperatures when they draw isotherms on maps. This ought to be misleading when we are trying to understand the distribution of organisms, because the rates of most metabolic processes increase with temperature exponentially (not linearly) over much of the normal range experienced. A temperature coefficient, Q_{10}, defines the response to a rise in temperature of 10°C and, in a typical example (Figure 2.5), for every 10°C rise in temperature the reaction rate increases 2.5 times.

However, when we plot the growth and development of whole organisms against temperature there is quite commonly an extended range over which there are only slight deviations from linearity (Figures 2.6a, b). It is easier to defend the use of arithmetic mean temperatures in biogeographical studies when the range of temperatures involved lies clearly within the range in which the organisms' responses are effectively linear.

When the relationship between growth and/or development *is* effectively linear the temperature experiences of an organism can be summarized in a single very useful value, the number of day-degrees. For instance, the development of the egg of the grasshopper, *Austroicetes cruciata*, to hatching takes 17.5 days at 20°C (4°C above a threshold of 16°C), but only 5 days at 30°C (14°C above the same threshold). At

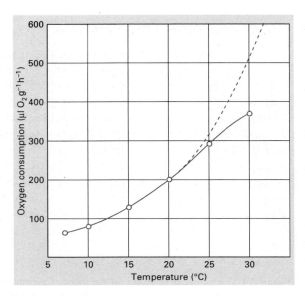

Figure 2.5 The rate of oxygen consumption of the Colorado beetle (*Leptinotarsa decemlineata*) increases with temperature (—). Over most of the range the rate increases roughly 2.5 times for a 10°C rise in temperature (Q_{10} = 2.5). (- - -) Shows the expected line had Q_{10} stayed constant rather than decreasing at the highest temperatures. (After Marzusch, 1952; from Schmidt-Nielsen, 1983.)

both temperatures, therefore, development requires 70 day-degrees (or, more properly, 'day-degrees above threshold'), i.e. $17.5 \times 4 = 70$, and $5 \times 14 = 70$. This is also the requirement for development in the grasshopper at other temperatures within the non-lethal range. In a similar way, the butterflies in Figure 2.6b require 174 day-degrees above their threshold to complete their development. Such organisms cannot be said to require a certain length of time for development. What they require is a combination of time and temperature: this combination is often referred to as physiological time.

time and temperature combine in 'physiological time'

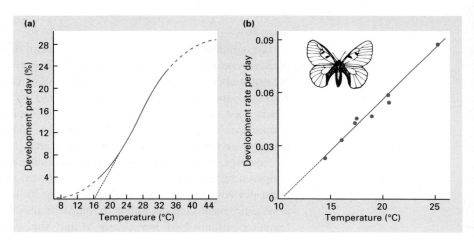

Figure 2.6 Development requires a given amount of 'physiological time', which is explained in the text. (a) Development of post-diapause eggs of the grasshopper, *Austroicetes cruciata*, requires 70 day-degrees above a developmental 'threshold' of 16°C; the relationship departs from linearity only at the lowest and highest temperature. (After Davidson, 1944.) (b) Development of the cabbage white butterfly, *Pieris rapae*, from egg-hatch to pupa requires 174 day-degrees above a threshold temperature of 10.5°C. (After Gilbert, 1984.)

54   CHAPTER 2

2.2.1 Distributions and extreme conditions

For most of the life of most organisms the temperature of their immediate environment is not its mean temperature—the average condition is a rare condition. In fact, the distributions of many species appear to be accounted for much more strongly by the occasional extremes of temperature than by the range commonly experienced. In some cases we can relate the distributional limits of a species to the occasional lethal temperature experience that precludes its existence. For instance, injury by frost is probably the single most important factor limiting plant distribution (and it is also a major cause of damage to crops). To take one example: the saguaro cactus is liable to be killed when temperatures remain below freezing for 36 h, but if there is a daily thaw it is under no threat. In Arizona, the northern and eastern edges of the cactus' distribution correspond to a line joining places where on occasional days it fails to thaw. Thus, the saguaro is absent where there are occasionally lethal conditions—an individual need only be killed once.

extreme conditions may be more important than average conditions ...

It is very rare for the climatic limits on the natural distribution of species to be deliberately tested by attempting to establish them outside their normal climatic range. But, a great number of species are deliberately cultivated outside their normal range in gardens and in agriculture. Indeed, there is scarcely any crop that is grown on a large commercial scale in the climatic conditions of its wild ancestors. In these cases we usually know what has caused failures and these usually turn out to be the risk of extreme events—especially frosts and drought, rather than the average prevailing conditions.

The climatic limit to the geographical range for the production of coffee (*Coffea arabica* and *C. robusta*) is defined by the 13 °C isotherm for the coldest month of the year. In most coffee-growing regions the mean annual temperature is approximately 21 °C. Much of the world's crop is produced in the highland microclimates of the São Paulo and Paraná districts of Brazil. Here, the average minimum temperature is 20 °C, but occasionally cold winds and just a few hours of temperature close to freezing are sufficient to kill or severely damage the trees (and play havoc with world coffee prices).

... as, for example, in coffee

Agriculturalists and foresters recognize that the length of the frost-free period limits when and where they can cultivate a particular species. The frost-free period is usually defined for a particular species as the time between the last lethal frost of spring and the first lethal frost of autumn—a crude measure of the potential growing season. (A more general measurement is of the interval between the last spring and first autumn temperature of −2 °C.) We have to take care how we interpret a frost-free period because what is lethal for a caterpillar or for the leaves and shoots of a plant may be a safe experience for a dormant pupa, seed, rhizome or the perennial shoots of a tree. However, for very many species the 'frost-free period' defines the maximum time available in 1 year in which active growth might be possible. A remarkable feature of the 'frost-free period' is its very high variation. For example, at Iowa Falls, Iowa, the average frost-free season is 150 days, but it has varied from as short as 111 days to as long as 188 days and over a 45-year period at Moscow, Idaho, the average was 149.7 days but varied from 83 to 192 days. Maps can be made of the mean length of the frost-free period, but they are of little predictive value for a farmer and little guide to what natural selection might favour in the natural world because it is the variance and extremes in the values that really matter.

Agricultural examples such as that of coffee carry three important messages: (i) we may most easily establish what limits the natural distributions of organisms if we try to establish them outside their normal range and wait to see what goes wrong; (ii) the crucial limiting conditions may occur only rarely and we may need to wait a long time before we can begin to measure the risks involved; and (iii) maps of average conditions are not very useful when it is the variance and extreme values that really matter.

If occasional and rare events are the important conditions that impose limits on species' distributions, there are real problems for ecologists because biostatistics has channelled them into thinking primarily about means and variances and not about the tips of the tails of probability distributions (Gaines & Denny, 1993). We return to this problem in Section 2.9.1.

2.2.2 Distributions and the interaction of temperature with other factors

We know enough about the intensity with which organisms interact with each other that it would be surprising if we could explain the influence of temperature on the distribution of a species by considering the response of that species alone. Temperature will affect the behaviour of all the organisms that it interacts with, its competitors, parasites, symbionts and all those other species that create or modify the physical environment in which it lives. For instance, many animals have a distribution which is correlated with temperature, but they are really limited by the occurrence or quality of their food. For example, the small moth, *Coleophora alticolella*, feeds over the Cumbrian moors in England on the seeds of the heath rush, *Juncus squarrosus*, but is absent above an altitude of approximately 600 m. *Juncus* grows above this height and produces flowers, and the adult female moths lay their eggs on the flowers as they do elsewhere; but the temperatures here are generally too low for successful seed production, and the newly hatched moth larvae therefore starve (Randall, 1982). Here, temperature is controlling the distribution of the moth by altering the nature of a biological interaction.

Another, more complicated example is provided by the relationship between temperature and the southern limit of the barnacle, *Balanus balanoides* (Barnes, 1957; see also Lewis, 1976). This Arctic species is killed by high air temperatures in summer, approaching 25 °C. On the other hand, gonad maturation and fertilization only take place when air temperatures drop below 10 °C and are maintained at that level for some time (perhaps 20 days). In Europe, it appears that south of the southern limit, the summer air temperatures are too high for adult survival and the winter air temperatures are too high for gonad maturation. On the Atlantic coast of North America, however, the seasonal variation in temperature is far greater, and in South Carolina for instance, 480 km south of the species' southern limit, the winter conditions are still suitable for breeding. Here, therefore, it is only the summers that are too hot, and the southern-most records are all from north-facing, shaded locations. On the Pacific coast of North America, the temperature conditions are such that *B. balanoides* might be expected to occur at least 1600 km further south than its present southern limit. That it does not do so is attributed to the adverse effects of competition from other local species of barnacle (*B. balanoides* is a relatively recent migrant to the Pacific via the Bering Strait). In other words, the temperature conditions in which *B. balanoides* fails to establish a population because it is out-competed are far less extreme than those in which it is simply unable to survive,

temperature may determine a
species' distribution through
interaction with other
factors ...

... such as its food supply ...

... its competitors

and it is therefore not limited by temperature alone but by the effects of temperature on its competitors.

There is abundant evidence that increasing temperature influences the development of disease in crop plants by accelerating the development of the pathogens and shortening the latent periods in their development (Rowell, 1984). For example, a rise of only a few degrees produces dramatic changes in the phenotypic expression of black stem rust infection in wheat (Burdon, 1987). This sort of evidence makes it likely that one effect of global warming (see Section 2.11.2) will be to increase levels of pest and disease in natural communities because of the rapid responsiveness of their life cycles and, as a consequence, to change the distribution and abundance of their hosts.

... or disease

Many of the interactions between temperature and other physical conditions are so strong that it is not sensible to consider them separately. The interactions between temperature and the concentration of dissolved gases in aquatic systems, and between temperature and humidity in terrestrial communities are particularly important.

In rivers the level of oxygen availability is set mainly by physical forces, being high in upstream regions where turbulence is great and air is effectively mixed with water, and low in downstream regions where higher temperatures mean lower gaseous concentrations because the solubility of oxygen in water decreases with a rise in temperature. Table 2.1 lists some fish typical of different regions in British rivers, and also their requirements in terms of temperature and oxygen concentration. An upstream species like the brown trout is limited in its distribution, at least in part, because as temperature increases downstream the trout's requirement for oxygen goes up whilst the oxygen concentration of the water goes down. Similar comments apply to the downstream limits of midstream species like the pike. In these cases it is impossible to isolate the effect of temperature from that of oxygen concentration: their physiological effects are interrelated and they affect distribution in concert (the upstream limits of these species must be related more to suboptimally low temperatures, to high-velocity flow or possibly to competition or predation).

The relative humidity of the atmosphere is an important condition in the life of terrestrial organisms because it plays a major part in determining the rate at which they lose water. Water is of course a resource rather than a condition in the life of organisms and is therefore considered further in the next chapter. However, animals and plants differ in their abilities to reduce and counteract water losses, and they

Table 2.1 There is a close correspondence in British rivers between the temperature and oxygen requirements of different fish species and the conditions where they are found. Temperatures are lowest and oxygen concentrations are highest in upstream waters; temperatures are highest and oxygen concentrations lowest in downstream waters. (After Varley, 1967.)

Exemplar	Distribution	Oxygen requirement for survival (ml l^{-1})	Upper lethal temperature limit (°C)	Optimum temperature for growth (°C)
Brown trout	Upstream	5–11	< 28	7–17
Pike	Midstream	4	28–34	14–23
Carp	Downstream	0.5	> 34	20–28

differ in how they respond to relative humidity which may permit or prevent their existence. In practice, it is rarely possible to make a clean distinction between the effects of relative humidity and of temperature. This is simply because a rise in temperature leads to an increased rate of evaporation. A relative humidity that is acceptable to an organism at a low temperature may therefore be unacceptable at a higher temperature. Moreover, both relative humidity and temperature may themselves interact with wind speed: the rapid movement of air across an evaporative surface maintains the moisture gradient, increases the rate of evaporation and the latent heat of evaporation lowers the temperature.

Microclimatic variations in relative humidity can be even more marked than those involving temperature. For instance, it is not unusual for the relative humidity to be almost 100% (i.e. almost saturated) at ground level amongst dense vegetation and within the soil, whilst the air immediately above, perhaps 40 cm away, has a relative humidity of only 50%. The organisms most obviously affected by humidity in their distribution are those 'terrestrial' animals that are actually, in terms of the way they control their water balance, 'aquatic'. Amphibians, terrestrial isopods, nematodes, earthworms and molluscs are all, at least in their active stages, confined to microenvironments where the relative humidity is at or very close to 100%. The major group of animals to escape such confinement are the terrestrial arthropods, especially insects. Even here though, the evaporative loss of water is often of crucial importance. This can be seen in the case of the fruit-fly *Drosophila subobscura* which, like other insects, has an impermeable exoskeleton; but it loses water across its respiratory surfaces when its spiracles are open, which they need to be when it is active. The species is largely confined to woodlands (high relative humidity), making only occasional forays into open areas (low relative humidity). The individuals fly (their most energetic activity) only at particular times of the day: just after dawn and just before dusk. These are the periods of daylight with the lowest temperature and the highest relative humidity. During the middle of the day a flying *Drosophila* would rapidly dehydrate. Moreover, its periods of activity are even more narrowly circumscribed in arid, open areas than they are in damper woodlands (Inglesfield & Begon, 1981).

2.2.3 Conditions as stimuli

We have seen that temperature as a condition affects the rate at which organisms develop. It may also act as a stimulus, determining whether or not the organism starts its development at all. For instance, for many species of temperate, Arctic and alpine herbs, a period of chilling or freezing (or even of alternating high and low temperatures) is necessary before germination will occur (see Chapter 5, pp. 195–198). A cold experience (physiological evidence that winter has passed) is required before the plant can start on its cycle of growth and development. Temperature may also interact with other stimuli (e.g. photoperiod) to break dormancy and so time the onset of growth. The seeds of the birch (*Betula pubescens*) require a photoperiodic stimulus (i.e. experience of a particular regime of day length) before they will germinate, but if the seed has been chilled it starts growth without a light stimulus. However, the temperatures and photoperiods that act as stimuli, for example in breaking dormancy (or setting in throw the process of acclimatization; see Section 2.3.3), are often quite different from those that control the subsequent rates of growth and development. All of these comments apply in a similar fashion to

chilling, freezing or a particular photoperiod may break seed dormancy …

or initiate acclimation

the effects of temperature and photoperiod on diapause (essentially dormancy) in insects, as discussed in Chapter 5, Section 5.5.1.

2.3 Life at low temperatures

The greater part of our planet is cold (below 5°C) and 'cold is the fiercest and most widespread enemy of life on earth' (Franks *et al.*, 1990). More than 70% of the planet is covered with seawater: mostly deep ocean with a remarkably constant temperature of about 2°C. If we include the polar ice-caps, more than 80% of earth's biosphere is permanently cold. We inhabit a cold planet, and should perhaps regard those organisms that are best able to cope with low temperatures as being its most successful colonizers (Russell, 1990).

By definition, all temperatures below the optimum are harmful but there is usually a wide range of such temperatures that cause no physical damage and over which any effects are fully reversible. There are, however, two quite distinct types of damage at low temperatures (chilling and freezing), that can be lethal, either to tissues or to whole organisms.

2.3.1 Injury from chilling

cold can sometimes be damaging even when no ice is formed

Cold itself can have lethal physical and chemical consequences even though ice may not be formed. Water may supercool to temperatures at least as low as −40°C, remaining in an unstable liquid form in which its physical properties change in ways that might be expected to be biologically significant: its viscosity increases, its diffusion rate decreases. Over the range from 25 to −25°C the degree of ionization of water decreases nearly 100-fold, and as hydrogen ions (H^+) and hydroxyl ions (OH^-) are involved in almost all biological processes this can scarcely fail to be important. Many organisms are damaged by exposure to temperatures that are low, but above freezing point—so-called 'chilling injury'. The fruits of the banana blacken and rot after exposure to chilling temperatures and many species of tropical rainforest are

chilling injury occurs above 0°C

sensitive to chilling. The nature of the injury is obscure, although it seems to be associated with the breakdown of membrane permeability and the leakage of specific ions such as calcium (Minorsky, 1985). Organisms are said to be 'chill tolerant' when they survive temperatures that are suboptimal but are never low enough for freezing to occur. Chilling injury and tolerance (like the response to supercooling) are quite different from injury by and tolerance to the formation of ice, i.e. freezing.

2.3.2 Injury from freezing

Ice seldom forms in an organism until the temperature has fallen several degrees below 0°C—it remains in a supercooled state until it solidifies suddenly around particles that act as nuclei. Pure water then separates out as the solid phase and the concentration of solutes in the remaining liquid phase rises as a consequence. When ice forms in plant or animal tissues it is almost always extracellular water that freezes . It is very rare for ice to form within cells and it is then inevitably lethal, but

ice formed within cells is inevitably lethal ...

the freezing of extracellular water is one of the factors that prevents ice forming within the cells themselves (Franks *et al.*, 1990). As extracellular ice forms, water is withdrawn from the cell, and solutes in the cytoplasm (and vacuoles) become more

... but, is usually intercellular ...

concentrated. The effects of freezing are therefore mainly osmoregulatory: the water balance of the cells is upset and cell membranes are destabilized. The effects are essentially similar to those of drought and salinity. Extreme water withdrawal from plant cells destroys the semi-permeability of the plasma membranes and may even cause the physical tearing of the cytoplasm away from the cell walls.

The risk of damage from freezing is greatest if supercooling has occurred and ice forms suddenly—it is then that there is the greatest chance that lethal intracellular ice will form. For this reason, mechanisms that encourage the formation of ice at temperatures only just below 0°C protect an organism from freezing damage. Certain bacteria (*Pseudomonas syringae* and *Erwinia herbicola*) are able to synthesize materials that catalyse the formation of ice at temperatures as high as −4°C (Schnell & Vali, 1976). Some species of bivalves and gastropods also produce ice-nucleating proteins that induce the formation of extracellular ice during the winter (Johnston, 1990).

When extracellular ice is formed, the withdrawal of water from the cells is resisted by the accumulated osmotically active molecules and ions. In addition, much of the ability of both plants and animals to tolerate freezing temperatures (and drought and salinity), appears to depend on antifreezes. For example, the blood of the fish, *Pagothenia borchgrevinki*, from McMurdo Sound, Antarctica, has a freezing point of −2.7°C compared with −0.8°C for a comparable but more typical marine fish, and this is due partly to a higher concentration of sodium chloride and partly to peptides and large glycopeptides which depress the freezing point 200–300 times more than expected from their concentration. Such compounds are particularly important in the process of 'frost hardening' by which organisms acquire a tolerance of low temperatures.

2.3.3 Acclimation to life at low temperatures

Perhaps what is most striking about the tolerances of organisms to low temperatures is that they are not fixed but are preconditioned by the experience of temperatures in their recent past. This process is called *acclimation* when it occurs in the laboratory and *acclimatization* when it occurs naturally. The exposure of an individual for several days to a relatively low temperature can shift its whole temperature response downwards along the temperature scale (Figure 2.7a). Similarly, exposure to a high temperature can shift the temperature response upwards.

In higher plants the development of cold tolerance is triggered by environmental cues. Most commonly a period of 4–6 weeks at temperatures of 0–5°C, usually accompanied by decreasing day length (photoperiod), provides the stimulus for acclimatization. The increase in freezing tolerance (frost hardening) that is gained by acclimatization ranges from a few degrees in herbaceous species to tens of degrees in winter cereals and some trees (Figure 2.7b), and to more than 100°C in some extreme examples amongst deciduous trees (Steponkus *et al.*, 1990).

During a period of frost hardening most species of plants and animals accumulate compounds that protect the cells (especially their membranes) from the effects of dehydration rather than preventing it. These compounds are mainly free amino acids (especially in plants) and polyhydroxy compounds (PHCs) of low molecular mass (e.g. glycerol and sorbitol). The sugars and sugar alcohols are particularly important compounds ecologically because they protect complex protein and lipid structures from quite extreme dehydration—whether produced by freezing or heat (Franks *et al.*, 1990).

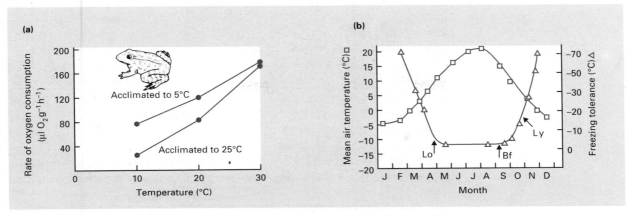

Figure 2.7 Acclimation to low temperatures. (a) The rate of oxygen consumption of frogs (*Rana pipiens*) at a given temperature depends on the temperature of acclimation. (After Rieck *et al.*, 1960.) (b) Seasonal changes in the freezing tolerance (i.e. the ability to withstand short periods at low temperatures) of the willow, *Salix sachalinensis*, (△) on Mt Kurodake, Japan and mean air temperature (□). Bf, buds forming; Lo, leaves opening; Ly, leaves yellowing. Freezing tolerance is much greater in winter when mean temperatures are far lower. (After Sakai & Otsuka, 1970; from Sutcliffe, 1977.)

damage from freezing is avoided in some insects ...

... by preventing intracellular ice formation ...

... and in others by encouraging extracellular ice formation

Insects have been shown to have two strategies that allow survival through the low temperatures of winter. A 'freeze-avoiding' strategy uses low-molecular-weight polyhydric alcohols (polyols, such as glycerol and sorbitol) that prevent the formation of intracellular ice by depressing both the freezing point and the supercooling point and uses specialized 'thermal hysteresis' proteins to prevent ice nuclei from forming (Figure 2.8). A contrasting 'freeze-tolerant' strategy, which also involves the formation of polyols, encourages the formation of extracellular ice, but protects the cell membranes from damage when water is withdrawn from the cells (Storey, 1990). Acclimatization starts as the weather becomes colder in the autumn and stimulates the conversion of almost the entire glycogen reserve of the animals into polyols. Such protection from frost damage during the winter is an energetically costly affair: about 16% of the carbohydrate reserve is consumed in the conversion of the glycogen reserves to polyols.

Acclimatization aside, individuals commonly vary in their temperature response depending on the stage in development that they have reached. Probably the most extreme form of this is when an organism has a dormant stage in its life cycle. Dormant stages are typically dehydrated, metabolically slow and tolerant of extremes of temperature.

2.3.4 Genetic variation and the evolution of cold tolerance

Even within species there are often differences in temperature response between populations from different locations, and these differences have frequently been found to be the result of genetic differences rather than being attributable solely to acclimatization (Figure 2.9). It can therefore be misleading on several counts to think of a species as having one temperature response.

some temperature tolerances are heritable ...

There are striking cases where the geographical range of a crop has been

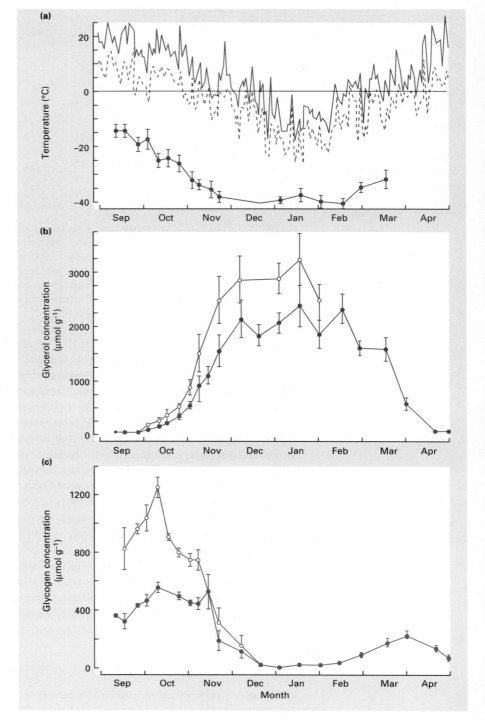

Figure 2.8 Changes in the (b) glycerol and (c) glycogen content of the freeze-avoiding larvae of the goldenrod gall moth, *Epiblema scudderiana*. Also plotted are (a) the daily temperature minima and maxima and whole larvae supercooling points. The metabolic contents are expressed relative both to wet mass (●) and dry mass (○) of the animal. (After Storey, 1990; from Rickards *et al.*, 1987.)

extended into colder regions by plant breeders. Programmes of deliberate selection applied to corn (*Zea mays*) have extended the area of the USA over which the crop can be profitably grown into colder states. Over the period 1920–1930 to

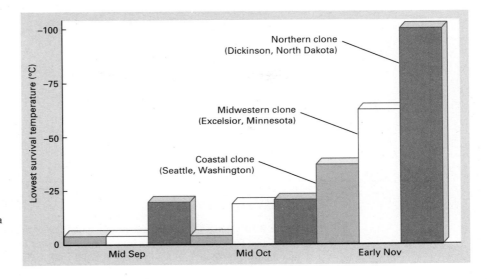

Figure 2.9 Variation in the seasonal patterns of cold resistance shown by three climatic races of red-osier dogwood, *Cornus stolonifera*. The histograms are for clones from North Dakota, Minnesota and Washington grown in Minnesota. (After Weiser, 1970.)

... and may vary within a species

1940–1949, the production of corn in Iowa and Illinois increased by 21.6 and 27.3%, respectively, whereas in the colder state of Wisconsin it increased by 54.3%. Indeed, in 1953 the average corn yield in Wisconsin exceeded that of any other state for the first time in history. However, most of the genetic change that has allowed the crop to spread into colder regions has involved the development of varieties with shorter growing seasons and a reduced risk that the seeds will rot when sown into cold soil rather than an increased tolerance of cold itself.

Powerful evidence that cold tolerance varies between geographical races of a species comes from a study of the cactus, *Opuntia fragilis*. Cacti are generally species of hot dry habitats, but *O. fragilis* extends as far north as 56°N and at one site the lowest extreme minimum temperature recorded was –49.4°C. Twenty populations were sampled from diverse localities in northern USA and Canada, and tested for freezing tolerance and ability to acclimate to cold. Individuals from the most freeze-tolerant population (from Manitoba) tolerated –49°C in laboratory tests and acclimated by 19.9°C, whereas plants from a population in the more equable climate of Hornby Island, British Columbia, tolerated only –19°C and acclimated by only 12.1°C. Even more significant evidence of local climatic races in the species was that freeze tolerance and cold acclimation were both positively correlated ($r^2 \cong 0.7$) with the minimum temperature at the 20 locations (Loik & Nobel, 1993).

If deliberate selection can change the tolerance and distribution of a domesticated plant we should expect natural selection to do the same thing in nature. It is difficult to test this unless populations of a species are deliberately grown outside their normal range in order to determine whether they change their behaviour over time. An attempt to do this was made with the plant *Umbilicus rupestris*, which lives on walls in mild maritime areas of Great Britain (Woodward, 1990). A population of plants and seeds was taken from a donor population in the mild wintered habitat of Cardiff (South Wales) and introduced onto a wall in a cooler environment at an altitude of 157 m in Sussex (South England). After 8 years the temperature response of seeds from the donor (Cardiff) and the introduced populations had diverged quite strikingly (Figure 2.10a), and subfreezing temperatures that have been observed to kill in South Wales (–12°C) were then tolerated by 50% of the Sussex population

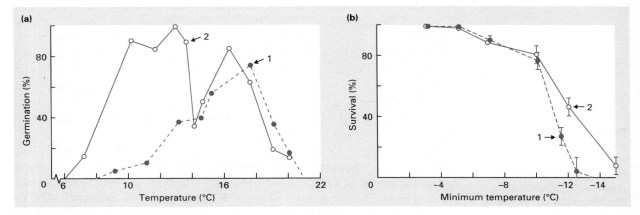

Figure 2.10 Changes in the behaviour of populations of the plant *Umbilicus rupestris*, established for a period of 8 years in a cool environment in Sussex from a donor population in a mild-wintered area in South Wales (Cardiff). (a) Temperature responses of seed germination; (1) Responses of samples from the donor population (Cardiff) in 1978, (2) responses from the Sussex population in 1987. (b) The low-temperature survival of the donor population at Cardiff, 1978 (1) and of the established population in Sussex, 1987 (2). (After Woodward, 1990.)

(Figure 2.10b). This experiment is very suggestive of rapid genetic change (although not tested by deliberate breeding experiments). It also strengthens the suspicion that past climatic changes, for example ice-ages, will have changed the temperature tolerance of species as well as forcing their migration.

2.4 Life at high temperatures

Perhaps the most important thing about dangerously high temperatures is that, for a given organism, they usually lie only a few degrees above the metabolic optimum. This is largely an unavoidable consequence of the physicochemical properties of most enzymes (see Heinrich, 1977; Privalov, 1983). High temperatures may be dangerous because they lead to the inactivation or even the denaturation of enzymes, but they may also have damaging indirect effects by leading to dehydration. All terrestrial organisms need to conserve water, and at high temperatures the rate of water loss by evaporation can be lethal, but they are caught between the devil and the deep blue sea because evaporation is an important means of reducing body temperature. If surfaces are protected from evaporation (e.g. by closing stomata in plants or spiracles in insects) the organisms may be killed by too high a body temperature, but if their surfaces are not protected they may die of desiccation.

high temperatures may cause enzyme inactivation ...

and cause dehydration

Death Valley, California, in the summer, is probably the hottest place on earth in which higher plants make active growth. Air temperatures during the daytime may approach 50°C and soil surface temperatures may be very much higher. The perennial plant, desert honeysweet (*Tidestromia oblongifolia*), grows vigorously in such an environment despite the fact that its leaves are killed if they reach the same temperature as the air. Very rapid transpiration keeps the temperature of the leaves at 40–45°C, and in this range they are capable of extremely rapid photosynthesis (Berry & Björkman, 1980).

64   CHAPTER 2

Most of the plant species that live in very hot environments suffer severe shortage of water and are therefore unable to use the latent heat of evaporation of water to keep leaf temperatures down. This is especially the case in desert succulents in which water loss is minimized by a low surface to volume ratio and a low frequency of stomata. In such plants the risk of overheating may be reduced by spines (which shade the surface of a cactus), or hairs or waxes, which reflect a high proportion of the incident radiation. Nevertheless, such species experience and tolerate temperatures in their tissues of more than 60°C when the air temperature is above 40°C (Nobel, 1983; Downton *et al.*, 1984; Smith *et al.*, 1984).

cysts, spores and seeds are commonly heat tolerant

There are stages in the life history of many organisms that are particularly tolerant of high temperature. This is especially true of dormant structures such as the resting spores of fungi, cysts of nematodes and seeds of plants, probably as a result of their naturally dehydrated state. Dry wheat grains can withstand 90°C for up to 10 min, but after being soaked for 24 h, 60°C kills them in about 1 min (Sutcliffe, 1977). As with temperature responses generally, the responses to high temperature are subject to acclimatization. The ability of plants to acclimatize to high temperatures appears to be less than to low temperatures, although desert cacti may raise their tolerance of heat by 10–20°C.

fire is an important ecological heat hazard ...

Fires are responsible for the highest temperature that organisms face on earth and, before the fire-raising activities of humans, were caused mainly be lightning strikes. The recurrent risk of fire has shaped the species composition of arid and semi-arid woodlands in many parts of the world. All plants are damaged by burning but it is remarkable powers of regrowth from protected meristems on shoots and seeds that allow a specialized subset of species to recover from damage and form characteristic fire floras. Fire may be a requirement for the natural regeneration of some species (e.g. *Eucalyptus regnans*; Gilbert, 1959). Some striking examples of fire-tolerant species of *Eucalyptus* depend on an accumulation of dormant buds on large 'lignotubers' to generate vigorous shrubby regrowth after fire. (See Hodgkinson & Griffin, 1982; Hodgkinson, 1991, 1992, for discussion of the role of fire in Australian forests, and Naveh, 1975, for an analysis of the evolutionary significance of fire in the Mediterranean region.)

... and can cause specialized fire floras to develop

Amongst eukaryotic organisms it is fungi such as species of *Mucor*, *Rhizopus* and *Humicola* that tolerate the highest temperatures (between 50 and 60°C). A few protozoa can tolerate temperatures between 50 and 55°C. Only a few eukaryotes tolerate temperatures higher than 60°C and there are very few habitats on earth where these high-temperatures occur. Decomposing organic matter in heaps of farmyard manure, compost heaps and damp hay provides appropriate, although almost always man-made, environments. Stacks of damp hay are heated to temperatures of 50–60°C by the metabolism of fungi such as *Aspergillus fumigatus* and *Geotrichum candidum*, carried further to approximately 65°C by thermophilic organisms such as *Mucor pusillus* and *Humicola lanuginosa* and then a little further by bacteria and actinomycetes. Biological activity stops well short of 100°C but autocombustible products are formed which cause further heating, drive off water and may even result in fire.

there are very few 'hot' environments in nature

Another hot environment is that of natural hot springs and in these *Thermus aquaticus* grows at temperatures of 67°C and tolerates temperatures up to 79°C. This organism has also been isolated from domestic hot water systems. Many (perhaps all) of the extremely thermophilic species are prokaryotes with very odd metabolism being, for example phototrophic (*Chloroflexus aurantiacus*), or capable of oxidizing ferrous iron (*Caldariella acidophila*), methane (*Methanobacterium*) or sulphur (*Sulfolobus acidocaldarius*). Many of them have other curious tolerances: *S. acidocalcarius* is not only thermophilic (optimum growth at 70–74°C, maximum at 85°C), but grows at pH 2–3! These data for thermophiles are given in Stolp (1988) with many others including two of the most extreme thermophiles, the archaebacteria *Pyrococcus furiosus* and *Pyrodictium occultum*, with optimal temperatures for growth of 100 and 105°C, respectively.

the most heat tolerant organisms are prokaryotes

In environments with very high temperatures the communities contain few species. In general, eukaryotes are the most sensitive to heat followed by fungi, and in turn by bacteria, actinomycetes and archaebacteria. This is essentially the same order as is found in response to many other extreme conditions, such as low temperature, salinity, metal toxicity and desiccation.

An ecologically very remarkable hot environment has been discovered only towards the end of the present century and its study presents a massive technological challenge. In 1979 a deep oceanic site was discovered in the eastern Pacific at which fluids at high temperatures ('smokers') were vented from the sea floor forming thin-walled 'chimneys' of mineral materials. Since that time many more vent sites have been discovered at midocean crests in both the Atlantic and Pacific oceans. The depth of the oceans at these sites is 2000–4000 m below sea-level, giving pressures of 200–400 bars (20–40 MPa). The boiling point of water is raised to 370°C at 200 bars (20 MPa) and 404°C at 400 bars (40 MPa). The superheated fluid emerges from the chimneys at temperatures as high as 350°C and, as it cools to the temperature of seawater at about 2°C, provides a continuum of environments at intermediate temperatures.

superheated water in deep oeanic vents ...

Environments at such extreme pressures and temperatures are obviously extraordinarily difficult to study *in situ* and in most respects impossible to maintain in the laboratory. Some thermophilic bacteria collected from vents have been cultured successfully at 100°C, or only slightly above normal barometric pressures (e.g. the sulphur-reducing *Staphylothermus marinus* and the methane-producing *Methanococcus jannaschii*), and growth observed even at 110°C at pressures that are sustainable in the laboratory (Jannasch & Mottl, 1985). But, there is much circumstantial evidence that some microbial activity occurs at much higher temperatures and may form the energy resource for the warm water communities outside the vents. For example, particulate DNA has been found in samples taken from within the 'smokers' at concentrations that point to intact bacteria being present at temperatures very much higher than those conventionally thought to place limits on life (Straube *et al.*, 1990; Deming & Baross 1993; Baross & Deming, 1995).

... associated with thermophilic chemotrophic bacteria ...

which may provide the energy source ...

The environment in the local neighbourhood of vents may have its temperature elevated far above that characteristic of deep oceans. There is a rich eukaryotic fauna that is quite atypical of the deep oceans in the large number of invertebrate species and the bulk of its biomass. A thermistor probed into the sediment at one of these sites (occupied by a population of giant clams) recorded 9.35°C at the sediment surface, 21.44°C at 20 cm depth and 33.63°C at 40 cm depth.The fauna in these warm areas around vent sites contains pogonophorans, massive bivalve molluscs, polychaete worms and crabs. Some of these local communities have been surveyed photographically and by video from deep dives: at least 55 taxa were documented near one vent in the Middle Valley, North-East Pacific, and of these 15 were new or probably new species (Juniper *et al.*, 1992) (Figure 2.11). There is strong evidence of mutualistic associations between many of these invertebrates and bacteria that

... for species-rich communities near the vents

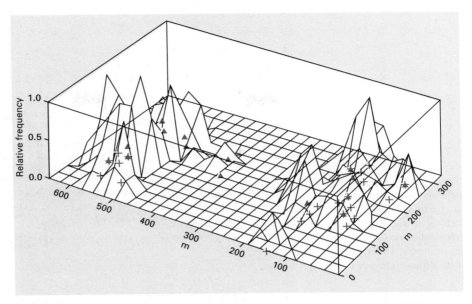

Figure 2.11 Map of an area of hydrothermal activity at a depth of approximately 2420 m at Middle Valley, northern Juan de Fuca ridge, North-East Pacific. The area was surveyed by photography and the plot shows the number of types of organism recorded in each photograph as a proportion of all the types of organism recorded in the photographs. (+), Photographs in which clams were seen; (▲), the sites of vent chimneys and (or) visible vent flows. (After Juniper *et al.*, 1992.)

appear to contain sulphur granules. The energetics of this very specialized community may depend, at least in part, on chemosynthetic microbial activity (Douglas, 1994).

There can be few environments in which so complex and specialized a community depends on so localized a special condition. The closest known vents with similar conditions are 2500 km distant. Such communities add a further list to the planet's record of species richness. They present tantalizing problems in evolution and daunting problems for the technology needed to observe, record and then study the physiology of organisms, some of which may be living at temperatures and pressures far beyond the range experienced by the typical biota of the planet.

2.5 Ectotherms and endotherms

Most of the organisms that we have discussed in this chapter have a body temperature that differs little, if at all, from that of their environment and there is little that they can do to change it. A parasitic worm in the gut of a mammal, a bacterial cell or fungal mycelium in the soil and a sponge, coral or kelp in the sea acquire the temperature of the medium in which they live.

Terrestrial organisms, exposed to the sun and the air, are different because they may acquire heat directly by absorbing solar radiation or be cooled by the latent heat of evaporation of water (typical pathways of heat exchange are shown in Figure 2.12). Various fixed properties may ensure that body temperatures are higher (or lower) than the ambient temperatures. For example, the reflective, shiny or silvery leaves of many desert plants reflect warming radiation that might otherwise heat the

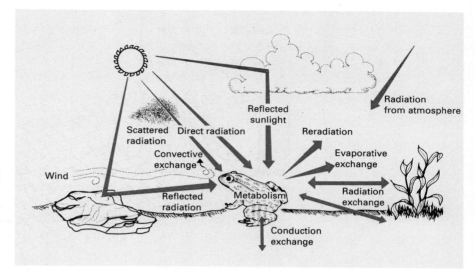

Figure 2.12 Schematic diagram of the avenues of heat exchange between an ectotherm and a variety of physical aspects of its environment. (After Tracy, 1976; from Hainsworth, 1981.)

leaves, and the wet skin of a frog will keep the body temperature lower than the ambient. Organisms that can move have some control over their body temperature because they can seek out warmer or cooler environments, as when a lizard chooses to warm itself by basking on a hot sunlit rock or escape from the heat by finding shade. Even fish, which live in environments where temperature gradients are usually very weak, congregate at characteristic preferred temperatures when they are placed in a laboratory temperature gradient (Figure 2.13).

Amongst insects there are examples of body temperatures raised by controlled muscular work, as when bumblebees raise their body temperature by shivering their flight muscles. Social insects such as bees and termites control the temperature of their colonies and regulate them with remarkable thermostatic precision by controlled metabolic activity. From late spring to autumn honeybees control the

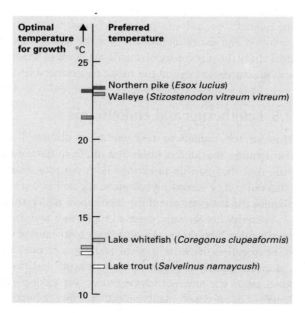

Figure 2.13 Fish regulate their temperature behaviourally. Four species select characteristic preferred temperatures in a laboratory temperature gradient, clearly related to their optimal temperature for growth. (After Christie & Regier, 1988.)

temperature of their hives between 34.5 and 35.5°C even when the surface temperature outside has been as high as 70°C, and in winter there is a record of 31°C maintained by a cluster of bees in a hive when the temperature outside was −28°C (observations by various authors reported in Wilson, 1975).

All of these organisms—plants, microorganisms, insects and other invertebrates and some vertebrates (reptiles, amphibians and fish)—have been grouped together as 'cold-blooded' organisms, distinguishing them from those like mammals and birds which are thought of as 'warm blooded'. But, the terms are clearly subjective and not very useful since many 'cold-blooded' organisms are as warm as we are (37°C), whenever their environment is at that temperature. A more useful classification divides organisms into two categories of: (i) *homeotherms*, which maintain an approximately constant body temperature; and (ii) *poikilotherms*, which have one that varies. But, like so many biological categories, this distinction breaks down when we remember that many classic 'homeothermic' animals, such as dormice, hedgehogs and bats, hibernate with body temperatures scarcely different from that of their surroundings.

A more satisfactory distinction is between endotherms and ectotherms. Endotherms regulate their temperature by the production of heat within their own bodies; ectotherms rely on external sources of heat. Very broadly, this is a distinction between the birds and mammals in the first case, and the other animals, plants, fungi and protists in the second case; but even here the distinction is not clear cut. There are a number of reptiles, fish and insects (e.g. certain bees, moths and dragonflies) that use heat generated in their own bodies in order to regulate body temperature for limited periods; in some plants, metabolic heat maintains a relatively constant temperature in the flowers (e.g. *Philodendron*); and there are some birds and mammals that relax or suspend their endothermic abilities at the most extreme temperatures (Bartholomew, 1982).

Despite all these reservations, the distinctions between endotherms and ectotherms and between homeotherms and poikilotherms can be useful.

2.5.1 Endotherms

Underlying the endotherm's homeothermy is a relationship like the one in Figure 2.14. Over a certain temperature range (the thermoneutral zone) an endotherm consumes energy at a basal rate. But, at environmental temperatures further and further above or below that zone, the endotherm consumes more and more energy in maintaining a constant body temperature. Moreover, even in the thermal neutral zone an endotherm typically consumes energy many times more rapidly than an ectotherm of comparable body size.

Endotherms produce heat at a rate controlled by a thermostat in the brain. They usually maintain a constant body temperature between 35 and 40°C, and they therefore tend to lose heat in most environments; but this loss is moderated by insulation in the form of fur, feathers and fat, by controlling blood flow near the skin surface. When it is necessary to increase the rate of heat loss, this too can be achieved by the control of surface blood flow, and by a number of other mechanisms shared with ectotherms like panting and the simple choice of an appropriate habitat (Bartholomew, 1982). Together, all these mechanisms and properties give endotherms a powerful (although not a perfect) capability for regulating their body temperature (Figure 2.14), and the benefits they obtain from this are a constancy of

organisms can be divided into homeotherms and poikilotherms ...

... or into endotherms and ectotherms

ectotherms modify their gain and loss of heat—but, only to a limited extent

endotherms regulate temperature effectively—but, they can expend large amounts of energy doing so

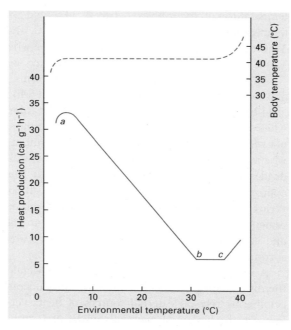

Figure 2.14 Thermostatic heat production by an endotherm is constant in the thermoneutral zone, i.e. between *b*, the lower critical temperature, and *c*, the upper critical temperature. Heat production rises, but body temperature remains constant, as environmental temperature declines below *b*, until heat production reaches a maximum possible rate at a low environmental temperature. Below *a*, heat production and body temperature both fall. Above *c*, metabolic rate, heat production and body temperature all rise. Hence, body temperature is constant at environmental temperatures between *a* and *c*. (After Hainsworth, 1981.)

performance and a much greater 'peak' or 'burst' performance. But, the price they pay for this capability is a large expenditure of energy (Figure 2.15), and thus, a correspondingly large requirement for food to provide that energy.

We may expect the minimization of costs of endothermy to be one of the forces operating in natural selection. In general, mammals from cold climates have shorter extremities (e.g. ears and limbs) than those with otherwise similar characteristics from warmer climates ('Allen's rule'); and foxes, deer and other mammals with a wide distribution are often larger in colder areas ('Bergmann's rule'). The apparent explanation in both cases is that the endotherms in colder climates should, relative to their volume, have a smaller surface area across which they lose heat, and animals of the same shape necessarily experience a decrease in the ratio of surface area to volume as they get larger. However, while Allen's rule, involving only a change in shape, appears to be very widely applicable, Bergmann's rule, involving body size which must be subject to very many other selective forces, is much less universally true.

Many endothermic animals escape from some of the costs of endothermy by hibernating during the coldest seasons: they behave almost like ectotherms. Indeed, '...hibernation is a practical and perhaps even enviable, solution to a mammalian problem' (Nedergaard & Cannon, 1990). Endothermy is energetically particularly costly for small mammals because their metabolic rate is high in proportion to their body mass:

Allen's rule, Bergmann's rule and geographical distribution generally

hibernation ...

70   CHAPTER 2

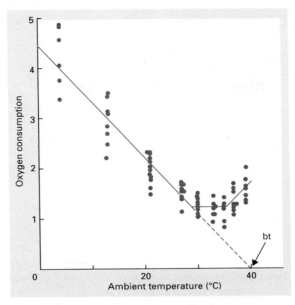

Figure 2.15 The effect of environmental temperature on the metabolic rate (rate of oxygen consumption) of the eastern chipmunk (*Tamias striatus*). bt, body temperature. Note that at temperatures between 0 and 30°C oxygen consumption decreases approximately linearly as the temperature increases. Above 30°C further increase in temperature has little effect until near the animal's body temperature when oxygen consumption increases again. (After Neumann, 1967; Nedergaard & Cannon, 1990.)

$$\text{metabolic rate } (W) \text{ of an animal} = 3.34 \times (\text{body mass (kg))}^{0.75}. \tag{2.1}$$

Thus, in order to maintain its body temperature for 100 days without food a mouse (0.01 kg) would need to carry three times as much fat as its whole lean body weight, whereas a bear (1000 kg) would need only a very small fraction (about 0.17 of lean body weight). Although bears are said to hibernate, they scarcely change their body temperature during the winter and remain active and alert in their dens. This is in sharp contrast with true hibernators such as marmots (the largest true hibernators), ground squirrels, hamsters, dormice and many bats. In these, the body temperature in winter falls to that of the surroundings (Figure 2.16a). Most mammals that truly hibernate are small and, just as their small weight leads them to lose heat rapidly, the relative cost of rewarming their bodies after hibernation is comparatively small (Table 2.2).

... has costs, such as that of rewarming

It is tempting to think of ectotherms as 'primitive' and endotherms as having gained 'advanced' control over their environment, but it is difficult to justify this view. Both ectotherms and endotherms are at risk of being killed by even short exposures to very low temperatures and by more prolonged exposure to moderately low temperatures (although the exact temperature depends on the stage in development and the species concerned). Both endotherms and ectotherms have an optimal environmental temperature and upper and lower lethal limits. There are costs to both ectotherms and endotherms when they live at temperatures that are not optimal. For the ectotherm the costs may be slower growth and reproduction, slow movement, failure to escape predators and a sluggish rate of search for food. But, for the endotherm the maintenance of body temperature costs energy that

71   CONDITIONS

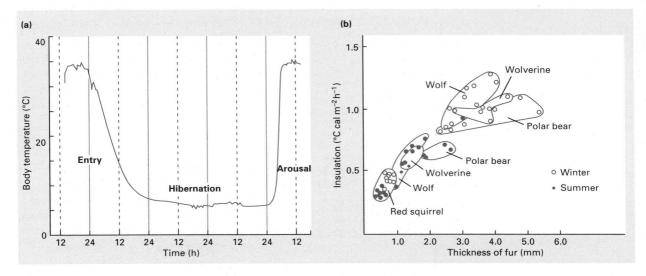

Figure 2.16 (a) Changes in the body temperature of a European hamster during a 3-day bout of hibernation. (After Nedergaard & Cannon, 1990; from Pévet *et al.*, 1989.) (b) Changes in insulation due to changes in the thickness of fur.

might have been used to catch more prey, produce and nurture more offspring or escape more predators. There are also costs of insulation (e.g. blubber in whales, fur in mammals) and costs of changing the insulation between seasons (Figure 2.16b). Temperatures only a few degrees higher than the metabolic optimum are liable to be lethal to endotherms as well as ectotherms.

Some of the hottest environments on earth are inhabited by mixed communities of endothermic and ectothermic animals, for example desert rodents and desert lizards. Some of the coldest environments on earth are also inhabited by a mixture of the two types, for example penguins and whales together with fish and krill at the edge of the Antarctic ice-sheet and polar bears and fish in the Arctic. It is clearly difficult for an organism to be both endothermic and to be smaller than the smallest

mixed communities of endotherms and ectotherms are found in many environments

Table 2.2 The energetic cost of rewarming after hibernation. The energetic cost of rewarming mammals of different sizes from 7 to 37°C. Body mass and metabolic rate are used to calculate the amount of time for which the animal could have maintained its normal active body temperature for the same energy cost as rewarming (rewarming equivalent). For a very small animal even short periods of hibernation (or torpor) are energetically meaningful because if the time spent at low temperature exceeds 3 h the animal will have gained energetically. (After Nedergaard & Cannon, 1990.)

Body mass (kg)	Animal	Metabolic rate (W)	Rewarming cost (kJ)	Rewarming equivalent (h)
0.01	Mouse	0.11	1.25	3
0.1	Rat	0.59	12.5	6
1	—	3.34	125	10
10	Marmot	18.8	1250	18
100	Human	106	12 500	33
1000	Bear	593	125 000	58

rodents or humming birds. It may also be difficult for a land animal to be ectothermic and to be as large as an elephant (although the largest sharks show that it is clearly possible to be a very large ectotherm in the sea). It is surprisingly difficult to make sweeping generalizations about the ecological conditions in which endotherms are 'superior to' ectotherms and might oust them in a struggle for existence. Indeed, debate and argument continues about whether dinosaurs, the most successful of all animals judged by their long domination of the planet, were endo- or ectotherms!

2.6 pH of soil and water

very high and very low pH may be lethal ...

The pH of soil in terrestrial environments or of water in aquatic ones is a condition which can exert a powerful influence on the distribution and abundance of organisms. The protoplasm of the root cells of most vascular plants is damaged as a direct result of toxic concentrations of H^+ or OH^- ions in soils below pH 3 or above pH 9, respectively. Further, indirect effects occur because soil pH influences the availability of nutrients and/or the concentration of toxins (Figure 2.17). Below pH 4.0–4.5, mineral soils contain such a high concentration of aluminium ions (Al^{3+}) as to be severely toxic to most plants. In addition, both manganese (Mn^{2+}) and iron (Fe^{3+}), which are essential plant nutrients, can be present at toxic concentrations at low pH. Increased acidity may act in three ways: (i) directly, by upsetting osmoregulation, enzyme activity or gaseous exchange across respiratory surfaces; (ii) indirectly, by increasing the concentration of toxic heavy metals, particularly Al^{3+}; and (iii) indirectly, by reducing the quality and range of food sources available to animals (e.g. fungal growth is reduced at low pH (Hildrew et al., 1984) and the aquatic flora is often absent or less diverse).

but most pH effects are indirect ...

Tolerance limits for pH vary amongst plant species, but only a minority are able to grow and reproduce at a pH below about 4.5. The situation is similar for animals inhabiting streams, ponds and lakes; typically the species richness of communities decreases in acid waters. There are important parallels in the role of pH in soil and aquatic systems; in both cases the effects may be direct (toxic concentrations of H^+ or OH^-) or indirect. Indirect effects involve either an interaction between two different conditions (pH and the concentration of a toxic chemical), or an interaction

... such as its influence on heavy metal toxicity ...

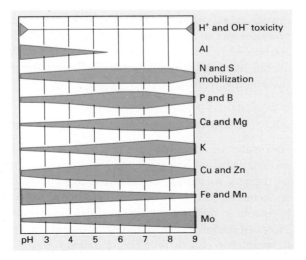

Figure 2.17 The toxicity of H^+ and OH^- to plants, and the availability to them of minerals (indicated by the widths of the bands) is influenced by soil pH. (After Larcher, 1980.)

between a *condition* (pH) and the availability of a *resource* (reduced fungal biomass as food for an invertebrate in a stream at low pH).

In alkaline soils, Fe^{3+} and phosphate (PO_4^{3+}), and certain trace elements such as Mn^{2+} are fixed in relatively insoluble compounds, and plants may then suffer because there is too little rather than too much of them. For example, calcifuge plants (those characteristic of acid soils) commonly show the symptoms of Fe^{3+} deficiency when they are transplanted to more alkaline soils. In general, however, soils and waters with pH above 7 tend to be hospitable to many more species than those that are more acid. Chalk and limestone grasslands carry a much richer flora (and associated fauna) than acid grasslands and the situation is similar for animals inhabiting streams, ponds and lakes.

Some prokaryotes, especially archaebacteria, can tolerate and even grow best in environments with pH far outside the range tolerated by eukaryotes. Such environments are rare, but occur in volcanic lakes and geothermal springs where they are dominated by chemolithotropic sulphur-oxidizing bacteria whose pH optima lie between 2 and 4 and which cannot grow at neutrality (Stolp, 1988). Examples are *Bacillus acidocaldarius* and *Cyanidium caldarius*. *Thiobacillus ferroxidans* occurs in the waste from industrial metal-leaching processes and tolerates pH 1.0; *T. thiooxidans* can not only tolerate but can grow at pH 0. Towards the other end of the pH range are the alkaline environments of soda lakes with pH values of 9–11, which are inhabited by cyanobacteria such as *Anabaenopsis arnoldii* and *Spirulina platensis; Plectonema nostocorum* can grow at pH 13.

... and on the availability of essential micronutrients

2.7 Salinity

For terrestrial plants the concentration of salts in the soil water offers osmotic resistance to water uptake. The most extreme saline conditions occur in arid zones where the predominant movement of soil water is towards the surface and cystalline salt accumulates. This occurs especially when crops have been grown in arid regions under irrigation; salt pans have then developed and the land is lost to agriculture. The main effect of salinity is to create the same kind of osmoregulatory problems as drought and freezing and the problems are countered in much the same ways. For example, many of the higher plants that live in saline environments (halophytes) accumulate electrolytes in their vacuoles, but maintain the concentration low in the cytoplasm and organelles (Robinson *et al.*, 1983). Such plants maintain high osmotic pressures and so remain turgid, and are protected from the damaging action of the accumulated electrolytes by polyols and membrane protectants (just like plants that tolerate freezing). A useful introduction to the literature of salt tolerance is given in Osmond *et al.* (1987).

salinity can cause problems in osmoregulation

2.7.1 The effects of salinity on the ecology of aquatic organisms

Freshwater environments present a set of specialized environmental conditions because water tends to move into organisms from the environment and this needs to be resisted. In marine habitats, the majority of organisms are isotonic to their environment so that there is no net flow of water: but, there are many that are hypotonic so that water flows out from the organism to the environment, putting them in a similar position to terrestrial organisms. Thus, for many aquatic organisms the regulation of body fluid concentration is a vital and sometimes an energetically

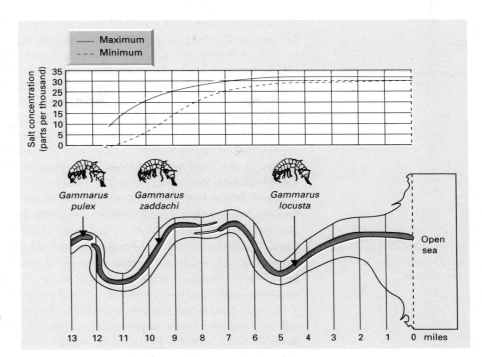

Figure 2.18 The distribution along British rivers of three closely related species of amphipod crustacean (relative abundance indicated by the width of the band), relative to the concentration of salt in the water. (After Spooner, 1947; from Cox *et al.*, 1976.)

expensive process. The salinity of an aquatic environment can have an important influence on distribution and abundance, especially in places like estuaries where there is a particularly sharp gradient between truly marine and freshwater habitats.

Figure 2.18 shows the typical distribution along British rivers of three closely related species of amphipod crustacean (Spooner, 1947). *Gammarus locusta* is an estuarine species and is found where the salt concentration never falls below 25 ppt (parts per 1000). *G. zaddachi* is moderately salt tolerant and is found in the region where the salt concentration exhibits considerable variation as a result of the tidal cycle, averaging 10–20 ppt. *G. pulex is* a truly freshwater species found only where the river shows no signs of tidal influence.

2.8 Conditions at the boundary between sea and land: interactions between salinity, exposure and substrate

Salinity has important effects on the distribution of organisms in intertidal areas but it does so through interactions with other conditions—notably exposure and the nature of the substrate. These interactions must have presented extreme ecophysiological problems in the evolution of land floras and faunas from the sea and land (Little, 1990).

Algae of all types have found suitable habitats permanently immersed in the sea, but permanently submerged higher plants are almost completely absent—except for one specialized group, the sea grasses. This is a striking contrast with submerged freshwater habitats where a variety of flowering plants have a conspicuous role. The main reason why most families of ferns and flowering plants have failed to colonize the sea seems to be that they require a substrate in which their roots can find anchorage. Large marine algae, like *Himanthalia* and *Laminaria*, which are

in intertidal areas salinity interacts with exposure and character of the substrate

75   CONDITIONS

continuously submerged except at extremely low tides, largely take the place of submerged flowering plants and ferns in marine communities. These algae do not have roots but attach to rocks by specialized 'holdfasts'. They are excluded from regions where the substrates are soft and 'holdfasts' cannot 'hold fast'. It is in such regions that the few truly marine flowering plants, for example sea grasses such as *Zostera* and *Posidonia*, form submerged communities, that produce abundant biomass and support complex animal communities.

algae dominate these areas

Most species of higher plant that root in seawater have leaves and shoots that are exposed to the atmosphere for a large part of the tidal cycle, such as mangroves, species of the grass genus *Spartina* and extreme halophytes such as species of *Salicornia* which have aerial shoots but whose roots are exposed to the full salinity of seawater. Where there is a stable substrate in which plants can root, communities of flowering plants may extend right through the intertidal zone in a continuum extending from those continuously immersed in full-strength seawater (like the sea grasses) through to totally non-saline conditions (see, e.g. Ranwell, 1972). Salt-marshes in particular encompass a range of salt concentrations running from full-strength seawater down to totally non-saline conditions (and support a range of associated species with different degrees of salt tolerance).

higher plants are absent unless there is a substrate in which they can root

Higher plants are absent from intertidal rocky sea shores except where pockets of soft substrate may have formed in crevices. Instead, such habitats are dominated by the algae which give way to lichens at and above high tide level where the exposure to desiccation is highest. The plants and animals that live on rocky seashores are influenced by environmental conditions in a very profound and often particularly obvious way by the extent to which they tolerate exposure to the aerial environment and the forces of wave and storm. This expresses itself in the *zonation* of the organisms, with different species at different heights up the shore; yet, as Figure 2.19 makes clear, the precise nature of this zonation is crucially dependent on the physical characteristics of the particular shore. On a very calm and sheltered shore (Figure 2.19a), variations in exposure are almost entirely the result of the twice-daily, tidal changes in sea-level. But, on a very exposed shore, waves and sea-spray change the pattern. They extend the shore environment upwards and they alter the detailed nature of zonation (although not its underlying character; Figure 2.19b).

waves and storms are limiting conditions for life on rocky shores

To talk of 'zonation as a result of exposure', however, is to over-simplify the matter greatly (see, e.g. Lewis, 1976, for a more extended discussion). In the first place, 'exposure' can mean a variety, or a combination of, many different things: desiccation, extremes of temperature, changes in salinity, excessive illumination and the sheer physical forces of pounding waves and storms (to which we return in Section 2.9). For example, amongst the seaweeds around British coasts the red algae of the lower shore appear in unusual luxuriance in the midshore wherever the heavier, blanketing layers of the brown fucoid algae, *Fucus serratus* and *Ascophyllum nodosum*, are experimentally removed (Lewis, 1976). Furthermore, 'exposure' only really explains the *upper* limits of these essentially marine species, and yet zonation depends on them having lower limits too. For some species there can be *too little exposure* in the lower zones. For instance, green algae would be starved of blue and especially red light if they were submerged for long periods too low down the shore. For many other species though, a lower limit to distribution is set by competition and predation (see, e.g. discussion in Paine, 1994). The seaweed *F. spiralis* will readily extend lower down the shore than usual in Great Britain whenever the competing midshore fucoids are scarce.

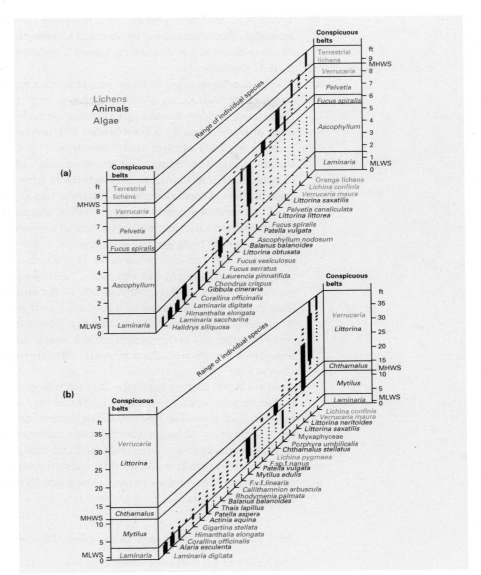

Figure 2.19 The seashore zonation of animals and plants around Great Britain: (a) on a calm and sheltered shore, Clachan Sound; and (b) on an exposed shore, Parkmor Point. (After Lewis, 1976.) MHWS, mean high water of spring tides; MLWS, mean low water of spring tides.

2.9 The physical forces of winds, waves and currents

In nature there are many forces of the environment that have their effect by virtue of the force of physical movement—wind and water are prime examples.

In streams and rivers, both plants and animals face the continual hazard of being washed away. The average velocity of flow generally increases in a downstream direction, but the greatest influence on the benthic (bottom-dwelling) community is in upstream regions, because the water here is turbulent and shallow, and the animals and plants are exposed to their greatest danger of being washed away. The only plants to be found in the most extreme flows are literally 'low-profile' species like encrusting and filamentous algae, mosses and liverworts. Where the flow is slightly less extreme there are plants like the water crowfoot (*Ranunculus fluitans*),

77   CONDITIONS

which is streamlined, offering little resistance to flow and which anchors itself around an immovable object by means of a dense development of adventitious roots. Plants such as the free-floating duckweed *(Lemna* spp.) are usually only found where there is negligible flow.

The conditions of exposure on seashores place severe limits on the life forms and habits of species that can tolerate repeated pounding and the suction of wave action. Algae, such as fucoids and kelps anchored on rocks, survive the repeated pull and push of wave action by a combination of powerful attachment by 'holdfasts' and extreme flexibility of their thallus structure. There is no woody tissue in an algal thallus. Animals in the same environment either move with the mass of water or, like the algae, rely on subtle mechanisms of firm adhesion such as the powerful organic glues of barnacles and the muscular feet of limpets.

A comparable diversity of morphological specializations is to be found amongst the invertebrates that tolerate the hazards of turbulent, freshwater streams. Extreme flattening of the body permits some species to live in the relatively still 'boundary layer' just above the substratum of the stream bed, or to live in crevices under stones and thus avoid the current altogether. Some species are able to maintain their position by means of hooks or suckers, whilst others, such as the mayfly nymph *(Baetis)*, have long tail cerci which keep their body facing into the current where the streamlined body shape reduces drag to a minimum (Townsend, 1980). Many fish can, of course, stay in the same place in a rapidly flowing current only by sheltering from it (e.g. loaches and eels) or by continual energy-expensive swimming against the current.

Wave action in the sea and the forces of water currents in rivers and streams are part of the everyday (or at least the every year) experience of the organisms that live in them. Such forces can be described as hazards, i.e. regular and reliable risks of life in these habitats. At longer intervals these same forces occur on a more dramatic scale as storms that can sweep rocks (and everything attached to them) down a river or stream, or overturn and strip rocks and stones on a maritime shore (rolling stones gather no mussels—neither do they allow a natural community of aquatic colonizers to progress very far).

2.9.1 Hazards, disasters and catastrophes: the ecology of extreme events

The wind and the tides are normal daily hazards in the life of many organisms. Their structure and behaviour bear some witness to the frequency and intensity that such frequent hazards have played in the evolutionary history of their species. Thus, most trees withstand the force of most storms without falling over or losing their living branches. Most limpets, barnacles and kelps hold fast to the rocks through the normal day to day forces of the waves and tides. We can also recognize a scale of more severely damaging forces (we might call them 'disasters'), that occur occasionally, but with sufficient frequency to have contributed repeatedly to the forces of natural selection. When such a force recurs it will meet a population that still has a genetic memory of the selection that acted on its ancestors—and may therefore suffer less than they did. In the woodlands and shrub communities of arid zones, fire has this quality, and tolerance of fire damage is a clearly evolved response (see Section 2.4 and Naveh, 1975). There are clearly also many species of plant and animal whose life depends on opportunistically exploiting 'disasters' that have affected others.

When disasters strike natural communities it is only rarely that they have been carefully studied before the event. One exception is cyclone 'Hugo' which struck the Caribbean island of Guadeloupe in 1994. Detailed accounts of the dense humid forests of the island had been published only 4 years before cyclone 'Hugo' (Ducrey & Labbé, 1985, 1986). The cyclone devastated the forests with mean maximum wind velocities of 270 km h^{-1} and gusts of 320 km h^{-1}. Up to 300 mm of rain fell in 40 h. Early stages of regeneration after the cyclone (Labbé, 1994) typify the responses of long-established communities on both land or sea to massive forces of destruction. Even in 'undisturbed' communities there is continual creation of gaps as individuals (e.g. trees in a forest, kelps on a seashore) die and the space they occupied is recolonized. After massive devastation by cyclones or other widespread disasters, much of the recolonization follows much the same course as the microsuccessions that were part of the natural regeneration cycle in the previously undisturbed community. Species that normally colonized only natural gaps in the vegetation come to dominate a continuous community. Whether the community is that of a rocky shore or a tropical rainforest the early succession after disasters tends to be of competition- (or shade-) intolerant, relatively short-lived species with high rates of precocious reproduction.

In contrast to conditions that we have called 'hazards' and 'disasters' there are natural occurrences that are enormously damaging, yet occur so rarely that they may have no lasting selective effect on the evolution of the species. We might call such events 'catastrophes', for example the volcanic eruption of Mt St Helens or of Krakatau. The next time that Krakatau erupts there are unlikely to be any genes persisting that were selected for volcano tolerance!

Of course, what we have called ecological hazards, disasters and catastrophes are arbitrary stages on a continuum. It would be interesting, and perhaps valuable to ecologists, to quantify this continuum—the equivalent of an ecological risk analysis. We might then hope to express the frequency of risk in terms that have relevance to the length of life of particular organisms, the length of periods of dormancy, the length of the juvenile non-reproductive phase of life cycles, etc. The bulk of statistical theory has, however, been mainly concerned with means and variances, i.e. with the 'normal' range of happenings. Ecologists require a statistical procedure that is appropriate for rare events and it is convenient that the probability structure of extreme values conforms to a generalized distribution that can be estimated by maximum likelihood techniques (Gaines & Denny, 1993). Such a statistical procedure has been applied to sea surface temperatures, wave forces, wind speeds and human life spans and (given that some statistical precautions, such as the removal of trends, are observed) accurate long-term predictions can apparently be made from a surprisingly small number of measurements. Figure 2.20 gives one example of the application of the technique to wind speeds to predict the time expected to elapse before a particular deviation from average weekly maximum wind speed is observed again—the 'return time'. We have emphasized that for the distribution of organisms in nature it is usually extremes rather than averages that matter. The statistics of extreme values is therefore likely to become increasingly valuable to ecologists.

2.10 Environmental pollution

A number of environmental conditions that are, regrettably, becoming increasingly important are due to the accumulation of toxic by-products of humans' activities.

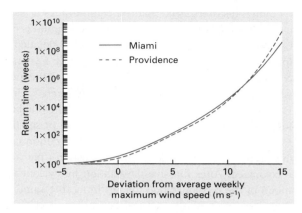

Figure 2.20 Predicted return times for deviations from the average weekly maximum wind speed (i.e. the predicted time interval before a deviation of particular magnitude occurs again). Calculations were based on measurements made at 3-hourly intervals for 7 years at Providence, Rhode Island and Miami, Florida, but the period did not include any hurricanes. (After Gaines & Denny, 1993.)

Sulphur dioxide emitted from power stations, and metals like copper, zinc and lead, dumped around mines or deposited around refineries, are just some of the pollutants that limit distributions, especially of plants. Many such pollutants are present naturally, but at low concentrations, and some are indeed essential nutrients for plants. But, in polluted areas their concentrations can rise to lethal levels and there is a succession of disappearances of one species after another. The loss of species is often the first indication that pollution has occurred and changes in the species richness of a river, lake or area of land provide bioassays of the extent of their pollution (see, e.g. Lovett Doust *et al.*, 1994).

Yet, it is rare to find even the most inhospitable polluted areas entirely devoid of species; there are usually at least a few individuals of a few species that can tolerate the conditions. Even natural populations from unpolluted areas often contain a low frequency of individuals that tolerate the pollutant (Figure 2.21); this is part of the genetic variability present in sexually reproducing populations. Such individuals may be the only ones to survive or colonize as pollutant levels rise, and then become the founders of a tolerant population to which they have passed on their 'tolerance' genes. Pollution, therefore, provides us with an ideal opportunity to observe evolution in action. However, sufficient genetic variability is not present in all populations; some species repeatedly give rise to tolerant populations, whilst others rarely if ever do so (Gartside & McNeilly, 1974; Bradshaw & McNeilly, 1981).

Thus, in very simple terms, a pollutant has a twofold effect. When it is newly arisen or is at extremely high concentrations, there will be few individuals of any species present (the exceptions being naturally tolerant variants or their immediate descendants). Subsequently, however, the polluted area is likely to support a much higher density of individuals, but these will be representatives of a much smaller range of species than would be present in the absence of the pollutant. Such newly evolved, species-poor communities are now an established part of human environments (Bradshaw, 1987).

genetic variation may sometimes allow the evolution of pollution-tolerant forms

80   CHAPTER 2

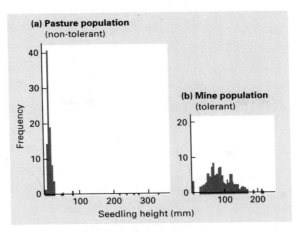

Figure 2.21 When seedlings of a generally non-tolerant pasture population of bent grass (*Agrostis tenuis*) are grown from seed on copper-contaminated soil (a), a few individuals exhibit copper tolerance, and grow successfully. These individuals might presumably give rise to a copper-tolerant population, since their tolerance is comparable with that shown by the most tolerant individuals from a population growing near a copper mine, where, however, mean tolerance is much higher (b). (After Walley *et al.*, 1974.)

some pollutants may have effects far from their source ...

... for example 'acid rain'

Pollution can of course have its effects far from the original source (Figure 2.22). Toxic effluents from a mine or a factory may enter a watercourse and affect its flora and fauna for its whole length downstream. Effluents from large industrial complexes can pollute and change the flora and fauna of many rivers and lakes in a region and cause international disputes. A striking example is the creation of 'acid rain' falling in Scotland and Scandinavia from industrial activities in other countries. Since the Industrial Revolution, the burning of fossil fuels and the consequent emission to the atmosphere of various pollutants, notably sulphur dioxide, has produced a deposition of dry acidic particles and rain that is essentially dilute sulphuric acid. Our knowledge of the pH tolerances of diatom species enables an approximate pH history of a lake to be constructed. The history of the acidification of lakes is often recorded in the succession of diatom species accumulated in lake sediments (Battarbee, 1984).

changing diatom populations reflect a history of environmental pollution

Figure 2.23 shows how diatom species composition has changed over the past 400 years or more in Round Loch of Glenhead, Scotland—far from major industrial sites. A sediment core containing the accumulated identifiable remains of diatoms has been analysed section by section, and these show a rapid and dramatic decline in those species that are rarely found below pH 5.5 (e.g. *Anomoeoneis vitrea* and *Tabellaria flocculosa*), and at the same time an increase in species typical of acid conditions (e.g. *Eunotia veneris* and *T. quadriseptata*). Since about 1850 the pH has declined from about 5.5 to about 4.6.

2.11 Global change

In Chapter 1 we discussed some of the ways in which global environments have changed over the long time scales involved in continental drift and the shorter time scales of the repeated ice-ages. Over these time scales some organisms have failed to accommodate to the changes and have become extinct, others have migrated so that

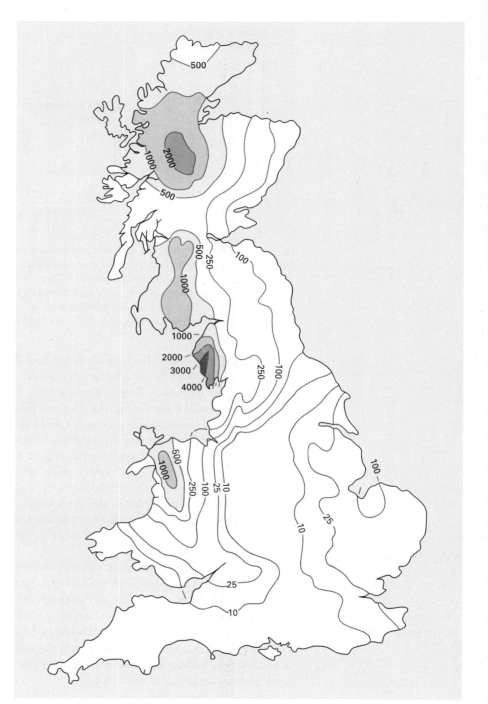

Figure 2.22 An example of long-distance environmental pollution. The distribution in Great Britain of fallout of radioactive caesium (Bq m^{-2}) from the Chernobyl nuclear accident in the Soviet Union in 1986. The map shows the persistence of the pollutant on acid-upland soils where it is recycled through soils, plants and animals. Sheep in the upland areas contained more caesium-137 in 1987 and 1988 (after recycling) than in 1986. Caesium-137 has a half-life of 30 years! On typical lowland soils it is more quickly immobilized and does not persist in the food chains. (After NERC, 1990.)

they continue to experience the same conditions but in a different place, and it is probable that others have changed their nature (evolved) and tolerated some of the changes. We now turn to consider global changes that are occurring in our own lifetimes—consequences of our own activities—and that are predicted, in most scenarios, to bring profound changes in the ecology of the planet.

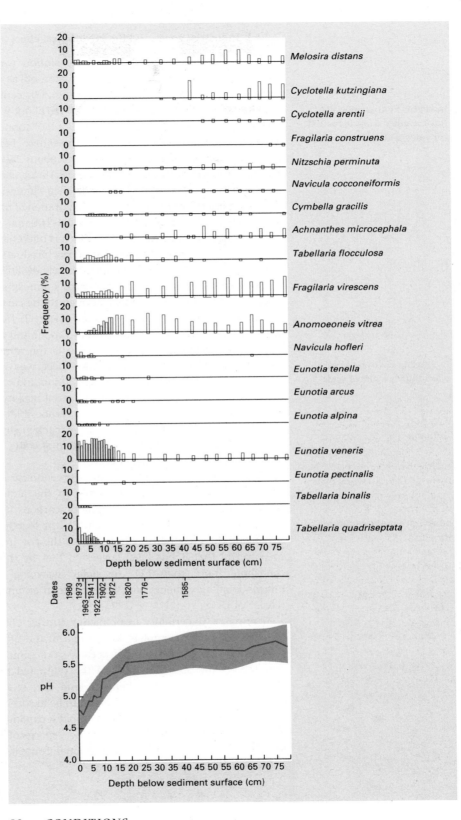

Figure 2.23 The percentage frequency of selected diatom species at various depths in a sediment core from Round Loch of Glenhead, Scotland. Dates corresponding to particular depths have been obtained by an isotope radioactive decay technique (lead-210). This has allowed the pH history to be reconstructed on the basis of knowledge of pH tolerance ranges of the species concerned (on the assumption that these have not changed over time). (After Flower & Báttarbee, 1983.)

2.11.1 Industrial gases and the greenhouse effect

increases in atmospheric CO_2 have resulted from industrial activities and deforestation

A major element of the Industrial Revolution was the switch from the use of sustainable fuels to the use of coal (and later, oil) as a source of power. Between the middle of the 19th and the middle of the 20th century the burning of fossil fuels, together with extensive deforestation, added about 9×10^{10} tonnes of carbon dioxide (CO_2) to the atmosphere and a further 9×10^{10} tonnes have been added since 1950. The concentration of CO_2 in the atmosphere before the Industrial Revolution (measured in gas trapped in ice cores) was about 280 ppm and had risen to 345 ppm by 1985 (Figure 2.24; see also Figure 18.15 for changes measured directly over a shorter period in Hawaii). There have been changes in other atmospheric gases in the same period. For example, the concentration of methane (CH_4) has risen from 0.7 to 1.7 ppm, and of ozone (O_3) from 0 to 10–100 ppv (parts per volume). The increase in CH_4 is not all explained but probably has microbial origin in intensive agriculture on anaerobic soils (especially increased rice production) and in the digestive process of ruminants (a cow produces approximately 40 litres of CH_4 each day).

... causing environmental warming on a global scale

Solar radiation incident on the earth's atmosphere is in part reflected, in part absorbed and part is transmitted through to the earth's surface, which absorbs and is warmed by it. Part of this absorbed energy is radiated back to the atmosphere where atmospheric gases, mainly water vapour and CO_2 absorb about 70% of it. It is this trapped reradiated energy that heats the atmosphere in what is called the 'greenhouse effect'. The greenhouse effect was of course part of the normal environment before the Industrial Revolution and carried responsibility for some of the environmental warmth before industrial activity started to enhance it. At that time, atmospheric water vapour was responsible for the greater portion of the greenhouse effect. There is a positive feedback in so far that warming of the earth's surface provides latent heat of evaporation of water which then contributes further to the trapping of reradiated energy.

other trace gases contribute to the effect

In addition to the enhancement of greenhouse effects by increased CO_2, other trace gases have increased markedly in the atmosphere particularly CH_4, O_3, nitrous oxide (N_2O) and the chlorofluorocarbons (CFCs, e.g. carbon tetrachloride (CCl_3) and dichlorodifluoromethane (CCl_2F_2)): together these contribute as much to enhancing the greenhouse effect as does the rise in CO_2. The effect of the CFCs is so powerful that it has been suggested that by the year 2050 they alone could contribute more than CO_2 to global warming (Dickson & Cicerone, 1986). International agreements have been made in an attempt to halt further rise in the release of CFCs.

The major uncertainty in predicting further changes in the greenhouse effect comes from very incomplete knowledge of the effect of the oceans in buffering atmospheric levels of CO_2. The oceans store some 50 times more CO_2 than the atmosphere and less than 60% of the estimated production of CO_2, due to the burning of fossil fuel and changing land use, is still present in the atmosphere (IGBP, 1990b). The oceans are thought to be absorbing most of the remainder—but, there is much ignorance. The very rapidly expanding research programmes and literature can be followed through the reports of the International Geosphere–Biosphere programme of the International Council of Scientific Unions (e.g. IGBP, 1990a, b).

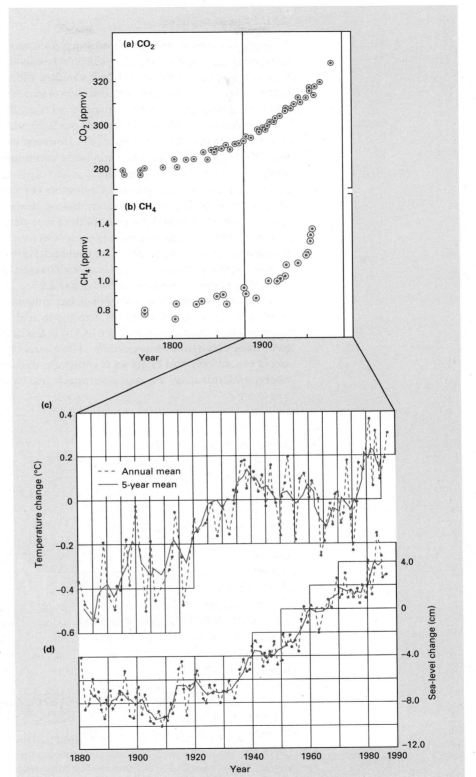

Figure 2.24 The record of global changes over 150 years until 1990. Curves (a) and (b) are reconstructions of atmospheric CO_2 and methane (CH_4) concentrations based on the analysis of gas trapped in ice collected at Siple Station, West Antarctica. (After Stauffer, 1988.) Concentrations in 1990 were 345 ppm (by volume) CO_2 and 1.7 ppm CH_4. Curves (c) and (d) are records of global mean annual temperature and sea-level changes. In both (c) and (d) the average value for 1951–1970 is set equal to zero, with data reported as deviations from zero: the lines drawn are 5-year running means. (After IGBP, 1990a.)

2.11.2 Global warming

Predictions of the extent of global warming consequent on the enhanced greenhouse effect come from two sources: (i) predictions based on sophisticated computer models ('General Circulation Models') which simulate the world's climate; and (ii) trends detected in measured data sets. A doubling of atmospheric CO_2 is widely expected to occur by the year 2100 and perhaps as soon as 2020. Not surprisingly, different global circulation models differ in their predictions of the rise in global temperature that will result from the doubling of CO_2. However, most model predictions vary only from 2.3 to 5.2°C (most of the variation is accounted for by the way in which the effects of cloud cover are modelled).

A diversity of direct and indirect indicators of recent past climates provides some evidence that global temperatures are indeed rising. Sources of relevant measurements include pollen analysis, the width of tree rings, carbon-14 analysis of dated tree rings, sea-level records, the gas content of ice-cores at polar sites, measures of the rate of retreat of glaciers, dated annual layers in lake and other terrestrial sediments, dated marine sedimentology with oxygen and carbon isotope ratios. Some examples are given in Figures 2.24c, d and 2.25.

Computer modelled projections of global temperatures (based on the greenhouse effect) and extrapolation of trends detected in real and surrogate climatic records taken together make a projected rise of 3°C in the next 100 years seem a reasonable value from which to make projections of ecological effects. But, temperature regimes are, of course, only part of the set of conditions that determine which organisms live where. Unfortunately, we can place much less faith in computer projections of

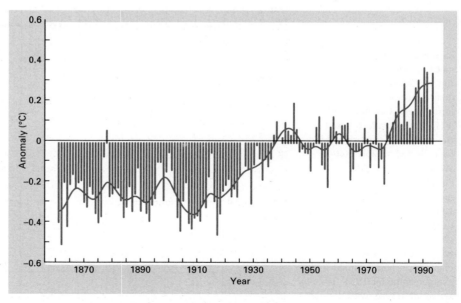

Figure 2.25 Annual global averages for combined land and sea temperatures from 1861 to 1993 (GER, 1993). The graph represents anomalies in degrees Celsius from the 1951 to 1980 average. The continuous line is a smoothed estimate designed to suppress temperature variations on a scale of less than about one decade.

rainfall and evaporation because it proves very hard to build good models of cloud behaviour into a general model of climate. If we consider only temperature as a relevant variable we would project a 3°C rise in temperature giving London (England) the climate of Lisbon (Portugal) (with an appropriate vegetation of olives, vines, *Bougainvillea* and semi-arid scrub). But, with more reliable rain it would be nearly subtropical, and with a little less might qualify for the status of an arid zone!

We have emphasized that the behaviour of many organisms is determined by the occasional extremes rather than by average conditions. Computer modelled projections imply that global climatic change will bring greater variance that, in Great Britain, implies a dramatic increase in the number of very warm years. The English summer in 1976 was exceptionally warm (mean 17.8°C)—a record achieved on average only once in 1000 years (0.1%). An increase in mean temperature of 2.1°C, *even without the increase in variance*, implies that such summers would occur on average every third year (HMSO, 1991).

A wide-ranging review of the ecological consequences of global change is given by Vitousek (1994).

In Chapter 1 we described some of the changes in the distribution of forests that accompanied the advance and retreat of the last ice-age—these indicate that the movement of isotherms was considerably faster than the movement of vegetation. Moreover, there was strong evidence that assemblages of species (community types) did not move as if they were entities—but, rather that different species migrated at different speeds and in different directions. There is every reason to expect that global climatic change in the coming centuries will follow similar patterns except that: (i) the projected speed of change under the conditions of the greenhouse effect is approximately 30 times faster than occurred during the retreat of the last glaciation; and (ii) large tracts of land over which vegetation might advance and retreat have been fragmented in the process of civilization, putting major barriers in the way of vegetational advance. It will be very surprising if many species do not get lost on the journey.

the predicted speed of global warming is much faster than that at the end of the last ice-age

2.12 The ecological niche

The list of environmental conditions that we have considered is not exhaustive, but general principles have emerged, supported by a range of different examples. We are now in a position to consider a concept that is central to much of ecological thinking, the definition of which depends crucially on the responses of organisms to environmental conditions. This concept is the ecological niche.

The term ecological niche has been part of the ecologist's vocabulary for more than half a century, but for the first 30 years its meaning was rather vague (see Vandermeer, 1972, for a historical review). The current, generally accepted definition (Hutchinson, 1957) is best illustrated by example.

Organisms of any given species can survive, grow, reproduce and maintain a viable population only within certain temperature limits. This range of temperature is the species' ecological niche in one dimension (i.e. the dimension 'temperature'; Figure 2.26a). Examples of optimum temperature ranges for a variety of plant species are illustrated in Figure 2.27a. (It should be noted that the niche of a species in any given dimension cannot always be conceived simply in terms of a maximum and a minimum value between which survival and reproduction are possible. Sometimes,

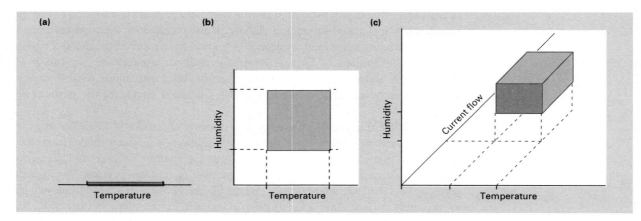

Figure 2.26 Ecological niches. (a) In one dimension (temperature); (b) in two dimensions (temperature and humidity); (c) in three dimensions (temperature, humidity and, say, pH).

there is no minimum (as in Figure 2.1c), or we may need to consider more complex 'regimes' which include, e.g. allowable daily or seasonal fluctuations.)

Of course, an organism is not affected by temperature in isolation, nor by any other single condition. Thus, organisms of the species in question may also be able to survive and reproduce only within certain limits of relative humidity. Taking temperature and humidity together, the niche becomes two dimensional and can be

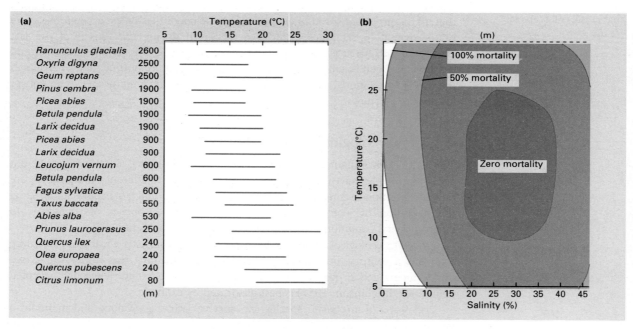

Figure 2.27 (a) Niches in one dimension: optimum temperature ranges for net photosynthesis (at low light intensity, 70 W m^{-2}) in species originating from various altitudes in the Alps. (After Pisek *et al.*, 1973.) (b) A niche in two dimensions for the sand shrimp (*Crangon septemspinosa*). Contours represent 0, 50 and 100% mortality in experiments in aerated water with egg-bearing females at a range of temperatures and salinities. (After Haefner, 1970.)

visualized as an *area* (see Figure 2.26b). A real example is given in Figure 2.27b, showing the niche of a sand shrimp in two dimensions—temperature and salinity.

If further conditions are brought into consideration (as they should be) then the next step is a three-dimensional description of the niche, a *volume* (see Figure 2.26c); but the incorporation of more than three dimensions is impossible to visualize. Nevertheless, the process can be continued in an abstract way, by analogy, and a species' true ecological niche can be thought of as an *n-dimensional hypervolume* within which it can maintain a viable population.

This, in essence, is the niche concept developed by Hutchinson (1957), except that in addition he proposed a separate niche dimension not only for each important environmental condition, but also for each of the resources that the organism requires (e.g. solar radiation, water, mineral nutrients for plants; food, nest sites, etc.: for animals, see Chapter 3). This view is now one of the cornerstones of ecological thought.

Provided that a location is characterized by conditions within acceptable limits for a given species, and provided also that it contains all necessary resources, then the species can, potentially, occur and persist there. Whether or not it does so depends on two further factors. First, it must be able to reach the location, and this depends in turn on its powers of colonization and the remoteness of the site. Second, its occurrence may be precluded by the action of individuals of other species which compete with it or prey on it.

Usually, a species has a larger ecological niche in the absence of competitors and predators than it has in their presence. In other words, there are certain combinations of conditions and resources which *can* allow a species to maintain a viable population, but only if it is not being adversely affected by enemies. This led Hutchinson to distinguish between the *fundamental* and the *realized* niche. The former describes the overall potentialities of a species; the latter describes the more limited spectrum of conditions and resources which allow it to persist, even in the presence of competitors and predators. Fundamental and realized niches will receive more attention in Chapter 7, when we look at interspecific competition.

In common usage the word 'niche' means a position in space. This is not the meaning in ecological science and it is important to recognize that an ecological niche is not something that can be seen. It is an abstract concept that brings together, in a single descriptive term, all of the environmental conditions and resources that are necessary for an organism to maintain a viable population. Nor is it necessary to make measurements along each and every niche dimension for the concept to be useful.

Habitats, by contrast, are actual places, and as such they provide a variety of conditions and resources which may satisfy the requirements (i.e. provide the niches) for warblers, oak trees, spiders and myriads of other species. The niches of the species that occur together in any one habitat are likely to be different, sometimes markedly so. We return to this point in Chapter 7.

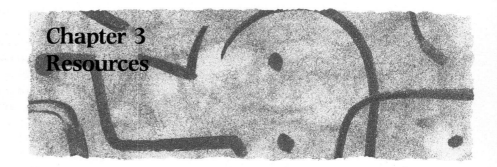

Chapter 3
Resources

3.1 Introduction

Science adopts many words from common speech, and they can carry with them a cloud of nuances that bear little relation to the science. This is true of words like 'conditions' and 'resources', which have become assimilated into the science of ecology. Yet, the very act of searching for what is meant by a word can deepen understanding.

what is a resource

According to Tilman (1982), *all things consumed by an organism* are resources for it:

> Just as nitrate, phosphate and light may be resources for a plant, so may nectar, pollen and a hole in a log be resources for a bee, and so may acorns, walnuts, other seeds, and a larger hole in a log be resources for a squirrel.

But, even Tilman's broad definition poses further questions. For example, what is meant by 'consumed'? It cannot simply mean *eaten* (although, nectar and acorns are eaten) nor yet *incorporated into biomass* (although, this happens to nitrate, nectar and acorns): bees and squirrels do not eat holes nor incorporate them into biomass. Nevertheless, a hole that is occupied is no longer available to another bee or squirrel, just as the atom of nitrogen, the sip of nectar or the mouthful of acorn is no longer available to other consumers. Similarly, females that have already mated may be unavailable to other mates. All these things have been consumed in the sense that the stock or supply has been reduced. Thus, they are resources, not conditions, because they represent *quantities* that can be reduced by the activity of the organism.

The resources of living organisms are mainly the stuffs of which their bodies are made, the energy that is involved in their activities and the places or spaces in which they act out the cycles of their lives. The body of a green plant is assembled from inorganic ions and molecules which represent its *food* resources, whilst solar radiation, trapped in photosynthesis, provides the *energy* resource. Green plants themselves represent packages of food resources for herbivores which, in turn, form packages of resources for carnivores. The bodies of organisms also represent food resources for parasites and, when dead, for microflora and detritivores. A large part of ecology is about the assembly of inorganic resources by green plants and the reassembly of these packages at each successive stage in a web of consumer interactions. Thinking about resources becomes particularly important when we come to consider how what one organism consumes affects what is available to others (of the same or of different species). We develop this theme in Chapters 6 and 7.

3.2 Radiation as a resource

Solar radiation is the only source of energy that can be used in metabolic activities by green plants. It differs in many ways from all other resources.

Radiant energy comes to the plant as a flux of radiation from the sun, either directly or after it has been diffused by the atmosphere or reflected or transmitted by other objects. The relative amounts of direct and diffused radiation arriving at an exposed leaf depend on the amount of dust in the air and, in particular, the thickness of the scattering air layer between the sun and the plant. The direct fraction is highest at low latitudes (Figure 3.1).

When a leaf intercepts radiant energy it may be reflected (with its wavelength unchanged), transmitted (after some wavebands have been filtered out) or absorbed. Part of the fraction that is absorbed may raise the leaf temperature and be reradiated (but now at much longer wavelengths), and part may contribute latent heat of evaporation of water and so power the transpiration stream. A small part may reach the chloroplasts and drive the process of photosynthesis (Figure 3.2).

radiant energy must be captured or is lost forever

Radiant energy is converted during photosynthesis into energy-rich chemical compounds of carbon which will subsequently be broken down in respiration (either by the plant itself or by those that consume it). But, unless the radiation is captured and chemically fixed at the instant it falls on the leaf, it is irretrievably lost for photosynthesis. Radiant energy that has been fixed in photosynthesis passes just once through the world. This is in complete contrast to an atom of nitrogen or carbon or a molecule of water that may *cycle* repeatedly through endless generations of organisms.

radiation is a spectrum of which photosynthesis uses only a part

Solar radiation is a resource continuum—a spectrum of different wavelengths— but, the photosynthetic apparatus is able to gain access to energy in only a restricted band of this spectrum. All green plants depend on chlorophyll pigments for the photosynthetic fixation of carbon, and these pigments fix radiation in a waveband between 380 and 710 nm (or in broad terms 400–700 nm). This is the band of 'photosynthetically active radiation' (PAR). It corresponds broadly with the range of

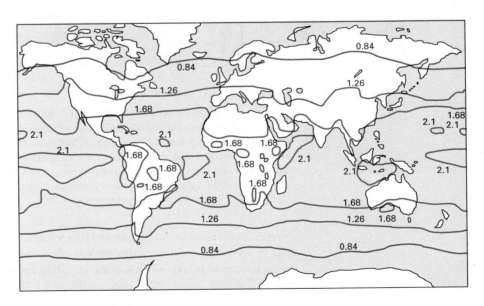

Figure 3.1 Global map of the solar radiation absorbed annually in the earth–atmosphere system: from data obtained with a radiometer on the Nimbus 3 meteorological satellite. The units are J cm^{-2} min^{-1}. (After Raushke et al., 1973.)

Figure 3.2 The reflection (R) and attenuation of solar radiation falling on various plant communities. The arrows show the percentage of incident radiation reaching various levels in the vegetation. (a) A boreal forest of mixed birch and spruce; (b) a pine forest; (c) a field of sunflowers; and (d) a field of corn (maize). These figures represent data obtained in particular communities and great variation will occur depending on the stage of growth of the forest or crop canopy, and on the time of day and season at which the measurements are taken. (After Larcher, 1980, and a variety of sources.)

the spectrum visible to the human eye that we call 'light'. It is not strictly correct to refer to light as a resource, although it is often convenient to do so—it is radiation that is the resource. About 56% of the radiation incident on the earth's surface lies outside the PAR range and is unavailable as a resource for green plants. In other organisms there are pigments that operate in photosynthesis outside the PAR range of green plants, e.g. bacteriochlorophyll has its peak absorption at 800, 850 and 870–890 nm. However, the chlorophyll pigments of green plants put fundamental constraints on the radiation that green plants can use as a resource, and so also constrain how much energy they contribute to the community in which they live.

3.2.1 Radiation is a diurnally and seasonally variable resource

A major reason why plants seldom achieve their intrinsic photosynthetic capacity is that the intensity of radiation varies continually (Figure 3.3). Plant morphology and leaf physiology that are optimal for photosynthesis at one intensity of radiation will usually be inappropriate at another. Most leaves in the real world live in a radiation regime that varies throughout the day and the year, and they live in an environment of other leaves that modifies the quantity and quality of radiation received. This illustrates two vital properties of all resources: their supply can vary both systematically and unsystematically.

systematic variation in the supply of radiation

Annual and diurnal rhythms are systematic variations in solar radiation (Figure 3.3). The green plant experiences periods of famine and glut in its radiation resource every 24 h (except near the poles) and seasons of famine and glut every year (except in the tropics). The ways in which an organism or an organ reacts to a systematic (predictable) supply of a resource reflects both its present physiology and its past evolution. The seasonal shedding of leaves by deciduous trees in temperate regions *in part* reflects the annual rhythm in the intensity of radiation—they are shed when they are least useful. In consequence, an evergreen leaf of an understorey species may experience a further systematic change, because the seasonal cycle of leaf production of overstorey species determines what radiation remains to penetrate to the understorey. The daily movement of leaves in many species also reflect the changing intensity and direction of incident radiation.

unsystematic variation in the supply of radiation

Less systematic variations in the radiation environment of a leaf are caused by the nature and position of neighbouring leaves and those that overtop it. Each canopy, each plant and each leaf, by intercepting radiation, creates a resource-depletion zone (RDZ)—a moving band of shadow over other leaves of the same plant, or of others.

Deep in a canopy, shadows (radiation-depletion zones) become less well defined because much of the radiation loses its original direction by diffusion and reflection. The composition of radiation that has passed through leaves in a canopy is also altered—it is photosynthetically less useful because the PAR component has been reduced.

strategic and tactical responses to variation in supply

Amongst animals and plants there are classic responses to varying environments. Where the environmental variation is systematic and repeated, there is commonly a determinate pattern of response—an evolved, genotypically fixed programme that allows little flexibility or plasticity. The analogy in military terms is with a rigid strategy, leaving little room for tactical manoeuvre. By contrast, the typical response to non-systematic, unpredictable variation in the environment is to make tactical changes.

The major 'strategic' differences between species in their reaction to the intensity of radiation are those evolved between 'sun species' and 'shade species'. In general, plant species that are characteristic of shaded habitats use radiation at low intensities more efficiently than sun species, but the reverse is true at high intensities (Figure 3.4).

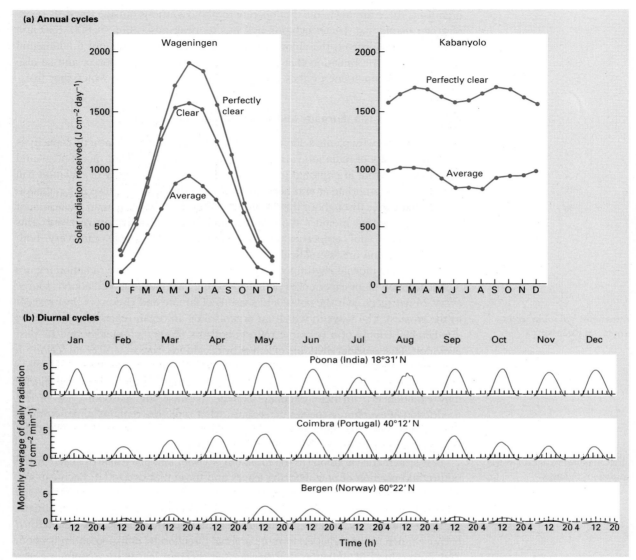

Figure 3.3 (a) The daily totals of solar radiation received throughout the year at Wageningen (Holland) and Kabanyolo (Equatorial Africa). (b) Monthly average of daily radiation recorded at Poona (India), Coimbra (Portugal) and Bergen (Norway). (After de Wit, 1965; and others.)

Part of the difference between sun and shade species lies in the physiology of the leaves, but the morphology of the plants also influences the efficiency with which radiation is captured. The leaves of sun plants are commonly exposed at acute angles to the midday sun (see, e.g. Poulson & DeLucia, 1993). This spreads an incident beam of radiation over a larger leaf area, and effectively reduces its intensity. An intensity of radiation that is superoptimal for photosynthesis when it strikes a leaf at 90° may therefore be optimal for a leaf inclined at an acute angle. The leaves of sun plants are often superimposed into a multilayered canopy. In bright sunshine even the shaded leaves in lower layers may have positive rates of net photosynthesis. Shade plants commonly have leaves held near to the horizontal and in a single-layered canopy.

leaves at angles and leaves in multilayered canopies

94   CHAPTER 3

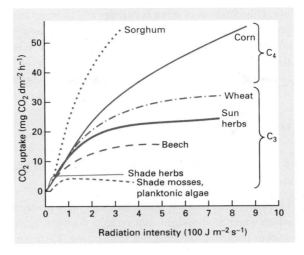

Figure 3.4 The response of photosynthesis to light intensity in various plants at optimal temperatures and with the natural supply of CO_2. Note that corn and sorghum are C_4 plants and the remainder are C_3. (After Larcher, 1980; and a variety of sources.)

In contrast to these 'strategic' differences, it may also happen that as a plant grows, its leaves develop differently as a tactical response to the radiation environment in which it developed. This often leads to the formation of 'sun leaves' and 'shade leaves' within the canopy of a single plant. Sun leaves are typically smaller, thicker, have more cells per unit area, denser veins, more densely packed chloroplasts and greater dry weight per unit area of leaf. The shade leaves have a correspondingly large area relative to their dry weight, are often more translucent and may have only 20% of the photosynthetic activity (Björkman, 1981). The lower shade leaves of a tree may not make a very great contribution to the energy budget of the plant of which they are a part, but, with their lower compensation points, they may at least pay for their own respiration.

sun and shade leaves

These 'tactical manoeuvres', then, tend to occur not at the level of the whole plant, but at the level of the individual leaf or even its parts. Nevertheless, they take time. To form sun or shade leaves as a tactical response, the plant, its bud or the developing leaf must sense the leaf's environment and respond by growing a leaf with appropriate structure. For example, it is impossible for the plant to change its form fast enough to track the changes in intensity of radiation between a cloudy and a clear day. It can, however, change its rate of photosynthesis extremely rapidly, reacting even to the passing of a fleck of sunlight.

The rate at which a leaf photosynthesizes also depends on the demands that are made on it by other vigorously growing parts. Photosynthesis may be reduced, even though conditions are otherwise ideal, if there is no demanding call on its products.

3.2.2 The value of radiation as a resource depends critically on the supply of water

It is not sensible to consider radiation as a resource independently of water, but the argument is complex. Intercepted radiation does not result in photosynthesis unless there is carbon dioxide (CO_2) available and the prime route of entry of CO_2 is through open stomata. If the stomata are open, water will evaporate through them and part of the incident radiation provides the latent heat for this evaporation. If water is lost faster than it can be gained, the leaf (and the plant) will sooner or later wilt and/or

95   RESOURCES

die. This determines what is perhaps the most crucial dilemma in the optimization of plant behaviour, because in most terrestrial communities water is sometimes in short supply. Should the 'ideal' plant conserve water at the expense of present photosynthesis or maximize photosynthesis at the risk of running out of water before it matures? Once again we meet the problem of whether the optimal solution involves a strict strategy of plant development or the ability to make local tactical responses. As we come to expect in nature, we can find good examples of both solutions and also compromises.

3.2.3 Strategic solutions to the reconciliation of photosynthetic activity with controlled water loss

1 Plants may have a short life and high photosynthetic activity during periods when water is abundant, but remain dormant as seeds during the rest of the year, neither photosynthesizing nor transpiring (e.g. many desert annuals, annual weeds and most annual crop plants).

2 Plants with long life may produce leaves during periods when water is abundant and shed them during droughts (e.g. many species of *Acacia*). Some shrubs of the Israeli desert (e.g. *Teucrium polium*) bear finely divided, thin-cuticled leaves during a season when soil water is freely available. These are replaced by undivided, small, thick-cuticled leaves in more drought-prone seasons, which in turn fall and may leave only green spines or thorns (Orshan, 1963). In these species, there is a sequential polymorphism of the leaf canopy through the season, each leaf morph being replaced in turn by a less photosynthetically active but more watertight structure.

3 Leaves may be produced that are long-lived, transpire only slowly and tolerate a water deficit, but which are unable to photosynthesize rapidly even when water is abundant (e.g. evergreen desert shrubs). Structural features such as hairs, sunken stomata and the restriction of stomata to specialized areas on the lower surface of a leaf, make the diffusion pathway for water vapour from the wet cell surfaces of the mesophyll to the outside atmosphere less steep, and so slow down water loss. But at the same time these morphological features make the diffusion gradient for CO_2 into the plant less steep and reduce its rate of entry. These features are not responsive from day to day or minute to minute like stomatal closure, and they simply slow down both water loss and photosynthesis. Waxy and hairy leaf surfaces may, however, reflect a greater proportion of radiation that is not in the PAR range and so keep the leaf temperature down and reduce water loss.

4 C_4 and crassulacean acid metabolism (CAM) physiology. We consider these in more detail in Sections 3.3.1–3.3.3. At this point we simply note that plants with 'normal' (i.e. C_3) photosynthesis are wasteful of water compared with plants that possess the modified C_4 and CAM physiology. The latter can achieve significantly higher photosynthetic rates at a given level of stomatal conductance, and so with less water loss (Table 3.1). The water-use efficiency of C_4 plants (the amount of carbon fixed per unit of water transpired) may be double that of C_3 plants.

3.2.4 Tactical solutions to the reconciliation of photosynthetic activity and controlled water loss

The major tactical control of the rates of both photosynthesis and water loss is of course through the changes in stomatal conductance that may occur rapidly during

structural features that make the diffusion gradient less steep

Table 3.1 Examples of plants with C_3, C_4 and CAM photosynthetic systems.

___

Typical C_3 plants
Triticum vulgare (wheat), *Secale cereale* (rye), *Lolium perenne* (rye grass), *Dactylis glomerata* (orchard grass), *Vicia faba* (field bean), *Phaseolus multiflorus* (French bean), *Trifolium repens* (white clover), *Medicago sativa* (alfalfa). All species of *Quercus* (oaks), *Fagus* (beech), *Betula* (birch), *Pinus* (pines)

Typical C_4 plants
Zea mays (corn), *Saccharum officinale* (sugar cane), *Panicum miliaceum* (millet), *Sorghum vulgare* (sorghum), *Echinochloa crus-galli* (cockspur), *Setaria italica* (Italian millet). This group includes several thousand species in at least 17 families, especially Amaranthaceae (amaranths), Portulacaceae and Chenopodiaceae.

Typical CAM (crassulacean acid metabolism) plants (see Section 3.3.3)
Echinocactus fendleri, *Ferocactus acanthoides* and *Opuntia polyacantha* (all desert succulents), *Tillandsia recurvata* (an epiphyte)

Mixed genera. Some genera contain both C_3 and C_4 species
Euphorbia maculata (C_4), *E. corollata* (C_3), *Cyperus rotundus* (C_4), *C. papyrus* (C_3), *Atriplex rosea* (C_4) and *A. hastata* (C_3)

___

the course of a day and allow very rapid tactical response to immediate water shortages.

stomatal opening

Rhythms of stomatal opening and closure may ensure that the aerial parts of the plant remain more or less watertight except during controlled periods of active photosynthesis. These rhythms may be diurnal; alternatively, the movements may be quickly responsive to the plant's internal water status. Stomatal movement may even be triggered directly by conditions at the leaf surface itself—the plant then responds to desiccating conditions at the very site, and at the same time, as the conditions are first sensed.

It is usual to find that the number of stomata per unit area of leaf is highest in plants that are likely to suffer drought, and that these same species have more heavily cuticularized (waterproofed) leaf surfaces. This has the effect of allowing ready exchanges of gas and water during photosynthesis, but making the plant more watertight at other times. Thus, the exposed sun leaves of trees usually have a higher stomatal density and thicker cuticle than shade leaves on the same plant.

The conductance of water by stomata and the rate of photosynthesis are strongly correlated even during short periods within a day; but which is cause and which is effect? There is now much evidence that stomatal conductance adjusts to match the intrinsic photosynthetic capacity rather than the other way round (Farquahar & Sharkey, 1982). In an ideal world, if there is a limited amount of water available to the plant, the stomata should open and close in such a way as to maximize the photosynthesis that might be achieved (Cowan & Farquhar, 1977). We do not yet know whether plants in nature achieve this best of all possible worlds and it may even be quite wrong to expect plants always to behave as highly efficient systems. Evolution by natural selection does not necessarily optimize—it occurs because some organisms leave more descendants (or descendant genes) than others. It is not difficult to envisage situations in which in a population of plants the individuals that will leave most descendants are those that are profligate with their resources if by doing so they reduce the number of descendants left by their neighbours.

3.2.5 Net photosynthesis is the gain from gross photosynthesis minus the losses due to respiration and the loss of plant parts

The rate of photosynthesis is a gross measure of the rate at which a plant captures radiant energy and fixes it in carbon compounds. However, it is often more important to consider, and very much easier to measure, the *net* gain. *Net photosynthesis* is the increase (or decrease) in dry matter that results from the difference between gross photosynthesis and the losses due to respiration and the death of plant parts (Figure 3.5).

Net photosynthesis is negative in darkness, when respiration exceeds photosynthesis, and increases with the intensity of PAR. The compensation point is the intensity of PAR at which the gain from gross photosynthesis exactly balances the respiratory and other losses. The leaves of shade species tend to respire at lower rates than those of sun species. Thus, when both are growing in the shade the *net photosynthesis* of shade species is greater than that of sun species.

In practice, most of what we know about the effects of radiation as a resource depends on measurements of net photosynthesis rather than on direct measures of gross photosynthesis. This is because it is exceedingly difficult to measure the rate of respiration in the light, and we commonly have to assume that it continues at the same rate in the light as in the dark. Growth of a plant involves not only the fixation of CO_2 in photosynthesis, but the subsequent assembly and maintenance of complex tissue components. These account for some of the respiratory costs and some of the differences between rates of gross and of net photosynthesis. However, this may not be the whole story. There is a pathway of respiratory metabolism in many leaves and roots that generates no adenosine triphosphate (ATP) and, hence, contributes no energy to the processes of synthesis and maintenance (Laties, 1982). It may have a role in burning off excess photosynthates (Lambers, 1985). This suggests that

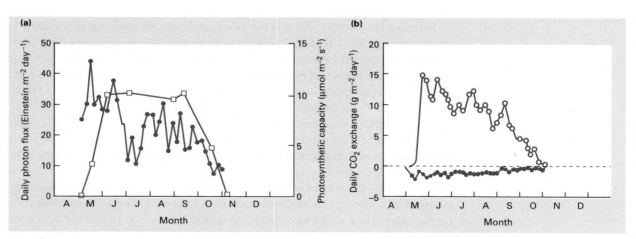

Figure 3.5 The annual course of events that determined the net photosynthetic rate of the foliage of maple (*Acer campestre*) in 1980. (a) Variations in the intensity of PAR (●), and changes in the photosynthetic capacity of the foliage (□) appearing in spring, rising to a plateau and then declining through late September and October. (b) The daily fixation of CO_2 (○) and its loss through respiration during the night (●). The annual total gross photosynthesis was $1342\,g\,CO_2\,m^{-2}$ and night respiration was $150\,g\,CO_2\,m^{-2}$, giving a balance of $1192\,g\,CO_2\,m^{-2}$ net photosynthesis. (After Pearcy *et al.*, 1987; from data of Küppers.)

photosynthesis, rather than being always a critical limiting process in plant growth, may sometimes be carried to excess. It may be that green plants are 'pathological overproducers of carbohydrates' and that some part of their respiration is concerned with disposing of the excess. If this is the case, it would help to explain why plants so often seem to photosynthesize at rates below those of which they are capable.

The photosynthetic capacity of leaves varies by nearly 100-fold (Mooney & Gulmon, 1979). This 'capacity' is defined as the rate of photosynthesis when incident radiation is saturating, temperature is optimal, relative humidity is high, and CO_2 and oxygen (O_2) concentrations are normal. When the leaves of different species are compared under these ideal conditions, the ones with the highest photosynthetic capacity are generally those from environments where nutrients, water and radiation are seldom limiting (at least during the growing season). These include many agricultural crops and their weeds. Species from resource-poor environments (e.g. shade plants, desert perennials, heathland species) usually have low photosynthetic capacity—even when abundant resources are provided.

Needless to say, ideal conditions in which plants may achieve their photosynthetic capacity are rarely present outside a physiologist's controlled environment chamber. In nature, the plants that approach their photosynthetic capacity most closely are annuals and grasses in deserts after rain. Then, the usually incompatible conditions of high intensity of radiation and abundant water supply coincide (Ehleringer & Mooney, 1984). They are the opportunists of such habitats. In the same deserts, evergreen perennials tolerate long periods of drought but have very low photosynthetic capacities. These two very different styles of physiology appear to have arisen as evolutionary alternatives. During periods when water is freely available, individuals may gain fitness by using water extravagantly and photosynthesizing fast, if by doing so they can ripen seeds quickly, and 'selfishly' use up the water needed by their slower growing neighbours. However, this leaves some tightly held soil water that can only be extracted slowly and with difficulty, providing a niche for longer lived 'lethargic' individuals or species.

The potential photosynthesis of plants in nature may also fall below their capacity because periods of radiation are short and temperatures low. The evergreen shrub, *Heteromeles arbutifolia*, loses at least 48% of its potential carbon gain because of low temperatures and short days in the autumn, winter and spring, and because cloudy days reduce incident radiation (Mooney *et al.*, 1975).

3.2.6 A resumé for radiation

The limitations of green plants as users of radiation as a resource can be summarized as follows.

1   For much of the year in temperate climates, and for the whole of the year in arid climates, the leaf canopy does not cover the land surface, so that most of the incident radiation falls on bare branches or on bare ground.

2   The rate at which photosynthesis may proceed is limited by conditions (e.g. temperature) and by the availability of resources other than radiant energy, especially water. The conflict between conserving water within the plant and maintaining high rates of photosynthesis is only partially resolved by metabolic, morphological and behavioural specializations.

3   About 56% of incident radiation lies outside the range of wavelengths available to the photosynthetic system.

4 Leaves seem to achieve their maximal photosynthetic rate only when the products are being actively withdrawn (to developing buds, tubers, etc.).

5 The rate of photosynthesis increases with the intensity of PAR, but in most species (C_3 plants) reaches a plateau at intensities of radiation well below that of full solar radiation.

6 Because radiation intensity varies from day to day and within a day, no particular leaf morphology, physiology or orientation can be optimal all the time.

7 The photosynthetic capacity of leaves is highly correlated with leaf nitrogen content, both between leaves on a single plant and between the leaves of different species. There is an even closer correlation between N-uptake and maximum photosynthetic capacity (Woodward, 1994). Around 75% of leaf nitrogen is invested in chloroplasts. This suggests that the availability of nitrogen as a resource may place strict limits on the ability of plants to garner CO_2 and energy in photosynthesis.

8 The highest efficiency of utilization of radiation by green plants is 3–4.5%, obtained from cultured microalgae at low intensities of PAR. In tropical forests values fall within the range 1–3%, and in temperate forests 0.6–1.2%. The approximate efficiency of temperate crops is only about 0.6%. It is on such levels of efficiency that the energetics of all communities depends.

We return to the issue of radiation as a resource in Chapter 18, when we consider the net photosynthesis of areas of vegetation instead of that of individual plants.

Very detailed reviews of radiation as a plant resource are to be found in the sequence of volumes of *Encyclopedia of Plant Physiology* (new series), Springer-Verlag, Berlin.

3.3 CO_2 as a resource

The photosynthetic fixation of atmospheric CO_2 is the unique process by which green plants capture and conserve the energy of solar radiation in chemical bonds. It is then available as the power that drives virtually all biological processes. The CO_2 used in photosynthesis is obtained almost entirely from the atmosphere, where its concentration has risen from approximately $280\,\mu l\,l^{-1}$ in 1750 to about $350\,\mu l\,l^{-1}$ in 1995 (see Chapter 19, Section 19.4.6) and is still increasing by 0.4–0.5% year^{-1} (see Figure 19.15).

In a terrestrial community the flux of CO_2 at night is upwards, from the soil and vegetation to the atmosphere; on sunny days above a photosynthesizing canopy there is a downward flux. In one study, the concentration difference between points 48 and 138 cm above a crop of corn was found to be of the order of only 2–12 ppm during the night and 2–4 ppm in the daytime (Wright & Lemon, 1966) because of the efficiency with which gases become mixed in the air column above a vegetation canopy.

CO$_2$ varies little in its availability

However the situation is quite different within and beneath canopies. Changes in CO_2 concentration in the air within a mixed deciduous forest in New England were measured at various heights above ground level during the year (Bazzaz & Williams, 1991) (Figure 3.6a). Highest concentrations, up to $\cong 1800\,\mu l\,l^{-1}$, were measured near the surface of the ground, tapering off to $\cong 400\,\mu l\,l^{-1}$ at 1 m above the ground. These high values near ground level were achieved in the summer when high temperatures allowed the rapid decomposition of litter and soil organic matter. At greater heights within the forest the CO_2 concentrations scarcely ever (even in winter) reached the value of $350\,\mu l\,l^{-1}$ which is the atmospheric concentration of bulk air

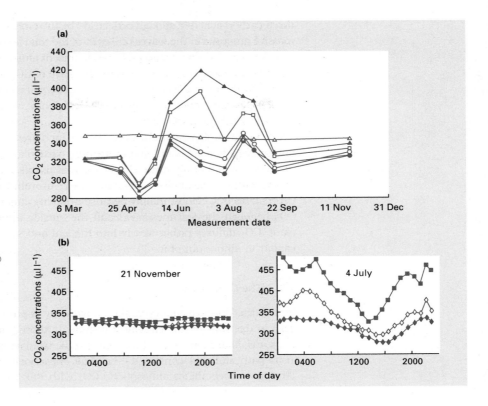

Figure 3.6 (a) CO_2 concentrations in a mixed deciduous forest (Harvard Forest, Massachusetts, USA) at various times of year at five heights above ground: ▲ 0.05 m; □ 0.20 m; ■ 3.00 m; ○ 6.00 m; ● 12.00 m. Data from the Mauna Loa CO_2 observatory (△) are given on the same axis for comparison. (b) CO_2 concentrations for each hour of the day (averaged over 3- to 7-day periods) on 21 November and 4 July. (After Bazzaz & Williams, 1991.)

measured at the Mauna Loa laboratory in Hawaii (see Chapter 19, Section 19.4.6). During the winter months the concentration of CO_2 remained virtually constant through day and night at all heights. But, in the summer months, major diurnal cycles of concentration developed which reflected the interaction between the production of CO_2 by decomposition and its consumption in photosynthesis (Figure 3.6b).

That CO_2 concentrations vary so widely within vegetation means that plants growing in different parts of a forest will experience quite different CO_2 environments. Indeed the lower leaves on a forest shrub will usually experience higher CO_2 concentrations than its upper leaves and seedlings will live in environments richer in CO_2 than mature trees.

A process as fundamental to life on earth as photosynthesis might be expected to be underpinned by a single unique biochemical pathway. It is therefore very remarkable that there are three such pathways (and variants within them) and just as remarkable that it was not until the 1960s that this was realized. The three pathways by which CO_2 is absorbed and fixed have important ecological consequences, especially as they affect the reconciliation of photosynthetic activity and controlled water loss (see Section 3.2.2). The physiology, ecology and evolution of the pathways are discussed in an elegant review by Ehleringer and Monson (1993), who also give a very full list of references.

3.3.1 The C_3 pathway

In this, the most common process (the Calvin–Benson cycle), CO_2 is fixed into a three-carbon acid (phosphoglyceric acid) by the enzyme Rubisco which is present in massive amounts in the leaves (25–30% of the total leaf nitrogen). This same enzyme can also act as an oxygenase and, under present atmospheric conditions, this activity (photorespiration) results in a wasteful release of CO_2—reducing by about one-third the net amounts of CO_2 that are fixed. Photorespiration increases with temperature with the consequence that the overall efficiency of carbon fixation declines with increasing temperature.

The rate of photosynthesis of C_3 plants increases with the intensity of radiation, but reaches a plateau. In many species, particularly shade species, this plateau occurs at radiation intensities far below that of full solar radiation (see Figure 3.4). Plants with C_3 metabolism have low water-use efficiency compared with C_4 and CAM plants (see below), mainly because Rubisco cannot maintain a steep gradient of CO_2 between the leaf mesophyll and the outside atmosphere. Therefore, in a C_3 plant, CO_2 diffuses rather slowly into the leaf and so allows time for a lot of water vapour to diffuse out of it.

3.3.2 The C$_4$ pathway

In this (the Hatch–Slack) cycle the C_3 pathway is present but it is confined to cells deep in the body of the leaf—most commonly those that form a sheath around the veins. CO_2 that diffuses into the leaves via the stomata meets mesophyll cells that lie in its path and that contain the enzyme phosphoenolpyruvate (PEP) carboxylase. This enzyme combines atmospheric CO_2 with PEP to produce a four-carbon acid. This diffuses, and releases CO_2 to the inner cells where it enters the traditional C_3 pathway. PEP carboxylase has a much greater affinity than Rubisco for CO_2. There are profound consequences.

1   C_4 plants can absorb atmospheric CO_2 much more effectively than C_3 plants, so that the gradient of CO_2 into the leaf in C_4 plants is much steeper than in C_3 plants, whilst the gradient of water out of the leaf is not changed.

2   As a result, C_4 plants may lose much less water per unit of carbon fixed.

3   The wasteful release of CO_2 by photorespiration is almost wholly prevented.

4   As a consequence, the efficiency of the overall process of carbon fixation does not change with temperature.

5   The concentration of Rubisco in the leaves is a third to a sixth of that in C_3 plants and the leaf nitrogen content is correspondingly lower.

6   As a consequence, C_4 plants are dietetically much less attractive to some (but not all) herbivores and also achieve more photosynthesis per unit of nitrogen absorbed.

One might wonder how C_4 plants, with such high water-use efficiency, have failed to dominate the vegetation of the world; but there are clear costs to set against the gains. The C_4 system has a high light-compensation point and is inefficient at low light intensities; C_4 species are therefore ineffective as shade plants. Moreover, C_4 plants have higher temperature optima for growth than C_3 species: most C_4 plants are found in arid regions or the tropics. In North America C_4 species of dicotyledons appear to be favoured in sites of limited water supply (Stowe & Teeri, 1978) (Figure 3.7), whereas the abundance of C_4 species of monocotyledons is strongly correlated with the maximum daily temperatures during the growing season (Teeri & Stowe, 1976). But, these correlations are not universal. More generally, where there are mixed populations of C_3 and C_4 plants, the proportion of C_4 species tends to fall with elevation on mountain ranges, and in seasonal climates it is C_4 species that tend to

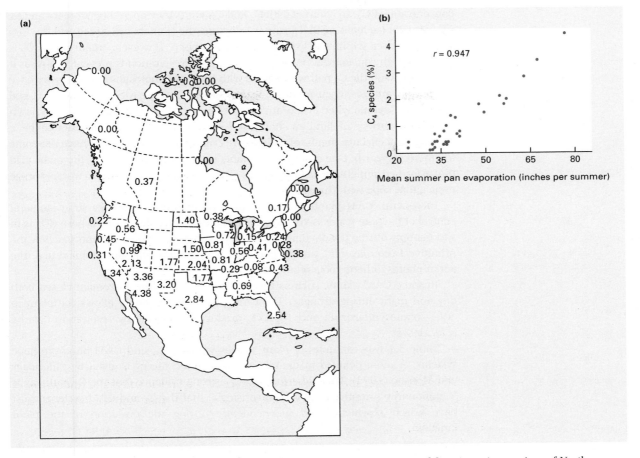

Figure 3.7 (a) The percentage of native C_4 species of dicot in various regions of North America. (b) The relationship between the percentage of native C_4 species in 31 geographical regions of North America, and the mean summer (May–October) pan evaporation—a climatic indicator of plant/water balance. Regions for which appropriate climatic data were unavailable were excluded, together with south Florida, where the peculiar geography and climate may explain the aberrant composition of the flora. (After Stowe & Teeri, 1978.)

dominate the vegetation in the hot dry seasons and C_3 species in the cooler wetter seasons.

The few C_4 species that extend into temperate regions (e.g. *Spartina* spp.) are found in marine or other saline environments where osmotic conditions may especially favour species with efficient water use. Perhaps the most remarkable feature of C_4 plants is that they do not seem to use their high water-use efficiency in faster shoot growth, but instead devote a greater fraction of the plant body to a well-developed root system. This is one of the hints that the rate of carbon assimilation is not the major limit to their growth, but that the shortage of water and/or nutrients matters more.

3.3.3 The CAM pathway

CAM plants

Plants with this CAM pathway also use PEP carboxylase with its strong power of

103   RESOURCES

concentrating CO_2. In contrast with C_3 and C_4 plants they open their stomata and fix CO_2 at night (as malic acid). During the daytime the stomata are closed and the CO_2 is then released within the leaf and fixed by Rubisco. However, because the CO_2 is then at a high concentration within the leaf, photorespiration is prevented, just as it is in plants with the C_4 pathway. Plants with the CAM photosynthetic pathway have obvious advantages when water is in short supply, because their stomata are closed during the daytime when evaporative forces are strongest. The system is now known in a wide variety of families, not just the Crassulaceae. This appears to be a marvellously effective means of water conservation, but CAM species have not come to inherit the earth. One cost to CAM plants is the problem of storing the malic acid that is formed at night: most CAM plants are succulents with extensive water-storage tissues that cope with this problem.

In general, CAM plants are found in arid environments where strict stomatal control of daytime water is vital for survival (desert succulents) and where CO_2 is in short supply during the daytime, for example submerged aquatic plants (such as the primitive *Isoetes howellii*), and in photosynthetic organs that lack stomata (e.g. the aerial photosynthetic roots of orchids).

In some CAM plants, such as *Opuntia basilaris*, the stomata remain closed both day and night during drought. The CAM process then simply allows the plant to 'idle'—photosynthesizing only the CO_2 produced internally by respiration (Szarek *et al.*, 1973).

Table 3.1 lists examples of plant species with C_3, C_4 and CAM photosynthetic systems. A more detailed taxonomic and systematic survey is given by Ehleringer and Monson (1993). They describe the very strong evidence that the C_3 pathway is evolutionarily primitive and, very surprisingly, that the C_4 and CAM systems must have arisen repeatedly and independently during the evolution of the plant kingdom.

3.3.4 The response of plants to changing atmospheric concentrations of CO_2

Of all the various resources required by plants, CO_2 is the only one that is increasing on a global scale. This rise is strongly correlated with the increased rate of consumption of fossil fuels and the clearing of forests. There is also evidence of large-scale changes in atmospheric CO_2 over much longer time scales. Carbon balance models suggest that during the Triassic, Jurassic and Cretaceous, atmospheric concentrations of CO_2 were four to eight times greater than at present, falling after the Cretaceous from 1400 to 2800 μl l^{-1} to below 1000 μl l^{-1} in the Eocene, Miocene and Pliocene, and fluctuating between 180 and 280 μl l^{-1} during subsequent glacial and interglacial periods (Berner, 1991; Cerling, 1991; Ehleringer & Monson, 1993).

changes in atmospheric CO_2 from the Triassic to the present

The declines in CO_2 concentration in the atmosphere after the Cretaceous may have been the primary force that favoured the evolution of plants with C_4 physiology (Ehleringer *et al.*, 1991), because at low concentrations of CO_2, photorespiration places C_3 plants at a particular disadvantage. The steady rise in CO_2 since the Industrial Revolution is therefore a partial return to pre-Pleistocene conditions. C_4 plants may begin to lose some of their advantage. Mechanistic models of leaf biochemistry can be linked with geochemical data about past atmospheres (not only CO_2 but also atmospheric oxygen which appears also to have varied widely in Phanerozoic times) to reconstruct the rates of photosynthesis of past vegetation.

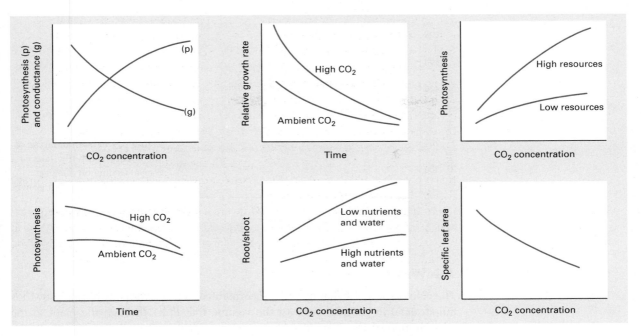

Figure 3.8 General trends in the responses of plants to the concentration of atmospheric CO_2. (After Bazzaz, 1990.)

Details of the ways in which this can be done, together with a very full list of references are given by Beerling (1994).

When other resources are present at adequate levels, additional CO_2 increases the rate of photosynthesis of C_3 (but, scarcely that of C_4) plants. Indeed, artificially increasing the CO_2 concentration in greenhouses is a commercial technique to increase crop yields. We might reasonably predict dramatic increases in the productivity of individual plants and of whole crops, forests and natural communities as atmospheric concentrations of CO_2 continue to increase. But, there is now much evidence that the responses may be very complicated, even when we consider only the response to rising CO_2 levels and ignore the global warming and changes in precipitation that may be its indirect effects (Bazzaz, 1990). General trends are illustrated in Figure 3.8, but there are other effects that appear sometimes to be species specific. For example, doubling CO_2 reduced the time to flowering of *Gaura pulchella* (Onagraceae), but increased it in *Lupinus texensis* (Leguminosae). The main effect of doubling CO_2 on the growth of some early successional species from the rainforest of Mexico was to change their architecture. It led to an increase in the height of the canopy of *Cecropia obtusifolia*, *Piper auritum* and *Trichospermum mexicanum*, but reduced it in *Senna multijuga* (Reekie & Bazzaz, 1989). When six species of temperate forest tree were grown for 3 years in a CO_2-enriched atmosphere in a glasshouse they were generally larger than controls, but the CO_2 enhancement of growth declined during the experiment at different rates for different species (Bazzaz *et al.*, 1993).

There may be indirect effects of CO_2 on plant–animal interactions because an increased concentration of CO_2 usually reduces the concentration of nitrogen in leaves. Insect herbivores may then eat 20–80% more foliage to maintain their nitrogen intake and fail to gain weight as fast (Figure 3.9) (Fajer *et al.*, 1991).

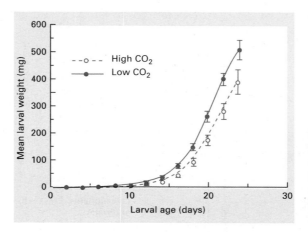

Figure 3.9 Growth of larvae of the buckeye butterfly (*Junonia coenia*) feeding on *Plantago lanceolata* that had been grown at ambient and elevated CO_2 concentrations. (After Fajer, 1989.)

3.4 Water

The volume of water that becomes incorporated in the plant body during growth is infinitesimal in comparison with the volume that flows through the plant in the transpiration stream. Nevertheless, water is a critical resource: it is a quantity that is reduced by the activity of the organism and its neighbours.

Hydration is a necessary condition for metabolic reactions to proceed. It is the medium in which the reactions occur. Because no organism is completely watertight its water content needs continual replenishment. Most terrestrial animals drink free water and/or obtain it from the food they eat. Some water is generated from the metabolism of food and body materials, and there are extreme cases in which animals of arid zones may obtain all their water in this way. However, the availability of water-holes may place strict limits on the distribution and abundance of the larger herbivores and, in turn, affect the pattern of vegetation at various distances from drinking holes (Goodall, 1967).

For most terrestrial plants the main source of water is the soil and they gain access to it through a root system. Roots are morphologically very odd structures and it is not easy to see how they evolved by the modification of any more primitive organ (Harper *et al.*, 1991), yet the colonization of the land by large terrestrial plants could not have happened without them. The evolution of the root was almost certainly the most influential event that made an extensive land flora and fauna possible. Once roots had evolved they provided secure anchorage for structures of the size of trees and a means for making intimate contact with mineral nutrients and water within the soil.

Water enters the soil as rain or melting snow and forms a reservoir in the pores between soil particles. What happens to it then depends on the size of the pores, which may hold it by capillary forces against gravity. If the pores are wide, as in a sandy soil, much of the water will drain away until it reaches some impediment and accumulates as a rising water-table or finds its way into streams or rivers.

the soil as a reservoir: the 'field capacity' and the 'permanent wilting point'

The water held by soil pores against the force of gravity is called the 'field capacity' of the soil. This is the upper limit of the water that a freely drained soil will retain against gravity. There is a less clearly defined lower limit on the water that is held in the soil and that can be used in plant growth (Figure 3.10). This lower limit is determined by the ability of plants to extract water from the narrower soil pores,

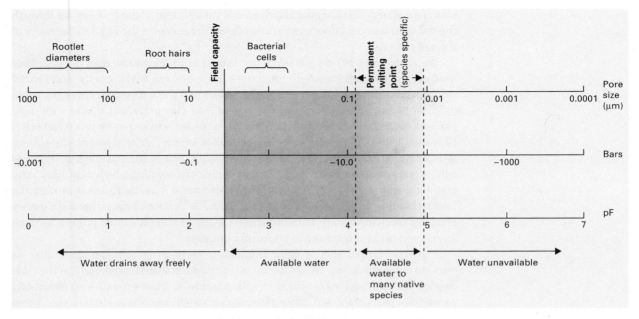

Figure 3.10 The status of water in the soil, showing the relationship between three measures of water status: (i) pF, the logarithm of the height (cm) of the column of water that the soil would support; (ii) water status expressed as atmospheres or bars; (iii) the diameter of soil pores that remain water filled. The size of water-filled pores may be compared in the figure with the sizes of rootlets, root hairs and bacterial cells. Note that for most species of crop plant the permanent wilting point is at approximately -15 bars (-1.5×10^6 Pa), but in many other species it reaches -80 bars (-8×10^6 Pa), depending on the osmotic potentials that the species can develop.

and is known as the 'permanent wilting point'—the soil water content at which plants wilt and are unable to recover. The soil water content at the permanent wilting point does not differ much between the plant species of mesic environments (or between species of crop plants), but many species native to arid regions can extract significantly more water from the soil. Solutes in the soil solution add osmotic forces to the capillary forces, which the plant must match when it extracts water from the soil. These osmotic forces become particularly important in the saline soils of arid areas. Here, much water movement is upwards from the soil to the atmosphere and salts rise to the surface, creating osmotically lethal salt pans.

the capture of water by roots

As a root withdraws water from the soil pores at its surface, it creates water-depletion zones around it. These determine gradients of water potential between the interconnected soil pores. Water flows in the capillary pores, along the gradient into the depleted zones, supplying further water to the root. This simple process is made much more complex because the more the soil around the roots is depleted of water, the more resistance there is to water flow. As the root starts to withdraw water from the soil, the first water that it obtains is from the wider pores because they hold the water with weaker capillary forces. This leaves only the narrower, more tortuous water-filled paths through which flow can occur, and so increases the resistance to water flow. Thus, when the root draws water from the soil very rapidly, the RDZ becomes sharply defined and water can move across it only slowly. For this reason, rapidly transpiring plants may wilt in a soil which contains

the distribution of water
in the soil

abundant water. The fineness and degree of ramification of the root system through the soil then become important in determining the access of the plant to the water in the soil reservoir.

Water that arrives on a soil surface as rain or as melting snow does not distribute itself evenly. Instead, it tends to bring the surface layer to field capacity, and further rain extends this layer further and further down into the soil profile. This means that different parts of the same plant root system may encounter water held with quite different forces, and indeed the roots can move water between soil layers (Caldwell & Richards, 1986). In arid areas, where rainfall is in rare, short showers, the surface layers may be brought to field capacity whilst the rest of the soil stays at or below wilting point. This is a potential hazard in the life of a seedling which may, after rain, germinate in the wet surface layers lying above a soil mass that cannot provide the water resource to support its further growth. A variety of specialized dormancy breaking mechanisms are found in species living in such habitats, protecting them against too quick a response to an insufficient rain.

The root system that a plant establishes early in its life can determine its responsiveness to future events. Many plants that are waterlogged early in their life develop a superficial root system which is inhibited from growth into anaerobic, water-filled parts of the soil. Later in the season, when water is in short supply, these same plants may suffer from drought because their root system has not developed into the deeper layers. In an environment in which most water is received as occasional showers on a dry substrate, a seedling with a developmental programme that puts its early energy into a deep taproot will gain little from subsequent showers. By contrast, a programme that determines that the taproot will be formed early in life may guarantee continual access to water in an environment in which heavy rains fill a soil reservoir to depth in the spring, but there is then a long period of drought. A review of the behaviour of water in the soil–plant–atmosphere continuum is given by Schulze *et al.* (1987).

There are quite profound differences in the ways in which different groups of ecologists analyse and explain the loss of water from plants to the atmosphere. Plant physiologists, since as early as the work of Brown and Escombe (1900), have emphasized the way in which the behaviour of stomata determines the rate at which a leaf loses water. It now seems obvious that it is the frequency and aperture of pores in an otherwise mainly waterproof surface that will control the rate at which water diffuses from a leaf to the outside atmosphere. A massive research effort has supported this view, ranging from theoretical studies of diffusion through single pores (e.g. Parlange & Waggoner, 1970), to laboratory studies of the behaviour of leaves in enclosed chambers and on the leaves of plants growing in the field (e.g. Burrows & Milthorpe, 1976).

But, there is a quite different interpretation due mainly to micrometeorologists and which focuses on vegetation as a whole rather than on the single stoma, leaf or plant (Penman, 1948; Thornthwaite, 1948). This approach goes even further back in history than theories of stomatal control, to Aristotle who asked whether the evaporation of water was driven by the heating of the sun or the movement of the wind. Micrometeorologists point out that water will be lost by evaporation only if the latent heat for this evaporation is provided. This may be by solar radiation received directly by the transpiring leaves or as advective energy (i.e. heat received as solar radiation elsewhere but transported in moving air). The latter will be particularly important at the edges of vegetation—an oasis effect.

The micrometeorologists have developed formulae for the rate of water loss that are based entirely on the weather (e.g. wind speed, solar radiation, saturation deficit and temperature); they wholly ignore both the species of plants and their physiology, but they nevertheless prove to be powerful predictors of the evaporation of water from vegetation that is not suffering from drought.

There are few cases in ecological science in which the importance of the scale of study has so dramatic an effect on interpretation. Both the plant physiologists and ecologists working with individual leaves and plants, and the meteorologists and agronomists working with extensive areas of vegetation are correct in their assessment of the control, or lack of control, of transpiration by the stomata. The conflict of opinion is not a conflict of scientific evidence but of interpretation. The results from either group are not applicable to the plant systems studied by the other, unless proper allowance is made for the change of scale (Jarvis & McNaughton, 1986; McNaughton & Jarvis, 1991). In particular, the dependence of transpiration on stomatal conductance decreases and its dependence on radiation increases as the spatial scale is increased from the individual leaf to an area of vegetation.

Climatically based models are likely to be the most relevant in predicting the evapotranspiration, photosynthesis and net primary production that might occur in areas of vegetation as a result of global warming and changes in precipitation. Aber and Federer (1992) give an example of such modelling applied to temperate and boreal forests.

3.5 Mineral nutrients

It takes more than light, CO_2 and water to make a plant. Mineral resources are also needed. The mineral resources that the plant must obtain from the soil (or, in the case of aquatic plants, from the surrounding water) include macronutrients (i.e. those needed in relatively large amounts)—nitrogen (N), phosphorus (P), sulphur (S), potassium (K), calcium (Ca), magnesium (Mg) and iron (Fe)—and a series of trace elements—for example; manganese (Mn), zinc (Zn), copper (Cu), boron (B) and molybdenum (Mo) (Figure 3.11). (Many of these elements are also essential to animals, although it is more common for animals to obtain them in organic form in their food than as inorganic chemicals.) Some plant groups have special require-ments. For example, aluminium is a necessary nutrient for some ferns, silicon for diatoms and selenium for certain planktonic algae. Cobalt is required in the mutualistic association between leguminous plants and the nitrogen-fixing bacteria of their root nodules.

all green plants need the same 'essential' elements, but may differ in the proportions they require

Green plants do not obtain their mineral resources as a single package. Each element enters the plant independently as an ion or a molecule, and each has its own characteristic properties of absorption in the soil and of diffusion, which affect its accessibility to the plant even before any selective processes of uptake occur at the root membranes. All green plants require all of the 'essential' elements listed in Figure 3.11, although not in the same proportion, and there are some quite striking differences between the mineral compositions of plant tissues of different species and between the different parts of a single plant (Figure 3.12).

water and minerals share characteristics as resources and they may interact as resources

Many of the points made about water as a resource, and about roots as extractors of this resource, apply equally to mineral nutrients. Coarse differences in develop-mental programmes can be recognized between the roots of different species (Figure 3.13a) and some of these are of major importance in the 'match between the

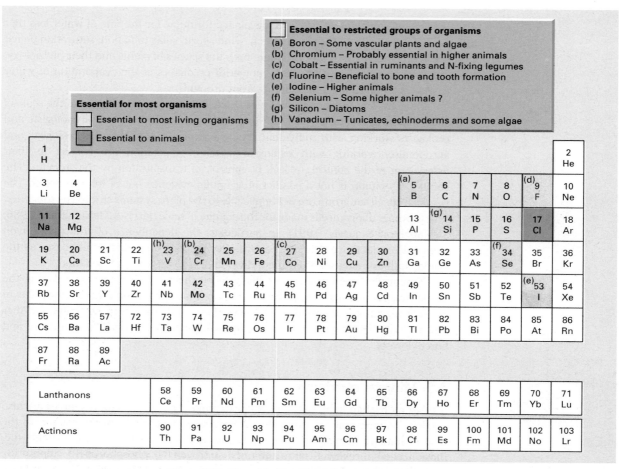

Figure 3.11 The periodic table of the elements showing those that are essential resources in the life of various organisms.

organism and its environment'. But, it is the ability of root systems to override strict programmes and be opportunistic that makes them effective exploiters of the soil. Most roots have features that ensure that they are explorers. Roots elongate before they produce laterals and this ensures that exploration precedes exploitation. Branch roots usually emerge on radii of the parent root, secondary roots radiate from these primaries and tertiaries from the secondaries. These rules of growth guide the exploration of a volume of soil, and reduce the chance that two branches of the same root will forage in the same soil particle and enter each other's depletion zones.

exploration of the soil and the capture of water

Roots pass through a medium in which they meet obstacles and encounter heterogeneity—patches of nutrient that vary on the same scale as the diameter of a root itself. In 1 cm of growth, a root may encounter a boulder, pebbles and sand grains, a dead or living root or the decomposing body of a worm. It may pass through layers of clay soil with fine pores and layers of sand or loam with much coarser pores (see Figure 3.13b). As a root passes through a heterogeneous soil (and all soils are heterogeneous seen from a 'root's-eye view'), it responds by branching freely in zones that supply resources, and scarcely branching in less rewarding patches. That it can do so depends on the individual rootlet's ability to react on an extremely local scale to the conditions that it meets.

roots as opportunists

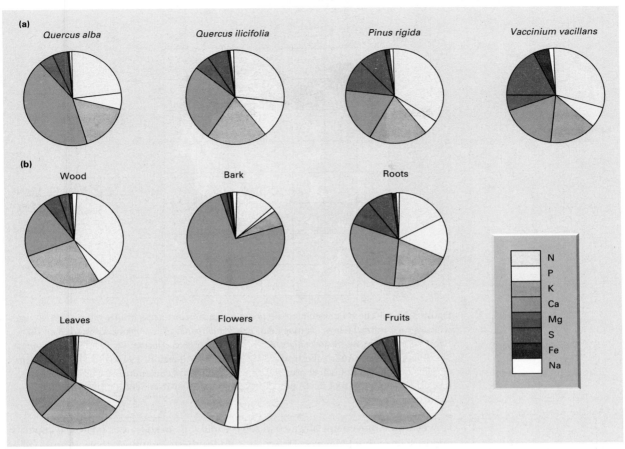

(a)

Quercus alba   Quercus ilicifolia   Pinus rigida   Vaccinium vacillans

(b)

Wood   Bark   Roots

N
P
K
Ca
Mg
S
Fe
Na

Leaves   Flowers   Fruits

Figure 3.12 (a) The relative concentration of various minerals in whole plants of four species in the Brookhaven Forest, New York. (b) The relative concentration of various minerals in different tissues of the white oak (*Quercus alba*) in the Brookhaven Forest. (After Woodwell *et al.*, 1975.) Note that the differences between species are much less than between the parts of a single species.

the architecture of a root determines access to mineral resources

The effectiveness of a root as a forager for resources must depend on its architecture. Root systems appear to be much less firmly programmed in their development than shoots, but we are still extraordinarily ignorant of how root growth and proliferation is determined. In particular, we need to know how a root responds when it starts to enter the RDZ of another root within soil.

There are strong interactions between water and nutrients as resources for plant growth. Roots will not grow freely into soil zones that lack available water, and so nutrients in these zones will not be exploited. Plants deprived of essential minerals make less growth and may then fail to reach volumes of soil that contain available water. There are similar interactions between mineral resources. A plant starved of nitrogen makes poor root growth and so may fail to 'forage' in areas that contain available phosphate or indeed contain more nitrogen.

Although there are strong *interactions* between the uptake of water and minerals from the soil, the *correlation* between the two is really strong only in the case of nitrate ions. Of all the major plant nutrients, nitrates move most freely in the soil solution and are carried from as far away from the root surface as water is carried.

111   RESOURCES

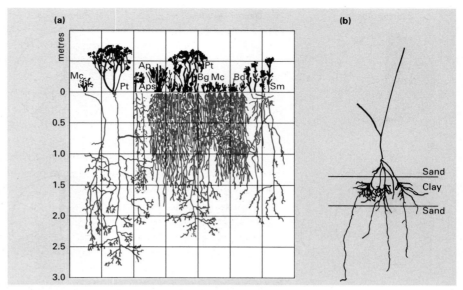

Figure 3.13 (a) The root systems of plants in a typical short-grass prairie after a run of years with average rainfall (Hays, Kansas). Ap, *Aristida purpurea*; Aps, *Ambrosia psilostachya*; Bd, *Buchloe dactyloides*; Bg, *Bouteloua gracilis*; Mc, *Malvastrum coccineum*; Pt, *Psoralia tenuiflora*; Sm, *Solidago mollis*. (After Albertson, 1937; Weaver & Albertson, 1943.) (b) The root system developed by a plant of wheat grown through a sandy soil containing a layer of clay. Note the responsiveness of root development to the localized environment that it encounters. (Courtesy of J.V. Lake.)

However, the movement of water in soil is rapid only in soils at or near field capacity, and in soils with wide pores. Hence, it is under these conditions that nitrates will be most mobile. The RDZ for nitrates will then be wide, and concentration gradients around the roots will be shallow. When RDZs are wide, those produced around neighbouring roots will be more likely to overlap. Competition can then occur (even between the roots of a single plant), because one organ begins to compete with another only when they begin to draw on resources accessible to both, i.e. when their RDZs overlap.

The concept of RDZs is important not only in visualizing how one organism influences the resources available to another, but also in understanding how the architecture of the root system affects the capture of these resources. For a plant growing in an environment in which water moves freely to the root surface, those nutrients that are freely in solution will move with the water. They will then be most effectively captured by wide-ranging, but not intimately branched, root systems. The less freely that water moves in the soil, the narrower will be the depletion zones, and the more it will 'pay' to explore the soil intensively rather than extensively.

the minerals 'available' are a biased sample of the minerals present

Just because a root system has access to the nutrient solution in the soil does not mean that it has equal access to all nutrients. The soil solution that flows through soil pores to the root surface has a biased mineral composition compared with what is potentially available. This is because different mineral ions are held by different forces in the soil. Ions such as nitrate, calcium and sodium may, in a fertile agricultural soil, be carried to the root surface faster than they are accumulated in the body of the plant. By contrast, the phosphate and potassium content of the soil solution will often fall far short of the plant's requirements. Phosphate is bound on soil colloids by surfaces that bear calcium, aluminium and ferric ions. The rate at which phosphate ions are released into the soil solution then depends on: (i) the rate at which the root withdraws them from the soil solution; and (ii) the rate at which the concentration is replenished by release from the colloids and by diffusion. The diffusion rate is the main factor that determines the width of an RDZ for ions that are bound on soil surfaces. In dilute solutions the diffusion coefficients of ions that are not absorbed, such as nitrate, are of the order of 10^{-5} cm^2 s^{-1} and for cations such as calcium, magnesium, ammonium and potassium they are 10^{-7} cm^2 s^{-1}. For strongly absorbed anions such as phosphate, the coefficients are as low as 10^{-9} cm^2 s^{-1}.

For resources like phosphate that have low diffusion coefficients, the RDZs will be narrow (Figure 3.14); roots or root hairs will only tap common pools of resource (i.e. will compete) if they are very close together. It has been estimated that more than 90% of the phosphate absorbed by a root hair in a 4-day period will have come from the soil within 0.1 mm of its surface. Two roots will therefore only draw on the same phosphate resource in this period if they are less than 0.2 mm apart. If phosphate is in short supply, the intimate branching of the root, and particularly its extension by root hairs and associated mycorrhizal fungi, will dramatically affect its potential for absorption of phosphate, compared with that for more mobile nutrients. A widely spaced, extensive root system tends to maximize access to nitrate, whilst a narrowly spaced, intensively branched root system tends to maximize access to phosphates (Nye & Tinker, 1977). Plants with different shapes of root system may therefore tolerate different levels of soil-mineral resources, and different species may deplete different mineral resources to different extents. This may be of great importance in allowing a variety of plant species to cohabit in the same area (see Chapter 20, Section 20.3.1).

mycorrhiza

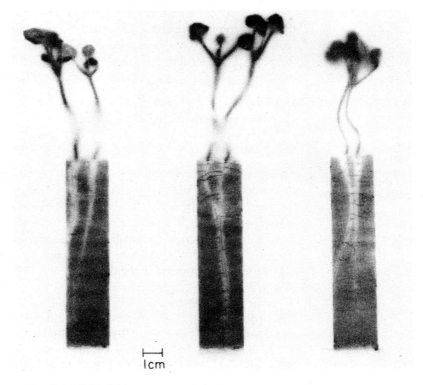

Figure 3.14 Radioautograph of soil in which seedlings of mustard have been grown. The soil was supplied with radioactively labelled phosphate ($^{32}PO_4^-$) and the zones that have been depleted by the activity of the roots show up clearly as white. (After Nye & Tinker, 1977.)

1cm

113 RESOURCES

The parts of this chapter that concern water and nutrient resources in the soil have been written on the assumption that plants have roots. In fact, most plants do not have roots—they have mycorrhiza. These are associations of fungal and root tissue which have resource-gathering properties differing greatly from those observed in laboratory studies of sterile roots. The properties of mycorrhiza are discussed in Chapter 13, Section 13.7.

3.6 O_2 as a resource

O_2 is a resource for both animals and plants. Only a few prokaryotes can do without it. Its diffusibility and solubility in water are very low and so it becomes limiting most quickly in aquatic and waterlogged environments. Its solubility in water decreases rapidly with increasing temperature. When organic matter decomposes in an aquatic environment, microbial respiration makes a demand for O_2 and this 'biological O_2 demand' may constrain the types of higher animal that can persist. High biological O_2 demands are particularly characteristic of still waters into which leaf litter or organic pollutants are deposited, and becomes most acute during periods of high temperature.

Because O_2 diffuses so slowly in water, aquatic animals must either maintain a continual flow of water over their respiratory surfaces (e.g. the gills of fish), or have very large surface areas relative to body volume (many aquatic crustacea have large feathery appendages, or have specialized respiratory pigments or a slow respiration rate, e.g. the midge larvae that live in still and nutrient-rich waters), or continually return to the surface to breath (e.g. whales, dolphins, turtles, newts).

The roots of many higher plants fail to grow into waterlogged soil, or die if the water table rises after they have penetrated deeply. These reactions may be direct responses to O_2 deficiency or responses to the accumulation of gases such as hydrogen sulphide, methane and ethylene, which are produced by microorganisms engaged in anaerobic decomposition. Even if roots do not die when starved of O_2, they may cease to absorb mineral nutrients so that the plants suffer from mineral deficiency.

3.7 Organisms as food resources

3.7.1 Introduction

autotrophs and heterotrophs

decomposers, parasites, predators and grazers

Autotrophic organisms (green plants and certain bacteria) assimilate inorganic resources into 'packages' of organic molecules (proteins, carbohydrates, etc.). These become the resources for heterotrophs (decomposers, parasites, predators and grazers—organisms that require resources in organic, energy-rich form), and take part in a chain of events in which each consumer of a resource becomes, in turn, a resource for another consumer. At each link in this *food chain* we can usually recognize three pathways to the next body of consumers.

1 *Decomposition*, in which the bodies (or parts of bodies) of organisms die, and together with waste and secretory products, become a food resource for 'decomposers' (bacteria, fungi and detritivorous animals)—a group which can use other organisms as food but only after they have died.

2 *Parasitism*, in which the living organism is used as a food resource whilst it is still alive. We define a parasite as a consumer that usually does not kill its food organism

and feeds from only one or a very few host organisms in its lifetime (see Chapter 12).

3 *Predation*, the final category, applies to those cases in which the consumer kills and eats another organism (or part of it) as food. Examples of predator–prey interactions are a seedling attacked by the fungus *Pythium*, a water flea consuming phytoplankton cells, an acorn eaten by a beetle or squirrel (consumers of seeds bring an end to the life of the embryos within them), a mountain lion consuming a rabbit, a whale eating krill and perhaps even a pitcher-plant drowning a mosquito. *Grazing* can be regarded as a type of predation, but the food (prey) organism is not killed; only part of the prey is taken, leaving the remainder with the potential to regenerate. Grazers, in contrast to parasites but like most other predators, are likely to feed on (or from) many prey during their lifetime.

ways of classifying organisms

The three categories are not neat and tidy and can be endlessly subdivided by the pedantic. Two more deserve mention here: (i) the degree to which a consumer's diet is specialized or generalized in nature; and (ii) the relationship between the life spans of the consumer and its resource. Consumers may be generalists (polyphagous), taking a wide variety of prey species, although often with a rank order of acceptability amongst the foods that they take. Alternatively, a consumer may specialize on specific parts of its prey but range over a number of species. This is most common amongst herbivores because, as we shall see, different parts of plants are quite different in their composition. Thus, a variety of birds specialize on seeds, although they are seldom restricted to a particular species; many grazing animals specialize on leaves and do not usually take roots; and certain species of eelworm (nematodes) and the larvae of some beetle species are specialized root feeders. Amongst consumers of animals, parasites (e.g. including liver flukes) often attack particular tissues or organs but are not necessarily restricted to a single host species. Cichlid fish of the species *Genychromis meuto* are unusual in their specialized feeding on scales from other fish. Finally, a consumer may specialize on a single species or a narrow range of closely related species (when it is said to be monophagous).

generalists and specialists

The more specialized the food resource required by an organism, the more it is constrained to live in patches of that resource *or* to spend time and energy in searching for it amongst a mixture. Such specialization may be fixed by peculiar structures, particularly mouthparts, which make it possible to deal efficiently with some resources, but of course make it more difficult to deal with others. The stylet of the aphid (Figure 3.15) can be interpreted as an exquisite product of the evolutionary process that has given the aphid access to a uniquely valuable food resource, *or* as an example of the ever-deepening rut of specialization that has evolved, narrowed and constrained how aphids can feed and what they can feed on. Figure 3.16 illustrates mouth-part specialization amongst insects and birds.

Many food resources are seasonal, and this is illustrated by reference to a population of wild raspberries in a temperate woodland. In winter, the plant is a mass of tiny twigs, but in spring it develops an abundance of young, protein-rich buds and juvenile leaves. Flowering brings a short period of nectar production offering a quite new type of resource, but only for the flowering period. As fruit is set, a new flush of resources is displayed in ripening and ripe fruit (Figure 3.17). Such seasonal structures may be reliable resources *either* to generalist herbivores that can turn to other foods when raspberry is out of season *or* to specialists that have an active life concentrated in the appropriate season and spend the rest of the year making no demands on food (dormant or in diapause). The bird species feeding on raspberry fruits are generalists, taking them only as a seasonal component of a diet

the seasonal availability of resources

115   RESOURCES

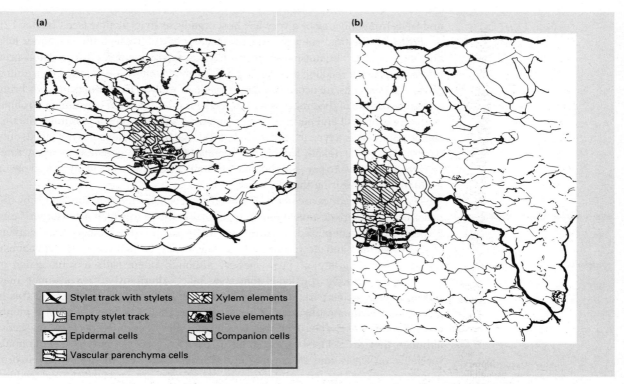

Figure 3.15 (a, b) The pathways of two stylets of the aphid, *Aphis fabae*, into the leaf of the broadbean (*Vicia faba*), determined by complete serial sectioning. The stylets thread their way to the phloem sieve tubes, but the mechanism of search is obscure. Note that there are branches in some of the pathways where a stylet has partially withdrawn and then followed a new path. (After Tjallingii & Hogen Esch, 1993.)

that varies continuously throughout the year. In complete contrast, the raspberry beetle (*Byturus tormentosus*) is a raspberry specialist. It lays eggs in the flower and the larva completes its life cycle within the developing fruit. It then remains in diapause as a pupa until the raspberry flowers again 10–11 months later. The larva of the raspberry moth (*Lampronia rubiella*) has a longer life on the pith within the woody stems: this resource is present throughout the year. This example illustrates both how a single resource (the raspberry plant) can be used by a variety of types of consumer and also, that a single species of plant can be many different resources.

3.7.2 The nutritional content of plants and animals as food

As a 'package' of resources, the body of a green plant is quite different from the body of an animal. This has a tremendous effect on the value of these resources as potential food (Figure 3.18). The most important contrast is that plant cells are bounded by walls of cellulose, lignin and/or other structural materials. It is these cell walls which give plant material its high-fibre content. The presence of cell walls is also largely responsible for the high fixed carbon content of plant tissues and the high ratio of carbon to other important elements. For example, the carbon : nitrogen (C : N) ratio of plant tissues commonly exceeds 40 : 1, in contrast to the ratios of approximately 8 : 1 to 10 : 1 in bacteria and fungi, detritivores, herbivores and

C : N ratios in animals and plants

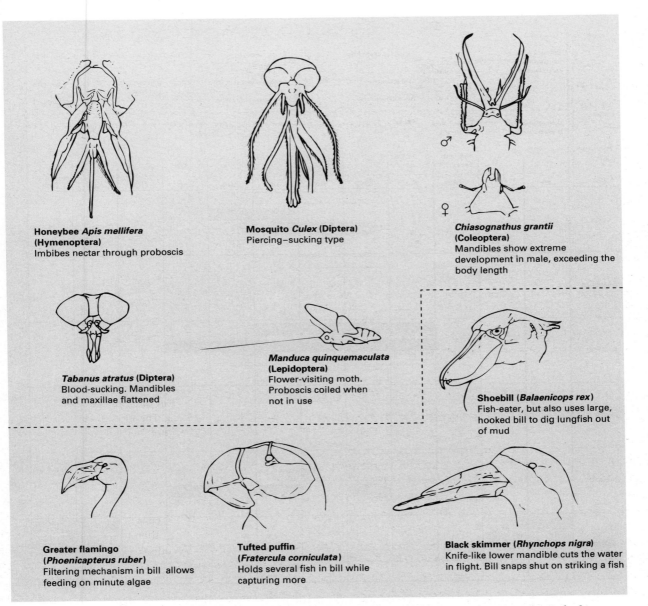

Honeybee *Apis mellifera* (Hymenoptera)
Imbibes nectar through proboscis

Mosquito *Culex* (Diptera)
Piercing–sucking type

♂ ♀ *Chiasognathus grantii* (Coleoptera)
Mandibles show extreme development in male, exceeding the body length

***Tabanus atratus* (Diptera)**
Blood-sucking. Mandibles and maxillae flattened

***Manduca quinquemaculata* (Lepidoptera)**
Flower-visiting moth. Proboscis coiled when not in use

Shoebill (*Balaenicops rex*)
Fish-eater, but also uses large, hooked bill to dig lungfish out of mud

Greater flamingo (*Phoenicapterus ruber*)
Filtering mechanism in bill allows feeding on minute algae

Tufted puffin (*Fratercula corniculata*)
Holds several fish in bill while capturing more

Black skimmer (*Rhynchops nigra*)
Knife-like lower mandible cuts the water in flight. Bill snaps shut on striking a fish

Figure 3.16 The specializations of organisms with respect to the nature of their food resources are illustrated by the variety of their mouth parts. Insects and aquatic birds, illustrated here, show such specialization to a marked degree. (After Snodgrass, 1944; Richards & Davies, 1977; Daly *et al.*, 1978; and the *Encyclopedia Britannica*.)

carnivores. Unlike plants, animal tissues contain no structural carbohydrate or fibre component but are rich in fat and, in particular, protein.

Herbivores that consume living plant material—and bacteria, fungi and detritivores which consume dead plant material—all utilize a food resource that is rich in carbon and poor in protein. The transition from plant to consumer involves a massive burning off of carbon as the C : N ratio is lowered. The main waste products of organisms that consume plants are therefore carbon-rich compounds (CO_2 and fibre). The greater part of the energy requirements of carnivores is obtained from the

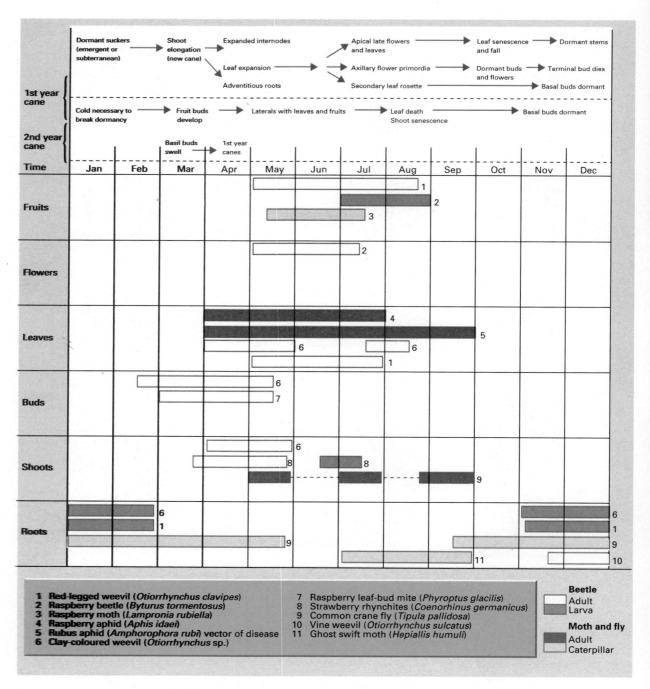

Figure 3.17 The life cycle and phenology of the raspberry plant (*Rubus idaeus*) and of some of the animals that eat it. (Data from a variety of sources.)

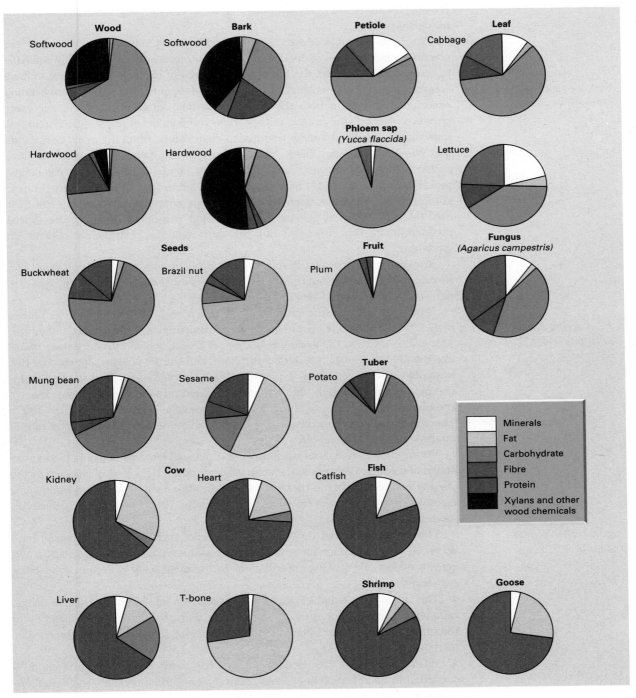

Figure 3.18 The composition of various plant parts and of the bodies of animals that serve as food resources for other organisms. (Data from a variety of sources.)

protein and fats of their prey, and their main excretory products are in consequence nitrogenous.

The large amounts of fixed carbon in plant materials means that they are potentially rich sources of energy. Yet, most of that energy is not directly available to consumers. To exploit the full energy resource of plants it is necessary to have enzymes capable of mobilizing cellulose and lignins. But, the overwhelming majority of species in both the plant and animal kingdoms lack the necessary enzymes. Hence, the major energy rich component of most plant tissue is unavailable as a direct energy resource to plant eaters. Of all the many constraints that put limits on what living organisms can do, the failure of so many to have evolved cellulolytic enzymes is an evolutionary puzzle. It may be that cellulolytic prokaryotes have so readily formed symbiotic relationships with herbivores (in their gut floras) that there has been little selection pressure to evolve cellulases of their own (Martin, 1991). It is now recognized that a number of insects do indeed produce their own cellulases but the vast majority nevertheless depend on symbionts.

It appears that for most herbivores and decomposers, the body of a plant is a superabundant energy and carbon resource; it is other components of the diet (e.g. nitrogen) that are more likely to be limiting.

Even if the cell wall fraction is excluded from a consideration of plants as articles of diet, the C : N ratio is high in green plants compared with most other organisms. This is illustrated by the feeding behaviour of aphids, which introduce their stylets into the phloem transport system of the plant, and extract phloem sap which is rich in soluble sugars (see Figures 3.15 and 3.18). The aphid uses only a fraction of this energy resource—indeed, the amino acid concentration, rather than that of carbohydrates is a measure of the quality of the sap as a resource (Douglas, 1993). The aphids convert the excess sugar into mellibiose which may drip as a rain of honeydew from an aphid-infested tree.

even excluding cell walls, plants have a high C : N ratio

Because most animals lack cellulases, the cell wall material of plants hinders the access of digestive enzymes to the contents of plant cells. The acts of chewing by the grazing mammal, cooking by humans and grinding in the gizzard of birds allow digestive enzymes to reach cell contents more easily. The carnivore, by contrast, can more safely gulp its food.

organisms that possess cellulases

Those organisms that do possess cellulases (of their own or vicariously) have access to a food resource for which they compete only amongst themselves. Their activity makes two striking contributions to the availability of resources to other organisms.

1 The alimentary canal of herbivores provides a miniature habitat in which cellulolytic bacteria gain especially easy access to fragmented cell wall material. In the ruminants, the major by-products of this bacterial fermentation are absorbed by the host (see Chapter 13).

2 When plant parts are decomposed, material with a high carbon content is converted to microbial bodies with a relatively low carbon content—the limitations on microbial growth and multiplication are resources other than carbon. Thus, when microbes multiply on a decaying plant part, they withdraw nitrogen and other mineral resources from their surroundings and build them into their own microbial bodies. For this reason, and because microbial tissue is more readily digested and assimilated, plant detritus which has been richly colonized by microorganisms is generally preferred by detritivorous animals.

a plant is a population of parts
that differ greatly in
composition and food value

The cellular packages of plants are aggregated into tissues of similar cells and into organs composed of heterogeneous assemblages of cell types. The highest concentrations of nitrogen and mineral elements are found in the young growing tips, in axillary buds and in seeds, and the highest concentrations of carbohydrate in phloem sieve tubes and in storage organs such as tubers and some seeds. The highest concentrations of cellulose and lignin are in old and dead tissues such as wood and bark. The dietary value of different tissues and organs is so different that it is no surprise to find that most small herbivores are specialists—not only on particular species or plant groups, but on particular plant parts: meristems, leaves, roots, stems, etc. The smaller the herbivore, the finer is the scale of heterogeneity of the plant on which it may specialize. Extreme examples can be found in the larvae of the oak gall wasps, some of which may specialize on young leaves, some on old, some on vegetative buds, some on male flowers and others on root tissues.

Larger herbivores also select what they eat but do so on a grosser scale. There are interesting behavioural conflicts between spending time searching for easily digested nutritious items of diet and spending time making indigestible material more digestible by chewing it and accommodating it in the gut whilst it is fermented (see Chapter 13, Section 13.5). In herbivorous vertebrates the rate of energy gain from different dietary resources is determined by the structure of the gut—in particular, the balance between: (i) a well-stirred anterior chamber in which microbial fermentation occurs; (ii) a connecting tube in which there is digestion, but no fermentation; and (iii) a posterior fermentation chamber, the colon and caecum. Models of such three-part digestive systems (Alexander, 1991) suggest that large (i), small (ii) and small (iii) (e.g. the ruminant) would give near-optimal gains from poor-quality food, and that large (iii), as in horses, is more appropriate for food with less cell wall material and more cell contents. For very high-quality food (very high proportion of cell contents and little cell wall material) the optimum gut has long (ii) and no (i) or (iii).

Elephants, lagomorphs and some rodents eat their own faeces and so double the distance travelled by the food resource through the digestive system. This allows further fermentation and digestion but may also allow time for dietary deficiencies (e.g. of vitamins) to be made good by microbial synthesis.

Seeds are the most nutritionally complete part of plants. They are particularly concentrated resources of carbohydrate, fat, protein and minerals, and provide the food resource for many small herbivores. A single seed may provide a lifetime's food supply for a bruchid beetle. The egg is laid in or on the seed and the larva may complete its life cycle to pupation within that single seed. The same seed might, however, have formed just part of 1 day's diet for a bird or have been cached in the winter stores of a rodent. Similarly, a leaf of clover in a pasture may form part of one bite for a sheep, a whole day's diet for a slug or snail or a lifetime's diet for a weevil, a leaf-mining caterpillar or a fungal leaf pathogen.

the composition of animal
bodies is less variable than
that of plants

Although plants and their parts may differ widely in the resources they offer to potential consumers, the composition of the bodies of different herbivores is remarkably similar. In terms of the content of protein, carbohydrate, fat, water and minerals per gram there is very little to choose between a diet of caterpillars or cod, or of earthworms, shrimps or venison. The packages may be differently parcelled (and the taste may be different), but the contents are essentially the same. Carnivores, then, are not faced with problems of digestion (and they vary rather little in their digestive apparatus), but rather with difficulties in finding, catching and handling their prey (see Chapter 8).

3.7.3 Food resources are often defended against consumers

All organisms are potentially food resources for others and it is in the interests of any organism to stay alive (when that makes it more likely to leave more genes to subsequent generations). Not surprisingly, many organisms have evolved physical, chemical, morphological and/or behavioural defences that reduce the chance of an encounter with a consumer and/or increase the chance of surviving such an encounter. But, the interaction does not necessarily stop there. A better defended food resource itself exerts a selection pressure on consumers. Those best able to deal with the defence will leave more descendants, and their traits are likely to spread through the consumer population. A continuing evolutionary interaction can be envisaged in which the evolution of each taxon is partially dependent upon the evolution of the other: such an interaction is referred to as coevolution (Ehrlich & Raven, 1964). Although an enticing idea, the role of coevolution as a widespread and important evolutionary force is by no means fully established (general reviews of the subject are collected in Nitecki, 1983; Futuyma & Slatkin, 1983 and for marine communities in Steneck, 1991; Hughes & Gliddon, 1991).

At this point it should be noted that the resources of green plants (and of autotrophs in general) do not grow, do not reproduce and do not evolve. With the exception of radioactive decay, atoms are immutable. The resources for plant growth cannot therefore evolve defences. Coevolution is also not possible between decomposer organisms and their dead food resources. However, bacteria, fungi and detritivorous animals will often have to contend with residual effects of physical and, in particular, chemical defences in their food. Strong interactions may of course occur between autotrophs or between decomposers in competition for these resources (see Chapters 6 and 7).

physical defences: spines

The spiny leaves of holly are not eaten by larvae of the oak eggar moth, but if the spines are removed the leaves are eaten quite readily. No doubt a similar result would be achieved in equivalent experiments with perch or foxes as predators and de-spined sticklebacks or hedgehogs as prey! Spines can be an effective deterrent.

Many small planktonic invertebrates that live in lakes, including rotifers and cladoceran water fleas, exhibit a remarkable degree of variation in the occurrence and size of spines and crests and other appendages which have been shown to reduce their vulnerability to predation. It may seem surprising that a predator, simply by its presence, can induce the development of such physical defences, but this is just what is observed amongst many of these species. Thus, for example, spine development in the offspring of an egg-bearing rotifer, *Keratella cochlearis*, is promoted if their mother was cultured in a medium conditioned by the predatory rotifer, *Asplachna priodonta* (Stemberger & Gilbert, 1984). Similarly, the development of the crest in the water flea, *Daphnia pulex*, is promoted in the presence of predatory phantom midge larvae (*Chaoborus*) (Krueger & Dodson, 1981).

protective coatings—perhaps utilizing what is most readily available

Many plant surfaces are clothed in epidermal hairs (trichomes) and in some species these develop thick secondary walls to form strong hooks or points which may trap or impale insects. The bean *Phaseolus vulgaris*, has hooked trichomes, 0.06–0.11 mm long on the undersurface of its leaves which trap both nymph and adult leafhoppers (*Empoasca fabae*). The trapped insects stop feeding and many dehydrate or starve. Likewise, the trichomes on *Passiflora* leaves are a very effective defence against most caterpillars (but, at least one species escapes the problem—caterpillars of *Mechanitis isthmia* cooperate to build a web suspended below the leaf and feed from the undefended edges (Rathcke & Poole, 1975)).

Any feature of an organism's lifestyle that increases the energy that a consumer spends in discovering or handling it, is a defence if, as a consequence, the consumer eats less of it. The thick shell of a nut or the fibrous cone on a pine increases the time spent by an animal in extracting a unit of effective food, and this may reduce the number of nuts or seeds in a cone that are eaten. The green plant uses none of its energetic resources in running away and so may have relatively more available to invest in energy rich defence structures. Moreover, it may be that most green plants are relatively overprovided with energy resources and that it is cheap to build husks and shells around seeds (and woody spines on stems) mainly out of cellulose and lignin, and so protect the real riches—the scarce resources of nitrogen, phosphorus, potassium, etc., in embryos and meristems.

A predator of mussels faces much the same problem as a predator of walnuts. Adult mussels do not move about, but they have shells which at least demand a substantial handling time and may protect the animal completely against some predators. Even a slow-moving snail has some of the same properties. The mussel may be carried up into the air by a crow and dropped onto rocks to break the shell; the snail may be broken on a stone by a thrush; the nut may be cracked by the teeth of a squirrel—but, in each case the shell wastes the time and energy of the predator. Janzen (1981) pointed out that our ancestors have probably known for the past 5 million years that truly hard biotic containers often contain otherwise undefended contents (turtles and hedgehogs are very nutritious!). Generally, the only seeds that can be eaten with confidence are those encased in thick, hard walls. In short, a castle wall can substitute for other defences such as running away or the presence of toxins.

As we might expect, the more valuable a part of an organism is to its fitness, the more strongly is it usually defended. The most concentrated food resources offered by plants are the growing points (root and shoot meristems) and the seeds. The meristems represent the potential for new growth; the seeds represent the potential descendants. Shoot meristems are commonly protected against small predators (those of the same order of size as the bud themselves) by the presence of cuticularized and often lignified scales. But, such protection will usually be no defence against larger predators like birds, for example the bullfinch concentrates on the developing flower buds of fruit trees in the spring. Buds may escape predation from larger herbivores too by being an integral part of a woody stem or twig, and may be further protected by spines intimately associated with the bud itself.

Seeds are most at risk to predators when they have just ripened and are still attached, in a cone or ovary, to the parent plant. Seeds massed together in a capsule during the ripening process offer a potentially marvellous food package to a predator, but its value is literally dissipated as soon as the capsule opens and the seeds are shed. The poppy weeds that live in cornfields and the cultivated oil-seed poppy illustrate this point. The seeds of the wild poppies are shed through a series of pores at the apex of the capsule as it waves in the wind. Two of the species, *Papaver rhoeas* and *P. dubium*, open these pores as soon as the seed is ripe and the capsules are often empty by the following day. Two other species, *P. argemone* and *P. hybridum*, have seeds that are large relative to the size of the capsule pores and dispersal is a slow process over the autumn and winter months. The capsules of these species are defended by spines. Occasionally, spineless capsuled forms of *P. argemone* have been found and these are the only ones which are attacked by birds. The cultivated poppy (*P. somniferum*), by contrast, has been selected by humans not to disperse its seeds—the capsule pores do not open. Birds can therefore be a serious pest on crops of the cultivated poppy; they tear open the capsules to reach an oil- and protein-rich reward. Humans have, of course, selected most of their crops to retain and not disperse their seeds. One consequence is that the seeds of most agricultural crops represent a sitting target for seed-eating birds.

chemical defences

In addition to the bewildering array of physical defences exhibited by different organisms, there is a battery of chemical defences. The plant kingdom is very rich in chemicals that apparently play no role in the normal pathways of plant biochemistry. These 'secondary' chemicals range from simple molecules like oxalic acid and cyanide to the more complex glucosinolates, alkaloids, terpenoids, saponins, flavonoids and tannins (Futuyma, 1983).

A defensive function is generally ascribed to these secondary chemicals and in some cases has been demonstrated unequivocally. Populations of white clover, *Trifolium repens*, are commonly polymorphic for the ability to release hydrogen cyanide when the tissues are attacked. The cyanogenic property is dependent on two pairs of alleles, one determining the presence or absence of linamarin (a glucosinolate) and the other of a β-glucosidase enzyme that releases hydrogen cyanide from the linamarin when the tissues are damaged, for instance by chewing. Plants that lack the glucosinolate or the enzyme, or both, are eaten by slugs and snails. The cyanogenic forms, however, are nibbled but then rejected (Table 3.2; see also Chapter 1, Figure 1.20).

However, many secondary chemicals appear not to have a defensive role and it is common to find that species contain a battery of related compounds rather than a single uniquely powerful poison or deterrent. It may be that compounds that are at present inactive have had an evolutionary history of activity, but that pests have evolved resistance to them. Alternatively, it has been argued that plants that produce a diversity of secondary chemicals will be evolutionarily favoured because there is a greater chance that they will, by chance, occasionally produce a biologically active compound (Jones & Firn, 1991).

Noxious plant chemicals have been broadly classified into two types: (i) toxic (or qualitative) chemicals, that are poisonous even in small quantities; and

Table 3.2 The grazing activity of slugs *(Agriolimax reticulatus)* on cyanogenic (AcLi) and acyanogenic (acli) plants of white clover *(Trifolium repens)*. Two plants, one of each genotype, were grown together in plastic containers and slugs were allowed to graze for seven successive nights. The figures show the numbers of leaves that suffered various categories of grazing damage. (+) and (–) indicate more or less than expected in this category by chance. The difference from expectation is significant at $P < 0.001$. (After Dirzo & Harper, 1982.)

	Conditions of leaves after grazing			
	Not damaged	Nibbled	Up to 50% of leaf removed	More than 50% of leaf removed
Cyanogenic plants (AcLi)	160 (+)	22 (+)	38 (–)	9 (–)
Acyanogenic plants (acli)	87 (–)	7 (–)	30 (+)	65 (+)

(ii) digestion-reducing (or quantitative) chemicals, which act in proportion to their concentration. Tannins are an example of the second type. They act by binding proteins, and render tissues, such as mature oak leaves, relatively indigestible. The growth of caterpillars of the winter moth (*Operophtera brumata*) decreases with an increase in the concentration of tannin included in the diet (Feeny, 1976).

Short-lived, ephemeral plants gain a measure of protection from consumers because of the unpredictability of their appearance in space and time (they are said to be 'unapparent'). It has been argued that they need to invest less in defence than predictable, long-lived species like forest trees (Rhoades & Cates, 1976). In addition, the former are predicted to contain 'qualitative' defences that protect only against generalist predators whereas the predictable, long-lived (apparent) species need to possess quantitative defences, relatively effective against all kinds of consumer and less susceptible to coevolution of a specialist predator (Feeny, 1976). It is likely that some plant species vary their investment in qualitative and quantitative defences as the season progresses. For example, in the bracken fern (*Pteridium aquilinum*), the young leaves that push up through the soil in spring are less apparent to potential herbivores than the luxuriant foliage in late summer. Intriguingly, the young leaves are rich in cyanogenic glucosinolates, whilst the tannin content steadily increases in concentration to its maximum in mature leaves (Figure 3.19).

chemical defences in animals ...

Animals have more options than plants when it comes to defending themselves, but some still make use of chemicals. For example, defensive secretions of sulphuric acid of pH 1 or 2 occur in some marine gastropod groups, including the cowries. The bombardier beetle (*Brachinus crepitans*) possesses a reservoir in its abdomen filled with hydroquinone and hydrogen peroxide. When threatened, these chemicals are ejected into an explosion chamber, and mixed with a peroxidase enzyme which allows the hydrogen peroxide to oxidize hydroquinone to noxious quinone. The accompanying release of O_2 gas causes the fluid to be ejected in an explosive spray.

often obtained from their poisonous food plants

Other animals, which can tolerate the chemical defences of their plant food, store and use them in their own defence. A classic example is that of the monarch butterfly (*Danaus plexippus*), whose caterpillars feed on milkweeds (*Asclepias* spp.). Milkweeds contain secondary chemicals, cardiac glycosides, that affect the vertebrate heartbeat and are poisonous to mammals and birds. Caterpillars of the monarch butterfly can store the poison, and it is still present in the adult, which in consequence is completely unacceptable to bird predators. A naive blue jay (i.e. one that has not tried a monarch butterfly before) will vomit violently after eating one, and once it recovers, will reject all others on sight. In contrast, monarchs reared on

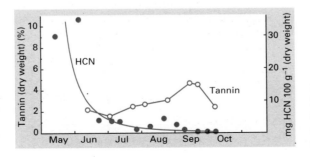

Figure 3.19 Seasonal variation in the concentration of cyanide and tannin in bracken, *P. aquilinum*. (After Rhoades & Cates, 1976.)

125 RESOURCES

cabbage, or on one of the few milkweed species that lack the glycoside, are edible (Brower & Corvinó, 1967).

Chemical defences are not equally effective against all consumers. Indeed, what is unacceptable to some animals may be the chosen, even a unique, diet of others. Many herbivores, particularly insects, have evolved to become specialist consumers of the same plant species whose particular defence they have overcome. An important step in the evolutionary interaction of plants and herbivores is the evolution of tolerance of, and then even attraction to and utilization of, the chemical defence and the plant producing it. For example, gravid females of the cabbage root fly (*Delia brassicae*) home in on a brassica crop from distances as far as 15 m downwind of the plants (Figure 3.20). It is probably hydrolysed glucosinolates (toxic to many other species) that provide the attractive odour.

An animal may be less obvious to a predator if it matches its background, or possesses a pattern that disrupts its outline, or resembles an inedible feature of its environment. Straightforward examples of such *crypsis* are the green coloration of many grasshoppers and caterpillars, and the transparency of many planktonic animals that inhabit the surface layers of oceans and lakes. More dramatic cases are the sargassum fish, whose body outline mimics the sargassum weed in which it is found, the many small invertebrates that closely resemble twigs, leaves and flower parts and the caterpillar of the viceroy butterfly that resembles a bird dropping. Cryptic animals may be highly palatable, but their morphology and colour (and their choice of the appropriate background) reduce the likelihood that they will be used as a resource.

Whilst crypsis may be a defence strategy for a palatable organism, noxious or dangerous animals often seem to advertise the fact by bright, conspicuous colours and patterns. This phenomenon is referred to as *aposematism*. The monarch butterfly, discussed above in the context of its chemical defence and distastefulness to birds, is aposematically coloured, as is its caterpillar which actually sequesters the defensive cardiac glucosinolates from its food. One attempt by a bird to eat such a butterfly is so memorable that others are subsequently avoided for some time. But, clearly the advantage will only accrue to other monarch butterflies if they *look like* those which have actually been tasted (and killed). Thus, there is a strong selection pressure for monarch butterflies to look like their conspecifics and for their form and colour to be easily memorized by a predator.

The adoption of memorable body patterns by distasteful prey immediately opens the door for deceit by other species because there will be a clear selection advantage to a palatable prey (the mimic) if it looks like an unpalatable species (the model). Such *Batesian mimicry* is a widespread phenomenon. In one sense it is a special case of

'one man's meat is another man's poison'

morphology and colour as defence: crypsis

warning coloration

Batesian mimicry

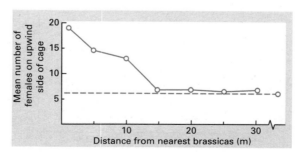

Figure 3.20 The mean numbers of female cabbage root flies (*Delia brassicae*) orientating upwind towards a brassica crop, in relation to their distance from the plants. (After Hawkes, 1974.)

126   CHAPTER 3

camouflage (crypsis), because the mimic resembles something else in the environment, but it differs because the mimic, rather than remaining inconspicuous, produces clear signals likely to be detected by the predator. Developing the story of the monarch butterfly a little further, we can note that the adult of the palatable viceroy butterfly mimics the distasteful monarch, and a blue jay that has learned to avoid monarchs will also avoid viceroys. Intriguingly, whilst the distasteful caterpillar of the monarch, like the adult, is aposematically coloured, the caterpillar of the palatable viceroy is cryptically coloured and resembles a bird dropping.

Mimicry of a distasteful organism appears not to be confined to animals. For instance, the mistletoes often show a remarkable resemblance to their evergreen, but not to their deciduous, host plants.

By living in holes, animals (millipedes, moles) may avoid stimulating the sensory receptors of predators, and by 'playing dead' (opossum, African ground squirrel, many beetles and grasshoppers), animals may fail to stimulate a killing response. Animals which withdraw to a prepared retreat (rabbits and prairie dogs to their burrows, snails to their shells), or which roll up and protect their vulnerable parts by a tough exterior (armadillos, hedgehogs, pill millipedes), reduce their chance of capture. However, once the rabbit or armadillo has withdrawn into its 'castle' it stakes its life on the chance that the attacker will not be able to breach its walls (a rabbit burrow is no defence against a weasel), and it has also sacrificed its ability to tell what is going on outside. Other animals seem to try to bluff themselves out of trouble by threat displays. The startle response of moths and butterflies that suddenly expose eye spots on their wings is one example. Others include the rattling of quills by the African porcupine and tail erection and foot stamping by the skunk.

behavioural defences

No doubt the most common behavioural response of an animal in danger of being preyed upon is to flee. For example, the small lake-dwelling copepod, *Cyclops vicinus*, when a young fish is near, exhibits a rapid, darting motion that brings it safely through an encounter with all but the most effective of predators. Figure 3.21 shows how a young bream, with its protrusible mouth and ability to develop strong suction pressure, can learn to deal successfully with *Cyclops*, whilst the less well-equipped roach cannot cope efficiently with such elusive prey.

This example illustrates three important points.

1   Like all defensive behaviours, 'flight' reduces the likelihood that the animal will end up as a food resource.

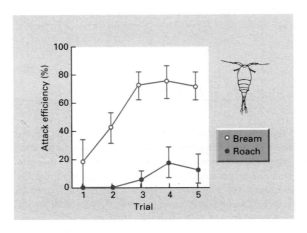

Figure 3.21 With increasing experience, young bream (*Abramis brama*) learn to capture the elusive copepod prey (*Cyclops vicinus*—shown in vignette), but roach (*Rutilus rutilus*) never achieve more than a 20% success rate. (After Winfield *et al.*, 1983.)

127   RESOURCES

2 Perhaps inevitably, one or more predators exist that can overcome the defence.

3 The defence costs energy, and the rewards (escape) must be seen as having to be paid for by the sacrifice of energy or other resources that might have been used in other activities.

3.8 A classification of resources

We have seen that every plant requires 20–30 distinct resources to complete its life cycle, and most plants require the same set of resources, although in subtly different proportions. Each of the resources has to be obtained independently of the others, and often by quite different uptake mechanisms—some as ions (potassium), some as molecules (CO_2), some in solution, some as gases. Carbon cannot be substituted by nitrogen, nor phosphorus by potassium. Only a very few of the resources needed by higher plants may be substituted in whole or in part. Nitrogen can be taken up by most plants either as nitrate or as ammonium ions, but there is no substitute for nitrogen itself. In complete contrast, for many carnivores, most prey of about the same size are wholly substitutable, one for another, as articles of diet. This contrast between resources that are individually *essential* for an organism, and those that are *substitutable*, can be extended into a classification of resources taken in pairs (Figure 3.22).

In this classification, the concentration or quantity of one resource is plotted on the x-axis, and that of the other resource on the y-axis. We know that different combinations of the two resources will support different growth rates for the

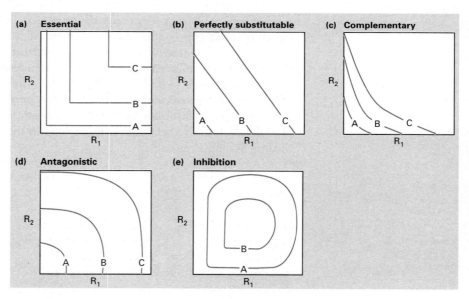

Figure 3.22 Resource-dependent growth isoclines. Each of the growth isoclines represents the amounts of two resources (R_1 and R_2) that would have to exist in a habitat for a population to have a given growth rate. Because this rate increases with resource availability, isoclines further from the origin represent higher population growth rates— isocline A has zero growth rate, isocline B an intermediate rate and isocline C the higher rate. (a) Essential resources; (b) perfectly substitutable; (c) complementary; (d) antagonistic; and (e) inhibition. (After Tilman, 1982.)

organism in question (this can be individual growth or population growth, i.e. survival and reproduction). Thus, we can join together points (i.e. combinations of resources) with the same growth rates, and these are therefore contours or 'isoclines' of equal growth. For example, in Figure 3.22 line B in each case is an isocline of zero net growth. In other words, each of the resource combinations on these lines allows the organism just to maintain itself, neither increasing nor decreasing. The A isoclines, therefore, with less resources than B, join combinations giving the same *negative* growth rate; whilst the C isoclines, with more resources than B, join combinations giving the same *positive* growth rate. As we shall see, the shapes of the isoclines vary with the nature of the resources.

3.8.1 Essential resources

Two resources are said to be *essential* when one is unable to substitute for another. Thus, the growth that can be supported on resource 1 is absolutely dependent on the amount available of resource 2 and vice versa. This is denoted in Figure 3.22a by the isoclines running parallel to both axes. They do so because the amount available of one resource defines a maximum possible growth rate, irrespective of the amount of the other resource. This growth rate is achieved unless the amount available of the other resource defines an even lower growth rate. It will be true for nitrogen and potassium as resources in the growth of green plants, and for the two obligate hosts in the life of a parasite or pathogen that is required to alternate in its life cycle (see Chapter 12). In the life of *Heliconius* butterflies, resource 1 might be the leaves of a particular species of *Passiflora* vine for the larval food, and resource 2 would be the pollen of the cucurbit plant *Gurania* for the adult butterfly.

3.8.2 Other categories of resource

Two resources are said to be *perfectly substitutable* when either can wholly replace the other. This will be true for seeds of wheat or barley in the diet of a farmyard chicken, or for zebra and gazelle in the diet of a lion. Note that we do not imply that the two resources are as good as each other. This feature (perfectly substitutable but not necessarily as good as each other) is included in Figure 3.22b by the isoclines having slopes that do not cut both axes at the same distance from the origin. Thus, in Figure 3.22b, in the absence of resource 2, the organism needs relatively little of resource 1, but in the absence of resource 1 it needs a relatively large amount of resource 2.

complementary resources   Substitutable resources are defined as *complementary* if the isoclines bow inwards towards the origin (Figure 3.22c). This shape means that a species requires less of two resources when taken together than when consumed separately. Humans eating certain kinds of beans together with rice can increase the usable protein content of their food by 40% (Lappe, 1971). The beans are rich in lysine, an essential amino acid poorly represented in rice, whilst rice is rich in sulphur-containing amino acids that are present only in low abundance in beans.

antagonistic resources   A pair of substitutable resources with isoclines that bow away from the origin are defined as *antagonistic* (Figure 3.22d). The shape indicates that a species requires proportionately more resource to maintain a given rate of increase when two resources are consumed together than when consumed separately. This could arise, for example, if the resources contain different toxic compounds which act synergistically (more than just additively on their consumer). For example, D, L-pipecolic acid and djenkolic acid (two secondary chemicals believed to have a defensive function in certain seeds) had no significant effect on the growth of the seed-eating larva of a bruchid beetle if consumed separately, but they had a pronounced effect if taken together (Janzen *et al.*, 1977). If the seed of one species contained one of these compounds, and another seed the second compound, then a mixed diet would produce an adverse effect on growth.

inhibition

Finally, Figure 3.22e illustrates the phenomenon of *inhibition* at high resource levels (for a pair of essential resources). It is not difficult to find examples of resources that are essential but become toxic or damaging when in excess (see Figure 2.1d). CO_2, water and mineral nutrients such as iron are all required for photosynthesis, but each is lethal in excess. Similarly, light leads to increased growth rates in plants through a broad range of intensities, but can inhibit growth at very high intensities. In such cases, the isoclines form closed curves because growth decreases with an increase in resources at very high levels. This is a situation in which what are *resources* at one level become limiting *conditions* at another.

3.9 Resource dimensions of the ecological niche

In Chapter 2 we developed the concept of the ecological niche as an *n*-dimensional hypervolume. This defines the limits within which a given species can survive and reproduce, for a large number (*n*) of environmental factors, including both conditions (see Chapter 2) and resources (this chapter). Note, therefore, that the zero growth isoclines in Figure 3.22 define niche boundaries in two dimensions. Resource combinations to one side of the A line allow the organisms to thrive—but, to the other side of the line the organisms decline.

The resource dimensions of a species' niche can sometimes be represented in a manner similar to that adopted for conditions, with lower and upper limits within which a species can thrive. Thus, for a size-selective predator, limits to its ability to detect and handle prey mean that it is able to exploit only a limited range within a continuum, from tiny to very large potential prey. For other resources, such as mineral nutrients for plants, a lower limit to nutrient concentration may be defined below which individuals cannot grow and reproduce—but, an upper limit may not exist (Figure 3.22a–d). However, many resources cannot be described as continuous variables; rather, they must be viewed as discrete entities. Larvae of butterflies in the genus *Heliconius* require *Passiflora* leaves to eat; those of the monarch butterfly specialize on plants in the milkweed family; and various species of animals require nest sites with particular specifications. These resource requirements cannot be arranged along a continuous graph axis labelled, for example, 'food plant species'. Instead, the food plant or nest-site dimension of their niches needs to be defined simply by a restricted list of the appropriate resources.

Resource dimensions, however they are defined, are crucial components of the *n*-dimensional niche. We will see that the niche concept is at its most powerful when we consider problems related to the use of limiting, or potentially limiting, resources. It will figure prominently in later chapters which consider the potential role of interspecific competition for resources in determining community composition and diversity (see Chapters 7, 18 and 20).

3.10 Space as a resource

space as a portmanteau term

All living organisms occupy space, and it has not normally been regarded as a resource. A position in space has been treated as defining where resources are to be found rather than as a resource in its own right. When authors have written about organisms competing for space they have usually used 'space' as a 'portmanteau' word to describe the resources that may be captured within it rather than as a resource itself. A plant may be said to compete with another for a space in a canopy, but this seems to say only that they are competing for the light (or CO_2) that might be captured in that space. Similarly, plants may be said to require root space, but this seems inevitably to be saying that they require the resources of water and nutrients that are in the space rather than the space itself. Even a dense forest will contain abundant space for many more trees and, above the surface of a pond, there is enough space for an infinite number of layers of duckweeds on a pond—in both cases, it will be the supply of light, water or mineral resources that will ultimately limit growth, not a shortage of space.

But, space itself can become a limiting resource if it is the physical packing of organisms that limits what they can do, even though food is abundant. Barnacles and mussels may pack on a rock surface at densities at which there is literally no physical room for others. Crocus corms may develop at such density that they physically force each other out of the ground. There are other cases in which it is abundantly clear that it is space itself that is the resource. Thus, we can envisage warm basking sites on rocks as a resource for lizards. Lizards can hardly be said to be consuming temperature (a condition), but they certainly use up favourable micro-sites, making them unavailable for others. By definition, such microsites are resources. In addition, we can identify nest sites and hiding places as potential resources for many kinds of animal. When they are occupied they are unavailable to others but, if the occupant deserts or dies, the hole or hiding place or warm rock becomes available for another occupant (consumption was only temporary!).

sites as resources

There is an important sense in which the territorial behaviour of some animals has made space itself into a resource. We shall see in Chapter 7 (when competition between organisms is considered) that it is possible to distinguish, on the one hand, situations in which individual A reduces the level of a resource and individual B reacts to this reduced level. The two organisms respond not to each other's presence, but to the level of resource depletion that each produces (exploitation competition). On the other hand, particularly in higher animals and birds, the level of conflict may be shifted so that A and B interact directly by capturing space (territory)—they then respond directly to each other's presence (interference competition), rather than to the level to which they have depleted a more conventional resource (e.g. food).

exploitation and interference

Territorial behaviour is almost always associated with mating behaviour. Typically, an individual establishes his (in a few species it is her) territory by aggressive behaviour (often by ritual display rather than actual fighting), and this in turn determines success in attracting a mate. The territory is in a sense a surrogate for the consumable resources of food or nesting sites within it. (The really crucial resource may of course be the number of available mates, and the establishment of a territory is a ticket to a share in the limited resource of available mates.) An elegant demonstration of the relationship between sexuality and the capture of resource space was made by Moss *et al.* (1994), in an experiment with red grouse (*Lagopus lagopus lagopus*), which have strong territorial behaviour (see Chapter 6, Section 6.12).

They determined the territories of cocks in two areas separated by a buffer zone and implanted testosterone into the cocks in one of these areas. The implanted cocks increased the size of their territories by capturing space from their neighbours and excluded some of them altogether (Figure 3.23).

Other examples of the capture of resource space by birds and plants are shown in Figures 6.36 and 6.37. Territorial animals are for the most part motile and can choose and defend a position in space and can also choose to desert it and move elsewhere. Plants have no such freedom, nor have most clonal and modular animals (or the few unitary animals such as mussels and barnacles, which are fixed and rooted for most of their lives). A successful egg or a larval stage is one that lands in a position in space where it is safe from major hazards and can obtain the nutrition and conditions needed for growth. As it grows it may exclude some of its neighbours in a jostling for space (e.g. mussels). In plants and other modular organisms (see Chapter 4), parts of individuals extend into the territory of others and suppress them (e.g. the branches of trees as they meet neighbours in a forest and the spreading tillers of a grass as it spreads in a grassland) (Harper, 1985).

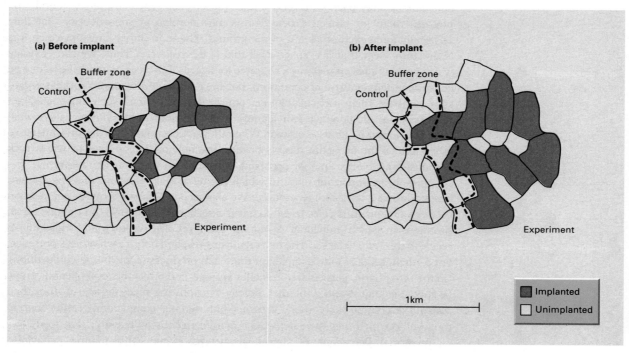

Figure 3.23 The effects on territories and populations of implanting cocks of the red grouse (*Lagopus lagopus lagopus*) with testosterone. (a) The territories of male grouse before implantation in 1991. Shaded areas are the territories of cocks chosen for implantation. A buffer zone was defined between the broken lines. (b) The territories established in spring 1992. Two cocks were killed by predators, one of them in the experimental and one in the control area. The cock in the control area was replaced by an unmarked cock, leaving the territories largely unchanged. Similar results were obtained when the treatments were crossed over in 1992–1993, except that two implanted cocks expanded their territories across the buffer zone, causing the loss of two cocks from the control area. (After Moss *et al.*, 1994.)

As soon as we begin to think of space itself as a resource, the growth habit, form and architecture of modular organisms take on new significance—for it is they that determine how space is explored and captured. This is just one of the reasons why the study of space and its exploitation has become so important in ecology in the 1990s—'the final frontier for ecological theory' (Kareiva, 1994). It is as though we come to regard the supermarket as the resource rather than the array of food items on its shelves.

In the study of modular and other non-motile organisms, the dynamic events that occur where individual organisms meet each other or invade open spaces are the key to understanding community and global-level changes. This has suggested to many modellers the value of cellular automata, which can describe the dynamical behaviour of systems by treating space and time in a discrete manner (Wolfram, 1983, 1984). Such automata consist of a regular lattice of cells, each of which may have one of a range of states, for example vacant or occupied by a particular species. Each cell in the lattice has four neighbouring cells which share its boundaries. The procedure is especially appropriate for studying the way in which the growth form of modular organisms determines how they capture space (Oborny, 1994) and has been used to interpret the spatial population dynamics of white clover in grasslands (Cain *et al.*, 1995). If we know the relative success of each of a set of species in invading and excluding the others we can write transition probabilities for each unit of time for each cell, and the process can be repeated (iterated) to predict changes in their distribution.

the partitioning of space between species

Five species of pasture grass were grown in an array of contiguous hexagonal plots so that the invasive properties of each species could be determined with respect to the others (Figure 3.24a). These experimental data were then used in a cellular automaton to predict the long-term partitioning of space between the species. In one automaton (Figure 3.24b) the individuals of the five species were initially distributed at random with respect to each other, thus minimizing the number of intraspecific contacts. One of the species rapidly monopolized almost all of the lattice. In other simulations (Figures 3.24c–e), the five species were initially concentrated in five bands across the lattice—thus, maximizing the number of intraspecific neighbour contacts. More species then persisted for longer and the way they partitioned the space between them, over the long term, depended on the initial arrangement of the bands (Figures 3.24b–d). The simulation makes two important points: (i) intraspecific aggregation slows down the rate at which space is partitioned between the species; and (ii) the way in which the space is partitioned between species is dependent on the initial conditions. Cellular automata have also been used to model the way in which space is partitioned between annual and perennial plants (Crawley & May, 1987).

The concept of space as a resource appears again later in this book in different contexts: (i) as patches within a community that are habitable but unoccupied and which may become inhabited if propagules are efficiently dispersed (in a discussion of metapopulation dynamics, see Chapter 15, Section 15.6.5); and (ii) as places in a community where space is occupied by an individual (or part of one), and is therefore a resource that might be captured by competition by one of the same or of a different species (see Chapters 6 and 7).

There are some striking differences in the ways in which plant and animal ecologists have studied resources. Much of the history of plant ecology has been dominated by 'ecophysiology' and particularly by the ways in which *individual* plant parts obtain light, water and nutrients. There has been a strong emphasis on

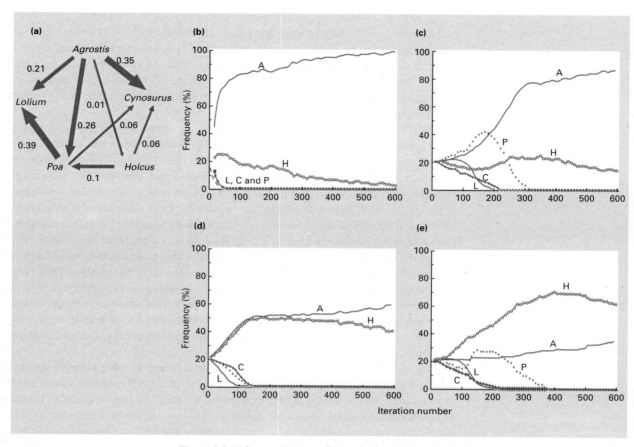

Figure 3.24 The partitioning of space between five species of grass over time as determined in a cellular automaton model using values obtained for the replacement rates of each of the species against the others in an experiment of Thórhallsdóttir (1990). (a) Observed net rates of invasion by five species of grass (A, *Agrostis capillaris*; C, *Cynosurus cristatus*; H, *Holcus lanatus*; L, *Lolium perenne*; P, *Poa trivialis*) into the territory of the others. The arrows point from the invading species to the invaded species. (b) The abundances of the five grass species over successive iterations of the cellular automaton model. At the start of the iterations the five species were distributed at random. (c) As (b), but at the start of the iterations the five species were concentrated in five horizontal bands across the lattice of cells in the order A–H–L–C–P (i.e. the population of A had neighbours of H on one side and none on the other, P had neighbours of C on one side and none on the other). (d) As (c), but the five horizontal bands were in the order A–L–C–H–P. (e) As (c), but the five horizontal bands were in the order A–H–P–C–L. (After Silvertown *et al.*, 1992a.)

laboratory studies. In contrast, the animal ecologist has more often concentrated on resources as the object of competition between organisms, or as the material of predator–prey interactions. Some of this historical difference in approach is apparent in the structure of this chapter, for example in the ecophysiological approach to the resources of light, water and nutrients. As ecology becomes an increasingly integrated science, these differences in approach may largely disappear. The realization that space itself can be studied and modelled as a resource is a powerful force for the integration of ecological disciplines (van Tongeren & Prentice, 1986; Iwasa *et al.*, 1991; Tilman, 1994; Holmes *et al.*, 1994).

Chapter 4
Life and Death in Unitary and Modular Organisms

4.1 Introduction: an ecological fact of life

In this chapter we change the emphasis of our approach. We will not be concerned so much with the interaction between individuals and their environment, but rather with the numbers of individuals and the processes leading to changes in the number of individuals.

In this regard, there is a fundamental and unalterable ecological fact of life:

$$N_{\text{now}} = N_{\text{then}} + B - D + I - E. \tag{4.1}$$

In other words, the number of a particular organism presently occupying a particular site of interest (N_{now}) is equal to the number previously there (N_{then}), plus the number of births between then and now (B), minus the number of deaths (D), plus the number of immigrants (I), minus the number of emigrants (E). Or, to put it a slightly different way:

$$N_{\text{future}} = N_{\text{now}} + B - D + I - E \tag{4.2}$$

where B, in this case, is the number of births between now and some time in the future, and so on.

These facts of life define the main aim of ecology: to describe, explain and understand the distribution and abundance of organisms. Ecologists studying the effects of an insect pest on a crop will probably be trying to find ways of increasing N_{future} for the crop by reducing N_{future} for the pest; and those studying the distribution of a rare plant in a protected site will monitor the variation in N_{now} from place to place and will probably try to find out whether particular microsites favour colonization and birth, or hasten death. Ecologists studying the animal and plant community of a polluted stream will probably catalogue N_{now} for a wide range of species, and compare these with data from similar but unpolluted streams. Even a study of the effects of temperature, light or a pollutant, such as mercury, on a particular organism has only ecological relevance insofar as the factor affects the birth, death or migration of the organism. In all cases, ecologists are interested in the number of individuals, the distributions of individuals, the demographic processes (birth, death and migration) which influence these and the ways in which these demographic processes are themselves influenced by environmental factors.

Thus, the present chapter lays foundations for ecology by examining patterns of birth and death, and to a lesser extent migration. (Migration will be considered in much greater detail in Chapter 5.) Particular attention will also be paid to the ways in which these patterns are quantified, and to discovering if there are generalizations linking apparently dissimilar types of organisms.

4.2 What is an individual? Unitary and modular organisms

First, however, an important assumption, implicit in what has been said so far, must be recognized and rejected. The 'ecological fact of life' was framed in terms of 'individuals' dying, migrating and so on. This implies that any one individual is just like any other, which is patently false on a number of counts.

individuals differ in their life-cycle stage and their condition

First, almost all species pass through a number of *stages* in their life cycle: insects metamorphose from eggs to larvae, sometimes to pupae and then to adults; plants pass from seeds to seedlings to photosynthesizing adults; and so on. In all such cases, the different stages are likely to be influenced by different factors and to have different rates of migration, death and of course reproduction. These stages need to be recognized and treated separately.

Second, even within a stage, or where there are no separate stages, individuals can differ in 'quality' or 'condition'. The most obvious aspect of this is size, but it is also common, for example, for individuals to differ in the amount of stored reserves they possess, as Figure 4.1 shows for the quantity of stored fat in brown trout.

The most important area in which the simplistic view of an individual breaks down, however, is when an organism is *modular* rather than *unitary*.

4.2.1 Unitary and modular organisms

In unitary organisms, form is highly determinate. Barring aberrations, all dogs have four legs, all locusts have six legs, all fish have one mouth and all squid have two eyes. Humans are perfect examples of unitary organisms. A life begins when a sperm fertilizes an egg to form a zygote. This implants in the wall of the uterus, and the complex processes of embryonic development commence. By 6 weeks the foetus has a recognizable nose, eyes, ears and limbs with digits, and accidents apart, it will remain in this form until it dies. The foetus continues to grow until birth, and then the infant grows until perhaps the 18th year of life; but the only changes in form (as opposed to size) are the relatively minor ones associated with sexual maturity. The reproductive phase lasts for perhaps 30 years in females and rather longer in males.

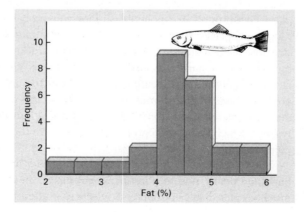

Figure 4.1 Variation in individual quality: fat content in a sample of small trout all weighing close to 11 g. (After Elliott, 1976.)

This is followed by, or may merge into, a phase of senescence. Death can intervene at any time, but for surviving individuals the succession of phases is, like form, entirely predictable.

modular organisms

In modular organisms, on the other hand, neither timing nor form is predictable. The zygote develops into a unit of construction (a module, e.g. a leaf with its attendant length of stem), which then produces further, similar modules. Individuals are composed of a highly variable number of such modules, and their programme of development is strongly dependent on their interaction with their environment. The product is almost always branched, and except for a juvenile phase, effectively immobile. Most plants are modular and are certainly the most obvious group of modular organisms. There are, however, many important groups of modular animals (indeed, some 19 phyla, including sponges, hydroids, corals, bryozoans and colonial ascidians), and many modular protists and fungi. Reviews of the growth, form, ecology and evolution of a wide range of modular organisms may be found in Jackson *et al.* (1985), Harper *et al.* (1986a), Hughes (1989), Andrews (1991) and Room *et al.* (1994). The essential components of modularity are summarized by Harper *et al.* (1986b).

growth in higher plants

In the growth of a higher plant, the fundamental module of construction above ground is the leaf with its axillary bud and the attendant internode of the stem. As the bud develops and grows, it produces further leaves, each bearing buds in their axils. The plant grows by accumulating these modules. At some stage in the development, a new sort of module appears, associated with reproduction (e.g. the flowers in a higher plant), ultimately giving rise to new zygotes. Modules that are specialized for reproduction usually cease to give rise to new modules (although this is not true of all modular animals). The roots of a plant are also modular, although the modules are quite different (Harper *et al.*, 1991). The programme of development in modular organisms is typically determined by the proportion of modules that are allocated to different roles (e.g. to reproduction or to continued growth).

Many ecological and evolutionary generalizations have been made in the past as if the unitary animal (such as a human or a mosquito) in some way typifies the living world. This is highly misleading. Modular organisms such as seaweeds, corals, forest trees and grasses dominate large parts of the terrestrial and aquatic environments. A variety of growth forms and architectures produced by modular growth in animals and plants is illustrated in Figure 4.2.

horizontal and vertical disposition of modules

Modular organisms may broadly be divided into those that concentrate on vertical growth, and those that spread their modules laterally, over or in a substrate. Many plants produce new root systems associated with a laterally extending stem: these are the rhizomatous and stoloniferous plants. The connections between the parts of such plants may die and rot away, so that the product of the original zygote becomes represented by physiologically separated parts. (Modules with the potential for separate existence are known as 'ramets'.) The most extreme examples of plants 'falling to pieces' as they grow are the many species of floating aquatics like duckweeds (*Lemna*), the water hyacinth (*Eichhornia*) and the water lettuce (*Pistia*). Whole ponds, lakes or rivers may be filled with the separate and independent parts produced by a single zygote.

Trees are the supreme example of plants whose growth is concentrated vertically. The peculiar feature distinguishing trees and shrubs from most herbs is the connecting system linking modules together and connecting them to the root system. This does not rot away, but thickens with wood, conferring perenniality. Most of the

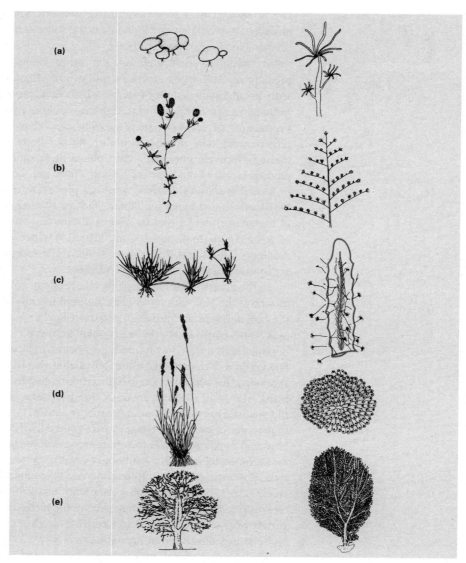

Figure 4.2 A range of modular organisms: plants to the left and animals to the right. (a) Modular organisms that fall to pieces as they grow: duckweed (*Lemna* spp.) and *Hydra* spp. (a non-colonial cnidarian). (b) Freely branching organisms that are relatively short lasting: the annual hare's foot clover (*Trifolium arvense*) and *Pennaria* spp. (Cnidaria: Hydrozoa). (c) Rhizomatous and stoloniferous organisms in which clones spread laterally: buffalo grass (*Buchloe dactyloides*) and *Campanularia* spp. (Cnidaria: Hydrozoa). (d) Tussock formers comprising tightly packed modules: 6-weeks fescue grass (*Festuca octoflora*) and *Cryptosula* spp. (an encrusting bryozoan). (e) Persistent, multiple-branched organisms: an oak tree (*Quercus* spp.) and *Gorgonia* spp., the common sea fan (a coral—Cnidaria: Anthozoa).

structure of such a woody tree is dead, with a thin layer of living material lying immediately below the bark. The living layer, however, continually regenerates new tissue, and adds further layers of dead material to the trunk of the tree. The majority of a tree is a 'cemetery' in which dead stem tissues of the past are interred, but the strength of the trunk solves the difficult morphological problem of obtaining water and nutrients below the ground, but also light perhaps 50 m away at the top of the canopy.

modularity can be apparent at more than one level

We can often recognize two or more levels of modular construction. The fundamental units of higher plants are assembled into clusters with a form that is itself continually repeated. The strawberry is a good example of this: leaves are repeatedly developed from a bud, but these leaves are arranged into rosettes. The strawberry plant grows: (i) by adding new leaves to a rosette, and (ii) by producing new rosettes on stolons grown from the axils of its rosette leaves. Trees also exhibit modularity at several levels: the leaf with its axillary bud, the whole shoot on which the leaves are arranged and the whole branch systems that repeat a characteristic pattern of shoots.

Obelia: a modular animal

The growth of modular animals can be illustrated by a hydrozoan, like *Obelia*. Development begins when a short-lived, free-swimming planula larva attaches itself to a solid object. It gives rise to a horizontal root-like structure that bears a number of branched stalks. The basic *Obelia* modules, the polyps (which are both feeding and defensive structures), are borne on these stalks. The terminal polyp of each branch is temporarily the youngest, but is overgrown by the next one to develop, which arises as a bud at its base. The branched stalks remain as an interconnecting network between all the polyps in a colony. Reproduction in *Obelia* begins when tiny, free-swimming jellyfish are budded off from modified polyps called gonophores; these jellyfish then reproduce sexually to produce the dispersing planula larvae. Thus, these and similar animals, despite variations in their precise method of growth and reproduction, are as 'modular' as any plant. Moreover, in corals, for example, just like many plants, the individual may exist as a physiologically integrated whole, or may be split into a number of colonies—all part of one individual, but physiologically independent (Hughes *et al.*, 1992).

Returning to the question of what is an individual, one could determine the number of rabbits in a field by counting their ears and dividing by two, or their legs and dividing by four. But, no such calculation is possible from the number of leaves on a tree, the number of fronds on a fern or the number of zooids of an ascidian or bryozoan. N_{now} may represent the current number of surviving zygotes, but it can give only a partial and misleading impression of the 'size' of the population if the organism is modular.

individual modular organisms: genets

Kays and Harper (1974) coined the word 'genet' to describe the 'genetic individual': the product of a zygote. In modular organisms, then, whilst the distribution and abundance of genets (individuals) is important, it is also necessary, and often more useful, to study the distribution and abundance of modules (ramets, shoots, tillers, zooids, polyps or whatever). For instance, the amount of grass in a field available to a herd of cattle is not determined by the number of grass genets but by the number of leaves, i.e. the number of modules. For modular organisms, therefore, the 'ecological fact of life' comprises at least two equations: the one already established (see Equation 4.1), plus another:

$$\text{modules}_{now} = \text{modules}_{then} + \text{modular birth} - \text{modular death}. \tag{4.3}$$

module abundance is often more important than genet abundance

The importance of this is emphasized by comparing two populations of creeping buttercup, *Ranunculus repens*—one growing in woodland, the other in grassland (Figure 4.3). Over a study period of 18 months, only four established seedlings (new genets) were recorded in the grassland plot, of which one was still alive at the end, and only seven in the woodland plot, of which three were still alive. There was no

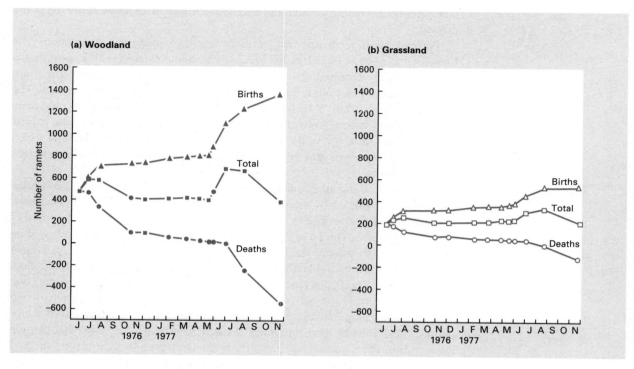

Figure 4.3 The net number of ramets and the cumulative births and deaths of ramets of the creeping buttercup, *Ranunculus repens*, in ten 0.5 m × 0.5 m quadrats in (a) woodland and (b) grassland in North Wales. (After Lovett Doust, 1981.)

question, however, of 'nothing going on' within the populations. In the woodland plots, especially, an overall approximate constancy in the total number of ramets was accompanied by marked declines in numbers in autumn (August–November), marked rises in summer (June–August), and especially by a very large number of modular births and deaths, which a simple monitoring of net numbers would fail to convey.

modularity can lead to extreme individual variability

Moreover, the potentialities for individual difference are far greater in modular than in unitary organisms. For example, an individual of the annual plant *Chenopodium album*, may, if grown in poor or crowded conditions, flower and set seed when only 50 mm high. Yet, given more ideal conditions, it may reach 1 m in height, and produce 50 000 times as many seeds as its depauperate counterpart. It is modularity and the differing birth and death rates of plant *parts*, which give rise to this plasticity.

whole modular individuals often do not senesce, but …

In fact, there is often no programmed senescence of the whole modular organism—they appear to have perpetual somatic youth. Even in trees that accumulate their dead stem tissues, or gorgonian corals that accumulate old calcified branches, death often results from becoming too big or succumbing to disease rather than from programmed senescence. This is illustrated for three types of coral in the Great Barrier Reef in Figure 4.4. Annual mortality declined sharply with increasing colony size (and hence, broadly, age) until, amongst the largest, oldest colonies, mortality was virtually zero, with no evidence of any increase in mortality at extreme old age (Hughes & Connell, 1987).

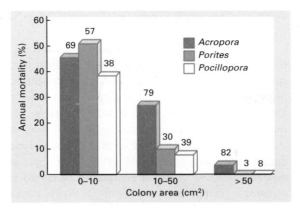

Figure 4.4 Mortality rate declines steadily with colony size (and hence, broadly, age) in three coral taxa from the reef crest at Heron Island, Great Barrier Reef (sample sizes are given above each bar). (After Hughes & Connell, 1987; Hughes *et al.*, 1992.)

... individual modules do: modular individuals have an age structure

At the modular level, things are quite different. The annual death of the leaves on a deciduous tree is the most dramatic example of senescence—but roots, buds, flowers and the modules of modular animals all pass through phases of youth, middle age, senescence and death. The growth of the individual genet is the combined result of these processes. Thus, the body of an individual modular organism has an age structure; it is composed of young and developing, actively functioning and senescent parts. (Moreover, a leaf or root, for instance, changes its activity as it ages, and may also change its dietary value and attraction to a herbivore. For the ecologist to treat all leaves in a pasture or forest canopy as equal is to ignore the fact that other organisms will discriminate between them.) Figure 4.5 shows that the age structure of shoots of the sedge *Carex arenaria* is changed dramatically by the application of NPK fertilizer, even when the total number of shoots present is scarcely affected by the treatment. The fertilized plots became

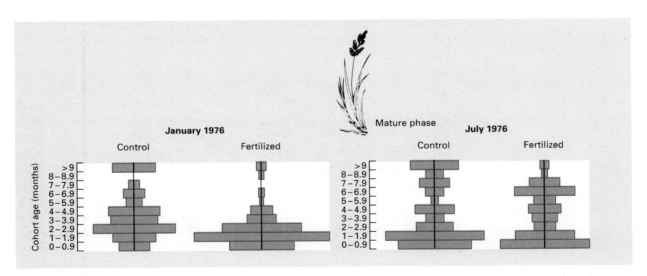

Figure 4.5 The age structure of shoots in clones of the sand sedge *Carex arenaria* growing on sand-dunes in North Wales. Clones are composed of shoots of different ages. The effect of applying fertilizer is to change this age structure. The clones become dominated by young shoots and the older shoots die. (After Noble *et al.*, 1979.)

141   LIFE AND DEATH

dominated by young shoots, as the older shoots which were common on control plots were forced into early death.

For many rhizomatous and stoloniferous species, this changing age structure is in turn associated with a changing level to which the connections between individual ramets remain intact. A young ramet may benefit from the nutrients flowing from an older ramet to which it is attached and from which it grew, especially as the 'parent' and 'daughter' ramets may, in other respects, have conflicting demands for soil nutrients and light (*intraspecific competition* is discussed fully in Chapter 6). However,

the pros and cons of module integration

the pros and cons of attachment will have changed markedly by the time the daughter is fully established in its own right and the parent has entered a post-reproductive phase of senescence (a comment equally applicable to unitary organisms with parental care—parents please note!) (see Caraco & Kelly, 1991).

The changing benefits and costs of integration have been studied experimentally in the pasture grass *Holcus lanatus*, by comparing the growth of (i) ramets that were left with a physiological connection to their parent plant, and in the same pot, so that parent and daughter might compete (unsevered, unmoved: UU); (ii) ramets that were left in the same pot (competition possible), but had their connection severed (SU); and (iii) ramets that had their connection to their parent severed and were repotted in their parent's soil, but after the parent had been removed, so no competiton was possible (SM) (Figure 4.6). These treatments were applied to daughter ramets of various ages, which were then examined after a further 8 weeks' growth. For the youngest daughters (Figure 4.6a) attachment to the parent significantly enhanced growth (UU > SU), but competition with the parent had no apparent effect (SU ≈ SM). For slightly older daughters, however (Figure 4.6b), growth could be depressed by the parent (SU < SM), but physiological connection effectively negated this (UU > SU; UU ≈ SM). For even older daughters, the balance shifted further still: physiological connection to the parent was either not enough to fully overcome the adverse effects of the parent's presence (Figure 4.6c; SM > UU > SU), or eventually appeared to represent a drain on the resources of the daughter (Figure 4.6d; SM > SU > UU). However, this tells us only about the pros and cons of connection for individual

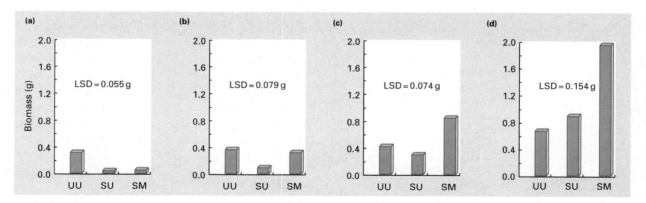

Figure 4.6 The growth of daughter ramets of the grass *Holcus lanatus* which were initially (a) 1 week, (b) 2 weeks, (c) 4 weeks and (d) 8 weeks old, and were then grown on for a further 8 weeks. LSD, least significant difference, which needs to be exceeded for two means to be significantly different from each other. For further discussion, see text. (After Bullock *et al.*, 1994a.)

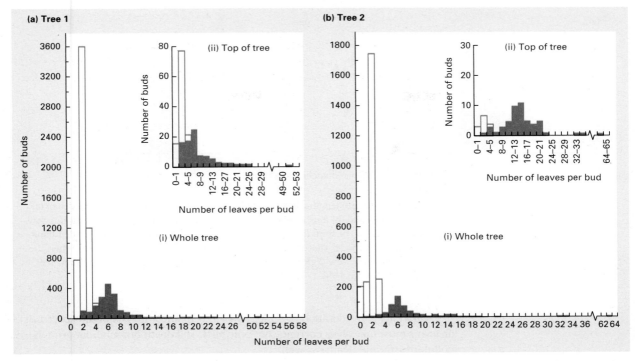

Figure 4.7 The frequency distributions, for two silver birch trees, of the number of leaves produced per bud on representative branches, for the whole tree (i) or for the top of the tree only (ii), where the top of the tree is taken to be the leader shoot or any branches produced by the leader shoot in the previous year. (□) short shoots; (■) long shoots. (After Maillette, 1982.)

daughter ramets. Determining the net benefits of various levels of connection to the plant as a whole will be a far more complex problem.

module behaviour varies with position on the genet

Thus, the way in which modular organisms interact with their environment is determined by their architecture—how their modules are placed in relation to other modules of the same organism and the modules of other organisms. There is little mobility, search or escape—only that which can be achieved by growing from one place to another or by releasing specialized dispersal units (or, in some aquatic plants, falling into pieces and floating away). Variation in module behaviour, dependent on location within the genet, is seen clearly in silver birch trees, *Betula pendula* (Figure 4.7). Most buds produce short shoots with one to three leaves, or very rarely four. At the top of the tree, however, a much higher proportion of buds gives rise to long shoots with many more leaves, and many fewer of the buds remain inactive (i.e. stay dormant or die without producing leaves).

In some cases, this type of variation in architecture has been shown to be more patchily distributed within a plant—not simply the consequence of being at the top or bottom of a tree. As stolons of *Glechoma hederacea* and *Lamiastrum galeobdolon* grow from high-light into low-light conditions, there is stolon elongation: internode lengths increase so that leaves in the low-light environment are more widely spaced. But, when the stolons grow in the reverse direction from low to high light, there is a corresponding shortening of internodes (Figure 4.8).

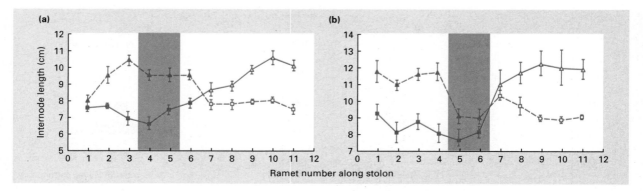

Figure 4.8 Mean (± SE) lengths of sequentially produced primary stolon internodes of (a) *Glechoma hederacea* and (b) *Lamiastrum galeobdolon*, growing either from high-light into low-light conditions (■—△) or from low-light to high-light conditions (▲—□). The shaded band indicates stolon internodes that were still elongating under 'old' light conditions when the stolon apex grew into 'new' light conditions. (After Slade & Hutchings, 1987; Dong, 1993; Hutchings & de Kroon, 1994.)

Clearly, any consideration of ecology which is to encompass a broad spectrum of life forms must take as its currency not only birth and death but modular growth (i.e. modular birth and death) as well.

4.3 Counting individuals

If we are going to study birth, death and modular growth seriously, we must quantify them. This means counting individuals and (where appropriate) modules. Often in modular organisms we cannot recognize individual genets—their parts can become separated, and clones (products of a single zygote) can intermingle. We can then count only modules.

numerical change is often monitored without monitoring demographic processes

The remainder of this chapter will examine and quantify patterns of birth, death and modular growth. However, many studies concern themselves not with birth and death but with their consequences, i.e. the total number of individuals present and the way these numbers vary with time. Such studies can often be useful nonetheless. For instance, Figure 4.9a shows the steady exponential increase since the early years of the 20th century in the number of pairs of black noddy birds on Heron Island, at the southern end of the Great Barrier Reef; Figure 4.9b shows that the number of pine looper moths in a Scottish woodland fluctuated repeatedly with a periodicity of around 6–8 years; whilst Figure 4.9c shows the decline in abundance to an apparently constant low level experienced in recent years by economically important fish, Cape hake, off the coast of South Africa. Of course, the best studies provide data not only on the numbers of individuals, but also the processes affecting numbers.

the meaning of 'population'

It is usual to use the term population to describe a group of individuals of one species under investigation. However, what actually constitutes a population will vary from species to species and from study to study. In some cases, the boundaries of a population are readily apparent: the sticklebacks occupying a small, homogeneous, isolated lake are the 'stickleback population of that lake'. In other cases, boundaries are determined by an investigator's purpose or convenience: it is possible to study the population of lime aphids inhabiting one leaf, one tree, one stand of trees

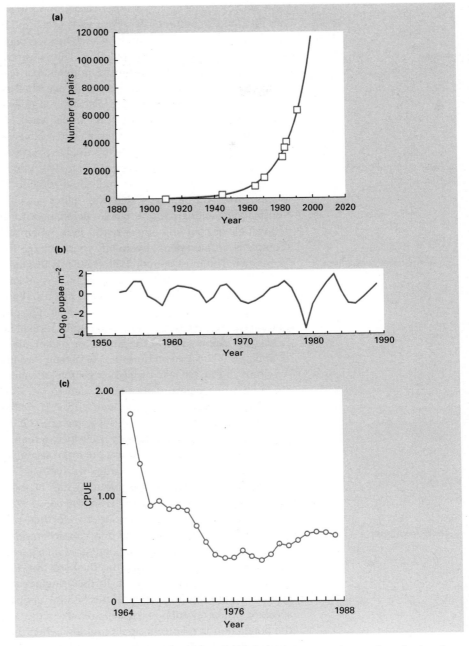

Figure 4.9 Numerical change in populations. (a) The steady rise in the number of pairs of black noddy birds (*Anous minutus*) on Heron Island. (After Ogden, 1993.) (b) Fluctuations in the size of a Scottish woodland population of the pine looper moth, *Bupalus piniaria*. (After Broekhuizen *et al.*, 1993.) (c) The decline in abundance to an approximately constant low level of Cape hake (*Merluccius capensis* and *M. paradoxus*) off the coast of South Africa, as estimated by an index of abundance: the number caught per unit of fishing effort (CPUE). (After Hilborn & Walters, 1992.)

or a whole woodland. In yet other cases—and there are many of these—individuals are distributed continuously over a wide area, and an investigator must define the limits of a population arbitrarily. In such cases, especially, it is often more convenient to consider the *density* of a population. This is usually defined as 'numbers per unit area', but in certain circumstances 'numbers per leaf', 'numbers per host' or some other measure may be appropriate. As we shall see in Chapter 6, Section 6.3, however, although calculating a density is usually treated as if it were a simple matter, there are actually at least three ways of calculating the density of a population—each of them different and conveying different information—and ecologists have by no means always used the most appropriate measure.

density

The most straightforward approach to determining the number of individuals in a population is to count every individual present. This is a possibility with many plants, and with animals that are either sessile or very slow, or large and conspicuous enough for a whole population to be censused without missing individuals or counting others twice. Even when they are possible, however, such 'complete enumerations' frequently require large investments of time and trained personnel. The most common alternative is to *sample* the population. Samples are taken from one or (much more sensibly) several small portions of the population. These portions are of known size and are usually chosen at random so as to be representative of the population as a whole. For plants and animals living on the ground surface, the sample unit is generally a small area known as a quadrat (which is also the name given to the square or rectangular device used to demarcate the boundaries of the area on the ground). For soil-dwelling organisms the unit is usually a volume of soil, for lake dwellers a volume of water; for many herbivorous insects the unit is one typical plant or leaf, and so on. In each case, every individual or module within the unit is counted, and it is therefore assumed that individuals remain within the unit whilst this is going on. The data from sampling lead directly to estimates of population density. Density can then be converted to population size if, as in the case of ground surface, the ratio of sampled area to total population area is known. Further details of sampling methods, and of methods for counting individuals generally, can be found in one of many texts devoted to ecological methodology (e.g. Kershaw, 1973; Southwood, 1978a; Krebs, 1989).

complete enumerations

the sampling of populations

For animals, especially, there are two further methods of estimating population size or density. The first is known as capture–recapture. Very simply, this involves catching a random sample of a population, marking individuals so that they can be recognized subsequently, releasing them so that they remix with the rest of the population and then catching a further random sample. Population size can be estimated from the proportion of this second sample which bear a mark. Roughly speaking, the proportion of marked animals in the second sample will be high when the population is relatively small, and low when the population is relatively large. Further details are provided by Begon (1979).

capture–recapture

The final method is to use an index of abundance. This can provide information on the relative size of a population, but by itself usually gives little indication of absolute size. The relative abundance of fruit-flies, for example, can be estimated by monitoring the numbers attracted each day to a standard bait. As another example, Figure 4.9c is based on an index of abundance for fish and whales: the numbers caught per unit fishing effort (where effort is measured in terms combining the number of ships, their size and the length of time they fish). Despite their shortcomings, even indices of abundance can provide valuable information.

indices of abundance

146 CHAPTER 4

4.4 Life cycles and the quantification of death and birth

semelparity and iteroparity

Having recognized the complexities of individuality and the necessities of counting, it is possible now to examine patterns of birth, death and growth. To a large extent, these are a reflection of the organism's life cycle, of which there are five main types (although there are many life cycles that defy this simple classification). Birth, death and growth in these various life cycles will be discussed in some detail in the remainder of this chapter. The life cycles are summarized diagrammatically in Figure 4.10. The species are said to be either *semelparous* or *iteroparous* (often referred to by plant scientists as monocarpic and polycarpic). Like so many dichotomies in ecology, this one, whilst useful, is not clear-cut: some species occupy the continuum between the two extremes. Nonetheless, we may say that in semelparous species, individuals have only a single, distinct period of reproductive output in their lives, prior to which they have largely ceased to grow, during which they invest little or nothing in survival to future reproductive events and after which they therefore die. In iteroparous species, on the other hand, an individual normally experiences several or many such reproductive events, which may in fact merge into a single extended period of reproductive activity. During each period of reproductive activity the individual continues to invest in future survival and possibly growth, and beyond each it therefore has a reasonable chance of surviving to reproduce again.

life tables, survivorship curves and fecundity schedules

Often, in order to monitor and examine changing patterns of mortality with age or stage, a *life table* is used. Frequently this allows a *survivorship curve* to be constructed, which traces the decline in numbers, over time, of a group of newly born or newly emerged individuals or modules—or it can be thought of as a plot of the probability, for a representative newly born individual, of surviving to various ages. Patterns of birth amongst individuals of different ages are often monitored at the same time as life tables are constructed. These patterns are displayed in *fecundity schedules*.

Life tables, survivorship curves and fecundity schedules are of the utmost importance because they contain the raw material of our 'ecological fact of life'. Without them we have little hope of understanding the N_{now} of the species that interest us, and still less hope of predicting the N_{future}.

4.5 Annual species

Annual life cycles take approximately 12 months or rather less to complete (Figure 4.10a–d). Usually, every individual in a population breeds during one particular season of the year, but then dies before the same season in the next year. Generations are therefore said to be discrete, in that each generation is distinguishable from every other; the only overlap of generations is between breeding adults and their offspring during and immediately after the breeding season. Species with discrete generations need not be annual, since generation lengths other than 1 year are conceivable. In practice, however, most are: the regular annual cycle of climates provides the major pressure in favour of synchrony.

4.5.1 Simple annuals: cohort life tables

The common field grasshopper, *Chorthippus brunneus*, is an annual species living throughout much of Europe. The first-instar nymphs emerge from their eggs in late

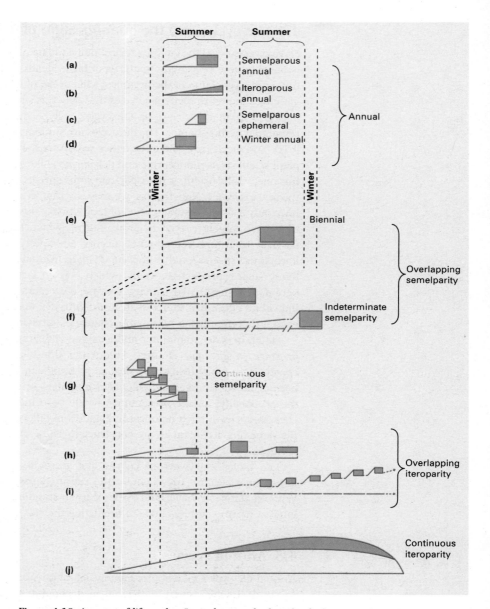

Figure 4.10 A range of life cycles. In each case, the length of a bar represents the length of an individual's life. The height of a bar represents individual size (although on a logarithmic scale, so that the exponential growth of an individual early in life appears linear). The shaded region indicates the proportion of available resources allocated to reproduction in illustrative rather than exact terms, i.e. semelparous species, when they reproduce, allocate a large proportion of resources to reproduction and die soon thereafter; iteroparous species tend to allocate less to current reproduction but may survive for further reproductive bouts. Most life cycles should be self-explanatory. In overlapping semelparity, species may have a strict 2-year ((e) biennial) or 3-year, etc. cycle, or they may have a cycle of indeterminate length (f). In overlapping iteroparity, species may be relatively short lived (h) or long lived (i), and, if the latter, may continue to grow, apparently indefinitely, as shown here.

spring, and then feed, grow and eventually metamorphose into larger but otherwise fairly similar second-instar nymphs. Two other nymphal instars follow, until in mid-summer the fourth instars moult into winged adults. The adult females lay eggs in the soil (in egg pods containing about 11 eggs each), but by mid-November all the adults will have died, so ending that particular generation. Meanwhile, however, the next generation is starting: the eggs begin development, which proceeds until the following spring, when first-instar nymphs emerge once again.

A life table and fecundity schedule for an isolated population of *C. brunneus* from near Ascot, England, is presented in Table 4.1. The life table is known as a *cohort* life table, because a single cohort of individuals (i.e. a group of individuals born within the same short interval of time) was followed from birth to the death of the last survivor. With an annual species like *C. brunneus*, there is no other way of constructing a life table.

The first column of Table 4.1 sets out the various stages of the life cycle that have been distinguished. The second column, a_x, then lists the major part of the raw data: it gives the total number of individuals observed in the population at each stage (a_0 individuals in the initial stage, a_1 in the following one, and so on). The problem with any a_x column is that the information in it is specific to one population in one year, making comparisons with other populations and other years very difficult. The data have therefore been standardized in a column of l_x-values. This is headed by an l_0-value of 1.000, and all succeeding figures have been brought into line accordingly (e.g. $l_2 = 1.000 \times 2529/44\,000 = 0.058$). Thus, whilst the a_0-value of 44 000 is peculiar to this set of data, *all* studies have an l_0-value of 1.000, making all studies comparable. The l_x-values are best thought of as the proportion of the original cohort surviving to the start of each stage.

To consider mortality more explicitly, the proportion of the original cohort dying during each stage (d_x) has been computed, being simply the difference between l_x and l_{x+1}, for example $d_1 = 0.080 - 0.058 = 0.022$. The stage-specific mortality rate, q_x, has also been computed. This considers d_x as a fraction of l_x, so that, for instance, q_3 (the fraction dying during the third nymphal instar), is $0.011/0.044 = 0.25$. Note too that q_x may be thought of as the average 'chance' or probability of an individual dying. It is therefore equivalent to $(1 - p_x)$ where p refers to the probability of survival (i.e. the fraction dying and the fraction surviving must always add up to 1).

The advantage of the d_x-values is that they can be summed: thus, the proportion of the cohort dying as nymphs was $d_1 + d_2 + d_3 + d_4$ ($= 0.050$). The disadvantage is that the individual values give no real idea of the intensity or importance of mortality during a particular stage. This is because the d_x-values are larger the more individuals there are, and hence the more there are available to die. The q_x-values, on the other hand, are an excellent measure of the intensity of mortality. For instance, in the present example it is clear from the q_x column that the mortality rate remained almost constant throughout the early nymphal stages; this is not clear from the d_x column. The q_x-values, however, have the disadvantage of not being liable to summation: $q_1 + q_2 + q_3 + q_4$ does not give the overall mortality rate for the nymphs.

The advantages are combined, however, in the last column of the life table, which contains k_x-values (Haldane, 1949; Varley & Gradwell, 1970). k_x is defined simply as $\log_{10} a_x - \log_{10} a_{x+1}$ (or, equivalently, $\log_{10} a_x / a_{x+1}$), and is sometimes referred to as a 'killing power'. Like q_x-values, k_x-values reflect the intensity or rate of mortality (as Table 4.1 shows); but unlike summing the q_x-values, summing k_x-values is a

Table 4.1 A cohort life table for the common field grasshopper, *Chorthippus brunneus*. The columns are explained in the text. (After Richards & Waloff, 1954.)

Stage (x)	Number observed at start of each stage a_x	Proportion of original cohort surviving to start of each stage l_x	Proportion of original cohort dying during each stage d_x	Mortality rate q_x	$\log_{10} a_x$	$\log_{10} l_x$	$\log_{10} a_x - \log_{10} a_{x+1} = k_x$	Eggs produced in each stage F_x	Eggs produced per surviving individual in each stage m_x	Eggs produced per original individual in each stage $l_x m_x$
Eggs (0)	44 000	1.000	0.920	0.92	4.64	0.00	1.09	—	—	—
Instar I (1)	3513	0.080	0.022	0.28	3.55	−1.09	0.15	—	—	—
Instar II (2)	2529	0.058	0.014	0.24	3.40	−1.24	0.12	—	—	—
Instar III (3)	1922	0.044	0.011	0.25	3.28	−1.36	0.12	—	—	—
Instar IV (4)	1461	0.033	0.003	0.11	3.16	−1.48	0.05	—	—	—
Adults (5)	1300	0.030	—	—	3.11	−1.53	—	22 617	17	0.51

$R_0 = \Sigma l_x m_x = 0.51.$

$\dfrac{\Sigma F_x}{a_0} = 0.51.$

legitimate procedure. Thus, the killing power or k-value for the nymphal period is $0.15 + 0.12 + 0.12 + 0.05 = 0.44$, which is also the value of $\log_{10} a_1 - \log_{10} a_5$. Note too that the k_x-values can be computed from the l_x-values as well as from the a_x-values (Table 4.1); and that, like l_x-values, k_x-values are standardized, and are therefore appropriate for comparing quite separate studies. In this and later chapters, k_x-values will be used repeatedly.

The fecundity schedule in Table 4.1 (the final three columns) begins with a column of raw data, F_x: the total number of eggs deposited during each stage. This is followed in the next column by m_x: the individual fecundity or birth rate, i.e. the mean number of eggs produced per surviving individual. (Note that the number of eggs produced per female would be roughly twice this value.) Because only the adults reproduced, there are, in the present case, entries in the fecundity schedule only for this final stage.

Perhaps the most important summary term that can be extracted from a life table and fecundity schedule is the basic reproductive rate, denoted by R_0. This is the mean number of offspring (of the first stage in the life cycle—in this case fertilized eggs) produced per original individual by the end of the cohort. It therefore indicates, in annual species, the overall extent by which the population has increased or decreased over that time. (As we shall see in Section 4.7, the situation becomes more complicated when generations overlap or species breed continuously.)

There are two ways in which R_0 can be computed. The first is from the formula:

$$R_0 = \frac{\Sigma F_x}{a_0} \qquad (4.4)$$

i.e. the total number of fertilized eggs produced during one generation divided by the original number of individuals (ΣF_x means the sum of the values in the F_x column). The more usual way of calculating R_0, however, is from the formula:

$$R_0 = \Sigma l_x m_x \qquad (4.5)$$

i.e. the sum of the number of fertilized eggs produced per original individual during each of the stages (the final column of the fecundity schedule). As Table 4.1 shows, the basic reproductive rate is the same, whichever formula is used.

Another cohort life table and fecundity schedule are set out in Table 4.2, this time for the annual plant *Phlox drummondii* in Nixon, Texas (Leverich & Levin, 1979). The most obvious difference between this and Table 4.1 is apparent from the first column. Leverich and Levin (1979) divided the life cycle of *Phlox* not into a number of stages but into a number of age classes. They recorded the size of the seed population on various occasions before germination, and then did so again at regular intervals until all individuals had flowered and died. The advantage of using age classes is that it allows an observer to look in detail at the patterns of birth and mortality within a stage. The disadvantage is that the age of an individual is not necessarily the best, nor even a satisfactory, measure of the individual's biological 'status'. In many long-lived plants, for instance (see below), individuals of the same age may be reproducing actively, or growing vegetatively but not reproducing, or doing neither. In such cases, a classification based on developmental stages (as opposed to ages) is clearly appropriate. Leverich and Levin's decision to use age classes in *Phlox* was based on the small number of stages, the demographic variation within each and the synchronous development of the whole population.

It is apparent from Table 4.2 that the use of age classes has exposed the detailed

Table 4.2 A cohort life table for *Phlox drummondii*. The columns are explained in the text. (After Leverich & Levin, 1979.)

Age interval (days) $x - x'$	Number surviving to day x a_x	Proportion of original cohort surviving to day x l_x	Proportion of original cohort dying during interval d_x	Mortality rate per day q_x	$\text{Log}_{10}\, l_x$	Daily killing power k_x	F_x	m_x	$l_x m_x$
0–63	996	1.000	0.329	0.006	0.00	0.003	—	—	—
63–124	668	0.671	0.375	0.013	− 0.17	0.006	—	—	—
124–184	295	0.296	0.105	0.007	− 0.53	0.003	—	—	—
184–215	190	0.191	0.014	0.003	− 0.72	0.001	—	—	—
215–264	176	0.177	0.004	0.002	− 0.75	0.001	—	—	—
264–278	172	0.173	0.005	0.002	− 0.76	0.001	—	—	—
278–292	167	0.168	0.008	0.004	− 0.78	0.002	—	—	—
292–306	159	0.160	0.005	0.002	− 0.80	0.001	53.0	0.33	0.05
306–320	154	0.155	0.007	0.003	− 0.81	0.001	485.0	3.13	0.49
320–334	147	0.148	0.043	0.025	− 0.83	0.011	802.7	5.42	0.80
334–348	105	0.105	0.083	0.106	− 0.98	0.049	972.7	9.26	0.97
348–362	22	0.022	0.022	1.000	− 1.66	—	94.8	4.31	0.10
362–	0	0.000	—	—	—	—	—	—	—
							2408.2		2.41

$$R_0 = \Sigma l_x m_x = \frac{\Sigma F_x}{a_0} = 2.41.$$

patterns of fecundity and mortality. (Note that the variable lengths of the age classes has necessitated the conversion of the q_x and k_x columns to 'daily' rates.) The age-specific fecundity, m_x (the fecundity per surviving individual), demonstrates the existence of a pre-productive period, a gradual rise to a peak and then a rapid decline. The reproductive output of the whole population, F_x, parallels this pattern to a large extent, but also takes into account the fact that whilst the age-specific fecundity was changing, the size of the population was gradually declining. This combination of fecundity and survivorship is an important property of F_x-values, shared by the basic reproductive rate (R_0). It reinforces the point that actual reproduction depends both on reproductive potential (m_x) and on survivorship (l_x).

In the case of the *Phlox*, R_0 was 2.4. This means that there was a 2.4-fold increase in the size of the population over one generation. For the grasshopper the value was 0.51 (the population declined to 0.51 of its former size). If such values were maintained from generation to generation, the *Phlox* population would grow ever larger and soon cover the globe, whilst the grasshopper population would quickly decline to extinction. Neither of these events has happened. The reasons why will be considered in later chapters. The important point to realize now is that life tables and fecundity schedules drawn up in different years would exhibit rather different patterns.Therefore, a balanced and realistic picture of life and death in the grasshopper, *Phlox*, or any other species can only emerge from several or many years' data.

The pattern of mortality in the *Phlox* population is illustrated in Figure 4.11a, using both q_x- and k_x-values. Mortality rate was fairly high at the beginning of the seed stage but became very low towards the end. Then, amongst the adults, there was a period where mortality rate fluctuated about a moderate level, followed finally

a realistic picture requires data for several years

patterns of mortality: survivorship curves

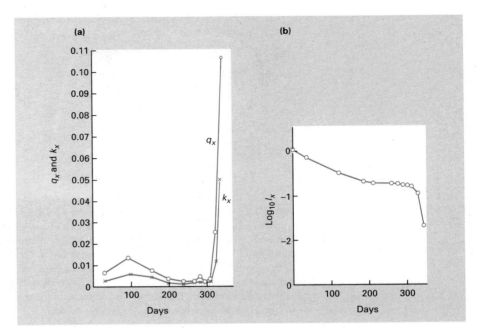

(a)

Figure 4.11 Mortality and survivorship in the life cycle of *Phlox drummondii*. (a) The age-specific daily mortality rate (q_x) and daily killing power (k_x). (b) The survivorship curve: $\log_{10} l_x$ plotted against age. (After Leverich & Levin, 1979.)

by a sharp increase to very high levels during the last weeks of the generation. The same pattern is shown in a different form in Figure 4.11b. This is a *survivorship curve*, and follows the decline of $\log_{10} l_x$ with age. When the mortality rate is roughly constant, the survivorship curve is more or less straight; when the rate increases, the curve is convex; and when the rate decreases, the curve is concave. Thus, the curve is concave towards the end of the seed stage, and convex towards the end of the generation. Survivorship curves are the most widely used way of depicting patterns of mortality.

a classification of survivorship curves

Pearl (1928), in a much-quoted attempt at classification (Figure 4.12), described convex, straight and concave survivorship curves as type I, type II and type III, respectively (also discussed in some depth by Deevey (1947)). In practice, however, most species, when observed throughout their whole life, do not display a single type of curve. Instead, like *Phlox*, they display a succession of shapes as stage follows stage.

'death before birth'

Most survivorship curves show changes in numbers of individuals from the time they are born until they die. This completely misses a stage at which massive and ecologically important mortality may occur. The life of an individual starts with the formation of the zygote, and in mammals and seed plants the first part of an individual's life is spent within the mother's body. Competition between the individuals may start at this stage, as developing embryos compete for limited maternal resources. Those failing to obtain sufficient resources may be resorbed by the mother or may simply cease growth and abort. There is extraordinarily little information about this part of the survivorship curve for most animals or plants, but Wiens (1984) gives data which suggest that abortion rates of ovules are of the order of 15% for annual and 50% for perennial plants. In humans, one estimate suggests that only about 31% of zygotes survive to birth (Biggers, 1981).

the use of logarithms

The y-axis in Figure 4.11b is logarithmic. The importance of using logarithms in survivorship curves can be seen by imagining two investigations of the same population. In the first, the whole population is censused: there is a decline in one

153   LIFE AND DEATH

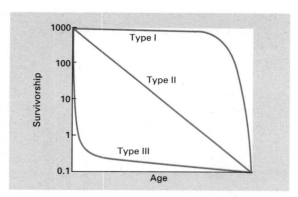

Figure 4.12 A classification of survivorship curves. (After Pearl, 1928; Deevey, 1947.) Type I (convex)—epitomized perhaps by humans in rich countries, cosseted animals in a zoo or leaves on a plant—describes the situation in which mortality is concentrated at the end of the maximum life span. Type II (straight) indicates that the probability of death remains constant with age, and may well apply to the buried seed banks of many plant populations. Type III (concave) indicates extensive early mortality, with those that remain having a high rate of survival subsequently. This is true, for example, of many marine fish which produce millions of eggs of which very few survive to become adults.

time interval from 1000 to 500 individuals. In the second, samples are taken, and over the same time interval this index of density declines from 100 to 50. The two cases are biologically identical, i.e. the rate or probability of death per individual over the time interval (the *per capita* rate) is the same. The slopes of the two logarithmic survivorship curves reflect this—both would be -0.301. But, on simple linear scales the slopes would differ. Logarithmic survivorship curves therefore have the advantage of being standardized from study to study, just like the 'rates' q_x, k_x and m_x. Plotting numbers on a logarithmic scale will also indicate when per capita rates of increase are identical. 'Log numbers' will therefore often be used in preference to 'numbers' when numerical change is being plotted.

Finally, there is another difference between the life cycle of the grasshopper and *Phlox* which is not apparent from Tables 4.1 and 4.2. Although the reproductive season for the *Phlox* population lasts for 56 days, each individual plant is semelparous. It has a single reproductive phase during which all of its seed develop synchronously (or nearly so). The extended reproductive season occurs because different individuals enter this phase at different times. The grasshopper, on the other hand, is iteroparous. Each adult female, assuming she survives for long enough, produces a number of egg pods, with a period of perhaps 1 week or more between each. During these intervals between pods, active maintenance of body tissues persists, and stored reserves may be laid down.

In fact, semelparity is the more common pattern for annual plants, whilst for animals, iteroparity is more usual. There are many exceptions, however. For instance, in the freshwater isopod *Asellus aquaticus*, females generally lay a single clutch of eggs which they hold, externally, beneath their abdomen, 'brooding' them there until the independent juveniles can fend for themselves. If for some reason these eggs are not fertilized, the semelparous female will die without ever laying a replacement clutch. Conversely, a number of iteroparous annual plants, unlike *Phlox*, start to flower when small, and continue to grow, flower and set seed until they die from some extrinsic cause, such as frost or drought. This is the case with

semelparity and iteroparity illustrated

many species of *Veronica* (speedwell), with *Poa annua* (annual meadow grass) and with *Senecio vulgaris* (groundsel).

4.5.2 Seed banks

Using *Phlox* as an example of an annual plant has, to a certain extent, been misleading, because the group of seedlings developing in one year is a true cohort: it derives entirely from seed set by adults in the previous year. Seeds which do not germinate in one year will not survive till the next. In most 'annual' plants this is not the case. Instead, seeds accumulate in the soil in a buried *seed bank*. At any one time, therefore, seeds of a variety of ages are likely to occur together in the seed bank; and when they germinate, the seedlings will also be of varying ages (age being the length of time since the seed was first produced). The formation of something comparable to a seed bank is rarer amongst animals, but there are examples to be seen amongst the eggs of nematodes, mosquitos and fairy shrimps, the gemmules of sponges and the statocysts of bryozoans. In the case of the fairy shrimp, *Streptocephalus vitreus*, living in temporary ponds in Kenya, resistant eggs are present in the dry sediment after a pool has dried out. Only a proportion hatch at the onset of the next rains, however—the others require at least one more drying and wetting sequence before they hatch. Even eggs produced by the same female in the same batch show variation in hatching—some hatch after one drying–wetting sequence, others only after two or more such episodes (Hildrew, 1985). This presumably serves to spread the risk of eggs hatching and failing to develop to the point of further reproduction before the pool dries again.

Figure 4.13 illustrates the results of an experiment designed to follow the fates of some annual weed seeds in horticultural soil (Roberts, 1964). No new seed was allowed to enter the soil after the start of the experiment (year 0) but, as Figure 4.13 shows, many of the seeds remained viable in the soil for periods greater than 1 year, and seedlings therefore developed from seeds of various ages.

different species make different contributions to the seed bank
As a general rule, dormant seeds, which enter and make a significant contribution to seed banks, are more common in annuals and other short-lived plant species than they are in longer-lived species. A notable consequence is that seeds of short-lived species tend to predominate in buried seed banks, even when most of the

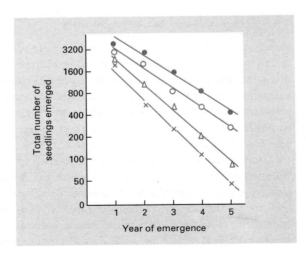

Figure 4.13 The consequence of a seed bank. The number of seedlings emerging in successive years from a horticultural soil into which no additional seed was allowed to enter after year zero. ●, *Thlaspi arvense*; △, *Stellaria media*; ○, *Capsella bursa-pastoris*; ×, *Senecio vulgaris*. (After Roberts, 1964.)

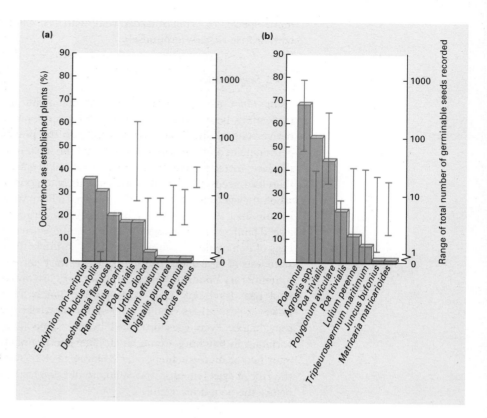

Figure 4.14 There are marked discrepancies between the relative abundances of mature plants (histograms) and of seed in the soil (vertical bars) for the major species of (a) a deciduous woodland and (b) an arable field. (After Thompson & Grime, 1979.)

established plants above them belong to much longer-lived species (Figure 4.14).

Note that species commonly referred to as 'annual', but with a seed bank, are not strictly annual species at all, even if they progress from germination to reproduction within 1 year, since some of the seeds destined to germinate each year will already be more than 12 months old. All we can do, though, is bear this fact in mind, and note that it is just one example of real organisms spoiling our attempts to fit them neatly into clear-cut categories.

4.5.3 Ephemerals and facultative annuals

Annual species with seed banks are not the only ones for which the term annual is, strictly speaking, inappropriate. For example, there are many annual plant species living in deserts that are far from seasonal in their appearance. They have a substantial buried seed bank, with germination occurring on rare occasions after substantial rainfall. Subsequent development is usually rapid, so that the period from germination to seed production is short. Such plants are best described as semelparous *ephemerals* (see Figure 4.10c).

Data concerning the appearance of one such species, *Eriophyllum wallacei*, were collected by Juhren *et al.* (1956). Some degree of consistency was apparent: *E. wallacei* germinated usually when temperatures were relatively low, and when there had been either heavy rainfall or a period of moderate rainfalls. Under these conditions, *E. wallacei*, like many desert annuals, developed rapidly with high survivorship and rather low individual fecundity. More apparent than the consis-

desert annuals

156   CHAPTER 4

tency, however, was the unpredictability. Over a period of 4 years, this opportunistic ephemeral germinated at various times between early September and late May.

facultative annuals

Another group of species to which a simple annual label does not apply is that in which the majority of individuals in each generation are annual, but in which a small number postpone reproduction until their second summer. For example, this is true of the terrestrial isopod *Philoscia muscorum* living in north-east England (Sunderland *et al.*, 1976). Approximately 90% of females bred only in the first summer after they were born; the other 10% bred only in their second summer. In some other species, the difference in numbers between those that reproduce in their first or second years is so slight that the description *annual–biennial* is most appropriate.

In short, it it clear that annual life cycles merge into more complex ones without any sharp discontinuity.

4.6 Individuals with repeated breeding seasons

Many species breed repeatedly (assuming they survive long enough), but nevertheless have a specific breeding season. Thus, they have overlapping generations (*overlapping iteroparity*—see Figure 4.10h,i). Amongst the more obvious examples are temperate-region birds living for more than 1 year, some corals, most trees and other iteroparous perennial plants. In all of these, individuals of a range of ages breed side by side. Nonetheless, some species in this category, some grasses for example, and many birds, live for relatively short periods (see Figure 4.10h).

4.6.1 Cohort life tables

following a cohort of red deer

Constructing a cohort life table for such species is even more difficult than constructing one for an annual species. A cohort must be recognized and followed (often for many years), even though the organisms within it are coexisting and intermingling with organisms from many other cohorts, older and younger. This was possible, though, as part of an extensive study of red deer (*Cervus elaphus*) on the small island of Rhum, Scotland (Lowe, 1969). The deer live for up to 16 years, and the females (hinds) are capable of breeding each year from their fourth summer onwards. In 1957, Lowe and his co-workers made a very careful count of the total number of deer on the island, including the total number of calves (less than 1 year old). Lowe's cohort consisted of the deer that were calves in 1957. Thus, each year from 1957 to 1966, every one of the deer that was discovered that had died from natural causes, or had been shot under the rigorously controlled conditions of this Nature Conservancy Council reserve, was examined and aged reliably by examining tooth replacement, eruption and wear. It was therefore possible to identify those dead deer that had been calves in 1957; and by 1966, 92% of this cohort had been observed dead and their age at death therefore determined. The life table for this cohort of hinds (or the 92% sample of it) is presented in Table 4.3; the survivorship curve is shown in Figure 4.15. There appears to be a fairly consistent increase in the risk of mortality with age (the curve is convex).

cohort life tables are often more easily constructed for sessile organisms

The difficulties of constructing a cohort life table for an organism with overlapping generations are eased somewhat when the organism is sessile. In such a case, newly arrived or newly emerged individuals can be mapped, photographed or even marked in some way, so that they (or their exact location) can be recognized whenever the site is revisited subsequently. A number of workers have followed

Table 4.3 Cohort life table for red deer hinds on the island of Rhum that were calves in 1957. (After Lowe, 1969.)

Age (years) x	Proportion of original cohort surviving to the beginning of age-class x l_x	Proportion of original cohort dying during age-class x d_x	Mortality rate q_x
1	1.000	0	0
2	1.000	0.061	0.061
3	0.939	0.185	0.197
4	0.754	0.249	0.330
5	0.505	0.200	0.396
6	0.305	0.119	0.390
7	0.186	0.054	0.290
8	0.132	0.107	0.810
9	0.025	0.025	1.000

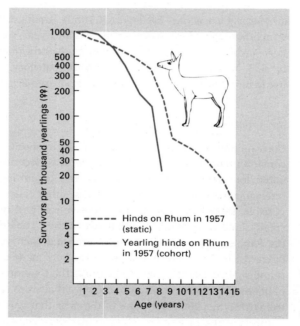

Figure 4.15 Two survivorship curves for red deer hinds on the island of Rhum. As explained in the text, one is based on the cohort life table for the 1957 calves and therefore applies to the post-1957 period; the other is based on the static life table of the 1957 population and therefore applies to the pre-1957 period. (After Lowe, 1969.)

cohorts of grasses in this way (e.g. see Begon & Mortimer, 1986). Connell (1970) constructed a cohort life table for the barnacle *Balanus glandula* over 8 years (see Table 4.6 below). Over a rather longer period, a cohort of the shrub *Acacia burkittii* was followed in permanently established plots in a vegetation reserve in South Australia, from germination in 1928 until 1970 (Figure 4.16).

4.6.2 Static life tables

using age structures for life tables: an imperfect but often unavoidable alternative

Taken overall, however, practical problems have tended to deter ecologists from constructing cohort life tables for long-lived iteroparous organisms with overlapping generations, even when the individuals are sessile. But, there is an alternative: the

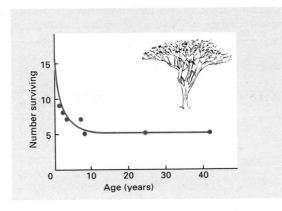

Figure 4.16 Survival of a cohort of *Acacia burkittii* individuals recorded regularly in permanent plots in South Australia since their germination in 1929. (After Crisp & Lange, 1976.)

construction of a static life table. As will become clear, this alternative is seriously flawed—but it is often better than nothing at all.

An interesting example of a static life table emerges from Lowe's (1969) study of red deer on Rhum. As has already been explained, a large proportion of the deer that died from 1957 to 1966 could be aged reliably. Thus, if, for example, a fresh corpse was examined in 1961 and found to be 6 years old, it was known that in 1957 the deer was alive and 2 years old. Lowe was therefore eventually able to reconstruct the age structure of the 1957 population: age structures are the basis for static life tables. Of course, the age structure of the 1957 population could have been ascertained by shooting and examining large numbers of deer in 1957; but since the ultimate aim of the project was the enlightened conservation of the deer, this method would have been somewhat inappropriate. (Note that Lowe's results did not represent the total numbers alive in 1957, because a few carcasses must have decomposed or been eaten before they could be discovered and examined.) Lowe's raw data for red deer hinds are presented in column 2 of Table 4.4.

Remember that the data in Table 4.4 refer to ages in 1957. They can be used as a basis for a life table, but only if it is assumed that there had been no year-to-year variation prior to 1957 in either the total number of births or the age-specific survival rates. In other words, it must be assumed that the 59 6-year-old deer alive in 1957 were the survivors of 78 5-year-old deer alive in 1956, who were themselves the survivors of 81 4-year olds in 1955, and so on; or, in short, that the data in Table 4.4 are the same as would have been obtained if a single cohort *had* been followed.

Having made these assumptions, l_x, d_x and q_x columns have been constructed. It is clear, however, that the assumptions are false. There were actually more animals in their seventh year than in their sixth year, and more in their 15th year than in their 14th year. There were therefore 'negative' deaths and meaningless mortality rates. The pitfalls of constructing such static life tables (and equating age structures with survivorship curves) are amply illustrated.

static life tables can be useful ...

Nevertheless, the data can be useful. Lowe's aim was to provide a *general* idea of the population's age-specific survival rate prior to 1957 (when culling of the population began). He could then compare this with the situation after 1957, as illustrated by the cohort life table previously discussed. He was more concerned with general trends than with the particular changes occurring from one year to the next. He therefore 'smoothed out' the variations in numbers between ages 2–8 and 10–16 years to give a steady decline during both of these periods. The results of this process

Table 4.4 A static life table for red deer hinds on the island of Rhum, based on the reconstructed age structure of the population in 1957. (After Lowe, 1969.)

Age (years) x	Number of individuals observed of age x a_x	l_x	d_x	q_x	Smoothed l_x	d_x	q_x
1	129	1.000	0.116	0.116	1.000	0.137	0.137
2	114	0.884	0.008	0.009	0.863	0.085	0.097
3	113	0.876	0.251	0.287	0.778	0.084	0.108
4	81	0.625	0.020	0.032	0.694	0.084	0.121
5	78	0.605	0.148	0.245	0.610	0.084	0.137
6	59	0.457	0.047	—	0.526	0.084	0.159
7	65	0.504	0.078	0.155	0.442	0.085	0.190
8	55	0.426	0.232	0.545	0.357	0.176	0.502
9	25	0.194	0.124	0.639	0.181	0.122	0.672
10	9	0.070	0.008	0.114	0.059	0.008	0.141
11	8	0.062	0.008	0.129	0.051	0.009	0.165
12	7	0.054	0.038	0.704	0.042	0.008	0.198
13	2	0.016	0.008	0.500	0.034	0.009	0.247
14	1	0.080	−0.023	—	0.025	0.008	0.329
15	4	0.031	0.015	0.484	0.017	0.008	0.492
16	2	0.016	—	—	0.009	0.009	1.000

are shown in the final three columns of Table 4.4, and the survivorship curve is plotted in Figure 4.15. A general picture does indeed emerge: the introduction of culling on the island appears to have decreased overall survivorship significantly, overcoming any possible compensatory decreases in natural mortality.

Notwithstanding this successful use of a static life table, the interpretation of static life tables generally, and the age structures from which they stem, is fraught with difficulty: usually, age structures offer no easy short cuts to understanding the dynamics of populations. Sometimes, however, background knowledge greatly facilitates their interpretation, making their construction worthwhile. For instance, Crisp and Lange (1976), apart from following a cohort of *A. burkittii*, were also able to examine the age structure of a number of populations (Figure 4.17). (The age of an individual could be estimated from a number of size measurements, since, in a sample of trees of known age, there were several highly significant age–size correlations.)

The two populations in Figure 4.17 are obviously different in age structure. But

... within very strict limits

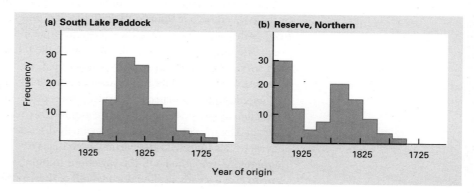

Figure 4.17 Age structures of *Acacia burkittii* populations at two sites in South Australia. (After Crisp & Lange, 1976.)

Table 4.5 Mean clutch size and age of great tits in Wytham Wood, near Oxford. (After Perrins, 1965.)

Age (years)	1961		1962		1963	
	Number of birds	Mean clutch size	Number of birds	Mean clutch size	Number of birds	Mean clutch size
Yearlings	128	7.7	54	8.5	54	9.4
2	18	8.5	43	9.0	33	10.0
3	14	8.3	12	8.8	29	9.7
4			5	8.2	9	9.7
5			1	8.0	2	9.5
6					1	9.0

it was also known that the South Lake Paddock population (Figure 4.17a) had been grazed by sheep from 1865 to 1970 (when the data were collected) and by rabbits from 1885 to 1970, whilst the reserve population (Figure 4.17b) had been fenced in 1925 to exclude sheep. It was therefore possible to draw a number of conclusions: (i) grazing from 1865 onwards led to drastic reductions in the number of successful new recruits to the populations; (ii) fencing in 1925 allowed more successful recruitment into the reserve population; and (iii) even after fencing, rabbit grazing led to levels of recruitment lower than those prior to 1865 (i.e. the 1920–1945 age class is much smaller than the 1845–1860 class, even though the latter has survived for an additional 75 years).

4.6.3 Fecundity schedules

Static fecundity schedules, i.e. age-specific variations in fecundity within a particular season, can also provide useful information, especially if they are available from successive breeding seasons. We can see this for a population of great tits (*Parus major*) in Wytham Wood, near Oxford (Table 4.5), where the data could be obtained only because individual birds could be aged (in this case, because they had been marked with individually recognizable leg-rings soon after hatching). The table shows that mean fecundity rose to a peak in 2-year-old birds and declined gradually thereafter. Indeed, most iteroparous species show an age- or stage-related pattern of fecundity. For instance, Figure 4.18 shows the size-dependent fecundity of three species of oak.

mast years in trees

Table 4.5 also shows that mean fecundity nevertheless varied considerably from year-to-year. Some variation in individual fecundity from season-to-season is found in all iteroparous species. The phenomenon is most marked, however, in a number of forest trees that exhibit *masting*. A mast year is a year in which a tree produces a massive crop of seeds, fruit or cones, and the effect is often amplified by the synchronous masting of many or most of the trees of that species in a forest. Mast years are generally interspersed amongst years in which trees are sterile or nearly so. The possible consequences of masting will be discussed in Chapter 8, Section 8.4.

4.6.4 The importance of modularity

The sedge *Carex bigelowii*, growing in a lichen heath in Norway, illustrates the

161   LIFE AND DEATH

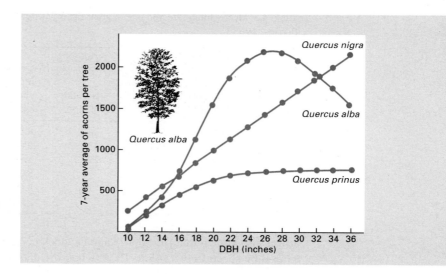

Figure 4.18 Size-specific variation in fecundity for three species of oak tree. DBH, diameter at breast height. (After Downs & McQuilkin, 1944.)

modular stage structure and a modular static life table

difficulties of constructing any sort of life table for organisms that are not only iteroparous with overlapping generations but also modular (Figure 4.19). *Carex bigelowii* has an extensive underground rhizome system which produces tillers (aerial shoots) at intervals along its length as it grows. It grows by producing a lateral meristem in the axil of a leaf belonging to a 'parent' tiller. This lateral is completely dependent on the parent tiller at first, but is potentially capable of developing into a vegetative parent tiller itself, and also of flowering, which it does when it has produced a total of 16 or more leaves. Flowering, however, is always followed by tiller death, i.e. the tillers are semelparous although the genets are iteroparous.

Callaghan took a number of well-separated young tillers, and excavated their rhizome systems through progressively older generations of parent tillers. This was made possible by the persistence of dead tillers. He excavated 23 such systems containing a total of 360 tillers, and was able to construct a type of static life table (and fecundity schedule) based on growth stages (Figure 4.19). There were, for example, 1.04 dead vegetative tillers (per m²) with 31–35 leaves. Thus, since there were also 0.26 tillers in the next (36–40 leaves) stage, it can be assumed that a total of 1.30 (i.e. 1.04 + 0.26) living vegetative tillers entered the 31–35 leaf stage. As there were 1.30 vegetative tillers and 1.56 flowering tillers in the 31–35 leaf stage, 2.86 tillers must have survived from the 26–30 stage. It is in this way that the life table—applicable not to individual genets but to tillers (i.e. modules)—was constructed.

There appeared to be no new establishment from seed in this particular population (no new genets); tiller numbers were being maintained by modular growth alone. However, a 'modular growth schedule' (*laterals*), analogous to a fecundity schedule, has been constructed.

the importance of stage not age in modular organisms

Note finally that stages rather than age classes have been used here—something which is almost always desirable and necessary when dealing with modular iteroparous organisms, because variability stemming from modular growth accumulates year-upon-year, making age a particularly poor measure of an individual's chances of death, reproduction or further modular growth.

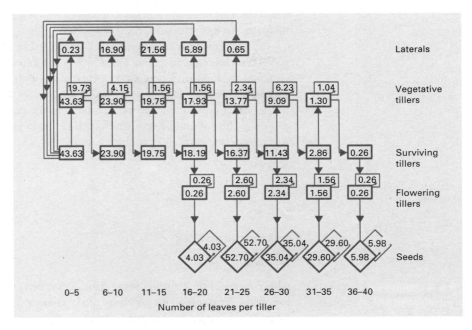

Figure 4.19 A reconstructed static life table for the modules (tillers) of a *Carex bigelowii* population. The densities per m² of tillers are shown in rectangular boxes, and those of seeds in diamond-shaped boxes. Rows represent tiller types, whilst columns depict size classes of tillers. Thin-walled boxes represent dead tiller (or seed) compartments, and arrows denote pathways between size classes, death or reproduction. (After Callaghan, 1976.)

4.7 Reproductive rates, generation lengths and rates of increase

4.7.1 The relationships between the variables

In the previous section we saw that the life tables and fecundity schedules drawn up for species with overlapping generations are at least superficially similar to those constructed for species with discrete generations. With discrete generations, we were able to compute the basic reproductive rate (R_0) as a summary term describing the overall outcome of the patterns of survivorship and fecundity (see Section 4.5.1). Can a comparable summary term be computed when generations overlap?

with overlapping generations, R_0 still refers to the average number of offspring produced by an individual

Note immediately that previously, for species with discrete generations, R_0 described two separate population parameters. It was the number of offspring produced on average by an individual over the course of its life; but it was also the multiplication factor which converted an original population size into a new population size, one generation hence. With overlapping generations, when a cohort life table is available, the basic reproductive rate can be calculated using the same formula:

$$R_0 = \Sigma \, l_x m_x, \qquad (4.6)$$

and it still refers to the average number of offspring produced by an individual. But further manipulations of the data are necessary before we can talk about the rate at which a population increases or decreases in size—or, for that matter, about the length of a generation. The difficulties are much greater still when only a static life table (i.e. an age structure) is available (see below).

We begin by deriving a general relationship that links population size, the rate of population increase and time, but which is not limited to measuring time in terms of generations. Imagine a population which starts with 10 individuals, and which, after successive intervals of time, rises to 20, 40, 80, 160 individuals and so on. We refer to the initial population size as N_0 (meaning the population size when no time has elapsed). The population size after one time interval is N_1, after two time intervals it is N_2, and in general after t time intervals it is N_t. In the present case:

$$N_0 = 10 \tag{4.7}$$

and:

$$N_1 = 20 \tag{4.8}$$

and, we can say that:

$$N_1 = N_0 R \tag{4.9}$$

the 'fundamental net reproductive rate' R

where R, which is 2 in the present case, is known as the *fundamental net reproductive rate* or the *fundamental net per capita rate of increase*. Clearly, populations will increase when $R > 1$, and decrease when $R < 1$.

general equations, based on R, for population increase over time

R combines the birth of new individuals with the survival of existing individuals. Thus, when $R = 2$, each individual could give rise to two offspring but die itself, or give rise to only one offspring and remain alive: in either case, R (birth plus survival) would be 2. Note too that in the present case R remains the same over the successive intervals of time, i.e.:

$$N_2 = 40 = N_1 R \tag{4.10}$$
$$N_3 = 80 = N_2 R \tag{4.11}$$

and so on.
Thus:

$$N_3 = N_1 R \times R = N_0 R \times R \times R = N_0 R^3 \tag{4.12}$$

and in general terms:

$$N_{t+1} = N_t R \tag{4.13}$$

and:

$$N_t = N_0 R^t. \tag{4.14}$$

the relationship between R_0, R and T, the generation length

Equations 4.13 and 4.14 link together population size, rate of increase and time; and we can now link these in turn with R_0, the basic reproductive rate, and with the generation length (defined as lasting T intervals of time). In Section 4.5.1, we saw that R_0 is the multiplication factor that converts one population size to another population size, one generation later, i.e. T time intervals later.
Thus:

$$N_T = N_0 R_0. \tag{4.15}$$

But we can see from Equation 4.14 that:

$$N_T = N_0 R^T. \tag{4.16}$$

Therefore:

$$R_0 = R^T \tag{4.17}$$

or, if we take natural logarithms of both sides:

$$\ln R_0 = T \ln R. \tag{4.18}$$

ln $R = r$, the 'intrinsic rate of natural increase'

The term $\ln R$ is usually denoted by r, *the intrinsic rate of natural increase*. It is the rate at which the population increases in size, i.e. the change in population size per individual per unit time. Clearly, populations will increase in size for $r > 0$, and decrease for $r < 0$; and we can note from the preceding equation that:

$$r = \frac{\ln R_0}{T}. \tag{4.19}$$

R_0, r and T

Summarizing so far, we have a relationship between the average number of offspring produced by an individual in its lifetime, R_0, the increase in population size per unit time, r ($= \ln R$), and the generation time, T. Previously, with discrete generations (see Section 4.5.1) the unit of time was a generation. It was for this reason that R_0 was the same as R.

4.7.2 Estimating the variables from life tables and fecundity schedules

when fecundity and survivorship schedules remain steady, a population achieves a rate of increase of r, and a stable age structure

In populations with overlapping generations (or continuous breeding), r is the intrinsic rate of natural increase that the population has the *potential* to achieve; but it will only actually achieve this rate of increase if the survivorship and fecundity schedules remain steady over a long period of time. If they do, r will be approached gradually (and thereafter maintained), and over the same period the population will gradually approach a stable age structure (i.e. one in which the proportion of the population in each age class remains constant over time; see below). If, on the other hand, the fecundity and survivorship schedules alter over time—as they almost always do—then the rate of increase will continually change, and it will be impossible to characterize in a single figure. Nevertheless, it can often be useful to characterize a population in terms of its potential, especially when the aim is to make a comparison, for instance comparing various populations of the same species in different environments, to see which environment appears to be the most favourable for the species.

an exact, but biologically obscure equation for r

The most precise way to calculate r is from the equation:

$$\Sigma e^{-rx} l_x m_x = 1 \tag{4.20}$$

where the l_x- and m_x-values are taken from a cohort life table, and e is the base of natural logarithms. However, this is a so-called 'implicit' equation, which cannot be solved directly (only by iteration, nowadays usually on a computer), and it is an equation without any clear biological meaning. It is therefore customary to use instead an approximation to Equation 4.19, namely:

an approximate, but biologically transparent equation for r based on T_c, the cohort generation time

$$r \approx \frac{\ln R_0}{T_c} \tag{4.21}$$

where T_c is the *cohort generation time* (see below). This equation shares with Equation 4.19 the advantage of making explicit the dependence of r on the reproductive output of individuals (R_0) and the length of a generation (T). Equation 4.21 is a good approximation when $R_0 \approx 1$ (i.e. population size stays approximately constant), or when there is little variation in generation length, or for some combination of these two things (May, 1976).

We can estimate r from Equation 4.21, if we know the value of the cohort generation time T_c, which is the average length of time between the birth of an individual and the birth of one of its own offspring. This, being an average, is the sum of all these birth-to-birth times, divided by the total number of offspring, i.e.:

$$T_c = \frac{\Sigma\, x l_x m_x}{\Sigma\, l_x m_x}$$

or $\quad T_c = \dfrac{\Sigma\, x l_x m_x}{R_0}.$ \hfill (4.22)

This is only approximately equal to the true generation time T, because it takes no account of the fact that some offspring may themselves develop and give birth during the reproductive life of the parent.

Thus, Equations 4.21 and 4.22 allow us to calculate T_c, and thus an approximate value for r, from a cohort life table of a population with either overlapping generations or continuous breeding. In short, they give us the summary terms we require. A worked example is set out in Table 4.6, using data for the barnacle *Balanus glandula*. Note that the precise value of r, from Equation 4.20, is 0.085, compared to the approximation 0.080; whilst T, calculated from Equation 4.19, is 2.9 years compared to $T_c = 3.1$ years. The simpler and biologically transparent approximations are clearly satisfactory in this case. They show that since r was somewhat greater than zero, the population would have increased in size, albeit rather slowly, if the schedules had remained steady. Alternatively, we may say that, as judged by this cohort life table, the barnacle population had a good chance of continued existence.

Table 4.6 A cohort life table and a fecundity schedule for the barnacle *Balanus glandula* at Pile Point, San Juan Island, Washington (Connell, 1970). The computations for R_0, T_c and the approximate value of r are explained in the text. Numbers marked with an asterisk were interpolated from the survivorship curve.

Age (years) x	a_x	l_x	m_x	$l_x m_x$	$x l_x m_x$
0	1 000 000	1.000	0	0	
1	62	0.0000620	4600	0.285	0.285
2	34	0.0000340	8700	0.296	0.592
3	20	0.0000200	11600	0.232	0.696
4	15.5*	0.0000155	12700	0.197	0.788
5	11	0.000110	12700	0.140	0.700
6	6.5*	0.0000065	12700	0.082	0.492
7	2	0.0000020	12700	0.025	0.175
8	2	0.0000020	12700	0.025	0.200
				1.282	3.928

$R_0 = 1.282.$ $\quad T_c = \dfrac{3.928}{1.282} = 3.1.$ $\quad r \approx \dfrac{\ln R_0}{T_c} = 0.08014.$

Table 4.7 A hypothetical cohort life table and fecundity schedule, and a model population growing under the influence of these. Values for R_0, T_c, approximate and exact r, and R have been calculated from the cohort data. Values for R and r have also been estimated from the rate of increase of the population between time intervals eight and nine.

Cohort life-table data			Age structure (with static life-table entries in brackets)				
				After one time	After two time	After eight time	After nine time
x	l_x	m_x	Initial population	interval	intervals	intervals	intervals
0	1.000	0	1600 (1.00)	1880 (1.00)	2560 (1.00)	12 305 (1.000)	15 990 (1.000)
1	0.100	5	100 (0.063)	160 (0.085)	188 (0.073)	948 (0.077)	1231 (0.077)
2	0.060	15	60 (0.038)	60 (0.032)	96 (0.038)	437 (0.036)	569 (0.036)
3	0.018	10	20 (0.013)	18 (0.010)	18 (0.007)	101 (0.008)	130 (0.008)
4	0	—	0	0	0	0	0
						13 791	17 920

$R_0 = 1.58$.

$T_c = 1.80$.

$r \approx \dfrac{\ln R_0}{T_c} = 0.254$.

$r = 0.262$.

$e^r = R = 1.30$.

$\dfrac{17\ 920}{13\ 791} = 1.30 = $ estimated R.

Estimated $r = \ln R = 0.262$.

What, though, of static life tables and the age structures from which they spring? Table 4.7 shows what happens in a model population where the survivorship schedule (l_x column of a cohort life table) and the fecundity schedule (m_x column) remain steady over long periods. Initially, there are 100, 60 and 20 individuals in age classes 1, 2 and 3, respectively. At the start of the first time interval these produce the 1600 individuals in age-class 0, as a result of the birth rates in the m_x column [$1600 = (100 \times 5) + (60 \times 15) + (20 \times 10)$]. During the first time interval, the numbers in each age-class are subjected to the survival rates in the l_x column, giving the numbers in age classes 1, 2 and 3 *after one time interval*. The numbers in age class 0 (i.e. 1880) are then arrived at in the same way as before. The process has been repeated to show the nature of the population after eight and nine time intervals.

a model population achieving a rate of increase of r and a stable age structure

why cohort and static life tables are usually different

Values for R_0, T_c, approximate and exact r, and R have all been computed from the cohort life table using Equations 4.20, 4.21 and 4.22. Table 4.7 confirms that after a sufficiently long period of constancy in the l_x and m_x columns (eight time intervals) the population did indeed increase at a rate equal to r, and the age structure of the population was stable (as shown by the entries of the static life tables). However, even after eight and nine time intervals, the static life tables give a misleading impression of the true survivorship schedule shown in the cohort life table. In fact, each entry in a static life table, after a stable age structure has been achieved, is given by the following formula: $l_x e^{-rx}$, where l_x is the entry from the cohort life table. Hence, the cohort and static life tables are the same only when the population remains of a constant size ($r = 0$, $e^{-rx} = 1$). In a population that increases steadily in size, as here ($r > 0$, $e^{-rx} < 1$), the older age classes are increasingly poorly represented in static life tables and age structures; whilst in a population declining in size ($r < 0$, $e^{-rx} > 1$) the older age classes are increasingly well represented. Thus, static life tables and age structures fail properly to reflect the survivorship and fecundity schedules of a population not only when these schedules vary, but also when they are constant but the population itself is not constant in size. Clearly, though, the age-specific survivorship and fecundity schedules of a population, its potential rate of increase and its stable age structure are all interlinked. We return to this point when we look at human populations in Section 4.9.

4.8 Biennial and related life cycles

white sweet clover: a strictly biennial plant

Returning to the various types of life cycle, 'overlapping semelparity' (see Figure 4.10e–f) refers to species which are semelparous and have a distinct breeding season, but in which not all individuals breed within that season—some continue developing. This is seen at its simplest in organisms that are strictly biennial, i.e. where each individual takes two summers and the intervening winter to develop, but has only a single reproductive phase, in its second summer. An example is the white sweet clover, *Melilotus alba*, growing in New York State (Klemow & Raynal, 1981). Two cohort survivorship curves for one clover population are shown in Figure 4.20. These were constructed by carefully mapping the position of every plant in a number of randomly selected permanent plots, and then up-dating these maps every 2 weeks from 1976 to 1978 (winter mapping was prevented by snow). The survivorship curves show that, as usual, the 2 years were different in detail (the summer of 1977 was hotter and drier than that of 1976). Nevertheless, the two cohorts displayed a similar pattern, typical of a strictly biennial life cycle: relatively high mortality during the first growing season (whilst seedlings were developing into established plants), followed by much lower mortality until the end of the second summer, when the plants flowered and survivorship decreased rapidly. No plants survived to a third summer. Thus, in *M. alba* there is an overlap of two generations at most—in the summer and autumn there are newly emerged, growing plants, and plants that are reproducing or about to do so.

a semelparous plant with a less determinate life cycle

A more typical example of a semelparous species with overlapping generations is provided by the composite plant *Grindelia lanceolata* (Figure 4.21). This figure also contains two cohort survivorship curves, obtained here by placing wire rings around all newly germinated seedlings at the ends of the germinating seasons in 1972 and 1973, and then following the marked cohorts carefully until all individuals had died.

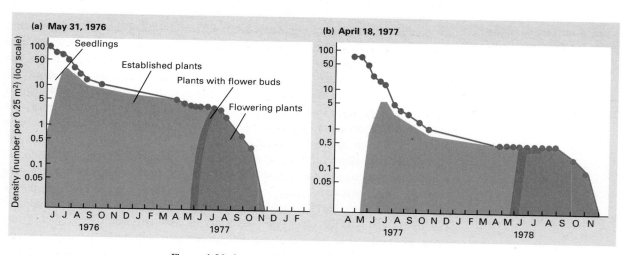

Figure 4.20 Survivorship curves for two cohorts of *Melilotus alba* (white sweet clover), a strict biennial. Survivorship was moderate in seedlings, good in established plants and very poor after flowering. (After Klemow & Raynal, 1981.)

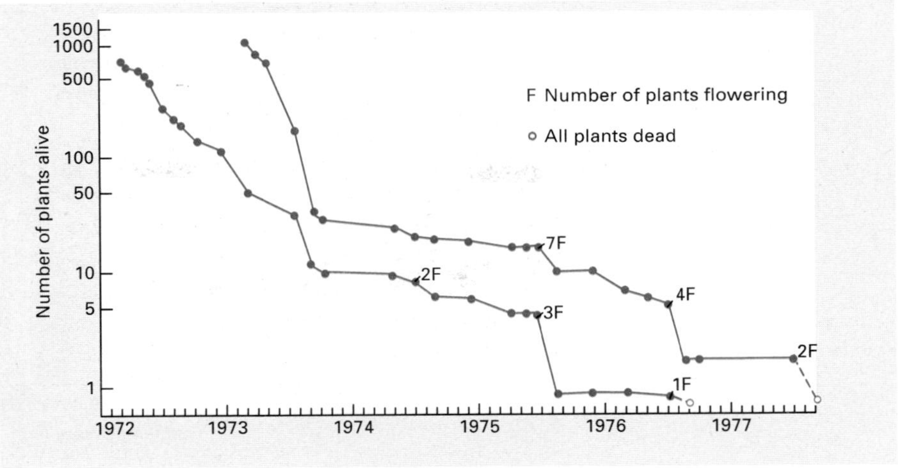

Figure 4.21 Survivorship curves of two cohorts of *Grindelia lanceolata* in Tennessee. Survivorship was moderate in young plants, good in established plants and very poor after flowering. (After Baskin & Baskin, 1979.)

(Despite the overlapping generations, cohort life tables are obviously easier to construct in these relatively short-lived species than they are in the iteroparous organisms considered in Section 4.6.) The major difference between Figures 4.21 and 4.20 is that *G. lanceolata*, rather than flowering in its second year, flowered in its third, fourth or fifth years. But, whenever an individual did flower, it died soon after. This has led to near-vertical drops down the survivorship curve associated with each flowering season. However, apart from these drops, the curves (as in Figure 4.20) indicate low survivorship during establishment, but high survivorship of established plants until they reproduce.

the importance of stage—and age

The behaviour of these indeterminate semelparous plants is, to a large extent, a reflection of their modularity and the fact that size (i.e. the extent of modular growth) is as important as age in determining an individual's fate. For instance, individuals of the semelparous teazel, *Dipsacus fullonum*, were sown in sites where they had previously been absent, and followed for a period of 5 years from emergence. During this time, rosette size and the vegetative or flowering condition of individuals were noted. The risks of pre-reproductive death and the chances of flowering were both clearly related to rosette size, but this relationship to size was essentially the same whether the rosettes were 2, 3 or 4 years old (Werner, 1975). On the other hand, within a given size class of rosettes of the semelparous *Lobelia telekii* growing on Mount Kenya, slower-growing (and hence, older) rosettes consistently exhibited higher rates of mortality than faster-growing, younger rosettes (Figure 4.22). Clearly, both age and stage can be important.

overlapping and continuous semelparity in animals

One well-known example of a semelparous animal with overlapping generations is provided by those species of salmon that grow without reproducing for several years before making a single determined effort to migrate up-river to spawn. Another is the common octopus, *Octopus vulgaris*, which has a life span in the Mediterranean of 15–24 months (Nixon, 1969). When reproduction is initiated in the females, protein synthesis in the muscles is suppressed, and there is rapid growth of the ovaries, with a concomitant loss of weight elsewhere (O'Dor & Wells, 1978). The final release of the single brood of offspring is followed rapidly by the death of the adult. In fact, the octopus *O. cyanea*, which lives in the Indo–Pacific region where breeding is possible at all times of the year (van Henkelem, 1973), is one example of

169 LIFE AND DEATH

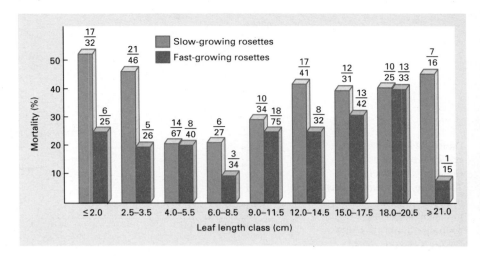

Figure 4.22 Between May 1979 and November 1981, for any given size class of rosette of *Lobelia telekii*, growing on Mount Kenya, mortality was greater for those that had grown less than the average between February 1978 and May 1979, and were older, than for those younger individuals that had grown more than the average. The fractions above the bars represent the number of deaths divided by the total number of plants observed in the class. (After Young, 1985.)

a species of semelparous individuals where breeding within a population goes on continuously (see Figure 4.10g). A much more widespread and perhaps more familiar group displaying *continuous semelparity* are those semelparous insect species living in environments kept constant by humans (e.g. many pests of warehouses).

4.9 Continuous iteroparity: human demography

The meaning of this final type of life cycle (see Figure 4.10j) should, by now, be clear: individuals reproduce repeatedly and can do so at any time throughout the year. This life cycle occurs, for instance, in environments which are not seasonal. This is the case in some tropical species, in many parasites and in many inhabitants of artificial environments kept constant by humans (e.g. grain stores). Such a life cycle can also occur in species that are essentially oblivious to seasonal changes in the environment because of their sophisticated physiologies. The most obvious example of this is *Homo sapiens*.

The quantitative study of the size, distribution, structure and dynamics of human populations is a mature discipline in its own right. Data of various sorts are essential for national and international planning and policy making, and life-table analysis was in fact originally developed for devising pension and insurance schemes on the basis of individuals' life expectancies. Thus, the data that are available for human populations have a history, a scope and an accuracy which cannot be matched by those for any other species.

the population pyramid

Figures 4.23 and 4.24a contain age-structure data in a form which is commonly used with human populations: the population pyramid. Each age class is represented by a horizontal bar, the length of which represents the size or relative size of the age class. The pyramid is constructed by arranging the age classes vertically with the youngest at the base and the oldest at the apex. Figure 4.23 shows how, with

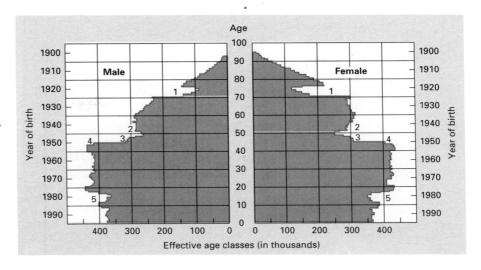

Figure 4.23 A population pyramid for France on January 1, 1992. 1, shortfall in births due to the 1914–1918 war (empty classes); 2, passage of empty classes to age of fecundity; 3, shortfall in births due to the 1939–1945 war; 4, 'baby boom'; 5, non-replacement of generations. (After Johnson *et al.*, 1994.)

humans, such data can be combined with historical and sociological information to provide an intelligible picture of a country's population (in this case France). Figure 4.24a contrasts the age distributions of the total populations of 'developing' (i.e. poor) countries and of developed countries in 1980. The age distribution for the developing nations tapers sharply from its base to its apex, whilst that for the developed nations has near-vertical or even overhanging sides until the older age classes. This contrast is partly due to the higher birth rates and lower survivorship rates in the poorer countries (Figure 4.24b); but as explained in Section 4.7, it also reflects the fact that the populations of the developing nations are expanding rapidly, whilst those of the developed nations are increasing slowly if at all (see Figure 4.24b, c).

prospects for the next century

The projections into the next century (Figure 4.24b, c), especially for the developing nations, are bound to cause concern when we think of the problems of feeding these multitudes. Indeed, even though the birth rates and survivorships are likely to alter and promote slower rates of increase over the next 20 years, the total population size will still increase rapidly because of the large number of people still to enter their reproductive years in developing countries. As May (1980) has discussed, this phenomenon—known as the *momentum of population growth*—can be expressed most dramatically by noting that even if an intrinsic *r*-value of zero could be achieved overnight, the developing countries would still be committed to a rough doubling of total population before coming to equilibrium (we saw in Section 4.7 that intrinsic rates of increase take some time to achieve). Thus, whilst we can derive some comfort (or complacency?) from the recent declines in birth rates in developing countries, we must still face up to a problem of quite frightening proportions.

4.10 Looking ahead

Having examined a wide variety of life cycles, it is natural to wonder how it comes about that some species exhibit one type of life cycle and other species another. It would be foolish to pretend that ecologists know the detailed answer to this question. Trying to understand the correspondence between life cycles and environments is one of the exciting challenges of modern ecology; but, before we take up this challenge here, we must recognize that for the majority of organisms the most

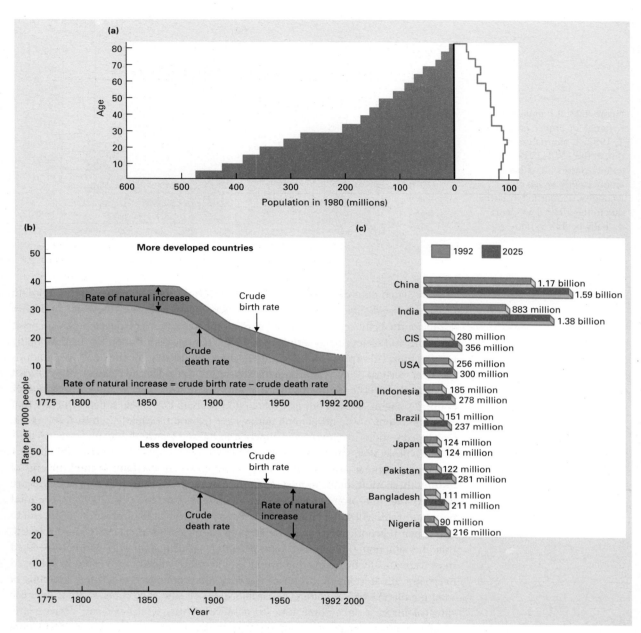

Figure 4.24 (a) The magnitude and the age composition of the total population in developing countries (shaded, on the left) and in developed countries (unshaded, on the right) for 1980. (After May, 1980.) (b) The changing per capita birth rates, death rates and net rates of increase for poor and developed countries in the past, and projected into the future. (c) The sizes and projected future sizes of the human populations in the 10 most populous countries, including examples from both the poor and the developed world. (After Miller, 1993.)

important aspects of the environment are the other organisms living alongside them. It is therefore necessary to postpone our attempts to understand life cycles until after we have examined the interactions that occur within and between species (see Chapters 6–13). We will return to life cycles in Chapter 14.

Chapter 5
Dispersal, Dispersion and Migration in Space and Time

5.1 Introduction

All organisms in nature are where we find them because they have moved there. This is true for even the most apparently sedentary of organisms, such as oysters and redwood trees. Their movements range from the passive transport that affects many plant seeds to the active movement of many mobile animals and their larvae. The effects of such movements are also varied. In some cases, they aggregate members of a population into clumps; in others they continually redistribute and shuffle them amongst each other and in others they spread the individuals out and 'dilute' their density.

The terms *dispersal* and *migration* are used to describe certain aspects of the movement of organisms. *Dispersal* is most often taken to mean a spreading of individuals away from others (e.g. their parents or siblings), and it may involve active (walking, swimming, flying) or passive movements (carriage in water or wind). Dispersal is therefore an appropriate description for several kinds of movements: (i) of plant seeds or starfish larvae away from each other and their parents; (ii) of voles from one area of grassland to another, usually leaving residents behind and being counterbalanced by the dispersal of other voles in the other direction; and (iii) of land birds amongst an archipelago of islands (or aphids amongst a mixed stand of plants) in the search for a suitable habitat.

Migration is most often taken to mean the mass directional movements of large numbers of a species from one location to another. The term therefore applies to classic migrations (the movements of locust swarms, the intercontinental journeys of birds, the transoceanic movements of eels), but also to less obvious examples like the to-and-fro movements of shore animals following the tidal cycle (see Section 5.7).

The terms dispersal and migration are both defined for *groups* of organisms. However, it is the individual that actually moves. Migration is mass movement, and an individual can only disperse, literally, if it separates into pieces (see Section 5.6 for dispersal of parts of a genet). Many dispersing organisms (especially plant seeds and many marine larvae) have little or no control over where or how far they travel. They are simply hazarded into the world at large to be carried at the mercy of winds and waves. At the level of the individual, there is no sharp distinction between migration and dispersal. Nevertheless, their effects are sufficiently different for them to be discussed separately in the course of this chapter.

5.2 Patterns of distribution: dispersion

The movements of organisms affect the spatial pattern of their distribution (their *dispersion*) and we can recognize three main patterns of dispersion, although they form part of a continuum.

Random dispersion occurs when there is an equal probability of an organism occupying any point in space (irrespective of the position of any others). The result is that individuals are unevenly distributed because of chance events.

Regular dispersion (also called *uniform, even* or *overdispersion*) occurs either when an individual has a tendency to avoid all other individuals, or when individuals that are especially close to others die. The result is that individuals are more evenly spaced than expected from chance.

Aggregated dispersion (also called *contagious, clumped* or *underdispersion*) occurs either when individuals tend to be attracted to (or are more likely to survive in) particular parts of the environment, or when the presence of one individual attracts, or gives rise to, another close to it. The result is that individuals are closer together than expected from chance.

These patterns are defined by the relative positions of the organisms to one another, but how they will appear to an observer or be relevant to the life of another organism depends on how the spatial scale is sampled. For example, consider the distribution of an aphid living on a particular species of tree in a woodland. If the area is sampled with large samples, for example acre or hectare quadrats, the aphids will appear to be aggregated in particular parts of the world, i.e. in woodlands, as opposed to other types of habitat. If our samples are smaller and taken only in woodlands, the aphids will still appear to be aggregated, but now on their host tree species rather than on the trees in general. However, if our samples were still smaller (25 cm², about the size of a leaf) and were taken within the canopy of a single tree, the aphids might appear to be randomly distributed over the tree as a whole. An even smaller quadrat (1 cm²) might detect a regular distribution because individual aphids on a leaf avoid one another.

In practice, the populations of all species are patchily distributed at some scale or another (e.g. Figure 5.1b) and ecologists have spent a lot of time designing ways of describing and measuring this patchiness (e.g. see Southwood, 1978a; Okubo, 1980; Greig-Smith, 1983; Murray, 1989). One of the most important consequences of the patchy distribution of organisms is that we can have quite misleading measures of their *effective* density. For example, the human density of a country or region is usually calculated as the total number of individuals divided by the total land area. For the contiguous 48 states of the USA, the 1960 census would give the population density as 59.94 persons per square mile. But, because most people live in patches (towns and cities), their *effective* density is really about 3000 persons per square mile (Lewontin & Levins, 1989). When we come to consider the role of density in regulating natural populations it becomes vitally important to distinguish *average density* from *effective density*: a population may have a very low average density, but the few individuals may be very close together.

It is immensely more difficult, but ecologically even more relevant, to describe the dispersion of organisms on scales that are relevant to the life style of motile organisms (a 'worms-eye' or 'birds-eye' view). MacArthur and Levins (1964) introduced the concept of environmental *grain* to make this point. For example, the

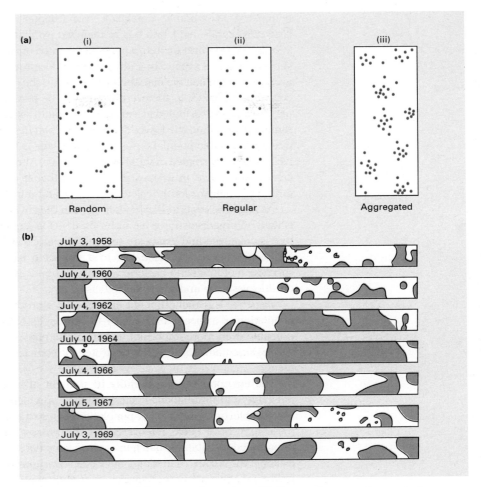

Figure 5.1 (a) Patterns of dispersion of organisms. (b) Changes from year to year in the patchiness of the grass *Festuca ovina* (shaded) on a transect through heathland (Breckland in eastern England). (After Watt, 1970.)

canopy of an oak hickory forest is *fine grained* (not patchy), from the point of view of a bird like the scarlet tanager which forages indiscriminately in both oaks and hickories, but is *coarse grained* (patchy) for defoliating insects which attack either oaks or hickories preferentially. Indeed, it has been argued that 'the problem of pattern and scale is the central problem in ecology, unifying population biology and ecosystem science, and marrying basic and applied ecology' (Levin, 1992).

Patchiness may be a feature of the physical environment: a localized distribution of resources or conditions is reflected in a localized distribution of organisms. Islands surrounded by water are obvious examples of a patchy distribution of physical conditions associated with patches of terrestrial organisms. Rocky outcrops in moorland make a patchy environment of 'islands' suitable for lichens and mosses, and patches of water (lakes) in a terrestrial landscape are 'islands' for fish.

Equally important patchiness is created by the activities of organisms themselves. The living cover of almost all land surfaces and much of tidal and subtidal areas is dominated by modular organisms (see Chapter 4, Section 4.2: trees, grasses, seaweeds, corals, etc.), and it is in their nature that an individual (the product of a zygote) develops into a patch, a population, patch or cluster of parts (leaves, branches, shoots, fronds, polyps). Thus, the trees in a forest may be distributed at random or evenly spaced when we map their trunks—but, the population of leaves on each tree contributes a patch to the overall canopy. The porcupine searching for tree trunks may find them distributed at random, but a moth searching for egg-laying sites in the same trees will find the leaves distributed as patches. Moreover, when an individual tree dies, a whole population or patch of foliage is lost from the canopy and a new well-lit island is created in the shady forest floor. Almost every activity of an organism creates patchiness in its environment, whether it is patchy grazing, deposition of dung, trampling or local depletion of water and mineral resources.

patches and gaps have life histories

As might be expected, patches or gaps in the environment that are created by the activity of organisms have life histories. Thus, a gap created in a forest by a falling tree is colonized and grows up to contain mature trees, whilst other trees fall and create new gaps. The dead leaf in a grassland is a patch for colonization by a succession of fungi and bacteria which eventually exhaust it as a resource, but new dead leaves arise and are colonized elsewhere.

Dispersal is usually thought of as part of a process of colonization in which the population of a species expands by increasing its occupied territory. However, it is probably even more important as part of the process by which the population of a species maintains itself as the environment changes, or as it changes the environment.

habitable sites or patches

The environment can usefully be thought of as composed of a patchwork of habitable and uninhabitable sites—patches in which the resources and conditions are suitable, or unsuitable, for populations of a particular species. However, not all of these may have been 'discovered' in the process of dispersal, and they remain unoccupied. A habitable patch may cease to be habitable over the course of time, for example the resources in it may have been consumed or the conditions in it changed by the invasion of another species. A well-lit gap in a forest may have provided a habitable site for short-lived, light-demanding species, but as the surrounding canopy closes, the patch ceases to support these species and becomes habitable only for those that tolerate shade. The light demanders then become locally extinct. New habitable sites arise elsewhere (e.g. a tree dies and creates a canopy gap, a storm strips the mussels from a seashore rock). The important feature of most environmental patchiness is that habitable patches are patchy in time (have a limited habitable life) and, by definition, patchy in space (more or less separated from one another). Dispersal provides the process by which new habitable patches are continually colonized in the spatial and temporal flux, which underlies even the most stable of populations. We return to these ideas and develop them further in Chapter 15, Sections 15.5 and 15.6.

5.2.1 Forces that favour aggregation

The simplest evolutionary explanation for the patchiness of populations is that organisms aggregate when and where they find resources and the conditions that favour reproduction and survival. These resources and conditions are usually patchily distributed in both space and time. It pays (and has paid in evolutionary time) to live in these patches when and where they occur. There are, however, other

aggregation can result from a shared habitat selection

specific ways in which organisms may gain from being close to neighbours in space and time. For example, warning cries from one individual may protect other members of a group, but bring no advantage to isolated individuals. But, it is difficult to explain how evolutionary forces act to favour living as a group, because, as we have pointed out earlier, evolution happens because some individuals leave more descendants than others and we are therefore forced to look for ways in which the individual (or at least its kin) gains from being part of a group (see the discussion in Dawkins' *The Selfish Gene* (1976) and *The Extended Phenotype* (1982), but also Wynne-Edwards (1993), for a vigorous defense of group selection).

aggregation and 'domains of danger'

An elegant theory identifying selective advantage to individuals that aggregate with others was suggested by Hamilton (1971) in his paper 'Geometry for the selfish herd'. He argued that the risk to an individual from a predator may be lessened if it places another potential-prey individual between itself and the predator. The consequence of many individuals doing this is bound to be an aggregation. The 'domain of danger' for individuals in a herd is at the edge, so that an individual would gain an advantage if its social status allowed it to assimilate into the centre of a herd. We might then expect subordinate individuals to be forced into the region of greater danger on the edge of the flock. This seems to be the case in reindeer (*Rangifer tarandus*) and in woodpigeons (*Columba palumbus*), where a newcomer may have to join the herd or flock at its risky perimeter and can only establish itself in a more protected position within the flock after social interaction (Murton *et al.*, 1966). Individuals may also gain from living in groups if this helps to locate food, give warning of predators or if it pays for individuals to join forces in fighting off a predator (Pulliam & Curaco, 1984).

The principle of the selfish herd as described for the aggregation of organisms in space is just as appropriate for the synchronous appearance of organisms in time. The individual that is precocious or delayed in its appearance, outside of the norm for its population, may be at greater risk from predators than those conformist individuals that take part in 'flooding the market' and thereby dilute their own risk.

predator satiation in time

Amongst the most remarkable examples of synchrony are the periodic cicadas, the adults of which emerge simultaneously after 13 or 17 years of life underground as nymphs. Williams *et al.* (1993) studied the mortality of populations of 13-year periodical cicadas that emerged in north-western Arkansas in 1985. Birds consumed almost all of the standing crop of cicadas when the density was low, but this became only 15–40% when the cicadas reached peak density. Predation then rose to near 100% as the cicada density fell again (Figure 5.2).

Many species of tree, especially in temperate regions, have synchronous 'mast' years (see Chapter 8, Section 8.4), in which they produce vast crops of seed followed by a run of years in which they are relatively barren. In many species of bamboo too, the individuals flower and set abundant seed synchronously, in some species after intervals of as much as 100 years. If the populations of predators cannot increase fast enough to consume the glut, more seeds may survive to germinate and grow than if seed production had been regular and reliable (Janzen, 1976). Synchronous behaviour that saturates the needs of predators in some years and starves them in others probably most reduces risks to the prey when the predators are specialists— and when the synchrony occurs throughout patchily dispersed populations. It may not work quite as well if the predators are generalists and the synchrony is just within patches (Ims, 1990), although the example of the cicadas (Figure 5.2) seems to defy this generalization.

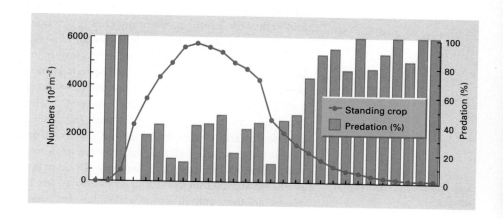

Figure 5.2 Changes in the density of a population of 13-year periodical cicadas in north-western Arkansas in 1985, and changes in the percentage eaten by birds. (After Williams et al., 1993.)

There are also strong selective force pressures that can act *against* aggregation in space or time. The foremost of these is certain to be the more rapid local depletion of resources, that arises from many individuals being in the same place at the same time (see Chapter 6). A group of individuals may actually concentrate a predator's attention (the opposite effect to the 'selfish herd'). Overall, the types of dispersion found in nature are bound to be compromises between opposing forces: the attractant forces of shared habitat requirements and the benefits of group living, and the repellent forces of overcrowding and the costs of group living.

5.3 Dispersal

5.3.1 Dispersal as escape and discovery

Dispersal is the term applied to the process by which individuals escape from the immediate environment of their parents and neighbours, and become less aggregated; dispersal may therefore relieve local congestion. But, dispersal can also often involve a large element of discovery. A useful distinction can be made between two types of such 'discovery dispersal' (Baker, 1978). First, there is dispersal in which individuals visit and 'explore' a large number of sites before finally returning and settling in a chosen one. Second, there is dispersal in which individuals visit a succession of locations, but then cease to move (with no element of 'return' to a site previously explored). In fact, this latter category can be split further into cases where the cessation of movement is under the dispersing organism's control, and cases where it is not.

The dispersal of plant seeds is non-exploratory and beyond the control of the seed itself. The discovery aspect of seed dispersal is therefore a matter of chance (although the chances of reaching a suitable site may be increased by the specializations for dispersal that the seeds possess). Animal dispersal, on the other hand, can fall into any of the three categories. Some animals have essentially the same type of dispersal as plant seeds (e.g. the simple freshwater invertebrates described in Section 5.37). Many other animals cannot be said to explore, but they certainly control their settlement, and cease movement only when an acceptable site has been found. For example, most aphids, even in their winged form, have powers of flight which are too weak to counteract the forces of prevailing winds. But, they control their take-off

margin note: exploratory and non-exploratory discovery

Table 5.1 Examples of species of insects able, or unable, to migrate or to fly, in relation to the permanence of their habitat. (Permanent: lakes, rivers, streams, canals, trees, bushes, salt-marshes, woods. Temporary: ponds, ditches, pools, annual plants, perennial plants but not climax vegetation, arable lands.) (After Johnson, 1969.)

	Permanent habitats	Temporary habitats
British Anisoptera		
Total number of species	20	23
Pronounced dispersive phase	6	13
British Macrolepidoptera		
Total number of species	594	181
Pronounced dispersive phase	14	42
British water beetles		
Total number of species	52	127
Able to fly	13	81
Unable to fly	20	9
Flight variable	19	37

from their site of origin, they control when they drop out of the windstream and they make additional, often small-scale flights if their original site of settlement is unsatisfactory. Their dispersal, therefore, involves 'discovery', over which they have some, albeit limited, control.

The extent of truly exploratory dispersal is difficult to judge: appropriate data are sparse. It is not sufficient to know where an animal started from and where it finished; it is also necessary to know where the animal went in between. However, exploratory dispersal may be quite widespread (Baker, 1982).

some species disperse more than others ...

All species disperse, but some are more dispersive than others. Insects living in habitats that are, by nature, temporary have a more pronounced dispersive phase than insects living in more permanent habitats (Table 5.1). In a similar way, 'supertramp' bird species (Diamond, 1973; Diamond & May, 1976) are good dispersers and good early colonizers of islands, but they do not persist for long when the habitat becomes colonized by other species. The pigeon *Macropygia macinlayi*, is a classic supertramp, occurring only on small and species-poor islands of the Bismarck archipelago (east of New Guinea) (Figure 5.3). It contrasts with the cuckoo *Centropus violaceus*, which is a late colonizer and then only of islands already well stocked with other species. It is hard to be a successful cuckoo unless other species are present to foster its young!

In general, dispersal is essential for the persistence of species that exploit temporary stages in a changing community (a community 'succession'—see Chapter 17, Section 17.4). The descendants of individuals of all successional species are doomed in their local habitats. Yet, even the species of so-called 'climax' communities (the relatively stable endpoints of successions) are doomed in the long run unless they colonize new areas. The movements of forests following the advance and retreat of ice-sheets (see Chapter 1, Section 1.22), or of tropical forests following arid periods, are on a different time scale to that usually associated with the dispersal of organisms. However, they make the same point, that in the life of all terrestrial organisms home is sooner or later a dangerous place.

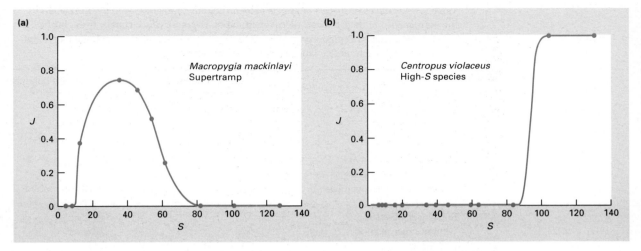

Figure 5.3 The incidence of (a) the 'supertramp' pigeon (*Macropygia mackinlayi*) and (b) the cuckoo (*Centropus violaceus*) on islands in the Bismarck archipelago. *S* is the number of bird species on an island and *J* is the fraction of the islands on which the pigeon or the cuckoo are present. The pigeon tends to be an early colonist of islands, but not to persist in species-rich communities. The cuckoo, in contrast, is found only on islands that have been colonized by a rich bird fauna. (After Diamond & May, 1976.)

5.3.2 The argument that all organisms should have some dispersive properties

Dispersal is usually risky, and there will always be a balance of risk between living longer in an already occupied habitat and hazarding resources in an act of colonization. As Gadgil (1971) has argued:

> In a very general way, the factor favouring evolution of dispersal would be the chance of colonizing a site more favourable than the one that is presently inhabited…. An organism should disperse if the chance of reaching a better site exceeds the loss from the risk of death during dispersal or the chance of reaching a poorer habitat.

Even in very stable habitats, it appears that in their evolutionary history all organisms will have been under selective pressure to disperse some of their progeny (Hamilton & May, 1977). Imagine a population in which the majority of organisms have a stay-at-home, non-dispersive genotype O, but in which a rare mutant genotype, X, keeps some offspring at home but commits others to dispersal. The disperser X is certain to replace itself in its own habitat. This is because the other, O-type individuals make no attempt to displace it. However, X also has at least some chance of displacing some of the non-dispersers. The disperser X will therefore increase in frequency in the population. On the other hand, if the majority of the population are type X, whilst O is the rare mutant, O will still do worse than X. This is because O can never displace any of the Xs, whilst O itself has to contend with several or many dispersers. Dispersal is therefore said to be an evolutionarily stable strategy (ESS) (Maynard Smith, 1972; Parker, 1984). A population of non-dispersers will evolve towards the ubiquitous possession of a dispersive strategy; a population of dispersers will show no tendency to lose that strategy. Even when efficient dispersal is achieved only at the cost of a reduced number of offspring, a smaller but still positive fraction of the progeny should be dispersed (Cohen & Mitro, 1989). A high degree of spatial variability in the environment favours (both within and between species) forms with high rates of dispersal (Cohen & Levin, 1987).

5.3.3 The demographic significance of dispersal

The ecological fact of life propounded in Chapter 4, Section 4.1 emphasized that dispersal can have a potentially profound effect on the dynamics of populations. In practice, however, many studies have paid little attention to dispersal. The reason often given is that emigration and immigration are approximately equal, and they therefore cancel one another out. One suspects, though, that the real reason is that dispersal is usually extremely difficult to quantify. The studies that *have* looked carefully at dispersal have tended to bear out what the ecological fact of life suggests. In a long-term and intensive study of a population of great tits in Wytham Wood, Oxford, England (Greenwood *et al.*, 1978; see Chapter 5, Section 5.4.4) it was observed that 57% of breeding birds were immigrants rather than born in the population. In a review of dispersal in mice and voles (Gaines & McClenaghan, 1980), there were a number of cases in which either more than 50% of a population emigrated each week, or more than 50% of a population were individuals that had immigrated during the previous week. Significantly, when populations of the voles *Microtus pennsylvanicus* and *M. ochrogaster* were enclosed to prevent emigration, densities rose to abnormally high levels and the vegetation was overgrazed in a way not normally observed (Krebs *et al.*, 1969).

In a study of the spruce budworm in Canada, estimates of adult dispersal prior to egg laying for the years 1949–1955 suggested values ranging from a loss of more than 90% by emigration to an increase of 1700% by immigration (Greenbank, 1957). In a population of the Colorado potato beetle in Canada, the average emigration rate of newly emerged adults was 97% (Harcourt, 1971). This can be interestingly compared with the spread of the beetle in Europe in the middle of this century (Figure 5.4; Johnson, 1967). Dispersal is undoubtedly a crucially important demographic process, although it is much neglected.

5.3.4 Passive dispersal on land and in the air: the seed rain

the blurred distinction between active and passive dispersal

Most biological categories are blurred at the edges. One example is the distinction between active and passive dispersers. For most mobile animals there is some element of behaviour involved in movement, whereas in sessile organisms it is forces outside the organism that control the distance and direction of dispersal. Yet, passive dispersal in air currents is not a phenomenon restricted to plants. Young spiders that climb to high places and then release a gossamer thread which carries them on the wind are as passively at the mercy of air currents as the winged fruits of a maple or the winged seeds of a pine. Nor is passive dispersal confined to organisms totally lacking control of their movement (Johnson, 1969). The wings of birds and insects are not only active sources of directed flight; they are also often aids to the passive movement of these organisms in air currents (Table 5.2).

Despite these blurred distinctions, however, the distance and direction that a passive unit is carried can be seen as being determined not by any mobile quality of its own, but by the force that carries it (whether this be wind, water or a moving organism that carries seed on its coat or in its gut). The higher plant has control not

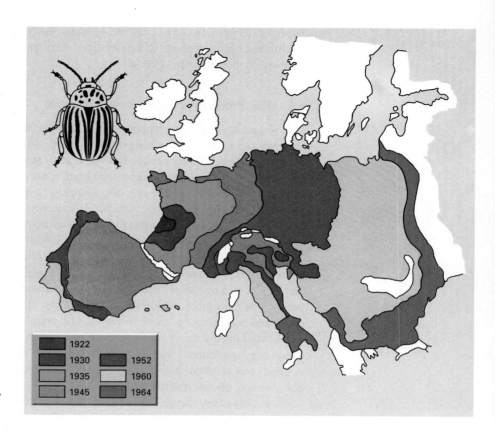

Figure 5.4 Spread of the Colorado beetle (*Leptinotarsa decemlineata*) in Europe. (After Johnson, 1967.)

1922	
1930	1952
1935	1960
1945	1964

of where its seeds will land, but of how they will interact with wind currents and eddies. Burrows (1986) gives an elegant analysis of the way in which the morphology of seeds interacts with aerodynamic forces.

The rain of seed that falls from a parent plant is probably never distributed at random. Most seeds fall close to the parent and their density declines with distance. This is the case for wind-dispersed seeds and also for those that are ejected actively by maternal tissue (e.g. many legumes, *Viola* and *Geranium*). The density of seeds is often low immediately under the parent, rises to a peak close by and then falls off

Table 5.2 Comparison of the relative abundance of total insects and selected groups of Diptera obtained by sampling for 6–7 days at different heights in the air. (After Glick, 1939.) The great heights attained in many of these examples are quite unnecessary for powered flight from one place to another but enormously important for long distance passive transport in air currents.

	Height (feet)					
	20	200	1000	2000	3000	5000
Total insects	26 512	13 396	4749	2357	1364	612
Total Diptera	11 018	5175	1979	1024	586	279
Chloropidae	3435	1010	459	212	104	74
Sarcophagidae	57	17	4	1	2	2
Muscidae	14	6	1	2	0	0
Calliphoridae	0	3	0	0		

steeply with distance. A number of seed dispersal curves are shown in Figure 5.5. The decline in density with distance from the parent often approaches the inverse square law for wind-dispersed seeds, but approaches or is even steeper than the inverse cube law in the case of species that ejaculate their seeds from explosive capsules, or in which the seeds have no obvious dispersal mechanism. Plant

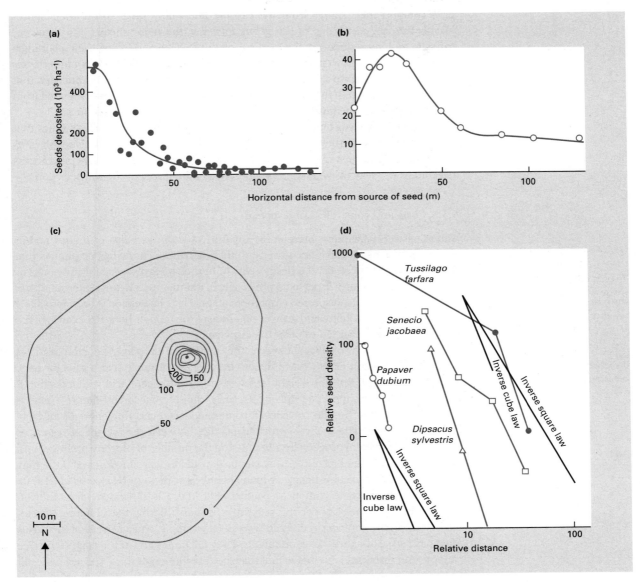

Figure 5.5 (a) The dispersal of seeds of *Eucalyptus regnans* from trees approximately 75 m high from the edge of a dense forest and (b) from an isolated tree. (c) Contour map of the distribution of one-seeded fruits of the tropical tree *Lonchocarpus pentaphylla* from the parent tree. Contour interval = 50 fruits m⁻². (After Augspurger & Hogan, 1983.) (d) The relationship between relative seed density and relative dispersal distance for a variety of plants: (i) *Senecio jacobaea* from the edge of a dense population in the direction of the prevailing wind; (ii) *Tussilago farfara* from the edge of a dense population; (iii) *Papaver dubium* from an isolated plant; and (iv) *Dipsacus sylvestris* from isolated plants. (After a variety of sources; Harper, 1977.)

pathologists have recognized that diseases of which the dispersal is approximated by the inverse square law tend to form isolated foci of infection at a distance, whereas diseases with dispersal following the inverse cube law tend to spread as an advancing and quite well-marked horizon or front. The same generalizations probably apply to the dispersal of seeds (Harper, 1977).

The explosive propulsion of dispersal units is not restricted to seed plants. Many ferns catapult their spores from the capsule and many fungi throw their spores forcibly into the air. Fungi of the genus *Pilobolus* (the name means 'hat thrower') grow in and reproduce on the surface of dung. They orientate their sporangia towards the light in the daytime and then, at night, explosively fire them into the air with a velocity of up to 27 m s^{-1}. Each sporangium carries a drop of mucilage that sticks it to vegetation. If this is eaten (e.g. by horses) the spores are then released from the sporangium in the gut of the animal and further dispersed in its dung.

Because most seeds tend to fall close to the parent, the majority of seedlings that develop are likely, if they compete, to do so with the parent or close relatives as neighbours. This is just one of the ways in which the nature of the dispersal process determines the way in which a population meets the forces of natural selection.

5.3.5 Costs and constraints of dispersal

Almost all specializations for seed dispersal involve the sacrifice of tissues and/or resources that might have been used in other activities. The wings or plumes that support wind-borne seeds, the fleshy coats of bird-dispersed seeds and the hooks or spines of seeds that are dispersed on the coats of animals are always maternal tissue, and are not part of the dispersed embryonic plant. Their possession is therefore likely to detract from the vitality and growth potential of the mother herself, because of the resources she has diverted to offspring dispersal.

the cost of dispersal to the 'mother'

There is also a conflict between the weight of a dispersed unit and its dispersibility. Weight clearly has a major effect on the distance that a seed (or baby spider) is dispersed. But this weight largely represents the capital of resources that it carries with it to support it during its early phases of establishment. There is therefore, as with a human traveller, a conflict between carrying enough 'luggage' and being mobile. There is yet a further conflict between the weight of resources allocated to individual progeny by a parent and the number of progeny produced. All organisms have limited resources that can be devoted to reproduction and these may be allocated either to a few heavy, or many lightweight progeny (Harper *et al.*, 1970).

dispersibility and capital investment

The evolutionary resolution of conflicts like these is discussed more fully in Chapter 14, Section 14.4. For the present, it is worth noting that different species have resolved the conflicts in different ways. For example, tropical trees of the genus *Mora* have seeds that are like cannonballs. They fall under and at the edge of the tree canopy, and dispersal is so ineffectual that the distribution of these species may still be limited by the dispersal that has occurred since they evolved! Such species are good examples of well-endowed, heavy progeny that appear to have sacrificed dispersibility. By contrast, many parasitic plants (see Chapter 12, Section 12.2.2) produce vast numbers of wind-dispersed 'dust seeds', which are almost entirely dependent on their host for the nutrition of their earliest developmental stages. The 'dust' seeds of orchids carry virtually no food reserves and are dependent on mutualism with fungi (mycorrhiza) for their early nutrition (see Chapter 13, Section 13.7.3).

5.3.6 Passive dispersal by an active agent

Much of the uncertainty concerning the fate of dispersed progeny is removed if an active agent of dispersal is involved. The seeds of many herbs of the woodland floor carry spines or prickles that increase their chance of being carried passively on the coats of animals. The seeds may then be concentrated in nests or burrows when the animal grooms itself (Sorensen, 1978). The seeds of many species, particularly of shrubs and lower canopy trees, are fleshy and attractive to birds, and the seed coats resist digestion in the gut (see Chapter 13, Figure 13.6). Such seeds are then deposited in faeces. Where the seed is dispersed to depends on the defecating behaviour of the bird.

birds and seeds

There are many examples of what might be coevolved relationships between particular plant species and specialist dispersing birds, although there seems to be no bird that is totally specialized on any one species of fruit. A close approach to such specialization, however, is found in the mistletoe bird of Australia (Figure 5.6), which feeds almost exclusively on mistletoe berries. It defecates frequently, and at each defecation it wriggles itself against the branch of a tree. This has the effect of lodging the seeds in bark crevices where they germinate and establish themselves as parasites on the host.

jays disperse and bury acorns . . .

An even more striking example of the dispersal of seeds by birds is that of acorns by jays (*Garrulus glandarius*) (Bossema, 1979). The jay collects acorns and buries them individually in the soil. It selects acorns that are ripe and undamaged, and has a preference for the larger acorns and for particular shapes. The jay carries between one and five acorns during a flight—one (usually the largest) in its beak and the rest held in the throat and oesophagus. The acorns may be carried distances from 100 m

Figure 5.6 The Australian mistletoe bird (*Dicaeum livundinaceum*) feeding its young on mistletoe berries.

185　DISPERSAL, DISPERSION AND MIGRATION

to a few kilometres, and each is then buried individually, not in caches. The further the jay flies to deposit the acorns the greater is the number that it carries (Figure 5.7). The jays usually choose to bury the seeds near a tree, or a hedge at the edge of a piece of disturbed ground. This means that the behaviour of the bird may determine the microsites that the oak colonizes. In the spring the birds recover the acorns, and of course, if the birds recovered all of them, the oak would simply suffer loss, as from a predator. In practice, however, many of the birds will have died during the winter and never recover their stores, and even when the birds rediscover the acorns, they have usually started to germinate. The bird then commonly removes only one of the cotyledons—and the young plant continues to grow.

... and remember where they are buried

The most remarkable part of the story is that the jays remember where they have planted the individual acorns—they do not search by smell, they simply remember where they are and can even locate them under a snow cover. This feat of memory is the more remarkable because an individual bird may collect and bury 4600 acorns in a season!

It is dangerous, however, to interpret all bird-feeding habits as in some way optimizing the dispersal of seeds, and as examples of coevolution. Figure 5.8 shows the seeds of oak (acorns) stored in holes in fencing posts by the acorn woodpecker (there may be 50 000 holes bored in a single tree). There is little in such behaviour that can be interpreted as increasing the fitness of the oaks!

Much of the ecology of seed dispersal is still in the realm of anecdotal natural history—beautifully described in the classic text of Ridley (1930) and by van der Pijl (1969).

There are also important examples in which animals are dispersed by an active agent. For instance, there are many species of mite that are taken very effectively and

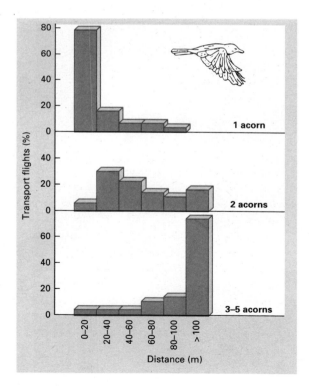

Figure 5.7 Distances travelled by jays (*Garrulus glandarius*) dispersing seeds (acorns) of oak (*Quercus*). Note that the more seeds they carry the further they take them. (After Bossema, 1979.)

Figure 5.8 Acorns stored in fencing posts by the acorn woodpecker (*Melanerpes formicovorus*). (Photograph courtesy of J.L. Harper.)

directly from dung pat to dung pat, or from one piece of carrion to another, by attaching themselves to dung beetles or carrion beetles. They usually attach to a newly emerging adult, and leave again when that adult reaches a new patch of dung or carrion. Often, the interaction is mutualistic (advantageous to both—see Chapter 13): the mites gain a dispersive agent, but many of them attack and eat the eggs of flies that would otherwise compete with the beetles.

5.3.7 Passive dispersal in water

Weight is less of a hindrance to the dispersal of propagules in water. It is striking, however, that the passive dispersal of seeds is rare in freshwater; most aquatic flowering plants produce their seeds out of water. In rivers the larvae of freshwater invertebrates make use of the flowing column of water to dispersing from hatching sites to appropriate microhabitats (their movements are referred to as 'invertebrate drift'—Townsend, 1980; Brittain & Eikeland, 1988), but most freshwater insects depend on flying adults for upstream dispersal and movement from stream to stream. Clearly, passive transport in the water is a useless means of spread between ponds and lakes. The dispersal of most freshwater organisms without a free-flying stage depends on resistant wind-blown structures (e.g. gemmules of sponges, cysts of brine shrimps, ephippia of cladocerans, eggs of *Hydra* and the desiccated bodies of tardigrades and rotifers).

In the sea the situation is quite different. Marine habitats are large and continuous, and are more or less accessible to pelagic larvae thanks to rapid diffusion by tidal and residual currents. In marine invertebrates, the pelagic, short-lived larvae are usually the dispersal units; the sedentary adult is usually the stage in the life cycle at which most feeding and growth is done. This is in complete contrast to the freshwater insects (Figure 5.9).

most freshwater insects disperse as flying adults ...

... but most marine invertebrates are dispersed passively as larvae

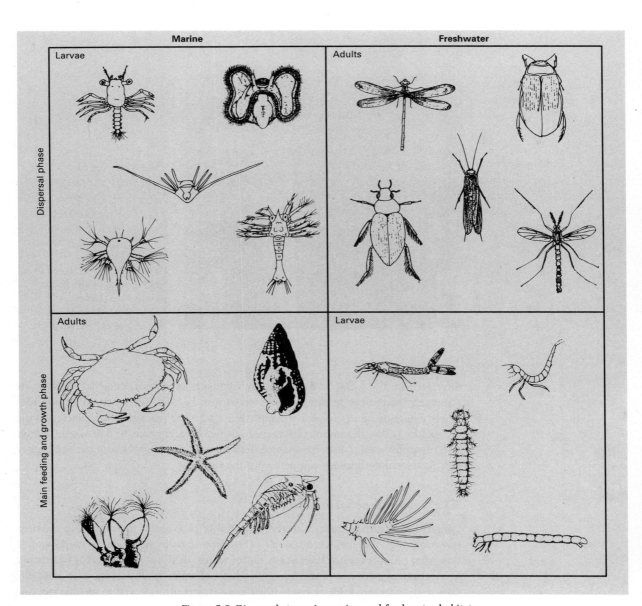

Figure 5.9 Dispersal stages in marine and freshwater habitats.

In discussing seed dispersal we made the point that there is a conflict between producing a few large, well-provisioned seeds or a larger number of small ones carrying few reserves. Amongst the pelagic larvae of the marine habitat there is an alternative—many fend for themselves, and do not depend wholly on yolk reserves provided by the parent. This has been called the planktotrophic strategy. Such pelagic larvae are active swimmers, using energy and actively consuming resources whilst they are dispersing—very different from a passively dispersed seed (Crisp, 1976). In such cases, settlement is also an active process. The larvae can make a choice of where to settle, and this contrasts dramatically with the impotence of a seed to determine where it will fall.

188 CHAPTER 5

5.4 Variation in dispersal within and amongst populations

5.4.1 The sexual dimension

dispersal of males and females is often different

Males and females often differ in their liability to disperse. Amongst species of birds it seems to be usually females that are the main dispersers, but amongst mammals it is the males (Table 5.3; Greenwood, 1980). This difference between the sexes in the two taxa is linked to a number of other trends: (i) the mating system in mammals is usually based on competition between males for mates rather than for territories; (ii) mammals are mainly polygamous whereas birds are mainly monogamous; and (iii) male mammals contribute less to the care of their progeny than is commonly the case with birds. Altruism also appears to be more common in the non-migrating sex, i.e. more common amongst male than amongst female birds, but more common amongst female than amongst male mammals.

Differences in dispersal between the sexes are especially strong in some insects, where it is the male that is usually the more active disperser. For example, in the winter moth (*Operophtera brumata*), the female is wingless whilst the male is free-flying. There are some exceptions to this general rule, such as the beet leafhopper (*Circulifer tenellus*), which makes long-distance migrations, crossing mountains thousands of feet high. In this case it is the females that disperse further, and the populations that the migrants leave behind therefore become male dominated.

5.4.2 Dispersal polymorphism

Another source of variability in dispersal within populations is a somatic polymorphism amongst the progeny of a single parent. Environments and habitats are patchy and change with time. Thus, the sites in which progeny establish themselves successfully may sometimes be close to home and sometimes far away. Also, the situation may change from season to season and from year to year. There is then no single perfect pattern of dispersal behaviour, and the fittest parents may be those that produce both dispersing ('high-risk') and non-dispersing ('low-risk') progeny (Figure 5.10) (Venable *et al.*, 1987; Venable & Brown, 1988).

high risk and low risk dispersal

seed polymorphism

A classic example is the desert annual plant *Gymnarrhena micrantha*. This bears a very few (one to three) large seeds (achenes) in flowers that remain unopened below the soil surface, and these seeds germinate in the original site of the parent. The root system of the seedling may even grow down through the dead parent's root channel. But, the same plants also produce above ground smaller seeds with a feathery pappus, and these are wind dispersed. In very dry years only the undispersed underground seeds are produced, but in wetter years the plants grow vigorously and produce a large number of seeds above ground, which are released to the hazards of dispersal (Koller & Roth, 1964).

Table 5.3 Number of species with predominant dispersal by one sex. (After Greenwood, 1980.)

	Mainly by males	Mainly by females	No sex difference
Birds	3	21	6
Mammals	45	5	15

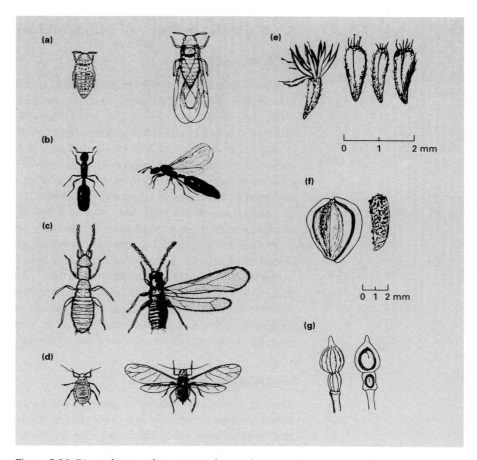

Figure 5.10 Dimorphism with respect to dispersal in animals and plants. Winged and wingless females of: (a) the grape phylloxera (*Viteus vitifoliae*); (b) the little black ant (*Monomorium minimum*); (c) the zorapteran (*Zorotypu hubbardi*); and (d) the apple aphid (*Aphis pomi*). Contrasting sizes and dispersal features of seeds and fruits: (e) *Galinsoga parviflora*, fruits from the centre (on left) and margin (on right) of the capitulum; (f) *Dimorphotheca pluvialis*, again showing central and marginal fruits; (g) *Rapistrum rugosum*, a two-seeded pod with marked difference in seed size. (After a variety of sources.)

There are very many examples of such seed dimorphism amongst the flowering plants, especially amongst grasses, composites and members of the Chenopodiaceae and Cruciferae families (all rather recently evolved families). In these, the difference between seed types is determined by the mother, not by genetic differences between the seeds. Both the dispersed and the 'stay-at-home' seeds will, in their turn, produce both dispersed and stay-at-home progeny. Commonly, in seed polymorphism, the 'stay-at-home' seed is produced from self-pollinated flowers below ground or from unopened flowers. The seeds that are dispersed are more often the product of cross-fertilization. Two phenomena are thus coupled: the tendency to disperse is coupled with the possession of new, recombinant ('experimental') genotypes, whereas the stay-at-home progeny are more likely to be the product of self-fertilization.

A dimorphism of dispersers and non-dispersers is also a common phenomenon amongst aphids (winged and wingless progeny). As this differentiation occurs during

dispersal polymorphism in insects

the phase of population growth when reproduction is parthenogenetic, the winged and wingless forms are genetically identical. Whether a mother produces winged or wingless progeny, or a mixture of the two, seems to be determined by the conditions of crowding and the quality of food that she receives (Harrison, 1980). This situation is closely parallel to that in *Gymnarrhena*.

Often the dispersal polymorphism in insect populations is of long-winged and short-winged forms. Amongst the water striders (*Gerris*), short-winged (stay-at-home) forms tend to predominate during phases of population growth, and long-winged (dispersive) forms are developed just before the end of the breeding season. Environmental factors seem to be of prime importance in determining the proportion of long-winged and short-winged individuals in a population. Temperature and photoperiod, the availability and quality of food and the effect of crowding have all been implicated (Harrison, 1980). These are presumably used as environmental cues that provide some information about the favourableness of the current environment and prospects for the future (e.g. if days are short and growing shorter, then winter is approaching). Plant hoppers (e.g. *Prokelisia* spp.) developed winged forms when they were experimentally subjected to crowding. This occurred even when they were crowded by members of a sympatric species, and this suggests that in nature the increasing abundance of one species might force members of the other to emigrate (Denno & Roderick, 1992).

5.4.3 Genetic and behavioural differences in dispersal behaviour within populations

There is increasing evidence that immigrant and emigrant individuals are not just random samples from the population as a whole, but are genetically or phenotypically different from each other and from the rest of the population. Sometimes, it is easy to show that those individuals that disperse from a population are a specialized subset, for example mainly members of one sex, or the winged forms in a dimorphic species. The Japanese geneticist Sakai (1958) studied dispersal and migration in *Drosophila melanogaster* by establishing populations in culture tubes which were connected to vacant tubes. He then counted the numbers that entered the vacant tubes—the migrants. There were six populations: one with a long history of laboratory culture, and the other five obtained from field samples captured on small islands not far from each other. Random migration (i.e. dispersal) continued at the same rate irrespective of the density of the population in the founder tube. It resulted from the random movements of individual flies and was more common amongst the wild flies than in those with a history of life in culture. There was also mass migration which was most pronounced at higher densities, and in this the strains differed. The laboratory strain did not make mass migrations until the density in the founder tubes exceeded 150, whereas in wild strains the critical density that stimulated migration was much lower: 40, 60 or 80 depending on the strain (Figure 5.11). Thus, there appeared to be genetic differences amongst strains suggesting that a tendency to migrate had been lost after a long period of culture in the laboratory.

There are many hypotheses that attempt to explain dispersal in small rodents. For example, the social subordination hypothesis (Christian, 1970) proposes that, as population density increases, the resulting shortage of resources leads to increased aggression which forces social subordinates to disperse. But, this only raises the question of what determines social subordination. Often, it appears to be the young

191   DISPERSAL, DISPERSION AND MIGRATION

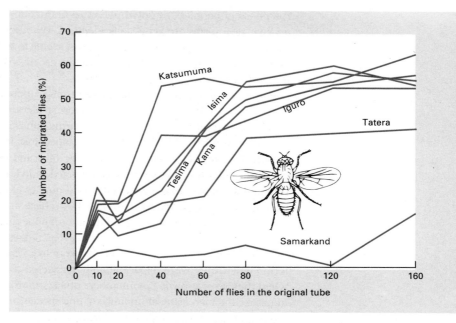

Figure 5.11 The percentage of flies (*Drosophila melanogaster*) that migrated from populations in culture tubes after they were connected to new, unoccupied tubes. The figure shows the effect of the density of the population on the number of flies that migrate. The names refer to islands off the coast of Japan from which the strains of *Drosophila* had been collected. Samarkand is a long-established laboratory strain. (After Sakai, 1958.)

that are forced to emigrate. In some species the dispersers are predominantly males and in others females (Johnson & Gaines, 1990), although this seems to be related to the nature of the mating system and whether males compete mainly for food or for mates (in monogamous mammals natal dispersal is generally similar between the sexes; Dobson, 1982). One view holds that individuals are either innately aggressive, *or* have an innately high reproductive output and that at high densities it is the individuals with high reproductive potential that become socially subordinate and disperse (Krebs, 1978). Another view is that the social subordinates are individuals that have not formed social ties—the asocial rather than the oppressed (Bekoff, 1977).

are dispersers genetically different?

A comparison of dispersing and non-dispersing voles (*Microtus pennsylvanicus*) in southern Indiana showed that they differed in at least two loci, and one rare homozygote was found only amongst dispersers (Myers & Krebs, 1971). Other studies have obtained similar results with different species of vole, and indicate that the dispersers are not a random selection from the whole population. They do not imply that the genetic differences have anything to do with the inheritance of dispersibility, although there is evidence from other species that a tendency to disperse may indeed be heritable (Waser & Jones, 1989).

The nature of the forces that determine dispersal in small mammals remains an exciting area of ignorance. It has been extremely difficult to identify dispersing individuals, yet we need to know the genealogical relationships between individuals, to have correct identification of individuals and to know their fate. Now that techniques of genetic fingerprinting can be applied to individuals in nature, some of the problems may disappear (Ribble, 1992). However, deep questions about proximal motivation remain: what stimuli act to make an individual decide whether to leave home or to stay there? There is a wonderful literature of scientific imagination, controversy and theory to explore (Lidicker & Patton, 1987; Anderson, 1989; Johnson & Gaines, 1990). The role of dispersal in the determination of population size in rodents is considered in Chapter 15, Section 15.4.4.

5.4.4 Dispersal and outbreeding

If organisms and their progeny remain aggregated together, then there are many opportunities for inbreeding between sibs and half-sibs, progeny and other relatives. One consequence of dispersal (although not necessarily of migration) is that it favours the mixing of the progeny from different parents and hence favours outbreeding. Whether and when outbreeding results in improved fitness is another question!

It is particularly amongst populations that normally outbreed that matings between close relatives often result in reduced viability and fertility in the offspring. This point has been beautifully illustrated by Darlington (1960), who examined family histories (e.g. of the Bach and Darwin families), comparing branches where there had and had not been a history of inbreeding. In general, families with a history of inbreeding suffer from outbreeding and outbreeders suffer from inbreeding. Genealogies are also known in some other long-studied animals, especially birds, because the progeny of ring-marked parents can themselves be marked in the nest.

inbreeding depression in great tits?

For instance, in a long-term study of the great tit in Wytham Woods near Oxford, England, there were 885 pairings between 1964 and 1975, in which the identities of both male and female were known. Pairs were formed both from birds born in the wood (residents) and from those born outside (immigrants). These 885 pairings consisted of 194 matings between resident males and resident females, 239 between resident males and immigrant females, 158 between immigrant males and resident females and 294 between immigrant males and immigrant females. There was no evidence that inbreeding pairs produced smaller clutches of eggs than outbreeding pairs, but nestling mortality was significantly higher amongst inbreeding (27.7%) than outbreeding (16.2%) pairs (Greenwood *et al.*, 1978). Females dispersed further than males in their first year (Figure 5.12), although subsequently the birds tended to stay in their territories. Significantly, many of the cases of inbreeding were associated with less dispersal than might be expected on average. The evidence from this and many other studies suggests that dispersal

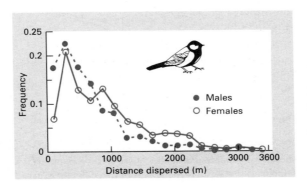

Figure 5.12 The distance dispersed by outbreeding individuals of the great tit (*Parus major*) from the site of birth to the site of first breeding. (After Greenwood *et al.*, 1978.)

contributes significantly to the extent of outbreeding, and that outbreeding produces fitter progeny (Greenwood, 1983).

If dispersal is poor, close neighbours are likely to be related and are also likely to have experienced much the same local selective forces. Outbreeding brings the risk that locally specialized gene combinations may be broken-up. Inbreeding might be expected to hold together and congeal locally selected gene combinations. However, inbreeding commonly produces 'inbreeding depression' (a reduction in offspring vitality), especially in populations that have a history of some outbreeding. There are then two forces that act against each other, and it might be expected that the healthiest offspring result from matings between individuals that are neither too similar in genetic make-up nor too dissimilar (Bateson, 1978, 1980; Price & Waser, 1979). An experiment was made to test this hypothesis by hand-pollinating flowers of *Delphinium nelsoni* at various distances from each other (Price & Waser, 1979). There was a clear peak in the number of seeds set per flower at an outcrossing distance of 10 m, compared with both the results of self-fertilization and of matings between close neighbours, and with the results of matings with very distant neighbours, 100 and 1000 m apart, respectively (Figure 5.13).

Thus, the studies of great tits and of delphiniums are both suggestive of ways in which natural selection might act to influence the distance that individuals disperse. Has natural selection adjusted the degree of dispersal in a population to optimize the degree of outbreeding, or has the degree of outbreeding been selected to a level appropriate for a particular pattern of dispersal?

Gametes and spores often move further than the individuals that produce them. This means that genes, and particular gene combinations of higher plants and some marine invertebrates may colonize populations that the adults would never meet. The consequence may be that precise local specialization of populations is hindered by genetic contamination from elsewhere—but, also that the evolution of local populations may be able to draw on genetic variants that were not present within the population itself. The enormous variety of specialized mechanisms that ensure the dispersal of pollen is some measure of the importance that such '*haploid*' dispersal at this scale has had in the evolution of the lives of higher plants (Levin & Kerster, 1974).

optimal outbreeding in delphiniums?

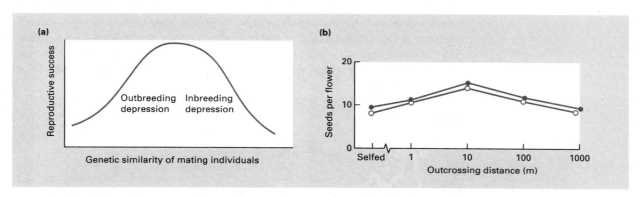

Figure 5.13 (a) A generalized diagram to show the relationship between the relatedness of mating individuals and the success of the mating (vigour or number of progeny). (b) The effect of crossing plants of *Delphinium nelsoni* by deliberate hand-pollination of individuals chosen from different distances apart in the community. (After Price & Waser, 1979.) The symbols ○ and ● show the results of two different experiments.

5.5 Dormancy: dispersal in time

An organism gains in fitness by dispersing its progeny, as long as the progeny are more likely to leave descendants than if they remained undispersed. Similarly, an organism gains in fitness by delaying its arrival on the scene, so long as the delay increases its chances of leaving descendants. This will often be the case when conditions in the future are likely to be better than those in the present. Thus, a delay in the recruitment of an individual to a population may be regarded as *migration in time*, and as an alternative to *migration in space*, and there are trade-offs between the two types of dispersal (Cohen & Levin, 1987).

Organisms generally spend their period of delay in a state of *dormancy*. This relatively inactive state has the benefit of conserving energy, which can then be used during the period following the delay. In addition, the dormant phase of an organism is often more tolerant of the adverse environmental conditions prevailing during the delay (i.e. tolerant of drought, extremes of temperature, lack of light, and so on). Dormancy can be either *predictive* or *consequential* (Müller, 1970). Predictive dormancy is initiated in advance of the adverse conditions, and is most often found in predictable, seasonal environments. It is generally referred to as 'diapause' in animals, and in plants as 'innate' or 'primary' dormancy (Harper, 1977). Consequential (or 'secondary') dormancy, on the other hand, is initiated in response to the adverse conditions themselves. As we shall see, it may be either 'enforced' or 'induced' (Harper, 1977). Note, though, that all of these distinctions are somewhat blurred at the edges.

predictive and consequential dormancy

5.5.1 Diapause: predictive dormancy in animals

Diapause has been most intensively studied in insects, where examples occur in all developmental stages. The common field grasshopper *Chorthippus brunneus* is a fairly typical example. This annual species passes through an *obligatory* diapause in its egg stage, where, in a state of arrested development, it is resistant to the cold winter conditions that would quickly kill the nymphs and adults. In fact, the eggs require a long cold period before development can start again (around 5 weeks at 0°C, or rather longer at a slightly higher temperature; Richards & Waloff, 1954). This ensures that the eggs are not affected by a short, freak period of warm winter weather which might then be followed by normal, dangerous, cold conditions. It also means that there is an enhanced synchronization of subsequent development in the population as a whole. The grasshoppers 'migrate in time' from late summer to the following spring.

obligatory diapause

Diapause is also common in species with more than one generation per year. For instance, the fruit-fly *Drosophila obscura*, passes through four generations per year in England, but enters diapause during only one of them (Begon, 1976). This *facultative* diapause shares important features with obligatory diapause; it enhances survivorship during a predictably adverse winter period, and it is experienced by *resistant* diapause adults with arrested gonadal development and large reserves of stored abdominal fat. In this case, synchronization is achieved not only during diapause but also prior to it. Emerging adults react to the short day lengths of autumn by laying down fat and entering the diapause state; they recommence development in response to the longer days of spring. Thus, by relying, like many species, on the utterly predictable *photoperiod* as a cue for seasonal development, *D. obscura* enters a

facultative diapause

the importance of photoperiod

state of predictive diapause which is confined to those generations that inevitably pass through the adverse conditions.

5.5.2 Seed dormancy in plants

Seed dormancy is an extremely widespread phenomenon in flowering plants. The young embryo ceases development whilst still attached to the mother plant, and enters a phase of suspended activity, usually losing much of its water and becoming dormant in a desiccated condition. In a few species of higher plants, such as some mangroves, a dormant period is absent, but this is very much the exception—almost all seeds are dormant when they are shed from the parent and require special stimuli to return them to an active state (germination).

We can recognize three types of dormancy.

innate dormancy

1 *Innate dormancy* is a state in which the embryo or the maternal tissues that enclose it have an absolute requirement for some special external stimulus to reactivate the process of growth and development. The stimulus may be the presence of water, low temperature, light, photoperiod or an appropriate balance of near- and far-red radiation. Such dormancy breaking requirements tend to synchronize germination with some phase of the seasons. Seedlings of species with this sort of dormancy tend to appear in sudden flushes of almost simultaneous germination.

enforced dormancy

2 *Enforced dormancy* is a state of the seed that has been imposed on it by external conditions (i.e. it is consequential dormancy). The absence of normal requirements for growth (water, an appropriate temperature, a supply of oxygen), or the presence of inhibiting elements (such as high concentrations of carbon dioxide), may hold a seed in a dormant state. It may, of course, then die, but in many groups of higher plants, seeds can be held in the dormant state for very long periods. Seeds of *Chenopodium album* collected from archaeological excavations have been shown to be viable when 1700 years old (Ødum, 1965). Seeds that are held in enforced dormancy may be caused to germinate simply by supplying the missing resource or conditions. Many of the seeds in a seedsman's packet are held in enforced dormancy only by the lack of water, and they germinate readily as soon as water is supplied. In nature, enforced dormancy may be broken by rain falling after a period of drought, or by exposure to the more favourable atmosphere near the soil surface (perhaps as a result of the activities of earthworms or other burrowing animals). The progeny of a single plant with enforced dormancy may be dispersed in time over years, decades or even centuries.

induced dormancy

3 *Induced dormancy* is a state produced in a seed during a period of enforced dormancy in which it acquires some new requirement before it can germinate. The seeds of many agricultural and horticultural weeds will germinate without a light stimulus when they are released from the parent; but after a period of enforced dormancy they require exposure to light before they will germinate. For a long time it was a puzzle that soil samples taken from the field to the laboratory would quickly generate huge crops of seedlings, although these same seeds had failed to germinate in the field. It was a simple idea of genius that prompted Wesson and Wareing (1969) to collect soil samples from the field at night and bring them to the laboratory in darkness. They obtained large crops of seedlings from the soil only when the samples were exposed to light. They also dug pits in the field at night, and covered some with slate to exclude light whilst they covered others with glass. Seedlings appeared in the pit only under the glass. This type of induced dormancy is responsible for the accumulation of large populations of seeds in the soil. In nature they germinate only when they are brought to the soil surface by earthworms, or by the exposure of soil after a tree falls, or when the soil is disturbed by burrowing animals such as moles or gophers or by human tillage.

seed dormancy and the light beneath a canopy

Seed dormancy may be induced by radiation that contains a relatively high ratio of far-red (730 nm) to near-red (approximately 660 nm) wavelengths. Light that has filtered through a leafy canopy has its spectral composition biased in this way, and the seeds of sensitive species can be brought into a state of dormancy by exposing them to light that has filtered through leaves. In the field, this must have the effect of holding sensitive seeds in the dormant state when they land on the ground under a canopy, whilst releasing them into germination only when the overtopping plants have died away. It may also disperse a population of seeds in time, ensuring that individual seeds germinate only when they cease to be overtopped by vegetation.

Seedlings that germinate early, if they survive, usually have an expectation of a longer growth period and a higher fecundity than those that germinate late. Every delay in the life cycle of an organism is a reduction in its potential fecundity and so of its fitness. On the other hand, delayed germination may mean that individuals escape lethal hazards in the environment. There is obviously a compromise involved in dispersal in time—a compromise between the risk of death by starting growth too early and a loss of fecundity from starting growth too late (Venable & Brown, 1988).

dormancy as a compromise

Effective dispersal in time depends not only on seeds being dormant for a period, but also on them being long lived. Such longevity appears to be under genetic control because it has been lost from most long-domesticated cultivars together with much of their innate dormancy. Most of the species of plant with seeds that persist for long in the soil are annuals and biennials, and they are mainly weedy species (which is cause and which is effect?). They are largely species that lack features that will disperse them extensively in space. The seeds of trees usually have a very short expectation of life in the soil, and many are extremely difficult to store artificially for more than 1 year. The seeds of many tropical trees are particularly short lived, a matter of weeks or even days. Amongst trees, the most striking longevity is seen in those that retain the seeds in cones or pods on the tree until they are released after fire (many species of *Eucalyptus* and *Pinus*). This phenomenon of *serotiny* protects the seeds against risks on the ground until fire creates an environment suitable for their rapid establishment.

the distribution of seed dormancy amongst plant types

5.5.3 Non-seed dormancy in plants

Dormancy in plants is not confined to seeds. For example, the sand sedge *Carex arenaria*, as it grows tends to accumulate dormant buds along the length of its predominantly linear rhizome. These may remain alive but dormant long after the shoots with which they were produced have died, and they have been found in numbers of up to 400–500 m^{-2} (Noble *et al.*, 1979). They play a role analogous to the bank of dormant seeds produced by other species. In a similar vein, Prescott chervil (*Chaerophyllum prescottii*), found in meadows in the USSR, has no detectable reserve of dormant seed, but can remain as dormant underground tubers in the soil for over 10 years. When meadows are ploughed, or areas of grass are killed, hundreds of dormant chervil tubers are activated into growth (Rabotnov, 1964).

Finally, the very widespread habit of deciduousness is a form of innate dormancy displayed by many perennial trees and shrubs. Established individuals pass through periods, usually of low temperatures and low light levels, in a leafless state of low metabolic activity.

5.5.4 Consequential dormancy in animals

Consequential dormancy in animals, like consequential (enforced and induced) dormancy in plants, may be expected to evolve in environments that are relatively unpredictable. In such circumstances, there will be a disadvantage in responding to adverse conditions only after they have appeared. But, this may very well be outweighed by the advantages: (i) of responding to favourable conditions *immediately* they reappear; and (ii) of entering a dormant state only if adverse conditions *do* appear. Thus, when many mammals enter hibernation, they do so (after an obligatory preparatory phase) in direct response to the adverse conditions. Having achieved 'resistance' by virtue of the energy they conserve at a lowered body temperature, and having periodically emerged and monitored their environment, they eventually cease hibernation whenever the adversity disappears. A particularly remarkable example of consequential dormancy in mammals is provided by those marsupials that are able to hold back the development of a foetus for many months during periods when the mother is short of resources. Such a delaying option is unavailable to placental mammals.

A consequential dormancy similar to that shown by plant seeds is apparent in the lives of many nematode worms, particularly parasites, where dormant cysts may remain in suspended animation for many years, breaking dormancy only after some special stimulus is received which signifies that the environment has become favourable for development (Sunderland, 1960). The signal may be that the cyst has been ingested by a host or, in the case of nematodes parasitic on higher plants, it may be that the growing host has released germination stimulants into the environment of the cyst. Protozoa too may encyst in a dormant state that is broken only by the return of specialized conditions such as the refilling of a dried-up pond.

5.6 Clonal dispersal

By definition, modular growth involves the repetition by a single genetic individual of units of its structure (see Section 4.2.1). Thus, in almost all clonal organisms, an individual genet branches and spreads its parts around it as it grows. There is a sense in which a developing tree or coral disperses its parts into, and therefore explores, the surrounding environment. When much of the modular growth is horizontal as a stolon or rhizome (in many higher plants), a laterally spreading clone is formed. Often, the interconnections of such a clone decay so that it becomes represented by a number of dispersed parts. In extreme cases, such clonal spread may result ultimately in the product of one zygote being represented by a clone of great age spread over great distances. The Finnish botanist Oinonen (1967) estimated that some clones of the rhizomatous bracken fern (*Pteridium aquilinum*) were more than 1400 years old and one extended over an area of 474×292 m (nearly 14 ha).

A few trees are also capable of clonal spread from root suckers, and the aspen (*Populus tremuloides*) may cover even greater areas than bracken from the product of a single original seed. Organisms that form clones in this way achieve a form that reflects the disposition of their modules. Amongst the forms that extend by rhizomes or stolons, we can recognize two extremes in a continuum. At one extreme the connections between modules are long (often also thin and short-lived), so that the shoots of a clone are widely spaced. These have been called 'guerilla' forms because they give the plant, hydroid or coral a character like that of the military guerilla force—constantly on the move, disappearing from some territories and penetrating into others. They are both fugitive and opportunist. At the other end of the continuum are 'phalanx' forms in which the connections between modules are short (often thick and long lasting) and the modules are tightly packed (Lovett Doust & Lovett Doust, 1982). The tussock grasses of arid regions are fine examples of phalanx growth forms. Plants and modular animals with phalanx growth forms expand their clones slowly, retain their original site occupancy for long periods, and neither penetrate readily amongst neighbouring plants nor are easily penetrated by them (Sackville Hamilton *et al.*, 1987). The term phalanx is used to describe this growth form by analogy with the units of a Roman army, tightly packed with their shields held around the group.

In military strategy it is well known that to organize an army of guerillas is no way to fight a phalanx, and to arrange an army in phalanges is no way to fight guerillas. There may well be ecological parallels in the outcome of competitive interactions between phalanx and guerilla species—neither may be wholly successful in eliminating the other from a mixed community.

Even amongst non-clonal trees, it is easy to see that the way in which the buds are placed gives them a guerrilla *or* a phalanx type of growth form. The dense packing of shoot modules in species like cypresses (*Cupressus*) produces a relatively undispersed and impenetrable foliage (phalanx canopy), whilst many loose-structured, broad-leaved trees (*Acacia*, *Betula*) can be seen as guerilla canopies, bearing buds that are widely dispersed and shoots that interweave with the buds and branches of neighbours. The twining or clambering lianes in a forest are guerilla growth forms *par excellence*, dispersing their foliage and buds over immense distances, both vertically and laterally.

The ways in which modular organisms disperse and display their modules affect the ways in which they interact with neighbours. Those with guerilla form will continually meet and interact with (compete with) other species and other genets of their own kind. With a phalanx structure, however, most meetings will be between modules of a single genet. For a tussock grass or a cypress tree, competition must occur very largely between the parts of itself.

Clonal growth is most effective as a means of dispersal in aquatic environments. Many aquatic plants fragment easily, and the parts of a single clone become independently dispersed because they are not dependent on the presence of roots to maintain their water relations. The major aquatic weed problems of the world are caused by plants that multiply as clones and fragment and fall to pieces as they grow (see Chapter 4). Duckweeds (Figure 4.2), the water hyacinth, Canadian pond weed and the water fern *Salvinia* (see Chapter 15, Figure 15.28) are all examples of organisms that disperse their modules, and in time, the product of a single zygote may become dispersed through the watercourses of an entire nation.

5.7 Patterns of migration

5.7.1 Diurnal and tidal movements

Individuals of many species move from one habitat to another and back again repeatedly during their life. The time scale involved may be hours, days, months or years. In some cases, these movements have the effect of maintaining the organism in the same type of environment. This is the case in the movement of crabs on a shoreline: they move with the advance and retreat of the tide. In other cases, diurnal migration may involve moving between two environments, each of which can supply a limiting resource. For example, planktonic algae both in the sea and in freshwater lakes descend to the depths at night but move to the surface during the day. It appears that they accumulate phosphorus and perhaps other nutrients in the deeper water (the hypolimnion) at night before returning to photosynthesize near the surface (the epilimnion) during daylight hours (Salonen *et al.*, 1984).

Birds, bats, slugs, snails and a wide variety of other mobile animals move from one habitat to another regularly within the course of a 24-h cycle of rest and activity. As a rule, they aggregate into tight populations during the resting period and separate from each other when feeding. For example, most snails rest in confined humid microhabitats by day, but range widely when they search for food by night. These are all species with fundamental niches that can only be satisfied by alternating life in two distinct habitats within each day of their lives.

5.7.2 Seasonal movement between habitats

Many motile organisms make seasonal moves between habitats. The patches of the environment in which resources are available change with the seasons, and populations move from one type of patch to another. The altitudinal migration of grazing animals in mountainous regions is one example, where, for instance the American elk and mule deer move up into high mountain areas in the summer and down to the valleys in the winter. (Such annual altitudinal migration has been conspicuous in the way humans have managed livestock in mountainous areas. Cattle, sheep, goats and even pigs were moved to high pastures during the summer—often with the women and younger children as shepherds—whilst the men harvested hay from valley pastures. The livestock returned to the lowlands during the winter to be fed on the stored hay.) By migrating seasonally the animals escape the major changes in food supply and climate that they would meet if they stayed in the same place.

This pattern can be contrasted with the 'migration' of amphibians (frogs, toads, newts) between an aquatic breeding habitat in spring and a terrestrial environment for the remainder of the year. The young develop (as tadpoles) in water with a different food resource from that which they will later eat on land. They will return to the same aquatic habitat for mating, aggregate into dense populations for a time and then separate to lead more isolated lives on land. An individual may make several return journeys in its lifetime (Figure 5.14).

5.7.3 Long-distance migration

The most remarkable habitat shifts are those that involve travelling very long distances. Most terrestrial birds in the Northern hemisphere move north in the spring when food supplies will become abundant during the warm summer period, and move south to savannahs in the autumn when food becomes abundant only after the rainy season. The migrations seem, in virtually every case, to involve transit between areas that supply abundant food, but only for a limited period. They are areas in which seasons of comparative glut and famine alternate, and which cannot

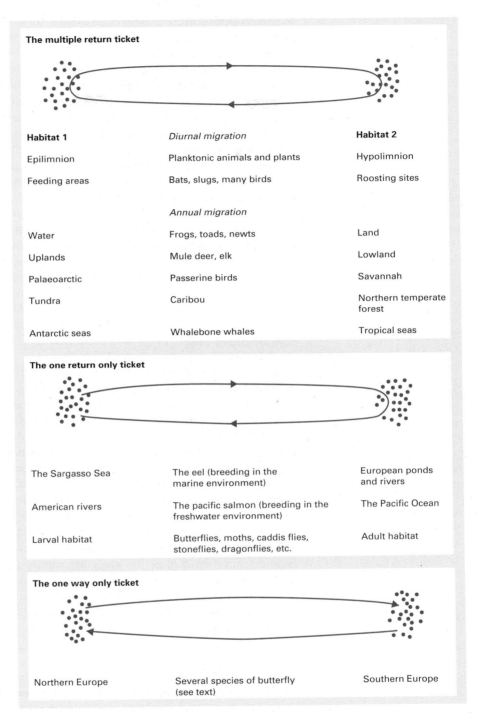

The multiple return ticket

Habitat 1	*Diurnal migration*	Habitat 2
Epilimnion	Planktonic animals and plants	Hypolimnion
Feeding areas	Bats, slugs, many birds	Roosting sites
	Annual migration	
Water	Frogs, toads, newts	Land
Uplands	Mule deer, elk	Lowland
Palaeoarctic	Passerine birds	Savannah
Tundra	Caribou	Northern temperate forest
Antarctic seas	Whalebone whales	Tropical seas

The one return only ticket

The Sargasso Sea	The eel (breeding in the marine environment)	European ponds and rivers
American rivers	The pacific salmon (breeding in the freshwater environment)	The Pacific Ocean
Larval habitat	Butterflies, moths, caddis flies, stoneflies, dragonflies, etc.	Adult habitat

The one way only ticket

Northern Europe	Several species of butterfly (see text)	Southern Europe

Figure 5.14 Migration patterns in relation to life cycles.

the costs and benefits of long-distance migration

support large resident populations all the year round. For example, swallows (*Hirundo rustica*) which migrate seasonally to South Africa, vastly outnumber a related resident species. The all-year-round food supply can support only a small population of residents, but a seasonal glut exceeds what the residents can consume.

201   DISPERSAL, DISPERSION AND MIGRATION

Migrants can then make a large contribution to the diversity of a local fauna. Moreau (1952) calculated that of the 589 species of birds (excluding seabirds) that breed in the Palaeoarctic, 40% spend the winter elsewhere.

Of those species that breed in the Palaeoarctic (temperate Europe and Asia) and leave for the winter, 98% travel south to Africa. They spend the winter in areas of thornbush and savannah (deciduous vegetation), and their arrival usually coincides with the ripening of an immense crop of seeds of the dominant grasses.

The Arctic tern (*Sterna paradisaea*) travels from its Arctic breeding ground to the Antarctic pack-ice and back each year—about 10 000 miles (16 100 km) each way (although unlike many other migrants it can feed on its way). The terns that breed in eastern Canada cross the Atlantic twice on their way south (Kullenberg, 1946).

subspecies may have different migration routes

The same species may behave in different ways in different places. All robins (*Erithacus rubecula*) leave Finland and Sweden in winter, but on the Canary Islands the species is resident the whole year round. In most of the intervening countries, a part of the population migrates and a part remains resident (Lack, 1954). Such variations in the pattern of migration are sometimes clear examples of evolutionary divergence. This is the case in the knots (*Calidris canutus*), a species of small wading bird most of which breed in remote areas of the Arctic tundras and 'winter' in the summers of the Southern hemisphere. At least five subspecies appear to have diverged in the late Pleistocene (genetic evidence from the sequencing of mitochondrial DNA; see Baker, 1992), and these now have strikingly different patterns of distribution and migration (Figure 5.15).

On the whole, the birds and mammals of tropical rainforest or evergreen montane forest do not migrate. In these communities production is less seasonal and all available resources are more likely to be consumed by the residents.

We know most about the migratory habits of birds because marking by leg rings is so easy and there are so many ornithologists. A survey of the science of bird migration is given by Berthold (1994). However, migration is a feature of many other groups too. Baleen whales (Mysticeti) in the Southern hemisphere move south in summer to feed in the food-rich waters of the Antarctic. In winter they move north to breed (but scarcely to feed) in tropical and subtropical waters. Caribou (*Rangifer tarandus*) travel several hundred miles per year from northern forests to the tundra and back; whilst tunny fish (*Thunnus thynnus*) breed in the Mediterranean in April and May, and then migrate to the northern part of the North Sea. In all of these examples, an individual of the migrating species may make the return journey several times. It is therefore at least possible that it learns a route, and uses landmarks, in combination with magnetic information about position and/or information from the sun and stars (see Baker, 1982).

5.7.4 'One-return journey' migration

Many migrant species make only one return journey during their lifetime. They are born in one habitat, make their major growth in another habitat, but then return to breed and die in the home of their infancy. The eel and the migratory salmon are

eels and salmon

classic examples. The European eel (*Anguilla anguilla*) travels from European rivers and ponds across the Atlantic to the Sargasso Sea, where it is thought to reproduce and die (although spawning adults and eggs have never actually been caught there). The American eel (*Anguilla rostrata*) makes a comparable journey from areas ranging between the Guianas in the south, to south-west Greenland in the north.

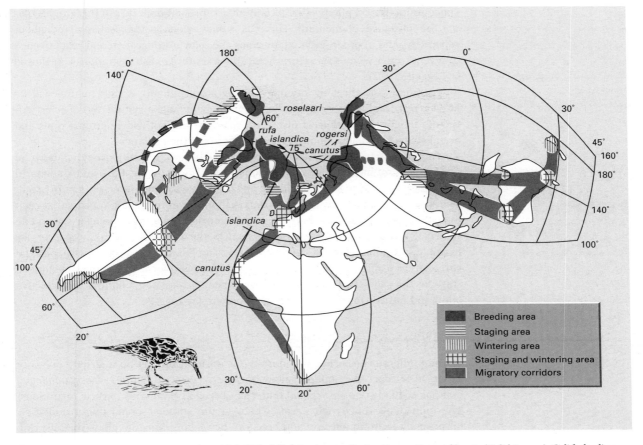

Figure 5.15 Global distribution and migration pattern of knots (*Calidris* spp.). Solid shading indicates the breeding areas; horizontally striped spots indicate the stop-over areas, used only during south- and northward migration; the cross-hatched spots indicate the areas used both as stop-over and wintering sites; and the vertically striped spots designate areas used only for wintering. The grey shaded corridors indicate proven migration routes; the broken-shaded corridors indicate tentative migration routes suggested in the literature. (After Piersma & Davidson, 1992b.)

The salmon makes a comparable transition, but from a freshwater egg and juvenile phase to mature as a marine adult. It then returns to freshwater sites to lay eggs. After spawning, all Pacific salmon (*Oncorhynchus nerka*) die without ever returning to the sea. Many Atlantic salmon (*Salmo salar*) also die after spawning, but some survive to return to the sea and then migrate back upstream to spawn again.

Migratory fish return, often with very high precision, to the home of their infancy and some are known to use information from the sun, polarized light and geomagnetic fields as generalized cues that increase the 'homeward-bound' direction of migration. But, the more sensitive control of the direction of long-distance migrations appears to be determined by olfactory cues and, when closer to home, memory of local topography. There is strong evidence that each river and tributary has a characteristic odour and that young fish imprint to this odour before they migrate to the sea. On the homeward journey they track the changing concentrations of the chemicals responsible for the odour. The eel can apparently detect

203 DISPERSAL, DISPERSION AND MIGRATION

concentrations of β-phenylethyl alcohol that represent only two to three molecules on the olfactory epithelium. There is some evidence that salmon remember separately, and in order, each odour along the seaward migration and react to each in reverse order when they return home (the literature of fish migrations is reviewed in Leggett, 1977).

An important effect of 'homing' in migratory animals is that mating will be largely restricted to geographically localized populations within the species. Although they may travel great distances, such fish are not widely promiscuous and are likely to find mates amongst the neighbours of their infancy.

The monarch butterfly (*Danaus plexippus*) has a 'return-journey' pattern of migration (Urquhart, 1960). Populations travel north in the USA and Canada to breed in summer, and south to Florida and California to overwinter, but not to breed. The same individuals may return to the north in the spring, but it appears that each individual makes no more than one return journey. On a much smaller scale, 'one return ticket' migrations can also be seen in the life cycles of many insects, for example butterflies, moths, caddis-flies and stoneflies. The individual spends the early part of its life in one habitat (the larval food plant for the butterfly, the aquatic stage for the caddis), migrates to another habitat as a sexually mature, reproductive adult and returns to the juvenile habitat only to lay eggs.

5.7.5 'One-way only' migration

In some migratory species, the journey for an individual is on a strictly one-way ticket. In Europe, the clouded yellow (*Colias croceus*), red admiral (*Vanessa atalanta*) and painted lady (*Vanessa cardui*) butterflies breed at both ends of their migrations. The individuals that reach Great Britain in the summer breed there, and their offspring fly south in autumn and breed in the Mediterranean region—the offspring of these in turn come north in the following summer.

Most migrations occur seasonally in the life of individuals or of populations. They usually seem to be triggered by some external seasonal phenomenon (e.g. changing day length), and sometimes also by an internal physiological clock. They are often preceded by quite profound physiological changes such as the accumulation of body fat. They represent strategies evolved in environments where seasonal events like rainfall and temperature cycles are reliably repeated from year to year. There is, however, a type of migration that is tactical, forced by events such as overcrowding, and appears to have no cycle or regularity and these are most common in environments where rainfall is not seasonally reliable. The economically disastrous migration-plagues of locusts in arid and semi-arid regions are the most striking examples.

Locusts, like a number of other insects, change their behaviour when they become crowded during periods of peak population growth. There is a 'phase' change that occurs over three generations, from a *solitaria* phase (when they behave as individuals), through a generation in a *transiens* phase, to a third generation in a *gregaria* phase when they start to band together and behave as members of a herd and they then fly as powerful flying machines, forming immense swarms in the daytime.

A plague of the desert locust *Schistocerca gregaria*, illustrates many of the features of large-scale plague migrations. The species has no built-in mechanism of dormancy or diapause that allows individuals to persist through periods of severe desiccation and so, when eggs are laid, there must be sufficient water to sustain the developmental stages and for the growth of sufficient vegetation for food. In areas of very uncertain rainfall the locusts survive in their arid environment when their flying swarms move from one area where rain has fallen to another—often more than 1000 km away. When there is a sequence of favourable rains, populations of the locust may grow by an order of magnitude per generation and change from the *solitaria* to the *gregaria* phase and form day-flying swarms. The next stage involves the chance that major air movements will transport the flying insects to another site at which rains have recently fallen and where the population can continue to grow.

A plague in 1986–1989 was followed in great detail and the stages in its development and subsidence are shown in Figure 5.16. From the summer of 1985 to that of 1988, there was successful breeding and gregarization in the central and western parts of its invasion area because suitably timed rains fell and there was almost continuous breeding in West and North-West Africa. It was not until the end of 1988 and early 1989 that adverse weather and pesticide control measures forced an end to the plague and a little, new eastern movement started again. During the peak of the migrations the populations swarmed over North Africa, the Sahel, the Sudan, the Near East, South-West Asia and swarms even reached the Cape Verde Islands and crossed the Atlantic to the Caribbean. Massive populations drowned in the Atlantic.

There is a profound contrast between the exquisitely timed seasonally predictable migrations of migratory birds, salmon, eels, etc. and the largely opportunistic plague migrations of locusts. Neither the time, place nor the direction of locust migrations appear to be seasonally programmed. This means that control programmes have to be pragmatic and respond to the same unpredictable features of the environment that affect the locusts. Immense resources are employed internationally through the Food and Agriculture Organization of the United Nations to detect where new migrating locust swarms are developing and to track the changes in wind and rainfall that determine where swarms will be carried. The successful control of such migratory pests is further bedevilled by two very different problems: (i) the main chemical control agents used in dusts, sprays and baits are regarded as environmentally undesirable and are often banned (especially dieldrin); and (ii) civil disturbance and wars prevent monitoring and control in areas that may be crucial for the early development of swarming populations (e.g. Eritrea, Chad, Mauritania and the western Sahara).

Accounts of recent migrations of other swarming pests (e.g. army-worm (*Spodoptera exempta*), Senegalese grasshopper (*Oedaleus senegalensis*), spruce budworm (*Choristoneura fumiferana*)), and details of monitoring and control measures can be found in Rainey *et al.* (1990).

5.8 A forward look

Phases of dispersal or migration are crucial in the life histories of all organisms. Every organism is doomed to extinction if it, or its descendants, stay in the same place. But, the extent of dispersal of individuals and populations in nature is usually the most difficult part of the life history to measure. Of all the propagules produced by a higher plant (e.g. a tree), it may sometimes be the odd, exceptional, furthest-flung seed that produces the 'inoculum' from which new colonies and populations develop. When this is the case, it is not the average distance of dispersal that matters, but the remote

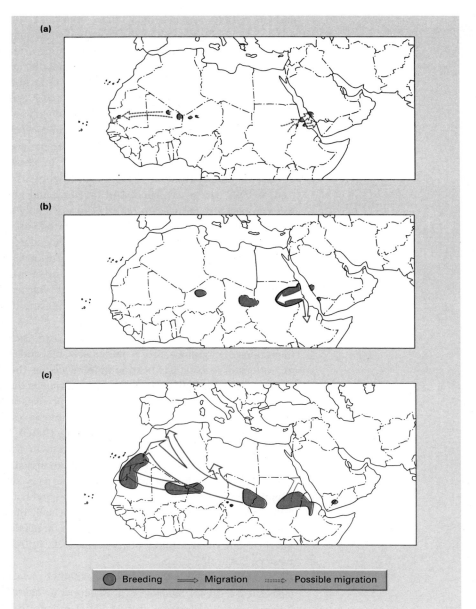

Figure 5.16 The development, migration and subsidence of an outbreak of the desert locust (*Schistocerca gregaria*) in Africa and Arabia from 1985 to 1989 after a 20-year period (1965–1985) when the populations were in recession. (After Skaf *et al.*, 1990.) (a) 1985. (b) 1986–1987 winter–spring breeding. (c) 1987 summer breeding. (d) (*facing page*) 1987–1988 winter–spring breeding. (e) Movement of swarms of the first summer generation (June–September 1988). (f) Movement of swarms of the second summer generation (October 1988–May 1989).

probabilities of extreme distances of dispersal. These are, by their nature, the most difficult to measure.

Much of what we know about dispersal relates to the spread of progeny from isolated individuals—because we can then recognize which were the parents. It may often be much more important to discover how far dispersal occurs within and between populations, for instance, of seeds within a forest, or of migrant voles within populations of voles. Sometimes, it may be more important to know about the dispersal of genes rather than of individuals—especially because the pollen of higher

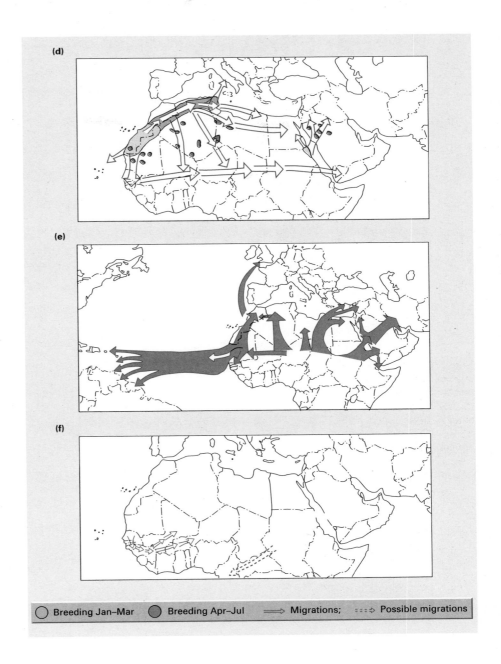

(d)

(e)

(f)

○ Breeding Jan–Mar ● Breeding Apr–Jul ⟹ Migrations; ⋯⋯⇨ Possible migrations

plants and the gametes of some aquatic invertebrates are dispersed much further than their parents. We will not be able to make much sense of this sort of dispersal until we can identify which offspring derive from which parents. Techniques of genetic engineering may allow neutral marker genes to be introduced into populations and so reveal the detailed patterns of dispersal and migration that occur within populations. It may well be that we need to understand the subtle aspects of the dispersal of individuals and their genes before we can really understand the dynamics of populations.

207   DISPERSAL, DISPERSION AND MIGRATION

PART 2
INTERACTIONS

Introduction

The activity of any organism changes the environment in which it lives. It may alter conditions—as when the movement of an earthworm through soil aerates the soil, or when the transpiration of a tree cools the atmosphere. Or it may add or substract resources from the environment that might have been available to other organisms—as when a tree shades what is beneath it, or a cow eats grass. In addition, though, organisms will actually interact when individuals enter into the lives of others. In the following chapters (6–13) we consider the variety of these interactions. We distinguish five main categories: competition, predation, parasitism, mutualism and detritivory. Like most biological categories, these five are not perfect pigeon-holes.

In very broad terms, 'competition' is an interaction in which one organism consumes a resource that would have been available to, and might have been consumed by, another. One organism deprives another, and, as a consequence, the other organism grows more slowly, leaves fewer progeny or is at greater risk of death. The act of deprivation can occur between two members of the same species or between individuals of different species. We choose to consider these situations in separate chapters (Chapters 6 and 7), partly because it seems a reasonable assumption that members of the same species are particularly likely to make the same resource demands on the environment (and to react symmetrically to the presence of each other) while members of different species are likely to differ (and so react asymmetrically to each other). But we need to be very careful not to take this distinction too literally—the very process of natural selection depends on differences between members of the same species, and conspecific individuals can differ, too, in their condition, their stage of development and so on. Moreover, members of two very different species may use and deplete the same resources: a rabbit in a field may be more effectively deprived of food by a sheep than by another rabbit.

Chapters 8, 9 and 10 deal with various aspects of 'predation', though we have defined predation very broadly. We have combined those situations in which one organism eats another and kills it (such as an owl preying on mice), and those in which the consumer takes only part of its prey, which may then regrow to provide another bite another day (grazing). We have also combined herbivory (animals eating plants) and carnivory (animals eating animals). In Chapter 8 we examine the nature of predation, i.e. what happens to the predator and what happens to the prey (be it animal or plant). Particular attention is paid to herbivory because of the subtleties that characterize the response of a plant to attack. In Chapter 9 we discuss the behaviour of predators—an area in which population ecology and behavioural ecology merge. The way in which individual predators behave clearly influences the dynamics of predator and prey populations, and these 'consequences of consumption' form the material of Chapter 10. This is the part of ecology that has the most obvious relevance to those concerned with the management of natural resources: the efficiency of harvesting (whether of fish, whales, grasslands or prairies) and the biological and chemical control of pests and weeds—themes that we take up later.

Four of the processes in this section—competition, predation, parasitism and mutualism—involve interactions between organisms that generate processes of natural selection. The activities of predators may be expected to influence the evolution of their prey and vice versa; and the activities of parasites may be expected to influence the evolution of their hosts and vice versa. If the process of evolution is largely the consequence of interactions between organisms (as Darwin supposed), it is among predators and prey, parasites and hosts, pairs of mutualists and between competitors that we might expect to find ecological and evolutionary processes most closely intertwined in action. By contrast, dead material leaves no descendants and cannot itself evolve, nor can it multiply. However, it is involved in complex interactions, and the processes by which it is decomposed involve competition, parasitism, predation and mutualism: microcosms of all the major ecological processes (except photosynthesis). Detritivory, the consumption of dead organic matter, is discussed in Chapter 11.

Chapter 12, Parasitism and Disease, deals with a subject that has rarely been represented by a full chapter in ecology texts. Yet more than half of the species on the face of the earth are parasites. The category of parasitism has blurred edges, particularly where it merges into predation. But whereas a predator usually takes all or part of many individual prey, a parasite normally takes its resources from one or a very few hosts, and (like many grazing predators) it rarely kills its hosts immediately if at all. The insect parasitoids (like ichneumon wasps which develop (often singly) within the bodies of their hosts and kill them) fit neatly into neither the predators nor the parasites. We have discussed them, along with predators and grazers, in Chapters 8–10.

Finally, Chapter 13 is concerned with mutualistic interactions between individuals of different species. These are interactions in which both organisms experience a net benefit. The ecology of mutualism has been neglected and, like parasitism, it rarely claims a chapter in an ecology textbook. Yet the greater part of the world's biomass is composed of mutualists.

Interactions between organisms have often been summarized by a simple code which represents each one of the pair of interacting organisms by a ' + ', '–' or '0', depending on how it is affected by the interaction. Thus, a predator–prey (including a herbivore–plant) interaction, in which the predator benefits and the prey is harmed, is denoted by + –; and the parasite–host interaction is also clearly + –. Another straightforward case is mutualism, which is obviously + +; whereas if organisms do not interact at all, we can denote this by 00 (sometimes called 'neutralism'). Detritivory must be denoted by + 0, since the detritivore itself benefits, while its food (dead already) is unaffected. The general term applied to + 0 interactions is 'commensalism', but paradoxically this term is not usually used for detritivores. Instead, it is reserved for cases, allied to parasitism, in which one organism ('the host') provides resources or a home for another organism, but in which the host itself suffers no tangible ill-effects. Competition is often described as a – – interaction, but, as we shall see in Chapters 6 and 7, it is often impossible to establish that both organisms are harmed. Such asymmetric interactions may then approximate to a –0 description, generally referred to as 'amensalism'. True cases of amensalism may occur when one organism produces its ill-effect (for instance a toxin) whether or not the potentially affected organism is present.

The eight chapters of this section are rather different in style. This reflects some very real differences in the various 'states of the art'. Studies of parasitism and mutualism have often concentrated on disentangling the intricacies of the relationships between the organisms concerned, rather than dealing with the role of the phenomena in the larger communities in which the organisms live. Thus we know a great deal about the ecology of rhizobial nitrogen-fixing bacteria in the root nodules of legume plants, but we know much less about the dynamics of nodulated legumes in natural vegetation. Likewise, we know a great deal about the ecology of parasitic worms in intestines but much less about the role of infection in determining the abundance of natural populations. With interspecific competition, on the other hand, the problem is slightly different. We have a very sophisticated science concerned with the experimental and theoretical study of competition, but we are still fearfully ignorant about just how often it occurs and how important it is as a force in natural communities. As another point of contrast, there has been a long history of mathematical modelling of predator–prey interactions, whereas it is only recently that modellers have turned their attention to mutualism. For reasons such as these, and also because of the special interests of the most influential investigators, ecological study proceeds in fits and starts. There are parts that have powerful theory but lack a good underpinning of large and reliable data sets, and there are other parts that rely heavily on natural history but have tended to lack any real theory to give them proper coherence. Ecology is a science that is full of gaps and areas of ignorance—much like any other science.

6.1 Introduction: the nature and features of intraspecific competition

Organisms grow, reproduce, die and migrate (see Chapters 4 and 5). They are affected by the conditions in which they live (see Chapter 2), and by the resources that they obtain (see Chapter 3). Yet, no organism lives in isolation. Each, for at least part of its life, is a member of a population composed of individuals of its own species.

Individuals of the same species have very similar requirements for survival, growth and reproduction; but their combined demand for a resource may exceed the immediate supply. The individuals then compete for the resource and, not surprisingly, at least some of them become deprived. This chapter is concerned with the nature of such intraspecific competition, its effects on the competing individuals and on populations of competing individuals. We begin, however, with a working definition: 'competition is an interaction between individuals, brought about by a shared requirement for a resource in limited supply, and leading to a reduction in the survivorship, growth and/or reproduction of at least some of the competing individuals concerned.' We can now look more closely at competition.

Consider, initially, a simple hypothetical community: a thriving population of grasshoppers (all of one species) feeding on a field of grass (also of one species). In order to provide themselves with energy and material for growth and reproduction, grasshoppers must eat grass; but, in order to find and consume that grass, they must use energy. Any grasshopper might find itself at a spot where there is no grass because some other grasshopper has eaten it. The grasshopper must then move on and expend more energy before it takes in food. The more grasshoppers there are competing for food, the more often this will happen. Yet, an increased energy expenditure and a decreased rate of food intake may all decrease a grasshopper's chances of survival, and also leave less energy available for development and reproduction. Survival and reproduction determine a grasshopper's contribution to the next generation. Hence, the more intraspecific competitors for food a grasshopper has, the less its likely contribution will be.

As far as the grass itself is concerned, an isolated seedling in fertile soil may have a very high chance of surviving to reproductive maturity. It will probably exhibit an extensive amount of modular growth, and will probably therefore eventually produce a large number of seeds. However, a seedling which is closely surrounded by neighbours (shading it with their leaves and depleting the water and nutrients of its soil with their roots) will be very unlikely to survive, and if it does, will almost certainly form few modules and set few seeds.

We can see immediately that the ultimate effect of competition on an individual is a decreased contribution to the next generation compared with what would have happened had there been no competitors. Intraspecific competition typically leads to decreased rates of resource intake per individual, and thus to decreased rates of individual growth or development, or perhaps to decreases in amounts of stored reserves or to increased risks of predation. These may lead, in turn, to decreases in survivorship and/or decreases in fecundity. As we saw in Chapter 4, survivorship and fecundity together determine an individual's reproductive output.

the ultimate effect is on fecundity and survivorship

In many cases, competing individuals do not interact with one another directly. Instead, individuals respond to the level of a resource, which has been depressed by the presence and activity of other individuals. Thus, grasshoppers competing for food are not directly affected by other grasshoppers, but by a reduction in food level and an increased difficulty in finding good food. Similarly, a competing grass plant is adversely affected by the presence of close neighbours because the zone from which it extracts resources (light, water, nutrients) has been overlapped by the 'resource depletion zones' of these neighbours, making it more difficult for the original plant to extract its resources. In all these cases, competition may be described as *exploitation*, in that each individual is affected by the amount of resource that remains after it has been exploited by others. Exploitation can only occur, therefore, if the resource in question is in limited supply.

exploitation and interference

In many other cases, however, competition takes a form known as *interference*. Here, individuals interact directly with each other, and one individual will actually prevent another from exploiting the resources within a portion of the habitat. For instance, this is seen amongst animals that defend territories (discussed in more detail in Section 6.12), and also amongst the sessile animals and plants that live on rocky shores and amongst terrestrial plants. For example, the presence of a barnacle on a rock prevents any other barnacle from occupying that same position, even though the supply of food at that position may exceed the requirements of several barnacles. In such cases, space can be seen as a resource in limited supply. Another type of interference competition occurs when, for instance, two red deer stags fight for access to a harem of hinds. In this case, the hinds are a resource in limited supply, because although either stag, alone, could readily mate with all the hinds, they cannot both do so since matings are limited to the 'owner' of the harem.

Thus, interference competition may occur for a resource of real value (e.g. space on a rocky shore for a barnacle), in which case the interference is accompanied by a degree of exploitation, or for a surrogate resource (a territory, or ownership of a harem), which is only valuable because of the access it provides to a real resource (food within a territory, the right to mate with all females). Interference for a surrogate can replace, in a sense, exploitative competition for a real resource. Hence, whereas with exploitation the intensity of competition is closely linked to the level of resource present and the level required, with interference, intensity may be high even when the level of the real resource is not limiting.

In practice, many examples of competition probably include elements of both exploitation and interference. For instance, our hypothetical grass plant, apart from suffering from the resource depletion of its competitors, would be physically excluded by them from the sites they already occupy. To take a more specific example, adult cave beetles, *Neapheanops tellkampfi*, in Great Onyx Cave, Kentucky, compete amongst themselves but with no other species and have only one type of food—cricket eggs, which they obtain by digging holes in the sandy floor of the cave.

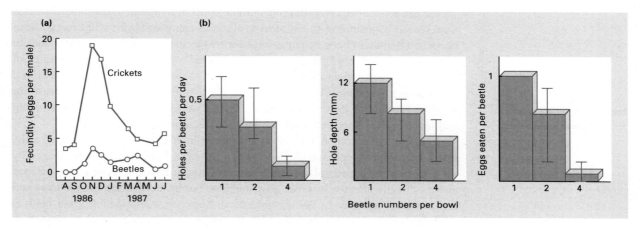

Figure 6.1 Intraspecific competition amongst cave beetles (*Neapheanops tellkampfi*). (a) Exploitation. Beetle fecundity is significantly correlated ($r = 0.86$) with cricket fecundity (itself a good measure of the availability of cricket eggs—the beetles' food). The beetles themselves reduce the density of cricket eggs. (b) Interference. As beetle density in experimental arenas with 10 cricket eggs increased from 1 to 2 to 4, individual beetles dug fewer and shallower holes in search of their food, and ultimately ate much less ($P < 0.001$ in each case), in spite of the fact that 10 cricket eggs was sufficient to satiate them all. Means and standard deviations are given in each case. (After Griffith & Poulson, 1993.)

On the one hand, they suffer indirectly from exploitation: beetles reduce the density of their resource (cricket eggs) and then have markedly lower fecundity when food availability is low (Figure 6.1a). But, they also suffer directly from interference: at higher beetle densities they fight more, forage less, dig fewer and shallower holes and eat far fewer eggs than could be accounted for by food depletion alone (Figure 6.1b).

Whether they compete through exploitation or interference, individuals within a species are in essence equivalent, having many fundamental features in common, using similar resources and reacting in much the same way to conditions. However, there are many occasions when intraspecific competition is very one sided: a strong, early seedling will shade a stunted, late one; and an older and larger bryozoan on the shore will grow over a smaller and younger one. One example is shown in Figure 6.2. The overwinter survival of red deer calves in the resource-limited population on the island of Rhum (see Chapter 4) declined sharply as the population became more crowded, but those that were smallest when born were by far the most likely to die.

Intrinsic, heritable differences between individuals may also ensure that competitive interactions are not reciprocal. For instance, tall corn plants will usually shade and suppress genetically distinct short plants of the same species.

This lack of exact equivalence means that the ultimate effect of competition is far from being the same on different individuals. Weak competitors may make only a small contribution to the next generation, or no contribution at all. Strong competitors may have their contribution only negligibly affected. Indeed, a strong competitor may actually make a larger *proportional* contribution when there is intense competition than when there is no competition at all (i.e. if they maintain their contribution whilst all around them are losing theirs). In other words, although the ultimate effect of competition is a decrease in reproductive output, this does not always mean a decrease in individual fitness (i.e. 'relative' contribution),

one-sided competition—which may increase fitness

216 CHAPTER 6

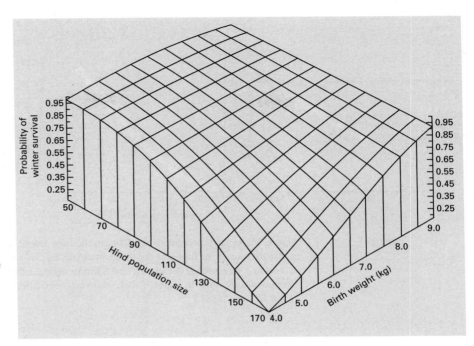

Figure 6.2 Those red deer that are smallest when born are the least likely to survive over winter when, at higher densities, survival declines. (After Clutton-Brock *et al.*, 1987.)

especially not for the strongest competitors. It would not be correct, therefore, to say that competition 'adversely affects' all competing individuals (Wall & Begon, 1985).

Finally, note that the likely effect of intraspecific competition on any individual is greater the more competitors there are. The effects of intraspecific competition are thus said to be density dependent, and we turn next to a more detailed look at the density dependent effects of intraspecific competition on death, birth and growth.

the effects of competition are density dependent

6.2 Intraspecific competition, and density dependent mortality and fecundity

Figure 6.3 shows the pattern of mortality in the flour beetle *Tribolium confusum* when cohorts were reared at a range of densities. Known numbers of eggs were placed in glass tubes with 0.5 g of a flour–yeast mixture, and the number of individuals that survived to become adults in each tube was then scored. The same data have been expressed in three complementary ways, and in each case the resultant curve has been divided into three regions. Figure 6.3a describes the relationship between density and the per capita mortality rate—literally, the mortality rate 'per head', i.e. the probability of an individual dying, or the proportion that died between the egg and adult stages. Figure 6.3b describes how the number that died prior to the adult stage changed with density, and Figure 6.3c describes the relationship between density and the numbers that survived.

density independent mortality

Throughout region 1 (low density) mortality rate remained constant as density was increased (Figure 6.3a). The numbers dying and the numbers surviving both rose (Figures 6.3b, c) (not surprising, given that the numbers 'available' to die and survive increased as density increased), but the proportion dying remained the same, which accounts for the straight lines in region 1 of Figures 6.3b and c. Mortality in this region is said to be density *in*dependent, and judged by this, there was

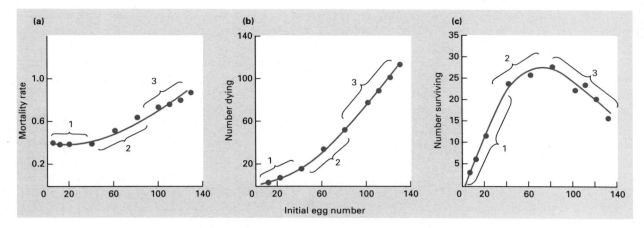

Figure 6.3 Density dependent mortality in the flour beetle *Tribolium confusum* (a) as it affects mortality rate, (b) as it affects the numbers dying, and (c) as it affects the numbers surviving. In region 1 mortality is density independent; in region 2 there is undercompensating density dependent mortality; in region 3 there is overcompensating density dependent mortality. (After Bellows, 1981.)

apparently no intraspecific competition between the beetles at these densities. Individuals died, but the chance of an individual surviving to become an adult was not changed. Such density independent deaths affect the population at all densities, and represent a baseline, which any *density dependent* mortality will exceed.

In region 2, the density dependent effects of intraspecific competition are apparent. The numbers dying continued to rise with density, but they did so more rapidly than in region 1 (Figure 6.3b). The numbers surviving also continued to rise, but they did so less rapidly than in region 1 (Figure 6.3c). The reason for this is that the mortality rate increased with density (Figure 6.3a): there was density dependent mortality. However, although the risk of death increased in region 2, it did so less rapidly than the increase in density. Over this range, increases in egg density continued to lead to increases in the total number of surviving adults. The mortality rate had increased, but it 'undercompensated' for increases in density.

In region 3, intraspecific competition was even more intense. The increasing mortality rate 'overcompensated' for any increase in density, i.e. an increase in the initial number of eggs led to an even greater proportional increase in the mortality rate. Thus, over this range, the more eggs there were present, the fewer adults survived. Indeed, if the range of densities had been extended, there would have been tubes with no survivors: the developing beetles would have eaten all the available food before any of them reached the adult stage.

A slightly different situation is shown in Figure 6.4. This illustrates the relationship between density and mortality in young trout. None of the densities examined were low enough to avoid intraspecific competition entirely; even at the lower densities there was undercompensating density dependence. At higher densities, however, mortality never overcompensated, but compensated exactly for any increase in density: any rise in the number of fry was matched by an exactly equivalent rise in the mortality rate. The number of survivors therefore approached and maintained a constant level, irrespective of initial density.

Another example, mortality of soybeans, is illustrated in Figure 6.5 (the same

undercompensating density dependence

overcompensating density dependence

exactly compensating density dependence

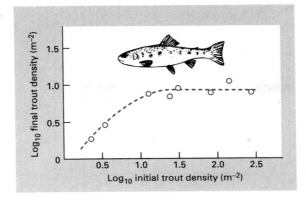

Figure 6.4 The relationship between density and mortality amongst trout fry. At high trout densities, the increasing mortality rate compensates exactly for increasing trout density and a constant number of trout survive. (After Le Cren, 1973; Hassell, 1976.)

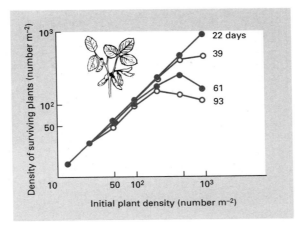

Figure 6.5 Density dependent mortality in the soybean (*Glycine soja*). After 61 and 93 days the increasing mortality rate overcompensates for increases in sowing density, and the number of surviving plants declines. (After Yoda *et al.*, 1963.)

a plant example

type of plot as in Figure 6.3a). After 22 days, almost every seed was represented by a plant; there had been scarcely any mortality and certainly no density dependent mortality. After 39 days, however, there was some evidence of undercompensating density dependent mortality; and after 61 and 93 days, the mortality at high density was overcompensating—the more seeds were sown, the fewer plants survived.

Thus, to summarize, irrespective of variations in over- and undercompensation, the essential point is a simple one: at appropriate densities, intraspecific competition can lead to density dependent mortality, which means that mortality rate increases as density increases.

intraspecific competition and fecundity

The patterns of density dependent fecundity that result from intraspecific competition are, in a sense, a mirror-image of those for density dependent mortality (Figure 6.6). Here, though, the per capita birth rate falls as intraspecific competition intensifies. At the lowest densities, birth rate is density independent (Figures 6.6a and b); but as density increases, and the effects of intraspecific competition become apparent, birth rate initially shows undercompensating density dependence (Figures 6.6a–c), and may then show exactly compensating (Figure 6.6c) or overcompensating density dependence (Figure 6.6a).

Density dependence and intraspecific competition are obviously bound closely together. Whenever there is intraspecific competition, its effect, whether on survival, fecundity or a combination of the two, is density dependent. However, as subsequent

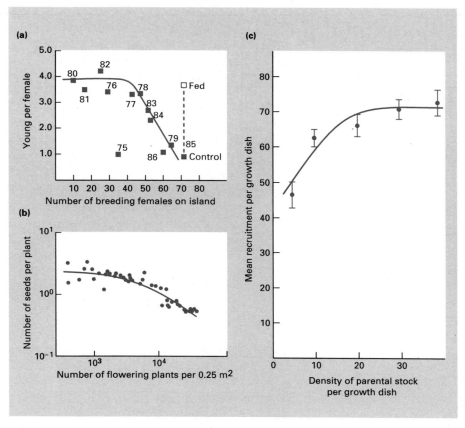

Figure 6.6 Density dependent fecundity. (a) In the population of song sparrows (*Melospiza melodia*) on Mandarte Island, British Columbia, Canada, fecundity changes from density independence to overcompensating density dependence as density increases, over the years 1975–1986. (For instance, at a density of 40, the total number of young produced was roughly $40 \times 4 = 160$, whereas at a density of 70 it was roughly $70 \times 1 = 70$.) Providing supplementary food in 1985 suggested that competition for food was the cause. (After Arcese & Smith, 1988.) (b) In the annual dune plant (*Vulpia fasciculata*), fecundity changes from approximate density independence to undercompensating density dependence. (After Watkinson & Harper, 1978; from Watkinson & Davy, 1985.) (c) In the fingernail clam (*Musculium securis*), fecundity changes from undercompensating density dependence to nearly exactly compensating density dependence. (After Mackie *et al.*, 1978.)

chapters will show, there are processes other than intraspecific competition which also have density dependent effects.

6.3 Density or crowding?

Of course, the intensity of intraspecific competition experienced by an individual is not really determined by the density of the population as a whole. The effect on an individual is determined, rather, by the extent to which it is crowded or inhibited by its immediate neighbours.

One way of emphasizing this is by noting that there are actually at least three different meanings of 'density' (see Lewontin & Levins, 1989, where details of calculations and terms can be found). Consider a population of insects, distributed over a population of plants on which they feed. This is a typical example of a very general (perhaps universal) phenomenon—a population (the insects in this case) being distributed amongst different patches of a resource (the plants). Density would usually be calculated as the number of insects (let us say 1000) divided by the number of plants (say 100), i.e. 10 insects per plant. This, which we would normally call simply 'the density', is actually the 'resource-weighted density'. However, it gives an accurate measure of the intensity of competition suffered by the insects (the extent to which they are crowded) only if there are exactly 10 insects on every plant.

Suppose, instead, that 10 of the plants support 91 insects each and the remaining 90 support just one insect. The resource-weighted density would still be 10 insects per plant. But, the average density experienced by the insects would be 82.9 insects per plant. That is, one adds up the densities experienced by each of the insects $(91 + 91 + 91 ... + 1 + 1)$ and divides by the total number of insects. This is the 'organism-weighted density', and it clearly gives a much more satisfactory measure of the intensity of competition the insects are likely to suffer.

However, there remains the further question of the average density of insects experienced by the plants. This, which may be referred to as the 'exploitation pressure', comes out at 1.1 insects per plant, reflecting the fact that most of the plants support only one insect.

What, then, is the density of the insect? Clearly, it depends on whether you answer from the perspective of the insect or the plant—but whichever way you look at it, the normal practice of calculating the resource-weighted density and calling it 'the density' looks highly suspect. The difference between resource- and organism-weighted densities is illustrated for the human population of a number of USA states in Table 6.1 (where the 'resource' is simply land area). The organism-weighted densities are so much larger than the usual, but rather unhelpful, resource-weighted densities essentially because the large counties within a state support small populations and the small counties support large populations—most people live, crowded, in cities (Lewontin & Levins, 1989).

The difficulties of relying on density to characterize the potential intensity of intraspecific competition are particularly acute with sessile, modular organisms, because, being sessile, they compete almost entirely only with their immediate

Table 6.1 A comparison of the resource- and organism-weighted densities of five states, based on the 1960 USA census, where the 'resource patches' are the counties within each state. (After Lewontin & Levins, 1989.)

State	Resource-weighted density (km^{-2})	Organism-weighted density (km^{-2})
Colorado	44	6252
Missouri	159	6525
New York	896	48 714
Utah	28	684
Virginia	207	13 824

neighbours, and being modular, competition is directed disproportionately at those modules that are closest to those neighbours. Thus, for instance, when silver birch trees (*Betula pendula*) are grown in small groups, the sides of individual trees that interface with neighbours typically have a lower 'birth' rate of buds and a higher death rate (see Chapter 4, Section 4.2); branches are shorter, and the growth form approaches that of a forest tree. However, on sides of the same tree with no interference, bud birth rate is higher, death rate lower, branches are longer and the form approaches that of an open-grown individual (Figure 6.7). Different modules within the same individual genet experience different intensities of competition, and quoting the density at which an individual was growing would be all but pointless.

These subtleties are further illustrated by the results of growing pairs of clover

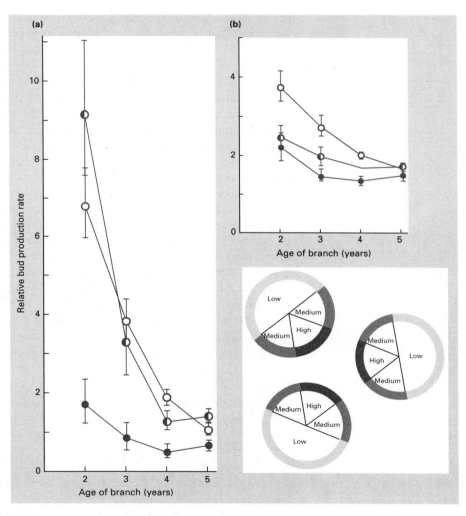

Figure 6.7 Mean relative bud production (new buds per existing bud) for silver birch trees (*Betula pendula*), expressed (a) as gross bud production and (b) as net bud production ('birth minus death'), in different interference zones. These zones are themselves explained in the inset. ●, High interference; ◑, medium; ○, low. Bars represent standard errors. (After Jones & Harper, 1987.)

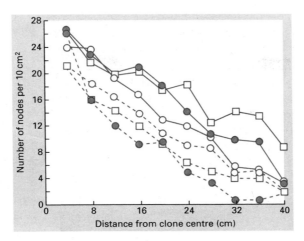

Figure 6.8 Variation in node density from the centre to the edge of paired clover plants (*Trifolium repens*) on the sides where the plants meet (- - -) and where there was no neighbour (—). Each point represents the mean of each six replicates, and the two sides are significantly different in each case ($P < 0.05$). ●, Pairs started at 12 cm distant; □, pairs started at 8 cm distant; ○, pairs started at 5 cm distant. (After Solangaarachchi & Harper, 1989.)

or at least, parts of them do

plants (*Trifolium repens*) at various distances apart (Figure 6.8). For all plants, the density of nodes (branching points) declined from the centre to the edge because the centre had been growing for longer, but this decline was significantly sharper (i.e. there was less growth) on the sides of plants facing the neighbour than on the 'open' side (rather like the birch branches). However, perhaps surprisingly, this competitive effect was most marked between the pairs grown furthest apart (12 cm). The explanation is that when plants are closer together, they grow into and through each other before they have become dense enough to impede greatly each others' development. When they start further apart, they have developed denser 'advancing fronts' and have correspondingly greater effects on each other when they meet.

As we have seen, the same problems in the use of 'density' to measure the potential for intraspecific competition are still present, although to a lesser extent, with mobile organisms. Different individuals meet or suffer from different numbers of competitors. Density, especially resource-weighted density, is an abstraction which applies to the population as a whole, but need not apply to any of the individuals within it. Nonetheless, density may often be the most convenient way of expressing the degree to which individuals are crowded—and it is certainly the way it usually has been expressed.

6.4 Intraspecific competition and the regulation of population size

Despite the variations from example to example, there are obviously typical patterns in the effects of intraspecific competition on death (see Figures 6.3–6.5) and birth (Figure 6.6). These generalized patterns are summarized in Figures 6.9 and 6.10.

Figures 6.9a–c reiterate the fact that as density increases, the per capita birth rate eventually falls and the per capita death rate eventually rises. There must, therefore, be a density at which these curves cross. At densities below this point, the birth rate exceeds the death rate and the population increases in size. At densities above the

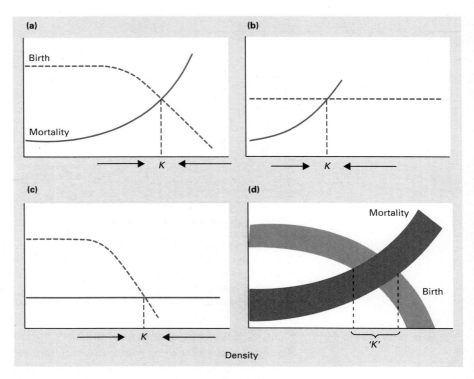

Figure 6.9 Density dependent birth and mortality rates lead to the regulation of population size. When both are density dependent (a), or when either of them is (b, c), their two curves cross. The density at which they do so is called the carrying capacity (*K*). Below this the population increases, above it the population decreases: *K* is a stable equilibrium. However, these figures are the grossest of caricatures. The situation is closer to that shown in (d), where mortality rate broadly increases, and birth rate broadly decreases, with density. It is possible, therefore, for the two rates to balance not at just one density, but over a broad range of densities, and it is towards this broad range that other densities tend to move.

competition may lead to a stable equilibrium

crossover point, the death rate exceeds the birth rate and the population declines. At the crossover density itself, the two rates are equal and there is no net change in population size. This density therefore represents a stable equilibrium, in that all other densities will tend to approach it. In other words, intraspecific competition, by acting on birth rates and death rates, can regulate populations at a stable density at which the birth rate equals the death rate. This density is known as the *carrying capacity* of the population and is usually denoted by *K* (Figure 6.9). It is called a carrying capacity because it represents the population size that the resources of the environment can just maintain ('carry') without a tendency to either increase or decrease.

natural populations lack simple carrying capacities

However, whilst hypothetical populations caricatured by line drawings like Figures 6.9a–c can be characterized by a simple carrying capacity, this is not true of any natural population. There are unpredictable environmental fluctuations; individuals are affected by a whole wealth of factors of which intraspecific competition is only one; and resources not only affect density but respond to density as well. Hence, the situation is likely to be closer to that depicted in Figure 6.9d. Intraspecific

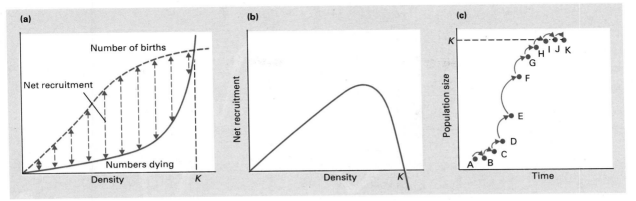

Figure 6.10 Some general aspects of intraspecific competition. (a) Density dependent effects on the numbers dying and the number of births in a population: net recruitment is 'births minus deaths'. Hence, as shown in (b), the density dependent effect of intraspecific competition on net recruitment is a domed or 'n'-shaped curve. (c) A population increasing in size under the influence of the relationships in (a) and (b). Each arrow represents the change in size of the population over one interval of time. Change (i.e. net recruitment) is small when density is low (i.e. at small population sizes: A to B, B to C) and is small close to the carrying capacity (I to J, J to K), but is large at intermediate densities (E to F). The result is an 'S'-shaped or sigmoidal pattern of population increase, approaching the carrying capacity.

competition does not hold natural populations to a predictable and unchanging level (the carrying capacity), but it may act upon a very wide range of starting densities and bring them to a much narrower range of final densities, and it therefore tends to keep density within certain limits. It is in this sense that intraspecific competition may be said typically to be capable of regulating population size. For instance, Figures 6.11a and b show the fluctuations within and between years in populations of the brown trout (*Salmo trutta*) and the grasshopper (*Chorthippus brunneus*). There are no simple carrying capacities in these examples, but there are clear tendencies for the 'final' density each year ('late summer numbers' in the first case, 'adults' in the second) to be relatively constant, despite the large fluctuations in density within each year and the obvious potential for increase which both populations possess. These figures illustrate another point too: it is often not the 'peak' size that a population can achieve that is most obviously regulated, but the 'trough' size to which it repeatedly falls.

In fact, the concept of a population settling at a stable carrying capacity, even in caricatured populations, is relevant only to situations in which density dependence is not strongly overcompensating. Where there is overcompensation, cycles or even chaotic changes in population size may be the result. We will return to this point later (see Sections 6.8 and 6.9).

the fastest rate of population increase is at intermediate densities

An alternative general view of intraspecific competition is shown in Figure 6.10a, which deals with numbers rather than rates. The difference between the two curves ('births minus deaths' or 'net recruitment') is the net number of additions expected in the population during the appropriate stage or over one interval of time: the amount by which the population changes in size. Because of the shapes of the birth and death curves, the net number of additions is small at the lowest densities, increases

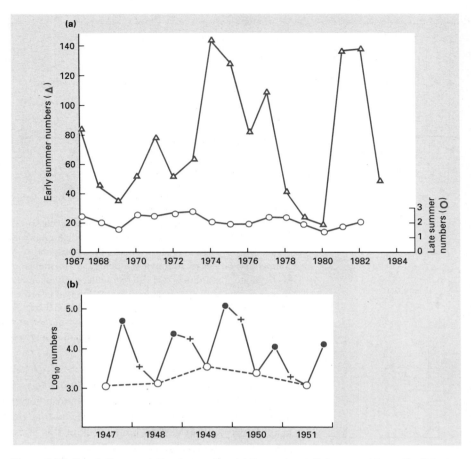

Figure 6.11 Population regulation in practice. (a) Brown trout (*Salmo trutta*) in an English Lake District stream. △, Numbers in early summer, including those newly hatched from eggs; ○, numbers in late summer. Note the difference in vertical scales. (After Elliott, 1984.) (b) The grasshopper (*Chorthippus brunneus*) in southern England. ●, Eggs; +, nymphs; ○, adults. Note the logarithmic scale. (After Richards & Waloff, 1954.) There are no definitive carrying capacities, but the 'final' densities each year ('late summer' and 'adults') are relatively constant despite large fluctuations within years.

as density rises, declines again as the carrying capacity is approached and is then negative (deaths exceed births) when the initial density exceeds K (Figure 6.10b). Thus, total recruitment into a population is small when there are few individuals available to give birth, and small when intraspecific competition is intense. It reaches a peak, i.e. the population increases in size most rapidly, at some intermediate density.

The precise nature of the relationship between a population's net rate of recruitment and its density varies with the detailed biology of the species concerned (e.g. the pheasants, flies and whales in Figures 6.12a–c). Moreover, because recruitment is affected by a whole multiplicity of factors, the data points never fall exactly on any single curve. Yet, in each case in Figures 6.12a–c, an 'n'-shaped curve is apparent. This reflects the general nature of density dependent birth and

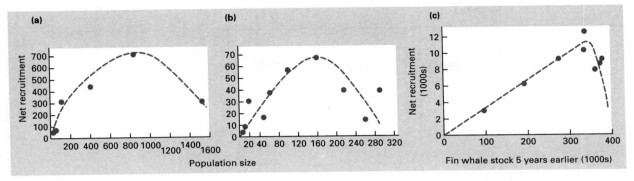

Figure 6.12 Some 'n'-shaped net-recruitment curves, drawn by eye through the data points shown. (a) The ring-necked pheasant on Protection Island following its introduction in 1937. (After Einarsen, 1945.) (b) An experimental population of the fruit-fly *Drosophila melanogaster*. (After Pearl, 1927.) (c) Estimates for the stock of Antarctic fin whales. (After Allen, 1972.)

death whenever there is intraspecific competition. Note, therefore, that an 'n'-shaped curve also describes the relationship between the leaf area index (LAI) of a plant population (the total leaf area being borne per unit area of ground) and the population's growth rate (modular birth minus modular death): growth rate is low when there are few leaves, peaks at an intermediate LAI, and is then low again at a high LAI, where there is much mutual shading and competition and many leaves may be consuming more in respiration than they contribute through photosynthesis (see Chapter 8, Figure 8.8).

In addition, curves of the type shown in Figures 6.10a and b may be used to suggest the pattern by which a population might increase from an initially very small size (e.g. when a species colonizes a previously unoccupied area). If a succession of time intervals are taken singly, then each final density can be treated as the initial density for the next time interval. This is illustrated in Figure 6.10c. Imagine a small population, well below the carrying capacity of its environment (Figure 6.10c, point A). Because the population is small, it increases in size only slightly during one time interval (Figure 6.10b), and only reaches point B (Figure 6.10c). Now, however, being larger, it increases in size more rapidly during the next time interval (to point C), and even more during the next (to point D). This process continues until the population passes beyond the peak of its net recruitment curve (Figure 6.10b). Thereafter, the population increases in size less and less with each time interval (points G, H, I and J), until the population reaches its carrying capacity (*K*) and ceases completely to increase in size. The population might therefore be expected to follow an S-shaped or 'sigmoidal' curve as it rises from a low density to its carrying capacity. This is a consequence of the hump in its recruitment rate curve, which is itself a consequence of intraspecific competition.

populations rise from low density following an 'S'-shaped curve

Of course, Figure 6.10c, like the rest of Figure 6.10, is a gross simplification, a caricature. It assumes, apart from anything else, that changes in population size are affected only by intraspecific competition. Nevertheless, something akin to sigmoidal population growth can be perceived in many natural and experimental situations (Figure 6.13).

Intraspecific competition will be obvious in certain cases (such as overgrowth competition between sessile organisms on a rocky shore), but this will not be true of

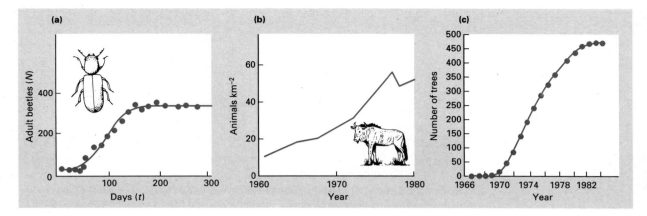

Figure 6.13 Real examples of S-shaped population increase. (a) The beetle (*Rhizopertha dominica*) in 10 g of wheat grains replenished each week. (After Crombie, 1945.) (b) The population of wildebeest (*Connochaetes taurinus*), of the Serengeti region of Tanzania and Kenya seems to be levelling off after rising from a low density caused by the disease rinderpest. (After Sinclair & Norton-Griffiths, 1982; Deshmukh, 1986.) (c) The population of the willow tree (*Salix cinerea*) in an area of land after myxomatosis had effectively prevented rabbit grazing. (After Alliende & Harper, 1989.)

intraspecific competition is widespread—but, by no means all-important

every population examined. Individuals are affected not only by intraspecific competitors, but also by predators, parasites and prey, competitors from other species, and the many facets of their physical and chemical environment. Any of these may outweigh or obscure the effects of intraspecific competition; or the effect of these other factors at one stage may reduce density to well below the carrying capacity for all subsequent stages. Nevertheless, intraspecific competition probably affects most populations at least sometimes during at least one stage of their life cycle.

6.5 Intraspecific competition and density dependent growth

Intraspecific competition, then, can have a profound effect on the number of individuals in a population; but it can have an equally profound effect on the individuals themselves. In populations of unitary organisms, rates of growth and rates of development are commonly influenced by intraspecific competition. This necessarily leads to density dependent effects on the composition of a population. For instance, Figures 6.14a and b show two examples in which the distribution of sizes within a population has been altered as a result of intraspecific competition. This, in turn, often means that although the numerical size of a population is regulated only approximately by intraspecific competition, the total biomass is regulated much more precisely. This is illustrated by observations on the limpet *Patella cochlear* (Figure 6.15). In the low-density populations the individuals were relatively large, whilst in the high-density populations they were relatively small. Overall, however, the total biomass was roughly the same at all densities in excess of 400 individuals per m^2.

modular organisms: constant final yield

Such effects are particularly marked in modular organisms. For example, as subterranean clover (*Trifolium subterraneum*) plants grow, the yield (i.e. the total weight) per unit area increases. At the first harvest (62 days) the yield is closely related to the density of seeds sown (Figure 6.16a). After 181 days, however, yield

competition affects growth and development in unitary organisms

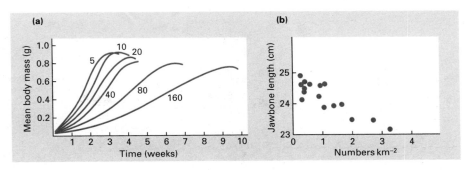

Figure 6.14 The effects of density on growth rate and size. (a) The density dependent growth rate of the frog *Rana tigrina*, where numbers refer to individuals per 2 l aquarium. (After Dash & Hota, 1980.) (b) Mean size (jawbone length) decreases with density in a reindeer population. (After Skogland, 1983.)

no longer reflects the numbers sown. Rather it becomes the same over a wide range of initial densities. This pattern has frequently been uncovered by plant ecologists and has been called 'the law of constant final yield' (Kira *et al.*, 1953). The yield is constant over a wide range of densities because individuals suffer density dependent reductions in growth rate and thus in individual plant size, and in particular because the reduction in mean plant weight *compensates exactly* for increases in density.

Yield is density (d) multiplied by mean weight per plant ($\bar{w}$). Thus, if yield is constant (c):

$$d\bar{w} = c \tag{6.1}$$

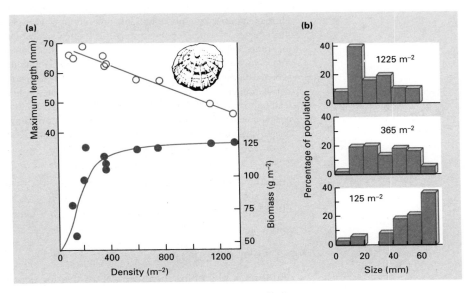

Figure 6.15 Intraspecific competition and growth in populations of the limpet *Patella cochlear*. (a) Individual size declines with density leading to an exact regulation of the population's total biomass. (b) High-density populations have many small and a few large individuals; low-density populations have many large and a few small individuals. (After Branch, 1975.)

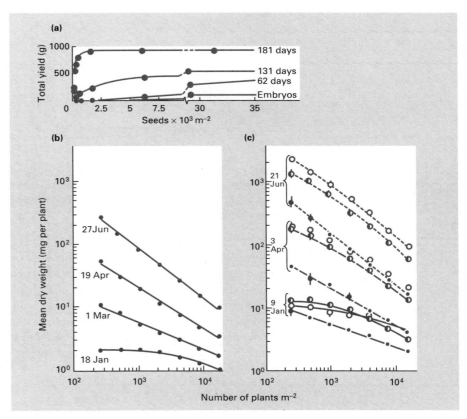

Figure 6.16 The 'constant final yield' of plants sown at a range of densities. This can be illustrated either by a horizontal line when total yield is plotted against density (a), or by a line of slope −1 when log mean weight is plotted against log density (b, c)—see text. (a) Subterranean clover, *Trifolium subterraneum*. (After Donald, 1951.) (b, c) The dune annual, *Vulpia fasciculata*. (After Watkinson, 1984.) (b) On 18 January, at low densities especially, growth and hence mean dry weight were roughly independent of density. But, by 27 June, density dependent reductions in growth compensated exactly for variations in density, leading to constant yield and a slope of −1 (see text). (c) Constant yield was exhibited sooner at low concentrations of nitrogen in soil (●) than at medium (◑) or high (○) since resources became limiting sooner; bars indicate standard errors.

and so:

$$\log d + \log \bar{w} = \log c \qquad (6.2)$$

and:

$$\log \bar{w} = \log c - 1 \times \log d \qquad (6.3)$$

and thus, a plot of log mean weight against log density should have a slope of −1.

Data on the effects of density on the growth of the grass *Vulpia fasciculata* are shown in Figure 6.16b, and the slope of the curve towards the end of the experiment does indeed approach a value of −1. Here too, as with the clover, individual plant weight at the first harvest was reduced only at very high densities—but, as the plants became larger, they interfered with each other at successively lower densities.

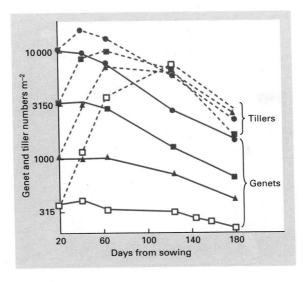

Figure 6.17 Intraspecific competition in plants often regulates the number of modules. When populations of rye grass (*Lolium perenne*) were sown at a range of densities, the range of final tiller (i.e. module) densities was far narrower than that of genets. (After Kays & Harper, 1974.)

The results of these experiments suggest that there may be limited resources available for plant growth, especially at high densities. If this is the case, we would expect the provision of extra resources to allow greater growth of individual plants and greater yield per unit area. This expectation was borne out when three levels of nitrogen fertilizer were given to populations of *V. fasciculata* growing at a range of densities (Figure 6.16c).

competition tends to regulate module number

The constancy of the final yield is a result, to a large extent, of the modularity of plants. This can be seen more explicitly from what happened when perennial rye grass (*Lolium perenne*) was sown at a 30-fold range of densities (Figure 6.17). After 180 days, some genets had died; but as a result of density dependent modular growth, the range of final tiller densities was far narrower than that of genets. In modular organisms, then, the regulatory powers of intraspecific competition frequently operate by affecting the number of modules per genet instead of, or as well as, affecting the number of genets themselves.

different plant parts are affected to different extents

It must not be imagined, however, that plants at high densities are typically scaled-down versions of those at low densities, differing only in the possession of fewer modules. Intraspecific competition affects not only rates of growth, but also rates of development and maturation, and it therefore affects the way in which biomass is distributed within individual plants. Figure 6.18 shows an example of a very common phenomenon. At high densities the maize plants were not only smaller but devoted a lower proportion of their biomass to seeds; the result was that seed output per unit area declined at the higher densities. Indeed, in many examples of density dependent fecundity, in both unitary and modular organisms, the immediate effect of density is on growth rate and size, and it is this that affects fecundity.

6.6 Quantifying intraspecific competition

Every population is unique. Nevertheless, we have already seen that there are general patterns in the action of intraspecific competition. In this section we take such generalizations a stage further. A method will be described, utilizing *k*-values

k-values can be used in quantification

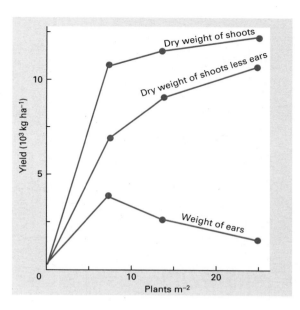

Figure 6.18 Competition in populations of maize (*Zea mays*) affects seed production (weight of ears) far more than it affects the growth of shoots. (After Harper, 1961.)

(see Chapter 4), which can be used to summarize the effects of intraspecific competition on mortality, fecundity and growth. Mortality will be dealt with first. The method will then be extended for use with fecundity and growth.

A *k*-value was defined by the formula:

$$k = \log(\text{initial density}) - \log(\text{final density}) \tag{6.4}$$

or, equivalently:

$$k = \log\left(\frac{\text{initial density}}{\text{final density}}\right). \tag{6.5}$$

For present purposes, 'initial density' may be denoted by *B*, standing for 'numbers *before* the action of intraspecific competition', whilst 'final density' may be denoted by *A*, standing for 'numbers *after* the action of intraspecific competition'. Thus:

$$k = \log(B/A). \tag{6.6}$$

Note that *k* increases as mortality rate increases, i.e. as the proportion surviving (*A/B*) decreases.

when *k* is plotted against the log of density, the slope quantifies competition

Some examples of the effects of intraspecific competition on mortality are shown in Figure 6.19, in which *k* is plotted against log *B*. In several cases, *k* is constant at the lowest densities. This is an indication of density independence: the proportion surviving is not correlated with initial density. At higher densities, *k* increases with initial density; this indicates density dependence. Most importantly, however, the way in which *k* varies with the logarithm of density indicates the precise nature of the density dependence. For example, Figures 6.19a and b describe, respectively, situations in which there is under- and exact compensation at higher densities. The exact compensation in Figure 6.19b is indicated by the slope of the curve (denoted by *b*) taking a constant value of 1. The undercompensation which preceded this at lower densities, and which is seen in Figure 6.19a even at higher densities, is indicated by the fact that *b* is less than 1.

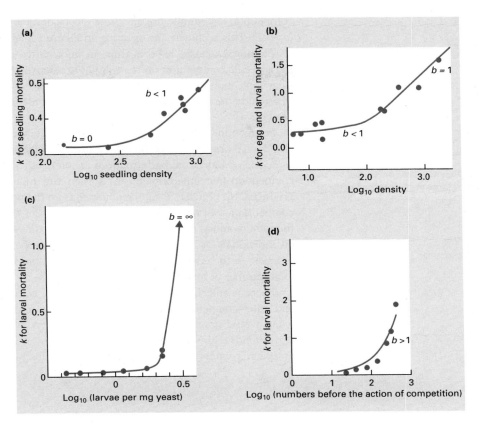

Figure 6.19 The use of *k*-values for describing patterns of density dependent mortality. (a) Seedling mortality in the dune annual, *Androsace septentrionalis*, in Poland. (After Symonides, 1979.) (b) Egg mortality and larval competition in the almond moth, *Ephestia cautella*. (After Benson, 1973a.) (c) Larval competition in the fruit-fly, *Drosophila melanogaster*. (After Bakker, 1961.) (d) Larval mortality in the moth, *Plodia interpunctella*. (After Snyman, 1949.)

For the mathematically inclined: exact compensation means that A is constant. The slope, b, is given by:

$$b = \frac{k_2 - k_1}{\log_{10} B_2 - \log_{10} B_1}$$

$$= \frac{\log_{10} B_2 - \log_{10} A - (\log_{10} B_1 - \log_{10} A)}{\log_{10} B_2 - \log_{10} B_1}$$

$$= \frac{\log_{10} B_2 - \log_{10} B_1}{\log_{10} B_2 - \log_{10} B_1} = 1.$$

(6.7)

contest and scramble competition

Exact compensation ($b = 1$) is often referred to as pure contest competition, because there are a constant number of winners (survivors) in the competitive process. The term was initially proposed by Nicholson (1954b), who contrasted it with what he called pure scramble competition. Pure scramble is the most extreme form of overcompensating density dependence, in which all competing individuals are so adversely affected that none of them survive, i.e. $A = 0$. This would be indicated in Figure 6.19 by a *b* value of infinity (a vertical line), and Figure 6.19c is an example in which this is the case. More common, however, are examples in which competition is scramble-*like*, i.e. there is considerable but not total overcompensation ($b \gg 1$). This is shown, for instance, in Figure 6.19d.

Plotting k against log B is thus an informative way of depicting the effects of intraspecific competition on mortality. Variations in the slope of the curve (b) give a clear indication of the manner in which density dependence changes with density. The real value of the method, however, is that it can be extended to fecundity and growth.

For fecundity, it is necessary to think of B as 'the total number of offspring that *would* have been produced had there been no intraspecific competition', i.e. if each reproducing individual had produced as many offspring as it would have done in a competition-free environment. A is then the total number of offspring *actually* produced. (In practice, B can be obtained from the population experiencing the least competition—not necessarily competition free.) For growth, B must be thought of as the total biomass, or total number of modules, that would have been produced had all individuals grown as if they were in a competition-free environment (or, in practice, in the environment with least competition). A is then the total biomass or total number of modules actually produced.

There is still, in these cases, an underlying comparison of 'before competition' and 'after competition'. Now, however, 'before' must be imagined on the basis of what would have happened had there been no competition. 'After' clearly corresponds with what actually happens. Figures 6.20 and 6.21 illustrate examples in which k-values are used to describe the effects of intraspecific competition on

k-values can also be used for fecundity and growth

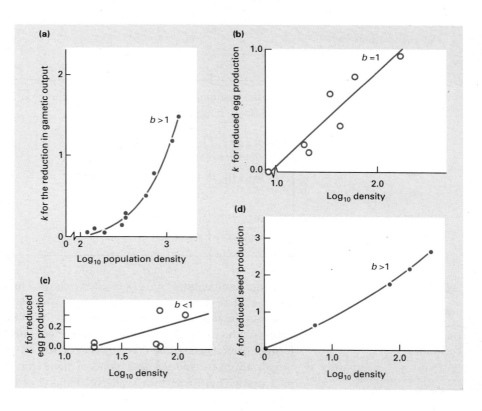

Figure 6.20 The use of *k*-values for describing density dependent reductions in fecundity as a consequence of competition. (a) The limpet *Patella cochlear* in South Africa. (After Branch, 1975.) (b) The cabbage root fly, *Eriosichia brassicae*. (After Benson, 1973b.) (c) The grass mirid *Leptoterna dolabrata*. (After McNeill, 1973.) (d) The plantain *Plantago major*. (After Palmblad, 1968.)

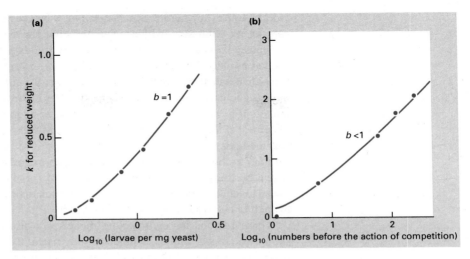

Figure 6.21 The use of *k*-values for describing density dependent reductions in growth as a consequence of competition. (a) The relative failure of adult females to gain weight, as a consequence of larval competition in *Drosophila melanogaster*. (After Bakker, 1961.) (b) Weight loss in the shepherd's purse, *Capsella bursa-pastoris*. (After Palmblad, 1968.)

fecundity and growth, respectively. The patterns are essentially similar to those in Figure 6.19. Each falls somewhere on the continuum ranging between density independence and pure scramble, and their position along that continuum is immediately apparent. Using *k*-values, all examples of intraspecific competition can be quantified in the same terms. With fecundity and growth, however, the terms 'scramble' and, especially, 'contest' are less appropriate, and it is usually preferable to talk in terms of exact, over- and undercompensation.

6.7 Mathematical models: introduction

The desire to formulate general rules in ecology often finds its expression in the construction of mathematical or graphical models. It may seem surprising that those interested in the natural living world should spend time reconstructing it in an artificial mathematical form; but there are several good reasons why this should be done. The first is that models can crystallize, or at least bring together in terms of a few parameters, the important, shared properties of a wealth of unique examples. This simply makes it easier for ecologists to think about the problem or process under consideration, by forcing us to try to extract the essentials from complex systems. Thus, a model can provide a 'common language' in which each unique example can be expressed; and if each can be expressed in a common language, then their properties relative to one another, and relative perhaps to some ideal standard, will be more apparent.

These ideas are more familiar, perhaps, in other contexts. Newton never laid hands on a perfectly frictionless body, and Boyle never saw an ideal gas—other than in their imaginations—but Newton's Laws of Motion and Boyle's Law have been of immeasureable value to us for centuries.

Perhaps more importantly, however, models can actually shed light on the real world, which they mimic. Specific examples below will make this apparent. Models can, as we shall see, exhibit properties that the system being modelled had not previously been known to possess. More commonly, models make it clear how the properties of the behaviour of a population, for example, depend on the properties of the individuals that comprise that population. To put this same point another way: models allow us to see the likely consequences of any assumptions that we choose to make—'If it were the case that only juveniles migrate, what would this do to the dynamics of their populations?'—and so on. Models can do this because mathematical methods are designed precisely to allow a set of assumptions to be followed through to their natural conclusions. As a consequence, models often suggest what would be the most profitable experiments to carry out or observations to make—'Since juvenile migration rates appear to be so important, these should be measured in each of our study populations.'

These reasons for constructing models are also criteria by which any model should be judged. Indeed, a model is only useful (i.e. worth constructing) if it does perform one or more of these functions. Of course, in order to perform them a model must adequately describe real situations and real sets of data, and this 'ability to describe' or 'ability to mimic' is itself a further criterion by which a model can be judged. However, the crucial word is 'adequate': models must *adequately* describe. The only perfect description of the real world is the real world itself. A model is an adequate description, ultimately, as long as it performs a useful function.

In the present case, some simple models of intraspecific competition will be described. They will be built up from a very elementary starting-point, and their properties (i.e. their ability to satisfy the criteria described above) will then be examined. Initially, a model will be constructed for a population with discrete breeding seasons. This will be followed by a model in which breeding is continuous.

6.8 A model with discrete breeding seasons

6.8.1 The basic equations

In Chapter 4, Section 4.7, we developed a simple model for species with discrete breeding seasons, in which the population size was N_t, at time t, and the population altered in size over time under the influence of a fundamental net reproductive rate, R. This model can be summarized in two equations, namely:

$$N_{t+1} = N_t R \tag{6.8}$$

and:

$$N_t = N_0 R^t. \tag{6.9}$$

no competition: an exponentially increasing population

The model, however, describes a population in which there is no competition. R is constant, and if $R > 1$, the population will continue to increase in size indefinitely ('exponential growth', shown in Figure 6.22). The first step is, therefore, to modify the equations by making the net reproductive rate subject to intraspecific competition. This is done in Figure 6.23, which has three components.

the incorporation of intraspecific competition

Point A describes a situation in which the population size is very small (N_t is virtually zero). Competition is therefore negligible, and the actual net reproductive rate is adequately defined by an unmodified R. Thus, at these very low densities, Equation 6.8 is still appropriate, or, rearranging the equation:

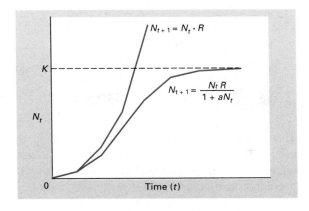

Figure 6.22 Mathematical models of population increase with time, in populations with discrete generations: exponential increase (left) and sigmoidal increase (right).

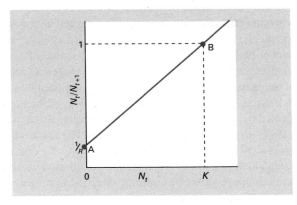

Figure 6.23 The simplest, straight-line way in which the inverse of generation increase (N_t/N_{t+1}) might rise with density (N_t). For further explanation, see text.

$$\frac{N_t}{N_{t+1}} = \frac{1}{R}. \tag{6.10}$$

Point B, by contrast, describes a situation in which the population size (N_t) is very much larger and there is a significant amount of intraspecific competition. In fact, at point B, the actual net reproductive rate has been so modified by competition that the population can collectively do no better than replace itself each generation, because 'births' equal 'deaths'. In other words, N_{t+1} is simply the same as N_t, and N_t/N_{t+1} equals 1. The population size at which this occurs is, by definition, the carrying capacity, K (see Figure 6.9).

The third component of Figure 6.23 is the straight line joining point A to point B and extending beyond it. This describes the progressive modification of the actual net reproductive rate as population size increases; but its straightness is simply an assumption made for the sake of expediency, since all straight lines are of the simple form: $y =$ (slope) $x +$ (intercept). In Figure 6.23, N_t/N_{t+1} is measured on the y-axis, N_t on the x-axis, the intercept is $1/R$ and the slope, based on the segment between points A and B, is $(1 - 1/R)/K$. Thus:

$$\frac{N_t}{N_{t+1}} = \frac{1 - \dfrac{1}{R}}{K} \cdot N_t + \frac{1}{R} \tag{6.11}$$

or, rearranging:

237   INTRASPECIFIC COMPETITION

$$N_{t+1} = \frac{N_t R}{1 + \frac{(R-1)N_t}{K}}.$$ (6.12)

a simple model of intraspecific competition

For further simplicity, $(R - 1)/K$ may be denoted by a giving:

$$N_{t+1} = \frac{N_t R}{(1 + aN_t)}.$$ (6.13)

This is a model of population increase limited by intraspecific competition. Its essence lies in the fact that the unrealistically constant R in Equation 6.8 has been replaced by an actual net reproductive rate, $R/(1 + aN_t)$, which decreases as population size (N_t) increases.

which comes first, a or K?

We, like many others, derived Equation 6.13 as if the behaviour of a population is jointly determined by R and K, the per capita rate of increase and the population's carrying capacity—a is then simply a particular combination of these. An alternative point of view is that a is meaningful in its own right, measuring the per capita susceptibility to crowding: the larger the value of a, the greater the effect of density on the actual rate of increase in the population (Kuno, 1991). Now the behaviour of a population is seen as being jointly determined by two properties of the individuals within it—their intrinsic per capita rate of increase and their susceptibility to crowding, R and a. The carrying capacity of the population ($K = (R - 1)/a$) is then simply an outcome of these properties. The great advantage of this viewpoint is that it places individuals and populations in a more realistic biological perspective. Individuals come first: individual birth rates, death rates and susceptibilities to crowding are subject to natural selection and evolve. Populations simply follow: a population's carrying capacity is just one of many features that reflect the values these individual properties take.

the properties of the simple model

The properties of the model in Equation 6.13 may be seen by reference to Figure 6.23 (from which the model was derived) and Figure 6.22 (which shows a hypothetical population increasing in size over time in conformity with the model). The population in Figure 6.22 increases exponentially when N_t is very low, but the rate of increase declines progressively as population size rises, until at the carrying capacity the rate is zero; the result is an 'S'-shaped curve. This is a desirable quality of the model, but it should be stressed that there are many other models which would also generate such a curve. The advantage of Equation 6.13 is its simplicity.

The behaviour of the model in the vicinity of the carrying capacity can best be seen by reference to Figure 6.23. At population sizes that are less than K the population will increase in size; at population sizes that are greater than K the population size will decline; and at K itself the population neither increases or decreases. The carrying capacity is therefore a stable equilibrium for the population, and the model exhibits the regulatory properties classically characteristic of intraspecific competition.

It is not yet clear, however, just exactly what type or range of competition this model is able to describe. This can be explored by tracing the relationship between k-values and $\log N$ (as in Section 6.6). Each generation, the potential number of individuals produced (i.e. the number that would be produced if there were no competition) is $N_t R$. The actual number produced (i.e. the number that survive the effects of competition) is $N_t R/(1 + aN_t)$.

Section 6.6 established that:

$$k = \log \text{(number produced)} - \log \text{(number surviving)}. \tag{6.14}$$

Thus, in the present case:

$$k = \log N_t R - \log N_t R / (1 + aN_t) \tag{6.15}$$

or, simplifying:

$$k = \log (1 + aN_t). \tag{6.16}$$

Figure 6.24 shows a number of plots of k against $\log_{10} N_t$ with a variety of values of a inserted into the model. In every case, the slope of the graph approaches and then attains a value of 1. In other words, the density dependence always begins by undercompensating and then compensates perfectly at higher values of N_t. The model is therefore limited in the type of competition that it can produce, and all we have been able to say so far is that *this type* of competition leads to very tightly controlled regulation of populations.

One simple modification that we can make is to relax the assumption that populations respond instantaneously to changes in their own density, i.e. that present density determines the amount of resource available to a population, and this in turn determines the net reproductive rate within the population. Suppose instead that the amount of resource available is determined by the density one time interval previously. To take a specific example, the amount of grass in a field in spring (the resource available to cattle) might be determined by the level of grazing (and hence, the density of cattle) in the previous year. In such a case, the reproductive rate itself will be dependent on the density one time interval ago. Thus, since in Equations 6.8 and 6.13:

time lags provoke population fluctuations

$$N_{t+1} = N_t \times \text{reproductive rate} \tag{6.17}$$

Equation 6.13 may be modified to:

$$N_{t+1} = \frac{N_t R}{1 + aN_{t-1}}. \tag{6.18}$$

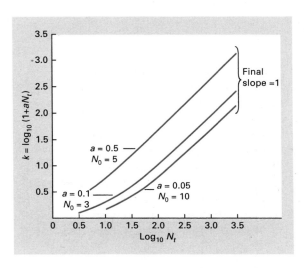

Figure 6.24 The intraspecific competition inherent in Equation 6.13. The final slope of k against $\log_{10} N_t$ is unity (exact compensation), irrespective of the starting density N_0, or the constant a ($= (R-1)/K$).

239 INTRASPECIFIC COMPETITION

There is a time lag in the population's response to its own density, caused by a time lag in the response of its resources. The behaviour of the modified model is as follows:

$R < 1.33$: direct approach to a stable equilibrium
$R > 1.33$: damped oscillations towards that equilibrium.

In comparison, the original Equation 6.13, without a time lag, gave rise to a direct approach to its equilibrium for all values of R. The time lag has provoked the fluctuations in the model, and it can be assumed to have similar, destabilizing effects on real populations.

6.8.2 Incorporating a range of competition

A simple modification of Equation 6.13 of far more general importance was originally suggested by Maynard Smith and Slatkin (1973) and discussed in detail by Bellows (1981). It alters the equation to:

$$N_{t+1} = \frac{N_t R}{1 + (aN_t)^b}. \tag{6.19}$$

the model provides a good description of real data, and a common language

The justification for this modification may be seen by examining some of the properties of the revised model. For example, Figure 6.25 shows plots of k against $\log N_t$, analogous to those in Figure 6.24: k is now $\log[1 + (aN_t)^b]$. It is apparent that the slope of the curve, instead of approaching 1 as it did previously, now approaches the value taken by b in Equation 6.19. Thus, by the choice of appropriate values, the model can portray undercompensation ($b < 1$), perfect compensation ($b = 1$), scramble-like overcompensation ($b \gg 1$) or even density independence ($b = 0$). The model has the generality that Equation 6.13 lacks, with the value of b determining the type of density dependence which is being incorporated.

Further evidence of the descriptive powers and generality of Equation 6.19 is displayed in Figures 6.26a–e. Each plot in this figure represents a set of field or laboratory data (the dots) to which a curve conforming to the equation has been fitted by computer; the computer program has selected the combination of values of a, b and R which gives the most closely fitting curve. Equation 6.19 can clearly model

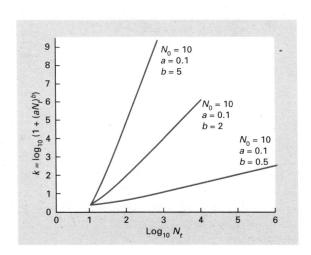

Figure 6.25 The intraspecific competition inherent in Equation 6.19. The final slope is equal to the value of b in the equation.

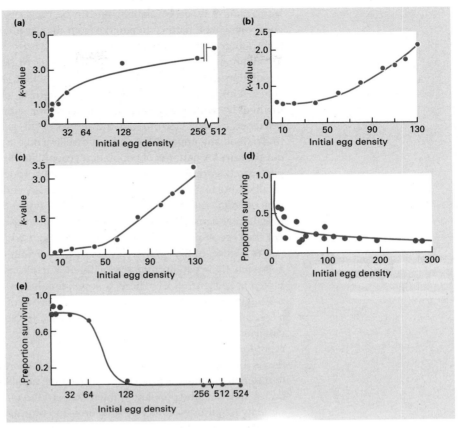

Figure 6.26 Equation 6.19 as a description of density dependent mortality (—) fitted to sets of data (●). (a) The beetle, *Stegobium panaceum*, in the laboratory. (b) The beetle, *Tribolium confusum*, in the laboratory. (c) The beetle, *T. castaneum*, in the laboratory. (d) The winter moth, *Operophtera brumata*, in the field. (After Varley & Gradwell, 1968.) (e) The beetle, *Lasioderma serricorne*, in the laboratory. (After Bellows, 1981.)

situations in which k tends to level off with density (Figure 6.26a), as well as those in which there is an upward sweep to the curve (Figures 6.26b, c); and it can model situations in which the proportion surviving declines monotonically with density (Figure 6.26d) as well as those in which the proportion remains roughly constant at low density, then declines sharply and then levels out again at higher densities (Figure 6.26e). The descriptive properties of Equation 6.19 are impressive, and this is a desirable quality for any model to have.

In addition, the model can aid in classifying the observed spectrum of density dependent relationships. Figure 6.26 indicated that a wide range of data sets could be described satisfactorily by assigning values to a, b and R. These values, therefore, immediately become a common language which can be used to compare and contrast different relationships. The advantages of this should not be underestimated. Ecological data sets come from an enormous variety of organisms in an enormous variety of circumstances; progress is only possible if the connections between them can be clearly and simply expressed. Without a common language, ecologists could, for instance, merely point to the general difference in shape between Figures 6.26a and b, and to the fact that Figures 6.26b and c are essentially similar. As it is, each relationship can be precisely encapsulated in a three-figure code, allowing a pattern to emerge from the variation.

Another desirable quality which Equation 6.19 shares with other good models is an ability to throw fresh light on the real world. By sensible analysis of the population dynamics generated by the equation, it is possible to draw guarded conclusions regarding the dynamics of natural populations. The mathematical method by which this and similar equations may be examined is set out and discussed by May (1975a), but the results of the analysis (Figures 6.27a, b) can be understood and appreciated without dwelling on the analysis itself. Figure 6.27b sets out the various patterns of population growth and dynamics that Equation 6.20 can generate. Figure 6.27a sets out the conditions under which each of these patterns occurs. Note first that the pattern of dynamics depends on two things: (i) b, the precise type of competition or density dependence; and (ii) R, the effective net reproductive rate (taking density independent mortality into account). By contrast, a determines not the type of pattern, but only the level about which any fluctuations occur.

the model indicates that intraspecific competition can lead to a wide range of population dynamics

As Figure 6.27a shows, low values of b and/or R lead to populations that approach their equilibrium size without fluctuating at all. This has already been hinted at in Figure 6.21. There, a population behaving in conformity with Equation 6.13 approached equilibrium directly, irrespective of the value of R. Equation 6.13 is a special case of Equation 6.19 in which $b = 1$ (perfect compensation); Figure 6.27a confirms that for $b = 1$, monotonic damping is the rule whatever the effective net reproductive rate.

As the values of b and/or R increase, the behaviour of the population changes first to damped oscillations gradually approaching equilibrium, and then to 'stable limit cycles' in which the population fluctuates around an equilibrium level, revisiting the same two, four or even more points time and time again. Finally, with large values of b and R, the population fluctuates in an apparently irregular and chaotic fashion.

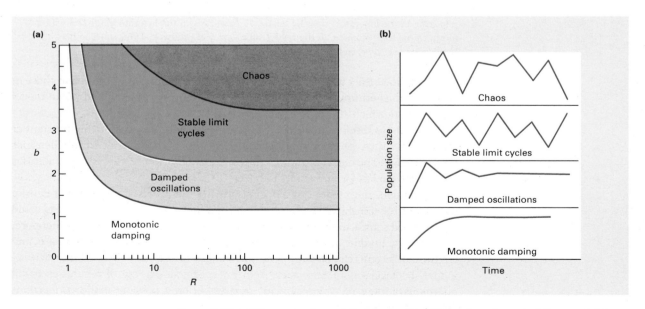

Figure 6.27 (a) The range of population fluctuations (themselves shown in (b)) generated by Equation 6.19 with various combinations of b and R inserted. (After May, 1975a; Bellows, 1981.)

6.9 Chaos

Thus, a model built around a density dependent, supposedly regulatory process (intraspecific competition) can lead to a very wide range of population dynamics. If a model population has even a moderate fundamental net reproductive rate (and an individual leaving 100 (= R) individuals in the next generation in a competition-free environment is not unreasonable), and if it has a density dependent reaction which even moderately overcompensates, then far from being stable, it may fluctuate widely in numbers without the action of any extrinsic factor. The biological significance of this is the strong suggestion that even in an environment that is wholly constant and predictable, the intrinsic qualities of a population and the individuals within it may, by themselves, give rise to population dynamics with large and perhaps even chaotic fluctuations. (A much fuller, very readable account of how 'chaos' may emerge in apparently simple systems is given by Gleick, 1987.) The consequences of intraspecific competition are clearly not limited to 'tightly controlled regulation'. There are, therefore, two important conclusions to be drawn. The first is that time lags, high reproductive rates and overcompensating density dependence are capable (either alone or in combination) of provoking all types of fluctuations in population density, without invoking any extrinsic cause. The second, equally important conclusion is that this has been made apparent by the analysis of mathematical models.

In fact, the recognition that even simple ecological systems may contain the seeds of chaos has led to chaos itself becoming a topic of considerable interest amongst ecologists (Schaffer & Kot, 1986; May, 1987; Logan & Allen, 1992; Godfray & Grenfell, 1993; Hastings et al., 1993; Perry et al., 1993). A detailed exposition of the nature of chaos is not appropriate here, but a few key points should be understood.

key characteristics of chaotic dynamics

1 The term 'chaos' may itself be misleading if it is taken to imply a fluctuation with absolutely no discernable pattern. Chaotic dynamics do not consist of a sequence of random numbers. On the contrary, there are tests (although they are not always easy to put into practice) designed to distinguish chaotic from random and other types of fluctuations.

2 Moreover, fluctuations in chaotic ecological systems occur between definable upper and lower densities. Thus, in the model of intraspecific competition that we have discussed, the idea of 'regulation' has not been lost altogether, even in the chaotic region.

3 Unlike the behaviour of truly regulated systems, however, two similar population trajectories in a chaotic system will not tend to converge ('be attracted to') the same equilibrium density or the same limit cycle. Rather, the behaviour of a chaotic system is governed by a 'strange attractor'. Initially, very similar trajectories will diverge from one another, exponentially, over time: chaotic systems exhibit 'extreme sensitivity to initial conditions'.

4 Mathematically, this difference between 'convergent' and 'divergent' systems may be characterized by the value of what is known as the system's largest *Lyapunov exponent*. The details of its calculation are not important here. Note simply that for systems that subsequently converge when initially slightly different, all Lyapunov exponents are negative, whereas for those that diverge at least one is positive—one of the hallmarks of chaos.

5 Hence, the long-term future behaviour of a chaotic system is effectively impossible to predict, and prediction becomes increasingly inaccurate as one moves

further into the future. Even if we appear to have seen the system in a particular state before—and know precisely what happened subsequently last time—tiny (perhaps unmeasurable) initial differences will be magnified progressively, and past experience will become of increasingly little value.

Ecology must aim to become a predictive science. Chaotic systems set us some of the sternest challenges in prediction. There has been an understandable interest, therefore, in the question 'How often, if ever, are ecological systems chaotic?' Attempts to answer this question, however, whilst illuminating, have certainly not been definitive.

are ecological dynamics chaotic? An initial, apparently negative answer by Hassell *et al.* (1976)

In the earliest study, Hassell *et al.* (1976) analysed data from 24 natural populations of insects and estimated values for R and b in an equation essentially the same as Equation 6.19, allowing them to position each population on a figure like Figure 6.27a. Of these populations, none lay in the region of chaos, only one lay in the region of limit cycles and one lay in the region of damped oscillations—the rest were in the region of monotonic approach to a stable equilibrium. This was influential for many years in suggesting to ecologists that, in practice, chaos is of little importance in ecological systems. However, as Hassell *et al.* made clear, such a conclusion is only valid insofar as the equation on which it is based is itself a valid representation of the populations' dynamics. Since the equation envisages a population interacting only within itself and limited only by intraspecific competition, its validity in this context is in fact doubtful, and the conclusions are questionable.

the more general approach of Turchin and Taylor (1992)...

Turchin and Taylor (1992), therefore, suggested the use of a more general and flexible equation, which may be written as follows:

$$r = \ln\{R\} = \ln\left\{\frac{N_{t+1}}{N_t}\right\} = c_0 + c_1 N_t^{d_1} + c_2 N_{t-1}^{d_2} + c_3 (N_t^{d_1})^2 + c_4 (N_{t-1}^{d_2})^2 + c_5 N_t^{d_1} N_{t-1}^{d_2} \qquad (6.20)$$

where c_0–c_5 and d_1, d_2 are constants. A full justification for the structure of what appears to be a complex model is not necessary here as long as the following two points are appreciated. (i) Abundance, N_{t+1}, is determined both by numbers one time interval ago, N_t, and by those one time interval before that, N_{t-1} (i.e. the model can take account of a time lag in the population's response). (ii) The model is especially flexible because of a mathematical theorem that says that the influence of a range of other species on a target species' abundance, N_{t+1}, can be taken into account, even when the abundances of those other species are not known, by using the abundances of the target species at various times in the past: N_t, N_{t-1}, N_{t-2}, etc. Moreover, Turchin and Taylor found that in their case, N_t and N_{t-1} were sufficient. Hence, their model is limited neither to interactions within the target species alone nor to intraspecific competition.

Each point in a time series of population abundances has a value for N_t (numbers now), a value for N_{t-1} (numbers at the previous point in the series), and a value for N_{t+1} (numbers at the next point). In essence, Turchin and Taylor's approach involves determining, statistically, the set of constants that generate the best predictions of the various values of N_{t+1}, given the various corresponding values of N_t and N_{t-1}. The end result is a 'response surface'—so called because it predicts the 'response' of N_{t+1} to various combinations of N_t and N_{t-1}. Having determined the most appropriate values for the constants in the model, the final step is then to insert them and run the model for many generations on a computer, to see how a population would behave if governed by these values. It might approach a single stable abundance, either directly or via damped oscillations, or a relatively simple cycle in which the same abundances are revisited repeatedly or follow an apparently 'strange' attractor (Figure 6.28), suggestive of chaos.

Of 36 time series (14 from insects, 22 from vertebrates), Turchin and Taylor found that four were apparently unregulated, nine showed a monotonic approach to a stable equilibrium, 17 showed damped oscillations, five showed persistent cycles and one was apparently chaotic (Figure 6.28). Whilst this last case has been challenged on the basis of further data (Perry *et al.*, 1993), and Turchin and Taylor were themselves careful to point out that they had not shown that the intrinsic dynamics of the species concerned (as opposed to the particular time series) were chaotic, these results do at least suggest that natural populations exhibit a wider and more interesting range of dynamics than had previously been supposed.

Moreover, a further analysis of 34 vole and lemming time series, in which not only response surfaces but also Lyapunov exponents were estimated, suggested that chaos itself may not be uncommon (Turchin, 1993; see also Hanski *et al.*, 1993). Of 11 'temperate' studies (39–56°N latitude), all had Lyapunov exponents that were all negative, implying relatively stable, convergent dynamics, in line with the general observation that vole abundances in these regions fluctuate rather little (see Chapters 10 and 15). But, of 23 'northern' studies (64–71°N latitude), 15 had positive Lyapunov exponents, implying divergent, chaotic dynamics. Violent fluctuations in the abundances of voles and lemmings in these regions have frequently been observed, and an apparent regularity in these fluctuations (a 3–5-year cycle) has been the focus of much attention (see Chapters 10 and 15). Turchin, too, found an underlying regularity in his northern studies—emphasizing that chaotic dynamics may themselves contain a significant periodic component.

Overall, the potential importance of chaos in ecological systems is clear. From a fundamental point of view, we need to appreciate that if we have a relatively simple system, it may nevertheless generate complex, chaotic dynamics; and that if we observe complex dynamics, the underlying explanation may nevertheless be simple. From an applied point of view, if ecology is to become a predictive and manipulative science, then we need to know the extent to which long-term prediction is threatened by one of the hallmarks of chaos—extreme sensitivity to initial conditions. As we have seen, however, the key practical question—how common is chaos?—remains largely unanswered. Further progress requires further advances in analytical techniques, especially those that can deal with the relatively short runs of data available for ecological systems (tens or, rarely, hundreds of data points rather than the thousands or more sometimes available in other spheres). But, more longer and

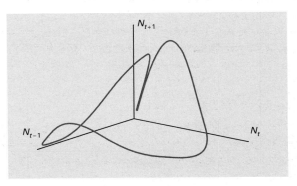

Figure 6.28 The apparently 'strange' attractor, suggestive of chaotic dynamics, generated by the time series of population sizes when parameter values for Equation 6.20 are estimated by response surface methodology for the aphid *Phyllaphis fagi*. (After Turchin & Taylor, 1992.)

more reliable runs of data are also required. These may emerge from studies of experimental, laboratory populations, but natural populations cannot be so readily rushed. Seeking answers to questions about chaos is likely to require great patience, and a willingness to fund and undertake long-term studies of population dynamics in the field.

6.10 Continuous breeding: the logistic equation

The model derived and discussed in Section 6.8 was appropriate for populations that have discrete breeding seasons and can therefore be described by equations growing in discrete steps, i.e. by 'difference' equations. Such models are not appropriate, however, for those populations in which birth and death are continuous. These are best described by models of continuous growth, or 'differential' equations, which will be considered next.

The net rate of increase of such a population will be denoted by dN/dt (referred to in speech as 'dN by dt'). This represents the 'speed' at which a population increases in size, N, as time, t, progresses. The increase in size of the whole population is the sum of the contributions of the various individuals within it. Thus, the average rate of increase per individual, or the 'per capita rate of increase' is given by $dN/dt(1/N)$. But, we have already seen in Chapter 4, Section 4.7, that in the absence of competition, this is the definition of the 'intrinsic rate of natural increase', r. Thus:

the intrinsic rate of natural increase: r

$$\frac{dN}{dt}\left(\frac{1}{N}\right) = r \qquad (6.21)$$

and:

$$\frac{dN}{dt} = rN. \qquad (6.22)$$

A population increasing in size under the influence of Equation 6.22, with $r > 0$, is shown in Figure 6.29. Not surprisingly, there is unlimited, 'exponential' increase. In fact, Equation 6.22 is the continuous form of the exponential difference Equation 6.9. Indeed, as discussed in Chapter 4, Section 4.7, r is simply $\log_e R$. (Mathematically adept readers will see that Equation 6.22 can be obtained by differentiating Equation 6.9.) R and r are clearly measures of the same commodity: 'birth plus

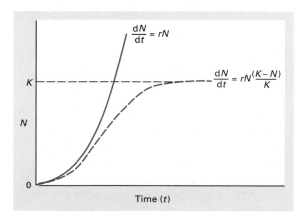

Figure 6.29 Exponential (—) and sigmoidal (− −) increase in density (N) with time for models of continuous breeding. The equation giving sigmoidal increase is the logistic equation.

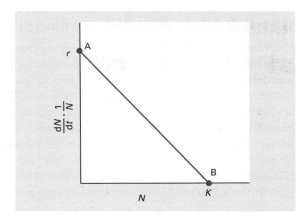

Figure 6.30 The simplest, straight-line way in which the rate of increase per individual $(dN/dt(1/N))$ might fall with density (N). For further explanation, see text.

survival' or 'birth minus death'; the difference between R and r is merely a change of currency.

For the sake of realism, intraspecific competition must obviously be added to Equation 6.22. This can be achieved most simply by the method set out in Figure 6.30, which is exactly equivalent to the one used in Figure 6.23. The net rate of increase per individual is unaffected by competition when N is very close to zero. It is still therefore given by r (point A). When N rises to K (the carrying capacity) the net rate of increase per individual is zero (point B). As before, a straight line between A and B is assumed. Thus:

$$\frac{dN}{dt}\left(\frac{1}{N}\right) = \left(\frac{-r}{K}\cdot N\right) + r \tag{6.23}$$

or, rearranging:

$$\frac{dN}{dt}\left(\frac{1}{N}\right) = r\left(1 - \frac{N}{K}\right) \tag{6.24}$$

and:

$$\frac{dN}{dt} = rN\left(\frac{K-N}{K}\right). \tag{6.25}$$

the logistic equation

This is known as the logistic equation (coined by Verhulst, 1838), and a population increasing in size under its influence is shown in Figure 6.29.

The logistic equation is the continuous equivalent of Equation 6.13, and it therefore has all the essential characteristics of Equation 6.13, and all of its shortcomings. It describes a sigmoidal growth curve approaching a stable carrying capacity, but it is only one of many reasonable equations which do this. Its major advantage is its simplicity. Moreover, whilst it was possible to incorporate a range of competitive intensities into Equation 6.13, this is by no means easy with the logistic. The logistic is therefore doomed to be a model of perfectly compensating density dependence. Nevertheless, in spite of these limitations, the equation will be an integral component of models in Chapters 7 and 10, and it has played a central role in the development of ecology. Note, finally, that like Equation 6.13, the behaviour of the logistic is conventionally seen as dependent on r and K, whereas there is much to be said for seeing it as dependent on r and q ($= r/K$), the individuals' susceptibility to crowding (see Section 6.8.1; Kuno, 1991).

6.11 Individual differences: asymmetric competition

Until now, we have focused on what happens to the whole population or the average individual within it. Intraspecific competition, however, acts directly on individuals, and different individuals may respond in very different ways. Figure 6.31 shows the results of an experiment in which flax (*Linum usitatissimum*) was sown at three densities, and harvested at three stages of development, recording the weight of each plant individually. This made it possible to monitor the effects of increasing amounts of competition not only as a result of variations in sowing density, but also as a result of plant growth (between the first and the last harvests). When intraspecific competition was at its least intense (at the lowest sowing density after only 2-weeks' growth) the individual plant weights were distributed symmetrically about the mean. When competition was at its most intense, however, the distribution was strongly skewed to the left: there were many very small individuals and a few large ones. As

competition can lead to skewed weight distributions within populations

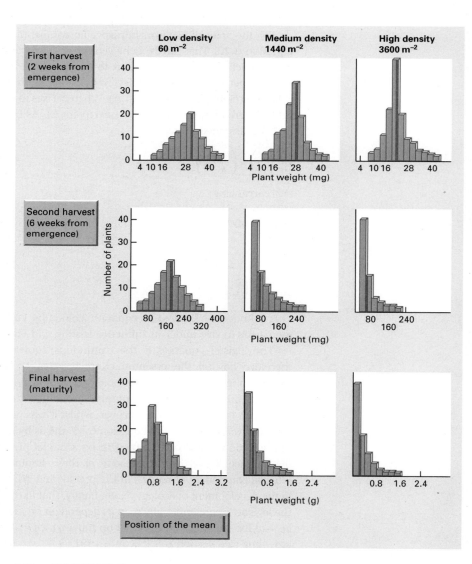

Figure 6.31 Competition and a skewed distribution of plant weights. Frequency distributions of individual plant weights in populations of flax (*Linum usitatissimum*), sown at three densities and harvested at three ages. (After Obeid *et al.*, 1967.)

the intensity of competition gradually increased, the degree of skewness increased as well. More generally, we may also say that increased competition increased the degree of size inequality within the population, i.e. the extent to which total biomass was unevenly distributed amongst the different individuals (Weiner, 1990). Rather similar results have been obtained from natural populations of limpets (see Figure 6.15) and from a number of other populations of animals (Uchmanski, 1985) and plants (Uchmanski, 1985; Weiner & Thomas, 1986). Typically, populations experiencing the most intense competition have the greatest size inequality and often have a size distribution in which there are many small and a few large individuals. Characterizing a population by an arbitrary 'average' individual can obviously be very misleading under such circumstances, and can divert attention from the fact that intraspecific competition is a force affecting individuals, even though its effects may often be detected in whole populations.

individual differences are exaggerated in plants by pre-emption of space

An indication of the way in which competition can exaggerate underlying inequalities in a population comes from the results of an experiment in which the grass *Dactylis glomerata*, was sown at random on a soil surface, and seedlings were marked as they emerged, so that their time of emergence would be known subsequently (Figure 6.32). Seven weeks after the start of the experiment, all plants were harvested and weighed. Not surprisingly, those that had emerged earliest (and had grown longest) were the largest. Yet, if weights had been determined by the length of the growing period alone, they would have fallen on the dashed line in Figure 6.32. Instead, the later a plant emerged, the further it fell below this expected line (the slower it grew). Thus, the later a plant emerged, the more it was affected by neighbours that had become established earlier. Plants which emerged early pre-empted or 'captured' space, and were little affected by intraspecific competition subsequently. Plants which emerged later entered a universe in which most of the available space had already been pre-empted; they were therefore greatly affected by intraspecific competition. Competition was asymmetric: there was a heirarchy. Some individuals were affected far more than others, and small initial differences were transformed by competition into a 1000-fold range of plant weights at the time of harvest.

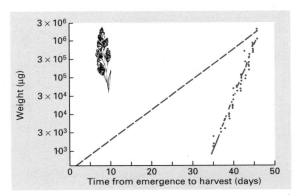

Figure 6.32 Space capture amongst individuals in a population of the grass *Dactylis glomerata*, which emerged over a period of around 10 days but were harvested simultaneously. Had weight been determined by the length of the growing period alone, the data would have fallen along the dashed line. Instead, those that emerged later (shorter period from emergence to harvest) grew less than expected, because space had been pre-empted or captured by those that emerged earlier. (After Ross & Harper, 1972.)

A similar result has been obtained for a natural, crowded population of the woodland annual *Impatiens pallida*, in south-eastern Pennsylvania. Over an 8-week period of very rapid plant growth, growth was very much faster in large than in small plants—in fact, small plants did not grow at all (Figure 6.33a). This increased significantly the size inequality within the population (Figure 6.33b).

If competition is asymmetric because superior competitors pre-empt resources, then competition is most likely to be asymmetric when it occurs for resources that are most liable to be pre-empted. Specifically, competition amongst plants for light, in which a superior competitor can overtop and shade an inferior, might be expected to lend itself far more readily to pre-emptive resource capture than competiton for soil nutrients or water, where the roots of even a very inferior competitor will have more immediate access to at least some of the available resources than the roots of its

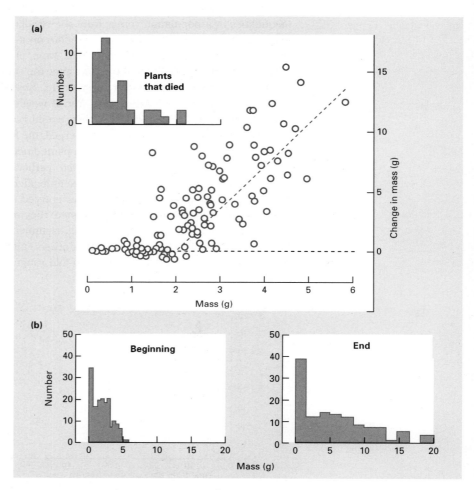

Figure 6.33 Asymmetric competition in a natural population of *Impatiens pallida*. (a) The increase in mass of survivors of different sizes over an 8-week period, and the distribution of initial sizes of those individuals that died over the same period. The horizontal axis is the same in each case. (b) The distribution of individual weights at the beginning (Gini coefficient, a measure of inequality, 0.39) and the end of this period (Gini coefficient, 0.48). (After Thomas & Weiner, 1989.)

superiors. This expectation is borne out by the results of an experiment in which morning glory vines (*Ipomoea tricolor*) were grown as single plants in pots ('no competition'), as several plants rooted in their own pots but with their stems intertwined on a single stake ('shoots competing'), as several plants rooted in the same pot, but with their stems growing up their own stakes ('roots competing') and as several plants rooted in the same pot with their stems intertwined on one stake ('shoots and roots competing') (Figure 6.34). Despite the fact that root competition was more intense than shoot competition, in the sense that it led to a far greater decrease in the mean weight of individual plants, it was shoot competition for light that led to a much greater increase in size inequality.

competitive asymmetries can be manifested in other ways too

Skewed distributions are one possible manifestation of hierarchical, asymmetric competition, but there are many others. For instance, Rubenstein (1981), studied competition in populations of the Everglades pygmy sunfish (*Elassomia evergladei*). An increase in density led to marked decreases in growth rate, in the fecundity of females and in the reproductive activity of males—or at least it did when 'average' individuals at high and low densities were compared. But, the 'best' individuals were only very slightly affected by competition. This emphasizes that intraspecific competition is not only capable of exaggerating individual differences, it is also greatly affected by individual differences. In particular, there is often a marked asymmetry between different stages or age classes in the same population. For example, when 'large' larvae (penultimate larval instar) of the damselfly, *Ischnura elegans*, compete with 'small' larvae (two instars earlier), with which they frequently

asymmetries between age classes or stages

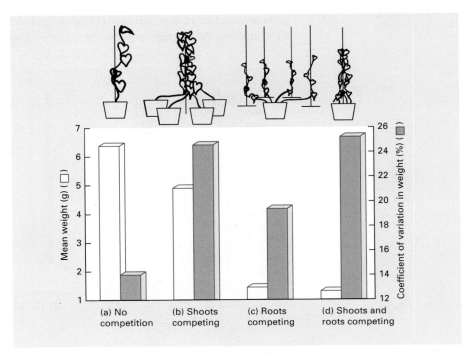

Figure 6.34 When morning glory vines competed, root competition was most effective in reducing mean plant weight (treatments significantly different, $P < 0.01$, for all comparisons except (c) with (d)), but shoot competition was most effective in increasing the degree of size inequality, as measured by the coefficient of variation in weight (significant differences between treatments (a) and (b), $P < 0.05$, and (a) and (d), $P < 0.01$). (After Weiner, 1986.)

251   INTRASPECIFIC COMPETITION

co-occur in the field, the growth and development rates of the small larvae are significantly depressed, whereas those of the large larvae are not affected (Gribbin & Thompson, 1990).

Asymmetric competition was observed on a much longer time scale in a population of the herbaceous perennial *Anemone hepatica*, growing in Sweden (Figure 6.35). Despite the crops of seedlings that entered the population between 1943 and 1952, it is quite clear that the most important factor determining which individuals survived to 1956 was whether or not they were established in 1943. Of the 30 individuals that had reached large or intermediate size by 1943, 28 survived until 1956, and some of these had branched. By contrast, of the 112 plants that were either small in 1943 or appeared as seedlings subsequently, only 26 survived to 1956, and not one of these was sufficiently well established to have flowered. Similar patterns can be observed in tree populations. The survival rates, the birth rates and thus the fitnesses of the few established adults are high; those of the many seedlings and saplings are comparatively low.

These considerations illustrate an important general point: asymmetries tend to reinforce the regulatory powers of intraspecific competition. Tamm's established plants were successful competitors year after year, but his small plants and seedlings were repeatedly unsuccessful. This guaranteed a near constancy in the number of established plants between 1943 and 1956. Each year there was a near-constant number of 'winners', accompanied by a variable number of 'losers' that not only failed to grow, but usually, in due course, died.

The interaction of asymmetries in growth and death is further illustrated by data from an experiment in which seeds of the annual plant *Lapsana communis* were sown by scattering them irregularly on the surface of sterilized compost. The seedlings

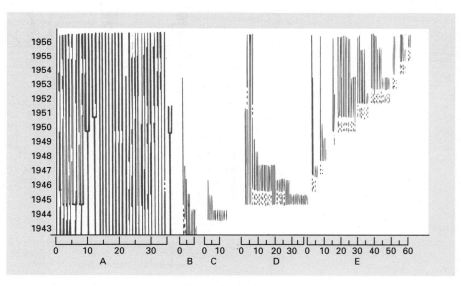

Figure 6.35 Space pre-emption in a perennial, *Anemone hepatica*, in a Swedish forest. Each line represents one individual: straight for unramified ones, branched where the plant has ramified, bold where the plant flowered and broken where the plant was not seen that year. Group A were alive and large in 1943, group B alive and small in 1943, group C appeared first in 1944, group D in 1945 and group E thereafter, presumably from seedlings. (After Tamm, 1956.)

emerged almost synchronously, and detailed maps were made of their distribution (Figure 6.36a). Then, the position of each individual with respect to its neighbours was described by constructing 'Thiessen polygons' from the perpendicular bisectors of the lines joining each plant to its neighbours (Figure 6.36b). After 15 weeks the population was mapped again, new polygons were drawn (Figure 6.36c), and the plants were dried and weighed. There were three major findings.

1 Plants with close neighbours (i.e. with small polygons) had the greatest risk of dying.

2 Survivors that had had close neighbours were smaller. Hence, there were both death and growth asymmetries, as there had been in the study of the *I. pallida* population in Figure 6.36a. In that case, the death rates were insufficient to prevent an increase in size inequality. In the present case, however, mortality was more influential.

asymmetries can also destroy skewness when they lead to the mortality of weaklings

3 Initially, therefore, the plant sizes and polygon areas had frequency distributions that were highly skewed (Figure 6.36d), but subsequently they became symmetrical (Figure 6.36e). Thus, whilst asymmetric competition can create a skewed distribution in a population, it can also destroy the skew by causing the mortality of the small weaklings.

6.12 Territoriality

Territoriality is one particularly important and widespread form of asymmetric intraspecific competition. It occurs when there is active interference between individuals, such that a more or less exclusive area, the territory, is defended against intruders by a recognizable pattern of behaviour.

Individuals of a territorial species that fail to obtain a territory often make no contribution whatsoever to future generations. Territoriality, then, is a 'contest'. There are winners (those that come to hold a territory) and losers (those that do not), and at any one time there can be only a limited number of winners. The exact number of territories (winners) is usually somewhat indeterminate in any one year, and certainly varies from year to year, depending on environmental conditions. Nevertheless, the contest nature of territoriality ensures, like asymmetric competition generally, a comparative constancy in the number of surviving, reproducing individuals. One important consequence of territoriality, therefore, is population regulation, or more particularly, the regulation of the number of territory holders. Thus, when territory owners die, or are experimentally removed, their places are often rapidly taken by newcomers. For instance, in great tit populations, vacated woodland territories are reoccupied by birds coming from hedgerows where reproductive success is noticeably lower (Krebs, 1971).

population regulation is a consequence of territoriality

Wynne-Edwards felt that the regulatory consequences of territoriality must themselves be the root causes underlying the evolution of territorial behaviour. He suggested that territoriality was favoured because the population as a whole benefited from the rationing effects of territoriality, which guaranteed that the population did not overexploit its resources (Wynne-Edwards, 1962; retracted by Wynne-Edwards, 1977; retraction retracted by Wynne-Edwards, 1986). However, there are powerful and fundamental reasons for rejecting this 'group-selectionist' explanation (essentially, it stretches evolutionary theory beyond reasonable limits): the ultimate cause of territoriality must be sought within the realms of natural selection, in some advantage accruing to the individual.

population regulation is not the cause of territoriality

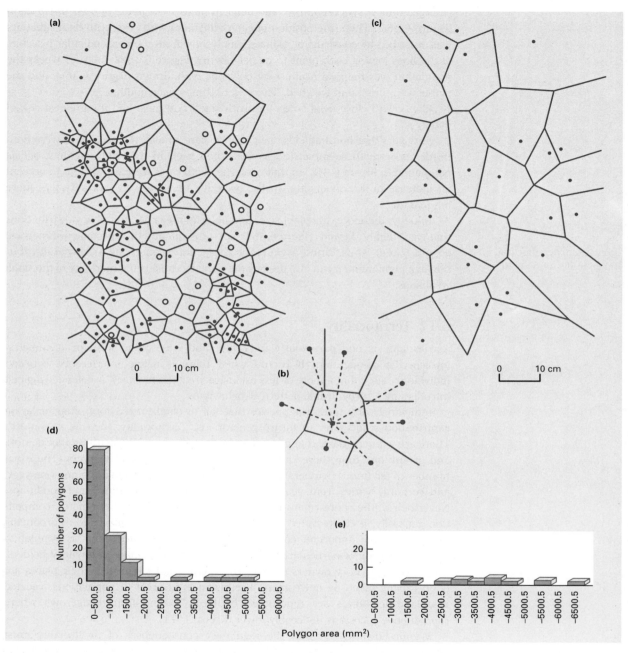

Figure 6.36 (a) The distribution of plants in a seedling population of the annual plant *Lapsana communis*, 3 days after seedling emergence. Each seedling is represented by a point, except for those that survived for at least 15 weeks, which are shown as circles. (b) The spatial distribution of sessile organisms (as in (a)) can be described by drawing 'Thiessen polygons', in which polygon sides are drawn at right-angles through the midpoint of lines joining neighbouring organisms: Theissen polygons are shown in (a), and also in (c) which shows the distribution of survivors after 15 weeks. (d) The frequency distribution of polygon areas 3 days after seedling emergence. (e) The frequency distribution of polygon areas 15 weeks after emergence, after self-thinning had occurred. (After Mithen *et al.*, 1984.)

Any benefit that an individual does gain from territoriality, of course, must be set against the costs of defending the territory. In some animals this defence involves fierce combat between competitors, whilst in others there is a more subtle mutual recognition by competitors of one another's keep-out signals (e.g. song or scent). Yet, even when the chances of physical injury are minimal, territorial animals typically expend energy in patrolling and advertising their territories, and these energetic costs must be exceeded by any benefits if territoriality is to be favoured by natural selection (see Davies & Houston, 1984).

In some cases, these benefits are clear and direct. For instance, in the great tit, territoriality greatly reduces the chances of eggs and incubating females being preyed upon (Krebs, 1971), especially by weasels (Dunn, 1977); and in Belding's ground squirrel (*Spermophilus beldingi*), larger territories lead to decreased chances of cannibalism of the young by other members of the same species (Sherman, 1981).

territorial individuals may gain
more energy (from food) than
they spend

It is probably more common, however, for the benefits to come in terms of an increased intake of energy (from food), which more than compensates for the energy expended in territorial defence. It has been possible, for instance, to estimate the time and energy budgets of golden-winged sunbirds (*Nectarinia reichenowi*) in East Africa where they defend patches of *Leonotis* flowers (Gill & Wolf, 1975). The nectar contents of territories were calculated along with the energetic costs of foraging for nectar (1000 cal h^{-1}), of sitting (400 cal h^{-1}) and of territory defence (3000 cal h^{-1}). By expending energy on territory defence (taking up about 0.28 h day^{-1}) when they would otherwise have been sitting, the birds were able to raise the average nectar content of their flowers (because no other birds were foraging from them). Therefore, because foraging brought a greater rate of return, they needed to spend around 1.3 h less per day foraging, in order to satisfy their energy needs. What they gained typically ($780 \text{ cal} = 1.3 \times (1000 - 400) \text{ cal}$) was greater than what they spent ($728 \text{ cal} = 0.28 \times (3000 - 400) \text{ cal}$).

On the other hand, explaining territoriality only in terms of a net benefit to the territory owner is rather like history always being written by the victors. There is another, possibly trickier question, which seems not to have been answered—could those individuals without a territory not do better by challenging the territory owners more often and with greater determination?

Of course, describing territoriality in terms of just 'winners' and 'losers' is an over-simplification. Generally, there are first, second and a range of consolation prizes—not all territories are equally valuable. This has been demonstrated in an unusually striking way in a study of oystercatchers (*Haematopus ostralegus*) on the coast of The Netherlands, where pairs of birds defend both nesting territories on the salt-marsh, and feeding territories on the mudflats (Ens *et al.*, 1992). For some birds ('residents'), the feeding territory is simply an extension of the nesting territory: they form one spatial unit. For other pairs, however ('leapfrogs'), the nesting territory is further inland and hence separated spatially from the feeding territory (Figure 6.37a). Residents fledge many more offspring than do leapfrogs (Figure 6.37b), because they deliver far more food to them (Figure 6.37c). From an early age, resident chicks follow their parents onto the mudflats, taking each prey item as soon as it is captured. Leapfrog chicks, however, are imprisoned on their nesting territory prior to fledging; all their food has to be flown in. It is far better to have a resident than a leapfrog territory.

Territoriality, then, can be seen as a flexible and subtle pattern of behaviour which has evolved as a result of the net advantages accruing to individual

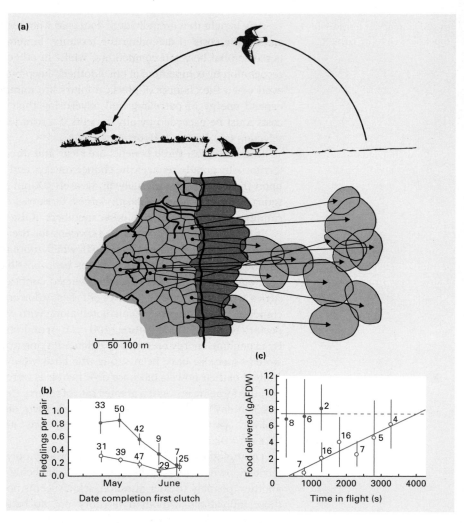

Figure 6.37 (a) A coastal area in The Netherlands providing both nesting and feeding territories for oystercatchers. In 'resident' territories (dark shading), nesting and feeding areas are adjacent and chicks can be taken from one to the other at an early age. 'Leapfrogs', however, have separate nesting and feeding territories (light shading) and food has to be flown in until the chicks fledge. (b) Residents (●) fledge more chicks than leapfrogs (○). (c) Residents (●) deliver more food per tide (grams of ash-free dry weight, with standard deviations) than leapfrogs (○). The latter deliver more, the more effort (in flying) they expend, but still cannot match the residents. (After Ens *et al.*, 1992.)

competitors. As an independent consequence of this, there is a particularly powerful regulatory influence on the populations concerned. This is just one (albeit extreme) example in which intraspecific competition is recognizably asymmetric.

6.13 Self-thinning

We have seen throughout this chapter that intraspecific competition can influence the number of deaths, the number of births and the amount of growth within a population. We have illustrated this largely by looking at the end results of competition. But, in practice, the effects are often progressive. As a cohort ages, the individuals grow in size, their requirements increase, and they therefore compete at a greater and greater intensity. This in turn tends gradually to increase their risk of dying. But, if some individuals die, then density and the intensity of competition are decreased—which affects growth, which affects competition, which affects survival, which affects density, and so on.

The patterns that emerge in ageing, crowded cohorts have been a focus of particular attention in plant populations. For example, perennial rye grass, *Lolium perenne*, was sown at a range of densities, and samples from each density were harvested after 14, 35, 76, 104 and 146 days (Figure 6.38a). Figure 6.38a has the same logarithmic axes—density and mean plant weight—as Figures 6.16b and c. It is most important to appreciate the difference between the two. In Figure 6.16, each line represented a separate yield–density relationship at different ages of a cohort. Successive points along a line represented different initial sowing densities. In Figure 6.38, each line itself represents a different sowing density, and successive points along a line represent populations of this initial sowing density at different ages. The lines are therefore trajectories that follow a cohort through time. This is indicated by arrows, pointing from many small, young individuals (bottom right) to fewer larger, older individuals (top left).

From 35 days onwards, plant growth was apparently slower in the higher density populations: mean plant weight (at a given age) was always greatest in the lowest density populations (Figure 6.38a). It is also clear that the highest density populations were the first to suffer substantial mortality. What is most noticeable in

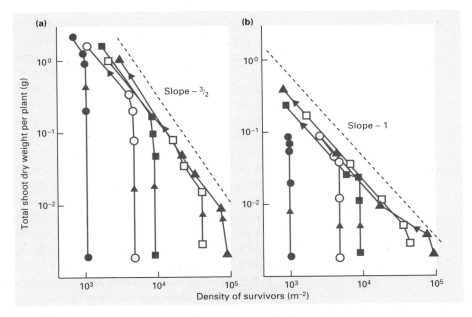

Figure 6.38 Self-thinning in *Lolium perenne* sown at five densities: 1000 (●), 5000 (○), 10 000 (■), 50 000 (□) and 100 000 (▲) 'seeds' m^{-2}. (a) 0% shade; (b) 83% shade. The lines join populations of the five sowing densities harvested on five successive occasions. They therefore indicate the trajectories, over time, that these populations would have followed. The arrows indicate the directions of the trajectories, i.e. the direction of self-thinning. For further discussion, see text. (After Lonsdale & Watkinson, 1982.)

these high-density populations, however, is that density declined and mean plant weight increased in unison: both populations progressed along a single straight line with a slope of approximately −3/2. The populations are said to have experienced *self-thinning* (i.e. a progressive decline in density in a population of growing individuals), and the line with a slope of roughly −3/2, which they approached and then followed, is known as a *dynamic thinning line* (Weller, 1990).

cohorts approach and then follow a dynamic thinning line

In fact, the lower the sowing density, the later was the onset of self-thinning. In all cases, though, the populations initially followed a trajectory that was almost vertical, i.e. there was little mortality. Then, as they neared the thinning line, the populations suffered increasing amounts of mortality, so that the slopes of all the self-thinning trajectories gradually approached −3/2. Finally, on reaching the thinning line, the populations progressed along it.

Self-thinning plant populations have repeatedly been found (if sown at sufficiently high densities) to approach and then follow a thinning line with a slope of roughly −3/2. The relationship has therefore often been referred to in the past as 'the −3/2 power law' (Yoda *et al.*, 1963; Hutchings, 1983), since density (d) is related to mean weight ($\bar{w}$) by the equation:

'the −3/2 power law'

$$\log \bar{w} = \log c - 3/2 \log d \qquad (6.26)$$

or:

$$\bar{w} = c d^{-3/2} \qquad (6.27)$$

where c is constant.

Note, in passing, that Figure 6.38 has been drawn, following convention, with log density on the *x*-axis and log mean weight on the *y*-axis. This is not meant to imply that density is the independent variable on which mean weight depends. Indeed, it can be argued that mean weight increases naturally during plant growth, and this determines the decrease in density. The most satisfactory view is that density and mean weight are wholly interdependent: neither is independent of the other.

A slope of −3/2 indicates that in a growing, self-thinning population, mean plant weight increases faster than density decreases. A population following a −3/2 thinning line will therefore steadily increase its total weight (or yield). Eventually, of course, this must stop: yield cannot increase indefinitely. Instead, the thinning line might be expected to change from a slope of −3/2 to a slope of −1, such that the total weight per unit area remains constant. This actually occurred when populations of *L. perenne* (Figure 6.38b) were grown at low light intensities (17% of full light). A slope of −1 implies that the further growth of survivors exactly balances the deaths of other individuals, i.e. it seems to occur when total yield reaches a maximum which cannot be exceeded by that species in that environment. In an environment with reduced light intensities, maximum yield is reduced; slopes of −1 are thus apparent at lower densities than they would otherwise be (Figure 6.38b).

thinning slopes of −1

In fact, in many cases where self-thinning relationships have been documented, it is not a single cohort that has been followed over time, but a series of crowded populations of different ages that have been compared. This has been especially true with populations of trees and other long-lived species. In such cases, it is more correct to speak of a *species boundary line*—a line beyond which combinations of density and mean weight are not possible for that species (Weller, 1990). It has often been assumed that the dynamic thinning line and the species boundary line are one and the same: self-thinning cohorts approach the boundary line and then thin along

species boundary lines

it. This may often be true—but, it is by no means invariably so. We have only to compare Figures 6.38a and b to see that the light regime can alter the dynamic thinning line. So too can soil fertility and even the spatial arrangement of seedlings at the time of establishment (Weller, 1990). Of course, the thinning line can never be above the boundary line.

Intriguingly, when the thinning and boundary lines of all sorts of plants are plotted on the same figure, they all appear to have approximately the same slope (roughly −3/2) and also to have intercepts (i.e. values of c in the equation) falling within a narrow range (Figure 6.39). To the lower right of the figure are high-density

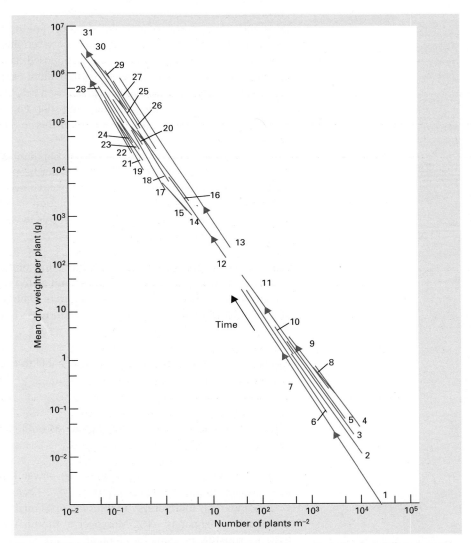

Figure 6.39 Self-thinning in a wide variety of herbs and trees. Each line is a different species, and the line itself indicates the range over which observations were made. The arrows, drawn on representative lines only, indicate the direction of self-thinning over time. The figure is based on Figure 2.9 of White (1980), which also gives the original sources and the species names for the 31 data sets.

259 INTRASPECIFIC COMPETITION

populations of small plants (annual herbs and perennials with short-lived shoots), whilst to the upper left are sparse populations of very large plants, including coastal redwoods (*Sequoia sempervirens*), the tallest known trees. Fashions change in science as in everything else. At one time, ecologists looked at Figure 6.39 and saw uniformity—all plants marching in −3/2 time (White, 1980; previous editions of this text), with variations from the norm seen as either 'noise' or as only of minor interest. More recently, however, it has become more popular to cast serious doubt on the conformity of individual slopes to −3/2, and on the whole idea of a single, ideal thinning line—whether it has a slope of −3/2 or anything else (Weller, 1987, 1990, 1991; Zeide, 1987; Lonsdale, 1990). There really is no contradiction, though. On the one hand, the lines in Figure 6.39 occupy a very much smaller portion of the graph than one would expect by chance alone, and hence there *is* apparently some fundamental phenomenon linking this whole spectrum of plant types. On the other hand, the variations between lines are real and important and in as much need of explanation as any supposed 'general' rule—which is probably inapplicable in detail to any one of the individual examples. We proceed, therefore, first by suggesting why all plants might have a self-thinning slope of approximately −3/2 and roughly the same intercept, and then enquiring why each species might display its own variation on this common theme (Long & Smith, 1984; Osawa & Allen, 1993).

different species share roughly the same thinning line—or do they?

explanations for the slopes and intercepts

In a growing cohort, as the mass of the population increases, the leaf area index (L, the leaf area per unit area of land) does not keep on increasing. Instead, beyond a certain point, it remains constant irrespective of plant density (N). It is, in fact, precisely beyond this point that the population is believed to follow the dynamic thinning line (Figure 6.40). We can express this by writing:

$$L = \lambda N = \text{constant} \tag{6.28}$$

where λ is the mean leaf area per surviving plant. However, the leaf area of individual plants increases as they grow, and so too therefore does their mean, λ. It is reasonable to expect λ, because it is an area, to be related to linear measurements of a plant, such as stem diameter, D, by a formula of the following type:

$$\lambda = a D^2 \tag{6.29}$$

where a is a constant. Similarly, it is reasonable to expect mean plant weight, $\bar{w}$, to be related to D by:

$$\bar{w} = b D^3 \tag{6.30}$$

where b is also a constant. Putting Equations 6.28–6.30 together, we obtain:

$$\bar{w} = b(L/a)^{3/2} \cdot N^{-3/2} \tag{6.31}$$

This is structurally equivalent to the −3/2 power law in Equation 6.27, with the intercept constant, c, given by $b(L/a)^{3/2}$.

It is apparent, therefore, why thinning lines might generally be expected to have slopes of approximately −3/2. Moreover, if the relationships in Equations 6.29 and 6.30 are roughly the same for all plant species, and if all plants support roughly the same leaf area per unit area of ground (L), then the constant c will be approximately the same for all species. On the other hand, suppose that L is not quite constant for some species (Equation 6.28), or that the powers in Equations 6.29 and 6.30 are not exactly 2 or 3, or that the constants in these equations (a and b) either vary between

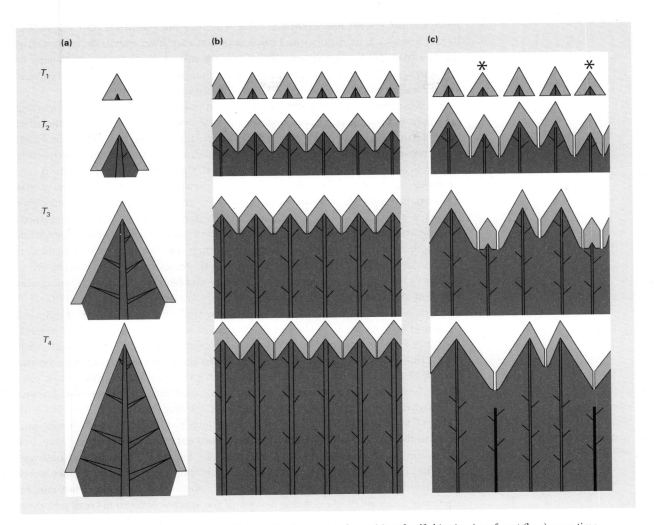

Figure 6.40 Growth of an isolated tree (a) and self-thinning in a forest (b, c), over time (T_1–T_4), in diagrammatic form. The outer zone in each diagram represents the canopy of leaves and the region in which buds are alive and active. Here the birth rate of buds exceeds their death rate and new additions are continually being made to the architectural skeleton of the canopy. In the shaded zone beneath the canopy, the death rate of buds (or their enforced dormancy) exceeds their birth rate, many of the smaller branches die, but those that extend into the canopy become secondarily thickened and persistent. In (b), a forest 'monoculture', all competition is symmetric. The canopy becomes continuous between T_1 and T_2, reaching its 'ceiling' LAI. Thereafter, the density of the canopy scarcely changes, but it becomes higher and higher, whilst the zone beneath becomes deeper and most of the branches within it die. The difference between (b) and (c) is that in (c), more realistically, competition is asymmetric. Individuals marked * may have germinated later, may have developed from smaller seeds or may be of a competitively inferior genotype (or species, in the case of interspecific competition—see Chapter 7). By T_2 these individuals are shorter and contribute proportionately less to the canopy. The difference has become exaggerated by T_3, and by T_4 these suppressed individuals are dead. Note that the trees whose neighbours have died have asymmetrical crowns, reflecting the different intensities of competition experienced on their sides.

species or are not actually constants at all. Thinning lines will then have slopes that depart from −3/2, and slopes and intercepts that vary from species to species. It is easy to see why, at first glance, there is a broad similarity in the behaviour of different species, but also why, on closer examination, there are variations between species and no such thing as a single, 'ideal' thinning line.

Furthermore, contrary to the implicit assumptions of the argument supporting Equations 6.28–6.31, it is not necessarily the case that the yield–density relationship in a growing cohort depends only on the numbers that die and the manner in which the survivors grow. We have seen (see Section 6.11) that competition is frequently highly asymmetric (see also Figure 6.40c). If those that die in a cohort are predominantly the very smallest individuals, then such mortality will have little effect on the total yield of the cohort. 'Asymmetric' mortality may therefore itself give rise to a significant increase in mean weight without there being any growth amongst the survivors. This cannot, of course, be the whole self-thinning story. Even if, hypothetically, those that died weighed nothing, the thinning line could have a slope no steeper than −1, in the absence of growth. Nevertheless, variations between species in the nature and extent of asymmetric competition will be another reason why they differ in their thinning lines.

Is it possible, though, to use departures from the assumptions built into Equations 6.28–6.30, to explain at least some of the variations from the 'general' −3/2 rule? One study suggests that the answer may be 'yes'. Osawa and Allen (1993) estimated a number of the parameters in these equations from data on the growth of individual plants of mountain beech (*Nothofagus solandri*) and red pine (*Pinus densiflora*). They estimated, for instance, that the exponents in Equations 6.29 and 6.30 were not 2 and 3, but 2.08 and 2.19 for mountain beech, and 1.63 and 2.41 for red pine. These suggest thinning slopes of −1.05 in the first case and −1.48 in the second, which compare quite remarkably well with observed slopes of −1.06 and −1.48 (Figure 6.41a). The similarities between estimates and observations for the intercept constants were equally impressive. These results show therefore that thinning lines with slopes other than −3/2 can occur, but can be explicable in terms of the detailed biology of the species concerned—and that even when slopes of −3/2 do occur, they may do so, as with red pine, for the 'wrong' reason (−2.41/1.63 rather than −3/2!).

The results also hold out the possibility of a classification of thinning lines (and species boundaries) in which particular groups of plant that share a common growth pattern also share a broadly similar combination of slope and intercept. Some simple intergroup patterns are apparent in any case. The trees are located to the left in Figure 6.39 because their canopies maintain a very considerable weight of tissue, which they can do because most of that tissue is dead (necromass rather than biomass—see Figure 6.40), and requires very little maintenance. Grasses, by contrast, to the right, can maintain relatively little tissue, because most of it does indeed require active maintenance. Beyond that, suggestions regarding the relative positions of, for example, coniferous and deciduous trees or grasses and dicotyledonous herbs on plots like Figure 6.39, have in the past been contradictory (Lonsdale & Watkinson, 1983; Lonsdale, 1990). Needless to say, there are no agreed reasons for any differences between groups. The resolution of these difficulties awaits future developments.

Animals, whether they are sessile or mobile, must also 'self-thin', insofar as growing individuals within a cohort increasingly compete with one another and

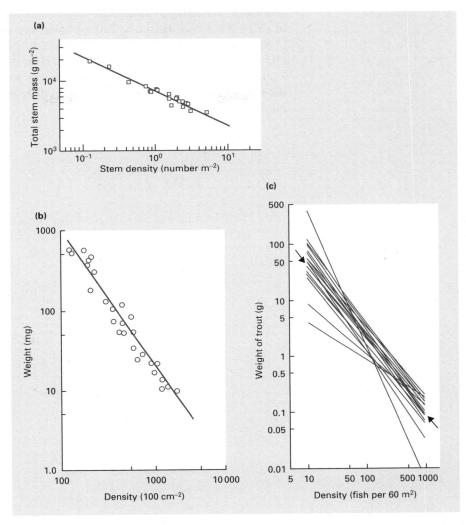

Figure 6.41 Further examples of self-thinning. (a) Species boundary line for populations of red pine, *Pinus densiflora* (slope = −1.48) from northern Japan. (After Osawa & Allen, 1993.) (b) Species boundary line for populations of the barnacle, *Semibalanus balanoides* (slope = −1.62) from North Wales. (After Hughes & Griffiths, 1988.) (c) Dynamic thinning lines for 23 year-classes of sea trout, *Salmo trutta*, from an English Lake District stream, with the position of the mean regression line (slope = −1.35) indicated by the arrows. (After Elliott, 1993.)

self-thinning in animals

reduce their own density. However, since there is nothing linking all animals quite like the shared need for light interception that links all plants, there is even less likelihood of a general self-thinning 'law' for animals. On the other hand, crowded sessile animals can, like plants, be seen as needing to pack 'volumes' beneath an approximately constant area, and it is therefore interesting that mussels have been found to follow a thinning line with a slope of −1.41 and barnacles a line with a slope of −1.62 (Hughes & Griffiths, 1988; Figure 6.41b). For mobile animals, it has been suggested that energetic constraints could generate thinning lines with slopes of −4/3 (−1.33) (Begon *et al.*, 1986). Whilst it is also suggested that this would be especially

263   INTRASPECIFIC COMPETITION

sensitive to the nature of the animal–resource interaction, an average thinning slope of −1.29 has indeed been obtained for laboratory populations of the grasshopper, *Chorthippus brunneus* (range for different cohorts:−1.03 to −1.58) (Begon *et al.*, 1986) and one of −1.35 obtained for field populations of the trout, *Salmo trutta* (range: −2.42 to −0.68) (Figure 6.41c). Plants are not so consistent in their pattern of self-thinning as was once thought. It may be that animals are not much less bound than plants by 'general' self-thinning rules.

Chapter 7
Interspecific Competition

7.1 Introduction

The essence of interspecific competition is that individuals of one species suffer a reduction in fecundity, survivorship or growth as a result of resource exploitation or interference by individuals of another species. This competition is likely to affect the population dynamics of the competing species, and the dynamics, in their turn, can influence the species' distributions and their evolution. Of course, evolution, in *its* turn, can influence the species' distributions and dynamics. Here, we concentrate on the effects of competition on species' populations, whilst Chapter 20 examines its role in shaping the structure of ecological communities. There are several themes introduced in this chapter that are taken up and discussed more fully in Chapter 20. The two chapters should be read together for a full coverage of interspecific competition.

7.2 Some examples of interspecific competition

There have been many studies of interspecific competition between species of all kinds. We have chosen six initially, to illustrate a number of important ideas.

7.2.1 Competition between salamanders

The first example concerns two species of terrestrial salamanders, *Plethodon glutinosus* and *P. jordani*, which live in the southern Appalachian Mountains of the USA. Generally, *P. jordani* lives at higher altitudes than *P. glutinosus*, but in certain areas their altitudinal distributions overlap. Hairston (1980) carried out an experiment at two sites, one in the Great Smoky Mountains and the other in the Balsam Mountains. These sites both had populations of the two salamander species, they had similar salamander faunas overall, they were at the same elevation and they faced the same direction. At both sites, seven experimental plots were established in 1974: two from which *P. jordani* was removed, two from which *P. glutinosus* was removed and three as controls. Then, six times in each of the next 5 years, the numbers of both species and their ages were estimated in all plots.

In the control plots, and naturally, *P. jordani* was by far the more abundant of the two species; and in the plots from which it was removed, there was a statistically significant increase in the abundance of *P. glutinosus* at both locations. In the plots from which *P. glutinosus* was removed, there was no significant reciprocal increase in the abundance of *P. jordani*. However, there was, at both sites, a statistically significant increase in the proportion of *P. jordani* in the 1- and 2-year-old age classes. This was presumably a result of increased fecundity and/or increased survival of young, both of which are crucial components of the basic reproductive rate.

The important point is that individuals of both species must, originally, have been adversely affected by individuals of the other species, since when one species was removed, the remaining species showed a significant increase in abundance and/or fecundity and/or survivorship. It appears, therefore, that in the control plots and in the other zones of overlap generally, these species competed with one another but still coexisted.

7.2.2 Competition between bedstraws (*Galium* spp.)

A.G. Tansley, one of the greatest of the 'founding fathers' of plant ecology, studied competition between two species of bedstraw (Tansley, 1917). *Galium hercynicum* is a species which grows in Great Britain at acidic sites, whilst *G. pumilum* is confined to more calcareous soils (in Tansley's time they were known as *G. saxatile* and *G. sylvestre*). Tansley found that as long as he grew them alone, both species would thrive on both the acidic soil from a *G. hercynicum* site and the calcareous soil from a *G. pumilum* site. Yet, if the species were grown together, only *G. hercynicum* grew successfully in the acidic soil and only *G. pumilum* grew successfully in the calcareous soil. It seems, therefore, that when they grow together the species compete, and that one species wins, whilst the other loses so badly that it is competitively excluded from the site. The outcome depends on the habitat in which the competition occurs.

7.2.3 Competition between barnacles

The third study concerns two species of barnacle in Scotland: *Chthamalus stellatus* and *Balanus balanoides* (Figure 7.1; Connell, 1961). These species are frequently found together on the same Atlantic rocky shores of north-west Europe. However, adult *Chthamalus* generally occur in an intertidal zone which is higher up the shore than that of adult *Balanus*, even though young *Chthamalus* settle in considerable numbers in the *Balanus* zone. In an attempt to understand this zonation, Connell monitored the survival of young *Chthamalus* in the *Balanus* zone. He took successive censuses of mapped individuals over the period of 1 year and, most important, he ensured at some sites that young *Chthamalus* that settled in the *Balanus* zone were kept free from contact with *Balanus*. In contrast with the normal pattern, such individuals survived well, irrespective of the intertidal level. Thus, it seemed that the usual cause of mortality in young *Chthamalus* was not the increased submergence times of the lower zones, but competition from *Balanus* in those zones. Direct observation confirmed that *Balanus* smothered, undercut or crushed *Chthamalus*, and the greatest *Chthamalus* mortality occurred during the seasons of most rapid *Balanus* growth. Moreover, the few *Chthamalus* individuals that survived 1 year of *Balanus* crowding were much smaller than uncrowded ones, showing, since smaller barnacles produce fewer offspring, that interspecific competition was also reducing fecundity.

Thus, *Balanus* and *Chthamalus* compete. They coexist on the same shore, but on a finer scale their distributions overlap very little. *Balanus* outcompetes and excludes *Chthamalus* from the lower zones; but *Chthamalus* can survive in the upper zones where *Balanus*, because of its comparative sensitivity to desiccation, cannot.

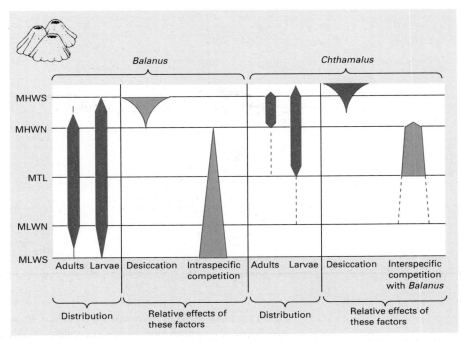

Figure 7.1 The intertidal distribution of adults and newly settled larvae of *Balanus balanoides* and *Chthamalus stellatus*, with a diagrammatic representation of the relative effects of desiccation and competition. Zones are indicated to the left: from MHWS (mean high water, spring) down to MLWS (mean low water, spring). (After Connell, 1961.)

7.2.4 Competition between *Paramecium* species

The fourth example comes from the classic work of the great Russian ecologist G.F. Gause, who studied competition in laboratory experiments using three species of the protozoan *Paramecium* (Gause, 1934, 1935). All three species grew well alone, reaching stable carrying capacities in tubes of liquid medium. There, *Paramecium* consumed bacteria or yeast cells, which themselves lived on regularly replenished oatmeal (Figure 7.2a).

When Gause grew *P. aurelia* and *P. caudatum* together, *P. caudatum* always declined to the point of extinction, leaving *P. aurelia* as the victor (Figure 7.2b). *P. caudatum* would not normally have starved to death as quickly as it did, but Gause's experimental procedure involved the daily removal of 10% of the culture and animals. Thus, *P. aurelia* was successful in competition because near the point where its population size levelled off, it was still increasing by 10% day^{-1} (and able to counteract the enforced mortality), whilst *P. caudatum* was only increasing by 1.5% day^{-1} (Williamson, 1972).

By contrast, when *P. caudatum* and *P. bursaria* were grown together, neither species suffered a decline to the point of extinction—they coexisted. But, their stable densities were much lower than when grown alone (Figure 7.2c), indicating that they were in competition with one another. A closer look, however, revealed that although they lived together in the same tubes, they were, like Connell's barnacles, spatially separated. *P. caudatum* tended to live and feed on the bacteria suspended in the medium, whilst *P. bursaria* was concentrated on the yeast cells at the bottom of the tubes.

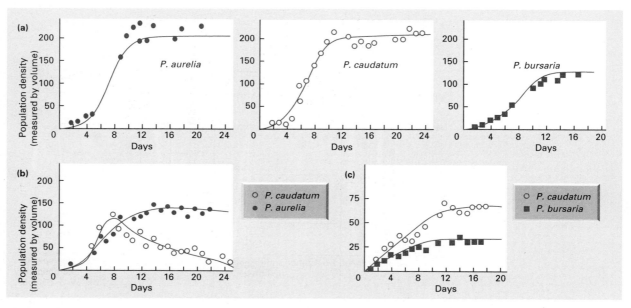

Figure 7.2 Competition in *Paramecium*. (a) *P. aurelia*, *P. caudatum* and *P. bursaria* all establish populations when grown alone in culture medium. (b) When grown together, *P. aurelia* drives *P. caudatum* towards extinction. (c) When grown together, *P. caudatum* and *P. bursaria* coexist, although at lower densities than when alone. (After Clapham, 1973; from Gause, 1934.)

7.2.5 Coexistence amongst tits

The next study considers the coexistence of five closely related bird species in English broad-leaved woodlands: the blue tit (*Parus caeruleus*), the great tit (*P. major*), the marsh tit (*P. palustris*), the willow tit (*P. montanus*) and the coal tit (*P. ater*) (Lack, 1971). Four of these species weigh between 9.3 g and 11.4 g on average (the great tit weighs 20.0 g); all have short beaks and hunt for food chiefly on leaves and twigs, but at times on the ground; all eat insects throughout the year, and also seeds in winter; and all nest in holes, normally in trees. Nevertheless, in Marley Wood, Oxford, where Lack studied them, all five species bred, and the blue, great and marsh tits were common. In short, they coexisted and seemed very similar in where they lived, where they bred and what they ate. The closer Lack examined these species, however, the more he concluded that they were separated, at most times of the year, by precisely where within the trees they fed, by the size of their insect prey and by the hardness of the seeds they took. He found, too, that this separation was associated with (often minor) differences in overall size, and in the size and the shape of the birds' beaks. Despite their similarities, these coexisting species exploited slightly different resources in slightly different ways.

7.2.6 Competition between diatoms

The final example is from a laboratory investigation of two species of freshwater diatom: *Asterionella formosa* and *Synedra ulna* (Tilman *et al.*, 1981). Both these algal species require silicate in the construction of their cell walls, and the investigation

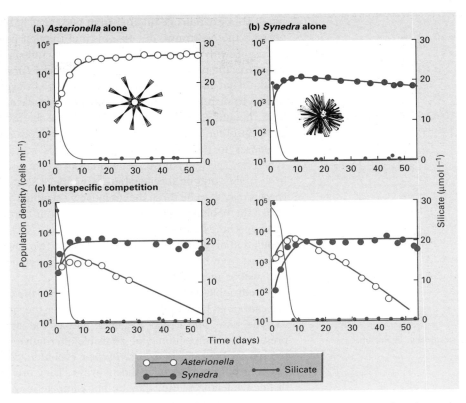

Figure 7.3 Competition between diatoms. (a) *Asterionella formosa*, when grown alone in a culture flask, establishes a stable population and maintains a resource, silicate, at a constant low level. (b) When *Synedra ulna* is grown alone it does the same, but maintains silicate at an even lower level. (c) When grown together, in two replicates, *Synedra* drives *Asterionella* to extinction. (After Tilman *et al.*, 1981.)

was unusual, because at the same time as population densities were being monitored, the impact of the species on their limiting resource (silicate) was being recorded. When either species was cultured alone in a liquid medium to which resources were continuously being added, it reached a stable carrying capacity whilst maintaining the silicate at a constant low concentration (Figures 7.3a, b). However, in exploiting this resource, *Synedra* reduced the silicate concentration to a lower level than did *Asterionella*. Hence, when the two species were grown together, *Synedra* maintained the concentration at a level which was too low for the survival and reproduction of *Asterionella*. *Synedra* therefore competitively excluded *Asterionella* from mixed cultures (Figure 7.3c).

7.3 Assessment: some general features of interspecific competition

These examples show that individuals of different species can compete. This is hardly surprising. The salamanders and barnacles also show that different species do compete in nature (i.e. there was a measurable interspecific reduction in abundance and/or fecundity and/or survivorship). It seems, moreover, that competing species may either exclude one another from particular habitats so that they do not coexist (as with the bedstraws, the diatoms and the first pair of *Paramecium* spp.), or may coexist (as with the salamanders), perhaps by utilizing the habitat in slightly different ways (e.g. the barnacles and the second pair of *Paramecium* spp.).

coexisting competitors or 'the ghost of competition past'?

But what about the tit study? Certainly the five species coexisted and utilized the habitat in slightly different ways. But did this have anything to do with competition? Lack thought so. In Connell's (1980) phrase, he invoked 'the ghost of competition past'. In other words, he believed that they coexisted as a result of evolutionary responses to interspecific competition. This requires some further explanation. When two species compete, individuals of one or both species may suffer reductions in fecundity and/or survivorship, as we have seen. The fittest individuals of each species may then be those that (relatively speaking) escape competition because they utilize the habitat in ways that differ most from those adopted by individuals of the other species. Natural selection will then favour such individuals, and eventually the population may consist entirely of them. The two species will evolve to become more different from one another than they were previously, they will compete less, and thus will be more likely to coexist.

or simply, evolution?

The trouble with this as an explanation for the tit data is that there is no proof. We cannot go back in time to check whether the species ever competed more than they do now, and it was not even part of the study to determine the extent of present-day competition. One plausible alternative interpretation, therefore, is that the species have, in the course of their evolution, responded to natural selection in different but entirely independent ways. They are distinct species, and they have distinctive features. But, they do not compete now, nor have they ever competed; they simply happen to be different. If all this were true, it might seem that the coexistence of the tits has nothing to do with competition. On the other hand, it may be that competition in the past eliminated a number of other species, leaving behind only those that were different in their utilization of the habitat. In other words, we can still see the hand of the ghost of competition past, but it acted as an ecological force (eliminating species), rather than an evolutionary one (changing them).

interspecific competition: an ecological and an evolutionary process

The tit study therefore, and the difficulties with it, illustrate two important general points. The first is that we must pay careful, and separate, attention to both the ecological and the evolutionary effects of interspecific competition. The ecological effects, as we have seen, are, broadly, that species may be eliminated from a habitat by competition from individuals of other species, or, if competing species coexist, that individuals of at least one of them suffer reductions in survival and/or fecundity. The evolutionary effects appear to be that species differ more from one another than they would otherwise do, and hence compete less (but see Section 7.9).

The second point, however, is that there are profound difficulties in invoking competition as an explanation for observed patterns, and especially in invoking it as an evolutionary explanation. An experimental manipulation (for instance, the removal of one or more species) can, as we have seen with the salamanders and barnacles, indicate the presence of current competition, if, say, it leads to an increase in the fecundity or survival or abundance of the remaining species. But negative results would be equally compatible with the past elimination of species by competition, the evolutionary avoidance of competition in the past, and independent evolution of non-competing species. In fact, for many sets of data, there are no easy or agreed methods of distinguishing between these explanations (see Chapter 20). Thus, in the remainder of this chapter and in Chapter 20, when we examine the ecological and especially the evolutionary effects of competition, both authors and readers will need to be more than usually cautious.

For now, though, what other general features emerge from our examples? As with intraspecific competition, a basic distinction can be made between interference and exploitation competition (although elements of both may be found in a single interaction). With exploitation, individuals interact with each other indirectly, responding to a resource level that has been depressed by the activity of competitors. The diatom work provides a clear example of this. By contrast, Connell's barnacles provide an equally clear example of interference competition. *Balanus*, in particular, directly and physically interfered with the occupation by *Chthamalus* of limited space on the rocky substratum.

Interference, on the other hand, is not always as direct as this. Amongst plants, it has often been claimed that interference occurs through the production of chemicals that are toxic to other species but not to the producer (known as allelopathy). There is no doubt that chemicals with such properties can be extracted from plants, but their role is controversial. Allelopathy has been proposed with enthusiasm (Muller, 1969), welcomed and supported with enthusiasm (Whittaker & Feeney, 1971), dismissed with great scepticism (Harper, 1977), treated as an established, uncontroversial fact (Rice, 1984) and suggested soberly as a subject for serious enquiry (Williamson, 1990). Amongst competing tadpole species, too, water-borne inhibitory products have been implicated as a means of interference (most notably, perhaps, an alga produced in the faeces of the common frog, *Rana temporaria*, inhibiting the natterjack toad, *Bufo calamita* (Beebee, 1991; Griffiths *et al.*, 1993)), but here again their importance in nature is unclear (Petranka, 1989). Of course, the production by fungi, actinomycetes and bacteria—especially those found in the soil—of chemicals that inhibit the growth of potentially competing micro-organisms is widely recognized—and exploited in the selection and production of antibiotics.

Interspecific competition (like intraspecific competition) is frequently highly asymmetric—the consequences are often not the same for both species. For instance, with Connell's barnacles, *Balanus* excluded *Chthamalus* from their zone of potential overlap, but any effect of *Chthamalus* on *Balanus* was negligible: *Balanus* was limited by its own sensitivity to desiccation. A closely analogous situation is provided by two species of cattail (reedmace) in ponds in Michigan, where one species, *Typha latifolia*, was mostly found in shallower water, whilst the other, *T. angustifolia*, occurred in deeper water (Figure 7.4a). Experimental manipulations suggested that amongst newly established plants, *T. latifolia* normally excludes *T. angustifolia* from shallower water. But the distribution of *T. latifolia* is unaffected by competition with *T. angustifolia* (Figure 7.4b). On the other hand, in the longer term, after establishment is complete, the asymmetry is largely reversed: at all but the shallowest depths, *T. angustifolia* expands at the expense of *T. latifolia* and is apparently unaffected by its presence (Figure 7.4c).

On a broader front, it seems that highly asymmetric cases of interspecific competition (where one species is little affected) outnumber symmetric cases by around two to one in insects (Lawton & Hassell, 1981) in herbaceous plants (Keddy & Shipley, 1989) and more generally, too (Connell, 1983). The more fundamental point, however, is that there is a continuum linking the perfectly symmetric competitive cases to the 'perfectly' asymmetric ones. For instance, Jackson (1979), showed this in a study of 'overgrowth competition' amongst bryozoan species

interference and exploitation

allelopathy

interspecific competition is frequently highly asymmetric

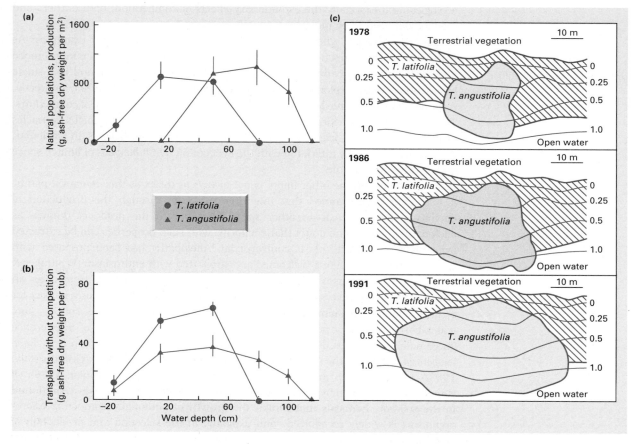

Figure 7.4 Asymmetric competition between cattail species. (a) The natural distributions in Michigan, USA, of coexisting populations of *Typha latifolia* (in shallower water) and *T. angustifolia* (in deeper water). (b) When transplants were grown alone, *T. angustifolia* grew over a much wider range of depths (suggesting it is normally excluded from shallower water), but *T. latifolia* grew at the same range of depths as it did when in competition. (After Grace & Wetzel, 1981.) (c) However, in the longer term, at the edge of a Swedish pond, a natural *T. angustifolia* stand has expanded at the expense of *T. latifolia* and, at all but the shallowest depths, is apparently unaffected by it. The lines are water-depth contours (cm). (After Weisner, 1993.)

(colonial, modular animals) living on the undersurfaces of corals off the coast of Jamaica. For the pair-wise interactions amongst the seven most commonly interacting species, he found that the 'percentage wins' varied more or less continuously from 50% (perfect symmetry) to 100% (perfect asymmetry).

competition for one resource affects competition for other resources …

Finally, it is worth noting that competition for one resource often affects the ability of an organism to exploit another resource. For example, Buss (1979), showed that in bryozoan overgrowth interactions, there appears to be an interdependence between competition for space and for food. When a colony of one species contacts a colony of another species, it interferes with the self-generated feeding currents on which bryozoans rely (competition for space affects feeding). But a colony short of food will, in turn, have a much reduced ability to compete for space (by overgrowth).

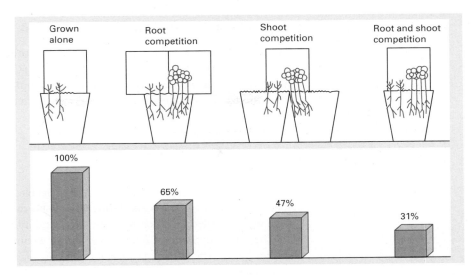

Figure 7.5 Root and shoot competition between subterranean clover (*Trifolium subterraneum*) and skeleton weed (*Chondrilla juncea*). Above are the experimental designs used; below are the dry weights of skeleton weed produced as a percentage of the production when grown alone. (After Groves & Williams, 1975.)

Comparable examples are found amongst rooted plants. If one species invades the canopy of another and deprives it of light, the suppressed species will suffer directly from the reduction in light energy that it obtains, but this will also reduce its rate of root growth, and it will therefore be less able to exploit the supply of water and nutrients in the soil. This in turn will reduce its rate of shoot and leaf growth. Thus, when plant species compete, repercussions flow backwards and forwards between roots and shoots (Wilson, 1988a). A number of workers have attempted to separate the effects of canopy and root competition by an experimental design in which two species are grown: (i) alone; (ii) together; (iii) in the same soil, but with their canopies separated; and (iv) in separate soil with their canopies intermingling. One example is a study of subterranean clover (*Trifolium subterraneum*) and skeleton weed (*Chondrilla juncea*). The clover was not significantly affected under any circumstances (another example of asymmetric competition). However, as Figure 7.5 shows, the skeleton weed was affected when the roots intermingled (reduced to 65% of the control value of dry weight) and when the canopies intermingled (47% of the control). When both intermingled, the effect was multiplicative, dry weight being reduced to 31% of the control, compared with the 30.6% (65 × 47%) that might have been expected. Clearly, both were important.

... root and shoot competition

7.4 Competitive exclusion or coexistence?

The results of experiments such as those described here highlight a critical question in the study of the ecological effects of interspecific competition: what are the general conditions that permit the coexistence of competitors, and what circumstances lead to competitive exclusion? Mathematical models have provided important insights into this question.

273 INTERSPECIFIC COMPETITION

7.4.1 A logistic model of interspecific competition

The 'Lotka–Volterra' model of interspecific competition (Volterra, 1926; Lotka, 1932) is an extension of the logistic equation described in Chapter 6, Section 6.10. As such, it incorporates all of the logistic's shortcomings, but a useful model can nonetheless be constructed, shedding light on the factors that determine the outcome of a competitive interaction.

The logistic equation:

$$\frac{dN}{dt} = rN \frac{(K - N)}{K} \tag{7.1}$$

contains, within the brackets, a term responsible for the incorporation of intraspecific competition. The basis of the Lotka–Volterra model is the replacement of this term by one which incorporates both intra- and interspecific competition.

The population size of one species can be denoted by N_1, and that of a second species by N_2. Their carrying capacities and intrinsic rates of increase are K_1, K_2, r_1 and r_2, respectively.

Suppose that 10 individuals of species 2 have, between them, the same competitive, inhibitory effect on species 1 as does a single individual of species 1. The total competitive effect on species 1 (intra- and interspecific) will then be equivalent to the effect of $(N_1 + N_2/10)$ species 1 individuals. The constant (1/10 in the present case) is called a competition coefficient and is denoted by α_{12}. It measures the per capita competitive effect on species 1 of species 2. Thus, multiplying N_2 by α_{12} converts it to a number of 'N_1-equivalents'. (Note that $\alpha_{12} < 1$ means that individuals of species 2 have less inhibitory effect on individuals of species 1 than individuals of species 1 have on others of their own species, whilst $\alpha_{12} > 1$ means that individuals of species 2 have a greater inhibitory effect on individuals of species 1 than do the species 1 individuals themselves.)

α: the competition coefficient

The crucial element in the model is the replacement of N_1 in the bracket of the logistic equation with a term signifying 'N_1 plus N_1-equivalents', i.e.:

the Lotka–Volterra model: a logistic model for two species

$$\frac{dN_1}{dt} = r_1 N_1 \frac{(K_1 - (N_1 + \alpha_{12} N_2))}{K_1} \tag{7.2}$$

or:

$$\frac{dN_1}{dt} = r_1 N_1 \frac{(K_1 - N_1 - \alpha_{12} N_2)}{K_1} \tag{7.3}$$

and in the case of the second species:

$$\frac{dN_2}{dt} = r_2 N_2 \frac{(K_2 - N_2 - \alpha_{21} N_1)}{K_2}. \tag{7.4}$$

These two equations constitute the Lotka–Volterra model.

To appreciate the properties of this model, we must ask the question: when (under what circumstances) does each species increase or decrease in abundance? In order to answer this, it is necessary to construct diagrams in which all possible combinations of species 1 and species 2 abundance can be displayed (i.e. all possible combinations of N_1 and N_2). These will be diagrams (Figures 7.6 and 7.8), with N_1 plotted on the horizontal axis and N_2 plotted on the vertical axis, such that there are low numbers of both species towards the bottom left, high numbers of both species

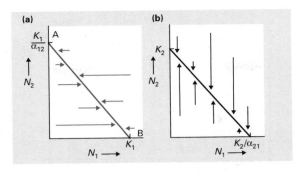

Figure 7.6 The zero isoclines generated by the Lotka–Volterra competition equations. (a) The N_1 zero isocline: species 1 increases below and to the left of it, and decreases above and to the right of it. (b) The equivalent N_2 zero isocline.

towards the top right, and so on. Certain combinations of N_1 and N_2 will give rise to increases in species 1 and/or species 2, whilst other combinations will give rise to decreases in species 1 and/or species 2. Crucially, there must also therefore be 'zero isoclines' for each species (lines along which there is neither increase nor decrease), dividing the combinations leading to increase from those leading to decrease. Moreover, if a zero isocline is drawn first, there will be combinations leading to increase on one side of it, and combinations leading to decrease on the other.

the behaviour of the Lotka–Volterra model is investigated using 'zero isoclines'

In order to draw a zero isocline for species 1, we can use the fact that on the zero isocline $dN_1/dt = 0$ (by definition), that is (from Equation 7.3):

$$r_1 N_1 (K_1 - N_1 - \alpha_{12} N_2) = 0. \tag{7.5}$$

This is true when the intrinsic rate of increase (r_1) is zero, and when the population size (N_1) is zero, but—much more importantly in the present context—it is also true when:

$$K_1 - N_1 - \alpha_{12} N_2 = 0 \tag{7.6}$$

which can be rearranged as:

$$N_1 = K_1 - \alpha_{12} N_2. \tag{7.7}$$

In other words, everywhere along the straight line which this equation represents, $dN_1/dt = 0$. The line is therefore the zero isocline for species 1; and since it is a straight line it can be drawn by finding two points on it and joining them. Thus, in Equation 7.7, when:

$$N_1 = 0, N_2 = \frac{K_1}{\alpha_{12}} \qquad \text{(point A, Figure 7.6a)} \tag{7.8}$$

when:

$$N_2 = 0, N_1 = K_1 \qquad \text{(point B, Figure 7.6a)} \tag{7.9}$$

and joining them gives the zero isocline for species 1. Below and to the left of this, numbers of both species are relatively low, and species 1, subjected to only weak competition, increases in abundance (arrows in the figure, representing this increase, point from left to right, since N_1 is on the horizontal axis). Above and to the right of the line, numbers are high, competition is strong and species 1 decreases in abundance (arrows from right to left). Based on an equivalent derivation, Figure 7.6b has combinations leading to increase and decrease in species 2, separated by a species 2 zero isocline, with arrows, like the N_2-axis, running vertically.

Finally, in order to determine the outcome of competition in this model, it is

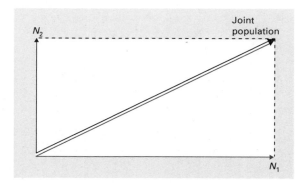

Figure 7.7 Vector addition. When species 1 and 2 increase in the manner indicated by the N_1 and N_2 arrows (vectors), the joint population increase is given by the vector along the diagonal of the rectangle, generated as shown by the N_1 and N_2 vectors.

necessary to fuse Figures 7.6a and b, allowing the behaviour of a joint population to be predicted. In doing this, it should be noted that the arrows in Figure 7.6 are actually vectors—with a strength as well as a direction—and that to determine the behaviour of a joint N_1, N_2 population the normal rules of vector addition should be applied (Figure 7.7).

there are four ways in which the two zero isoclines can be arranged

Figures 7.8a–d show that there are, in fact, four different ways in which the two zero isoclines can be arranged relative to one another, and the outcome of competition will be different in each case. The different cases can be defined and distinguished by the intercepts of the zero isoclines. For instance, in Figure 7.8a:

$$\frac{K_1}{a_{12}} > K_2 \quad \text{and} \quad K_1 > \frac{K_2}{a_{21}} \tag{7.10}$$

i.e.:

$$K_1 > K_2 a_{12} \quad \text{and} \quad K_1 a_{21} > K_2. \tag{7.11}$$

The first inequality ($K_1 > K_2 a_{12}$) indicates that the inhibitory intraspecific effects that species 1 can exert on itself are greater than the interspecific effects that species 2 can exert on species 1. The second inequality, however, indicates that species 1 can exert more of an effect on species 2 than species 2 can on itself. Species 1 is thus a

strong interspecific competitors outcompete weak interspecific competitors

strong interspecific competitor, whilst species 2 is a weak interspecific competitor; and as the vectors in Figure 7.8a show, species 1 drives species 2 to extinction and attains its own carrying capacity. The situation is reversed in Figure 7.8b. Hence, Figures 7.8a and b describe cases in which the environment is such that one species invariably outcompetes the other.

In Figure 7.8c:

$$K_2 > \frac{K_1}{a_{12}} \quad \text{and} \quad K_1 > \frac{K_2}{a_{21}} \tag{7.12}$$

i.e.:

$$K_2 a_{12} > K_1 \quad \text{and} \quad K_1 a_{21} > K_2. \tag{7.13}$$

when interspecific competition is more important than intraspecific, the outcome depends on the species' densities

Thus, individuals of both species compete more strongly with individuals of the other species than they do amongst themselves. This will occur, for example, when each species produces a substance that is toxic to the other species but harmless to itself, or when each species is aggressive towards or even preys upon individuals of the other species, more than individuals of its own species. The consequence, as the figure shows, is an unstable equilibrium combination of N_1 and N_2 (where the

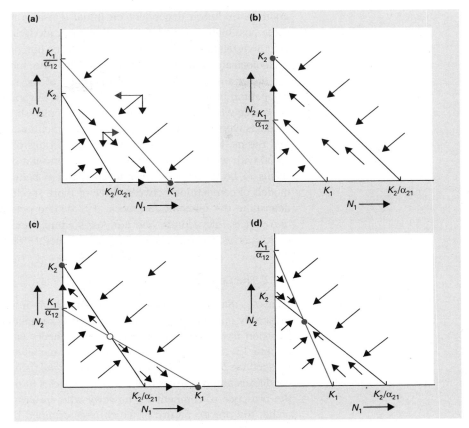

Figure 7.8 The outcomes of competition generated by the Lotka–Volterra competition equations for the four possible arrangements of the N_1 and N_2 zero isoclines. Vectors, generally, refer to joint populations, and are derived as indicated in (a). The solid circles show stable equilibrium points. The open circle in (c) is an unstable equilibrium point. For further discussion, see text.

isoclines cross), and two stable points. At the first of these stable points, species 1 reaches its carrying capacity with species 2 extinct; whilst at the second, species 2 reaches its carrying capacity with species 1 extinct. Which of these two outcomes is actually attained is determined by the initial densities: the species which has the initial advantage will drive the other species to extinction.

Finally, in Figure 7.8d:

$$\frac{K_1}{\alpha_{12}} > K_2 \quad \text{and} \quad \frac{K_2}{\alpha_{21}} > K_1 \tag{7.14}$$

i.e.:

$$K_1 > K_2\alpha_{12} \quad \text{and} \quad K_2 > K_1\alpha_{21}. \tag{7.15}$$

when inter- is less significant than intra-, the species coexist

In this case, both species have less competitive effect on the other species than they have on themselves. The outcome, as Figure 7.8d shows, is a stable-equilibrium combination of the two species, which all joint populations tend to approach.

Overall, therefore, the Lotka–Volterra model of interspecific competition is able to generate a range of possible outcomes: the predictable exclusion of one species by

277 INTERSPECIFIC COMPETITION

another, exclusion dependent on initial densities and stable coexistence. Each of these possibilities will be discussed in turn, alongside the results of laboratory and field investigations. We will see that the three outcomes from the model correspond to biologically reasonable circumstances. The model, therefore, in spite of its simplicity and its failure to address many of the complexities of the dynamics of competiton in the real world, serves a useful purpose.

Ks, *α*s ... and *r*s?

Before we move on, however, one particular shortcoming of the Lotka–Volterra model is worth noting. The outcome of competition in the model depends on the *K*s and the *α*s, but not on the *r*s, the intrinsic rates of increase. These determine the speed with which the outcome is achieved but not the outcome itself. This, though, seems to be a result peculiar to competition between only two species, since in models of competition between three or more species, the *K*s, *α*s and *r*s combine to determine the outcome (Strobeck, 1973). The potential inadequacies of studying two-species interactions, and the possible improvements arising from extensions to three species, are themes taken up in Chapter 15 on Abundance.

7.4.2 The Competitive Exclusion Principle

Figures 7.8a and b describe cases in which a strong interspecific competitor invariably outcompetes a weak interspecific competitor. It is useful to consider this situation from the point of view of niche theory (see Chapter 2, Section 2.12 and Chapter 3, Section 3.8). Recall that the niche of a species in the absence of competitors from other species is its *fundamental* niche (defined by the combination of conditions and resources that allow the species to maintain a viable population). In the presence of competitors, however, the species may be restricted to a *realized* niche, the precise nature of which is determined by which competing species are present. This distinction stresses that interspecific competition reduces fecundity and survival, and that there may be parts of a species' fundamental niche in which, as a result of interspecific competition, the species can no longer survive and reproduce successfully. These parts of its fundamental niche are absent from its realized niche. Thus, returning to Figures 7.8a and b, we can say that the weak interspecific competitor lacks a realized niche when in competition with the stronger competitor. The real examples of interspecific competition previously discussed can now be re-examined in terms of niches.

fundamental and realized niches

coexisting competitors often exhibit a differentiation of realized niches

In the case of the diatom species, the fundamental niches of both species were provided by the laboratory regime (they both thrived when alone). Yet, when *Synedra* and *Asterionella* competed, *Synedra* had a realized niche whilst *Asterionella* did not: there was competitive exclusion of *Asterionella*. The same outcome was recorded when Gause's *P. aurelia* and *P. caudatum* competed; *P. caudatum* lacked a realized niche and was competitively excluded by *P. aurelia*. When *P. caudatum* and *P. bursaria* competed, on the other hand, both species had realized niches, but these niches were noticeably different: *P. caudatum* living and feeding on the bacteria in the medium, *P. bursaria* concentrating on the yeast cells on the bottom of the tube. Coexistence was therefore associated with a differentiation of realized niches, or a 'partitioning' of resources.

In the *Galium* experiments, the fundamental niches of both species included both acidic and calcareous soils. In competition with one another, however, the realized niche of *G. hercynicum* was restricted to acidic soils, whilst that of *G. pumilum* was restricted to calcareous ones: there was reciprocal competitive exclusion. Neither habitat allowed niche differentiation, and neither habitat fostered coexistence.

Amongst Connell's barnacles, the fundamental niche of *Chthamalus* extended down into the *Balanus* zone; but competition from *Balanus* restricted *Chthamalus* to a realized niche higher up the shore. In other words, *Balanus* competitively excluded *Chthamalus* from the lower zones, but for *Balanus* itself, even its fundamental niche did not extend up into the *Chthamalus* zone: its sensitivity to desiccation prevented it surviving even in the absence of *Chthamalus*. Hence, overall, the coexistence of these species was also associated with a differentiation of realized niches.

the 'Competitive Exclusion Principle'

The pattern that has emerged from these examples has also been uncovered in many others, and has been elevated to the status of a principle: the 'Competitive Exclusion Principle' or 'Gause's Principle'. It can be stated as follows: if two competing species coexist in a stable environment, then they do so as a result of niche differentiation, i.e. differentiation of their realized niches. If, however, there is no such differentiation, or if it is precluded by the habitat, then one competing species will eliminate or exclude the other. Thus, exclusion occurs when the realized niche of the superior competitor completely fills those parts of the inferior competitor's fundamental niche which the habitat provides.

When there *is* coexistence of competitors, a differentiation of realized niches is sometimes seen to arise from current competition (an 'ecological' effect), as with the barnacles. Often, however, the niche differentiation is believed to have arisen either as a result of the past elimination of those species without realized niches (leaving behind only those exhibiting niche differentiation—another ecological effect), or as an *evolutionary* effect of competition. In either case, present competition may be negligible or at least impossible to detect. Thus, we return to the difficulties discussed earlier (see p. 270). Consider again the coexisting tits. The species coexisted and exhibited differentiation of their realized niches. But it was not established that they compete now, or that they had ever done so in the past, or that other species had been competitively excluded in the past. It is impossible to say with certainty whether the Competitive Exclusion Principle was relevant. If the species do actually compete currently, or if other species are being or have been competitively excluded, then the Principle is relevant in the strictest sense. If they competed only in the past, and that competition has led to their niche differentiation, then the Principle is relevant, but only if it is extended from applying to the coexistence of 'competitors' to the coexistence of 'species that are *or have ever been* competitors'. Of course, if the species have never competed, then the Principle is of no relevance here. Clearly, interspecific competition cannot be studied by the mere documentation of present interspecific differences.

difficult methodological problems in proving, and especially disproving, the Principle

With Hairston's salamanders, on the other hand, the two species competed and coexisted, and the Competitive Exclusion Principle would suggest that this was a result of niche differentiation. But, whilst reasonable, this is by no means proven, since such differentiation was neither observed nor shown to be effective. Thus, when two competitors coexist, it is often difficult to establish positively that there is niche differentiation. Worse still, it is impossible to prove the absence of it. When ecologists fail to find differentiation, this might simply mean that they have looked in the wrong place or in the wrong way. Clearly, there can be very real methodological problems in establishing the pertinence of the Competitive Exclusion Principle in any particular case.

niche differentiation and interspecific competition: a pattern and a process that are not always linked

The Competitive Exclusion Principle has become widely accepted: (i) because there is much good evidence in its favour; (ii) because it makes intuitive good sense; and (iii) because there are theoretical grounds for believing in it (the Lotka–Volterra model). But there will always be cases in which it has not been positively established; and as Section 7.5 will make plain, there are many other cases in which it simply does not apply. In short, interspecific competition is a process that is often associated, ecologically and evolutionarily, with a particular pattern (niche differentiation), but interspecific competition and niche differentiation (the process and the pattern) are not inextricably linked. Niche differentiation can arise through other processes, and interspecific competition need not lead to a differentiation of niches.

7.4.3 Mutual antagonism

Figure 7.8c, derived from the Lotka–Volterra model, describes a situation in which interspecific competition is, for both species, a more powerful force than intraspecific competition. This is known as mutual antagonism.

An extreme example of such a situation is provided by work on two species of flour beetle: *Tribolium confusum* and *T. castaneum* (Park, 1962). Park's experiments in the 1940s, 1950s and 1960s were amongst the most influential in shaping ideas about interspecific competition. He reared the beetles in simple containers of flour, which provided fundamental and often realized niches for the eggs, larvae, pupae and adults of both species. There was certainly exploitation of common resources by the two species; but in addition, the beetles preyed upon each other. Larvae and adults ate eggs and pupae, cannibalizing their own species as well as attacking the other species, and their propensity for doing so is summarized in Table 7.1. The important point is that taken overall, beetles of both species ate more individuals of the other species than they did of their own. Thus, a crucial mechanism in the interaction of these competing species was reciprocal predation (i.e. mutual antagonism), and it is easy to see that both species were more affected by inter- than intraspecific predation.

reciprocal predation in flour beetles

Table 7.1 Reciprocal predation (a form of mutual antagonism) between two species of flour beetle, *Tribolium confusum* and *T. castaneum*. Both adults and larvae eat both eggs and pupae. In each case, and overall, the preference of each species for its own or the other species is indicated. Interspecific predation is more marked than intraspecific predation. (After Park *et al.*, 1965.)

	'Predator'	'Shows a preference for . . .'
Adults eating eggs	*T. confusum*	*T. confusum*
	T. castaneum	*T. confusum*
Adults eating pupae	*T. confusum*	*T. castaneum*
	T. castaneum	*T. confusum*
Larvae eating eggs	*T. confusum*	*T. castaneum*
	T. castaneum	*T. castaneum*
Larvae eating pupae	*T. confusum*	*T. castaneum*
	T. castaneum	*T. confusum*
Overall	*T. confusum*	*T. castaneum*
	T. castaneum	*T. confusum*

to persist at a site. It does so by attaching to the bare rock, usually in gaps in the mussel bed created by wave action. However, the mussels themselves slowly encroach on these gaps, gradually filling them and precluding colonization by *Postelsia*. Paine found that these species coexisted only at sites in which there was a relatively high average rate of gap formation (about 7% of surface area per year), and in which this rate was approximately the same each year. Where the average rate was lower, or where it varied considerably from year to year, there was (either regularly or occasionally) a lack of bare rock for colonization. This led to the overall exclusion of *Postelsia*. At the sites of coexistence, on the other hand, although *Postelsia* was eventually excluded from each gap, these were created with sufficient frequency and regularity for there to be coexistence in the site as a whole.

7.5.2 Unpredictable gaps: the pre-emption of space

When two species compete on equal terms, the result is usually predictable. But in the colonization of unoccupied space, competition is rarely even handed. Individuals of one species are likely to arrive, or germinate from the seedbank, in advance of individuals of another species. This, in itself, may be enough to tip the competitive balance in favour of the first species. If space is pre-empted by different species in different gaps, then this may allow coexistence, even though one species would always exclude the other if they competed 'on equal terms'.

For instance, Figure 7.10 shows the results of a competition experiment between the annual grasses *Bromus madritensis* and *B. rigidus* which occur together in Californian rangelands (Harper, 1961). When they were sown simultaneously in an equiproportional mixture, *B. rigidus* contributed overwhelmingly to the biomass of the mixed population. But, by delaying the introduction of *B. rigidus* into the mixtures, the balance was tipped decisively in favour of *B. madritensis*. It is therefore quite wrong to think of the outcome of competition as being always determined by

first come, best served

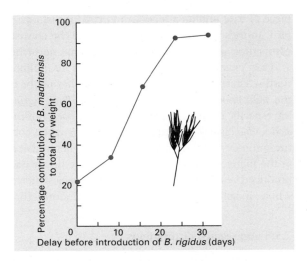

Figure 7.10 The effect of timing on competition. *Bromus rigidus* makes an overwhelming contribution to the total dry weight per pot after 126 days growth when sown at the same time as *B. madritensis*. But, as the introduction of *B. rigidus* is delayed, its contribution declines. Total yield per pot was unaffected by delaying the introduction of *B. rigidus*. (After Harper, 1961.)

the inherent competitive abilities of the competing species. Even an 'inferior' competitor can exclude its superior if it has enough of a head start. This can foster coexistence when repeated colonization occurs in a changing or unpredictable environment.

7.5.3 Fluctuating environments

'the paradox of the plankton'

The balance between competing species can be shifted repeatedly, in fact, and coexistence therefore fostered, simply as a result of environmental change. This was the argument used by Hutchinson (1961) to explain 'the paradox of the plankton'— the paradox being that numerous species of planktonic algae frequently coexist in simple environments with little apparent scope for niche differentiation. Hutchinson suggested that the environment, although simple, was continually changing, particularly on a seasonal basis. Thus, although the environment at any one time would tend to promote the exclusion of certain species, it would alter and perhaps even favour these same species before exclusion occurred. In other words, the equilibrium outcome of a competitive interaction may not be of paramount importance if the environment typically changes long before the equilibrium can be reached. Since all environments keep changing, competitive balances must be forever shifting, and coexistence must commonly be associated with niche differences that would promote exclusion in an unvarying world.

7.5.4 Ephemeral patches with unpredictable life spans

Many environments, by their very nature, are not simply variable but ephemeral. Amongst the more obvious examples are decaying corpses (carrion), dung, rotting fruit and fungi, and temporary ponds. But note too that a leaf or an annual plant can be seen as an ephemeral patch, especially if it is palatable to its consumer for only a limited period. Often, these ephemeral patches have an unpredictable life span—a piece of fruit and its attendant insects, for instance, may be eaten at any time by a bird. In these cases, it is easy to imagine the coexistence of two species: a superior competitor, and an inferior competitor that nevertheless reproduces early.

coexistence of the good with the fast

One example concerns two species of pulmonate snail living in ponds in north-eastern Indiana, USA. Artificially altering the density of one or other species in the field showed that the fecundity of *Physa gyrina* was significantly reduced by interspecific competition from *Lymnaea elodes*, but the effect was not reciprocated. *L. elodes* was clearly the superior competitor when competition continued throughout the summer. Yet *P. gyrina* reproduced earlier and at a smaller size than *L. elodes*, and in the many ponds that dried up by early July it was often the only species to have produced resistant eggs in time. The species therefore coexisted in the area as a whole, in spite of *P. gyrina*'s apparent inferiority (Brown, 1982).

7.5.5 Aggregated distributions

A more subtle, but more generally applicable path to the coexistence of a superior and an inferior competitor on a patchy and ephemeral resource is based on the idea that the two species may have independent, aggregated (i.e. clumped) distributions over the available patches. This would mean that the powers of the superior competitor were mostly directed against members of its own species (in the

a clumped superior competitor adversely affects itself and leaves gaps for its inferior

high-density clumps), but that this aggregated superior competitor would be absent from many patches—within which the inferior competitor could escape competition. An inferior competitor may then be able to coexist with a superior competitor that would rapidly exclude it from a continuous, homogeneous environment. Certainly it can do so in a series of models (Shorrocks *et al.*, 1979; Hanski, 1981; Atkinson & Shorrocks, 1981; de Jong, 1982; Ives & May, 1985; Kreitman *et al.*, 1992). For instance, a simulation model (Figure 7.11) shows that the persistence of such coexistence between competitors increases with the degree of aggregation (as measured by the parameter k of the 'negative binomial' distribution) until, at high levels of aggregation, coexistence is apparently permanent, although this has nothing to do with any niche differentiation. Since many species have aggregated distributions in nature, these results may be applicable widely.

again, though, coexistence occurs when intra- is more powerful than inter-

Note, however, that whilst such coexistence of competitors has nothing to do with niche differentiation, it is linked to it by a common theme—that of species competing more frequently and intensively intraspecifically than they do interspecifically. Niche differentiation is one means by which this can occur; but temporary aggregations can give rise to the same phenomenon—even for the inferior competitor.

Hanski and Kuusela (1977) provide an example of this from a study of carrion flies in Finland. They established 50 small patches of carrion in a very limited area (approximately 5 m²) and allowed the naturally occurring flies to lay eggs in (i.e. colonize) the patches. Overall, nine species of carrion fly emerged from the patches. Yet the average number of species per patch was only 2.7. This was because all of the species were very highly aggregated in their distribution (with many offspring emerging from a few patches, and a few offspring or none emerging from many others). As a result, particular pairs of species came into contact only rarely, and this is bound to have fostered coexistence overall, even though there may have been exclusion where the species did come into contact.

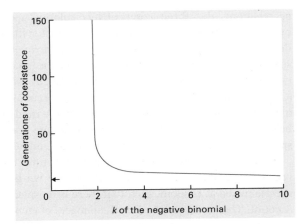

Figure 7.11 When two species compete on a continuously distributed resource, one species would exclude the other in approximately 10 generations (as indicated by the arrow). However, with these same species on a patchy and ephemeral resource, the number of generations of coexistence increases with the degree of aggregation of the competitors, as measured by the parameter k of the 'negative binomial' distribution. Values above 5 are effectively random distributions; values below 5 represent increasingly aggregated distributions. (After Atkinson & Shorrocks, 1981.)

In seeking to justify the applicability of these models to the real world, however, two questions arise: (i) are species really aggregated so that, at all densities, their dispersion can be represented by the same parameter value in the same distribution (k of the negative binomial); and especially (ii) are two similar species really likely to have independent distributions over available patches of resource? The questions have been addressed through an examination of a large number of data sets from Diptera, especially drosophilid flies—where eggs are laid, and larvae develop, in ephemeral patches (fruits, fungi, flowers, etc.). In answer to (i), most species are highly aggregated and the negative binomial is usually a good description of this; but, the degree of aggregation typically declines (k increases) at higher densities (Rosewell *et al.*, 1990). However, in answer to (ii), there is no good evidence for a lack of independence in the aggregations of coexisting species (Shorrocks *et al.*, 1990; see also Worthen & McGuire, 1988). Furthermore, computer simulations suggest that whilst a positive association between species (i.e. a tendency to aggregate in the same patches) does make coexistence more difficult, the degrees of association and aggregation actually found would still generally lead to coexistence, whereas there would be exclusion in a homogeneous environment (Shorrocks & Rosewell, 1987).

grasses in a cellular automaton

The importance of aggregation for coexistence has been further supported by another spatially explicit model based on a two-dimensional lattice of cells (see Section 7.5.1), each of which could be occupied by one of five species of grass: *Agrostis stolonifera*, *Cynosurus cristatus*, *Holcus lanatus*, *Lolium perenne* and *Poa trivialis* (Silvertown *et al.*, 1992a). The model was a 'cellular automaton', in which each cell can exist in a limited number of discrete states (in this case, which species was in occupancy), with the state of each cell determined at each time step by a set of rules. In this case, the rules were based on the cell's current state, the state of neighbouring cells and the probability that a species in a neighbouring cell would replace its current occupant. These replacement rates of each species by each other species were themselves based on field observations (Thórhallsdóttir, 1990).

If the initial arrangement of the species over the grid was random (no aggregation), the three competitively inferior species were quickly driven to extinction, and of the survivors, *Agrostis* (greater than 80% cell occupancy) rapidly dominated *Holcus*. If, however, the initial arrangement was five equally broad single-species bands across the landscape, the outcome changed dramatically: (i) competitive exclusion was markedly delayed even for the worst competitors (*Cynosurus* and *Lolium*); (ii) *Holcus* sometimes occupied more than 60% of the cells, at a time (600 time steps) where, with an initially random arrangement, it would have been close to extinction; and (iii) the outcome itself depended largely on which species started next to each other, and hence, initially competed with each other.

There is no suggestion, of course, that natural communities of grasses exist as broad single-species bands—but, neither are we likely to find communities with species mixed at random, such that there is no spatial organization to be taken into account. The model emphasizes the dangers of ignoring aggregations (because they shift the balance towards intra- rather than interspecific competition, and hence promote coexistence), but also the dangers of ignoring the juxtaposition of aggregations, since these too may serve to keep competitive subordinates away from their superiors.

in general: heterogeneity often stabilizes

Repeatedly in this section, then, the heterogeneous nature of the environment can be seen to have fostered coexistence without there being a marked differentiation of niches. A realistic view of interspecific competition, therefore, must acknowledge that it often proceeds not in isolation, but under the influence of, and within the constraints of, a patchy, impermanent or unpredictable world. These ideas are reviewed from a more mathematical point of view by Abrams (1983) and Chesson and Case (1986). Furthermore, the heterogeneity need not be in the temporal or spatial dimensions that we have discussed so far. Individual variation in competitive ability within species can also foster stable coexistence in cases where a superior non-variable competitor would otherwise exclude an inferior non-variable species (Begon & Wall, 1987). This reinforces a point which recurs throughout this text: heterogeneity (spatial, temporal or individual) can have a stabilizing influence on ecological interactions.

7.6 Apparent competition: enemy-free space

Another reason for being cautious in our discussion of competition is the existence of what Holt (1977, 1984) has called 'apparent competition', and what others have called 'competition for enemy-free space' (Jeffries & Lawton, 1984, 1985).

two prey species being attacked by a predator are, in essence, indistinguishable from two consumer species competing for a resource

Imagine a single species of predator that attacks two species of prey. Both prey species are harmed by the predator, and the predator benefits from both species of prey. Hence, the increase in abundance that the predator achieves by consuming prey 1 increases the harm it does to prey 2. Indirectly, therefore, prey 1 adversely affects prey 2 and vice versa. These interactions are summarized in Figure 7.12c, which shows that from the point of view of the two prey species, the signs of the interactions are indistinguishable from those that would apply in the indirect interaction of two species competing for a single resource (exploitation competition; Figure 7.12b). In the present case, there appears to be no limiting resource. Hence, the term 'apparent competition'. On the other hand, we may choose to think of 'enemy-free space' as the limiting resource.

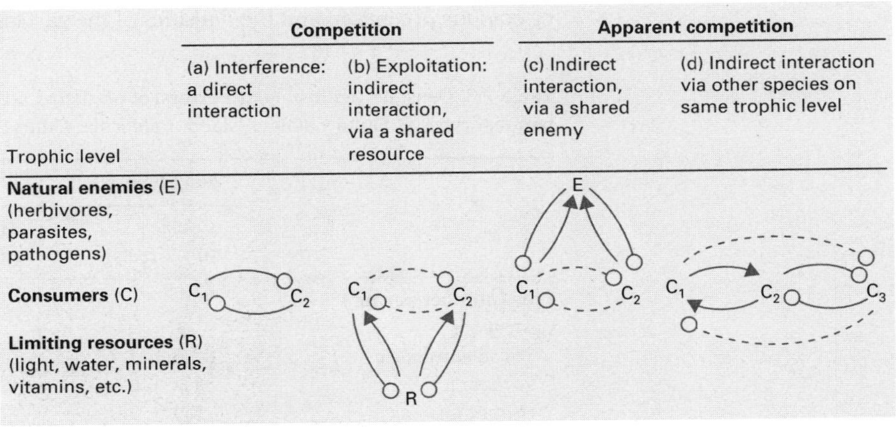

Figure 7.12 In terms of the signs of their interactions, all of the following are indistinguishable from one another: (a) two species interfering directly (interference competition); (b) two species consuming a common resource (exploitation competition); (c) two species being attacked by a common predator ('apparent competition' for 'enemy-free space'); and (d) two species linked by a third which is a competitor of one and a mutualist of the other. (—), Direct interactions; (- - -), indirect interactions; arrows indicate positive influences, circles indicate negative influences. (After Holt, 1984; Connell, 1990.)

Furthermore, the persistence of prey species 1 will be favoured by avoiding attacks from the predator, which we know also attacks prey 2. Clearly, prey 1 can achieve this by occupying a habitat, or adopting a form or a behavioural pattern, that is sufficiently different from that of prey 2. In short, 'being different' (i.e. niche differentiation) will once again favour coexistence—but, it will do so because it diminishes apparent competition or competition for enemy-free space.

In fact, there is at least one other indirect interaction between two species that qualifies for the term 'apparent competition' (Figure 7.12d), where species 1 and 2 have negative impacts on one another, and species 2 and 3 have positive (mutualistic) impacts (see Chapter 13). Species 1 and 3 then have indirect negative impacts on one another without sharing a common resource or, for that matter, a common predator. They exhibit apparent competition, although not for enemy-free space (Connell, 1990).

evidence for apparent
competition on rocky shores

A rare experimental demonstration of apparent competition for enemy-free space involves two groups of prey living on subtidal rocky reefs at Santa Catalina Island, California. The first comprises three species of mobile gastropods, *Tegula aureotincta*, *T. eiseni* and *Astraea undosa*; the second comprises sessile bivalves, dominated by the clam, *Chama arcana*. Both groups were preyed upon by a lobster (*Panulirus interruptus*), an octopus (*Octopus bimaculatus*) and a whelk (*Kelletia kelletii*), although these predators showed a marked preference for the bivalves. In areas characterized by large boulders and much crevice space ('high relief'), there were high densities of bivalves and predators but only moderate densities of gastropods; whereas in low-relief areas largely lacking crevice space ('cobble fields'), there were apparently no bivalves, only a few predators but high densities of gastropods (Table 7.3).

The densities of the two prey groups were inversely correlated, but there was little in their feeding biology to suggest that they were competing for a shared food resource. On the other hand, when bivalves were experimentally introduced into cobble-field areas, the number of predators congregating there increased, the mortality rates of the gastropods increased (often observably associated with lobster or octopus predation) and the densities of the gastropods declined (Figures 7.13a, b).

Table 7.3 Densities (with standard errors) of predators and two contrasting types of prey on two reef types at Santa Catalina Island, California. (After Schmitt, 1987.)

	Reef type	
	Cobble (*n* = 7 reefs)	High relief (*n* = 11 reefs)
Prey (number per m²)		
Gastropods		
Tegula aureotincta	47.6 ± 8.2	1.7 ± 0.8
T. eiseni	51.2 ± 8.6	21.9 ± 6.0
Astraea undosa	7.9 ± 2.0	2.6 ± 0.9
Sessile bivalves*	0.0	11.7 ± 5.1
Predators (number per m²)		
Panulirus interruptus	0.01 ± 0.01	0.16 ± 0.04
Octopus bimaculatus	0.12 ± 0.04	0.28 ± 0.05
Kelletia kelletii	0.02 ± 0.01	0.40 ± 0.24

*Data for these groups gathered from seven of the 11 high relief reefs.

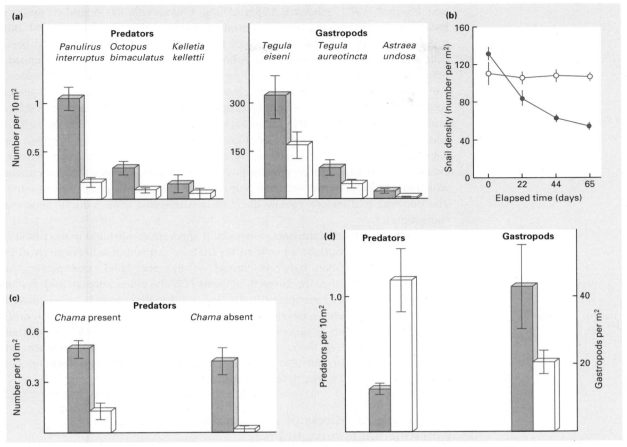

Figure 7.13 Evidence for apparent competition for predator-free space at Santa Catalina Island. (a) Predator density (number per 10 m², with standard errors) and gastropod mortality increased (number of 'newly dead' shells per site, with standard errors) when bivalves were added to gastropod-dominated cobble sites (coloured bars) relative to controls (white bars). (b) This led to a decline in gastropod density (standard error bars shown). (c) Predator density was higher (number per 10 m², with standard errors) at high (coloured bars) than at low (white bars) gastropod-density cobble sites, both in the presence and absence of *Chama*. (d) Densities of predators were lower (number per 10 m², with standard errors) and densities of gastropods higher (number per m², with standard errors) at high-relief sites without *Chama* (coloured bars) than at those with (white bars). (After Schmitt, 1987.)

Experimental manipulation of the (mobile) gastropods proved impossible, but cobble sites with high densities of gastropods supported higher densities of predators, and had higher mortality rates of experimentally added bivalves than did sites with relatively low densities of gastropods (Figure 7.13c). On the rare high-relief sites without *Chama* bivalves, predator densities were lower, and gastropod densities higher, than was normally the case (Figure 7.13d). It seems clear that each prey group adversely affected the other through an increased number of predators, and hence increased predator-induced mortality.

Perhaps more revealing, however, is a reappraisal (Connell, 1990) of the wide range of plant examples in two extensive surveys of field experiments on 'competition' (Connell, 1983; Schoener, 1983; see also Chapter 20). Of 54 studies, authors had claimed to have demonstrated conventional interspecific competition in 50. But a closer look revealed that in many of these, insufficient information had been collected to distinguish between conventional and either form of apparent competition; and in a number of others the information was available—but was ambiguous. For example, one study showed that removal of *Artemisia* bushes from a large site in Arizona led to much better growth of 22 species of herb than was observed in either undisturbed sites or sites where *Artemisia* was removed from narrow 3 m strips. This was originally interpreted in terms of greatly reduced exploitative competition for water in the former case (Robertson, 1947). However, the herbs in the larger site also experienced greatly reduced grazing pressure from deer, rodents and insects, for which the *Artemisia* bushes were not only a source of food but a place of shelter, too. The outcome is therefore equally likely to have resulted from reduced apparent competition.

This emphasizes that the relative neglect of apparent competition in the past has been unwarranted, but also re-emphasizes that the distinction is important within interspecific competition between pattern on the one hand, and process or mechanism on the other. In the past, patterns of niche differentiation, and also of increased abundance of one species in the absence of another, have been interpreted as evidence of competition too readily. Now we can see that such patterns can arise through a wide variety of processes, and that a proper understanding requires that we distinguish between them—not only discriminating between conventional and apparent competition, but also specifying mechanisms within, say, conventional competition—a point to which we return in Section 7.10.

7.7 Ecological effects of interspecific competition: experimental approaches

Notwithstanding the important interactions between competition and environmental heterogeneity, and the complications of apparent competition, a great deal of attention has been focused on conventional competition itself. We have already noted the difficulties in interpreting merely observational evidence, and it is for this reason that many studies of the ecological effects of interspecific competition have taken an experimental approach (see Schoener, 1983; Connell, 1983; see also Chapter 20). For example, we have seen manipulative field experiments involving salamanders (see Section 7.2.1), barnacles (see Section 7.2.3), cattails (see Section 7.3) and freshwater pulmonates (see Section 7.5.4), where the density of one or both species was altered (usually reduced). The fecundity, the survivorship, the abundance or the resource utilization of the remaining species was subsequently monitored. It was then compared either with the situation prior to the manipulation, or, far better, with a comparable control plot in which no manipulation had occurred. Such experiments have consistently provided valuable information, but they are typically easier to perform on some types of organism (e.g. sessile organisms) than they are on others.

The second type of experimental evidence has come from work carried out under artificial, controlled (often laboratory) conditions. Again, the crucial element has usually been a comparison between the responses of species living alone and their responses when in combination. Such experiments have the advantage of being comparatively easy to perform and control, but they have two major disadvantages.

The first is that species are examined in environments that are different from those they experience naturally. The second is the simplicity of the environment: it may preclude niche differentiation because niche dimensions are missing that would otherwise be important. Nevertheless, these experiments can provide useful clues to the likely effects of competition in nature.

7.7.1 Longer term experiments

The most direct way of discovering the outcome of competition between two species in the laboratory, or under other controlled conditions, is to put them together and leave them to it. However, since even the most one-sided competition is likely to take a few generations (or a reasonable period of modular growth) before it is completed, this direct approach is easier, and has been more frequently used, in some species than in others. It has most frequently been applied to insects (such as the flour beetle example in Section 7.4.3) and microorganisms (such as the *Paramecium* example in Section 7.2.4), and further invertebrate and microorganism examples are shown in Figure 7.14. The figure also contains a rare higher plant example, but neither higher plants, nor vertebrates nor large invertebrates lend themselves readily to this approach. We must be aware that this may bias our view of the nature of interspecific competition.

7.7.2 Single-generation experiments

Given these problems, the alternative 'laboratory' approach, with plants especially (although the methods have occasionally been used with animals) has generally been to follow populations over just a single generation, comparing 'inputs' and 'outputs'. A number of experimental designs have been used.

substitutive experiments: the de Wit replacement series

In 'substitutive' experiments, the effect of varying the proportion of each of two species is explored whilst keeping overall density constant (de Wit, 1960). Thus, at an overall density of say 200 plants, a series of mixtures would be set up: 100 of

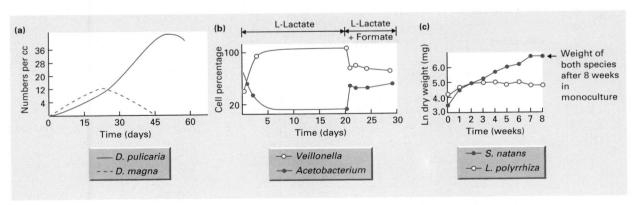

Figure 7.14 Experiments to demonstrate the outcome when two species are left together to compete. (a) Populations of the cladocerans, *Daphnia pulicaria* and *D. magna*. (After Frank, 1957.) (b) Populations of the bacteria *Veillonella alcalescens* and *Acetobacterium* NS.L40 in a chemostat with either L-lactate or L-lactate plus formate as the energy source. (After Veldkamp *et al.*, 1984.) (c) The growth in mixed culture over 8 weeks of the water plants, *Lemna polyrrhiza* and *Salvinia natans*. (After Clatworthy & Harper, 1962.)

291 INTERSPECIFIC COMPETITION

species A with 100 of species B, 150 A and 50 B, 0 A and 200 B, and so on. At the end of the experimental period, the amount of seed or the biomass of each species in each mixture would be monitored. Such replacement series may then be established at a range of total densities. In practice, however, most workers have used only a single total density, and this has led to considerable criticism of the design, since it means that the effect of competition over several generations—when total density would inevitably alter—cannot even be predicted (see Firbank & Watkinson, 1990).

Nonetheless, replacement series (the analysis of which is reviewed, for example, by Harper, 1977; Begon & Mortimer, 1986) have provided valuable insights into the nature of interspecific competition and the factors influencing its intensity (Firbank & Watkinson, 1990). An early, influential study was that of de Wit *et al.* (1966) on competition between the grass *Panicum maximum* and the legume *Glycine javanica*, which often form mixtures in Australian pastures. *Panicum* acquires its nitrogen only from the soil, but *Glycine* acquires part of its nitrogen from the air, by nitrogen fixation, through its root association with the bacterium, *Rhizobium* (see Chapter 13, Section 13.10.1). The competitors were grown in replacement series with and without an inoculation of *Rhizobium*, and the results (Figure 7.15) are given both as replacement diagrams and as 'relative yield totals'. The relative yield of a species in a mixture is the ratio of its yield in mixture to its yield alone in the replacement series, removing any absolute yield differences between species and referring both to the same scale. The relative yield total of a particular mixture is then the sum of the two relative yields. It is fairly clear from the replacement series (Figure 7.15a) that both species, but especially *Glycine*, fared better (were less affected by interspecific competition) in the presence than in the absence of *Rhizobium*. This is clearer still, however, from the relative yield totals (Figure 7.15b), which never departed significantly from 1 in the absence of *Rhizobium*, but consistently exceeded 1 in its presence, suggesting that niche differentiation was not possible without *Rhizobium* (a second species could only be accommodated by a compensatory reduction in the

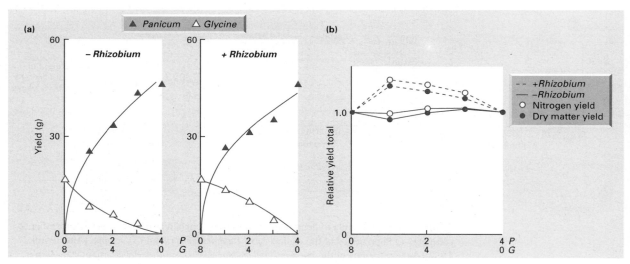

Figure 7.15 A substitutive experiment on interspecific competition between *Panicum maximum* (P), and *Glycine javanica* (G), in the presence and absence of *Rhizobium*: (a) replacement diagrams; (b) relative yield totals. (After de Wit *et al.*, 1966.)

output of the first), but that niche differentiation occurred in its presence (the species yielded more between them than either could alone).

additive experiments

A second popular approach in the past has been the use of an 'additive' design, in which one species (typically a crop) is sown at a constant density, along with a range of densities of a second species (typically a weed). The justification for this is that it mimics the natural situation of a crop infested by a weed, and it therefore provides information on the likely effect on the crop of various levels of infestation (see Cousens, 1985; Firbank & Watkinson, 1990). A problem with additive experiments, however, is that overall density and species proportion are changed simultaneously. It has therefore proved difficult to separate the effect of the weed itself on crop yield from the simple effect of increasing total density (crop plus weed). An example is shown in Figure 7.16, describing the effects of two weeds, sicklepod (*Cassia obtusifolia*) and redroot pigweed (*Amaranthus retroflexus*), on the yield of cotton grown in Alabama, USA (Buchanan *et al.*, 1980). As weed density increased, so cotton yield decreased, and this effect of interspecific competition was always more pronounced with sicklepod than with redroot pigweed.

response surface analyses

In substitutive designs the proportions of competitors are varied but total density held constant, whilst in additive designs proportions are varied but the density of one competitor is held constant. It is perhaps not surprising, therefore, and certainly welcome, that a 'response surface analysis' has been proposed and applied, in which two species are grown alone and in mixtures at a wide range of densities and proportions (Firbank & Watkinson, 1985; Law & Watkinson, 1987 (Figure 7.17); Bullock *et al.*, 1994b (although the last of these deals with clones of the same species)). Overall, these studies suggest that good equations for describing the competitive effect of one species (A) on another (B) are, for mortality:

$$N_A = N_{i,A} (1 + m (N_{i,A} + \beta N_{i,B}))^{-1} \tag{7.16}$$

and for fecundity:

$$Y_A = N_A R_A (1 + a (N_A + \alpha N_B))^{-b} \tag{7.17}$$

which can both be seen to be related to Equation 6.18 (see Chapter 6, Section

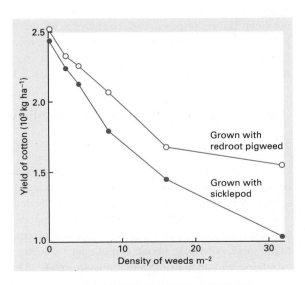

Figure 7.16 An 'additive design' competition experiment: the yield of cotton produced from stands planted at constant density, infested with weeds (either sicklepod or redroot pigweed) at a range of densities. (After Buchanan *et al.*, 1980.)

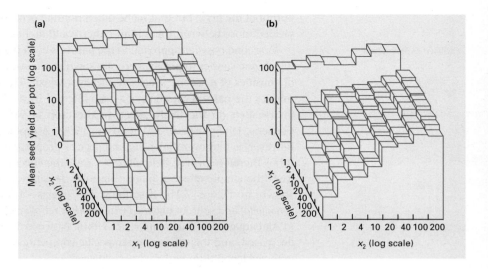

Figure 7.17 The response surface of competition, as indicated by seed production per pot, between (a) *Phleum arenarium* and (b) *Vulpia fasciculata* sown alone and in mixtures over a range of densities and frequencies. (After Law & Watkinson, 1987.)

6.8.1—basic model of intraspecific competition) and Equation 7.3 (see Section 7.4.1—incorporation of interspecific competition). Thus, $N_{i,A}$ and $N_{i,B}$ are the initial numbers of species A and B; N_A and N_B are the numbers of species A and B after mortality; Y_A is the yield (seeds or biomass) of species A; *m* and *a* are susceptibilities to crowding; β and α are competition coefficients; R_A is the basic reproductive rate of species A (and hence, $N_A R_A$ is the yield in the absence of competition); and *b* determines the type of density dependence (assumed equal to 1 for mortality—perfect compensation). Data like those shown in Figure 7.17, obtained over a single generation, can thus be used to fit values (by computer program) to the parameters in Equations 7.16 and 7.17, and the equations in their turn can be used to predict the outcome of competition between the species over many generations—not possible with either substitutive or additive designs.

On the other hand, Law and Watkinson (1987) found that they could obtain an improved fit to their response surfaces, especially for one of their species, if they used an equation in which competition coefficients were not fixed, but varied with both frequency and density—although the meaning of this in terms of 'plant behaviour' is not clear. Hence, response surface analyses, in revealing the potential complexities in interactions between competing species, also reveal that knowing or predicting dynamical outcomes may be only part of the story. It may also be necessary to understand underlying mechanisms (see Section 7.10).

7.8 Evolutionary effects of interspecific competition

7.8.1 Natural experiments

We have seen that interspecific competition is commonly studied by an experimenter comparing species alone and in combination. Nature, too, often provides information of this sort: the distribution of certain potentially competing species is such that they sometimes occur together (sympatry) and sometimes occur alone (allopatry). These 'natural experiments' can provide additional information about interspecific competition, and especially about evolutionary effects, since the differences between sympatric and allopatric populations are often of long standing. The attractions of

the pros and cons of natural experiments

natural experiments are first that they *are* natural—they are concerned with organisms living in their natural habitats—and second, that they can be 'carried out' simply by observation—no difficult or impracticable experimental manipulations are required. They have the disadvantage, however, of lacking truly 'experimental' and 'control' populations. Ideally, there should be only one difference between the populations: the presence or absence of a competitor species. In practice, though, populations typically differ in other ways too, simply because they exist in different locations. Natural experiments should therefore always be interpreted cautiously.

competitive release and character displacement

Evidence for competition from natural experiments usually comes either from niche expansion in the absence of a competitor (known as *competitive release*), or simply from a difference in the realized niche of a species between sympatric and allopatric populations. If this difference is accompanied by morphological changes, then the effect is referred to as *character displacement*. On the other hand, physiological, behavioural and morphological traits are all equally likely to be involved in competitive interactions and to be reflections of a species' realized niche. One difference may be that morphological distinctions are most obviously the result of evolutionary change, but as we shall see, physiological and behavioural 'characters' are also liable to 'competitive displacement'.

gerbils in Israel: competitive release

One example of natural competitive release is provided by work on two gerbilline rodents living in the coastal sand-dunes of Israel (Abramsky & Sellah, 1982). In northern Israel, the protrusion of the Mt Carmel ridge towards the sea separates the narrow coastal strip into two isolated areas, north and south. *Meriones tristrami* is a gerbil that has colonized Israel from the north. It now occurs, associated with the dunes, throughout the length of the coast, including the areas both north and south of Mt Carmel. *Gerbillus allenbyi* is another gerbil, also associated with the dunes and feeding on similar seeds to *M. tristrami*; but this species has colonized Israel from the south and has not crossed the Mt Carmel ridge. To the north of Mt Carmel, where *M. tristrami* lives alone, it is found on sand as well as other soil types. However, south of Mt Carmel it occupies several soil types but not the coastal sand-dunes. Here, only *G. allenbyi* occurs on dunes.

invoking the ghost of competition past

This appears to be a case of competitive exclusion and competitive release: exclusion of *M. tristrami* by *G. allenbyi* from the sand to the south of Mt Carmel; release of *M. tristrami* to the north. Is this present-day competitive exclusion, however, or an evolutionary effect? Abramsky and Sellah set up a number of plots south of Mt Carmel from which *G. allenbyi* was removed, and they compared the densities of *M. tristrami* in these plots with those in a number of similar control plots. They monitored the plots for 1 year; but the abundance of *M. tristrami* remained essentially unchanged. It seems that south of Mt Carmel, *M. tristrami* has evolved to select those habitats in which it avoids competition with *G. allenbyi*, and that even in the absence of *G. allenbyi* it retains this genetically fixed preference. Note, though, as ever, that this interpretation, because it invokes the ghost of competition past, may be sound and sensible—but it is not established fact.

worker ants: morphological character displacement

A possible case of morphological character displacement comes from work on the seed-eating harvester ant, *Veromessor pergandei*, collected at various desert sites in the south-western USA (Davidson, 1978). Food (i.e. seeds) can be an important limiting resource for these ants (Brown & Davidson, 1977), and different ant species specialize on seeds of different sizes depending on their own size (Davidson, 1977). In the case of *V. pergandei*, Davidson looked at mandible length and variability in foraging workers; and as Figure 7.18a shows, variability decreased significantly as

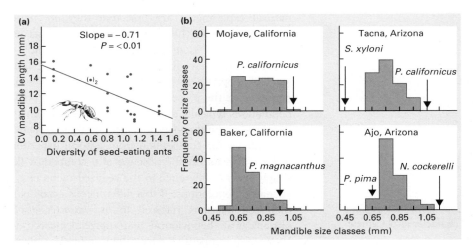

Figure 7.18 Character displacement. (a) The relationship between the within-colony coefficient of variation (CV) in mandible length for *Veromessor pergandei* and the species diversity of seed-eating ants in the community. *V. pergandei* is most variable where there are fewest competitors. (b) Some frequency distributions of mandible size classes for *V. pergandei* at different sites, with the mean mandible lengths of competitors most similar in size indicated by arrows. *V. pergandei* varies in average size from site to site, such that it is always different from the competitors with which it coexists. (After Davidson, 1978.)

the diversity of potential competitors at a site increased. In other words, *V. pergandei* is more of a size specialist at those sites in which interspecific competition is most likely. The same picture emerges from Figure 7.18b, which suggests, in addition, that worker size itself varies from site to site in a way which tends to make *V. pergandei* different in size from the species with which it coexists. It is this morphological differentiation in the presence of potential competitors that is known as character displacement. The inference is that *V. pergandei* workers have been selected that are significantly different in size from individuals of coexisting species, thus allowing resource partitioning (different-sized ants eat different-sized seeds) and the coexistence of potential competitors.

A further example concerns populations of the originally marine three-spined stickleback, *Gasterosteus aculeatus*, living in freshwater lakes in British Columbia, Canada, having apparently been left behind either following uplifting of the land after deglaciation, around 12 500 years ago, or after the subsequent rise and fall of sea levels around 11 000 years ago (Schluter & McPhail, 1992, 1993). As a result of this 'double invasion', some lakes now support two species of *G. aculeatus* (although they have not, as yet, been given their own specific names), whilst others support only one. Wherever there are two species, one is always 'limnetic', the other 'benthic'. The first concentrates its feeding on plankton in the open water and has correspondingly long (and closely spaced) gill rakers which seive the plankton from the stream of ingested water. The second, with much shorter gill rakers, concentrates on larger prey which it consumes largely from vegetation or sediments (Figure 7.19). Wherever there is only one species in a lake, however, this species exploits both food resources and is morphologically intermediate (Figure 7.19). Presumably, either ecological character displacement has evolved since the second invasion, and has promoted the coexistence of the species-pairs, or it was a

morphological and ecological displacement in sticklebacks

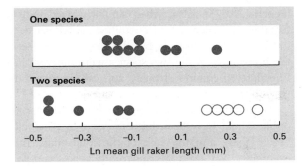

Figure 7.19 Character displacement in three-spined sticklebacks (*Gasterosteus aculeatus*). In small lakes in coastal British Columbia supporting two stickleback species (lower graph), the gill rakers of the benthic species (●) are significantly shorter than those of the limnetic species (○), whilst those species of sticklebacks that occupy comparable lakes alone (upper graph) are intermediate in length. Lengths of gill rakers have been adjusted to take account of species differences in overall size. (After Schluter & McPhail, 1993.)

necessary prerequisite for the second invasion to be successful (see also Chapter 20, Section 20.4.2).

Another plausible example of character displacement is provided by work on the mud snails *Hydrobia ulvae* and *H. ventrosa*. When the two species live apart, their sizes are more or less identical; but when they coexist they are always different, both in Denmark and in Finland (Figures 7.20a, b). When the similarly sized species lived apart they consumed similarly sized food particles; but when they coexisted, *H. ulvae*, which was then larger, tended to consume larger food particles than *H. ventrosa*

mud snails: a classic example of character displacement ...

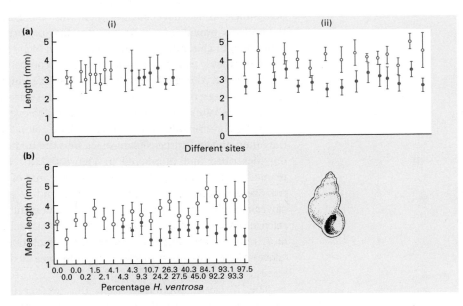

Figure 7.20 (a) Character displacement: average lengths (plus standard deviations) of *Hydrobia ulvae* (○) and *H. ventrosa* (●) at various sites in Denmark. When they live alone (i) their sizes are similar; but when they live together (ii) their sizes differ. (After Fenchel, 1975.) (b) A similar pattern in Finland: the sites are arranged in order of increasing percentage of *H. ventrosa* (horizontal scale). (After Saloniemi, 1993.)

(Fenchel, 1975). Also, when similarly sized individuals of the two species grew together in experimental containers, interspecific competition was as intense as intraspecific competition; but this depressant effect was significantly reduced when the individuals of the two species were different in size (Fenchel & Kofoed, 1976). The data, therefore, strongly suggest character displacement, allowing resource partitioning and coexistence.

... or are they?

Despite its appeal and plausibilty, good examples of morphological character displacement have been hard to come by (Connell, 1980; Arthur, 1982; Taper & Case, 1994), and the case of the mud snails has been considered to be near-exemplary. Yet, even this apparently good example is now open to serious question (e.g. Cherrill & James, 1987; Cherrill, 1988; Fenchel, 1988; Saloniemi, 1993). In particular, in both Great Britain and Finland, the sympatric and allopatric habitats were found not to be the same: *H. ulvae* and *H. ventrosa* coexisted in sheltered water bodies rarely affected by tidal action, *H. ulvae* was found alone in relatively exposed tidal mudflats and salt marshes, and *H. ventrosa* was found alone in non-tidal lagoons and pools. Moreover, *H. ulvae* simply grows larger in less tidal habitats, and *H. ventrosa* may grow less well in this habitat, at least in Finland. This alone could account for the size differences between sympatry and allopatry in these species. This emphasizes the major problem with natural experiments such as those that seem to demonstrate character displacement: sympatric and allopatric populations can occur in different environmental conditions over which the observer has no control. It may be these environmental differences, rather than competition, that has led to the character displacement.

7.8.2 Experimenting with natural experiments

Sometimes, as we have already seen with the gerbils, natural experiments may themselves provide an opportunity for a further—and more informative—experimental manipulation. In one such case, niche divergence was sought in clover, *Trifolium repens*, as a result of its having to compete with the grass *Lolium perenne* (Turkington & Mehrhoff, 1990). Clover was examined from two sites: one in which it achieved a ground coverage of 48%, but the grass achieved a coverage of 96% (a 'two-species' site); and another in which it achieved 40% coverage, but *L. perenne* covered only 4% (effectively a 'clover-alone' site). A total of three transplant (into the other site) and three replant (back into the home site) experiments were carried out (described and numbered in Figure 7.21a). *T. repens*, from both sites, was planted: (i) in plots at the two-species site cleared of *T. repens* only; (ii) in plots at the two-species site cleared of both *T. repens* and *L. perenne*; and (iii) in plots at the clover-alone site cleared of *T. repens*. The extent of competitive suppression or release was assessed from the amount of growth achieved by the different plantings of *T. repens*. From this, the extent of the evolution of niche divergence between 'clover-alone' and 'two-species' *T. repens* was deduced, as was that between *T. repens* and *L. perenne*.

The *T. repens* population from the two-species site had indeed apparently diverged from the *L. perenne* population with which it was coexisting (and with which it may otherwise have competed strongly), and had diverged too from the clover-alone population (Figure 7.21b). When the two-species site was cleared of *T. repens* only, the *re*planted *T. repens* grew better than the *trans*planted clover-alone plants (treatments 1 and 4, Figure 7.21b; $P = 0.086$, close to significance), suggesting

niche divergence in clover–grass competition

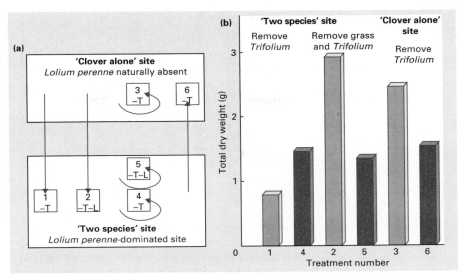

Figure 7.21 (a) Experimental design to test for the evolution of *Trifolium repens* (T) in competition with *Lolium perenne* (L). Indigenous populations of *T. repens*, and sometimes also *L. perenne*, were removed. *Trifolium repens* was removed from the base of the arrow and transplanted, or replanted, at the head of the arrow. Treatment numbers are consistent with the usage of Connell (1980). (b) The results of this experiment is in terms of the total plot dry weight achieved by *T. repens* in the various treatments. Significance levels for comparisons between pairs of treatments are given in the text. (After Turkington & Mehrhoff, 1990.)

that the clover-alone plants were competing more with the resident *L. perenne*. Moreover, when *L. perenne* was also removed, this made effectively no difference to the two-species *T. repens* (treatments 4 and 5; $P > 0.9$), but led to a large increase in the growth of the clover-alone plants (treatments 1 and 2; $P < 0.005$). Also, when *L. perenne* was removed, the clover-alone plants grew better than the two-species ones (treatments 2 and 5; $P < 0.05$)—all of which suggests that only the clover-alone plants were released from competition by the absence of *L. perenne*. Finally, at the clover-alone site, the two-species clover plants grew no better than they had at their home site (treatments 4 and 6; $P > 0.7$), whereas the clover-alone plants grew far better than they had at the two-species site in the presence of the grass (treatments 1 and 3; $P < 0.05$). Thus, the clover from the two-species population hardly competes with the *L. perenne* with which it coexists, whereas the clover-alone population would do—and does so if transplanted to the two-species site.

7.8.3 Selection experiments

direct demonstration of the evolutionary effects of competition has been rare

The most direct way of demonstrating the evolutionary effects of competition within a pair of competing species is for the experimenter to induce these effects—impose the selection pressure (competition) and observe the outcome. Surprisingly perhaps, there have been very few successful experiments of this type (Arthur, 1982). In some cases, a species has responded to the selection pressure applied by a second, competitor species by apparently increasing its 'competitive ability', in the sense of increasing its frequency within a joint population. An example of this with two species of *Drosophila* is shown in Figure 7.22. Such results, however, tell us nothing

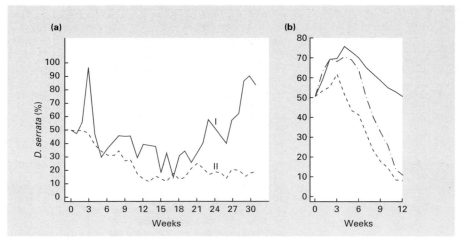

Figure 7.22 Apparent evolution of competitive ability in *Drosophila serrata*. (a) Of two experimental populations coexisting (and competing) with *D. nebulosa*, one (I) increased markedly in frequency after around week 20. (b) Individuals from this population did better in further competition with *D. nebulosa* ((—), mean of five populations) than did individuals from population II ((- - -), mean of five), or individuals from a stock not previously subjected to interspecific competition ((·—·—·), mean of five). (After Ayala, 1969.)

about the means by which such apparent increases were achieved (e.g. whether it was as a result of niche differentiation).

Tribolium castaneum individuals show a greater tendency to attack *T. confusum* rather than members of their own species (see Section 7.4.3), the greater the proportion of *T. confusum* in the stocks in which they had been reared (the more they had been subject to inter- rather than intraspecific competition) (Dawson, 1979). However, to find an example of a demonstrable increase in competitive ability associated with a demonstrable differentiation of niches in a selection experiment, we must turn away from interspecific competition in the strict sense to competition between two types of the same species which were not able to interbreed: the 'wild type' and the 'dumpy' mutant of *D. melanogaster* (Figure 7.23). After five generations of 'selection' (being reared in competition with the other type), the results of a replacement series experiment showed not only that the performance of each type against its competitor had improved, but also that the total productivity of mixed populations had increased—indicative of increased niche differentiation (Figure 7.23). Overall, however, the experimental selection of increased niche differentiation amongst competing species appears to be either frustratingly elusive or sadly neglected.

7.9 Niche differentiation and similarity amongst coexisting competitors

It might be imagined that scientific progress is made by providing answers to questions. In fact, progress often consists of replacing one question with another, more pertinent, more challenging question. In this section, we deal with an area where this is the case: the questions of *how* different coexisting competitors are, and how different coexisting species need to be if competition is not to eliminate one of them.

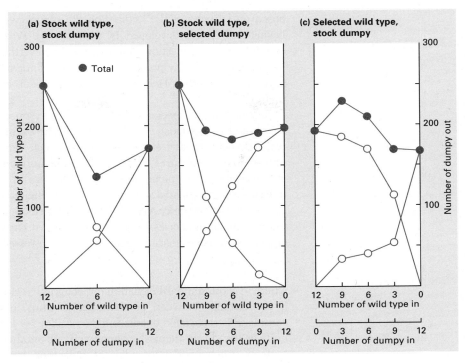

Figure 7.23 Three replacement series examining competition between 'wild type' and 'dumpy' *Drosophila melanogaster*. (a) Unselected stocks. (b) Unselected wild type but dumpy subjected to five generations of selection against wild type. (c) Unselected dumpy but wild type subjected to five generations of selection against dumpy. In (b) and (c) both types, when selected, are more productive at any given proportion than they are in (a), but in addition, the total productivities of mixtures in (b) and (c) are increased—in (c) such that productivity is higher than for either type alone. Selection has apparently also led to niche differentiation. (After Seaton & Antonovics, 1967.)

The Lotka–Volterra model predicts the stable coexistence of competitors in situations where interspecific competition is, for both species, less significant than intraspecific competition. Niche differentiation will obviously tend to concentrate competitive effects more within species than between them. The Lotka–Volterra model, and the Competitive Exclusion Principle, therefore imply that *any* amount of niche differentiation will allow the stable coexistence of competitors. Hence, in an attempt to discover whether this was 'true', the question 'do competing species need to be different in order to coexist stably?' greatly exercised the minds of ecologists during the 1940s (Kingsland, 1985).

It is easy to see now, however, that the question is badly put, since it leaves the precise meaning of 'different' undefined. We have seen examples in which the coexistence of competitors is apparently associated with some degree of niche differentiation, but it seems that if we look closely enough, all coexisting species will be found to be different—without this having anything to do with competition. A more pertinent question, therefore, would be 'is there a minimum amount of niche differentiation that has to be exceeded for stable coexistence?' That is, is there a limit to the similarity of coexisting species?

One influential attempt to answer this question for exploitative competition,

how much?

a simple model developed to
provide an answer

based on variants of the Lotka–Volterra model, was initiated by MacArthur and
Levins (1967) and developed by May (1973). With hindsight, their approach is
certainly open to question (Abrams, 1983). Nevertheless, we can learn most about
'the limiting similarity problem' by first examining their approach and then looking
at the objections to it. Here, as so often, the models can be instructive without being
'right'.

Imagine three species competing for a resource that is unidimensional and
distributed continuously; food size is a clear example. Each species has its own
realized niche in this single dimension which can be visualized as a resource-
utilization curve (Figures 7.24a, b). The consumption rate of each species is highest
at the centre of its niche and tails off to zero at either end, and the more the
utilization curves of adjacent species overlap, the more the species compete. Indeed,
by assuming that the curves are 'normal' distributions (in the statistical sense), and
that the different species have similarly shaped curves, the competition coefficient
(applicable to both adjacent species) can be expressed by the following formula:

$$\alpha = e^{-d^2/4w^2} \tag{7.18}$$

where w is the standard deviation (or, roughly, 'relative width') of the curves, and d
is the distance between adjacent peaks. Thus, α is very small when there is
considerable separation of adjacent curves ($d/w \gg 1$; Figure 7.24a), and approaches
unity as the curves themselves approach one another ($d/w < 1$; Figure 7.24b).

How much overlap of adjacent utilization curves is compatible with stable
coexistence? Assume that the two peripheral species have the same carrying capacity
(K_1, representing the suitability of the available resources for species 1 and 3), and
consider the coexistence, in between them on the resource axis, of another species
(carrying capacity K_2). When d/w is low (α is high and the species are similar) the
conditions for coexistence are extremely restrictive in terms of the $K_1 : K_2$ ratio; but
these restrictions lift rapidly as d/w approaches and exceeds unity (Figure 7.25). In
other words, coexistence is possible when d/w is low, but only if the suitabilities of
the environment for the different species are extremely finely balanced.
... which it does ...
Furthermore, if the environment is assumed to vary, then the fluctuations will lead

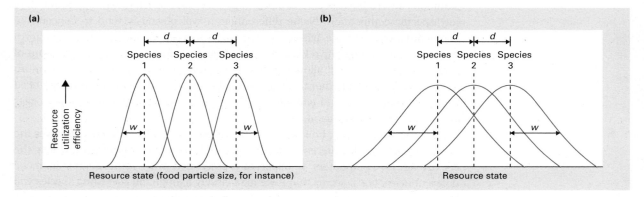

Figure 7.24 Resource-utilization curves for three species coexisting along a one-dimensional
resource spectrum. d is the distance between adjacent curve peaks, w is the standard
deviation of the curves. (a) Narrow niches with little overlap ($d > w$), i.e. relatively little
interspecific competition. (b) Broader niches with greater overlap ($d < w$), i.e. relatively
intense interspecific competition.

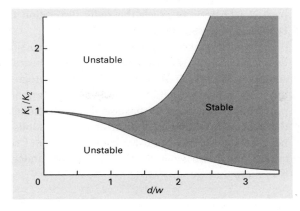

Figure 7.25 The range of habitat favourabilities (indicated by the carrying capacities K_1 and K_2, where $K_1 = K_3$) that permit a three-species equilibrium community with various degrees of niche overlap (d/w). (After May, 1973.)

to variations in the $K_1 : K_2$ ratio, and coexistence will now only be possible if there is a broad range of $K_1 : K_2$ ratios leading to stability, i.e. if, roughly, $d/w > 1$.

This model, then, suggests that there *is* a limit to the similarity of coexisting competitors, and that the limit is given by the condition $d/w > 1$. Are these the correct answers? In fact, it seems most unlikely that there is a universal limit to similarity, or even a widely applicable one that we could express in such a simple way as $d/w > 1$. Abrams (1976, 1983), amongst others, has emphasized that models with competition in several dimensions, with alternative utilization curves, and so on, all lead to alternative limits to similarity, and often to much lower values of d/w being compatible with robust, stable coexistence. In other words, '$d/w > 1$' is a property of one type of model analysis, but not of others, and thus, almost certainly, **... but, the answer is almost certainly wrong in detail** not of nature as a whole. Furthermore, we have already seen that because of environmental heterogeneity, apparent competition and so on, exploitative competition and any niche differentiation associated with it are not necessarily the whole story when it comes to the coexistence of competitors. This too argues against the idea of a universal limit.

On the other hand, the most general messages from the early models still seem valid, namely: (i) in the real world, with all its intrinsic variability, there *are* likely to be limits to the similarity of coexisting exploitative competitors; and (ii) these limits will reflect not only the differences between species, but also the variability within them, the nature of the resource, the nature of the utilization curve, and so on.

But is even the limiting similarity question the best question to ask? We want to understand the extent of niche differentiation amongst coexisting species. If species were always packed as tightly together as they could be, then presumably they **ecological and evolutionary effects, again** would differ by the minimum (limiting) amount. But why should they be? We return, once again, to the distinction between the ecological and the evolutionary consequences of competition (Abrams, 1990). The ecological effects are that species with 'inappropriate' niches will be eliminated (or repelled if they try to invade), and the limiting similarity question implicitly concerned itself with this: how many species can be 'packed in'? But coexisting competitors may also evolve. Do we generally observe the ecological, or the combined ecological and evolutionary effects? Do they differ? We cannot attempt to answer the first question without answering the second, and the answer to that seems to be, perhaps inevitably, 'it depends'. Different models, based on different underlying mechanisms in the competitive process, predict that evolution will lead to more widely spaced niches, or to more closely packed niches or to much the same disposition of niches as those predicted by ecological processes alone (Abrams, 1990).

Two points, therefore, emerge from this discussion. The first, which will not have escaped notice, is that it has been entirely theoretical. This is a reflection of the second point, which is that we have seen progress—but, in terms of successive questions superceding their predecessors rather than actually answering them. Data provide answers—what we have seen is a refinement of questions. The latest stage in this appears to be that attempts to answer questions regarding 'niche similarity' may need to be postponed until we know more about resource distributions, utilization curves, and, more generally, the mechanisms underlying exploitative competition. It is to these that we now turn.

the answer?—It depends

7.10 Niche differentiation and mechanisms of exploitation

In spite of all the difficulties of making a direct connection between interspecific competition and niche differentiation, there is no doubt that niche differentiation is often the basis for the coexistence of species.

differential resource utilization: easy to imagine in animals— less easy in plants

There are a number of ways in which niches can be differentiated. One is resource partitioning, or, more generally, differential resource utilization. This can be observed when species living in precisely the same habitat nevertheless utilize different resources. Since the majority of resources for animals are individuals of other species (of which there are literally millions of types), or parts of individuals, there is no difficulty, in principle, in imagining how competing animals might partition resources amongst themselves. Plants, on the other hand, all have very similar requirements for the same potentially limited resources (see Chapter 3), and there is much less apparent scope for resource partitioning (but see below).

niche differentiation based on resources and conditions

In many cases, the resources used by ecologically similar species are separated spatially. Differential resource utilization will then express itself as either microhabitat differentiation between the species (e.g. different species of fish feeding at different depths), or even a difference in geographical distribution. Alternatively, the availability of the different resources may be separated in time, for example, different resources may become available at different times of the day or in different seasons. Differential resource utilization may then express itself as a temporal separation between the species.

The other major way in which niches can be differentiated is on the basis of conditions. Two species may use precisely the same resources; but if their ability to do so is influenced by environmental conditions (as it is bound to be), and if they respond differently to those conditions, then each may be competitively superior in different environments. This too can express itself as a microhabitat differentiation, or a difference in geographical distribution or a temporal separation, depending on whether the appropriate conditions vary on a small spatial scale, a large spatial scale or over time. Of course, in a number of cases (especially with plants) it is not easy to distinguish between conditions and resources (see Chapter 3). Niches may then be differentiated on the basis of a factor (such as water) which is both a resource and a condition.

There are many examples of the separation of competing species in space or time involving both animals and plants. For example, tadpoles of two anuran species in New Jersey, USA (*Hyla crucifer* and *Bufo woodhousii*), have their feeding periods offset by around 4–6 weeks each year, apparently, though not certainly, associated with differential responses to environmental conditions rather than seasonal changes in resources (Lawler & Morin, 1993). The gerbils, *Gerbillus allenbyi* and *G. pyramidus* in Israeli deserts, typically forage at different times of the night and in slightly different habitats, apparently as a combined result of the higher energy efficiency of the former and the greater aggression of the latter (Ziv *et al.*, 1993). Two phloem-feeding bark beetles, *Ips duplicatus* and *I. typographus*, on Norway spruce trees, in Norway, are separated in their feeding sites on a small spatial scale by trunk diameter, although the reason for this is not at all clear (Schlyter & Anderbrandt, 1993). But, it is amongst plants and other sessile organisms, because of their limited scope for differential resource utilization at the same location and instant, that spatial and temporal separation are likely to be of particular significance (see Harper, 1977), although as ever, it is one thing to show that species differ in their spatial or temporal distribution—it is quite another to prove that this has anything to do with competition. The cattails in Section 7.3 provide one example of competing plants separated spatially. Another is shown in Figure 7.26: the annuals *Sedum smallii* and *Minuartia uniflora* which dominate the vegetation growing on granite outcrops in the south-eastern USA. The adult plants exhibit an especially clear spatial zonation associated with soil depth (itself strongly correlated with soil moisture), and further experimental results reinforce the idea that it is competition rather than mere differences in tolerance that gives rise to this zonation.

Describing the outcome of competition, however—'one species coexists with or excludes another'—and even associating this with niche differentiation, whether based on resources themselves, or conditions or merely differences in space or time,

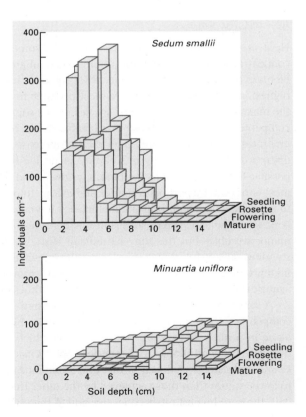

Figure 7.26 The zonation of individuals, according to soil depth, of two annual plants, *Sedium smallii* and *Minuartia uniflora* at four stages of the life cycle. (After Sharitz & McCormick, 1973.)

actually provides us with rather little understanding of the competitive process. For this, as we have seen repeatedly in this chapter, we may need to focus more on the mechanisms of exploitation. *How*, precisely, does one species outexploit and outcompete another? *How* can two consumers coexist on two limiting resources, when both resources are absolutely essential to both consumers?

Furthermore, as Tilman (1990) has pointed out, whilst monitoring the population dynamics of two competing species may give us some powers of prediction for the next time they compete, it will give us very little help in predicting how each would fare against a third species—whereas, if we understood the dynamics of the interaction of all the species with their shared limiting resources, then we might be able to predict the outcome of exploitative competition between any given pair of them. We therefore turn now to some attempts to explain the coexistence of species competing for limiting resources that explicitly consider not only the dynamics of the competing species but also the dynamics of the resources themselves. Rather than going into details, we examine the outlines of models and some major conclusions.

the need to consider resource dynamics

7.10.1 Exploitation of a single resource

successful exploitation and the reduction in resource concentration

Tilman (1990) shows, for a number of models, what we have already seen demonstrated empirically in Section 7.2.6, that when two species compete exploitatively for a single limiting resource, the outcome is determined by which species is able, in its exploitation, to reduce the resource to the lower equilibrium concentration, R^*. Different models, based on varying details in the mechanism of exploitation, give rise to different formulae for R^*, but even the simplest model is revealing, giving:

$$R_i^* = m_i C_i/(g_i - m_i). \tag{7.19}$$

but how is this achieved?

Here m_i is the mortality or loss rate of consumer species i; C_i is the resource concentration at which species i attains a rate of growth and reproduction per unit biomass (relative rate of increase, RRI) equal to half its maximal RRI (C_i is thus highest in consumers that require most resource in order to grow rapidly); and g_i is the maximum RRI achievable by species i. This suggests that successful exploitative competitors (low R_i^*) are those that combine resource-utilization efficiency (low C_i), low rates of loss (low m_i) and high rates of increase (high g_i). On the other hand, as discussed much more fully in Chapter 14 on life-history strategies, it may not be possible for an organism to combine, say, low C_i and high g_i. A plant's growth will be most enhanced by putting its matter and energy into leaves and photosynthesis—but to enhance its nutrient-utilization efficiency it would have to put these into roots. A lioness will be best able to subsist at low densities of prey by being fleet-footed and manoeuverable—but this may be difficult if she is often heavily pregnant. Understanding successful exploitative competitiveness, therefore, may require us ultimately to understand how organisms trade-off features giving rise to low values of R^* against features that enhance other aspects of fitness.

answers from grasses

A rare test of these ideas is provided by Tilman's own work on terrestrial plants competing for nitrogen (Tilman & Wedin, 1991a, b). Five grass species were grown alone in a range of experimental conditions which gave rise in turn to a range of nitrogen concentrations. Two species, *Schizachyrium scoparium* and *Andropogon gerardi*, consistently reduced nitrate and ammonium concentrations in soil solutions to lower values than those achieved by the other three species (in all soils but those with the very highest nitrogen levels), and of these three other species, one, *Agrostis scabra*, left behind higher concentrations than the other two, *Agropyron repens* and *Poa pratensis*. Then, when *A. scabra* was grown with *A. repens*, *S. scoparium* and *A. gerardi*, the results, especially at low nitrogen concentrations where nitrogen was most likely to be limiting, were very much in line with the exploitative competition theory (Figure 7.27). The species that could reduce nitrogen to the lowest concentration always won—*A. scabra* was always competitively displaced. Moreover, the displacement of *A. scabra* was more rapid against the species that reduced nitrogen to the the lowest concentrations, *S. scoparium* and *A. gerardi*, than it was against *A. repens*.

In fact, these five species were chosen from various points in a typical old-field successional sequence in Minnesota, USA (Figure 7.28a), and it is clear that the better competitors for nitrogen are found later in the sequence. These species, and *S. scoparium* and *A. gerardi* in particular, had higher root allocations, but lower above-ground vegetative growth rates and reproductive allocations (e.g. Figure 7.28b). In other words, they achieved their low values of R^* by the high resource-utilization efficiency given to them by their roots (low C_i; Equation 7.19), even though they appeared to have paid for this through a reduction in growth and

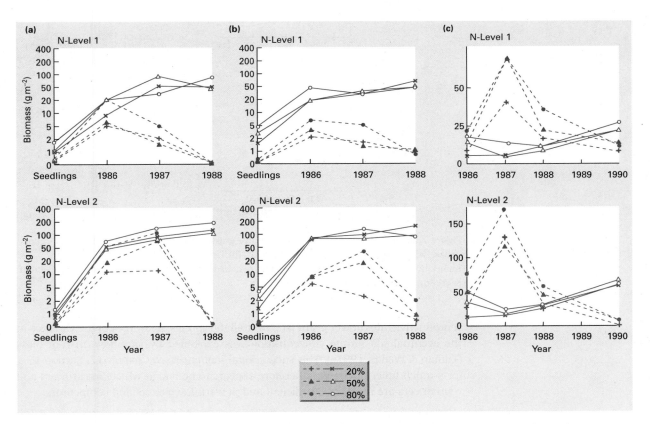

Figure 7.27 The results of competition experiments in which *Agrostis scabra* (- -) was competitively displaced by (a) *Schizachyrium scoparium*, (b) *Andropogon gerardi* and (c) *Agropyron repens* (—), at each of two nitrogen levels (both low) and whether it represented 20%, 50% or 80% of the initial seed sown. In each case, *A. scabra* had lower values of R^* for nitrate and ammonium (see text). Displacement was least rapid in (c) where the differential was least marked. (After Tilman & Wedin, 1991b.)

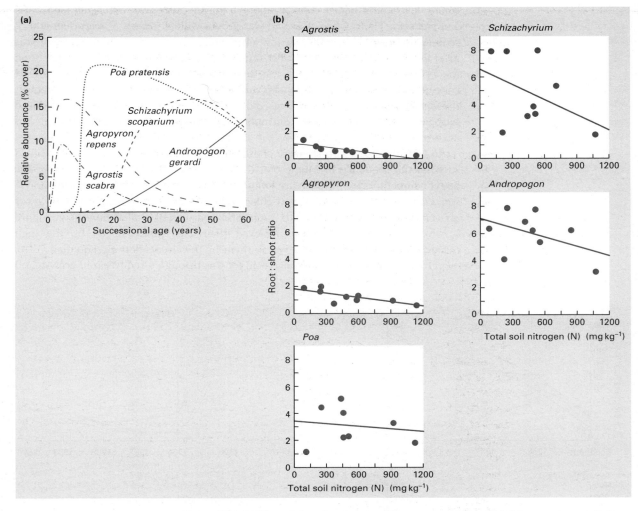

Figure 7.28 (a) The relative abundances of five grasses during old-field successions at Cedar Creek Natural History Area, Minnesota, USA. (b) The root : shoot ratios were generally higher in the later successional species and declined as soil nitrogen increased. (After Tilman & Wedin, 1991a.)

reproductive rates (lower g_i). In fact, over all the species, a full 73% of the variance in the eventual soil nitrate concentration was explained by variations in root mass (Tilman & Wedin, 1991a). This successional sequence (see Chapter 17, Section 17.4 for a much fuller discussion) therefore appears to be one in which fast growers and reproducers are replaced by efficient and powerful exploiters and competitors.

7.10.2 Exploitation of two resources

a model based on the dynamics of competitors and their resources

Tilman (1982, 1986; see also Chapter 3, Section 3.8) has also considered what happens when two competitors compete for two resources. Beginning with intraspecific competition, we can define a zero net growth isocline for a single species utilizing two essential resources (see Chapter 3, Section 3.8). This isocline is the boundary between resource combinations that allow the species to survive and reproduce, and resource combinations that do not (Figure 7.29), and therefore represents the boundary of the species' niche in these two dimensions. For present purposes, we can ignore the complications of overcompensation, chaos, etc. and assume that intraspecific competition brings the population to a stable equilibrium. Here, however, the equilibrium has two components: both population size and the resource levels should remain constant. Population size is constant (by definition) at all points on the isocline, and Tilman established that there is only one point on the isocline where resource levels are also constant (point S^* in Figure 7.29). This point, which is the two-resource equivalent of R^* for one resource, represents a balance between the consumption of the resources by the consumer (taking the resource concentrations towards the bottom left of the figure) and the natural renewal of the resources (taking the concentrations towards the top right). Indeed, in the absence of the consumer, resource renewal would take the resource concentrations to the 'supply point', shown in the figure.

To move from intra- to interspecific competition, it is necessary to superimpose the isoclines of two species on the same diagram (Figures 7.30a, b). The two species will presumably have different consumption rates, but there will still be a single supply point. The outcome depends on the position of this supply point.

In Figure 7.30a, the isocline of species A is closer to both axes than that of species B. There are three regions in which the supply point might be found. If it was in region 1, below the isoclines of both species, then there would never be sufficient resources for either species and neither would survive. If it was in region 2, below the isocline of species B but above that of species A, then species B would be unable to survive and the system would equilibrate on the isocline of species A. If the supply point was in region 3, then this system too would equilibrate on the isocline of species A. Analogous to the one-resource case, species A would competitively exclude species B because of its ability to exploit both resources down to levels at

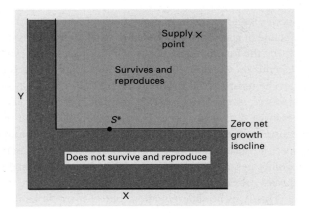

Figure 7.29 The zero net growth isocline of a species potentially limited by two resources (X and Y), divides resource combinations on which the species can survive and reproduce, from those on which it cannot. The isocline is rectangular in this case because X and Y are essential resources (see Chapter 3, Section 3.8). Point S^* is the only point on the isocline at which there is also no net change in resource concentrations (consumption and resource renewal are equal and opposite). In the absence of the consumer, resource renewal would take the resource concentrations to the 'supply point' shown.

309   INTERSPECIFIC COMPETITION

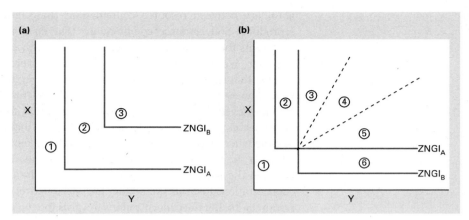

Figure 7.30 (a) Competitive exclusion: the isocline of species A lies closer to the resource axes than the isocline of species B. If the resource supply point is in region 1, then neither species can exist. But if the resource supply point is in regions 2 or 3, then species A reduces the resource concentrations to a point on its own isocline (where species B cannot survive and reproduce): species A excludes species B. (b) Potential coexistence of two competitors limited by two essential resources. The isoclines of species A and B overlap, leading to six regions of interest. With supply points in region 1 neither species can exist; with points in regions 2 and 3, species A excludes species B; and with points in regions 5 and 6, species B excludes species A. Region 4 contains supply points lying between the limits defined by the two dashed lines. With supply points in region 4 the two species coexist. For further discussion, see text.

which species B could not survive. Of course, the outcome would be reversed if the positions of the isoclines were reversed.

coexistence—dependent on the ratio of resource levels at the supply point

In Figure 7.30b the isoclines of the two species overlap, and there are six regions in which the supply point might be found. Points in region 1 are below both isoclines and would allow neither species to exist; those in region 2 are below the isocline of species B and would only allow species A to exist; and those in region 6 are below the isocline of species A and would only allow species B to exist. Regions 3, 4 and 5 lie within the fundamental niches of both species. However, the outcome of competition depends on which of these regions the supply point is located in.

The most crucial region in Figure 7.30b is region 4. For supply points here, the resource levels are such that species A is more limited by resource X than by resource Y, whilst species B is more limited by Y than X. However, species A consumes more X than Y, whilst species B consumes more Y than X. Because each species consumes more of the resource that more limits its own growth, the system equilibrates at the intersection of the two isoclines, and this equilibrium is stable: the species coexist.

a subtle niche differentiation— each species consumes more of the resource that more limits its own growth

This is niche differentiation, but of a subtle kind. Rather than the two species exploiting different resources, species A disproportionately limits itself by its exploitation of resource X, whilst species B disproportionately limits itself by its exploitation of resource Y. The result is the coexistence of competitors. By contrast, for supply points in region 3, both species are more limited by Y than X. But species A can reduce the level of Y to a point on its own isocline below species B's isocline, where species B cannot exist. Conversely, for supply points in region 5, both species are more limited by X than Y, but species B depresses X to a point below species A's isocline. Thus, in regions 3 and 5, the supply of resources favours one species or the other, and there is competitive exclusion.

It seems then that two species can compete for two resources and coexist as long as two conditions are met.

1 The habitat (i.e. the supply point) must be such that one species is more limited by one resource, and the other species more limited by the other resource.

2 Each species must consume more of the resource that more limits its own growth. Thus, it is possible, in principle, to understand coexistence in competing plants on the basis of differential resource utilization. The key seems to be an explicit consideration of the dynamics of the resources as well as the dynamics of the competing species. As with other cases of coexistence by niche differentiation, the essence is that intraspecific competition is, for both species, a more powerful force than interspecific competition.

The best evidence for the validity of the model comes from Tilman's own experimental laboratory work on competition between the diatoms *Asterionella formosa* and *Cyclotella meneghiniana* (Tilman, 1977). For both species, Tilman observed directly the consumption rates and the isoclines for both phosphate and silicate. He then used these to predict the outcome of competition with a range of resource supply points (Figure 7.31). Finally, he ran a number of competition experiments with a variety of supply points, and the results of these are also illustrated in Figure 7.30. In most cases, the results confirmed the predictions. In the two that did not, the supply points were very close to the regional boundary. The

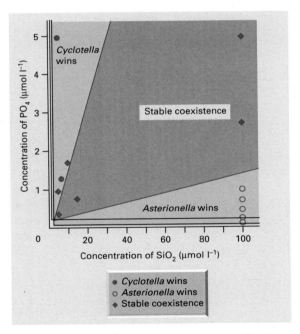

Figure 7.31 The observed isoclines and consumption vectors of two diatom species, *Asterionella formosa* and *Cyclotella meneghiniana*, were used to predict the outcome of competition between them for silicate and phosphate. The predictions were then tested in a series of experiments, the outcomes of which are depicted by the symbols explained in the key. Most experiments confirmed the predictions, with the exception of two lying close to the regional boundary. (After Tilman, 1977, 1982.)

results are therefore encouraging. However, it has proved extremely difficult to transfer this approach from the laboratory, where supply points can be manipulated, to natural populations, where they cannot, and where even estimation of supply points has proved practically impossible (Sommer, 1990). There is considerable need for consolidation and extension from work on other types of plants and animals.

Our survey of interspecific competition has concluded with a realization that we need to understand much more about the mechanisms underlying interactions between consumers and their resources. If these resources are alive, then we normally refer to such interactions as predation; and if they were alive once but are now dead, we refer to them as detritivory. It would seem, therefore, that the distinction normally made between competition and predation is, in a very real sense, an artificial one (Tilman, 1990). Nonetheless, having dealt with competition in the past two chapters, we turn next, in a separate series of chapters, to predation and detritivory.

Chapter 8
The Nature of Predation

8.1 Introduction: the types of predators

Consumers affect the distribution and abundance of the things they consume and vice versa, and these effects are of central importance in ecology. Yet, it is never an easy task to determine what the effects are, how they vary and why they vary. These topics will be dealt with in this and the next few chapters. We begin here by asking 'What is the nature of predation?' and 'What are the effects of predation on the predators themselves and on their prey?' Then, in the next chapter, we examine the behaviour of predators and the way this affects what they consume. In Chapter 10, we turn to the consequences of predation for the dynamics of predator and prey populations.

a definition of predation

Predation, put simply, is consumption of one organism (the prey) by another organism (the predator), in which the prey is alive when the predator first attacks it. This excludes detritivory, the consumption of dead organic matter, which is discussed in its own right in Chapter 11. Nevertheless, it is a definition that encompasses a wide variety of interactions and a wide variety of 'predators'.

There are two main ways in which predators can be classified. Neither is perfect, but both can be useful. These classifications will not be pursued with the aim of constructing a neat and tidy catalogue, or of settling any issues in semantics. However, by distinguishing various types of predator, and establishing what characteristics they share and how they differ, we will be able to develop a fuller understanding of the precise nature of predation. The most obvious classification, perhaps, is 'taxonomic': carnivores consume animals, herbivores consume plants and omnivores consume both. An alternative, however, is a 'functional' classification (Thompson, 1982) of the type already outlined in Chapter 3. Here, there are four main types of predator: true predators, grazers, parasitoids and parasites (the last divisible further into microparasites and macroparasites as explained in Chapter 12).

taxonomic and functional classifications of predators

True predators kill their prey more or less immediately after attacking them; during their lifetime they kill several or many different prey individuals. Often, they consume prey in their entirety, but some true predators consume only parts of their prey. Most of the more obvious carnivores like tigers, eagles, coccinellid beetles and carnivorous plants are true predators, but so too are seed-eating rodents and ants, plankton-consuming whales, and so on.

true predators

Grazers also attack large numbers of prey, one after the other, during their lifetime, but they remove only a part of each prey individual rather than the whole. Their effect on a prey individual, although typically harmful, is rarely lethal in the

grazers

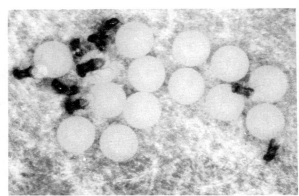

Figure 8.1 (a) *Trichogramma* spp.: egg parasitoids of moth pests (here, the cabbage moth, *Mamestra brassicae*). The most widely used parasitoid in biological control today, it is mass produced and released to control various moth pests around the world. (b) *Exerastes cintipes*, an ichneumonid parasitoid of larvae of *M. brassicae*. Eggs are laid in young larvae (as here). (Courtesy of CAB International Institute of Biological Control.)

short term, and certainly never predictably lethal (in which case they would be true predators). Amongst the more obvious examples are the large vertebrate herbivores like sheep and cattle, but the flies that bite vertebrates, and leeches that suck blood, are also undoubtedly grazers by this definition.

parasites

Parasites, like grazers, consume parts of their prey (their 'host') rather than the whole. Also like grazers, their attacks are typically harmful but rarely lethal in the short term. Unlike grazers, however, their attacks are concentrated on one or a very few individuals during the course of their life. There is, therefore, an intimacy of association between parasites and their hosts which is not seen in true predators and grazers. Tapeworms, liver flukes, the measles virus, the tuberculosis bacterium and the flies and wasps that form mines and galls on plants are all obvious examples of parasites. There are also considerable numbers of plants, fungi and microorganisms that are parasitic on plants (often called 'plant pathogens'), for instance, the tobacco mosaic virus, the rusts and smuts and the mistletoes. Moreover, there are also many herbivores that may readily be thought of as parasites. For example, aphids extract sap from one or a very few individual plants with which they enter into intimate contact. Even caterpillars often rely on a single plant for their development (although the association here is not so intimate). Caterpillars may therefore be parasites, grazers (where they take small parts of several plants) or even true predators (where they destroy several plants—perhaps young seedlings)—and many undoubtedly sit somewhere on a continuum between these categories. Plant pathogens, and animals parasitic on animals, are dealt with together in Chapter 12. 'Parasitic' herbivores, like aphids and caterpillars, are dealt with here and in the two chapters that follow. In these three chapters they will be grouped together with true predators, grazers and parasitoids under the umbrella term 'predator'.

parasitoids

The parasitoids (Figure 8.1) are a group of insects that are classified as such on the basis of the egg-laying behaviour of the adult female and the subsequent developmental pattern of the larva. They belong mainly to the order Hymenoptera, but also include many Diptera. They are free-living as adults, but they lay their eggs in, on or near other insects (or, more rarely, in spiders or woodlice). The larval parasitoid then develops inside (or more rarely, on) its host individual, which is itself usually a pre-adult. Initially, it does little apparent harm to the host, but eventually it almost totally consumes the host and therefore kills it before or during the pupal stage. An adult parasitoid, rather than an adult host, emerges from what is apparently a host pupa. Often, just one parasitoid develops from each host, but in some cases several individuals share a host. Thus, parasitoids are intimately associated with a single host individual (like parasites); they do not cause immediate death of the host (like parasites and grazers), but their eventual lethality is inevitable (like predators). For parasitoids, and also for the many herbivorous insects that feed as larvae on plants, the rate of 'predation' is determined very largely by the rate at which adult females oviposit. Each egg laid by the female is an 'attack' on the prey or host, even though it is the larva that develops from the egg that actually does the eating.

Parasitoids may seem to be an unusual group of limited general importance. However, it has been estimated that they account for 10% or more of the world's species (Godfray, 1994). This is not surprising when we consider that there are so many species of insects, that most of these are attacked by at least one parasitoid and that even parasitoids themselves may be attacked by parasitoids. Moreover, a number of parasitoid species have been intensively studied by ecologists, and they have provided a wealth of information relevant to predation generally.

In the remainder of this chapter, we examine the nature of predation. We will look at the effects of predation on the prey individual (see Section 8.2), the effects on the prey population as a whole (see Section 8.3) and the effects on the predatory individual itself (see Section 8.4). In the cases of attacks by true predators and parasitoids, the effects on prey individuals are very straightforward: the prey are killed. Attention will therefore be focused in Section 8.2 on prey subject to grazing and parasitic attack. In fact, attention will be mainly focused on herbivory, since, apart from being important in its own right, it serves as a useful vehicle for discussing the subtleties and variations in the effects that predators can have on their prey.

8.2 The effects of herbivory on individual plants

The effects of herbivory on a plant depend on precisely which parts are affected, and on the timing of the attack relative to the plant's development. Leaf biting, sap sucking, mining, meristem consumption, flower and fruit damage and root pruning are all likely to differ in the effect they have on the plant. The consequences of defoliating a germinating seedling are unlikely to be the same as those of defoliating a plant that is setting its own seed. Moreover, because the plant usually remains alive in the short term, the effects of herbivory are crucially dependent on the response of the plant. Minerals or nutrients may be diverted from one part to another, the overall rate of metabolism may change, the relative rates of root growth, shoot growth and reproduction may alter and special protective chemicals or tissues may be produced. Overall, the effect of a herbivore may be more drastic than it appears, or less drastic. It is only rarely what it seems.

8.2.1 Plant compensation

herbivory can reduce self-shading

In a variety of ways, individual plants can compensate for the effects of herbivory (Trumble et al., 1993). In the first place, the removal of leaves from a plant may decrease the shading of other leaves and thereby increase their rate of photosynthesis.

Alternatively, if shaded leaves are removed (with normal rates of respiration but low rates of photosynthesis; see Chapter 3), then this may improve the balance between photosynthesis and respiration in the plant as a whole. Thus, the weevil *Phyllobius argentatus* has been observed to feed mainly on the lower, shaded leaves towards the centre of a beech tree, and it has little impact on a tree's overall productivity (Nielsen & Ejlerson, 1977).

Second, in the immediate aftermath of an attack from a herbivore, many plants compensate by utilizing reserves stored in a variety of tissues and organs. For example, when two varieties of Italian rye grass (*Lolium multiflorum*) were completely defoliated, the variety 'Liscate', which had higher levels of stored carbohydrates in its roots and stubble, exhibited higher initial rates of leaf regrowth than the variety 'S.22' (Figure 8.2). However, within a short time of attack, the role of supplying new tissues usually passes from stored reserves to current photosynthesis.

Herbivory also frequently alters the distribution of photosynthate within the plant, the general rule being apparently that a balanced root : shoot ratio is maintained. When shoots are defoliated, an increased fraction of net production is channelled to the shoots themselves; and when roots are destroyed, the switch is towards the roots (Crawley, 1983). In fact, the defoliation of grasses often stops root growth altogether, and it may even lead to a reduction in root weight when roots that die naturally are not replaced by fresh growth (Ryle, 1970). The redistribution of photosynthate can also compensate for the effects of patchy attack within a plant. Sometimes when tillers are differentially defoliated, the plant responds by redirecting carbohydrates to the damaged tillers themselves (e.g. Italian rye grass—Marshall &

herbivory can lead to the mobilization of stored carbohydrates

herbivory can lead to an altered pattern of photosynthate distribution

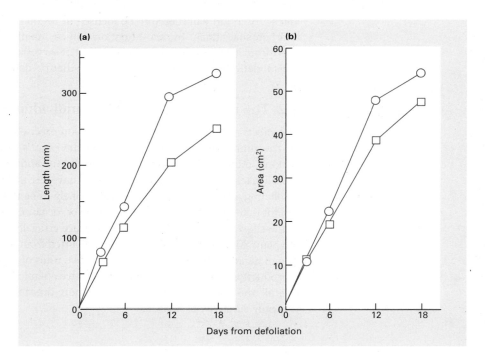

Figure 8.2 After defoliation, the regrowth of the variety 'Liscate' of *Lolium multiflorum* (○) is better than that of the variety 'S.22' (□) in terms of (a) leaf elongation and (b) leaf area expansion. The fraction of water-soluble carbohydrates in the roots and stubble at the time of defoliation was 27% higher in Liscate than in S.22. (After Kigel, 1980.)

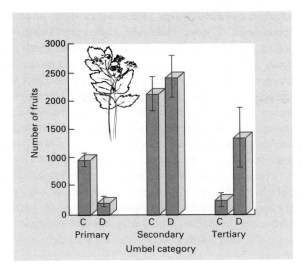

Figure 8.3 Compensation via reduced death rate of flowers. Although most of the flowers and fruits of primary umbels of *Pastinaca sativa* are destroyed by parsnip webworm, damaged plants (D) produce similar numbers of fruits from their secondary umbels and many more fruits from their tertiary umbels than do control plants (C) (means plus the standard error). (After Hendrix, 1979; from Crawley, 1983.)

Sagar, 1968); but there are some grass species that cut their losses by redirecting carbohydrates to the less badly damaged tillers (Ong *et al.*, 1978). Similarly, pod loss in soybeans (*Glycine max*) is compensated for by an increase in the weight of the seeds produced by the remaining pods (Smith & Bass, 1972).

herbivory can lead to an increase in unit leaf rate (ULR)

Herbivore attack may also lead to an increase in the rate of photosynthesis per unit area of surviving leaf (the 'unit leaf rate', ULR). Thus, when *Agropyron smithii* was partially defoliated experimentally, there was a 10% increase in the ULR of the remaining leaves over the succeeding 10 days, at a time when control plants showed a 10% decrease (Painter & Detling, 1981). ULR can also show a compensatory increase when sap-sucking insects like aphids consume vast quantities of carbohydrate.

herbivory can lead to a reduced death rate of plant parts

Often, there is compensatory regrowth of defoliated plants when buds that would otherwise remain dormant are stimulated to develop. There is also, commonly, a reduced subsequent death rate of surviving plant parts. This is especially prevalent in plants that exhibit a high natural rate of flower abortion prior to the production of fruit or seed. For example, the wild parsnip (*Pastinaca sativa*), produces a moderate number of seeds on its primary umbels, a larger number on its secondary umbels, but only a small number on its tertiary umbels because of flower abortion prior to seed production (Figure 8.3). However, when it is attacked by larvae of the parsnip webworm (*Depressaria pastinacella*—a moth), although most of the flowers and fruits on the primary umbels are destroyed, there is little effect on the secondary umbels, and a greatly reduced rate of abortion on the tertiary umbels. Overall, therefore, the quantity of seed produced is largely unaltered.

despite compensation, herbivores harm plants

Clearly, then, there are a number of ways in which individual plants compensate for the effects of herbivory. But perfect compensation is rare. Plants are usually harmed by herbivores even though the compensatory reactions tend to counteract the harmful effects.

8.2.2 Defensive responses of plants

plants make defensive responses …

Aside from compensating for the attacks of herbivores, plants may also respond by initiating or increasing their production of defensive structures or chemicals. Pines

317 NATURE OF PREDATION

attacked by wood wasps and sawflies show altered phenol metabolism and the appearance of novel defensive chemicals (Thiegles, 1968); artificially wounded potato and tomato plants produce increased levels of protease inhibitors (Green & Ryan, 1972); and the prickles of dewberries on cattle-grazed plants are longer and sharper than those on ungrazed plants nearby (Abrahamson, 1975). Particular attention has been paid to 'rapidly inducible' defences, mostly the production of chemicals within the plant that inhibit the protease enzymes of the herbivores. These changes can occur within individual leaves, within branches or throughout whole tree canopies, and they may be detectable within a few hours, days or weeks, and last a few hours, days, weeks or years (Fowler & Lawton, 1985; Schultz, 1988).

... or do they? ...

There are, however, a number of problems in interpreting these responses (Schultz, 1988). First, are they 'responses' at all?—or merely an incidental consequence of regrowth tissue having different properties from that removed by the herbivores (properties which may nonetheless make them less susceptible to those particular herbivores). Second, are they actually defensive in the sense of having an ecologically significant effect on the herbivores that seem to have induced them? Finally, are they truly defensive in the sense of having a measurable, positive impact on the plant making them (and perhaps even on the plant population), especially after the costs of mounting the response have been taken into account?

Fowler and Lawton (1985) reviewed the effects of rapidly inducible plant defences and found little clear-cut evidence for their being effective against insect herbivores, despite a widespread belief that they were so. (In fact, many studies simply compared undamaged and recently damaged leaves without specifying the nature of the plant's response.) Even amongst laboratory studies, they found that most revealed only small adverse effects (less than 11%) on such characters as larval development time and pupal weight, with many of the studies claiming larger effects being flawed statistically. They argue that such effects may have negligible consequences for field populations.

The first problem—are responses defensive or incidental?—may in large part be semantic. If the metabolic responses of a plant to tissue removal happen to be defensive, then natural selection will favour them and reinforce their use. Haukioja and Neuvonen (1985) addressed the question directly using fertilized and unfertilized white birch trees (*Betula pubescens*), arguing that if responses to herbivory were only incidental, then fertilized trees, with much less pressure on resources, would make less of a response. In fact, using growth of a geometrid moth, *Epirrita autumnata*, the most important defoliator of birch trees in Finnish Lapland, as a measure of the plants' response, they found that the response was no less profound in the fertilized trees, and indeed that it was greater in trees defoliated by the moths than in trees defoliated artificially. The response seems genuinely defensive.

... are herbivores really affected adversely? ...

There are also a number of cases in which the plants' responses do seem to be genuinely harmful to the herbivores. When larch trees were defoliated by the larch budmoth, *Zeiraphera diniana*, the survival and adult fecundity of the moths were reduced throughout the succeeding 4–5 years as a combined result of delayed leaf production, tougher leaves, higher fibre and resin concentration and lower nitrogen levels (Baltensweiler *et al.*, 1977). Similarly, snowshoe hares (*Lepus americanus*) showed all the adverse signs normally associated with crowding (see p. 377) when fed birch leaves that had regenerated after severe defoliation (Bryant & Kuropat, 1980). When 25% of the leaf area of oak leaves (*Quercus robur*) containing mines of *Phyllonorycter* spp. were artificially removed, larval deaths in the mines were

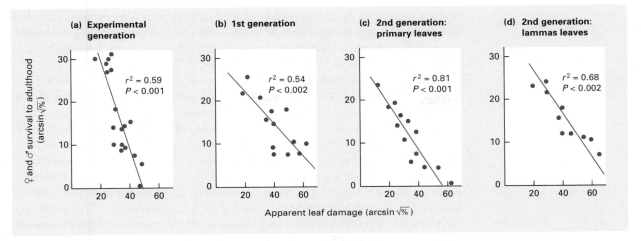

Figure 8.4 Larval survival to adulthood for both sexes of the leaf-mining *Phyllonorycter* spp. declined with the level of natural leaf damage to pedunculate oak, *Quercus robur*, in Wytham Woods near Oxford, England. The experimental generation emerged early in the season as a result of artificial warming, and was then followed by two natural generations. Primary leaves are those produced at budburst in April; lammas leaves are those produced during refoliation in mid-June. (After West, 1985.)

50–100% greater than those of controls (West, 1985), suggesting that rapid damage-induced changes in the leaves could affect the population dynamics of herbivores in the field. West also found a strong negative correlation between natural damage levels and larval survival to adulthood (Figure 8.4). Moreover, a further study of a leaf-mining *Phyllonorycter* spp., but on the willow *Salix lasiolepis*, found that another common response to leaf damage, early abscission ('dropping off') of damaged leaves, was induced by the leaf mining of the moths, but also that early abscission of mined leaves was an important mortality factor for the moths (Preszler & Price, 1993)—that is, they were harmed by the response.

... and do the plants really benefit?

The final question—do plants benefit from their 'defensive' responses?—has proved the most difficult to answer. Conifers seem to be protected by insect-stimulated terpene defences (Raffa & Berryman, 1983), but if the potential costs to plants of chemical production are included in calculations, there is apparently no evidence that plant losses are actually prevented by induced responses (Schultz, 1988). It is a lesson worth learning that there is often in ecology—and doubtless elsewhere—a big difference between a widely held belief ('induced plant defences harm herbivores and benefit the plants') and an established, demonstrated fact.

8.2.3 Herbivory, defoliation and plant growth

Herbivory can stop plant growth; it can have a negligible effect on growth rate; and it can do just about anything in between. When leaves are produced synchronously, the effects of defoliation depend crucially on timing: removal of 75% of the foliage of mature oaks early in the season leads to a 50% loss in wood production, but similar removal later has no noticeable effect on growth (Franklin, 1970; Rafes, 1970). By contrast, in plants where leaf production is continuous, the loss of young leaves may be compensated for by the production of new leaves, such that earlier defoliation

timing is crucial

319 NATURE OF PREDATION

(leaving more time for compensation) is less harmful than later defoliation. Thus, 25% defoliation of cabbages early in the growing season led to merely a 3% loss of final yield—which could also arise from only a 5% defoliation much later (at the time of heading) (Wit, 1985).

In cases like this, in fact, there is considerable difficulty in even assessing the extents of defoliation, *re*foliation, and hence net growth. 'Snapshots' taken at one point in time can give no real impression of the previous dynamics of these processes, but most studies have relied on such snapshots. By contrast, it was clear when waterlily leaf beetles (*Pyrrhalta nymphaeae*) were closely monitored grazing on waterlilies (*Nuphar luteum*) that leaves were rapidly removed, but new leaves were also rapidly produced. Marked leaves on grazed plants had all disappeared within 17 days, compared with leaves on ungrazed plants (at another site) which were 100% intact after 17 days and still 65% intact after 6 weeks (Figure 8.5). Yet snapshots throughout this period suggested only around a grossly underestimated 13% loss of leaf to the beetles. Apart from simply ignoring dynamics, snapshots may also underestimate the defoliating effects of herbivory because of the role in this of increased rates of leaf abscission, noted earlier.

The plants that seem to be most tolerant of grazing, especially vertebrate grazing, are the grasses. In most species, the meristem is almost at ground level amongst the basal leaf sheaths, and this major point of growth (and regrowth) is therefore usually protected from grazers' bites. Following defoliation, new leaves are produced using either stored carbohydrates or the photosynthates of surviving leaves, and new tillers are also often produced. The grasses do not benefit directly from the grazers' attentions. But it is likely that they are aided by the grazers in their competitive interactions with other plants (see below), accounting for the predominance of grasses in so many natural habitats that suffer intense vertebrate grazing.

The effects of herbivory on plant growth, however, may be underestimated if the herbivores remove sap or xylem without altering the obvious physical structure of the plant. For example, lime aphids (*Eucallipterus tiliae*) live on the leaves of limes (*Tilia vulgaris*) and extract phloem sap through their piercing stylets. They can build up extremely heavy infestations: a 14 m lime tree with 58 000 leaves may support more than one million aphids. Yet, when uninfested saplings were compared with saplings infested from the time their leaves were half-grown, there were no differences in tree girth, height increment, leaf number or leaf size. The aphid infestations, though, led to a virtual cessation of root growth; and this in turn led,

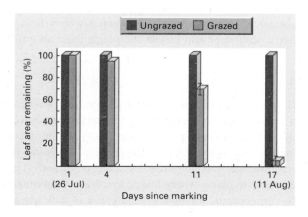

Figure 8.5 The survivorship of leaves on waterlily plants grazed by the waterlily leaf beetle was much lower than that on ungrazed plants. Effectively, all leaves had disappeared at the end of 17 days, despite the fact that 'snapshot' estimates of loss rates to grazing on grazed plants during this period suggested only around a 13% loss. (After Wallace & O'Hop, 1985.)

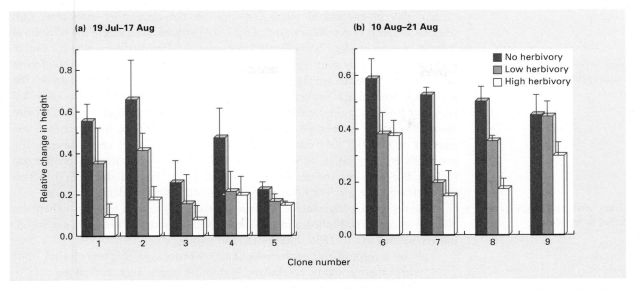

(a) 19 Jul–17 Aug

(b) 10 Aug–21 Aug

No herbivory
Low herbivory
High herbivory

Relative change in height

Clone number

Figure 8.6 Relative growth rates (changes in height, with standard errors) of a number of different clones of the sand-dune willow, *Salix cordata*, (a) in 1990 and (b) in 1991, subjected either to no herbivory, low herbivory (four flea beetles per plant) or high herbivory (eight beetles per plant). (After Bach, 1994.)

over the season, to a total weight increase of infested saplings which was only 8% that of uninfested saplings (Dixon, 1971). Hence, the visible effects of the aphids were misleading; infested trees grew normally in volume above the ground, but grew hardly any new roots and put on very little weight. This would doubtless have had repercussions subsequently.

8.2.4 Herbivory and plant survival

Generally, it is more usual for herbivores to increase a plant's susceptibility to mortality than to kill it outright. For example, defoliation by the cinnabar moth (*Tyria jacobaeae*) of the ragwort (*Senecio jacobaea*) gives rise to plants that are particularly susceptible to mortality through frost damage. This has led to successful use of the moth as a biological control agent of ragwort in eastern Canada, because the frosts there are characteristically early (Harris *et al.*, 1978). In other regions, where frosts occur later, there is sufficient time for ragwort regrowth after moth pupation, and weed control is much less successful (Crawley, 1989). Similarly, the flea beetle, *Altica subplicata*, reduced the growth rate of the sand-dune willow, *Salix cordata*, in both 1990 and 1991 (Figures 8.6a, b), but there was significant mortality as a result of drought stress in these plants only in 1991. Then, however, susceptibility was strongly influenced by the herbivore: 80% of plants died in a high herbivory treatment (eight beetles per plant), 40% died at four beetles per plant, but none of the control plants without any beetles on them died (Bach, 1994).

> mortality is often the result of an interaction with another factor

Repeated defoliation, especially, can have a drastic effect. A single defoliation of oak trees by the gypsy moth (*Lymantria dispar*) led to only a 5% mortality rate (no different from the natural rate in unattacked but crowded woodlands); but three successive heavy defoliations led to mortality rates of up to 80% (Stephens, 1971).

> repeated defoliation can kill plants

Mortality of established plants, however, is not necessarily associated with massive amounts of defoliation. One of the most extreme cases where the removal of a small amount of plant has a disproportionately profound effect is ring-barking of trees, for example by squirrels or porcupines. The cambial tissues and the phloem are torn away from the woody xylem, and the carbohydrate supply link between the leaves and the roots is broken. Thus, these pests of forestry plantations often kill young trees whilst removing very little tissue. Surface-feeding slugs can also do more damage to newly established grass populations than might be expected from the quantity of material they consume (Harper, 1977). The slugs chew through the young shoots at ground level. They leave the felled leaves uneaten on the soil surface, but consume the meristematic region at the base of shoots from which regrowth would occur. They therefore effectively destroy the plant.

However, even a single attack is frequently sufficient to kill a seedling—where the plant is scarcely established and its powers of compensation are least developed. Indeed, as long ago as 1859, Charles Darwin wrote:

> ... on a piece of ground three feet long and two wide, dug and cleared, and where there could be no choking from other plants, I marked all the seedlings of our native weeds as they came up, and out of the 357 no less than 295 were destroyed, chiefly by slugs and insects (Darwin, 1859).

Often, however, vertebrate predators of seedlings may be of greatest importance. For example, Mills (1983) looked at first-year establishment of seedlings following fire in southern California chaparral, especially of the most important plants, ceanothus (*Ceanothus greggii*) and chamise (*Adenostoma fasciculatum*). A series of caging experiments (excluding the herbivores), backed up by direct observation of marked seedlings, showed that insects had no significant effect, but that vertebrates, especially the brush rabbit (*Sylvilagus bachmani*), had an important impact on seedling abundance, particularly of ceanothus.

Predation of seeds, not surprisingly, has an even more predictably harmful effect on individual plants (i.e. the seeds themselves). For example, Davidson and co-workers (e.g. Davidson *et al.*, 1985) have demonstrated dramatic impacts of granivorous ants and rodents on the seed banks of 'annual' plants in the deserts of the south-western USA—and also a significant effect on the composition of the plant community (see Section 8.3), at least partly by affecting interspecific plant competition (see Section 8.2.6).

8.2.5 Herbivory and plant fecundity

The effects of herbivory on plant fecundity are, to a considerable extent, reflections of the effects on plant growth: smaller plants bear fewer seeds. Plants may also be affected more directly—by the removal, or destruction, of flowers, flower buds or seeds prior to dispersal. Thus, in a North Wales pasture heavily grazed by cattle, only 15% of buttercups (*Ranunculus* spp.) that flowered set seed, whereas 48% set seed in lightly grazed swards (Sarukhán, 1974).

It is important to realize, however, that many cases of 'herbivory' of reproductive tissues are actually mutualistic, i.e. beneficial to both the herbivore and the plant (see Chapter 13). Animals that 'consume' pollen and nectar usually transfer pollen inadvertently from plant to plant in the process; and there are many fruit-eating animals that also confer a net benefit on both the parent plant and the individual seed within the fruit. Most vertebrate fruit-eaters, in particular, either eat the fruit

ring-barking and meristem consumption can kill plants

many seeds and seedlings are killed by herbivores

smaller plants bear fewer seeds

herbivores often destroy reproductive structures directly ...

... but, much pollen and fruit 'herbivory' is mutualistic

but discard the seed, or eat the fruit but expel the seed in the faeces. This disperses the seed, rarely harms it and frequently enhances its ability to germinate.

Insects that attack fruit, on the other hand, are very unlikely to have a beneficial effect on the plant. They do nothing to enhance dispersal, and they may even make the fruit less palatable or unpalatable to vertebrates. However, some large animals that normally kill seeds can also play a part in dispersing them, and they may therefore have at least a partially beneficial effect. There are some 'scatter-hoarding' species, like certain squirrels, that take nuts and bury them at scattered locations; and there are other 'seed-caching' species, like some mice and voles, that collect scattered seeds into a number of hidden caches (see also Chapter 5; p. 185). In both cases, although many seeds are eaten, the seeds are dispersed, they are hidden from other seed predators and a number are never relocated by the hoarder or cacher (Crawley, 1983).

even some seed-eaters benefit plants

Herbivores also influence fecundity in a number of other ways. One of the most common responses to herbivore attack is a delay in flowering. For instance, in longer-lived semelparous species, herbivory frequently delays flowering for 1 year or more; and this typically increases the longevity of such plants since death almost invariably follows their single burst of reproduction (see Chapter 4). *Poa annua* on a lawn can be made almost immortal by mowing it at weekly intervals, whereas in natural habitats, where it is allowed to flower, it is commonly an annual as its name implies (Crawley, 1983).

herbivory can lead to delayed flowering ...

... which can increase plant longevity

Generally, the timing of defoliation is critical in determining the effect on plant fecundity. If leaves are removed before inflorescences are formed, then the extent to which fecundity is depressed clearly depends on the extent to which the plant is able to compensate. Early defoliation of a plant with sequential leaf production may have a negligible effect on fecundity; but where defoliation takes place later, or where leaf production is synchronous, flowering may be reduced or even inhibited completely. If leaves are removed after the inflorescence has been formed, the effect is usually to increase seed abortion or to reduce the size of individual seeds.

8.2.6 Interactive effects of herbivory

The effects of herbivory, as we have seen, are often most important because of their interaction with other factors affecting the plant. Herbivores can have severe effects on plants, for example when they act as vectors for plant pathogens: bacteria, fungi and especially viruses—what the herbivores take from a plant is far less important than what they give it! For instance, scolytid beetles feeding on the growing twigs of elm trees act as vectors for the fungus that causes Dutch elm disease. This killed vast numbers of elms in the north-eastern USA in the 1960s, and virtually eradicated the elms in southern England in the 1970s and early 1980s (Strobel & Lanier, 1981). Likewise, the *Cactoblastis* caterpillars that control *Opuntia* cacti in Australia (see pp. 395 and 618) exert a large part of their ill-effect by creating feeding scars which are invaded by bacteria that destroy the cactus tissues (Dodd, 1940).

herbivores can act as vectors for disease

Probably the most widespread reason for herbivory having a more drastic effect than is initially apparent is the interaction between herbivory and plant competition (the range of possible consequences of which are discussed by Pacala & Crawley, 1992). An interesting example concerns the chrysomelid dock beetle, *Gastrophysa viridula*, and two species of dock plant, *Rumex obtusifolius* and *R. crispus*. *Rumex crispus* is only moderately affected by competition from *R. obtusifolius* and little

herbivory and plant competition often combine to produce a severe effect

affected by light beetle grazing. Between the two plants, the beetle feeds preferentially on *R. obtusifolius*. Light grazing and competition together, however, have a considerable impact on *R. crispus* (Figure 8.7). On the other hand, in the perhaps more usual competitive setting of *R. obtusifolius* pitted against various grasses, it is *R. obtusifolius* that suffers most from the combination of dock beetle grazing and competition (Cottam *et al.*, 1986).

8.2.7 Air pollution and insect herbivores

herbivores and pollutants may combine to harm plants

One particular class of interaction in which plants are disproportionately affected by herbivores deserves special mention: those in which the plants are simultaneously affected by air pollutants—particularly sulphur dioxide (SO_2), ozone and the oxides of nitrogen (Riemer & Whittaker, 1989). Controversy surrounds the possible effects on plants, especially trees, of 'acid rain' (rain with these oxides of nitrogen and sulphur dissolved in it—see Chapter 19). The resolution of the controversy will require careful experimentation to link cause unequivocally with effect. Yet experiments with plants and air pollutants alone, however careful, may be misleading if a three-way interaction between plant, pollutant and herbivore is at the heart of the matter.

There is certainly evidence that insect herbivores increase in abundance in the presence of pollutants (Riemer & Whittaker, 1989), i.e. indirect evidence that plants are likely to be harmed by the combined effects of pollutants and herbivores. For example, a series of surveys in Finland of the numbers of the pine bark bug (*Aradus cinnamomeus*) along gradients from factories emitting nitrogen oxides and SO_2 showed that the density of the bugs was lowest right next to the factories, but rose quickly to a peak density 1–2 km away from them. At a greater distance, density slowly reverted to background levels (roughly 0.01% of the density in the heavily infested zone). Heavy attack by the pine bark bug in Finland is only seen in industrial areas (Heliövaara & Väisänen, 1986).

herbivores concentrated near factories

Furthermore, a significant increase has been demonstrated under experimental conditions in the growth rate of aphids, *Aphis fabae*, feeding on beans, *Vicia faba*, that had been prefumigated for 7 days with either 148 ppb SO_2 or 204 ppb nitrogen dioxide (NO_2). Fumigation of aphids living on artificial (non-plant) diets showed no direct effect of the gases on their growth rates. In a related experiment, the system

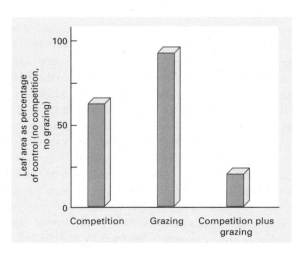

Figure 8.7 The effects of grazing by the beetle *Gastrophysa viridula* on the dock plant *Rumex crispus* are increased significantly when combined with competition from another dock, *R. obtusifolius*. Plant performance is measured as leaf area, relative to control plants (no grazing, no interspecific competition). (After Bentley & Whittaker, 1979.)

was fumigated with ambient London air (13–54 ppb SO_2 and 10–50 ppb NO_2). The aphids had a significantly higher growth rate than with charcoal-filtered air (Dohmen *et al.*, 1984).

herbivores in polluted atmospheres

There is also some direct evidence, albeit under artificial, experimental conditions, that the interaction between herbivores and pollutants affects plant growth, survival or fecundity. Mexican bean beetles (*Epilachna varivestis*) damaged 45% of the leaf area of control soybean plants (*Glycine soja*), whereas 75% was damaged on plants treated with 135 ppb SO_2 (Hughes, 1983); and an increase in the population of aphids (*Acyrthosiphon pisum*) on pea plants (*Pisum sativum*), attributable to a fumigation of the plants with 50 or 80 ppb SO_2, resulted in a 10% reduction in the yield of peas compared with aphid-infested control plants (Warrington *et al.*, 1978).

It would be wrong, however, to suppose that the typical interaction is simply 'pollutant affects plant, which benefits herbivore, which further harms plant'. For instance, the damage caused by the leafhopper *Ossiannilssonola callosa* to seedlings of sycamore (*Acer pseudoplatanus*), significantly increased the rate of uptake of SO_2 by the plants overnight (Warrington *et al.*, 1989). Clearly, there is the potential for a truly synergistic interaction between herbivores and pollutants in their adverse effects on plants.

Examples like these support the view that plants may readily be disproportionately affected by herbivores in the presence of pollutants. The weight of evidence is very heavily against two alternative explanations: (i) that the insects benefit directly from the pollutants (see, for instance, the *A. fabae* example above); or (ii) that the plants benefit and the insects are simply responding to an improved quantity of resource (Riemer & Whittaker, 1989). White (1984, 1993), in particular, has stressed the possible continuity between the effects of airborne pollutants on plants and the effects of other adverse influences such as herbicides, irradiation, extremes of temperature, disease, drought and waterlogging. The suggestion is that pollutants, like other factors (see Section 8.4), increase the concentration of available nitrogen in the plant tissue as nitrogen is translocated from senescent, damaged or diseased tissue to areas of new growth. The insect herbivores, which are typically most limited by the availability of nitrogen (see Chapter 3), then benefit from the greatly improved quality (although perhaps reduced quantity) of their resource.

the possible importance of nitrogen availability

There appear, however, to have been virtually no field studies simultaneously investigating the effects of pollution on plant quality, herbivore performance and herbivore density—and one that did so was unable to demonstrate any clear connections (Koricheva & Haukioja, 1995). Pollution-induced changes in chemical characteristics of birch leaves (*Betula pubescens* and *B. pendula*) in the vicinity of a copper–nickel smelter in south-western Finland showed no consistent relationship with either the larval performance or the densities of *Eriocrania* leaf-mining moths.

8.3 The effect of predation on prey populations

Returning now to predators in general, it may seem that since the effects of predators are harmful to individual prey, the immediate effect of predation on a population of prey must also be predictably harmful. However, these effects are not always so predictable, for one or both of two important reasons. In the first place, the individuals that are killed (or harmed) are not always a random sample of the population as a whole, and may be those with the lowest potential to contribute to the population's future. Second, there may be compensatory changes in the growth, survival or reproduction of the surviving prey: they may experience reduced competition for a limiting resource, or produce more offspring, or other predators may take fewer of the prey. In other words, whilst predation is bad for the prey that get eaten, it may be good for those that do not. Moreover, predation is least likely to affect prey dynamics if it occurs at a stage of the prey's life cycle that does not have a significant effect, ultimately, on prey abundance.

To deal with this last point first, if, for example, plant recruitment is not limited by the number of seeds produced, then insects that reduce seed production are unlikely to have an important effect on plant population dynamics (Crawley, 1989). For instance, the weevil, *Rhinocyllus conicus*, does not reduce recruitment of the nodding thistle, *Carduus nutans*, in southern France despite inflicting seed losses of over 90%. Indeed, sowing 1000 thistle seeds per m^2 also led to no observable increase in the number of thistle rosettes. Hence, recruitment appears not to be limited by the number of seeds produced; although whether it is limited by subsequent predation of seeds or early seedlings, or the availability of germination sites, is not clear (Crawley, 1989). On the other hand, *R. conicus* has been used successfully to control nodding thistle in Virginia, USA, reducing thistle density by 95% (Kok & Surles, 1975)—there, presumably, the thistle is recruitment-limited.

The same point is illustrated by a study of two shrubs, *Haplopappus venetus* and *H. squarrosus*, along a 60 km gradient inland from the coast in California (Louda, 1982, 1983). *Haplopappus venetus* decreased steadily in abundance away from the coast, whereas *H. squarrosus* increased. For *H. venetus*, the level of insect damage to developing flowers and seeds by a range of species was high. Experimental exclusion of flower and seed predators therefore caused an 11% decrease in the abortion of flower-heads, a 19% increase in pollination success and a 104% increase in the number of developing seeds escaping damage. This led to an increase in the number of seedlings established. But subsequently, this was followed by a much greater loss of seedlings, probably to vertebrate herbivores, especially inland. As a consequence, the original abundance gradient was re-established in spite of the short-term importance of the seed predators. For *H. squarrosus*, on the other hand, fortnightly insecticide applications reduced the loss of fruit to tephritid flies, moths and chalcicoid wasps from 94 to 41%, leading to significant increases in the number of viable seeds, the density of seedlings and the number of young plants recruited after 1 year. In this case, therefore, seed predation did indeed seem to limit recruitment into the adult population, and, being most intense at the coast, it seemed to be instrumental in establishing the abundance gradient. There is, nonetheless, a big difference between showing that individual seeds are killed and showing that this actually alters plant abundance.

The impact of predation is most commonly limited by compensatory reactions amongst the survivors as a result of reduced intraspecific competition. Thus, in an experiment in which large numbers of woodpigeons (*Columba palumbus*) were shot, the overall level of winter mortality was not increased, and stopping the shooting led to no increase in pigeon abundance (Murton *et al.*, 1974). This was because the number of surviving pigeons was determined ultimately not by shooting but by food availability, and so when shooting reduced density, there were compensatory reductions in intraspecific competition and in natural mortality, as well as density dependent immigration of birds moving in to take advantage of unexploited food.

Indeed, whenever density is high enough for intraspecific competition to occur, the effects of predation on the population will be ameliorated by the consequent reductions in intraspecific competition. This can be seen very clearly by reconsidering the n-shaped curves of net recruitment or net productivity against density that were discussed in Chapter 6, Section 6.4. Net recruitment is low when there are few reproducing individuals, just as net plant productivity is low when there are few leaves (a small leaf area index). But net recruitment is also low where there are many, crowded individuals; and plant productivity is low where there is a large leaf area index and a great deal of shading (Figure 8.8). Thus, if a predator or a herbivore attacks a population that lies on the right-hand limb of such a curve, density falls but net recruitment or net productivity rises. This hastens the rate at which the population recovers.

The effect is probably most important in populations of plants (and especially grasses) where it is surviving plant parts and not only surviving individuals that compensate. Thus, even when defoliation has a disastrous effect on an individual tiller or even a whole individual genet, it may have no serious effect on the yield of the sward as a whole.

... but compensation is usually
not perfect

Compensation, however, is by no means always perfect. When 75% of emerging adults were removed each day from experimental populations of blowflies (*Lucilia cuprina*), there was some compensation, but nonetheless a 40% reduction in population size (Nicholson, 1954b). When a subterranean clover population was defoliated below a leaf area index of 4.5 (on the left-hand limb of Figure 8.8), there was drastic reduction in the rate of leaf production. Typically, therefore, predation leads to compensatory decreases in intraspecific competition. But equally typically, those powers of compensation are limited (especially in low-density prey or plant populations). These topics will be considered further in Chapter 16 in the context of harvesting. For now, though, it is worth noting that humans rely on the compensatory powers of populations to allow them to take repeated harvests; but the limitations in these compensatory powers may bring a heavily harvested population to (or beyond) the brink of extinction.

predatory attacks are often
directed at the weakest prey

Turning to the non-random distribution of predators' attention within a population of prey, it is likely, for example, that predation by many large carnivores is focused on the old (and infirm), on the young (and naive) or on the sick. For instance, a study in the Serengeti found that cheetahs and wild dogs killed a disproportionate number from the younger age classes of Thomson's gazelles (Figure 8.9a), because: (i) these young animals were easier to catch (Figure 8.9b); (ii) they had lower stamina and running speeds; (iii) they were less good at outmanoeuvering the predators (Figure 8.9c); and (iv) they may even have failed to recognize the predators

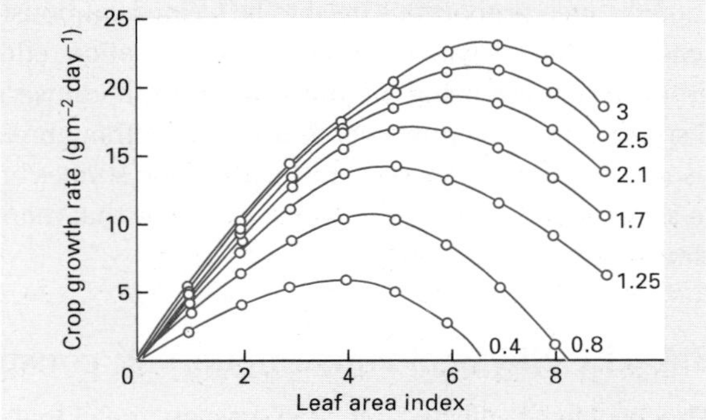

Figure 8.8 The relationship between crop growth rate of subterranean clover (*Trifolium subterraneum*) and leaf area index at various intensities of radiation (kJ cm^{-2} day^{-1}). (After Black, 1963.)

327   NATURE OF PREDATION

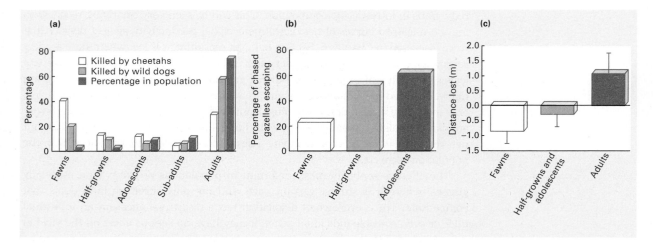

Figure 8.9 (a) The proportions of different age classes (determined by tooth wear) of Thomson's gazelles in cheetah and wild dog kills is quite different from their proportions in the population as a whole. (b) Age influences the probability for Thomson's gazelles of escaping when chased by cheetahs. (c) When prey (Thomson's gazelles) 'zigzag' to escape chasing cheetahs, prey age influences the mean distance lost by the cheetahs. (After FitzGibbon & Fanshawe, 1989; FitzGibbon, 1990.)

(FitzGibbon & Fanshawe, 1989; FitzGibbon, 1990). Yet these young gazelles will also have been making no reproductive contribution to the population, and the effects of this level of predation on the prey population will therefore have been less than would otherwise have been the case.

Similar patterns may also be found in plant populations. Mortality of mature *Eucalyptus* trees in Australia, resulting from defoliation by the sawfly *Perga affinis affinis*, was restricted almost entirely to weakened trees on poor sites, or to trees that had suffered from root damage or from altered drainage following cultivation (Carne, 1969).

Taken overall, then, it is clear that the step from noting that individual prey are harmed by individual predators to demonstrating that prey abundance is adversely affected is not an easy one to take. To do so it is necessary either to carry out an artificial experiment or to observe a natural one (see Chapter 15). However, of 28 available studies in which herbivorous insects were experimentally excluded from plant communities with the use of insecticides, only around 50% provided evidence of an effect on plants at the population level (Crawley, 1989). As Crawley notes, though, such proportions need to be treated cautiously. There is an almost inevitable tendency for 'negative' results (no population effect) to go unreported, on the grounds of there being 'nothing' to report; and insecticide application is worryingly disruptive to the community as a whole—although picking the insects off by hand is usually impossible. Moreover, the exclusion studies have often taken 7 years or more to show any impact on the plants: it may be that many of the 'negative' studies were simply given up too early.

the difficulties of demonstrating effects on prey populations

8.4 The effects of consumption on consumers

The beneficial effects that food has on individual predators are not difficult to imagine. Generally speaking, an increase in the amount of food consumed leads to increased rates of growth, development and birth, and decreased rates of mortality. This, after all, is implicit in any discussion of intraspecific competition amongst consumers (see Chapter 6): high densities, implying small amounts of food per individual, lead to low growth rates, high death rates, and so on. Similarly, many of the effects of migration previously considered (see Chapter 5) reflect the responses of individual consumers to the distribution of food availability. However, there are a number of ways in which the relationships between consumption rate and consumer benefit can be more complicated than they appear. In the first place, all animals require a certain amount of food simply for maintenance (Figure 8.10); and unless this threshold is exceeded the animal will be unable to grow or reproduce, and will therefore be unable to contribute to future generations. In other words, low consumption rates, rather than leading to a small benefit to the consumer, simply alter the rate at which the consumer starves to death.

At the other extreme, the birth, growth and survival rates of individual consumers cannot be expected to rise indefinitely as food availability is increased. Rather, the consumers become satiated. Consumption rate eventually reaches a plateau, where it becomes independent of the amount of food available, and benefit to consumers therefore also reaches a plateau. Thus, there is a limit to the amount that a particular consumer population can eat, a limit to the amount of harm that it can do its prey population at that time, and a limit to the extent by which the consumer population can increase in size. This is discussed more fully under 'Functional responses' in Chapter 9, Section 9.5.

The most striking example of whole populations of consumers being satiated simultaneously is provided by the many plant (especially tree) species that have mast years—occasional years in which there is the synchronous production of a large volume of seed by most conspecific trees in the same geographical area, with a dearth of seeds produced in the years in between (Figure 8.11; but see also Figure 5.2). This is seen particularly often in tree species that suffer generally high

consumers often need to exceed a threshold of consumption

consumers may become satiated

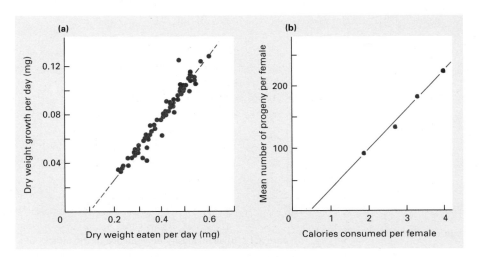

Figure 8.10 Prey availability has to exceed a certain maintenance threshold before predators derive a measurable benefit in terms of growth or reproduction. (a) Growth in the spider *Linyphia triangularis* (Turnbull, 1962). (b) Reproduction in the water flea *Daphnia pulex* var. *pulicaria* (Richman, 1958). (After Hassell, 1978.)

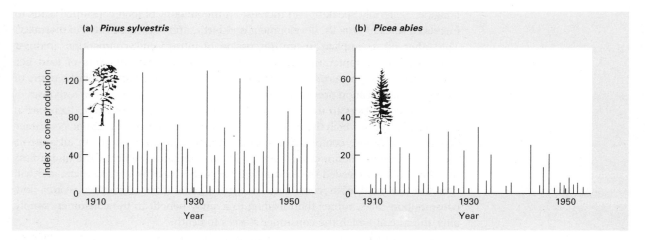

Figure 8.11 Masting: fluctuations in the cone crop of Scots pine (*Pinus sylvestris*) and Norway spruce (*Picea abies*) in northern Sweden. Based on counts made from reports by rangers. (After Hagner, 1965.)

mast years and the satiation of seed predators

intensities of seed predation (Silvertown, 1980), and it is therefore especially significant that the chances of escaping seed predation are typically much higher in mast years than in other years. The individual predators of seeds are satiated in mast years, and the populations of predators cannot increase in abundance rapidly enough to exploit the glut. By the time they have increased (usually the following year), glut has given way to famine. Figure 8.12 shows an example of such inverse density dependence (higher rates of plant survival at higher prey densities) for pre-dispersal seed predation by a weevil of seeds of witch-hazel in Michigan.

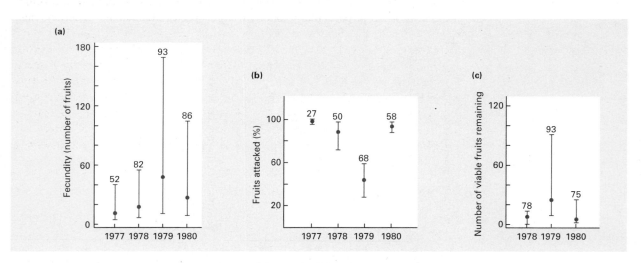

Figure 8.12 Inverse density dependence in the attacks of a weevil, *Pseudanthonomus hamamelidis*, on fruits of the witch-hazel (*Hamamelis virginiana*), such that as (a) plant fecundity varied greatly from 1977 to 1980, (b) the percentage of fruits attacked was inversely related to fecundity, so that (c) significant numbers of viable fruit remained only in the year of high fecundity (medians and 25th and 75th percentiles and sample sizes are shown in each case). (After De Steven, 1983.)

330   CHAPTER 8

On the other hand, the production of a mast crop makes great demands on the internal resources of a plant. In a mast year a spruce tree averages 38% less annual growth than in other years, and the annual ring increment in forest trees may be reduced by as much during a mast year as by a heavy attack of defoliating caterpillars. The years of seed famine are therefore essentially years of plant recovery.

Apart from illustrating the potential importance of predator satiation, the example of masting highlights a further point which relates to time scales. The seed predators are unable to extract the maximum benefit from (or do the maximum harm to) the mast crop because their generation times are too long. A hypothetical seed predator population that could pass through several generations during a season would be able to increase exponentially and explosively on the mast crop and destroy it. Generally speaking, consumers with relatively short generation times tend to track closely fluctuations in the quantity or abundance of their food or prey, whereas consumers with relatively long generation times take longer to respond to increases in prey abundance, and longer to recover when reduced to low densities.

a consumer's numerical response is limited by its generation time ...

This is further illustrated in desert communities, where year-to-year variations in precipitation, often falling in a short rainy season, are both considerable and unpredictable—as, therefore, are the variations in the productivities of many desert plants. In the rare years of high productivity, herbivores are typically at low abundance following 1 or more years of low plant productivity. Thus, they are frequently satiated, allowing plant populations to add considerably to their reserves in those years—either their buried seed banks ('annual' plants especially) or perhaps their underground storage organs (Ayal, 1994). The example of fruit production by *Asphodelus ramosus* in the Negev desert in Israel in shown in Figure 8.13. The mirid bug, *Capsodes infuscatus*, feeds on *Asphodelus*, exhibiting a particular preference for the developing flowers and young fruits. Potentially, therefore, it can have a profoundly harmful effect on the plant's fruit production. But it only passes through one generation per year. Hence, its abundance tends never to match that of its host plant (Figure 8.13). In 1988 and 1991, fruit production was high but mirid abundance was relatively low: the reproductive output of the mirids was therefore high (3.7 and 3.5 nymphs per adult, respectively), but the proportion of fruits damaged relatively low (0.78 and 0.66). In 1989 and 1992, on the other hand, when fruit production had dropped to much lower levels, the proportion of fruits damaged was much higher (0.98 and 0.87), but the reproductive output lower (0.30 nymphs per adult in 1989; unknown in 1992). This suggests that herbivorous insects, at least, may have a limited ability to affect plant population dynamics in desert

... as illustrated by desert interactions

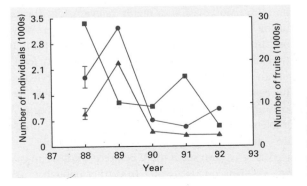

Figure 8.13 Fluctuations in the fruit production of *Asphodelus* (■) and the number of *Capsodes* nymphs (●) and adults (▲) at a study site in the Negev desert, Israel. (After Ayal, 1994.)

331   NATURE OF PREDATION

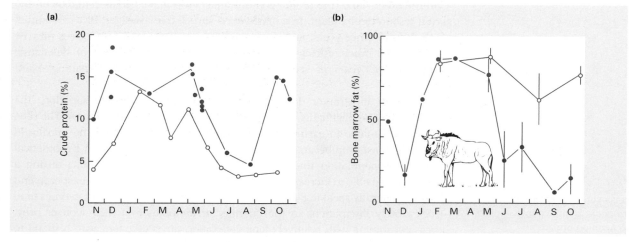

Figure 8.14 (a) The quality of food measured as percentage crude protein available to (○) and eaten by (●) wildebeest in the Serengeti during 1971. Despite selection ('eaten' > 'available'), the quality of food eaten fell during the dry season below the level necessary for the maintenance of nitrogen balance (5–6% of crude protein). (b) The fat content of the bone marrow of the live male population (○) and those found dead from natural causes (●). Vertical lines, where present, show 95% confidence limits. (After Sinclair, 1975.)

communities, but that the potential is much greater for the dynamics of herbivorous insects to be affected by their food plants (Ayal, 1994).

Chapter 3 stressed that the abundance or quantity of prey or food consumed may

food quality rather than quantity can be of paramount importance

be less important than its quality. In fact, food quality, which has both positive aspects (like the concentrations of nutrients) and negative aspects (like the concentrations of toxins), can only sensibly be defined in terms of the effects of the food on the animal that eats it; and this is particularly pertinent in the case of herbivores. For instance, it was mentioned in Section 8.2.2, that the survival and fecundity of larch budmoths and snowshoe hares is as much dependent on the quality of their food as on how much of it they consume. Along similar lines, Sinclair (1975) examined the effects of grass quality (protein content) on the survival of wildebeest in the Serengeti of Tanzania. Despite selecting protein-rich plants and plant parts (Figure 8.14a), the wildebeest consumed food in the dry season that contained well below the level of protein necessary even for maintenance (5–6% of crude protein); and to judge by the depleted fat reserves of dead males (Figure 8.14b), this was an important cause of mortality. Moreover, it is highly relevant that the protein requirements of females during late pregnancy and lactation (December–May in the wildebeest) are three to four times higher than the normal (Agricultural Research Council, 1965). It is therefore clear that the shortage of high-quality food (and not just food shortage per se) can have a drastic effect on the growth, survival and fecundity of a consumer. In the case of herbivores especially, it is possible for an animal to be apparently surrounded by its food whilst still experiencing a food shortage. We can see the problem if we imagine that we ourselves are provided with a perfectly balanced diet—diluted in an enormous swimming-pool. The pool contains everything we need, and we can see it there before us, but we may very well starve to death before we can drink enough water to extract enough nutrients to sustain ourselves. In a similar fashion, herbivores may frequently be confronted with a pool of available nitrogen that is so dilute that they have difficulty processing enough material to extract what they need. Outbreaks of herbivorous insects may then be associated with rare elevations in the concentration of available nitrogen in their food plants (see Chapter 3, Section 3.7.2), perhaps associated with unusually dry or, conversely, unusually waterlogged conditions (White, 1978, 1984, 1993). Consumers obviously need to acquire resources—but, to benefit from them fully they need to acquire them in appropriate quantities and in an appropriate form. The behavioural strategies that have evolved in the face of the pressures to do this are the main topic of the next chapter.

Chapter 9
The Behaviour of Predators

9.1 Introduction

In this chapter we discuss the behaviour of predators. We examine where they feed, what they feed on, how they are affected by other predators and how they are affected by the density of their prey. These topics are of interest in their own right, but they are also relevant in two other, broader contexts. First, foraging is an aspect of animal behaviour that is subject to the scrutiny of evolutionary biologists, within the general field of 'behavioural ecology'. The aim, put simply, is to try to understand how natural selection has favoured particular patterns of behaviour in particular circumstances (how, behaviourally, organisms match their environment). In this chapter we deal directly with this problem in the sections on 'optimal foraging' (see Sections 9.3 and 9.9), but the general evolutionary approach will underlie much of what is said throughout the chapter.

predatory behaviour as behavioural ecology

Second, the various aspects of predatory behaviour can be seen as components that combine to influence the population dynamics of both the predator itself and its prey. The population ecology of predation is dealt with much more fully in the next chapter. Nonetheless, it will be useful at various points in the present chapter to suggest the ways in which individual behaviour may affect population dynamics. As Chapter 6 made clear, if predation is such that one or other of the populations is subjected to density dependent increases in mortality (or decreases in birth rate), then this will tend to regulate the size of that population within certain limits, i.e. it will tend to stabilize the dynamics of the population. This in turn will tend to stabilize the population dynamics of the interaction as a whole (the populations will be persistent and show relatively little variation in abundance). Conversely, if there is 'inverse density dependence', such that mortality decreases (or birth rate increases) with increasing density, then this will tend to destabilize the dynamics of the interaction. Predatory behaviour can clearly have a significance beyond its effects on the individuals concerned.

predatory behaviour and population dynamics

9.2 The widths and compositions of diets

Consumers can be classified as either monophagous (feeding on a single prey-type), oligophagous (few prey-types) or polyphagous (many prey-types). Often, an equally useful distinction is between specialists (broadly, monophages and oligophages) and generalists (polyphages). Herbivores, parasitoids and true predators can all provide examples of monophagous, oligophagous and polyphagous species. But the distribution of diet widths differs amongst the various types of consumer. True predators with specialized diets do exist; for instance the Everglades kite (*Rostrahamus sociabilis*) feeds almost entirely on snails of the genus *Pomacea*—but most true predators have relatively broad diets. Parasitoids, on the other hand, are typically specialized and may even be monophagous. Herbivores are well represented in all categories, but whilst grazing and 'predatory' herbivores typically have broad diets, 'parasitic' herbivores are very often highly specialized. For instance, Janzen (1980) examined 110 species of beetle that feed as larvae inside the seeds of dicotyledonous plants in Costa Rica ('parasitizing' them) and found that 83 attacked only one plant species, 14 attacked only two, nine attacked three, two attacked four, one attacked six and one attacked eight. This was in spite of there being 975 plant species present in the area.

9.2.1 Food preferences

It must not be imagined that polyphagous and oligophagous species are indiscriminate in what they choose from their acceptable range. On the contrary, some degree of preference is almost always apparent. An animal is said to exhibit a preference for a particular type of food when the proportion of that type in the animal's diet is higher than its proportion in the animal's environment. To measure food preference in nature, therefore, it is necessary not only to examine the animal's diet (usually by the analysis of gut contents) but also to assess the 'availability' of different food types. Ideally, this should be done not through the eyes of the observer (i.e. not by simply sampling the environment), but through the eyes of the animal itself.

preference is defined by
comparing diet with
'availability'

The exact quantification of preference is therefore fraught with difficulties—but it is generally much easier at least to establish that a preference does exist. For instance, the results of an accidental field experiment, in which deer broke into a tree plantation, provide a nice example. The plantation contained equal numbers of four species arranged at random: white pine, red pine, jack pine and white spruce. As Table 9.1 shows, the deer, with free access to all four species, exhibited a fairly consistent preference for jack pine, followed by white pine, with red pine being only lightly browsed and white spruce ignored.

9.2.2 Ranked and balanced preferences

A food preference can be expressed in two rather different contexts. There can be a preference for items that are the most valuable amongst those available, or for items that provide an integral part of a mixed and balanced diet. These will be referred to as ranked and balanced preferences, respectively. In the terms of Chapter 3, Section 3.8, where resources were classified, individuals exhibit ranked preferences in

Table 9.1 A feeding preference: the percentages of various planted trees browsed by deer when they broke into a plantation in which the four species of tree were equally abundant and arranged at random. (After Horton, 1964.)

	White pine	Red pine	Jack pine	White spruce
Winter 1956–1957	31	19	84	0
Winter 1958–1959	9	1	48	0
Winter 1960–1961	17	0	70	0

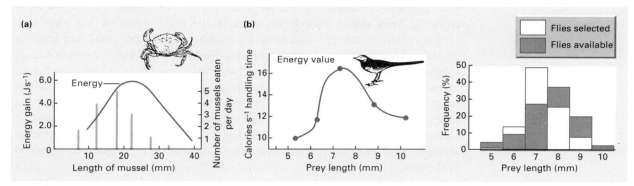

Figure 9.1 Predators eating 'profitable' prey, i.e. predators showing a preponderance in their diet for those prey items that provide them with the most energy. (a) When crabs (*Carcinus maenas*) were presented with equal quantities of six size classes of mussels (*Mytilus edulis*), they tended to show a preference for those providing the greatest energy gain (energy per unit handling time). (After Elner & Hughes, 1978.) (b) Pied wagtails (*Motacilla alba yarrellii*) tended to select, from scatophagid flies available, those providing the greatest energy gain per unit handling time. (After Davies, 1977; Krebs, 1978.)

discriminating between resource types that are 'perfectly substitutable', and exhibit balanced preferences between resource types that are 'complementary'.

Ranked preferences are usually seen most clearly amongst carnivores. For instance, Figure 9.1 shows two examples in which carnivores actively selected prey items that were the most profitable in terms of energy intake per unit time spent dealing with (or 'handling') prey. Results such as these reflect the fact that a carnivore's food often varies little in composition (see Chapter 3, p. 121), but may vary in size or accessibility. This allows a single measure (like 'energy gained per unit handling time') to be used to characterize food items, and it therefore allows food items to be ranked. In other words, Figure 9.1 shows consumers exhibiting an active preference for food of a high rank.

For many consumers, however, especially herbivores and omnivores, no simple ranking is appropriate, since none of the available food items matches the nutritional requirements of the consumer. These requirements can therefore only be satisfied either by eating large quantities of food, and eliminating much of it in order to get a sufficient quantity of the nutrient in most limited supply (for example aphids excrete vast amounts of carbon in honeydew to get sufficient nitrogen from plant sap), or by eating a combination of food items that between them match the consumer's requirements. In fact, many animals exhibit both sorts of response. They select food that is of generally high quality (so the proportion eliminated is minimized), but they also select items to meet specific requirements. For instance, sheep and cattle show a preference for high-quality food. They select leaves in preference to stems, and green matter in preference to dry or old material; and compared to the forage as a whole, the selected material is usually higher in nitrogen, phosphorus, sugars and gross energy, and lower in fibres (Arnold, 1964). In fact, all generalist herbivores appear to show rankings in the rate at which they eat different food plants when given a free choice in experimental tests (Crawley, 1983).

On the other hand, a balanced preference is also quite common. For instance, the plate limpet, *Acmaea scutum*, selects a diet of two species of encrusting microalgae

ranked preferences predominate when food items can be classified on a single scale

but many consumers exhibit a combination of ranked and balanced preferences

that contains 60% of one species and 40% of the other, almost irrespective of the proportions in which they are available (Kitting, 1980); whilst caribou, which survive on lichen through the winter, develop a sodium deficiency by the spring which they overcome by drinking seawater, eating urine-contaminated snow and gnawing shed antlers (Staaland *et al.*, 1980). We have only to look at ourselves to see an example in which 'performance' is far better on a mixed diet than on a pure diet of even the 'best' food.

There are two other important reasons why a mixed diet may be favoured. First, consumers may accept low-quality items simply because, having encountered them, they have more to gain by eating them (poor as they are) than by ignoring them and continuing to search. This is discussed in detail in Section 9.3. Second, consumers may benefit from a mixed diet because each food type contains a different undesirable toxic chemical. A mixed diet would then keep the concentrations of all of these chemicals within acceptable limits. It is certainly the case that toxins can play an important role in food preference. For instance, one study examined the winter food of a variety of Arctic animals: three species of ptarmigan, three grouse, capercaillie, two kinds of hare and the moose (Bryant & Kuropat, 1980). In each case, the conclusions were the same: animals ranked their foods on neither energy

nor nutrient content. Instead, preference was strongly and negatively correlated with the concentrations of certain toxins. Overall, however, it would be quite wrong to give the impression that all preferences have been clearly linked with one explanation or another. For example, Thompson (1988) has reviewed the relationship beween the oviposition preferences of phytophagous insects and the performance of their offspring on these foodplants in terms of growth, survival and reproduction. A number of studies have shown a good association (i.e. females preferentially oviposit on plants where their offspring perform best), but in many others the association is poor. In such cases, there is generally no shortage of explanations for the apparently unsuitable behaviour, but these explanations are, as yet, often just untested hypotheses.

9.2.3 Switching

The preferences of many consumers are fixed, i.e. they are maintained irrespective of the relative availabilities of alternative food types. But many others switch their preference, such that food items are eaten disproportionately often when they are common and are disproportionately ignored when they are rare. The two types of

switching involves a preference
for food types that are
common

preference are contrasted in Figure 9.2. Figure 9.2a shows the fixed preference exhibited by predatory shore snails when they were presented with two species of mussel prey at a range of proportions. The line in Figure 9.2a has been drawn on the assumption that they exhibited the same preference at all proportions. This assumption is clearly justified: irrespective of availability, the predatory snails showed the same marked preference for the thin-shelled, less protected *Mytilus edulis*, which they could exploit more effectively. By contrast, Figure 9.2b shows what happened when guppies (a species of fish) were offered a choice between fruit-flies and tubificid worms as prey. The guppies clearly switched their preference, and consumed a disproportionate number of the more abundant prey-type.

There are a number of situations in which switching can arise. Probably the most common is where different types of prey are found in different microhabitats, and the consumers concentrate on the most profitable microhabitat. This was the case

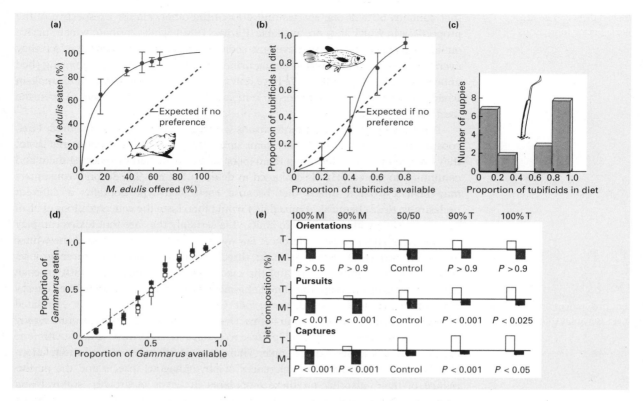

Figure 9.2 Switching. (a) A lack of switching: snails exhibit a consistent preference amongst the mussels *Mytilus edulis* and *M. californianus*, irrespective of their relative abundance (means plus standard errors). (After Murdoch & Stewart-Oaten, 1975.) (b) Switching by guppies fed on tubificids and fruit-flies: they take a disproportionate amount of whichever prey type is the more available (means and total ranges). (After Murdoch *et al.*, 1975.) (c) Preferences shown by the individual guppies in (b) when offered equal amounts of the two prey types: individuals were mostly specialists on one or other type. (d) Switching by sticklebacks fed mixtures of *Gammarus* and *Artemia*: overall they take a disproportionate amount of whichever is more available. However, in the first series of trials, *Gammarus* availability decreasing (closed symbols), first-day trialists (■) tended to take more *Gammarus* than third-day trialists (●), whereas with *Gammarus* availability increasing, firsts (□) tended to take less *Gammarus* than thirds (○). The effects of learning are apparent. (After Hughes & Croy, 1993.) (e) The behaviour of dragonfly nymphs presented with 50% mayflies (M), 50% tubifex (T), having previously been trained at 100% M, 90% M, 50% M–50% T, 90% T or 100% T. The 50 : 50 training represents a control: the significance levels refer to departures from control proportions. Pursuit and capture probabilities reflected past experience (and were the basis for switching by the predator): orientation probability did not. (After Bergelson, 1985.)

for the guppies in Figure 9.2b: the fruit-flies floated at the water surface, whilst the tubificids were found at the bottom. Switching can also occur (Bergelson, 1985) when there is:

the circumstances when switching arises

1 an increased probability of orientating toward a common prey-type, i.e. consumers develop a 'search image' for abundant food (Tinbergen, 1960) and concentrate on their 'image' prey to the relative exclusion of non-image prey;

2 an increased probability of pursuing a common prey-type;

3 an increased probability of capturing a common prey-type;

4 an increased efficiency in handling a common prey-type.

For instance, switching occurred in the 15-spined stickleback, *Spinachia spinachia*, feeding on the crustaceans *Gammarus* and *Artemia* as alternative prey (Figure 9.2d) as a result of learned improvements in capturing and handling efficiencies, especially of *Gammarus*. Fish were fed *Gammarus* for 7 days, which was then replaced in the diet, in 10% steps, with *Artemia* until the diet was 100% *Artemia*. This diet was then maintained for a further 7 days, when the process was reversed back down to 100% *Gammarus*. Each 'step' itself lasted 3 days, on each of which the fish were tested. The learning process is apparent in Figure 9.2d in the tendency for first-day trialists to be more influenced than third-day trialists by the previous dietary mix.

Learning is also apparent in the switching by dragonfly nymphs, *Anax junius*, between mayfly nymphs and tubifex worms (Figure 9.2e). All predators were presented with a 50 : 50 mixture of the prey-types, but their responses reflected their past experience—pursuing and capturing with a significantly enhanced probability the prey-type that they knew best. On the other hand, experience did not affect, and switching was thus not affected by, the orientation of the predators towards the different types of prey (Figure 9.2e).

switching and the individual

Interestingly, switching in a population often seems to be a consequence not of individual consumers gradually changing their preference, but of the proportion of specialists changing. This is illustrated by a study of switching in woodpigeons (*Columba palumbus*) feeding on maple peas and tic beans (Murton, 1971). When the two were equally abundant there was a slight preference for maple peas; but, when there were 82% tic beans on offer, the birds switched to an average of 91% tic beans in their diet. This average, however, included two birds that specialized on the rarer maple peas, taking only 5% and 0% tic beans. Figure 9.2c shows a comparable pattern for the guppies fed on fruit-flies and tubificids: when the prey-types were equally abundant, individual guppies were not generalists. Rather, there were approximately equal numbers of fruit-fly and tubificid specialists.

9.3 The optimal foraging approach to diet width

diet width and evolution

Predators and prey have undoubtedly influenced one another's evolution. This can be seen in the distasteful or poisonous leaves of many plants, in the spines of hedgehogs and in the camouflage coloration of many insect prey; and in the stout ovipositors of wood wasps, the multichambered stomachs of cattle and the silent approach and sensory excellence of owls. Such specialization makes it clear, though, that no predator can possibly be capable of consuming all types of prey. Simple design constraints prevent shrews from eating owls (even though shrews are carnivores) and prevent humming-birds from eating seeds.

Even within their constraints, however, most animals consume a narrower range of food types than they are morphologically capable of consuming. In trying to understand what determines a consumer's actual diet within its wide potential range, ecologists have increasingly turned to optimal foraging theory. The aim of optimal foraging theory is to predict the foraging strategy to be expected under specified conditions. It generally makes such predictions on the basis of a number of assumptions.

certain assumptions are inherent in optimal foraging theory

1 The foraging behaviour that is exhibited by present-day animals is the one that has been favoured by natural selection in the past but also most enhances an animal's fitness at present.

2 High fitness is achieved by a high net rate of energy intake (i.e. gross energy intake minus the energetic costs of obtaining that energy).

3 Experimental animals are observed in an environment to which their foraging behaviour is suited, i.e. it is a natural environment very similar to that in which they evolved, or an experimental arena similar in essential respects to the natural environment.

These assumptions will not always be justified. First, other aspects of an organism's behaviour may influence fitness more than optimal foraging does. For example, there may be such a premium on the avoidance of predators that animals forage at a place and time where the risk from predators is lower, and in consequence gather their food less efficiently than is theoretically possible (see Section 9.4). Second, and just as important, for many consumers (particularly herbivores and omnivores) the efficient gathering of energy may be less critical than some other dietary constituent (e.g. nitrogen), or it may be of prime importance for the forager to consume a mixed and balanced diet. In such cases, the value of existing optimal foraging theory is limited. However, in circumstances where the energy maximization premise can be expected to apply, optimal foraging theory offers a powerful insight into the significance of the foraging 'decisions' that predators make (for reviews see Krebs, 1978; Townsend & Hughes, 1981; Krebs *et al.*, 1983; Stephens & Krebs, 1986; Krebs & Kacelnik, 1991).

the theoreticians are omniscient mathematicians— but the foragers need not be

Typically, optimal foraging theory makes predictions about foraging behaviour based on mathematical models constructed by ecological theoreticians who are omniscient ('all-knowing') as far as their model systems are concerned. The question therefore arises: is it necessary for a real forager to be equally omniscient and mathematical, if it is to adopt the appropriate, optimal strategy? The answer is 'no'. The theory simply says that if there is a forager that in some way (in any way) manages to do the right thing in the right circumstances, then this forager will be favoured by natural selection; and if its abilities are inherited, these should spread, in evolutionary time, throughout the population.

mechanistic models complement optimal foraging theory

Optimal foraging theory does not specify precisely how the forager should make the right decisions, and it does not require the forager to carry out the same calculations as the modeller. Instead, there is another group of 'mechanistic' models (see Section 9.9.3) that attempt to show how a forager, given that it is not omniscient, might nevertheless manage to respond by 'rules of thumb' to limited environmental information and thereby exhibit a strategy which is favoured by natural selection. But it is optimal foraging theory that predicts the nature of the strategy that should be so favoured.

9.3.1 The diet-width model: 'searching and handling'

The first paper on optimal foraging theory (MacArthur & Pianka, 1966) sought to understand the determination of diet 'width' (the range of food types eaten by an animal) within a habitat. Subsequently, the model has been developed into a more rigorous algebraic form, notably by Charnov (1976a). MacArthur and Pianka argued that to obtain food, any predator must expend time and energy, first in searching for its prey and then in handling it (i.e. pursuing, subduing and consuming it). Whilst searching, a predator is likely to encounter a wide variety of food items. MacArthur and Pianka therefore saw diet width as depending on the responses of predators once they had encountered prey. Generalists pursue (and may then subdue and consume) a large proportion of the prey they encounter; specialists continue searching except when they encounter prey of their specifically preferred type.

The 'problem' for any forager is this: if it is a specialist, then it will only pursue profitable prey items, but it may expend a great deal of time and energy searching for them; whereas if it is a generalist, it will spend relatively little time searching, but it will pursue both more and less profitable types of prey. An optimal forager should balance the pros and cons so as to maximize its overall rate of energy intake. MacArthur and Pianka expressed the problem as follows. Given that a predator already includes a certain number of profitable items in its diet, should it expand its diet (and thereby decrease its search time) by including the next most-profitable item as well?

We can refer to this 'next most-profitable' item as the ith item. E_i / h_i is then the profitability of the item, where E_i is its energy content, and h_i its handling time. In addition, $\bar{E} / \bar{h}$ is the average profitability of the 'present' diet (i.e. one that includes all prey-types that are more profitable than i, but does not include prey-type i itself), and $\bar{s}$ is the average search time for the present diet. If a predator does pursue a prey item of type i, then its expected rate of energy intake is E_i / h_i. But if it ignores this prey item, whilst pursuing all those that are more profitable, then it can expect to search for a further $\bar{s}$, following which its expected rate of energy intake is $\bar{E} / \bar{h}$. The total time spent in this latter case is $\bar{s} + \bar{h}$, and so the overall expected rate of energy intake is $\bar{E} / (\bar{s} + \bar{h})$. The most profitable, optimal strategy for a predator will be to pursue the ith item if, and only if:

$$E_i / h_i \geqslant \bar{E} / (\bar{s} + \bar{h}). \tag{9.1}$$

In other words, a predator should continue to add increasingly less profitable items to its diet as long as Equation 9.1 is satisfied (i.e. as long as this increases its overall rate of energy intake). This will serve to maximize its overall rate of energy intake, $\bar{E} / (\bar{s} + \bar{h})$.

This optimal diet model leads to a number of predictions.

1 Predators with handling times that are typically short compared to their search times should be generalists, because in the short time it takes them to handle a prey item that has already been found, they can barely begin to search for another prey item. (In terms of Equation 9.1: E_i / h_i is large (h_i small) for a wide range of prey-types, whereas $\bar{E} / (\bar{s} + \bar{h})$ is small ($\bar{s}$ large) even for broad diets.) This prediction seems to be supported by the broad diets of many insectivorous birds feeding in trees and shrubs. Searching is always moderately time-consuming: but handling the minute, stationary insects takes negligible time and is almost always successful. A bird, therefore, has something to gain and virtually nothing to lose by consuming an item once found, and overall profitability is maximized by a broad diet.

2 By contrast, predators with handling times that are long relative to their search times should be specialists. That is, if $\bar{s}$ is always small, then $\bar{E} / (\bar{s} + \bar{h})$ is similar to $\bar{E} / \bar{h}$. Thus, maximizing $\bar{E} / (\bar{s} + \bar{h})$ is much the same as maximizing $\bar{E} / \bar{h}$, which is achieved, clearly, by including only the most profitable items in the diet. For instance, lions live more or less constantly in sight of their prey so that search time is negligible: handling time, on the other hand, and particularly pursuit time, can be long (and very energy consuming). Lions consequently specialize on those prey that can be pursued most profitably: the immature, the lame and the old.

to pursue or not to pursue?

searchers should be generalists

handlers should be specialists

3 Other things being equal, a predator should have a broader diet in an unproductive environment (where prey items are relatively rare and $\bar{s}$ is relatively large) than in a productive environment (where $\bar{s}$ is smaller). This prediction is broadly supported by the two examples shown in Figure 9.3: in experimental arenas, both bluegill sunfish (*Lepomis macrochirus*) and great tits (*Parus major*) had more specialized diets when prey density was higher.

specialization should be greater in more productive environments

4 Equation 9.1 depends on the profitability of the *i*th item (E_i / h_i), depends on the profitabilities of the items already in the diet ($\bar{E} / \bar{h}$), and depends on the search times for items already in the diet ($\bar{s}$) and thus on their abundance. But it does not depend on the search time for the *i*th item, s_i. In other words, predators should ignore insufficiently profitable food types irrespective of their abundance. Re-examining the examples in Figure 9.3, we can see that these both refer to cases in which the optimal diet model does indeed predict that the least profitable items should be ignored completely. The foraging behaviour was very similar to this prediction, but in both cases the animals consistently took slightly more than expected of the less profitable food types. In fact, this sort of discrepancy has been uncovered repeatedly, and there are a number of reasons why it may occur (Krebs & McCleery, 1984), which can be summarized crudely by noting that the animals are not omniscient. The optimal diet model, however, does not predict a perfect

the abundance of unprofitable prey-types is irrelevant

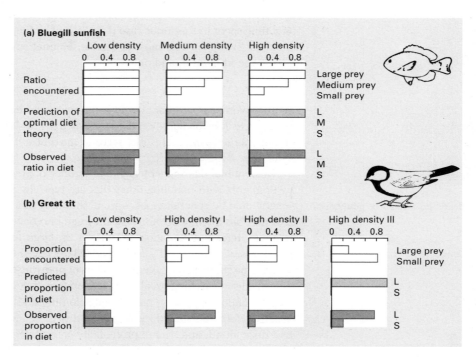

Figure 9.3 Two studies of optimal diet choice that show a clear but limited correspondence with the predictions of Charnov's (1976a) optimal diet model. Diets are more specialized at high prey densities; but more low profitability items are included than predicted by the theory. (a) Bluegill sunfish preying on different size classes of *Daphnia*: the histograms show ratios of encounter rates with each size class at three different densities, together with the predicted and observed ratios in the diet. (After Werner & Hall, 1974.) (b) Great tits preying on large and small pieces of mealworm. (After Krebs *et al.*, 1977.) The histograms in this case refer to the proportions of the two types of item taken. (After Krebs, 1978.)

correspondence between observation and expectation. It predicts the sort of strategy that will be favoured by natural selection, and says that the animals that come closest to this strategy will be most favoured. From this point of view, the correspondence between data and theory in Figure 9.3 seems much more satisfactory.

5   Equation 9.1 also provides a context for understanding the narrow specialization of predators that live in intimate association with their prey, especially where an individual predator is linked to an individual prey (e.g. many parasitoids and parasitic herbivores—and many parasites (see Chapter 12)). Since their whole life style and life cycle are finely tuned to those of their prey (or host), handling time ($\bar{h}$) is low; but this precludes their being finely tuned to other prey species, for which, therefore, handling time is very high. Equation 9.1 will thus only apply within the specialist group, but not to any food item outside it.

On the other hand, polyphagy has definite advantages. Search costs ($\bar{s}$) are typically low—food is easy to find—and an individual is unlikely to starve because of fluctuations in the abundance of one type of food. In addition, polyphagous consumers can, of course, construct a balanced diet, and maintain this balance by varying preferences to suit altered circumstances, and can avoid consuming large quantities of a toxin produced by one of its food types. These are considerations ignored by Equation 9.1.

Overall, then, evolution may broaden or restrict diets. Where prey exert evolutionary pressures demanding specialized morphological or physiological responses from the consumer, restriction is often taken to extremes. But where consumers feed on items that are individually inaccessible or unpredictable or lacking in certain nutrients, the diet often remains broad. An appealing and much-discussed idea is that particular pairs of predator and prey species have not only evolved but have coevolved. In other words, there has been an evolutionary 'arms race', whereby each improvement in predatory ability has been followed by an improvement in the prey's ability to avoid or resist the predator, which has been followed by a further improvement in predatory ability, and so on. This may itself be accompanied, on a long-term, evolutionary time scale, by speciation, so that, for example, related species of butterfly are associated with related species of plants—all the species of the Heliconiini feed on members of the Passifloracaea (Ehrlich & Raven, 1964; Futuyma & May, 1992). To the extent that coevolution occurs, it may certainly be an additional force in favour of diet restriction. At present, however, hard evidence for predator–prey or plant–herbivore coevolution is proving difficult to come by (Futuyma & Slatkin, 1983; Futuyma & May, 1992).

coevolution: predator–prey arms races?

9.3.2 Switching and optimal diets

There may seem, at first sight, to be a contradiction between the predictions of the optimal diet model and switching. In the latter, a consumer switches from one prey-type to another as their relative densities change. But the optimal diet model suggests that the more profitable prey-type should always be taken, irrespective of its density or the density of any alternative. Switching is presumed to occur, however, in circumstances to which the optimal diet model does not strictly apply. Specifically, switching often occurs when the different prey-types occupy different microhabitats, whereas the optimal diet model predicts behaviour within a microhabitat. Moreover, most other cases of switching involve a change in the profitabilities of items of prey

switching complements the optimal diet model

as their density changes, whereas in the optimal diet model these are constants. Indeed, in cases of switching, the more abundant prey-type is the more profitable, and in such a case the optimal diet model predicts specialization on whichever prey-type is more profitable, i.e. whichever is more abundant, i.e. switching.

9.4 Foraging in a broader context

It is worth stressing that foraging strategies will not always be strategies for simply maximizing feeding efficiency. On the contrary, natural selection will favour foragers that maximize their net benefits, and strategies will therefore often be modified by other, conflicting demands on the individuals concerned. In particular, the need to avoid predators will frequently affect an animal's foraging behaviour.

This has been shown in work on foraging by nymphs of an aquatic insect predator, the backswimmer *Notonecta hoffmanni* (Sih, 1982). These animals pass through five nymphal instars (with 'I' being the smallest and youngest, and 'V' the oldest), and in the laboratory the first three instars are liable to be preyed upon by adults of the same species, such that the relative risk of predation from adults was:

I > II > III > IV = V $\cong$ no risk.

backswimmers forage 'suboptimally' but avoid being preyed upon ...

These risks appear to modify the behaviour of the nymphs, in that they tend (both in the laboratory and in the field) to avoid the central areas of water bodies, where the concentration of adults is greatest. In fact, the relative degree of avoidance was the same as the relative risk of predation from adults:

I > II > III > IV = V $\cong$ no avoidance.

Yet these central areas also contain the greatest concentration of prey items for the nymphs, and so, by avoiding predators, nymphs of instars I and II showed a reduction in feeding rate in the presence of adults (although those of instar III did not). The young nymphs displayed a less than maximal feeding rate as a result of their avoidance of predation, but an increased survivorship.

... as do certain fish ...

The modifying influence of predators on foraging behaviour has also been studied by Werner *et al.* (1983a) working on bluegill sunfish. They estimated the net energy returns from foraging in three contrasting laboratory habitats—in open water, amongst water weeds and on bare sediment—and they examined how prey densities varied in comparable natural habitats in a lake through the seasons. They were then able to predict the time at which the sunfish should switch between different lake habitats so as to maximize their overall net energy returns. In the absence of predators, three sizes of sunfish behaved as predicted (Figure 9.4). But in a further field experiment, this time in the presence of predatory largemouth bass, the small sunfish restricted their foraging to the water-weed habitat (Figure 9.5; Werner *et al.*, 1983b). Here, they were relatively safe from predation, although they could only achieve a markedly submaximal rate of energy intake. By contrast, the larger sunfish are more or less safe from predation by bass, and they continue to forage according to the optimal foraging predictions. In a similar vein, several species of zooplanktivorous fish largely restrict their feeding to the hours of darkness, a time when their feeding rates are likely to be relatively low but when the risk of predation by piscivorous fish and birds is much reduced (Townsend & Winfield, 1985).

Taken together, the work on *Notonecta* and fish emphasizes that a foraging strategy is an integral part of an animal's overall pattern of behaviour. It is strongly

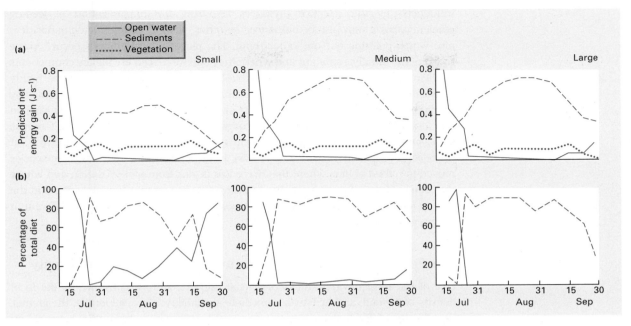

Figure 9.4 Seasonal patterns in (a) the predicted habitat profitabilities (net rate of energy gain) and (b) the actual percentage of the diet originating from each habitat, for three size classes of bluegill sunfish (*Lepomis macrochirus*). Piscivores were absent. (The 'vegetation' habitat is omitted from (b) for the sake of clarity—only 8–13% of the diet originated from this habitat for all size classes of fish.) There is good correspondence between the patterns in (a) and (b). (After Werner *et al.*, 1983a.)

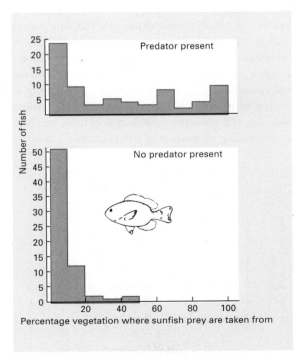

Figure 9.5 In contrast to Figure 9.4 and to the lower figure here, when largemouth bass (which prey on small bluegill sunfish) are present (upper figure) many sunfish take prey from areas where the percentage vegetation is high and where they are relatively protected from predation. (After Werner *et al.*, 1983b.)

influenced by the selective pressures favouring the maximization of feeding efficiency, but it may also be influenced by other, possibly conflicting demands. It is also worth pointing out one other thing. The places where animals occur, where they are maximally abundant and where they choose to feed are all key components of their 'realized niches'. We saw in Chapter 7 that realized niches can be highly constrained by competitors. Here, we see that they can also be highly constrained by predators. This is also seen in the effects of predation by the barn owl (*Tyto alba*) on the foraging behaviour of three heteromyid rodents, the Arizona pocket mouse (*Perognathus amplus*), Bailey's pocket mouse (*P. baileyi*) and Merriam's kangaroo rat (*Dipodomys merriami*) (Brown *et al.*, 1988). In the presence of owls, all three species moved to microhabitats where they were less at risk from owl predation and where they reduced their foraging activity; but they did so to varying extents, such that the way in which the microhabitat was partitioned between them was quite different in the presence and absence of owls.

predation and the realized niche

9.5 Functional responses: consumption rate and food density

One of the most obvious things of crucial importance to a consumer is the local density of its food, and hence its immediate availability, since, generally, the greater the density of food, the more the consumer eats. The relationship between an individual's consumption rate and local food density is known as the consumer's functional response (Solomon, 1949). Not surprisingly, the detailed nature of the response varies. It has been classified into three 'types' by Holling (1959).

a functional response relates consumption rate to local food density

9.5.1 The type 2 functional response

The most frequently observed functional response is the 'type 2' response, in which consumption rate rises with prey density, but gradually decelerates, until a plateau is reached at which consumption rate remains constant irrespective of prey density. This is shown for a carnivore and a herbivore in Figure 9.6. Holling's (1959) explanation for the type 2 response can be stated as follows. A consumer has to devote a certain handling time to each prey item it consumes (i.e. pursuing, subduing and consuming the prey item, and then preparing itself for further search). As prey density increases, finding prey becomes increasingly easy. Handling a prey item, however, still takes the same length of time, and handling overall therefore takes up an increasing proportion of the consumer's time—until at high prey densities the

type 2 responses and handling times

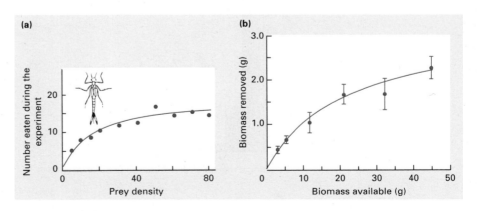

Figure 9.6 Type 2 functional responses. (a) Tenth-instar damselfly nymphs eating *Daphnia* of approximately constant size. (After Thompson, 1975.) (b) Bank voles, *Clethrionomys glareolus*, eating willow shoots, *Salix myrsinifolia*. (After Lundberg, 1988.)

consumer is effectively spending all of its time handling prey. Consumption rate therefore approaches and then reaches a maximum (the plateau), determined by the maximum number of handling times that can be fitted into the total time available.

This view of the type 2 functional response can be examined through a study of the ichneumonid parasitoid *Pleolophus basizonus*, attacking cocoons of the European pine sawfly, *Neodiprion sertifer* (Griffiths, 1969). Griffiths plotted the number of ovipositions per parasitoid over a range of host densities, dealing separately with parasitoids of different ages. He also calculated the actual maximum oviposition rate by presenting other parasitoids of the same ages with a superabundant supply of host cocoons. Figure 9.7 shows that the type 2 curves did indeed approach their appropriate maxima. However, whilst these maxima (of around 3.5 ovipositions per day) suggest a handling time of about 7 h, further direct observation appeared to indicate that oviposition takes, on average, only 0.36 h. In fact, this type of discrepancy is extremely common. For example, when oystercatchers (*Haematopus ostralegus*) fed on cockles in North Wales, the observed handling time of a cockle varied from 19 to 29 s; but the value estimated from a functional response curve was 75 s (Sutherland, 1982).

In the case of Griffiths's work, the discrepancy is accounted for by the existence of a 'refractory period' following oviposition, during which there are no eggs ready to be laid. 'Handling time', therefore, includes not only the time actually taken in oviposition, but also the time taken preparing for the next oviposition. Likewise, the true handling times of Sutherland's oystercatchers, and of the animals in Figure 9.6, include time devoted to feeding-related activities other than the direct manipulation of food items. For example, the damselfly larvae in Figure 9.6a are affected by a

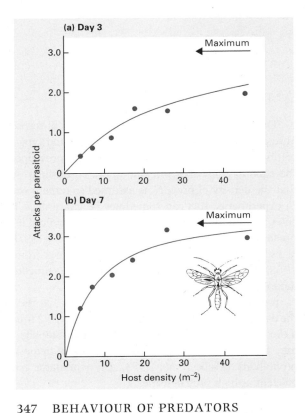

Figure 9.7 The type 2 functional responses of the ichneumonid parasitoid *Pleolophus basizonus* to changes in the density of its host *Neodiprion sertifer*. Arrows indicate maxima observed in the presence of excess hosts. (a) Parasitoids on their third day of adult life. (b) Parasitoids on their seventh day. (After Griffiths, 1969.)

factor known as 'gut limitation', whereby their maximum rate of consumption is determined by the capacity of their gut, and by the speed with which it can make space available for further food items. 'Handling time' includes all such phenomena.

A further point to note from Figure 9.7 is that whilst the plateau level (and thus, the handling time) was approximately the same for parasitoids of different ages, the rate of approach to that plateau was much more gradual in the younger parasitoids. This indicates that the younger parasitoids have a lower 'searching efficiency', or, synonymously, a lower 'attack rate': at low host densities they oviposit less often than the older parasitoids. But at high host densities there is such a ready supply of hosts that even they are limited only by their handling time. The form taken by a type 2 functional response curve, therefore, can be characterized simply in terms of a handling time (determining the level of the plateau) and an attack rate (determining the rate of approach to the plateau).

a type 2 response can be defined by a handling time and a searching efficiency

More specifically, we can derive a relationship between P_e (the number of prey items eaten by a predator during a period of searching time, T_s) and N, the density of those prey items (Holling, 1959). In fact, P_e increases with the time available for searching, it increases with prey density and it increases with the searching efficiency or attack rate of the predator, a'. Thus:

$$P_e = a'T_s N. \tag{9.2}$$

However, the time available for searching will be less than the total time, T, because of time spent handling prey. Hence, if T_h is the handling time of each prey item, then $T_h P_e$ is the total time spent handling prey, and:

$$T_s = T - T_h P_e. \tag{9.3}$$

Substituting this into Equation 9.2 we have:

$$P_e = a'(T - T_h P_e)N \tag{9.4}$$

or, rearranging:

Holling's disc equation

$$P_e = \frac{a'NT}{1 + a'T_h N}. \tag{9.5}$$

This equation describes a type 2 functional response, and is known as Holling's 'disc equation' because Holling first generated type 2 responses experimentally by getting a blindfolded assistant to pick up ('prey upon') sandpaper discs. Note that the equation describes the amount eaten during a specified period of time, T; and that the density of prey, N, is assumed to remain constant throughout that period. In experiments, this can sometimes be guaranteed by replacing any prey that are eaten; but more sophisticated models are required if prey density is depleted by the predator. Such models are described by Hassell (1978), who also discusses methods of estimating attack rates and handling times from a set of data. (Trexler et al., 1988, discuss the general problem of fitting functional response curves to sets of data.)

there are alternative reasons for a type 2 response

It would be wrong, however, to imagine that the existence of a handling time is the only or the complete explanation for all type 2 functional responses. For instance, if the prey items are actually of variable profitability, then at high densities the diet may tend towards a decelerating number of items which are nevertheless of high profitability (Krebs et al., 1983). Or a predator may become confused and less efficient at high prey densities.

348   CHAPTER 9

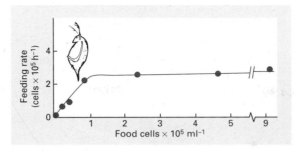

Figure 9.8 The type 1 functional response of *Daphnia magna* to different concentrations of the yeast *Saccharomyces cerevisiae*. (After Rigler, 1961.)

9.5.2 The type 1 functional response

An example of a 'type 1' functional response is illustrated in Figure 9.8, which shows the rate at which *Daphnia magna* consumed yeast cells when the density of cells varied. The consumption rate rose linearly to a maximum as density increased, and then remained at that maximum irrespective of further increases. This occurred because the yeast cells were extracted by the *Daphnia* from a constant volume of water washed over their filtering apparatus, and the amount extracted therefore rose in line with food concentration. Above 10^5 cells ml^{-1}, however, the *Daphnia* were unable to swallow (i.e. handle) all the food they filtered, and they therefore ingested food at their maximum (plateau) rate irrespective of its concentration. In other words, in a type 1 functional response, below the plateau, the handling time is zero and $T_s = T$. Equation 9.2 therefore applies, and the slope of the response is a' (the attack rate or searching efficiency), i.e. the proportion of prey eaten per unit time.

Fully described type 1 responses with a linear rise and a plateau like that in Figure 9.8 are rather rare. But, there are many examples, especially amongst herbivores, where, over the range of food densities examined, the consumption rate rises linearly (Crawley, 1983). In other words, there is no evidence in these cases for the deceleration which characterizes type 2 responses.

9.5.3 The type 3 functional response

Type 3 functional responses are illustrated in Figures 9.9a–c. At high food densities they are similar to a type 2 response, and the explanations for the two are the same. At low food densities, however, the type 3 response has an accelerating phase where an increase in density leads to a more than linear increase in consumption rate. Overall, therefore, a type 3 response is 'S-shaped' or 'sigmoidal'.

type 3 responses—and switching …

One important way in which a type 3 response can be generated is by switching on the part of the consumer (see Section 9.2.3). The similarities between Figures 9.2 and 9.9 are readily apparent. The difference is that discussions of switching focus on the density of a prey-type relative to the densities of alternatives, whereas functional responses are based on only the absolute density of the single prey-type being considered. In practice, though, absolute and relative densities are likely to be closely correlated, and switching is therefore likely to lead frequently to a type 3 functional response.

… and changes in handling time or searching efficiency

More generally, a type 3 functional response will arise whenever an increase in food density leads to an increase in the consumer's searching efficiency or a decrease in its handling time. Between them, these two determine consumption rate, and so

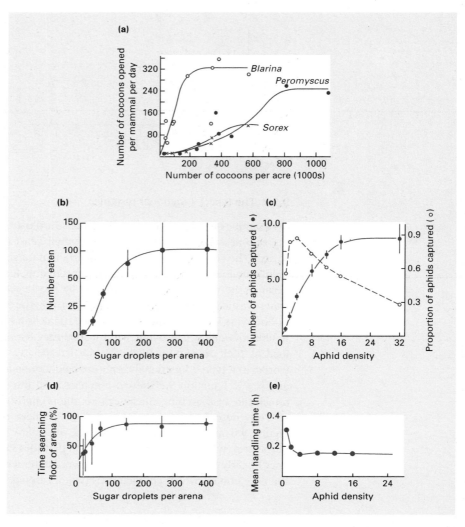

Figure 9.9 Type 3 (sigmoidal) functional responses. (a) The shrews *Sorex* and *Blarina* and the deer mouse *Peromyscus* responding to changing field densities of cocoons of the European pine sawfly, *Neodiprion sertifer*, in Ontario, Canada. (After Holling, 1959.) (b) The bluebottle fly, *Calliphora vomitoria*, feeding on sugar droplets. (After Murdie & Hassell, 1973.) (c) The wasp, *Aphelinus thomsoni*, attacking sycamore aphids, *Drepanosiphum platanoidis*: note the density dependent increase in prey mortality rate at low prey densities (– – –) giving rise to the accelerating phase of the response curve (——). (After Collins *et al.*, 1981.) (d) The basis of the response in (b): searching efficiency of *C. vomitoria* increases with 'prey' (sugar droplet) density. (After Murdie & Hassell, 1973.) (e) The basis of the response in (c): handling time in *A. thomsoni* decreases with aphid density. (After Collins *et al.*, 1981.)

an increase in a' or a decrease in T_h will make consumption rate rise faster than would be expected from the increase in food density alone. Thus, the small mammals in Figure 9.9a appear to develop a search image for sawfly cocoons (increasing efficiency); the bluebottle fly, *Calliphora vomitoria* (Figure 9.9b) spends an increasing proportion of its time searching for 'prey', as prey density increases (Figure 9.9d), also increasing efficiency; whilst the wasp *Aphelinus thomsoni* (Figure 9.9c) exhibits a reduction in mean handling time as the density of its sycamore aphid prey increases (Figure 9.9e). In each case, a type 3 functional response is the result.

At one time, type 2 functional responses were referred to as invertebrate functional responses, and type 3 responses as vertebrate responses, the implication being that the alterations in behaviour associated with type 3 responses were limited largely to vertebrates, as in the classic example shown in Figure 9.9a. Now, as Figures 9.2, 9.9b and c suggest, this view appears to do less than justice to invertebrate predators. On the other hand, good examples of type 3 responses generally are apparently rare. It seems likely that most functional responses are type 2.

9.5.4 The consequences of functional responses for the dynamics of populations

Different types of functional response have different effects on the dynamics of the populations concerned. With type 2 responses, with type 1 responses as the plateau is reached and with type 3 responses at higher densities, prey in higher-density populations have less chance of being affected than prey in lower-density populations. This inverse density dependence tends to have a destabilizing effect on the dynamics of the populations. With type 3 responses at lower densities, on the other hand, individuals at lower densities have less chance of being affected than individuals at higher densities, and this density dependent effect tends to stabilize the population dynamics.

It is important to realize, however, that these are only tendencies, which may be overriden (or supported) by forces arising from other components of the interaction. Their role in population dynamics is discussed fully in Chapter 10.

9.6 The effects of consumer density: mutual interference

mutual interference leads to reductions in consumption rate

No consumer lives in isolation: all are affected by other consumers. The most obvious effects are competitive; many consumers experience exploitation competition for limited amounts of food when their density is high or the amount of food is small, and this results in a reduction in the consumption rate per individual as consumer density increases. However, even when food is not limited, the consumption rate per individual can be reduced by increases in consumer density by a number of processes known collectively as mutual interference. For example, many consumers interact behaviourally with other members of their population, leaving less time for feeding and therefore depressing the overall feeding rate. For instance, hummingbirds actively and aggressively defend rich sources of nectar; badgers patrol and visit the 'latrines' around the boundaries between their territories and those of their neighbours; and females of *Rhyssa persuasoria* (a parasitoid of wood wasp larvae) will threaten and, if need be, fiercely drive away an intruding female from their own area of tree trunk (Spradbery, 1970). Alternatively, an increase in consumer density may lead to an increased rate of emigration, or of consumers stealing food from one another (as do many gulls) or the prey themselves may respond to the presence of consumers and become less available for capture.

Hassell and Varley (1969; see also Hassell, 1978) have shown how various cases of mutual interference can all be reduced to a common form by calculating the searching efficiency of the consumer and plotting this against consumer density on logarithmic scales (Figure 9.10). (The searching efficiency of the consumer, a', is the

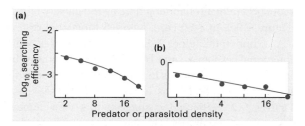

Figure 9.10 Mutual interference: the relationship between searching efficiency and the density of searching parasitoids or predators (log scales). (a) *Encarsia formosa* parasitizing the whitefly *Trialeurodes vaporariorum*. (After Burnett, 1958.) (b) *Phytoseiulus persimilis* feeding on nymphs of the mite *Tetranchyus urticae*. (After Fernando, 1977; Hassell, 1978.)

same as that in Holling's disc equation.) As expected, the slopes of the graphs are negative: searching efficiency, and thus the consumption rate per individual, declines with density. The slope in each case is said to take the value $-m$, and m is known as the coefficient of interference. At the very lowest densities (e.g. Figure 9.10a) the effect of interference is likely to be negligible. But, at moderate and high densities m tends to remain constant (Figures 9.10a, b); and m is actually often constant over the entire range of densities examined (e.g. the constant slope in Figure 9.10b). The biologically realistic range of values for m appears to be between 0 and 1 (Hassell, 1978).

the coefficient of interference

It is important to remember of course that the consumption rate per individual does not always decline with increasing consumer density. Sometimes, especially at low densities, there can be an increase in consumption rate which occurs because of social facilitation amongst the consumers. For instance, in many herds, flocks or schools of vertebrates an increase in flock size can lead to a reduction in the time each individual spends on the look-out for its own predators, and this leaves more time for feeding (see Barnard & Thompson, 1985, for examples of this in birds).

the reverse of interference: facilitation

Without doubt, though, the more general pattern is for individual consumption rate to decrease with consumer density. This reduction is likely to have an adverse effect on the fecundity, growth and mortality of individual consumers, which intensifies as consumer density increases. The consumer population is thus subject to density dependent control, and mutual interference therefore tends to stabilize the dynamics of predatory populations, and to stabilize predator–prey dynamics generally (see Chapter 10, Section 10.3).

mutual interference tends to stabilize predator–prey dynamics

9.7 Consumers and food patches

For all consumers, food is distributed patchily. The patches may be natural and discrete physical objects: a bush laden with berries is a patch for a fruit-eating bird; a leaf covered with aphids is a patch for a predatory ladybird. Alternatively, a 'patch' may only exist as an arbitrarily defined area in an apparently uniform environment; for a wading bird feeding on a sandy beach, different 10 m^2 areas may be thought of as patches that contain different densities of worms. In all cases though, a patch must be defined with a particular consumer in mind. One leaf is an appropriate patch for a ladybird, but for a larger and more active insectivorous bird, 1 m^2 of canopy or even a whole tree may represent a more appropriate patch.

the meaning of 'patch'

Consumers select those habitats that contain their food. But even within these broad habitats, food is distributed patchily, and consumers typically show a preference for particular patches. The basis for this preference may be the quality of the food; for instance mountain hares preferentially graze patches of heather that are rich in nitrogen (Moss *et al.*, 1981). Alternatively, food patches may be chosen so as to avoid predators (see Section 9.4), or parasites or severe weather conditions.

9.7.1 Aggregative responses to prey density

Because of its potential consequences for population dynamics, however, ecologists have been most interested in patch preferences where patches vary in the density of the food or prey items they contain. At one time, it appeared, and was widely believed that: (i) predators generally spent most time in patches containing high densities of prey (because these were the most profitable patches); (ii) most predators were therefore to be found in such patches; and (iii) prey in those patches were therefore most vulnerable to predation whereas those in low-density patches were relatively protected and most likely to survive. Examples certainly exist to support the first two of these propositions (Figure 9.11). These show an 'aggregative response' by the predators that is directly density dependent: predators spending most time in high-prey-density patches such that prey and predator densities are

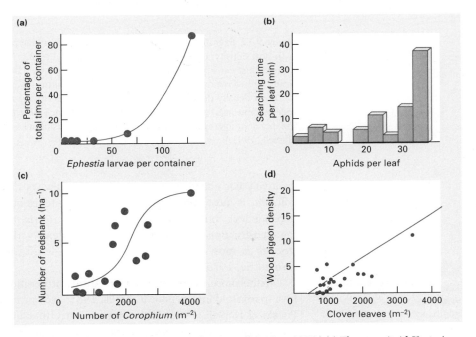

Figure 9.11 Aggregative responses. (After Hassell & May, 1974.) (a) The parasitoid *Venturia canescens* spends more time in containers with its host *Ephestia cautella* at high densities. (After Hassell, 1971.) (b) Coccinellid larvae (*Coccinella septempunctata*) spend more time on leaves with high densities of their aphid prey (*Brevicoryne brassicae*). (c) Redshank (*Tringa totanus*) aggregate in patches with higher densities of their amphipod prey *Corophium volutator*. (After Goss-Custard, 1970.) (d) Woodpigeons (*Columba palumbus*) aggregate in areas with higher densities of clover leaves. (After Murton *et al.*, 1966.)

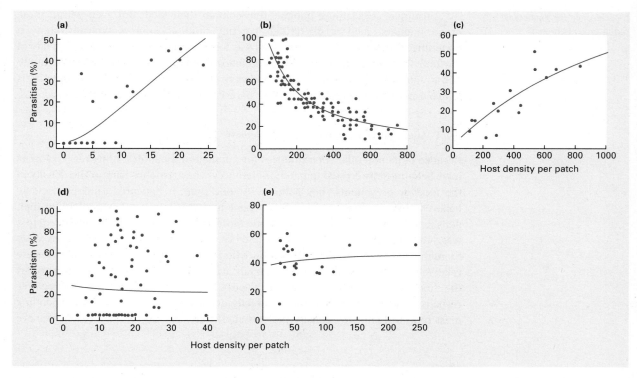

Figure 9.12 Examples of field studies showing percentage parasitism by parasitoids varying with host density; curves are derived from a model relationship of the response of parasitoids to local host density. (a) Direct density dependence when *Delia radicum* attacks *Trybliographa rapae*. (b) Inverse density dependence when *Ooencyrtus kuwanai* attacks *Lymantria dispar*. (c) Direct density dependence when *Aspidiotiphagus citrinus* attacks *Fiorinia externa*. (d) Density independent aggregation when *Tetrasticus* spp. attack *Rhopalomyia californica*. (e) Density independent aggregation when *Coccophagoides utilis* attacks *Parlatoria oleae*. (After Pacala & Hassell, 1991.)

positively correlated. However, this is not always the case. The beetle *Phyllotreta cruciferae* is most abundant in the densest patches of *Brassica* plants (direct density dependence), but larvae of the butterfly *Pieris rapae* are most abundant in low-density *Brassica* populations (inverse density dependence), whilst *Phyllotreta striolata* is most abundant in intermediate-density plots (a 'domed' relationship) (Cromartie, 1975).

responses may be directly or inversely density dependent, 'domed', or density independent

Furthermore, contrary to the third proposition, wide-ranging reviews of host–parasitoid interactions (Lessells, 1985; Stiling, 1987; Walde & Murdoch, 1988; Pacala & Hassell, 1991) have shown clearly that the prey (hosts) in high-density patches are not necessarily the most vulnerable to attack (direct density dependence): percentage parasitism may also be inversely density dependent or density independent between patches (Figure 9.12). Indeed, the reviews suggest that only around 50% of the studies examined show evidence of density dependence, and in only around 50% of these is the density dependence direct, as opposed to inverse. Nonetheless, despite this variation in pattern, it remains true that the risk of predation often varies greatly between patches, and hence between individual prey.

9.7.2 Aggregations of herbivores

herbivores often aggregate
without this being an
'aggregative response'

Many herbivores display a marked tendency to aggregate and many plants show marked variation in their risk of being attacked, without this being an aggregative response to 'prey density'. The cabbage aphid (*Brevicoryne brassicae*) forms aggregates at two separate levels (Way & Cammell, 1970). Nymphs quickly form large groups when isolated on the surface of a single leaf; and populations on a single plant tend to be restricted to particular leaves. At the level of the whole plant, aggregations can be seen when the moth *Cactoblastis cactorum*, attacks the prickly pear cacti *Opuntia inermis* and *O. stricta* (Monro, 1967). The female moths deposit egg-sticks on the plants, each containing 70–90 eggs; and as Table 9.2 shows, the distribution of these egg-sticks over the plants is highly aggregated.

aggregation and protection

When cabbage aphids attack only one leaf of a four-leaved cabbage plant (as they do naturally), the other three leaves survive; but if the same number of aphids are evenly spread over the four leaves, then all four leaves are destroyed (Way & Cammell, 1970). When *Cactoblastis* attacks a prickly pear population, an 'unexpectedly' large number of plants escape attack altogether (Table 9.2). The aggregative behaviour of the herbivores affords protection to a number of cabbage leaves and a number of prickly pear plants.

Such patterns, and the similar patterns seen in the distribution of risk amongst prey generally, have been widely believed in the past necessarily to lend stability to the population dynamics of interactions (Begon *et al.*, 1990). Now the picture appears to be far more complex (see Chapter 10, Section 10.5). Note, however, that any stabilizing effect that there might be is a by-product of the consumers' preference for high-profitability patches. The predators are not behaving in this way 'in order to' stabilize the predator–prey interaction!

9.7.3 Patchiness and time: hide-and-seek

In fact, the responses of consumers to their food patches often have not just a spatial component but a temporal component as well. When this is the case, the

Table 9.2 The observed distribution of *Cactoblastis* egg-sticks on *Opuntia* plants is aggregated, compared to a random Poisson distribution (in which the variance and mean are equal). (After Monro, 1967.)

Site	Mean density (egg-sticks per segment)	Egg-sticks per plant		Comparison by χ^2 test of distributions with Poisson distributions of the same mean
		Mean	Variance	
1	0.398	2.42	6.1	$*P < 0.001$
2	0.265	2.09	22.40	$*P < 0.001$
3	0.084	1.24	5.09	$*P < 0.001$
4	0.031	0.167	0.247	$\dagger P > 0.05$
5	0.112	0.53	1.47	$*P < 0.01$
6	0.137	1.97	18.76	$*P < 0.001$
7	0.175	0.62	3.55	$*P < 0.001$
8	0.112	0.34	0.90	$*P < 0.05$

*Egg-sticks more clumped than expected for random oviposition.
†Egg-sticks not distributed differently from random.

protagonists can appear to play 'hide-and-seek'. The most famous example is the experimental work of Huffaker (Huffaker, 1958; Huffaker *et al.*, 1963), who studied a system in which the predatory mite *Typhlodromus occidentalis* fed on the herbivorous mite *Eotetranychus sexmaculatus*, which fed on oranges interspersed amongst rubber balls in a tray. In the absence of its predator, *Eotetranychus* maintained a fluctuating but persistent population (Figure 9.13a); but if *Typhlodromus* was added during the early stages of prey population growth, it rapidly increased its own population size, consumed all of its prey and then became extinct itself (Figure 9.13b).

The interaction was altered, however, when Huffaker made his microcosm more 'patchy'. He spread the oranges further apart, and partially isolated each one by placing a complex arrangement of vaseline barriers in the tray, which the mites could not cross. But he facilitated the dispersal of *Eotetranychus* by inserting a number of upright sticks from which they could launch themselves on silken strands carried by air currents. Dispersal between patches was therefore much easier for prey than it was for predators. In a patch occupied by both *Eotetranychus* and *Typhlodromus*, the predators consumed all the prey and then either became extinct themselves or dispersed (with a low rate of success) to a new patch. But in patches occupied by prey alone, there was rapid, unhampered growth accompanied by

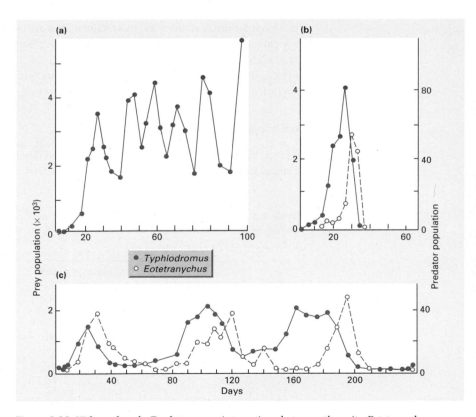

Figure 9.13 Hide-and-seek. Predator–prey interactions between the mite *Eotetranychus sexmaculatus* and its predator, the mite *Typhlodromus occidentalis*. (a) Population fluctuations of *Eotetranychus* without its predator. (b) A single oscillation of the predator and prey in a simple system. (c) Sustained oscillations in a more complex system. (After Huffaker, 1958.)

successful dispersal to new patches. In a patch occupied by predators alone, there was usually death of the predators before their food arrived. Each patch was therefore ultimately doomed to the extinction of both predators and prey; but overall, at any one time, there was a mosaic of unoccupied patches, prey–predator patches heading for extinction and thriving prey patches; and this mosaic was capable of maintaining persistent populations of both predators and prey (Figure 9.13c).

aggregation, hide-and-seek and stability

A similar example, from a natural population, is provided by work off the coast of southern California on the predation by starfish of clumps of mussels Murdoch & Stewart-Oaten, 1975). Clumps that are heavily preyed upon are liable to be dislodged by heavy seas so that the mussels die; the starfish are continually driving patches of their mussel prey to extinction. The mussels, however, have planktonic larvae which are continually colonizing new locations and initiating new clumps, whereas the starfish disperse much less readily. They aggregate at the larger clumps, but there is a time lag before they leave an area when the food is gone. Thus, patches of mussels are continually becoming extinct, but other clumps are growing prior to the arrival of the starfish. As with Huffaker's mites, the combination of patchiness, the aggregation of predators in particular patches and a lack of synchrony between the behaviour of different patches appears capable of stabilizing the dynamics of a predator–prey interaction.

9.8 Behaviour that leads to aggregated distributions

There are various types of behaviour underlying the aggregative responses of consumers (and their responses to patches generally), but they fall into two broad categories: those involved with the location of profitable patches, and the responses of consumers once within a patch. The first category includes all examples in which consumers perceive, at a distance, the existence of heterogeneity in the distribution of their prey. For instance, the parasitoid *Callaspidia defonscolombei* is attracted to concentrations of its syrphid (fly larva) host by the odours produced by the syrphid's own prey—various species of aphid (Rotheray, 1979).

patch location

Within the second category—responses of consumers within patches—there are two main aspects of behaviour. The first is a change in the consumer's pattern of searching in response to encounters with items of food. In particular, there is often a slowing down of movement and an increased rate of turning immediately following the intake of food, both of which lead to the consumer remaining in the vicinity of its last food item ('area-restricted search'). Consumers therefore tend to remain in high-density patches of food (where the high rate of encounter leads to a slowing down and a high turning rate), and to leave low-density patches (where they move faster and turn less): Figure 9.14 illustrates this sort of behaviour for birds feeding on a lawn. Alternatively, consumers may simply abandon unprofitable patches more rapidly than they abandon profitable ones. This is what happens when the carnivorous, net-spinning larva of the caddis-fly *Plectrocnemia conspersa* feeds on chironomid larvae in a laboratory stream. Caddis in their nets were provided with one prey item at the beginning of the experiment and then fed daily rations of zero, one or three prey. The tendency to abandon the net was lowest at the high feeding rates (Hildrew & Townsend, 1980; Townsend & Hildrew, 1980).

area-restricted search

Plectrocnemia's behaviour in relation to prey patches also has an element of area-restricted search; the likelihood that it will spin a net in the first place depends

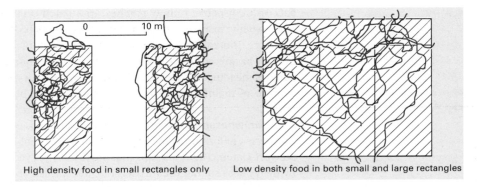

High density food in small rectangles only   Low density food in both small and large rectangles

Figure 9.14 The search paths of thrushes (the song thrush, *Turdus philomelos*, and the European blackbird, *T. merula*). The birds tend to remain within areas of high prey density (left), because in these, compared with low-density areas (right), they turn more often and more sharply. (After Smith, 1974.)

on whether it happens to encounter a food item (which it can consume even without a net). Figure 9.15a shows that larvae that have fed begin net-building immediately, whereas unfed larvae continue wandering and are more likely to leave the patch. Overall, therefore, a net is more likely to be constructed, and less likely to be abandoned, in a rich patch. These two behaviours account for a directly density dependent aggregative response in the natural stream environment observed for much of the year (Figure 9.15b).

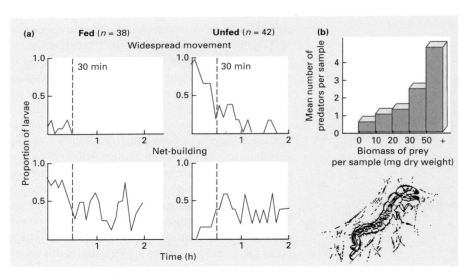

Figure 9.15 (a) On arrival in a patch, fifth-instar *Plectrocnemia conspersa* larvae that encounter and eat a chironomid prey item at the beginning of the experiment ('fed') quickly cease wandering and commence net-building. Predators that fail to encounter a prey item ('unfed') exhibit much more widespread movement during the first 30 min of the experiment, and are significantly more likely to move out of the patch. (b) Directly density dependent aggregative response of fifth-instar larvae in a natural environment expressed as mean number of predators against combined biomass of chironomid and stonefly prey per 0.0625 m^2 sample of streambed (*n* = 40). (After Hildrew & Townsend, 1980; Townsend & Hildrew, 1980.)

The difference in the rates of abandonment of patches of high and low profitability can be achieved in a number of ways, but two are especially easy to envisage. A consumer might leave a patch when its feeding rate drops below a threshold level, or a consumer might have a giving-up time—it might abandon a patch whenever a particular time interval passes without the successful capture of food. Whichever mechanism is used, or indeed if the consumer simply uses area-restricted search, the consequences will be the same: individuals will spend longer in more profitable patches, and these patches will therefore contain more consumers (but see Section 9.10).

9.9 The optimal foraging approach to patch use

The advantages to a consumer of spending more time in higher profitability patches are easy to see. However, the detailed allocation of time to different patches is a subtle problem, since it depends on the precise differentials in profitability, the average profitability of the environment as a whole, the distance between patches, and so on. The problem has been a particular focus of attention for optimal foraging theory. In particular, a great deal of interest has been directed at the very common situation in which foragers themselves deplete the resources of a patch, causing its profitability to decline with time. Amongst the many examples of this are insectivorous insects removing prey from a leaf, and bees consuming nectar from a flower.

9.9.1 The marginal value theorem

Charnov (1976b) and Parker and Stuart (1976) produced similar models to predict the behaviour of an optimal forager in such situations. They found that the optimal stay-time in a patch should be defined in terms of the rate of energy extraction experienced by the forager at the moment it leaves a patch (the 'marginal value' of the patch). Charnov called the results the 'marginal value theorem'. The models were formulated mathematically, but their salient features are shown in graphical form in Figure 9.16.

The primary assumption of the model is that an optimal forager will maximize its overall intake of a resource (usually energy) during a bout of foraging, taken as a whole. Energy will, in fact, be extracted in bursts because the food is distributed patchily; the forager will sometimes move between patches, during which time its intake of energy will be zero. But once in a patch, the forager will extract energy in a manner described by the curves in Figure 9.16a. Its initial rate of extraction will be high; but as time progresses and the resources are depleted, the rate of extraction will steadily decline. Of course, the rate will itself depend on the initial contents of the patch (i.e. its productivity or profitability) and on the forager's efficiency and motivation (Figure 9.16a).

The problem under consideration is this: at what point should a forager leave a patch? If it left all patches immediately after reaching them, then it would spend most of its time travelling between patches, and its overall rate of intake would be low. If it stayed in all patches for considerable lengths of time, then it would spend little time travelling, but it would spend extended periods in depleted patches, and its overall rate of intake would again be low. Some intermediate stay-time is therefore optimal. In addition, though, the optimal stay-time must clearly be greater for

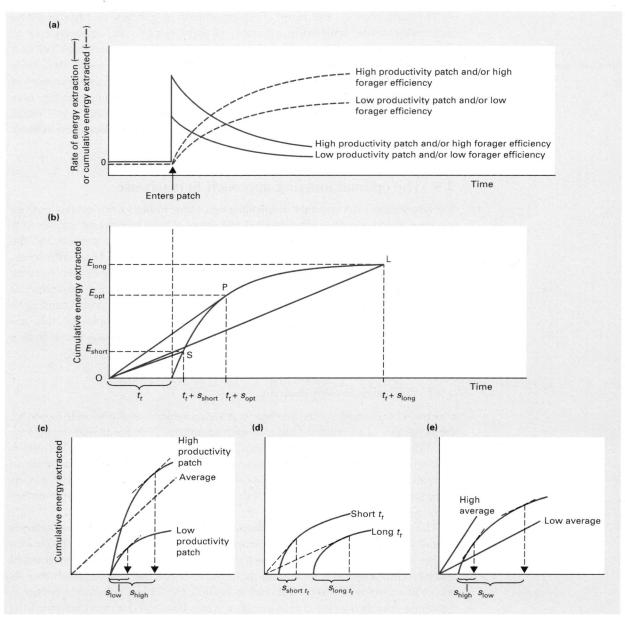

Figure 9.16 The marginal value theorem. (a) When a forager enters a patch, its rate of energy extraction is initially high (especially in a highly productive patch or where the forager has a high foraging efficiency), but this rate declines with time as the patch becomes depleted. The cumulative energy intake approaches an asymptote. (b) The options for a forager. The solid curve is cumulative energy extracted from an average patch, and t_t is the average travelling time between patches. The rate of energy extraction (which should be maximized) is energy extracted divided by total time, i.e. the slope of a straight line from the origin to the curve. Short stays in the patch (slope = $E_{short} / (t_t + s_{short})$) and

long stays (slope = $E_{long} / (t_t + s_{long})$) both have lower rates of energy extraction (shallower slopes) than a stay (s_{opt}) which leads to a line just tangential to the curve. s_{opt} is therefore the optimum stay-time, giving the maximum overall rate of energy extraction. *All* patches should be abandoned at the *same* rate of energy extraction (the slope of the line OP). (c) Low productivity patches should be abandoned after shorter stays than high productivity patches. (d) Patches should be abandoned more quickly when travelling time is short than when it is long. (e) Patches should be abandoned more quickly when the average overall productivity is high than when it is low.

360   CHAPTER 9

profitable patches than for unprofitable ones, and it must depend on the profitability of the environment as a whole.

Consider, in particular, the forager in Figure 9.16b. It is foraging in an environment where food is distributed patchily and where some patches are more valuable than others. The average travelling time between patches is t_t. This is therefore the length of time the forager can expect to spend on average after leaving one patch before it finds another. The forager in Figure 9.16b has arrived at an average patch for its particular environment, and it therefore follows an average extraction curve. In order to forage optimally it must maximize its rate of energy intake not merely for its period in the patch, but for the whole period since its departure from the last patch (i.e. for the period $t_t + s$, where s is the stay-time in the patch).

If it leaves the patch rapidly then this period will be short ($t_t + s_{short}$ in Figure 9.16b). But by the same token, little energy will be extracted (E_{short}). The rate of extraction (for the whole period $t_t + s$) will be given by the slope of the line OS (i.e. $E_{short} / (t_t + s_{short})$). On the other hand, if the forager remains for a long period (s_{long}) then far more energy will be extracted (E_{long}); but, the overall rate of extraction (the slope of OL) will be little changed. To maximize the rate of extraction over the period $t_t + s$, it is necessary to maximize the slope of the line from O to the extraction curve. This is achieved simply by making the line a tangent to the curve (OP in Figure 9.16b). No line from O to the curve can be steeper, and the stay-time associated with it is therefore optimal (s_{opt}).

The optimal solution for the forager in Figure 9.16b, therefore, is to leave that patch when its extraction rate is equal to (tangential to) the slope of OP, i.e. it should leave at point P. In fact, Charnov, and Parker and Stuart, found that the optimal solution for the forager is to leave all patches, irrespective of their profitability, at the same extraction rate (i.e. the same 'marginal value'). This extraction rate is given by the slope of the tangent to the average extraction curve (e.g. in Figure 9.16b), and it is therefore the maximum average overall rate for that environment as a whole.

The model therefore confirms that the optimal stay-time should be greater in more productive patches than in less productive patches (Figure 9.16c). Moreover, for the least productive patches (where the extraction rate is never as high as OP) the stay-time should be zero. The model also predicts that all patches should be depleted such that the final extraction rate from each is the same (i.e. the 'marginal value' of each is the same); and it predicts that stay-times should be longer in environments where the travelling time between patches is longer (Figure 9.16d), and that stay-times should be longer where the environment as a whole is less profitable (Figure 9.16e).

9.9.2 Experimental tests of the marginal value theorem

Encouragingly, there is evidence from a number of cases that lends support to the marginal value theorem. For instance, Cowie (1977) has tested quantitatively the prediction set out in Figure 9.16d: that a forager should spend longer in each patch when the travelling time is longer. He used captive great tits in a large indoor aviary, and got the birds to forage for small pieces of mealworm hidden in sawdust-filled plastic cups—the cups were 'patches'. All patches on all occasions contained the same number of prey; but travelling time was manipulated by covering the cups with cardboard lids that varied in their tightness and therefore varied in the time needed

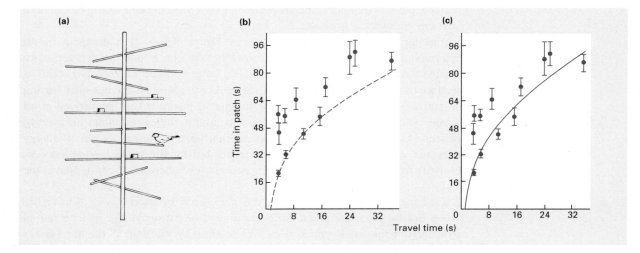

Figure 9.17 (a) An experimental 'tree' for great tits, with three patches. (b) Predicted optimal time in a patch plotted against travelling time (– – –), together with the observed mean points (± standard error) for six birds, each in two environments. (c) The same data points, and the predicted time taking into account the energetic costs of travelling between patches. (After Cowie, 1977; from Krebs, 1978.)

Cowie's experiments with great tits

to prize them off. Birds foraged alone, and Cowie used six in all, subjecting each to two habitats. One of these habitats always had longer travelling times (tighter lids) than the other.

For each bird in each habitat (12 in all) Cowie measured the average travelling time and the curve of cumulative food intake within a patch. He then used the marginal value theorem to predict the optimal stay-time in habitats with different travelling times, and compared these predictions with the stay-times he actually observed. As Figure 9.17 shows, the correspondence was quite close. It was closer still when he took account of the fact that there was a net loss of energy when the birds were travelling between patches.

Hubbard and Cook's experiments with *Venturia*

The predictions of the marginal value theorem have also been examined using the ichneumonid parasitoid *Venturia canescens*, attacking the flour moth *Ephestia cautella* (Hubbard & Cook, 1978). Individual *Venturia* searched in an arena containing patches of prey concealed in plastic dishes covered with wheat bran, and patches contained varying numbers of prey. Hubbard and Cook made a number of observations broadly consistent with the marginal value theorem, amongst which was the one shown in Figure 9.18. This indicates that the encounter rate with unparasitized hosts at the end of the experiment (the marginal value) was very similar in patches of different initial density. By contrast, the initial encounter rate was, as expected, higher in the more profitable patches. The *Venturia* appeared to deplete the patches until all profitabilities were similar.

9.9.3 Mechanistic explanations for 'marginal value behaviour'

Much fuller reviews of the tests of the marginal value theorem are provided by Townsend and Hughes (1981), Krebs and McCleery (1984), Stephens and Krebs (1986) and Krebs and Kacelnik (1991). The picture these convey is one of

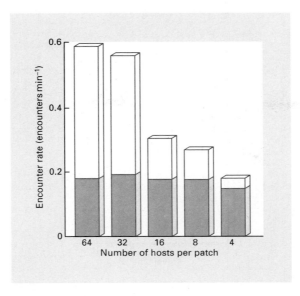

Figure 9.18 The estimated terminal encounter rates of *Venturia canescens* with its host in patches of different host density (shaded areas) compared with the initial rates in these patches (open areas). (After Hubbard & Cook, 1978.)

encouraging but not perfect correspondence. The main reason for the imperfection is that the animals, unlike the modellers, are not omniscient. They may need to spend time learning about and sampling their environment, and they may need to spend time doing things other than foraging (e.g. looking for predators). Nevertheless, they often seem to come remarkably close to the predicted strategy. Ollason (1980) has developed a mechanistic model to account for this in the great tits studied by Cowie, whilst Waage (1979) has developed a mechanistic model for the foraging behaviour of *Venturia*.

Ollason's mechanistic model for the great tit

Ollason's is a memory model. It assumes that an animal has a 'remembrance of past food' which Ollason likens to a bath of water without a plug. Fresh remembrance flows in every time the animal feeds. But remembrance is also draining away continuously. The rate of input depends on the animal's feeding efficiency and the productivity of the current feeding area. The rate of out-flow depends on the animal's ability to memorize and the amount of remembrance. Remembrance drains away quickly, for example, when the amount is large (high water level) or the memorizing ability is poor (tall, narrow bath). Ollason's model simply proposes that an animal should stay in a patch until remembrance ceases to rise; an animal should leave a patch when its rate of input from feeding is slower than its rate of declining remembrance.

An animal foraging consistently with Ollason's model behaves in a way very similar to that predicted by the marginal value theorem. This is shown for the case of Cowie's great tits in Figure 9.19 (and Ollason argues that the discrepancy can be explained in terms of Cowie's experimental procedure). As Ollason himself remarks, this shows that to forage in a patchy environment in a way that approximates closely to optimality, an animal need not be omniscient, it does not need to sample and it does not need to perform numerical analyses to find the maxima of functions of many variables. All it needs to do is to remember, and to leave each patch if it is not feeding as fast as it remembers doing (although, as Krebs & Davies, 1993, point out, this is no more surprising than the observation that the same birds can fly without any formal qualification in aerodynamics).

363   BEHAVIOUR OF PREDATORS

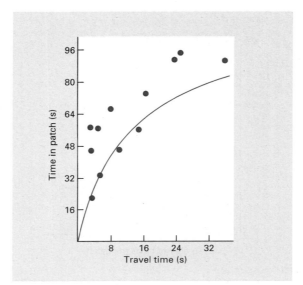

Figure 9.19 Cowie's (1977) great tit data (see Figure 9.17) compared to the predictions of Ollason's (1980) mechanistic memory model.

Waage's mechanistic model for *Venturia*

Waage's (1979) model for *Venturia* is based very largely on his own and others' observations of the parasitoid's behaviour. Host larvae produce a contact chemical that elicits a sharp U-turn by *Venturia* when contact with hosts is lost (i.e. when the edge of the patch is reached). This effect is enhanced by increases in the concentration of the chemical (i.e. with increases in host density). However, *Venturia* gradually becomes habituated to this chemical and will eventually leave a patch (no U-turn) despite the loss of contact. A successful oviposition in a patch, on the other hand, temporarily reverses the effects of habituation. Frequent oviposition will therefore prolong the stay-time in a patch. In addition, it seems that low oviposition rates over several days may increase responsiveness to patch stimuli, thereby increasing average stay-time in low-productivity habitats. Waage argues that these patterns of behaviour, taken together, can account for the observed foraging pattern of the parasitoids.

For both the great tit and *Venturia*, therefore, optimal foraging and mechanistic models are seen to be compatible and complementary in explaining how a predator has achieved its observed foraging pattern, and why that pattern has been favoured by natural selection. Between them they provide grounds for hoping that a comprehensive understanding of patch use by foragers may emerge.

optimal foraging in plants

Finally, although this is not strictly an aspect of predation, the principles of optimal foraging are now being applied to investigations of the foraging strategies of plants for nutrients (reviewed by Hutchings & de Kroon, 1994). When does it pay to produce long stolons moving rapidly from patch to patch? When does it pay to concentrate root growth within a limited volume, foraging from a patch until it is close to depletion? Certainly, it is good to see such intellectual cross-fertilization across the taxonomic divide.

9.10 Ideal free and related distributions: aggregation and interference

We can see, then, that consumers tend to aggregate in profitable (often high prey density) patches where their expected rate of food consumption is highest. Yet we

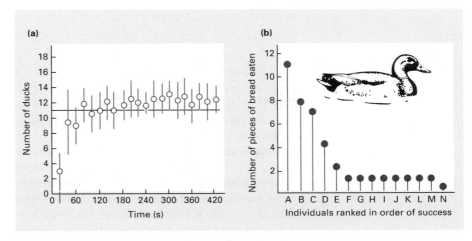

Figure 9.20 (a) When 33 ducks were fed pieces of bread at two stations around a pond (with a profitability ratio of 2 : 1), the number of ducks at the poorer station rapidly approached one-third the total, in apparent conformity with the predictions of ideal free theory. (b) However, contrary to the assumptions and other predictions of simple theory, the ducks were not all equal. (After Harper, 1982; from Milinski & Parker, 1991.)

have also seen (see Section 9.6) that consumers tend to compete and interfere with one another, thereby reducing their per capita consumption rate. It follows from this that patches that are initially most profitable become immediately less profitable because they attract most consumers. We might therefore expect the consumers to redistribute themselves, and it is perhaps not surprising that the observed patterns of predator distributions across prey patches (e.g. Figures 9.11 and 9.12) are varied. But can we make some sense of this variation in pattern?

In an early attempt to do so, it was proposed that if a consumer forages optimally, the process of redistribution will continue until the profitabilities of all patches are equal (Fretwell & Lucas, 1970; Parker, 1970). This will happen because as long as there are dissimilar profitabilities, consumers should leave less profitable patches and be attracted to more profitable ones. Fretwell and Lucas called the consequent
distribution the ideal free distribution: the consumers are 'ideal' in their judgement of profitability, and 'free' to move from patch to patch. Consumers were also assumed to be equal. Hence, with an ideal free distribution, because all patches come to have the same profitability, all consumers have the same consumption rate. There are some simple cases where consumers appear to conform to an ideal free distribution insofar as they distribute themselves in proportion to the profitabilities of different patches (e.g. Figure 9.20a), but even in such cases, one of the underlying assumptions is likely to have been violated (e.g. Figure 9.20b—all consumers are not equal).

The early ideas have been much modified taking account, for example, of unequal competitors (see Milinski & Parker, 1991; Tregenza, 1995, for reviews). In particular, ideal free theory was put in a more ecological context by Sutherland (1983) who explicitly incorporated predator handling times and mutual interference amongst the predators. He found that predators should be distributed such that the proportion of predators in site i, p_i, is related to the proportion of prey (or hosts) in site i, h_i, by the equation:

the ideal free distribution: a balance between attractive and repellent forces

incorporating a range of interference coefficients

$$p_i = k(h_i^{1/m}) \tag{9.6}$$

where m is the coefficient of interference (see Section 9.6), and k is a 'normalizing constant' such that the proportions, p_i, add up to 1. It is now possible to see how the patch to patch distribution of predators might be determined jointly by interference and the selection by the predators of intrinsically profitable patches.

If there is no interference amongst the predators, then $m = 0$. All should exploit only the patch with the highest prey density (Figure 9.21), leaving lower density patches devoid of predators.

If there is a small or moderate amount of interference (i.e. $m > 0$, but $m < 1$—the range, as already noted, that is most realistic biologically), then high-density prey patches should still attract a disproportionate number of predators (Figure 9.21). In other words, there should be an aggregative response by the predators, which is not only directly density dependent, but actually accelerates with increasing prey density in a patch (like, for example, the data in Figure 9.11a). Hence, the prey's risk of predation might itself be expected to be directly density dependent: the greatest risk of predation in the highest prey density patches (like the examples in Figures 9.12a, c).

With a little more interference—$m \approx 1$—the proportion of the predator population in a patch should still increase with the proportion of prey, but now it should do so more or less linearly rather than accelerating, such that the ratio of predators : prey is roughly the same in all patches (Figure 9.21, and, for example, Figure 9.11b). Here, therefore, the risk of predation might be expected to be the same in all patches, and hence independent of prey density (like the examples in Figures 9.12d, e).

Finally, with a great deal of interference—$m > 1$—the highest density prey patches should have the lowest ratio of predators : prey (Figure 9.21). The risk of predation might therefore be expected to be greatest in the lowest prey density patches, and hence inversely density dependent (like the data in Figure 9.12b).

It is clear, therefore, that the range of patterns amongst the data in Figures 9.11 and 9.12 reflects a shifting balance between the forces of attraction and of repulsion. Predators are attracted to highly profitable patches; but they are repelled by the presence of other predators that have been attracted in the same way.

This description, however, of the relationship between the distribution of predators and the distribution of predation risk has been peppered with 'might be expected to's. The reason is that the relationship also depends on a range of factors not so far considered. For example, Figure 9.22 shows a case where the parasitoid *Trichogramma pretiosum* aggregates in high-density patches of its moth host, but the

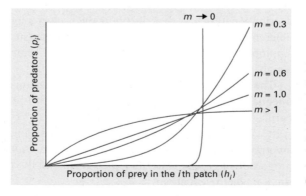

Figure 9.21 The effect of the interference coefficient, m, on the expected distribution of predators amongst patches of prey varying in the proportion of the total prey population they contain (and hence, in their 'intrinsic' profitability). (After Sutherland, 1983.)

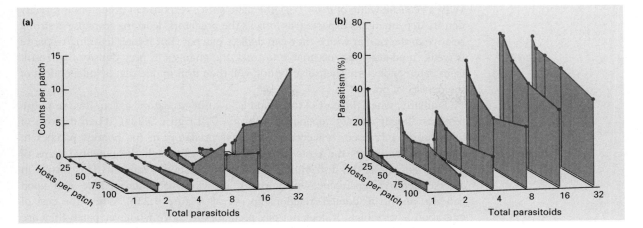

Figure 9.22 (a) The aggregative response of the egg parasitoid, *Trichogramma pretiosum*, which aggregates on patches with high densities of its host *Plodia interpunctella*. (b) The resultant distribution of ill-effects: hosts on high-density patches are least likely to be parasitized. (After Hassell, 1982.)

pseudo-interference

learning and migration

risk of parasitism to the moth is greatest in low-host-density patches. The explanation probably lies in time wasted by parasitoids in high-density host patches, dealing with already parasitized hosts that may still attract parasitoids because they are not physically removed from a patch (unlike preyed-upon prey) (Morrison & Strong, 1981; Hassell, 1982). Thus, earlier parasitoids in a patch may interfere indirectly with later arrivals, in that the previous presence of a parasitoid in a patch may reduce the effective rate at which later arrivals attack unparasitized hosts. This effect has been termed 'pseudo-interference' (Free *et al.*, 1977); its potentially important effects on population dynamics are discussed in Chapter 10, Section 10.5.3.

Expected patterns are modified further still if we incorporate learning by the predators, or the costs of migration between patches (Bernstein *et al.*, 1988, 1991).

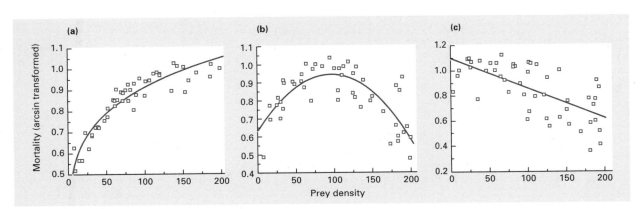

Figure 9.23 The effect of a cost to migration in predators distributing themselves across prey patches in a simulation model. The interference coefficient, *m*, is 0.3 and would lead to direct density dependence in the absence of a migration cost. (a) Low migration cost: direct density dependence maintained. (b) Intermediate cost: a 'domed' relationship. (c) High cost: inverse density dependence. (After Bernstein *et al.*, 1991.)

367   BEHAVIOUR OF PREDATORS

With a realistic value of m (= 0.3), the aggregative response of predators is directly density dependent (as expected) as long as the predators' learning response is strong relative to the rate at which they can deplete patches. But if their learning response is weak, predators may be unable to track the changes in prey density that result from patch depletion. Their distribution will then drift to one that is independent of the density of prey.

Similarly, when the cost of migration is low, the predators' aggregative response remains directly density dependent (with m = 0.3; Figure 9.23a). When the cost of migration is increased, however, it still pays predators in the poorest patches to move, but for others the costs of migration can outweigh the potential gains of moving. For these, the distribution across prey patches is random. This results in inverse density dependence in mortality rate between intermediate and good patches, and in a 'domed' relationship overall (Figure 9.23b). When the cost of migration is very high, it does not pay predators to move whatever patch they are in—mortality is inversely density dependent across all patches (Figure 9.23c).

Clearly, there is no shortage of potential causes for the wide range of types of distributions of predators, and of mortality rates, across prey patches (see Figures 9.11, 9.12 and 9.22). Their consequences, in terms of population dynamics, are one of the topics dealt with in the chapter that follows. This highlights the crucial importance of forging links between behavioural and population ecology.

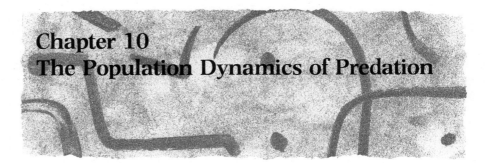

Chapter 10
The Population Dynamics of Predation

10.1 Introduction: patterns of abundance and the need for their explanation

We turn now to the effects of predation on the population dynamics of the predator and its prey, where even a limited survey of the data reveals a varied array of patterns. There are certainly cases where predation has a profoundly detrimental effect on a prey population. For example, the 'vedalia' ladybird beetle (*Rodolia cardinalis*) is famous for having virtually eradicated the cottony cushion-scale insect (*Icerya purchasi*), a pest that threatened the California citrus industry in the late 1880s (DeBach, 1964). On the other hand, there are many cases where predators and herbivores have no apparent effect on their prey's dynamics or abundance. For example, the weevil *Apion ulicis* was introduced into New Zealand in an attempt to control the abundance of gorse bushes (*Ulex europaeus*), and it has become one of the most abundant insects in New Zealand. Yet, despite eating up to 95% of the gorse seeds every year, it has had no appreciable impact on the numbers of the plant (Miller, 1970).

There are also many examples in which a predator retains a fairly constant density in spite of fluctuations in the abundance of its prey (tawny owls and small mammals in Figure 10.1a); and there are cases in which a predator or herbivore population tracks the abundance of its prey, although the prey itself varies in density as a result of some other factor (cinnabar moth larvae and ragwort plants in Figure 10.1b). There are studies, too, that appear to show predator and prey populations linked together by coupled oscillations in abundance (Figure 10.1c). Finally, there are many examples in which predator and prey populations fluctuate in abundance apparently independently of one another.

It is clearly a major task for ecologists to develop an understanding of the patterns of predator–prey abundance, and to account for the differences from one example to the next. It is equally clear, though, that none of these predator and prey populations exist as isolated pairs, but rather as parts of multispecies systems, and that all these species are affected by environmental conditions. These more complex problems are addressed in Chapter 15. However, as with any complex process in science, we are unlikely to be able to understand the full complexity without a reasonable understanding of the sensibly identified, conceptually isolated components—in this case, populations of predators and prey. Hence, this chapter deals with the consequences of predator–prey interactions for the dynamics of the populations concerned.

The approach will be firstly to use simple models to deduce the effects produced

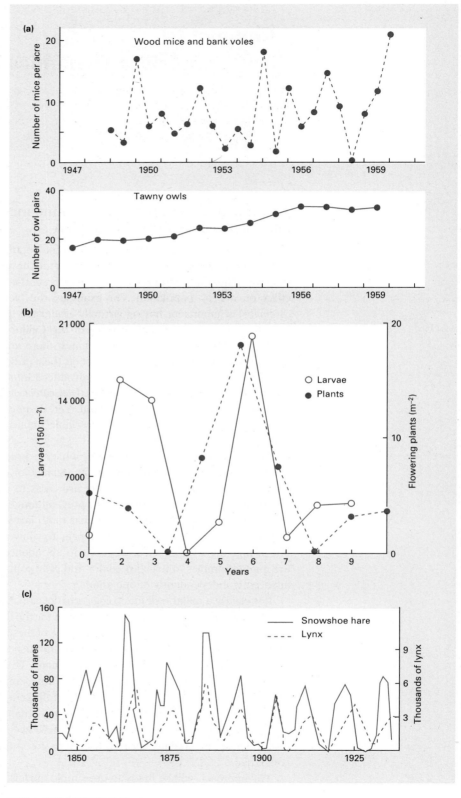

by different components of the interactions. (Models are, indeed, normally indispensable in teasing out the separate effects of components in a complex process, before seeking to understand their effects in combination.) Then, in each case, field and experimental data will be examined to see whether the deductions appear to be supported or refuted. In fact, simple models are most useful when their predictions are *not* supported by real data—as long as the reason for the discrepancy can subsequently be discovered. Confirmation of a model's predictions constitutes consolidation; refutation with subsequent explanation constitutes progress.

10.2 The basic dynamics of predator–prey and plant–herbivore systems: a tendency towards cycles

There have been two main series of models developed as attempts to understand predator–prey dynamics. Both will be examined here. The first (see Section 10.2.1) is based on differential equations (and hence, applies most readily to populations in which breeding is continuous), but relies heavily on simple graphical models (Rosenzweig & MacArthur, 1963). The second (see Section 10.2.2) uses difference equations to model host–parasitoid interactions with discrete generations. Despite this taxonomic limitation, these models have the advantage of having been subject to rigorous mathematical exploration (and, as we noted in Chapter 8, Section 8.1, there are many important parasitoid species).

10.2.1 The Lotka–Volterra model

The simplest differential equation model is known (like the model of interspecific competition) by the name of its originators: Lotka–Volterra (Volterra, 1926; Lotka, 1932). This will serve as a useful point of departure. The model has two components: P, the numbers present in a predator (or consumer) population, and N, the numbers or biomass present in a prey or plant population.

It can be assumed initially that in the absence of consumers the prey population increases exponentially:

$$\frac{dN}{dt} = rN. \tag{10.1}$$

But prey individuals are removed by predators, and this occurs at a rate that depends on the frequency of predator–prey encounters. Encounters will increase as the numbers of predators (P) increase and as the numbers of prey (N) increase. However, the exact number encountered and successfully consumed will depend on the searching and attacking efficiencies of the predator, i.e. on a', the 'searching

---

Figure 10.1 *(facing page)* The various patterns of predator–prey abundance. (a) Tawny owls (*Strix aluco*) near Oxford, England, maintain a constancy of abundance despite fluctuations in the numbers of their small mammal prey. (After Southern, 1970.) (b) The number of cinnabar moth larvae (*Tyria jacobaeae*) in 1 year at an East Anglian site is determined largely by the number of flowering ragwort plants (*Senecio jacobaea*) in the previous year—but changes in plant abundance are due mainly to changes in germination conditions. The insect is food limited, but the plant is not herbivore limited. (After Dempster & Lakhani, 1979; from Crawley, 1983.). (c) The apparently coupled oscillations in abundance of the snowshoe hare (*Lepus americanus*) and Canada lynx (*Lynx canadensis*) as determined from numbers of pelts lodged with the Hudson Bay Company. (After MacLulick, 1937.)

efficiency' or 'attack rate' (see Chapter 9). The frequency of 'successful' predator–prey encounters, and hence, the consumption rate of prey, will thus be $a'PN$, and overall:

the Lotka–Volterra prey equation

$$\frac{dN}{dt} = rN - a'PN.$$ (10.2)

As we saw in Chapter 8, in the absence of food, individual predators lose weight and starve to death. Thus, in the model, predator numbers are assumed to decline exponentially through starvation in the absence of prey:

$$\frac{dP}{dt} = -qP$$ (10.3)

where q is their mortality rate. This is counteracted by predator birth, the rate of which is assumed to depend on only two things: (i) the rate at which food is consumed, $a'PN$; and (ii) the predator's efficiency, f, at turning this food into predator offspring. Predator birth rate is therefore $fa'PN$, and overall:

the Lotka–Volterra predator equation

$$\frac{dP}{dt} = fa'PN - qP.$$ (10.4)

Equations 10.2 and 10.4 constitute the Lotka–Volterra model.

The properties of this model can be investigated by finding zero isoclines. Zero isoclines were described for models of two-species competition on pp. 274–277. They are lines along which a population just maintains itself, neither increasing nor decreasing. Here, there are separate predator and prey zero isoclines, both of which are drawn on a graph of prey density (x-axis) against predator density (y-axis). Each is a line joining those combinations of predator and prey density that lead either to an unchanging prey population ($dN / dt = 0$; prey zero isocline) or an unchanging predator population ($dP / dt = 0$; predator zero isocline). Having drawn, say, a prey zero isocline, we know that combinations to one side of it lead to prey decrease, and combinations to the other to prey increase. Thus, as we shall see, if we plot the prey and predator zero isoclines on the same figure, we can begin to determine the pattern of dynamics of the joint predator–prey populations.

In the case of the prey (Equation 10.2), when:

the properties of the model are revealed by zero isoclines

$$\frac{dN}{dt} = 0, \quad rN = a'PN$$ (10.5)

or:

$$P = \frac{r}{a'}.$$ (10.6)

Thus, since r and a' are constants, the prey zero isocline is a line for which P itself is a constant (Figure 10.2a).

Likewise, for the predators (Equation 10.4), when:

$$\frac{dP}{dt} = 0, \quad fa'PN = qP$$ (10.7)

or:

$$N = \frac{q}{fa'}.$$ (10.8)

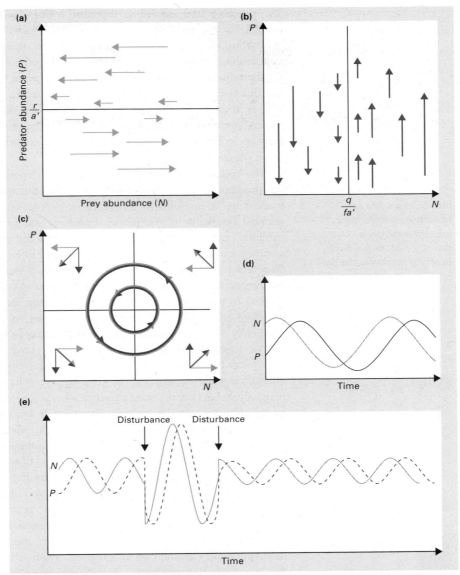

Figure 10.2 The Lotka–Volterra predator–prey model. (a) The prey zero isocline, with prey (N) increasing in abundance (arrows left to right) at lower predator densities (low P) and decreasing at higher predator densities. (b) The predator zero isocline, with predators increasing in abundance (arrows pointing upwards) at higher prey densities and decreasing at lower prey densities. (c) When the zero isoclines are combined, the arrows can also be combined, and these joint arrows progress in anticlockwise circles. In other words, the joint population moves with time from low predator/low prey (bottom left in (c)), to low predator/high prey (bottom right), to high predator/high prey, to high predator/low prey and back to low predator/low prey. Note, however, that the lowest prey abundance ('9 o'clock') comes one-quarter-cycle before the lowest predator abundance ('6 o'clock'—anticlockwise movement). These coupled cycles of predator–prey abundance, continuing indefinitely, are shown as numbers against time in (d). However, as shown in (e), these cycles exhibit neutral stability: they continue indefinitely if undisturbed, but each disturbance to a new abundance initiates a new, different series of neutrally stable cycles, around the same means but with a different amplitude.

The predator zero isocline is therefore a line along which N is constant (Figure 10.2b).

Putting the two isoclines together (Figure 10.2c) shows the behaviour of joint populations: they undergo coupled oscillations in abundance, which continue indefinitely. Predators increase in abundance when there are large numbers of prey. But this leads to an increased predation pressure on the prey, and thus to a decrease in prey abundance. This then leads to a food shortage for predators and a decrease in predator abundance, which leads to a relaxation of predation pressure and an increase in prey abundance, which leads to an increase in predator abundance, and so on (Figure 10.2d).

the model reveals an
underlying tendency for
coupled oscillations—but
actually exhibits indefinite,
neutrally stable fluctuations

The Lotka–Volterra model, then, is useful in pointing to an underlying tendency for predator–prey interactions to generate fluctuations in the prey population tracked by fluctuations in the predator population (i.e. coupled oscillations). The detailed behaviour of the model, however, should not be taken too seriously. It exhibits 'neutral stability', which means that the populations follow precisely the same cycles indefinitely unless some external influence shifts them to new values, after which they follow new cycles indefinitely (Figure 10.2e). In practice, of course, environments are continually changing, and populations would continually be 'shifted to new values'. A population following the Lotka–Volterra model would, therefore, not exhibit regular cycles, but, because of repeated disturbance, fluctuate erratically. No sooner would it start one cycle than it would be diverted to a new one.

For a population to exhibit regular and recognizable cycles, the cycles must themselves be stable: when an external influence changes the population level, there must be a tendency to return to the original cycle. Such cycles, in contrast to the neutrally stable Lotka–Volterra fluctuations, are known as stable limit cycles. In fact, as we shall see, predator–prey models (once we move beyond the very limiting assumptions of Lotka–Volterra) are capable of generating a whole range of abundance patterns: stable-point equilibria, multigeneration cycles, one-generation cycles, chaos, etc.—a range repeated in surveys of real populations. The challenge is to discover what light the models can throw on the real populations' behaviour.

time delays

The basic mechanism generating the coupled oscillations is a series of repeated time delays. The first is between 'many prey' and 'many predators' (a time delay indicating that predator abundance cannot respond instantaneously to prey abundance). However, the increase in predator abundance necessarily leads to a reduction in prey abundance to 'few prey'. There is then another time delay between 'few prey' and 'few predators', and then between 'few predators' and 'many prey', and so on. Broadly speaking, therefore, if we imagine that the cycle is characterized by a series of time delays all of at least approximately the same duration, then the total length of the cycle is four times the length of the time delay. The same general picture emerges from a wide range of broadly appropriate equations (May, 1981a).

10.2.2 The Nicholson–Bailey model

Turning now to parasitoids, the basic model (Nicholson & Bailey, 1935) is again not so much realistic as a reasonable basis from which to start. Let H_t be the number of hosts, and P_t the number of parasitoids in generation t; r is the intrinsic rate of natural increase of the host (see Chapter 4, Section 4.7). If H_a is the number of hosts attacked by parasitoids (in generation t), then, assuming no intraspecific competition, and that each host can support only one parasitoid (commonly the case):

$$H_{t+1} = e^r(H_t - H_a) \tag{10.9}$$

$$P_{t+1} = H_a. \tag{10.10}$$

In other words, hosts that are not attacked reproduce, and those that are attacked yield not hosts but parasitoids.

To derive a simple formulation for H_a, let E_t be the number of host–parasitoid encounters in generation t. Then, if A is the parasitoid's searching efficiency:

$$E_t = AH_t P_t \tag{10.11}$$

and:

$$\frac{E_t}{H_t} = AP_t. \tag{10.12}$$

Note the similarity to the formulation in Equation 10.2. Remember, though, that we are dealing with parasitoids and hence, a single host can be encountered several times, although only the first encounter leads to successful parasitization. Predators, by contrast, would remove their prey and prevent re-encounters. Thus, Equation 10.2 dealt with instantaneous rates, rather than numbers.

a model based on random encounters …

If encounters are assumed to occur more or less at random, then the proportions of hosts encountered zero, one, two or more times are given by the successive terms in the appropriate 'Poisson distribution' (see any basic statistics textbook). The proportion not encountered at all, p_0, would be given by e^{-E_t/H_t}, and thus the proportion that is encountered (one or more times) is $1 - e^{-E_t/H_t}$. The number encountered (or attacked) is then:

$$H_a = H_t(1 - e^{-E_t/H_t}) \tag{10.13}$$

and using this and Equation 10.12 to substitute into Equations 10.9 and 10.10 gives us:

$$H_{t+1} = H_t(1 - e^{(r - AP_t)}) \tag{10.14}$$

$$P_{t+1} = H_t(1 - e^{(-AP_t)}). \tag{10.15}$$

… giving rise to coupled oscillations

This is the basic Nicholson–Bailey model of a host–parasitoid interaction. Its behaviour is reminiscent of the Lotka–Volterra model. An equilibrium combination of the two populations is a possibility, but even the slightest disturbance from this equilibrium leads to divergent coupled oscillations.

10.2.3 One-generation cycles

The coupled oscillations generated by the basic Lotka–Volterra and Nicholson–Bailey models are multigeneration cycles, i.e. there are several generations between successive peaks (or troughs). Other models of host–parasitoid systems, however, are able to generate coupled oscillations just one host generation in length, even in completely or largely aseasonal environments (such as the tropics or man-made environments like grain stores), where there is continuous breeding of both species (Godfray & Hassell, 1989; Gordon *et al.*, 1991). They occur essentially when the generation length of the parasitoid is half that of its host—as it often is. Any small peak in host abundance tends to generate a peak in host abundance one host-generation later. But any associated peak in parasitoid abundance is generated half a host-generation length later, creating a trough in host abundance between the twin peaks. Moreover, the parasitoid peak fails to coincide with the second host peak. Thus, the parasitoids have alternate 'feasts' and 'famines' which accentuate the natural peaks and troughs in host abundance, and hence promote one-generation cycles.

10.2.4 Predator–prey cycles in nature: or are they?

despite the underlying tendencies, predator–prey cycles are not necessarily seen—nor are they to be 'expected'

The inherent tendency for predator–prey interactions to generate coupled oscillations in abundance has, at times in the past, produced an 'expectation' of such oscillations in real populations. This expectation, however, should immediately be tempered by two thoughts. First, there are many important aspects of predator and prey ecology that have not been considered in the models derived so far; and as subsequent sections will show, these can greatly modify our expectations. Second, even if a population exhibits regular oscillations, this does not necessarily provide support for the Lotka–Volterra, Nicholson–Bailey or any other simple model. If a herbivore population fluctuates in abundance, that may reflect its interaction with its food or with its predators. The abundance of a predator may cycle as it tracks cycles in prey abundance, even though the predator–prey interaction itself did not generate those cycles in the first place. Predator–prey interactions *can* generate regular cycles in the abundance of both interacting populations, and they can reinforce such cycles if they exist for some other reason; but attributing a cause to regular cycles in nature is generally a difficult task (see Chapter 15, Section 15.4).

cycles in the laboratory

It has been possible in some cases to generate coupled predator–prey oscillations, several generations in length, in the laboratory. For instance, Figure 10.3 shows this for a parasitoid–host system—the azuki bean weevil, *Callosobruchus chinensis*, and a braconid parasitoid, *Heterospilus prosopidis* (Utida, 1957). A more typical example, however, was discussed in Chapter 9, Section 9.7.3 (Figure 9.13): Huffaker's predatory mites and their mite prey were only able to sustain coupled oscillations in a heterogeneous environment. In the homogeneous type of environment that the simple models envisage, both mite populations rapidly fluctuated to extinction. Even under controlled laboratory conditions, therefore, it is necessary to take account of many aspects of predator and prey ecology before cycles of abundance can be understood.

Figure 10.4a shows another laboratory host–parasitoid study (the Indian meal moth *Plodia interpunctella*, and the parasitoid, *Venturia canescens*), apparently exhibiting coupled oscillations one host generation in length. Figure 10.4b, however,

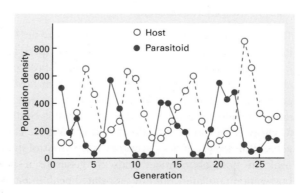

Figure 10.3 Coupled predator–prey oscillations in the laboratory: the azuki bean weevil (*Callosobruchus chinensis*) and its braconid parasitoid (*Heterospilus prosopidis*). (After Utida, 1957.)

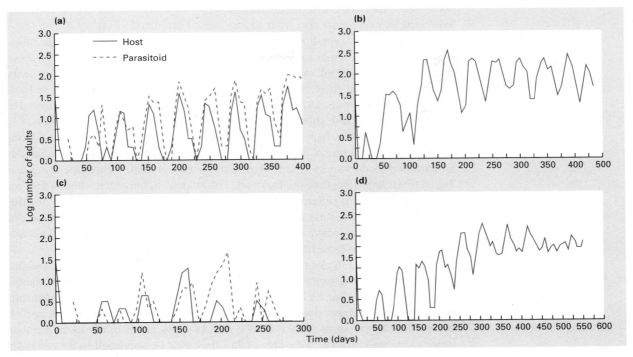

Figure 10.4 Long-term population dynamics in laboratory population cages of a host (*Plodia interpunctella*), with and without its parasitoid (*Venturia canescens*). (a) Host and parasitoid in deep medium, exhibiting coupled cycles in abundance, approximately one host generation in length. (b) The host alone in deep medium, exhibiting similar cycles. (c) Host and parasitoid in shallow medium, unable to persist. (d) The host alone in shallow medium, able to persist. The deep medium provides a refuge from attack for a proportion of the host population that is not present in the shallow medium (see Section 10.5.5). All data sets are selected from several replicates showing the same pattern. (After Begon *et al.*, 1995a.)

shows that moth abundance cycled in essentially the same way even in the absence of the parasitoid. These were coupled predator–prey cycles, but they may not have been generated simply by the predator–prey interaction itself—a conclusion that could only be drawn because the original observation was accompanied by a 'control': host-alone dynamics were also examined.

Amongst field populations, there are a number of examples in which regular cycles of prey and predator abundance can be discerned. These are discussed in Chapter 15, Section 15.4, but for now we examine a single example. Cycles in hare populations have been discussed by ecologists since the 1920s, and were recognized by fur trappers more than 100 years earlier (Keith, 1983; Krebs *et al.*, 1992). Most famous of all is the snowshoe hare, *Lepus americanus*, which in the boreal forests of North America follows a '10-year cycle' (although in reality this varies in length between 8 and 11 years; see Figure 10.1c). The snowshoe hare is the dominant herbivore of the region, feeding on the terminal twigs of numerous shrubs and small trees. A number of predators, including the Canada lynx (*Lynx canadensis*), have associated cycles of similar length. There are also 10-year cycles of certain other herbivores, notably the ruffed grouse and the spruce grouse. The hare cycles often involve 10- to 30-fold changes in abundance, and 100-fold changes can occur in

the complex natural plant–hare–lynx–grouse cycles …

some habitats. They are made all the more spectacular by being virtually synchronous over a vast area from Alaska to Newfoundland.

The declines in snowshoe hare abundance are accompanied by low birth rates, low juvenile survivorship and low growth rates or even weight loss. All of these features can be induced experimentally by food shortages. Direct measurements have suggested that there is often a shortage of food during periods of peak hare abundance (Pease *et al.*, 1979; Keith *et al.*, 1984), but it seems likely that these are frequently not absolute shortages, leading to starvation, but relative shortages, i.e. shortages of accessible, high-quality food (Keith *et al.*, 1984; Smith *et al.*, 1988). More specifically, the quantity of willow and birch twigs declines as hare density increases, so that total food quantity is low when hare density is high. Hence, the hares are forced to eat high-fibre (low-quality) food. This is likely to have some direct effect on their body condition, perhaps making them more susceptible to predation. But the hares also spend more time searching for food, exposing them more to predation. As a consequence, the predators concentrate, behaviourally, on hares, and also increase numerically, and so drive hare numbers down again.

After the decline in plant and hare abundance, the regrowth shoots of the plants contain high levels of toxic, secondary chemicals which greatly reduce the plants' attractiveness and digestibility (Sinclair *et al.*, 1988). The quantity of relatively palatable, mature shoots does not then recover until after a time lag of another 2–3 years. Thus, there seems to be little of a role for these toxins in the hares' decline (since their concentration rises after it), but they may be important in delaying the start of the next phase of rapid increase.

... which shows both the value and the inadequacies of the simple models

Is this, then, a hare–plant cycle or predator–hare cycle? Experiments, as ever, are instructive in providing an answer (Krebs *et al.*, 1992). Normally, as we have seen, with both plant and predator effects present, there are cycles. But if food is added and predators excluded from experimental areas (neither effects active), then hare numbers increase 10-fold and stay there—the cycles are lost. However, if either predators are excluded but no food added (food effect alone), or food is added in the presence of predators (predator effect alone), numbers of hares double but then drop again—the 'cycle' is retained. Thus, both the hare–plant and the predator–hare interaction have some propensity to cycle on their own, but in practice the cycle seems normally to be generated by the interaction between the two (Figure 10.5). Moreover, the predators eat a large number of grouse when the predator : hare ratio is high, but only a small number when the ratio is low. It appears to be this that generates the cycles in these subsidiary herbivores (Figure 10.5).

It is tempting to suggest that the dynamics are cyclic, and the cycles last around 10 years, because of the time lag in plant palatability lasting 25% of the cycle's length (in line with the simple models) (May, 1981a). This requires us to think of the plants' low point being directly after the hare decline, as plant toxin concentration starts to rise, and of the hares' low point as being at the end of this period of high toxin concentration, after the hares have suffered firstly from low-quantity and high-fibre content and then from the toxins. However, cycles need not necessarily be four times the length of a time lag. For example, Trostel *et al.* (1987) generated an 8–11-year hare cycle in a predator–hare model with a 1-year lag in the predator's numerical response, a type 2 functional response in the predators and realistic field values for their model's parameters. Thus, the simple models are clearly helpful in providing a framework for understanding natural cycles in abundance, but they are far from being able to provide a complete explanation.

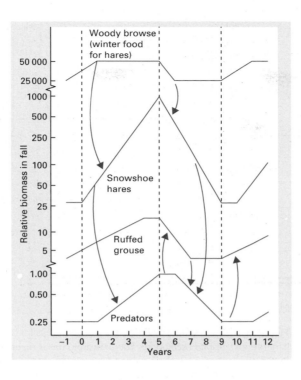

Figure 10.5 Fluctuations in the relative biomass of the major components of 'wildlife's 10-year cycle' in Alberta, Canada. The arrows indicate the major causative influences. (After Keith, 1983.)

10.2.5 Delayed density dependence

Varley (1947) used the term 'delayed density dependence' to describe the time delayed regulatory effect that a predator has on a prey population when their abundances are linked closely together. There are two crucial points that should be noted about delayed density dependence.

The first is that compared to other regulatory effects, those of delayed density dependence are relatively difficult to demonstrate. To see this, note the coupled oscillations produced by a particular parasitoid–host model, shown in Figure 10.6a (Hassell, 1985). The model population is more stable than those of previous sections, in that the oscillations are damped; but the details of the model need not concern us. What is important is that the prey population, subject to delayed density dependence, is regulated in size by the predator. Yet, when we plot the k-values of predator-induced mortality over a generation against the log of prey density in that generation (Figure 10.6b), no clear relationship is apparent, in spite of the fact that this is the conventional way of revealing density dependence (see Chapter 6). On the other hand, when the same points are linked together, each generation to the next ('serial linking': Figure 10.6c), they can be seen to describe a spiral travelling in an anticlockwise direction. This spiralling is characteristic of delayed density dependence. Here, because the oscillations are damped, the points actually spiral inwards. Moreover, when we plot the k-values of predator-induced mortality against the log of prey density *two generations previously* (Figure 10.6d), the delayed density dependence is also clearly revealed.

It is relatively straightforward to reveal the regulatory effects of delayed density dependence for the model population of Figure 10.6, because it is not subject to the fluctuations of a natural environment, it is not subject to the density dependent

the regulatory effects of predators are frequently difficult to demonstrate

379 POPULATION DYNAMICS OF PREDATION

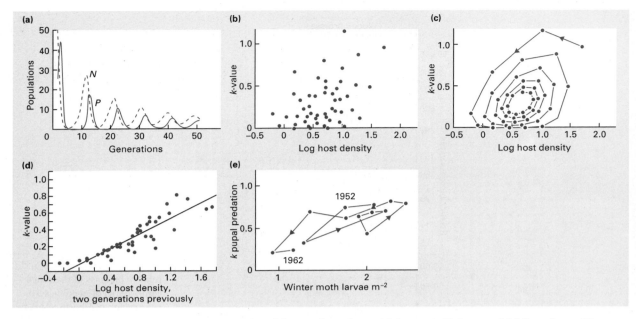

Figure 10.6 Delayed density dependence. (a) A parasitoid–host model followed over 50 generations: despite oscillations, the parasitoid has a regulatory effect on the host population. (b) For the same model, the *k*-value of generation mortality plotted against the log of host density: no clear density dependent relationship is apparent. (c) The points from (b) linked serially from generation to generation: they spiral in an anticlockwise direction—a characteristic of delayed density dependence. (After Hassell, 1985). (d) The *k*-value of generation mortality plotted against the log of host density two generations previously: a clear delayed density dependent relationship is again apparent. (e) Field data: the *k*-value for predation (mainly by beetles) of pupae of the winter moth (*Operophtera brumata*) plotted against the log of density. This emerges unequivocally as a density dependent factor in spite of a certain amount of delayed density dependence (anticlockwise spiralling). (After Varley *et al.*, 1973.)

attacks of any other predator, it is not subject to the inaccuracies of sampling error, and the length of the delay can be deduced, roughly, from a clear and extended set of data (Figure 10.6a). Data of this quality, however, are rarely if ever available for natural or even experimental populations. It is not surprising, therefore, that there have been difficulties in determining the role of predators in the regulation of natural populations (Dempster, 1983; Hassell, 1985; Dempster & Pollard, 1986; Hassell, 1987). In some cases, the oscillations are damped rapidly enough for a positive density dependent relationship to be apparent in spite of a tendency to spiral (e.g. Figure 10.6e). But we must certainly be wary of underestimating the regulatory effects of predators simply because intrinsic time delays and natural variability combine to cloud clear density dependent relationships.

delayed density dependence may regulate but certainly need not stabilize

The second, related point concerns the meanings of the terms 'regulation' and 'stability' in predator–prey interactions. Most natural predator and prey populations exhibit apparently less violent, and certainly less regular fluctuations than those we have seen generated by the simplest models. Most of the rest of this chapter describes the search for explanations: (i) for these more stable patterns; and (ii) for the variations in dynamical pattern from case to case. A population that remains roughly constant in size provides evidence for the effects of both regulatory and stabilizing forces. The delayed density dependence of a predator–prey interaction 'regulates' in the sense of acting strongly on large populations and making them smaller, and acting only weakly on small populations and allowing them to grow larger; but, as we have already seen, it can hardly be said, typically, to stabilize either population. What follows in this chapter is, therefore, in large part, a search for stabilizing forces that might complement the (delayed) regulatory forces that occur inherently in predator–prey interactions.

10.3 The effects of crowding

Most predator–prey interactions are influenced by crowding in either one or both of the populations, i.e. by intraspecific competition or mutual interference.

10.3.1 Crowding in the Lotka–Volterra model

The effects of these processes can be investigated by modifying the Lotka–Volterra predator and prey isoclines. The predator zero isocline in the Lotka–Volterra model is vertical. This implies that the ability of a predator population to increase in abundance is determined by the absolute abundance of prey (and not, for example, by the prey : predator ratio), and that the abundance of the predators themselves is unimportant. Often, however, as predator density increases, so too will mutual interference amongst the predators (see Chapter 9, Section 9.6). If this happens, individual consumption rates will go down and additional prey are required to maintain a predator population of any given size. The predator zero isocline can thus be expected to depart increasingly from the vertical (Figure 10.7a). Moreover, at high densities, even in the presence of excess food, most predator populations will be limited by the availability of some other resource: resting sites, perhaps, or safe refuges of their own. This will put an upper limit on the predator population irrespective of prey numbers. Overall, it seems reasonable to assume that consumers generally have a zero isocline resembling that in Figure 10.7a.

Note that this is a line along which the predators neither increase nor decrease in abundance. At predator–prey combinations above and to the left of it (relatively few prey, relatively many predators), the predators decrease, whilst at combinations below and to the right of it they increase. There will, though, be considerable variation from predator to predator in the location of the isocline on the axes, and in the densities at which particular effects become apparent.

The detailed methods of incorporating intraspecific competition into the prey zero isoclines are described by Begon *et al.* (1990). The end result shown in Figure 10.7b, however, can be understood without reference to these details. At low prey densities there is no intraspecific competition, and the prey isocline is the same as in the Lotka–Volterra model. But as density and intraspecific competition increase, the isocline is increasingly depressed until at the carrying capacity (K_N) it actually reaches the prey axis; in other words, at a density K_N, the prey population can only just maintain itself even in the absence of predators.

The likely effects of crowding in either population can now be deduced by combining the predator and prey isoclines (Figure 10.7c). Oscillations are still apparent for the most part, but these are no longer neutrally stable. Instead, they are damped so that they converge to a stable equilibrium. Predator–prey interactions in which either or both populations are limited substantially by crowding, therefore,

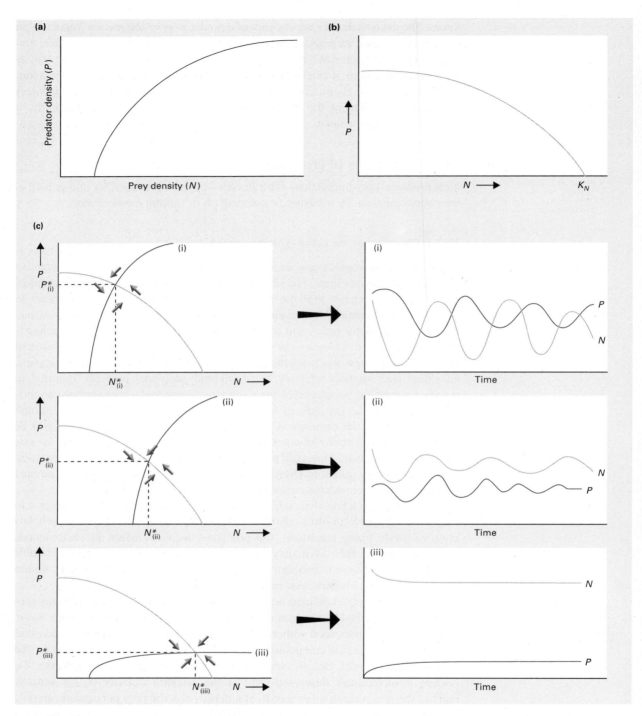

Figure 10.7 (a) A predator zero isocline subject to crowding (see the text). (b) A prey zero isocline subject to crowding. At the lowest prey densities this is the same as the Lotka–Volterra isocline, but when the density reaches the carrying capacity (K_N) the population can only just maintain itself even in the complete absence of predators. (c) The prey zero isocline combined with predator zero isoclines with increasing levels of crowding: (i), (ii) and (iii). P^* is the equilibrium abundance of predators, and N^* the equilibrium abundance of prey. Combination (i) is least stable (most persistent oscillations), and has most predators and least prey: the predators are relatively efficient. Less efficient predators, (ii), give rise to a lowered predator abundance, an increased prey abundance and less persistent oscillations. Strong predator self-limitation, (iii), can eliminate oscillations altogether, but P^* is low and N^* is close to K_N.

are likely to exhibit patterns of abundance which are relatively stable, i.e. in which fluctuations in abundance are relatively slight.

More particularly, when the predator is relatively inefficient, i.e. when many prey are needed to maintain a population of predators (curve (ii) in Figure 10.7c), the oscillations are damped quickly but the equilibrium prey abundance (N^*) is not much less than the equilibrium in the absence of predators (K_N). By contrast, when the predators are more efficient (curve (i)) N^* is lower and the equilibrium density of predators, P^*, is higher—but the interaction is less stable (the oscillations are more persistent). Moreover, if the predators are very strongly self-limited, perhaps by mutual interference, then stability is relatively high and abundance may not oscillate at all (curve (iii)); but P^* will tend to be low, whilst N^* will tend to be not much less than K_N. Hence, for interactions where there is crowding, there appears to be a contrast between those in which predator density is low, prey abundance is little affected and the patterns of abundance are stable, and those in which predator density is higher and prey abundance is more drastically reduced, but the patterns of abundance are less stable.

... but this is most marked for inefficient predators where prey abundance is therefore little affected

Essentially similar conclusions emerge from modifications of the Nicholson–Bailey model that incorporate either simple (logistic) crowding effects amongst the hosts or mutual interference amongst the predators (Hassell, 1978). However, as we describe below, more interesting conclusions emerge when we use the Nicholson–Bailey framework to incorporate a wider range of types of host self-limitation and combine these with other important modifications with which they can interact.

10.3.2 Crowding in practice

There are certainly examples that appear to confirm the stabilizing effects of crowding in predator–prey interactions. For instance, there are two groups of primarily herbivorous rodents that are widespread in the Arctic: the microtine rodents (lemmings and voles) and the ground squirrels. The microtines are renowned for their dramatic, cyclic fluctuations in abundance (see Chapter 15, Section 15.4.2), but the ground squirrels have populations that remain remarkably constant from year to year. Significantly, the ground squirrels are strongly self-limited by their aggressive territorial defence of burrows used for breeding and hibernating, and it is to this that their stability has been convincingly attributed (Batzli, 1983).

ground squirrels provide support for the stabilizing effects of self-limitation ...

An example that lends more specific support to Figure 10.7c is provided by work on an avian herbivore, the red grouse (*Lagopus lagopus scoticus*), feeding on heather (*Calluna vulgaris*) on Scottish moorlands (Watson & Moss, 1972; see also Caughley & Lawton, 1981). Heather comprises at least 90% of the red grouse's diet over most of the year, and it is the dominant (and sometimes virtually the only) higher plant on the moors where the grouse live. The grouse themselves are strongly territorial, with the size of the spring breeding population being determined by the number of territories established by cocks in the previous autumn. The 'surplus' birds are then forced from the moor, and account for the bulk of the overwinter mortality. Some of the grouse populations are fairly constant in size, although others fluctuate over perhaps a threefold range of densities (Jenkins *et al.*, 1967); but in line with Figure 10.7 (curves (ii) and (iii)), the moors support a great deal of heather with relatively few red grouse. In fact, the grouse eat only about 2% of the total annual heather production. This may, of course, be a higher percentage of the nutritious heather, but the predicted effects of mutual

as does the grouse–heather interaction

interference on an interaction are, at the very least, not refuted by these populations.

10.4 Functional responses and the Allee effect

the 'stabilizing' effects of type 3 responses are probably of little importance in practice ...

Prey zero isoclines can be further modified to take account of the various types of functional response (see Chapter 9, Section 9.5). A type 3 response will give rise to a low predation rate where prey density is low. We might therefore suppose that prey at low densities can increase in abundance virtually irrespective of predator density, and that the prey zero isocline will therefore rise vertically at low prey densities (Figure 10.8). Although in principle this might seem capable of lending a considerable degree of stability to an interaction (Figure 10.8, curve (i)), that would require the predator to be highly efficient (readily capable of maintaining itself) at low prey densities—which contradicts the whole idea of a type 3 response (ignoring prey at low densities). Hence, curve (ii) in Figure 10.8 is likely to apply, and the stabilizing influence of the type 3 response may in practice be of little importance.

... but switching type 3 responses can stabilize prey abundance ...

On the other hand, a predator is likely to have a type 3 functional response to one particular type of prey because it switches its attacks amongst various prey types depending on which are most abundant. In such a case, the population dynamics of the predator would be independent of the abundance of any particular prey type, and the position of its zero isocline would therefore be the same at all prey densities. As Figure 10.9 shows, this can lead potentially to the predators regulating the prey at a low and stable level of abundance.

An apparent example of this is provided by studies of vole cycles in Europe

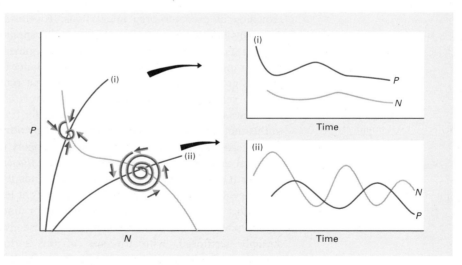

Figure 10.8 The prey zero isocline is that which is appropriate when consumption rate is particularly low at low prey densities because of a type 3 functional response, an aggregative response (and partial refuge), an actual refuge or because of a reserve of plant material that is not palatable. With a relatively inefficient predator, predator zero isocline (ii) is appropriate and the outcome is not dissimilar from Figure 10.7. However, a relatively efficient predator will still be able to maintain itself at low prey densities. Predator zero isocline (i) will therefore be appropriate, leading to a stable pattern of abundance in which prey density is well below the carrying capacity and predator density is relatively high.

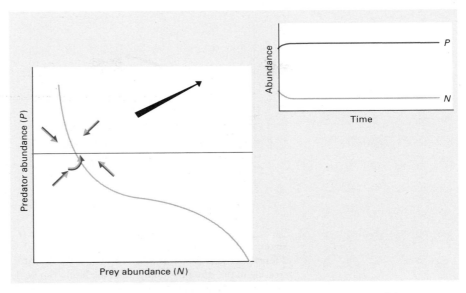

Figure 10.9 When a type 3 functional response arises because the predator exhibits switching behaviour, the predator's abundance may be independent of the density of any particular prey type (main figure), and the predator zero isocline may therefore be horizontal (unchanging with prey density). This can lead to a stable pattern of abundance (inset) with prey density well below the carrying capacity.

... as seems to be the case for northern European voles

(Hanski *et al.*, 1991; see also Chapter 15, Section 15.4). In subarctic Finnish Lapland, there are regular 4- or 5-year cycles, with a ratio of maximum : minimum rodent densities generally exceeding 100. In southern Sweden (and in most of Central Europe) small rodents show no regular multi-annual cycles. But between the two, moving north to south in Fennoscandia, there is a gradient of decreasing regularity, amplitude, length and interspecific synchrony of the cycle. Hanski *et al.* argue that this gradient is itself correlated with a gradient of increasing densities of generalist predators (especially red foxes, badgers, domestic cats, buzzards, tawny owls and crows) and specialist bird predators (especially other owl species and kestrels). The suggestion is that the former switch between alternative prey as relative densities change, whilst the latter, being wide ranging in their activity, switch between alternative areas as relative densities change. In both cases, the effect is to add stability. In fact, Hanski *et al.* were able to go further by constructing a simple model of prey (microtines) interacting with specialist predators (mustelids: stoats and weasels) and generalist (switching) predators. They were able to support their general contention. As the number of generalist predators increased, oscillations in microtine and mustelid abundance (which may or may not be the basis for the microtine cycle) decreased in length and amplitude; large enough densities of generalists stabilized the cycle entirely.

the 'destabilizing' effects of type 2 responses may also be of little importance in practice

If the predator has a type 2 response (see Chapter 9, Section 9.5) that reaches its plateau at relatively low prey densities (well below K_N), then the prey zero isocline has a hump, because there is a range of intermediate prey densities where the functional responses have reached saturation but the effects of competition amongst the prey are not intense. A hump will also arise here if the prey are subject to an 'Allee effect'. An Allee effect (Allee, 1931) is said to occur where individuals in a

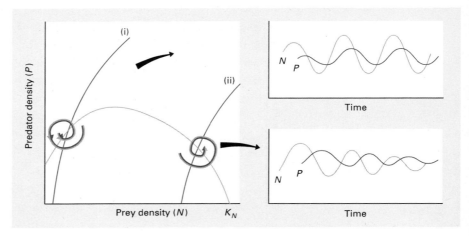

Figure 10.10 The possible effects of a prey isocline with a 'hump', either as a result of a type 2 functional response or an Allee effect. (i) If the predator is highly efficient, with its isocline crossing to the left of the hump, then the hump can be destabilizing, leading to the persistent oscillations of a limit cycle (inset). (ii) But if the predator is less efficient, crossing to the right of the hump, then the hump has little effect on the dynamics: oscillations converge (inset).

population have a disproportionately low rate of recruitment when their own density is low, perhaps because mates are difficult to find or because a 'critical number' must be exceeded before a resource can be properly exploited, i.e. there is inverse density dependence at low population densities. If the predator zero isocline crosses to the left of the hump (Figure 10.10), then the outcome will be the persistent oscillations of a limit cycle, rather than convergent oscillations, i.e. the interaction will be destabilized. However, for a type 2 response to have this effect, handling time must be very long, whereas for insects at least, handling times may be relatively short (Hassell, 1981). The potentially destabilizing effects of type 2 responses (see Chapter 9, Section 9.5.4) may therefore be of little practical importance.

The destabilizing role of the Allee effect has apparently been neither supported nor refuted quantitatively. However, the idea of a prey population declining towards extinction because it is 'too small' (i.e. below some critical density) is familiar, at least, in the context of exploited whale and fish populations (see Chapter 16, Section 16.12). Decline to extinction is, perhaps, the ultimate in ecological instability.

10.5 Heterogeneity, aggregation and spatial variation

Until now in this chapter, environmental heterogeneities, and the variable responses of predators and prey to such heterogeneities—all of which we saw in the previous chapter to be commonplace—have been ignored. We can ignore them no longer.

10.5.1 Heterogeneity in the graphical model

We can start by incorporating into the Lotka–Volterra isoclines some relatively simple types of heterogeneity. Suppose that a portion of the prey population exists in a refuge: for example shore snails packed into the limited space available in cracks in the cliff-face, away from marauding birds, or plants that maintain a reserve of material that cannot be grazed, possibly close to or under the ground (see Chapter 8, Section 8.2.1). In such cases, the prey zero isocline again rises vertically at low prey densities (see Figure 10.8), since prey at low densities can increase in abundance irrespective of predator density, unaffected by the predators because they are hidden in a refuge.

Now, suppose instead that predators tend always to ignore prey in low-density patches, as we appear to have seen in a number of aggregative responses (see Chapter 9, Section 9.7.1). This comes close to the prey at low density being in a refuge, in the sense that the predators do not (rather than cannot) attack them. The prey may therefore be said to have a partial refuge, and this time the prey isocline can be expected to rise almost vertically at low prey densities (see Figure 10.8).

We saw above, when discussing type 3 functional responses, that such isoclines have a tendency to stabilize interactions, and early analyses of both the Lotka–Volterra and the Nicholson–Bailey systems (and earlier editions of this text) certainly agreed with this conclusion: that spatial heterogeneities, and the responses of predators and prey to them, stabilize predator–prey dynamics, often at low prey densities (Beddington *et al.*, 1978). However, as we shall see next, subsequent developments have shown that the effects of heterogeneity are more complex than was previously supposed: the effects of heterogeneity vary depending on the type of predator, the type of heterogeneity, and so on.

refuges, partial refuges and their *apparently* stabilizing effects

10.5.2 Heterogeneity in the Nicholson–Bailey model

Most progress has been made in untangling these effects in host–parasitoid systems. A good starting point is the model constructed by May (1978), in which he ignored the precise nature of host and parasitoid distributions and efforts and argued simply that the distribution of host–parasitoid encounters was not random but aggregated. In particular, he assumed that this distribution could be described by a particular statistical model, the negative binomial. In this case (in contrast to Section 10.2.2), the proportion of hosts not encountered at all is given by:

May's detail-independent model of aggregation

$$p_0 = \left[1 + \frac{AP_t}{k}\right]^{-k} \tag{10.16}$$

where k is a measure of the degree of aggregation: maximal aggregation at $k = 0$, but a random distribution (recovery of the Nicholson–Bailey model) at $k = \infty$. If this is incorporated into the Nicholson–Bailey model (Equations 10.14 and 10.15), then we have:

$$H_{t+1} = H_t e^r \left[1 + \frac{AP_t}{k}\right]^{-k} \tag{10.17}$$

$$P_{t+1} = H_t \left\{1 - \left[1 + \frac{AP_t}{k}\right]^{-k}\right\}. \tag{10.18}$$

The behaviour of a version of this model, which also includes a density dependent host rate of increase, is illustrated in Figure 10.11, from which it is clear that the system is given a marked boost in stability by the incorporation of significant levels of aggregation ($k \leqslant 1$). Of particular importance is the existence of stable systems with low values of H^* / K; i.e. aggregation appears capable of generating stable host abundances well below the host's normal carrying capacity. This coincides with the conclusion drawn from Figure 10.8.

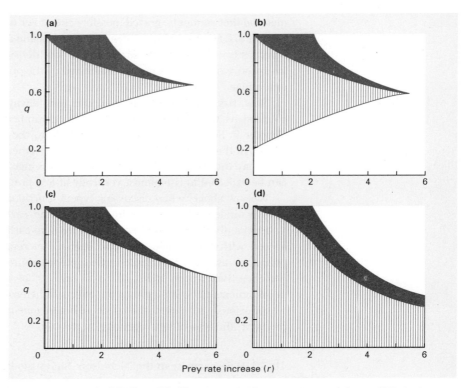

Figure 10.11 May's (1978) model of host–parasitoid aggregation, with host self-limitation incorporated, illustrates that aggregation can enhance stability and give rise to stability at low values of $q = H^* / K$. In the shaded area there is an exponential approach to equilibrium; in the hatched area there is an oscillatory approach to equilibrium; outside these there is instability (the oscillations either diverge or are sustained). The four figures are for four values of k, the exponent of the negative binomial distribution in the model. (a) $k = \infty$: no aggregation, least stability; (b) $k = 2$; (c) $k = 1$; (d) $k = 0.1$: most aggregation. (After Hassell, 1978.)

10.5.3 Aggregation of risk and spatial density dependence

aggregation of risk can increase pseudo-interference . . .

How does this stability arise out of aggregation? The answer lies in what has been called 'pseudo-interference' (Free *et al.*, 1977; see also Chapter 9, Section 9.10). The term was coined to describe a relationship reminiscent of mutual interference amongst predators. With mutual interference, as predator density increases, predators spend an increased amount of time interacting with one another rather than searching for and attacking prey, and the attack rate of predators therefore declines. With pseudo-interference, attack rate also declines with predator (or parasitoid) density, but as a result of an increasing fraction of encounters being wasted on prey or hosts that have already been attacked. The crucial point is that 'aggregation of risk' amongst hosts (see Chapter 9, Section 9.7) tends to increase the amount of pseudo-interference. At low parasitoid densities, a parasitoid is unlikely to have its attack rate reduced as a result of aggregation. But at higher parasitoid densities, parasitoids in aggregations (where most of them are) will increasingly be faced with host patches in which most or all of the hosts have already been parasitized, so that their 'attacks' must be either aborted or wasted. As parasitoid density increases, therefore, their effective attack rate rapidly declines. Increasing

388   CHAPTER 10

pseudo-interference thus makes parasitoid birth rate decline more rapidly with parasitoid density than it would otherwise do—a *directly* density dependent effect. This dampens both the natural oscillations in parasitoid density, and their impact on host mortality.

... a temporal, directly density dependent effect that increases stability

To summarize, aggregation of risk stabilizes host–parasitoid interactions by strengthening direct (not delayed) density dependencies that already exist (Taylor, 1993). The stabilizing powers of this spatial phenomenon, aggregation of risk, therefore arise not from any spatial density dependencies, but from its translation into direct, temporal density dependence.

Note also, though, that pseudo-interference implies reduced parasitoid effectiveness. Hence, although it generates stability, it necessarily also gives rise to increased equilibrium host abundance. Thus, even where aggregation of risk is capable of generating stable equilibria at low host abundance, increased aggregation, whilst increasing stability, actually increases host abundance.

But how does aggregation of risk actually relate to the aggregative responses of parasitoids? And do aggregative responses and aggregation of risk necessarily lead to enhanced stability? We can address these questions by examining Figure 10.12, bearing in mind from Chapter 9, Section 9.7 that aggregated predators do not necessarily spend most time foraging in patches of high host density (spatial density dependence); foraging time can also be negatively correlated with host density or independent of host density. Start with Figure 10.12a. The distribution of parasitoids over host patches follows a perfect, straight-line density dependent relationship. But, since the host : parasitoid ratio is therefore the same in each host patch, the risk is likely to be the same in each host patch, too. Immediately we can see, therefore, that positive spatial density dependence does not necessarily lead to aggregation (i.e. variation) in risk and does not necessarily enhance stability. On the other hand, with a perfect density dependent relationship that accelerates (Figure 10.12b), there does appear to be aggregation of risk, and this might well enhance stability (Hassell & May, 1973); but it turns out that whether or not it does so depends on the parasitoids' functional response (Ives, 1992a). With a type 1 response, assumed by most analyses, stability is enhanced. But with a more realistic type 2 response, initial increases in density dependent aggregation from zero aggregation actually decrease aggregation of risk and are destabilizing (because the effects of type 2 responses— lower attack rates in low host-density patches where the parasitoids are satiated— counteract the effects of aggregation). Only high levels of density dependent aggregation (outstripping the effects of the type 2 response) are stabilizing.

positive spatial density dependence does not necessarily enhance stability ...

Moreover, it is clear from Figures 10.12c and d that there can be considerable aggregation of risk with either inverse spatial density dependence or no spatial density dependence of any sort—and these would not be counteracted by a type 2 functional response. In partial answer to our two questions above, therefore, aggregative responses that are spatially density dependent are actually least likely to lead to aggregation of risk, and therefore least likely to enhance stability.

... and inverse spatial density dependence and density independence seem more likely to do so

In practice, of course, with real sets of data (like those in Chapter 9, Figure 9.12), aggregation of risk will often arise from a combination of spatially density dependent (direct or inverse) and density independent responses (Chesson & Murdoch, 1986; Hassell & Pacala, 1990; Pacala et al., 1990; Hassell et al., 1991; Pacala & Hassell, 1991). Pacala, Hassell and co-workers have called the former the 'host-density dependent' (HDD) component, and the latter the 'host-density independent' (HDI) component, and have described methods by which, in real data sets like Figure 9.12,

the division of spatial variation into HDD and HDI components

389 POPULATION DYNAMICS OF PREDATION

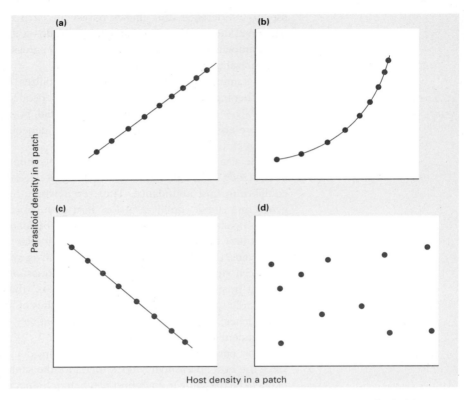

Figure 10.12 The aggregative responses of parasitoids and the aggregation of risk. (a) Parasitoids aggregate in high, host-density patches, but the parasitoid : host ratio is the same in all patches (a perfect straight-line relationship), and hence the risk to hosts is apparently the same in all patches. (b) Parasitoid aggregation to high, host-density patches now accelerates with increasing host density, and hosts in high-density patches are thus apparently at greater risk of parasitization: there is aggregation of risk. (c) With perfect inverse density dependence (i.e. parasitoid aggregation in *low* host-density patches) the hosts in the low-density patches are apparently at a much greater risk of parasitization: again there is aggregation of risk. (d) Even with no aggregative response (density independence) hosts in some patches are apparently at a greater risk of parasitization (are subject to a higher parasitoid : host ratio) than others: here too there is aggregation of risk.

the aggregation of risk can be split into these HDD and HDI components. To do so, they quantify the aggregation of risk by its 'coefficient of variation' (CV). The CV, described in most statistics texts, is the standard deviation in risk divided by the mean risk. A number of conclusions emerge from their analysis of CV.

the CV² > 1 rule: useful if treated cautiously

First, they suggest a 'CV² > 1 rule': stability requires that the CV^2 of risk exceeds 1. They derive the rule from an analysis of mathematical models. It holds exactly for May's (1978) negative binomial model (which, it turns out, is purely HDI; Chesson & Murdoch, 1986), and it holds approximately for a number of other models. However, there are other plausible models for which the rule does not hold (Taylor, 1993). For example, in models where host self-regulation is important, stability may have little to do with aggregation of risk (Hochberg & Lawton, 1990); and where several parasitoid species attack a host, the rule may not hold for any one of them, nor for all of them combined (Taylor, 1993). Clearly, in spite of (or perhaps because of) its endearing simplicity, the rule needs to be treated with some caution. Nonetheless, Pacala and Hassell's (1991) finding is interesting, that of 65 data sets they examined, representing 26 different host–parasitoid combinations, 18 had $CV^2 > 1$. This does at least provide us with a crude measure of the extent to which aggregation of risk, acting alone, is likely to be capable of stabilizing interactions.

In addition, however, and of particular interest, they found that in 14 of these 18 cases, it was HDI variation that contributed most to the total, further weakening any imagined link between spatial density dependence and stability.

HDI seems more important than HDD variation in giving rise to stability

10.5.4 Heterogeneity in some continuous-time models

We have been pursuing parasitoids and hosts; and in doing so we have been retaining certain structural features in our analysis which it is now appropriate to reconsider. In particular, our parasitoids have been assumed, in effect, to arrange themselves over host patches at the beginning of a generation (or whatever the time interval is between t and $t + 1$), and then to have to suffer the consequences of that arrangement until the beginning of the next generation. But suppose we move into continuous time—as appropriate for many parasitoids as it is for many other predators. Now, aggregation should be assumed to occur on a continuous basis, too. Predators in a depleted, or even a depleting patch, should leave and redistribute themselves (see Chapter 9, Section 9.10). The whole basis of pseudo-interference and hence stability, namely wasted predator attacks in high predator-density patches, tends to disappear.

continuous redistribution can eliminate pseudo-interference ...

Murdoch and Stewart-Oaten (1989) went to, perhaps, the opposite extreme to the one we have been concentrating on, by constructing a continuous-time model in which prey moved instantly into patches to replace other prey that had been consumed, and predators moved instantly into patches to maintain a consistent pattern of predator–prey covariation over space, unaltered by any prey depletion. The effects on their otherwise neutrally stable Lotka–Volterra model contrast strongly with those we have seen previously. First, predator aggregation that is independent of local prey density now has *no* effect on either stability or prey density. However, predator aggregation that is directly dependent on local prey density has an effect which depends on the strength of this dependence—although it always lowers prey density (because predator efficiency is increased). If such density dependence is relatively weak (as Murdoch & Stewart-Oaten, 1989, argue it usually is in practice), then stability is *de*creased. Only if it is stronger than seems typical in nature is stability increased.

... and radically alters the effects on stability

Other, less 'extreme' continuous-time formulations (Ives, 1992b), or those that combine discrete generations with redistribution within generations (Rohani *et al.*, 1994a) produce results which are themselves intermediate between the 'Nicholson–Bailey extreme' and the 'Murdoch–Stewart-Oaten extreme'. It seems certain, however, that a preoccupation with models lacking within-generation movement has, in the past, led to a serious overestimation of the significance of aggregation to patches of high host density in stabilizing host–parasitoid interactions.

10.5.5 The metapopulation perspective

The continuous- and discrete-time approaches clearly differ, but they share a common perspective in seeing predator–prey interactions occurring within single populations, albeit populations with in-built variability. An alternative is a 'metapopulation' perspective (see Chapter 15, Section 15.6), in which environmental patches support subpopulations that have their own internal dynamics, but are linked to other subpopulations by movement between patches, which follows specified, although often very simple, rules.

A number of studies have investigated predator–prey metapopulation models, usually with unstable or neutrally stable dynamics within patches. Mathematical difficulties have often limited analysis to two-patch models, where, if the patches are the same, and dispersal is uniform, stability is unaffected: patchiness and dispersal have no effect in their own right (Murdoch *et al.*, 1992; Holt & Hassell, 1993).

Differences between the patches, however, tend, in themselves, to stabilize the interaction (Ives, 1992b; Murdoch *et al.*, 1992; Holt & Hassell, 1993). The reason is that any difference in parameter values between patches leads to asynchrony in the fluctuations in the patches. Inevitably, therefore, a population at the peak of its cycle tends to lose more by dispersal than it gains, a population at a trough tends to gain more than it loses, and so on. Dispersal and asynchrony together give rise to stabilizing temporal density dependence in net migration rates.

dispersal between asynchronous patches gives rise to stabilizing temporal density dependence

Not surprisingly, perhaps, the situation becomes much more complex with the inclusion of aggregative behaviour in the prey and/or the predator, since dispersal rates themselves become a much more complex function of both prey and predator densities in all of the patches. Aggregation appears to have two opposing effects (Murdoch *et al.*, 1992). It tends to increase the asynchrony between fluctuations in predator abundance (enhancing stability) but to reduce the asynchrony between prey fluctuations (decreasing stability). The balance between these forces appears sensitive to the strength of aggregation, but perhaps even more sensitive to the assumptions built into the models (Ives, 1992b; Godfray & Pacala, 1992; Murdoch *et al.*, 1992). Aggregation may either stabilize or destabilize. In contrast to previous analyses, it has no clear effect on prey density since its stabilizing powers are not linked to predator efficiency.

an explicitly, and visually, spatial model

The treatment of a spatially heterogeneous predator–prey interaction as a problem in metapopulation dynamics was taken a stage further by Comins *et al.* (1992). They constructed computer models of an environment consisting of a patchwork of squares, that could actually be visualized as such (Figure 10.13). In each generation, two processes occurred in sequence. First, a fraction, μ_P, of predators, and a fraction, μ_N, of prey, dispersed from each square to the eight neighbouring squares. At the same time, predators and prey from the eight neighbouring squares were dispersing into the first square. Thus, for example, the dynamics for the density of prey, $N_{i,t+1}$, in square i in generation $t+1$, was given by:

$$N_{i,t+1} = N_{i,t}(1 - \mu_N) + \mu_N \bar{N}_{i,t} \tag{10.19}$$

or:

$$N_{i,t+1} = N_{i,t} + \mu_N(\bar{N}_{i,t} - N_{i,t}) \tag{10.20}$$

where $\bar{N}_{i,t}$ is the mean density in the eight squares neighbouring square i in generation t. The second phase then consisted of one generation of standard predator–prey dynamics, either following the Nicholson–Bailey equations or a discrete-time version of the Lotka–Volterra equations (May, 1973). Simulations were started with random prey and predator populations in a single patch, with all other patches empty.

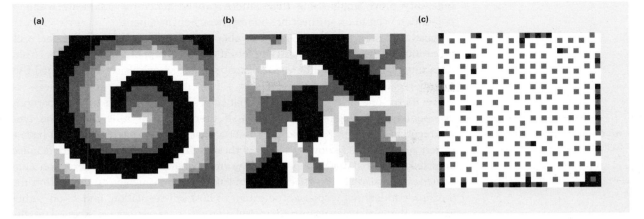

Figure 10.13 Instantaneous maps of population density for simulations of the dispersal model of Comins *et al.* (1992) with Nicholson–Bailey local dynamics. Different levels of shading represent different densities of hosts and parasitoids. Black squares represent empty patches; dark shades becoming paler represent patches with increasing host densities; light shades to white represent patches with hosts and increasing parasitoid densities. (a) Spirals: $\mu_N = 1$, $\mu_P = 0.89$; (b) spatial chaos: $\mu_i = 0.2$, $\mu_P = 0.89$; (c) a 'crystalline lattice': $\mu_N = 0.05$; $\mu_P = 1$. (The maps are single frames from simulations over many generations, which take about 0.2 s per generation on a 33 MHz 80486 personal computer.) (After Comins *et al.*, 1992.)

We know that within individual squares, if they existed in isolation, the dynamics would be unstable. But within the patchwork of squares as a whole, stable or at least highly persistent patterns can readily be generated (Figure 10.13). The general message is similar to the results that we have already seen: that stability can be generated by dispersal in metapopulations in which different patches are fluctuating asynchronously. Note especially, in this case, that a patch experiences a net gain in migrants when its density is lower than the mean of the eight patches with which it connects (Equation 10.20) but experiences a net loss when its density is higher—a kind of density dependence. Note, too, that the asynchrony arises in the present case because the population has spread from a single initial patch (all patches are, in principle, the same) and that it is maintained by dispersal being limited to neighbouring patches (rather than being a powerful equalizing force over all patches).

Moreover, the explicitly spatial aspects of this model have, quite literally, added another dimension to the results. Depending on the dispersal fractions and the host reproductive rate, a number of quite different spatial structures can be generated (although they tend to blur into one another) (Figures 10.13a–c). 'Spatial chaos' can occur, in which a complex set of interacting wave fronts are established, each one persisting only briefly. With somewhat different parameter values, and especially when both predator and prey are highly mobile, the patterns are more structured than chaotic, with 'spiral waves' rotating around almost immobile focal points. The model, therefore, makes the point very graphically that persistence at the level of a whole population does not necessarily imply either uniformity across the population or stability in individual parts of it. Static 'crystalline lattices' can even occur within a narrow range of parameter values, with highly mobile predators and rather

persistence, again, from
dispersal between
asynchronous patches

spatial chaos, spiral waves and
crystalline lattices

sedentary prey, emphasizing that pattern can be generated internally within a population even in an intrinsically homogeneous environment.

Lastly, Comins *et al.* (1992) were able to compute values of CV^2 for each generation of their simulations. They found qualified support for the $CV^2 > 1$ rule. Associations that persisted by virtue of spatial chaos or spirals consistently had CV^2 slightly greater than 1.

what does aggregation do?— well, it depends ...

Is there one general message that can be taken from this apparently exponentially-growing body of theory? If there is, then it is not of the type 'aggregation does *x* to predator–prey interactions'. Rather, aggregation can have a variety of effects, and knowing which of these is likely to apply will require detailed knowledge of predator and prey biology and behaviour for the interaction concerned. In particular, the effects of aggregation have been seen to depend on the predator's functional response, the extent of host self-regulation, and so on—other features that we have examined in isolation. It is necessary, as stressed at the beginning of this chapter, in seeking to understand complex processes, to isolate conceptually the different components. But it is also necessary, ultimately, to recombine those components.

10.5.6 Aggregation, heterogeneity and spatial variation in practice

metapopulation effects

What, then, can be said about the role of heterogeneity in practice? The stabilizing effects of heterogeneity have already been described for the case of Huffaker's mites (see Chapter 9, Section 9.7.3). In a similar vein, it is significant that populations of the 'cyclic' snowshoe hare (p. 377) never show cyclic behaviour in habitats that are a mosaic of habitable and uninhabitable areas. In mountainous regions, and in areas fragmented by the incursions of agriculture, the snowshoe hare maintains relatively stable, non-cyclic populations (Keith, 1983). These examples support the idea of a stabilizing role for patchiness (and dispersal) in a metapopulation context.

heterogeneity and biological control: the holly leaf-miner ...

The effects of aggregative responses, on the other hand, have been the subject of lively debate in considering the properties and nature of biological control agents. A biological control agent is a natural enemy of a pest species that is imported into an area, or otherwise aided and abetted, in order to control that pest (see Chapter 16). What is required of a good biological control agent is the ability to depress prey (pest) abundance to a stable level well below its 'normal', harmful level, and we have seen that some theoretical analyses suggest that this is precisely what aggregative responses help to generate. Arising directly from this work, Heads and Lawton (1983) examined some of the natural enemies of the holly leaf-miner (*Phytomyza ilicis*, a small agromyzid fly), which is a pest species in Canada because larvae of the fly tunnel within the leaves of the holly and spoil it for ornamental purposes. Of the parasitoids and predators considered by Heads and Lawton, one, the larval parasitoid *Chrysocharis gemma*, produced an aggregated distribution of host mortality. Significantly, this aggregation became most marked at the lowest host densities. On the other hand, there was no evidence of aggregative responses from any other of the natural enemies, including the pupal parasitoids *Sphegigaster pallicornis* and *C. pubicornis*. Heads and Lawton were working not in Canada but in northern England. Nonetheless, it is noteworthy that whereas neither *S. pallicornis* nor *C. pubicornis* was apparently successful in combating the leaf-miner in Canada, *C. gemma* was reported by the entomologists working on the project to be an effective control agent.

... the olive scale ...

On the other hand, a case of successful biological control (of the olive scale,

Parlatoria oleae) has been described in which no aggregative responses were apparent even though they were searched for (Murdoch *et al.*, 1984); and Murdoch *et al.* (1985) have argued that, in general, pest populations persist after successful biological control not as a result of aggregative responses but because of the stochastic creation of host patches by colonization and their subsequent extinction when discovered by the agent: essentially, a metapopulation effect. Waage and Greathead (1988), however, suggest that a broader perspective can incorporate both the importance of aggregative responses and the views of Murdoch and his colleagues. They propose that scale insects and other homopterans, and mites (like Huffaker's), which may reproduce to have many generations within a patch, are often stabilized by asynchronies in the dynamics of different patches; whereas lepidopterans and hymenopterans, which typically occupy a patch for only part of a single generation, may often be stabilized by an aggregative response.

... and prickly pear cacti

The importance of heterogeneity in biological control appears to be reaffirmed by a plant–herbivore interaction in the control of the prickly pear cacti, *Opuntia inermis* and *O. stricta*, in Australia (Monro, 1967; see p. 355). Prior to the introduction of the moth *Cactoblastis cactorum* in 1925, vast areas of otherwise useful land in Australia were covered with dense stands of these cacti. *Cactoblastis* was then introduced precisely because it was a natural consumer of the prickly pear in South America, where they both originated. As the moths spread to more and more areas between 1928 and 1935, the dense stands of cacti were virtually wiped out. A population crash usually took about 2 years, and the prickly pear has remained at low (and economically acceptable) levels ever since ($N^* / K_N = 0.002$, according to Caughley & Lawton, 1981).

The *Cactoblastis* larvae are doubly aggregated: eggs are laid 80 at a time in egg-sticks, and the egg-sticks themselves have a clumped distribution amongst the plants. In addition, the larvae have a very limited mobility, and they are likely to perish along with the plant itself when the plant carries too many egg-sticks (roughly, more than two per plant). Thus, in this case, the interaction is apparently stabilized at low densities by the pseudo-interference resulting from the high death rate of larvae on overloaded plants at high larval densities.

One major problem in making pronouncements about the stabilizing role of aggregation is that although, as we have seen, there have been wide-ranging surveys of the data on spatial distributions of attacks (Lessells, 1985; Stiling, 1987; Walde & Murdoch, 1988; Pacala & Hassell, 1991), these data generally come from studies of very short duration—often of only one generation. We do not know if the observed spatial patterns are typical for that interaction; nor do we know if the population dynamics show the degree of stability that the spatial patterns might seem to predict.

real data confirm the complexity of natural systems

One investigation that, by contrast, has examined several generations of population dynamics and spatial distributions is that of Redfern *et al.* (1992), who made a 7-year (seven-generation) study of two tephritid fly species (the prey) that attack thistle flower-heads, and their associated parasitoid guilds. For one, *Terellia serratulae*, there was evidence of year-to-year density dependence in the overall rate of parasitism (Figure 10.14a), but no strong evidence of significant levels of aggregation within generations, either overall (Figure 10.14c) or for parasitoid species individually. For the other species, *Urophora stylata*, the data seemed to tell the opposite story (Figures 10.14b, d), and to repeat a pattern we have seen before, most heterogeneity was contributed by the HDI component. The study is equally informative, however, as a cautionary tale. Both hosts were attacked by several

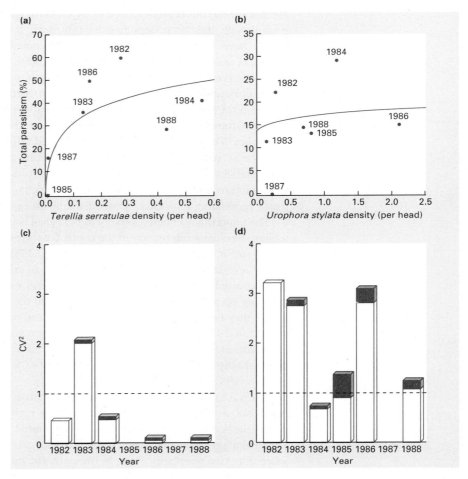

Figure 10.14 Attacks by parasitoids on tephrytid flies (*Terellia serratulae* and *Urophora stylata*) that attack thistle flower-heads. Temporal density dependence of parasitoid attacks on *T. serratulae* (a) is significant ($r^2 = 0.75$; $P < 0.05$), but for *U. stylata* (b) it is not ($r^2 = 0.44$; $P > 0.05$); both fitted lines take the form $y = a + b\log_{10}x$. However, whereas for *T. serratulae* (c) there is little aggregation of risk of parasitoid attack within years (measured as CV^2), with *U. stylata* (d) there is far more, most of which is HDI (no shading) rather than HDD (dark shading). (After Redfern *et al.*, 1992.)

parasitoid species—not one, as assumed by most models. Levels of aggregation (and to a lesser extent the HDI or HDD contributions) varied considerably and apparently randomly from year to year (Figures 10.14c, d): no one year was typical (or informative). Finally, there was no clear reflection of the differing patterns of temporal density dependence and spatial aggregation in the hosts' population dynamics (Figures 10.14a, b).

Another multigeneration study which provides support for the stabilizing powers of a particular type of heterogeneity, a physical refuge (see Section 10.5.1), was illustrated in Figure 10.4. A proportion of a prey population may be in a refuge because they are beyond the reach of their predator's foraging apparatus. In this case, hosts (*Plodia interpunctella*) living deeper in their food are beyond the reach of the parasitoids (*Venturia canescens*) attempting to lay their eggs in them. In the absence of this refuge, in a shallow food medium, this host–parasitoid interaction is unable to persist (see Figure 10.4c), although the host alone does so readily (see Figure 10.4d). With a refuge present, however, in a deeper food medium, the host and parasitoid can apparently persist together indefinitely (see Figure 10.4a).

Clearly, spatial and other heterogeneities *can* exert important regulatory effects on prey populations. However, as we have seen, there are a number of roles that heterogeneity can play, and its effectiveness in these roles depends on its interaction with other factors. We have also seen that the combination of delayed density dependence and the fluctuations characteristic of all natural environments can lead to patterns of generation mortality which do not immediately suggest population regulation at all, even though powerful regulatory forces may be acting within each generation. Hassell (1985) has remarked that the solution lies with 'two-dimensional' sampling programmes that follow populations over several generations but also document the spatial and other variations occurring within generations. Certainly, without such a dual approach we are unlikely to identify the true reasons for the relative stability of many natural populations. But recent advances have emphasized the range of within-generation information required, and have suggested that for 'several' generations we may have to read 'many'.

10.6 Multiple equilibria: an explanation for outbreaks

Ecologists working in a variety of fields have come to realize that when predator and prey populations interact, there is not necessarily just one equilibrium combination of the two (about which there may or may not be oscillations). There can, instead, be 'multiple equilibria', or 'alternative stable states'. The idea has been pursued most often in cases where the predator or herbivore numbers are not determined by the abundance of their prey (see May, 1977, for a review). Good examples of this would be a herd or flock of grazing mammals (where the flock size is determined by the farmer; see Noy-Meir, 1975), or a fleet of trawlers exploiting fish populations (where the fleet size is determined by the trawler owners—or perhaps by international treaty). Situations of this type will be considered in the sections on harvesting (see Chapter 16), but multiple equilibria can also emerge when predators and prey interact with one another.

Figure 10.15 is a model with multiple equilibria. The prey zero isocline has both a vertical section at low densities and a hump. This could reflect a type 3 functional response of a predator that also has a long handling time, or perhaps the combination of an aggregative response and a decline in prey recruitment at low density. As a consequence, the predator zero isocline crosses the prey zero isocline three times. The strengths and directions of the arrows in Figure 10.15a indicate that two of these points (X and Z) are fairly stable equilibria (although some oscillations around each are to be expected). The third point however (point Y) is unstable: populations near here will move towards either point X or point Z. Moreover, there are joint populations close to point X where the arrows lead to the zone around point Z, and joint populations close to point Z where the arrows lead back to the zone around point X. Even small environmental perturbations could put a population near point X on a path towards point Z, and vice versa.

The behaviour of a hypothetical population, consistent with the arrows in Figure 10.15a, is plotted in Figure 10.15b on a joint abundance diagram, and in Figure 10.15c as a graph of numbers against time. The prey population, in particular,

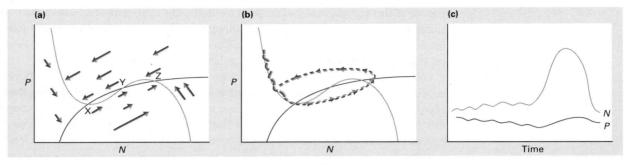

Figure 10.15 A predator–prey zero isocline model with multiple equilibria. (a) The prey zero isocline has a vertical section at low densities and a hump; the predator zero isocline can therefore cross it three times. Intersections X and Z are stable equilibria, but intersection Y is an unstable 'breakpoint' from which the joint abundances move towards either intersection X or intersection Z. (b) A feasible path that the joint abundances might take when subject to the forces shown in (a). (c) The same joint abundances plotted as numbers against time, showing that an interaction with characteristics that do not change can lead to apparent 'outbreaks' in abundance.

... leading to 'outbreak' or 'eruptive' patterns of abundance

displays an 'outbreak' or 'eruption' in abundance, as it moves from a low-density equilibrium to a high-density equilibrium and back again. This eruption is in no sense a reflection of an equally marked change in the environment. It is, on the contrary, a pattern of abundance generated by the interaction itself (plus a small amount of environmental 'noise'), and in particular it reflects the existence of multiple equilibria. Similar explanations may be invoked to explain apparently complicated patterns of abundance in nature.

10.6.1 Multiple equilibria in nature?

There are certainly examples of natural populations exhibiting outbreaks of abundance from levels that are otherwise low and apparently stable (Figures 10.16a–c); and there are other examples in which populations appear to alternate between two stable densities (Figure 10.16d). But it does not follow that each of these examples is necessarily an interaction with multiple equilibria.

In some cases, a plausible argument for multiple equilibria can be put forward. This is true, for instance, of Clark's (1964) work in Australia on the eucalyptus psyllid (*Cardiaspina albitextura*), a homopteran bug (Figure 10.16a). These insects appear to have a low-density equilibrium maintained by their natural predators (especially birds), and a much less stable high-density equilibrium reflecting intraspecific competition (the destruction of host tree foliage leading to reductions in fecundity and survivorship). Outbreaks from one to the other can occur when there is just a short-term failure of the predators to react to an increase in the density of adult psyllids.

Spruce budworm populations (*Choristoneura fumiferana*; Figure 10.16b) also appear to exhibit multiple equilibria, although probably not of the exact type described in Figure 10.15 (see Peterman *et al.*, 1979). In an immature balsam fir and white spruce forest, or in a mixed spruce-fir–hardwood stand, the quality of food for the budworm is low and their rate of recruitment is low. It has therefore been hypothesized that they are held in check at a low-density stable equilibrium by their

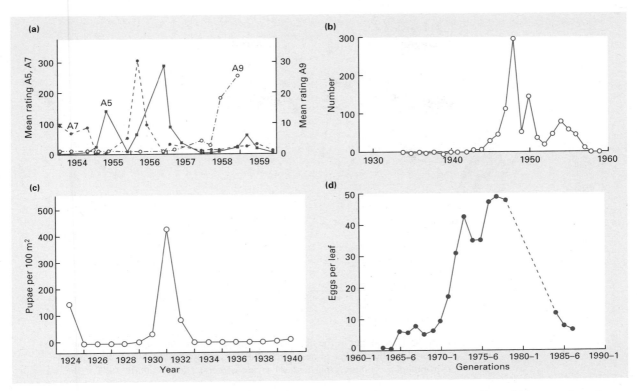

Figure 10.16 Possible examples of outbreaks and multiple equilibria. (a) Mean ratings of relative abundance of the eucalyptus psyllid, *Cardiaspina albitextura*, in three study areas in Australia (A5, A7 and A9). (After Clark, 1962.) (b) Observed changes in numbers of the spruce budworm, *Choristoneura fumiferana*, in Canada. (After Morris, 1963.) (c) Counts of pupae of the pine beauty moth, *Panolis flammaea*, whose larvae eat needles of the Scots pine in Germany. (After Schwerdtfeger, 1941.) (d) The mean number of eggs per leaf of the viburnum whitefly, *Aleurotrachelus jelinekii*, on a viburnum bush in Silwood Park, Berkshire, UK. No samples were taken between 1978 and 1979, and 1984 and 1985. (After Southwood *et al.*, 1989.)

predators (as in Figure 10.8, curve (i)), although there is no experimental proof of this important assumption. However, as a spruce and fir forest gradually matures, food suitability improves, and there is an increase in the rate of budworm recruitment and in the carrying capacity of the forest. The spruce budworm zero isocline may then rise above that of its predators such that the isoclines only cross at a high-density equilibrium (as in Figure 10.7). This outbreak, however, quickly destroys the mature trees. The forest therefore reverts to relative immaturity, and the isoclines and equilibria revert to those in Figure 10.8.

The important point is that there can be a large change in abundance (i.e. an outbreak and subsequent crash) as a result of either a small change in carrying capacity (i.e. maturity) or a small environmental perturbation when maturity is close to a critical threshold value.

It is particularly interesting that the empirical observation of two alternative equilibria in Figure 10.16d for the viburnum whitefly, *Aleurotrachelus jelinekii*, is reinforced by a model for that population which predicts the same pattern (Southwood *et al.*, 1989). The basis for this model is that the shortfall between the

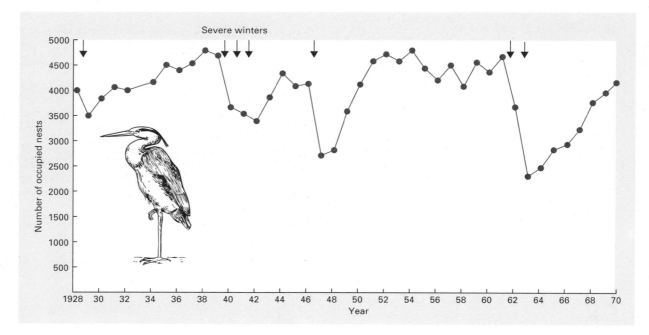

Figure 10.17 Changes in the abundance of herons (*Ardea cinerea*) in England and Wales (measured by the number of nests occupied) are readily attributable to changes in environmental conditions (particularly severe winters). (After Stafford, 1971.)

... but sudden changes in abundance can also result from sudden changes in the environment

observed and the potential number of eggs laid in any generation is density dependent, but specifically falls into two quite distinct relationships depending on the density of potential eggs per leaf.

On the other hand, there are many cases in which sudden changes in abundance are fairly accurate reflections of sudden changes in the environment or a food source. For instance, the number of herons nesting in England and Wales normally fluctuates around 4000–4500 pairs, but the population declines markedly after particularly severe winters (Figure 10.17; Stafford, 1971). This fish-eating bird is unable to find sufficient food when inland waters become frozen for long periods, but there is no suggestion that the lower population levels (2000–3000 pairs) are an alternative equilibrium. The population crashes are simply the result of density independent mortality from which the herons rapidly recover.

10.7 Conclusion: beyond predator–prey

The simplest mathematical models of predator–prey interactions produce coupled oscillations that are highly unstable. However, by adding various elements of realism to these models it is possible to reveal the features of real predator–prey relationships that are likely to contribute to their stability. Crowding effects in either population (intraspecific competition amongst prey, mutual interference amongst predators), limited dispersal in spatially explicit models and heterogeneity/aggregative responses can be particularly powerful stabilizing influences. A further insight provided by models is that predator–prey systems may exist in more than one stable state.

We have seen that a variety of patterns in the abundance of predators and prey, both in nature and in the laboratory, are consistent with the conclusions derived from models. Unfortunately, we are rarely in a position to apply specific explanations to particular sets of data, because the critical experiments and observations to test the models have rarely been made. Natural populations are affected not just by their predators or their prey, but also by many other environmental factors that serve to 'muddy the waters' when direct comparisons are made with simple models.

Moreover, the attention of both modellers and data-gatherers (not that the two need be different) is increasingly being directed away from single- or two-species systems, towards those in which three species interact: for example, a pathogen attacking a predator which attacks a prey, or a parasitoid and a pathogen both attacking a prey/host. Interestingly, in several of these systems, unexpected dynamical properties emerge that are not just the expected blend of the component two-species interactions (Hassell & Anderson, 1989; Begon *et al.*, 1995b). We return to the problems of 'abundance' in a broader context in Chapter 15.

Chapter 11
Decomposers and Detritivores

11.1 Introduction

When plants and animals die, their bodies become resources for other organisms. Of course, in a sense, most consumers live on dead material—the carnivore catches and kills its prey, and the living leaf taken by a herbivore is dead by the time digestion starts. The critical distinction between the organisms in this chapter, and herbivores, carnivores and parasites, is that the latter all directly affect the rate at which their resources are produced. Whether it is lions eating gazelles, gazelles eating grass or grass parasitized by a rust fungus, the act of taking the resource harms its ability to regenerate new resource (more gazelles or grass leaves). Some categories of mutualist, on the other hand, can *increase* the supply of resources provided by their partner (see Chapter 13). In contrast with these groups, decomposers (bacteria and fungi) and detritivores (animal consumers of dead matter) do not control the rate at which their resources are made available or regenerate; they are dependent on the rate at which some other force (senescence, illness, fighting, the shedding of leaves by trees) releases the resource on which they live. Exceptions exist amongst necrotrophic parasites (see Chapter 12) that kill and then continue to extract resources from the dead host. Thus, *Botrytis cinerea* attacks living bean leaves but continues its life, after the host's death, as a decomposer. Similarly, maggots of the sheep blowfly *Lucilia cuprina*, may parasitize and kill their host, whereupon they become detritivores on the corpse. In these cases the decomposer/detritivore can be said to have a measure of control over the supply of its food resource.

Pimm (1982) describes the relationship that generally exists between decomposers or detritivores and their food as *donor-controlled*: the donor (prey) controls the density of the recipient (predator), but not the reverse. The dynamics of donor-controlled models differ in a number of ways from the traditional Lotka–Volterra models of predator–prey interactions (see Chapter 10). One important distinction is that interacting assemblages of species that possess donor-controlled dynamics are expected to be particularly stable, and furthermore, that stability is independent of, or actually increases with, increased species richness and food-web complexity. This is in complete contrast to the situation where Lotka–Volterra dynamics apply. Detailed discussion of the important topic of food-web complexity and community stability is reserved for Chapter 22.

Whilst there is generally no direct feedback between the consumer population and the resource, nevertheless it is possible to see an indirect effect, for example via the release of nutrients from decomposing litter, that may ultimately affect the rate at

decomposers and detritivores do not generally control their supply of resources ...

... their interactions are 'donor-controlled'

which trees produce more litter. In fact, it is in nutrient recycling that decomposers and detritivores play their most fundamental role.

Immobilization occurs when an inorganic nutrient element is incorporated into organic form—primarily during the growth of green plants. Conversely, decomposition involves the release of energy and the *mineralization* of chemical nutrients—conversion of elements from organic to inorganic form. Decomposition is defined as the gradual disintegration of dead organic matter and is brought about by both physical and biological agencies. It culminates with complex energy rich molecules being broken down by their consumers (decomposers and detritivores) into carbon dioxide, water and inorganic nutrients. Some of the chemical elements will have been locked up for a time as part of the body structure of the decomposer organisms, and the energy present in the organic matter will have been used to do work and is eventually lost as heat. Ultimately, the incorporation of solar energy in photosynthesis, and the immobilization of inorganic nutrients into biomass, is balanced by the loss of heat energy and organic nutrients when the organic matter is mineralized. Thus, a given nutrient molecule may be successively immobilized and mineralized in a repeated round of nutrient cycling. We discuss the overall role played by decomposers and detritivores in the fluxes of energy and nutrients at the community level in Chapters 18 and 19. In the present chapter, we introduce the organisms involved and look in detail at the ways they deal with their resources.

It is not only the bodies of dead animals and plants that serve as resources for decomposers and detritivores. Dead organic matter is continually produced during the life of both animals and plants and can be a major resource. Unitary organisms shed dead parts as they develop and grow—the larval skins of arthropods, the skins of snakes, the skin, hair, feathers and horn of other vertebrates. Specialist feeders are often associated with these cast-off resources. Amongst the fungi there are specialist decomposers of feathers and of horn, and there are arthropods that specialize on sloughed-off skin. Human skin is a resource for the household mites that are omnipresent inhabitants of house dust, and cause problems for many allergy sufferers.

The continual shedding of dead parts is even more characteristic of modular organisms. The older polyps on a colonial hydroid or coral die and decompose, whilst other parts of the same genet continue to regenerate new polyps. Most plants shed old leaves and grow new ones; the seasonal litter-fall onto a forest floor is the most dramatic of all the sources of resource for decomposers and detritivores, but the producers do not die in the process. Higher plants also continually slough off cells from the root caps, and root cortical cells die as a root grows through the soil. This supply of organic material produces the very resource-rich *rhizosphere*. Plant tissues are generally leaky, and soluble sugars and nitrogenous compounds also become available on the surface of leaves, supporting the growth of bacteria and fungi in the *phyllosphere*.

Finally, animal faeces, whether produced by detritivores, herbivores, carnivores or parasites, are a further category of resource for decomposers and detritivores. They are composed of dead organic material that is chemically related to what their producers have been eating.

The remainder of this chapter is in two parts. In Section 11.2, we describe the 'actors' in the decomposition 'play', and consider the relative roles of the bacteria and fungi on the one hand, and the detritivores on the other. Then, in Section 11.3, we consider, in turn, problems and processes involved in the consumption by detritivores of plant detritus, faeces and carrion.

decomposition defined

decomposition of dead bodies

decomposition of shed parts of organisms

decomposition of faeces

11.2 The organisms

11.2.1 The decomposers: bacteria and fungi

If a scavenger does not take a dead resource immediately, the process of decomposition usually starts with colonization by bacteria and fungi. Other changes may occur at the same time: enzymes in the dead tissue may start to autolyse it and break down carbohydrates and proteins to simpler, soluble forms. The dead material may also become leached by rainfall or, in an aquatic environment, may lose minerals and soluble organic compounds as they are washed out in solution.

bacteria and fungi are early colonists of newly dead material

Bacteria and fungal spores are omnipresent in the air and the water, and are usually present on (and often in) dead material before it is dead. They usually have first access to a resource, together with the necrotrophic parasites (see Chapter 12). These early colonists tend to use soluble materials, mainly amino acids and sugars, that are freely diffusible. They lack the array of enzymes necessary for digesting structural materials such as cellulose, lignin, chitin and keratin. Many species of *Penicillium*, *Mucor* and *Rhizopus*, the so-called 'sugar fungi' in soil, grow fast in the early phases of decomposition; together with bacteria having similar opportunistic physiologies, they tend to undergo population explosions on newly dead substrates. As the freely available resources are consumed, these populations collapse, leaving very high densities of resting stages from which new population explosions may develop when another freshly dead resource becomes available. They may be thought of as the opportunist '*r*-selected species' amongst the decomposers (see Chapter 14, Section 14.9), and can be compared with the 'tramps' and 'supertramps' amongst the bird colonists of islands (see Chapter 5, Section 5.3.1).

The early phases of the decomposition process are encountered in a variety of domestic and industrial situations.

domestic and industrial decomposition

1   The early colonizers of staling bread are the opportunistic starch and sugar fungi such as *Penicillium* and *Mucor*.

2   The early colonizers of the nectar in flowers are predominantly yeasts (simple sugar fungi), and these may spread to the ripe fruit, or explode as populations in juice to give the well-known staling products that are wine and beer.

3   Making silage from green fodder for conservation as a livestock feed, or sauerkraut from cabbage for human consumption, depends on primary colonizers that metabolize soluble or readily solubilized carbohydrates. The organisms involved are bacteria, mainly lactobacilli, that respire soluble sugars and produce organic acids (mainly lactic acid) when the fermentation is anaerobic. The acidity may 'pickle' the plant material by reducing the pH below a level at which other decomposers can act.

4   Plants that are used as sources of stem fibres (e.g. flax and jute) are 'retted' microbially to free the fibres from the rest of the plant tissue. Bacteria are primarily responsible for this process, and they enter through cuts and damage points and grow up through the soft, easily decomposed tissues, attacking cell walls that have not yet become thickened with cellulose and lignins. Again we see simple, opportunistic invaders as the first decomposers.

aerobic and anaerobic decomposition in nature

In nature, as in the various industrial and domestic processes, the activity of the early colonizers is dominated by the metabolism of sugars and is strongly influenced by aeration. When oxygen is in free supply, sugars are metabolized to carbon dioxide by growing microbes. Under anaerobic conditions, fermentations produce a less

404   CHAPTER 11

complete breakdown of sugars to by-products such as alcohol and organic acids that change the nature of the environment for subsequent colonizers. In particular, the lowering of the pH by the production of acids has the effect of favouring fungal as opposed to bacterial activity.

The most widespread class of anoxic habitats occurs in sediments of oceans and lakes that receive a continuous supply of dead organic matter from the water column above. Aerobic decomposition (mainly by bacteria) quickly exhausts the available oxygen because this can only be supplied from the surface of the sediment by diffusion. Thus, at some depth, from zero to a few centimetres below the surface, depending mainly on the load of organic material, sediments are completely anoxic. Below this level are found a variety of bacterial types that employ different forms of anaerobic respiration, i.e. they use terminal inorganic electron acceptors other than oxygen in their respiratory process. The bacterial types occur in a predictable pattern with denitrifying bacteria at the top, sulphate-reducing bacteria next and methanogenic bacteria in the deepest zone. Sulphate is comparatively abundant in seawater and so the zone of sulphate-reducing bacteria is particularly wide. In contrast, freshwater sediments contain much less sulphate, and methanogenesis plays a correspondingly larger role (Fenchel, 1987b).

A strong element of chance determines which species are the first to colonize newly dead material, but in some environments there are specialists with properties that enhance their chances of arriving early. Litter that falls into streams or ponds is often colonized by aquatic fungi (e.g. Hyphomycetes), which bear spores with sticky tips (Figure 11.1a), and are often of a curious form that seems to maximize their chance of being carried to and sticking to leaf litter. They may spread by growing from cell to cell within the tissues (Figure 11.1b).

After colonization of dead material by the 'sugar' fungi and bacteria, and perhaps also after leaching by rain or in the water, the residual resources are not diffusible and are more resistant to attack. In broad terms, the major components of dead organic matter are, in a sequence of increasing resistance to decomposition: sugars < (less resistant than) starch < hemicelluloses, pectins and proteins < cellulose < lignins < suberins < cutins. Hence, after an initial rapid breakdown of sugar, decomposition proceeds more slowly, and involves microbial specialists that can use celluloses and lignins and break down the more complex proteins, suberin (cork) and cuticle. These are structural compounds, and their breakdown and

zonation of anaerobic bacteria in aquatic sediments

after the initial phase, decomposition of more resistant tissues proceeds more slowly

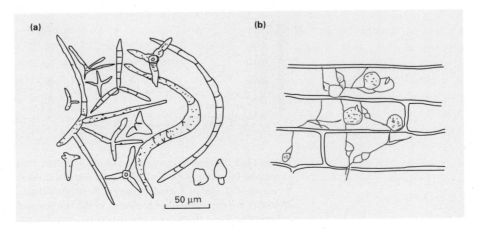

Figure 11.1 (a) Spores (conidia) of aquatic hyphomycete fungi from river foam. (After Webster, 1970.)
(b) Rhizomycelium of the aquatic fungus *Cladochytrium replicatum* within the epidermis of an aquatic plant. The circular bodies are zoosporangia. (After Webster, 1970.)

50 μm

metabolism depend on very intimate contact with the decomposers (most cellulases are surface enzymes requiring actual physical contact between the decomposer organism and its resource). The processes of decomposition may now depend on the rate at which fungal hyphae can penetrate from cell to cell through lignified cell walls. Few, if any, organisms are able to attack both lignin and cellulose. In the decomposition of wood by fungi, two major categories of specialist decomposers can be recognized: the brown rots which can decompose cellulose but leave a predominantly lignin-based brown residue, and the white rots which decompose mainly the lignin and leave a white cellulosic residue.

there is a natural succession of decomposing microorganisms

The organisms capable of dealing with progressively more refractory compounds represent a natural succession starting with simple sugar fungi (mainly Phycomycetes and Fungi Imperfecti), and usually followed by septate fungi (Basidiomycetes and Actinomycetes) and Ascomycetes which are slower growing, spore less freely, make intimate contact with their substrate and have more specialized metabolism (Pugh, 1980). The diversity of the microflora that decomposes a fallen leaf tends to decrease, as fewer but more highly specialized species are concerned with the last and most resistant remains. An example of the succession of fungi involved in the decomposition of forest litter is given in Chapter 17, Section 17.4.1.

the changing chemical nature of the resource

The nature of a decomposing resource changes during the course of the succession; indeed, the changes are largely responsible for driving the succession. The changing nature of a resource is illustrated in Figure 11.2 for oak-leaf litter on a woodland floor and in a small stream running through woodland. The patterns are remarkably similar. Soluble carbohydrates quickly disappeared, mainly as a result of leaching in water (and comparatively more quickly in the aquatic environment). The resistant structural cellulose and hemicellulose components were reduced through enzymatic degradation more slowly, and lignin (only measured in the woodland floor study) most slowly of all. Lipid (only measured in the stream study) was degraded at an intermediate rate.

most microbial decomposer species are relatively specialized

Individual species of microbial decomposer are not biochemically very versatile; most of them can cope with only a limited number of substrates. It is the diversity of species involved that allows the structurally and chemically complex tissues of a

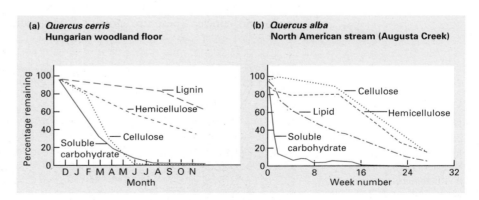

Figure 11.2 Changes in composition of oak-leaf litter during decomposition in contrasting situations: (a) leaves of *Quercus cerris* on a woodland floor in Hungary, through the year; (b) leaves of *Q. alba* in a small stream in North America, during a 28-week experiment. Amounts are expressed as percentages of the starting quantities. (After Toth *et al.*, 1975; Suberkropp *et al.*, 1976, respectively.)

plant or animal corpse to be decomposed. Between them, a varied flora of bacteria and fungi can accomplish the complete degradation of dead material of both plants and animals; but in practice they seldom act alone, and the process would be much slower and, moreover, incomplete, if they did so. The major factor that delays the decomposition of organic residues is the compartmentalized cellular structure of plant remains—an invading decomposer meets far fewer barriers in an animal body. The process of plant decomposition is enormously speeded up by any activity that grinds up and fragments the tissues, such as the chewing action of detritivores. This breaks open cells and exposes the contents and the surfaces of cell walls to attack.

11.2.2 The detritivores and specialist microbivores

specialist bacterial and fungal microbivores

The microbivores are a group of animals that operate alongside the detritivores, and which can be difficult to distinguish from them. The name microbivore is reserved for the minute animals that specialize at feeding on microflora, and are able to ingest bacteria or fungi but to exclude detritus from their guts. Exploitation of the two major groups of microflora requires quite different feeding techniques, principally because of differences in growth form. Bacteria (and yeasts) show a colonial growth form arising by the division of unicells, usually on the surface of small particles. Specialist consumers of bacteria are inevitably very small; they include free-living protozoans such as amoebae, in both soil and aquatic environments, and the terrestrial nematode worm *Pelodera*, which does not consume whole sediment particles but grazes amongst them consuming the bacteria on their surfaces. Fungi, in contrast to bacteria, produce extensively branching, filamentous hyphae, that in many species are capable of penetrating organic matter. Some specialist consumers of fungi possess piercing, sucking stylets (e.g. the nematode *Ditylenchus*) that they insert into individual fungal hyphae. However, most fungivorous animals graze on the hyphae and consume them whole. In some cases, close mutualistic relationships exist between fungivorous beetles, ants and termites and characteristic species of fungi. These mutualisms are discussed in Chapter 13. Note that microbivores consume a living resource and cannot be described as being donor-controlled.

most detritivores consume both detritus and its microflora

The larger the animal, the less able it is to distinguish between microflora as food and the plant or animal detritus on which these are growing. In fact, the majority of the detritivorous animals involved in the decomposition of dead organic matter are generalist consumers, of both the detritus itself and the associated microfloral populations.

terrestrial detritivores are usually classified according to their size

The invertebrates that take part in the decomposition of dead plant and animal materials are a taxonomically diverse group. In terrestrial environments they are usually classified according to their size. This is not an arbitrary basis for classification, because size is an important feature for organisms that reach their resources by burrowing or crawling amongst cracks and crevices of litter or soil. The *microfauna* (including the specialist microbivores) includes protozoans, nematode worms and rotifers (Figure 11.3). The principal groups of the *mesofauna* (animals with a body width between 100 μm and 2 mm) are litter mites (Acari), springtails (Collembola) and pot worms (Enchytraeidae). The *macrofauna* (2–20 mm body width) and, lastly, the *megafauna* (greater than 20 mm) include woodlice (Isopoda), millipedes (Diplopoda), earthworms (Megadrili), snails and slugs (Mollusca) and the larvae of certain flies (Diptera) and beetles (Coleoptera). These animals are mainly responsible for the initial shredding of plant remains. By their action, they may bring

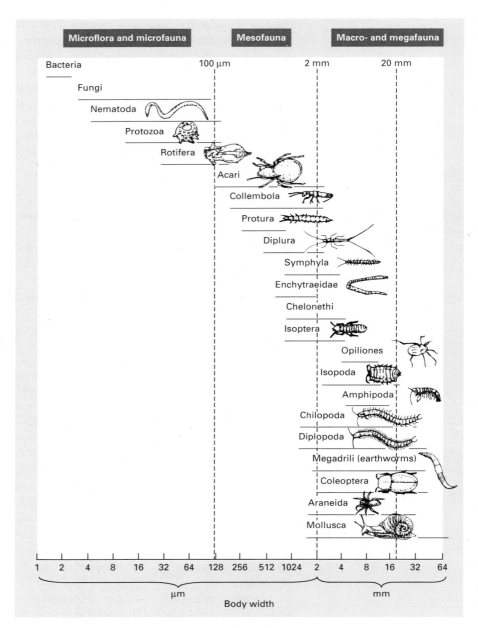

Figure 11.3 Size classification by body width of organisms in terrestrial decomposer food webs. The following groups are wholly carnivorous: Opiliones (harvest spiders), Chilopoda (centipedes) and Araneida (spiders). (After Swift *et al.*, 1979.)

about a large-scale redistribution of detritus, and thus contribute directly to the development of soil structure.

the important role of earthworms

Darwin (1888) estimated that earthworms in some pastures close to his house formed a new layer of soil 18 cm deep in 30 years, bringing about 50 tons ha^{-1} (5.1 kg m^{-2}) to the soil surface each year as worm casts. Figures of this order of magnitude have since been confirmed on a number of occasions. Even higher values have been reported from a pasture in western Nigeria: 170 tons ha^{-1} (17.3 kg m^{-2}), all in the space of 2–6 months of the rainy season (Madge, 1969). Moreover, not all species of earthworm put their casts above ground, so the total amount of soil and organic matter that they move may be much greater than this. Where earthworms are abundant, they bury litter, mix it with the soil (and so expose it to other decomposers and detritivores), create burrows (so increasing soil aeration and drainage) and deposit faeces rich in organic matter. It is not surprising that agricultural ecologists become worried about practices that reduce worm populations.

Detritivores occur in all types of terrestrial habitat and are often found at remarkable species richness and in very great numbers. Thus, for example, 1 m² of temperate woodland soil may contain 1000 species of animals, in populations exceeding 10 million for nematode worms and protozoans, 100 000 for springtails (Collembola) and soil mites (Acari) and 50 000 or so for other invertebrates (Anderson, 1978). The relative importance of microfauna, mesofauna and macrofauna in terrestrial communities varies along a latitudinal gradient (Figure 11.4). The microfauna is relatively more important in the organic soils in boreal forest, tundra and polar desert. Here, the plentiful organic matter stabilizes the moisture regime in the soil and provides suitable microhabitats for protozoans, nematodes and rotifers

the relative importance of micro-, meso- and macrofauna varies latitudinally

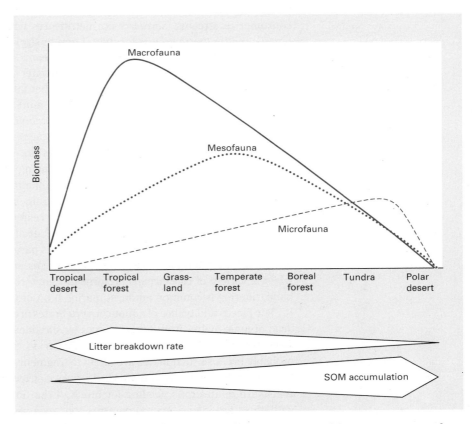

Figure 11.4 Patterns of latitudinal variation in the contribution of the macro-, meso- and microfauna to decomposition in terrestrial ecosystems. Soil organic matter (SOM) accumulation (inversely related to litter breakdown rate) is promoted by low temperatures and waterlogging, where microbial activity is impaired. (After Swift *et al.*, 1979.)

409 DECOMPOSERS AND DETRITIVORES

that live in interstitial water films. The hot, dry, mineral soils of the tropics have few of these animals. The deep organic soils of temperate forests are intermediate in character; they maintain the highest mesofaunal populations of litter mites, springtails and pot worms. The majority of the other soil animal groups decline in numbers towards the drier tropics, where they are replaced by termites.

the terrestrial detritivore community also varies with dryness

On a more local scale too, the nature and activity of the decomposer community depends on the conditions in which the organisms live. Temperature has a fundamental role in determining the rate of decomposition (Swift *et al.*, 1979), and moreover, the thickness of water films on decomposing material places absolute limits on mobile microfauna and microflora (protozoa, nematode worms, rotifers and those fungi that have motile stages in their life cycles; see Chapter 3, Figure 3.10). In dry soils, such organisms are virtually absent. A continuum can be recognized from dry conditions through waterlogged soils to true aquatic environments. In the former, the amount of water and thickness of water films are of paramount importance, but as we move along the continuum, conditions change to resemble more and more closely those of the bed of an open-water community, where oxygen shortage, rather than water availability, may dominate the lives of the organisms.

aquatic detritivores are usually classified according to their feeding mode

In freshwater ecology the study of detritivores has been concerned less with the size of the organisms than with the ways in which they obtain their food. Cummins (1974) devised a scheme that recognizes four main categories of invertebrate consumer in streams. *Shredders* are detritivores which feed on coarse particulate organic matter (particles greater than 2 mm in size), and during feeding these serve to fragment the material. Very often in streams, the shredders, such as cased caddis-fly larvae of *Stenophylax* spp., freshwater shrimps *(Gammarus* spp.) and isopods (e.g. *Asellus* spp.), feed on tree leaves that fall into the stream. *Collectors* feed on fine particulate organic matter (less than 2 mm). Two subcategories of collectors are defined. *Collector–gatherers* obtain dead organic particles from the debris and sediments on the bed of the stream, whereas *collector–filterers* sift small particles from the flowing column of water. Some examples are shown in Figure 11.5. *Grazer–scrapers* have mouth parts appropriate for scraping off and consuming the organic layer attached to rocks and stones, comprised of attached algae, bacteria, fungi and dead organic matter adsorbed to the substrate surface. The final invertebrate category is *carnivores*. Figure 11.6 shows the relationships amongst these invertebrate feeding groups and three categories of dead organic matter. This scheme, developed for stream communities, has obvious parallels in terrestrial ecosystems as well as in other aquatic ecosystems. Earthworms are important shredders in soils, whilst a variety of crustaceans perform the same role on the seabed. On the other hand, filtering is common amongst marine but not terrestrial organisms.

most aquatic detritivores are markedly generalized

The faeces and bodies of aquatic invertebrates are generally processed along with dead organic matter from other sources by shredders and collectors. Even the large faeces of aquatic vertebrates do not appear to possess a characteristic fauna, probably because such faeces are likely to fragment and disperse quickly as a result of water movement. Carrion also lacks a specialized fauna—many aquatic invertebrates are omnivorous, feeding for much of the time on plant detritus and faeces with their associated microorganisms, but ever ready to tackle a piece of dead invertebrate or fish when this is available. This contrasts very strongly with the situation in the terrestrial environment, where both faeces and carrion have specialized detritivore faunas (see later, pp. 420–427).

Some animal communities are composed almost exclusively of detritivores and

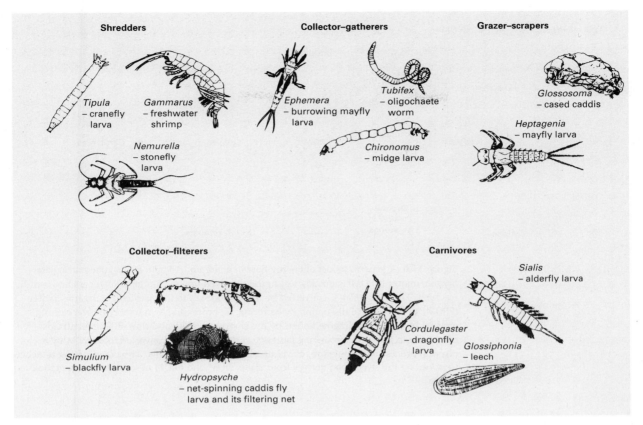

Figure 11.5 Examples of the various categories of invertebrate consumer in freshwater environments.

detritivore-dominated
communities

their predators. This is true not only of the forest floor, but also of shaded streams, the depths of oceans and lakes and the permanent residents of caves: in short, wherever there is insufficient light for appreciable photosynthesis, but nevertheless, an input of organic matter from nearby plant communities. The forest floor and shaded streams receive most of their organic matter as dead leaves from the trees. The beds of oceans and lakes are subject to a continuous settlement of detritus from above. Caves receive dissolved and particulate organic matter percolating down through soil and rock, together with windblown material and debris of migrating animals.

11.2.3 The relative roles of microflora and detritivores

The roles of the microflora and the detritivores in decomposing dead organic matter can be compared in a variety of ways. A comparison of numbers will reveal a predominance of bacteria. This is almost inevitable because we are counting individual cells. A comparison of biomass gives a quite different picture. Figure 11.7 shows the relative amounts of biomass represented in different groups involved in the decomposition of litter on a forest floor (expressed as the relative amounts of nitrogen present). For most of the year, decomposers (microorganisms) accounted for

411 DECOMPOSERS AND DETRITIVORES

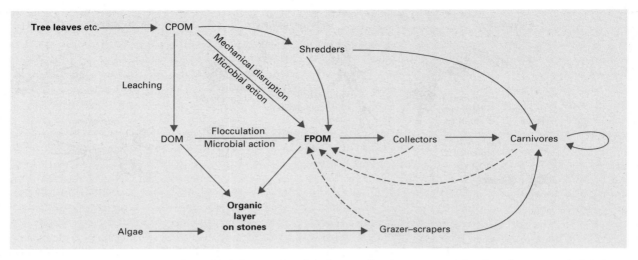

Figure 11.6 A general model of energy flow in a stream. A fraction of coarse particulate organic matter (CPOM) is quickly lost to the dissolved organic matter (DOM) compartment by leaching. The remainder is converted by three processes to fine particulate organic matter (FPOM): (i) mechanical disruption by battering; (ii) processing by microorganisms causing gradual break-up; (iii) fragmentation by the shredders. Note also that all animal groups contribute to FPOM by producing faeces (_ _ _). DOM is also converted into FPOM by a physical process of flocculation, or via uptake by microorganisms. The organic layer attached to stones on the streambed derives from algae, DOM and FPOM adsorbed onto an organic matrix.

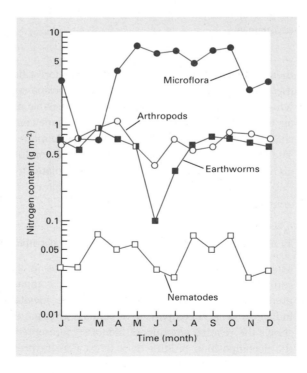

Figure 11.7 The relative importance in forest litter decomposition of microflora in comparison to arthropods, earthworms and nematodes, expressed in terms of their relative content of nitrogen, a measure of their biomass. Microbial activity is much greater than that of detritivores, but the latter is more constant through the year. (After Ausmus *et al.*, 1976.)

five to ten times as much of the biomass as the detritivores. The biomass of detritivores varied less through the year because these are less sensitive to climatic change, and they were actually predominant during a period in the winter.

biomass: an interesting, but imperfect measure of importance

Unfortunately, the biomass present in different groups of decomposers is itself a poor measure of their relative importance in the process of decomposition. Populations of organisms with short lives and high activity may contribute more to the activities in the community than larger, long-lived, sluggish species (e.g. slugs!) that make a greater contribution to biomass. All in all, it is very difficult to assess the relative contribution of microorganisms and detritivores to the process of decomposition. Their lives are intimately interwoven.

the crucial importance of interactions between detritivores and microbes in plant decomposition

The decomposition of dead material is not simply due to the sum of the activities of microbes and detritivores: it is largely the result of interaction between the two (Lussenhop, 1992). The nature of such interaction is seen in Figure 11.8, which gives the results of a very simple experiment in which up to six isopods were added to microcosms containing 1 g of oak-leaf litter. The presence of the isopods caused increased microbial respiration (an index of rate of decomposition), which continued for more than 40 days. The shredding action of these detritivores produced smaller particles with a larger surface area (per unit volume of litter), and thus increased the area of substrate available for microorganism growth. The activity of fungi may have been stimulated by the disruption, through grazing, of competing hyphal networks. In addition, the activity of both fungi and bacteria may have been enhanced by the addition of mineral nutrients in urine and faeces (Lussenhop, 1992). A higher density of 10 animals per experiment also stimulated microbial respiration, at least initially. However, there were further, more subtle consequences, since counts of bacterial cells and lengths of fungal hyphae indicated that the animals had stimulated bacterial growth (probably by mixing cells with fresh substrate) but inhibited the fungi (perhaps by overgrazing). In another experiment, also involving oak-leaf litter, microbial respiration was significantly higher on the faeces produced by the terrestrial larva of the caddis-fly (*Enoicyla pusilla*) than on its food, dried oak leaves (Figure 11.9). Close similarity in rates of respiration on faeces and on oak leaves, ground to the same fragment size as faeces, indicates that it is the surface area of the material that determines how fast it decomposes.

The ways in which the microflora and the detritivores interact might be studied

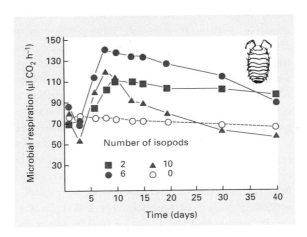

Figure 11.8 The influence of woodlice (Isopoda) activity on microbial breakdown of leaf litter in laboratory microcosms. Their presence caused increased microbial respiration over an extended period. (After Hanlon & Anderson, 1980.)

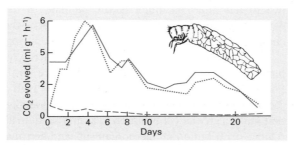

Figure 11.9 Microbial respiration on food and faeces of the terrestrial caddis (*Enoicyla pusilla*). The similarity in activity on faecal pellets (——) and on ground oak leaves of similar fragment size (· · · ·) contrasts with that on intact leaves (– – –). (After van der Drift & Witkamp, 1959.)

a reductionist view of guts, microflora and detritus—the life (after death) of a plant cell wall

by following a leaf fragment through the process of decomposition, focusing attention on a part of the wall of a single cell. Initially, when the leaf falls to the ground, the piece of cell wall is protected from microbial attack because it lies within the plant tissue. The leaf is now chewed and the fragment enters the gut of an isopod. Here, it meets a new microbial flora in the gut and is acted on by the digestive enzymes of the isopod. The fragment emerges, changed by passage through the gut. It is now part of the isopod's faeces and is much more easily attacked by microorganisms, because it has been fragmented and partially digested. Whilst microorganisms are colonizing, it may again be eaten, perhaps by a coprophagous springtail, and pass through the new environment of the springtail's gut. Incompletely digested fragments may again appear, this time in springtail faeces, yet more easily accessible to microorganisms. The fragment may pass through several other guts in its progress from being a piece of dead tissue to its inevitable fate of becoming carbon dioxide and minerals. This 'reductionist' way of thinking about the process of decomposition emphasizes the manner in which microflora and detritivores interact. There is, however, an even more dramatic way of visualizing the process, when we remember that the gut of the animal is really a tube of the external environment that passes through its body. The particle of dead leaf passes through a succession of different environments in the course of its decomposition; some of these environments are the controlled environments of the gut, others are those within the faeces, or on the soil particles with which they become mixed.

the role of invertebrate fragmentation varies in terrestrial, freshwater and marine environments

Fragmentation by detritivores plays a key role in terrestrial situations because of the tough cell walls characteristic of vascular plant detritus. The same is true in many freshwater environments where terrestrial litter makes up most of the available detritus. In contrast, plant detritus in marine environments consists of phytoplankton cells and seaweeds; the former present a high surface area without the need for physical disruption, and the latter, lacking the structural polymers of vascular plant cell walls, are prone to fragmentation by physical factors. Rapid decomposition of marine detritus is probably less dependent on invertebrate comminution; shredders are rare in the marine environment compared to its terrestrial and freshwater counterparts (Plante *et al.*, 1990).

interactions between detritivores and microbes in animal decomposition

Enhancement of microbial respiration by the action of detritivores has also been reported in the decomposition of small mammal carcasses. Two sets of insect-free rodent carcasses weighing 25 g were exposed under experimental conditions in the field. In one set, the carcasses were left intact. In the other, the bodies were artificially riddled with tunnels by repeated piercing of the material with a dissecting needle to simulate the action of blowfly larvae in the carcass. The results of the experiment, carried out in an English grassland in autumn, showed that the tunnels

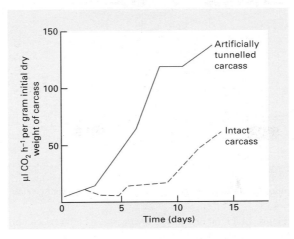

Figure 11.10 The evolution of carbon dioxide, a measure of microbial activity, from carcasses of small mammals placed in 'respiration' cylinders and screened from insect attack. One set of carcasses was left intact, whilst the second set was pierced repeatedly with a dissecting needle to simulate the action of tunnelling by blowfly larvae. (After Putman, 1978a.)

enhanced microbial activity (Figure 11.10) by disseminating the microflora as well as increasing the aeration of the carcass.

11.2.4 The chemical composition of decomposers, detritivores and their resources

There is a great contrast between the composition of dead plant tissue (primary resource) and that of the tissues of the heterotrophic organisms that consume and decompose it. Whilst the major components of plant tissues, particularly cell walls, are structural polysaccharides, these are only of minor significance in the bodies of microorganisms and detritivores. However, being harder to digest than storage carbohydrates and protein, the structural chemicals still form a significant component of detritivore faeces. Detritivore faeces and plant tissue have much in common chemically, but the protein and lipid contents of detritivores and microflora are significantly higher than those of plants and faeces.

decomposition rate is influenced by the availability of nitrogen and phosphorus in dead tissue ...

The nutrient element composition of dead plant material, decomposers and detritivores is illustrated in Table 11.1. The rate at which dead organic matter decomposes is strongly dependent on its biochemical composition. This is because microbial tissue has very high nitrogen and phosphorus contents, indicative of high requirements for these nutrients. Roughly speaking, the ratios of carbon : nitrogen (C : N) and carbon : phosphorus (C : P) in decomposers are 10 : 1 and 100 : 1, respectively (e.g. Goldman *et al.*, 1987). In other words, a microbial population of 111 g can only develop if there is 10 g of nitrogen and 1 g of phosphorus available. Terrestrial plant material has much higher ratios, ranging from 19 to 315 : 1 for

Table 11.1 Nutrient element composition of resources, decomposers and detritivores (percentage dry weight). (After Swift *et al.*, 1979.)

	Nitrogen	Phosphorus	Potassium	Calcium	Magnesium
Deciduous leaf	0.56	0.15	0.60	2.35	?
Deciduous wood	0.1–0.5	0.1–0.17	0.06–0.4	0.06–1.3	0.01–0.15
Bacteria	4.0–15	1.0–6.0	1.0–2.0	1.0	0.15–1.0
Fungi	1.3–3.6	0.1–0.7	0.1–2.9	0.1–3.3	0.1–0.2
Invertebrate detritivores	5.8–10.5	0.8–6.9	0.1–0.7	0.3–10.3	0.2–0.3

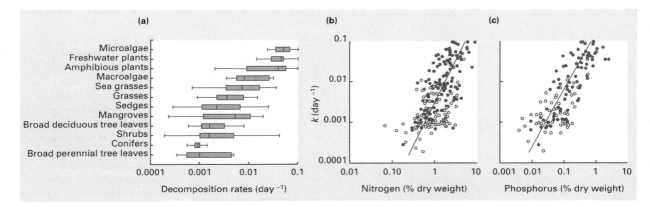

Figure 11.11 (a) Box plots showing the recorded decomposition rates of detritus from different sources. Decomposition rate is expressed as k (in log units per day), derived from the equation $W_t = W_0 e^{-kt}$, which describes the loss in plant dry weight (W) with time (t) since initiation of measurements. Boxes encompass the 25 and 75% quartiles of all data from the literature for each plant type. The central line represents the median, and bars extend to the 95% confidence limit. The relationships between decomposition rate and the initial concentrations in the tissues (percentage of dry weight) of (b) nitrogen and (c) phosphorus are also shown. Solid lines represent fitted regression lines and open and closed circles represent detritus decomposing on land and submersed, respectively. (After Enriquez *et al.*, 1993.)

C : N, and from 700 to 7000 : 1 for C : P (Enriquez *et al.*, 1993). Consequently, this material can support only a limited biomass of decomposer organisms and the whole pace of the decomposition process will itself be limited by nutrient availability. Marine and freshwater plants tend to have ratios more similar to the decomposers (Duarte, 1992), and their rates of decomposition are correspondingly faster (Figure 11.11a). Figures 11.11b and c illustrate the strong relationships between initial nitrogen and phosphorus concentration in plant tissue and its decomposition rate for a wide range of plant detritus from terrestrial, freshwater and marine species.

... and in the surrounding environment

The rate at which dead organic matter decomposes is also influenced by mineral nutrients, especially nitrogen (ammonium or nitrate), that are available from the environment. Thus, greater microbial biomass can be supported, and decomposition proceeds faster, if nitrogen is absorbed from outside. For example, grass litter decomposes faster in streams running through tussock grassland in New Zealand that has been improved for pasture (where the water is, in consequence, richer in nitrate) than in 'unimproved' streams (Young *et al.*, 1994). Another consequence of the capacity of microflora to use environmental nutrients is that after plant material is added to soil, the level of soil nitrogen tends to fall rapidly as it is incorporated into microbial biomass. The effect is particularly evident in agriculture, where the ploughing in of stubble can result in nitrogen deficiency of the subsequent crop. The same process occurs when leaf litter falls on a forest floor. When nitrogen is not available, the whole process of decomposition is slowed down, and undecomposed litter accumulates.

In contrast to terrestrial plants, the bodies of animals have nutrient ratios that are of the same order as those of microbial biomass; thus, their decomposition is not limited by the availability of nutrients, and animal bodies tend to decompose much faster than plant material.

When dead organisms or their parts decompose in or on soil, they begin to acquire the C : N ratio of the decomposers. On the whole, if material of nitrogen content less than 1.2–1.3% is added to soil, any available ammonium ions are absorbed. If the material has a nitrogen content greater than 1.8%, ammonium ions tend to be released. One consequence is that the C : N ratios of soils tend to be rather constant around values of 10; the decomposer system is in general remarkably homeostatic. However, in extreme situations, where the soil is very acid or waterlogged, the ratio may rise to 17 (an indication that decomposition is slow).

It should not be thought that the only activity of the microbial decomposers of dead material is to respire away the carbon and mineralize the remainder. A major consequence of microbial growth is the accumulation of microbial by-products, particularly fungal cellulose and microbial polysaccharides, which may themselves be slow to decompose and contribute to maintaining soil structure.

11.3 Detritivore–resource interactions

11.3.1 Consumption of plant detritus

Two of the major organic components of dead leaves and wood are cellulose and lignin. These pose considerable digestive problems for animal consumers, most of which are not capable of manufacturing the enzymatic machinery to deal with them. Cellulose catabolism (cellulolysis) requires *cellulase* enzymes. Without these, detritivores are unable to digest the cellulose component of detritus, and so cannot derive from it either energy to do work or the simpler chemical modules to use in their own tissue synthesis.

only a few detritivores have
their own cellulase

Cellulases of animal origin have been definitely identified in remarkably few species, be they detritivorous or herbivorous invertebrates. In fact, strong cases for the occurrence of animal cellulases have been made in only one species of cockroach and a few species of higher termites in the subfamily Nasutitermitinae (Martin, 1991). In these organisms, cellulolysis poses no special problems.

most detritivores rely on
microbial cellulases

The majority of detritivores, lacking their own cellulases, rely on the production of cellulases by associated microflora or, in some cases, protozoa. The interactions range from *obligate mutualism* between a detritivore and a specific and permanent gut microflora or microfauna, through *facultative mutualism*, where the animals make use of cellulases produced by a microflora that is ingested with detritus as it passes through an unspecialized gut, to the 'external rumen', where animals assimilate the metabolic products of the cellulase-producing microflora associated with decomposing plant remains or faeces (Figure 11.12).

termites rely on protozoa and
bacteria

Clear examples of obligate mutualism are found amongst certain species of cockroach and termite that rely on symbiotic bacteria or protozoa for digestion of structural plant polysaccharides. In lower termites, such as *Eutermes*, protozoa may make up more than 60% of the insect's body weight. The protozoa are located in the hindgut, which is dilated to form a rectal pouch. They ingest fine particles of wood, and are responsible for extensive cellulolytic activity, although bacteria are also implicated. Termites feeding on wood made up of 54.6% cellulose, 18.0% pentosans and 27.4% lignin, produce faeces containing 18% cellulose, 8.5% pentosans and 75.5% lignin, indicating effective digestion of cellulose but not of lignin (Lee & Wood, 1971). This pattern is apparently generally the case for termites, but *Reticulitermes* has been reported to digest 80% or more of the lignin present in its food. This topic is dealt with in detail in Chapter 13.

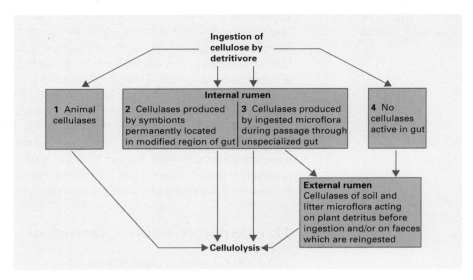

Figure 11.12 The range of mechanisms that detritivores adopt for digesting cellulose (cellulolysis). (After Swift *et al.*, 1979.)

woodlice rely on an ingested microflora

Woodlice provide a good example of detritivores that indulge in a facultative symbiotic association in an unspecialized gut. Cellulase activity has been recorded in isopod guts, but experimentation has shown this to be of microbial origin (Hassall & Jennings, 1975). When excised guts of *Philoscia muscorum* were incubated in contact with a film of carboxymethyl-cellulose and then stained to determine whether cellulolysis had occurred, it was possible to demonstrate the presence of cellulase at particular places in the gut. After the addition of antimicrobial agents to kill any microflora in the gut, no cellulolytic activity could be detected, but this was restored when the animal was allowed to ingest leaf litter, with its associated microflora.

many species rely on an exogenous microflora

Finally, a wide range of detritivores appear to have to rely on the exogenous microflora to digest cellulose. The invertebrates then consume the partially digested plant detritus along with its associated bacteria and fungi, no doubt obtaining a significant proportion of the necessary energy and nutrients by digesting the microflora itself. These animals, such as the springtail *Tomocerus*, can be said to be making use of an 'external rumen' in the provision of assimilable materials from indigestible plant remains. The 'external rumen' phenomenon reaches a pinnacle of specialization in 'ambrosia' beetles and in certain species of ants and termites that 'farm' fungus in specially excavated gardens (see Chapter 13).

why no cellulases?

Given the versatility apparent in the evolutionary process, it seems surprising that so few animals that consume plants can produce their own cellulase enzymes. Janzen (1981) has argued that cellulose is the master construction material of plants 'for the same reason that we construct houses of concrete in areas of high termite activity'. He views the use of cellulose, therefore, as a defence against attack, since higher organisms can rarely digest it unaided. From a different perspective, it has been suggested that cellulolytic capacity is uncommon simply because it is a trait that is rarely advantageous for animals to possess (Martin, 1991). For one thing, diverse bacterial communities are commonly found in hindguts and this may have facilitated the evolution of symbiont-mediated cellulolysis. For another, the diets of plant-eaters generally suffer from a limited supply of critical nutrients, such as nitrogen and phosphorus, rather than of energy. This imposes the need for processing large volumes of material to extract the required quantities of nutrients. In fact, the energy yield from cellulolysis, if this were routinely employed by animals, could conceivably create serious problems in disposal of waste energy (Swift *et al.*, 1979).

fruit-flies and rotten fruit

Of course, not all plant detritus is so difficult for detritivores to digest. For example, fallen fruit is readily exploited by many kinds of opportunist feeders, including insects, birds and mammals. However, like all detritus, decaying fruits have associated with them a microflora, in this case mainly dominated by yeasts. Fruit-flies (*Drosophila* spp.) specialize at feeding on these yeasts and their by-products; and in fruit-laden domestic compost heaps in Australia, five species of fruit-fly show preferences for particular categories of rotting fruit and vegetables (Oakeshott *et al.*, 1982). *Drosophila hydei* and *D. immigrans* prefer melons, *D. busckii* specializes on rotting vegetables, whilst *D. simulans* is catholic in its tastes for a variety of fruits. The common *D. melanogaster*, however, shows a clear preference for rotting grapes and pears. Note that rotting fruits can be highly alcoholic. Yeasts are commonly the early colonists and the fruit sugars are fermented to alcohol which is normally toxic, eventually even to the yeasts themselves. *Drosophila melanogaster* tolerates such high levels of alcohol because it produces large quantities of alcohol dehydrogenase (ADH), an enzyme that breaks down ethanol to harmless metabolites. More surprisingly, it has been shown that this fruit-fly can use ethanol as a resource (Parsons & Spence, 1981). Decaying vegetables produce little alcohol, and *D. busckii*, which is associated with them, produces very little ADH. Intermediate levels of ADH were produced by the species preferring moderately alcoholic melons. The boozy *D. melanogaster* is also associated with winery wastes!

plant detritus and the microflora are typically consumed together ...

Because of the intimate intermeshing of microfloral and plant detrital tissues there are inevitably many generalist consumers that ingest both resources. These animals simply cannot manage to take a mouthful of one without the other. Thus, for example, Figures 11.13a and b illustrate the various components of the diet of several oribatid mites occurring in the litter of a woodland in Kent, England. There are clear interspecific differences in the relative importance of fungal material, higher plant material and amorphous organic matter, and some species are more specialized feeders than others, at least at some times of the year. All consume more than one of the potential diet components and most are remarkably generalist.

Microfloral biomass seems to constitute a better food resource than plant detritus. A clue to its relative importance in diet has been provided by studies on the efficiency with which ingested pure fungus or unconditioned tree leaf is assimilated. For example, a freshwater shredder, *Gammarus pseudolimnaeus*, assimilated pure sterile leaf discs of elm or maple with an efficiency of only 17–19% of the energy content, the remainder being lost in faeces. On the other hand, when the detritivores were provided with a pure culture of aquatic fungi, normally associated with tree leaves in streams, the assimilation efficiency was 74–83% (Bärlocher & Kendrick, 1975).

... but, the microflora is the more nutritious

Such results have led Cummins (1974) to liken microbial tissue to peanut butter—a nutritious food that occurs on a nutritionally unsuitable cracker biscuit, the plant detritus. Whilst this analogy is quite apt, it perhaps overstresses the role of microbial biomass. By far the largest organic component ingested by detritivores is generally plant detritus, so although it is assimilated with low efficiency, it may still provide an appreciable portion of the animal's requirements. It is perhaps more accurate to compare what detritivores ingest to a very thick cracker biscuit spread

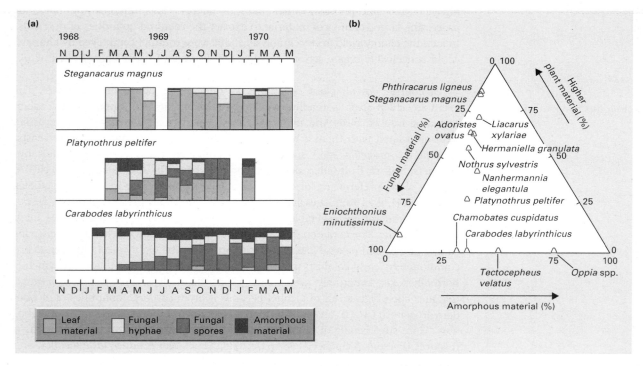

Figure 11.13 Gut contents of oribatid mites from a woodland in Kent, England. (a) Temporal variation in feeding by three species from *Castanea* litter showing the average composition of gut contents on each sampling occasion. (b) Mean proportions of the three major food items in the species that were most abundant during a 20-month study. Specialized feeders occur near the apices of the triangle, with more generalist species nearer the centre. (After Anderson, 1975.)

only thinly with peanut butter. Nevertheless, there is no doubt that an important part of the diet of terrestrial and aquatic detritivores derives from fungi and bacteria, and there is considerable evidence, for example amongst freshwater detritivores, that the animals prefer microbially conditioned detritus and grow better when fed on it.

11.3.2 Feeding on faeces

Invertebrate faeces

millipedes do best when they can eat their own faeces

A large proportion of dead organic matter in soils and aquatic sediments may consist of invertebrate faeces. Detritivores with generalist feeding habits will usually include these in their diet. It many cases, reingestion of faeces is critically important. In the laboratory, the litter-dwelling millipede *Apheloria montana* showed only a small weight increase if it was fed solely on leaf detritus (1.1% in 30 days). A second group of animals, which were allowed access to their faeces, increased their ingestion rates by more than 60% and their growth rates by 16% compared to the controls (McBrayer, 1973). It seems that the conversion of plant remains to digestible microbial biomass and also to partially degraded plant compounds is of great importance to many detritivores. In one sense, the detritivores' guts act as factory conveyor belts producing cracker biscuits (faeces) to be spread with peanut butter (microflora) over and over again.

A remarkable story of coprophagy (eating faeces) has been unravelled in some small bog lakes in north-east England (MacLachlan *et al.*, 1979). These murky water bodies have restricted light penetration because of dissolved humic substances derived from the surrounding sphagnum peat, and they are characteristically poor in plant nutrients. Primary production is insignificant. The main organic input consists of poor-quality peat particles resulting from wave erosion of the banks. By the time the peat has settled from suspension it has been colonized, mainly by bacteria, and its caloric and protein contents have increased by 23 and 200%, respectively. These small particles are consumed by *Chironomus lugubris* larvae, the detritivorous young of a non-biting chironomid midge. The faeces that the larvae produce become quite richly colonized by fungi, microbial activity is enhanced and they would seem to constitute a high-quality food resource. But, they are not reingested by *Chironomus* larvae, mainly because they are too large and too tough for its mouth parts to deal with. However, another common inhabitant of the lake, the small crustacean *Chydorus sphaericus*, finds chironomid faeces very attractive. It seems always to be associated with them and probably depends on them for food. *Chydorus* clasps the chironomid faecal pellet just inside the valve of its carapace and rotates it whilst grazing the surface, causing gradual disintegration. In the laboratory, the presence of chydorids has been shown to speed up dramatically the breakdown of large *Chironomus* pellets to smaller particles. The final and most intriguing twist to the story is that the fragmented chironomid faeces (mixed probably with chydorid faeces) are now small enough to be used again by *Chironomus*. It is probable that *C. lugubris* larvae grow faster when in the presence of *C. sphaericus* because of the availability of suitable faecal material to eat. The interaction benefits both participants.

Vertebrate faeces

The dung of carnivorous vertebrates is relatively poor-quality stuff. Carnivores assimilate their food with high efficiency (usually 80% or more is digested), and their faeces retain only the least digestible components. In addition, carnivores are necessarily much less common than herbivores, and their dung is probably not sufficiently abundant to support a specialist detritivore fauna. What little research has been done suggests that decay is effected almost entirely by bacteria and fungi (Putman, 1983).

In contrast, herbivore dung still contains an abundance of organic matter and is sufficiently thickly spread in the environment to support its own characteristic fauna, consisting of many occasional visitors but with several specific dung feeders. A good example is provided by elephant dung. Two main patterns of decay can be recognized, related to the wet and dry seasons in the tropics (Kingston, 1977). During the rains, within a few minutes of dung deposition the area is alive with beetles. The adult dung beetles feed on the dung but they also bury large quantities along with their eggs to provide food for developing larvae. For example, the large African dung beetle, *Heliocopris dilloni*, carves a lump out of fresh dung and rolls this away for burying several metres from the original dung pile. Each beetle buries sufficient dung for several eggs. Once underground, a small quantity of dung is shaped into a cup and lined with soil; a single egg is laid and then more dung is added to produce a sphere that is almost entirely covered with a thin layer of soil. A

a midge and a cladoceran eat one another's faeces

carnivore dung is attacked mainly by microbes

the highly efficient decomposition of elephant dung by dung beetles in the wet season ...

421   DECOMPOSERS AND DETRITIVORES

Figure 11.14 (a) African dung beetle rolling a ball of dung. (Courtesy of Heather Angel.) (b) The larva of the dung beetle *Heliocopris* excavates a hollow as it feeds within the dung ball. (After Kingston & Coe, 1977.)

small area at the top of the ball, close to the location of the egg, is left clear of soil, possibly to facilitate gas exchange. After hatching, the larva feeds by a rotating action in the dung ball, excavating a hollow, and, incidentally, feeding on its own faeces as well as the elephant's (Figure 11.14). When all the food supplied by its parents is used up, the larva covers the inside of its cell with a paste of its own faeces, and pupates.

The full range of tropical dung beetles in the family Scarabeidae vary in size from a few millimetres in length up to the 6 cm long *Heliocopris.* Not all remove dung and bury it at a distance from the dung pile. Some excavate their nests at various depths immediately below the pile, whilst others build nest chambers within the dung pile itself. Species in other families do not construct chambers, but simply lay their eggs in the dung, and their larvae feed and grow within the dung mass until fully developed, when they move away to pupate in the soil. The beetles associated with elephant dung in the wet season may remove 100% of the dung pile. Any left may be processed by other detritivores such as flies and termites, as well as by microflora.

Dung that is deposited in the dry season is colonized by relatively few beetles (adults emerge only in the rains). Some microbial activity is evident but this soon

... but, a very different picture emerges in the dry season

declines as the faeces dry out. Rewetting during the rains stimulates more microbial activity but beetles do not exploit old dung. In fact, a dung pile deposited in the dry season may persist for longer than 2 years, compared with 24 h or less for one deposited during the rains.

The breakdown of ungulate dung in temperate environments has also been much studied. Cow dung in Great Britain is attacked by many invertebrates (amongst which earthworms are dominant), as well as by bacteria and fungi (Denholm-Young, 1978). Very few species of dung-burying beetles operate in Great Britain, but one, *Geotrupes spiniger*, may remove up to 13% by dry weight of a cow pat. The maggots of many species of fly, including the specialist yellow dungfly *Scatophaga stercoraria*, also play a role. Adult *Scatophaga* are predatory or nectar-feeding and take only small quantities of liquid dung. However, they congregate to mate around fresh droppings and the female lays her eggs on the dung pile. Larvae hatch out on the surface and immediately burrow down into the dung. A new generation of adults emerges within 3–4 weeks.

An Australian dilemma: bovine dung, but no bovine dung beetles

Australian cow dung poses a problem

Bovine dung has provided an extraordinary and economically very important problem in Australia. During the past 200 years the cow population has risen from just seven individuals (brought over by the first English colonists in 1788) to 30 million or so. These produce some 300 million cow pats per day, covering as much as 6 million acres per year with dung. Deposition of bovine dung poses no particular problem elsewhere in the world, where bovines have existed for millions of years and have an associated fauna which exploits the faecal resources. However, the largest herbivorous animals in Australia until European colonization were marsupials such as kangaroos. The native detritivores which deal with the dry, fibrous dung pellets that these leave cannot cope with cow dung, and the loss of pasture under dung has imposed a huge economic burden on Australian agriculture. The decision was therefore made in 1963 to establish in Australia beetles of African origin, able to dispose of bovine dung in the most important places and under the most prevalent conditions where cattle are raised (Waterhouse, 1974); so far, 21 species have been introduced (Doube *et al.*, 1991).

Adding to the problem, Australia is plagued by native bushflies (*Musca vetustissima*) and buffalo flies (*Haematobia irritans exigua*), which deposit eggs on dung pats. The larvae fail to survive in dung that has been buried by beetles, and the presence of beetles has been shown to be effective at reducing fly abundance (Ridsdill-Smith, 1991). Success depends on dung being buried within about 6 days of production, the time it takes for the fly egg (laid on fresh dung) to hatch and develop to the pupal stage. Edwards and Aschenborn (1987) surveyed the nesting behaviour in southern Africa of 12 species of dung beetles in the genus *Onitis*. They concluded that *O. uncinatus* was a prime candidate for introduction to Australia for fly-control purposes, since substantial amounts of dung were buried on the first night after pat colonization. The least suitable species, *O. viridualus*, spent several days constructing a tunnel and did not commence burying until 6–9 days had elapsed.

11.3.3 Consumption of carrion

many carnivores are opportunistic carrion-feeders

The chemical composition of the diet of carrion-feeders is quite distinct from that of other detritivores, and this is reflected in their complement of enzymes. Carbohydrase activity is weak or absent, but protease and lipase activity is vigorous. Carrion-feeding detritivores possess basically the same enzymatic machinery as carnivores, reflecting the chemical identity of their food. In fact, many species of carnivore are also opportunistic carrion-feeders.

When considering the decomposition of dead bodies, it is helpful to distinguish three categories of organisms that attack carcasses. As before, both microflora and invertebrate detritivores have a role to play, but, in addition, scavenging vertebrates are often of considerable importance. Many carcasses of a size to make a single meal for one of a few of these scavenging detritivores will be removed completely within a very short time of death, leaving nothing for bacteria, fungi or invertebrates. This role is played, for example, by Arctic foxes and skuas in polar regions, by crows, gluttons and badgers in temperate areas and by a wide variety of birds and mammals, including kites, jackals and hyenas, in the tropics.

scavenging vertebrates often remove whole carcasses ...

The relative roles played by microflora, invertebrates and vertebrates are influenced by factors that affect the speed with which carcasses are discovered by scavengers in relation to the rate at which they disappear through microbial and invertebrate activity. This is illustrated graphically for small rodent carcasses whose disappearance/decomposition was monitored in the Oxfordshire countryside in both summer–autumn and winter–spring periods (Figure 11.15). There are two points to note.

... although this can vary seasonally in a subtle way

1 The rate at which carcasses were removed was faster during summer and autumn, reflecting greater scavenger activity at this time (presumably because of higher scavenger population densities and/or higher feeding rates—these were not monitored in the study).

2 A greater percentage of the rodent bodies was removed in the winter–spring period, albeit over a longer time scale. At a time when microbial decay proceeds most slowly, all the carcasses persisted for long enough to be found by scavengers. During summer and autumn, decomposition was much more rapid and any carcass that was undiscovered for 7 or 8 days would have been largely decomposed and removed by bacteria, fungi and invertebrate detritivores.

seasonal variations in invertebrate and microbial activity

Larger carcasses, such as those of dogs, sheep and antelopes, may also be dealt with predominantly by vertebrate scavengers. Sheep carcasses in temperate regions are often stripped completely by foxes or eagles within a short time of death, whilst vultures and hyenas remove as many as 90% of the bodies of ungulates that die in

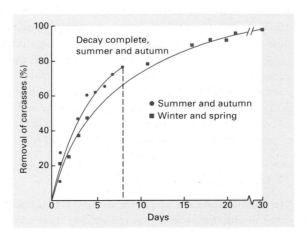

Figure 11.15 Rate of removal of small mammal corpses in the Oxfordshire countryside in two periods: summer–autumn and winter–spring. (After Putman, 1983.)

southern Africa. In a study of decomposition of the largest terrestrial carcasses of all, those of elephants, Coe (1978) found that although scavengers played a major role, some carcasses went unexploited by vertebrates, and in these cases invertebrate detritivores and particularly the microflora played an important role. However, this study was carried out in a period of unusually high elephant mortality in Kenya, and the scavenger populations were unable to exploit the exceptional availability of food as fully as usual.

Carrion undiscovered by large scavengers is available for processing by microflora and invertebrates, and their relative roles are influenced by prevailing conditions. Invertebrates such as the larvae of blowflies in the genus *Lucilia* (which are characteristic of decaying carrion), may remove a significant portion of carrion whilst conditions are favourable. For example, Putman (1978b) estimates that in Great Britain blowfly maggots consume as much as 80% of small mammal carcasses available during the summer. In the winter, however, decomposition is largely a microbial process. The activity of blowflies is seasonal in tropical environments too. In wet seasons they play an important role in decomposition, but they are unable to colonize in dry seasons, and carrion is likely to shrivel and dry, any decomposition being attributable to microorganisms. Tropical dry-season mummification parallels the winter mummification of carrion often observed in temperate regions.

the specialist consumers of bone, hair and feathers

Certain components of animal corpses are particularly resistant to attack and are the slowest to disappear. However, some consumer species possess the enzymes to deal with them. For example, the blowfly larvae of *Lucilia* spp. produce a collagenase that can digest the collagen and elastin present in tendons and soft bones. The chief constituent of hair and feathers, keratin, forms the basis of the diet of species characteristic of the later stages of carrion decomposition, in particular tineid moths and dermestid beetles. The midgut of these insects secretes strong reducing agents which break the resistant covalent links binding together peptide chains in the keratin. Hydrolytic enzymes then deal with the residues. Fungi in the family Onygenaceae are specialist consumers of horn and feathers. It is the corpses of larger animals that generally provide the widest variety of resources, and thus attract the greatest diversity of carrion consumers (Doube, 1987). In contrast, the carrion community associated with dead snails and slugs consists of a relatively small number of sarcophagid and calliphorid flies (Kneidel, 1984).

the remarkable burying beetles: *Necrophorus* spp.

One group of carrion-feeding invertebrates deserves special attention—the burying beetles (*Necrophorus* spp.). These species live exclusively on carrion on which they play out their extraordinary life history. Adult *Necrophorus*, presumably orientating by smell, arrive at the carcass of a small mammal or bird within 1–2 h of death. The beetle may tear flesh from the corpse and eat it or, if decomposition is sufficiently advanced, consume blowfly larvae instead. However, should a burying beetle arrive at a completely fresh corpse it sets about burying it where it lies, or may drag the body (many times its own weight) for several metres before starting to dig. It works beneath the corpse, painstakingly excavating and dragging the small mammal down little by little until it is completely underground (Figure 11.16). Some species, such as *N. vespilloides*, only just cover the corpse, whilst others, including *N. germanicus*, may bury it to a depth of 20 cm.

During the excavation, other burying beetles are likely to arrive. Competing individuals of the same or other species are fiercely repulsed, sometimes leading to the death of one combatant. A prospective mate, on the other hand, is accepted and the male and female work on together.

425   DECOMPOSERS AND DETRITIVORES

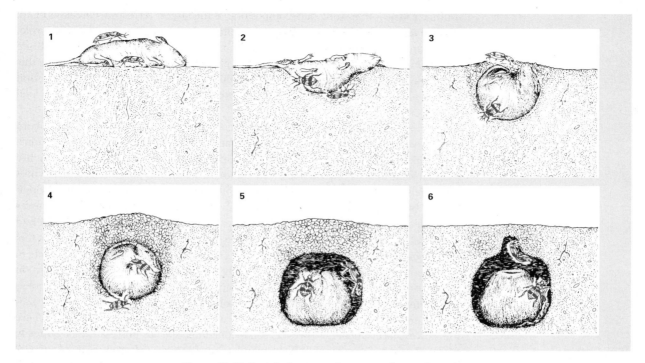

Figure 11.16 Burial of a mouse by a pair of *Necrophorus* beetles. (After Milne & Milne, 1976.)

The buried corpse is much less susceptible to attack by other invertebrates than it was whilst on the surface. Additional protection is provided, under some circumstances, by virtue of a mutualistic relationship between *Necrophorus* and a species of mite, *Poecilochirus necrophori*, which invariably infests adult burying beetles, hitching a ride to a suitable carrion source. When the carcass is first buried the beetle systematically removes its hair and this clears it of virtually all the eggs of blowflies. However, if the carcass is buried only shallowly, flies will often lay more eggs and maggots would compete with the beetle larvae. It is now that the presence of mites has a beneficial effect (Figure 11.17). By piercing and consuming fly eggs,

the presence of mites can improve beetle breeding success

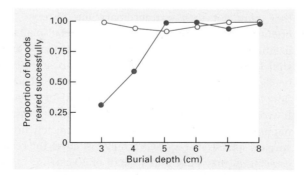

Figure 11.17 When mites, *Poecilochirus necrophori*, are present on the buried corpse (○), brood success by the burying beetle, *Necrophorus tomentosus*, is high. Experimental removal of mites (●) has no effect on beetle brood success for deeply buried carcasses, but results in low survival of beetle broods in carcasses buried to a depth of 4 cm or less. (After Wilson, 1986.)

the mites keep the carcass free of the beetle's competitors and dramatically improve beetle brood success (Wilson, 1986).

When excavation of the burial chamber is complete, the beetles set about removing fur or feathers, and they work the mass of flesh into a compact ball. Now the beetles copulate, and eggs are laid in a blind passage leading to the roof of the chamber. The eggs are not abandoned, but rather both adults, or sometimes just the female, remain in the chamber. If the parents are removed, a dense mould grows over the surface of the carcass and brood success is low. Chemicals in the faeces or saliva of the beetles are thought to be responsible for this antibiotic action. The beetles go on to provide parental care for the larvae. A conical depression is prepared in the top of the meat ball, into which droplets of partially digested meat are regurgitated. When the eggs hatch, the adult calls the larvae to the feeding location by stridulation, and the larvae are fed droplets from the pool of liquid food by the parent. Older larvae are able to feed themselves from the pool or to remove strips of meat from the carcass. Only when their offspring are ready to pupate do the adults force their way out through the soil and fly away.

We have already noted that in freshwater environments carrion lack a specialized fauna. However, specialist carrion-feeders are found on the seabed in very deep parts of the oceans. As the detritus sinks through very deep water all but the largest particles of organic matter are completely decomposed before they reach the bottom. In contrast, the occasional body of a fish, mammal or large invertebrate does settle on the seabed. A remarkable diversity of scavengers exist there, although at low density, and these possess several characteristics that match a way of life in which meals are well spread out in space and time. For example, Dahl (1979) describes several genera of deep-sea gammarid crustaceans that, unlike their relatives at shallower depths and in freshwater, possess dense bundles of exposed chemosensory hairs that sense food, and sharp mandibles that can take large bites from carrion. These animals also have the capacity to gorge themselves far beyond what is normal in amphipods. Thus, *Paralicella* possesses a soft body wall that can be stretched when feeding on a large meal, so that the animal swells to two or three times its normal size, and *Hirondella* has a midgut that expands to fill almost the entire abdominal cavity and in which it can store meat.

11.4 Conclusion

Decomposer communities are, in their composition and activities, as diverse as or more diverse than any of the communities more commonly studied by ecologists. Generalizing about them is unusually difficult because the range of conditions experienced in their lives is so varied. As in all natural communities, the inhabitants not only have specialized requirements for resources and conditions, but their activities change the resources and conditions available for others. Most of this happens hidden from the view of the observer, in the crevices and recesses of soil and litter, and in the depths of water bodies.

Despite these difficulties, some broad generalizations may be made.

1 Decomposers and detritivores tend to have low levels of activity when temperatures are low, aeration is poor, soil water is scarce and conditions are acid.

2 The structure and porosity of the environment (soil or litter) is of crucial importance, not only because it affects the factors listed in (1), but because most of the organisms responsible for decomposition must swim, creep, grow or force their

way through the medium in which their resources are dispersed.

3 The activities of the microflora and detritivores are intimately interlocked and tend to be synergistic. For this reason, it is very difficult, and perhaps not sensible, to try to discover their relative importance.

4 Many of the decomposers and detritivores are specialists, and the decay of dead organic matter results from the combined activities of organisms with widely different structure, form and feeding habits.

5 Organic matter may cycle repeatedly through a succession of microhabitats within and outside the guts and faeces of different organisms, as they are degraded from highly organized structures to their eventual fate as carbon dioxide and mineral nutrients.

6 The activity of decomposers unlocks the mineral resources such as phosphorus and nitrogen that are fixed in dead organic matter. The speed of decomposition will determine the rate at which such resources are released to growing plants (or become free to diffuse and thus to be lost from the ecosystem). This topic is taken up and discussed in Chapter 19.

7 Many dead resources are patchily distributed in space and time (Doube, 1987). An element of chance operates in the process of their colonization; the first to arrive have a rich resource to exploit, but the successful species may vary from dung pat to dung pat, and from corpse to corpse. The dynamics of competition between exploiters of such patchy resources require their own particular mathematical models (Atkinson & Shorrocks, 1981; see Chapter 7). Because detritus is often an 'island' in a sea of quite different habitat, its study is conceptually similar to that discussed under the heading of island biology in Chapter 23.

8 Finally, it may be instructive at this point to switch the emphasis away from the success with which decomposers and detritivores deal with their resources. It is, after all, the failure of organisms to decompose wood rapidly that makes the existence of forests possible! Moreover, deposits of peat, coal and oil are testaments to the failures of decomposition.

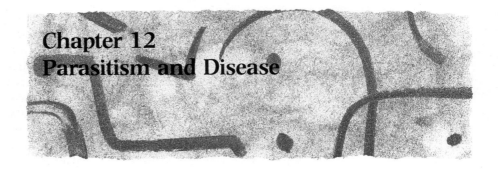

Chapter 12
Parasitism and Disease

12.1 Introduction

definitions of parasitism: harm, intimacy and dependence

In Chapter 8 we defined a parasite as an organism that obtains its nutrients from one or a very few host individuals, normally causing harm but not causing death immediately.

parasites cause substantial economic loss and misery ...

Parasites are an extraordinarily important group of organisms. Millions of people are killed each year by various types of infection, and many millions more are debilitated or deformed (250 million cases of elephantiasis at present, over 200 million cases of bilharzia, and the list goes on). When the effects of parasites on domesticated animals and crops are added to this, the cost in terms of human misery and economic loss becomes incalculable. Of course, humans make things easy for the parasites by living in dense and aggregated populations and forcing their domesticated animals and crops to do the same. But, it is nonetheless reasonable to believe that animals and plants *in general* are harmed and killed in vast numbers by parasitism and disease. Parasites are certain to represent an important source of mortality and unfulfilled fecundity in many, probably most, natural populations.

... they affect populations generally ...

A free-living organism that does *not* harbour several parasitic individuals of a number of species is a rarity. Moreover, many parasites and pathogens are host-specific or at least have a limited range of hosts. Together, these two points would seem to make the conclusion unavoidable that more than 50% of the species on the earth, and many more than 50% of individuals, are parasites. Most bacterial and viral parasites probably still remain unidentified!

... and they are numerically of great importance

In the light of this richness, a survey of the diversity of parasites might seem impracticable. On the other hand, an outline survey, especially of the types that have been most intensively studied and are of most importance to humans, will provide a valuable background for the remainder of this chapter.

12.2 The diversity of parasites

The language and jargon used by plant pathologists and animal parasitologists are often so different as to give the impression that the subjects have nothing in common. For the ecologist, however, such distinction is superficial. There are important differences in the ways in which living animals and plants serve as habitats for parasites, and important differences in the way they respond to infection, but the differences are less striking than the resemblances. As ecologists, therefore, we will deal with the two together.

Parasites have been classified in many different ways, but a distinction that is particularly useful to ecologists recognizes a difference between macroparasites and microparasites (May & Anderson, 1979).

Microparasites are small, often extremely numerous, and they multiply directly within their host. Because most are intracellular they become deeply caught up in the intimacies of cell metabolism and antibody reactions. It is always difficult, and usually impossible, to count the number of microparasites in a host, and the number of infected hosts is the parameter usually studied, rather than the number of parasites. For example, a study of epidemics of measles will involve counting the number of cases of the disease, rather than the number of particles of the measles virus.

Macroparasites have a quite different biology: they grow but do not multiply in their host. They produce infective stages that are released to infect new hosts. The macroparasites of animals mostly live on the body or in the body cavities (e.g. the gut), rather than within the host cells and do not multiply in the host. In plants, they are generally intercellular. It is usually possible to count or at least estimate the numbers of macroparasites in or on a host (e.g. worms in an intestine or lesions on a leaf), and the numbers of parasites as well as the numbers of infected hosts can both be studied by the epidemiologist.

Parasites with more complex life cycles present special problems. For example, an epidemiologist studying infection by schistosome parasites may use the infected snail as one unit of study (because the parasite multiplies in the snail, and so behaves as a microparasite), but, in an infected human (where the parasite behaves as a macroparasite: it grows but does not multiply), he/she may be able to estimate the number of parasites directly.

12.2.1 Microparasites

the range of microparasites

Probably the most obvious microparasites are the bacteria and viruses that infect animals (such as the measles virus and the typhoid bacterium) and plants (e.g. the yellow net viruses of beet and tomato and the bacterial crown gall disease). The other major group of microparasites affecting animals are the protozoa (e.g. the trypanosomes that cause sleeping sickness and the *Plasmodium* spp. that cause malaria; Figure 12.1a). In plant hosts some of the simpler fungi behave as microparasites (e.g. *Physoderma zeae-maydis*, Figure 12.1b).

directly transmitted microparasites

The transmission of a microparasite from one host to another can in some cases be almost instantaneous, as in venereal disease and the short-lived infective agents carried in the water-droplets of coughs and sneezes (influenza, measles, etc.). In other species the parasite may spend an extended dormant period 'waiting' for its new host. This is the case with the ingestion of food or water contaminated with the protozoan *Entamoeba histolytica*, which causes amoebic dysentery, and with the plant parasite *Plasmodiophora brassicae*, which causes 'club-root disease' of crucifers. Alternatively, a microparasite may depend for its spread from host to host on indirect transmission by a vector, usually some other organism. Microparasites have no special mechanisms for dispersal in space and have no special features that allow them to do so. Their host or a vector always does it for them and plays a direct role in their population dynamics.

vector-transmitted microparasites: sleeping sickness and malaria ...

The two most economically important groups of vector-transmitted protozoan parasites of animals are: (i) the trypanosomes, which cause sleeping sickness in humans and Nagana in domesticated (and wild) mammals and are transmitted by

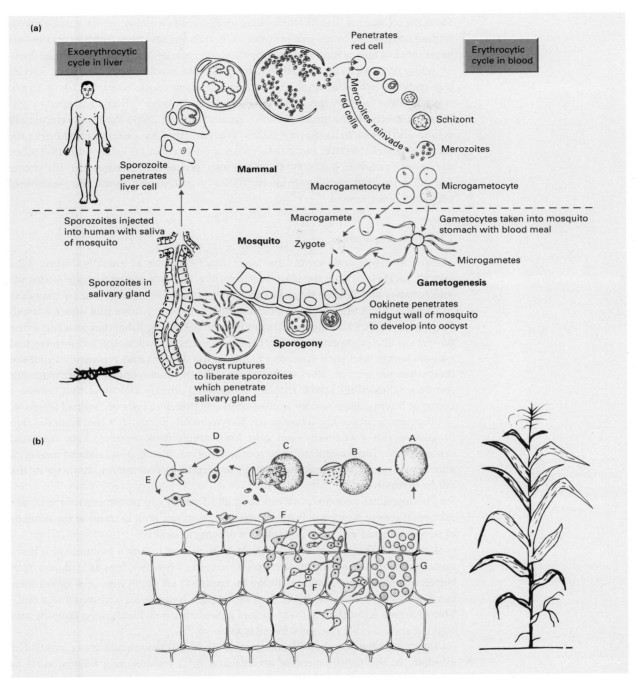

Figure 12.1 Microparasites of plants and animals. These are usually intracellular and multiply within the host or hosts. (a) The life history of the malarial parasite *Plasmodium*. The parasite is injected directly into the bloodstream of the host by a mosquito. Multiplication occurs in the liver cells of the host and the products enter the red blood cells where they multiply again and form sexual stages which are taken up by a feeding mosquito. Sexual fusion occurs in the mosquito gut and after yet a further phase of multiplication the salivary glands of the mosquito become infected and the cycle is ready to start again. (After Vickerman & Cox, 1967.) (b) The life history (A–G) of *Physoderma zeae-maydis*, a parasite that infects the leaves of corn (*Zea mays*). Dispersed spores (A) germinate to release motile infective zoospores (B–D) which infect the host (E). Resistant-dispersal spores are formed on the death of the host or its parts. (After Tisdale, 1919.)

tsetse flies (*Glossina*); and (ii) the various species of *Plasmodium*, which cause malaria and are transmitted by anopheline mosquitos. In both these cases, the flies act not only as vectors but also as intermediate hosts, i.e. the parasite also multiplies within them.

... plant viruses

Many plant viruses are transmitted by aphids. In some 'non-persistent' species (e.g. cauliflower mosaic virus), the virus is only viable in the vector for 1 h or so and is often borne only *on* the aphid's mouth parts. In other 'circulative' species (e.g. lettuce necrotic yellow virus), the virus passes from the aphid's gut to its circulatory system and thence to its salivary glands. There is therefore a latent period before the vector becomes infective, but it then remains infective for an extended period (often the aphid's entire life). Finally, there are also 'propagative' viruses (e.g. the potato leaf roll virus), that multiply within the aphid. Nematode worms are also widespread vectors of plant viruses.

12.2.2 Macroparasites

the range of macroparasites

The parasitic helminth worms are major macroparasites of animals (Figures 12.2 and 12.3). They include the platyhelminths (e.g. the tapeworms and the trematode schistosomes and flukes), the acanthocephalans and the roundworms (nematodes). In addition, there are lice, fleas, ticks and mites and some fungi that attack animals (see Figure 12.2). Plant macroparasites include the higher fungi that give rise to the powdery mildews, downy mildews, rusts and smuts, as well as the gall-forming and mining insects, and such flowering plants as the dodders and broomrapes that are themselves parasitic on other higher plants. Like the microparasites, macroparasites can be subdivided into those that are transmitted directly and those that require a vector or intermediate host for transmission and therefore have an indirect life cycle.

directly transmitted macroparasites: monogeneans ...

The monogeneans are ectoparasitic platyhelminth worms that feed from the skin or gills of fish particularly, but also from amphibians, reptiles, cetaceans and cephalopods. They maintain their position by means of a specialized posterior attachment organ. New hosts are actively located by free-swimming larvae or by the adults themselves.

... intestinal nematodes

The intestinal nematodes of humans, all of which are transmitted directly, are perhaps the most important human intestinal parasites, both in terms of the number of people infected and their potential for causing ill health.

... lice and fleas ...

Lice spend all stages of their life cycle on their host (either a mammal or a bird), and transmission is usually by direct physical contact between host individuals, often between mother and offspring. Fleas, by contrast, lay their eggs and spend their larval lives in the 'home' (usually the nest) of their host (again, a mammal or a bird). The emerging adult then actively locates a new host individual, often jumping and walking considerable distances in order to do so.

... mildews and other plant-infecting fungi ...

Direct transmission is common amongst the fungal macroparasites of plants. For example, in the development of an infection by a leaf-infecting fungus, such as mildew on a crop of wheat, infection involves contact between a spore (usually wind dispersed) and a leaf surface, followed by penetration of the fungus into or between host cells, where it begins to grow. Sooner or later this infection, if successful, will become apparent as a lesion of altered host tissue. This phase of invasion and colonization (the lag or juvenile phase) precedes an infective stage when the lesion matures and starts to produce spores. Subsequently, the lesion may cease to produce spores, and that part of the host tissue is then either dead or incapable of being reinfected.

(a)

(b)

(c)

Figure 12.2 Macroparasites of animals.
(a) Hedgehog ticks (*Ixodes hexagonus*) *in situ* with heads buried in the host's skin.
(b) Fungal parasite of *Campanotus* ant. (Courtesy of BPS.) (c) Three-spined stickleback with cestode worm parasites (the pleurocercoid stage of *Schistocephalus solidus*). (Courtesy of Heather Angel.)

... parasitic flowering plants ...

Most of the macroparasites of higher plants are fungi or insects, such as the gall formers. Flowering plants in a number of families have, however, become specialized as parasites on other flowering plants. These are of two quite distinct types: (i) *holoparasites*, which lack chlorophyll and are wholly dependent on the host plant for the supply of water, nutrients and fixed carbon; and (ii) *hemiparasites*, which are photosynthetic but have a poorly developed root system of their own, or none at all. The hemiparasites form connections with the roots or stems of other species and draw most or all of their water and mineral nutrient resources from the host. A few of the hemiparasites, such as *Odontites verna*, *Rhinanthus minor* and *Euphrasia* spp., can grow to maturity without a host, although they are then severely stunted.

The most extreme holoparasitism is found in *Rafflesia arnoldii*, in which the entire vegetative body of the plant is a network within the host. Only the flower develops externally and appears on the surface of the host roots. It adds to its idiosyncrasy by

433 PARASITISM AND DISEASE

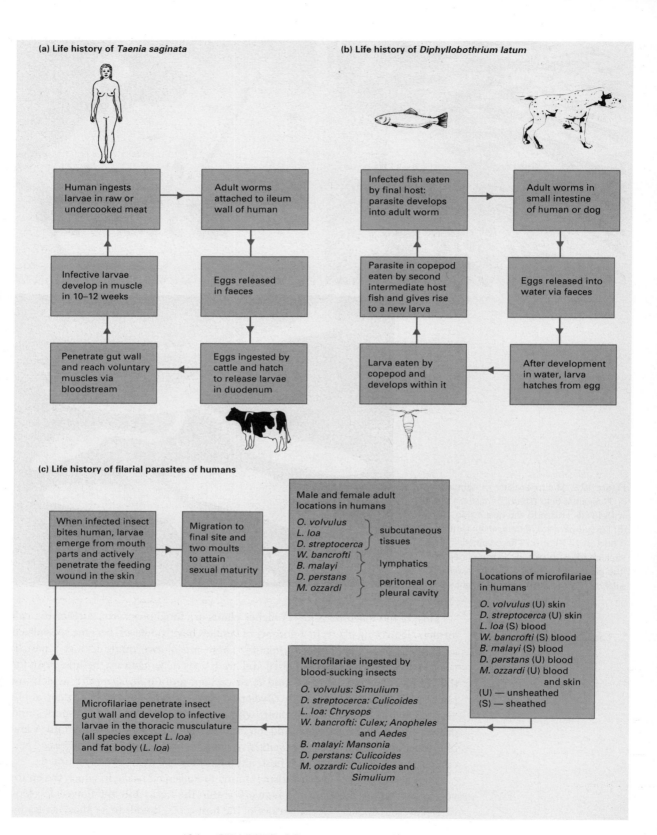

(a) Life history of *Taenia saginata*

Human ingests larvae in raw or undercooked meat

Adult worms attached to ileum wall of human

Infective larvae develop in muscle in 10–12 weeks

Eggs released in faeces

Penetrate gut wall and reach voluntary muscles via bloodstream

Eggs ingested by cattle and hatch to release larvae in duodenum

(b) Life history of *Diphyllobothrium latum*

Infected fish eaten by final host: parasite develops into adult worm

Adult worms in small intestine of human or dog

Parasite in copepod eaten by second intermediate host fish and gives rise to a new larva

Eggs released into water via faeces

Larva eaten by copepod and develops within it

After development in water, larva hatches from egg

(c) Life history of filarial parasites of humans

When infected insect bites human, larvae emerge from mouth parts and actively penetrate the feeding wound in the skin

Migration to final site and two moults to attain sexual maturity

Male and female adult locations in humans

O. volvulus
L. loa
D. streptocerca } subcutaneous tissues
W. bancrofti
B. malayi } lymphatics
D. perstans
M. ozzardi } peritoneal or pleural cavity

Locations of microfilariae in humans

O. volvulus (U) skin
D. streptocerca (U) skin
L. loa (S) blood
W. bancrofti (S) blood
B. malayi (S) blood
D. perstans (U) blood
M. ozzardi (U) blood and skin
(U) — unsheathed
(S) — sheathed

Microfilariae ingested by blood-sucking insects

O. volvulus: *Simulium*
D. streptocerca: *Culicoides*
L. loa: *Chrysops*
W. bancrofti: *Culex*; *Anopheles* and *Aedes*
B. malayi: *Mansonia*
D. perstans: *Culicoides*
M. ozzardi: *Culicoides* and *Simulium*

Microfilariae penetrate insect gut wall and develop to infective larvae in the thoracic musculature (all species except *L. loa*) and fat body (*L. loa*)

being the biggest known flower, reaching 1 m in diameter. All of the resources used in its growth are derived directly from the host.

The hemiparasites on trees tap the sapwood of their host either by sending in one primary haustorium that branches in the cortex of the host and may encircle a branch from within, or by developing a root system on the outside of the branch from which haustoria tap into the sapwood of the host at intervals. There is clear evidence that the presence of branch hemiparasites reduces the growth of the host. Stands of silver fir in France that carried mistletoe populations were reduced in stem volume by 19% (Klepac, 1955), and dwarf mistletoe has caused losses of one-third in extractable timber in stands of lodgepole pine in Colorado (Gill, 1957).

The more complete the dependence of plant parasites on their hosts, the more limited is the range of hosts that they parasitize. Thus, the hemiparasitic *Odontites verna* has a wide range of potential hosts, but the holoparasitic species of the genus *Orobanche* (the broomrapes) have a range which is often restricted to a single host species. Also, the more wholly parasitic the parasite, the smaller tend to be its seeds, and the more numerous. It is as if these species have gained access to a food resource for their early development that is an alternative to parentally supplied seed reserves and so have been able to increase the numbers of their seeds at the expense of seed size.

There are many types of medically important animal macroparasites with indirect life cycles. For example, the tapeworms are intestinal parasites as adults, absorbing host nutrients directly across their body wall, and consisting of a large number of segments (or 'proglottids') proliferating behind an attached head. The most distal proglottids contain masses of eggs and are voided *via* the host's faeces. The larval stages of the life cycle then proceed through one or two intermediate hosts before the definitive host (in these cases, the human) is reinfected. Figure 12.3 outlines the life cycles of two human intestinal tapeworms.

The schistosomes causing human schistosomiasis (bilharzia) are representative of a much larger group of trematode worms, in that their life cycle involves sexual reproduction in a terrestrial vertebrate host and clonal multiplication in a snail (Figure 12.4). They are therefore macroparasites of humans but microparasites of snails. Schistosomiasis affects the gut wall where eggs become lodged, and also affects the blood vessels of the liver and lungs when eggs become trapped there too.

Filarial nematodes are long-lived parasites of humans that all require a period of larval development in a blood-sucking insect. *Wucheria bancrofti*, which causes Bancroftian filariasis, does its damage by the accumulation of adults in the lymphatic system (classically, but only rarely, leading to elephantiasis). Larvae (microfilariae) are released into the blood and are ingested by mosquitos which also transmit more developed, infective larvae back into the host. *Onchocerca volvulus*, which causes 'river blindness' is transmitted by blackflies (the larvae of which live in rivers, hence the name). Here, though, it is the microfilariae that do the major damage when they are released into skin tissue and reach the eyes.

Figure 12.3 (*facing page*) (a, b) The life histories of two cestode parasites that infect humans: (a) *Taenia saginata* which is transmitted in raw or undercooked beef and (b) *Diphyllobothrium latum* which is transmitted in incompletely cooked fish. (c) The life histories of the seven most important filarial parasites of humans. The filariae are long-lived nematode worms that are transmitted by blood-sucking hosts. B, *Brugia*; D, *Dipetalonema*; L, *Loa*; M, *Mansonella*; O, *Onchocerca*; W, *Wuchereria*. (After Whitfield, 1982.)

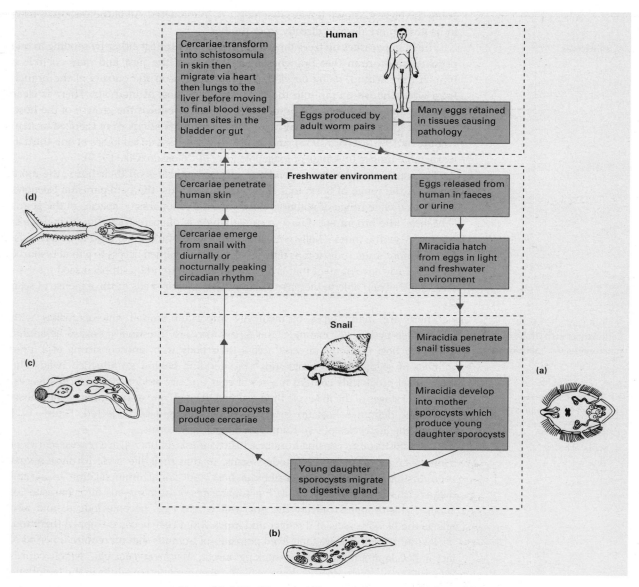

Figure 12.4 The life cycle of human schistosomes (a class of digenean worms): (a) the ciliated larva, which infects molluscs (miracidia); (b) the sporocysts, which develops in the mollusc; (c) rediae, which migrate to the mollusc's digestive gland and may multiply there eventually releasing (d) larval stages that leave the mollusc—these cercariae may in some cases infect the host directly or may encyst as a dormant stage that is eaten by the host. (After Whitfield, 1982.)

... rust fungi

Indirect transmission of plant macroparasites via an intermediate host is common amongst the rust fungi. For example, in black stem rust (Figure 12.5), infection is transmitted from an annual grass host (especially the cultivated cereals such as wheat) to the barberry shrub (*Berberis vulgaris*) and from the barberry back to wheat. Infections on the cereal are polycyclic, i.e. within a season spores may infect, form lesions on the leaves and stem, and release spores which disperse and in turn infect

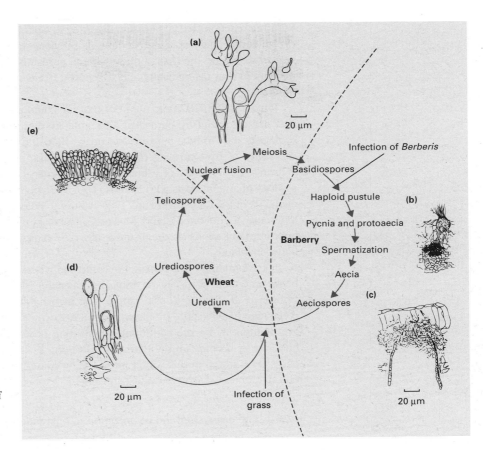

Figure 12.5 The life history of the black stem rust disease of cereals (*Puccinia graminis*—see text). (After Webster, 1970.)

further cereal plants. This cycle of repeated reinfection may continue throughout most of the season as a phase of intense clonal multiplication by the parasite, and it is this phase that is responsible for epidemic outbreaks of disease. At the end of the season the parasite produces a different type of spore, a thick-walled resistant 'teliospore' which overwinters. At the start of the next growing season the spores germinate and (without at this stage infecting a host) almost immediately produce haploid spores (basidiospores) that can infect only the barberry. The infection spreads as a haploid mycelium within the host and becomes 'spermatized' by insect-transmitted spores that are, in effect, gametes. The parasite on the barberry now forms lesions from which yet a further (new) type of spore is produced (aeciospore), that can infect only the grass host. A proper sexual process occurs only on the barberry. The barberry is a long-lived shrub and the rust is persistent within it. Infected barberry plants may therefore serve as primary and persistent foci for the spread of the rust into crops of the much shorter-lived annual cereal.

12.3 Hosts as habitats

the habitat of a parasite is alive

The essential difference between the ecology of parasites and that of free-living organisms is that the habitats of parasites are themselves alive. A living habitat is capable of growth (in numbers and/or size); it is potentially reactive, i.e. it can respond *actively* to the presence of a parasite by changing its nature, developing

437 PARASITISM AND DISEASE

immune reactions to the parasite, digesting it, isolating or imprisoning it; it is able to evolve (every host is a descendant of previous hosts and a product of natural selection); and in the case of many animal parasites it is mobile and has patterns of movement that dramatically affect dispersal (transmission) from one habitable host to another.

Janzen (1968, 1973) pointed out that we could usefully think of hosts as islands that are colonized by parasites. This analogy was one of the catalysts that made ecologists start to take parasites seriously. By using the same vocabulary it brought the study of host/parasite relationships into the same arena as MacArthur and Wilson's (1967) study of island biogeography (see Chapter 15, Section 15.6). The analogy of animal hosts with islands is not so obvious (islands are rarely mobile), but elements of the analogy are still real. A human colonized by the malarial parasite is in a sense an inhabited island. Mosquitos can disperse the parasite from one island to another. Limitations on the distance flown by mosquitos correspond to the distance between islands.

The island analogy causes some confusion when, for example, a study is made of hosts and their parasites on real (e.g. oceanic) islands (Dobson & Pacala, 1992; Dobson et al., 1992). Considering a host as a habitable patch in the environment conveniently avoids the problem (Dobson & Keymer, 1990). Populations of parasites are maintained by the continual colonization of new host patches as old infected hosts die or become immune to new infection. As soon as the population biology of parasites is stated in this way it immediately comes into the context of metapopulation dynamics (which we discuss in Chapter 15, Section 15.6).

12.3.1 The dispersion and colonization of host patches

Not surprisingly, different species of parasite are transmitted in different ways between host patches. For directly transmitted microparasites, in which infection is by physical contact between hosts or by a very short-lived infective agent, the net rate of transmission is usually directly proportional to how often infected hosts meet susceptible (uninfected) hosts. The net rate of parasite transmission is then always greatest in populations that contain high densities of susceptible individuals, or high densities of infecteds and especially in those containing high densities of both. The transmission rate of many parasites varies with the season because of the tendencies of hosts to aggregate or disperse in tune with variations in the climate, and so increase or relax the frequency of contacts between infecteds and susceptibles. Seasonal variations in aggregation are especially important in the epidemiology of many common viral infections of humans, such as measles, mumps and chickenpox (Anderson, 1982).

transmission is affected by contact rates and therefore densities

Where hosts are infected by long-lived infective agents (e.g. dormant and persistent spores), the rate at which hosts acquire parasites is usually directly proportional to the frequency of contact between hosts and infective stages (and is therefore dependent on the densities of both). For vector-transmitted microparasites, on the other hand, the rate of transmission from infected vectors to susceptible hosts is proportional to the 'host-biting rate'. This itself is dependent on the overall frequency of vector 'bites' and the fraction of the host population that is susceptible. The transmission rate from infected hosts back to susceptible vectors depends on the frequency of vector bites and the fraction of hosts that are infected.

Many soil-borne fungal diseases of plants are spread from one host plant to

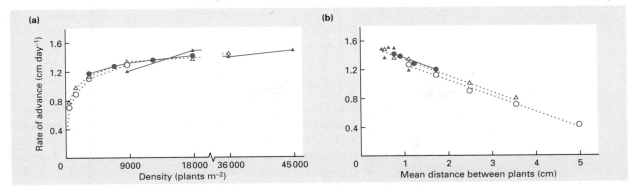

Figure 12.6 The rate of advance of the pathogen *Pythium irregulare* into populations of cress (*Lepidium sativum*), from infected seedlings placed at the edge of the plant populations. The spread of the disease is shown (a) in relation to the density of cress plants, and (b) in relation to the mean distance between plants. ○ , ● , △ and ▲ represent replicates. (After Burdon & Chilvers, 1975.)

the spread of a disease is affected by the distance between hosts

another by root contacts, or by the growth of the fungus through the soil from a 'food base' established on one plant which gives it the resources from which to attack another. For example, the honey fungus *Armillaria mellea* spreads through soil as a bootlace-like 'rhizomorph' and can infect another host (usually a woody tree or shrub) where their roots meet. In effect, the parasite spreads from one host patch to another only along corridors that link the patches. Such a parasite can spread through a plantation or along a hedge, and its spread can be controlled by digging a trench across its path. A somewhat similar type of corridor transmission occurs in the swollen shoot virus disease of cacao. The virus is spread by mealy bugs, but these are flightless and can transmit the disease only between trees that make physical contact with each other. In the past this would presumably have prevented epidemics of the disease in naturally diverse forests, but when cacao is cultivated as 'continents' of continuous interplant contacts there are greatly increased opportunities for infection to spread. In the period from 1922 to 1930 this disease spread and destroyed plantations over wide areas of West Africa.

On a much finer scale, fungi of the genus *Pythium*, which parasitize and kill the seedlings of many higher plants, also depend for their rapid spread on the seedlings being close to each other (Figure 12.6). It is very characteristic of such soil-borne diseases that epidemics spread as 'invading fronts' or 'horizons of infection'.

wind-borne diseases

For diseases that are spread by wind, foci of infection may become established at great distances from the origin; but the rate at which an epidemic develops locally is strongly dependent on the distance between individuals. It is characteristic of wind-dispersed propagules (spores, pollen and seeds) that the distribution achieved by dispersal is usually strongly 'leptokurtic': a few propagules go a very long way but the majority are deposited close to the origin. An extreme example of a wind-borne parasite of animals is the virus causing foot-and-mouth disease in farm livestock (Donaldson, 1983). The disease is normally absent from Great Britain, but occasionally a case is reported involving long-distance spread, for instance from The Netherlands to eastern England. Foci of infection are very rarely formed from this sort of distance, but if they establish they can spread extremely fast within herds, especially in the crowded conditions of cattle markets. Because Great Britain is an

leptokurtic transmission

island, the effective control of the disease is by the compulsory slaughter of all infected animals and likely contacts. On the continent of Europe such a policy could not be sustained, and vaccination is adopted.

12.3.2 The specificity of hosts as habitats

Much of ecology is about the ways in which organisms are constrained and limited to life in specialized environments. Much of evolutionary theory is about how such constraints and limits have arisen. We have seen, for example, in the chapters on the interactions between predators and their prey that there is often a high degree of specialization of a particular predator species on a particular species of prey (monophagy). The specialization of particular parasites on one, or a restricted range of, host species is even more striking. For any species of parasite (be it tapeworm, virus, protozoan or fungus) the potential hosts are a tiny subset of the available flora and fauna. The overwhelming majority of other organisms are quite unable to serve as hosts: usually we do not have the faintest idea why!

Not only is a species of parasite commonly restricted to one species of host, but there may be detailed correspondence between specific genotypes of parasite and specific genotypes of the host. In its most extreme form there is a strict one-for-one correspondence between genes that confer pathogenicity on a parasite, and genes conferring resistance on a host.

The study of specialization, whether of predators on prey, or parasites on hosts (or even specialized matches between species and patches of the physical environment), forms an essential part of ecology. It poses both the *proximal* questions about what it is in the present biology of individuals that constrains where they can live, and the *ultimate* questions of how these specialization's and constraints have evolved. We make the implicit assumption that when a predator chooses its food or a parasite discriminates between potential hosts the range selected is, or is close to, an evolutionarily stable strategy. In other words we assume that the existence of a narrow and restricted range of hosts (or prey) represents some sort of evolutionarily optimal condition. We presume that natural enemies, chemical defences or some other force would reduce the fitness of any parasites that extended the range (Ward, 1992). These are rash assumptions that beg to be tested.

Studies of parasitoid wasps offer the chance to test some of the theories of host/parasite specialization. Their larvae develop in or on another insect—the host, killing it in the process. The way in which host preference may change was tested by reducing the life expectancy of parasitoids (*Venturia canescens*) by starving them. As predicted by theory, starved parasitoids were less choosy. They were willing to lay their eggs in low-quality hosts more frequently than unstarved controls (Fletcher *et al.*, 1994).

12.3.3 The specificity of habitats within hosts

hosts are heterogeneous environments

Although for many purposes we may regard hosts as habitable patches of the environment, they are by no means homogeneous patches. Every organism is a heterogeneous environment—a matrix of potential habitats. Even a bacterial cell is a heterogeneous environment to an infective particle of phage. Gut, blood, liver, brain, pancreas and eyes are different environments for the parasites of a mammal. Leaves, stems, roots, fruits and seeds offer different resources, are differently protected and provide different physical environments for the parasites of plants (Figure 12.7).

440 CHAPTER 12

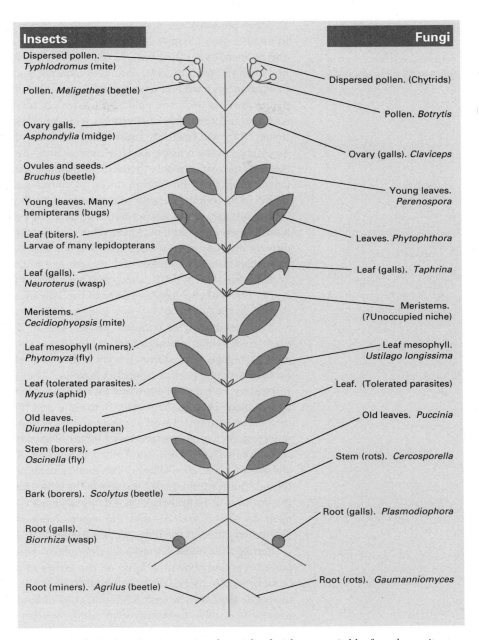

Insects

Dispersed pollen. *Typhlodromus* (mite)

Pollen. *Meligethes* (beetle)

Ovary galls. *Asphondylia* (midge)

Ovules and seeds. *Bruchus* (beetle)

Young leaves. Many hemipterans (bugs)

Leaf (biters). Larvae of many lepidopterans

Leaf (galls). *Neuroterus* (wasp)

Meristems. *Cecidiophyopsis* (mite)

Leaf mesophyll (miners). *Phytomyza* (fly)

Leaf (tolerated parasites). *Myzus* (aphid)

Old leaves. *Diurnea* (lepidopteran)

Stem (borers). *Oscinella* (fly)

Bark (borers). *Scolytus* (beetle)

Root (galls). *Biorrhiza* (wasp)

Root (miners). *Agrilus* (beetle)

Fungi

Dispersed pollen. (Chytrids)

Pollen. *Botrytis*

Ovary (galls). *Claviceps*

Young leaves. *Perenospora*

Leaves. *Phytophthora*

Leaf (galls). *Taphrina*

Meristems. (?Unoccupied niche)

Leaf mesophyll. *Ustilago longissima*

Leaf. (Tolerated parasites)

Old leaves. *Puccinia*

Stem (rots). *Cercosporella*

Root (galls). *Plasmodiophora*

Root (rots). *Gaumanniomyces*

Figure 12.7 The higher plant is a series of specialized niches occupied by fungal parasites and by insect predators and parasites. Comparable, although rather less diverse, diagrams could be drawn for nematodes and for pathogenic bacteria. The subtlety of niche specialization is even greater than is implied by the diagram, for instance species that produce leaf galls on oak are believed to specialize at a subspecific level with respect to the tip, middle and base of a leaf, as well as to leaf age. (After Askew, 1961.) Some species may alternate between two highly specific niches on the same plant, for example *Biorhiza pallida* forms galls on the roots and shoot meristems in different seasons. (After Harper, 1977.)

Most parasites are specialized to live in particular parts of their host. Like mutualists (see Chapter 13) they vary in how intimately they enter into the environment that is their host. At one extreme there are ectoparasites, which feed on host material from outside. Examples amongst animal parasites include fleas, lice, ticks, monogeneans and even some fungi; whilst amongst the ectoparasites that use plants as hosts there are aphids and powdery mildews which live exposed on leaf surfaces and send only their stylets (aphids) or haustoria (mildews) into the host tissue. They are relatively uninsulated from the elements and share some of the experience of the outside world with their host. An intermediate group of parasites develops in cavities within the host; for example nematode worms in the alimentary canal, lungs and Eustachian tubes of animals, and fungi growing as mycelia between, but not in, the cells of a plant host. Such parasites may not enter the host cells at all; they are in a strict sense 'outside' the cellular body of the host, although it forms their environment. At the other end of the spectrum are parasites that enter directly into host tissues and cells. The malarial parasites and the piroplasms that cause babesiosis in domesticated animals live in the red blood cells of vertebrates. *Theileria* parasites of cattle, sheep and goats live in the lymphocytes of the mammal and in the epithelial cells and later in the salivary gland cells of the tick that is the disease vector.

The distribution of parasites on the gills of fishes is a particularly beautiful example of habitat specialization within a host. The teleost fishes have four pairs of gills. Each gill is formed from a bony arch which carries primary lamellae, which in turn carry tightly packed secondary lamellae. They provide an enormous surface for respiratory exchange of gases and for occupancy by parasites. There is a continuous flow of blood close to the surface of the gills and the flow of water over them keeps the microenvironment well oxygenated. The gills are parasitized by fungi, protozoa, monogenean worms, metacercarial larvae of trematodes, copepods, glochidia larvae of bivalve molluscs, leeches and mites. In a survey of the parasites on the fish *Lepomis globosus* in West Lake, Ontario, the monogenean parasites were most abundant on the anterior face and medial sections of the gills. The copepods were concentrated on the dorsal and ventral regions. The monogeneans and glochidia were found most often on gills II and III, and the copepod *Ergasilus caeruleus* was rather evenly distributed amongst the different gills (Hanek, 1972).

Perhaps the most remarkable example of habitat specificity known amongst parasites is that shown by fungi on the bodies of insects. Within the fungal genus *Laboulbenia*, are species that are specialists on beetles. On one particular host species, the beetle *Bembidion picipes*, different species of fungus colonize different parts of the body. It is possible to identify many of the species of fungus with precision simply from the part and the sex of the beetle that they parasitize (Benjamin & Shorer, 1952).

Many parasites are limited in their growth to sites within a host that they reach first, but in others the specific sites are sought out by migration after they have entered the host. Helminth worms are particularly good examples of parasites that search for specific habitats within their hosts. For example, the nematode *Angiostrongylus cantonensis* has a remarkable circuit of habitats in the rat (Mackerras & Saunders, 1955; Alicata & Jindrak, 1970). The parasites penetrate the wall of the gut, enter the small veins and pass to the liver in the hepatic portal system. They then travel to the right atrium of the heart and then to the lungs through the pulmonary artery and back to the main arterial circulation from the heart. The

young adults then pass into cerebral veins, return to the heart and lungs and become established in the pulmonary arteries. There are no free-living organisms with patterns of migration that approach this complexity.

By experimentally transplanting parasites from one part of the host's body to another it can be shown that they home in on target habitats. When nematode worms (*Nippostrongylus brasiliensis*) were transplanted from the jejunum into the anterior and posterior parts of the small intestine of rats, they migrated back to their original habitat (Alphey, 1970). Another nematode (*Spirocerca lupi*), when transplanted into the thoracic cavity of a dog, migrated back to its normal site on the wall of the oesophagus (Bailey, 1972).

The examples above are of parasites that move bodily, but in other cases habitat search may involve growth (e.g. of a fungus) from the point at which infection occurs to a quite different part of the body where spore production takes place. Loose smut of wheat, *Ustilago tritici*, infects the exposed stigmas of the flowers of wheat, and then grows as an extending filamentous system into the young embryo. Growth continues in the seedling, and the fungus mycelium keeps pace with the growth of the shoot. Ultimately, the fungus grows rapidly into the developing flowers and converts them into masses of spores.

12.3.4 The dynamics of parasite populations within hosts

competition between parasites increases with density ...

Because populations of parasites occupy specific sites within their host, it is to be expected that individuals will sometimes interfere with each other's activities, and that there will be intraspecific competition between parasites and density dependence in their growth and in their birth and/or death rates. It is extremely difficult to determine the way in which populations of microparasites are regulated within the cells of their host. However, it is relatively easy to manipulate populations of the larger parasites like the helminths, and show how they respond to their own density (Figure 12.8). In experimental infections of mice with the tapeworm *Hymenolepis microstoma*, the total weight of worm per mouse rises asymptotically, as if to some ceiling value, as the number of parasites is increased. The weight of individual parasites decreases correspondingly. The total number of eggs produced *per host* is much the same at all levels of infection, but the number of eggs produced *per parasite* falls in a strongly density dependent fashion. These patterns are clearly reminiscent of many of those in Chapter 6, especially the 'constant final yield' found in many plant monocultures (e.g. Figure 6.16).

A similar crowding effect can be seen in the liver fluke (*Fasciola hepatica*), and two additional effects now come into play. As density increases, the percentage of infections that develop into established worms declines, and, even more important, the development of a worm to the egg-producing stage is greatly delayed. This has the direct effect of slowing down the potential rate of increase of the population of worms.

... but, so may the host's immune response

We need to be cautious how we interpret what appear to be density dependent effects within parasite populations as the result of intraspecific competition. The intensity of the immune reaction elicited from the host (see below) invariably depends on the density of parasites. Part of the effect of density on the performance of parasites then reflects active reaction by the parasite's environment—its host.

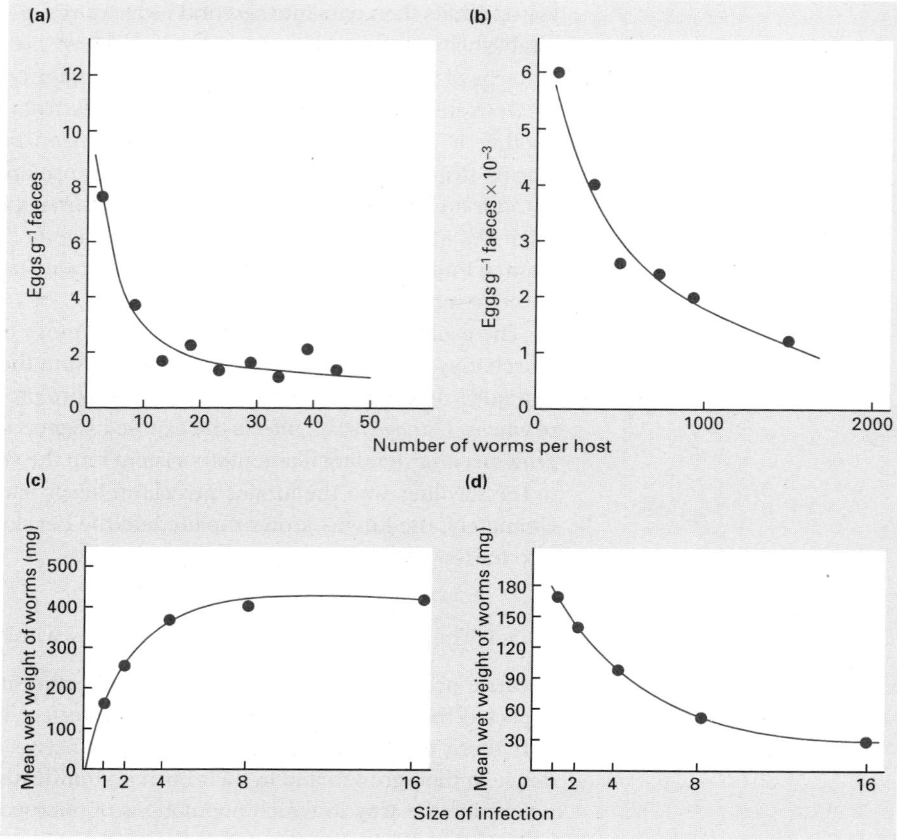

Figure 12.8 Density dependent responses in populations of parasites. (a) Egg production by the roundworm *Ascaris* in humans. (After Croll *et al.*, 1982.) (b) Egg production in *Ancylostoma* (a tapeworm) in humans. (After Hill, 1926.) (c) The mean weight of worms per infected mouse after deliberate infection with different levels of infection by the tapeworm *Hymenolepis microstoma*, and (d) the mean weight of individual worms from the same experiment. (After Moss, 1971.)

12.3.5 Hosts as reactive environments

The presence of a parasite in a host appears always to elicit a response. This must be so as a matter of definition—if there were no response we ought to describe the organism as a commensal, not a parasite. The most obvious response is for the whole host to die, or just its infected parts. We recognize a major difference amongst

necrotrophs and biotrophs

parasites between those that kill and can then continue life saprophytically on the dead host (*necrotrophic parasites*) and those for which the host must be alive (*biotrophic parasites*). This tidy distinction is somewhat spoiled by the responses of modular organisms, such as higher plants. In these, a part (e.g. a leaf) may be killed by a necroparasite but the remainder of the plant may remain uninfected. For a biotroph, the death of its host (or, in a modular organism, the death of the infected part) spells the end of its active life. A dead host or the dead part of a host is an uninhabitable patch in the environment of a biotroph. Most parasitic worms, lice and fleas, protozoa and parasitoids and the gall-forming parasites of higher plants are biotrophic, as are the rusts, smuts and mildews amongst the fungal parasites of plants. Biotrophs have specialized nutritional requirements, and until recently few could be cultured outside their host.

Any reaction by an organism to the presence of another depends on it recognizing a difference between what is 'self' and what is 'not-self'. In invertebrate animals, populations of phagocytic cells are responsible for most of a host's response to invaders, even to inanimate particles. Phagocytes may engulf and digest small alien bodies, and encapsulate and isolate larger ones. In vertebrates there is also a phagocytic response to material that is 'not-self', but their armoury is considerably extended by a much more elaborate process: the immune response.

The ability of vertebrates to recognize and reject tissue grafts from unrelated members of their own species is just one example of the immune response in action (Figure 12.9). For the ecology of parasites, an immune response has two vital features: (i) it may enable a host to recover from infection; and (ii) it can give a once-infected host (= habitat) a 'memory' that changes its reaction if the parasite (= colonist) strikes again, i.e. the host has become immune to reinfection. In mammals, the transmission of immunoglobulins to the offspring can sometimes even extend protection to the next generation. The response of vertebrates to infection therefore protects the individual host and increases its potential for survival to reproduction. The response of invertebrates gives poorer protection to the individual, and the recovery of their populations after disease depends more on the high reproductive potential of the survivors than on the recovery of those that have been infected (Tripp, 1974).

For most viral and bacterial infections of higher animals, the colonization of the host is a brief and transient episode in the host's life. The parasites multiply within the host and elicit a strong immunological response. By contrast, the immune responses elicited by many of the macroparasites and protozoan microparasites tend to be rather short lived. The infections themselves, therefore, tend to be persistent, and hosts are subject to continual reinfection.

transient and persistent infections

The modular structure of plants, the presence of cell walls and the absence of a true circulating system (such as blood or lymph) would all make any form of

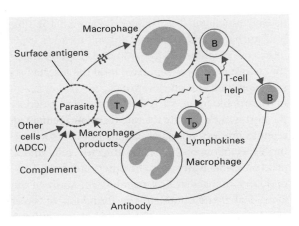

Figure 12.9 A vertebrate host is a reactive habitat in which the immune response of the host converts it from being habitable into one that is uninhabitable. This sort of reaction distinguishes living organisms as habitats from the sorts of environment more normally considered by ecologists. The response begins when the immune system is stimulated by an antigen that is taken up and processed by a macrophage. The antigen is a part of the parasite, such as a surface molecule. The processed antigen is presented to T and B lymphocytes. T lymphocytes respond by stimulating various clones of cells, some of which are cytotoxic, whilst others secrete lymphokines and yet others stimulate B lymphocytes to produce antibodies. The parasite that bears the antigen can now be attacked in a variety of ways. (After Cox, 1982.)

immunological response an inefficient protection. There is no migratory population of phagocytes in plants that can be mobilized to deal with invaders. There is however growing evidence that higher plants possess elegant and complex systems of defense against pathogens (and in some cases also herbivores). These defenses have been classified as 'constitutive' or 'induced'. Constitutive defenses are physical or biological barriers against invading organisms that are present whether the pathogen is present or not. Inducible defenses are those that arise in response to pathogenic attack and, in these, signal and receptor mechanisms are present that rival the complexity and sophistication of signalling systems in animals (Ryan & Jagendorf, 1995).

systematic acquired resistance

After a plant has survived a pathogenic attack, it may become immune to subsequent attacks—a response called 'systematic acquired resistance' is elicited from the host. For example, tobacco plants infected on one leaf with tobacco mosaic virus can produce local lesions that restrict the virus infection locally: the plants then become resistant to new infections not only by the same virus but to other pathogens as well. In some cases the process involves the production of 'elicitins' which have been purified and shown to induce vigorous defense responses by the host (Yu, 1995) which may include the production of extracellular pathogenesis related (PR) proteins. One signal that leads to the response by the whole plant appears to be salicylic acid (Lee *et al.*, 1995). The detailed chemistry of signalling and transcription is discussed in Ryan *et al.* (1995).

Many specialized parasites (especially parasites of some insects) kill their host quickly, but there are many that are tolerated by their host that shows little response to infection. A most remarkable example of a parasite that uses host resources without stimulating a strong host reaction is seen in the parasitism of higher plants by aphids. Aphids gain access to plant resources by penetrating with their stylets into the phloem and drawing phloem sap from the sieve tubes whilst these are still actively functional (see Chapter 3, Figure 3.15). Phloem cells are extremely sensitive to disturbance and extremely difficult to study experimentally. Yet, aphid stylets are inserted into phloem with so little disruption that plant physiologists, interested in the nature of phloem sap, have used the inserted stylets as the most effective way of sampling undisturbed sieve tubes.

many biotrophs are tolerated

Biotrophic fungi, such as the rusts and powdery mildews, penetrate the cells of the host, which usually tolerate them and remain alive. In some cases, the infections seem to produce no major change in the behaviour of the host apart from reducing the expectation of life of the infected cells and changing the pace of metabolic processes and water loss. In other situations, the presence of a parasite induces profound changes in host metabolism. Many of the biotrophic parasites that infect living leaf tissue stimulate the formation of 'green islands' around the infection which are not only sites of increased photosynthetic activity but also serve as 'sinks' to which metabolites flow from uninfected parts of the same plant.

many biotrophs alter the pattern of host morphogenesis

Many biotrophic parasites induce a new *programmed* change in the development of the host. The agromyzid flies and cecidomyid and cynipid wasps that form galls on higher plants are remarkable examples. The insects lay eggs in host tissue, which responds by renewed growth. The galls that are produced are the result of a morphogenetic response that is quite different from any structure that the plant normally produces. Just the presence, for a time, of the parasite egg may be sufficient to start the host tissue into a morphogenetic sequence that can continue even if the developing larva is removed. There is clearly a process of recognition in which specific alien material (the insect egg) cues the process of gall formation. Amongst the gall-formers that attack the common English oak (*Quercus robur*), each elicits a unique morphogenetic response from the host (Askew, 1961, 1971). Many parasitic aphids also induce specialized galls on higher plants and there is evidence that it is the aphids that determine the details of gall morphology—they may be said to extend their phenotype to include plant material (Stern, 1995)!

Fungal parasites can also induce morphogenetic change. Parasites such as *Plasmodiophora brassicae* stimulate cell multiplication and enlargement, increasing the number of habitable sites (root cells) in which the parasite multiplies. Nematode parasites may also force plant hosts into morphogenetic responses such as enormous cell enlargement and the formation of nodules and other 'deformations'. One such intervention by a parasite in controlling the growth of the host is that caused by the bacterium *Agrobacterium tumefaciens*, which induces the formation of galls on plant tissues. After infection, gall tissue can be recovered that lacks the parasite but has now been set in its new morphogenetic pattern of behaviour; it continues to produce gall tissue. In this case, the parasite has induced a genetic transformation of the host cells. The process involves plasmid transmission, and something similar may be involved in the formation of the insect-induced galls and the *Rhizobium*-induced nodules of leguminous plants (see Chapter 13, Section 13.10.1).

A particular type of control of host growth is accomplished by parasitoids (see Chapter 8, Section 8.1). The host remains 'alive' as its body becomes increasingly transformed into the body of its parasitoid, but it has no continued life after the parasitic larvae mature. Such parasitoids kill the host, but use the resources of its body only whilst it is alive. They kill but are not necroparasites. Some of the more specialized parasitic fungi also 'take control' of their host plant and 'castrate' or 'sterilize' it. The fungus *Epichloe typhina*, which parasitizes grasses, prevents them from flowering and setting seed—the grass remains a vegetatively vigorous 'eunuch' leaving descendant parasites but no descendants of its own. This 'take-over' of the host is a common phenomenon amongst systemic parasites of plants, and these vegetative eunuchs may actually have greater vegetative vigour than unparasitized plants (Bradshaw, 1959).

some parasites alter host behaviour

Most of the responses of modular organisms to parasites (and other environmental stimuli) involve changes in growth and form, but in unitary organisms the response of hosts to infection more often involves a change in behaviour: this often increases the chance of transmission of the parasite. Irritation of the anus of worm-infected hosts stimulates scratching, and parasite eggs are then carried from the fingers or claws to the mouth. Sometimes, the behaviour of infected hosts seems to maximize the chance of the parasite reaching a secondary host or vector. Praying mantises have been observed walking to the edge of a river and apparently throwing themselves in. Such suicidal individuals were seen (J.L. Harper, personal observation) in two successive years by the River Hérault in southern France, and within a minute of entering the water a gordian worm (*Gordius*) had emerged from the anus. This worm is a parasite of terrestrial insects but depends on an aquatic host for part of its life cycle. It seems that an infected host develops a 'hydrophilia' that ensures that the parasite reaches a watery habitat. Suicidal mantises that had been rescued immediately returned to the river bank and threw themselves in again!

12.3.6 Necrotrophic parasites

Lucilia cuprina, the blowfly of sheep, is a necroparasite on an animal host. The fly lays eggs on the living host and the larvae (maggots) eat into its flesh and may kill it. The maggots continue to exploit the carcass after death but they are now detritivores rather than parasites. Necroparasites on plants include many that attack the vulnerable seedling stage (especially species of the genus *Pythium*) and cause the 'damping-off' of seedlings. They are rarely host-specific and attack a wide range of species. *Botrytis fabi* is a typical necroparasite. It develops in the leaves of the bean *Vicia faba*, and the cells are killed, usually in advance of penetration. Spots and blotches of dead tissues form on the leaves and the pods. The fungus continues to develop as a decomposer (saprophyte), and forms and disperses spores from the dead tissue. It does not do so when the host tissue is alive.

necrotrophs as pioneering decomposers

When modular organisms are the host, a necroparasite will often kill a part rather than the whole host. Modular organisms grow by continually producing new parts as older parts die. A necroparasite may simply speed up the natural death rate of the modules of its host (e.g. the leaves of a plant) leaving the remainder alive and iterating new parts. Most necroparasites can be regarded as pioneer decomposers. They are 'one jump ahead' of competitors because they can kill the host (or its parts) and so gain first access to the resources of its dead body. Their resource requirements are typically simple; they can usually be grown on simple artificial media. The response of the host to necroparasites is never very subtle. Amongst plant hosts, the most common response is to shed the infected leaves (e.g. alfalfa sheds leaves that are attacked by *Pseudopeziza medicaginis*) or to form specialized barriers that isolate the infection (potatoes form corky scabs on the tuber surface which isolate infections by *Actinomyces scabies*).

12.3.7 The distribution of parasites within host populations

The distribution of parasites within populations of hosts is rarely random. For any particular species of parasite it is usually the case that many hosts harbour few or no parasites, and a few hosts harbour many, i.e. the distributions are usually aggregated or clumped (Figure 12.10). Some of the aggregation results when microparasites multiply within randomly infected hosts, and quickly generate gross differences between the zero populations of parasites in uninfected hosts and dense populations in others. When dispersal is weak or slow, very dense, very local aggregations of parasites can occur. Aggregations also arise because individual hosts vary in their susceptibility to infection (whether due to genetic, behavioural or environmental factors).

parasites are usually aggregated

In such locally aggregated populations, the mean density of parasites has rather little useful meaning. In a population of humans in which only one person is infected with anthrax, the mean density of *Bacillus anthracis* is a particularly useless piece of information! The most widely used epidemiological statistic for microparasites is the *prevalence* of infection: the proportion or percentage of a host population that is infected with a specific parasite. This is a particularly important measure in the case of microparasites, where the number of individual parasites cannot be counted anyway (the infected host is the unit of study). On the other hand, infection may often vary in severity between individuals and is often clearly related to the number of parasites that they harbour. The number of parasites in or on a particular host is

the prevalence, intensity and mean intensity of infection

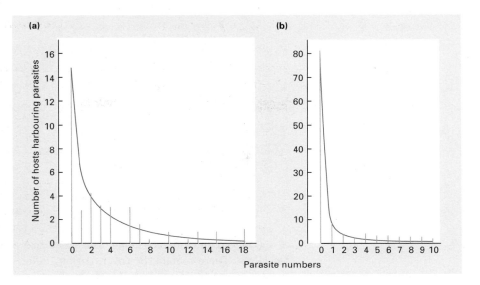

Figure 12.10 Examples of aggregated distributions of parasite numbers per host for which the negative binomial is a good empirical model. The vertical bars represent the observed frequencies and the solid line is the fit of the negative binomial distribution. The distributions of (a) *Toxicara canis*, a gut nematode parasite of foxes and (b) lice (*Pediculus humanis capitis*) on a population of humans. (After Watkins & Harvey, 1942; Williams, 1944.)

distribution, prevalence and mean intensity

referred to as the *intensity* of infection. The *mean intensity* of infection is then the mean number of parasites per host in a population (including those hosts that are not infected).

The relationships between prevalence and mean intensity for various types of frequency distribution are shown in Figure 12.11. If parasites are evenly distributed, prevalence tends to be relatively high and mean intensity relatively low. This could arise because of density dependent mortality amongst the parasites (see below), or acquired resistance to reinfection or because high intensities of infection kill the hosts. In the more usual case of aggregation, mean intensity is relatively high but prevalence is relatively low. The epidemiological consequences of aggregation are discussed in subsequent sections; and we return to the concepts of prevalence and intensity in the more general context of 'abundance' and 'rarity' in Chapter 15.

12.3.8 The diversity of parasite communities

Most organisms are host to a diverse community of parasites and this raises all the classical ecological questions about the ways in which communities are structured and organized. We have pointed out (see Section 12.3.3) that a host offers a variety

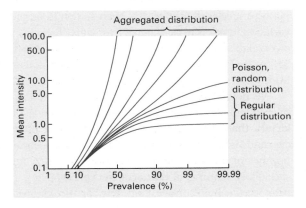

Figure 12.11 The relationship between the prevalence and mean intensity of infection for aggregated, random and regular distributions of parasites per host. (After Anderson, 1982.)

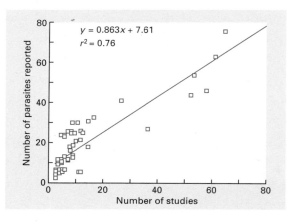

Figure 12.12 The relationship between the total number of parasite species reported and the number of published studies of 60 species of Canadian freshwater fish. (After Poulin, 1991.)

of specialized microhabitats and that these are exploited as niches by parasite specialists. There remain, however, curious differences in the richness of parasite communities in different types of host and various studies have sought to uncover the explanations.

Parasites are particularly likely to spread rapidly in species of host that form local aggregations, and it might be expected that, in this respect, what favours one species might be expected to favour many. Searches for ecological correlations amongst data sets of this sort are fraught with problems, not least that the number of species recorded in community samples (e.g. of parasite numbers) is a function of the research effort put into the study (Figure 12.12). We might predict that species that live in groups should carry the most diverse populations of parasites. The ectoparasite communities of fish that form shoals (schools) can be compared with the communities on species that are solitary as a test of this theory. It might also be predicted that internal parasites, which require vectors or other intermediate steps for transmission, should be much less affected by group living. But, a study of the parasites of 60 species of freshwater fish in Canada detected no such relationships (Poulin, 1991). A different analysis of essentially the same data set came to quite different conclusions (Ranta, 1992), namely the following.

1 Solitary species of salmonids carry a smaller diversity of parasites than salmonids that school outside the breeding season (Figure 12.13). Very surprisingly, most of the species in the analysis were internal helminth parasites with intermediate hosts—just the situation in which the effect of schooling had been predicted to be low.

2 Host species with a wider geographical range harboured a greater diversity of parasite species (this effect has been shown in other studies, for example 68% of the variation in parasite richness of freshwater fish in Great Britain was accounted for by a correlation with host range (Price & Clancy, 1983)).

3 In the Percidae the diversity of parasites increased with the size and age of the host.

We noted earlier (see Section 12.3.3) that the gills of fish offer a specialized microhabitat for ectoparasites, especially monogeneans. In a statistical analysis of factors that might account for the diversity of these parasites in West African cyprinid fishes, the size of the host species accounted for 77% of the variation in number of parasites per host (Figure 12.14). (A further 8% was accounted for by specialized features of host ecology: species living in turbulent waters or rapids, or near the water surface had few or no parasites.) A comparable, but quite

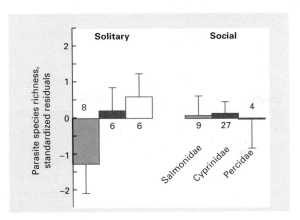

Figure 12.13 The species richness of the parasite fauna in solitary and social salmonid, cyprinid and percid fish in Canadian freshwaters. The values shown are the residuals after a regression model has taken account of the relationship between the number of parasite species reported and the number of published studies, shown in Figure 12.12. The number of species in each class is shown. (After Ranta, 1992.)

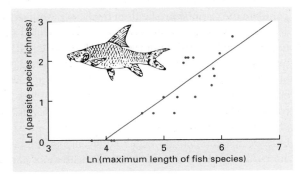

Figure 12.14 The relationship between the species richness of the parasite fauna and the maximal length (mm) of the host species in 19 West African cyprinid fishes. (After Guégan *et al.*, 1992.)

independent study was made of the species richness of internal helminth parasites in Canadian freshwater fish (Bell & Burt, 1991). They also found that the diversity of the parasite fauna was positively correlated with host size (or age). Size and diet accounted for 40% of the variance amongst host species. But, as we stress repeatedly in this book, correlations are no proof of causation, and cross-correlations can have wonderfully dangerous effects. For example, the size of an animal species must be the resultant of a host of evolutionary selective forces and itself is often highly correlated with species' environmental range. Of course, if we continue with our vision of hosts as islands it is not surprising to find that the larger the fish the greater the number of parasite species that it harbours. The exponent relating diversity of parasites to the body area of the fish is 0.22, which is indeed consistent with the mean exponent of 0.31 (with a standard deviation of 0.23), relating the number of species to the area of 90 true, i.e. maritime, islands in a survey made by Connor and McCoy (1979).

The papers of Poulin, Ranta, Guégan *et al.* and Bell and Burt are fascinating examples of what ecologists can achieve by using subtle statistics on large data sets, and are highly relevant to understanding the nature of global biodiversity. But these studies suffer from one common problem: they did not control for potential phylogenetic effects (see Harvey & Pagel (1991) for a description of ways in which this can be done). Throughout evolutionary time, host lineages undergo speciation and changes in ecological traits (body size, geographical range, etc.). This leads to differences among host lineages in the probability of acquiring (e.g. parasites colonizing new hosts) or losing (e.g. through parasite extinction) parasite species and to differences in the species richness of parasite communities. However, extant hosts have inherited some parasite species from their ancestors, independently of whether or not their ecological traits have remained the same. Therefore, contemporary parasite communities are made up of inherited parasites (historical legacy) and acquired parasites (depending on host ecology). Any attempt to establish the relationship between certain ecological variables and parasite diversity among several host species should control for these phylogenetic influences. For example, among 54 genera of birds, there is a strong correlation between host body size and the richness of the gastrointestinal parasite fauna ($P < 0.01$). But when phylogenetic effects are removed, i.e. when phylogenetically independent contrasts are used rather than genus values, the correlation disappears, suggesting that the apparent link between host body size and parasite richness in birds may be a phylogenetic artefact (Poulin, 1995).

There has been a long-continued programme of research on the lizards that populate the Caribbean islands of the Lesser Antilles (see, for example, Roughgarden, 1992). Each island has one or two closely related species of the genus *Anolis*, and these occur throughout all the natural habitats of the islands and at an abundance often of more than one lizard per m². As expected, the lizards harbour populations of parasites and these are represented by nine genera: six nematodes, two digeneans and an acanthocephalan (Dobson *et al.*, 1992). The diversity of the parasite community poses questions at a variety of levels. At one level the number and diversity of parasites per lizard poses the question whether intraspecific and interspecific competition occur between parasites within the individual host. At a quite different level the number and diversity of parasites per island, or per habitat within an island, poses questions about evolution, colonization and extinction. An analysis of extensive field samples answers some of these questions (Dobson & Pacala, 1992).

In places where the burden of parasites per host and the diversity of species was low, the number of species per host did not differ from random. But, where there was a high diversity of parasite species in a location, there were more single-species infections and fewer infections by a mixture of species than expected by chance. This suggests some degree of competitive exclusion occurring amongst the parasites. If this were the case, the populations of particular parasite species could be expected to be lower in hosts harbouring mixed populations than in those harbouring only a single species. Many of the species of parasite were too rare to make useful comparisons, but data for the two most common species are shown in Figure 12.15: they do not give tidy answers. In the case of one common species, *Thelandros cubensis*, there was no difference between its density in hosts in which it was the only species present and its density in mixed-species populations. The burden of the population of the other common parasite, *Skrjabinodon* spp., in its hosts was reduced in mixed-species infections at two of the more heavily infected sites. This provides some, but not very strong, evidence of a degree of mutual exclusion between parasites within hosts.

The analysis suggests that those features that make a parasite a good colonizer of

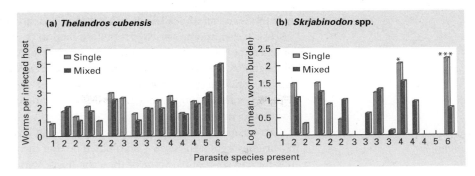

Figure 12.15 The number of worms per infected host (the worm burden) in single- and mixed-species infections. In both (a) *Thelandros cubensis* and (b) *Skrjabinodon* spp., the samples are arranged in order of increasing richness of parasites per host (1–6). Asterisks indicate a significant difference between the worm burden: * $P < 0.1$ and *** $P < 0.01$. (After Dobson & Pacala, 1992.)

susceptible hosts are the same as those that allow it to colonize new habitats and islands. *T. cubensis*, in particular, was not only the most abundant species in most hosts but was also the species common to all the habitats and islands that were sampled. Perhaps the most significant finding from this remarkable survey and analysis is how important are the unpredictable extinction and colonization events that have determined what is absent from particular islands (and indeed from particular hosts on those islands).

12.3.9 The effects of diversity in host populations

Much of what we know about the spread of infections within populations comes from the rather specialized monocultures of agriculture and forestry and the local aggregations of humans. Natural populations of hosts are usually far more diverse, and it is normal for individuals of susceptible species to be dispersed within a milieu of other species or forms that are immune or resistant. Under these circumstances, the effective density of a host (to a particular race of parasite) is the density of (really the distance between) susceptible hosts, not the total density of the community or population.

The further that individual hosts (animal, plant or individual leaves) are isolated from others of their own sort, the more remote are the chances that a parasite will spread between them. If hosts are isolated from others of their own kind by a mixture of other species, the transmission stages of parasites are likely to end up stranded on alien 'non-hosts'. This is probably one reason why epidemics of plant disease are uncommon in nature since monotonous stands of a single-host species are rare. The major disease epidemics known amongst plants have occurred in crops that are not islands in a sea of other vegetation, but 'continents'—large areas of land occupied by one single species (and often by one single variety of that species).

mixtures dilute densities

In agriculture and forestry, some control of disease may be obtained by growing mixtures of species, or mixture of genotypes of a single species. In mixtures, the effective density of each species is reduced—diluted amongst the others. Each disease-resistant form in a mixture may serve as a barrier to the spread of infection from one susceptible individual to another.

In agricultural practice, resistant cultivars offer a challenge to evolving parasites:

mutants that can attack the resistant strain have an immediate gain in fitness. New, disease-resistant crop varieties therefore tend to be widely adopted into commercial practice; but they then often succumb, rather suddenly, to a different race of the pathogen. A new resistant strain of crop is then used, and in due course a new race of pathogen emerges. There is a 'boom and bust' cycle which is repeated endlessly. This keeps the pathogen in a continually evolving condition, and plant breeders in continual employment.

mixtures of susceptibles and immunes

An escape from this boom and bust cycle can be gained by the deliberate mixing of varieties so that the crop is dominated neither by one virulent race of the pathogen nor by one susceptible form of the crop itself. Various types of mixture have been used, including random mixtures of different genotypes, and multilines (populations that differ within themselves only with respect to the resistance genes that they carry). These procedures have indeed been shown to control the rate of disease development, and Figure 12.16 illustrates the effect of varietal mixing on the development of epidemics of late blight of potatoes.

In nature there may be a particular risk of disease spreading from perennial plants to seedlings of the same species growing close to them. If this were commonly the case it could contribute to the species richness of communities by preventing the development of monocultures (this has been called the Janzen–Connell effect). Augspurger (1983) followed the fate of seedlings of the neotropical tree *Platypodium elegans*, that had established at different distances from isolated mature trees of the same species. Most of the seedlings died within 3 months and 75% of the deaths were caused by damping-off disease. The closer a seedling was to a mature tree the greater was the risk that it would die. The risk of death was less for seedlings that had established in light gaps than for those under a closed canopy. However, not all the species that she studied behaved in this way, and in some the distance of a seedling from a mature plant of the same species did not appear to affect the risk of damping-off, although pathogenic attack was the largest single cause of seedling mortality in six out of seven species that she studied (Augspurger, 1984, 1990).

After the spread of an epidemic through a population of higher animals, those

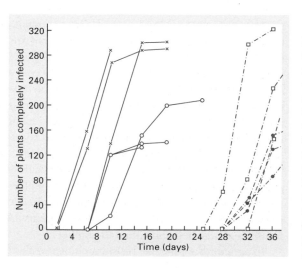

Figure 12.16 The rate of spread of infections by the parasite that causes late blight of potatoes (*Phytophthora infestans*) in pure populations of the potato varieties Pentland Crown and Pentland Dell, and in a mixed population of the two varieties. ×, pure populations of Pentland Crown; □, pure populations of Pentland Dell; ○, Pentland Crown in mixed populations; ●, Pentland Dell in mixed populations. (After Skidmore, 1983.)

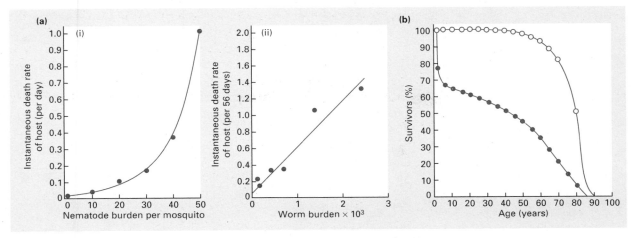

Figure 12.17 (a) The influence of the level of infection by parasites on the death rate of the host. (i) The influence of the burden of the nematode parasite *Dirofilaria immitis*, on the death rate of its host mosquito (*Aedes trivittatus*). (After Christensen, 1978.) (ii) The influence of the worm burden of the digenean liver fluke *Fasciola hepatica*, on the death rate of sheep. (After Boray, 1969.) (b) The percentage of individuals surviving to a given age in a prosperous industrial country (○) and in a poor developing country (●). (After Bradley, 1977.)

that have contracted the disease but recovered may have acquired immunity (see Section 12.3.5). They then slow down the multiplication of the parasite in the same way as resistant forms in a population. They effectively dilute the density (or contact rate) of susceptibles, and reduce the effectiveness of vectors.

12.4 The effects of parasites on the survivorship, growth and fecundity of hosts

The success of a biotrophic parasite must depend on it being an effective competitor for resources against the parts of the host that it parasitizes. It is such internal competition that is responsible, at least in part, for the reduced survival, fecundity, growth or competitive ability of the host. The influence of parasite burden on host death rate is shown in Figure 12.17 for mosquitos infected with nematodes, sheep infected with liver flukes and humans infected with a variety of diseases in relatively poorly developed countries. Infection with the gut-inhabiting nematode *Trichostrongylus tenuis*, has a significantly adverse effect on the reproductive output of female red grouse (*Lagopus lagopus scoticus*) (Potts *et al.*, 1984); whilst infection of the water bug *Hydrometra myrae*, with the mite *Hydraphantes tenuabilis*, reduces the

parasitism and other factors often interact to harm the host

survival, reproduction and rate of development of the host (Figure 12.18). More subtly, infection may make hosts more susceptible to predation (Anderson, 1979). For example, cormorants capture a disproportionately large number of roach infected with the tapeworm *Ligula intestinalis*, compared with the prevalence of infected fish in the population as a whole (van Dobben, 1952). The same effect occurs in very different species and environments. For example, post-mortem examination of red grouse showed that birds killed by predators in the spring and summer carried significantly greater burdens of the parasite *T. tenuis* than birds shot during the autumn (Hudson *et al.*, 1992a).

455   PARASITISM AND DISEASE

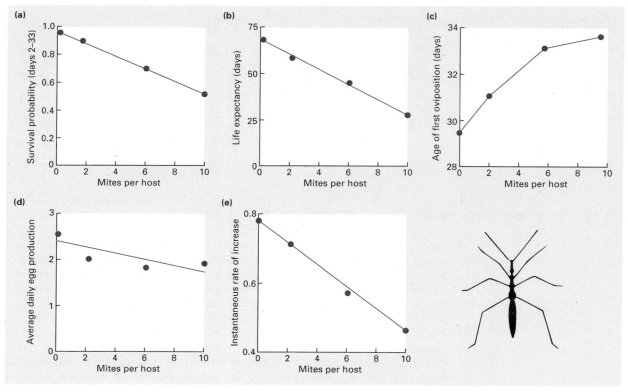

Figure 12.18 The relationships between the rate of infection of the water bug *Hydrometra myrae*, by the parasitic mite *Hydraphantes tenuabilis*, and (a) host survival, (b) the life expectancy of the host, (c) host maturity, (d) fecundity and (e) the rate of increase of the host population. (After Lanciani, 1975.)

Of these examples, the ones most relevant to populations in general are probably the humans affected by disease, the roach and grouse infected with tapeworms and the oats affected by nematodes. With the oats there was an interaction between nematode infection and competition from barley; with the roach the interaction was between infection and predation, whilst with the humans there was an interplay between disease and nutritional deficiency. Such interactions between diseases and other factors are certain to be widespread: parasites may be most harmful when the hosts are weakened by some other factor. The majority of infant deaths in underdeveloped countries appear to be due to childhood infections that are not lethal in the better nourished populations of developed countries (Figure 12.17b).

We know immensely more about the effects of parasites on human populations than we do about wild species, but as our interest in nature conservation grows it becomes important to know how often disease is responsible for the abundance of species in nature. A valuable compendium and review of the literature up to 1993 is given by Gulland (1995).

One important effect of parasitism may be to weaken an aggressive competitor and so allow weaker associated species to persist. The best evidence comes from experiments under controlled conditions such as Park's (1948) classic study of competition between two species of flour beetle (*Tribolium*), in which unexpected

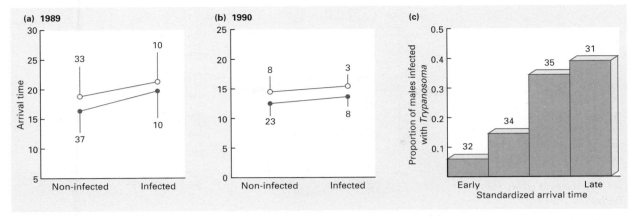

Figure 12.19 The mean date of arrival (1 = 1 May) in Finland of male pied flycatchers (*Ficedula hypoleuca*) infected and uninfected with *Trypanosoma*. ●, adult males; ○, yearling males. (a) 1989, (b) 1990. Sample sizes are indicated near the standard deviation bars. (c) The proportion of males infected with *Trypanosoma* amongst groups of migrants arriving in Finland at different times. (After Rätti *et al.*, 1993.)

changes in the competitive relationships of two species were found to be due to infection by a protozoan parasite (*Adelina triboli*).

Evidence that parasites may disturb competitive relationships in nature comes from a study of two *Anolis* lizards that live on the Caribbean island of St Maarten. *A. gingivinus* is the stronger competitor and appears to exclude *A. wattsi* from most of the island. The malarial parasite *Plasmodium azurophilum*, very commonly affects *A. gingivinus* but rarely *A. wattsi*. Wherever the parasite infects *A. gingivinus*, *A. wattsi* is present and wherever the parasite is absent, even in very local communities, only *A. gingivinus* occurs (Schall, 1992).

It is part of the definition of parasites that they cause harm to their host. But, it is not always easy to demonstrate this harm (Toft & Karter, 1990), which may be detectable only at some peculiarly sensitive stage of the host's life history. For example, the pied flycatcher (*Ficedula hypoleuca*) migrates from tropical West Africa to Finland to breed, and males that arrive early are particularly successful in finding mates. Males infected with the blood parasite *Trypanosoma*, have shorter tails, tend to have shorter wings and arrive in Finland late and so presumably suffer fitness costs (Figure 12.19).

It appears that many plants and animals may tolerate 'parasites' without showing any harmful consequences—although, by definition they then really cease to be parasites and should perhaps be redefined as commensals or as having 'guest status'. But, like guests under other circumstances, they may have unwelcome effects when the hosts are ill or distressed! Lice that feed on the feathers of birds are commonly regarded as 'benign' parasites, with little or no effects on the fitness of their hosts. But, a long-term comparison of the effects of the lice on feral rock doves (*Columba livia*), showed that the lice reduced the thermal protection given by the feathers and that heavily infected birds incurred the costs of maintaining the higher metabolic rates that maintained their body temperatures (Booth *et al.*, 1993) and in the time that the birds spent in preening to keep the lice population under control. A further

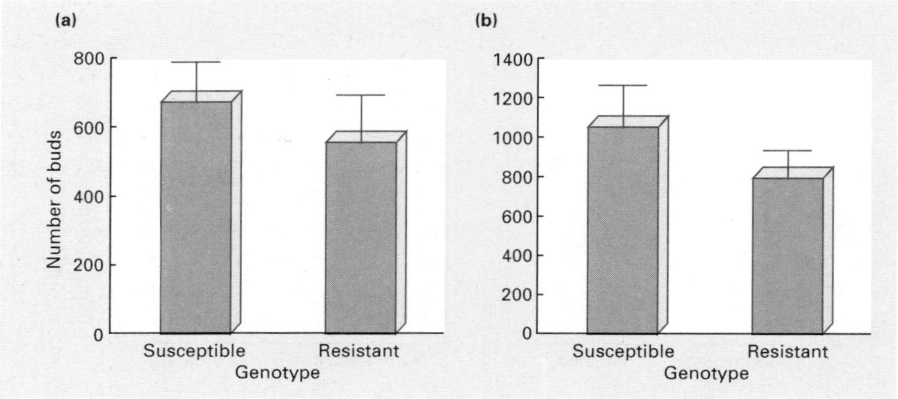

Figure 12.20 The number of buds produced by resistant and susceptible genotypes of two cultivars of lettuce. (a) Cultivar AS, (b) cultivar AA. Error bars are ± 2 standard errors. (After Bergelson, 1994.)

cost of feather parasitism is that heavily parasitized males engaged in less metabolically expensive courtship display and obtained fewer mates than less parasitized males (Clayton, 1990). Indeed, it may be that the influence of parasites on the quality of male sexual characteristics can sometimes be the most important cost of parasitism because it translates so directly into fitness costs (see, for example, the discussion of parasites and sexual selection in Hamilton & Zuk, 1982; Folstad & Karter, 1992).

There are examples of studies of parasites which feed on a host but appear to do it no harm. For example, in natural populations of the sleepy lizard *Tiliqua rugosa*, in Australia their longevity was either not correlated or was positively correlated with the load of ectoparasitic ticks (*Aponomma hydrosauri* and *Amblyomma limbatum*), and lizards in mating pairs carried a higher tick burden than those never found in pairs! There was no evidence that the ticks reduced host fitness (Bull & Burzacott, 1993). Researchers and editors are usually loth to publish such negative results and, if a search for a cost of parasitism fails to find any, the data are likely to be ignored or lost. We need to remember this in any discussion of the costs of parasitism and also that it is not logically possible to prove a negative. There will always be room for exceptions and when we fail to find a cost of what looks like parasitism it may simply mean that we have not yet looked in the right place.

Organisms that are resistant to parasites avoid the costs of parasitism but, as with resistance to other natural enemies, resistance itself may carry a cost. This was tested in a horticultural context with two cultivars of lettuce (*Lactuca sativa*) resistant or susceptible, by virtue of two tightly linked genes, to leaf root aphid (*Pemphigus bursarius*) and downy mildew (*Bremia lactucae*). Pests and diseases were controlled by weekly applications of insecticides and fungicides. Resistant forms of lettuce bore fewer axillary buds than susceptibles (Figure 12.20) and this cost of resistance was most marked when the plants were making poor growth because of nutrient deficiency. In nature, hosts must always be caught between the costs of susceptibility and the costs of resistance.

12.5 The population dynamics of infection

In principle, the sorts of conclusions that were drawn in Chapter 10 regarding the population dynamics of predator–prey and herbivore–plant interactions can be extended to parasites and hosts. Parasites harm individual hosts, which they use as a resource. The way in which this affects their populations varies with the densities of both and with the details of the interaction. In particular, infected and uninfected hosts can exhibit compensatory reactions which may greatly reduce the effects on the host population as a whole. Theoretically, a range of outcomes can be predicted: varying degrees of reduction in host-population density, varying levels of parasite prevalence, various fluctuations in abundance and even multiple equilibria (Anderson, 1981; Grenfell & Dobson, 1995).

The theory of the dynamics of infectious disease is the oldest branch of biomathematics and is now one of the most sophisticated parts of ecological science. Medical records provide some of the most extensive data sets, and there are comparable records of epidemics in diseases of crop plants (Figure 12.21). There are, however, plenty of problems.

One difficulty is that parasites often cause a reduction in the 'health' of their host rather than its immediate death, and it is therefore usually difficult to disentangle the effects of the parasites from those of some other factor or factors with which they interact (see Section 12.4). Another problem is that even when parasites cause a death, this may not be obvious without a detailed post-mortem examination (especially in the case of microparasites). Also, the biologists who describe themselves as parasitologists have in the past tended to study the biology of their chosen parasite without much consideration of the effects on whole host populations, and ecologists have tended to ignore parasites; whilst plant pathologists and medical and veterinary parasitologists have, for obvious reasons, chosen to study parasites with known severe effects, living typically in dense and aggregated populations of hosts. They too have therefore paid little attention to the more typical effects of parasites in populations of hosts less strongly influenced by humans. Perhaps the most certain

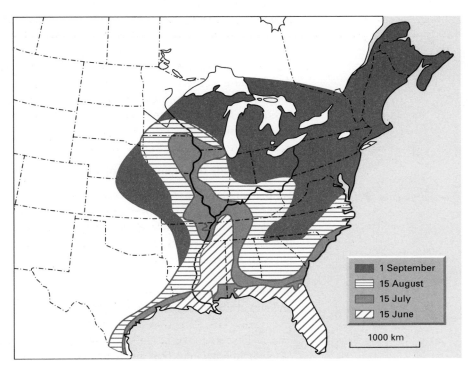

Figure 12.21 The development of an epidemic of southern corn leaf blight (*Helminthosporium maydis*) in the USA in 1970. (After Zadoks & Schein, 1979.)

thing that can be said about the role of parasites in host-population dynamics is that its elucidation is one of the major challenges facing ecology.

epidemiological and ecological approaches

Here, we begin by looking at the dynamics of disease within host populations without considering any possible effects on the total abundance of hosts. This 'epidemiological' approach (Anderson, 1991) has especially dominated the study of human disease, where total abundance is usually considered to be determined by a whole spectrum of factors and is thus effectively independent of the prevalence of any one disease. Disease only affects the partitioning of this population into susceptible (uninfected), infected and other classes. We then take a more 'ecological' approach by considering the effects of parasites on host abundance in a manner much more akin to conventional predator–prey dynamics.

12.5.1 The basic reproductive rate and the transmission threshold

In all studies of the dynamics of parasite populations or the spread of a disease, there are a number of particularly key concepts. The first is the *basic reproductive rate*, R_p (see Anderson & May, 1991, for a review). We have already met essentially the same concept, R_0, in Chapter 4, Section 4.5.1 in a more general context, although for parasites the term is usually formulated slightly differently. In particular, in the case of microparasites, where it is the number of infected hosts that is the unit of study (rather than the number of parasites), R_p is the average number of new cases of the disease (secondary infections) that would arise from a single infectious host introduced into a population of susceptible hosts. For macroparasites, the basic reproductive rate is, more straightforwardly, the average number of established, reproductively mature offspring produced by a mature parasite throughout its life in a population of uninfected hosts.

The *transmission threshold*, which must be crossed if a disease is to spread, is then given by the condition $R_p = 1$. An infection will eventually die out for $R_p < 1$ (each present infection or parasite leads to less than one infection or parasite in the future), but an infection will spread for $R_p > 1$.

Insights into the dynamics of infection can be gained by considering the various determinants of the basic reproductive rate. We do this in some detail for directly transmitted microparasites, and then deal more briefly with related issues for indirectly transmitted microparasites, and directly and indirectly transmitted macroparasites.

12.5.2 Directly transmitted microparasites

the determinants of R_p

For many microparasites with direct transmission, R_p can be seen to increase: (i) with the average period of time over which an infected host remains infectious, L; (ii) with the density of susceptible individuals in the host population, S, because higher densities offer more opportunities for transmission of the parasite or pathogen; and (iii) with the transmission rate of the disease, β, which itself depends first on the infectiousness of the disease—the probability that contact leads to transmission—but also on the pattern of host behaviour insofar as this affects the likelihood of infectious and susceptible hosts coming into contact (Anderson, 1982). Thus, overall:

$$R_p = S\beta L. \tag{12.1}$$

Note immediately that by this definition, R_p is host-density dependent: the greater the density of susceptible hosts, the higher the basic reproductive rate of the disease. Note too that for most biotrophic parasites, L is the period of the host's life when it is infectious; but for necrotrophic parasites and some biotrophs, the hosts may remain infectious long after they are dead (and indeed decomposed). They may leave a residue of resting spores where a diseased corpse or root system had been, and these may then germinate and produce a new infection when they make contact with a new host (resting spores of *Plasmodiophora* after a plant has suffered from club-root, or the resting spores of *Bacillus anthracis* after a host has died from anthrax).

critical threshold densities; epidemics and endemic infections

The transmission threshold can now be expressed in terms of a critical 'threshold density' S_T, where, because $R_p = 1$ at that threshold:

$$S_T = \frac{1}{\beta L}. \tag{12.2}$$

In populations with densities of susceptibles lower than this, the disease will die out ($R_p < 1$). With densities higher than this the disease will spread ($R_p > 1$).

However, if a disease spreads then the number of susceptible hosts is likely to decline—either because they die or because they become immune. The value of R_p itself will then decline, until, if some equilibrium or balance is achieved between the parasite and its host, $R_p = 1$. Recognizing this allows us to make an operational distinction between the *epidemic* phase of an infection and its *endemic* phase (Bailey, 1975). In an epidemic, there is a rapid increase in the incidence of a disease because R_p clearly exceeds 1: the dynamics are determined essentially by the components of R_p itself and often largely by the density of susceptibles. In an endemic infection, on the other hand, R_p remains close to 1 and the dynamics of the disease are controlled more by the pattern of recruitment of new susceptibles into the population, relative to the rate at which they are being removed by the disease.

patterns in the dynamics of infection

If diseases are highly infectious (large βs), or give rise to long periods of infectiousness (large Ls), then they will have relatively high R_p values even in small populations (S_T is small) and will therefore be able to persist there. Conversely, if diseases are of low infectivity or have short periods of infectiousness, they will have relatively small R_p values and will only be able to persist in large populations (Anderson & May, 1978). This provides an explanation for a number of patterns (Anderson, 1982; Anderson & May, 1991).

Many protozoan diseases of vertebrates, and also some virus diseases such as herpes, are persistent within individual hosts (large L), often because the immune response to them is either ineffective or short lived. A number of plant diseases, too, like club-root, have very long periods of infectiousness. In each case, the threshold density is therefore small, explaining why they can and do survive endemically even in small or low-density host populations. On the other hand, the immune responses to many other human viral and bacterial infections are powerful enough to ensure that they are only very transient in individual hosts (small L), and they often induce lasting immunity. Thus, for example, a disease like measles has a threshold density of around 300 000 individuals, and is unlikely to have been of great importance until quite recently in human biology. However, it has generated major epidemics in the growing cities of the industrialized world in the 18th and 19th centuries, and in the growing concentrations of population in the developing world in the 20th century. Current estimates suggest that around 900 000 deaths occur each year from measles infection in the developing world (Walsh, 1983).

461   PARASITISM AND DISEASE

We have already noted that the immunity induced by many bacterial and viral infections reduces S, reduces R_p and therefore tends to lead to a decline in the incidence of the disease itself. However, in due course there will be an influx of new susceptibles into the population (as a result of new births or perhaps immigration). There is thus a marked tendency with such diseases to generate a sequence from 'high incidence', to 'few susceptibles', to 'low incidence', to 'many susceptibles', to 'high incidence', etc.—just like any other predator–prey cycle. This undoubtedly underlies the observed cyclic incidence of many human diseases, with the differing lengths of cycle reflecting the differing characteristics of the diseases: measles with peaks every 1 or 2 years (Figure 12.22a), pertussis (whooping cough) 3–4 years (Figure 12.22b), diphtheria 4–6 years, and so on (Anderson & May, 1991). Protozoan diseases, by contrast, are much less variable in their prevalence.

Recognizing the importance of threshold densities also throws light on vaccination (injection) and other immunization programmes, in which susceptible hosts are rendered non-susceptible without ever becoming diseased (showing clinical symptoms), usually through exposure to a killed or attenuated pathogen. The direct effects here are obvious: the immunized individual is protected. But, by reducing the number of susceptibles, such programmes also have the indirect effect of reducing R_p. Indeed, seen in these terms, the fundamental aim of an immunization programme is clear—to hold the number of susceptibles below S_T so that R_p remains less than 1. To do so is said to provide 'herd immunity'.

immunization, herd immunity and critical proportions immunized

In fact, a simple manipulation of Equation 12.1 gives rise to a formula for the critical proportion of the population, p_c, that needs to be immunized in order to provide herd immunity (bring R_p down to 1). If we define S_0 as the typical number of susceptibles prior to any immunization and note that S_T is the number still susceptible (not immunized) once the programme to achieve $R_p = 1$ has become fully established, then the proportion immunized is:

$$p_c = 1 - \frac{S_T}{S_0}.$$

(12.3)

The formula for S_T is given in Equation 12.2, whilst that for S_0, from Equation 12.1,

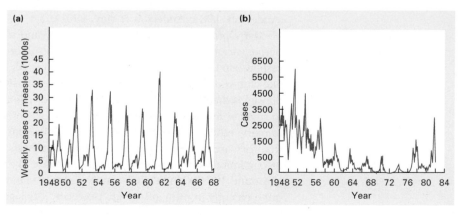

Figure 12.22 (a) Reported cases of measles in England and Wales from 1948 to 1968, prior to the introduction of mass vaccination. (b) Reported cases of pertussis (whooping cough) in England and Wales from 1948 to 1982. Mass vaccination was introduced in 1956. (After Anderson & May, 1991.)

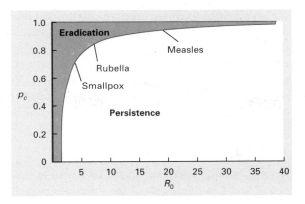

Figure 12.23 The dependence of the critical level of vaccination coverage required to halt transmission, p_c, on the basic reproductive rate, R_0, with values for some common human diseases indicated. (After Anderson & May, 1991.)

is simply $R_0 / \beta L$, where R_0 is the basic reproductive rate of the disease prior to immunization. Hence:

$$p_c = 1 - \frac{1}{R_0}. \tag{12.4}$$

This reiterates the point that in order to eradicate a disease, it is not necessary to immunize the whole population—just a proportion sufficient to bring R_p below 1. It also shows that this proportion will be higher the greater the 'natural' basic reproductive rate of the disease (without immunization). This general dependence of p_c on R_0 is illustrated in Figure 12.23, with the estimated values for a number of human diseases indicated on it. Note that smallpox, the only disease where in practice immunization seems to have led to eradication, has unusually low values of R_0 and p_c.

Not all directly transmitted microparasites are transmitted at a rate that depends simply on the total density of susceptible hosts, S—as if the rate of transmission depended on the probability of infectious and susceptible individuals 'bumping into one another'. For example, with sexually transmitted diseases, where transmission occurs after an infected individual 'seeks out' (or is sought out by) a susceptible individual, it is more accurate to think of transmission depending on the *proportion* of hosts that are susceptible, S / N (i.e. the *probability* that an infected individual's contact is with a susceptible). In such 'frequency dependent' cases, the basic reproductive rate is given by:

frequency dependent transmission: sexually transmitted diseases

$$R_p = \beta L. \tag{12.5}$$

Here, clearly, R_p is not host-density dependent. There is, essentially, no threshold density and such diseases can therefore persist even at extremely low densities (where, to a first approximation, the chances of sexual contact for an infected host are the same as at high densities). In fact, as we shall see shortly when plant macroparasites are discussed, frequency dependent transmission can also occur outside the context of conventional sexually transmitted diseases.

12.5.3 Crop pathogens: macroparasites viewed as microparasites

Most of plant pathology has been concerned with the dynamics of diseases within crops, and hence with the spread of a disease *within* a generation. Moreover, although most commonly-studied plant pathogens are macroparasites in the sense

463 PARASITISM AND DISEASE

we have defined them, they are typically treated like microparasites in that disease is monitored on the basis of some measure of disease severity—often, the proportion of the population infected (i.e. prevalence). It is also usually necessary with plant pathogens to take explicit account of the latent period between the time when a lesion is initiated and the time when it becomes infective (spore-forming) itself. The rate of increase in the proportion of a plant population affected by lesions (Vanderplank, 1963; Zadoks & Schein, 1979; Gilligan, 1990) may thus be given by:

$$\frac{\mathrm{d}y_T}{\mathrm{d}t} = D(y_{T-P} - y_{T-P-L})(1 - y_T). \tag{12.6}$$

Here, y_T is the proportion affected by lesions at time T, P is the length of the latent period and L is the length of the infectious period. The rate of increase therefore depends on the rate of disease increase per unit of infectious disease (D), on the proportion of the population affected by *infectious* lesions ($y_{T-P} - y_{T-P-L}$) and on the proportion of the population susceptible to infection ($1 - y_T$). This essentially logistic formulation gives rise to S-shaped curves for the progress of a disease within a crop that broadly match the data derived from many crop–pathogen systems (Figure 12.24).

In the progress of such diseases, plant pathologists recognize three phases.
1 The exponential phase, when the most rapid acceleration of parasite prevalence occurs. The disease is rarely detectable in this phase: 'epidemiologically significant multiplication escapes observation'. This is also the phase in which chemical control would be most effective, but in practice it is usually applied in the next phase. The exponential phase is usually considered arbitrarily to end at $y = 0.05$, and this is about the level of infection at which a non-specialist might detect that an epidemic was developing (the perception threshold).
2 The second phase, which extends to $y = 0.5$. (This is sometimes confusingly called the 'logistic' phase, although the whole curve is logistic.)
3 The terminal phase, which continues until y approaches 1.0. In this phase chemical treatment is virtually useless—yet, it is at this stage that the greatest damage is done to the yield of a crop.

The population dynamics of epidemic foliage parasites tends to be dominated by three factors (Burdon, 1987). The first is the initial dispersal of parasites to the host population. Epidemics may be initiated by a single spore landing within a stand of a crop, or a spore cloud being deposited from a heavily infected area elsewhere (Figure 12.25). In a crop, or in any other static host population, the spatial and temporal dynamics of a disease are especially clearly related. The second factor is the speed with which infectivity develops on infected hosts (i.e. the value of P in Equation 12.6). The last factor is the proximity of susceptible hosts to each other in relation to the direction of air movement.

The basic reproductive rate of crop diseases like those described by Equation 12.6 can be shown to be essentially equivalent to that already described: $R_p = \beta SL$ (May, 1990b). Here, however, since the focus is on progress within a generation, a disease can appear to 'spread' from an initial inoculum even if $R_p < 1$, in the sense that even with low R_p values, an increasing fraction of the population will become infected as diseased individuals 'accumulate'.

On the other hand, some crop diseases are not simply transmitted by the passive spread of infective particles from one host to another. For example, the anther smut fungus, *Ustilago violacea*, is spread between host plants of white campion, *Silene*

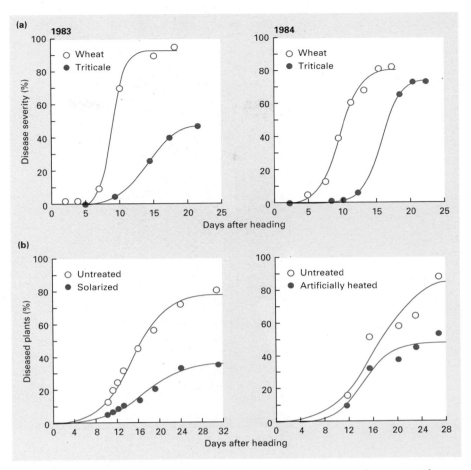

Figure 12.24 'S-shaped' curves of the progress of diseases through crops from an initial inoculum to an asymptotic proportion of the total population infected. (a) *Puccinia recondita* attacking wheat (cultivar Morocco) and triticale (a crop derived from hybridization of wheat and rye) in 1983 and 1984. (b) *Fusarium oxysporum* attacking tomato in experiments comparing untreated and sterilized soil and untreated and artificially heated soil. (After Gilligan, 1990, in which original data sources and methods of curve-fitting may be found.)

alba, by pollinating insects that adjust their flight distances to compensate for changes in plant density, such that the rate of transmission is effectively independent of host density (Figure 12.26a). However, this rate decreases significantly with the proportion of the population that are susceptible: transmission is frequency dependent (Figure 12.26b), favouring, as we have seen, persistence of the disease even in low-density populations. Of course, this is really just another case of frequency dependent transmission in a sexually transmitted disease—except that sexual contact here is indirect rather than intimate.

12.5.4 Other classes of parasite

For microparasites that are spread from one host to another by a vector more generally (where the vector does not compensate for changes in host density), the life-cycle characteristics of both host and vector enter into the calculation of R_p. In

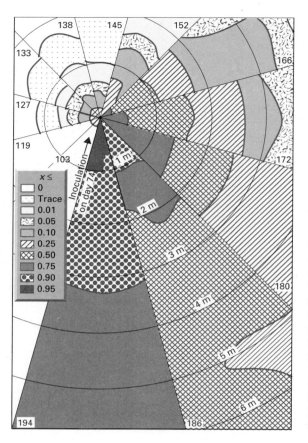

Figure 12.25 The development of an epidemic of the yellow stripe rust of wheat (*Puccinia striiformis*) from artificially established foci of infection. The circles show the distance of infection from the original focus and the shading indicates the intensity of infection. The progress of the epidemic can be traced by following the 'spider's web' clockwise. The figures around the margin indicate time in days from 1 January. (After Zadoks & Schein, 1979.)

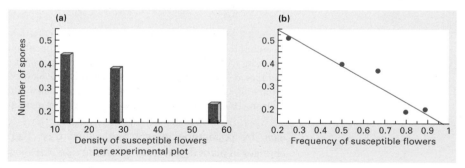

Figure 12.26 Frequency dependent transmission of a sexually transmitted disease. The number of spores of *Ustilago violacea* deposited per flower of *Silene alba* ($\log_{10} (x + 1)$ transformed) where spores are transferred by pollinating insects. The number is (a) independent of the density of susceptible (healthy) flowers in experimental plots ($P > 0.05$) (and shows signs of decreasing rather than increasing with density, perhaps as the number of pollinators becomes limiting). However, (b) the number decreases with the frequency of susceptibles ($P = 0.015$). (After Antonovics & Alexander, 1992.)

vector-transmitted microparasites: the importance of the vector : host ratio

particular, the transmission threshold ($R_p = 1$) is dependent on a ratio of vector : host densities. For a disease to establish itself and spread, that ratio must exceed a critical level—hence, disease control measures are usually aimed directly at reducing the numbers of vectors, and are aimed only indirectly at the parasite. Many virus diseases of crops, and vector-transmitted diseases of humans and their livestock (malaria, onchocerciasis, etc.), are controlled by insecticides rather than chemicals directed at the parasite; and the control of all such diseases is of course crucially dependent on a thorough understanding of the vector's ecology.

directly transmitted
macroparasites: density
dependence within the host

The effective reproductive rate of a directly transmitted macroparasite is directly related to the length of its reproductive period within the host (i.e. again, to L) and to its rate of reproduction (rate of production of infective stages). Both of these are subject to density dependent constraints that can arise either because of competition between the parasites, or commonly because of the host's immune response (see Section 12.3.4). Their intensity varies with the distribution of the parasite population between its hosts, and as we have seen, aggregation of the parasites is the most common condition. This means that a very large proportion of the parasites exist at high densities where the constraints are most intense, and this tightly controlled density dependence undoubtedly goes a long way towards explaining the observed stability of many helminth infections (such as hookworms and roundworms) even in the face of perturbations induced by climatic change or humans' intervention (Anderson, 1982).

Most directly transmitted helminths have an enormous reproductive capability. For instance, the female of the human hookworm *Necator*, produces roughly 15 000 eggs per worm per day, whilst the roundworm *Ascaris*, can produce in excess of 200 000 eggs per worm per day. The critical threshold densities for these parasites are therefore very low, and they occur and persist endemically in low-density human populations, such as hunter–gatherer communities.

density dependence and
indirectly transmitted
macroparasites:
schistosomiasis

Density dependence within hosts also plays a crucial role in the epidemiology of indirectly transmitted macroparasites, such as schistosomes (see Figure 12.4). In this case, however, the regulatory constraints can occur in either or both of the hosts: adult worm survival and egg production are influenced in a density dependent manner in the human host, but production of cercariae is virtually independent of the number of miracidia that penetrate the snail. Thus, levels of schistosome prevalence tend to be stable and resistant to perturbations from outside influences.

The threshold density for the spread of infection depends directly on the abundance of both humans and snails (i.e. a product as opposed to the ratio which was appropriate for vector-transmitted microparasites). This is because transmission in both directions is by means of free-living infective stages. Thus, since it is inappropriate to reduce human densities, schistosomiasis is often controlled by reducing snail densities with molluscicides in an attempt to depress R_p below unity (the transmission threshold). The difficulty with this approach, however, is that the snails have an enormous reproductive capacity, and they rapidly recolonize aquatic habitats once molluscicide treatment ceases. The limitations imposed by low snail densities, moreover, are offset to an important extent by the long life span of the parasite in humans (L is large): the disease can remain endemic despite wide fluctuations in snail density.

the complex determination of a
basic reproductive rate for
black stem rust

The determinants of the basic reproductive rate for a plant macroparasite like black stem rust (see Section 12.2 and Figure 12.5) are even more complex. Forgetting for the moment the existence of the sexual stage on the barberry, the parasite has a basic reproductive rate during the growing season of the cereal. Lesions give rise to daughter lesions, and there is a local multiplication of lesions within a crop. But, spore clouds are carried for long distances, and epidemics often develop far more quickly than would be possible from purely local multiplication. A proper calculation of R_p for black stem rust would therefore need to include the number of daughter lesions produced per lesion along such a path of disease migration. R_p calculated in this way would be the basic reproductive rate of clones, spreading and multiplying without sexual recombination. But there is a quite different R_p that could be calculated—the basic sexual reproductive rate that takes into account the fact that new clones (new recombinants) arise only on the barberry. For this we would need to measure the number of daughter lesions produced per mother lesion on the barberry.

The control of a disease with such a complex life history and mass long-distance dispersal poses interesting problems. If the barberry was the crop that required protection, the eradication of the cereal would eliminate the disease on a time scale determined solely by the length of life of the barberry (spores released from the barberry cannot infect barberry). Similarly, the elimination of the barberry on a local scale would prevent the local perpetuation of infections on the cereal where it does not overwinter; but epidemics could still arise from the mass influx of spores in the annual 'migration' of spore clouds. In milder climates, spore production continues on grasses and cereals with overwintering foliage. Some control may then be achieved by preventative spraying with fungicides, but this is expensive. Thus, the major means of control of such a disease, whether or not there is an alternative host in the life cycle, is by the use of genetic resistance bred into the crop. An alternative host on which sexual stages occur then has a quite different significance in epidemiology. It is the place where new 'races' of the parasite may arise through recombination, breaking through the resistance barriers developed by the plant breeder.

12.6 Parasites and the population dynamics of hosts

There are certainly cases where a parasite or pathogen can be seen to reduce the population density of its host. The widespread and intensive use of sprays, injections and medicines in agricultural and veterinary practice all bear witness to the disease-induced loss of yield that would result in their absence. (In human medicine, there is often more emphasis on saving the individual rather than increasing the size or yield of the population as a whole.) Reductions of host density by parasites have also been seen in controlled environments (Figures 12.27a, b). However, such direct evidence is rare in laboratory environments and extraordinarily difficult to obtain from natural populations. Even when a parasite is present in one population but absent in another, the parasite-free population is certain to live in an environment that is different from that of the infected population; and it is likely also to be infected with some other parasite that is absent from or of low prevalence in the first population. There are nevertheless some sets of field data in which a parasite is implicated in the detailed dynamics of its host, either because the dynamics of parasite and host appear to be strongly correlated (the larch budmoth, *Zeiraphera diniana*, and its granulosis virus; Figure 12.27c) or because the dynamics of a host are so different in the presence and absence of a parasite (the red grouse, *Lagopus lagopus scoticus*, and its parasitic nematode *Trichostrongylus tenuis*; Figure 12.27d). We therefore proceed in a manner similar to that used in Chapter 10, when the population dynamics of predator–prey interactions generally were discussed. A number of simple models are examined, in tandem either with the data that inspired them or with data on which they cast light.

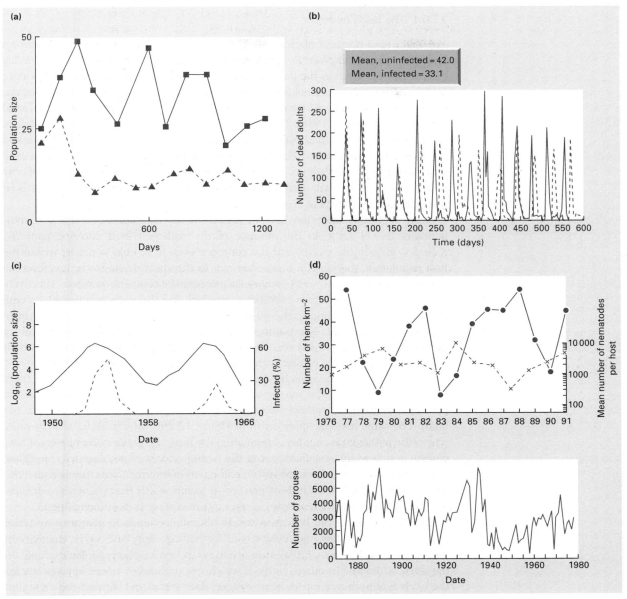

Figure 12.27 (a) The depression of the population size of the flour beetle *Tribolium castaneum*, infected with the protozoan parasite *Adelina triboli*: (———), uninfected; (- - -), infected. (After Park, 1948; see also Section 12.4.) (b) The depression of the population size of the Indian meal moth *Plodia interpunctella*, infected with a granulosis virus: (———), uninfected; (- - -), infected; means are derived from three replicate populations in each case, of which only one is shown in the figure. (After Sait *et al.*, 1994.) (c) Apparent linkage over two cycles between the abundance (numbers per 7.5 kg of branches) of the larch budmoth, *Zeiraphera diniana* (———) and the prevalence of its granulosis virus (- - -) in the European Alps. (After Auer, 1971; Anderson & May, 1980.) (d) (Above) Regular cycles in the abundance (breeding hens per km²) of red grouse, *Lagopus lagopus scoticus* (———) and the mean number of nematodes (*Trichostrongylus tenuis*) per host at Gunnerside, North Yorkshire, UK. (Below) A pattern of grouse abundance at Kerloch, Scotland without regular cycles and where the nematode is not considered important. (After Potts *et al.*, 1984; Dobson & Hudson, 1992.)

12.6.1 The larch budmoth

We begin with a model appropriate to the interaction between a host without an immune response (a plant or an invertebrate) and a microparasite with free-living infective stages, such as the larch budmoth and its granulosis virus shown in Figure 12.27c. Many insect pests of forestry (especially moths and sawflies) exhibit regular cycles of abundance peaking around every 10 years, and many (if not all) are affected by pathogens (Myers, 1988). Do the pathogens cause, or are they instrumental in, the cycles? Could the pathogens be manipulated to interfere with the cycles and perhaps control the pests? The model is described in Figure 12.28. For this, and those that follow, we shall not concern ourselves with the model's analysis, but only with the results of that analysis—which can themselves be understood without absorbing all the details of Figure 12.28.

$S_T < K$—a threshold condition for self-regulated hosts

Note first that since the host is subject to self-regulation (intraspecific competition), it will settle in the absence of the pathogen at a carrying capacity, K (= $(a - b) / q$ in the model). For the pathogen even to be able to persist within the host population, therefore, it is necessary for its threshold density to be less than the carrying capacity (i.e. $S_T < K$)—otherwise intraspecific competition would effectively prevent that threshold from ever being reached and the pathogen would become extinct. In fact, depending on the values of the various biological parameters, a range of outcomes is possible. The pathogen, as we have noted, may be absent (allowing the host to settle at K), it may regulate the host to a stable density less than K, or both host and pathogen may exhibit regular cycles of abundance (Figure 12.29a).

maximum host depression at intermediate pathogenicity—a message for pest controllers

Broadly, the pathogen will be absent ($S_T < K$) when the rate at which it kills hosts is too great relative to the rate at which infected hosts produce new pathogens: a pathogen may be too pathogenic to persist (disease-induced mortality rate, α, too high). The same message emerges from Figure 12.29b, which shows that in cases where the pathogen is capable of regulating the host, the greatest depression of host density occurs at intermediate, not at the highest values of pathogenicity, α. Other models produce much the same result: pathogens of intermediate virulence give rise to the greatest reductions in host density—a lesson worth learning when parasites are used, as they are increasingly, in attempts to control populations of pests.

the model suggests that what the larch budmoth exhibits are not simply host–pathogen cycles

The regulation of the host to a steady abundance tends to give way to cycles when the pathogen has high values both for pathogenicity and for productivity in individual hosts (Figure 12.29a). But, are these the cycles observed, for example, in the larch budmoth? In other words, if we choose parameter values appropriate for the larch budmoth–granulosis virus system, does the model support the view that the pathogen is responsible for the cycles? In fact, cycles *are* produced when the best estimates of appropriate parameter values are used (Figure 12.29c), and they bear a superficial resemblance to those in nature (see Figure 12.27c) with a period (7.9 years) encouragingly close to that observed (8.6 years). However, in other respects the fit is by no means as good. The prevalence of the disease peaks at around 65%, compared to observations of 25–50%; whilst the abundance of the moths peaks well below the carrying capacity (around $0.13 K$), whereas in nature, to judge by the massive defoliation of larch trees, the carrying capacity is regularly exceeded. Hence, in this system, and in forest insects generally, the model undermines any suggestion that the cycles observed are host–pathogen cycles in the simplest sense. On the other hand, this does not eliminate the possibility that the cycles are the combined result of host–pathogen interactions and some form of self-regulation more complex than that

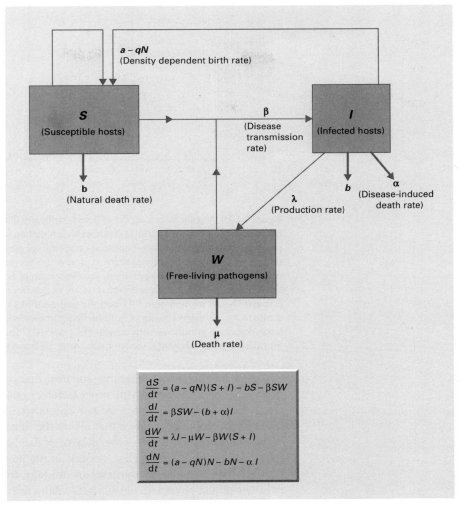

Figure 12.28 Flow diagram (above) depicting the dynamics of a microparasitic infection of a non-vertebrate host (no immune response) where the pathogen has free-living infective stages; and (below) the model equations describing those dynamics. In the flow diagram, boxes refer to densities either of the pathogen or of subclasses of the host. Arrows refer to rates of gain into or loss from those densities, or to transfers from one density to another. Taking the equations in order, they describe: (i) susceptible hosts (S) increasing as a result of density dependent birth from both the susceptible and infected classes, but decreasing both as a result of natural (density independent) death and also by becoming infected through contact with pathogens; (ii) infected hosts (I) increasing as a result of susceptibles becoming infected, and decreasing as a result of both natural and disease-induced death; and (iii) free-living pathogen particles (W) increasing as a result of being produced by infected hosts, but decreasing both as a result of death and also by being 'consumed' by hosts. (Note that $\beta W(S + I)$ describes the removal of particles because they are consumed by both susceptibles and infecteds, but βSW describes the process of infection since it only occurs when susceptibles contact pathogen particles.) Finally, the equation for the total host population ($N = S + I$) is derived by summing the equations for S and I. (After Anderson & May, 1981; Bowers *et al.*, 1993.)

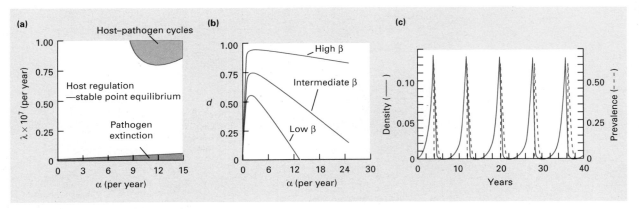

Figure 12.29 Results from a model of insect–pathogen (larch budmoth–granulosis virus) interactions (see Figure 12.28). (a) Pathogen elimination, stable regulation of the host by the pathogen, and persistent host–pathogen cycles are all possible, depending (amongst other things) on the values of pathogenicity, α, and pathogen productivity, λ. (b) For parameter values giving rise to stable regulation, the depression of host abundance is greatest at intermediate values of α. Depression, d, takes its maximum value (1) when the host is eliminated, and its minimum (0) when the pathogen has no effect and the host therefore settles at its carrying capacity, K. (c) The cycles generated when parameter values from the larch budmoth–granulosis virus system are employed. The cycle period is similar, but disease prevalence and moth abundance are not. (After Bowers *et al.*, 1993.)

envisaged by the model—reiterating the view that parasites exert their influences on host populations in concert with other factors, so that the resulting dynamics can be attributed to neither the parasite nor any other factor alone. Indeed, a model of lepidopteran population cycles that allows the 'quality' of offspring to be a delayed density dependent function of the 'quality' of their mothers, gives good prediction of the period of population cycles. 'Quality' in the model is a measure of fecundity or survivorship, or could be the level of ovarially transmitted viruses or protozoan parasites or just physiological vigour (Ginzburg & Taneyhill, 1994).

12.6.2 Fox-rabies

We turn next to rabies: a directly transmitted (no free-living stages) disease of vertebrates, including humans, that attacks the central nervous system and is much feared both for the unpleasantness of its symptoms and the high probability of death once it has taken hold. In Europe, recent interest has focused on the interaction between rabies and the red fox (*Vulpes vulpes*). An epidemic in foxes has been spreading westwards and southwards since the 1940s, and whilst the direct threat to humans that this poses is questionable, there is economically-significant transmission of rabies from foxes to cattle. The authorities in Great Britain are especially worried about rabies: the disease has yet to cross the Channel from mainland Europe (Macdonald, 1980).

Do we know enough about fox-rabies population dynamics to suggest how further spread of the disease might be prevented and how it might even be eliminated where it already exists? A simple model of fox-rabies dynamics is described in Figure 12.30, which does indeed seem to capture the essence of the interaction successfully, since, with values for the various biological parameters

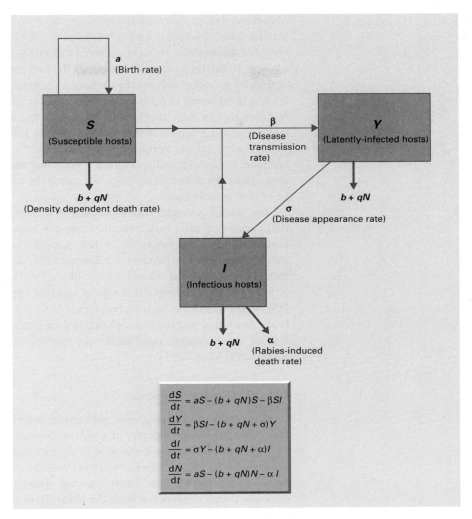

Figure 12.30 Flow diagram (above) depicting the dynamics of a rabies infection of a vertebrate host (such as the fox); and (below) the model equations describing those dynamics. Taking the equations in order, they describe: (i) susceptible hosts (S) increasing as a result of (density independent) birth from the susceptible class only, but decreasing both as a result of natural (density dependent) death and also by becoming infected through contact with infectious hosts; (ii) latently infected (non-infectious) hosts (Y) increasing as a result of susceptibles becoming infected, and decreasing both as a result of natural (density dependent) death and also (as the rabies appears) by becoming converted into infectious hosts; and (iii) infectious hosts (I) increasing as a result of disease development in latently infected hosts, but decreasing as a result of natural and disease-induced mortality. Finally, the equation for the total host population (N = S + Y + I) is derived by summing the equations for S, Y and I. (After Anderson *et al.*, 1981.)

taken from field data, the model predicts regular cycles of fox abundance and rabies prevalence, around 4 years in length—just like those found in a number of areas where rabies is established (Anderson *et al.*, 1981).

There are two methods that have a realistic chance of controlling rabies in foxes. The first is to kill numbers of them on a continuing basis, so as to hold their

control by culling foxes?

473 PARASITISM AND DISEASE

abundance below the rabies transmission threshold. The model suggests that this is around 1 km^{-2}, which is itself a helpful piece of information, given credence by the ability of the model to recreate observed dynamics. As discussed much more fully in Chapter 16 (in the context of harvesting), the problem with repeated culls of this type is that by reducing density they relieve the pressure of intraspecific competition, leading to increases in birth rates and declines in natural death rates. Thus, culling becomes rapidly more problematic the greater the discrepancy between the normal density and the target density (in this case, 1 km^{-2}). Culling may, therefore, be feasible with natural densities of around only 2 km^{-2}, but since densities in, for instance, Great Britain often average 5 km^{-2} and may reach 50 km^{-2} in some urban areas, culls of a sufficient intensity will usually be unattainable. Culling will typically be of little practical use.

control by vaccination?   The second potential control method is vaccination—in this case, placement of oral vaccine in baits to which the foxes are attracted. Such methods can reach around 80% of a fox population. Is that enough? The formula for answering this has already been given as Equation 12.3, application of which suggests that vaccination should be successful at natural fox densities up to 5 km^{-2}. Vaccination may therefore be successful, for example throughout much of Great Britain, but appears to offer little hope of control in many urban areas. At this point, of course, the simple models break down and much more sophisticated analyses are required to design vaccination or other control programmes for strongly heterogeneous populations (see Anderson & May, 1991).

12.6.3 Red grouse and nematodes

Lastly we look at the red grouse—of interest both because it is a 'game' bird, and hence, the focus of an industry in which landowners charge for the right to shoot at it on their land, and also because it is another species that often, although by no means always, exhibits regular cycles of abundance (Figure 12.27d). The underlying cause of these cycles has been disputed (Lawton, 1990), but one mechanism receiving strong support has been the influence of the parasitic nematode, *T. tenuis*, occupying the birds' gut caeca and reducing survival and breeding production (Figures 12.31a, b).

A model for this type of host–macroparasite interaction is described in Figure 12.32. Its analysis suggests that regular cycles both of host abundance and of mean number of parasites per host will be generated if:

$$\frac{\delta}{\alpha} > k. \tag{12.7}$$

Here, δ is the parasite-induced reduction in host fecundity, α is the pathogenicity or parasite-induced host death rate and k is the 'aggregation parameter' for the (assumed) negative binomial distribution of parasites amongst hosts. Cycles arise when the destabilizing effects of the relatively delayed density dependence of reduced fecundity overcome the stabilizing effects of both the relatively direct density dependence of increased mortality and the aggregation of parasites (providing a 'partial refuge' for the hosts) (see Chapter 10).

Data from a study population in the north of England that does show regular cycles (see Figure 12.27d (above)) indicate that α is 3×10^{-4} per worm per year, that

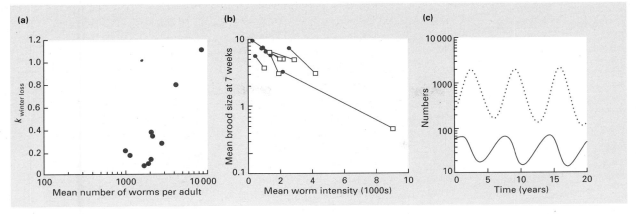

Figure 12.31 (a) *Trichostrongylus tenuis* reduces survival in the red grouse: over 10 years (1980–1989) winter loss (measured as a *k*-value) increased significantly ($P < 0.05$) with the mean number of worms per adult. (b) *T. tenuis* reduces fecundity in the red grouse: in each of 8 years, females treated with a drug to kill nematodes (●, representing mean values) had fewer worms and larger brood sizes (at 7 weeks) than untreated females (□). (c) The pattern of host–parasite dynamics generated by the model in Figure 12.32, with parameter values from the grouse–nematode system, closely resemble those observed. (After Dobson & Hudson, 1992; Hudson *et al.*, 1992a.)

δ is 5×10^{-4} per worm per year and that k is 1 (rather high compared to many other systems—relatively little aggregation) (Hudson *et al.*, 1992a; Dobson & Hudson, 1992). Hence, Equation 12.7 is satisfied: the parasite has more of an effect on host fecundity than on host mortality, and the parasites exhibit a relatively low degree of aggregation. In fact, with field values inserted for all of the model's parameters, cycles are generated that closely resemble those observed (see Figure 12.31c). By contrast, grouse populations that fail to show regular cycles (see Figure 12.27d (ii)), or show them only very sporadically, are typically those in which the nematode cannot properly establish ($S_T > K$) (Dobson & Hudson, 1992).

These examples, therefore, illustrate the similarities and parallels between host–parasite population dynamics and those of predators and prey more generally. Indeed, the importance of disease in population dynamics is increasingly being suggested and confirmed (see, for example, Grenfell & Dobson, 1995). However, it is also becoming apparent that it is dangerous to consider (and to model) the interaction between parasites and hosts as though there are only these two components to the interaction. The influence of a parasite on its host may well be to change its response to competitors and mutualists, alter its reaction to physical conditions of the environment or damage its ability to garner resources. Indeed Price *et al.* (1986) argue that:

> The evidence suggests that parasites may modify the ecology and evolution of every kind of interaction, from the evolution of molecules such as haemoglobin and immunoglobulins to the evolution and extinction of biota on a global scale.

Studies of host–parasite interactions are therefore increasingly having to incorporate the behaviour of at least a third species in order to increase their realism and predictive value (Begon & Bowers, 1995).

host–parasite dynamics—two species are not enough

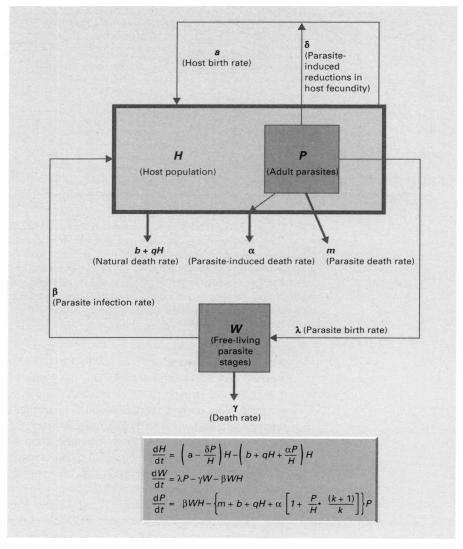

Figure 12.32 Flow diagram (above) depicting the dynamics of a macroparasitic infection such as the nematode *Trichostrongylus tenuis* in red grouse, where the parasite has free-living infective stages; and (below) the model equations describing those dynamics. Taking the equations in order, they describe: (i) hosts (H) increasing as a result of (density independent) births (which, however, are reduced at a rate dependent on the average number of parasites per host, P/H), but decreasing as a result of deaths—both natural (density dependent) and induced by the parasite (again dependent on P/H); (ii) free-living parasite stages (W) increasing as a result of being produced by parasites in infected hosts, but decreasing both as a result of death and by being consumed by hosts; and (iii) parasites within hosts (P) increasing as a result of being consumed by hosts, but decreasing as a result of their own death within hosts, of the natural death of the hosts themselves and of disease-induced death of hosts. This final term is dependent on the distribution of parasites amongst hosts—here assumed to follow a negative binomial distribution, parameter k, accounting for the term in square brackets. (After Anderson & May, 1978; Dobson & Hudson, 1992.)

12.7 Polymorphism and genetic change in parasites and their hosts

When populations of host and parasite evolve together without the intervention of humans as a plant breeder, the multiplication of a particular race of pathogen is followed by an increase in the proportion of resistant hosts. This increase in the representation of resistance genes leads to a decline in the representation of the appropriate virulence gene in the pathogen. A further race of the pathogen then gains ascendancy, and this in turn leads to an increase in the representation of genes for resistance to that race of pathogens. Race after race of the pathogen becomes ascendant and declines, and the result is a maintained dynamic polymorphism in the genetic structure of both pathogen and host. In fact, the polymorphisms that develop naturally between interacting and evolving hosts and parasites produce effectively the same result as, for example, the deliberate mixing of strains of a crop that is being developed by agriculturists for control of disease (see Figure 12.16). A much fuller account of the evolutionary interactions between plant hosts and pathogens is given by Burdon (1987).

Genetic polymorphisms are actually only one of the evolved consequences of the reciprocating selection pressures that can occur between parasites and their hosts. A cyclical phenotypic response that does not involve genetic change occurs during recurrent bouts of sleeping sickness in an infected human host. The trypanosome parasite has a changing repertoire of antigens that are activated in turn by subtle molecular mechanisms as the host mounts successive immune defences (Figure 12.33). The result is violent fluctuations in the trypanosome population *within* a host.

A most dramatic example of the ecological and evolutionary interaction of a parasite and its host involves the rabbit and the myxoma virus which causes the disease myxomatosis. The disease originated in the South American jungle rabbit *Sylvilagus brasiliensis*, where it is a mild disease that only rarely kills the host. It is,

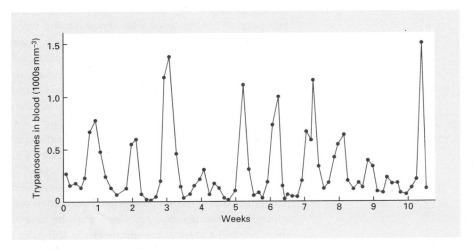

Figure 12.33 Fluctuations in the populations of trypanosomes in a patient with trypanosomiasis (sleeping sickness). The trypanosome population at each peak is antigenically different from those at preceding or succeeding peaks. (After Ross & Thomson, 1910.)

however, usually fatal when it infects the European rabbit *Oryctolagus cuniculus*. This behaviour fits the commonly held view that parasites evolve toward becoming benign to their hosts and that the most dramatic epidemics of lethal diseases occur when the parasite and host are new to each other and have therefore had no period of coevolution (but, see caveats later in this section).

In one of the greatest examples of the biological control of a pest, the myxoma virus was introduced into Australia in the 1950s to control the European rabbit which had become a pest of grazing lands. The disease spread rapidly in 1950–1951 and rabbit populations were greatly reduced—by more than 90% in some places. At the same time, the virus was introduced to England and France and there also it resulted in huge reductions in the rabbit populations. The evolutionary changes that then occurred in Australia were followed in detail by Fenner and his associates (Fenner & Ratcliffe, 1965; Fenner, 1983) who had the brilliant research foresight to establish baseline genetic strains of both rabbits and virus. They used these baseline strains to measure subsequent changes in the virulence of the virus and in resistance of the host as they evolved in the field.

When the disease was first introduced to Australia it killed more than 99% of infected rabbits. This 'case mortality' fell to 90% within 1 year and then declined further (Fenner & Ratcliffe, 1965). The virulence of isolates of the virus sampled from the field was graded according to the survival time and the case mortality of control rabbits. The original introduced virus (1950–1951) was Grade I which killed more than 99% of infected laboratory rabbits. Already by 1952 most of the virus isolates from the field were the less virulent Grades III and IV. At the same time, the rabbit population in the field was increasing in resistance so that when injected with a standard Grade III strain of the virus, field samples of rabbits declined by nearly 90% in 1950–1951, but less than 30% only 8 years later (Marshall & Douglas, 1961) (Figure 12.34).

In Australia in the first 20 years after introduction of the virus, its main vectors were mosquitos (especially *Anopheles annulipes*), and one reason why Grades I and II of the virus failed to maintain themselves was that they killed the host so quickly that there was only a very short time in which the vector could transmit them. At the other end of the virulence scale the mosquitos were unlikely to transmit Grade V of the virus because it produced very few infective particles in the host skin that could contaminate the vectors' mouth parts. The situation was complicated in the late 1960s when an alternative vector of the disease, the rabbit flea *Spilopsyllus cuniculi* (the main vector in England), was introduced to Australia. There is some evidence that more virulent strains of the virus may be favoured when the flea is the main vector (see discussion in Dwyer *et al.*, 1990).

Various mathematical models have been devised to predict the future of the interaction between myxoma and the rabbit (see especially Dwyer *et al.*, 1990; Nowak & May, 1994, for reviews of these). A very detailed model that takes account of demographic parameters of rabbits and changing virus levels in infected hosts concludes that strains of the virus with intermediate virulence will continue to control Australian rabbit populations in the near future (Dwyer *et al.*, 1990). But, there are still a great many unknowns in this most meticulously documented story. In particular, we do not know for how long the rabbit genome can continue to generate new resistant genotypes in a long drawn-out arms race with the virus.

It is convenient to think of natural populations of hosts as environmental patches, each of which carries one strain of parasite, but in the real world the environmental

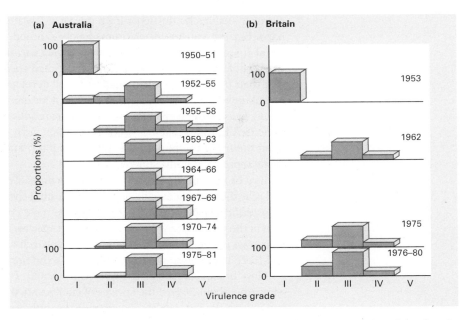

Figure 12.34 (a) The proportions in which various grades of myxoma virus have been found in wild populations of rabbits in Australia at different times from 1951 to 1981. Data showing the percentage of all field samples falling into virulence Grades I–V. (After Fenner, 1983.) Grade I is the most and Grade V the least virulent. (b) Similar data for wild populations of rabbits in Great Britain from 1953 to 1980. (After May & Anderson, 1983; from Fenner, 1983.)

patch that is a host may support two or more strains of a parasite. There may indeed be an ecological succession as one parasite strain infects a host that has already been infected by another (superinfection). There may then be competition between strains within a host or, under certain conditions, they may coexist within it (Levin, 1983). Clearly, this way of thinking about parasites and hosts has many formal analogies in the dynamics of metapopulations (see Chapter 15, Section 15.6.5; and see Nee & May, 1992). One important conclusion is that superinfection appears to allow the evolution of higher levels of virulence than are predicted by simpler models. 'The conventional wisdom that successful parasites have to become benign is not based on exact evolutionary thinking' (Nowak & May, 1994). This leaves the predicted state of myxomatosis and other host–parasite interactions in parlous uncertainty!

Reviews of the role of genetic diversity in host–parasite interactions are given in Hamilton and Howard (1994).

12.8 Brood and social parasitism

At first sight the presence of a section about cuckoos might seem out of place in this chapter. In all the examples that we have discussed involving animals, the host and its parasite have come from very distant systematic groups (mammals and bacteria, fish and tapeworms, rabbits and viruses). In contrast, brood parasitism usually occurs between quite closely related species and even between members of the same species! Yet, the phenomenon falls clearly within the definition of parasitism (a brood parasite 'obtains its nutrients from one or a few host individuals, *normally* causing harm but not causing death immediately'). Brood parasitism is well developed in social insects (sometimes then called 'social parasitism'). Here, the parasites use workers of another, usually very closely related species (Choudhary *et al.*, 1994), to rear their progeny. The phenomenon is well developed amongst birds but appears to be unknown in any other vertebrates except for one extraordinary example amongst fishes. Eggs of the catfish *Synodontis multipunctatus* are incubated in the mouths of some cichlid fishes together with the cichlids' own eggs. As they develop they feed upon the fry of the host whilst still in its mouth and so almost entirely destroy the host's reproductive investment.

Bird brood parasites lay their eggs in the nests of other birds, which then incubate and rear them. They usually depress the nesting success of the host. Amongst ducks *intraspecific* brood parasitism appears to be most common. Females may lay a few eggs in the nests of neighbours of their own species, although more usually they rear their own young. Young ducklings are largely self-supporting soon after they hatch and there is therefore probably little cost to the host bird.

Most brood parasitism is *interspecific*. About 1% of all bird species are brood parasites—about 50% of the species of cuckoos, two genera of finches, five cowbirds and a duck (Payne, 1977). They usually lay only a single egg in the host's nest and may adjust the host's clutch size by removing one of its eggs. The developing parasite may evict the host's eggs or nestlings and harm any survivors by monopolizing parental care. There is therefore the potential for brood parasites to have profound effects on the population dynamics of the host species. However, the frequency of parasitized nests is usually very low (less than 3%), and Lack (1963) concluded that 'the cuckoo is an almost negligible cause of egg and nestling losses amongst English breeding birds.'

Although the brood parasites appear in so many ways to be exceptional amongst parasites, many aspects of their behaviour reinforce the generalizations that emerge from studies of more conventional parasites. In particular, they are model systems for the coevolution of parasite and host (Rothstein, 1990). Highly host-specific relationships, polymorphisms and fine-scale specializations have developed within the parasites. For instance, the cuckoo *Cuculus canorum*, parasitizes many different host species, but there are strains ('gentes') within the species. Individual females of a strain are thought to favour just one host species and they lay eggs that match quite closely the colour and markings of the eggs of the preferred host. The host species differ in their ability to distinguish 'mimic' eggs from their own. Those with this ability tend to eject bad mimics or desert a nest that has been parasitized. The dunnock (*Prunella modularis*), on the other hand, makes no distinction between its own eggs, those of cuckoos or white or black model eggs, and the cuckoos that parasitize dunnocks do not produce eggs that mimic the dunnock's eggs (except in the darkness of their colour) (Brooke & Davies, 1988).

The match between the eggs of cuckoos and their hosts is strongest with host species that are good at recognizing mimics. How the cuckoo populations become divided up into separate, but coexisting strains is puzzling—there must be barriers to breeding between strains. One possible explanation is that the female cuckoos learn, as nestlings, to recognize their host species and seek out members of the same species when they, in turn, are looking for hosts for their eggs. It is more difficult to explain how male cuckoos mate assortatively with females of their own strain

(Harvey & Partridge, 1988), or why dunnocks have failed to evolve the ability to detect egg mimics.

The development of specialized host-specific races within species of brood parasitic cuckoos closely resembles that in biotrophic pathogens on plants, such as in the rusts. It emphasizes the great contrast between parasitism, which leads so often to a fine scale of host specificity, and mutualism, in which such specialization appears to be rather rare (see Chapter 13). The evolutionary consequences of battle and of partnership are quite different.

Chapter 13
Symbiosis and Mutualism

13.1 Introduction

Species do not live in isolation from each other: many form close associations. Many species form the habitats of others as in the intimate relationship between parasites and their hosts, the growth of rhizobial bacteria in the nodules of leguminous plants and the growth of lichens and mosses on the trunks of trees. Anton de Bary coined the word 'symbiosis'—living together—for such close physical associations between species (although in modern usage parasites are usually excluded from the category of symbionts). But, there are other types of association in which two species may come into physical contact only briefly in each others lives, as when trees are used as nesting sites by birds or insects pay fleeting visits to a flower in the search for nectar or pollen. The species in such cases can scarcely be regarded as 'living together' (symbionts), although one may play an essential part in the life of another.

Amongst the various associations between species we recognize a special category, 'mutualism', in which an association between species (usually a pair) brings mutual benefit. Mutualism need not involve close physical association. Mutualists need not be symbionts—although many are. For example, many plants gain dispersal of their seeds by offering a reward in the form of edible fleshy fruits, or assure cross-pollination by offering a resource of nectar in their flowers to visiting insects. These are mutualistic interactions but we would not consider them to be symbioses.

In this chapter we consider mutualisms as any association between species that confers mutual advantage or benefit. We usually recognize a mutualism when two (or conceivably more) species confer on each other higher birth rates, and/or lower death rates or higher carrying capacities.

Ecology textbooks have generally underemphasized or even ignored symbionts and mutualists, yet they compose most of the world's biomass. Almost all the plants that dominate the world's grasslands, heaths and forests have roots intimately associated with fungi. The polyps of most corals contain unicellular algae, many flowering plants depend on insect pollinators and a very great number of animals carry communities of microorganisms within their digestive systems.

the world's biomass is largely composed of mutualists

Mutualisms are represented in a much more varied range of species interactions than competition, predation and parasitism, and so this chapter contains much more natural history than others in this book.

We consider first, examples of mutualism in which there is little physical integration between the partners, and it is patterns of behaviour that cause their life cycles to intertwine. We then give examples of increasingly tight physical association in which the mutualisms are more truly symbioses. We then consider associations in which the lives (and bodies) of two species have become so integrated that the symbiosis is obligate, and finally, situations in which one partner has assimilated genetic information from the other and it has ceased to be sensible to regard them as distinct species. More detailed reviews are given by Boucher *et al.* (1984) and Douglas (1994).

13.2 Mutualisms that involve reciprocal links in behaviour

It is not easy to classify the variety of symbioses that involve intricate behavioural links or to prove mutualism in every case. The following examples illustrate the diversity.

13.2.1 Ant–plant mutualisms

The idea that there were mutualistic relationships between plants and ants was put forward by Belt (1874) after observing the behaviour of aggressive ants on species of *Acacia* with swollen thorns in Central America. This relationship, which might be described as 'cleaner/customer', was later described by Janzen (1967) for the Bull's horn acacia (*Acacia cornigera*) and its associated ant, *Pseudomyrmex ferruginea*. The plant bears hollow thorns which are used by the ants as nesting sites; it has finely pinnate leaves bearing protein-rich 'Beltian bodies' at their tips (Figure 13.1), which the ants collect and use for food; and it bears sugar-secreting nectaries on its vegetative parts which attract the ants. The ants, for their part, protect these small trees from competitors by actively snipping off shoots of other species that enter the acacia canopy, and also protect the plant from herbivores.

This pioneer study by Janzen stimulated a large body of research which reveals that ant–plant symbioses have evolved many times (even repeatedly in the same family of plants) and on different continents. There is evidence of competition between species of ant for host plants in which the smaller less aggressive species have been driven to occupy the most protected positions on the plants. The curious pruning activity in which ants cut off the shoots of neighbouring plants may have evolved primarily as a programme of defence, removing the bridges used by marauding ants. This raises the question whether the ant–plant symbioses are truly mutualisms or have evolved mainly as the consequence of competition between species of ant with the plant as a passive partner. However, the fact that species of *Acacia* provide a specialized and balanced food reward (the Beltian bodies) implies that at least in this case the symbiosis has evolved as a mutualism.

The demonstration of ant–plant mutualisms opens the way for new interpretations of other plant structures. Nectaries are present on the vegetative parts of plants of at least 39 families and in many communities throughout the world. Whilst nectaries on or in flowers are easily interpreted as attractants for pollinators, the role of extrafloral nectaries on vegetative parts is less obvious. They clearly attract ants, sometimes in vast numbers, but there has been a controversy between the 'protectionists' (who have supported the idea that the ants visiting the nectaries protect the plants from herbivorous animals), and the 'exploitationists' who have felt that the plants have no more use of ants than dogs have of fleas—a sceptical view that is proper in all scientific enquiry (Bentley, 1977). Such disagreement needs to be resolved by carefully designed and controlled experiments such as one

(a)

(b)

Figure 13.1 Structures of the Bull's horn acacia (*Acacia cornigera*) that attract its ant mutualist. (a) Protein-rich Beltian bodies at the tips of the leaflets. (b) Hollow thorns used by the ants as nesting sites. (Courtesy of L.E. Gilbert.)

made by Inouye and Taylor (1979). They excluded ants from plants of the aspen sunflower (*Helianthella quinquenervis*), which bears extrafloral nectaries. This plant can suffer heavy attacks from a terphritid fly whose larvae may destroy more than 85% of the developing seeds. The plants carrying ants escaped much of this predation.

Direct evidence that close association with ants does indeed defend higher plants against attack by phytophagous insects comes from a study of an Amazonian canopy tree *Tachigali myrmecophila*, which harbours the stinging ant *Pseudomyrmex concolor*, in its hollow leaf rachis and petioles. The ants were removed from selected plants and these then bore 4.3 times as many phytophagous insects as control plants and suffered much greater herbivory. Leaves on plants that carried a population of ants lived more than twice as long as those on unoccupied plants and nearly 1.8 times as long as those on plants from which ants had been deliberately removed (Figure 13.2).

Reviews of ant–plant interactions can be found in Beattie (1989), Huxley and Cutler (1991) and Davidson and McKey (1993).

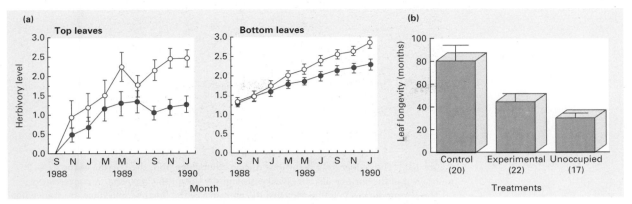

Figure 13.2 (a) The intensity of leaf herbivory on plants of *Tachigali myrmecophila* naturally occupied by the ant *Pseudomyrmex concolor* (●; *n* = 22) and on plants from which the ants had been experimentally removed (○; *n* = 23). Bottom leaves are those present at the start of the experiment and top leaves are those emerging subsequently. (b) The longevity of leaves on plants of *T. myrmecophila* occupied by *P. concolor* (control) and from which ants were experimentally removed or from which ants were naturally absent. Error bars ± the standard error. (After Fonseca, 1994.)

13.2.2 Cleaner fish and customers

In the mutualistic relationship between 'cleaner' and 'customer' fish, the former feed on ectoparasites, bacteria and necrotic tissue from the body surface of the latter. At least 45 species of cleaner fish (and a number of cleaner shrimps) have been recognized and these often hold territories with 'cleaning stations' which their customers visit. Customers visit cleaning stations more often when they carry many parasites. The cleaners gain a food source and the customers remain clean. In an experiment in the Bahamas, Limbaugh (1961) removed all the cleaner fish from patches of reef. The 'customers' developed skin diseases and their populations declined within 2 weeks.

13.2.3 The honey guide and the honey badger

An African bird, the honey guide (*Indicator indicator*), has formed a remarkable relationship with the ratel or honey badger (*Melliovora capensis*). A honey guide that has located a bees' nest leads the honey badger to it. The mammal tears open the nest and feeds on honey and bee larvae, and later the honey guide gets a meal of beeswax and larvae. The honey guide can locate a bees' nest but not break it open, whilst the honey badger is in just the opposite situation. The reciprocal link in their behaviour brings mutual benefit.

13.2.4 Shrimps and gobiid fish

Shrimps of the genus *Alpheus* dig burrows, and goby fish (*Cryptocentrus*) use these as safe sites in an environment that otherwise provides little or no shelter. The shrimp is almost completely blind and, when it leaves the burrow, keeps one antenna in contact with the fish, thereby getting warning of any disturbance (Figure 13.3; see also Chapter 1, Figure 1.1 for this symbiosis as one element in the ecological divergence

485   SYMBIOSIS AND MUTUALISM

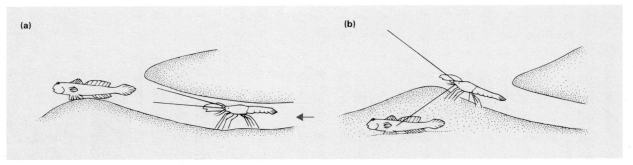

Figure 13.3 The sequence of behaviour during emergence from the burrow in the mutualism of the shrimp *Alpheus djiboutensis* and the goby fish *Cryptocentrus cryptocentrus*. (a) The shrimp moves head-first toward the entrance of the burrow; antennae are somewhat raised at an oblique forward angle with the tips 2–3 cm apart. (b) Outside the burrow the shrimp keeps one antenna in contact with the fish. (After Karplus *et al.*, 1972.)

in goby fishes). The goby gains a place to live in an environment of sediment containing abundant food. The shrimp gains an optical warning system that allows it to leave its burrow safely for short periods to feed on sediment outside (Fricke, 1975).

13.2.5 Clown fish and anemones

A variety of behavioural symbioses is found amongst the inhabitants of tropical coral reefs (where the corals themselves are mutualists—see Section 13.8). The clown fish (*Amphiprion*) lives close to a sea anemone (e.g. *Physobrachia*, *Radianthus*) and retreats amongst the anemone's tentacles whenever danger threatens. Whilst within the anemone, the fish gains from it a covering of mucus that protects it from the anemone's stinging nematocysts (the normal function of the anemone slime is to prevent discharge of nematocysts when neighbouring tentacles touch). The fish derives protection from this relationship, and the anemone also benefits because clown fish attack other fish that come near, including species that normally eat sea anemones (Fricke, 1975).

13.3 Mutualisms that involve the culture of crops or livestock

13.3.1 *Homo sapiens* is a mutualist in relation to crops and livestock

The most dramatic ecological mutualisms are those of human agriculture. The numbers of individual plants of wheat, barley, oats, corn and rice, and the areas these crops occupy, vastly exceed what would have been present if they had not been brought into cultivation. The increase in the human population since the time of hunter–gatherers is some measure of the reciprocal advantage to *Homo sapiens*. We may not do the experiment, but we can easily imagine the effect that the extinction of humans would have on the world population of rice plants or the effect the extinction of rice plants would have on the population of humans. The domestication of cattle, sheep and other mammals also involves a mutualism—the numbers of such animals would fall sharply in the absence of humans and human diet would change dramatically in the absence of livestock.

Similar 'farming' mutualisms have developed in termite and ant societies.

13.3.2 The 'farming' of caterpillars by ants

A classic example of livestock mutualism is that between the large blue butterflies (*Maculinea* spp.) and ants. All species of *Maculinea* lay their eggs on one or two specific species of plant and the young caterpillars feed on the flowers until their third instar, when they fall to the ground and remain within a few centimetres of the plant until they are discovered by a worker ant of the genus *Myrmica* and carried into the nest, where they remain for 10 months before they pupate in the upper chambers of the nest. Although many species of *Myrmica* will collect and carry away the caterpillars, each species of *Maculinea* is only reared successfully by a particular species of *Myrmica*. The caterpillars mimic ant larvae and those species that are the best mimics induce their ant host to feed them (e.g. *Maculinea rebeli* which feeds on the crossleaved gentian, *Gentiana cruciata*, and whose caterpillars mimic the larvae of the ant *Myrmica schenkii*). Other 'more primitive' species of *Maculinea* (such as *M. arion* which feeds on wild thyme) are predators on the ant larvae—indeed, an ant colony can be annihilated by the caterpillars of these species which then also die (Hochberg *et al.*, 1992).

The life histories of the *Maculinea* butterflies are so bizarre that it might seem incredible if they were not part of a complex series of ant–butterfly interactions found amongst many other members of the same family Lycaenidae. Other species of this family are also farmed by ants (although not all are carried into the colony), and the larvae of most species in the family bear active honey glands that are milked by ants. The interactions between species of the Lycaenidae and ants include examples from the whole continuum of species interactions from mutualism and commensalism, to parasitism and predation (Pierce & Young, 1986).

The large blue butterfly *M. arion*, is a rare and endangered species in Europe. It recently became extinct in Great Britain. Its life cycle and behaviour trap it into total dependence on flowering thyme (*Thymus serpyllum*) and subsequently on ant larvae for food. Only two species of ant appear to be acceptable as its hosts (*Myrmica scabrinoides* and *M. laevonoides*), and the calcareous grasslands that serve as habitat for both the thyme and the ants are rapidly disappearing. This life cycle is thus a classic example of species caught in 'the ever-narrowing rut of specialization', with their particular risk of extinction.

A theoretical model of the population dynamics of *Maculinea rebeli* has been constructed from data collected in 1990 at a 4 ha (4×10^4 m^2) site near Jaca in the Spanish Pyrenees, part of one of the largest known populations of the butterfly that extends over 100 ha (1×10^6 m^2) (Hochberg *et al.*, 1992). The model assumes that ant and plant populations remain constant over the short time scale of 5–10 years, but takes into account the role of specialist ichneumonid parasitoids which add further hazards to the survival of this seriously endangered species. The model indicates that the persistence of the butterfly population is much more sensitive to survival of larvae in the ants' nests than to competition between larvae on the food plant. At this site, changes in the population of the plant had virtually no impact on the butterfly population.

Ants 'farm' many species of Homoptera in return for sugary secretions. The ants defend their 'flocks' against predators and parasitoids in the same way that some species of ant, attracted by extrafloral nectaries, defend the plant that produces them against herbivores (Strong *et al.*, 1984).

13.3.3 The farming of fungi by beetles and ants

Much of plant tissue, including wood, is unavailable as a direct source of food to animals because they lack the enzymes that can digest cellulose and lignins (see

(a)

(b)

Figure 13.4 (a) Partially excavated nest of the leaf-cutting ant *Atta vollenweideri* in the Chaco of Paraguay. The above-ground spoil heap excavated by the ants extended at least 1 m below the bottom of the excavation. (b) Queen of *A. cephalotes* (with an attendant worker on her abdomen) on a young fungus garden in the laboratory, showing the cell-like structure of the garden with its small leaf fragments and binding fungal hyphae. (Courtesy of J.M. Cherrett.)

Chapters 3 and 11). However, many fungi possess these enzymes and an animal that can eat fungi gains indirect access to an energy-rich food. Some very specialized mutualisms have developed between animal and fungal decomposers. Beetles in the group *Scolytidae* tunnel deeply in the wood of dead and dying trees, and fungi that are specific for particular species of beetle grow in these burrows and are continually grazed by the beetle larvae. These 'ambrosia' beetles may carry inocula of the fungus in their digestive tract and some species bear specialized brushes of hairs on their heads which carry the spores. The fungi serve as food for the beetle and depend on it for dispersal to new tunnels—the relationship is wholly mutualistic.

Leaf-cutting ants of the genera *Atta* and *Acromyrmex* are the most remarkable of the fungus-farming ants. They excavate 2–3 litre cavities in the soil, and in these a basidiomycete fungus is cultured on leaves that are cut from neighbouring vegetation (Figures 13.4a, b). The ant colony may depend absolutely on the fungus for the nutrition of the larvae. Workers lick the fungus colonies and remove specialized swollen hyphae (gongylidia), which are aggregated into bite-sized 'staphylae'. These are fed to the larvae and this 'pruning' of the fungus may stimulate further fungal growth. The fungus gardens become more attractive to the workers if they have been temporarily excluded from them (Figure 13.5).

The fungus gains from the association: it is both fed and dispersed by leaf-cutting ants and has never been found outside their nests. The reproductive female carries her last meal as a culture when she leaves one colony to found another (some authors have claimed that the fungus culture is carried in a specialized pouch which the female deliberately fills before she leaves the nest).

Most phytophagous insects have very narrow dietary tolerances—indeed, the vast majority of insect herbivores are strict monophages. The leaf-cutting ants are remarkable amongst insect herbivores in their polyphagy. Ants from a nest of *A. cephalotes* harvest from 50 to 77% of the plant species in its neighbourhood (although they have clear preferences if given a choice). Leaf-cutting ants may harvest 17% of total leaf production in tropical rainforest and be the ecologically dominant herbivores in the community: it is their polyphagy that gives them this remarkable ecological status (Cherrett *et al.*, 1989). In contrast to the polyphagy of the adults the larvae appear to be extreme dietary specialists being restricted to nutritive bodies (gongylidia) produced by the fungus *Attamyces bromatificus*, which the adults cultivate and which decompose the leaf fragments (Cherrett *et al.*, 1989).

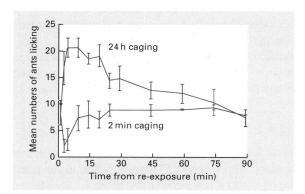

Figure 13.5 The numbers (means ± the standard error) of the ant *Atta sexdens* licking areas of fungus garden from which they have been excluded for 2 min or 24 h. (After Bass & Cherrett, 1994.)

13.4 Mutualisms involved in the dispersal of seeds and pollen

13.4.1 Seed dispersal mutualisms

Very many species use animals to disperse their seeds and pollen. About 10% of all flowering plants possess seeds or fruits that bear hooks, claws or barbs that become attached on the hairs, bristles or feathers of any animal that comes into contact with them. They are frequently an irritation to the animal which usually cleans itself and removes them if it can, but usually after carrying them some distance. Another group of flowering plants contains species with sticky seeds (e.g. *Salvia glutinosa*, *Plumbago capensis*). In all of these cases the benefit is to the plant (which has invested resources in attachment mechanisms) and there is no reward to the animal. This is not mutualism. Quite different is the true mutualism between higher plants and the birds and other animals that feed on fleshy fruits and disperse the seeds.

hooked and sticky seeds are dispersed by animals, but this is not mutualism

The plant kingdom has exploited a splendid array of morphological variations in the evolution of fleshy fruits (Figure 13.6) that offer different nutritional rewards. A detailed statistical survey of 910 plant species with fleshy fruits representing 392 genera and 94 families revealed some clearly different patterns of mutualism (Jordano, 1995). Bird-dispersed species bear smaller fruits than those that are mammal dispersed and the total energy reward per fruit follows the same trend. The fruits of bird-dispersed species have a higher content of lipids than mammal-dispersed species. This particular analysis was largely designed to determine how far the characteristics of the fruits of any particular species were determined by its ancestry (i.e. phylogenetic constraint), and how far they represented species-specific selective processes. It appears that variation in fruit morphology is largely an intrinsic and relatively stable property of higher taxonomic levels that reflects their phylogeny rather than the forces of recent natural selection.

Of course, for a fruit-eating bird and a higher plant to be in a mutualistic relationship it is essential that the bird digests only the fleshy fruit and not the seeds, which must remain viable when regurgitated or excreted. Thick strong defences that protect plant embryos are usually part of the price paid by the plant for dispersal by fruit-eating birds.

Mutualisms that involve animals that eat fleshy fruits and disperse seeds are seldom very specific to the species of animal involved. It is usually birds or mammals that are involved in these mutualisms and even in the tropics there are few plant species that fruit throughout the year and form a reliable food supply for any one specialist. Various tropical species of fig (*Ficus*) are unusual in being found in flower and fruit throughout the year, but one consequence is that, rather than allowing specialist feeders to evolve, they are eaten by a very wide range of both obligate and facultative frugivores (Figure 13.7). The Australian mistletoe bird (*Dicaeum livundinaceum*) (see Chapter 5, Figure 5.6) is one of the few fruit-eating birds that is relatively highly specialized to one species or genus of plants.

13.4.2 Pollination mutualisms

Most animal-pollinated flowers offer nectar, pollen or both as a reward to their visitors. Nectar seems to have no value to the plant other than as an attractant to animals and it has a cost to the plant, because the nectar carbohydrates might have been used in growth or some other activity.

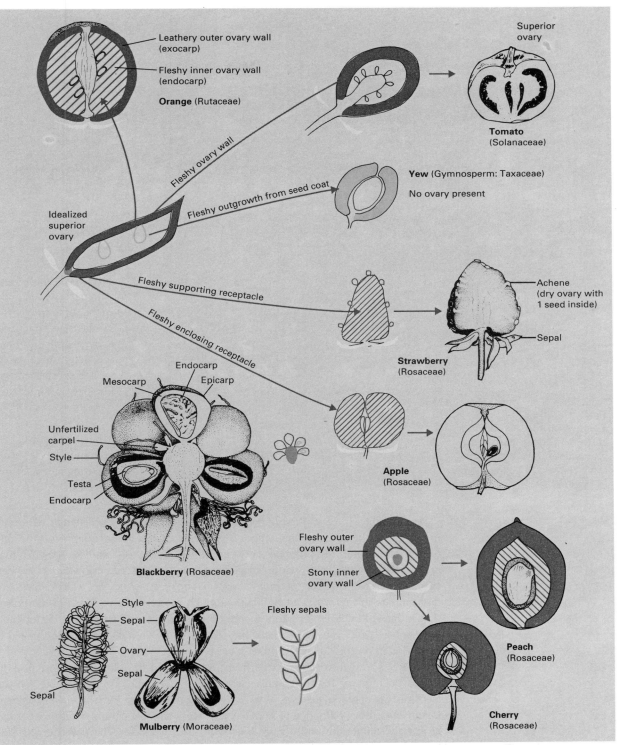

Figure 13.6 A variety of fleshy fruits involved in seed dispersal mutualisms illustrating morphological specializations that have been involved in the evolution of attractive fleshy structures.

491 SYMBIOSIS AND MUTUALISM

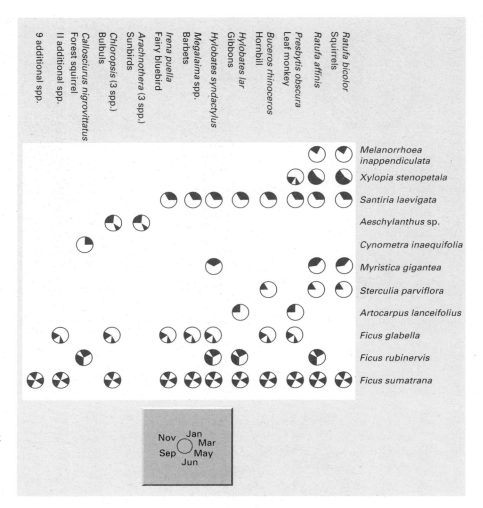

Figure 13.7 Birds and mammals that feed on the fruit of trees at various times of the year at Selangore, Malaysia. (After Harper, 1977; from data of McClure, 1966.)

insect-pollinated flowers may be generalists or specialists

Presumably, the evolution of specialized flowers and the involvement of animal mutualists in the act of pollination has been favoured because an animal may be able to recognize and discriminate between different flowers and so ensure that outbreeding is encouraged within species, and yet be hindered or prevented between species. Passive transfer of pollen, for example by wind or water, does not discriminate in this way and is of course much more wasteful. Much of the pollen that is carried by animal vectors reaches other flowers and where the vectors and flowers are highly specialized, as is the case in many orchids, virtually no pollen is wasted even on the flowers of other species.

There is a hazard that derives from the adoption of animals as mutualists in flower pollination. If the animals carry pollen they may be responsible for the transmission of sexual diseases as well. The fungal pathogen *Microbotryum violaceum*, is transmitted by pollinating visitors to the flowers of white campion (*Silene alba*) and in infected plants the anthers are filled with fungal spores. The pathogen *Botrytis anthophila*, is transmitted in a rather similar fashion as a venereal disease of clover plants (*Trifolium* spp.). That pollen visitors may also serve as disease vectors imparts a cost to the use of animals as agents of cross-pollination (Shykoff & Bucheli, 1995).

Figure 13.8 Pollinators. (a) Honeybee (*Apis mellifera*) on raspberry flowers. (b) Cape sugarbird (*Promerops cafer*) feeding on *Protea eximia*. (Courtesy of H. Angel.)

Many different kinds of animals have entered into pollination liaisons with flowering plants, including humming-birds, bats and even small rodents and marsupials (Figure 13.8). However, the pollinators *par excellence* are, without doubt, the insects. Pollen is a nutritionally-rich food resource and in the simplest insect-pollinated flowers pollen is offered in abundance and freely exposed to all and sundry. In more complex flowers, nectar (not much more than a solution of sugars) is produced as an additional or alternative reward. In simple flowers, the nectaries are unprotected; the bramble (*Rubus fruticosus*) openly exposes nectar in such abundance that it provides a local glut for a variety of nectar seekers. The mutualism in such a case is very unspecialized, but increasing specialization can be seen in families such as Ranunculaceae, in which the nectaries are enclosed in structures that restrict access to the nectar to just a few species of insect visitor. In the simple flower of *Ranunculus ficaria* the nectaries are exposed to all visitors, but in the more specialized flower of *R. bulbosus* there is a flap over the nectary and in *Aquilegia* the nectaries have developed into long tubes and only visitors with long tongues can reach the nectar. In the related *Aconitum* the whole flower is structured so that the nectaries are accessible only to insects of the right shape and size that are forced to brush against the anthers and pick up pollen.

An extreme example of highly specialized floral structure was reported by Darwin (1859) in the Madagascar star orchid (*Angraecum sesquipedale*), which produces its nectar in tubes approximately 30 cm long. He suggested that a pollinator would be found with an appropriate length of proboscis, and 40 years later a hawkmoth was indeed found with a proboscis 25 cm long. Darwin suggested that the long nectary forced the insect into close contact with the pollen and that natural selection would therefore favour even longer nectaries. As an evolutionary reaction the tongues of the pollinator would be selected for increasing length. In this way, the mutualism between plant and insect would become more and more obligate—a

the benefits of floral specialization

493 SYMBIOSIS AND MUTUALISM

reciprocal and escalating process of specialization. Nilsson (1988) deliberately shortened the nectary tubes of another long-tubed orchid (*Platanthera* spp.) and showed that the flowers then produced many fewer seeds—presumably because the pollinator was not forced into the right position. The orchids, more than any other group of flowering plants have been caught up in precise floral mutualisms that have overwhelmingly dominated evolutionary radiation in the family. The ecology of pollination in orchids is reviewed by Nilsson (1992).

Flowering is a seasonal event in most plants and this places strict limits on the degree to which a pollinator can become an obligate specialist. A pollinator can only become completely dependent on specific flowers as a source of food if its life cycle matches the flowering season of the plant. This is feasible for insects that use flowers for only a short period, for instance as a maintenance diet for adult butterflies and moths. Longer-lived pollinators such as bats and rodents, or bees with their long-lived colonies, are more likely to be generalists, turning from one relatively unspecialized flower to another through the seasons or to quite different foods when nectar is unavailable.

the pollination mutualism of the fig and the fig wasp

Pollination mutualisms exist between many species of fig (*Ficus*; family Moraceae), and their pollinators, the fig wasps, in which the interaction has become obligate for both partners (Janzen, 1979; Wiebes, 1979; Bronstein, 1988). The flowers of a typical hermaphrodite fig are borne within swollen receptacles (which become the fleshy fig) and male and female flowers are borne in separate regions within the receptacle. The female flowers are of two types: long styled and short styled. Female wasps enter the fig through an apical pore and lay their eggs in the short-styled ovaries (their ovipositors cannot reach down the long-styled ovaries). The parasitized short-styled ovaries develop into galls in which the fig wasp larvae complete their development. Male wasps bore their way out of the fig and the females follow them, but only after passing the male flowers and collecting pollen from these. The pollen is carried in specialized sacs, and when the females enter a new fig they have been observed shaking pollen out of these sacs and placing it on the stigmas of long-styled ovaries which, after pollination, develop seeds (Figure 13.9). Each species of fig has its own more-or-less specific species of wasp.

Such a specialized sequence of mutualistic events must have involved reciprocating evolutionary pressures in which each stage in the evolution of the fig has evoked its counterpart in the behaviour of the wasp (Wiebes, 1982). The wasp has become not only pollinator but also parasite.

In most species of fig there are two different species of wasp involved, each specialized on particular flowers within the fruit (in addition, there are usually two species of parasitic wasp which specialize on the two pollinating wasp species). A further wasp is often present which lays eggs in the female flowers of the fig but plays no part in the pollination process. It is not a mutualist but is in a real sense a 'cuckoo in the nest', reared by the fig but paying no price or fee for the privilege.

This pattern of interactions is repeated time and again in the various wild fig species, with different species of wasps as gall-former, parasite and cuckoo. The situation represents one of the great classics of coevolution, *and* of parallel evolution of mutualists *and* of parasitism. New variations and paradoxes are continually being discovered in fig–wasp symbioses (see, for example, Grafen & Godfray, 1991). (Note that the cultivated fig is parthenogenetic, requiring no act of pollination for the production of a ripened fruit. It has escaped from dependence on the wasp and also from dependence on outbreeding.)

Darwin realized that pollination biology posed some of the most profound questions in evolutionary theory, particularly in respect of the evolution of sex and the force of sexual selection. The study of pollination after Darwin was dominated by a vast effort at description and classification (Willson, 1994), and later by analysis of the genetic consequences of in- and outbreeding. Ecological aspects were largely ignored until the 1970s when a science of pollination biology with a strong ecological component suddenly developed and blossomed. It has its own sophisticated language (Inouye *et al.*, 1994). The subject is now too vast to compress into this chapter, but for the ecologist who wishes to explore this area of present excitement the following references provide an entrée to the literature (Howe & Westley, 1986; Lovett Doust & Lovett Doust, 1988; Kearns & Inouye, 1993; Arnold, 1994; Stanton, 1994; Willson, 1994). A detailed review of both seed dispersal and pollination mutualisms is also given by Thompson (1995), who gives a good account of the processes that may lead to the evolution of such mutualisms.

13.5 Mutualisms involving gut inhabitants

Most of the mutualisms so far discussed have depended on the patterns of behaviour of the animals involved. In all cases they involve acts of search, and usually (but not always) they have food as a reward. Both partners in such mutualisms are highly evolved complex organisms, and in most cases each spends a significant part of its life on its own. In many other mutualisms one of the partners is a unicellular eukaryote or prokaryote and is integrated more or less permanently into the body of its multicellular partner. The variety and intimacy of such close symbioses (which are not necessarily all mutualisms) is illustrated in Figure 13.10. The microbial floras of parts of the alimentary canal are the best known extracellular microbial symbionts.

the gut as a culture chamber

In most animals the gut is a microcosm of microbial life, and in many herbivores this microflora is crucial for the digestion of cellulose and perhaps also for the

495 SYMBIOSIS AND MUTUALISM

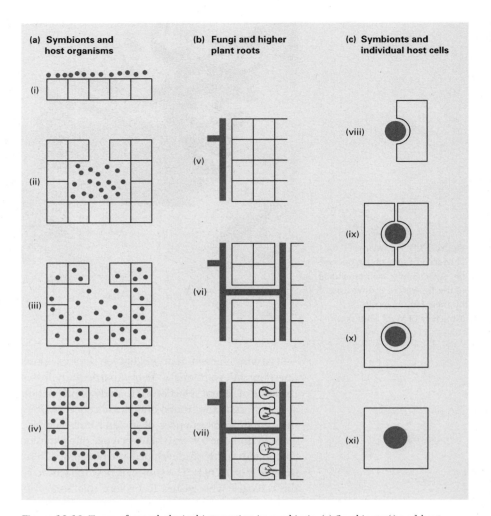

Figure 13.10 Types of morphological integration in symbiosis. (a) Symbionts (•) and host organisms (□): (i) symbionts entirely on surface (e.g. epiphytic or epizootic microbes); (ii) symbionts within organism (e.g. digestive-tract microorganisms such as rumen microbes); (iii) symbionts within organism, partly extracellular, partly intracellular, as occurs with cyanobacteria (blue–green algae) in tropical marine sponges; (iv) symbionts entirely intracellular. (b) Fungal hyphae and higher plant roots: (v) rhizosphere fungi which do not penetrate roots; (vi) ectotrophic mycorrhizal fungi of forest trees, which penetrate between but not into root cells; (vii) fungi which penetrate the root and into cells by haustoria, as in vesicular arbuscular mycorrhiza—the haustoria are enclosed by a host membrane. (c) Symbionts and individual host cells: (viii) symbionts in close physical contact with surface of host cells, but also in contact with external environment (e.g. lichens, *Prochloron* on ascidians); (ix) symbionts enclosed by host cells, but not within them; (x) symbionts within host cells, but enclosed in a host vacuole (most intracellular symbionts); (xi) symbionts lying free in the cytoplasm. (After Smith, 1979.)

synthesis of vitamins. The gut receives a steady flow of substrates for microbial growth in the form of food that has been eaten, chewed and partly homogenized. In the gut the mixture is stirred, its pH is regulated and anaerobic conditions are maintained. In warm-blooded animals the temperature, of what is effectively a culture chamber, is closely regulated. The contents of the culture chamber do not grow stale because waste substrate continually flows out of it. The system is closely analogous to the continuous fermentation vat used in the mass production of beer, where precisely the same set of environmental conditions is regulated by the brewer.

In small mammals (e.g. rodents, rabbits and hares) the caecum is the main fermentation chamber, and in larger non-ruminant mammals such as horses the caecum is the main site of fermentation. In elephants, which, like rabbits, refecate (consume their own faeces), all fermentation occurs in the caecum. In ruminants fermentation occurs in a specialized stomach (see Chapter 3, Section 3.7.2).

13.5.1 The rumen

The stomach of ruminants (which includes deer, cattle and antelope) is four chambered and food that is swallowed passes first into the rumen chamber. The first chewing reduces the size of food particles to volumes varying from 1 to 1000 µl and the particles may be as long as 10 cm. Only particles with a volume of about 5 µl or less can pass from the second chamber, the reticulum, into the next compartment, the omasum; the animal regurgitates and rechews the larger particles (the process of rumination). Dense populations of bacteria (10^{10}–10^{11} ml^{-1}) and of protozoa (10^5–10^6 ml^{-1}) are present in the rumen, the pH of which is regulated through secretion of 100–140 mmol l^{-1} bicarbonate and 10–50 mmol l^{-1} phosphate from the salivary glands. It is the activities of the host that provide the continual supply of substrates *and* the controlled conditions for microbial fermentation. The fermentation products form the principal food of the host (Figure 13.11).

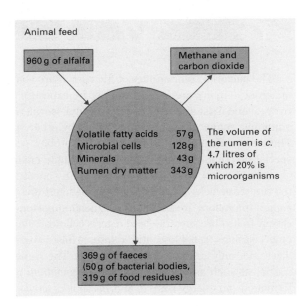

Figure 13.11 Microbial digestion within the rumen of the sheep. The data were obtained by continual feeding of alfalfa pellets. (After Hungate *et al.*, 1971.) The empirical formula for materials used and produced daily by the rumen system is:

$$C_{20.03}H_{36.99}O_{17.40}N_{1.34} + 5.6\,H_2O \rightarrow C_{12}H_{24}O_{10.1} + 0.83\,CH_4 + 2.76\,CO_2 + 0.50\,NH_3 + C_{4.44}H_{8.88}O_{2.35}N_{0.78}$$

difference between   + water   → volatile   + methane + carbon   + ammonia + microbial cells
food and faeces   fatty acids   dioxide

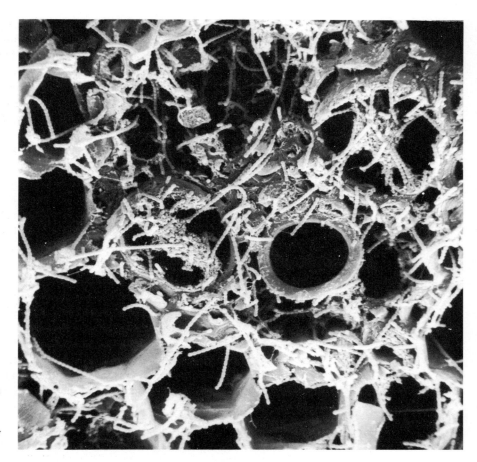

Figure 13.12 Attachment of *Ruminococcus flavefaciens* to the cell walls of perennial rye grass (*Lolium perenne*). Scanning electron micrograph, magnification × 1000. (Courtesy of B.E. Booker.)

The bacterial communities of the rumen are composed almost wholly of obligate anaerobes—many are killed instantly by exposure to oxygen. They require carbohydrate and most also need acetic, isobutyl, isovaleric and 2-methylbutyric acids and ammonia, many of which are provided by other species of bacteria in the rumen. The rumen flora is not a mixture of microbial generalists. Many have metabolic specialisms that constrain them to this unique environment.

the flora and fauna of the rumen ...

Cellulose and other fibres are the main constituents of the ruminant's diet, but the ruminant lacks the enzymes that can digest these. The cellulolytic activities of the rumen microflora are therefore of crucial importance. *Bacteroides succinogenes* sticks closely to its fibre substrate and digests cellulose; *Ruminococcus* (Figure 13.12) digests only cellulose, cellobiose and xylose; whilst others, like *Clostridium locheadii*, can digest not only cellulose but also starch. The rumen flora contains other resource specialists such as *Methanobacterium ruminantium* which can use only hydrogen or formate as an energy source, and *Bacteroides amylophilus* which can use only starch and its derivatives.

... which attack cellulose

the rumen contains a complete ecological community

The protozoa in the gut are also a complex mixture of specialists. Most are holotrich ciliates and entodiniomorphs (a group that is known only in rumen communities). The free-living protozoa of the rumen exist in a constant environment in which they are probably subject to intense competition from myriads of accompanying microbes. Their morphological diversity and complexity has been compared to that of tropical forests, where relatively constant environments also support great productivity and species diversity (Hungate, 1975).

A few of the protozoa can digest cellulose. The cellulolytic ciliates have intrinsic cellulases, although some other protozoa may use bacterial symbionts. Many of the protozoa ingest bacteria and if the protozoa are removed the numbers of bacteria rise. On the other hand, some of the entodiniomorphs are carnivores, preying on other protozoa. Thus, the diverse processes of competition, predation and mutualism, and the food chains that characterize terrestrial and aquatic communities in nature, are all present within the rumen microcosm. To add to the complexity, the species composition of the rumen microflora varies from one host species to another and it can change dramatically if the diet is suddenly altered.

The microbial population of the rumen is continually multiplying and being depleted as the rumen contents pass into the intestine. The ruminant's own enzymes are responsible for further digestive processes in the intestine, where some microbial bodies are themselves broken down. Within the rumen the main products of digestion are volatile fatty acids (acetic, propionic and butyric), ammonia, carbon dioxide and methane. The fatty acids are absorbed by the ruminant, and are its primary source of carbon nutrition. Propionic acid is particularly important since this is the only one of the volatile fatty acids that can be converted into carbohydrate by the ruminant, and is essential for its metabolism, especially during lactation.

The mutualistic character of the association of ruminants with a rumen microflora is clear: the microbial population gains a continuous supply of food and a rather stable environment; the ruminant gains digestible resources from a diet which its own enzymes cannot handle.

13.5.2 The termite gut

Termites are colonial social insects of the order Isoptera, many of which depend on mutualists for the digestion of wood. Primitive termites feed directly on wood, and most of the cellulose, hemicelluloses and lignins are digested by mutualists in the gut where the paunch (part of the segmented caecum) forms a microbial fermentation chamber (Figure 13.13). However, more advanced species of termite (75% of all the species) produce their own cellulase (Hogan *et al.*, 1988). A third group of termites (Macrotermitineae) cultivate fungi that digest wood and they then eat the fungus rather than the wood.

cellulose decomposition mainly by protozoa in the termite paunch

Termites refecate, i.e. they eat their own faeces, so that food material passes at least twice through the gut, and microbes that are reproduced during the first passage may be digested the second time round. The major group of microorganisms in the paunch of the primitive termites is of protozoans, consisting of anaerobic flagellates, such as *Trichomonas termopsidis*, and representing unique genera that are found only in termites and in a closely related species of wood-eating cockroach (*Cryptocercus*). Bacteria are also present, but cannot digest wood. The protozoa engulf particles of wood and ferment the cellulose within their cells, releasing carbon dioxide and hydrogen. The principal products are volatile fatty acids (as in the rumen), but in termites it is primarily acetic acid and this is absorbed through the hindgut.

The bacterial population of the termite gut is less conspicuous than that of the rumen, but appears to play a part in two distinct mutualisms.

1 Spirochaetes are important members of the bacterial flora and they, together

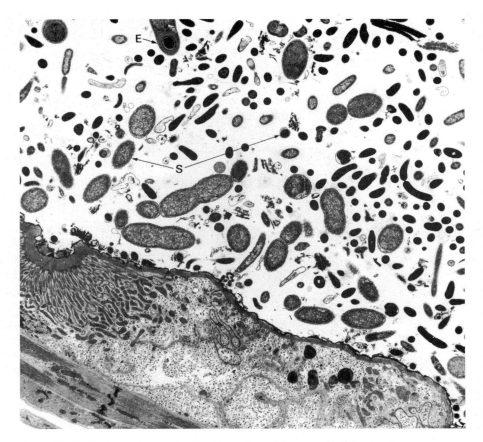

Figure 13.13 Electron micrograph of a thin section of the paunch of the termite *Reticulitermes flavipes.* Much of the flora is composed of aggregates of bacteria. Amongst them can be seen endospore-forming bacteria (E), spirochaetes (S) and protozoa. (After Breznak, 1975.)

with rod-shaped bacteria, tend to be concentrated at the surface of the flagellates. In the guts of one species (*Mastoterma paradoxa*) the spirochaetes have been observed in synchronized movement actually propelling the flagellates. The spirochaetes are so conspicuous that the flagellates were once thought to be ciliates (Figure 13.14). The association of spirochaete and flagellate is mutualistic—the spirochaete receiving nutrient from the protozoan and the protozoan gaining mobility from the spirochaete; so here we have a pair of mutualists living mutualistically within a third species.

nitrogen is fixed in the termite gut

2   Some bacteria in the termite gut are capable of fixing gaseous nitrogen (e.g. *Citrobacter freundii* and *Enterobacter*). This appears to be the only clearly established example of nitrogen-fixing symbionts in insects (Douglas, 1992). Nitrogen fixation stops when antibacterial antibiotics are fed (Breznak, 1975), and the rate of nitrogen fixation falls off sharply if the nitrogen content of the diet is increased.

13.6 Symbiosis and mutualism within animal tissues or cells

There are a great many careful records that establish the presence of symbionts (especially prokaryotes) within the cells of animals (especially insects)—intracellular

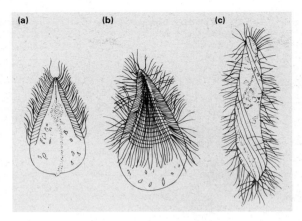

Figure 13.14 Symbiotic protozoa from the intestinal tract of termites. These drawings were made at a time when these organisms were thought to be ciliates. The cilia are now recognized as spirochaetes living in a mutualistic association with their protozoan host. (After Cupp & Kirby.)

symbionts. The majority of these 'guest' microorganisms are present in many or all the cells of a host insect and have no detectable effect on its growth, survival or fecundity: they are neither parasites nor mutualists. In contrast, mycetocyte symbionts are bacteria restricted to a single highly specialized insect cell type (mycetocytes) which may be scattered within a particular tissue or localized in discrete organs (mycetomes). Mycetocyte hosts include cockroaches, sucking lice, beetles, plant-hoppers and aphids but not most other insect groups. There is no doubt that the mycetocyte symbiosis is a mutualism.

In aphids the mycetocyte bacteria may represent 2–5% of the total biomass. If they are eliminated (e.g. by treating with antibiotics), the hosts grow and develop only slowly and die prematurely, often without reproducing. In most cases it has proved impossible to culture the microbial symbionts, and their metabolic activities are not understood. However, most mycetocyte insects feed on nutritionally poor or unbalanced diets and there are very suggestive specializations, for example those Homoptera that suck plant sap (e.g. aphids) have mycetocyte symbionts but those that feed on intact cells do not. The benefit of the bacterial symbionts to aphids appears to be that they synthesize and supply the host with essential amino acids that are lacking in phloem sap. They may also provide essential lipids (Douglas, 1989, 1993).

The bacteria of the aphid symbiosis (*Buchnera aphidicola*) have no free-living phase (i.e. they are obligate symbionts). Their very existence, therefore, depends on the mother host passing them to her offspring. This involves an often complex passage from mycetocyte to developing egg. One result of this transovarial transmission is that the symbiosis is maternally inherited and is just one example of the extremely tight integration of the two partners into each other's lives. A review of the mycetocyte symbiosis in insects is given by Douglas (1989, 1994).

When a mutualism involves such obligate and tight integration between symbiotic partners, it implies that it is the result of a long-established coevolutionary interaction. It has been possible to put a time scale on the evolution of the aphid–bacterial symbiosis by bringing together a phylogeny of the aphids based on their morphology (Heie, 1987) with a bacterial phylogeny based on sequences of 16S recombinant DNA (Moran *et al.*, 1993) (Figure 13.15). The minimum age of the association (and the age of aphids as a whole) is estimated at 160–280 million years.

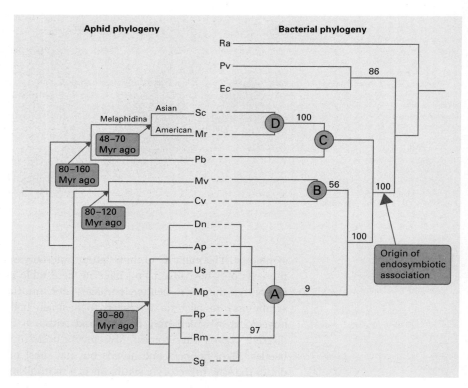

Figure 13.15 The phylogeny of selected aphids and their corresponding primary endosymbionts. Other bacteria are shown for comparison. The aphid phylogeny (after Heie, 1987) is shown on the left and the bacterial phylogeny on the right. Broken lines connect the associated aphids and bacteria. Three species of bacteria that are not endosymbionts are also shown in the phylogeny: Ec, *Escherichia coli*; Pv, *Proteus vulgaris*; Ra, *Ruminobacter amylophilus* (a rumen symbiont). The distances along branches are drawn to be roughly proportionate to time. (After Moran *et al.*, 1993.) Aphid species: Ap, *Acyrthosiphon pisum*; Cv, *Chaitophorus viminalis*; Dn, *Diuraphis noxia*; Mp, *Myzus persicae*; Mr, *Melaphis rhois*; Mv, *Mindarus victoriae*; Pb, *Pemphigus betae*; Rm, *Rhopalosiphum maidis*; Rp, *Rhodalosiphon padi*; Sc, *Schlectendalia chinensis*; Sg, *Schizaphis graminum*; Us, *Uroleucon sonchi*.

13.7 Mutualisms involving higher plants and fungi

A wide variety of symbiotic associations are formed between higher plants and fungi and many of them, but by no means all, have been shown to be truly mutualistic, i.e. bringing 'advantage' to both partners. It is usually easy to isolate species of fungi from plant tissues. At least 200 species of fungi have been isolated from the leaves of healthy winter wheat (Riesen & Sieber, quoted in Clay, 1990)! The majority probably have no more than 'guest' status in the living tissues. Many have very slow growth rates within their living hosts and produce no symptoms on them but they grow actively and often reproduce when the infected tissue dies (Carroll, 1988).

A very remarkable group of Ascomycete fungi, the Clavicipitaceae, grow in the tissues of many species of grass and a few species of sedge. The family includes species that are easily recognized as parasites (e.g. *Claviceps*, the ergot fungus, and *Epichloe*, the Choke disease of grasses), others which are clearly mutualistic and a large number occupying an uncertain 'no-man's land'. The fungal mycelium characteristically grows as sparsely branched filaments running through intercellular spaces along the axis of leaves and stems—but, they are not found in roots. In the genus *Acremonium* the mycelium grows up into the flowers and the infection is transmitted through the seeds. Some species of the tribe *Balansieae* grow only on the surfaces of leaves and inflorescences—really epiphytes not endophytes: they abort the host inflorescences. Many of the symbiotic fungi produce powerful toxic alkaloids that confer some protection from grazing animals (the evidence is reviewed in Clay, 1990) and, perhaps even more important, deter seed predators (Knoch *et al.*, 1993).

A quite different mutualism of fungi with higher plants occurs in roots. Most higher plants do not have roots, they have mycorrhiza—an intimate mutualism of fungus and root tissue. Plants of only a few families like the Cruciferae are the exception and do not form mycorrhiza. Most mosses, ferns, lycopods, gymnosperms and angiosperms contain tissues that are more or less intimately interwoven with fungal mycelium and all the dominant plants of the major vegetation types of the world—forest trees, grasses and heaths—are conspicuous mycorrhiza-formers. (The fossil record of the earliest land plants suggests that they too were heavily infected. These species lacked root hairs, even roots in some cases, and the early colonization of the land may have depended on the presence of the fungi to make the necessary intimate contact between plant and substrate.)

We can recognize distinct forms of mycorrhiza.

13.7.1 Sheathing mycorrhiza

A number of Basidiomycete and Ascomycete fungi form tightly matted sheaths (ectomycorrhiza) on the roots of trees. The association of fungus and host is facultative as many of the fungi involved can be cultured in isolation and their hosts can complete their life cycles in the absence of the fungi. Infected roots are usually concentrated in the litter layer of the soil. The fungal mycelium extends outwards from the sheath through the litter and produces large fruiting bodies which release enormous numbers of wind-borne spores. The fungal mycelium extends inwards from the sheath, penetrating between the cells of the root cortex to give intimate cell to cell contact with the host. The fungus usually induces morphogenetic changes in the host roots which cease to grow apically and remain stubby and are often dichotomously branched (Figure 13.16). Host roots that penetrate into the deeper, less organically rich layers of the soil remain uninfected: they retain normal apical growth and continue to elongate.

mycorrhizal sheathing fungi require soluble carbohydrates

The fungi that form mycorrhizal sheaths require soluble carbohydrates as their carbon resource, and in this respect differ from most of their non-symbiotic, free-living relatives which are cellulose decomposers. The mycorrhizal fungi gain part at least of their carbon resources from the host and absorb mineral nutrients from the soil. There is now no question that the fungus actively supplies mineral resources to the host (see Harley & Smith, 1983). Studies with radioactive tracers have shown that phosphorus, nitrogen and calcium may all move along the fungal hyphae into the host roots and then into the shoot system. Surprisingly, the mycorrhiza appear to function equally effectively without the hyphae that extend out from the sheath. The sheath itself must have some very remarkable properties as an absorber of nutrients and, subsequently, equally remarkable properties as a releaser of them to the host.

Figure 13.16 Mycorrhiza of pine (*Pinus sylvestris*). The swollen, much branched structure is the modified rootlet enveloped in a thick sheath of fungal tissue. (Courtesy of J. Whiting; photograph by S. Barber.)

13.7.2 Vesicular arbuscular mycorrhiza

Vesicular arbuscular (VA) mycorrhiza have been reported from an exceedingly wide range of host plants. They do not form a sheath but penetrate *within* the cells of the host, and do not alter the host's morphology. These features distinguish them sharply from the sheath-formers. The fungi responsible have been assigned to several genera (*Glomus*, *Gigaspora*, *Entrophosphora*, *Scutellispora*) and nearly 150 species have been described (Schenck & Pérez, 1990), based on the morphology of the asexual spores which are very large and produced in small numbers: a striking contrast with the sheath-forming fungi. There are some subtle specificities in the ecology of VA mycorrhiza. After five species of grass has been grown in monoculture on a gradient of soil mixtures the VA species present in the soil differed both between grass species and soil type (Johnson *et al.*, 1992).

No sexual phase has been observed in any of the species of VA mycorrhiza and (like the mycetocyte bacteria in insects) they have defied attempts to culture them outside the host.

Roots become infected from mycelium present in the soil or from germ tubes that develop from the large spores (Figure 13.17). Initially, the fungus grows between host cells but then enters them and forms a finely branched intracellular 'arbuscule'. The main advantage to the host of VA mycorrhiza seems to be that the fungal mycelium can obtain phosphate from the soil at distances greater than can be **phosphate is transported by the fungus** reached by an uninfected root and root hairs. Experiments using tracer phosphorus-32 show that the fungus can transport phosphate over distances of 1–2.7 cm. When phosphate is deficient in the soil, infection with VA mycorrhiza may increase the growth rate and phosphate content of the host (Figure 13.18).

The host plant certainly provides some carbon resources to the fungus (carbon-14 assimilated by the host can be found in the fungal hyphae within 24 h). How much of a cost this represents to the plant is difficult to measure. In onions the volume of fungus present in the roots is only 1–2% of the total root volume. There

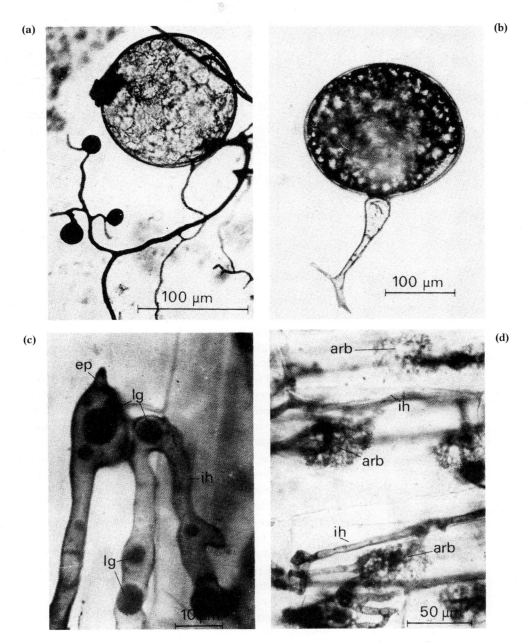

Figure 13.17 Vesicular arbuscular mycorrhiza. (a) A spore of the 'yellow vacuolate endophyte' *Glomus mosseae*. The endophytes have never been cultured successfully and their complete life history is unknown. The various forms are described as species based on the different forms of their spores. (b) A spore of the 'bulbous reticulate' type, *Gigaspora calospora*. (c) The entry point of the fungal filament into the host tissues. (d) A squashed preparation of an onion root showing the internal filaments of the fungus and the arbuscules (arb). (After Tinker, 1975.)

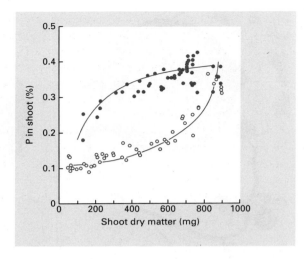

Figure 13.18 The relationship between phosphate concentration in the shoots of leeks and the dry weight of the shoots in plants infected with the mycorrhizal endophyte *Glomus mosseae* (●) and in non-infected plants (○). (After Stribley *et al.*, 1980.)

are, however, intriguing reports that the presence of VA mycorrhiza reduces the growth of the host when phosphate is readily available and this suggests that the fungus makes real demands on the host.

A factorial set of experiments was made on the growth of the annual grass *Vulpia ciliata* ssp. *ambigua* at sites in eastern England where there were large differences in the intensity of natural mycorrhizal infection (West *et al.*, 1993). In one treatment phosphate was applied, and in another the fungicide benomyl was used to control the fungal infection. Fecundity of the grass was scarcely affected by any of the treatments and although the performance of the plants was positively correlated with the extent of mycorrhizal infection at two of the sites, it was unrelated at one site and negatively correlated at another. In further experiments the fecundity of *Vulpia* plants was negatively correlated with the abundance of a group of root pathogenic fungi (*Fusarium oxysporum*, *Embellisia chlamydospora* and a species of *Phoma*), even though the plants developed no other symptoms of infection. The main benefit supplied by the arbuscular mycorrhizal fungi to the plant was not phosphate uptake but protection against pathogenic attack (Newsham *et al.*, 1994).

It is quite possible that what we observe in VA mycorrhiza is infection by specialized parasitic root fungi in which the virulence of the parasite has been largely controlled. The occasionally observed beneficial transfer of phosphate to the host under conditions of deficiency may be a simple consequence of the juxtaposition of fungal tissue between host and soil, with no special coevolved mutualism.

13.7.3 Other mycorrhizal symbioses

Some specialized associations of fungi with host roots occur without the formation of sheaths or vesicular arbuscules. For example, there is a characteristic form of mycorrhiza in the heaths (Ericaceae) and some evidence that nitrogen may be transferred to the host by the fungus. The orchids also have characteristic fungal associations, but the way in which these interact with the host plant is still obscure. The fungi are clearly involved in the germination of orchid seeds, and mycorrhizal fungi are used in commercial practice to ensure good seedling establishment. The seeds of orchids are minute, little more than packets of dried cells almost wholly lacking in food reserves. It seems highly probable that fungal infection provides fixed carbon and other resources to the developing seedlings. What it does afterwards is still a puzzle although various authors have described what looks like digestion of the fungus by the host. (This rather improbable process was even given a splendid name—*thamniscophysalodophagy*—which crudely translates as 'eating the little knobs from the ends of branches'.)

Some orchids (e.g. *Neottia*) and members of other families (e.g. *Monotropa*, Ericaceae) are wholly lacking in chlorophyll. They are heavily infected with mycorrhizal fungi which act as a link transporting materials from the roots of a second, photosynthetic host plant. There is no evidence that such relationships are mutualistic. Indeed, we could regard the plants as true parasites on the fungi (Lewis, 1974). The gametophyte generation in the life cycle of some ferns and lycopods and horsetails (*Equisetum*) also lacks chlorophyll. The prothalli live underground and are heavily infected with mycorrhiza. Again, the plants may be parasitic on the fungi.

13.8 Photosynthetic and chemosynthetic symbionts with invertebrates

Algae are found within the tissues of a variety of animals, particularly coelenterates. An especially detailed study of an animal–algae mutualism has been made on *Hydra viridis*, a species that can easily be cultured in the laboratory. In this animal, algal cells of the genus *Chlorella* are present in large numbers (1.5×10^5 per hydroid) within the digestive cells of the endoderm. Both the animal and the algae can be cultured without their symbiont (they are then called aposymbiotic). When a suspension of algae is injected into the coelenteron of an aposymbiont, some of the cells are ingested but there is a recognition process involved. Free-living *Chlorella* cells are treated like food particles—only cells of *Chlorella* isolated from symbiotic *Hydra* are retained. These are sequestered in individual vacuoles, and migrate to specialized positions at the base of the digestive cells, where they multiply. There must then be some regulating processes that harmonize the growth of the endosymbiont and its host (Douglas & Smith, 1984). If this were not the case, the symbionts would either overgrow and kill the host or fail to keep pace and become diluted as the host multiplied. Some such regulation presumably occurs in all mutualisms, maintaining an equilibrium between the partners.

regulatory processes must harmonize the growth of the endosymbiont and its host

Even when *Hydra* are maintained in darkness and fed daily with organic food, a population of algae is maintained in the cells for at least 6 months and quickly returns to normal within 2 days of exposure to light (Muscatine & Pool, 1979). There is no doubt that, in the light, *Hydra* receives fixed carbon products of photosynthesis from the algae and also 50–100% of its oxygen needs. It can, however, also use organic food. The dual nutrition of the symbiotic *Hydra* is remarkable. They can behave both as autotrophs and as heterotrophs.

an experimental demonstration of an alga–animal mutualism

In freshwater symbioses the algal symbiont is usually *Chlorella* and in marine environments it is usually a dinoflagellate *Symbiodinium* (of the algal family Dinophyceae). The Dinophyceae have a very distinctive morphology and physiology and most of them are nutritional opportunists, capable of photosynthesis but also of using organic foods. In both freshwater and marine environments the symbiotic algae are associated with protists and with sponges, molluscs and flatworms and

mutualisms with dinoflagellates

especially Cnidaria (e.g. hydras in freshwater and sea anemones, some jellyfish, stinging and stony corals and others in the sea).

There are many records of close associations between algae and protozoa (especially Foraminifera and Radiolaria) in the plankton of tropical seas. Some of the associations are intracellular and the alga is tightly integrated into the morphology of its host. For example, in the ciliate *Mesodinum rubrum*, chloroplasts are present which appear to be symbiotic algae. The mutualistic consortium of animals and algae forms dense populations, called blooms, in up-welling water currents and can fix carbon dioxide and take up mineral nutrients. Extraordinarily high production rates have been recorded from such populations (in excess of $2 \, \mathrm{g \, m^{-3} \, h^{-1}}$ of carbon), and these appear to be the highest levels of primary productivity ever recorded for populations of microorganisms. Other, less integrated associations are common, involving radiolarians carrying algal cells within their radiating pseudopods, and protozoa with individual algal cells (often dinophyceae but also cyanobacteria and diatoms) growing on their surface (Figure 13.19).

A very odd situation is found in relationships between algae and some invertebrates, falling somewhere between mutualism, parasitism and predation. In quite separate groups of invertebrates there has developed a system in which algae are digested but their chloroplasts are retained in a photosynthetically active form within the tissues of the animal. The photosynthetic activity is in a sense 'kidnapped' from the algae. A variety of algal taxa (especially members of the Siphonales) can provide chloroplasts used in this way, and it is principally a group of gastropod molluscs that use them (Figure 13.20). The chloroplasts may remain active in the

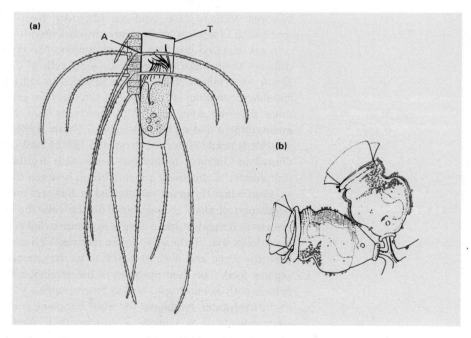

Figure 13.19 Symbiotic associations (presumed mutualisms) from tropical marine microplankton. (a) The filamentous diatom *Chaetoceros tetrastichon* (A) attached to the tintinnid *Eutintinnus pinguis* (T). (b) Two zooids of *Zoothamnium pelagicum* bearing epiphytic cyanobacteria. (After Taylor, 1982.)

508 CHAPTER 13

blooms of ciliates with mutualistic algae

kidnapping of chloroplasts

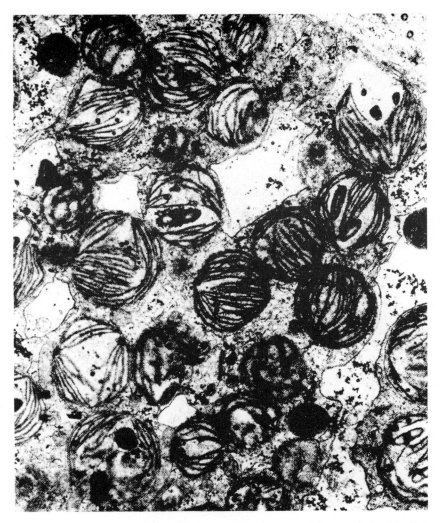

Figure 13.20 Electron micrograph showing the chloroplasts of the alga *Codium fragile* in a digestive cell of the gastropod mollusc *Elysia atroviridis*. (After Trench, 1975.)

animal for more than 2 months and are protected against digestion by the host until they lose activity. Within the host cells they fix carbon and evolve oxygen and the host uses the products.

In Chapter 2 we discussed the remarkable marine environments formed by hydrothermal vents and characterized by high temperatures and sulphide concentrations and with a diverse fauna of pogonophorans, polychaete worms, crabs and very large molluscs. Many of these have symbiotic chemosynthetic bacteria which are found in the gills of the molluscs and in the 'trunk' of the worm-like Pogonophorans. It seems probable that these peculiar symbioses are mutualisms in which the invertebrates benefit from carbon compounds fixed in a chemosynthesis powered by the oxidation of sulphides. However, the technical problems of establishing whether or not this symbiosis is a mutualism are immense in this curious environment.

13.9 The mutualism of fungus and algae: the lichens

A number of fungi have escaped from their normal life habit into a mutualistic association with algae. These 'lichenized' fungi include within their body a thin layer of algal cells near the surface (Figure 13.21). The algae form only 3–10% by weight of the thallus body. Of the 70 000 or so species of fungus that are known, approximately 25% are lichenized. Fungi that have become lichenized belong to diverse taxonomic groups and the mutualistic algae to 27 different genera. Presumably the lichen habit has evolved many times. Fungi that are not lichenized occupy a restricted range of habitats, either as parasites of plants or animals, or as decomposers of dead materials. Lichenization extends the ecological range of both fungus and alga enormously, both onto substrates (rock surfaces, tree trunks) and

the 'lichen habit' seems to have evolved many times

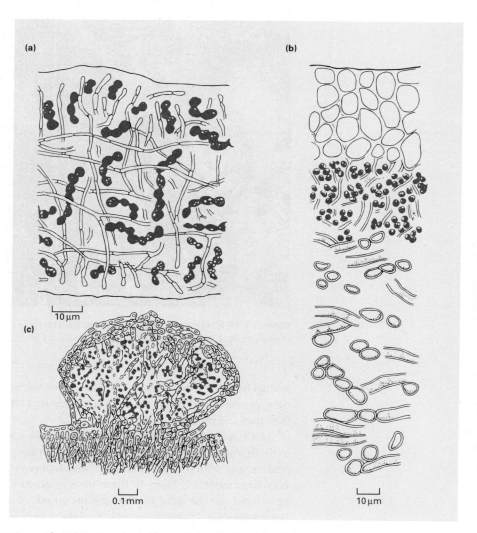

Figure 13.21 The structure of some lichen bodies. (a) A simple two-membered lichen, for example *Collema*, with algal cells distributed throughout the fungal body. (b) A more structured system in which the algae are confined to a central layer. (c) A section through a cephalodium, a specialized part of the thallus containing cyanobacteria.

into arid deserts and arctic and alpine regions that are barred to many other life forms.

The fungi gain photosynthates (and in a few cases fixed nitrogen) from their algal symbiont. The advantage to the algae, if any, has not been established so clearly. The fact that isolated algal cells are commonly found exposed on the surface of the lichen suggests that in some environments they do not need the protection of the fungal thallus. It may be that the algae are 'captured' by the fungus and exploited without any recompense. However, some of the species (e.g. of the genus *Trebouxia*) are rare in a free-living form, which suggests that there is something special about life in a fungal thallus that they need for comfort!

lichen mutualisms produce 'new species'

The most remarkable feature in the life of the lichenized fungi is that the growth form of the fungus is profoundly changed when the alga is present. When the fungi are cultured in isolation from the algae they grow slowly in compact colonies, much like related free-living fungi; but, in the presence of the algal symbionts they take on a variety of morphologies that are characteristic of specific algae—fungus partnerships. In fact, the alga stimulate morphological responses in the fungus that are so precise that the lichens are classified as distinct species, and different species of alga elicit quite different morphologies from the same fungus.

lichens are slow growers ...

All lichens are slow growing; the colonizers of rock surfaces rarely extend faster than 1–5 mm per year. They are very efficient accumulators of mineral cations, which they presumably extract from rainwater and from the flow and drip down the branches of trees. Their ability to accumulate minerals makes them particularly sensitive to environmental contamination from pollution by heavy metals and fluoride, and for this reason they are amongst the most sensitive indicators of environmental pollution. The 'quality' of an environment in humid regions can be judged rather accurately from the presence or absence of lichen growth on tombstones and tree trunks.

... and sensitive indicators of environmental pollution

Ninety per cent of the lichenized fungi are associated with green algae and these cannot fix atmospheric nitrogen. The remaining 10% contain cyanobacteria of the genera *Nostoc*, *Scytonema*, *Stigonema*, *Dichothrix* and *Calothrix*, which are nitrogen fixers. Surprisingly, these lichens are not especially characteristic of nitrogen-deficient habitats, although the algae certainly fix nitrogen within the host and leak substantial quantities to it.

13.10 The fixation of atmospheric nitrogen in mutualisms

nitrogen fixation is of crucial importance because nitrogen is often in limiting supply

The inability of most plants and animals to fix atmospheric nitrogen (conventionally referred to as dinitrogen, N_2) is one of the great puzzles in the process of evolution. The ability to fix nitrogen is widely but irregularly distributed amongst both bacteria and archaebacteria. Many of these have been caught up in tight mutualisms with systematically quite different groups of eukaryotes. Presumably such symbioses have evolved a number of times independently. They are of enormous ecological importance because nitrogen is in crucial limiting supply in many habitats. Gibson and Jordan (1983), Sprent (1987) and Sprent and Sprent (1990) review the ecology of nitrogen-fixing systems.

The nitrogen-fixing bacteria (excluding cyanobacteria) that have been found in symbiosis (not necessarily mutualistic) are members of the following taxa.

1 Azotobacteriaceae, which can fix nitrogen aerobically and are commonly found on leaf and root surfaces, for instance *Azotobacter*, *Azotococcus*.

2 Rhizobia, which fix nitrogen in the root nodules of most leguminous plants and just one non-legume *Parasponia* (a member of the family Ulmaceae, the elms). Three genera are recognized: *Rhizobium*, *Bradirhizobium* and *Azorhizobium* which are so distinct that they should perhaps be in different families (Sprent & Sprent, 1990). Rhizobia have been persuaded to fix nitrogen in anaerobic laboratory culture.

3 Bacillaceae, such as *Clostridium* spp., which occur in ruminant faeces, and *Desulfotomaculum* spp., which fix nitrogen weakly in mammalian guts.

4 Enterobacteriaceae, such as *Enterobacter* and *Citrobacter*, which occur in intestinal floras (e.g. of termites), although occasionally on leaf surfaces and on root nodules.

5 Spirillaceae, such as *Spirillum lipiferum*, which is an obligate aerobe found on grass roots.

6 Actinomycetes of the genus *Frankia*, which fix nitrogen in the nodules (actino-rhiza) of a number of non-leguminous and mainly woody plants, such as the alder (*Alnus*) and sweet gale (*Myrica*).

7 Cyanobacteria of the family Nostocaceae, which are found in association with a remarkable range of flowering and non-flowering plants (see Section 13.10.2).

Of all these symbioses, that of the rhizobia with legumes is the most thoroughly studied, because of the huge agricultural importance of legume crops.

13.10.1 The mutualism of rhizobia and leguminous plants

The establishment of the liaison between rhizobia and legume plants proceeds by a series of reciprocating steps. The bacteria occur in a free-living state in the soil and are stimulated to multiply by root exudates and cells that have been sloughed from roots as they develop. These exudates are also responsible for switching on a complex set of genes in the rhizobia (*nod* genes) that control the processs that induce nodulation in the roots of the host. In a typical case a bacterial colony develops on the root hair which then begins to curl, and the cell is penetrated by the bacteria. The host responds by laying down a wall that encloses the multiplying bacteria and forms an infection thread that is part host, part bacteria. This may then grow from cell to cell of the host root cortex, and in advance of it the host cells start to divide, beginning to form a nodule. As the host cells become colonized the bacteria change their form to become non-dividing swollen bacteroids.

A special vascular system develops in the host and supplies the products of photosynthesis to the nodule tissue and carries away fixed-nitrogen compounds (mainly asparagine) to other parts of the plant (Figure 13.22). The nitrogenase enzyme accounts for up to 40% of the protein in the nodules and depends for its activity on a very low oxygen tension (it is inactivated at oxygen concentrations of 1–$10 \, \mu\text{mol l}^{-1}$). A boundary layer of tightly packed cells within the nodule serves as a barrier to oxygen diffusion. Haemoglobin is formed within the nodules (leghaemo-globin) giving the active nodules a pink colour. It has a high affinity for oxygen and allows the symbiotic bacteria to respire aerobically in the virtually anaerobic environment of the nodule.

Wherever nitrogen-fixing symbioses occur, at least one of the partners has special structural (and usually also biochemical) properties that protect the obligately anaerobic nitrogenase enzyme from oxygen, yet allow normal aerobic respiration to occur around it.

The costs and benefits of this mutualism are not entirely clear. We need to compare the energetic costs of alternative processes by which a plant might obtain its

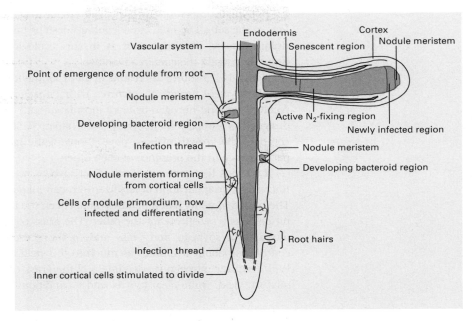

Figure 13.22 The development of the root nodule structure during the course of development of infection of a legume root by *Rhizobium*. (After Sprent, 1979.)

the energy costs of nitrogen fixation

supplies of fixed nitrogen. The route for most plants is direct from the soil as nitrate or ammonium ions. The metabolically cheapest route is the use of ammonium ions, which do not require to be reduced before they are incorporated into proteins, etc. But, in most soils ammonium ions are rapidly converted to nitrates by microbial activity (nitrification) and it is mainly in anaerobic and acid soils that ammonium ions are readily available.

The costs of the mutualistic process (including the maintenance costs of the bacteroids) are about 13.5 mol of adenosine triphosphate (ATP) consumed per mole of ammonia formed. This is energetically slightly more expensive than the 12 mol of ATP involved in reducing nitrate to ammonia. To the cost of nitrogen fixation must be added the costs of forming and maintaining the nodules themselves. This cost may be about 12% of the plant's total photosynthetic output. It is this enormous cost of nodule construction and maintenance that makes nitrogen fixation energetically inefficient. But, for green plants energy may be a very cheap currency (they may be 'pathological overproducers of carbohydrate'; see Chapter 3). A rare and valuable commodity bought with a cheap currency may be no bad bargain.

Nitrogen fixation declines rapidly when a nodulated legume is provided with nitrates, and it is in environments where fixed nitrogen is a rare commodity that the pay-off from a nitrogen-fixing mutualism is likely to be greatest. Sprent and Sprent (1990) gives an overview of the process of both free-living and symbiotic nitrogen fixation especially of the biochemistry of the process and the phylogeny of the organisms involved.

nitrogen-fixing mutualisms in their ecological context

The mutualism of rhizobia and legumes must not be seen just as an isolated interaction of bacteria with a host plant. The ecological status of the legumes themselves is profoundly changed when they carry nitrogen-fixing symbionts. In nature, legumes normally form mixed stands in association with non-legumes. These are potential competitors with the legumes for fixed nitrogen (nitrates or ammonium ions in the soil). The nodulated legume sidesteps this competition by its access to its unique source of nitrogen. It is in this ecological context that nitrogen-fixing mutualisms gain their main advantage.

Figure 13.23 shows the results of an experiment in which soybeans (*Glycine soja*, a legume) were grown in mixtures with *Paspalum*, a grass. The mixtures either received mineral nitrogen or were inoculated with *Rhizobium*. The experiment was designed as a 'replacement series' (see Chapter 7, Section 7.7.2), which allows us to compare the growth of pure populations of the grass and the legume with their performance in the presence of each other.

If we look first at the pure stands of soybean we see that *either* inoculation with *Rhizobium*, or application of fertilizer nitrogen increased its yield by at least 10-fold. The two treatments had much the same effect: the legumes can use either source of nitrogen as a substitute for the other. The grass responded only to the fertilizer. In mixtures of soybean and grass and in the presence of *Rhizobium*, the legume contributed about five times as much to the yield of the mixture as did the grass. With fertilizer nitrogen the legume and the grass contributed about equally to the mixture yield. Quite clearly, it is under conditions of resource deficiency that the

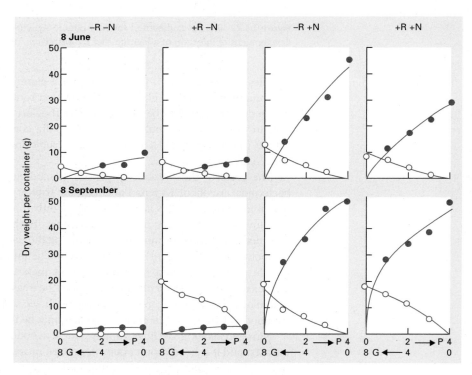

Figure 13.23 The growth of soybeans (*Glycine soja*, G) and a grass (*Paspalum*, P) grown alone and in mixtures with and without nitrogen fertilizer and with and without inoculation with nitrogen-fixing *Rhizobium*. The plants were grown in pots containing zero to four plants of the grass and zero to eight plants of *Glycine*. The horizontal scale on each figure shows the proportions of plants of the two species in each container. −R −N, no *Rhizobium*, no fertilizer; +R −N, inoculated with *Rhizobium* but no fertilizer; −R +N, no *Rhizobium* but nitrate fertilizer was applied; +R +N, inoculated with *Rhizobium* and nitrate fertilizer supplied. (After de Wit *et al.*, 1966.)

pay-offs from the mutualism are strongest—the creators of deficiency (the depleters) will be the species (e.g. *Paspalum*) that lack the mutualism.

In environments deficient in nitrogen, nodulated legumes have a great advantage over other species. But, as their activity raises the level of fixed nitrogen in the environment the advantage is lost—they have improved the environment of their competitors. In this way, organisms that can fix atmospheric nitrogen can be thought of as locally suicidal. This is one reason why it is very difficult to grow repeated crops of pure legumes in agricultural practice without aggressive grass weeds invading the nitrogen-enriched environment. It may also explain why leguminous herbs or trees usually fail to form dominant stands in nature.

Legume tissues usually have a high nitrogen content and in mixed clover/grass pastures the legumes make a large contribution to the protein intake of grazing herbivores. Moreover, after death, legumes augment the level of soil nitrogen on a very local scale with a 6–12 month delay as they decompose. A major consequence of this is that the growth of associated grasses will be favoured in these local patches. The grazing animal continually removes the grass foliage and the nitrogen status of the patches again declines to a stage at which the legume may once more be at a competitive advantage. In a stoloniferous legume, such as white clover, the plant is continually 'wandering' through the sward, leaving behind it local grass-dominated patches, whilst invading and enriching with nitrogen new patches where the nitrogen status has become low. (The grazing animal itself adds to the process by which patchiness is generated. It grazes over extensive areas of a sward and its dung and urine concentrate excreted nitrogen in local patches.) The symbiotic legume in such a community not only drives its nitrogen economy but also some of the cycles that occur within its patchwork (Cain *et al.*, 1995).

13.10.2 Nitrogen fixation in mutualisms with plants that are not legumes

Some other organisms that can fix dinitrogen are involved in symbioses with non-legumes. Their distribution in higher plants is very odd, and does not seem to make much evolutionary sense. Cyanobacteria form symbioses with three genera of liverwort (*Anthoceros*, *Blasia* and *Clavicularia*), with one fern (the free-floating aquatic *Azolla*), with many cycads (e.g. *Encephalartos*) and with all 40 species of the flowering plant genus *Gunnera* but with no other flowering plants. In the liverworts the cyanobacteria (*Nostoc*) live in mucilaginous cavities in the thallus and the plant reacts to their presence by developing fine filaments that maximize contact with it. *Nostoc* is found in specialized cavities at the base of the leaves of *Gunnera*, in the lateral roots of many cycads and in pouches in the leaves of the water fern *Azolla*. Cyanobacteria of the genus *Richelia* occur as symbionts under the cell wall of some marine diatoms.

Frankia is a genus of actinomycetes which form symbioses (actinorhiza) with members of at least eight families of flowering plants, almost all of which are shrubs or trees. The nodules are usually hard and woody. The best known hosts are the alder (*Alnus*), sea buckthorn (*Hippophaë*), sweet gale (*Myrica*), she-oak (*Casuarina*) and the Arctic–alpine subshrubs *Arctostaphylos* and *Dryas*. Plants of *Ceonothus* which form extensive stands in Californian chaparral also develop *Frankia* nodules.

Unlike rhizobia, the species of *Frankia* are filamentous and produce specialized vesicles and sporangia which release spores. The walls of the vesicles are massively thickened with as many as 50 monolayers of lipids which protect the nitrogenase from inactivation by oxygen. Thus, whilst the rhizobia rely on their host plant to protect their nitrogenase from oxygen, *Frankia* provides its own protection (many cyanobacteria such as *Nostoc* have somewhat similarly protected cells called heterocysts in which nitrogen fixation occurs).

13.10.3 Nitrogen-fixing mutualists in vegetational succession

A shortage of fixed nitrogen commonly hinders the earliest stages of the colonization of land by vegetation. Some fixed nitrogen will be contributed in rain after thunderstorms and some may be blown in as particles from other more established areas, but nitrogen-fixing organisms such as bacteria, cyanobacteria and lichens are important pioneers. Surprisingly, however, higher plants with nitrogen-fixing symbionts are rarely pioneers in vegetational succession. It is not obvious why this should be so. Of course, a seedling legume depends on nitrogen in its seed reserves and external sources of fixed nitrogen before it can grow to a stage at which it can nodulate and start to fix nitrogen for itself. It may be that only large seeded legumes carry enough fixed nitrogen in their seed reserves to carry them through the establishment phase and that species with such large seeds will not have the dispersibility needed to be pioneer colonizers (Grubb, 1986; see also the discussion of this problem in Sprent & Sprent, 1990).

the role of legumes in vegetational successions

All the examples of mutualism with *Frankia* allow the host to enter habitats that are nitrogen deficient, for example the peaty soils of *Alnus* and *Myrica*, the sand-dunes of *Hippophaë* and the thin nutrient-impoverished habitats of *Dryas* and *Arctostaphylos*. Here again the mutualists usually contribute to secondary or tertiary phases in vegetational succession rather than being true pioneers, only *Dryas* seems able to be a true pioneer.

Symbiotic nitrogen fixation is energetically demanding and it is not surprising that most of the species of higher plant that support nitrogen-fixing mutualists (rhizobia or *Frankia*) are intolerant of shade. This must be one of the features that usually excludes nitrogen-fixers from late stages of successions: they are seldom in at the beginning and seldom persist to the end.

13.11 The evolution of subcellular structures from symbioses

We have seen that there is remarkable variety in the types of association that may be regarded as symbioses, many of them clearly shown to be mutualisms, in both plant and animal kingdoms. They extend from patterns of behaviour that link two very different organisms that spend parts of their lives apart and retain their individual identities (gobiid fish with shrimps, large blue butterflies with ants), through the chemostat communities (strictly external to the body tissues) of the rumen and the caecum, to the intercellular sheathing mycorrhiza and the intracellular zooxanthellae of coelenterates and mycetocyte bacteria of insects. The stages can be read as a sequence from the integration of separate parts of a community to integration within what are, in most respects, the parts of a single 'organism'.

It is now generally accepted that the evolutionary process has in some cases progressed to the inextricable merging of the partners in a symbiosis. This view was developed particularly by Margulis (1975). She argued that critical steps in the evolution of the major groups of higher plants and animals have involved the incorporation of prokaryotes. This is the 'serial endosymbiosis theory of the evolution of the eukaryote cell'.

mutualism may have led to the evolution of the eukaryotes

In this theory the first step in the origin of eukaryotes from prokaryotes occurred when a fermentative anaerobe was captured by (or invaded) Krebs cycle-containing eubacteria (promitochondria) giving the mitochondria containing amoeboids from which all other eukaryotes were derived. These in turn acquired motile surface bacteria giving flagellates and ciliates. (We have already seen that the flagellates in the termite gut acquire motility from spirochaetes that become attached to them.) At a later stage the ingestion of cyanobacteria provides chloroplasts for the autotrophic 'organism' from which the plant kingdom might evolve. A review of such evolutionary theory is given by Douglas (1994). She points out that some elements in Margulis' original view have since been modified (for instance it is now clear that eukaryotes had evolved before mitochondria were acquired), but the symbiotic origin of mitochondria and plastids is still seen to lie at the heart of the evolutionary game. In this game intracellular symbionts (e.g. eubacteria) which have their own DNA, merge their identity within that of their host by transfer of some genes to a host nucleus. It has then become a host organelle and ceased to be a symbiont. Such evolution seems to be the mechanism by which many organisms, in their evolutionary history, have gained access to complex metabolic capabilities.

13.12 Models of mutualisms

A classic pair of simple equations defines the essence of competition between two species: members of two species (or populations) each suffer from association with the other, expressed in the code $--$ (p. 212). In Chapter 7, Section 7.4.1 we discussed the Lotka–Volterra equation which describes the population dynamics of two competing species and the terms $-\alpha_{12}N_2$ and $-\alpha_{21}N_1$ represent the negative effect of each species on the population of the other:

$$\frac{dN_1}{dt} = \frac{r_1 N_1 (K_1 - N_1 - \alpha_{12} N_2)}{K_1} \tag{13.1}$$

and for the second species:

$$\frac{dN_2}{dt} = \frac{r_2 N_2 (K_2 - N_2 - \alpha_{21} N_1)}{K_2} \tag{13.2}$$

At first sight we might imagine that an appropriate model for a mutualistic interaction would simply replace the negative contribution from the associated species with a positive contribution so that the presence of each had a positive effect on the growth of the other:

$$\frac{dN_1}{dt} = \frac{r_1 N_1 (K_1 - N_1 + \alpha_{12} N_2)}{K_2} \tag{13.3}$$

and for the second species:

$$\frac{dN_2}{dt} = \frac{r_2 N_2 (K_2 - N_2 + \alpha_{21} N_1)}{K_2} \tag{13.4}$$

However, this model leads to absurd solutions in which both populations explode to unlimited size (May, 1981b) because it places no limits on the carrying capacity of either species which would therefore increase indefinitely! In practice, intraspecific competition for limiting resources must eventually determine a maximum carrying capacity for any mutualist population, even if the population of the partner mutualist is present in excess (Dean, 1983). As an example, consider the shrimp–goby mutualism (see Section 13.2.4). The shrimp population alone is apparently limited by predation, and mutualism with the goby allows it to reduce this risk and enhance its survivorship. But, the resulting increase in the shrimp population must sooner or later expose it to some other density dependent constraint (e.g. food shortage, a limited number of refuges, parasitic infection) that prevents its population from expanding without check.

Similarly, a plant whose growth is limited by a shortage of fixed nitrogen may be released into faster growth by mutualism with a nitrogen-fixing partner, but its faster growth must soon become constrained by shortage of some other limiting resource (e.g. water, phosphate, radiant energy).

An interesting approach to the modelling of mutualisms comes from recognizing that some of them have properties more usually associated with predator–prey or parasite–host interactions (Wright, 1989). This is especially true in mutualisms in which one partner is a forager and spends its time partitioned between searching for items and handling them when they are found. Examples are pollinating insects, cleaner fish and ants tending aphids. In such situations the rate of foraging typically levels off as the object of the foraging becomes more abundant. This is the essence of the type 2 functional response in Holling's models of predator–prey interactions which we discussed in Chapter 9, Section 9.5.

A model for the population dynamics of a mutualist that incorporates a type 2 functional response is:

$$\frac{dN}{dt} = rN\left(1 - \frac{N}{K}\right) + \frac{fa'M}{1 + a'T_h M} \tag{13.5}$$

or, letting:

$$\chi = \frac{1}{a'T_h} \quad \text{and} \quad \beta = \frac{f}{T_h} \tag{13.6}$$

$$\frac{dN}{dt} = rN\left(1 - \frac{N}{K}\right) + \frac{\beta M}{(\chi + M)} \tag{13.7}$$

where N and M are the densities of the mutualists, K is the carrying capacity of mutualist N, r is the intrinsic rate for increase of N, a' is the attack rate of N on M, T_h is the handling time and f is a coefficient converting encounters with M into new units of N.

Isoclines for such mutualistic interactions are shown in Figure 13.24. Figure 13.24a shows the isoclines for the situation in which both partners to the mutualism are facultative and are represented by equations of the same form; r is positive. There is then always a stable non-negative equilibrium when the parameters are biologically realistic.

Obligate mutualisms are represented in the equation by negative values of r (N declines in the absence of M). In a mutualism in which both species are obligate mutualists, both species become extinct (Figure 13.24b) or there is both a stable equilibrium and also an unstable equilibrium in which both mutualists become extinct if either falls below some critical density (Figure 13.24c).

Experimental evidence that supports the type 2 functional response model comes from studies of the mutualism between ants and the aphids that they tend. The effect

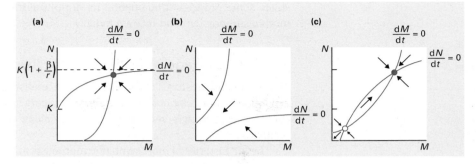

Figure 13.24 Model isoclines for mutualistic interactions in which both mutualists have a positive effect on their partners at low densities, but the effects are density dependent and become saturated at high densities; further increases in the mutualist populations have little effect. (a) A model for mutualists that are both facultative, giving one stable equilibrium point. (b, c) These represent the model for obligate mutualists. In (b) neither species can reach a population size high enough for a stable mutualism to form, and in (c) there is a critical population density above which the mutualists form a stable equilibrium but also a low density below which both species move to extinction. (After Wright, 1989.)

of tending appears to be lost when the aphids are most abundant (Addicott, 1979), i.e. the advantage of the mutualism is inversely density dependent. This interpretation was strongly confirmed in a series of experiments in which the densities of the aphid *Aphis varians* were varied in the presence of the ant *Formica cinerea*. The per capita growth rate of the aphids was increased by the presence of the ants at low aphid densities but their influence was less or disappeared entirely at high densities (Breton & Addicott, 1992).

There remains a big question whether any single mathematical model can capture the essence of all mutualisms (Vandermeer & Boucher, 1978). The examples given in this chapter represent a very diverse array. For example, they include mutualisms that are obligate (e.g. mycetocyte bacteria in aphids) and others that are facultative (seed dispersal from fleshy fruits by birds), but also mutualisms that are obligate in some habitats but facultative in others (the legume–*Rhizobium* mutualism is apparently obligate in some soils although facultative in most (Nutman, 1963)).

Even more important (and daunting!) is that it may not be sensible to bring just two mutualists into the model. The benefits of mutualism, in many of the examples, depend on the presence of one or more other species. For example, the legume–*Rhizobium* mutualism brings its great advantage to the legume when it is competing with some other plant (e.g. a grass) for limited nitrogen from the soil (see Figure 13.23). The ant–acacia mutualism brings advantage to the acacia if it is in competition with other shrubs (the ants prune the competitor), and the interaction is yet more complex because the ants also protect the acacia from herbivores. The dynamics of at least four species (probably more) would need to be built into a model that captured the essence of this situation.

should models of mutualism involve more than two species?

Realistic models of mutualisms may, then, need to involve the dynamics of three or more species and attempts to do this (Rai *et al.*, 1983; Addicott & Freedman, 1984) suggest that the presence of a third species (predator or competitor) may stabilize some mutualisms. However, it still seems unlikely that any single model will encapsulate some property fundamental to all mutualisms. Sadly, each different class of mutualism may demand its own model.

13.13 Some general features of the lives of mutualists

There are many features of the biology of mutualists (particularly those that involve a close symbiotic relationship) that set them apart from most other organisms. They contrast very strongly with aspects of the biology of parasites and with free-living relatives.

1   The life histories of most symbiotic mutualists are remarkably simple (contrasting particularly with the life histories of most parasites).

2   Sexuality appears to be suppressed in endosymbiotic mutualists, especially in comparison with parasites and with free-living close relatives (Law & Lewis, 1983).

3   There is no conspicuous dispersal phase in endosymbionts. Spores from the fruiting bodies of the sheath-forming mycorrhizal fungi are exceptions to this general rule, but these fungi may spend most of their lives in a free-living condition (or as parasites). This contrast between mutualists and parasites is particularly strong: dispersal rules in the population dynamics of most parasites.

4   It might be expected that coevolution of a mutualism would lead to mechanisms that disperse the two partners together. This happens when:

    (a) young queen ants or scolytid beetles take fungal inoculum with them to found a new colony;

    (b) fungus and algae are combined in the dispersal unit of many lichens;

    (c) mycetocyte bacteria are transmitted to the egg in mycetocyte bearing insects (see Section 13.6);

    (d) fungi such as the symbiotic (perhaps mutualistic Balansieae; see Section 13.7) invade and disperse in the seeds of grasses.

However, it is very striking that the root-infecting mycorrhizal fungi, the nitrogen-fixing and nodule-forming rhizobia and *Frankia* and mutualistic cyanobacteria are not dispersed with the seeds of their host. Like the mycorrhizal symbionts of orchids, each generation of seedlings needs to be newly reinfected. Of course, the integration of one partner in the morphology and genetics of the other in the early evolution of eukaryotes (see Section 13.11) was an ideal insurance that they would for ever be dispersed together.

5   There seems to be nothing in the life of mutualistic organisms comparable to epidemics amongst parasites. Populations of mutualists seem to have great stability when compared with those of parasites.

6   The numbers of endosymbionts per host seem to be remarkably constant and their population dynamics must have elements of density dependence.

7   The ecological range (and niche breadth) of organisms in mutualistic symbioses appears usually (perhaps always) to be greater than that of either species when living alone. This again contrasts with parasitic symbioses where the host's ecological range is probably usually reduced by the presence of parasites.

8   Surprisingly, species specificity of both partners in mutualisms is often quite flexible—ants and nectaries, algae and fungi in lichens, plants and pollinators, etc., often involve pairs of species that can live mutualistically with several, sometimes many, other partners.

There is no doubt that mutualism and other forms of symbiosis (even commensalism) have been seriously neglected by ecologists—even more so than parasitisms.

Too much attention may have been concentrated on the ecology of competition and the interactions of predators and prey. It is also clear that in the study of symbioses a great deal of energy has been put into detecting which symbioses are mutualisms. In many symbioses, however, the benefits are probably purely one sided (epiphytes on trees, encrusting animals on seaweeds). They are just as interesting and ecologically important as mutualisms.

Mutualisms lend themselves easily to the story-telling approach to the explanation of nature, in which everything is seen as a piece of evolved perfection. A more jaundiced (but, perhaps equally justified view) is that most so-called mutualisms are examples of the capture, slavery and exploitation of one species by another. The battery farming of poultry by humans may be the most appropriate analogy to the enslavement of rhizobia by legumes, of mycetobacteria by insects and even of mitochondria and plastids by eukaryotes.

Much of the literature concerning symbiosis is a stronghold of anecdotes—the structure of this chapter largely mirrors this truth. But, the study of mutualisms attacks one of the must fundamental problems in ecology. Do whole communities of organisms in nature represent more or less tightly coevolved relationships? Do 'holist' properties emerge from evolutionary interactions? Many of the examples in this chapter seem to support the view that there is a tendency for groups of two or more species to become tied together in associations that may sometimes be loose but may sometimes integrate into systems within systems—superorganisms. But, it is just as striking that the evolutionary process has generated so many examples of highly specialized harmful interactions. Without doubt it is the evolution of parasites, parasitoids and predators that has contributed most to the diversity of the world's flora and fauna, although it may be mutualists that have contributed most to its mass.

PART 3
THREE OVERVIEWS

Introduction

This section has three chapters. All deal with topics that could, in principle, have been discussed earlier in the book; but in all cases, delay has brought with it advantages. In Chapter 14 we survey, and seek to understand, the range of life-history patterns or 'life-history strategies' exhibited by living organisms. In Chapter 15 we survey, and seek to understand, the range of patterns of abundance exhibited by living organisms. In Chapter 16 we examine the ways in which our understanding of abundance has been utilized (or ignored!) in the applied contexts of pest control and the exploitation of natural populations. Hence, the chapters are, in a sense, overviews of the two levels of the ecological hierarchy that we have covered so far: Chapter 14 deals with individual organisms, and Chapters 15 and 16 with populations.

The chapters have been placed at this point in the book because we will need to draw on information from the preceding chapters in order to make sense of the topics addressed in the present section. In Chapter 14 we demonstrate that the life history of an organism can often be seen to reflect the environment in which it lives—yet this environment comprises, to a very large extent, the many other organisms with which it interacts. Likewise, the abundance of an organism is, essentially, a reflection of the combined effects of the many interactions in which it takes part. Thus, if we wish to disentangle these effects, we must first develop an understanding of them individually. Hence, Parts 1 and 2 of the book can be seen as a necessary preparation for Chapters 14, 15 and 16.

Notwithstanding this argument, it may seem regrettable that the present section interrupts the apparently logical flow from individual to population to interaction to community. But an alternative point of view would be to find it remarkable that this is the *only* interruption in the sequence. Every aspect of ecology can be fully understood only by reference to every other aspect—above it, below it and beside it in the hierarchy. To take just one example: we dealt with the responses of individuals to conditions very early in the book, and these responses set the scene for the interactions of the organisms with other organisms, and for the construction of the communities of which they are part. But the response of an organism to environmental conditions is itself crucially influenced by the presence of other, interacting organisms, and by the physical and biological structure of the surrounding community. Thus the 'logical' structure of the book is artificial. It has been imposed *by us*, for convenience, on a science that would more properly be described as a multidimensional array of topics with arrows of influence flowing backwards and forwards in every direction. In the present section, we collect together some of the arrows that have been passsing in the opposite direction to the one we have been following up the hierarchy.

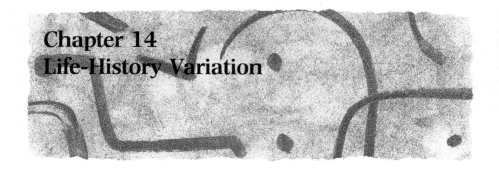

Chapter 14
Life-History Variation

14.1 Introduction

what is a life history?

Most natural history museums present us with a collection of mature adult organisms, as if the variety of nature could be properly seen in this one selected moment from the life of each species. But there is no such characteristic moment. Every organism can be truly represented only by its whole life history. An organism's life history is its lifetime pattern of growth, differentiation, storage and, especially, reproduction. But how does it come about—why is it—that orchids produce vast numbers of tiny seeds when tropical *Mora* trees produce just a few enormous ones? Why is it that swifts usually produce clutches of three eggs, when other birds produce larger clutches and the swifts themselves are physiologically capable of doing so? And why is it that the ratio of age-at-maturity : average life span is often roughly constant within a group of organisms but markedly different between groups (e.g. mammals 1.3, fish 0.45)? Trying to understand the similarities and differences in life histories is one of the fundamental challenges of modern ecology (see Southwood, 1988; Caswell, 1989; Lessells, 1991; Smith, 1991; Stearns, 1992, for further reviews).

life-history patterns: three types of question

In fact, there are at least three different, although related, types of question that are commonly asked about life histories.

1   The first is exemplified by our question about swift clutch size and is concerned with individual life-history traits: 'Can we establish that *this* clutch size is ultimately the most productive, i.e. the fittest in evolutionary terms, and what is it about this particular clutch size that makes it so?'

2   The second is exemplified by our question about maturity : life-span ratios and is concerned with links between life-history traits: 'Are age-at-maturity and average life span linked? What is the basis for the link within a group of related organisms? What is the basis for differences amongst groups?'

3   The final type is exemplified by our question about orchids and *Mora* trees and is concerned with links between life histories and habitats: 'Can the seed-size differences between orchids and *Mora* trees be related directly to differences in the habitats that they occupy, or to any other differences between them?'

In short, the study of life histories is a search for patterns—and for explanations for those patterns. Since we believe all organisms have evolved through natural selection, the explanations we seek are evolutionary ones: 'Why has natural selection favoured this particular pattern, this particular trait or this particular combination of traits?'

The search for pattern itself can be seen as an attempt by ecologists to construct their equivalent of the chemists' Periodic Table. For the 18th century chemist '... each fact had to be discovered by itself and each had to be remembered in isolation' (Southwood, 1977). A similar fate awaits ecologists unless we can discern, in outline at least, some patterns in the ways that organisms live out their lives.

We must remember, however, that every life history, and every habitat, is unique. In order to find ways in which life histories might be grouped, classified and compared, therefore, we must find ways of describing them that apply to all life histories and all habitats. Only then can we search for associations between one life-history trait and another or between life-history traits and features of the habitats in which the life histories are found.

life-history patterns reflect rules imposed within constraints

We will try to discern a set of rules describing when particular traits might be favoured. It is important to realize, however, that these rules work within constraints: the possession of one life-history trait may limit the possible range of some other trait, and the general morphology of an organism may limit the possible range of all its life-history traits. The most that natural selection can do is to favour, in a particular environment, with its many, various and often conflicting demands, the life history that has been most (not 'perfectly') successful, overall, at leaving descendants in the past.

life histories reflect the genotype, the environment and the interaction between the two

Finally, remember that the life history of an organism is not immutable. It is fixed within limits by the genotype of the individual, but within these limits there is plasticity: the interactions of an individual with its environment during its life may modify its life history. Thus, in dealing with individuals—products of the evolutionary process—and with the ways in which individuals interact with their immediate environment, we may also need to consider whether the interaction between the evolutionary product and the present environment has itself evolved. For instance, in the life histories of annual plants we may find that: (i) there is a number of seeds per pod that is characteristic of the species (genotypic); (ii) the number is reduced during drought (plasticity); but (iii) the extent to which seed number is reduced by drought is itself under genotypic control (some species may reduce the number of seeds set during drought, whilst others reduce the size of the seeds but keep the number the same).

14.2 The components of life histories

In order to recognize life-history patterns, it is necessary first to specify the various components of life histories, and to be aware of the benefits that these components can bring. To be worthy of our attention here, all such components must affect an organism's reproductive output and/or its survival, since it is these, ultimately, that determine its fitness.

size

Individual size is perhaps the most apparent aspect of an organism's life history. It varies from taxon to taxon, from population to population and from individual to individual. As Chapter 4 made clear, it is particularly variable in organisms with a modular construction. Large size may increase an organism's competitive ability, or increase its success as a predator or decrease its vulnerability to predation. Large organisms are also often better able to maintain a constancy of body function in the face of environmental variation (because their smaller surface : volume ratio makes them less 'exposed' to the environment). All of these factors tend to increase the survival of larger organisms. In addition, larger individuals within a species usually produce more offspring (Figure 14.1). Size, however, can increase some risks; a larger tree is more likely to be felled in a gale or struck by lightning. There have also been many studies in which predators exhibit a preference for larger prey (see Chapter 9); and larger individuals typically require more energy for maintenance, growth and reproduction, and may therefore be more prone to a shortage of resources. Hence, it is easy to see why the largest individuals may suffer a reduction in survival, and detailed studies are increasingly confirming an intermediate, not a maximum, size to be optimal (e.g. Thompson, 1989 (Figure 14.2); see also Roff, 1981).

growth and development

A particular size may be achieved by starting large, growing fast, growing for a long time or any combination of these. Development is the progressive differentiation of parts, enabling an organism to do different things (e.g. reproduce) at different stages in its life history. In many organisms, growth and development occur simultaneously. Development is quite separate from growth, however, in the sense that a given stage of development can be represented by a range of sizes. Rapid development can increase fitness because it leads to the rapid initiation of reproduction, to short generation lengths, and thus to high rates of increase (see Chapter 4). On the other hand, arrested development (as in dormancy or diapause) may improve an organism's chances of survival when resources are rare or conditions far from optimal (see Chapter 5).

reproductive precociousness and delay

A great deal of the variation amongst life histories relates to aspects of reproduction. To start with, as we have just discussed, organisms vary in the length of the pre-reproductive period, i.e. in the extent to which their reproductive maturity is precocious or delayed.

itero- and semelparity; clutch size and clutch number

Organisms can produce all of their offspring over one relatively short period in a single reproductive event (semelparity), or produce them in a series of separate events, during and after each of which the organisms maintain themselves in a condition that favours survival to reproduce again subsequently (iteroparity; see Chapter 4). Amongst iteroparous organisms, variation is possible in the number of separate clutches of offspring and in the number of offspring in a clutch.

offspring size and parental care

Individual offspring can vary in size, i.e. in the quantity of resources allocated to them by their parents. This variation may result from the provisioning of eggs with food reserves, or it can arise during the early growth and development of embryos, from resources supplied directly by the mother via a placenta or its equivalent. This is the situation in humans, other mammals and all seed plants. Parents may also vary in the amount of resources allocated to the care of offspring after birth:

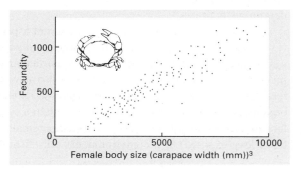

Figure 14.1 Larger females produce a greater number of young in the big-handed crab, *Heterozius rohendifrons*, from Kaikoura, New Zealand. (After Jones, 1978; see also Sibly & Calow, 1986.)

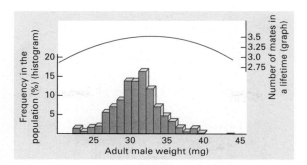

Figure 14.2 For adult male damselflies, *Coenagrion puella*, the predicted optimum size (weight) is intermediate (upper graph), and corresponds closely to the modal size class in the population (histogram below). The upper graph takes this form because mating rate decreases with weight, whereas life span increases with weight (mating rate = 1.15 – 0.018 weight, $P < 0.05$; life span = 0.21 – 0.44 weight, $P < 0.05$; n = 186). (After Thompson, 1989.)

protecting them, suckling them or foraging on their behalf. Large newly emerged or newly germinated offspring are often better competitors, better at obtaining nutrients and better at surviving in extreme environments. Hence, they often have a better chance of surviving to reproduce themselves.

Combining all of this detail, life histories are often described in terms of a composite measure of reproductive activity known as 'reproductive allocation' (also often called 'reproductive effort'). This is best defined as the proportion of the available resource input that is allocated to reproduction over a defined period of time; but it is far easier to define than it is to measure. Ideally, the appropriate, limited resource should be identified, and its allocation over a period of time to the various parts and physiological processes of an organism followed. The resource that matters most, however, varies between organisms and between organs—the anther of a flower contains different proportions of nitrogen, carbon, phosphorus and potassium from the pistil. In practice, though, even the better studies usually monitor only the allocation of energy (Figure 14.3a), or just dry weight (Figure 14.3b) to various structures at a number of stages in the organism's life cycle (Harper & Ogden, 1970; Howarth & Williams, 1972). Yet the justification for this is usually a faint-hearted one: measuring mass or energy is technically much easier than measuring most other resources.

There are, on the other hand, clear cases in which a particular resource can be identified as being crucial. For instance, the growth and reproduction of moose may be limited by the amount of sodium they can collect during a short season of aquatic grazing (Botkin *et al.*, 1973). Their life history is therefore likely to be influenced by the optimal allocation of sodium between parent and offspring and between various sodium-demanding activities (Harper, 1977). Fortunately, though, the search for life-history patterns and their determinants can usually proceed by assuming that organisms partition 'limited resources'—but without specifying what those resources are. Hence, the most common way of assessing reproductive allocation has been to measure such ratios as 'gonad weight : body weight', 'seed crop weight : plant weight' or 'clutch volume : body volume'. These obviously fall far short of the ideal, but they can nonetheless provide useful pointers to reproductive allocations when the intention is only to compare different individuals, populations or species (see discussion in Hart & Begon, 1982; Bazzaz & Reekie, 1985).

reproductive allocation—but which resources are allocated?

529 LIFE-HISTORY VARIATION

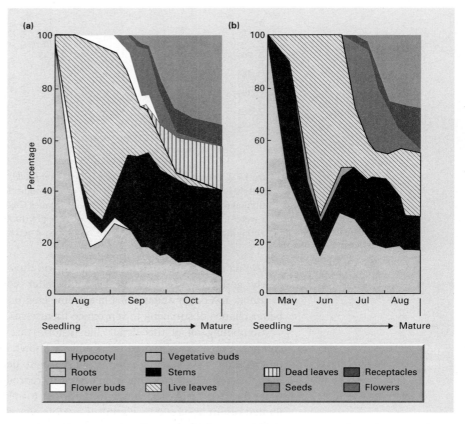

Figure 14.3 (a) Percentage allocation of calories to different structures throughout the life cycle of the annual weed *Senecio vulgaris*. (After Harper & Ogden, 1970.) (b) Percentage allocation of dry weight to different structures throughout the life cycle of another annual, *Chrysanthemum segetum*. (After Howarth & Williams, 1972.)

Direct measures of reproduction are not the only components of a life history. Most fundamentally, organisms will also generally benefit from the allocation of energy and/or resources to structures or activities—including size, already mentioned—that increase their own rate of resource capture or protect them from their enemies. The reason, of course, is that these in turn may enhance survivorship, growth and fecundity.

resource capture and protection

storage and dispersal

Storage of energy and/or resources will be of benefit to those organisms that pass through periods of reduced or irregular nutrient supply (probably most organisms at some time). It may be utilized later for metabolism, growth, defence or reproduction, and in general, for enhanced future survivorship and fecundity.

Finally, dispersal can affect both fecundity and survival (see Chapter 5) and is an integral part of any organism's life history.

14.3 Reproductive value

Natural selection favours those individuals that make the greatest proportionate contribution to the future of the population to which they belong. All life-history components affect this contribution, ultimately through their effects on fecundity

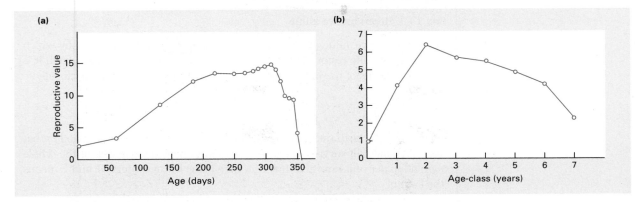

Figure 14.4 Reproductive value generally rises and then falls with age, as explained in the text. (a) The annual plant *Phlox drummondii*. (After Leverich & Levin, 1979.) (b) Female grey squirrels. (After Barkalow *et al.*, 1970; from Charlesworth, 1980.) Note that in both cases the vertical scale is arbitrary, in the sense that the rate of increase (R) for the whole population was not known, and a value had therefore to be assumed.

the need for a measure of fitness

and survival. It is necessary, though, to combine these effects into a single currency so that different life histories may be judged and compared. A number of measures of fitness have been used. All the better ones have made use of both fecundity and survival schedules, but they have done so in different ways, and there has often been marked disagreement as to which of them is the most appropriate. The intrinsic rate of natural increase, r, and the basic reproductive rate, R_0 (see Chapter 4) have had their advocates (Charlesworth, 1980; Sibly & Calow, 1986; Caswell, 1989; Smith, 1991), as has 'reproductive value' (Fisher, 1930; Williams, 1966; Lessells, 1991; see below), especially reproductive value at birth (Kozlowski, 1993). For an exploration of the basic patterns in life histories, however, the similarities between these various measures are far more important than the minor differences between them. We concentrate here on reproductive value.

reproductive value described in words

Reproductive value is described in some detail in Box 14.1. For most purposes though, these details can be ignored as long as it is remembered that: (i) reproductive value at a given age or stage is the sum of the current reproductive output and the residual (i.e. future) reproductive value (RRV); (ii) RRV combines expected future survival and expected future fecundity; (iii) this is done in a way which takes account of the contribution of an individual to future generations, relative to the contributions of others; and (iv) the life history favoured by natural selection from amongst those available in the population will be the one for which the sum of contemporary output and RRV is highest.

The way in which reproductive value changes with age in two contrasting populations is illustrated in Figure 14.4. It is low for young individuals when they have only a low probability of surviving to reproductive maturity, but then increases steadily as the age of first reproduction is approached, as it becomes more and more certain that surviving individuals will reach reproductive maturity. Reproductive value is then low again for old individuals, both because of the reproductive opportunities that have been left behind and the declines in either fecundity or survival in old age. The detailed rise and fall, of course, varies with the detailed age- or stage-specific birth or mortality schedules of the species concerned.

531 LIFE-HISTORY VARIATION

Box 14.1 Reproductive value

The reproductive value of an individual of age x (RV_x) is the currency by which the worth of a life history in the hands of natural selection may be judged. It is defined in terms of the life-table statistics discussed in Chapter 4. Specifically:

$$RV_x = \sum_{y=x}^{y=y_{max}} \left(\frac{l_y}{l_x} \cdot m_y \cdot R^{x-y} \right)$$

where m_x is the birth rate of the individual in age-class x; l_x is the probability that the individual will survive to age x; R is the net reproductive rate of the whole population per unit time (the time unit here being the age interval); and Σ means 'the sum of'.

To understand this equation, it is easiest to split RV_x into its two components:

$$RV_x = m_x + \sum_{y=x+1}^{y=y_{max}} \left(\frac{l_y}{l_x} \cdot m_y \cdot R^{x-y} \right).$$

Here, m_x, the individual's birth rate at its current age, can be thought of as its *contemporary reproductive output*. What remains is then the *residual reproductive value* (Williams, 1966): the sum of the 'expectations of reproduction' at all subsequent ages, modified in each case by R^{x-y} for reasons described below. The 'expectation of reproduction' for age-class y is $(l_y / l_x \cdot (m_y))$, i.e. it is the birth rate of the individual should it reach that age (m_y), discounted by the probability of it doing so given that it has already reached stage x (l_y / l_x).

Reproductive value takes on its simplest form where the overall population size remains approximately constant. In such cases, $R = 1$ and can be ignored. The reproductive value of an individual is then simply its total lifetime expectation of reproductive output (from its current age class and from all subsequent age classes).

However, when the population consistently increases or decreases, this must be taken into account. If the population increases, then $R > 1$ and $R^{x-y} < 1$ (because $x < y$). Hence, the terms in the equation are modified accordingly (reduced) the larger the value of y (the further into the future we go), signifying that future (i.e. 'residual') reproduction adds relatively little to RV_x, because the proportionate contribution made by a given reproductive output in the future is relatively small—whereas the offspring from present or early reproduction themselves have an early opportunity to contribute to the growing population. Conversely, if the population decreases, then $R < 1$ and $R^{x-y} > 1$, and the terms in the equation are successively increased, reflecting the greater proportionate contribution of future reproduction. Moreover, if it is wrongly assumed that the population does not change in size (i.e. $R = 1$), then this may lead, for example if $R > 1$, to the unwarranted undervaluation of the early reproducer, since no account is taken of the possible reproduction of the early reproducer's offspring.

In any life history, the reproductive values at different ages are intimately connected, in the sense that when natural selection acts to maximize reproductive value at one age, it constrains the values of the life-table parameters—and thus reproductive value itself—for subsequent ages. Hence, strictly speaking, natural selection acts ultimately to maximize reproductive value *at birth*, RV_0 (Kozlowski, 1993). (Note that there is no contradiction between this and the fact that reproductive value is typically low at birth (Figure 14.4). Natural selection can discriminate only between those options available at that stage.)

14.4 Trade-offs

It is not difficult to describe a hypothetical organism with all the attributes that confer high reproductive value. It reproduces immediately after its own birth; produces large clutches of large, protected offspring on which it lavishes parental care; it does this repeatedly and frequently throughout an infinitely long life; and it outcompetes its competitors, escapes predation and catches its prey with ease. But, whilst easy to describe, this organism is difficult to imagine, simply because if it puts all of its resources into reproduction it cannot also put all of them into survival; and if it spends all of its time caring for its offspring it cannot also spend all of its time searching for food. In short, common sense alone suggests that a real organism's life style and life history must be a compromise allocation of the resources that are available to it.

trade-offs: negative correlations between life-history traits

A 'trade-off' is a relationship between two life-history characteristics in which increased benefits from one are associated with decreased benefits from the other because of limitations in the total resource available to be allocated to them. For instance, female fruit-flies benefit both from finding egg-laying sites (which they do by flying in search of them) and from laying eggs in those sites, but if they are subjected to a period of induced flight, their subsequent fecundity is reduced (Figure 14.5a). Douglas fir trees (*Pseudotsuga menziesii*) benefit both from reproducing and from growing (since, amongst other things, this enhances future reproduction), but the more cones they produce the less they grow (Figure 14.5b). Male fruit-flies benefit both from a long period of reproductive activity and from a high frequency of matings, but the higher their level of reproductive activity earlier in life the sooner they die (Figure 14.5c). Each change in these life histories that might be expected to increase reproductive value is offset by an associated change—a trade-off—that tends to reduce reproductive value. Clearly, then, an investigation of trade-offs is central to one of our three types of life-history questions: what are the patterns that link life-history traits together, and what are the causes of these patterns?

... which cannot generally be revealed by simple observation of phenotypes

Yet it would be quite wrong to think that negative correlations between traits abound in nature, only waiting to be observed. On the contrary, we cannot generally expect to see trade-offs by simply observing correlations in natural populations (Lessells, 1991). In the first place, if there is just one clearly optimal way of combining, say, growth and reproductive output, then all individuals may approximate closely to this optimum and a population would then lack the variation in these traits necessary to generate a trade-off curve. Moreover, if there is variation between individuals in the amount of resource they have at their disposal, then there is likely to be a positive, not a negative correlation between two apparently alternative processes—some individuals will be good at everything, others consistently awful. For instance, in Figure 14.6, the grasshoppers that reach the adult stage first are also the largest. This occurs because the grasshoppers comprise a competitive hierarchy, in which the best competitors develop rapidly and grow large because they consume large amounts of food. The worst competitors obtain little food and are therefore small and slow to mature.

genetic comparisons: breeding and selection experiments

Two approaches have sought to overcome these problems and hence allow the investigation of the nature of trade-off curves. The first is based on comparisons of individuals differing *genetically*, where different genotypes are thought likely to give rise to different allocations of resources to alternative traits. Genotypes can be compared in two ways: (i) by a breeding experiment, in which groups of genetically

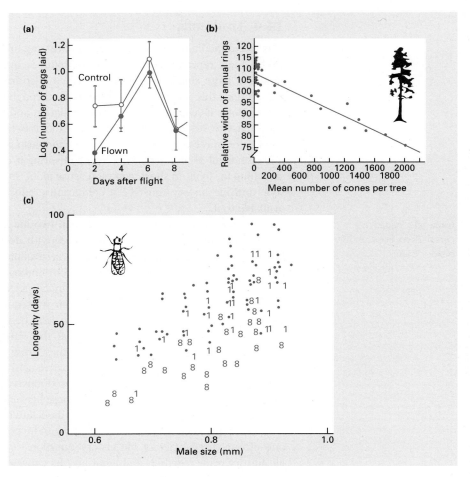

Figure 14.5 Life-history trade-offs. (a) Induced flight substantially reduces fecundity initially in female fruit-flies, *Drosophila subobscura*, and this never quite returns to the levels exhibited by non-migrant controls (standard errors shown). (After Inglesfield & Begon, 1983.) (b) The negative correlation between cone crop size and annual growth increment for a population of Douglas fir trees, *Pseudotsuga menziesii*. (After Eis *et al.*, 1965.) (c) The longevity of male fruit-flies (*D. melanogaster*) generally increases with size (thorax length). However, longevity was reduced in males provided with one virgin and seven mated females per day (1) compared with those provided with eight mated females (●), because of the increase in courtship activity, and reduced still further in males provided with eight virgins per day (8). (After Partridge & Farquhar, 1981.)

identical or similar individuals are bred and then compared; or (ii) by a selection experiment, in which a population is subjected to a selection pressure to alter one trait, and associated changes in other traits are then monitored. For example, in one selection experiment, populations of the Indian meal moth, *Plodia interpunctella*, that evolved increased resistance to a virus having been infected with it for a number of generations, exhibited an associated decrease (negative correlation) in their rate of development (Boots & Begon, 1993). Overall, however, the search for genetic correlations has generated more zero and positive than negative correlations (Lessells, 1991), and it has therefore had only limited success to date in measuring

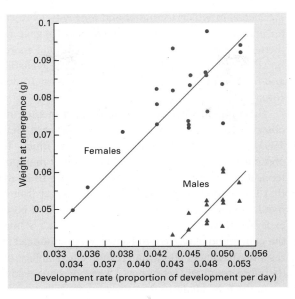

Figure 14.6 Differences in resource input obscuring possible trade-offs: the individuals that are largest at adult emergence are also first to emerge in male and female grasshoppers, *Chorthippus brunneus*, reared in the laboratory. (After Wall & Begon, 1987.)

trade-offs, despite receiving strong support from its adherents by virtue of its direct approach to the underlying basis for selective differentials between life histories (Reznick, 1985; Rose *et al.*, 1987).

phenotypic correlations: experimental manipulations

The alternative approach is to use experimental manipulation to reveal a trade-off directly from a negative *phenotypic* correlation. The two *Drosophila* studies in Figure 14.5 are examples of this. The great advantage of experimental manipulation over the simple observation of phenotypes is that individuals are assigned to experimental groups at random rather than differing from one another, for instance, in the quantity of resource that they have at their disposal. This contrast is illustrated in Figure 14.7, which shows two sets of data for the bruchid beetle, *Callosobruchus*

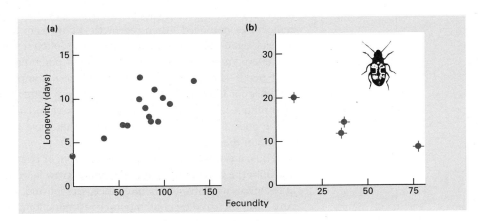

Figure 14.7 (a) The (positive) phenotypic correlation in an unmanipulated population between adult longevity and fecundity in female bruchid beetles, *Callosobruchus maculatus*. (b) The (negative) trade-off between the same traits when access to mates and/or egg-laying sites was manipulated. Points are means of four treatments with standard errors. (After Lessells, 1991; from K. Wilson, unpublished data.)

535 LIFE-HISTORY VARIATION

maculatus, in which fecundity and longevity were correlated. Simple observation of an unmanipulated population (Figure 14.7a) gave rise to a positive correlation: the 'better' individuals both lived longer and laid more eggs. When fecundity varied, however, not as a result of differing resource availability, but because access to mates and/or egg-laying sites was manipulated, a trade-off (negative correlation) was revealed (Figure 14.7b).

However, this contrast between experimental manipulation ('good') and simple observation ('bad') is not always straightforward (Bell & Koufopanou, 1986; Lessells, 1991). Some manipulations suffer from much the same problems as simple observations. For instance, if clutch size is manipulated by giving supplementary food, then improvements in other traits are to be expected as well. It is important that the manipulation should alter the target trait and nothing else. On the other hand, simple observation may be acceptable if based on the results of a 'natural' experiment. For example, it is likely, as a result of 'mast seeding' (see Chapter 8), that the population of fir trees in Figure 14.5b produced large and small crops of cones in response to factors other than resource availability, and that the negative correlation therefore genuinely represented an underlying trade-off.

In the final analysis, though, it is generally agreed that trade-offs are widespread and important. The problems arise in revealing and hence quantifying them.

14.4.1 The cost of reproduction

Most attention has been directed at trade-offs that reveal an apparent 'cost of reproduction' (CR). Here, 'cost' is used in a particular way to indicate that an individual, by increasing its current allocation to reproduction, is likely to decrease its survival and/or its rate of growth, and therefore decrease its potential for reproduction in future. This is shown by the fir trees and male fruit-flies in Figure 14.5 and by the beetles in Figure 14.7b.

current reproduction often leads to reductions in survival, growth or future

The costs of reproduction can be shown even more easily with plants. Semelparous plants such as the foxglove (*Digitalis purpurea*) flower 1 year or more after germination, when they reach a critical size. They normally die after they have set seed. However, if seed set is interfered with—for example if the inflorescence is removed or damaged—then regrowth occurs from the basal rosette, and the plant flowers again the next year, when it is often even bigger. Similarly, all good gardeners know that to prolong the life of perennial flowering herbs, the ripening seed heads should be removed, since these compete for resources that may be available for improved survivorship and even better flowering next year. At a more quantitative level, when ragwort plants (*Senecio jacobaea*) of a given size are compared at the end of a season, it is only those that have made the smallest reproductive allocation that survive (Figure 14.8).

Thus, individuals that delay reproduction, or restrain their reproduction to a level less than the maximum, may grow faster, grow larger or have an increased quantity of resources available for maintenance, storage and, ultimately, future reproduction. Any 'cost' incurred by contemporary reproduction, therefore, is likely to contribute to a decrease in RRV. Yet, as we have noted, natural selection favours the life history with the highest available *total* reproductive value. It therefore favours the life history with the greatest sum of two quantities, one of which (contemporary reproductive output) tends to go up as the other (RRV) goes down. Clearly, trade-offs involving the cost of reproduction are at the heart of the evolution of any life history.

natural selection favours life histories with the greatest sum of contemporary reproductive output plus residual reproductive rate

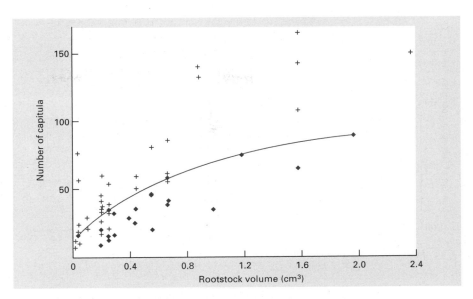

Figure 14.8 The cost of reproduction in ragwort (*Senecio jacobaea*). The line divides plants that survive, from those that have died by the end of the season. There are no surviving plants above and to the left of the line. For a given size (rootstock volume), only those that have made the smallest reproductive allocation (number of capitula) survive, although larger plants are able to make a larger allocation and still survive. (After Gilman & Crawley, 1990.)

14.4.2 The number and fitness of offspring

A second trade-off which has received considerable attention is that between the number of offspring and their individual fitness. At its simplest, this is a trade-off between the size and number of offspring, within a given total reproductive investment. That is, a reproductive allocation can be divided into fewer, larger offspring or more, smaller offspring. Parents, however, may also invest 'parental care' in their offspring after they have been produced, and the trade-off may therefore be, for example, between the number of offspring and the amount of food that each is provided with. Moreover, the size of an egg or seed is only an index of its likely fitness. It may be more appropriate to look for a trade-off between the number of offspring and, say, their individual survivorship or developmental rate.

Of the few genetic correlations between egg size and number that have so far been examined (dominated by domestic poultry), the majority have been negative, as expected (Lessells, 1991). Negative correlations have also been observed in simple cross-species or cross-population comparisons (e.g. see Figures 14.9a, b), although it is unlikely in such cases that the individuals from different species or populations are making precisely the same total reproductive allocation. Moreover, this type of trade-off is especially difficult to observe through experimental manipulation. To see why, note that we need to ask the following kind of question. Given that a plant, say, produces 100 seeds each weighing 10 mg and each with a 5% chance of developing to reproductive maturity, what would be the seed size, and what would be the chance of developing to maturity, if an identical plant receiving identical resources produced only 80 seeds? Clearly, it would be invalid to manipulate seed number by altering the provision of resources; and even if 20 seeds were removed at or close to their point of production, the plant would be limited in its ability to alter the size of

evidence for the trade-off—and the difficulties in finding it

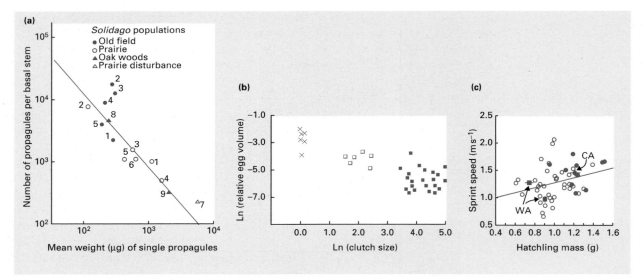

Figure 14.9 Evidence for a trade-off between the number of offspring produced in a clutch by a parent and the individual fitness of those offspring. (a) A negative correlation between the number of propagules per stem produced by goldenrod plants (*Solidago*) and the weight of single propagules. The species are: 1, *S. nemoralis*; 2, *S. graminifolia*; 3, *S. canadensis*; 4, *S. speciosa*; 5, *S. missouriensis*; 6, *S. gigantea*; 7, *S. rigida*; 8, *S. caesia*; and 9, *S. rugosa*, taken from a variety of habitats as shown. (After Werner & Platt, 1976.) (b) A negative correlation between clutch size and egg volume amongst Hawaiian species of *Drosophila*, developing either on a relatively poor and restricted resource, flower pollen (×), or on bacteria within decaying leaves (□) or on especially unpredictable but productive resources—yeasts within rotten fruits, barks and stems (■). (After Montague *et al.*, 1981; Stearns, 1992.) (c) The mass and sprint speed of California hatchlings of the lizard, *Sceloporus occidentalis*, are lower from eggs with some yolk removed (○) than from unmanipulated control eggs (●). Also shown are the means for the control California hatchlings (CA) (fewer, larger eggs) and those for two samples from Washington (WA) (more, smaller eggs). (After Sinervo, 1990.)

the remaining seeds, and their subsequent survivorship would not really address the question originally posed.

Nonetheless, Sinervo (1990) has manipulated the size of the eggs of an iguanid lizard (*Sceloporus occidentalis*) by removing yolk from them after they have been produced, giving rise to healthy but smaller offspring than unmanipulated eggs. These smaller hatchlings had a slower sprint speed (Figure 14.9c)—probably an indication of a reduced ability to avoid predators, and hence of a lower fitness. Within natural populations, this species produces smaller clutches of larger eggs in California than in Washington (typically seven to eight eggs with an average weight of 0.65 g, as against around 12 eggs weighing 0.4 g; see also Figure 14.9c). Thus, in the light of the experimental manipulations, the comparison between the two populations does indeed appear to reflect a trade-off between the number of offspring and their individual fitness.

14.5 Options sets and fitness contours

We turn next to another of our original life-history questions—are there patterns linking particular types of life history to particular types of organism or particular types of habitat? Given that trade-offs lie at the heart of any life history, the question: 'when and where will particular life histories be found?' can be thought of as: 'when and where will natural selection favour particular points on a trade-off curve?' To address this question, we introduce two further concepts. We do so in the context of the cost of reproduction, since the trade-offs associated with it are the most fundamental—but the same principles apply to any trade-off.

First, an *options set* describes the whole range of combinations of two life-history traits that an organism is capable of exhibiting. Hence, it reflects the organism's underlying physiology. Here, for purposes of illustration, we use the traits present reproduction, m_x, and growth (as a potentially important indicator of RRV) (Figure 14.10). The options set therefore describes, for any given level of present reproduction, the range of growth increments that the organism can achieve, and for any given growth increment, the range of levels of present reproduction that the organism can achieve. The outer boundary of the options set represents the trade-off curve. For any point on that boundary, the organism can only increase m_x by making a compensatory reduction in growth, in conformity with the underlying trade-off between the two.

options sets

An options set may be convex-outwards (Figure 14.10a), implying in the present case that a level of present reproduction only slightly less than the maximum nonetheless allows a considerable amount of growth. Alternatively, the set may be concave-outwards (Figure 14.10b), implying that substantial growth can only be achieved with a level of present reproduction considerably less than the maximum.

fitness contours

Second, a *fitness contour* is a line joining, in this case, combinations of m_x and growth for which fitness (reproductive value) is constant (Figure 14.10c). Contours further away from the origin therefore represent combinations with greater fitness (more growth and/or more reproduction). The shapes of fitness contours reflect the habitat in which an organism lives rather than the organism's intrinsic properties. For example, an organism may be capable of combining 75% of maximum present reproduction with 75% of maximum present growth. This would be reflected in its options set. However, for reasons described below, its habitat may be such that its '75% present reproduction' contributes very significantly to fitness, whereas its '75% growth' hardly contributes at all. This would be reflected in the shape of its fitness contours.

putting the two together

The combination of traits that, amongst those available, has the highest fitness determines the direction of natural selection. Natural selection therefore favours the point in the options set (on the trade-off curve) that reaches the highest fitness contour (indicated by the various asterisks in Figures 14.10d, 14.12a, 14.13a, 14.17a). Since different options sets imply different types of organism, and different shapes of fitness contour imply different types of habitat, they can be used together as a guide to where and when different types of life history might be found. We begin by classifying habitats.

14.6 Habitats: a classification

habitats must be seen from the organism's point of view

Each organism's habitat, like each organism's life history, is unique. Thus, if a pattern linking habitats and life histories is to be established, habitats must be classified in terms that apply to them all. Moreover, they must be described and classified from the point of view of the organism concerned. For instance, whether an annual plant will encounter an unusually cold winter or hot summer is difficult to

539 LIFE-HISTORY VARIATION

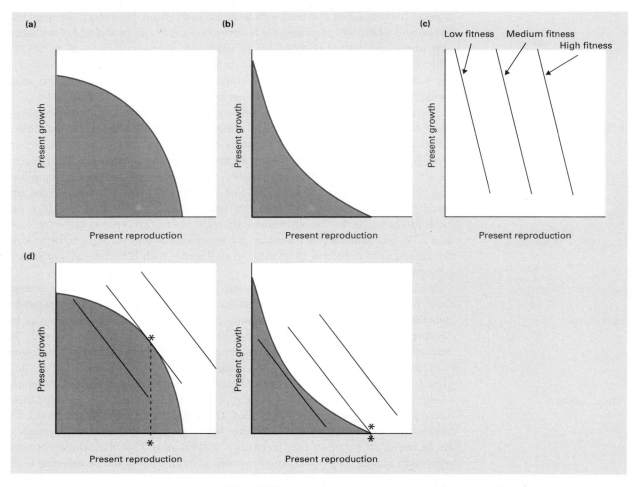

Figure 14.10 (a, b) Illustrate options sets—the combinations available to an organism of, in this case, present reproduction and present growth. As explained in the text, the outer boundary of the options set is a trade-off curve. (a) is convex-outwards, (b) concave-outwards. (c) Fitness contours linking combinations of present reproduction and present growth that have equal fitness in a given habitat. Hence, contours further from the origin have greater fitness. (d) The point in an options set with the greatest fitness is the one that reaches the highest fitness contour. This point, and the (optimal) value for present reproduction giving rise to it, are marked with an asterisk. (After Sibly & Calow, 1983.)

predict, since it lives through only 1 year. But the number of cool, moderate and hot summers experienced by a long-lived tree is much more predictable. Similarly, a small woodland might be a heterogeneous habitat for a beetle attacking aggregations of aphids, but a much more homogeneous habitat for a rodent collecting seeds from the woodland floor, and part of a heterogeneous habitat for a large, predatory buzzard ranging over a wider area of ground. Moreover, it is dangerous to assume too readily that particular habitats are 'benign' or 'hostile'. Even the most extreme habitats (polar ice, hot springs) may be benign for the organisms that live and have evolved there.

Habitat-type, therefore, is in the ecological eye of the beholder. Thus, when we say that the shapes of fitness contours reflect an organism's habitat, we mean that

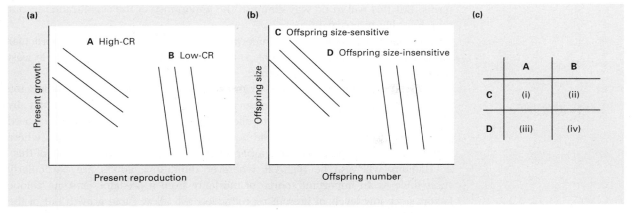

Figure 14.11 A demographic classification of habitats. (a) Habitats of established individuals can be either, (A), relatively high-CR (fitness contours indicate that residual reproductive value rises sharply with increased growth resulting from decreased present reproduction), or (B), relatively low-CR (fitness contours largely reflect the level of present reproduction). (b) Habitats of recently produced offspring can be either, (C), relatively offspring size-sensitive, or (D), relatively offspring size-insensitive. Larger offspring size is assumed to imply a smaller number of them (for a given reproductive allocation). Hence, for example, in (D), fitness largely reflects the number of offspring—not their individual size. (c) By combining these two contrasting pairs, the habitat of one organism over its whole life, compared with another organism, can be of four basic types, arbitrarily referred to as (i)–(iv) in the figure.

they reflect the effect of the habitat on that particular organism or the response of that organism to the habitat.

A number of classifications of habitat types have been proposed, to some of which we return later. Others have been suggested, for example, by Southwood (1977), Sibly and Calow (1985) and Townsend and Hildrew (1994). The way we classify habitats first here, however, is to focus on fitness contours, and hence on the ways in which present reproduction and growth combine to determine fitness in different types of habitat (combining the approaches of Sibly & Calow, 1983; following Levins, 1968; and Begon, 1985).

habitats of established individuals can be classified by the slopes of their fitness contours …

For established individuals (i.e. not newly emerged or newly born offspring) two contrasting habitat types can be recognized.

1 High-CR (high-cost-of-reproduction) habitats, in which any reduced growth that results from present reproduction has a significant negative effect on RRV, and hence on fitness. Thus, similar fitness can be achieved by combining high reproduction with low growth or low reproduction with high growth. Fitness contours therefore run diagonally with a negative slope (Figure 14.11a).

2 Low-CR habitats, in which RRV is little affected by the level of present growth. Fitness is thus essentially determined by the level of present reproduction alone and will be much the same whatever the level of present growth. The fitness contours therefore run approximately vertically (parallel to the 'growth' axis; Figure 14.11a).

… although the classification is comparative …

This classification, like many other aspects of the study of life histories, is comparative. In practice, a habitat can only be described as 'high-CR' relative to some other habitat which is, comparatively, low-CR. The purpose of the classification is to contrast habitats with one another, rather than to describe them in absolute

541   LIFE-HISTORY VARIATION

terms. But this will in no way diminish the importance of the classification in our search for life-history patterns.

A second point, also of crucial importance, is that a habitat can be of a particular type for a variety of different reasons. Habitats can be relatively high-CR for at least two reasons (although there are others—see below).

... and habitats of different types can arise for a variety of reasons

1 When there is intense competition amongst established individuals with only the best competitors surviving and reproducing, reduced present reproduction may, by virtue of the increased growth it allows, give rise to substantially increased competitive ability in the future, and hence increased RRV. Red deer stags, where only the best competitors can hold a harem of females, are a good example of this.

2 Habitats will also be high-CR wherever diminutive adults are particularly susceptible to an important source of mortality from a predator or some abiotic factor, since low levels of present reproduction will allow rapid growth out of the vulnerable size classes. For instance, mussels on the seashore may, through reproductive restraint, outgrow predation by both crabs and eider ducks.

Habitats can be relatively low-CR for at least three different reasons.

1 Much mortality may be indiscriminate and unavoidable, so that any increase in size caused by reproductive restraint is likely to be worthless in future. For instance, when temporary ponds dry out, most individuals die irrespective of their size or condition.

2 The habitat may be so benign and competition-free for established individuals that all of them have a high probability of surviving, and a large future reproductive output, irrespective of any present lack of reproductive restraint. This is true, at least temporarily, for the first colonists to arrive in a newly arisen habitat.

3 A habitat may be low-CR simply because there are important sources of mortality to which the largest individuals are especially prone. Thus, restrained present reproduction, by leading to greater size, may give rise to *lowered* survival in future. For instance, in the Amazon, avian predators preferentially prey upon the largest individuals of certain cyprinodont fish species.

In summary, then, a wide variety of habitats may be classified and compared on the basis of contrasts in the slopes of the fitness contours of the organisms occupying those habitats. These contrasts may then help us understand differences in life histories (see below).

a related classification for the habitats of newly emerged offspring

A related classification of habitats for newly born offspring can also be constructed. Again, there are two contrasting types (Figure 14.11b), assuming that, for a given reproductive allocation, larger offspring can be produced only if there are fewer of them.

1 'Offspring size-sensitive' habitats, in which the reproductive value of individual offspring rises significantly with size (as above, either because of competition amongst offspring, or because of important sources of mortality to which small offspring are especially vulnerable). An increase in size implies a significant rise up the fitness contours.

2 'Offspring size-insensitive' habitats in which the reproductive value of individual offspring is little affected by their size (as above, because of indiscriminate mortality, or because of superabundant resources or because there are sources of mortality to which larger individuals are more prone). An increase in size implies a negligible move up the fitness contours.

Together, clearly, the two contrasting pairs can be combined into four types of habitat (Figure 14.11c).

14.7 Reproductive allocation and its timing

14.7.1 Reproductive allocation

If we assume initially that all options sets are convex outwards, then we can see that relatively low-CR habitats should favour a higher reproductive allocation, whilst relatively high-CR habitats favour a lower reproductive allocation (Figure 14.12a). This pattern can be seen in three populations of the dandelion *Taraxacum officinale*. The populations were composed of a number of distinct clones which could be identified by electrophoresis and were found to belong to one or other of four biotypes (A–D). The habitats of the populations varied from a footpath (the habitat in which adult mortality was most indiscriminate—'lowest CR') to an old, stable pasture (the habitat with most adult competition—'highest CR'); the third site was intermediate between the other two. In line with predictions, the biotype that predominated in the footpath site (A), made the greatest reproductive allocation (whichever site it was obtained from), whilst the biotype that predominated in the old pasture (D), made the lowest reproductive allocation (Figures 14.12b, c). Biotypes B and C were

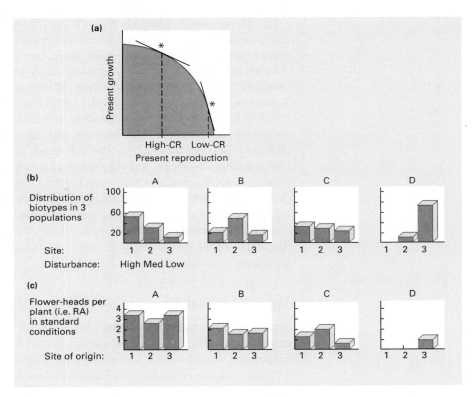

Figure 14.12 (a) Options sets and fitness contours (see Figure 14.11) suggest that relatively high-CR habitats should favour relatively small reproductive allocations. (b) The distribution of four biotypes (A–D) of the dandelion *Taraxacum officinale*, amongst three populations subject to low, medium and high levels of disturbance (i.e. habitats ranging from relatively high-CR to relatively low-CR). (c) The reproductive allocations (RAs) of the different biotypes from the different sites of origin, showing that biotype A, which predominates in the relatively low-CR habitat, has a relatively large RA, and so on. ((b, c) After Solbrig & Simpson, 1974.)

appropriately intermediate with respect both to their site occupancies and their reproductive allocations.

Another set of habitats that are presumably relatively low-CR (this time in the sense of being indiscriminately benign), are those of tapeworms and other gut parasites once the adults are established in their hosts, lying bathed in nutrients. This may account for their exceedingly high reproductive allocations (Calow, 1981).

A habitat may also be relatively high-CR because reproduction involves 'risks' over and above the simple diversion of resources. For instance, reproducing individuals may be more vulnerable to predation or less efficient at food capture because of the increased bulk of their reproductive organs. Hence, the extra costs rise as reproductive allocation rises. Whereas low-CR fitness contours are close to vertical, this additional cost brings the contours further round to the horizontal, suggesting that natural selection should favour a relatively low reproductive allocation. In support of this, lizards that chase their prey and flee from their predators, and thus pay an extra cost for reproduction in terms of decreased speed and mobility, have a relatively low reproductive allocation, whereas those that sit and wait for their prey, and use crypsis or armour to avoid their predators, have a higher allocation (Vitt & Congdon, 1978).

A habitat may be relatively high-CR, too, because it poses particularly severe physiological problems to an organism that require energy for their solution. This energy must be saved rather than spent on present reproduction—otherwise survival, and hence RRV, would be dangerously low. For example, Stearns (1980) examined two neighbouring populations of the mosquito fish (*Gambusia affinis*) in Texas, one of which lived in freshwater whilst the other lived in brackish water. He found a higher reproductive allocation in individuals from the brackish water population, but concluded that selection was not acting on reproductive allocation directly. It seemed, instead, that the individuals in the freshwater population had osmoregulatory problems, and they used extra energy to maintain their body fluids in an acceptable condition. They therefore had less 'disposable' energy, and it was this that led to their smaller reproductive allocation, emphasizing that a life history is a response to the whole of an organism's environment.

Thus, as judged by the dandelion, tapeworm, lizard and mosquito fish examples, the shapes of the fitness contours have a valuable role to play in understanding life-history patterns.

14.7.2 Age at maturity

Since relatively high-CR habitats should favour low reproductive allocations, maturity (the onset of sexual reproduction) should be relatively delayed in such habitats but should occur at a relatively large size (in deferring maturity at any given time, an organism is making a reproductive allocation of zero; Figure 14.13a). These ideas are supported by a comparison of populations of 'annual' meadow grass (*Poa annua*) from two types of habitat in north-west England and North Wales. One type ('opportunist') contained individuals at low density with large areas of bare ground between them, and had been subject to repeated disturbance (indiscriminate mortality). These habitats were therefore relatively low-CR. The other type had generally existed as permanent 'pasture' for some time and therefore contained individuals at high density, subject to intense competition (high-CR). Samples from each type were grown from seed under controlled, uncrowded conditions, and the

... tapeworms ...

... lizards ...

... and mosquito fish

high- and low-CR, and meadow grass ...

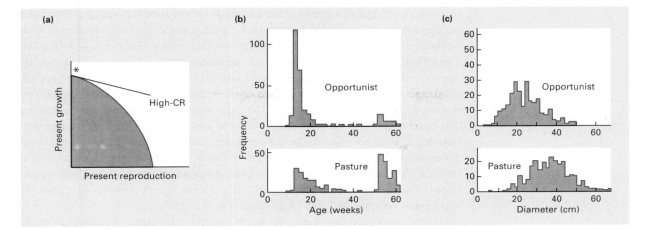

Figure 14.13 (a) A habitat may favour a reproductive allocation of zero if it is relatively high-CR, and will thus, at that time, favour deferred maturity and further growth. (b, c) *Poa annua* plants in high-CR habitats defer maturity to a relatively large size. (b) The distribution of pre-reproductive periods in plants from opportunistic (relatively low-CR) and pasture habitats. (c) The distribution of sizes in plants from opportunistic and pasture habitats. ((b, c) After Law *et al.*, 1977.)

... guppies ...

life histories of individuals were monitored. As predicted, individuals from the low-CR habitats matured earlier (Figure 14.13b) and were smaller (Figure 14.13c).

Another example, showing a similar pattern, comes from a study of guppies, *Poecilia reticulata*, in Trinidad (Table 14.1). The same work also provides support for

Table 14.1 A comparison of guppies (*Poecilia reticulata*) from relatively low-CR, offspring size-insensitive sites (*Crenicichla*—predation concentrated on larger, adult fish) and relatively high-CR, offspring size-sensitive sites (*Rivulus*—predation concentrated on small, juvenile fish). In the former, the guppies (male and female) mature earlier and smaller, make a larger reproductive allocation (shorter interlitter interval, higher percentage effort) and produce smaller offspring (and more of them). This is true both for natural populations from contrasting sites (left—after Reznick, 1982) and comparing a population introduced to a *Rivulus* site with its unmanipulated control (right—after Reznick *et al.*, 1990.)

	Reznick (1982)			Reznick *et al.* (1990)		
	Crenicichla		*Rivulus*	Control (*Crenicichla*)		Introduction (*Rivulus*)
Male age at maturity (days)	51.8	$P < 0.01$	58.8	48.5	$P < 0.01$	58.2
Male size at maturity (mg wet)	87.7	$P < 0.01$	99.7	67.5	$P < 0.01$	76.1
Female age at first birth (days)	71.5	$P < 0.01$	81.9	85.7	$P < 0.05$	92.3
Female size at first birth (mg wet)	218.0	$P < 0.01$	270.0	161.5	$P < 0.01$	185.6
Size of litter 1	5.2	$P < 0.01$	3.2	4.5	$P < 0.05$	3.3
Size of litter 2	10.9	NS	10.2	8.1	NS	7.5
Size of litter 3	16.1	NS	16.0	11.4	NS	11.5
Offspring weight (mg dry) litter 1	0.84	$P < 0.01$	0.99	0.87	$P < 0.10$	0.95
Offspring weight litter 2	0.95	$P < 0.05$	1.05	0.90	$P < 0.05$	1.02
Offspring weight litter 3	1.03	$P < 0.01$	1.17	1.10	NS	1.17
Interlitter interval (days)	22.8	NS	25.0	24.5	NS	25.2
Reproductive effort (%)	25.1	$P < 0.05$	19.2	22.0	NS	18.5

NS, not significant.

the patterns of reproductive allocation discussed above, and for patterns of variation in offspring size, discussed in Section 14.8. The guppies live in small streams that can be divided into two contrasting types. In one, their main predator is a cichlid fish, *Crenicichla alta*, which eats mostly large, sexually mature guppies. In the other, the main predator is a killifish, *Rivulus hartii*, which prefers small, juvenile guppies. The *Crenicichla* sites are therefore relatively low-CR, and, as predicted, the guppies there mature earlier and at a smaller size. They also make a larger reproductive allocation (left-hand columns in Table 14.1; Reznick, 1982). Moreover, when 200 guppies were introduced from a *Crenicichla* to a *Rivulus* site, and lived there for 11 years (30–60 generations), not only did the phenotypes in the field come to resemble those of other high-CR (*Rivulus*) sites, but it was also clear that these differences had evolved and were heritable, since they were also discernable under laboratory conditions (right-hand columns in Table 14.1; Reznick *et al.*, 1990).

... and sexual differences

Further support comes from comparing the ages of maturity of males and females within a species. As stressed previously, since the habitat must be seen from the individual organism's perspective, there can be important sex-related differences in habitat. Specifically, with polygynous mating systems (where a limited number of males get all the matings, perhaps by fighting or at least threatening other males), males should have most to gain from deferred maturity, because of the benefits they may get, for example, from increased size. With promiscuous mating, on the other hand, and especially with external fertilization, males do not control access to females. Hence, females, because of increased fecundity, should gain more from the increased size that comes with deferred maturity than males do. As predicted, in mammals and birds, where many species are polygynous, males tend to mature later and larger than females (Figure 14.14a), whereas in fishes, where external fertilization is common, the opposite is true (Figure 14.14b).

It is necessary here, however, to move beyond our simple classification of habitats. For example, the favoured age and size at maturity can be thought of as being governed by a trade-off between juvenile (pre-maturity) survival and

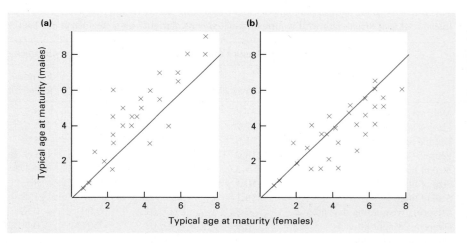

Figure 14.14 (a) In a sample of birds and mammals (containing some dramatically polygynous species—horses, dolphins, elephant seals) males tend to mature later and larger than females. (b) In North American freshwater fishes, females tend to mature later and larger than males. (After Bell, 1980; Stearns, 1992.)

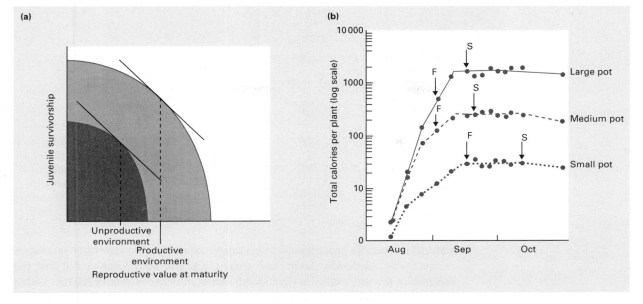

Figure 14.15 Age and size at maturity in productive and unproductive environments. (a) When reproductive value at maturity is traded off against juvenile survivorship, the trade-off curve at the edge of the options set in the productive environment lies beyond that in the unproductive environment, and earlier maturity at a larger size is predicted. (b) Growth curves for individuals of *Senecio vulgaris*, grown in large, medium and small pots. The first appearance of flowers (F) and seeds (S) is indicated in each case. With more limited resources, sexual maturity, as predicted, was deferred and occurred at a smaller size. (After Harper & Ogden, 1970.)

the trade-off between juvenile survivorship and reproductive value at maturity

reproductive value at maturity (for a review, see Stearns, 1992). Deferring maturity to a larger size increases the reproductive value at maturity, but it does so at the expense of decreased juvenile survival, since the juvenile phase is, by definition, extended when maturity is deferred. With this trade-off in mind, we can ask, for example, how age and size at maturity might differ between a 'productive' environment, with abundant food, and an 'unproductive' one in which individuals are poorly nourished. If increased food availability increases both the rate of growth (i.e. size at a given age) and juvenile survival (i.e. the probability of reaching a given age), then the options set in the productive environment will extend beyond that in the unproductive environment, whatever the shape of the trade-off curve (Figure 14.15a). Organisms in more productive environments should then mature both earlier and at a larger size. This can be seen for isolated individuals of *Senecio vulgaris*, grown in pots of various sizes—a larger pot being a more productive environment (Figure 14.15b), and has also been observed commonly in *Drosophila melanogaster*: flies growing at 27°C with abundant food at moderate densities start to reproduce at 11 days, weighing 1.0 mg, whereas crowded, poorly nourished flies start after 15 days or more, weighing 0.5 mg (Stearns, 1992). Note in these examples that we are comparing the immediate responses of individuals to their environments, rather than comparing two quite independent populations or species. We return to this point in Section 14.11.

Another approach to understanding age at maturity has been taken by Charnov and Berrigan (1991; see also Beverton & Holt, 1959). They note that female age at

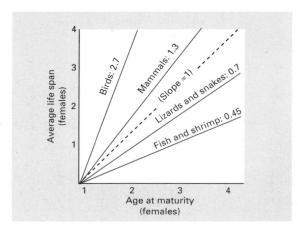

Figure 14.16 The relationship between female adult life span and female age at maturity is roughly constant within a number of taxa but very different between them. (After Charnov & Berrigan, 1991.)

the proportionality between
age at maturity and life
expectancy

maturity is roughly proportional to average female adult life span in a number of groups (birds, mammals, lizards and snakes, and fish and shrimps), but that this constant of proportionality is significantly different between groups (Figure 14.16). Reasons for both the constancy within and the differences between groups remain speculative, but what seems clear is that the taxonomic group to which an organism belongs sets limits to the way it lives its life, and hence to the whole pattern of its life history. Trying to understand age and size at maturity (or anything else) without regard to these limits can be misleading. This is another point to which we return later (see Section 14.12).

14.7.3 Semelparity

Returning to our comparison of high- and low-CR habitats, it is clear that semelparity is most likely to evolve in the latter (Figure 14.17a). This is borne out in a detailed way by work on two species of *Lobelia* living on Mount Kenya (Figure 14.17b). These are long-lived species: both take from 40 to 60 years even to mature, following which the semelparous *L. telekii* dies, whereas the iteroparous *L. keniensis* breeds only every 7–14 years. Young (1990) and Young and Augspurger (1991) showed that in drier *Lobelia* sites, probabilities of adult survival are smaller and periods between reproductive events longer, i.e. the drier sites are lower CR. Semelparity would only actually be favoured there, however, if semelparous plants also gained a sufficient reproductive advantage by diverting more resources to reproduction and less to future survival. In fact, there appears to be a close correspondence between the geographical boundary between the semelparous and the iteroparous species and the boundary where the balance of advantage swings from one to the other reproductive strategy (Figure 14.17b).

semelparous *Lobelia*s in low-
CR habitats

semelparity and synchronous
breeding

Semelparity is associated with synchronous breeding amongst individuals in a number of long-lived species of both animals (e.g. cicadas) and plants (e.g. bamboos). The main evolutionary force favouring such synchrony is probably predator satiation (see Chapter 8, Section 8.4), i.e. offspring produced in the year of synchronous breeding have a much higher chance of survival than any that might be produced in a year where there is little breeding. Thus, in the breeding year, fitness is determined very largely by the level of current reproduction—the deferred reproduction of an iteroparous breeder would be of little value because the offspring would be unlikely

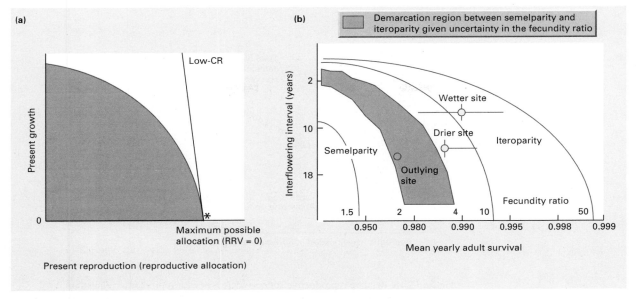

Figure 14.17 (a) Relatively low-CR habitats (near-vertical fitness contours) are more likely to give rise to semelparity (maximum reproductive allocation: nothing kept in reserve). (b) For *Lobelia* spp. on Mount Kenya, habitats become lower CR as interflowering interval increases and mean yearly adult survival decreases. Given that the semelparous *L. telekii* sets approximately four times the weight of seed set by the iteroparous *L. keniensis*, habitats can be predicted to favour either semelparity (bottom left) or iteroparity (upper right), with a region of uncertainty between the two. Three study populations of *L. keniensis* have, as predicted, either habitat characteristics favouring iteroparity, or, in the case of an outlying site, characteristics in the region of uncertainty. (After Young, 1990; Stearns, 1992.)

to survive. Fitness contours, in other words, are likely to be close to vertical, and semelparity is likely to be favoured.

If we now relax our assumption that all options sets are convex-outwards, it is apparent that semelparity is especially likely to evolve in organisms with options sets **concave options sets: salmon** that are concave-outwards, i.e. where even a low level of present reproduction leads to a considerable drop in, say, future survival, but increases from low to moderate levels have little influence on survival (see Figure 14.10). This is the likely explanation for why many species of salmon display suicidal semelparity. Reproduction for them demands a dangerous and effortful upstream migration from the sea to their spawning grounds, but the risks and extra costs are associated with the 'act' of reproduction and are largely independent of the magnitude of the reproductive allocation.

Another approach to the evolution of semelparity is to consider the relationship **reproductive allocation and** between reproductive allocation and reproductive success per unit allocation **success rate** (Schaffer & Rosenzweig, 1977). Semelparity will tend to be favoured where unit success rate increases with total allocation. This might occur, for example, where plants with larger inflorescences are more attractive to pollinators (leading to a higher proportion of fruit set). Certainly, there are several plant genera where this relationship exists amongst the semelparous but not amongst the iteroparous species (Figure 14.18).

549   LIFE-HISTORY VARIATION

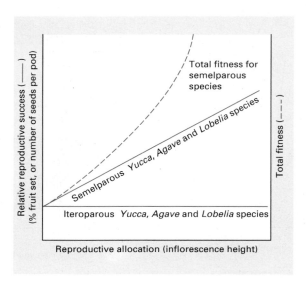

Figure 14.18 Relative reproductive success increases with reproductive allocation (inflorescence height) amongst the semelparous species of three genera of long-lived plants, but not amongst the iteroparous species. The positive relationship in the former itself suggests that total fitness (the product of allocation and relative success) will increase exponentially with increasing allocation—favouring the reproductive recklessness of semelparity. (After Young & Augspurger, 1991.)

14.8 The size and number of offspring

a smaller number of larger offspring in offspring size-sensitive habitats

To begin with the classification of Section 14.6, the division of a given reproductive allocation into a smaller number of larger offspring is expected in relatively offspring size-sensitive habitats. In support of this, Salthe (1969) demonstrated a relationship between egg size and clutch size in urodele amphibians (newts and salamanders). Limiting himself to species that were within a small size range, he identified three reproductive modes.

1 Those abandoning their eggs in the open in still waters, where, presumably, mortality was most indiscriminate and the habitat most offspring size-insensitive, produced the largest number of smallest eggs.

2 Those with direct development, where eggs were deposited in nests beneath objects on land, where there was parental care and newly hatched individuals needed to compete for relatively large items of prey (and where, presumably, mortality was most offspring size-sensitive), produced the smallest number of largest eggs.

3 Finally, those depositing their eggs in relatively simple nests, for example beneath stones in running water, and able to feed on relatively small prey on hatching, produced eggs and clutches of an intermediate size. Further support is provided by the observations and experiments on guppies described previously (Table 14.1): offspring size was larger where predation was most concentrated on the smaller juveniles; and also by examples in Figure 14.9: offspring size was larger in habitats where competition was likely to be most intense—goldenrods in prairies (as opposed to successional old field habitats) and *Drosophila* on pollen (as opposed to rich but unpredictable sources of yeast).

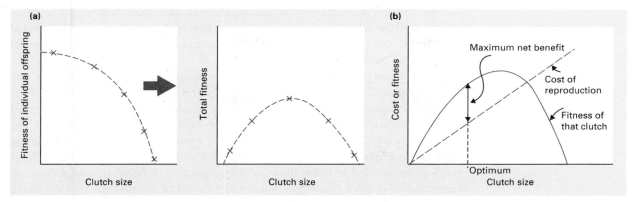

Figure 14.19 (a) The 'Lack clutch size'. If the fitness of each individual offspring decreases as total clutch size increases, then the total fitness of a clutch (the product of number and individual fitness) must be maximized at some intermediate ('Lack') clutch size. (b) However, if there is also a cost of reproduction, then the 'optimum' clutch size is that where the net fitness is greatest, i.e. here, where the distance between cost line and the 'benefit' (total clutch) curve is greatest. (After Charnov & Krebs, 1974.)

14.8.1 The number of offspring: clutch size

The offspring number and fitness trade-off, however, is perhaps best not viewed in isolation. Rather, if we combine it with the CR trade-off, we can turn to another of our types of life-history question and ask: 'How is it that particular clutch sizes, or particular sizes of seed crop, have been favoured?'

Lack (1947b) concentrated on the offspring number and fitness trade-off and proposed that natural selection will favour not the largest clutch size but a compromise clutch size, which, by balancing the number produced against their subsequent survival, leads to the maximum number surviving to maturity. This has come to be known as the 'Lack clutch size' (Figure 14.19a). A number of attempts, especially with birds and to a lesser extent with insects, have been made to test the validity of this proposal by adding eggs to or removing them from natural clutches or broods, determining which clutch size is ultimately the most productive, and comparing this with the normal clutch size. Most of these have suggested that Lack's proposal is wrong: the clutch size most commonly observed 'naturally' is not the most productive. Experimental increases in clutch size, in particular, often lead to apparent increases in productivity (Godfray, 1987; Dijkstra *et al.*, 1990; Lessells, 1991; Stearns, 1992). Nevertheless, as is so often the case, Lack's proposal, whilst wrong in detail, has been immensely important in directing ecologists towards an understanding of clutch size. A number of reasons for the lack of fit are now apparent (Lessells, 1991; Stearns, 1992) and two are particularly important.

In the first place, many of the studies are likely to have made an inadequate assessment of the fitness of individual offspring. It is not enough to add two eggs to a bird's normal clutch of four and note that six apparently healthy birds hatch, develop and fledge from the nest. How well do they survive the following winter? How many chicks do they have themselves? For example, in the egg parasitoid *Trichogramma evanescens*, where the clutch size is the number of eggs deposited in an individual host, larger than normal clutches appeared to be more productive—but when the effects of brood size on the daughters' own fecundity were taken into account, the

the 'Lack clutch size': probably wrong but certainly useful

551 LIFE-HISTORY VARIATION

ultimately most productive brood size was that actually observed (Waage & Godfray, 1985). On the other hand, this adjustment fails to bring observed and expected results into line in three other species of insect and does so in only two out of eight studies with birds (Lessells, 1991).

Second, perhaps the most important omission from Lack's proposal is any consideration of the cost of reproduction. Natural selection will favour a lifetime pattern of reproduction that gives rise to the greatest fitness overall. A large and apparently productive clutch may extract too high a price in terms of RRV. The favoured clutch size will then be less than what appears to be most productive in the short term (Figure 14.19b). On the other hand, of the few studies that have been sufficiently detailed to allow the cost of reproduction to be taken into account in assessing an optimal clutch size, fewer still (two out of seven—Lessells, 1991) have confirmed a correspondence between optimal and observed. (One of these, on kestrels, is discussed below—see Section 14.11.) It remains to be seen whether there are serious shortcomings in our understanding of clutch size, or whether the need is simply for more detailed and comprehensive sets of data. Either way, the study of clutch size illustrates the difficulties to be encountered when life-history investigations move from the comparative ('bigger?', 'slower?') to the quantitative ('three?', 'seventeen?').

combining with CR trade-offs

14.9 *r* and *K* selection

Some of the predictions of the previous sections can be brought together in a concept which has been particularly influential in the search for life-history patterns. This is the concept of *r* and *K* selection, originally propounded by MacArthur and Wilson (1967; MacArthur, 1962) and elaborated by Pianka (1970) (but, see Boyce, 1984). The letters refer to parameters of the logistic equation (see Chapter 6, Section 6.10). They are used to indicate that *r*-selected individuals have been favoured for their ability to reproduce rapidly (i.e. have a high *r*-value), whilst *K*-selected individuals have been favoured for their ability to make a large proportional contribution to a population which remains at its carrying capacity (*K*). The concept is therefore based on there being two contrasting types of habitat: *r*-selecting and *K*-selecting. Like all generalizations, this dichotomy is an oversimplification—but one that has been immensely productive.

K-selected individuals in K-selecting environments

A *K*-selected population lives in a habitat that experiences very little by way of random environmental fluctuations. As a consequence, a crowded population of fairly constant size is established. There is intense competition amongst the adults, and the results of this competition largely determine the adults' rates of survival and fecundity. The young also have to compete for survival in this crowded environment, and there are few opportunities for the young to become established as breeding adults themselves. In short, the population lives in a habitat that, because of intense competition, is both high-CR and offspring size-sensitive.

The predicted characteristics of these *K*-selected individuals are therefore larger size, deferred reproduction, iteroparity (i.e. more extended reproduction), a lower reproductive allocation and larger (and thus fewer) offspring with more parental care. The individuals will generally invest in increased survival (as opposed to reproduction); but in practice (because of the intense competition) many of them will have very short lives.

By contrast, an *r*-selected population lives in a habitat which is either unpredictable in time or short lived. Intermittently, the population experiences benign periods of rapid population growth, free from competition (either when the environment fluctuates into a favourable period, or when a site has been newly colonized). But these benign periods are interspersed with malevolent periods of unavoidable mortality (either in an unpredictable, unfavourable phase, or when an ephemeral site has been fully exploited or disappears). The mortality rates of both adults and juveniles are therefore highly variable and unpredictable, and they are frequently independent of population density and of the size and condition of the individuals concerned. In short, the habitat is both low-CR and offspring size-insensitive.

The predicted characteristics of *r*-selected individuals are therefore smaller size, earlier maturity, possibly semelparity, a larger reproductive allocation and smaller (and thus more) offspring. The individuals will invest little in survivorship, but their actual survival will vary considerably depending on the (unpredictable) environment in which they find themselves.

The '*r* / *K* concept', then, envisages two contrasting types of individual (or population or species), and predicts the association of *r*-type individuals with *r*-selecting environments, and *K*-type individuals with *K*-selecting environments. The concept originally emerged (MacArthur & Wilson, 1967) from the contrast between species that were good at rapidly colonizing relatively 'empty' islands (*r* species), and species that were good at maintaining themselves on islands once many colonizers had reached there (*K* species). Subsequently, the concept was applied much more generally.

The scheme is in fact a special case of the general classification of habitats in Figure 14.11c. Note, therefore, first, that adult and offspring habitats need not be linked in the way the *r* / *K* scheme envisages, and second, that the life-history characteristics associated with the *r* / *K* scheme can arise for all sorts of reasons beyond its scope (e.g. predation of diminutive adults as opposed to intense competition amongst adults; mortality of large offspring as opposed to a fluctuation between benign periods and indiscriminate mortality; and so on).

14.9.1 Evidence for the *r* / *K* concept

The *r* / *K* concept can certainly be useful in describing some of the general differences amongst taxa. For instance, amongst plants it is possible to draw up a number of very broad and general relationships (Figures 14.20a–c). Trees in relatively *K*-selecting, woodland habitats (relatively constant and predictable), exhibit long life, delayed maturity, large seed size, low reproductive allocation, large individual size and a very high frequency of iteroparity; whilst in more disturbed, open, *r*-selecting habitats, the plants tend to conform to the general syndrome of *r* characteristics. There is, however, one important respect in which plants and other modular organisms stand apart. Modular growth gives the genet the potential to increase in size exponentially by increasing its number of modules (branching corals, trees or clonal herbs). Delayed reproduction does not then necessarily delay population growth. Indeed, the number of descendant progeny produced by a zygote may well be greater when there is exponential modular growth followed by a burst of reproduction than when there are repeated short generations. This may account for the rather high frequency of species with clonal growth (and potentially infinite life) in *r*-selecting habitats.

There are also many cases in which populations of a species, or of closely related species, have been compared, and the correspondence with the *r* / *K* scheme has

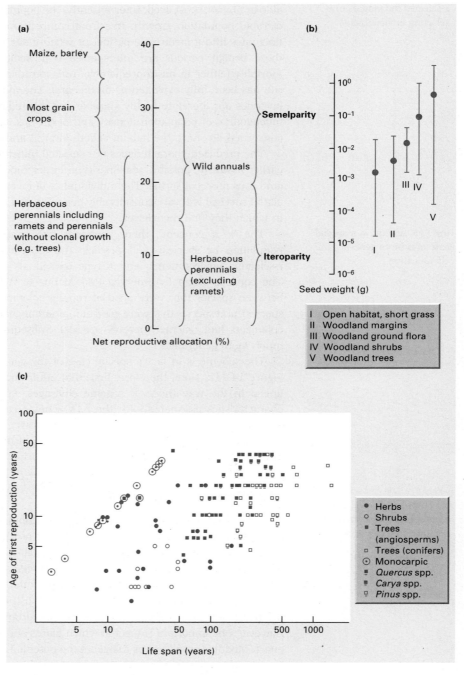

Figure 14.20 Broadly speaking, plants show some conformity with the r/K scheme. For example, trees in relatively K-selecting woodland habitats: (a) have a relatively high probability of being iteroparous and a relatively small reproductive allocation; (b) have relatively large seeds; and (c) are relatively long lived with relatively delayed reproduction. (After Harper, 1977; from Salisbury, 1942; Ogden, 1968; Harper & White, 1974.)

Table 14.2 Life-history traits of two *Typha* (cattail) species, along with properties of the habitats in which they grow. '$s^2 / \bar{x}$' refers to the variance : mean ratio, a measure of variability. The cattails conform to the r / K scheme. (After McNaughton, 1975.)

Habitat property	Measured by	Growing season	
		Short	Long
Climate variability	$s^2 / \bar{x}$ frost-free days per year	3.05	1.56
Competition	Biomass above ground (g m^{-2})	404	1336
Annual recolonization	Winter rhizome mortality (%)	74	5
Annual density variation	$s^2 / \bar{x}$ shoot numbers m^{-2}	2.75	1.51

Plant traits	*T. angustifolia*	*T. domingensis*
Days before flowering	44	70
Mean foliage height (cm)	162	186
Mean genet weight (g)	12.64	14.34
Mean number of fruits per genet	41	8
Mean weights of fruits (g)	11.8	21.4
Mean total weight of fruits (g)	483	171

been good. For instance, this is true of a study of *Typha* (cattail or reed mace) populations (Table 14.2). Individuals of a southerly species, *T. domingensis*, and a northerly species, *T. angustifolia*, were taken from sites in Texas and North Dakota, respectively, and grown side-by-side under the same conditions. In addition, certain aspects of the habitats with long and short growing seasons in which these species are found were quantified. It is clear from Table 14.2 that the former were relatively *K*-selecting and the latter relatively *r*-selecting. It is equally clear that the species inhabitating these sites conform to the r / K scheme. *Typha angustifolia* (which naturally has a short growing season) matures earlier (trait 1), is smaller (traits 2 and 3), makes a larger reproductive allocation (traits 3 and 6) and produces more and smaller offspring (traits 4 and 5) than does *T. domingensis* (long growing season).

the r / K scheme explains much, but leaves as much unexplained

There are, then, examples that fit the r / K scheme. Stearns (1977), however, in an extensive review of the data available at that time, found that of 35 thorough studies, 18 conformed to the scheme whilst 17 did not. We might regard this as a damning criticism of the r / K concept, since it undoubtedly shows that the explanatory powers of the scheme are limited. On the other hand, a 50% success rate is hardly surprising given the number of additional factors already described (or to be described) which further our understanding of life-history patterns. It is therefore equally possible to regard it as very satisfactory that a relatively simple concept can help make sense of a large proportion of the multiplicity of life histories. Nobody, though, can regard the r / K scheme as the whole story.

14.10 Further classifications of habitat

14.10.1 Triangular classifications for plants

Another way of relating types of life history to types of habitat is to begin with the organisms—grouping species together on the basis of similarities in their life histories and then asking what (if anything) the habitats of these species have in common. This approach has been taken with plant life histories by Silvertown *et al.* (1992b,

1993). The typical plant life cycle was divided into six 'transitions':

1 recruitment of seeds to the seed pool;
2 recruitment of seedlings from current seed production;
3 clonal growth;
4 retrogression (e.g. reverting from a flowering to a vegetative state);
5 stasis (i.e. survival from one year to the next in the same stage);
6 progression to later stages.

Then, for a number of species, ignoring technical details, the proportional 'importance' of each transition to the overall reproductive rate, R, was determined. In fact, (1) and (2) were combined as fecundity, F; (3) and (6) as growth, G; and (4) and (5) as survival, L. Individual species could then be placed on triangular diagrams with F, G and L as axes, from which it is apparent (Figure 14.21) that species of particular life forms (herbs, woody plants) and from particular types of environment (open habitats, forests) tend to be found together on the diagrams.

In fact, there is a tradition of triangular classifications in the study of plant life histories. Grime (1974; 1979; Grime *et al.*, 1988) developed a classification of habitats and plant life histories in which habitats, in the established phase of the plant, were seen as varying: (i) in their level of disturbance (brought about by herbivores,

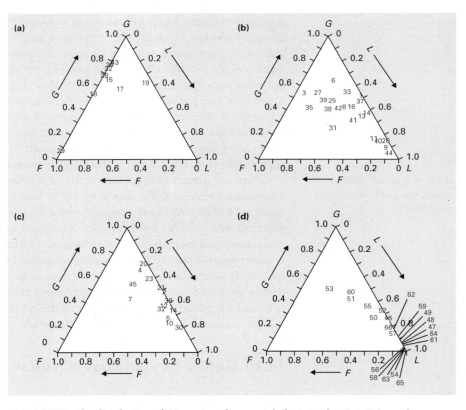

Figure 14.21 The distribution of 66 species of perennial plants in the G–L–F (growth, survival, fecundity) triangle. Note that G, L and F values are defined such that they always sum to 1. (a) Semelparous herbs, (b) iteroparous herbs of open habitats, (c) iteroparous forest herbs, (d) woody plants. (After Silvertown *et al.*, 1993, where the names of each of the numbered species may be found.)

pathogens, trampling, environmental disasters, etc.); and (ii) in the extent to which they experience what Grime called 'stress'—shortages of light, water, minerals, etc. which limit the rate of photosynthesis (although a habitat's level of disturbance or 'stress' will depend on which species is experiencing them). Grime suggests that a stress-tolerant strategy (S) is appropriate when 'stress' is severe but disturbance is uncommon. Conversely, a ruderal strategy (R) is appropriate when disturbance levels are high but conditions are benign and resources abundant. Finally, when disturbance is rare and resources are abundant, a crowded population is established and a competitive strategy (C) is appropriate. Plants were classified on the basis of indices composed of characteristics thought to suit the plants for one or other of these strategies, and each plant was placed in the CSR triangle. Then, like Silvertown, Grime was able to note the positions of groups of plants of different types, or different stages in succession, and so on, in different parts of the triangle. Unfortunately, reconciling the two triangular classifications may not be straightforward (Silvertown *et al.*, 1992b).

14.10.2 'Bet-hedging'

Our previous classifications of habitats, and a number of others, have contrasted the average conditions in one habitat with the average conditions in another. However, it is indisputable that all environments fluctuate and all populations suffer some random variations in birth rates, mortality rates, and so on. A number of modelling studies have therefore contrasted habitats in terms not of means but of variances (Orzack & Tuljapurkar, 1989; Phillipi & Seger, 1989), although as Stearns (1992) has remarked, the approach has reached a level of sophistication that not only restricts the potential audience but also goes well beyond the available sets of data.

A general idea of the approach can be gained by considering Schaffer's (1974) distinction between habitats in which adult mortality rates are the most variable and those in which juvenile rates are most variable (and adults might therefore be expected to 'hedge their bets' and not release all of their offspring at the same time). Schaffer's mathematical calculations predict high reproductive allocations where adult mortality rates fluctuate most, and lower allocations and iteroparity where juvenile rates fluctuate most. This, though, complements rather than contradicts our previous considerations, since unpredictable, unavoidable mortality of established adults is one way in which a habitat might be relatively low-CR.

14.11 Phenotypic plasticity

A life history is not a fixed property that an organism exhibits irrespective of the prevailing environmental conditions. An observed life history is the result of long-term evolutionary forces, but also of the more immediate responses of an organism to the environment in which it is and has been living. We have already seen from several examples in Chapter 6, for instance, that the relative allocations to different plant parts, and to different reproductive parts, frequently changes as population density increases. This ability of a single genotype to express itself in different ways in different environments is known as phenotypic plasticity.

polyphenism: seed dormancy

Phenotypic plasticity may express itself as variation in the proportion of discrete, alternative types in different environments. This is known as *polyphenism*. For example, this is true for the production of yellow (dormant) and green (non-dormant) seeds by the desert annual, *Ononis sicula*. The switch is determined by the relative rates of development of the seed pod and the seeds within it. Under long days, the development of the seeds is rapid so that by the time they are shed, the seed coat is hard, impermeable to water and the seed itself is dormant. Under short days, however, seed development is slower and seeds are shed whilst their coats are still soft, green, permeable to water, and hence not dormant (Evenari *et al.*, 1966; Silvertown, 1984).

reaction norms ... and their
genetic variation in *Drosophila*

Alternatively, phenotypic plasticity may express itself as continuous variation, in which case the relationship between the values of life-history traits and an environmental variable is known as a *reaction norm*. For instance, Figure 14.22 illustrates a number of reaction norms for the way in which the combination of the age and weight at which adult *Drosophila mercatorum* hatch from their pupae varies as the concentration of yeast in their food is altered. Each reaction norm refers to the offspring from a single cross (manipulated such that there was genetic variation between, but only minimal variation within, the results of each cross), and each is depicted as a polygon representing the 95% confidence envelope of the reaction norm. The trend in each was the same: as yeast concentration decreased, individuals hatched later and smaller. There was, however, considerable genetic variation (between crosses) in this trend. As a consequence, the relationship between age and size was positive at high yeast concentrations but negative at low concentrations. Not only are individual life-history traits subject to the influence of the immediate environment, but so too are relationships between traits.

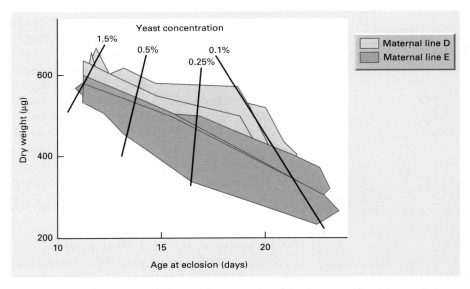

Figure 14.22 Reaction norms (95% confidence envelopes) for the age and weight at eclosion of *Drosophila mercatorum*, as the concentration of yeast in its food is altered. Six males, each from a different isofemale line, were crossed with females from one of two maternal lines that had been maintained parthenogenetically for several years (giving rise to genetic variation between, but only minimal variation within, the results of each cross). One of the 12 crosses failed. For clarity, not all of the remaining crosses are depicted. Also shown are the relationships, at each yeast concentration, between the average weights and ages of each cross: the sign of the slope itself alters with yeast concentration. (After Gebhardt & Stearns, 1988; Stearns, 1992.)

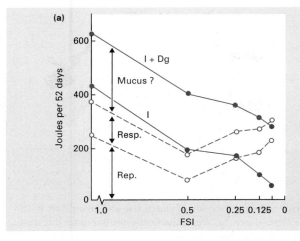
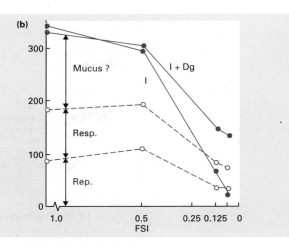

Figure 14.23 (a) Energy allocation by the flatworm *Bdellocephala punctata*, to respiration (Resp.), reproduction (Rep.) and secretion/excretion (Mucus ?). The energy available (I + Dg) comprises the food ingested (I) at different levels (indices) of food supply (FSI), plus the amount of degrowth (Dg). Degrowth is the mobilization of somatic tissue, making energy available for other processes. (b) Comparable data for the flatworm *Dugesia lugubris*. (After Woollhead, 1983.) The life histories are affected by the food supply, but the effect is different in the two species, as discussed in the text.

One of the most important questions we need to ask about phenotypic plasticity is the extent to which it represents a response by which an organism allocates resources differently in different environments such that it maximizes its fitness in each. The alternative would be that the response represented a degree of inevitable or uncontrolled damage or stunting by the environment (Lessells, 1991). Note, especially, that if phenotypic plasticity is governed by natural selection, then it is just as valid to seek patterns linking different environments and the different responses to them by a single individual, as it is to seek patterns linking habitats and the life histories of genetically different individuals. It seems likely that phenotypic plasticity is governed by natural selection. Understanding the details of its evolution, however, has proved difficult (Via & Lande, 1985; de Jong, 1990; Scheiner & Lyman, 1991; Stearns, 1992).

phenotypic plasticity and optimization in flatworms …

In some cases at least, the appropriateness of the plastic response seems clear. For example, Figure 14.23 illustrates the contrasting responses of two flatworm species to various levels of food supply (Woollhead, 1983). Of these, *Bdellocephala punctata* makes a high reproductive allocation at low food levels. Its hatchlings cope well with resource shortage. *Dugesia lugubris*, on the other hand, makes a low reproductive allocation at low food levels. Its hatchlings are far more susceptible to resource shortage.

… and kestrels

At a more detailed level, the apparent appropriateness of a plastic response has been demonstrated for kestrels with territories of differing quality in The Netherlands (Daan *et al.*, 1990). The kestrels vary in the size of their clutch and the date on which they lay it. This combination itself varies between territories, and since the differences appear not to be genetically determined, this seems to be an example of phenotypic plasticity. Is each combination optimal in its own territory?

The optimal combination of clutch size and laying date is, as usual, the one with

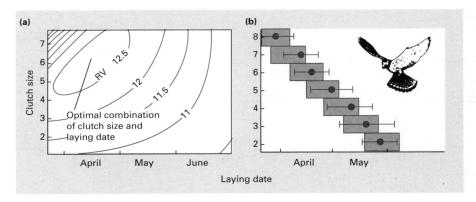

Figure 14.24 Phenotypic plasticity in the combination of clutch size and laying date in the kestrel, *Falco tinnunculus*, in The Netherlands. (a) Within particular territories (in this case, one of high quality) the expected, optimal combination is that with the highest total reproductive value (for calculation, see text). (b) Predicted (rectangles) and observed (points with standard deviations) combinations for territories varying in quality from high (left) to low (right). (After Daan *et al.*, 1990; Lessells, 1991.)

the highest total reproductive value—the value of the present clutch plus the parent's RRV. The value of the present clutch clearly increases with clutch size, and the value of each egg also varies with laying date. What, though, of RRV? This declines with increases in 'parental effort', i.e. number of hours per day in flight spent hunting in order to raise a clutch of chicks; and parental effort, in turn, decreases with increases in the 'quality' of a territory: the number of prey caught per hunting hour. Thus, RRV is lower: (i) with larger clutches; (ii) at particular, less productive times of the year; and (iii) in lower quality territories. On this basis, the total reproductive value of each combination of clutch size and date in each territory could be computed, the optimal combination predicted (Figure 14.24a) and the predicted and actual combinations compared in territories of different quality (Figure 14.24b). The correspondence is impressive. Each individual apparently comes close to optimizing its clutch size and laying date as an immediate response to the environment (territory) in which it finds itself.

14.12 Phylogenetic and allometric constraints

The life histories that natural selection favours (and we observe) are not selected from an unlimited supply. The favoured life history for a buttercup is selected from amongst those available to buttercups. An ecologist might, quite reasonably, try to explain how it comes about that buttercup species differ in their life histories—but it is no part of an ecologist's job to explain why buttercups are not butterflies. In other words, an organism's life history is constrained by the phylogenetic or taxonomic position that it occupies. For example, in the entire order Procellariiformes (albatrosses, petrels, fulmars) the clutch size is one, and the birds are 'prepared' for this morphologically by having only a single brood patch with which they can incubate this one egg (Ashmole, 1971). A bird might produce a larger clutch, but this is bound to be a waste unless it exhibits concurrent changes in all the processes in the development of the brood patch. Albatrosses are therefore prisoners of their evolutionary past, as are all organisms. Their life histories can evolve to only a

organisms are, to some extent, prisoners of their evolutionary past

limited number of options, and the organisms are therefore confined to a limited range of habitats.

It follows from these 'phylogenetic' constraints that caution must be exercised when life histories are compared. The albatrosses, as a group, may be compared with other types of birds in an attempt to discern a link between the typical albatross life history and the typical albatross habitat. The life histories and habitats of two albatross species might reasonably be compared. But if an albatross species is compared with a distantly related bird species, then care must be taken to distinguish between differences attributable to habitat (if any) and those attributable to phylogenetic constraints.

life histories reflect both habitat and phylogeny

14.12.1 The effects of size and allometry

phylogeny affects size, affects life history

One element of phylogenetic constraint is the constraint of size. Figure 14.25a shows the relationship between time to maturity and size (weight) in a wide range of organisms from viruses to whales. Note first that particular groups of organisms are confined to particular size ranges. For instance, unicellular organisms cannot exceed a certain size because of their reliance on simple diffusion for the transfer of oxygen from their cell surface to their internal organelles. Insects cannot exceed a certain size because of their reliance on unventilated tracheae for the transfer of gases to and from their interiors. Mammals, being endothermic, must exceed a certain size, because at smaller sizes the relatively large body surface would dissipate heat faster than the animal could produce it. Other groups are also confined to their own size range in other ways.

The second point to note is that time to maturity and size are strongly correlated. In fact, as Figures 14.25a–c illustrate, size is strongly correlated with many life-history components. Since the sizes of organisms are constrained within limits, these other life-history components will be constrained too. The life history of an organism can therefore once again be seen to be constrained by the phylogenetic position of that organism.

allometry defined

An allometric relationship (see Gould, 1966) is one in which a physical or physiological property of an organism alters relative to the size of the organism. For example, in Figure 14.26a an increase in size (actually volume) amongst salamander species leads to a decrease in the *proportion* of that volume which is allocated to a clutch of young. Likewise, in Figure 14.25b an increase in weight amongst bird species is associated with a decrease in the time spent brooding eggs *per unit body weight*. Such allometric relationships can be ontogenetic (changes occurring as an organism develops) or phylogenetic (changes which are apparent when related taxa of different size are compared), and it is the latter that are particularly important in the study of life histories (Figures 14.25 and 14.26).

Why are there allometric relationships? Briefly, if similar organisms that differed in size retained a geometric similarity (i.e. if they were *iso*metric), then all surface areas would increase as the square of linear size, whilst all volumes and weights increased as the cube. An increase in size would then lead to decreases in length : area ratios, decreases in length : volume ratios and, most important, decreases in area : volume ratios. Almost every bodily function depends for its efficiency on one of these ratios (or a ratio related to them). A change in size amongst isometric organisms would therefore lead to a change in efficiency.

without allometry, physiological efficiency would change with size

For example, the transfer of heat, or water, or gases or nutrients, either within an

561   LIFE-HISTORY VARIATION

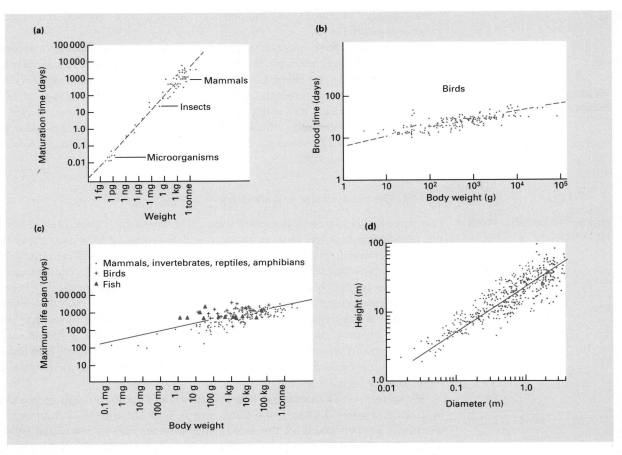

Figure 14.25 Allometric relationships, all plotted on log scales. (a) Maturation time as a function of body weight for a broad range of animals. (b) Brood time as a function of maternal body weight in birds. (c) Maximum life span as a function of adult body weight for a broad range of animals. (After Blueweiss *et al.*, 1978.) (d) The allometric relationship between tree height and trunk diameter 1.525 m from the ground, for 576 individual 'record' trees representing nearly every American species. (After McMahon, 1973.)

organism or between an organism and its environment, takes place across a surface, which has an area. The amount of heat produced or water required, however, depends on the volume of the organ or organism concerned. Hence, changes in area : volume ratios resulting from changes in size are bound to lead to changes in the efficiency of transfer per unit volume. Thus, if efficiency is to be maintained, this must be done by allometric alterations. Similarly, the weight of any part of an organism depends on its volume, whilst the strength of a structure supporting that weight depends on its cross-sectional area. Increases in size will therefore lead to problems of strength unless there are allometric alterations to compensate (as there are in the trees in Figure 14.25d).

In most examples, these simple area : volume relationships interact with one another and depend on the detailed morphology and physiology of the organisms concerned. Allometric slopes therefore vary from system to system and from taxon to taxon (for further discussion see Gould, 1966; Schmidt-Nielsen, 1984; and, in a more

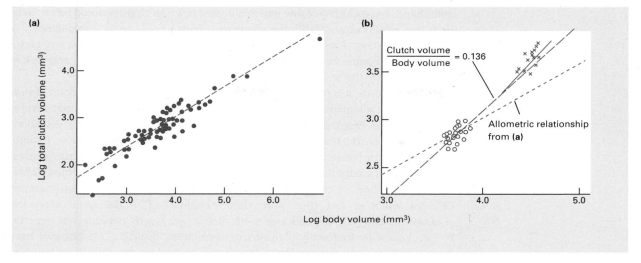

Figure 14.26 Allometric relationships between total clutch volume and body volume in female salamanders. (After Kaplan & Salthe, 1979.) (a) The overall relationship for 74 salamander species, using one mean value per species ($P < 0.01$). (b) The relationships within a population of *Ambystoma tigrinum* ($\times$) ($P < 0.01$), and within a population of *A. opacum* ($\bigcirc$) ($P < 0.05$). The allometric relationship from (a) is shown as a dashed line: *A. opacum* conforms closely to it; *A. tigrinum* does not. However, they both lie on an isometric line along which clutch volume is 13.6% of body volume (— — —).

ecological context, Peters, 1983). What, though, is the significance of allometry in the study of life histories?

The usual approach to the ecological study of life histories has been to compare the life histories of two or more populations (or species or groups), and to seek to **the properties of allometries and their consequences for life histories** understand the differences between them by reference to their environments. It must be clear by now, however, that taxa can also differ because they lie at different points on the same allometric relationship, or because they are subject to different phylogenetic constraints generally. It is therefore important to disentangle 'ecological' differences from allometric and phylogenetic differences (see, particularly, Pagel & Harvey, 1988; Harvey & Pagel, 1991; and also the summary in Stearns, 1992). This is not because the former are 'adaptive' whereas the latter are not. Indeed, we have seen, for example, that the basis for allometric relationships is a matching of organisms of different sizes to their respective environments. Rather, it is a question of the range of evolutionary responses of a species, say, to its habitat being limited by constraints that have themselves evolved.

These ideas are illustrated in Figure 14.26a, which shows the allometric relationship between clutch volume and body volume for salamanders generally. Figure 14.26b then shows the same relationships in outline; but superimposed upon it are the allometric relationships within populations for two salamander species, *Ambystoma tigrinum* and *A. opacum* (Kaplan & Salthe, 1979). If the species' means are compared without reference to the general salamander allometry, then the species are seen to have the same ratio of clutch volume : body volume (0.136). This seems **comparing salamanders: dangerous if the allometry is ignored** to suggest that the species' life histories 'do not differ', and that there is therefore 'nothing to explain'—but any such suggestion would be wrong. *Ambystoma opacum* conforms closely to the general salamander relationship. *Ambystoma tigrinum*, on the

563   LIFE-HISTORY VARIATION

other hand, has a clutch volume which is almost twice as large as would be expected from that relationship. Within the allometric constraints of being a salamander, *A. tigrinum* is making a much greater reproductive allocation than *A. opacum*; and it would be reasonable for an ecologist to look at their respective habitats and seek to understand why this might be so.

In other words, it is reasonable to compare taxa from an 'ecological' point of view as long as the allometric relationship linking them at a higher taxonomic level is known (Clutton-Brock & Harvey, 1979). It will then be their respective deviations from the relationship that form the basis for the comparison. Problems arise, though, when allometric relationships are unknown (or ignored). Without the general salamander allometry in Figure 14.26b, the two species would have seemed similar when in fact they are different. Conversely, two other species might have seemed different when in fact they were simply conforming to the same allometric relationship. Comparisons oblivious to allometries are clearly perilous, but regrettably, ecologists *are* frequently oblivious to allometries. Typically, life histories have been compared, and attempts have been made to explain the differences between them in terms of habitat differences. As previous sections have shown, these attempts have often been successful. But they have also often been unsuccessful, and unrecognized allometries undoubtedly go some way towards explaining this.

14.12.2 The effects of phylogeny

the influence of phylogeny after the effects of size have been removed

The approach used with the salamanders, of comparing species or other groups in terms of their deviations from an allometric relationship that links them, has been applied successfully to a number of larger assemblages. Note that because the effects of size are removed, the approach searches for phylogenetic relationships (reflecting constraints) beyond those associated with size or allometry. For example, Figure 14.27a shows, for a number of mammal species, that 'relative' age at first reproduction increases as 'relative' life expectancy increases (i.e. relative to a value expected on the basis of an underlying allometry). This reveals underlying similarities between species of very different sizes: elephants and otters, and mice and warthogs. It also shows a powerful relationship between these two life-history characters once the confounding effects of size have been removed. Another example (Figure 14.27b) shows that, amongst orders of eutherian mammals, once the effects of body size have been removed, the period of maternal investment decreases as annual fecundity increases. Bats (Chiroptera) have the longest period of investment but the lowest fecundity (closely followed by the primates): rabbits (Lagomorpha) lie at the opposite extreme.

The relationship in Figure 14.27a is clearly related to that previously discussed in Figure 14.16: not only are life span and age at maturity closely related in mammals once the effects of size have been removed (Figure 14.27a), they are also related when the effects of size are ignored, and this relationship differs significantly from that found in other groups (Figure 14.16). Together, these results stress the powerful influence that an organism's phylogenetic position can have on its life history.

Some impression of the strength of this influence can be gained from analyses like those in Table 14.3 (Read & Harvey, 1989). A nested analysis of variance has been applied to the variation in seven life-history traits amongst a large number of mammal species. This has led to the determination of the percentage of the total variance attributable to: (i) differences between species within genera; (ii) differences

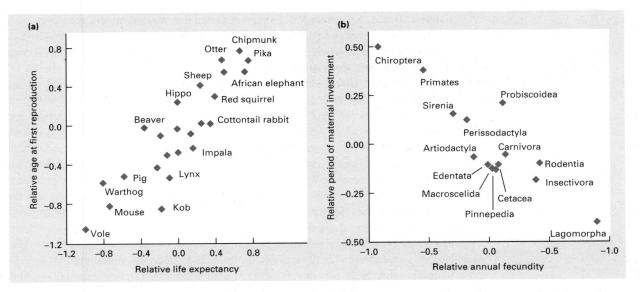

Figure 14.27 (a) After the effects of size have been removed, age at first reproduction increases with life expectancy at birth for 24 species of mammals. 'Relative' refers to the deviation from the underlying allometric relationship linking the character in question to organism size. (After Harvey & Zammuto, 1985.) (b) After the effects of size have been removed, the period of maternal investment decreases as annual fecundity increases for the orders of eutherian mammals. (After Read & Harvey, 1989.)

between genera within families; and so on. Species vary very little, independently of their phylogenetic background; the same is true of genera. Far and away the largest part of the variance, for all the traits, is accounted for by differences between orders within the mammalian class as a whole. This emphasizes that in simply comparing two species from different orders, we are in essence comparing those orders (which probably diverged many millions of years ago) rather than the species themselves. It does not mean, however, that because most of the life-history variation to be explained resides at the higher taxonomic levels, comparing species is only 'scratching the surface'. There is much to be gained, as we have seen, from comparisons. Even when two species are very similar in their life histories and their

Table 14.3 When nested analyses of variance are performed on data sets for a number of life-history traits from a large number of mammal species, the percentage of the variance is greatest at the highest taxonomic level (orders within the class) and least at the lowest level (species within genera). (After Read & Harvey, 1989.)

Trait	Species within genera	Genera within families	Families within orders	Orders within the class
Gestation length	2.4	5.8	21.1	70.7
Age at weaning	8.4	11.5	18.9	61.6
Age at maturity	10.7	7.2	26.7	55.4
Interlitter interval	6.6	13.5	16.1	63.8
Maximum life span	9.7	10.1	12.4	67.8
Neonatal weight	2.9	5.5	26.6	64.9
Adult weight	2.9	7.5	21.0	68.5

565 LIFE-HISTORY VARIATION

habitats, if one, say, makes a greater reproductive allocation and also lives in a habitat which is lower-CR, then this allows us to build a pattern linking the two.

Moreover, the strength of these relationships at the higher taxonomic levels does not mean that attempts to relate life histories to life styles and habitats should be abandoned even there, since life styles and habitats too are constrained by an organism's size and phylogenetic position. There may still, therefore, at these higher levels, be patterns linking habitats and life histories—rooted in natural selection. For example, insects (small size, many offspring, high reproductive allocation, frequent semelparity) have been described as relatively r-selected, compared to mammals (large size, few offspring, etc.—relatively K-selected) (Pianka, 1970). Such differences could be dismissed as being 'no more' than the product of an ancient evolutionary divergence (Stearns, 1992). But, as we have stressed, an organism's habitat reflects its own responses to its environment and is best captured by some general relationship such as those discussed in Section 14.6. Hence, a mammal and an insect living side by side are also almost certain to experience very different habitats. The larger, homeothermic, behaviourally sophisticated, longer-lived mammal is likely to maintain a relatively constant population size, subject to frequent competition, and relatively immune from environmental catastrophes and uncertainties. The smaller, poikilothermic, behaviourally unsophisticated, shorter-lived insect, by contrast, is likely to live a relatively opportunistic life, with a high probability of unavoidable death. Insects and mammals are prisoners of their evolutionary past in their range of habitats just as they are in their range of life histories—and the r/K scheme provides a reasonable (although certainly not perfect) summary of the patterns linking the two.

The same point is illustrated in a more quantitative way by an application of the 'phylogenetic-subtraction' method (Harvey & Pagel, 1991) to patterns of covariation in 10 life-history traits of mammals (Stearns, 1983). In the unmanipulated data, the pattern to be expected under the influence of r/K selection was pronounced: it accounted for 68% of the covariation. This r/K influence was reduced to about 42% when the effects of weight were removed, and was reduced still further when the trait values themselves were replaced in the analysis by their deviations from the mean value for the family to which the species belonged (33%), or their order (32%). In the first place, this reaffirms the importance of both size and phylogeny, since each clearly accounted for much of the interspecific variation. It is also significant that the r/K pattern remained clearly visible even after these other effects had been removed. But the strength of the pattern in the unmanipulated data set cannot simply be dismissed as an artefact arising out of phylogenetic naivety. It may well be that important differences in habitat, too, are associated with an organism's size, or its order or family.

It is undoubtedly true that life-history ecology cannot proceed oblivious to phylogenetic and allometric constraints. Yet it would be unhelpful to see phylogeny as an alternative explanation to habitat in seeking to understand life histories. Phylogeny sets limits to an organism's life history and to its habitat. But the essentially ecological task of relating life histories to habitats remains the most fundamental challenge.

Chapter 15
Abundance

15.1 Introduction: the interpretation of census data

Every species of plant and animal is always absent from almost everywhere. But, a large part of the science of ecology is concerned with trying to understand what determines the abundance of species in the restricted areas where they do occur. Why are some species rare and others common? Why does a species occur at low population densities in some places and at high densities in others? What factors cause fluctuations in a species' abundance? These are crucial questions. To provide complete answers for even a single species in a single location, we might need, ideally, a knowledge of physicochemical conditions, the level of resources available, the organism's life cycle and the influence of competitors, predators, parasites, etc., and an understanding of how all these things influence abundance through effects on rates of birth, death, dispersal and migration. In previous chapters, we have touched on each of these topics separately. We now try to bring them together and see how we might discover which factors actually matter in particular examples.

The raw material for the study of abundance is usually a census. In its crudest form, this consists of a list of presences and absences in defined sample areas. The most detailed censuses, on the other hand, involve counting individuals (and their parts, in the case of modular organisms), recognizing individuals of different age, sex, size and dominance and even distinguishing genetic variants.

If a census (over an area or over time) simply records the number of individuals present, we can attempt to 'explain' variations by correlating them with factors outside the population (weather, soil conditions, numbers of predators, and so on). But, the fact that two variables are correlated does not, of course, mean that it is the variation in one that causes the variation in the other! Moreover, a simple record of the numbers of individuals in a population can hide vital information. As an example, picture three human populations, shown by census to contain identical numbers of individuals. One of these is an old people's residential area, the second is a population of young children and the third is a population of mixed age and sex. No amount of attempted correlation with factors outside the population would reveal that the first was doomed to extinction (unless maintained by immigration), the second would grow fast but only after a delay and the third would continue to grow steadily.

The ecologist usually has to deal with census data that are grossly deficient in detail and there are many reasons for this.

1 It is usually a technically formidable task to follow individuals in a population throughout their lives, although it may be relatively easy at some stages (it is often

easy to count birds at nesting time or frogs at mating sites, but harder at other seasons). Often, a crucial stage in the life cycle is hidden from view—baby rabbits within their warrens, butterflies and moths as buried pupae and seeds in the soil. It is possible to mark birds with numbered leg-rings, young fish with dyes or metal tags, roving carnivores with radiotransmitters or seeds with radioactive isotopes, and to recognize marked individuals on resampling; but, the species and the numbers that can be censused in this way are severely limited. Only plants and sessile animals conveniently stay put and wait to be counted: even then, censusing the dispersal phases in their life histories poses enormous problems.

2   The results of a census will be misleading unless sampling is adequate over both space and time, and adequacy of either usually requires great commitment of time and money. The lifetime of investigators, the hurry to produce publishable work and the short tenure of most research programmes all deter individuals from even starting to make a census over an extended period of time. But, there are a few remarkable examples of long-term censuses and a large part of population theory depends on these few. Most of them are listed and described in Likens (1987), and particularly by Taylor (1987) in a survey of long-term studies in ecology.

3   As knowledge about populations grows, so the number of attributes that we hope to find recorded in a census grows and changes. Every census procedure is likely to be out of date almost as soon as it begins. Again and again, the analysis of census records shows that the next time it is done it should be done differently. Until the 1980s it was convenient to assume for census purposes that immigration and emigration balanced each other and could usually be ignored. This was until it was realized that the numbers of individuals in populations in nature might actually be determined by the balance between those leaving and those that stayed behind.

All the censuses described in this chapter are in one way or another less than ideal, judged in the light of what we now know we need to understand them fully. Furthermore, most of the really long-term or geographically extensive censuses have been made of organisms of economic importance such as fur-bearing animals, game birds, pest and disease organisms—or the furry and feathered favourites of amateur naturalists. Insofar as generalizations emerge, we should treat them with great caution.

The data that accumulate from a census may be used to establish correlations with *external* factors (e.g. the weather), or correlations between features within the census data (e.g. correlating numbers present in the spring with those present in the autumn). The correlations may then be used in either of two ways.

1   Correlations may be used to predict the future. For example, high intensities of the disease 'late blight' in the canopy of potato crops usually occur 15–22 days after a period in which the minimum temperature is not less than 10°C and the relative humidity is more than 75% for 2 consecutive days. This correlation may be used to predict outbreaks of the disease and alert the grower to the need for protective spraying. But, this does not imply that one is the cause of the other. A warm, humid environment permits but does not cause the disease.

2   Correlations may be used to suggest, although not to prove, causal relationships. For example, low populations of a pest may regularly follow cold winters. Such a correlation might suggest that cold winters kill the overwintering stage of the pest, but might just as well mean that cold winters favour its enemies or deplete part of its food resource. When we analyse variations in census records we need to be sceptical

about all claims to have discovered causes that are based only on the evidence of correlations.

The data obtained from a census are often used to suggest that *internal factors* are responsible for fluctuations in size of a population, for instance that it grew fastest when numbers were low and more slowly (or even began to fall) when numbers were high. Again, a correlation may be demonstrated (if so it is usually negative) between the size of a population and its growth rate (or its components: birth rates, death rates and migration rates). The correlation may hint that it is the size of the population itself that causes it to change but, like correlations with the climate or the weather, it does not prove that it is the cause. It may be that when the population is high many individuals starve to death, or fail to reproduce or become aggressive and drive out the weaker members. Only observing what is happening to the individuals will reliably reveal why the population changes its behaviour.

We need to be wary when we use words like 'abundance', 'population' and 'density' to describe the results of a census. We define the 'population' as a group of individuals of one species in an area—it may be the number of aphids living on one leaf, all the herons in the British Isles, all the human beings in China, all the buttercups in $1\ m^2$ of grassland or all the fruit-flies in one culture bottle. It is a word used quite pragmatically and often means simply the group of organisms that it is convenient and practicable to count. For geneticists the 'population' may mean those individuals of a species that are close enough to each other for there to be at least occasional mating between them.

Almost every population in nature is patchily distributed and the numbers of individuals vary from patch to patch (we have discussed some of the problems of describing patchiness in Chapter 5, Section 5.2). The term 'density' must be one of the most frequently used words in this book. But, it can mean very different things and again it is the patchiness of populations that causes problems. 'Density' is easy to define: it is the number of individuals per unit area (or, in some aquatic populations, per unit volume), but in a patchy community this does not tell us how crowded they are (see Lewontin & Levins, 1989; and Chapter 5, Section 5.2, for a discussion of this problem). Yet, it is often the degree of crowding, or the proximity of neighbours that we need to know about. We lose this information when we use the word 'density'. When organisms hold fixed positions, like most rooted plants, or barnacles, mussels and sea anemones on a rock, the distance between individuals is often a much more useful measure of their crowding than an abstraction into terms of numbers per unit area. The same may also be true in the case of animals that hold territories: the distance between them may again be a more useful measure than their 'density'. Information about the proximity of neighbours is particularly useful when we consider the ways in which individuals influence each other's activities—through competition or mutual exclusion.

Many of the studies that we discuss in this and other chapters have been concerned to detect 'density dependent' processes, as if density itself is the cause of changes in birth rates and death rates in a population. But, of course, this will rarely (if ever) be the case: organisms do not detect and respond to the density of their populations. They will usually respond to a shortage of resources caused by neighbours or mutual aggression. Of course, when we are considering motile organisms we can rarely identify which individuals have been responsible for the harm done to others, for example, which individuals ate the food items that then deprived another. We need continually to remember, when we use the word

populations are patchy

individuals do not respond to density

'density', that it is an abstraction that conceals what the real world is like as experienced in the lives of real organisms.

15.2 Fluctuation or stability

Perhaps the population census that covers the greatest time span is that of the swifts (*Micropus apus*) in the village of Selborne in southern England (Lawton & May, 1984). In one of the earliest published works on ecology, Gilbert White, who lived in the village, wrote (in 1778) of the swifts:

> I am now confirmed in the opinion that we have every year the same number of pairs invariably; at least, the result of my inquiry has been exactly the same for a long time past. The number that I constantly find are eight pairs, about half of which reside in the church, and the rest in some of the lowest and meanest thatched cottages. Now, as these eight pairs— allowance being made for accidents—breed yearly eight pairs more, what becomes annually of this increase?

Lawton and May visited the village in 1983, and found major changes in the 200 years since White described it. It is unlikely that swifts have nested in the church tower for 50 years, and the thatched cottages have disappeared or been covered with wire. Yet, the number of breeding pairs of swifts regularly to be found in the village is now 12. In view of the many changes that have taken place in the intervening centuries, this number is remarkably close to the eight pairs so consistently found by White.

many populations are very stable

A long-term study of nesting herons in the British Isles (where there is a large population of devoted ornithologists), reveals the same picture of a population that has remained remarkably constant over a long period (see Chapter 10, Figure 10.17). This census included seasons of severe weather when the population declined temporarily, but subsequently recovered. But, note that in this case the censused population was that of the whole of the British Isles and will have included the establishment of new colonies and the extinction of others.

Populations of plants can be mapped as part of a census. In a census of a population of creeping buttercups (*Ranunculus repens*) in an old permanent pasture in North Wales, Sarukhán and Harper (1973; Sarukhán, 1974) made detailed maps of the distribution of plants and seedlings and were able to follow the fate of every individual (an approach that is rarely possible for mobile animals). Deaths occurred throughout the year, with seasonal lethal periods occurring at the same time as the survivors were growing most rapidly. Additions to the population were made through two distinct processes: (i) the germination of seeds in annual flushes; and (ii) the clonal multiplication of rosettes. In clonal growth, the plants form rosettes of leaves; buds then grow out from the leaf axils to form elongated stolons and a new rooted rosette (or module) is formed at the stolon tip. A rosette may form several clonal daughters and these may, in turn, produce further stolons and rosettes in the course of a single season. In this way, a single genet (arising from a seedling) may come to be represented by a family of rosettes all of the same genotype. Each plant is now a subpopulation of parts which may lose their interconnections but remain modules of the same genetic individual.

The population dynamics of the creeping buttercup were sampled in a 1 ha (1×10^4 m^2) field of grassland at three separate 1 m^2 sites. The data are summarized in Table 15.1. On site A the population halved over 2 years, on site B it scarcely

Table 15.1 Population flux of *Ranunculus repens* in three sites of 1 m² in a grazed grassland.

	A	B	C
(a) Number of plants per m², April 1969	385	117	148
(b) Number of plants per m², April 1971	157	139	222
(c) Net change (b − a)	−228	+22	+74
(d) Rate of increase (b / a)	0.41	1.19	1.50
(e) Number of plants arrived between April 1969 and April 1971	344	244	466
(f) Total number of plants lost between April 1969 and April 1971	577	222	390
(g) Plants present April 1969, alive by April 1971	25	13	22
(h) Percentage survival of plants in (a) (g / a × 100)	6.5	11.1	7.4
(i) Expected time for complete turnover (years) 2 / (100 − h) × 100	2.14	2.25	2.16
(j) Total plants recorded during study	729	361	612
(k) Percentage annual mortality of all individuals (f / j) × 100	79.1	61.5	63.7

... but, the stability of the whole may conceal varied dynamics in its parts

changed and on site C it increased by 50%. Such a census reveals in detail how the births and deaths of individuals contribute to the overall changes in the size of populations (Figure 15.1). Over the 2 years of the study the number of buttercups on the three sites fell from 650 to 518. But, in this period 1054 new individuals were gained by the population and 1189 individuals were lost from it. The main lessons from this study are: (i) that different patches in the same 'population' showed quite different dynamics; (ii) that this was detected only because, for logistic reasons, the census was concentrated on three small plots; and (iii) that relatively little change may occur in populations despite very rapid flux of births and deaths.

Another example of a population showing relatively little change in adult numbers from year to year is seen in an 8-year study in Poland of the small, annual sand-dune plant *Androsace septentrionalis* (Symonides, 1979; Figure 15.2). Each year there were between 150 and 1000 seedlings per m², and each year mortality

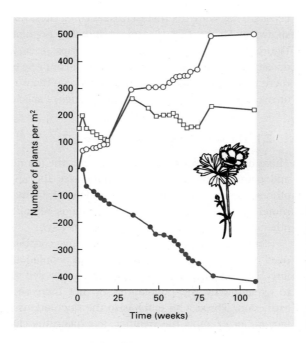

Figure 15.1 Changes in population size of the creeping buttercup (*Ranunculus repens*) at site C (see Table 15.1). ○, Cumulative gains from seed germination and clonal growth; ●, cumulative losses; □ net population size. (After Sarukhán & Harper, 1973.)

571 ABUNDANCE

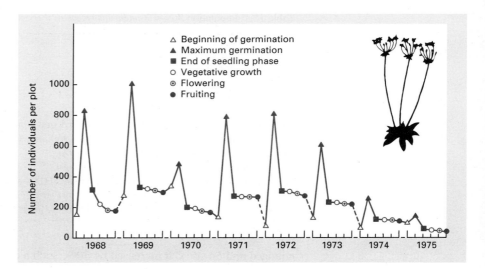

Figure 15.2 The population dynamics of *Androsace septentrionalis* during an 8-year study. (After Symonides, 1979. A more detailed analysis of these data is given by Silvertown, 1982.)

Legend:
△ Beginning of germination
▲ Maximum germination
■ End of seedling phase
○ Vegetative growth
⊙ Flowering
● Fruiting

reduced the population by between 30 and 70%. However, the population appears to be kept within bounds. At least 50 plants always survived to fruit and produce seeds for the next season.

As a final example of the relationship between population flux and relative constancy we take the classic study by Davidson and Andrewartha (1948a, b) of the apple-blossom thrips (*Thrips imaginis*), a small insect (approximately 1 mm long) found in the flowers of rose bushes, fruit trees, garden plants and weeds in southern Australia. They censused the thrips population in 20 roses picked at random from a long hedge every day, except Sundays and certain holidays, for 81 consecutive months. The variety of rose sampled (Cecile Brunner) lacks stamens and the thrips require pollen if they are to breed. The roses therefore served as 'traps' for the census. They were not themselves breeding sites. Examples of the census records are shown in Figure 15.3 and mean monthly population counts are summarized in Figure 15.4. Like many insects the thrips underwent very large fluctuations in density. Are these fluctuations determined entirely by randomly acting factors or are there processes at work which tend to keep them within bounds? We can begin to answer this question by looking at theories of species abundance in general.

a census of thrips

15.2.1 Theories of species abundance

There are contrasting theories to explain the abundance of populations of animals and plants. Some investigators emphasize the apparent stability of populations; others emphasize the fluctuations. Indeed, the same census data have been used by some authors to indicate that populations remain within narrow limits, and by others to stress that they change dramatically!

Those who have emphasized the relative constancy of populations argue that we need to look for stabilizing forces within populations (so-called 'density dependent' forces, for instance competition between crowded individuals for limited resources), to explain why the populations do not increase without bounds or decline to extinction. Those who emphasize the fluctuations in populations look to external factors, for example the weather, to explain the changes. The interest has been so

572   CHAPTER 15

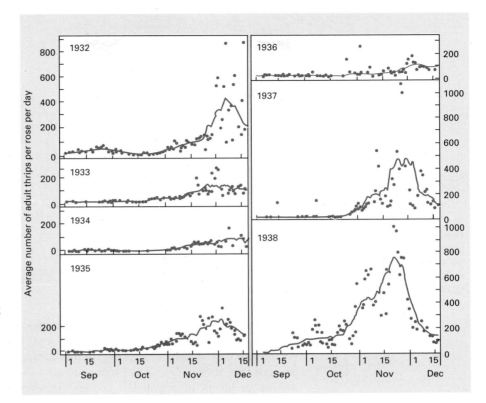

Figure 15.3 The numbers of *Thrips imaginis* per rose during the spring each year for 7 consecutive years. The points represent daily records; the curve is a 15-point moving average. (After Davidson & Andrewartha, 1948a.)

great, and the disagreement often so marked, that the subject has dominated much of ecology in this century.

The disagreements centred on the question of whether populations are regulated, at least at their upper limits, by forces that prevent their intrinsic rate of increase from being realized indefinitely. The controversy has not been resolved, but is now recognized to be in part the product of protagonists taking up extreme positions and arguing at cross purposes, and in part a consequence of loose usage of the concept of 'the population'. Most of the arguments were made between 1933 and 1958, when understanding of population ecology was less sophisticated than now, but considering the old viewpoints makes it easier to appreciate the details of an emerging modern consensus.

15.2.2 The view that population sizes are determined mainly by forces acting on the population from outside

Two Australian ecologists, Andrewartha and Birch (1954, and in a more recent restatement, 1984), made a major contribution to the debate. Their research was concerned mainly with the control of insect pests in the wild and they concluded that the numbers of animals in a natural population may be limited in three ways: (i) by shortage of material resources, such as food, places in which to make nests, etc.; (ii) by inaccessibility of these material resources relative to the animals' capacities for dispersal and searching; and (iii) by shortage of time when the rate of increase, r, is positive. They argued that of these three ways, the first is probably the least, and the

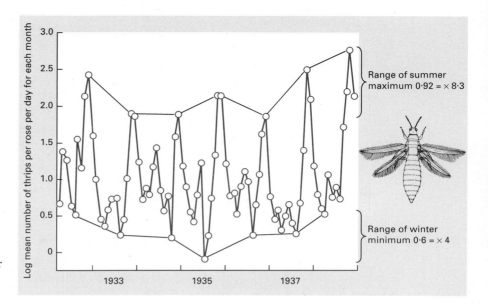

Figure 15.4 Mean monthly population counts per rose of *Thrips imaginis* (log scale). (After Davidson & Andrewartha, 1948a; Varley *et al.*, 1973.)

last is probably the most important in nature. They rejected the traditional subdivision of environment into physical and biotic factors (external to the population) and 'density dependent' factors (acting within the population) on the grounds that 'these were neither a precise nor a useful framework within which to discuss problems of population ecology' (Andrewartha & Birch, 1960). They attached great weight to the census of *T. imaginis* already referred to in Section 15.2 and in Figures 15.3 and 15.4. Estimates of abundance of the thrips were obtained for 81 consecutive months. For a further 7 years, estimates were also obtained for spring and early summer only, and local temperature and rainfall were monitored throughout the period. The census data were analysed by a multiple regression technique (see, for example, Poole, 1978) to determine how much of the population's variation from year to year could be 'explained' by a relationship with some aspect of the weather. For the analysis, the peak population each year was taken to be the mean logarithm of the numbers in the 30 days preceding the maximum. These peak values became the dependent variate in the regression analysis (log Y).

In seeking quantities that might be associated with the numbers of thrips in the spring, they looked first for one which would represent the opportunity for growth during autumn and winter by the annual plants which were chiefly important in the ecology of the thrips. They took, as the start of the season, the date on which the seeds of these annuals started to germinate. As a crude measure of the temperatures experienced in the following period, they determined the 'accumulated temperature' from the start of the season until 31 August (see Chapter 2, Section 2.2 for a discussion of accumulated temperatures). For this they summed the values of maximal daily temperature (in °F) for every day, having first subtracted 48°F from each record to represent a threshold below which activity was minimal. They then divided the result by 2—thus:

$$T = (\text{maximal daily temperature} - 48°F) / 2. \tag{15.1}$$

The values of T became the first chosen variate (x_1) for the regression. The second variate, x_2, was the total rainfall in September and October (Australian spring), and

x_3 was the daily 'effective temperature' in September and October. They recognized that the size of the population that carried over from the previous winter might be determined by conditions in the previous year. A further variate was introduced to allow for this: the value of x_1 from the previous year was taken as x_4. The regression equation obtained from this analysis is:

$$\log Y = -2.39 + 0.12x_1 + 0.20x_2 + 0.19x_3 + 0.08x_4. \tag{15.2}$$

The real values and those calculated from the regressions are shown in Figure 15.5. By far the most important variate was x_1 ($P < 0.001$), and x_2 was the next. Of the variance in population maxima, 78% was accounted for by the regression 'calculated entirely from meteorological records'.

correlations with the weather

This left virtually no chance of finding any other systematic cause for variation, because 22% is a rather small residuum to be left as due to random sampling errors. All the variation in maximal numbers from year to year may therefore be attributed to causes that are not density related. Not only did we fail to find a 'density dependent factor' but we also showed that there was no room for one (Andrewartha & Birch, 1954).

The census of *T. imaginis* is very remarkable both for its intensity and its duration. Some 6 000 000 thrips had been recorded by the end. The work also provides an excellent example of identifying correlations. Clearly, the weather (as represented by the four factors) plays a central and crucial role amongst the factors correlated with the peak numbers of thrips.

The same data have, however, been used by others to show that the thrips populations were regulated by some factor internal to the populations. The multiple regression model is not designed directly to reveal the presence of a density dependent factor (Varley *et al.*, 1973). When methods that were so designed were applied to the same data, there was strong evidence of density dependent population growth prior to the peaks (Smith, 1961). There were significant negative correlations between population *change* and the population size immediately preceding the spring peak. In addition, a fairly strong density dependent factor acting between the summer peak and the winter trough can be inferred from Figure 15.3, since the maxima are spread over an eightfold range but the minima cover only a fourfold range. In fact, Davidson and Andrewartha themselves (1948b) felt that weather acted as a density dependent component of the environment during the winter, by killing the proportion of the population inhabiting less-favourable 'situations'. (If the

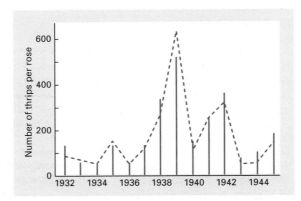

Figure 15.5 Observed and predicted population sizes of *Thrips imaginis*. The columns represent the geometric means of the daily counts of thrips per rose during the population peak. The dashed line represents the theoretical values calculated from the expression given above. (After Davidson & Andrewartha, 1948b.)

575 ABUNDANCE

number of safe sites is limited and remains roughly constant from year to year, then the proportion of individuals outside these sites that is killed by the weather will increase with the size of the whole population, and total mortality will appear to be dependent on density.)

15.2.3 The view that abundance is regulated by the effects of crowding

regulation and abundance

It is of critical importance to distinguish clearly between the ways in which the abundance of individuals is *determined* and the extent to which the abundance is *regulated*. *Regulation* refers to the tendency of a population to decrease in size when it is above a particular level, but to increase in size when below that level. In other words, regulation of a population can, by definition, occur only as a result of one or more features of the population itself that act on the rates of birth (and/or immigration) and/or death (and/or emigration) (Figure 15.6a). Various potentially density dependent processes have been discussed in earlier chapters on competition, predation and parasitism. On the other hand, abundance will be *determined* by the combined effects of all the factors and all the processes that impinge on a population, be they dependent or independent of density.

Figure 15.6b shows diagrammatically how an equilibrium level of abundance may be very greatly modified by the intensity of a density independent process. In this model the birth rate is density dependent, whilst the death rate is density

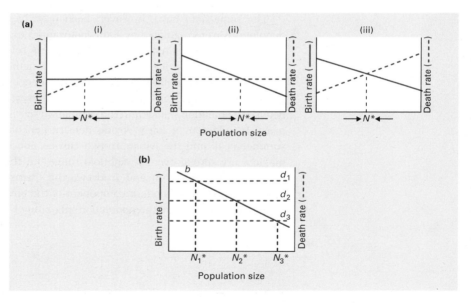

Figure 15.6 (a) Population regulation with: (i) density independent birth and density dependent death; (ii) density dependent birth and density independent death; and (iii) density dependent birth and death. Population size increases when birth rate exceeds death rate, and decreases when death rate exceeds birth rate. N* is therefore a stable equilibrium population size. The actual value of the equilibrium population size is seen to depend on both the magnitude of the density independent rate *and* the magnitude and slope of any density dependent process. (b) Population regulation with density dependent birth (*b*) and density independent death (*d*). Death rates are determined by physical conditions which differ in three sites (death rates d_1, d_2 and d_3). Equilibrium population size varies as a result (N_1^*, N_2^*, N_3^*).

independent but depends on physical conditions which differ in three locations. The figure shows three equilibrium populations (N_1, N_2, N_3), which correspond to the three death rates which in turn correspond to the physical conditions in the three environments. Such variations in density independent mortality were primarily responsible for differences in the abundance of the annual grass *Vulpia fasciculata*, on different parts of a sand-dune environment in North Wales. Reproduction was density dependent and regulatory, but varied little with physical conditions from site to site. However, physical conditions had strong density independent effects on mortality (Watkinson & Harper, 1978).

Clearly, a population can be changed only by birth (or germination or hatching), death or migration. The argument that there must be regulating factors that explain the relative stability of natural populations was well put by Haldane (1953):

Haldane's bordered whites: regulated but highly variable

Haldane (1953) argued that since no population increases without limit, there must be some regulating factors which, on the whole, cause the density of an animal or plant species in a given area to increase when it is small, and to decrease when it is large.

The example that Haldane used to illustrate his argument is very different from those of the herons and swifts that we have quoted. His illustration was a census of the numbers of the bordered white moth in a German pine forest from 1882 to 1940. During this period the highest density (in 1928) was at least 30 000 times the lowest density (in 1911). Clearly, regulation was not very precise! However, during the period of the study there were phases of extremely rapid population growth (e.g. 1922–1928), but these were not sustained. The population had the potential for extremely rapid growth but this was never realized for long. Hence, it can be argued that some regulating factor must have operated that prevented the populations from growing beyond bounds.

... but, may account for little of the variation in population size

A. J. Nicholson, an Australian theoretical and laboratory animal ecologist, is usually credited as the major proponent of the view that *density dependent*, biotic interactions (which Nicholson called 'density governing reactions') play the main role in determining population size (Nicholson, 1933, 1954a, b, 1957, 1958). In his own words 'Governing reaction induced by density change holds populations in a state of balance in their environments,' and 'The mechanism of density governance is almost always intraspecific competition, either amongst the animals for a critically important requisite, or amongst natural enemies for which the animals concerned are requisites'. He recognized that 'factors which are uninfluenced by density may produce profound effects upon density', but he thought that they did so only by 'modifying the properties of the animals, or those of their environment, so influencing the level at which governing reaction adjusts population densities'. Even under the extreme influence of density independent factors 'density governance is merely relaxed from time to time and subsequently resumed, and it remains the influence which adjusts population densities in relation to environmental favourability' (Nicholson, 1954b).

A very large number of censuses have been made of various kinds of animals, especially of insects. Density dependence has been detected in some but not in others. A survey of 24 sets of life-table data (see Chapter 4) for temperate Lepidoptera, plotting k-values against the logarithm of density on which mortality acts (see Chapter 6, Section 6.6), found density dependence due to intraspecific competition in only 13, and density dependence due to natural enemies in just three (Dempster, 1983). A further survey of 63 life tables from 49 separate insect studies failed to identify a density dependent process in around 47% of cases and found that no one density dependent process was of paramount importance in the remainder (Stiling, 1988). If these findings apply more generally, they might suggest that density dependent processes often play only a very minor role in population dynamics. On the other hand, they may indicate that the methods used to detect density dependence were either inadequate or inappropriate.

In the first place many of these studies may simply have been too short. In Stiling's data sets, the percentage of studies in which density dependence can be detected increases sharply with the length of the study (in generations). For instance, density dependence was detected in seven of the 10 studies that lasted for more than 12 generations, and the percentage rose to 80% of the studies of univoltine insects lasting more than 10 years (Hassell *et al.*, 1989). This emphasizes the enormous importance of long-term censuses, a point reinforced by analyses of the classic Rothamsted insect survey data (Woiwod & Hanski, 1992). The survey includes time-series data for hundreds of species of moths and aphids collected simultaneously from light traps (moths) or suction traps (aphids) throughout the British Isles. It does not suffer from the same degree of bias as many other data sets, which focus on pest or outbreak species because of their special interest for applied ecologists. Here, the species selected themselves by getting caught in standard traps. Using a range of alternative analyses, applied to 5715 time series of annual abundance from 447 species of aphids and moths, the influence of detectable density dependence increased significantly as time-series length rose from 10 to 24 years (Woiwod & Hanski, 1992). It was detected in 79% of the moth and 88% of the aphid series that lasted longer than 20 years. Moreover, density dependence appeared to be a species-specific characteristic: it was consistently detected in some species of noctuid moths and aphids, but not in others.

A second problem is that conventional life-table analyses may fail to detect *delayed* density dependence (e.g. brought about by the action of a natural enemy—see Chapter 10, Section 10.2.5), simply because they are not designed to do so (Turchin, 1990). An analysis of population time series for 14 species of forest insects detected direct density dependence clearly in only five, but it revealed delayed density dependence in seven of the remaining nine (Turchin, 1990). It may be that a similar proportion of the populations classified, from their life tables, as lacking density dependence were actually subject to the delayed density dependence of a natural enemy.

Lastly, there is a question of whether temporal runs of data alone, based on averages for each generation (or year) can always detect density dependence even when it exists. It may also be necessary to compare separate units *within* a generation in order to establish whether a population is subject to regulatory

density dependent processes have sometimes seemed rare. . .

. . . but, they are commonly detected in long data sets

delayed density dependence may be more difficult to detect

spatial variation may also need to be sampled

phenomena (May, 1989). When these 'regulatory' forces are capable of producing oscillatory or chaotic dynamics, perhaps by overcompensating density dependence (see Chapter 6, Section 6.9), then this, combined with the 'density independent noise' generated by environmental variations in a patchy world, is likely to make density dependence very difficult to detect in temporal runs of data—whereas, it may be readily apparent by sampling on an appropriate scale (Hassell, 1987). However, if the regulatory phenomena are contest-like, incapable of producing oscillatory dynamics, spatial heterogeneity may actually enhance our ability to detect density dependence in temporal runs of averages (Mountford, 1988).

There is a very strong bias towards insects in the data sets available for the analysis of the regulation and determination of population size, and amongst these there is a preponderance of studies of pest species. The limited information from other groups of animals suggests that terrestrial vertebrates may have significantly less variable populations than those of arthropods, that populations of lizards are especially constant and that populations of birds are more constant than those of mammals. Large terrestrial mammals seem to be regulated most often by their food supply, whereas in small mammals the single biggest cause of regulation seems to be the density dependent exclusion of juveniles from breeding (Sinclair, 1989). For birds, food shortage and competition for territories and/or nest sites seem to be most important. Such generalizations, however, may be as much a reflection of biases in the species selected for study and of the neglect of their predators and parasites, as they are of any underlying pattern.

Great efforts have been made to detect evidence of density dependence in animal populations and these have often succeeded only after great difficulty and argument. It could be argued that if it is so difficult to detect it is perhaps not often very important! Perhaps in giving so much emphasis to the ways in which populations are regulated we are liable to forget how they are determined.

Neither the bordered white nor the thrip populations suffered extinction or unrestrained growth, and the fluctuations in almost all populations are at least limited enough for us to be able to describe the species as 'common', 'rare', and so on. On the other hand, we must remember that in Davidson and Andrewartha's work, the weather accounted for 78% of the variation in peak number of thrips. If we wished to predict their abundance, or to explain why, in a particular year, one level of abundance was attained rather than another, we would be foolish to disregard the role of the weather. Although density dependent processes are an absolute necessity as a means of *regulating* populations, their importance in *determining* abundance depends very much on the species and environment in question.

the regulation and determination of abundance

It is logically unreasonable to suppose that any population is absolutely free from regulation; but, we need not expect that regulation is happening all the time (or even very often). Moreover, if the numbers of an organism fluctuate this does not necessarily imply that regulation is weak. It may be that the amount of resources fluctuate and a tightly regulated population dependent on those resources would also fluctuate. It seems reasonable to expect to find some populations in nature that are almost always recovering from the last disaster (Figure 15.7a), others that are usually limited by an abundant resource (Figure 15.7b), or a scarce resource (Figure 15.7c) and others that are usually in decline after sudden episodes of colonization (Figure 15.7d). We may indeed find parts of a large population that are in different phases at the same time.

We have quite a clear idea of what determines the abundance of *Vulpia* plants *and*

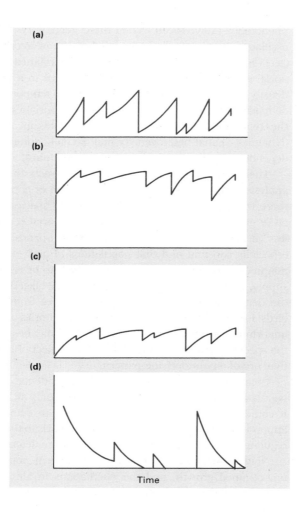

Figure 15.7 Idealized diagrams of population dynamics: (a) dynamics dominated by phases of population growth after disasters (*r* species); (b) dynamics dominated by limitations on environmental carrying capacity—carrying capacity high (*K* species); (c) same as (b) but carrying capacity low (*K* species); (d) dynamics within a habitable site dominated by population decay after more or less sudden episodes of colonization or recruitment, for example from the seed bank in the soil. (After Harper, 1981.)

what regulates their population about its mean density. In the case of *Thrips*, we know much about what determines its abundance but only have a hazy idea of what regulates it because the study was not designed to answer both questions and the data are therefore inadequate in this respect. In Section 15.3 we describe a technique that can throw light both on determination and regulation of abundance.

15.2.4 Chaos

Most of the excitement and satisfaction in studying population censuses comes from the discovery of regularities within them. These may be that numbers are more constant than might be expected from organisms' ability to multiply, or that the variations contain periodisms such as the repeating cycles of the kind we will discuss in Section 15.4. When we fail to detect regularities in census records the temptation has been to attribute the irregularities to external random influences such as climatic factors. It came as a shock when it was realized that quite realistic deterministic models of population behaviour could generate apparently irregular 'chaotic' variation (see Chapter 6, Section 6.9 for a discussion of the characteristics of models that can behave in this way, and references to the relevant literature). The question immediately arises of whether irregularities in the behaviour of populations *in nature* may sometimes be the chaotic consequences of such deterministic processes operating within the populations themselves. If so, how might we analyse time sequences of census data in order to recognize such complex dynamical behaviour in nature? In particular, how might we distinguish chaos that is determined by deterministic properties of the populations themselves from purely random effects of external features of the environment? A wide-ranging review of the methods available for detecting chaos in natural populations is given by Hastings *et al.* (1993), and see also Rohani *et al.* (1994b) and Rand and Wilson (1995).

15.3 Key-factor analysis

The number of individuals of a species present in an area at a point in time is the number present at some previous time plus births, minus deaths, plus immigrants and minus emigrants. All fluctuations in a population must be accountable in terms of these four processes. If we could break down the contributions of each to the changes that occur, we might be able to focus research more precisely on the stages that are responsible for changes in both density and stability.

The more frequently we can make a census and the more completely we can expose the details of a life cycle, the more likely we are to discover the crucial phases that determine population size. The approach, known as *key-factor analysis*, was developed for this purpose and applied to many univoltine insects and some other animals and plants. It is an approach based on the use of *k*-values (see also Chapter 4, Section 4.5.1 and Chapter 6, Section 6.6).

15.3.1 The Colorado potato beetle

In key-factor analysis, data from a series of censuses are compiled in the form of a life table, such as that for a Canadian population of the Colorado potato beetle (*Leptinotarsa decemlineata*) in Table 15.2 (Harcourt, 1971). In this species, 'spring adults' emerge from hibernation around the middle of June, when potato plants are breaking through the ground. Within 3 or 4 days oviposition begins, continuing for about 1 month and reaching its peak in early July. The eggs are laid in clusters (approximately 34 eggs) on the lower leaf surface, and the larvae crawl to the top of the plant where they feed throughout their development, passing through four instars. When mature, they drop to the ground and form pupal cells in the soil. The 'summer adults' emerge in early August, feed, and then re-enter the soil at the beginning of September to hibernate and become the next season's 'spring adults'.

determining *k*-values

... which indicate the average strengths of various mortality factors ...

The sampling programme provided estimates of the population at seven stages: eggs, early larvae, late larvae, pupal cells, summer adults, hibernating adults and spring adults. One further category was included, 'females × 2', to take account of any unequal sex ratios amongst the summer adults. Table 15.2 lists these estimates for a single season, and also gives the main cause of deaths in each stage of the life cycle. The mean *k*-values, determined for a single population over 10 seasons, are presented in the first column of Table 15.3. These indicate the relative strengths of the various factors that contribute to the total rate of mortality within a generation. Thus, the emigration of summer adults has by far the greatest proportional effect ($k_6 = 1.543$), whilst the starvation of older larvae, the frost-induced mortality of

Table 15.2 Typical set of life-table data collected by Harcourt (1971) for the Colorado potato beetle (in this case for Merivale, Canada, 1961–1962).

Age interval	Numbers per 96 potato hills	Numbers 'dying'	'Mortality factor'	$\text{Log}_{10}\,N$	k-value	
Eggs	11 799	2531	Not deposited	4.072	0.105	(k_{1a})
	9268	445	Infertile	3.967	0.021	(k_{1b})
	8823	408	Rainfall	3.946	0.021	(k_{1c})
	8415	1147	Cannibalism	3.925	0.064	(k_{1d})
	7268	376	Predators	3.861	0.024	(k_{1e})
Early larvae	6892	0	Rainfall	3.838	0	(k_2)
Late larvae	6892	3722	Starvation	3.838	0.337	(k_3)
Pupal cells	3170	16	D. doryphorae	3.501	0.002	(k_4)
Summer adults	3154	126	Sex (52% ♀)	3.499	−0.017	(k_5)
♀ × 2	3280	3264	Emigration	3.516	2.312	(k_6)
Hibernating adults	16	2	Frost	1.204	0.058	(k_7)
Spring adults	14			1.146		
					2.926	(k_{total})

Table 15.3 Summary of the life-table analysis for Canadian Colorado beetle populations. (After Harcourt, 1971.) b and a are, respectively, the slope and intercept of the regression of each k-factor on the logarithm of the numbers preceding its action; r^2 is the coefficient of determination. (See text for further explanation.)

		Mean	Coefficient of regression on k_{total}	b	a	r^2
Eggs not deposited	k_{1a}	0.095	−0.020	−0.05	0.27	0.27
Eggs infertile	k_{1b}	0.026	−0.005	−0.01	0.07	0.86
Rainfall on eggs	k_{1c}	0.006	0.000	0.00	0.00	0.00
Eggs cannibalized	k_{1d}	0.090	−0.002	−0.01	0.12	0.02
Eggs predation	k_{1e}	0.036	−0.011	−0.03	0.15	0.41
Larvae 1 (rainfall)	k_2	0.091	0.010	0.03	−0.02	0.05
Larvae 2 (starvation)	k_3	0.185	0.136	0.37	−1.05	0.66
Pupae (D. doryphorae)	k_4	0.033	−0.029	−0.11	0.37	0.83
Unequal sex ratio	k_5	−0.012	0.004	0.01	−0.04	0.04
Emigration	k_6	1.543	0.906	2.65	−6.79	0.89
Frost	k_7	0.170	0.010	0.002	0.13	0.02
	$k_{\text{total}} =$	2.263				

hibernating adults, the 'non-deposition' of eggs, the effects of rainfall on young larvae and the cannibalization of eggs all play substantial roles.

What the first column of Table 15.3 does not tell us, however, is the relative importance of these factors as determinants of the year-to-year fluctuations in mortality. For instance, we can easily imagine a factor that repeatedly takes a significant toll from a population, but which, by remaining constant in its effects, plays little part in determining the particular rate of mortality (and thus, the particular population size) in any 1 year. In other words, such a factor may, in a sense, be important in determining population size, but it is certainly not important in determining *changes* in population size, and it cannot help us understand why the population is of a particular size in a particular year. This can be assessed, however, from the second column of Table 15.3, which gives the regression coefficient of each individual k-value on the total generation value, k_{total}.

A mortality factor that is important in determining population changes will have a regression coefficient close to unity, because its k-value will tend to fluctuate in line with k_{total} in terms of both size and direction (Podoler & Rogers, 1975). A mortality factor with a k-value that varies quite randomly with respect to k_{total}, however, will have a regression coefficient close to zero. Moreover, the sum of all the regression coefficients within a generation will always be unity. The values of the regression coefficients will, therefore, indicate the relative strength of the association between different factors and the fluctuations in mortality. The largest regression coefficient will be associated with the key factor causing population change (Morris, 1959; Varley & Gradwell, 1968).

In the present example, it is clear that the emigration of summer adults, with a regression coefficient of 0.906, is the key factor; and other factors (with the possible exception of larval starvation) have a negligible effect on the changes in generation mortality, even though some have reasonably high mean k-values. A similar conclusion can be drawn by simply examining graphs of the fluctuations in k-values with time (Figure 15.8). (Note that Podoler and Roger's method, even though it is less arbitrary than this graphical alternative, still does not allow us to assess the statistical significance of the regression coefficients, because the two variables are not independent of one another.)

Thus, whilst mean k-values indicate the average strengths of various factors as causes of mortality each generation, key-factor analysis indicates their relative contribution to the yearly *changes* in generation mortality, and thus measures their importance as determinants of population size.

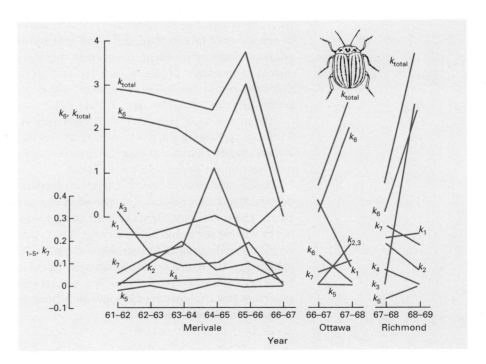

Figure 15.8 The changes with time of the various k-values of Colorado beetle populations at three sites in Canada. (After Harcourt, 1971.)

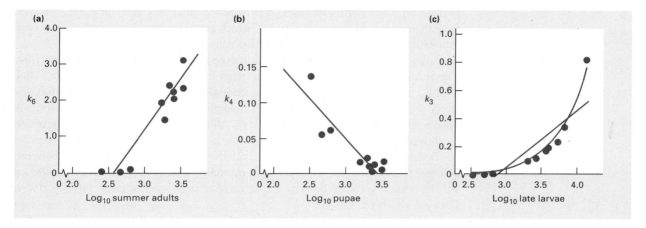

Figure 15.9 (a) Density dependent emigration by Colorado beetle 'summer' adults (slope 2.65). (b) Inversely density dependent parasitization of pupae (slope = 0.11). (c) Density dependent starvation of larvae (straight line slope = 0.37, final slope of curve = 30.95). (After Harcourt, 1971.)

... and which factors regulate rather than simply determine abundance

We must now consider the role of these factors in the *regulation* of the Colorado beetle population. In other words, we must examine the density dependence of each. This can be achieved most easily by plotting k-values for each factor against $\log_{10}$ of the numbers present before the factor acted. Thus, the last three columns in Table 15.3 contain, respectively, the slopes (b), intercepts (a) and coefficients of determination r^2, of the various regressions of k-values on their appropriate '$\log_{10}$ initial densities'. Three factors seem worthy of close examination.

The emigration of summer adults (the key factor) appears to act in an overcompensating density dependent fashion, since the slope of the regression (2.65) is considerably in excess of unity (see also Figure 15.9a). Thus, the key factor, although density dependent, does not so much regulate the population as lead to violent fluctuations in abundance (because of overcompensation). Indeed, the Colorado potato beetle–potato system would go extinct if potatoes were not continually replanted (Harcourt, 1971).

The rate of pupal parasitism by *Doryphorophaga doryphorae*, a tabanid fly (Figure 15.9b), is apparently inversely density dependent (although not significantly so, statistically), but because the mortality rates are small, any destabilizing effects this may have on the population are negligible.

Finally, the rate of larval starvation appears to exhibit undercompensating density dependence (although statistically this is not significant). An examination of Figure 15.9c, however, shows that the relationship would be far better represented not by a linear regression but by a curve. If such a curve is fitted to the data, then the coefficient of determination rises from 0.66 to 0.97, and the slope (b-value) achieved at high densities would be 30.95 (although it is, of course, much less than this in the range of densities observed). Hence, it is quite possible that larval starvation plays an important part in regulating the population, prior to the destabilizing effects of pupal parasitism and adult emigration.

15.3.2 Further examples of key-factor analysis

Key-factor analysis has been applied to a great many insect populations, but to far fewer bird, mammal or plant populations. Examples of these are shown in Figure 15.10, which in each case summarizes the changes through time of the total population mortality (k_{total}) and of the k-values of various other factors. Shown alongside these diagrams are factors that act in a density dependent manner.

The winter moth

The key factor in the life cycle of the winter moth (*Operophtera brumata*) in Wytham Wood near Oxford is easily identified as k_1, the overwintering loss of eggs and larvae before the first larval census each spring. This loss occurs mainly amongst larvae that have hatched before the oak leaves, on which they feed, have developed. They emigrate by spinning a length of silk on which they are blown away from the trees. In this study, the key factor (responsible for determining year to year changes in population size) was not the regulating factor that maintains the population within bounds. This is very often the case. In fact, predation of winter moth pupae in the soil (k_5) by beetles and small mammals was found to be density dependent (Figure 15.10a). The slope of the relationship between k_5 and pupal density is 0.35, showing that this density dependent factor undercompensates (it is less than 1.0) for changes in density. Note also that k_3, parasitism by insects (other than *Cyzenis*), acted in a slightly, but statistically significant, negatively density dependent way.

Tawny owls

A long-term study of the tawny owl (*Strix aluco*) population, again in Wytham Wood, revealed that the failure of birds to breed each year (k_1) was the key factor. Although the number of territory holding adults did not vary much (17–32 pairs), there was a great deal of variation in the number that attempted to breed, ranging from 22 pairs in 1959 to none in 1958. Years in which few attempted to breed were those when numbers of the owl's prey of mice and voles were particularly low. Prey availability did not depend on owl density, and thus k_1 did not operate as a density dependent factor. In contrast, k_5, which represents death or migration of young owlets after they leave the nest and before the start of the next season, was found to be density dependent (Figure 15.10b). Losses became density dependent only above a certain density (when the slope is 1.6, and overcompensating); they were probably brought about via intraspecific competition for territories.

African buffalo

Juvenile mortality (k_j) is the obvious key factor in the population dynamics of an African buffalo (*Syncerus caffer*) in the Serengeti region of East Africa, and once again it does not operate in a density dependent way. Juveniles suffered more than adults from a variety of endemic diseases and parasites, but calf mortality, although heavy, was random with respect to density. On the other hand, adult mortality (k_a) was found to be density dependent (undercompensating slope of 0.24; Figure 15.10c). Undernutrition appears to have been a primary agent.

Can any generalizations be made about density dependence in different groups of

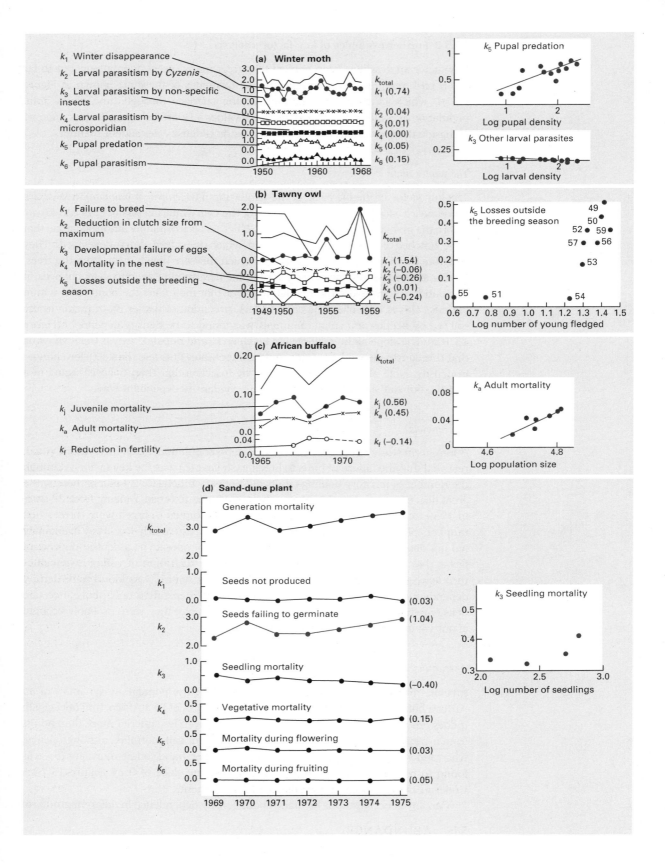

(a) Winter moth

k_1 Winter disappearance
k_2 Larval parasitism by *Cyzenis*
k_3 Larval parasitism by non-specific insects
k_4 Larval parasitism by microsporidian
k_5 Pupal predation
k_6 Pupal parasitism

k_{total}
k_1 (0.74)
k_2 (0.04)
k_3 (0.01)
k_4 (0.00)
k_5 (0.05)
k_6 (0.15)

1950   1960   1968

k_5 Pupal predation
Log pupal density

k_3 Other larval parasites
Log larval density

(b) Tawny owl

k_1 Failure to breed
k_2 Reduction in clutch size from maximum
k_3 Developmental failure of eggs
k_4 Mortality in the nest
k_5 Losses outside the breeding season

k_{total}
k_1 (1.54)
k_2 (−0.06)
k_3 (−0.26)
k_4 (0.01)
k_5 (−0.24)

1949 1950   1955   1959

k_5 Losses outside the breeding season
49 50 52 59 57 56 53 55 51 54
Log number of young fledged

(c) African buffalo

k_{total}
k_j (0.56)
k_a (0.45)
k_f (−0.14)

k_j Juvenile mortality
k_a Adult mortality
k_f Reduction in fertility

1965   1970

k_a Adult mortality
Log population size

(d) Sand-dune plant

k_{total} Generation mortality

k_1 Seeds not produced   (0.03)

k_2 Seeds failing to germinate   (1.04)

k_3 Seedling mortality   (−0.40)

k_4 Vegetative mortality   (0.15)

k_5 Mortality during flowering   (0.03)

k_6 Mortality during fruiting   (0.05)

1969 1970 1971 1972 1973 1974 1975

k_3 Seedling mortality
Log number of seedlings

animals? An analysis of 30 published accounts reached the following conclusions (Stubbs, 1977). Animals from more permanent habitats (vertebrates and some insects, including winter moth) tend to show undercompensating or exactly compensating mortalities (slope less than or equal to 1, never much more than 1). On the other hand, animals from more temporary habitats (many insects, including the Colorado potato beetle) tend to have small, undercompensating mortalities at low densities, rising sharply to overcompensating density dependence as numbers increase. In addition, 86% of density dependent factors acted on the young stages of animals of temporary habitats (*r* species), whereas the comparable figure for species of more permanent habitats (*K* species) was only 15%. Instead, parasitism and predation (30%) and reduced fecundity (35%) seem to act more frequently in animals of permanent habitats.

An annual plant

Finally, the key factor acting on a Polish population of the sand-dune annual plant *Androsace septentrionalis* (dynamics shown in Figure 15.10d) was found to be k_2, i.e. seed mortality in the soil. Again, this key factor did not operate in a density dependent manner, whereas the seedling mortality (k_3), which was not the key factor, was found to be density dependent (undercompensating slope of 0.2). Seedlings that emerge first in the season stand a much greater chance of surviving (Symonides, 1977), which suggests that competition for resources may be intense (and density dependent) and/or that the later emerging, smaller plants are more susceptible to various mortality factors.

All the animal studies referred to above have been performed on unitary organisms, and the only plant, *Androsace*, was treated from the point of view of the whole plant, rather than its parts. However, higher plants, and modular organisms in general, grow and die in parts, rarely as wholes: it may then be more useful to census the parts than the wholes. For example, predation on corals by *Acanthaster* (a brittle star) may be more sensibly censused by recording the number of polyps eaten than the number of individual corals destroyed, and a student of defoliating forest insects will have reason to be more interested in the number of leaves than in the number of trees in a forest. A beginning has been made to the study of the population dynamics of plant parts; they can, like whole organisms, be individually marked and life tables can be constructed, survivorship curves drawn (Figure 15.11a) and age structures determined (Figure 15.11b).

the 'factors' are actually phases

It is necessary to sound warning notes about key-factor analysis. In fact, it picks out *phases* in the life cycle when mortality (or failure to achieve reproductive

---

Figure 15.10 (*facing page*) Key-factor analysis of four contrasting populations. In each case, a graph of total generation mortality (k_{total}) and of various *k*-factors is presented. The values of regression coefficient of each individual *k*-factor on k_{total} are given in brackets. (After Podoler & Rogers, 1975; (d) from Silvertown, 1982.) The *k*-factor with the largest regression coefficient is the key factor and is shown as a coloured line. Alongside each key-factor graph are shown *k*-factors that act in a statistically significant, density dependent manner. (a) Winter moth, *Operophtera brumata*. (After Varley & Gradwell, 1968.) (b) Tawny owl, *Strix aluco*. (After Southern, 1970.) (c) Buffalo, *Syncerus caffer*. (After Sinclair, 1973.) (d) The sand-dune annual plant, *Androsace septentrionalis*. (After Symonides, 1979; analysis in Silvertown, 1982.)

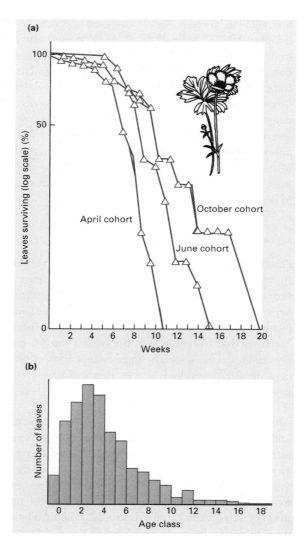

Figure 15.11 (a) Survivorship curves and (b) age structure of populations of individual leaves of the creeping buttercup, *Ranunculus repens*. (After Peters, 1980.)

potential) is important. It cannot by itself indicate what factors are responsible (e.g. competition for nitrate or breeding sites, a specific predator or parasite or the role of territoriality), unless only one factor acts in each phase. Moreover, key-factor analysis is not designed to detect delayed density dependence (Turchin, 1990), yet there is much evidence that this is a common phenomenon.

If the parts of a patchy population (or whole populations in different years) have different dynamics (some uncrowded and expanding, some crowded and stable or contracting) a key-factor analysis of the population *as a whole* is unlikely to reveal what is going on. For example, a key-factor analysis of the dynamics of mussel (*Mytilus edulis*) populations in 12 mussel beds in a single English estuary found the difference between egg production and spat settlement in spring to be the key factor on eight of the beds but net recruitment in the first summer was the key factor on the remaining four (McGrorty & Goss-Custard, 1993). This rare example of the application of key-factor analysis to marine invertebrates provides an elegant link between key-factor analysis and the population dynamics of patchy populations

key-factor analysis may not detect patchy dynamics

which we turn to in Section 15.5. It also warns against arguing too forcefully from the behaviour of a whole when the important events are happening in its parts.

Despite its limitations, however, key-factor analysis has clearly been useful as a technique for unravelling census and life-table data: in particular, it makes clear the crucial distinction between the *determination* and the *regulation* of abundance.

15.4 Population cycles and their analysis

Regular cycles in animal abundance were first observed in the long-term records of fur-trading companies, and of gamekeepers. Cycles have also been reported from many studies of voles and lemmings and in certain forest Lepidoptera (Myers, 1988). Cycles might be driven by periodic fluctuations in the environment. Alternatively, they might arise as a result of internal demographic processes that involve delayed, or overcompensating density dependence (see Chapter 10, Section 10.2).

15.4.1 Cycles and quasi-cycles

The great classic amongst long-term censuses is the record of furs received by the Hudson Bay Company from 1821 to 1934. Figure 15.12a shows the number of Canadian lynx trapped for the company and the peaks appear to the eye to be rather evenly spaced and to have much the same amplitude (see also Chapter 10, Section 10.2.4).

A statistical procedure was developed by Moran (1952) to check whether such cycles are real or simply a series of random fluctuations. This method correlates the number of animals counted in each year with the number counted in each succeeding year at increasing time intervals. In a cyclic series, high correlations occur when the intervals in years match corresponding phases of the cycle. For example, in a time series which peaks every 4 years, high positive correlations occur at years 4, 8, 12, etc., with negative correlations (but of similar strength) at 2, 6, 10, etc. The correlations for each time series are known as *autocorrelations*, and a graph

showing the strengths of correlations at different time intervals is a *correlogram*. If a population does not oscillate in a regular periodic fashion, the correlogram damps down quickly to low and insignificant levels of correlation (although purely chance fluctuations prevent it from ever damping down completely). A description of the procedure is given by Poole (1978).

The correlogram for lynx (Figure 15.12b) shows little damping down. The fluctuations in the population are truly cyclic with a period of 10 years. The cycles are now believed to arise as a result of cyclic interactions between snowshoe hares (the lynx's food), the plant food of the hares and the lynx itself (see also the discusssion of this cycle in Chapter 10, Section 10.2.4).

Other populations do not show such perfect cycling, but may demonstrate a *tendency* towards a cyclic type of population change: the correlogram does not damp down as fast as it would do in the complete absence of periodic fluctuation. Such *quasi-cycles* behave like a wave that dies away after a disturbance (or a shout that leaves dying echoes).

In an analysis of the records of red grouse (*Lagopus lagopus scoticus*) shot on a large number of moors in the north of England (each record was for at least 20 consecutive years during the period 1870–1977), the majority showed significant *negative* autocorrelation peaks at 2 or 3 years, indicating a cycling period of between 4 and 6

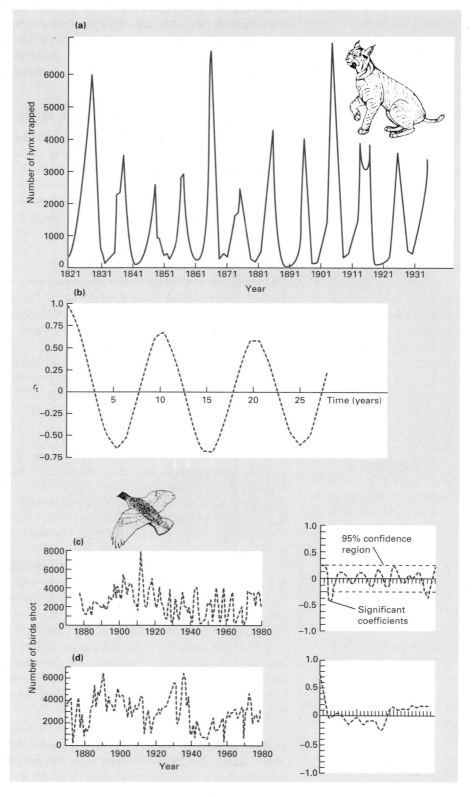

years (Potts *et al.*, 1984). Figures 15.12c and d show the original population data and correlograms for two of the moors, one with and the other without a quasi-cycle. Using a special statistical procedure, Potts *et al.* determined whether any of the autocorrelation peaks were statistically significant. Two autocorrelation coefficients were significant in the data from the moor shown in Figure 15.12c, but none in Figure 15.12d.

The causes of quasi-cycles in red grouse are still obscure and disputed (Watson & Moss, 1980; Potts *et al.*, 1984). The cycle periods appear to be different on different moors, and to be significantly longer in Scotland than in England. Very elegant experiments support quite different theories. One theory has it that the quasi-cycles result from an interaction between grouse and a parasitic nematode, *Trichostrongylus tenuis*. The key factor in the grouse cycles was winter loss; breeding losses were also significant: both were reduced when the parasite burdens were reduced by treating birds with nematocides. A model that incorporates an interaction between the grouse and the parasite using the long-term records to estimate the model's parameters gives a good fit to the cycles observed in the field. The analysis is consistent with the parasite being the cause of the cycles in northern England (Dobson & Hudson, 1992; Hudson *et al.*, 1992b). Alternative theories based on studies in Scotland emphasize the spacing behaviour of territorial cocks (Moss *et al.*, 1994).

The best documented cycles and strong quasi-cycles occur in northern high-latitude vegetation zones. On open tundra, lemmings and willow ptarmigan undergo a 4-year cycle. In the boreal forest, species such as snowshoe hare, spruce grouse, ruffed grouse and lynx follow a 10-year cycle (see Chapter 10, Section 10.2.4). The other important vertebrate herbivores in these environments are the larger caribou or reindeer, which make long-distance annual migrations. Their populations appear not to cycle.

15.4.2 Population changes in microtine rodents

A very great research effort has been made to understand the behaviour of small microtine rodents (particularly voles (*Microtus* spp.) and lemmings (*Lemmus* spp.) e.g. Figure 15.13) (Krebs & Myers, 1974; Krebs, 1985; Boyce & Boyce, 1988a; Lambin *et al.*,1992). The cycles have a typical periodicity of 3 or 4 years, although there are some populations that regularly or occasionally display a 2- or a 5-year cycle. There are also many populations and many species, particularly in less northerly regions, that fluctuate strongly but do not cycle). No clear explanation has yet appeared for these patterns of abundance. Indeed, it is tempting to abandon the search for a single unifying hypothesis (Lidicker, 1988). We will consider these studies in some detail, because they illustrate so well the joys, frustrations and difficulties of wrestling with

an important focus of interest, without a clear explanation ...

Figure 15.12 (*facing page*) (a) Number of Canadian lynx (*Lynx canadense*) trapped for the Hudson Bay Company. (After Elton & Nicholson, 1942.) (b) Correlogram for Canadian lynx. The correlogram does not damp down, showing clearly a true cycle with a period of 10 years. (After Moran, 1953.) (c) Analysis of serial correlations in red grouse (*Lagopus lagopus scoticus*) records from a typical quasi-cyclic moor and (d) a non-quasi-cyclic English moor. Beside the time series of original data of birds shot in each year is the correlogram produced by the autocorrelation technique. Significant negative coefficients occur in (c) at 2 and 3 years, indicating a quasi-cycle corresponding to between 4 and 6 years. No significant correlations occur in (d). (After Potts *et al.*, 1984.)

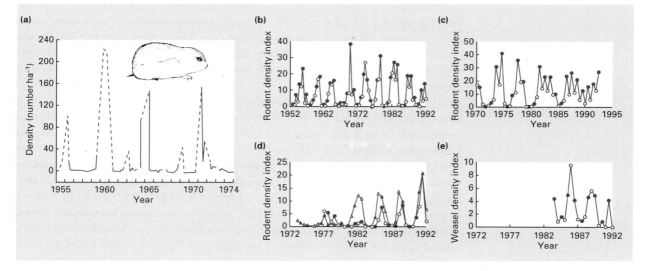

Figure 15.13 (a) Estimated lemming densities in the coastal tundra at Barrow, Alaska, for a 20-year period. (After Batzli *et al.*, 1980.) (b) The pooled density of five vole species, Kilpisjärvi, Finnish Lapland. (c) The pooled density of five vole species in old taiga forest, Pallasjärvi, Finnish Lapland. (d) *Microtus* densities at two study sites 14 km apart, Alajoki, western Finland. (e) Densities of the least weasel (*Mustela nivalis*) also at Alajoki, western Finland. ((b–e) After Hanski *et al.*, 1993.) Open symbols, spring densities; filled symbols, autumn densities.

a conspicuous empirical phenomenon that we ought to be able to explain, but cannot.

... posing questions pertinent to populations generally

Microtine cycles pose, in a particularly striking way, the more general question of how the sizes of populations are determined. The voles and lemmings, like other species, are affected by their food, by their predators, by disease and by the weather, and, as with many species, the characteristics of the individuals themselves vary with crowding and with the phase of the cycle. There are also genetic differences within and between populations. There are therefore problems of disentangling cause from effect, of distinguishing factors that change density from those that merely vary with density and of distinguishing those that affect density from those that actually impose a pattern of cycles.

the biology of lemmings in Alaska

An impression of microtine biology can be gained by considering an intensive study of the brown lemming (*L. sibericus*) in the coastal tundra at Barrow, Alaska, where it is the dominant herbivore (Batzli *et al.*, 1980). A second species, *Dicrostonyx torquatus*, is present but usually scarce (0.1 ha^{-1} ($1 \times 10^{-5} \text{ m}^{-2}$)), whereas the population of the brown lemming may reach mean densities of 225 ha^{-1} ($2.225 \times 10^{-2} \text{ m}^{-2}$) averaged over the entire tundra. Much of their life (in particular, their breeding) takes place underground or under snow. Censuses can be made only with the very greatest difficulty, except when the animals are moving above ground and can be trapped. Their behaviour does not help the researcher!

The distribution of the lemmings is patchy in both space and time. Much of the studied area is composed of a repeated pattern of polygons, formed by deeply penetrating wedges of ice. The patterns of snow accumulation, drainage and vegetation vary across a polygon and between polygons (Brown *et al.*, 1980). The life of the lemmings is strongly related to this pattern and changes with the season (Table 15.4). Mean values of lemming density greatly underestimate how crowded they are. For example, in 1974, 93% of nests were in polygon troughs, and it was here also that the lemmings clipped the greatest amount of vegetation.

a lemming increase and crash

One of the periodic increases in the lemming population starts when the lemmings reproduce in nests of grass and sedges at the base of the snow pack, and the population grows rapidly during the winter to reach a peak in late spring. Breeding ceases in May. Signs of overcrowding appear before the snow melts, when many lemmings burrow to the surface and wander about, some to die. During snow-melt, massive clippings of grasses and sedges, and the disruption of moss and lichen carpets are revealed, and lemmings scurry everywhere. Large numbers of predators attack the exposed lemmings. During the summer, lemming survival is low, and the population crashes to a low level, where it remains for 1–3 years (Batzli et al., 1980). This is a general picture, but each irruption of a population has its own characteristics. In 1963 and 1969 there was particularly heavy predation by weasels under the snow. In 1964 very high densities preceded the 1965 peak (the population doubled from 1964 to 1965), whereas in 1970 a very low density preceded the 1971 peak (the population increased 250-fold during the intervening winter).

It may be that there is no single unifying model that describes microtine dynamics. If there is such a theory it would need to take into account the following features (Warkowska-Dratnal & Stenseth, 1985):

1   a species is not necessarily cyclic throughout its range;
2   not all microtine species in a region are necessarily cyclic;
3   the cyclic species in a region need not necessarily be in phase with each other;
4   the ratio of maximum : minimum density can be in the order of thousands, although usually much less;
5   the period of the cycle seems to be more regular than its amplitude;
6   during periods of early population increase, most individuals have a high

Table 15.4 Summary of indicators of brown lemming activity in four habitats at Barrow. Note that densities are seasonal extremes. (After Brown et al., 1980.)

	1972	1973	1974
Winter nest density (number per ha)			
High-centred polygons	—	24.9	18.5
Low-centred polygons	—	18.2	12.9
Meadow	—	17.2	4.2
Polygons and ponds	—	2.7	0.4
Summer population density (number per ha)			
High-centred polygons	3.4–14.7	3.1–4.0	0.3–0.6
Low-centred polygons	12.9–46.1	3.1–4.9	0.3–0.6
Meadow	—	2.5–3.1	0.9–1.2
Polygons and ponds	1.6–7.7	0.3–1.5	0.0–0.6
Percentage of graminoid tillers clipped			
High-centred polygons	14.8	12.3	2.4
Low-centred polygons	—	25.1	6.7
Meadow	—	24.3	1.9
Polygons and ponds	—	5.4	0.3

reproductive rate and typically a high dispersal rate, whereas during periods of high but declining densities, most individuals have a low reproductive rate and a low dispersal rate;

7   the level of individual aggressiveness varies throughout the cycle (see below);

8   the specific dispersal rate depends on population density as well as on the rate of density change (i.e. the position during the cycle).

As has so often been the case in the history of population biology there are those who try to explain population behaviour by invoking extrinsic factors (weather, predators and parasites), and those who look for intrinsic causes (changes in the individual animals such as hormonal or behavioural changes, aggression, an urge to disperse). In the case of extrinsic factors, cycles may simply emerge from an interaction with or without the individual organisms altering in any marked way (see Chapter 6 for time lags and overcompensation; Chapter 10 for predation and alternative stable states; and Chapter 12 for parasitism). By contrast, the theories that rely on intrinsic causes involve genotypic change in the population (one type of individual being replaced by another), or phenotypic change (a given individual changing in response to the behaviour of its neighbours, for example becoming aggressive or submissive). Some authors combine intrinsic and extrinsic factors in their hypotheses.

15.4.3 Extrinsic theories of microtine abundance

The quantity or the quality of available food have both been invoked to explain microtine cycles. In Alaska, the vegetation can be devastated by a peak population of lemmings and many dead individuals are seen, but in most studies extreme starvation seems to be rare. However, the quality of the available food (e.g. the concentration of nutrients and toxins) might be more important to the rodents than its quantity (see, for example, Batzli, 1983).

Food quality has traditionally been seen as important by Finnish students of microtine cycles. Kalela (1962) argued that the nutritional state of food plants (as indicated by flowering intensity in particular) varies from year to year with meteorological conditions, and that the microtines respond to this nutritional state. Rodents increase to a peak density on high-quality food. A combination of flowering and microtine consumption, however, exhausts the nutrient and energy resources of the plants, their quality as food collapses and the microtine population declines (the decline being then reinforced by predators and disease). The plants typically take a number of years to recover in the northern latitude habitats with their short summer, and it is only when they have done so, and when meteorological conditions are favourable, that their nutritional state and the microtine density rise once again.

the equivocal results of food supplementation experiments

This model has not been favoured by many microtine ecologists partly because, in a number of studies, supplying extra food has not prevented population declines, especially in less northerly habitats. Yet, this may simply be due to the 'pantry effect': that predators are attracted from nearby non-supplemented populations and their activities counteract any microtine increase or lack of decline that there might otherwise have been. Certainly, when Ford and Pitelka (1984) provided supplementary food and water to penned populations of voles in California, *and controlled the number of predators*, their experimental populations declined only modestly during a summer when control populations crashed. There is also some evidence that *short* periods of food shortage can induce hypoglycaemic shock and that this might be critical during phases of population decline (Frank, 1953; Boyce & Boyce 1988a, b, c).

At least in theory, predators and parasites might be responsible for microtine cycles. Parasites have been known to cause considerable microtine mortality, but they seem mainly to affect individuals in poor condition that were probably susceptible to many other causes of mortality as well.

Predators are more obvious candidates to cause microtine cycles. The Alaskan lemmings are again a source of relevant data (Batzli *et al.*, 1980). At the time of year when the major increases in lemming population take place (a 'high' winter), the main predators are the Arctic fox (*Alopax lagopus*) and two species of weasel, the least weasel (*Mustela nivalis*) and the ermine (*M. erminea*). These were absent from the study site at Barrow during periods when the lemming populations were low, but they recolonized and multiplied fast during lemming peaks, reaching densities of up to 25 km^{-2}.

The other major predators are migrant birds, especially snowy owls, pomarine skuas (jaegars), glaucous gulls and sometimes short-eared owls. Their arrival precedes (snowy owl) or coincides with the onset of snow-melt, when the lemming populations are exposed. If the density of lemmings is low, the jaegars and owls move on, but if lemming density is high the birds stay and breed. The birds have smaller territories (occur at a higher density), lay larger clutches and rear more young in years when lemmings are abundant. In other words, the predators increase, both by reproduction and immigration, during lemming population peaks (Figure 15.14).

Populations of lemmings may be seriously reduced by the various predators (Figure 15.15), and these may contribute to population declines. Their activities may be sufficient to prevent increases at low densities and account for declines in summer—although there is a disagreement about how responsible they are for population crashes. Furthermore, relaxation of winter predation will not necessarily lead to population increases (Batzli *et al.*, 1980). The predator populations may track

a secondary role for parasites

the role of predators: Alaskan lemmings as a typical example

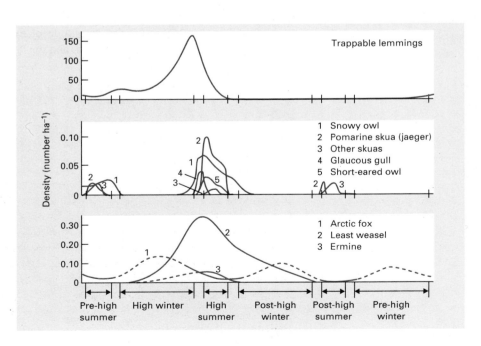

Figure 15.14 The estimated densities of predators during the course of a standard lemming cycle at Barrow. Periods of snow-melt and freeze are indicated between summer and winter. (After Batzli *et al.*, 1980.)

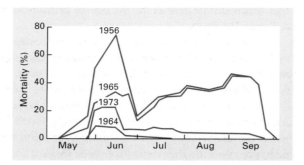

Figure 15.15 The impact of avian predators on lemming populations in 4 years as indicated by the percentage of mortality accounted for by predators. (After Osborn, 1975.)

the cycles in the populations of their prey, and account for a large number of deaths; but, many of those that are killed are probably doomed anyway. There has been a widely held view that, although predators may have important effects on microtine abundance, they play little part in generating microtine cycles. But, this view may now need to be revised.

It is a striking feature of microtine dynamics that the most marked cycles have been recorded in far northern environments of Alaska and Fennoscandia. Examples

the biology of voles in Finland

of cycles in populations of voles in northern Finland are illustrated in Figures 15.13b–e. Even for the same species both the strength and regularity of cycles become less marked in milder climatic regions. The nature of predator populations also changes from north to south in Finland (Figure 15.16), and it has been suggested that specialist predators (small weasels—mustelids) are wholly or largely responsible for the regular cycles in the northern populations. Further south, generalist predators (e.g. foxes, buzzards and cats) are more abundant. The role of generalist predators in vole dynamics is also discussed in Chapter 10, Section 10.4.

When the two kinds of predation were combined in the same predator–prey model, the generalist predators stabilized the cycles that were driven by the specialists. Indeed, strong predation by generalists could convert a limit cycle to a stable equilibrium. Moreover, the amplitude of the vole cycles decreased as the number of generalist predators increased, and the length of the cycle decreased from

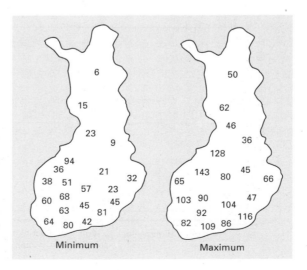

Figure 15.16 The density of birds of prey in different parts of Finland. The figure is derived from minimum and maximum estimates of numbers from Saurola (1985) multiplied by $W^{0.75}$, where W is the average weight of the species and the exponent 0.75 corrects for changing ingestion rate (Peters, 1983). The values obtained from these corrections are multiplied by 0.032 to give the values shown in the figure in units of predator weighing 100 g. (After Hanski *et al.*, 1991.)

5 to 3 years from 70 to 60°N. There was no clear multiannual cycle below 60°N (Hanski *et al.*, 1991).

The idea that weasels are instrumental in the maintenance of the regular vole cycle was supported by the following mathematical predator–prey model, which was parameterized with empirical data for voles and weasels (Hanski *et al.*, 1993). Key features of this model are a seasonal difference in the breeding of voles in summer and winter, no winter breeding at all in weasels (never recorded in Finland) and a threshold vole density for breeding by weasels.

Equation 15.3: prey dynamics in the 6 months of summer when the voles are breeding continuously. In the winter r and K are replaced by lower values r' and K'.

$$\frac{dN}{dt} = rN\left(1 - \frac{N}{K}\right) - \frac{cPN}{N+D} \tag{15.3}$$

Equation 15.4: predator dynamics in summer when $N > N_{crit}$ (i.e. when vole densities exceed a critical threshold).

$$\frac{dP}{dt} = vP\left(1 - \frac{qP}{N}\right) \tag{15.4}$$

Equation 15.5: predator–prey dynamics when $N \leq N_{crit}$ (i.e. when vole densities are at or below a critical threshold). When voles are abundant in winter ($N > N_{crit}$), d_{high} is replaced by a smaller value d_{low} (i.e. lower rate of exponential decline in the predator population).

$$\frac{dP}{dt} = -d_{high}P \tag{15.5}$$

An example of the dynamics of voles and weasels generated from the model is shown in Figure 15.17, which should be compared with the real census data in Figures 15.13b and c. Both the predicted and observed dynamics are chaotic (see Chapter 6, Section 6.9) with a significant periodic component. The comparison strongly supports the hypothesis that delayed density dependent predation explains multiannual microtine cycles in Finland. It may be that other explanations are relevant elsewhere.

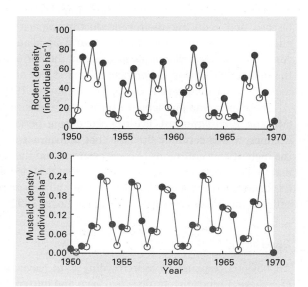

Figure 15.17 The dynamics of vole–weasel populations generated from the model described above with time shown in years since the beginning of the simulation. ○, Spring densities; ●, autumn densities. These results were obtained by using parameter values estimated from field data. (After Hanski *et al.*, 1993.)

Other attempts to model the behaviour of populations of small rodents are described in a series of papers edited by Hansson and Stenseth (1988).

15.4.4 Intrinsic theories of microtine abundance

intrinsic theories: aggression, dispersal and relatedness

We can now turn to the 'intrinsic' theories. It is not surprising if species like voles and lemmings that can achieve extremely high potential rates of population growth should experience periods of overcrowding. It might be predicted that the overcrowding might then produce changes in physiology and behaviour. Mutual aggression (even fighting) might become more common and have consequences in the physiology—especially hormonal balance of the individuals. There might be increased pressure on some individuals to defend territories and on others to escape. Powerful local forces of natural selection might be generated that favour particular genotypes (e.g. aggressors or escapists). There might also be family conflicts in which kin and non-kin behave differently to neighbours when they are crowded. These are responses that we easily recognize in crowded human societies and it is not surprising that ecologists have looked for the same phenomena when they try to explain the population behaviour of rodents. All these effects have been found or claimed by rodent ecologists. But, it remains an open question whether any of them gives a satisfactory explanation of the behaviour of rodent populations in nature.

Hormonal changes

Aggressiveness varies throughout the microtine cycle, although there is disagreement whether it is greatest in the increase phase and low during the decline (Krebs, 1985), or greatest when densities are high but declining and lowest during periods of early population increase (Warkowska-Dratnal & Stenseth, 1985). Strife certainly occurs in crowded rodent populations and can be detected in an increased proportion of males with wounds. Such strife might be expected to have direct effects on the pituitary–adrenocortical–endocrine system and increase the production of, for example, corticosterone. Laboratory experiments support the view that this occurs and suppresses the individuals' immune system, increasing their liability to disease and reducing their reproductive capacity (Christian, 1950, 1980). An alternative view (based on studies of an Australian marsupial (*Antechinus*) and the Australian bush rat (*Rattus fuscipes*)) is that the hormonal responses actually increase the individuals' ability to compete for mates but at the expense of very high male mortality after mating (see, e.g. Lee & Cockburn, 1985a, b; McDonald *et al.*, 1988).

There are some problems with this hypothesis. In populations of meadow voles (*Microtus pennsylvanicus*), live-trapped during periods of severe spring population declines, it was certainly the case that individuals in high density populations had the highest concentrations of corticosterone and the highest proportion of wounded males—but, these were the populations that declined the least. The population with the lowest density, lowest corticosterone and lowest proportion of wounded males suffered the greatest decline (Boonstra & Boag, 1992).

Dispersal

An obvious response to overcrowding is to leave the crowds and look for a life elsewhere. Unfortunately, dispersal from and into colonies of microtines is extremely difficult to measure. It can, however, be controlled experimentally by erecting fences to prevent entry and exodus, and there is much evidence that this upsets the population dynamics. A vole population enclosed with a mouse-proof fence to prevent emigration increased to a high density and suffered starvation. There was less evidence of aggression, and less decline in individual condition (Krebs *et al.*, 1969). When immigration was prevented there was also less aggression, but a much lower population peak, and the population neither crashed nor cycled (Gaines *et al.*, 1979).

If immigration and emigration play a crucial role in microtine dynamics we need to know what is the difference between the individuals that disperse and those that remain behind. There are several hypotheses.

1 The social subordination hypothesis proposes that as population density increases, the resulting shortage of resources leads to increased levels of aggression which in turn forces social subordinates (predominantly young individuals) to leave (Christian, 1970). The proportion of dispersers appears to be highest when the population is rising rather than when it is highest. It has also been suggested that aggression and a consequent urge to escape is exaggerated when individuals are more likely to meet others that are not close kin (Charnov & Finnerty, 1980). It is interesting that these theories place so much emphasis on the nature of encounters between individuals rather than on their response to some generalized holistic measure of 'population density'.

2 The genetic–behavioural polymorphism hypothesis (Krebs, 1978) proposes that individuals are *innately*, either aggressive or capable of high reproductive output. The aggressive types are favoured in competitive populations and force emigration of the rapid reproducers (*r* types), which are favoured in populations at low density. There is evidence that in many small mammals (although not all; see Gaines & McClenaghan, 1980) dispersers are not a random genetic sample of the population that they escape from. The technique of electrophoretic analysis of isoenzymes was used to show that dispersing and non-dispersing individuals of *M. pennsylvanicus* in Indiana differed at at least two loci and one rare homozygote was found only amongst dispersers (Myers & Krebs, 1971). It is not known whether these genetic differences have anything to do with the tendency to disperse as such, although in some other small mammals dispersibility has a clear heritable component (e.g. Waser & Jones, 1989). It is exciting that the techniques of genetic fingerprinting can now be applied to the small rodents and should allow much more detailed analysis of family structures and genetic traits (e.g. Ribble, 1992).

3 The saturation–pre-saturation dispersal hypothesis suggests that there are two distinct categories of disperse: social outcasts (juveniles, very old individuals, those in poor condition and in general those least able to cope) and pre-saturation dispersers which are in relatively good condition, but particularly sensitive to the early phases of population growth (Lidicker, 1975). In principle, we should expect individuals to disperse when they have more to gain by leaving than staying. This will frequently be the case for subordinate individuals with a future (e.g. young but healthy individuals): they will be capable of finding and exploiting new habitats and can be expected to leave expanding populations. In overcrowded populations, on the other hand, fewer of the subordinates will be healthy. Perhaps the only ones to leave will be those that are so oppressed that they are doomed if they remain.

4 The social cohesion hypothesis proposes that dispersers are individuals that have not formed social ties by interacting with their sibs, i.e. dispersers are asocial rather

than oppressed individuals (Bekoff, 1977). There are obvious analogies with human populations where family ties may hold an individual back from adventurous exploration.

Despite the variety of nuances in the different theories, there is much agreement that changes in the proportions of aggressive and docile individuals, strong and weak dispersers and strong and weak reproducers occur during population cycles. There is less support for the view that such behavioural changes cause the size of populations to change.

Sex, mating and territories

There may be real dangers in assuming that some one general theory can be applied to all small mammals—or, even to all species of one genus such as *Microtus*. It may usually be the case that it is the more aggressive individuals that do not disperse, but in *M. pennsylvanicus* dispersing males were more aggressive than the residents. Moreover, the different species of *Microtus* have different mating systems and this affects the way in which they compete and disperse. In some species of *Microtus* it is predominantly males that disperse and in other species it is the females. Much seems to depend on whether males compete mainly for food or for mates (Johnson & Gaines, 1990), and particularly whether females are territorial.

It is now well established that there is spacing behaviour in several species of vole with territorial behaviour in one or both sexes, and that this limits the number of individuals able to reproduce and has a strong influence on demography (e.g. Krebs, 1985; Lambin & Krebs, 1991a, b). Males of the Townsend's vole, *M. townsendii*, compete mainly for females and the females compete for breeding space. This makes it sensible to examine the behaviour of individuals in relation to their specific neighbours, rather than generalizing a study over a whole population. In a very subtle experimental design, Lambin and Krebs (1993) removed selected individuals from populations of the Townsend's vole so that the degree of relatedness of the females could be varied. This had involved a systematic marking of the new-born young and the radiotagging of adults. Predation was prevented. It was found that individual females that had mothers, sisters or daughters as a neighbour were more likely to survive the spring decline and reproduce than females without close relatives as neighbours. Moreover, closely related females lived closer to each other than more distantly related or unrelated females (Figure 15.18) (see also Kawata, 1987, for a very similar experiment). It is not yet clear whether such studies of intimate sociobiological behaviour will ultimately answer the still burning questions about cycles, peaks and crashes in microtine populations. They undoubtedly contribute new insights into the way in which the behaviour of the individual animals may interact to influence the abundance of populations.

15.4.5 Microtine cycles: a postscript

The forces that determine the population dynamics of small mammals remain an exciting area of ignorance. It has been extremely difficult to identify dispersing individuals, yet we need to know more about the genetics and the genealogy of individuals, and be able to identify accurately those that disperse and to know their fate. Now that techniques of genetic fingerprinting can be applied to individuals in nature some of the problems may be solved. But, deep questions about proximal

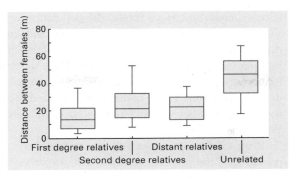

Figure 15.18 The distance between the nest (or the centre of activity) of female voles of different degrees of relatedness. First degree relatives are mother–daughters and littermate sisters. Second degree relatives are non-littermate sisters and aunt–nieces. Distant relatives are cousins, grandcousins, grandnieces and aunts. The box plots show median, interquartile ranges and the 95% confidence intervals. (After Lambin & Krebs, 1993.)

motivation remain: what stimuli make an individual decide whether to leave home or not and what does the future hold for those that decide one way or the other?

Most of the theories of microtine behaviour were already being aired more than 50 years ago and the popularity of different theories has waxed and waned. Theories about the roles of predators, territoriality and the social interactions of individuals now appear to be firmly back in fashion! Questions are much more interesting than answers. There is a wonderful literature of scientific imagination, controversy and theory to explore (e.g. Lidicker & Patton, 1987; Anderson, 1989; Johnson & Gaines, 1990; Stenseth, 1993).

15.5 Abundance determined by dispersal

One element in the population dynamics of microtines is undoubtedly the extent to which individuals disperse. We have already referred to this in Chapter 5, and we turn to it again now. In many studies of abundance, the assumption has been made that the major events occur within the census area, and that immigrants and emigrants can be safely ignored. But, it is becoming realized that migration can be a vital factor in determining and/or regulating abundance. We have already seen that emigration of summer adults of the Colorado potato beetle is both the key factor in determining population fluctuations and an overcompensating density dependent factor (see Section 15.3.1). Moreover, the winter moth population (see Section 15.3.2) was subject to an overwintering loss, due partly to larval emigration, which acted as the key factor but was not density dependent; whilst in the life cycle of the tawny owl (see Section 15.3.2), emigration (and death) of young owlets was shown to be density dependent, although not, in this case, the key factor.

When dispersal is a major factor determining population size, it can make life difficult for the investigator. A census made solely *within* an area will usually miss such events entirely, and a far more subtle censusing procedure is needed to account accurately for emigration losses and immigration gains. Even in species of which the individuals themselves cannot move there is always a dispersal phase amongst their progeny (planktonic larvae, seeds) and this is usually difficult to monitor.

In a study of *Cakile edentula,* a summer annual plant growing on the sand-dunes at Martinique Bay, Nova Scotia, the population was concentrated in the middle of the dunes and declined at both seaward and landward ends of its distribution (Keddy, 1981). Data gathered on seed production and mortality in the three parts of its range show convincingly that, in theory, an equilibrium population should be found only towards the seaward limit, where seed production is high and strongly density dependent. In the middle and landward sites, mortality exceeds fecundity so that the populations should rapidly go extinct (as they do in simulation models; Watkinson, 1984).

the crucial effects of dispersal on the dynamics of *Cakile* populations

If we take into account only the *in situ* births and deaths of plants, we would predict that the population would not persist in the middle area, but in fact the species is most abundant there. The natural abundance of this plant can only be explained when a landward migration of seeds, caused by both wave and wind action, is taken into account (Keddy, 1982; Watkinson, 1984). Plants that are found at the middle and landward ends of the dune gradient are there because of the high annual dispersal of seeds landward from the beach (indeed, the rate at which new individuals are added in this region exceeds the number produced by the resident

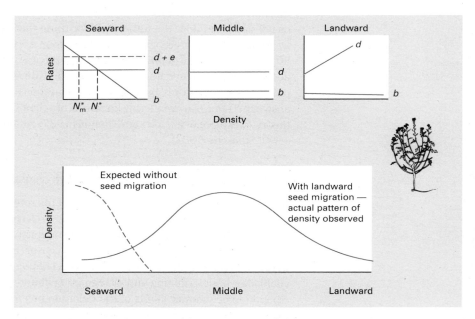

Figure 15.19 Idealized diagrammatic representation of variations in mortality (*d*) and seed production (births, *b*) of *Cakile edentula* in three areas along an environmental gradient from open sand beach (seaward) to densely vegetated dunes (landward). In contrast to other areas, seed production was prolific and density dependent at the seaward site. Mortality was only density dependent at the landward site. Births *in situ* exceeded deaths at the seaward site, in such a way that an equilibrium population density can be envisaged (N^*). Deaths always exceeded *in situ* births (new seeds) in the middle and landward sites so that populations should not, in theory, have persisted there ($---$ in lower graph). Persistence of populations in the middle and landward site occurs only because of the landward drift of the majority of seed produced by plants on the beach (seaward site). Thus, the sum of *in situ* births (*b*) plus immigrating seeds (*i*) exceeds mortality (*d*) in the middle and landward sites. Note also that losses from the population at the seaward site actually include both deaths (*d*) and emigrating seeds (*e*) so that the equilibrium density there (N^*_m) is less than it would be in the absence of seed dispersal (— in lower graph).

plants) (Figure 15.19). Nevertheless, it is the effect of crowding on seed production by plants at the seaward end of the dunes that regulates the overall population densities on the whole dune system.

15.6 Patches, islands and metapopulations

In Chapter 5 we emphasized that dispersal plays a critical role in determining the abundance of species because most populations are fragmented and patchy and may include patches that have quite different dynamics. The floor of a mature forest is just such a patchwork of dynamics. It includes the patches created where old trees have recently died (each has become locally extinct in its unique patch) and patches in which new individuals are colonizing and are in their turn on the way to becoming locally extinct. The process of vegetational succession (which is discussed in greater detail in Chapter 17) involves repeated episodes of local immigrations and extinctions that may all be occurring in different patches in a forest at the same time. In the population census of buttercups in a pasture (see Table 15.1) an overall relatively stable population was represented by three samples, in one of which the population increased and in another decreased. Such patchworks of successions affect not only the plants but the complexes of animal and microbial species associated with them. A census of the individuals in a whole forest or in a whole pasture will miss all these demographic niceties.

On a different scale, the aphids on one leaf may be crowded, but on other leaves on the same plant new colonists may be in the early phases of unimpeded population growth. On other, older leaves, the aphids may be becoming extinct as the leaf dies. Yet other leaves may still be unoccupied: 'habitable sites' that dispersers have not yet reached. A census of the dynamics of the population of aphids on a single plant would conceal these underlying dynamics. They would be totally lost in a census of the aphids present in a whole community of such plants.

15.6.1 Colonization and extinction in local populations

In his classic argument for the necessity for regulating factors, Haldane (1953) (the paper quoted above) used as part of his case that 'no population increases without limits and species only occasionally become extinct'. But the latter is only partly true. Local populations do indeed frequently become extinct. This point had already been made by Andrewartha and Birch (1954). They wrote 'spots that are occupied today may become vacant tomorrow and reoccupied next week or next year'. Indeed, it can be argued that there is a scale (albeit perhaps sometimes very small) on which every species is bound sooner or later to become locally extinct.

local extinction is common

The type of population dynamics that is revealed by a census may therefore depend almost entirely on how its patchwork is sampled (Figure 15.20) and see Wiens, 1989, for a description of sampling problems in patchy populations of birds).

patchiness is everywhere

Patchiness appears as the essence of quite different ecological processes in other chapters of this book. For example, islands are patches of land in a continuum of water and much of the theory of island biogeography places species' distributions and abundances in the context of the colonization of and extinction from these patches (see Chapter 23). Likewise, plant and animal hosts are patches from the point of view of a parasite or a pathogen. A population of a disease organism persists

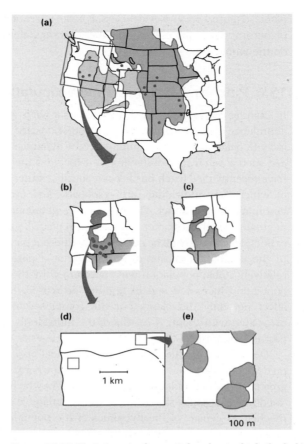

Figure 15.20 Variations in the spatial scale at which the habitat occupancy patterns of shrub–steppe birds were examined. At a biogeographical scale (a), variations in the densities of several species that commonly breed in subshrub environments were significantly correlated with several features of habitat structure. When the scale was focused on 14 sites within the subshrub region (b), these associations disappeared and instead the birds' distribution showed significant covariance with other features of habitat structure or flora. In an analysis of eight plots at four locations (c), two of the species showed no association at all with habitat patterns, and in a more detailed comparison of two study plots at a single site (d), patterns were again detected, but they differed from those found on the broader scales of sampling. In an even more detailed study of a single plot (e) (shading shows areas occupied by breeding territories), the patterns of habitat occupancy partly paralleled the patterns found at the regional scale (a), but not at the other scales. (After Wiens *et al.*, 1987; Wiens, 1989, gives a discussion of the wider problems of choosing appropriate scales for sampling populations.)

if there are patches (i.e. hosts) of which some are successfully occupied (infected), and remain occupied (i.e. do not become immune) for long enough to become infective, and close enough to disperse to (infect) uncolonized hosts (see Chapter 12). The same processes underpin the persistence of symbiotic and mutualistic associations (see Chapter 13). Similarly, when we consider the interactions between predator and prey populations we need to recognize that the prey are inevitably patchily distributed and this interacts profoundly with the searching behaviour of predators (see Chapters 8–10).

A population may be small because (Gadgil, 1971):

1 there are few patches (habitable sites) that provide the conditions and resources that it requires;

2 the habitable sites are small;

3 the habitable sites are far apart relative to the dispersibility of the species;

4 the habitable sites can support few individuals (their carrying capacity is low);

5 the habitable sites remain habitable for only a short time;

6 the population grows only slowly after colonizing a site and is slow to attain carrying capacity.

15.6.2 Dispersal between local populations

It is tidy to link so many of the processes that determine the abundance of organisms into the two concepts of habitable patch and dispersal distance (see Godfray & Pacala, 1992; and a review by Taylor, 1990). Sadly, the concepts are hard to translate into real measurements in nature. The dispersal distance for a species is not some simple parameter: dispersing individuals or their propagules travel widely different distances and the frequency distributions of the distances are often quite complex (see Chapter 5, Figure 5.5). Moreover, the occasional extreme events of long-distance dispersal that result in discovery of an unoccupied patch may have much greater demographic consequences than dispersal over common or average distances which may be more relevant to the way a colonist spreads within a patch.

The ways in which dispersal contributes to the colonization of habitable sites may be envisaged in three different ways, depending on how the spatial environment is characterized. This may be as 'islands', as 'stepping stones' or as a continuum (a masterly review of models and real data can be found in Kareiva, 1990). The key feature of island models is that a fraction of the individuals leave patches and enter a 'bath' of dispersers and are then redistributed amongst patches, usually at random. These models usually have no spatial dimension. In contrast, stepping-stone models have a spatial dimension and are useful for comparison of the effects of long-range and short-range dispersal. 'Stepping-stone' dispersal is probably most common in terrestrial communities of both plants and animals. In the sea, however, the larvae of invertebrates are typically dispersed for great distances and are thought to form a common larval pool which 'bathes' the community as in the island model (see, for example, Roughgarden *et al.*, 1984; Possingham & Roughgarden, 1990). However, even in the marine environment, the stepping-stone model appears to give the best fit to dispersal and patch colonization by an ascidian on a coral reef (Stoner, 1992).

In continuum models the environment is usually assumed to be homogeneous and dispersal is represented by a diffusion process. The latter might be thought to be very dull in the context of dispersal and patchiness, but it transpires that where there are species interactions, a system that is stable in the absence of dispersal can generate its own patchiness and produce regular or even bizarre spatial variations when dispersal occurs (Meinhardt, 1982; Kareiva & Odell, 1987; Kareiva, 1990).

The different models have different roles. The island models are mathematically elegant, stepping-stone models offer more realism and continuum models suggest intuitively unlikely situations worth looking for in the real world.

dispersal by 'stepping stones' or as a 'bath'

15.6.3 The habitability of local sites

habitable patches may be uninhabited

'Habitable patch' is another concept that is fraught with problems for both theorist and empiricist. Andrewartha and Birch emphasized (1954, 1960, 1984; see Section 15.2.2) that habitable sites might remain uninhabited because individuals failed to disperse into them. If we are to discover the limitations that the number of habitable sites place on species abundance, we need to be able to identify and measure habitable sites that are not inhabited. Only very rarely has this been attempted. One method involves identifying characteristics of habitat patches to which a species is restricted and then determining the distribution and abundance of similar patches in which the species might be expected to occur. The water vole (*Arvicola terrestris*) lives in river banks and in a survey of 39 sections of river bank, 100–300 m long in North Yorkshire, 10 contained breeding colonies of voles (core sites), 15 were visited by voles but they did not breed there (peripheral sites) and 14 were apparently never used or visited. A principle component analysis was used to characterize the 'core' sites and on the basis of these characteristics a further 12 unoccupied or peripheral sites were identified that should have been suitable for breeding voles (i.e. habitable sites). About 30% of habitable sites were uninhabited by voles because they were too isolated to be colonized or in some cases suffered high levels of predation by mink (Lawton & Woodroffe, 1991).

Habitable patches can be identified for a number of rare butterflies because the larvae feed on only one or a few patchily distributed plant species. Thomas *et al.* (1992) found that small patches and those isolated from sources of dispersal remained uninhabited. In such 'metapopulations', the butterfly *Phlebejus argus* was able to colonize virtually all habitable sites less than 1 km from existing populations. In a pretty test of the habitability of some of the isolated sites the butterfly was successfully introduced (Thomas & Harrison, 1992). This is presumably the crucial test of whether a site is really habitable or not! An example of the deliberate introduction of seed to identify habitable patches within plant communities is described in Section 15.7.1.

15.6.4 The development of metapopulation theory

Three main forces were responsible for changing the way we think about the role of patchiness and dispersal in determining abundance.

1　Students of genetics and evolution have long recognized that the individuals of many species are not grouped into isolated populations but are composed of local populations, connected to each other by a recurrent exchange of migrants. Migration of individuals (or, in the case of plants, the migration of pollen, spores, gametes or seeds, and in many marine invertebrates the migration of gametes) links local populations genetically and hinders evolutionary divergence from natural selection or genetic drift. Geneticists were therefore forced to model patchy populations in the context of gene flow and genetic isolation. Roughgarden (1979) is a good source for much of the relevant literature of that period (although he does not use the word 'metapopulation') and Gilpin (1991) for the evolutionary questions set in the context of metapopulation theory.

geneticists have recognized that populations are subdivided

2　A classic book *The Theory of Island Biogeography* by MacArthur and Wilson (1967) was an important catalyst in radically changing ecological theory. They focused attention on the extinction and colonization of species on maritime islands. They emphasized that populations differ in behaviour; that some species (or local populations) spend most of their time recovering from past crashes or in phases of invasion of new territories, and that others spend much of their time 'bumping up against' the limits of their environmental resources, or in other ways suffering from the effects of overcrowding (see Figure 15.7). These two conditions represent the ends of a continuum. At one extreme the size of a population usually reflects:

(a) the level to which it had last been reduced;

(b) the time elapsed for it to regrow;

(c) its intrinsic rate of population increase during that time.

two types of 'island' populations

At the other end of the continuum, the size of a population reflects some limiting resource that constrains its further growth by reducing the birth rate, increasing the death rate or stimulating the emigration of its members. (The ends of this continuum are MacArthur and Wilson's r selection and K selection, discussed in Chapter 14).

the r / K continuum joins the opposing poles

MacArthur and Wilson (1967) developed their ideas in the context of the floras and faunas of real (i.e. maritime) islands which they interpreted as a balance between the opposing forces of extinctions and colonizations (see Chapter 23). This thinking has been rapidly assimilated into much wider contexts with the realization that patches in terrestrial landscapes also have many of the properties of true islands, and indeed, that even isolated trees of one species in a forest of other species have some of the properties of islands for the insects that depend on them.

It is not unrealistic to think of individual organisms as islands that pathogens and parasites may invade, colonize and on which they may then become locally extinct. For example, the dynamics of an epidemic of measles involves dispersal from colonized (i.e. infected) host individuals to susceptible (i.e. 'habitable') but previously uninfected hosts (islands). The measles virus reveals quite different dynamics depending on whether it is studied within an individual patch, i.e. host (as its population increases, causes disease and then becomes extinct), or in a metapopulation of humans (an archipelago of islands) in which the population of infected individuals may show fascinating cycles of abundance and rarity (see Chapter 12, Figure 12.22a).

there are many kinds of 'islands'

metapopulations

3   At about the same time as MacArthur and Wilson's book was published, a simple model of 'metapopulation' dynamics was proposed by Levins (1969, 1970).

15.6.5 The dynamics of metapopulations

The concept of 'metapopulation' was introduced to refer to a subdivided and patchy population in which the population dynamics operates at two levels, within patches and between patches (local events of colonization and of extinction). For an environment that was patchy, Levins introduced the variable $p(t)$, the fraction of habitat patches occupied at time t. Note that the use of this single variable carries the profound notion that not all habitable patches are always inhabited.

The rate of change in the fraction of occupied habitat (patches, p) is given in Levins' model as:

$$\frac{dp}{dt} = mp(1-p) - \mu p \qquad (15.6)$$

in which μ is the rate of local extinction of patches and m the rate of recolonization of empty patches. This reflects the idea that the amount of recolonization increases

607   ABUNDANCE

both with the number of empty patches prone to recolonization $(1 - p)$ and with the number of occupied patches able to provide colonizers, p, but that extinctions increase simply with the number of patches prone to extinction, p. Rewriting this equation Hanski (1994a) shows that it is structurally identical to the logistic equation (see Chapter 6, Section 6.10):

$$\frac{dp}{dt} = (m - \mu)p \left(1 - \frac{p}{1 - \left(\frac{\mu}{m}\right)} \right). \tag{15.7}$$

Hence, as long as the intrinsic rate of recolonization exceeds the intrinsic rate of extinction $(m - \mu > 0)$, the total metapopulation will reach a stable equilibrium, with a fraction, $1 - (\mu / m)$ of the patches occupied.

The essence of this interpretation is that the metapopulation persists, stably, as a result of the balance between random extinctions and recolonizations, even though none of the local populations are stable in their own right. An example of this is shown in Figure 15.21, where within a persistent highly fragmented metapopulation of the Glanville fritillary butterfly (*Melitaea cinxia*) in Finland, even the largest local populations had a high probability of declining to extinction within 2 years.

On the other hand, whilst stable equilibria can readily be generated in simple metapopulation models, the observable dynamics of a species may often have more to do with the 'transient' behaviour of its metapopulations, far from equilibrium. For example, the silver-spotted skipper butterfly (*Hesperia comma*) declined steadily in Great Britain from a widespread distribution over most of the calcareous hills in 1900, to 46 or fewer refuge localities (local populations) in 10 regions by the early 1960s (Thomas & Jones, 1993). The probable reasons were changes in land use—increased ploughing of unimproved grasslands and reduced stocking with domestic grazing animals—and finally the virtual elimination of rabbits by myxomatosis with its consequent profound vegetational changes. Throughout this non-equilbrium period, rates of local extinction generally exceeded those of recolonization. In the 1970s and 80s, however, reintroduction of livestock and recovery of the rabbits led to increased grazing and suitable habitats increased again. Recolonization exceeded local extinction—but the spread of the skipper remained slow, especially into localities isolated from the 1960s refugia. Even in south-east England, where the density of refugia was greatest, it is predicted that the abundance

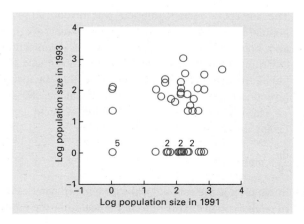

Figure 15.21 Comparison of the local population sizes in June 1991 (adults) and August 1993 (larvae) of the Glanville fritillary butterfly (*Melitaea cinxia*) on Åland island in Finland. Multiple data points are indicated by numbers. Many 1991 populations, including many of the largest, had become extinct by 1993. (After Hanski *et al.*, 1995.)

of the butterfly will increase only slowly—and remain far from equilibrium—for at least 100 years.

Levins' simple model does not take into account the variation in size of patches, their spatial locations, nor the dynamics of populations within individual patches. Not surprisingly, models that do take all these highly relevant variables into account become mathematically complex (Hanski, 1994a, b). Nevertheless, the nature and consequences of some of these modifications can be understood without going into the details of the mathematics.

patch-size variation, sources and sinks

For example, imagine that the habitat patches occupied by a metapopulation vary in size and that large patches support larger local populations. This allows persistence of the metapopulation, with lower rates of colonization than would otherwise be the case, as a result of the lowered rates of extinction on the larger patches (Hanski & Gyllenberg, 1993). Indeed, the greater the variation in patch size, the more likely it is that the metapopulation will persist, other things being equal. Variations in the size of local populations may, alternatively, be the result of variations in patch quality rather than patch size: the consequences would be broadly the same.

Taking variation in the quality of patches a stage further, patches may be divided into higher quality 'sources' (donor patches) and lower quality 'sinks' (receiver patches) (Pulliam, 1988). In source patches at equilibrium, the number of births exceeds the number of deaths, whereas in sink patches the reverse is true. Hence, source populations support one or more sink populations within a metapopulation as a whole: the persistence of the metapopulation depends not only on the overall balance between extinction and recolonization, as in the simple model, but also on the balance between sources and sinks. The dynamics of *Cakile edentula* (see Section 15.5) is a good example of a source–sink system: a seaward source, with middle and landward sinks.

a continuum of metapopulation types

There is likely to be a continuum of types of metapopulation: from collections of nearly identical local populations, all equally prone to extinction, to metapopulations in which there is great inequality between local populations, some of which are effectively stable in their own right. This contrast is illustrated in Figure 15.22 for the silver-studded blue butterfly (*Plejebus argus*) in North Wales.

pseudo-sinks

The identification of sink patches, however, is not straightforward. In a genuine sink patch, deaths exceed births but immigrants exceed emigrants when the population is at equilibrium; and deaths would also exceed births if the patch was isolated from all other patches. The sink patch would not be able to persist without immigrants from other (mainly source) patches. There are also likely, however, to be 'pseudo-sink' patches, in which deaths exceed births at the metapopulation equilibrium—but only because the inflow of migrants has led to density dependent decreases in birth rate or increases in mortality (Watkinson & Sutherland, 1995).

extinction rate declines as p increases: bimodal patch-occupancy distributions

The probability of extinction of local populations typically declines as local population size increases (Hanski, 1991). Moreover, as the fraction of patches occupied by the metapopulation, p, increases, there should, on average, be more migrants, more immigration into patches, and hence larger local populations (confirmed, for example, for the Glanville fritillary—Hanski *et al.*, 1995). Thus, the extinction rate, μ, should arguably not be constant as it is in the simple model, but should decline as p increases. Models incorporating this effect (Hanski, 1991; Hanski & Gyllenberg, 1993) often give rise to an intermediate unstable threshold value of p. Above the threshold, the sizes of local populations are sufficiently large, and their

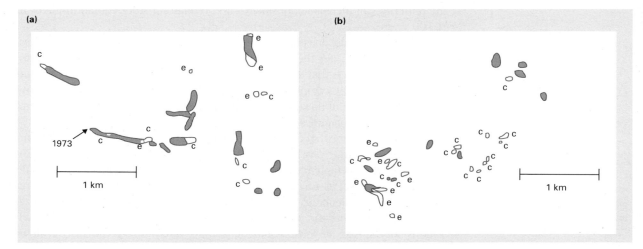

Figure 15.22 Two metapopulations of the silver-studded blue butterfly (*Plejebus argus*) in North Wales: (a) in a limestone habitat in the Dulas Valley, where there was a large number of persistent (often larger) local populations amongst smaller, much more ephemeral local populations; (b) in a heathland habitat at South Stack Cliffs, where the proportion of smaller and ephemeral populations was much greater. Filled outlines, present in both 1983 and 1990; open outlines, not present at both times; e, present only in 1983 (presumed extinction); c, present only in 1990 (presumed colonization). (After Thomas & Harrison, 1992.)

rate of extinction sufficiently low, for the metapopulation to persist at a relatively high fraction of patches, as in the simple model. Below the threshold, however, the average size of local populations is too low and their rate of extinction hence too high; the metapopulation declines either to an alternative stable equilibrium at $p = 0$ (extinction of the whole metapopulation) or to one in which p is low, where essentially only the most favourable patches are occupied.

Different metapopulations of the same species might therefore be expected to occupy either a high or a low fraction of their habitable patches (the alternative stable equilibria) but not an intermediate fraction (close to the threshold). Such a bimodal distribution is indeed apparent for the Glanville fritillary in Finland (Figure 15.23). In addition, these alternative equilibria have potentially profound implications for conservation (see Chapter 25), especially when the lower equilibrium occurs at $p = 0$, suggesting that the threat of extinction for any metapopulation may increase or decline quite suddenly as the fraction of habitable patches occupied moves below or above some threshold value.

15.7 The experimental perturbation of populations

the necessity for experimentation

The observation of nature can generate hypotheses about how it works. In ecology, field observation, theoretical models and the analysis of census data can all be used to suggest what forces are acting in the field, ocean or forest. But, just as we can only test what holds a pendulum at rest by setting it swinging, so we can only test ecological hypotheses by perturbing real ecological systems. If some correlations lead us to suspect that predators or competitors determine the size of a population, we can ask what happens if we remove them. If they are absent, we may ask what

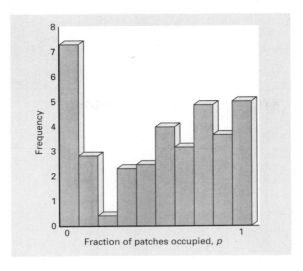

Figure 15.23 The bimodal frequency distribution of patch occupancy (proportion of habitable patches occupied, *p*) amongst different metapopulations of the Glanville fritillary (*Melitaea cinxia*) on Åland island in Finland. (After Hanski *et al.*, 1995.)

would happen if we add them. If we suspect that a resource limits the size of a population we may add more of it. Besides indicating the adequacy of our hypotheses, the results of such experiments may show that we ourselves have the power to determine a population's size, to reduce the density of a pest or weed or to increase the density of an endangered species. Ecology becomes a predictive science when it can forecast the future: it becomes a management science when it can determine the future.

15.7.1 Introduction of new species

There are many examples of species that have been accidentally introduced into alien habitats, but although the consequences have often been ecologically dramatic, they have not been part of controlled experiments. On the other hand, a deliberate introduction of species into a habitat can be a powerful means of beginning to understand the causes of commonness, rarity and absence. Such introductions are often frowned on by conservation-minded ecologists who dislike any disturbance of the *status quo*, and by those who rightly fear that experimental introduction of a species might lead to its uncontrolled spread. Such experiments are, however, probably the most powerful tools available to the ecologist.

plantain seeds sown in normal and 'abnormal' habitats

Seeds of three species of plantain, *Plantago lanceolata*, *P. media* and *P. major*, were sown into a variety of habitats in the neighbourhood of Oxford, England (Sagar & Harper, 1960). The chosen habitats included some from which the species was naturally absent and others in which it was rare or abundant (Figure 15.24). In three of the habitats (11, 12 and 13) the seeds germinated and the radicle emerged but blackened and died almost immediately and no seedlings emerged at all. These were sites on acid heath or in a bog dominated by *Sphagnum* moss, in which the plantains are normally absent from the natural vegetation. Apparently, the environments offered physical conditions outside the fundamental niches of all three species of plantain. In a patch of limestone grassland within a woodland (5), all three species of plantain had been absent and all three formed established populations from the sown seed; the plants persisted through the year. Apparently, the absence of the species from this site was due to the failure of seed to be dispersed naturally into it.

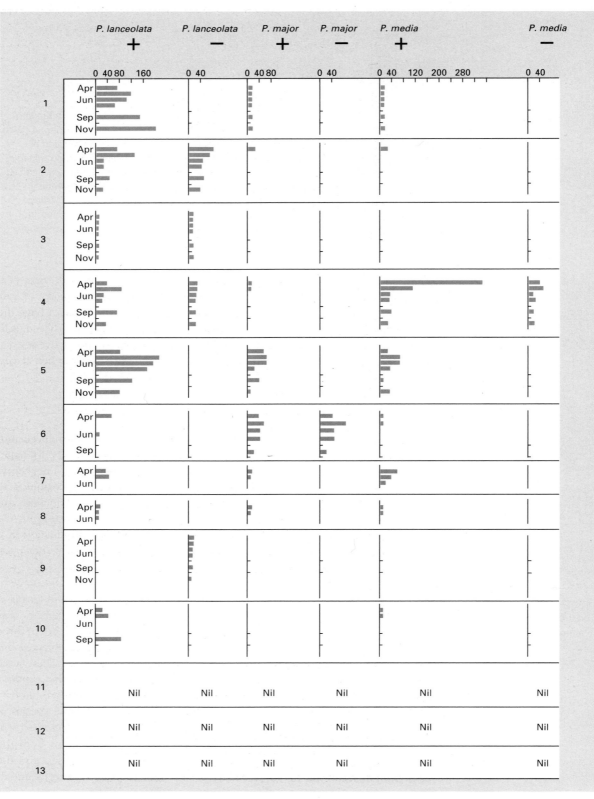

The conditions in the site clearly lay within the fundamental niche of all three species and there was 'room' (habitable sites) in which plants could become established. It is difficult to see how such dispersal-limited abundance could have been demonstrated without the experiment. On bare arable land (1), where there were no competitors, all three species of plantain again formed established populations.

On a trampled grassland (6) only *P. major* was originally present, and adding extra seed did not change the size of its population. The other two species were absent initially, and although seedlings appeared from the sowings, the plants did not survive. This was the commonest situation in a number of habitats: the addition of seed of a species resulted in a flush of seedlings and, if the species was already a part of the community, the population settled down quite quickly to the same population size as in unseeded areas. In these habitats it was very difficult to change the *status quo*: both the species and their populations had stabilized at densities that could not easily be increased.

15.7.2 Augmenting resources

If a population is limited by a resource that is in short supply, the addition of this resource should increase the abundance of the species. Tree squirrels of the genus *Tamiasciurus* are found in much of the boreal and temperate coniferous forest of North America. There is indirect evidence that the squirrel populations are limited by food; for example it has been shown that the size of territories is inversely related to the availability of food. Populations of the Douglas squirrel (*T. douglasii*) were estimated by live trapping in three areas of forest in British Columbia. Control areas received no additional food, but in experimental areas sunflower seeds and whole oats were distributed around each trap station every week from March to October 1977 and in March 1979. In the areas receiving food there was a five- to 10-fold increase in the number of squirrels trapped. Control densities generally varied from three to 10 squirrels per trapping area, but the experimental population increased to 65 animals during the winter feeding (Figure 15.25). This irruption was produced by a combination of immigration, a greater rate of reproduction in females and increased survival. After the food was withdrawn, the population declined to a level comparable with the controls (Sullivan & Sullivan, 1982).

That the augmentation of a resource increases population size is not in itself proof that competition for that resource had been occurring. In the food-rich environment the animals may simply have spent less energy in foraging and had more to spare for

food supplementation in tree squirrels

Figure 15.24 (*facing page*) The numbers of plants and seedlings of three species of plantain (*Plantago lanceolata, P. major* and *P. media*) in 30 cm² plots in a variety of habitats near Oxford, England. Five hundred and twelve seeds of each of the plantain species were sown in each of four plots in each habitat (+), and four controls were left unsown (–). The bar graphs show counts made in the 9 months after the seed was sown. A short narrow bar in each figure refers to a month in which no records were made. 1, recently ploughed bare arable land; 2, flood meadow grassland grazed by cattle; 3, as (2), but grazing prevented by fencing; 4, tightly grazed calcareous grassland; 5, open patch of limestone grassland enclosed in woodland; 6, heavily trampled grassland on light sand; 7, beech (*Fagus*) woodland with no ground flora; 8, oak (*Quercus*) woodland with *Mercurialis* ground flora; 9, ungrazed grassland (mainly *Festuca rubra*); 10, close grazed *Agrostis tenuis* turf with *F. rubra*; 11, lichen mat with *Rumex acetosella*; 12, *Calluna* heath with lichen mat; 13, *Sphagnum*-dominated valley bog.

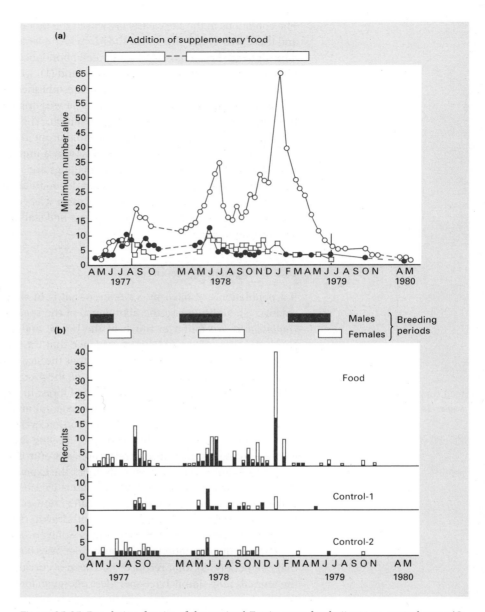

Figure 15.25 Population density of the squirrel *Tamiascurus douglasii* on two control areas (● and □) and one experimental area (supplementary food, ○) during 1977–1980. (b) Number of males (shaded) and females (unshaded) recruited into each population, and reproductive periods of males (scrotal testes) and females (lactation). (After Sullivan & Sullivan, 1982.)

reproduction (Reynoldson & Bellamy, 1971). The experiment did, however, demonstrate that food supply was limiting population size.

Other resources than food may limit population size, and again controlled experiments are likely to be more revealing of causes than simple correlations between abundance and some natural environmental variate. Nesting sites are thought to limit the populations of some woodland birds and providing nesting boxes is a convenient experimental test (Minot & Perrins, 1986).

15.7.3 Removal of possible competitors

A population may be limited because potential resources are shared with individuals of other species. In such a situation we could expect that the removal of the competitors would allow the population to increase (farmers control weeds mainly because the crop then grows better; several other examples are given in Chapter 7). It is easy to hypothesize that in natural communities of plants the abundance of some species is constrained by the presence of others. Only very rarely has such a hypothesis been tested experimentally, but when experimental removals are made this can reveal more about the community than any number of attempts to establish correlations within the *status quo*. Silander and Antonovics (1982) asked whether populations of plants of different species would increase if their neighbours were removed. A series of closely adjacent and floristically diverse coastal communities along a gradient across Core Banks, a barrier island on the coast of North Carolina, USA, were chosen for an attempt to answer this question. Figure 15.26 shows the distribution of species along the environmental gradient. At each of the sites, the dominant and subdominant species were removed singly or in groups, within $1\,m^2$ plots, by weeding and by the action of selective herbicides. At each site, an additional treatment involved the complete removal of all the vegetation, and finally there was an untreated control. Changes in vegetation were measured by using optical point cover. Individual plants (tillers, shoots) were not counted, except for the grasses, but counts and point cover gave very similar measures, so the experiment can be viewed as a study of changes in relative abundance. These changes are shown in Figure 15.27.

In the high marsh, the removal of *Spartina patens* allowed a massive expansion of *Fimbristylis*, and the removal of *Fimbristylis* allowed an even more dramatic increase in *S. patens*. This implies that each had been playing a major part in limiting the population size of the other, that they held habitable areas in common and shared the same fundamental niche. On the other hand, *S. patens* and *Uniola*, which have similar patterns of distribution on the dune, scarcely changed in abundance when either was removed from the neighbourhood of the other. This suggests that their 'habitable sites' (microhabitats) are species-specific and that they do not share a fundamental niche. Silander and Antonovics suggest that this perturbation approach might be applied to a whole community in which all the species were reciprocally removed. Sets of species that react to each other reciprocally could then be identified and might be considered as belonging to the same guilds.

In the experimental treatments in which all the vegetation was removed, four species (*Sabetia*, *Salicornia*, *Triglochin* and *Setaria*) that had been absent or in low frequency, exploded into abundance. These can be regarded as fugitive species in the area. Their abundance appears to depend on the frequency with which disasters strike the rest of the vegetation and free their populations to expand. The habitable sites for these species are apparently normally occupied (their niches are realized) by other species.

15.7.4 Removal of predators

It is sometimes possible to set up experiments in which predators can be excluded from a habitat by fencing. The preferred prey species of such an organism may now increase its population and occupy more of its habitable area. Repercussions may

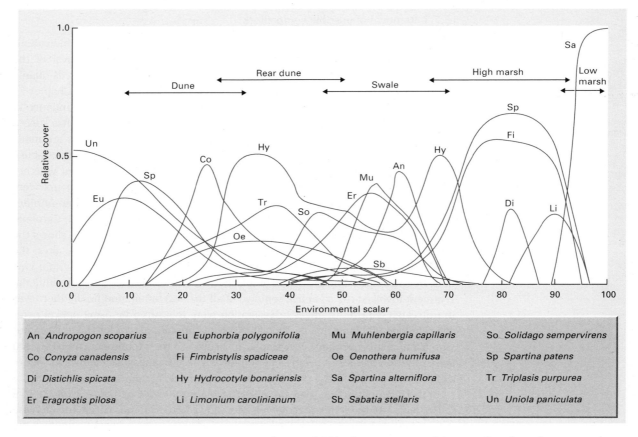

Figure 15.26 Distribution of 16 herbaceous species along a gradient from dune to marsh at Core Banks, North Carolina. The 'environmental scalar' is directly related to the distance from the beach and inversely related to the depth of the water table. (After Silander & Antonovics, 1982.)

then extend to other parts of the community.

the removal of rabbits from grassland

A striking example is provided by the changes in vegetation that followed the deliberate exclusion of rabbits from areas of the floristically very rich chalk grassland of southern Great Britain. The community rapidly degenerated to monotonous stands dominated by species of grass that had previously been a popular food of the rabbits (Tansley & Adamson, 1925). This experiment proved predictive because, after the experiment, the myxoma virus dramatically reduced the populations of rabbits and the same domination by grasses, with its associated loss of a diverse flora, followed over very large areas, to the consternation of conservationists.

Other examples of the effects of removing predators are discussed in Chapter 21.

Figure 15.27 (*facing page*) Perturbation response diagrams for species removal treatments applied to five sites from Core Banks, North Carolina. In each figure the area of the coloured central circle represents the relative abundance (percentage cover) of the species removed, *before* its removal. The areas of the outer circles represent the abundances of species in the control. The areas of the inner circles projecting into each coloured circle represent the increase in abundance of species following a removal. (After Silander & Antonovics, 1982.)

(a) Primary dune

Spartina patens removal

Hydrocotyle — Uniola
Conyza
Oenothera — Euphorbia
Triplasis
Eragrostis

Uniola removal

Conyza — Oenothera
Hydrocotyle — Eragrostis
Spartina patens — Triplasis
Euphorbia

(b) Rear dune

Spartina patens removal

Hydrocotyle
Conyza
Eragrostis
Triplasis — Euphorbia

Hydrocotyle removal

Triplasis
Conyza — Euphorbia
Eragrostis
Spartina patens

(c) Swale

Spartina patens removal

Muhlenbergia
Hydrocotyle
Solidago — Paspalum
Andropogon — Eragrostis

Muhlenbergia removal

Solidago — Andropogon
Hydrocotyle — Eragrostis
Paspalum
Spartina patens

Andropogon and *Eragrostis* removal

Muhlenbergia
Spartina patens
Solidago
Hydrocotyle

Hydrocotyle and *Solidago* removal

Muhlenbergia
Spartina patens
Andropogon — Paspalum
Eragrostis

(d) High marsh

Spartina patens removal

Fimbristylis
Aster
Limonium — Agalinis
Sabatia — Borrichia

Fimbristylis removal

Limonium
Sabatia — Aster
Spartina patens

617   ABUNDANCE

15.7.5 Experimental modification of territories

The local abundance of a species may be determined by territorial behaviour in which the males hold and defend territory, and females compete between each other for territorial males. It has been possible to show experimentally that territories are closely tied to the availability of resources, by for example varying the number of artificial feeders provided for hummingbirds (the literature is reviewed by Wiens, 1989). Hummingbirds desert territories when food supplies are low and cease to defend them when food is abundant. See also the discussion of changes in territory size (and consequent changes in population size) brought about in grouse by testosterone treatment in Chapter 3, Section 3.10 and Figure 3.23).

15.7.6 Introduction of a predator

the introduction of insects to control pest aquatic plants

There have been many occasions when aquatic plants have undergone massive population explosions after their introduction to new habitats, creating significant economic problems by blocking navigation channels and irrigation pumps and upsetting local fisheries. Notable amongst such species are Canadian pondweed (*Elodea canadensis*), water hyacinth (*Eichornia crassipes*) and the aquatic fern (*Salvinia molesta*). The population explosions occur as the plants grow clonally, break up into fragments and become dispersed.

Salvinia molesta, which originated in south-eastern Brazil, has appeared since 1930 in various tropical and subtropical regions. It was first recorded in Australia in 1952 and spread very rapidly: under optimal conditions *Salvinia* has a doubling time of 2.5 days. Significant pests and parasites appear to have been absent. In 1978, Lake Moon Darra (northern Queensland) carried an infestation of 50 000 tonnes fresh weight of *Salvinia* covering an area of 400 ha (Figure 15.28). Amongst possible control agents collected from *Salvinia*'s native range in Brazil, the black long-snouted weevil (*Cyrtobagous* sp.) was known to feed only on *Salvinia*. On June 3, 1980, 1500 adults were released in cages at an inlet to the lake and a further release was made on January 20, 1981. The weevil was free of any parasites or predators that might reduce its density and, by April 18, 1981, *Salvinia* throughout the lake had become dark brown. Samples of the dying weed taken 2 km apart contained 60 and 80 adult weevils per m^2 (suggesting a total population of 1 000 000 000 beetles on the lake). By August 1981, there was estimated to be less than 1 tonne of *Salvinia* left on the lake (Room *et al.*, 1981). This has been the most rapid success of any attempted biological control of one organism by the introduction of another. It was a controlled experiment to the extent that other lakes continued to bear large populations of *Salvinia*.

the introduction of *Cactoblastis* to control prickly pear

The interaction of *Cyrtobagous* and *Salvinia* is just one example of a deliberate experiment in which an abundant plant has been brought to a state of rarity by the action of an introduced insect. Perhaps the greatest success story in the history of biological control was the virtual elimination of the prickly pear cactuses *Opuntia inermis* and *O. stricta* from many parts of Australia. Huge tracts of country, useless because of *Opuntia* infestation, were brought back into production after introduction of the moth *Cactoblastis cactorum* between 1928 and 1930. The moth multiplied rapidly, and by 1932 the original stands of the cactus had collapsed. By 1940 there was virtually complete control, and still today the cactus and moth exist only at a low, stable equilibrium (Dodd, 1940; Monro, 1967). Caughley and Lawton (1981)

(a)

(b)

Figure 15.28 Lake Moon Darra (North Queensland, Australia). (a) Covered by dense populations of the water fern (*Salvinia molesta*). (b) After introduction of weevil (*Cyrtobagous* spp.). (Courtesy of P.M. Room.)

devised a model, based on the data of Dodd and Monro, that incorporates logistic plant growth, a type-2 functional response of the moth to cactus density, exponential herbivore population growth and intense competition between the larvae at high densities (see also Chapter 10, Section 10.5.6). Their simple model simulated, in a gratifyingly precise way, the sequence of actually observed events (Figure 15.29).

Almost without exception, the plant populations whose density has been dramatically reduced by an introduced predator had themselves been introduced, accidentally or deliberately, by humans, usually from transoceanic origins. The

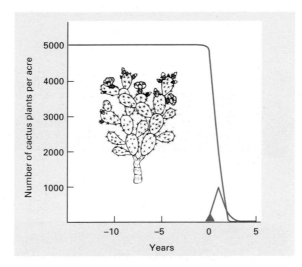

Figure 15.29 The control of *Opuntia* by *Cactoblastis*. The model derived by Caughley and Lawton (1981) accurately describes the crash of the cactus population from its pre-release level of 5000 plants per acre to a stable density of only 11 plants per acre within 2 years. Upper curve indicates cactus abundance; lower curve indicates moth numbers after introduction at year 0.

animal intended to control the 'weed' has normally been brought deliberately from the natural home of the plant, having been passed through a strict quarantine procedure to ensure that it is free from its own predators, parasites and diseases, which might have been determining its abundance in the native habitat. Ideally, its population size is now determined by its interaction with its food supply.

Plant species that have been introduced to new habitats and become vigorous invaders rarely if ever experience the same sort of population explosions in their native habitats. It is interesting to note that in successful cases of biological control, the plant in its exotic environment does not become extinct: its populations settle down to new but low densities at which it continues to support the introduced control organism. There are many cases in which a potential agent of biological control has been introduced without becoming established, but no cases in which a successful introduction has led to the extinction of the host plant. Finally, the successful biological control of plant infestations has almost always involved a perennial weed and an insect predator. Annual plants may too easily shrug off a specialist pest by disappearing as adult plants and reappearing from viable buried seed after the insect predator has become extinct.

The literature of biological control is a record of some remarkable induced changes in species abundance (see Chapter 16, Section 16.8). Nature may be unwilling to reveal its secrets unless we prod it.

Chapter 16
Manipulating Abundance:
Killing and Culling

16.1 Introduction

Arguments about which science, which subject or which discipline is the most important are all ultimately sterile. However, nobody can doubt the immediate and fundamental importance of the many applied problems addressed by ecologists, two of which we examine in this chapter. Both of them involve the manipulation of a population's abundance. The first, pest control, requires the abundance of undesirable species—pests—to be reduced, often drastically reduced, in order to allow the hungry to be fed and prevent the ravages of disease. The second, the exploitation of natural populations by culling or harvesting, requires the restrained reduction of abundance, feeding today's hungry mouths but preserving the exploited population for future harvests. A third area of applied ecology, the conservation of endangered species, which requires increases in abundance, or at least the prevention of decreases, is mentioned in this chapter but dealt with in detail in Chapter 25. Each of these topics touches on a different aspect of 'sustainability', which is dealt with briefly in its own right at the close of the present chapter.

16.2 Pest and weed control: introduction

what is a pest?

Put simply, a pest species is any species that we, as humans, consider undesirable. Giving a species pest status is therefore subjective: without humans there would be no pests. A more explicit definition is that pests compete with humans for food and other crops, transmit pathogens, feed on people or their domesticated animals or otherwise threaten human health, comfort or welfare. Weeds are included in this definition, but they are, more specifically, plants that threaten human welfare either by competing with other plants that have food, timber or amenity value, or by spoiling and thus diminishing the value of a product, for example *Allium* spp. have been weeds because they make bread taste of onions.

Thus, since we humans are a demanding lot, there are many species of pest, drawn from a great diversity of taxonomic groups and entering into a wide range of ecological relationships. From amongst these, we shall concentrate on the insect pests that threaten the production and storage of food and timber, on insect vectors of disease and on weeds. Our justification for doing so rests on the massive economic importance of each of these types of pest, on the correspondingly large amount of work and attention that has been devoted to them in the past, and on the general principles that they illustrate. We do not mean to imply that they are more important than, say, nematodes or fungal pests. Indeed, recent estimates suggest that there are around 67 000 species of pests that attack agricultural crops world-wide: 8000 weeds, 9000 insects and mites—but, 50 000 species of plant pathogens (Pimentel, 1993).

what makes a species a pest?

What, though, are the characteristics that make particular species pests? Applied to pests as a whole, the question is probably unanswerable: since our needs and desires are so various, pests could have almost any type of life history (especially if we include species that are pests because, rationally or irrationally, people fear them). Conway (1981) points out that the 'classic' pest is an *r* species (see Chapter 14), with explosive bursts of population increase, rapidly achieving levels at which immense damage is caused—rats, locusts and the rusts of wheat all conform to this description. However, *K* species can also be pests, particularly where the amount of damage is disproportionately large compared with the size of the pest population. For example, the codling moth (*Cydia pomonella*) lays relatively few eggs for an insect (up to 40–50) and often has only a single generation each year. Nevertheless, it is perhaps the single most important pest of apples world-wide (Hill, 1987). The damage it causes is insignificant in terms of biomass, but this is very important commercially as a result of the high value placed on unblemished fruit. Conway's conclusion, though, is that pests are frequently species that have 'escaped' the normal control of their enemies—often because they have been imported to new regions of the world having left their natural regulating agents in their country of origin, or because man-made changes have eliminated or reduced the efficiency of their natural enemies. Without the enemies, the species become pests.

what makes a plant a weed?

The question of what makes a plant a weed requires a different answer because weeds are pre-eminently successful competitors (against food or amenity plants, etc.), rather than successful predators. Weeds arise out of the mismatch between the habitats we create and the plants we choose to grow in them. When we open up a patch of bare soil in a flower-bed, or even a whole field of bare soil, we create an ideal habitat for an opportunistic, colonizing, *r*-type species: But the pretty flowers and the crops that we plant there are by no means ideally suited to such a habitat, and they are therefore likely to be plagued by *r*-type weeds. On the other hand, when we create the closed and crowded sward of a lawn, we are creating a habitat best exploited by dogged, perhaps *K*-type species, able to persist in the face of intense competition—the dandelions (*Taraxacum officinale*), daisies (*Bellis perennis*) and other lawn weeds. Thus, as many of us know to our cost, gardeners constantly struggle as hard to give the grasses an upper hand on their lawns as they do to eliminate grasses from their borders. A fuller discussion of which attributes contribute to 'weediness' is given by Mortimer (1989).

pests—and control agents—as invaders

One characteristic that may predispose a species to be a pest is a pronounced ability to invade ecological communities. Brown rats have become pests even on isolated islands because of their ability to jump ship and establish themselves. The Dutch elm disease fungus has caused devastation to elm trees in both North America and Great Britain because of its ability to invade and spread from an initially small bridgehead population. Of course, this is not so much answering a question as changing it: 'what makes a species a good invader?' In fact, the question is important not only in characterizing pests, but also in identifying potential control agents for those pests. As we shall see later in the chapter, 'biological control' frequently involves the release of a natural enemy of a pest which, it is hoped, will establish itself and control the pest's abundance. A successful biological control agent must also be a good invader. The question of what makes a good invader has been addressed (see the volume edited by Kornberg & Williamson, 1986), but it seems likely that the routes to success here are as various as the routes to successful pest status generally.

16.3 Aims of pest control: economic injury and action thresholds

One might imagine that the aim of pest control is total eradication of the pest—and at times in the past this has been quite widely believed. With some pathogens (where even a single human death is deplorable), and in certain amenity situations (where even a single ant in the kitchen is deemed deplorable) such an aim might be justified, but this is not the general rule. Rather, the aim is an economic one: to reduce the pest population to a level below which no further reductions are profitable, i.e. below which the extra costs of control exceed any additional revenue (or other benefits). This is known as the economic injury level (EIL) for the pest—or if social and amenity benefits are incorporated, the aesthetic injury level. (In the case of disease, therefore, total eradication *can* be justified, on the grounds that the benefits of even a single life saved are so high that they still exceed the possibly enormous costs of eliminating the final few pest organisms.)

the economic injury level (EIL) ...

The EIL for a hypothetical pest is illustrated in Figure 16.1. It is greater than zero, and hence eradication (or control to any level below the EIL) is not profitable; but it is also below the typical, average abundance of the species, and it is this that makes the species a pest. If the EIL was greater than the carrying capacity (see Chapter 6), and the species was therefore naturally self-limited to a density below the EIL, then the costs of control would always exceed the benefits gained, and the species could not be a pest. There are other species that have a carrying capacity in excess of their EIL, but have a typical abundance which is kept below their EIL by their natural enemies. These are potential pests. They can become actual pests if their enemies are removed.

... which defines actual and potential pests

The essence of the EIL concept is shown in Figure 16.2. At low pest densities, the value of a crop (or any other commodity) is affected to a negligible extent. However, above a threshold pest density, the crop loses value at an accelerating rate until a pest density is reached beyond which the crop is valueless. This value (or benefit)

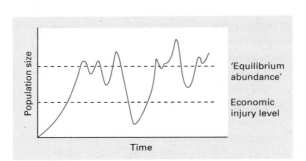

Figure 16.1 The population fluctuations of a hypothetical pest. Abundance increases rapidly when population size is small, and generally fluctuates around an 'equilibrium abundance' set by the pest's interactions with its food, predators, etc. It makes economic sense to control the pest when its abundance exceeds the economic injury level (EIL). Being a pest, its abundance exceeds the EIL most of the time (assuming it is not being controlled).

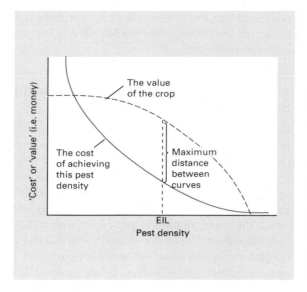

Figure 16.2 The definition of the economic injury level (EIL). The value of a crop declines with increasing pest density until the pests make the crop effectively valueless. The cost of achieving a particular pest density increases as that pest density gets smaller (i.e. more stringent 'control' is required). The EIL is the pest density at which value exceeds cost by the greatest amount. (After Headley, 1975.)

curve can now be matched against the cost of achieving, by control or otherwise, any particular pest density. Clearly, the cost of achieving a very high pest density is small—one simply provides the pest with a crop to eat and then does nothing. However, the cost of controlling the pest to lower and lower densities (moving right to left on the graph) increases at an accelerating rate, until at a density of zero (eradication) the cost is extremely high. The EIL is the density at which the difference between the two curves in Figure 16.2 (benefit minus cost) is greatest, and the optimal strategy is therefore to adopt measures designed to reduce pest abundance to the EIL but no further. Below the EIL, the cost of additional control rises more rapidly than the value of additional crop, and additional control is thus unprofitable. Above the EIL, the loss of crop value falls more rapidly than the savings on reduced control measures, and additional control is therefore called for.

Figure 16.2, however, is an over-simplification. In particular, EILs are likely to vary both within and between years. For example, they may change as the life cycle of the pest or its food source progresses, or because the weather especially favours or hinders the pest. Thus, the EIL for the two-spotted spider mite, *Tetranchyus urticae*, a major pest of cotton in Australia, depends not only on its abundance, nor only on the developmental stage of the crop, but also on the prior rate of population growth that the mite has been exhibiting (Wilson, 1993).

A more profound shortcoming in the concept of an EIL is the implied idea that control measures take effect immediately: action is taken against populations above the EIL, but as soon as populations fall below the EIL, control ceases because it is unprofitable. A glance at Figure 16.1, however, makes the point that if we wish to keep a population below its EIL, then action must be taken at some density less than the EIL, simply because, in the time it takes for the control measure to be put into operation and take effect, the population will have risen above the EIL and caused economic damage. Hence, in practical pest control, the EIL is often not as important as the economic threshold, or control action threshold (CAT)—the pest density at which action should be taken in order to prevent an impending pest outbreak (i.e. prevent the pest population exceeding its EIL). Again, there is never in reality a single

the EIL maximizes 'benefits minus costs'

the control action threshold (CAT) ...

CAT: it varies with time, with commodity prices and so on, and depends on the population levels of the pest's natural enemies. For example, the following are CATs for the spotted alfalfa aphid (*Therioaphis trifolii*) on hay alfalfa in California (Flint & van den Bosch, 1981).

1 Spring: when the population level reaches 40 aphids per stem.

2 Summer and autumn: when the population level reaches 20 aphids per stem, but do not treat the first three cuttings if the ratio of ladybirds (coccinellid beetles) to aphids is as follows:

> (a) on standing hay—one adult per five to 10 aphids or three larvae per 40 aphids;
> (b) on stubble—one larva per 50 aphids.

3 Overwintering populations: 50–70 aphids per stem.

… which is easier to define than to determine or operate

Such rules are easily stated. But, they cannot be formulated or put into operation without: (i) a sufficiently detailed understanding of the population ecology of the pest and its associated species to allow the future progress of any particular population to be predicted with confidence (thus determining the CAT); and (ii) a sufficiently accurate and frequent monitoring programme to keep a careful check on the pest and its associated species (thus determining when the CAT has been reached). Both economic and environmental considerations (see below) dictate that control measures should only be used when it is necessary to do so, but such a policy can only be adopted if time and effort have previously been invested in understanding and monitoring the pest's ecology.

16.4 A brief history of pest control

Our knowledge of the early history of pest control (see Flint & van den Bosch, 1981), after the development of agriculture approximately 10 000 years ago, is of course limited by the evidence that has survived to the present day. Nevertheless, we know that more than 4500 years ago the Sumerians were using sulphur compounds to control insects and mites, that 3200 years ago the Chinese were using insecticides derived from plants and by 2500 years ago had appreciated the beneficial role of natural enemies and the value of adjusting crop-planting times to avoid pest outbreaks, and that the Greeks and Romans understood the use of fumigants, mosquito nets, granaries on stilts, sticky bands on trees and pesticidal sprays and ointments (although throughout this period and long beyond, such sophisticated practices were accompanied by a widespread reliance on offerings to the gods and other superstitions).

ancient Chinese, Greeks and Romans

The Chinese continued to develop their pest-control technology, and by AD 300 they were using biological control, establishing nests of predatory ants in citrus orchards to control caterpillars and large boring beetles. Meanwhile, the Europeans, after the fall of the Roman Empire, relied increasingly on religious faith rather than biological knowledge. This decline was reversed by the Renaissance, and the 17th century saw an awakening of interest in biological control and the rediscovery and/or introduction into Europe of a variety of natural pesticides.

the Renaissance

The period from 1750 to 1880 in Europe was a time of the agricultural revolution, but this brought in its wake some of the greatest pest-driven agricultural disasters ever recorded: the potato blight in Ireland, England and Belgium (1840s) (see Large, 1958); the epidemic of powdery mildew in the grape-growing areas of Europe (1850s); the outbreak of fungus leaf spot disease of coffee, after which Ceylon switched from coffee to tea production; and the invasion from the Americas of an

agricultural revolution

insect, the grape phylloxera (*Viteus vitifoliae*), which almost destroyed the wine industry in France (1848–1878).

the sound and varied practices of the 19th century ...

Not surprisingly, over this period there was a surge of interest in developing pest-control techniques. After the appearance of the first books devoted entirely to pest control in the early 1800s, the 19th century saw a series of breakthroughs. Powdery mildew was controlled by two simple fungicides (Bordeaux mixture and Paris Green) and the grape phylloxera by the use of grafting onto resistant rootstocks (1870–1890); the first commercial pesticide spraying machine was produced (1880); the first major success in biological control by importation occurred (1888), when the vedalia beetle (*Rodolia cardinalis*) was imported to California from Australia and New Zealand to control a devastating citrus pest, the cottony cushion scale (*Icerya purchasi*); arthropods were recognized as vectors of human disease (1896); the first selective herbicide, iron sulphate, was introduced (1896); the first successful breeding programmes for resistant varieties were instituted (1899–1909), when strains of cotton, cowpeas and watermelon resistant to *Fusarium* wilt were developed; and 1901 saw the first successful biological control of a weed (*Lantana* in Hawaii).

The successful control of weeds throughout this period depended on the operation of three measures, all broadly aspects of 'cultural' control (Lockhart *et al.*, 1982): (i) the rotation of crops, which not only conserves fertility but also provides a succession of different environments so that no single weed species is favoured repeatedly; (ii) the disturbance of the soil by cultivation, which can kill weeds by burial, cutting, stimulation (of dormant seeds or buds to an active and therefore more vulnerable state), desiccation (when brought to the surface) or exhaustion (where repeated stimulation of perennials is coupled with a denial of photosynthetic activity); and (iii) the removal of weed seed from crop seed by winnowing, etc.

... continued into the early 20th

Thus, by the turn of the 19th century there were five main approaches to the control of pests which were well established and in common use: (i) biological control; (ii) chemical control (inorganic chemicals, especially sulphur and copper compounds, or plant-derived alkaloids, especially nicotine); (iii) mechanical and physical control (e.g. tree-banding with sticky substances); (iv) cultural and sanitation controls (like crop rotation); and (v) the use of resistant varieties. The first 40 years of the 20th century was then a period of steady progress in pest control, in which all of these five approaches played an important part.

the chemical pesticide revolution

Pest control was revolutionized, however (especially insect pest control), by the Second World War. Driven by the need to control insect vectors of human disease in the tropics, hundreds of manufactured chemicals were screened for insecticidal properties. In the USA the breakthrough came with dichlorodiphenyltrichloroethane (DDT), manufactured in Switzerland, followed by other chlorinated hydrocarbons. In Germany, another equally toxic group of compounds was developed, the organophosphates, whilst a third group of synthetic organic insecticides, the carbamates, was also discovered in the 1940s, by Swiss workers. The initial targets of the organic insecticides were the vectors of human disease, but after the war there was a rapid expansion into agriculture. 'Their success was immediate. They were cheap, effective in small quantities, easy to apply, and widely toxic. They seemed to be truly 'miracle insecticides' (Flint & van den Bosch, 1981).

During the 1930s and 1940s, too, the first organic selective herbicides were being developed, originally as a result of work on chemicals to regulate plant growth: 2-methyl-4-chlorophenoxyacetic acid (MCPA), 2,4-dichlorophenoxyacetic acid (2,4-D) and 4,6-dinitro-*o*-cresol (DNOC). The immediate impact was on cereal crops and, to a much lesser extent, on peas and grassland. An expansion into a much wider range of crops occurred from the 1960s onwards (Lockhart *et al.*, 1982).

The period from 1946 onwards has been described as the 'Age of Pesticides', divided into three phases: the Era of Optimism (1946–1962), the Era of Doubt (1962–1976) and the Era of Integrated Pest Management (IPM) (1976–) (Metcalf, 1980). It would be easy to quibble over the details of Metcalf's dating and terminology, but this division certainly captures the changes in the scientific climate. In the heady days of the 1940s and early 1950s, it was widely believed that generous doses of simple organics could entirely eradicate pests and the problems they caused. Then, the publication of Rachel Carson's *Silent Spring* in 1962—asserting that 'the chemical barrage, as crude a weapon as the cave man's club, has been hurled against the fabric of life'—heralded a public challenge to the idea that these chemicals brought enormous benefits but negligible risks. These doubts, which had been voiced much earlier by a few far-sighted entomologists (Strickland, 1945; Wigglesworth, 1945), grew steadily, until the XVth International Congress of Entomology in 1976 firmly rejected the widespread use of broad-spectrum and persistent pesticides in favour of an IPM approach. The nature of these chemicals, the problems with them, alternatives to them, and the nature of IPM are dealt with in the following sections.

optimism gives way to doubt and fears

16.5 Chemical pesticides

We will confine our description here to insecticides and herbicides (although we emphasize again that chemicals for the control of other types of pest are of comparable importance). The former can be divided into broad-spectrum insecticides (inorganics, botanicals, chlorinated hydrocarbons (organochlorines), organophosphates and carbamates) and narrow-spectrum or biorational insecticides (microbials (dealt with under biological control), insect growth regulators and semiochemicals). Herbicides comprise a much more varied range and defy so straightforward a classification. The following descriptions are based on more detailed expositions by Horn (1988) for insecticides and by Ware (1983) for herbicides.

16.5.1 Insecticides

The use of *inorganics* goes back to the dawn of pest control and, along with the botanicals, they were the chemical weapons of the expanding army of insect pest managers of the 19th and early 20th century. They are usually metallic compounds or salts of copper, sulphur, arsenic or lead, primarily stomach poisons (i.e. they are ineffective as contact poisons), and they are therefore effective only against insects with chewing mouth parts. This, coupled with their legacy of persistent, broadly toxic metallic residues, has led now to their virtual abandonment.

Naturally occurring insecticidal plant products, or *botanicals*, such as nicotine from tobacco and pyrethrum from chrysanthemums, having run a course similar to the inorganics, have now also been largely superceded, particularly because of their instability on exposure to light and air. However, a range of *synthetic pyrethroids*, with much greater stability, such as permethrin (Figure 16.3a) and deltamethrin, are now quickly replacing other types of organic insecticide (described below) because of their relative selectivity against pests as opposed to beneficial species (Pickett, 1988).

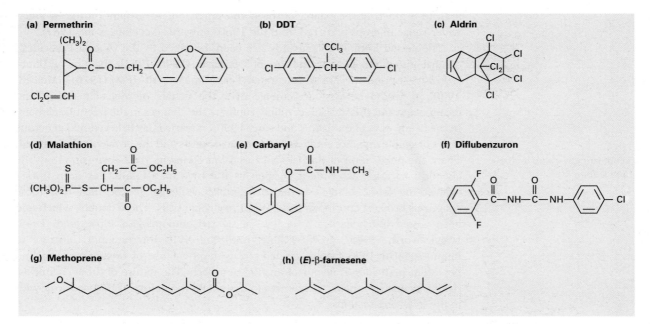

Figure 16.3 The chemical structure of selected insecticides.

Chlorinated hydrocarbons are contact poisons that affect nerve-impulse transmission. They are insoluble in water but show a high affinity for fats, thus tending to become concentrated in animal fatty tissue. The most notorious is DDT (Figure 16.3b): a Nobel Prize was awarded for its rediscovery in 1948, but it was suspended from all but emergency uses in the USA in 1973 (although it is still being used in poorer countries because it is so cheap). Others in use are toxaphene, aldrin (Figure 16.3c), dieldrin, lindane, methoxychlor and chlordane.

Organophosphates are also nerve poisons. They are much more toxic (to both insects and mammals) than the chlorinated hydrocarbons, but generally less persistent in the environment. Examples are malathion (Figure 16.3d), parathion and diazinon.

Carbamates have a mode of action similar to the organophosphates, but some have a much lower mammalian toxicity. However, most are extremely toxic to bees (necessary for pollination) and parasitic wasps (likely natural enemies of insect pests). The best-known carbamate is carbaryl (Figure 16.3e).

Insect growth regulators are chemicals of various sorts that mimic natural insect hormones and enzymes, and hence interfere with normal insect growth and development. As such, they are generally harmless to vertebrates and plants, although they may be as effective against a pest's natural insect enemies as against the pest itself. The two main types that have been used effectively to date are chitin-synthesis inhibitors such as diflubenzuron (Figure 16.3f), which prevent the formation of a proper exoskeleton when the insect moults, and juvenile hormone analogues such as methoprene (Figure 16.3g), which prevent pest insects from moulting into their adult stage, and hence reduce population size in the next generation.

Semiochemicals are not toxins but chemicals that elicit a change in the behaviour of the pest (literally 'chemical signs'). They are all based on naturally occurring substances, although in a number of cases it has been possible to synthesize either the semiochemicals themselves or analogues of them. Pheromones act on members of the same species; allelochemicals on members of another species. Sex-attractant pheromones are currently being used commercially to control pest moth populations by interfering with mating (Reece, 1985), whilst the aphid alarm pheromone (E)-β-farnesene (Figure 16.3h) is used to enhance the effectiveness of a fungal pathogen against pest aphids in glasshouses in Great Britain by increasing the mobility of the aphids, and hence, their rate of contact with fungal spores (Hockland *et al.*, 1986). These semiochemicals, along with the insect growth regulators, are sometimes referred to as 'third-generation' insecticides (following the inorganics and the organic toxins). Their development is relatively recent and is still proceeding rapidly (Forrester, 1993).

16.5.2 Herbicides

Here, too, the traditional agents were *inorganics*, although they have mostly been replaced, largely owing to the combined problems of persistence and non-specificity. However, for these very reasons, borates for example, absorbed by plant roots and translocated to above-ground parts, are still sometimes used to provide semi-permanent sterility to areas where no vegetation of any sort is wanted. Others include a range of arsenicals, ammonium sulphamate and sodium chlorate.

More widely used are the *organic arsenicals*, for instance DSMA (Figure 16.4a). These are usually applied as spot treatments (since they are non-selective) after which they are translocated to underground tubers and rhizomes where they disrupt growth.

By contrast, the highly successful *phenoxy* or *hormone* weed killers, translocated throughout the plant, tend to be very much more selective. For instance, 2,4-D (Figure 16.4b) is highly selective against broad-leaved weeds, whilst 2,4,5-trichlorophenoxyethanoic acid (2,4,5-T) (Figure 16.4c) is used mainly to control woody perennials. They appear to act by inhibiting the production of enzymes needed for coordinated plant growth, leading ultimately to plant death.

The *substituted amides* have diverse biological properties. For example, diphenamid (Figure 16.4d) is largely effective against seedlings rather than established plants, and is therefore applied to the soil around established plants as a 'pre-emergence' herbicide, preventing the subsequent appearance of weeds. Propanil, on the other hand, has been used extensively on rice fields as a selective post-emergence agent.

The *nitroanilines* (e.g. trifluralin, Figure 16.4e) are another group of soil-incorporated pre-emergence herbicides in very widespread use. They act, selectively, by inhibiting the growth of both roots and shoots.

The *substituted ureas* (e.g. monuron, Figure 16.4f) are mostly rather non-selective pre-emergence herbicides, although some have post-emergence uses. Their mode of action is to block electron transport.

The *carbamates* were described amongst the insecticides, but some are herbicides, killing plants by stopping cell division and plant tissue growth. They are primarily selective pre-emergence weed killers. One example, asulam (Figure 16.4g), is used mostly for grass control amongst crops, and is also effective in reforestation and Christmas tree plantings.

The *thiocarbamates* (e.g. EPTC, Figure 16.4h) are another group of soil-

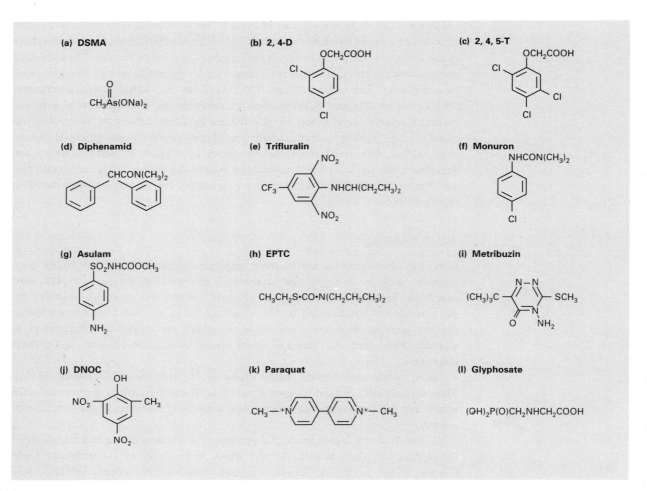

Figure 16.4 The chemical structure of selected herbicides.

incorporated pre-emergence herbicides, selectively inhibiting the growth of roots and shoots that emerge from weed seeds.

Amongst the *heterocyclic nitrogen* herbicides, probably the most important are the *triazines* (e.g. metribuzin, Figure 16.4i). These are effective blockers of electron transport, mostly used for their post-emergence activity.

The *phenol derivatives*, particularly the nitrophenols such as DNOC (Figure 16.4j), are contact chemicals with broad-spectrum toxicity extending beyond plants to fungi, insects and mammals. They act by uncoupling oxidative phosphorylation.

The *bipyridyliums* contain two important herbicides, diquat and paraquat (Figure 16.4k). These are powerful, very fast-acting contact chemicals of widespread toxicity that act by the destruction of cell membranes.

Finally worth mentioning is glyphosate (Figure 16.4l), a *glyphosphate* herbicide: a non-selective, non-residual, translocated, foliar-applied chemical, popular for its activity at any stage of plant growth and at any time of the year.

16.6 The problems with chemical pesticides

16.6.1 Widespread toxicity

mortality of non-target species

Chemical insecticides are generally intended for particular insect pests at a particular site. Nevertheless, problems often arise because these chemicals are usually toxic to a much broader range of organisms and also persist in the environment. As Table 16.1 shows, this is a problem to some extent with all the broad-spectrum insecticides. Problems have also arisen because insecticides have been applied carelessly and have spread far beyond the species and site of concern.

The potential for devastation is illustrated by the consequences of applying massive doses of dieldrin to large areas of Illinois farmland from 1954 to 1958, in order to 'eradicate' a grassland pest, the Japanese beetle *Popillia japonica*. Cattle and sheep on the farms were poisoned; 90% of the cats on the farms and a number of dogs were killed; and amongst the wildlife, 12 species of mammals and 19 species of birds suffered losses, with meadowlarks, robins, brown thrashers, starlings, common grackles and ringnecked pheasants being virtually eliminated from the area (Luckmann & Decker, 1960).

biomagnification of organochlorines

The problem is made more difficult with chlorinated hydrocarbons especially, because of their susceptibility to biomagnification: an increasing concentration of insecticide in organisms at higher trophic levels, as a result of a repeated cycle of concentration of the insecticide in particular tissues in a lower trophic level, consumption by the trophic level above, further concentration, further consumption, and so on, until top predators which were never intended as targets, suffer extraordinarily high doses. One example is illustrated in Figure 16.5. The nuisance caused by a non-biting gnat (*Chaoborus astictopus*) around Clear Lake, California prompted, in 1949, the spraying of the lake with dichlorodiphenyldichloroethane (DDD) (less toxic to fish than DDT) at around 0.02 ppm (parts per million). The initial disappearance of the insect was spectacular, but they returned in 1951, and spraying was therefore repeated regularly until 1954, when large numbers of carcasses of a diving bird, the western grebe, were found at the lake. The reason,

Table 16.1 The toxicity to non-target organisms, and the persistence of selected insecticides. Possible ratings range from a minimum of 1 (which may, therefore, include zero toxicity) to a maximum of 5. Most damage is done by insecticides that combine persistence with acute toxicity to non-target organisms. This clearly applies, to an extent, to each of the first six (broad-spectrum) insecticides. (After Metcalf, 1982; Horn, 1988.)

| | Toxicity | | | | |
	Rat	Fish	Bird	Honeybee	Persistence
Permethrin (pyrethroid)	2	4	2	5	2
DDT (organochlorine)	3	4	2	2	5
Lindane (organochlorine)	3	3	2	4	4
Ethyl parathion (organophosphate)	5	2	5	5	2
Malathion (organophosphate)	2	2	1	4	1
Carbaryl (carbamate)	2	1	1	4	1
Diflubenzuron (chitin-synthesis inhibitor)	1	1	1	1	4
Methoprene (juvenile hormone analogue)	1	1	1	2	2
Bacillus thuringiensis	1	1	1	1	1

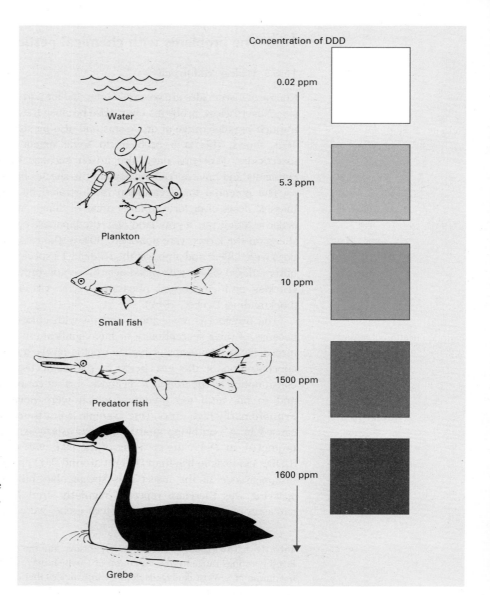

Concentration of DDD

Water   0.02 ppm

Plankton   5.3 ppm

Small fish   10 ppm

Predator fish   1500 ppm

Grebe   1600 ppm

Figure 16.5 The biomagnification of dichlorodiphenyldichloroethane (DDD) in Clear Lake, California. Added to the water to control gnats at a level not believed to be harmful, it created havoc further up the food chain. (After Flint & van den Bosch, 1981.)

depression of plant yield by insecticides

subsequently discovered, is illustrated in Figure 16.5: biomagnification. The grebes' body fat contained 1600 ppm DDD, and their diet of fish was so loaded with poison that their survival in Clear Lake was impossible in 1954 and for many years afterwards (Flint & van den Bosch, 1981).

Insecticides may also depress the growth and yield of the plants they are supposed to be protecting. For example, permethrin is known to reduce photosynthesis in lettuce seedlings by up to 80% (Sances *et al.*, 1981). (Similarly, herbicides— e.g. 2,4-D—can harm the plants they are supposed to protect by improving the performance of insect pests when they alter crop physiology (Oka & Pimentel, 1976).)

Of course, the most predictable problems stemming from the widespread toxicity of insecticides are those that arise out of adverse effects on insects, and arthropods generally, other than the pest. For instance, Table 16.1 shows that most broad-spectrum insecticides are highly toxic to honeybees, threatening the pollination of crops (e.g. almonds, apples, apricots, melons, alfalfa seed) and the livelihood of beekeepers. More than 50% of the very substantial number of losses of bee colonies in California are attributed to pesticides (Flint & van den Bosch, 1981).

The broadly toxic effects of herbicides have generally not been considered as great a problem as those of insecticides. One important reason is that many herbicides have a quite specific effect on plant physiology which has no exact equivalent in animal physiology. There are, though, a number of herbicides, for instance diquat and paraquat, which have high mammalian toxicity and where great care is therefore required in handling (especially as in these cases there are no known antidotes). Furthermore, in the 1960s, a controversy began over the possible effects to human health of 2,4,5-T and 2,4-D, which were used in combination ('Agent Orange') between 1962 and 1970 to defoliate swamps and forests in South Vietnam. Studies showed that low levels of 2,4,5-T caused high rates of birth defects in laboratory animals, probably as a result of very low levels of a manufacturing contaminant, a dioxin, TCCD. Subsequently, a large number of veterans from the Vietnam War complained of various symptoms, including birth defects in their children and certain rare cancers. (These veterans have been compensated in an out-of-court settlement by the manufacturers of Agent Orange, although no guilt has been admitted.) Later, further studies strongly suggested carcinogenic effects of 2,4-D, the active ingredient in more than 1500 herbicide products used by farmers and home gardeners (Miller, 1988).

Of course, most weed-control practices affect a wider range of plants than the target species. The result has been the disappearance of many attractive plants that have never been serious weeds. Species such as the wild delphinium (*Delphinium ajacis*), pheasant's eye (*Adonis annua*), corn cockle (*Agrostemma githago*) and the cornflower (*Centaurea cyanus*) have come to the verge of extinction in Great Britain and can now be found in abundance only in the peasant agricultures of eastern Europe.

16.6.2 Target pest resurgence

Of particular importance are the effects of insecticide on the (arthropod) natural enemies of an insect pest. This, in itself, may not seem too serious, apart from the regrettable loss in the natural diversity of harmless species. However, it can—and often has—had two extremely serious consequences. The first, target pest resurgence, refers to a rapid increase in pest numbers following some time after the initial drop in pest abundance caused by an application of insecticide. This rebound effect occurs when treatment kills not only large numbers of the pest, but large numbers of their natural enemies too (with any survivors likely to starve to death because there are insufficient pests on which to feed). Then, any pest individuals that survive (either because of resistance or good luck) or that migrate into the area, find themselves with a plentiful food resource but few if any natural enemies. A population explosion is the likely outcome.

Why, though, should it be the pest that bounces back? There are two main reasons. The first is that the enemies need the pest to support population growth, but the pest certainly does not need the enemies: the enemies are bound always to follow at least one step behind. The second, less obvious reason relates to the characteristics that make a species a pest in the first place. Notwithstanding the qualifications voiced in Section 16.2, these are often likely to include an ability to reproduce rapidly when a food resource becomes available, and an ability to find such food resources as are available (i.e. good dispersal ability). Hence, pests, naturally, are likely to be good at 'resurging'. For example, when the cabbage aphid, *Brevicoryne brassicae*, was first sprayed commercially in England with para-oxon, the initial kill was extremely high. However, the destruction of natural enemies resulted within 2 weeks in the largest outbreak of cabbage aphid ever seen in England (Ripper, 1956).

16.6.3 Secondary pest outbreaks

non-pests become pests when
their enemies and competitors
are killed

When natural enemies are destroyed, it is not only the target pest that might resurge. Alongside any actual pest are likely to be a number of potential pest species, which are not pests only because they are kept in check by their natural enemies. Thus, if a primary pest is treated with an insecticide that destroys a wide range of predators and parasitoids, other species may realize their potential and become 'secondary' pests.

A dramatic example of this concerns the insect pests of cotton in Central America (Flint & van den Bosch, 1981). In 1950, when mass dissemination of organic insecticides began, there were two primary pests: the boll weevil (*Anthonomus grandis*) and the Alabama leafworm (*Alabama argillacea*). Organochlorines and organophosphates, applied fewer than five times a year, initially had apparently miraculous results and yields soared. By 1955, however, three secondary pests had emerged: the cotton bollworm (*Heliothis zea*), the cotton aphid (*Aphis gossypii*) and the false pink bollworm (*Sacadodes pyralis*). The application rate rose to 8–10 per year. This reduced the problem of the aphid and the false pink bollworm, but led to the emergence of five further secondary pests. By the 1960s, the original two-pest species had become eight. There were, on average, 28 applications of insecticide per year.

On a broader scale, changes in the overall pattern of weed infestation can be seen as an example of the outbreak of secondary pests. The herbicides in use until the 1960s, when they were selective at all, tended to be most active against dicotyledonous weeds. The result was an upsurge in the importance of grass weeds (monocots), and the 1970s therefore saw the beginnings of a new drive towards the production of herbicides selective against grasses (Lockhart *et al.*, 1982).

16.6.4 Evolved resistance

This final problem is in many ways the most serious of all. Even before the advent of the organics, occasional examples of resistance to an insecticide had been found. For instance, in 1914 populations of the San Jose scale insect (*Quadraspidiotus perniciosus*), a very serious pest of fruit trees throughout the world, were found to be resistant to lime–sulphur sprays. Nevertheless, in the early years of the Age of Pesticides, the enormity of the problem was not appreciated and its very existence often ignored—even though the first reported case of DDT resistance was reported as early as 1946 (house flies, *Musca domestica*, in Sweden).

The evolution of pesticide resistance is simply natural selection occurring more rapidly than usual and on a particularly obvious character. Indeed, pesticide resistance provides us with some of our best examples of evolution in action (Roush & McKenzie, 1987). Within a large population subjected to a pesticide, one or a few individuals may be unusually resistant (perhaps because they possess an enzyme that can detoxify the pesticide). If such individuals exist at the outset, resistance can begin to spread in the population immediately; if they arise subsequently by mutation, then there will be a lag in the evolutionary response before this chance event occurs. In either case, the resistant individuals have an improved chance of surviving and breeding and, if the pesticide is applied repeatedly, each successive generation will contain a larger proportion of resistant individuals. Subsequent mutations may enhance resistance still further. Over succeeding generations, therefore, the resistant proportion and the degree of resistance may both increase. Because pests so often have a high intrinsic rate of reproduction, a few individuals in one generation may give rise to hundreds or thousands in the next, and resistance may therefore spread very rapidly in a population.

the problem is serious, growing and compounded by cross-resistance ...

The scale of the problem is illustrated in Figure 16.6, which shows the exponential increases in the number of insect species resistant to insecticides. World-wide, by 1987, nearly 500 species of insects, 150 species of plant pathogens, 50 species of fungi, more than 50 species of weeds and 10 species of small rodents (mostly rats) had strains resistant to one or more pesticides (Miller, 1988). The problems are further compounded by cross-resistance, which provides species resistant to one pesticide with resistance to other, related pesticides, because they share a common mode of action and can therefore be detoxified or otherwise resisted in the same way. Indeed, the insecticides that have been used extensively are all nerve poisons, acting on only three targets: the voltage-dependent sodium channel (DDT-type organochlorines, pyrethroids), the γ-aminobutyric acid-gated chlorine channel (other organochlorines, e.g. toxaphene, dieldrin, lindane) and the inhibition of acetylcholinesterase (organophosphates like parathion and malathion, carbamates) (Forrester, 1993). Thus, for example, lindane-resistant insects tend to tolerate dieldrin, and parathion-resistant insects tolerate malathion. On the other hand, even when they share a target, insecticides may differ in their mode of action, such that

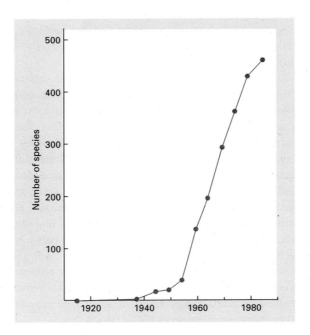

Figure 16.6 The growth in the number of insect species reported to be resistant to at least one insecticide. (After Metcalf, 1980; Miller, 1988.)

635   MANIPULATING ABUNDANCE

cross-resistance does not arise. A recently discovered group, for instance the pyrazolines, act on the voltage-dependent sodium channel, but at a site clearly different from the DDT–pyrethroid recognition site (Deecher & Soderlund, 1991).

... and multiple resistance

Far more serious, however, is multiple resistance: the resistance of species to a variety of pesticides with quite different modes of action. For example, the house fly, *M. domestica*, has developed resistance world-wide to virtually every chemical employed against it, including many of the modern synthetic pyrethroids and the juvenile hormone analogue methoprene; whilst the cotton bollworm (*H. zea*) and the tobacco budworm (*H. virescens*) have successively evolved resistance to DDT, toxaphene, endrin, methyl parathion, EPN, carbaryl and now synthetic pyrethroids (Metcalf, 1980).

herbicide resistance started slowly ...

Herbicide resistance was much slower to evolve than insecticide resistance (reviewed by Putwain, 1989). The first confirmed occurrence was in the state of Washington, USA, in 1968 when a triazine-resistant strain of groundsel was found in a tree nursery. Within a few years, reports of triazine-resistant weeds were common in the USA, Canada and western Europe. Subsequently, triazine-resistant weeds have become common in eastern Europe and Israel and also occur in Australia and New Zealand. In almost all cases, persistent 1,3,5-triazine herbicides have been used repeatedly in the same areas for 5–10 years or longer.

Multiple resistances involving triazines and other herbicides inhibiting the same biochemical pathway have recently been found in a few places. For instance, in Israel, triazine-resistant awned canary grass (*Phalaris paradoxa*) has been found also to be resistant to methabenzthiazuron and to diclofopmethyl. The latter compound is chemically unrelated to the triazines, and the multiple resistance was therefore unexpected.

... but, is catching up

Cases of resistance to herbicides other than the triazines have been slower to emerge. However, during the 1980s the evolution of resistance to a wider range of chemicals became more common in a variety of locations world-wide. For example, resistance to paraquat has evolved in at least eight species. Although paraquat is a non-persistent herbicide, growers caused a high selection pressure in favour of resistance by repeated application of the herbicide through each growing season. Thus, paraquat resistance has evolved despite the original expectation that it was unlikely to occur.

resistance management

One answer to the problem of pesticide resistance is to develop strategies of 'resistance management'. To date, these have consisted largely of two elements (Roush & McKenzie, 1987). The first is to reduce the frequency with which a particular pesticide is used, thus depriving the pest of a series of generations over which resistance may evolve. This may be done by using a range of pesticides in a repeated sequence, especially when they have different target sites or modes of action. Second, pesticides should be used at a concentration high enough to kill individuals heterozygous for the resistance gene, since this is where all resistance genes are likely to reside when resistance is rare. However, a further complication arises out of a tendency for non-target, but nonetheless harmful species, to evolve resistance as a by-product of treating the target pest, because these species, at the ecological periphery of the treatment zone, will generally receive a reduced dose which may allow heterozygotes to survive. For instance, in the case of cotton pests in Central America (described above), the malaria vector mosquito *Anopheles albimanus*, exhibited much higher levels of insecticide resistance in cotton-growing than in non-cotton-growing areas (Table 16.2).

Table 16.2 Insecticide resistance in the malarial vector mosquito *Anopheles albimanus* in Central America, which is highest in cotton-growing areas as a by-product of the use of the insecticides against cotton pests. (After Flint & van den Bosch, 1981.)

	Insecticide (country)	
	DDT (Guatemala) (%)	Propoxur (OMS-33) (El Salvador) (%)
Cotton-growing areas	79	35
Non-cotton-growing areas	54	7

Thus, together, the problems of resistance, target pest resurgence and secondary pest outbreaks have frequently met with a predictable but in many ways regrettable response: the application of more and other pesticides, leading to further resistance, further resurgence and further secondary pests, and so to more pesticide, more problems and more expense, as we have seen with Central American cotton. This has been called the 'pesticide treadmill'—once farmers (or whoever it may be) get on, they find it difficult to get off (Flint & van den Bosch, 1981).

the 'pesticide treadmill'

16.7 The virtues of chemical pesticides

Of course, if pesticides brought nothing but problems they would have fallen out of widespread use long ago—whatever the advertising and other pressures exerted by the manufacturing companies. They have not. Instead, their rate of production has increased rapidly in spite of the problems (Figure 16.7). There are four main reasons for this.

pesticide production has nonetheless increased rapidly

1 Until now, the chemical companies have managed, broadly speaking, to keep at least one step ahead of the pests. Indeed, more recent developments can be seen as direct responses to the problems of the past, with the emphasis amongst insecticides on, for example, synthetic pyrethroids, growth regulators and semiochemicals.

... by keeping one jump ahead ...

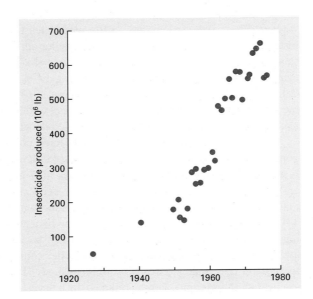

Figure 16.7 Various estimates for the quantity of insecticide produced in the USA, showing the rapid increase over time. (After Metcalf, 1980.)

2 The pesticides themselves have been used with increasing care. Many are now used as an integral part of a more varied armoury wielded against the pest, and are used not to eradicate the pest, nor according to a preset plan irrespective of pest density, but only when CATs are breached and EILs are therefore threatened. These approaches are discussed below under 'IPM'.

In addition, the methods by which pesticides are formulated and applied have themselves become increasingly designed to direct the action specifically against the target pest. This is important. For example, with insecticides a typical estimate for conventional application methods would be that less than 1% of the insecticide reaches the target pest (Metcalf, 1980). However, in the control of fire ants (*Solenopsis geminata*) on tobacco in the southern USA, if mirex (an organochlorine) is impregnated on ground corn-cob grits, then effective control can be achieved with mirex with as little as $1.7 \, \text{g acre}^{-1}$ ($4.2 \times 10^{-4} \, \text{g m}^{-2}$) per application (three applications per year), and since the grits are unattractive to non-target species, this method of delivery to the target is both ecologically and economically sound.

3 In spite of the steadily rising costs of pesticides—the result of increasing complexity (raising development and production costs) and of oil price rises—the benefit : cost ratio for the individual producer has remained in favour of pesticide use. For example, Pimentel *et al.* (1978) estimated that the ratio for the use of insecticides in USA agriculture was around $5 benefit for every $1 spent; whilst Metcalf (1980), on a more specific note, estimated that the ratio for using carbofuran to control corn rootworms (*Diabrotica* spp.) in USA corn (*Zea mays*) production was around $3.12 : $1. The picture changes when attempts are made to place an economic value on the social costs of pesticide use (e.g. effects on public health and the costs of pesticide regulation). Then the ratio for USA agriculture is around $2.40 : $1 (Pimentel, 1979). Moreover, to put this in perspective, the savings from biological control (see Section 16.8) in the USA are estimated to be around $30 for every $1 invested (Miller, 1988), and from cultural control (crop rotation, etc.—see Section 16.9) between $30 and $300 for every $1 (Pimentel, 1993). Of course, many millions more dollars have been invested in chemical than in biological or cultural control. Nevertheless, the rate of return on using chemicals is healthy. In short, as far as the economics of food production are concerned, pesticides work.

Pesticides have also worked, in the past at least, as disease control agents. For instance, the chlorinated hydrocarbons, despite all their attendant problems, have saved at least seven million lives from disease since 1947; or to take one specific example, more than one billion people have been freed from the risk of malaria (Miller, 1988). (Although, since 1970, malaria has made a remarkable come-back, owing, in large part, to the evolution of insecticide-resistant strains of mosquitos, leading to an increased emphasis on possible biological control measures.)

With herbicides, the economic benefits have not only been direct, i.e. smaller populations of weeds and hence smaller losses in crop yield. The replacement of the hoe as a major means of weed control has also made it possible to grow crops without soil disturbance, and there has been a consequent economizing on manpower, tractor power, fuel, capital investment and moisture conservation.

4 There are particular situations where the benefit : cost ratio is loaded especially heavily in favour of chemical pesticides by social circumstances and human attitudes. For instance, consumers in the richer countries have become conditioned to demand that their foodstuffs be unblemished. This means that pests have to be eradicated rather than simply reduced to a level where nutritional, rather than aesthetic or cosmetic harm is negligible. Chemicals generally act in a density independent manner, whereas most natural enemies are likely to be density dependent in their action. Hence, chemicals will usually be the only means of achieving eradication—and unblemished food.

Also, in many poorer countries, the prospect of imminent mass starvation, or of an epidemic of an insect-vectored disease, has often meant that the human costs of not producing food, or of letting disease go unchecked, are so great that the social and health costs of chemical use have been ignored, even when they are considerable. Thus, there is widespread use in many poor countries of toxic but relatively cheap chemicals that are banned, or whose use is restricted, in the home country of the company producing them. On the other hand, it has been estimated that the world death rate from pesticide poisoning is around 21 000 annually (Copplestone, 1977).

thus, chemicals 'work', but pesticides have inherent disadvantages

In summary, then, the case in favour of chemical pesticides is that they have worked in the past, as judged by objective measures such as 'lives saved', 'total food produced' and 'economic efficiency of food production', and that they are continuing to do so as a result of advances in the types of pesticide produced and the manner in which they are used. We have also seen, though, that many of the disadvantages of chemicals—widespread toxicity, secondary pests, resistance, escalating costs—are undeniable, such that the case for chemical control can only ever be one in which the advantages are shown to outweigh the regrettable disadvantages. These disadvantages are also, and perhaps most significantly of all, inherent, that is, newer and better chemicals may postpone or reduce the effects of these disadvantages (although probably at increased cost), but they are most unlikely to overcome them. The question, therefore, is whether or not there are alternatives to replace or use alongside chemicals. We turn to these next.

16.8 Biological control

biological control: the use of natural enemies in a variety of ways

The term biological control has been applied, at times, to virtually all pest control measures except the non-selective application of chemical pesticides, but it is best restricted to the use of natural enemies in pest control (Waage & Greathead, 1988). There are four types of biological control.

1 The *introduction* of a natural enemy from another geographical area—very often the area in which the pest originated prior to achieving pest status—in order that the control agent should persist and thus maintain the pest, long term, below its economic threshold. This is often called *classical biological control* or *importation*.

2 *Inoculation* is similar, but requires the periodic release of a control agent where it is unable to persist throughout the year, with the aim of providing control for only one or perhaps a few generations.

3 *Augmentation* involves the release of an indigenous natural enemy in order to supplement an existing population, and is also therefore carried out repeatedly, typically so as to intercept a period of particularly rapid pest-population growth.

4 Finally, *inundation* is the release of large numbers of a natural enemy, with the aim of killing those pests present at the time, but with no expectation of providing long-term control as a result of the control agent's population increasing or maintaining itself. By analogy with the use of chemicals, agents used in this way are referred to as biological pesticides.

Until now, insects have undoubtedly been the main agents of biological control against both insect pests (where parasitoids have been particularly useful) and weeds. Table 16.3 summarizes the extent to which they have been used and their success rate, i.e. the proportion of cases where the establishment of an agent has greatly reduced or eliminated the need for other control measures (Waage & Greathead, 1988).

the cottony cushion scale: a classic case of importation ...

Probably the best example of 'classical' biological control is itself a classic. Its success marked the start of biological control in a modern sense. The cottony cushion scale insect, *Icerya purchasi*, was first discovered as a pest of Californian citrus orchards in 1868. By 1886 it had brought the citrus industry close to the point of destruction. Would-be pest controllers initiated a world-wide correspondence to try and discover the natural home and natural enemies of the scale. In 1887 they received a report from Adelaide, Australia, of apparently natural populations of the scale along with a dipteran parasitoid, *Cryptochaetum* spp. Thus, in 1888, an expert insect taxonomist, Koebele, was despatched to collect parasitoids for importation into California. In spite of his expertize, however, and in spite of a widespread tour of Australia, he discovered that both scale and parasitoid were hard to find. Eventually he was successful—but he also found a ladybird beetle, the vedalia, *Rodolia cardinalis*, feeding on the scale insect. Then, on his way home, he visited New Zealand and found the vedalia doing very well on the scales. So, in all, he despatched around 12 000 *Cryptochaetum* and around 500 vedalia back to California. Initially, the parasitoids seemed simply to have disappeared, but the predatory beetles underwent such a population explosion that all infestations of the scale insects in California were controlled by the end of 1890. In fact, although the beetles have usually taken most or all of the credit, the long-term outcome has been that the vedalia has been largely instrumental in keeping the scale in check inland, but *Cryptochaetum* has been, and is, the main agent of control on the coast (Flint & van den Bosch, 1981).

... illustrating several general points

This example illustrates a number of important general points. Species often become pests simply because, by colonization of a new area, they escape the control of their natural enemies. The situation is typically as follows. Most crops have been introduced to most of the areas in which they are grown (see Chapter 2). Their pests have commonly followed them. We therefore tend to go back to where the crop came from to find the predators of the pest. Biological control by importation is thus, in an important sense, restoration of the status quo. It requires for its success the classical, perhaps unfashionable skills of the taxonomist in finding the pest in its native habitat, and particularly in identifying and isolating its natural enemies. This may often be a difficult task—especially if the natural enemy or enemies are potentially valuable by virtue of their efficiency, since in that case they are likely to keep the pest

Table 16.3 The record of insects as biological control agents against insect pests and weeds. (After Waage & Greathead, 1988.)

	Insect pests	Weeds
Control agent species	563	126
Pest species	292	70
Countries	168	55
Cases where agent has become established	1063	367
Substantial successes	421	113
Successes as a percentage of establishments	40	31

(and themselves) rare in their natural habitat. Nevertheless, the rate of return on investment can be highly favourable. In the case of the cottony cushion scale, biological control has subsequently been transferred to 50 other countries and savings have been immense: the outlay, including salaries, was around $5000. In addition, this example illustrates the importance of establishing several, hopefully complementary, enemies to control a pest. Finally, classical biological control, like natural control, can be destabilized by chemicals. The first use of DDT in Californian citrus orchards in 1946–1947 against the citricola scale *Coccus pseudomagnoliarum*, led to an outbreak of the (by then) rarely seen cottony cushion scale when the DDT almost eliminated the vedalia. The use of DDT was terminated (Metcalf, 1980).

'inoculation' against glasshouse pests

'Inoculation' as a means of biological control is widely practised in the control of arthropod pests in glasshouses—a means of crop production concentrated in western Europe—where crops are removed, along with the pests and their natural enemies, at the end of the growing season (van Lenteren & Woets, 1988). The two species of natural enemy that are used most widely in this way are *Phytoseiulus persimilis*, a mite which preys on the spider mite *Tetranychus urticae*, a pest of cucumbers and other vegetables, and *Encarsia formosa*, a chalcid parasitoid wasp of the whitefly *Trialeurodes vaporariorum*, a pest in particular of tomatoes and cucumbers. *Encarsia formosa* was widely used in the 1930s, but distribution was discontinued when organic insecticides were first introduced. However, as the problems of chemicals became increasingly apparent in the late 1960s and early 1970s, its value was rediscovered along with that of *P. persimilis*. In western Europe by 1985, around 500 million individuals of each species were being produced for release each year. *Encarsia formosa* is mass reared on tobacco and 'released' as a pupa glued to cardboard; *P. persimilis* is mass reared on beans and a mixture of all moving life-cycle stages released. Releases are generally delayed until the pests are actually observed. The growth in the use to which these two species have been put (Figure 16.8) is the soundest possible testimony to their efficacy (van Lenteren & Woets, 1988).

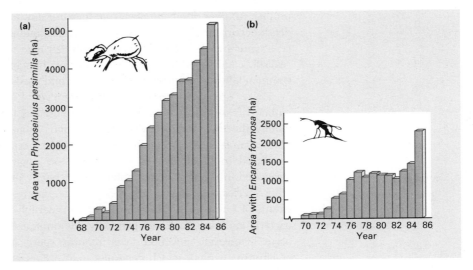

Figure 16.8 The growth world-wide in the use of biological control agents in glasshouses. (a) The area in which *Phytoseiulus persimilis* is used to control *Tetranychus urticae*. (b) The area in which *Encarsia formosa* is used to control *Trialeurodes vaporariorum*. (After van Lenteren & Woets, 1988.)

Recently, increasing attention in the control of insect pests has focused on the use of insect pathogens, largely as microbial insecticides (i.e. through inundation). The following description of their use is based on that by Payne (1988).

Around 100 species of insect-pathogenic bacteria have been identified, but only three have received a significant amount of attention, which has been concentrated very largely on just one: *Bacillus thuringiensis*. *Bacillus thuringiensis* is the only microbial pest control agent that has been commercialized world-wide on a large scale. It is easily produced on artificial media. After being ingested by insect larvae, gut juices release powerful toxins. Death occurs 30 min to 3 days after ingestion. Significantly, there is a range of varieties (or 'pathotypes') of *B. thuringiensis*, including ones specific against lepidoptera (many agricultural pests), diptera, especially mosquitos and blackflies (vectors of malaria and onchocerciasis) and beetles (many agricultural and stored product pests). *Bacillus thuringiensis* is used inundatively as a microbial insecticide. Its advantages are its powerful toxicity against target insects, but its lack of toxicity against organisms outside this narrow group (including ourselves and most of the pest's natural enemies).

More than 1600 virus isolates (species) have been recorded, causing disease in more than 1100 species of insects and mites. From amongst these, attention has been focused predominantly on the baculoviruses (essentially the nuclear polyhedrosis and granulosis viruses), a group with the advantage of having no apparent taxonomic affinity with any viruses known to attack plants or vertebrates. Indeed, these often highly virulent viruses, used mainly as microbial insecticides, are restricted in their host range principally to lepidoptera and hymenoptera, whilst individual isolates are in many cases genus- or species-specific. These are great advantages in terms of human safety, natural enemy protection, and so on. But there are also considerable disadvantages, in that: (i) it would often be necessary, where there are several pests, to use several different viruses (as opposed to perhaps one chemical); and (ii) a single-target species means a small potential market for a virus, making commercialization unattractive (the bigger the market, the bigger the profits). Another disadvantage is that they are slow acting relative to chemicals, with death occurring 3–4 days or more after infection (although feeding may effectively cease much earlier). Thus, Payne (1988) felt able to list only 14 insect viruses as realistic candidates for insect pest control, of which two were being produced commercially and a further five produced and used locally.

There are approximately 100 genera of fungi containing many species pathogenic to insects. Again, though, interest has focused on only a few of these, mainly the three in commercial production: *Beauveria bassiana* (produced in the USSR against the Colorado beetle, *Leptinotarsa decemlineata*), *Metarhizium anisopliae* (produced in Brazil against spittle bugs on sugar cane) and *Verticillium lecanii* (produced in the UK against aphids and whitefly on glasshouse crops). These and other species often have a broad host range amongst the insects (although different strains of a fungal 'species' can differ considerably in host range), and unlike the bacteria and viruses, do not generally rely on ingestion by the host for infection to occur. However, infection does require spore germination (giving rise to a penetrative hypha), which generally only occurs at very high relative humidities (greater than 92%). This limits the range of habitats in which fungi could be effective.

The last group of importance here are the nematodes. Their inclusion amongst 'microbial' control agents arises out of the mutualistic association of some with insect-pathogenic bacteria. In this association, the nematode first infects an insect host. The bacteria then escape from the nematodes, multiply and kill the insect host (usually within 48 h of invasion). The nematodes feed, develop and reproduce on the bacteria and decomposing host tissue, and a new generation of nematodes, complete with bacteria, is then released to infect new hosts. The nematodes can survive in damp soil for several months. This, their speed of action and their wide insect host range have made such nematodes and their bacteria attractive selective-control agents. They are the first commercially available biological product for use against insect pests in the soil.

Pathogens, particularly fungi, have also been used in the biological control of weeds, both for classical control and as mycoherbicides (see Cullen & Hasan, 1988, who list six examples in the former category and three in the latter, plus 35 other projects in both areas in the course of investigation). The most important has been the introduction from Europe in 1971 of the rust fungus, *Puccinia chondrillina*, for the control of the skeleton weed, *Chondrilla juncea*, in Australia, which had itself been introduced to Australia from Europe in the absence of any specific pathogens. The rust has gradually reduced the density of skeleton weed to its normal level in Europe: around 0.01% of its previous density (Cullen & Hasan, 1988). Significantly, *C. juncea* exists as several distinct genetic forms: it has been necessary to search for appropriately specialized strains of the pathogen.

Overall, there have been documented biological control programmes against around 86 naturalized weeds (i.e. introduced from another area) and 25 native weeds (Julien, 1982)—an expected pattern given that pest status is most likely to arise in the absence of natural enemies. Of these, the great majority have involved herbivorous insects. We have already discussed one of the most important, and certainly the most quoted example: the control of *Opuntia inermis* and *O. stricta* in Australia by *Cactoblastis cactorum* (see p. 395). Aquatic weeds provide a number of further examples of classical biological control. For instance, the free-floating *Salvinia molesta* (see Chapter 15, Section 15.7.6) has, since the 1930s, been introduced by humans from South America to Africa, India, Sri Lanka, South-East Asia, the Pacific Islands and Australia. Prior to 1972, *S. molesta* was confused with the related *S. auriculata*, and attempts at biological control with insects collected from *S. auriculata* were generally unsuccessful. Subsequently, following the discovery of the native range of the plant in south-east Brazil, a previously unknown and specific species of weevil, *Cyrtobagous salviniae*, has been introduced and is now controlling *S. molesta* at an increasing number of sites in Australia (Room *et al.*, 1985) and elsewhere (Thomas & Room, 1985).

The water hyacinth, *Eichhornia crassipes*, was introduced into the USA from South America at the 1884 Cotton States Exposition in New Orleans for its decorative blue flowers. It can infest a wide range of freshwater habitats—lakes, rivers, canals, swamps—and it has now spread throughout a large portion of the southern USA, particularly the south-east. A weevil discovered in South America, *Neochetina eichhorniae*, was first released in the state of Louisiana for biological control in 1974 and was distributed throughout Louisiana in 1976. The consequences are shown in Figure 16.9. The total area of water hyacinth in Louisiana remained high until the insect population increased in 1979. A sharp drop in area then followed. However, from 1980 to 1983 the water hyacinth recovered as its low abundance led to a reduction in the host-specific *N. eichhorniae* population (Cofrancesco *et al.*, 1985). Whether this is the first of a series of damped, persistent, or even escalating

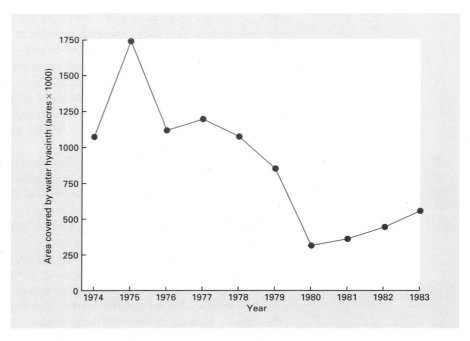

Figure 16.9 The decline in the area in the state of Louisiana occupied by the water hyacinth, *Eichhornia crassipes*, after the introduction for its control of the weevil, *Neochetina eichhorniae*, in 1974, followed by a subsequent increase as the abundance of the weevil itself declined. (After Cofrancesco *et al.*, 1985.)

oscillations, and what will be the eventual equilibrium level, if any, remains to be seen.

Aquatic weeds also provide an interesting example of biological control by inoculation. The grass carp, *Ctenopharyngodon idella*, is a fish native to northern China. It is almost completely herbivorous, and is attracted to a wide variety of troublesome aquatic weeds, both submerged and floating. It has been introduced for biological control in a number of regions, including Great Britain (Fowler, 1985). The fish survive well in the British climate, but they are most unlikely to breed since they require a simultaneous rapid rise in water flow and temperature—conditions not found in Great Britain. Hence, control, still in the early stages of development, requires the repeated release of fish reared in hatcheries. This increases the necessary expenditure, in economic and effort terms, required in using the grass carp for biological control. On the other hand, the risk of deleterious side-effects from the introduction of an exotic species are minimized.

The examples of biological control discussed here and in other chapters come from a large catalogue of successes, as Table 16.3 (confined to insects) has already indicated. (Reviews of many of these may be found in Huffaker, 1971; Huffaker & Messenger, 1976; Wood & Way, 1988; Corey *et al.*, 1993.) On the other hand, Table 16.3 makes it plain that biological control is not guaranteed to succeed. The success rate is, we hope, likely to improve in line with our understanding of what constitutes a good agent (see Chapter 10). For now, though, if we wish to assess the overall value of biological control (or any other measure), it is necessary to add the cost of the unsuccessful introductions to that of the successes. Yet, as we have already noted,

... and by inoculation with a fish

the value of biological control

compared with chemicals, biological control is cheap. Hence, even when incorporating the unsuccessful attempts, Table 16.3 indicates a substantial number of successes, economically achieved. Biological control is not easy: detailed ecological and taxonomic understanding, trained manpower and perseverence are all required. But the rates of return are generally favourable. Even when biological control alone is insufficient as a total control strategy, it may have an integral place in a larger scheme, as we shall see below.

16.9 Cultural control

Closely related to biological control is cultural control, i.e. the adoption of practices that make the environment unfavourable for the pest, often by making the environment favourable for the pest's natural enemies (Pimentel, 1993). Amongst the most obvious aspects of this are *crop rotation* (denying pests repeated access to their food) and *tillage* of the soil, burying crop residues that frequently harbour pests, both of which have been shown, for example, to provide effective means of control of insect pests of corn in the USA (Pimentel *et al.*, 1991). Other examples include the following.

1   *Polyculture* is the planting of several crops together in the same field. For instance, intercropping cowpeas with cassava reduces the abundance of whiteflies on cassava (Gold *et al.*, 1989).

2   *Trap crops* are plants grown amongst a main crop simply because they are more attractive to pests. For instance, Indian mustard is more attractive than cabbage for oviposition of diamondback moths (*Trichoplusia ni*) and leafrollers (*Crocidolomia binotalis*). Thus, when, in India, one row of mustard is planted between every five rows of cabbage, caterpillar populations build up on the mustard and can be treated by handsprayer with a high dosage of insecticide (Srinivasan & Moorthy, 1991).

3   *Sanitation* refers simply to the removal of crop residues and unharvestable (perhaps pest-infected) plants that might harbour pests. For instance, the destruction of abandoned apple trees prevents the invasion of pest insects from outside the crop area (Prokopy *et al.*, 1990).

4   Finally, in some cases, *planting times* can be delayed such that a crop is planted after a pest has emerged and died off. For instance, Hessian fly (*Mayetiola destructor*) populations are monitored in Georgia, in the USA, and farmers are advised when it is safe to plant their wheat crop (Buntin *et al.*, 1990).

16.10 Genetic control and resistance

There are several techniques that use genetic manipulation to kill pests, three of which we shall discuss here. The first, *autocidal control*, involves using the pest itself to increase its own rate of 'mortality'. In fact, the predominant method of autocidal control, the release of sterile males, leads to a decrease in birth rate. The principle is simple: large numbers of sterile males (usually irradiated) are released, these mate with females, which then lay infertile eggs. However, the method is expensive, and in order for it to work: (i) the females must mate only infrequently (preferably once); (ii) the males must be competitive with natural males (in spite of being irradiated); (iii) they must be amenable to mass rearing in the laboratory (they should far outnumber the natural males); and (iv) the target area should be isolated so that the programme is not undermined by the immigration of fertile males and females with

sterile male release: the screw-worm fly

fertile eggs. Nevertheless, the method has at least been successful once, in the virtual eradication from the southern USA and northern Mexico of the screw-worm fly, *Cochliomyia hominivorax* (Horn, 1988), which meets the criteria outlined above particularly well. The female screw-worm lays larvae directly onto fresh wounds in wildlife and domestic stock. Untreated, the victim dies. The EIL is very low: individual cattle are sufficiently valuable to make eradication desirable, justifying considerable expense. The females mate only once, and density is rather low (hence, released males can greatly outnumber natural males). Significantly, there is no alternative economical, chemical or cultural means of control.

the natural selection of resistant plant varieties

A more straightforward 'genetic manipulation' is the selection or breeding of plant varieties resistant to pests. (More recently, there has also been use of crop plants resistant to herbicides—allowing freer use of the herbicide in the control of competing weeds—and biological control agents resistant to insecticides). There can be little doubt that a great deal of selection has occurred naturally as crops and pests interact, and this has sometimes been brought to light in dramatic fashion as plants have been moved around the world. For example, the grape phylloxera, mentioned above, is an insect native to North America; the devastation caused by its attacks on western European vines was brought to an end only when American root stocks, naturally resistant to the phylloxera, were imported to the affected regions. The potato provides an interesting contrast. When it was introduced into Europe in the 16th century, the fungus *Phytophthora infestans*, which causes potato late blight, was not introduced with it (Barrett, 1983). The potato flourished and became a staple food for the northern European peasantry, but in the 1840s *P. infestans* reached Europe and gave rise to the Great Potato Blight of 1845, leading to the death of possibly millions of people. In 1833, a consignment of potatoes had been taken, in the absence of the fungus, from Europe to the area now known as Lesotho in Africa. Much later, these representatives of pre-blight European potatoes were compared experimentally with early 20th century varieties that had survived the Blight, but had not otherwise been subjected to any artificial selection: the post-Blight varieties were much the more resistant to the fungus.

... their selection and artificial breeding

In numerous other cases, throughout the history of agriculture, resistant varieties have been selected for widespread use by farmers from amongst those generally available. This continues up to the present. For instance, after the spotted alfalfa aphid, *Therioaphis trifolii*, devastated the Californian alfalfa crop in the 1950s, a variety of alfalfa resistant to the aphid was discovered quite fortuitously (Howe & Smith, 1957) and widely introduced. However, increasingly a more methodical approach has been taken to the selective breeding of resistant varieties. This, though, can be a lengthy process: several years to breed the variety, and several years more of testing before it can be released to growers. Even when this has been done, the pest itself may evolve a biotype that is virulent against the new variety. Hence, it is preferable, although more difficult, to breed varieties in which resistance is determined by more than one gene, since this seems to make it more difficult for the pest to make a rapid evolutionary response (Burdon, 1987).

gene-for-gene effects

A particularly well-understood, although probably fairly typical example of the relationship between different strains of pest and crop is shown in Table 16.4, illustrating data for a number of varieties of wheat and biotypes of the Hessian fly, *M. destructor* (Gallun, 1977; Gallun & Khush, 1980). Virulence in the fly is associated with recessive genes at four loci, designated k (virulent against wheat variety Knox 62), m (against Monon), s (against Seneca) and t (against Turkey). Resistance in wheat

Table 16.4 Different varieties of wheat are susceptible (S) or resistant (R) to different biotypes of the Hessian fly, depending on the resistance genes carried by the wheat and the virulence genes carried by the fly. (After Gallun, 1977.)

Hessian fly biotype (and virulence genes)	Wheat variety (and resistance genes)				
	Turkey (none)	Seneca (H_7 and H_8)	Monon (H_3)	Knox 62 (H_6)	Abe (H_5)
GP (tt)	S	R	R	R	R
A (tt, ss)	S	S	R	R	R
B (tt, ss, mm)	S	S	S	R	R
C (tt, ss, kk)	S	S	R	S	R
D (tt, ss, mm, kk)	S	S	S	S	R
E (tt, mm)	S	R	S	R	R
F (tt, kk)	S	R	R	S	R
G (tt, mm, kk)	S	R	S	S	R

is controlled by five dominant genes: H_3, H_5, H_6, H_7 and H_8. A fly that is homozygous for one of the virulence loci will thus be unaffected by the resistance arising from the corresponding gene (or genes) in wheat and can injure the crop. For example, the variety Seneca is resistant to all biotypes that are not homozygous for the gene s, namely biotypes GP, E, F and G.

A further problem is that varieties of a crop resistant to one pest may have an enhanced susceptibility to some other pest—simply another manifestation of the concept of a trade-off discussed in Chapter 14. For instance, DaCosta and Jones (1971) reported that a variety of cucumber resistant to chrysomelid cucumber beetles in the laboratory was devastated by red spider mites (*Tetranchyus urticae*) in the field.

genetically engineered plants ...

Lastly in this section, we discuss an area in which the phrase 'genetically manipulating crops' can now be taken literally (Meeusen & Warren, 1989). In 1987, the first success was reported of inserting a gene into a crop that conferred resistance against pests—the δ-endotoxin gene of *B. thuringiensis* was inserted into tobacco plants (Vaeck *et al.*, 1987), so that the δ-endotoxin was produced by the plant itself, conferring resistance potentially, by its toxicity, against a range of lepidopterous pests. This was quickly followed by other reports for the δ-endotoxin with tobacco (Barton *et al.*, 1987) and tomato (Fischhoff *et al.*, 1987) and for a protease inhibitor with the cowpea, *Vigna unguiculata* (Hilder *et al.*, 1987). Central to all this work was the common crown gall bacterium, *Agrobacterium tumefaciens*, a ubiquitous soil organism with the remarkable ability to move a portion of its own DNA into a plant cell during infection (Meeusen & Warren, 1989).

... and the consequent problem of continuous pesticide application

Clearly, the potential value of the development and use of such transgenic plants is immense, in terms of environmental and safety benefits, performance and the reduced costs of both development and application. But there are also serious potential problems. In particular, by incorporating a pesticide directly into a plant, that pesticide is being applied continuously, irrespective of need, subjecting the pest to an unrelenting selection pressure to which it is most likely to respond—not at all what 'resistance management' would recommend. This is especially worrying as there are now at least eight reported cases of resistance evolving to the *B. thuringiensis* δ-endotoxin, even when applied directly rather than in transgenic plants (McGaughey & Whalon, 1992). At the very least, resistance management programmes will have to become more sophisticated if transgenically toxic plants are not to be confronted rapidly with conventionally evolved resistant pests (May, 1993). On the other hand, other transgenic manipulations may not be so problematic—for instance, the engineering of herbicide resistance into a crop so that the herbicide can be used more effectively on its weeds, particularly as the early evidence suggests, for example, that the genetically engineered crop is not likely to become a weed itself, invading natural habitats (Crawley *et al.*, 1993).

There is always a danger that new technologies will be adopted and applied simply because they are novel and exciting. It may be, however, that the more straightforward application of ecological principles, to which we turn next, has more to offer pest control in the long term. Certainly, the new technologies have little chance of success in isolation. They will inevitably rely on a concurrent application of ecological insights.

16.11 Integrated pest management

the IPM philosophy: minimize disruption using anything or nothing, *as appropriate,* aiming at EILs

Integrated pest management (IPM) is a philosophy of pest management—although a very practical philosophy—rather than a specific, defined strategy. It has been practised, in effect, for a century or more in that it combines physical, cultural, biological and chemical control and the use of resistant varieties. However, propelled by a formal definition and statement of principles by Stern *et al.* (1959) (who called it 'integrated control'), it has come of age as part of the reaction against the unthinking use of chemical pesticides in the 1940s and 1950s.

... and the consequent difficulties of putting it into practice

IPM is ecologically rather than chemically based, although it utilizes all methods of control—including chemicals—where appropriate. It relies heavily on natural mortality factors, such as enemies and weather, and seeks to disrupt them as little as possible, or even to augment them. Crucially, its aim is not pest eradication but control below EILs, and it is therefore based on careful monitoring of pest and natural enemy abundance and the associated tailoring of control methods to present needs. The IPM approach views the various control methods we have discussed not so much as alternatives but as complementary parts of an overall programme. It is committed to considering the widest possible range of methods before the most appropriate combination is chosen. Thus, broad-spectrum pesticides in particular, although not excluded, are used only very sparingly in IPM because of their unavoidable tendency, as we have seen, to interfere with other methods of control. Similarly, there is an implicit commitment to invest time, effort and money in seeking efficient means of biological and/or cultural control because of the lack of disruption such methods cause. If chemicals are used at all, then the resistance of the pest is 'managed' so as to minimize costs and the quantities used.

the importance of the IPM adviser

More subtly, there is implicit in the IPM approach a need for specialist pest managers or advisers. The chemical approach asks the individual farmer (or whoever it may be) to do no more than 'read the instructions on the packet'. But it would be quite unrealistic to expect most individual farmers to implement an IPM programme unaided. After all, the essence of the IPM approach is to make the control measures fit the pest problem, and no two pest problems are the same—even in adjacent fields.

We can take as an example of a successful IPM programme, one developed to control the pests of cotton in the San Joaquin Valley of California where more than 90% of California's cotton is grown (van den Bosch *et al.*, 1971; Flint & van den Bosch, 1981). Prior to its establishment, the system had been afflicted by target pest resurgence (Figure 16.10a), secondary pest outbreaks (Figures 16.10b, c) and the evolution of resistance (Figure 16.10d), leading to an expansion in the range of pesticides used and an escalation in their frequency of application.

The key pest is the lygus bug *Lygus hesperus*, which feeds primarily on fruiting cotton buds, delaying fruit (cotton boll) formation and ultimately suppressing yield. Another major pest has been the cotton bollworm (*Heliothis zea*). However, in the San Joaquin Valley this seems to be largely a secondary pest induced by insecticides, as are two other lepidopterous pests, the beet army-worm (*Spodoptera exigua*) and the cabbage looper (*Trichoplusia ni*). An IPM programme therefore needed to reduce insecticide usage so as to prevent these secondary outbreaks, but it also needed specifically to avoid certain commonly used insecticides (e.g. methyl parathion, carbaryl) since these were found themselves to decrease the yield of cotton plants.

Prior to the development of the programme, a CAT had been established for the lygus bug—but from little information. Hence, the threshold remained the same throughout the year—a serious error. Further research showed that lygus could inflict serious damage only during the budding season (broadly, early June to mid-July). Thus, in the IPM programme the threshold was set at 10 bugs per 50 sweep-net sweeps on two consecutive sampling dates between 1 June and 15 July. After that time, insecticide application for lygus control was avoided. In fact, the threshold was subsequently refined to one based on a bug : bud ratio, which is more dynamic and more accurate.

Cultural control within the system has centred on manipulation of alfalfa, a favoured host plant of the lygus bug. In particular, the interplanting of narrow alfalfa strips (5–10 m) in the cotton fields has proved effective in attracting the bugs out of the cotton. In addition, attempts have been made to augment the populations of natural enemies by providing them with supplementary nutrition. For example, when the green lacewing *Chrysopa carnea*, is attracted by supplementary nutrients, there is an increase in the abundance of its own eggs and larvae, a significant impact on bollworm populations and a significant reduction in the injury to cotton bolls.

The key to a successful IPM programme is a good field monitoring system. Here, each field is sampled twice a week from the beginning of budding (roughly mid-May) to the end of August. Plant development, pest and natural enemy population data are all collected. Additional sampling for the cotton bollworm begins around 1 August and continues until mid-September. Overall, the programme represents a harmonious integration of naturally occurring (but augmented) biological controls and carefully timed insecticide treatments. The success of the programme is illustrated in Table 16.5 (see below).

Integration, however, does not stop at IPM. It has increasingly become apparent, in an agricultural context at least, that implicit in the philosophy of IPM is the idea that pest control cannot be isolated from other aspects of food production and is especially bound up with the means by which soil fertility is maintained and improved. By this argument, the management of pests and fertility should themselves be integrated into practices that aim to ensure the *sustainability* of food production methods. In contrast, the excessive and unthinking use of chemical pesticides is *un*sustainable in the sense that further problems are created—the evolution of resistance, secondary pests, death of non-targets—that require an escalating use of a wider range of ever more costly chemicals. Similarly, an overreliance on applications of simple inorganic chemical fertilizers (at the expense of soil and fertility conservation measures like crop rotation and the use of manures) is

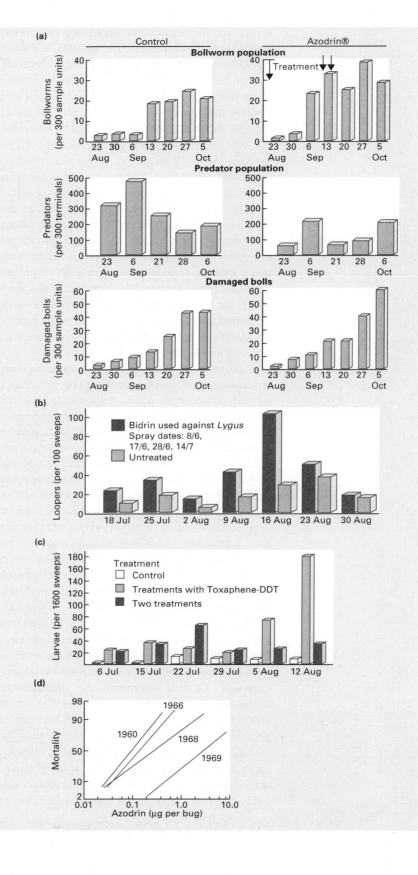

Table 16.5 The success of integrated pest management (IPM) in San Joaquin Valley, California, 1970–1971. With IPM, outlay on insecticides was cut drastically. Nevertheless, yield was increased. (Figures in brackets are standard deviations.) (After Hall *et al.*, 1975; Flint & van den Bosch, 1981.)

	IPM users	Non-users
Average insecticide cost per acre of cotton (US $)	4.9 (± 3.9)	12.0 (± 7.4)
Average yield per acre of cotton (US $)	270.2 (± 27.5)	247.8 (± 55.1)
Average insecticide cost per acre of citrus (US $)	20.5 (± 15.0)	42.4 (± 18.3)
Average yield per acre of citrus (US $)	515.8 (± 260.6)	502.9 (± 157.0)

unsustainable in the sense that more and more chemicals are needed to counter declining levels of natural soil fertility, water-holding capacity, and so on. Furthermore, the replacement, for example, of crop rotations by artificial chemical fertilizers, as the means of maintaining soil nitrogen levels, also makes life easier for the pest, by giving them an uninterrupted resource on which population levels can build up. This in itself generates a further need for pest control measures. By contrast, an avoidance of broad-spectrum pesticides allows, for instance, a thriving community of detritivores to remineralize dead organic matter efficiently, enhancing fertility.

Thus, a number of programmes have been initiated in order to develop and put into practice sustainable food production methods that incorporate IPM. These themselves, have spawned a series of further abbreviations and acronyms: LISA (low input sustainable agriculture) in the USA, and IFS (integrated farming systems), EFP (ecological food production) and LIFE (lower input farming and environment) in Europe (International Organisation for Biological Control, 1989; Begon, 1990; National Research Council, 1990). All share a committment to the development of sustainable agricultural systems. Sustainability is dealt with briefly again, as a concept in its own right, at the end of this chapter.

Of course, irrespective of the advantages of IPM and these other broader approaches in terms of reduced environmental hazard and so on, it is unreasonable to suppose that they will be adopted widely unless they are also sound in economic terms. Mounting evidence suggests that they frequently are. For example, Table 16.5, based on the cotton and citrus industries of the San Joaquin Valley, California (see above), gives details of: (i) the expenditure on insecticides; and (ii) the economic yields, for users and non-users of IPM consultancies (Hall *et al.*, 1975; in Flint & van den Bosch, 1981). IPM made economic as well as environmental good sense—partly, but not entirely, as a result of the reduction in pesticide costs.

More broadly, a number of studies of 'sustainable' agricultural systems, both in Europe and in the USA, support the economic claims of the integrated approach (International Organisation for Biological Control, 1989; National Research Council, 1990). For example, in West Germany the 'Lautenbach project' has, since 1978, compared the performance of an 'integrated' farm with a matched, adjoining

IPM makes economic sense

Figure 16.10 (*facing page*) Pesticide problems amongst cotton pests in San Joaquin Valley, California. (a) Target pest resurgence: bollworms resurge because the abundance of their natural predators is reduced—the number of damaged bolls is higher. (b, c) Secondary pest outbreaks. (b) An increase in cabbage loopers and (c) in beet army-worm when plots are sprayed against lygus bugs. (d) Increasing resistance of lygus bugs to Azodrin®. (After van den Bosch *et al.*, 1971.)

651   MANIPULATING ABUNDANCE

conventional farm, both growing cereals, sugar beet and legumes (El Titi, 1989). The integrated farm used significantly less mineral fertilizer and pesticide (36% less) and suffered significantly less soil erosion. Thus, it fared substantially better 'environmentally'. In addition, however, it produced greater yields of oats, sugar beet and beans (although less of wheat), achieved an increased net profit on winter wheat, sugar beet and beans (although not on spring wheat and oats) and managed an increased net profit of 4% overall. Hence, it even fared better than the conventional farm in narrow economic terms.

So, the 'ecological' advantages of IPM are clear and the potential economic advantages are apparent. Yet, certain disadvantages are apparent too, namely: (i) the difficulty of training the necessary army of advisers; (ii) the difficulty of developing a sufficiently detailed ecological understanding of the community of which the pest or pests are part; and (iii) the low rate of return on investment initially whilst training and ecological investigation are taking place. When we add to this the commercial pressures exerted on producers to use simpler, chemical methods of control, and the more immediate benefits that this approach brings, it is easy to see why, in spite of a widespread acceptance amongst ecologists that IPM is the ideal objective, actual examples of fully implemented IPM programmes have accumulated only slowly. On the other hand, the 1990s seems certain to be a decade in which the numbers of examples accelerates (e.g. Corey *et al.*, 1993). This is to be welcomed. In the USA, before 1945 and the extensive use of synthetic pesticides, crop losses to insects averaged around 7%. By 1991, after the use of insecticides in agriculture had increased 10-fold, crop losses to insects averaged 13% (Pimentel, 1993).

16.12 Harvesting, fishing, shooting and culling: introduction

harvesting aims to avoid both overexploitation and underexploitation

When a natural population is exploited by culling or harvesting—whether this involves the removal of whales or fish from the sea, the removal of deer from a moorland or the removal of timber from a forest—it is much easier to say what we want to avoid than precisely what we might wish to achieve. On the one hand, we want to avoid overexploitation, where too many individuals are removed and the population is driven into biological jeopardy, or economic insignificance or perhaps even to extinction. But we also want to avoid underexploitation, where far fewer individuals are removed than the population can bear, and a crop of food, for example, is produced which is smaller than necessary, threatening both the health of potential consumers and the livelihood of all those employed in the harvesting operation. However, as we shall see, the best position to occupy between these two extremes is not easy to determine, since it needs to combine considerations that are not only biological (the well-being of the exploited population) and economic (the profits being made from the operation), but also social (local levels of employment and the maintenance of traditional lifestyles and human communities) (Clark, 1976, 1981; Hilborn & Walters, 1992). We begin, though, with the biology.

blowflies stress the importance of intraspecific competition in harvesting

The effects of harvesting can be seen in laboratory cultures of Australian sheep blowflies, *Lucilia cuprina* (Table 16.6). As the harvesting (or exploitation) rate of emerging adults increased, both pupal production and adult emergence increased, and the rate of natural adult mortality decreased. Thus, by reducing density, harvesting reduced intraspecific competition, increasing the survivorship and fecundity of those that remained. Although the adult population size (878) was

Table 16.6 Effects produced in populations of the blowfly *Lucilia cuprina* by the destruction of different constant percentages of emerging adults. (After Nicholson, 1954b.)

Exploitation rate of emerging adults (%)	Pupae produced per day (a)	Adults emerged per day (b)	Mean adult population (c)	Mean birth rate (per individual per day) (a / c)	Natural adult deaths per day	Adults destroyed per day (d)	Accessions of adults per day (e = b − d)	Mean adult life span (days) (c / e)
0	624	573	2520	0.25	573	0	573	4.4
50	782	712	2335	0.33	356	356	356	6.6
75	948	878	1588	0.60	220	658	229	7.2
90	1361	1260	878	1.55	125	1134	126	7.0

depressed to well below the apparent carrying capacity (2520 at a zero exploitation rate), the yield from the harvest (column (*d*) in Table 16.6) increased as the exploitation rate rose to 90%. High yields, therefore, are obtained from populations held below (and in this case, well below) the carrying capacity.

16.12.1 Maximum sustainable yield

This fundamental pattern is captured by the model population in Figure 16.11. There, the natural net recruitment (or net productivity) of the population is described by an n-shaped curve (see Chapter 6, Section 6.4). Recruitment rate is low when there are few individuals and low when there is intense intraspecific competition. It is zero at the carrying capacity (*K*). The density giving the highest net recruitment rate depends on the exact form of intraspecific competition. This density is $K / 2$ in the logistic equation (see Chapter 6, Section 6.10), much less than $K / 2$ in Nicholson's blowflies (Table 16.6) and only slightly less than *K* in many large mammals (see Chapter 6, Figure 6.12c). Always, though, the rate of net recruitment is highest at an 'intermediate' density, less than *K*.

Figure 16.11 also illustrates three possible harvesting 'strategies', although in each case there is a fixed harvesting *rate*, i.e. a fixed number of individuals removed during a given period of time, or '*fixed quota*'. When the harvesting and recruitment lines cross, the harvesting and recruitment rates are equal and opposite; the number removed per unit time by the harvester equals the number recruited per unit time by the population. Of particular interest is the harvesting rate h_m, the line which crosses (or, in fact, just touches) the recruitment rate curve at its peak. This is the highest harvesting rate that the population can match with its own recruitment. It is known as the maximum sustainable yield (MSY), and as the name implies, it is the largest harvest that can be removed from the population on a regular and repeated (indeed indefinite) basis. It is equal to the maximum rate of recruitment, and it is obtained from the population by depressing it to the density at which the recruitment rate curve peaks (always below *K*).

the 'MSY' is obtained from the peak of the net recruitment curve

Another perspective on the same principle comes from asking: at what density can a harvest of a given size be taken most frequently? This is the sort of question that a farmer might ask who wishes to harvest grass from a field. The answer is illustrated in Figure 16.12, by assuming that the population follows an 'S'-shaped curve of population growth (see Chapter 6, Section 6.4). Clearly, harvests can be

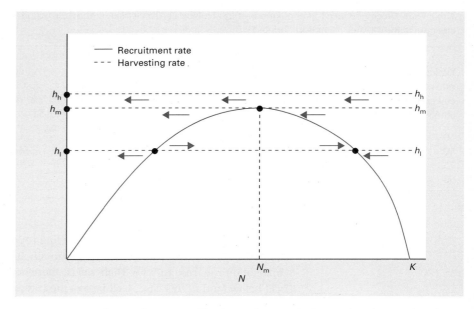

Figure 16.11 Fixed-quota harvesting. The figure shows a single recruitment curve (——) and three fixed-quota harvesting curves (- - -); high quota (h_h), medium quota (h_m) and low quota (h_l). Arrows in the figure refer to changes to be expected in abundance under the influence of the harvesting rate to which the arrows are closest. ● = equilibria. At h_h the only 'equilibrium' is when the population is driven to extinction. At h_l there is a stable equilibrium at a relatively high density, and also an unstable breakpoint at a relatively low density. The MSY is obtained at h_m because it just touches the peak of the recruitment curve (at a density N_m): populations greater than N_m are reduced to N_m, but populations smaller than N_m are driven to extinction.

taken most frequently at intermediate densities where the net recruitment rate (and hence the rate of population regrowth) is highest.

the MSY concept has severe shortcomings ...

The MSY concept is central to much of the theory and practice of harvesting. This makes the recognition of the following shortcomings in the concept all the more essential.

1 By treating the population as a number of similar individuals, or as an undifferentiated 'biomass', it ignores all aspects of population structure such as size or age classes and their differential rates of growth, survival and reproduction; alternatives that incorporate structure are considered below.

2 By being based on a single recruitment curve, it treats the environment as unvarying. This, however, is a problem common to virtually all harvesting models (but, see Iles, 1973).

3 In practice, it may be impossible to obtain a reliable estimate of the MSY.

4 Achieving an MSY is by no means the only, nor necessarily the best, criterion by which success in the management of a harvesting operation should be judged.

... but, it is frequently used

Despite all these difficulties, the MSY concept has dominated resource management for many years in fisheries, forestry and wildlife exploitation. Prior to recent changes, for example, there were 39 agencies for the management of marine fisheries, every one of which was required by its establishing convention to manage on the basis of an MSY objective (Clark, 1981). In many other areas, the MSY concept is still the guiding principle. Moreover, by assuming that MSYs are both

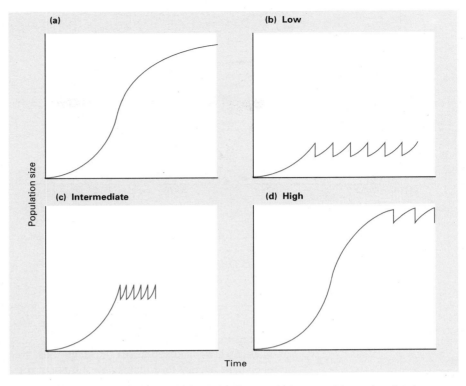

Figure 16.12 For a population exhibiting 'S'-shaped growth in size (a), at what density can a harvest of a given size (the vertical lines in (b–d)) be taken most frequently? The answer is 'at an intermediate density', shown in (c). At both low (b) and high (d) densities, the rate of growth and thus the frequency of harvest is lower.

desirable and attainable, a number of the basic principles of harvesting can be explained. Therefore, we begin here by exploring what can be learnt from analyses based on MSYs, but then look more deeply at management strategies for exploited populations by examining the various shortcomings of MSY in more detail.

16.13 Simple MSY models of harvesting

16.13.1 Fixed quotas

fixed-quota MSY harvesting is extremely risky

Returning to Figure 16.11, we can see that the MSY density (N_m) is an equilibrium (gains = losses). But when harvesting is based on the removal of a fixed quota, as it is in Figure 16.11, N_m is a very fragile equilibrium. If the density exceeds the MSY density, then h_m exceeds the recruitment rate and the population declines towards N_m. This, in itself, is satisfactory. But if, by chance, the density is even slightly less than N_m, then h_m will once again exceed the recruitment rate. Density will then decline even further, and if a fixed quota at the MSY level is maintained, the population will decline until it is extinct. Furthermore, if the MSY is even slightly overestimated, the harvesting rate will always exceed the recruitment rate (h_h in Figure 16.11). Extinction will then follow, whatever the initial density. In short, a fixed quota at the MSY level might be desirable and reasonable in a wholly

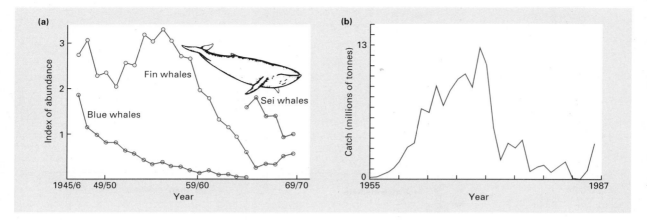

Figure 16.13 (a) The declines in the abundance of Antarctic baleen whales under the influence of human harvesting. (After Gulland, 1971.) (b) Catch history of the Peruvian anchoveta fishery. (After Hilborn & Walters, 1992.)

predictable world about which we had perfect knowledge. But in the real world of fluctuating environments and imperfect data sets, these fixed quotas are open invitations to disaster (Clark, 1981).

Nevertheless, a fixed-quota strategy has frequently been used, where a management agency formulates an estimate of the MSY, which is then adopted as the annual quota. On a specified day in the year, the fishery (or hunting season or whatever) is opened and the accumulated catch is logged. Then, when the quota has been taken, the fishery is closed for the rest of the year. This obviously encourages fishermen to compete. When annual quotas were used by the International Whaling Commission from 1949 to 1960, the ensuing scramble was commonly referred to as 'the whaling Olympics' (Clark, 1981). The whaling nations eventually agreed to allocate quotas to one another before the season opened; but even so, the effects of fixed quotas on whale populations have not been encouraging (Figure 16.13a).

Another, fairly typical example of the use of fixed quotas is provided by the Peruvian anchoveta (*Engraulis ringens*) fishery (Figure 16.13b). From 1960 to 1972 this was the world's largest single fishery, and it constituted a major sector of the Peruvian economy. Fisheries experts advised that the MSY was around 10 million tonnes annually, and catches were limited accordingly. But, the fishing capacity of the fleet expanded, and in 1972 the catch crashed. Overfishing seems at least to have been a major cause of the collapse, although its effects were compounded with the influences of profound environmental fluctuations, discussed below (Section 16.13.4). A moratorium on fishing might have allowed the stocks to recover, but this was not politically feasible: 20 000 people were dependent on the anchoveta industry for employment. The Peruvian government has therefore allowed fishing to continue each year. The catches have never recovered (Hilborn & Walters, 1992).

the dangers of fixed-quota MSY harvesting are illustrated by whaling …

… and by the Peruvian anchoveta fishery

16.13.2 Fixed harvesting effort

The risk associated with fixed quotas can be reduced if instead there is regulation of the harvesting *effort*. The yield from a harvest (*H*) can be thought of, simply, as being

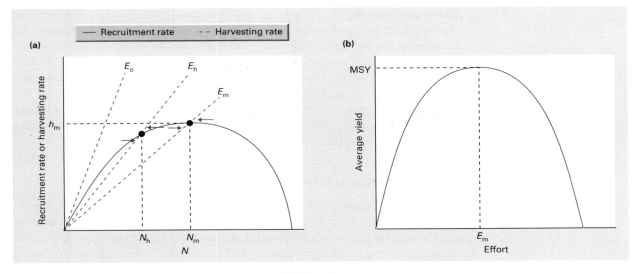

Figure 16.14 Fixed-effort harvesting. (a) Curves, arrows and dots as in Figure 16.11. The maximum sustainable yield (MSY) is obtained with an effort of E_m, leading to a stable equilibrium at a density of N_m with a yield of h_m. At a somewhat higher effort (E_h), the equilibrium density and the yield are both lower than with E_m but the equilibrium is still stable. Only at a much higher effort (E_o) is the population driven to extinction. (b) The overall relationship between the level of the fixed effort and average yield.

dependent on three things:

$$H = qEN. \tag{16.1}$$

regulating harvesting effort is less risky, but it leads to a more variable catch

Yield, H, increases with the size of the harvested population, N; it increases with the level of harvesting effort, E (e.g. the number of 'trawler-days' in a fishery or the number of 'gun-days' with a hunted population); and it increases with harvesting efficiency, q. On the assumption that this efficiency remains constant, Figure 16.14a depicts an exploited population subjected to three potential harvesting strategies differing in harvesting effort. Figure 16.14b then illustrates the overall relationship to be expected, in a simple case like this, between effort and average yield: there is an apparently 'optimum' effort giving rise to the MSY, E_m, with efforts both greater and less than this giving rise to smaller yields.

Adopting E_m is a much safer strategy than fixing an MSY quota. Now, in contrast to Figure 16.11, if density drops below N_m (Figure 16.14a), recruitment exceeds the harvesting rate and the population recovers. In fact, there needs to be a considerable overestimate of E_m before the population is driven to extinction (E_o in Figure 16.14a). However, because there is a fixed effort, the yield varies with population size. In particular, the yield will be less than the MSY whenever the population size, as a result of natural fluctuations, drops below N_m. The appropriate reaction would be to reduce effort slightly or at least hold it steady whilst the population recovers. But an understandable (albeit misguided) reaction might be to compensate by increasing the effort. This, however, might depress population size further (E_h in Figure 16.14a); and it is therefore easy to imagine the population being driven to extinction as very gradual increases in effort chase an ever-diminishing yield.

There are many examples of harvests being managed by legislative regulation of effort, and this occurs in spite of the fact that effort usually defies precise measurement and control. For instance, issuing a number of gun licenses leaves the accuracy of the hunters uncontrolled; and regulating the size and composition of a fishing fleet leaves the weather to chance. Nevertheless, the harvesting of mule deer, pronghorn antelope and elks in Colorado are all controlled by issuing a limited but varying number of hunting permits (Pojar, 1981). In the management of the important Pacific halibut stock, effort is limited by seasonal closures and sanctuary zones—although a heavy investment in fishery protection vessels is needed to make this work (Pitcher & Hart, 1982).

16.13.3 The instability of harvested populations: multiple equilibria

Even with the regulation of effort, however, harvesting near the MSY level may be courting disaster. Recruitment rate may be particularly low in the smallest populations (Figure 16.15a); for instance, recruitment of young salmon is low at low densities because of intense predation from larger fish, and recruitment of young whales may be low at low densities simply because of the reduced chances of males and females meeting to mate. Alternatively, harvesting efficiency may decline in large populations (Figure 16.15b). For instance, many clupeids (sardines, anchovies, herring) are especially prone to capture at low densities, because they form a small number of large schools that follow stereotyped migratory paths that the trawlers can intercept. In either case, small overestimates of E_m are liable to lead to overexploitation or even eventual extinction (Figure 16.15).

many harvesting operations have multiple equilibria, and are therefore susceptible to dramatic, irreversible crashes

Even more important, however, is the fact that these interactions may have crucial 'multiple equilibria' (see Chapter 10, Section 10.6), which arise here because the harvesting curve crosses the recruitment curve at three points. Of these, two are stable equilibria, but the central one is an unstable 'breakpoint'. If the population is driven slightly below the MSY density, or even to a level slightly above N_u, a

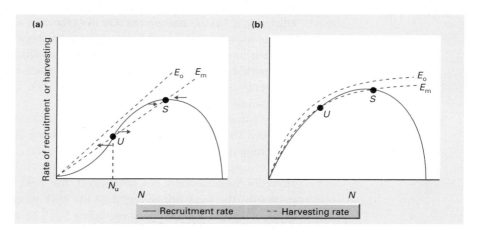

Figure 16.15 Multiple equilibria in harvesting. (a) When recruitment rate is particularly low at low densities, the harvesting effort giving the maximum sustainable yield (MSY) (E_m) has not only a stable equilibrium (S) but also an unstable breakpoint (U) at a density below which the population declines to extinction. The population can also be driven to extinction by harvesting efforts (E_0) not much greater than E_m. (b) When harvesting efficiency declines at high densities, comments similar to those in (a) are appropriate.

658   CHAPTER 16

breakpoint, it returns to the MSY density (Figure 16.15). But a marginally increased depression in density, to a level slightly below N_u, perhaps resulting from only a marginal increase in effort, would make the harvesting rate greater than the recruitment rate. The population would be *en route* to extinction. Moreover, once the population is on this slippery slope, much more than a marginal reduction in effort is required to reverse the process. This is the crucial, practical point about multiple equilibria: a very slight change in behaviour can lead to a wholly disproportionate change in outcome as the point of attraction in the system shifts from one stable state to another. Drastic changes in stock abundance can result from only small changes in harvesting strategy or small changes in the environment.

Regrettably, there are a number of examples where stocks have 'crashed' suddenly, where they have failed to recover even when harvesting efforts have been reduced and where patterns like those in Figure 16.15 have been implicated. This has been seen most importantly in a number of clupeid fisheries, including the Peruvian anchoveta, the North Sea herring (Figure 16.16) and the Pacific sardine (Murphy, 1977); and it has also been seen in certain Pacific salmon fisheries (Peterman *et al.*, 1979).

16.13.4 The instability of harvested populations: environmental fluctuations

It is tempting to attribute all fisheries' collapses simply to overfishing and human greed. Doing so, however, would be an unhelpful over-simplification. There is no doubt that fishing pressure often exerts a great strain on the ability of natural populations to sustain levels of recruitment that counteract overall rates of loss. But the immediate cause of a collapse—in one year rather than any other—is often the occurrence of unusually unfavourable environmental conditions. Moreover, when this is the case, the population is more likely to recover (once conditions have returned to a more favourable state) than it would be if the crash was the result of overfishing alone.

the anchoveta and El Niño

The Peruvian anchoveta (see Section 16.13.1; Figure 16.13b), prior to its major collapse from 1972 to 1973, had already suffered a dip in the upward rise in catches in the mid-1960s as a result of an 'El Niño event': the incursion of warm tropical water from the north severely reducing the upwelling, and hence the productivity, within the cold Peruvian current coming from the south. By 1973, however, because fishing intensity had so greatly increased, the effects of a subsequent El Niño event

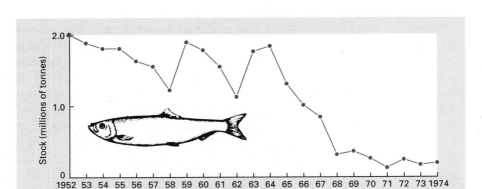

Figure 16.16 The decline in the stock of North Sea herring, *Clupea harengus*. (After Iles, 1981.)

were much more severe. Moreover, whilst the fishery showed some signs of recovery from 1973 to 1982, in spite of largely unabated fishing pressure, a further collapse occurred in 1983 associated with yet another El Niño event. Clearly, it is unlikely that the consequences of these natural perturbations to the usual patterns of current flow would have been so severe if the anchoveta had not been exploited or had been only lightly fished. It is equally clear, though, that the history of the Peruvian anchoveta fishery cannot be understood properly in terms simply of fishing, as opposed to natural, events.

the herring and cold water

The three Norwegian and Icelandic herring fisheries, all living in the Norwegian Sea, also collapsed in the early 1970s and had certainly been subjected to increasing fishing intensities prior to that. Once again, however, an oceanic anomaly is implicated (Beverton, 1993). In the mid-1960s, a mass of cold, low-salinity water from the Arctic Basin formed north of Iceland. It drifted south until it became entrained in the Gulf Stream several years later, and then moved north again—although well to the east of its southward track. It eventually disappeared off Norway in 1982 (Figure 16.17a). Data for the number of 'recruits per spawner', essentially the birth rate, are illustrated in Figure 16.17b for the Norwegian spring-spawning and the Icelandic spring- and summer-spawning herring between 1947 and 1990, in terms of the difference each year between that year's value and the overall average. Also illustrated are the corresponding yearly temperature differentials in the Norwegian Sea, reflecting the southward and northward passage of the anomalous cold water body. There was a good correspondence between the cold water and poor recruitment in both the Icelandic and Norwegian stocks in the late 1960s and in the Norwegian stocks in 1979–1981—the Icelandic stocks being extinct (spring spawners) or too far west. It seems likely that the anomalous cold water led to unusually low recruitment, which was strongly instrumental in the crashes experienced by each of these fisheries.

This cannot, however, account for all the details in Figure 16.17b—especially the succession of poor recruitment years in the Norwegian stocks in the 1980s. For this, a more complex explanation is required, probably involving other species of fish and perhaps alternative stable states (Beverton, 1993). Nonetheless, it remains clear that whilst the dangers of overfishing should not be denied, these must be seen within the context of marked and often unpredictable natural variations.

16.14 Recognizing structure in harvested populations: dynamic pool models

The simple models of harvesting that have been described so far are known as 'surplus yield' models. They are useful as a means of establishing some basic principles (like MSY), and they are good for investigating the possible consequences of different types of harvesting strategy. But they ignore population structure, and this is a bad fault for two reasons. The first is that 'recruitment' is, in practice, a complex process incorporating adult survival, adult fecundity, juvenile survival, juvenile growth, and so on, each of which may respond in its own way to changes in density and harvesting strategy. The second reason is that most harvesting practices are primarily interested in only a portion of the harvested population (e.g. mature trees, or fish that are large enough to be saleable). The approach that attempts to take these complications into account involves the construction of what are called 'dynamic pool' models.

'dynamic pool' (as opposed to 'surplus yield') models recognize population structure

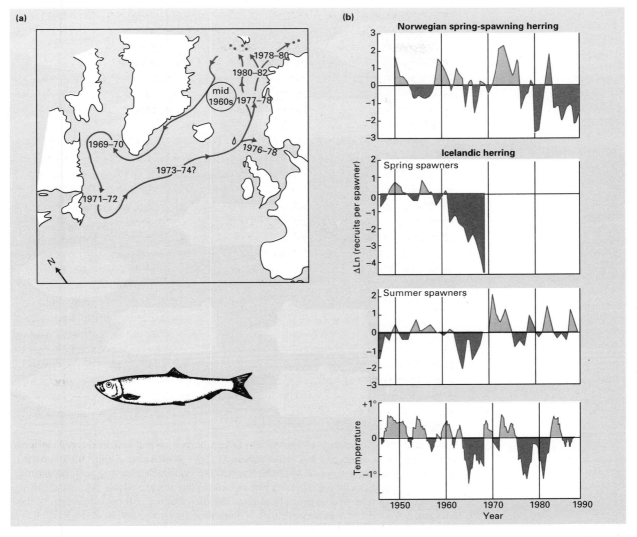

Figure 16.17 (a) The track of a large mass of cold, low-salinity water in the 1960s and 1970s, showing its presence in the Norwegian Sea both in the mid-1960s and the period 1977–1982. (b) Annual differentials between overall averages and (above) ln (recruits per spawner) for three herring stocks in the Norwegian Sea, and (below) the temperature in the Norwegian Sea. The Icelandic spring-spawning stock never recovered from its collapse in the early 1970s, preceded by low recruitment in the 1960s. (After Beverton, 1993.)

The general structure of a dynamic pool model is illustrated in Figure 16.18. Based on this ground plan, a wide spectrum of possible models has been proposed, varying in particular in the manner and detail in which the four fundamental components of the model are described (Beverton & Holt, 1957). These components or 'submodels' are: (i) the recruitment rate; (ii) the growth rate; (iii) the natural mortality rate; and (iv) the fishing rate of the exploited stock. They combine to determine the exploitable biomass of the stock and the way this translates into a yield to the fishing community. In contrast to the surplus yield models, this biomass yield depends not only on the number of individuals caught but also on their size (past

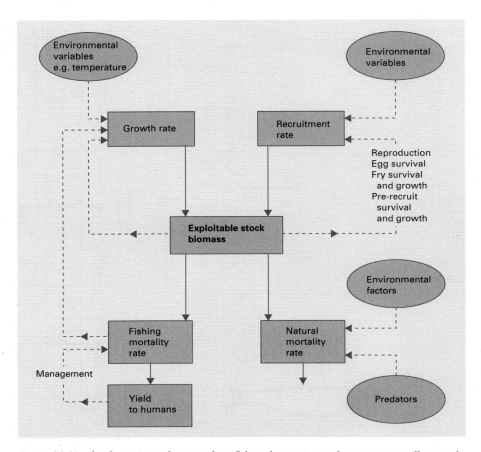

Figure 16.18 The dynamic pool approach to fishery harvesting and management, illustrated as a flow diagram. There are four main 'submodels': the growth rate of individuals and the recruitment rate into the population (which add to the exploitable biomass), and the natural mortality rate and the fishing mortality rate (which deplete the exploitable biomass). Solid lines and arrows refer to changes in biomass under the influence of these submodels. Dashed lines and arrows refer to influences either of one submodel on another, or of the level of biomass on a submodel or of environmental factors on a submodel. Each of the submodels can itself be broken down into more complex and realistic systems. Yield to humans is estimated under various regimes characterized by particular values inserted into the submodels. These values may be derived theoretically (in which case they are 'assumptions') or from field data. (After Pitcher & Hart, 1982.)

growth); whilst the quantity of exploitable (i.e. catchable) biomass depends not just on 'net recruitment' but on an explicit combination of natural mortality, harvesting mortality, individual growth and recruitment into the catchable age classes.

To generate the different variants on the ground plan, growth, natural mortality, susceptibility to capture, etc., can be dealt with separately in each of the age classes. For example, recruitment into the catchable part of the population does not need to be sudden: there can be non-catchable age classes, gradations of poorly catchable age classes, and so on. Furthermore, the four 'submodels' may, in different cases, incorporate as much or as little information as is available or desirable. Thus, growth curves may be obtained by deriving models from first principles or directly from field samples (e.g. ageing fish by examining their scales or their otoliths); and density dependence in either recruitment or natural mortality may be assumed and modelled or (better still) actually monitored in the field (Beverton & Holt, 1957).

In all cases, though, the basic approach is the same. Available information (both theoretical and empirical) is incorporated into a form that reflects the dynamics of the structured population. This then allows the yield and the response of the population to different harvesting strategies to be estimated. This in turn should allow a recommendation to the stock-manager to be formulated. The crucial point is that in the case of the dynamic pool approach, a harvesting strategy can include not only a harvesting intensity, but also a decision as to how effort should be partitioned amongst the various age classes.

dynamic pool models can lead to valuable recommendations ...

One commercially exploited population for which a dynamic pool model has been produced is the Arcto-Norwegian cod fishery, the most northerly of the Atlantic stocks (Garrod & Jones, 1974; see also Pitcher & Hart, 1982). Garrod and Jones took the age-class structure of the late 1960s, and used this to predict the medium-term effects on yield of different fishing intensities and different mesh sizes in the trawl. Some of their results are shown in Figure 16.19. The temporary peak after 5 or so years is a result of the very large 1969 year-class working through the population. Overall, however, it is clear that the best longer term prospects are predicted with a low fishing intensity and a large mesh size. Both of these give the fish more opportunity to grow (and reproduce) before they are caught, which is important because yield is measured in biomass, not simply in numbers. Higher fishing

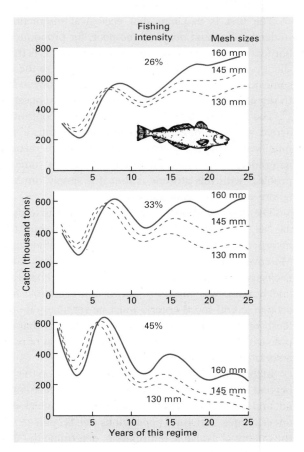

Figure 16.19 Garrod and Jones' (1974) predictions for the Arctic cod stock under three fishing intensities and with three different mesh sizes. (After Pitcher & Hart, 1982.)

663   MANIPULATING ABUNDANCE

intensities and mesh sizes of 130 mm were predicted to lead to overexploitation of the stock.

Sadly, Garrod and Jones' clear recommendations were ignored by those with the power to determine fishing strategies. Mesh sizes were not increased until 1979, and then only from 120 to 125 mm. Fishing intensity never dropped below 45% and catches of 900 000 tonnes were taken in the late 1970s. Not surprisingly perhaps, surveys late in 1980 showed that these and other North Atlantic cod stocks were very seriously depleted as a result of overfishing. Garrod and Jones' work provides encouragement for the use of dynamic pool models. But the response to their work has been somewhat less encouraging.

... but these may still be ignored

16.15 The objectives for managing harvestable resources

If we treat this last example as typical, then we might conclude that the biologist proposes—but the manager disposes. This is therefore an appropriate point at which to reconsider not only the objectives of harvesting programmes, but also the criteria by which successful management should be judged and the role of ecologists in management overall. As Hilborn and Walters (1992) have pointed out there are three alternative attitudes that ecologists can take, each of which has been popular but only one of which is sensible. Indeed, these are increasingly important considerations that apply not just to fisheries management but to every entry of ecologists into the public arena.

three attitudes for ecologists towards managers in the real world ...

The first is to claim that ecological interactions are too complex, and our understanding and our data too poor, for pronouncements of any kind to be made (for fear of being wrong). The problem with this is that if ecologists choose to remain silent because of some heightened sensitivity to the difficulties, there will always be some other, probably less qualified 'expert' ready to step in with straightforward, not to say glib, answers to probably inappropriate questions.

The second possibility is for ecologists to concentrate exclusively on ecology and arrive at a recommendation designed to satisfy purely ecological criteria. Any modification by managers or politicians of this recommendation is then ascribed to ignorance, inhumanity, political corruption or some other sin or human foible. The problem with this attitude is that it is simply unrealistic in any human activity to ignore social and economic factors.

The third alternative, then, is for ecologists to make ecological assessments that are as accurate and realistic as possible, but to assume that these will be incorporated with a broader range of factors when management decisions are made. Moreover, these assessments should themselves take account of the fact that the ecological interactions they address include humans as one of the interacting species, and humans are subject to social and economic forces. Finally, since ecological, economic and social criteria must be set alongside one another, choosing a single, 'best' option is a matter of opinion. It follows that a single recommendation is, in practice, far less useful than laying out a series of possible plans of action with their associated consequences.

only one of them sensible

In the present context, therefore, we develop this third alternative by first looking beyond MSY to criteria that incorporate risk, economics, social consequences, and so on, and then briefly examining the means by which crucial parameters and variables are estimated in natural populations (see Section 16.16), since these, by determining the quality of available information, determine the degree of confidence with which recommendations can be made. In doing this we have drawn heavily on the much more detailed account in Hilborn and Walters (1992) (see also Clark, 1985). Partly as a result of this, but partly as a shorthand, we refer frequently to 'fish' and 'fisheries'—but, the principles apply to any natural resource. Aspects have been selected with a view to conveying the most important messages—not because they necessarily represent the most important topics in the management of natural resources, nor those closest to the state-of-the-art for practising managers.

16.15.1 Economic and social factors

Perhaps the most obvious shortcoming of a purely ecological approach is its failure to recognize that the exploitation of a natural resource is usually a business enterprise, in which the value of the harvest must be set against the costs of obtaining that harvest. Even if we distance ourselves from any preoccupation with 'profit', it makes no sense to struggle to obtain the last few tonnes of an MSY if the money spent in doing so could be much more effectively invested in some other means of food production. The basic idea is illustrated in Figure 16.20. We seek to maximize not total yield but net value—the difference between the gross value of the harvest and the sum of the fixed costs (interest payments on ships or factories, insurance, etc.) and the variable costs, which increase with harvesting effort (fuel, crew's expenses, etc.). This immediately suggests that the economically optimum yield (EOY) is less than the MSY, and is obtained through a smaller effort or quota. However, the difference between the EOY and the MSY is least in enterprises where most costs are fixed ('total-cost' line virtually flat). This is especially the case in high-investment, highly technological operations such as deep-sea fisheries, which are therefore most prone to overfishing even with management aimed at economic optima.

the economically optimum yield (EOY)—typically less than the MSY

A second important economic consideration concerns 'discounting'. This refers to the fact that in economic terms, each bird in the hand now (or each fish in the hold) is worth more than an equivalent bird or fish some time in the future. The reason is basically that the value of the current catch can be placed in the bank to accrue interest, so that its total value increases. In fact, a commonly used discount rate for natural resources is 10% per annum (90 fish now are as valuable as 100 fish in 1 year's time) despite the fact that the difference between the interest rates in the banks and the rate of inflation is usually only 2–5%. The economists' justification for this is a desire to incorporate 'risk'. A fish caught now has already been caught; one

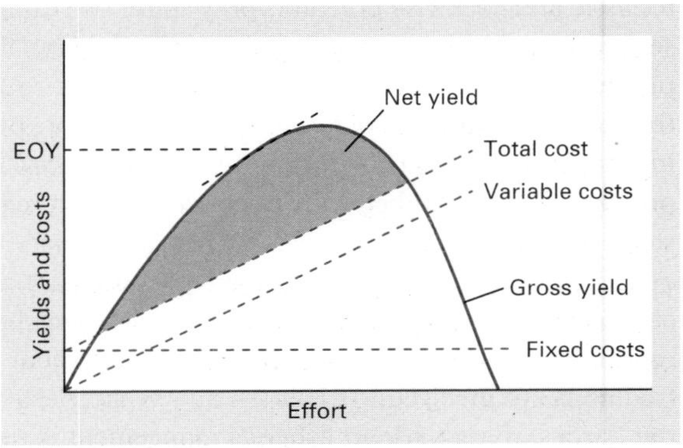

discounting—liquidating stocks, or leaving them to grow?

Figure 16.20 The economically optimum yield (EOY), that which maximizes 'profit', is obtained to the left of the peak of the yield-against-effort curve, where the difference between gross yield and total cost (fixed costs plus variable costs) is greatest. At this point, the gross yield and total cost lines have the same slope. (After Hilborn & Walters, 1992.)

still in the water might or might not be caught—a bird in the hand really is worth two in the bush.

On the other hand, the caught fish is dead, whereas the fish still in the water can grow and breed (although it may also die). In a very real sense, therefore, each uncaught fish will be worth more than 'one fish' in the future. In particular, if the stock left in the water grows faster than the discount rate, as is commonly the case, then a fish put on deposit in the bank is not so sound an investment as a fish left on deposit in the sea. Nevertheless, even in cases like this, discounting provides an economic argument for taking larger harvests from a stock than would otherwise be desirable.

Moreover, in cases where the stock is less productive than the discount rate—for example, many whales and a number of long-lived fish—it seems to make sense, in purely economic terms, not only to overfish the stock, but actually to catch every fish ('liquidate the stock'). The reasons for not doing so are partly ethical—it would clearly be ecologically short-sighted and a distainful (and distasteful) way of treating the hungry mouths to be fed in the future. But there are also practical reasons: jobs must be found for those previously employed in the fishery (or their families otherwise provided for), alternative sources of food must be found, and so on. This emphasizes, first, that a 'new economics' must be forged in which value is assigned not only to things that can be bought and sold—like fish and boats—but also to more abstract entities, like the continued existence of a species of whale (Pearce & Turner, 1990). It also stresses the danger of an economic perspective that is too narrowly focused. The profitability of a fishery cannot sensibly be isolated from the implications that the management of that fishery has in a wider sphere.

social repercussions

'Social' factors enter in two rather separate ways into plans for the management of natural resources. First, allied to the previous discussion of why it may be inappropriate to liquidate a stock even when it is apparently economically sensible to do so, the forces of both practical politics and sheer human decency may dictate, for instance, that a large fleet of small, individually inefficient boats is maintained in an area where there are no alternative means of employment. In addition, though, and of much more widespread importance, it is necessary for management plans to take full account of the way fishermen and harvesters will behave and respond to changing circumstances, rather than assuming that they will simply conform to the requirements for achieving either ecological or economic optima. Harvesting involves a predator–prey interaction: it makes no sense to base plans on the dynamics of the prey alone whilst simply ignoring those of the predator (us!).

the harvester as predator: human behaviour

The idea of the harvester as predator is reinforced in Figure 16.21, which shows a classic anticlockwise predator–prey spiral (see Chapter 10) for the North Pacific fur seal fishery in the last years of the 19th century. The figure therefore illustrates a numerical response on the part of the predator—extra vessels enter the fleet when the stock is abundant, but leave when it is poor. But the figure also illustrates the inevitable time lag in this response. Thus, whatever a modeller or manager might propose, there is unlikely ever to be some perfect match, at an equilibrium, between stock size and effort. Moreover, whilst the sealers in the figure left the fishery as quickly as they had entered it, this is by no means a general rule. The sealers were able to switch to fishing for halibut, but such switches are often not easy to achieve, especially where there has been heavy investment in equipment or long-standing traditions are involved. As Hilborn and Walters (1992) dramatically state 'Principle: the hardest thing to do in fisheries mangement is reduce fishing pressure'.

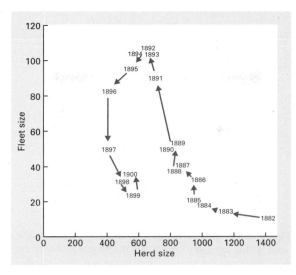

Figure 16.21 The fleet size of the North Pacific fur seal fishery (predators) responded to the size of the seal herd (prey) between 1882 and 1900 by exhibiting an anticlockwise predator–prey spiral. (After Hilborn & Walters, 1992; from data of Wilen, 1976, unpublished observations.)

Switching is one aspect of a harvester's predatory behaviour—its functional response (see Chapter 9). Harvesters will also generally 'learn': there is an inevitable trend towards technological improvement, and even without this, harvesters usually improve their efficiency the more they learn about their stock—notwithstanding the assumptions of simple fixed-effort models (see Section 16.14.2).

16.15.2 'Biological' criteria beyond MSY

$F_{0.1}$: avoiding risk in yield-per-recruit harvesting

Even when economic and social considerations are set to one side, MSY is not the only ecological management criterion of importance. Much attention has also been focused not on total-yield relationships (the basis of MSY) but on yield-per-recruit patterns. This may be especially relevant where the age structure of a population is being taken into account, since the age class of a 'recruit' is likely to be different from that from which 'yield' is calculated. The general relationship between yield per recruit and fishing mortality rate (proportion caught) or fishing effort is illustrated in Figure 16.22. Not surprisingly, yield per recruit is zero when there is no fishing, rises to a peak at some intermediate mortality rate but then declines again when the stock is overfished. The equivalent of an MSY would be to aim for a fishing mortality rate, F_{max}, which maximized the yield per recruit. However, most recent interest has centred on achieving an $F_{0.1}$ (Figure 16.22): the fishing mortality rate for which the slope of the yield per recruit against mortality rate curve is 0.1 times that at the origin. Clearly, $F_{0.1}$ is less than F_{max} and gives rise to a lower yield per recruit. The idea behind its use is simple: there is less risk of overfishing, and perhaps more chance of achieving an economic optimum, than in aiming for a maximum rate. Indeed, an analysis of three fish stocks, Pacific halibut (*Hippoglossus stenolepis*), western Lake Erie walleye (*Stizostenodon vitreum vitreum*) and Bering Sea Pacific cod (*Gaddus macrocephalus*), demonstrated that whereas an $F_{0.1}$ would remove 0.16, 0.29 and 0.46 of the total stocks, respectively, an MSY would remove 0.22, 0.42 and 0.69—in each case, around 50% as much again (Deriso, 1987). However, the choice of '0.1', as opposed to any other value, appears, despite its widespread acceptance, to be entirely arbitrary.

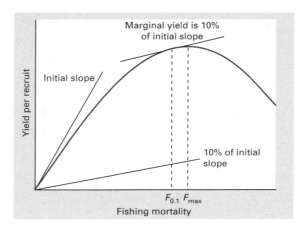

Figure 16.22 $F_{0.1}$ harvesting. Yield per recruit generally exhibits a dome-shaped relationship with fishing mortality rate. $F_{0.1}$ is the fishing mortality rate for which the slope of the yield per recruit against mortality rate curve is 0.1 or 10% that at the origin. This mortality rate must lie to the left of the peak of the curve and thus gives rise to a less than maximum yield per recruit (obtained at F_{max}). (After Deriso, 1987; Hilborn & Walters, 1992.)

Of course, similar risk-reducing modifications of the MSY criterion can, and have, been used. For instance, in 1975 the International Whaling Commission instituted a management procedure, basing the regulation of each stock on the current size of each stock relative to the MSY level, and then fixing quotas as a proportion of the MSY.

Pursuing further this theme of minimizing risk, an MSY, or a maximum (or any other) yield per recruit, is an average value to be expected over the course of many years. A number of studies have shown, however, that as average yield increases as a result of increased fishing intensity, so too does the variability in yield: there is a trade-off between size and predictability (see, for example, Beddington & May, 1977). Unpredictability creates immense difficulties for resource management: the sizes of the workforce and the harvesting and processing machinery are difficult to plan and each is liable to frequent bouts of both overstretching and underemployment. In short, 'maximization' is not an inevitable guiding principle—'risk minimization' may be just as important, not only ecologically (because of the risks of destroying stocks) but also socially and economically, because of the difficulties of managing investment and employment.

the trade-off between size and predictability of harvest

16.16 Estimates from data: putting management into practice

The role of the ecologist in the management of a natural resource is in *stock assessment*: making quantitative predictions about the response of the biological population to alternative management choices—addressing questions like whether a given fishing intensity will lead to a decline in the size of the stock, whether nets of a given mesh size will allow the recruitment rate of a stock to recover, and so on. In the past, it has often been assumed that this can be done simply by careful monitoring. For example, as effort and yield increase in an expanding fishery, both are monitored, and the relationship between the two is plotted until it seems that the top of a curve like that in Figure 16.14b has been reached or just exceeded,

monitoring effort and yield—
the difficulties of 'finding the
top'

identifying the MSY. This approach, however, is deeply flawed, as can be seen from Figure 16.23. In 1975, the International Commission for the Conservation of Atlantic Tunas (ICCAT) used the available data (1964–1973) to plot the yield–effort relationship for the yellowfin tuna (*Thunnus albacares*) in the eastern Atlantic. They felt that they had reached the top of the curve: a sustainable yield of around 50 000 tons (5.1×10^7 kg) and an optimum effort of about 60 000 fishing days. However, ICCAT were unable to prevent effort (and yield) rising further, and it soon became clear that the top of the curve had not been reached. A reanalysis using data up to 1983 suggested a sustainable yield of around 110 000 tons (1.1×10^8 kg) and an effort of 240 000 fishing days. Of course, the 1975 experience gives no grounds for confidence in even these estimates. Effort may need to be increased by another 20 or 30% (giving rise to a significant drop in yield) to be sure that this time the top has been reached.

This illustrates what Hilborn and Walters (1992) describe as another principle 'You cannot determine the potential yield from a fish stock without overexploiting it'. At least part of the reason for this is the tendency, already noted, for the variability in yield to increase as a MSY is approached. Furthermore, if we also recall the previously described difficulty in reducing fishing pressure, it is clear that in practice, management decisions are likely to have to wrestle with the combined challenge of estimation difficulties, ecological relationships (here, between yield and predictability) and socio-economic factors (here, concerning the regulation and reduction in effort). We are a long way from the simple fixed-effort models of Section 16.13.2.

The practical difficulties of parameter estimation are further illustrated in Figure 16.24, which displays the time series for total catch, fishing effort and catch per unit effort (CPUE) between 1969 and 1982 for yellowfin tuna for the whole Atlantic Ocean. As effort increased, CPUE declined—presumably, a reflection of a diminishing stock of fish. On the other hand, the catch continued to rise over this period, suggesting that perhaps the stock was not yet being overfished (i.e. the MSY had not yet been reached). These, then, are the data, and they come in probably the most commonly available form—a so-called 'one-way trip' time series. But can they suggest a MSY? Can they suggest the effort required to achieve that MSY? Certainly, methods exist for performing the necessary calculations, but these methods require assumptions to be made about the underlying dynamics of the population.

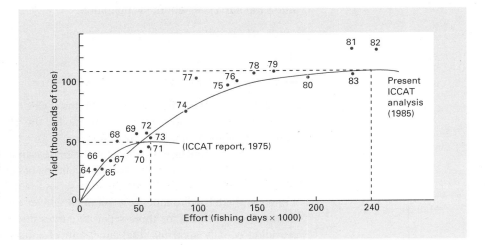

Figure 16.23 Estimated yield–effort relationships for the eastern Atlantic yellowfin tuna (*Thunnus albacares*) on the basis of the data for 1964–1973 (ICCAT, 1975) and the data for 1964–1983 (ICCAT, 1985). (After Hunter *et al.*, 1986; Hilborn & Walters, 1992.)

669   MANIPULATING ABUNDANCE

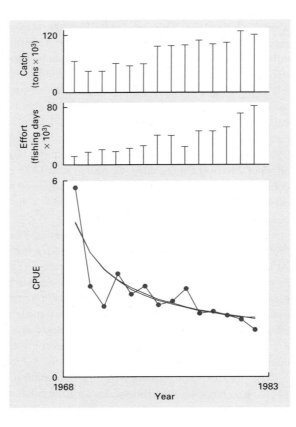

Figure 16.24 Changes in total catch, fishing effort and catch per unit effort (CPUE) between 1969 and 1982 for the yellowfin tuna (*Thunnus albacares*) in the Atlantic Ocean. Also shown are three separate curves fitted to the CPUE time series, by methods outlined in the text, the parameters of which are given in Table 16.7. (After Hilborn & Walters, 1992.)

The most frequently used assumption describes the dynamics of the stock biomass, *B*, by:

estimates from catch and effort data—applying the Schaefer model

$$\frac{dB}{dt} = rB\left(1 - \frac{B}{K}\right) - H \tag{16.2}$$

(Schaefer, 1954), which is simply the logistic equation of Chapter 6 (intrinsic rate of increase, *r*, carrying capacity, *K*) with a harvesting rate incorporated, which may itself be given, following Equation 16.1, by $H = qEB$, where *q* is harvesting efficiency and *E* the harvesting effort. By definition:

$$CPUE = H / E = qB. \tag{16.3}$$

Hence:

$$B = CPUE / q \tag{16.4}$$

and Equation 16.2 can be rewritten in terms of CPUE and either *H* or *E* as variables, with *r*, *q* and *K* as parameters. For this model, the MSY is given by $rK / 4$ and the effort required to achieve this by $r / 2q$.

There are a number of methods of obtaining estimates of these parameters from field data, perhaps the best of which is the fitting of curves to time series (Hilborn & Walters, 1992). However, when the time series is a one-way trip, as we have noted it often is, there is no unique 'best' set of parameter values. Table 16.7, for instance, shows the parameters for three separate curves fitted to the data in Figure 16.24, providing equally good fits (the same sum of squares), but with widely differing

even time-series analysis can provide equivocal answers ...

Table 16.7 Parameter estimates from three fits to the catch per unit effort (CPUE) time series for yellowfin tuna shown in Figure 16.23. r is the intrinsic rate of increase, K is the carrying capacity (equilibrium abundance in the absence of harvesting) and q is the harvesting efficiency. Effort is measured in fishing days; K and maximum sustainable yield (MSY) in tons. (After Hilborn & Walters, 1992.)

Fit number	r	K ($\times 1000$)	q ($\times 10^{-7}$)	MSY ($\times 1000$)	Effort at MSY ($\times 1000$)	Sum of squares
1	0.18	2103	9.8	98	92	3.8
2	0.15	4000	4.5	148	167	3.8
3	0.13	8000	2.1	261	310	3.8

parameter values. There are, in effect, a large number of equally good alternative explanations for the data in Figure 16.24, in some of which, for example, the population has a low carrying capacity but a high intrinsic rate of increase and is being harvested efficiently, whereas in others it has a high carrying capacity, a low rate of increase and is being harvested less efficiently. In the first case, the MSY had probably already been reached in 1980; in the second, catches could probably be doubled with impunity. Moreover, in each of these cases, the population is assumed to be behaving in conformity with Equation 16.2, which may itself be wide of the mark.

It is clear, therefore, even from this limited range of rather arbitrarily chosen examples, that there are immense limitations placed on stock assessments and management plans by inadequacies in both the available data and the means of analysing them. This is not meant, though, nor should it be taken as, a council of despair. Management decisions must be made; and the best possible stock assessments must form the basis—although not the sole basis—for these decisions. It is regrettable that we do not know more, but the problem would be compounded if we pretended that we did. Moreover, the ecological, economic and human behavioural analyses are important—as all analyses are—for identifying what we do *not* know, since, armed with this knowledge, we can set about obtaining whatever information is most useful. This has been formalized, in fact, in an 'adaptive management' approach, where, in an 'actively adaptive' strategy, a policy is sought which offers some balance between, on the one hand, probing for information (directed experimentation) and on the other, exercising caution about losses in short-term yield and long-term overfishing (Hilborn & Walters, 1992). Indeed, there is a strong argument that says that the inadequacies in data and theory make the need for ecologists all the more profound: who else can appreciate the uncertainties and provide appropriately enlightened interpretations?

... but difficulties and uncertainties make ecologists all the more invaluable

16.17 Conclusion: sustainability

We have dealt in this chapter, as separate topics, with the control of pests and the exploitation by harvesting of natural populations. We conclude by reiterating the interrelationships between these subjects and conservation, and the relationships between them and the notion of sustainability. One of the twin aims in harvesting is to avoid levels of overexploitation that would drive a natural population towards extinction. This is an aim that has not always been satisfied. For example, the blue whale (*Balaenoptera musculus*), by virtue of repeated overexploitation, has become an object of concern for all those with an interest in conservation; whereas for the passenger pigeon, repeated overexploitation in the last century has taken them beyond the interests of conservationists and into the biological history books.

Pest control could also be described as having twin aims: controlling the pest and minimizing the disturbance to other species in the community. Many of these disturbances pose problems in conservation. For instance, resurgences of target pests and outbreaks of secondary pests occur because natural enemies are destroyed. These enemies may not be aesthetically attractive to conservationists, and their disappearance may only be local, yet this need not disqualify them from conservationist interest. More obviously, when pesticides are liable to biomagnification, the worst affected species are top predators, usually vertebrates. Here, disappearance or decline on even a local scale is certain to excite conservationist interest. Even when there are plans to release natural pathogens as means of biological control, one of the major concerns is always that these will be pathogenic, too, to species related to the target.

Sustainability has rightly become one of the core concepts—perhaps *the* core concept—in an ever-broadening concern for the fate of the earth and the ecological communities that occupy it. For example, in 1991 the Ecological Society of America published 'The sustainable biosphere initiative: an ecological research agenda' (Lubchenko *et al.*, 1991), 'a call-to-arms for all ecologists', whilst The World Conservation Union, the United Nations Environment Programme and the World Wide Fund for Nature jointly published *Caring for the Earth. A Strategy for Sustainable Living* (IUCN/UNEP/WWF, 1991). The detailed contents and proposals of these documents are perhaps less important than their existence, indicating as they do a growing preoccupation with sustainability, shared by scientists and pressure groups, and a recognition that things cannot go on the way they are.

Both of these documents define sustainability. The former uses it '… to imply management practices that will not degrade the exploited systems or any adjacent systems.' The latter goes further:

> … if an activity is sustainable, for all practical purposes it can continue for ever. When people define an activity as sustainable, however, it is on the basis of what they know at the time. There can be no long-term guarantee of sustainability, because many factors remain unknown or unpredictable.

The document goes on to note that:

> … confusion has been caused because 'sustainable development', 'sustainable growth' and 'sustainable use' have been used interchangeably, as if their meanings were the same. They are not. 'Sustainable growth' is a contradiction in terms: nothing physical can grow indefinitely. 'Sustainable use' is applicable only to renewable resources: it means using them at rates within their capacity for renewal. 'Sustainable development' is used to mean: improving the quality of human life while living within the carrying capacity of supporting ecosystems. A 'sustainable economy' is the product of sustainable development. It maintains its natural resource base. It can continue to develop by adapting, and through improvements in knowledge, organization, technical efficiency, and wisdom.

The harvesting of natural resources fits most readily under 'sustainable use', although we have seen: (i) that conventional economics does not provide unquestioning support for sustainability; (ii) that sustainable use may be difficult to justify if it leads to measurable and significant human suffering at present (starvation, unemployment); but (iii) that the future must lie with sustainable development, fuelled by improvements in ecological knowledge and understanding and also by an improved ability to integrate this understanding into a wider human social and economic context. Sustainability in pest control, and in agriculture generally, refers first to a diminished reliance on non-renewable resources which therefore cannot be used sustainably (chemical pesticides and fertilizers), but also to the nurturing of resources that have the potential to be renewable, like the natural enemies of pests, or which will deplete only very slowly so long as unnecessary wastage is minimized, like the organic content of soils.

Claiming that an activity is sustainable requires us to predict the future. Reliable prediction of the future requires an especially profound understanding of the past and present. No qualifications are required in order to argue in favour of sustainability. But achieving sustainability will require the advances made by ecologists in years to come.

PART 4
COMMUNITIES

Introduction

In nature, areas of land and volumes of water contain assemblages of different species, in different proportions and doing different things. These communities of organisms have properties that are the sum of the properties of the individual denizens plus their interactions. It is the interactions that make the community more than the sum of its parts. Just as it is a reasonable aim for a physiologist to study the behaviour of different sorts of cells and tissues and then attempt to use a knowledge of their interactions to explain the behaviour of the whole organism, so an ecologist may use his or her knowledge of interactions between organisms in an attempt to explain the behaviour and structure of a whole community. Community ecology is the study of patterns in the structure and behaviour of multispecies biological assemblages.

We consider first the nature of the community—what we mean by the term. Usually we mean some unit of the natural world that we (human investigators) can categorize according to features that mean something to us. We impose an anthropogenic process of selection in deciding what will be regarded as a community. It is vitally important to be aware that the categories erected may lack any relevance to the lives of individual organisms within the communities. An oakwood or an estuary or the rumen of a cow may be recognized as communities by us, but they may be on a quite irrelevant scale in the life of a caterpillar, a shrimp or a rumen protozoan. To a caterpillar in an oakwood, the only community that matters in its life may consist of a few leaves on a branch, together with a handful of competitors and predators that visit its cluster of leaves. The organism's-eye view of the community in which it lives will differ from species to species. In Chapter 17 we impose our human perspective and describe patterns in the composition of communities in space and time.

In Chapters 18 and 19 we examine the ways in which arrays of feeders and their food may bind the inhabitants of a community into a web of interacting elements, through which energy and matter is moved. We explore the ways that human activities are upsetting energy and nutrient pathways, paying particular attention to the predicted impacts of global climate change.

Chapters 20 and 21 return to the topics of earlier chapters in the book and examine the extent to which competition, predation and disturbance influence the patterns we recognize at the community level. In Chapter 22 we integrate the wide variety of population interactions that occur in communities in our consideration of the structure of whole food webs and, in particular, those aspects of structure that contribute to stability.

Chapter 23 is concerned with an area of ecology that has been extraordinarily fruitful in producing new ecological insights—islands and island communities—while in Chapter 24 we attempt an overall synthesis of the factors that determine species richness. Finally, in Chapter 25, we address the question of how the richness of the world's biota can be maintained in the face of a variety of human influences that increase the risk of extinctions of species. To deal with this question we need to integrate information from every section of this text, drawing on knowledge and principles at the individual, population and community levels of ecological organization.

To pursue an analogy we introduced earlier, the study of ecology at the community level is a little like making a study of watches and clocks. A collection can be made and the contents of each timepiece classified. We can recognize characteristics that they have in common in the way they are constructed and what they do. We can recognize patterns and hierarchy in the collection as a whole. But to understand how they work, they must be taken to pieces, studied and put back together again. We will have understood the nature of natural communities when we have taken them to pieces and *know* how to recreate them.

Chapter 17
The Nature of the Community

17.1 Introduction

Physiological and behavioural ecologists are concerned primarily with individual *organisms*. Coexisting individuals of a single species possess characteristics such as density, sex ratio, age-class structure, rates of natality and immigration, mortality and emigration, that are unique to *populations*. We explain the behaviour of a population in terms of the behaviour of the individuals that comprise it. Finally, activities at the population level have consequences for the next level up—that of the *community*. The community is an assemblage of species populations that occur together in space and time. The principal focus of the community ecologist is the manner in which groupings of species are distributed in nature, the ways these groupings can be influenced by interactions between species and by the physical forces of their environment and the properties of these groupings.

A community is composed of individuals and populations, and we can identify and study straightforward *collective* properties, such as species diversity, community biomass and productivity. However, we have already seen that organisms of the same and different species interact with each other in processes of mutualism, parasitism, predation and competition. The nature of the community is obviously more than just the sum of its constituent species. It is their sum plus the interactions between them. Thus, there are *emergent* properties that appear when the community is the focus of attention, as there are in many other cases in which we are concerned with the behaviour of complex mixtures. For example, a cake has emergent properties of texture and flavour that are not apparent simply from a survey of the ingredients. In the case of ecological communities, the limits to similarity of competing species and the stability of the food web in the face of disturbance are examples of emergent properties. A primary aim of community ecology is to determine whether repeating patterns in collective and emergent properties exist, even when there are great differences in the particular species that happen to be assembled together.

Traditionally, another category of ecological study has been set apart: the *ecosystem* (first defined by Tansley, 1935). This comprises the biological community together with its physical environment. However, whilst the distinction between community and ecosystem may be helpful in some ways, the implication that communities and ecosystems can be studied as separate entities is wrong. No ecological system, whether individual, population or community, can be studied in isolation from the environment in which it exists. Thus, we will not distinguish a separate ecosystem level of organization. Nor will discussions of ecological energetics

communities have collective properties ...

... and emergent properties not possessed by the individual populations that comprise them

the relationship between community ecology and ecosystem ecology

and nutrient dynamics be segregated as ecosystem rather than community topics (as has usually been the practice in ecology textbooks). It is true that these phenomena depend explicitly on fluxes between living and non-living components of ecosystems, but the fundamental point is that they represent complementary means of approaching an understanding of community structure and functioning. This is not to say that the tasks set by ecosystem ecologists are unimportant. On the contrary, knowledge of the role that communities play in biogeochemical cycling is essential if we are to understand and combat the effects of acid rain, or increasing levels of atmospheric carbon dioxide or radioactivity on forests, streams, lakes and other communities (see Chapter 19).

Doing science at the community level presents daunting problems because the database may be enormous and complex. A first step is usually to search for patterns in community structure and composition. The need to develop procedures for describing and comparing communities has dominated the development of community ecology. In essence, this has been a search for simple ways to describe complex systems.

the search for pattern—a fundamental objective of any science

The recognition of patterns represents an important step in the development of all sciences (the periodic table in chemistry, movements of the heavenly bodies in astronomy, and so on). Patterns are repeated consistencies, such as the repeated grouping of the same species in different places (or the same growth forms, the same productivities, the same rates of nutrient turnover, etc.). Recognition of patterns leads in turn to the forming of hypotheses about the causes of these patterns. The hypotheses may then be tested by making further observations or by doing experiments.

A community can be defined at any size, scale or level within a hierarchy of habitats. At one extreme, broad patterns in the distribution of community types can be recognized on a global scale. The temperate forest biome is one example; its range in North America is shown in Figure 17.1. At this scale, ecologists usually recognize climate as the overwhelming factor that determines the limits of vegetation types. At a finer scale, the temperate forest biome in parts of New Jersey is represented by communities of two species of tree in particular, beech and maple, together with a very large number of other, less conspicuous species of plants, animals and microorganisms. Study of the community may be focused at this level. On an even finer habitat scale, the characteristic invertebrate community that inhabits water-filled holes in beech trees may be studied, or the flora and fauna in the gut of a deer in the forest. Amongst these various levels of community study, no one is more legitimate than another. The level appropriate for investigation depends on the sorts of questions that are being asked.

communities can be recognized at a variety of levels—all equally legitimate

Community ecologists sometimes consider all of the organisms existing together in one area, although it is rarely possible to do this without a large team of taxonomists. Others restrict their attention within the community to a single taxonomic group (e.g. birds, insects or trees), or a group with a particular activity (e.g. herbivores, detritivores). Thus, we may refer to the bird community of a forest or the detritivore community of a stream.

The rest of this chapter is in three sections. We start by explaining how the structure of communities can be measured and described (Section 17.2). Then we focus on patterns in community structure, in space (Section 17.3) and in time (Section 17.4).

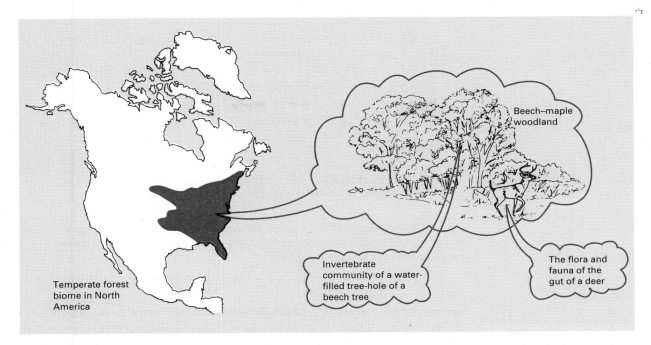

Figure 17.1 We can identify a hierarchy of habitats, nesting one into the other: temperate forest biome in North America; beech–maple woodland in New Jersey; water-filled tree-hole; or mammalian gut. The ecologist may choose to study the community that exists on any of these scales.

17.2 The description of community composition

species richness: the number of species present in a community

One way to characterize a community is simply to count or list the species that are present. This sounds a straightforward procedure that enables us to describe and compare communities by their species richness. In practice, though, it is often surprisingly difficult, partly because of taxonomic problems, but also because only a subsample of the organisms in an area can usually be counted. The number of species recorded then depends on the number of samples that have been taken, or on the volume of the habitat that has been explored. The most common species are likely to be represented in the first few samples, and as more samples are taken, rarer species will be added to the list. At what point does one cease to take further samples? Ideally, the investigator should continue to sample until the number of species reaches a plateau (Figure 17.2). At the very least, the species richnesses of different communities should be compared only if they are based on the same sample sizes (in terms of area of habitat explored, time devoted to sampling or, best of all, number of individuals or modules included in the samples). The analysis of species richness in contrasting situations figures prominently in Chapter 23 'Islands, areas and colonization' and Chapter 24 'Patterns in species richness' (see, for example, Chapter 23, Figure 23.1 and Chapter 24, Figure 24.3).

17.2.1 Diversity indices

An important aspect of the numerical structure of communities is completely ignored when the composition of the community is described simply in terms of the

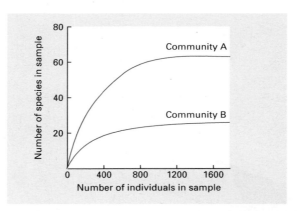

Figure 17.2 Relationship between species richness and number of individual organisms from two contrasting hypothetical communities. Community A has a total species richness considerably in excess of community B.

diversity incorporates richness, commonness and rarity

number of species present. It misses the information that some species are rare and others common. Intuitively, a community of 10 species with equal numbers in each seems more diverse than another, again consisting of 10 species, with more than 50% of the individuals belonging to the most common species and less than 5% in each of the other nine. Yet, each community has the same species richness.

If the community we are interested in is closely defined (e.g. the warbler community of a woodland), counts of the number of individuals in each species may suffice for many purposes. However, if we are interested in all the animals in the woodland, it makes very little sense to use the same sort of quantification for protozoa, woodlice, birds and deer. Their enormous disparity in size means that counts would be very misleading. We are also in great difficulty if we try to count plants (and other modular organisms). Do we count the number of shoots, leaves, stems, ramets or genets? One way round this problem is to describe the community in terms of the biomass (or the rate of production of biomass) per species per unit area.

Simpson's diversity index

The simplest measure of the character of a community that takes into account both the abundance (or biomass) patterns and the species richness, is Simpson's diversity index. This is calculated by determining, for each species, the proportion of individuals or biomass that it contributes to the total in the sample, i.e. the proportion is P_i for the ith species:

$$\text{Simpson's index, } D = \frac{1}{\sum\limits_{i=1}^{S} P_i^2} \tag{17.1}$$

where S is the total number of species in the community (i.e. the richness). As required, the value of the index depends on both the species richness and the evenness (equitability) with which individuals are distributed amongst the species. Thus, for a given richness, D increases with equitability, and for a given equitability, D increases with richness.

'equitability' or 'evenness'

Equitability can itself be quantified by expressing Simpson's index, D, as a proportion of the maximum possible value D would assume if individuals were

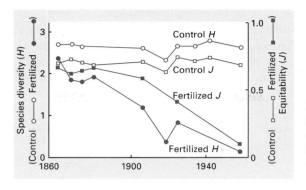

Figure 17.3 Species diversity (*H*) and equitability (*J*) of a control plot and a fertilized plot in the Rothamstead 'Parkgrass' experiment. (After Tilman, 1982.)

completely evenly distributed amongst the species. In fact, $D_{\max} = S$. Thus:

$$\text{equitability, } E = \frac{D}{D_{\max}} = \frac{1}{\sum_{i=1}^{S} P_i^{\ 2}} \times \frac{1}{S}.$$ (17.2)

Equitability assumes a value between 0 and 1.

the Shannon diversity index   Another index that is frequently used is the Shannon diversity index, *H*. This again depends on an array of P_i values. Thus:

$$\text{diversity, } H = -\sum_{i=1}^{S} P_i \ln P_i$$ (17.3)

and:

$$\text{equitability, } J = \frac{H}{H_{\max}} = \frac{-\sum_{i=1}^{S} P_i \ln P_i}{\ln S}.$$ (17.4)

An example of this kind of analysis is provided by an unusually long-term study that has been running since 1856 in an area of pasture at Rothamstead in England. Experimental plots have received a fertilizer treatment once every year, whilst control plots have not. Figure 17.3 shows how species diversity (*H*) and equitability (*J*) of grass species changed between 1856 and 1949. Whilst the unfertilized area remained essentially unchanged, the fertilized area has shown a progressive decline in diversity and equitability. The explanation may be that high nutrient availability leads to high rates of population growth and a greater chance of the most productive species coming to dominate and, perhaps, competitively exclude others. (This phenomenon is discussed further in Chapter 24, Section 24.3.1.)

17.2.2 Rank–abundance diagrams

Of course, attempts to describe a complex community structure by one single attribute, such as richness, diversity or equitability, can be criticized because so much valuable information is lost. A more complete picture of the distribution of species abundances in a community makes use of the full array of P_i values by plotting P_i against rank. Thus, the P_i for the most abundant species is plotted first, then the next most common, and so on until the array is completed by the rarest species of all. A rank–abundance diagram can be drawn for the number of individuals, or for the area of ground covered by different sessile species or for the biomass contributed to a community by the various species.

A range of the many forms that rank–abundance diagrams can take is shown in Figure 17.4. Two of these are basically statistical in origin (the log series and

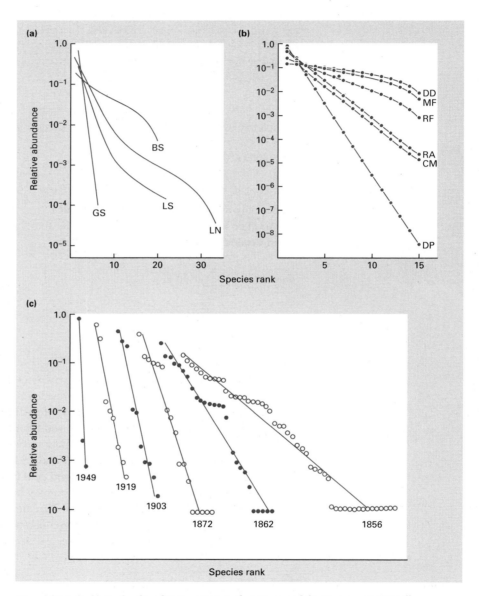

Figure 17.4 (a, b) Rank–abundance patterns of various models. Two are statistically orientated (LS and LN), whilst the rest can be described as niche orientated. (a) BS, broken stick; GS, geometrical series; LN, log-normal; LS, log series. (b) CM, composite; DD, dominance decay; DP, dominance pre-emption; MF, MacArthur fraction; RA, random assortment; RF, random fraction. (c) Change in the relative abundance pattern (geometrical series fitted) of plant species in an experimental grassland subjected to continuous fertilizer from 1856 to 1949. (After Tokeshi, 1993.)

rank–abundance models may
be based on statistical or
biological arguments

log-normal). However, we need to take care lest the esoteric delights of statistics lead us to forget the biology. The other models take into account the relationships between conditions, resources and species-abundance patterns (niche-orientated models) and are more likely to help us understand the mechanisms underlying community organization (Tokeshi, 1993). We illustrate the diversity of approaches by describing the basis of three of Tokeshi's niche-orientated models (see Tokeshi, 1993, for a complete treatment). The *dominance–pre-emption model*, which produces the least equitable species distribution, has successive species pre-empting a dominant portion (50% or more) of the remaining niche space; the first, most dominant species takes more than 50% of the total niche space, the next more than 50% of what remains, and so on. A somewhat more equitable distribution is represented by the *random-fraction model*, in which successive species invade and take over an arbitrary portion of the niche space of any species previously present. Finally, the *dominance–decay model* is the inverse of the dominance–pre-emption model, in that the largest niche in an existing assemblage is always subject to a subsequent (random) division. Thus, in this model the next invading species is supposed to colonize the niche space of the species currently most abundant, yielding more equitable species abundances than either of the other models.

community indices are
abstractions that may be
useful when making
comparisons

Rank–abundance diagrams, like indices of richness, diversity and equitability, should be viewed as abstractions of the highly complex structure of communities that may be useful when making comparisons. Progress so far has been limited, both because of problems of interpretation and the practical difficulty of testing for the best fit between model and data (Tokeshi, 1993). However, some studies have successfully focused attention on a change in dominance/evenness relationships in relation to environmental change, assuming that the *geometrical series model* can be appropriately applied. Figure 17.4c shows how dominance steadily increased, whilst species richness decreased, during the Rothamstead long-term grassland experiment described above. (A comparable pattern is evident in Chapter 24, Figure 24.20, but in this case dominance steadily decreased through a 40-year period of natural succession from an abandoned field to the development of woodland.)

Taxonomic composition and species diversity are just two of many possible ways of describing a community. Historically, most of the emphasis on diversity has been in terms of species, perhaps reflecting the dominance of taxonomists. It is important to note that there are other aspects of diversity that may be just as important, or even more important, when considering community structure. Thus, many species have life cycles that make them diverse contributors to a community (tadpoles/frogs, caterpillars/butterflies), or have structure that offers a diversity of resources (a tree compared with a herb, a cow compared with a nematode worm).

the energetics approach: an
alternative to taxonomic
description

Another alternative (not necessarily better but quite different) is to describe communities in terms of their standing crop and the rate of production of biomass by plants, and its use and conversion by heterotrophic microorganisms and animals. Studies that are orientated in this way may begin by describing the food web, and then define the biomasses at each trophic level and the flow of energy and matter from the physical environment through the living organisms and back to the physical environment. Such an approach can allow patterns to be detected amongst communities that may have no taxonomic features in common. This approach will be discussed in Chapters 18 and 19.

17.3 Community patterns in space

17.3.1 Gradient analysis

Figure 17.5 shows a variety of ways of describing the distribution of vegetation on the Great Smoky Mountains (Tennessee), USA, where the species of tree present give the vegetation its main character. Figure 17.5a shows the characteristic associations of the dominant trees on the mountainside, drawn as if the communities had sharp boundaries. The mountainside itself provides a range of conditions for plant growth, and two of these, altitude and moisture, may be particularly important in determining the distribution of the various tree species. Figure 17.5b shows the dominant associations graphed in terms of these two environmental dimensions. Finally, Figure 17.5c shows the abundance of each individual tree species (expressed as a percentage of all tree stems present) plotted against the single gradient of moisture.

Figure 17.5a is a subjective analysis which acknowledges that the vegetation of particular areas differs in a characteristic way from that of other areas. It could be taken to imply that the various communities are sharply delimited. Figure 17.5b gives the same impression. Note that both Figures 17.5a and b are based on descriptions of the *vegetation*. However, Figure 17.5c sharpens the focus by concentrating on the pattern of distribution of the individual *species*. It is then immediately obvious that there is considerable overlap in their abundance—there are no sharp boundaries. The various tree species are now revealed as being strung out along the gradient with the tails of their distributions overlapping. The results of this 'gradient analysis' show that the limits of the distributions of each species 'end not with a bang but with a whimper'.

Many other gradient studies have produced similar results. Figure 17.6a shows ecological response curves for selected grass species along a pH gradient in Britain, and Figure 17.6b shows the distribution of large invertebrates along an intertidal beach in Canada (here a gradient in particle size may be the key condition that determines the pattern of distribution). Chapter 15, Figure 15.26, provides a further example.

Perhaps the major criticism of gradient analysis as a way of detecting pattern in communities is that the choice of the gradient is almost always subjective. The investigator searches for some feature of the environment that appears to matter to the organisms and then organizes the data about the species concerned along a gradient of that factor. It is not necessarily the most appropriate factor to have chosen. The fact that the species from a community can be arranged in a sequence along a gradient of some environmental factor does not prove that this factor is the most important. It may only imply that the factor chosen is more or less loosely correlated with whatever really matters in the lives of the species involved. Gradient analysis is only a small step on the way to the objective description of communities.

17.3.2 The ordination and classification of communities

Formal statistical techniques have been defined to take the subjectivity out of community description. These techniques allow the data from community studies to sort themselves, without the investigator putting in any preconceived ideas about which species tend to be associated with each other or which environmental

the distributions of species along gradients end not with a bang but with a whimper

choice of gradient is almost always subjective

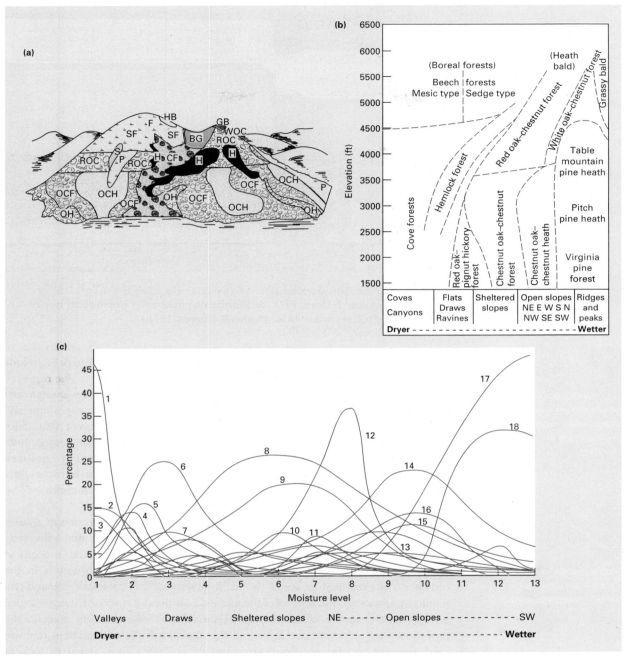

Figure 17.5 Three contrasting descriptions of distributions of the characteristic dominant tree species of the Great Smoky Mountains, Tennessee. (a) Topographical distribution of vegetation types on an idealized west-facing mountain and valley. (b) Idealized graphical arrangement of vegetation types according to elevation and aspect. (c) Distributions of individual tree populations (percentage of stems present) along the moisture gradient. Vegetation types: BG, beech gap; CF, cove forest; F, Fraser fir forest; GB, grassy bald; H, hemlock forest; HB, heath bald; OCF, chestnut oak–chestnut forest; OCH, chestnut oak–chestnut heath; OH, oak–hickory; P, pine forest and heath; ROC, red oak–chestnut forest; S, spruce forest; SF, spruce–fir forest; WOC, white oak–chestnut forest. Major species: 1, *Halesia monticola*; 2, *Aesculus octandra*; 3, *Tilia heterophylla*; 4, *Betula alleghaniensis*; 5, *Liriodendron tulipifera*; 6, *Tsuga canadensis*; 7, *B. lenta*; 8, *Acer rubrum*; 9, *Cornus florida*; 10, *Carya alba*; 11, *Hamamelis virginiana*; 12, *Quercus montana*; 13, *Q. alba*; 14, *Oxydendrum arboreum*; 15, *Pinus strobus*; 16, *Q. coccinea*; 17, *P. virginiana*; 18, *P. rigida*. (After Whittaker, 1956.)

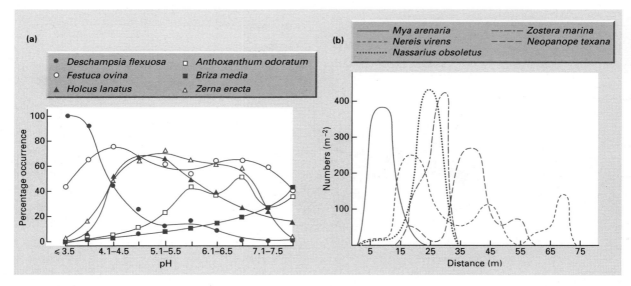

Figure 17.6 (a) Distribution gradients for selected grass species along a pH gradient in England. (After Grime & Lloyd, 1973.) (b) Distribution gradients for macrofaunal species along an oyster bed in Canada. (After Hughes & Thomas, 1971.)

variables correlate most strongly with the species distributions. One such technique is *ordination*.

in ordination, communities are displayed on a graph so that those most similar in composition are closest together ...

Ordination is a mathematical treatment that allows communities to be organized on a graph so that those that are most similar in both species composition and relative abundance will appear closest together, whilst communities that differ greatly in the relative importance of a similar set of species, or that possess quite different species, appear far apart. Details of the methods are provided by Ter Braak and Prentice (1988). Figure 17.7a shows the application of ordination to community data from 50 separate stands (communities) of vegetation on Welsh sand-dunes. The data consisted of lists of the species present and their abundances.

subsequently, it is necessary to ask what varies along the axes of an ordination graph

The axes of the graphs are derived mathematically solely from the species compositions of the various stands; they represent dimensions that effectively summarize community patterns. The interpretation of these patterns in terms of environmental variables is a second step, in which the scatter of points in the ordination is examined to see if the axes correspond to ecologically meaningful gradients. Obviously, the success of the method now depends on our having sampled an appropriate variety of environmental variables. This is a major snag in the procedure—we may not have measured the qualities in the environment that are most relevant. In the plant-ecologist's use of ordination it is usual to look at environmental factors such as moisture, nutrient levels, pH, rates of oxygen diffusion, and so on. However, gradients in community composition may be produced by grazing pressure, disease and a host of other interactions. There is no way in which the ecologist can determine in advance what factors will be most relevant, and there is no way in retrospect of discovering if a vital factor has been forgotten.

The sand-dune study (Figure 17.7a) was quite successful in pin-pointing environmental conditions that were closely related to the species composition of the

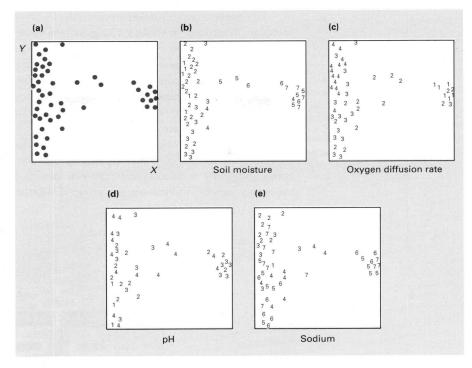

Figure 17.7 (a) Ordination of 50 stands (communities) of vegetation in Welsh sand-dunes; (b) ordination with scores for soil moisture from each stand superimposed; (c) ordination with scores for oxygen diffusion rate from each stand superimposed; (d) ordination with scores for pH from each stand superimposed; (e) ordination with scores for sodium concentration from each stand superimposed. (After Pemadasa *et al.*, 1974.)

communities. On to Figures 17.7b–e have been superimposed values corresponding to the soil moisture, oxygen diffusion rate, pH and sodium concentration at each site. The *x*-axis of the ordination is positively related to soil moisture; the lowest values for moisture tend to be to the left and the highest values to the right of the graph. In contrast, there is a tendency for the lowest values of oxygen diffusion rate to be towards the right (i.e. a negative relationship exists between the *x*-axis and diffusion rate). Neither pH nor sodium concentration exhibit any clear relationship with community structure in these sand-dunes.

What do these results tell us? First, and most specifically, the correlations with environmental factors, revealed by the analysis, give us some specific hypotheses to test about the relationship between community composition and underlying environmental factors. (Remember that correlation does not necessarily imply causation. For example, soil moisture and community composition may vary together because of a common response to another environmental factor. A direct causal link can only be proved by controlled experimentation.)

A second, more general point is relevant to the discussion of the nature of the community. The results of this ordination emphasize that under a particular set of environmental conditions, a predictable association of species is likely to occur. It shows that community ecologists have more than just a totally arbitrary and ill-defined set of species to study.

ordination can generate hypotheses for subsequent testing

689 NATURE OF THE COMMUNITY

The results of an ordination study on the invertebrate communities existing at 34 locations in streams in southern England are shown in Figure 17.8a. In this case, clear relationships were revealed between pH and the position of communities along the x-axis, and between average water temperature and their position along the y-axis. Once again, communities with predictable compositions occurred under specific sets of environmental conditions. If we knew the pH of a new stream in the area we could use the ordination data to predict the invertebrate fauna, and if we knew only the fauna we might predict the pH.

Classification, as opposed to ordination, begins with the assumption that communities consist of relatively discrete entities. It produces groups of related communities by a process conceptually similar to taxonomic classification. In taxonomy, similar individuals are grouped together in species, similar species in

classification involves grouping similar communities together in clusters

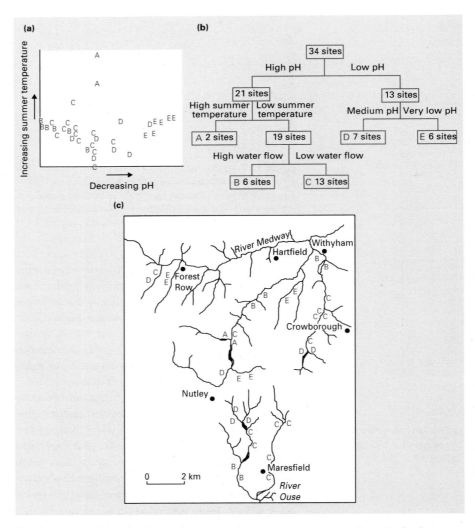

Figure 17.8 Analysis of 34 invertebrate communities in streams in southern England. (a) Ordination (letters A to E indicate community classes derived from classification). (b) Classification. (c) Distribution of community classes A to E in the stream catchment. (After Townsend *et al.*, 1983.)

genera, and so on. In community classification, communities with similar species-compositions are grouped together in subsets, and similar subsets may be further combined if desired (see Ter Braak & Prentice, 1988, for details of the procedure).

ordination and classification can complement one another

The stream invertebrate communities ordinated in Figure 17.8a have also been subjected to classification (Figure 17.8b) to produce five groupings. At each division, which is based solely on species composition, a significant difference in environmental conditions exists and the important factors are indicated. The relationship between ordination and classification can be gauged by noting that communities falling into classes A–E, derived from classification, are fairly distinctly separated on the ordination graph. In both cases, pH was implicated as a critical environmental condition (although the most influential factor may have been aluminium toxicity, often found at low pH), and communities in class E (associated with very low stream pH) were characterized particularly by the presence of larvae of the stoneflies *Leuctra nigra* and *Nemurella picteti*, the caddis-fly *Plectrocnemia conspersa* and the alderfly *Sialis fuliginosa*, all of which are tolerant of these extreme conditions.

The spatial distribution of each class of community in the stream system itself is shown in Figure 17.8c. Note that there is little consistent spatial relationship; communities in each class are dotted about the stream catchment. This illustrates one of the strengths of classification and/or ordination. The methods show the order or structure in a series of communities without the necessity of picking out some supposedly relevant environmental variable in advance, a procedure that is necessary for gradient analysis.

classification in an applied context

Classification was also applied to the benthic invertebrate communities of Oslofjord in Scandinavia (Gray, 1981). Seven reasonably distinct classes of communities were derived, and in this case there was a consistent spatial arrangement in the fjord, with community structure varying according to position in relation to the city of Oslo and its output of pollutants. Thus, one community class was characterized particularly by *Capitella*, *Polydora* and *Heteromastus* spp., which are invertebrates that can tolerate organically highly enriched sediments. This approach to community analysis shows its role in applied ecology, by revealing the extent of pollution. Further analyses of the communities through time could indicate whether antipollution measures were having the desired effects, or whether communities characteristic of polluted waters were becoming more common.

17.3.3 The problems of boundaries in community ecology

are communities discrete entities with sharp boundaries?

There may be communities that are separated by clear, sharp boundaries, where groups of species lie adjacent to, but do not intergrade into, each other. If they exist, they are exceedingly rare and exceptional. The meeting of terrestrial and aquatic environments might appear to be a sharp boundary but its ecological unreality is emphasized by the otters and frogs that regularly cross it and the many aquatic insects that spend their larval lives in the water but their adult lives as winged stages on land or in the air. On land, quite sharp boundaries occur between the vegetation types on acidic and basic rocks where outcrops meet, or where serpentine (a term applied to a mineral rich in magnesium silicate) and non-serpentine rocks are juxtaposed. However, even in such situations, minerals are leached across the boundaries, which become increasingly blurred. The safest statement we can make about community boundaries is probably that they do not exist, but that some communities are much more sharply defined than others. The ecologist is usually better employed looking at the ways in which communities grade into each other, than in searching for sharp cartographical boundaries.

In the first quarter of this century there was considerable debate about the nature of the community. Clements (1916) conceived of the community as a sort of *superorganism* whose member species were tightly bound together both now and in their common evolutionary history. Thus, individuals, populations and communities bore a relationship to each other which resembled that between cells, tissues and organisms.

In contrast, the *individualistic* concept devised by Gleason (1926) and others saw the relationship of coexisting species as simply the results of similarities in their requirements and tolerances (and partly the result of chance). Taking this view, community boundaries need not be sharp, and associations of species would be much less predictable than one would expect from the superorganism concept.

The current view is close to the individualistic concept. Results of direct gradient analysis, ordination and classification all indicate that a given location, by virtue mainly of its physical characteristics, possesses a reasonably predictable association of species. However, a given species that occurs in one predictable association is also quite likely to occur with another group of species under different conditions elsewhere.

Many ecologists have been preoccupied with the idea of community boundaries. Indeed, there has been much debate and concern over whether community ecology can legitimately be studied at all if communities do not exist as definable units. This problem has perhaps arisen because of a psychological need to be dealing with a readily defined entity. Whether or not communities have more or less clear boundaries is an important question, but it is not the fundamental consideration. Community ecology is the study of the *community level of organization* rather than of a spatially and temporally definable unit. It is concerned with the nature of the interactions between species and their environment, and with the structure and activities of the multispecies assemblage, usually at one point in space and time. It is not necessary to have discrete boundaries between communities to do community ecology.

the community: not so much a superorganism ...

... more a level of organization

17.4 Community patterns in time: succession

Just as the relative importance of species varies in space, so their patterns of abundance may change with time. In either case, a species will occur only where and when: (i) it is capable of reaching a location; (ii) appropriate conditions and resources exist there; and (iii) competitors, predators and parasites do not preclude it. A temporal sequence in the appearance and disappearance of species therefore seems to require that conditions, resources and/or the influence of enemies themselves vary with time.

For many organisms, and particularly short-lived ones, their relative importance in the community changes with time of year as the individuals act out their life cycles against a background of seasonal change. The explanation for such temporal patterns is straightforward and will not concern us here. Nor will we dwell on the variations in abundance of species from year to year as individual populations respond to a multitude of factors that influence their reproduction and survival (dealt with in Chapters 6–13 and 15). Rather, our focus will be on the process of *succession*, defined as *the non-seasonal, directional and continuous pattern of colonization and extinction on a site by species populations.* This general definition encompasses a range of successional sequences that occur over widely varying time scales and often as a result of quite different underlying mechanisms.

17.4.1 Degradative succession

some successions occur when a degradable resource is utilized successively by a number of species

One class of serial replacements may be termed *degradative successions*, and these occur over a relatively short time scale of months or years (Doube, 1987; Swift, 1987). Any packet of dead organic matter, whether the body of an animal or plant, a shed skin of a snake or arthropod or a faecal deposit, is exploited by microorganisms and detritivorous animals (see Chapter 11). Usually, different species invade and disappear in turn, as degradation of the organic matter uses up some resources and makes others available, whilst changes in the physical conditions of the detritus favour first one species, then another. (Since heterotrophic organisms are involved in such sequences they have often been referred to as *heterotrophic successions*.) Ultimately, degradative successions terminate because the resource is completely metabolized and mineralized.

a degradative succession on pine needles

Figure 17.9 shows the sequence of fungal species colonizing pine needles that enter the litter beneath a forest of Scots pine (*Pinus sylvestris*). In this environment, litter is continually being deposited on the surface of the ground, and it accumulates in layers because there is no earthworm activity to incorporate it. The surface litter is therefore young, and the deeper layers are old—time and depth are directly related.

The sequence begins even before the needles have been shed. *Coniosporium* is present on approximately 50% of living pine needles, but by the time the needles senesce and die this species can no longer be found. However, in August, when most needles fall, *Lophodermium* occurs on approximately 40% of needles, and *Fusicoccum*

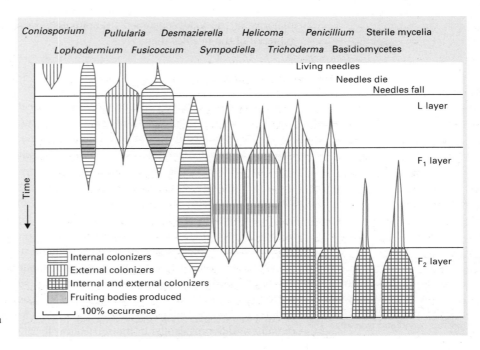

Figure 17.9 Temporal and spatial changes in fungal populations colonizing pine needles in litter layers beneath Scots pine (*Pinus sylvestris*) forest in England. (After Kendrick & Burges, 1962; from Richards, 1974.)

and *Pullularia* infect more than 80%. The needles fall to the ground to become part of the L layer ('litter' consisting of tough, light-brown needles which lay loose and uncompacted on the surface). During their 6 months in the L layer many of the needles are invaded by *Desmazierella*, and several other species also enter the sequence, gradually digesting and softening up the tissue.

The next stage of the succession is represented spatially by the F1 layer (the upper subhorizon of the 'fermentation' layer of the soil). The grey–black needles in this layer are more closely packed and have softened tissue with low tensile strength. The interior of the needle, particularly the phloem, is extensively attacked by *Desmazierella*, whilst its exterior, now surrounded by a much more humid atmosphere, is colonized by *Sympodiella* and *Helicoma*. Regions of internal attack are penetrated and further broken down by burrowing mites.

The character of the needles changes once more. After about 2 years in the F1 layer they are tightly compressed, and fragments that have been attacked by *Lophodermium*, *Fusicoccum* and *Desmazierella* are invaded by various soil animals, mainly collembolans, mites and enchytraeid worms. These eventually bring about the complete physical breakdown of the needles. During this time, the needle fragments are slowly attacked by basidiomycetes which can decompose cellulose and lignin. After about 7 years in the F2 layer, the pine needles are no longer recognizable, and biological activity in the highly acid conditions of the H layer ('humus') falls to a very low level. Humus typically comprises no more than 2% of the dry weight of soil, but it has a major role in determining physical structure, aeration and water-holding capacity. There is a profound link between the succession of organisms that decompose plant litter and the structure and properties of the soil.

Most of the fungi encountered in this study do not show strong antagonisms, and the timing of different stages of the succession is probably controlled mainly by changing environmental conditions and modification of the nutritional status of the organic material in the needles. The fungi perform an initial softening and digestion of the needle tissues, facilitating attack and entry of the detritivorous animals. The majority of the food consumed by animals is probably fungal tissue rather than material of the needles themselves.

the link between succession on plant litter and soil formation

17.4.2 Allogenic succession

A new habitat is created when an elephant defaecates, a tree sheds a leaf or a small mammal dies (see Chapter 11). Of more general interest to ecologists are those cases where the new habitat is an area of substrate opened up for invasion by green plants (referred to as *autotrophic successions*) or other sessile organisms. In these cases, the new habitat does not become degraded and disappear, but is merely occupied.

not all newly-created habitats degrade: some are simply occupied

It is important to distinguish between successions that occur as a result of biological processes—such as the accumulation of litter in a forest, or peat in a bog, or an increase in shading by the canopy—that modify conditions and resources (known as *autogenic* successions) and other serial replacements of species, occurring as a result of changing external geophysicochemical forces (*allogenic* successions).

allogenic successions are driven by external influences which alter conditions: a salt-marsh–woodland transition

There have been a number of reports of an apparent allogenic transition between salt-marsh and woodland caused primarily by silt deposition. One of the most thorough of these studies garnered evidence from historical maps, the present pattern of distribution of species in space and a stratigraphical record of plant remains at different depths in the soil (Ranwell, 1974). The estuary of the River Fal in Cornwall, England, like many other estuaries, is subject to a quite rapid deposition of silt (increased somewhat due to china-clay workings in its catchment). This accretion can occur at a rate of 1 cm per year on the mud flats which are found 15 km into the estuary. As a result, salt-marsh has extended 800 m seawards during the last century, whilst valley woodland has kept pace by invading the landward limits of the marshland (Figure 17.10a).

The vertical zonation (above sea-level) of selected plant species is shown in Figure 17.10b. Species colonize at a particular height above sea-level according to their tolerance of inundation by brackish water at high tides. Since the highest vertical levels correspond both to the landward edge of the study area and to the oldest stage in succession, Figure 17.10b illustrates both the horizontal distribution of species and the time sequence of their appearance in the succession. The lowest-lying pioneer species on these brackish mud-flats are the sedge *Scirpus* and the grass *Agrostis*. These span the entire range of the marsh, right up to the seaward limit of oak trees (*Quercus*), a vertical range of only 2.26 m on the Fal estuary. The range of two typical salt-marsh plants, *Puccinellia* and *Triglochin*, actually overlaps locally with the lower limits of the willow (*Salix*)–alder (*Alnus*)–oak tidal woodland. Clearly, differences in level of as little as 0.2 m may separate salt-marsh from woodland.

That this transition from salt-marsh to woodland truly represents a succession in time, rather than simply a fixed zonation of species, is proved by analysis of the plant remains in a soil core taken within the woodland. The species sequence described in space is faithfully recapitulated in time (= depth in the profile).

The external physical agency of silt deposition is no doubt primarily responsible for the succession observed. However, two biological processes are also capable of raising the level locally and may play some role: plants such as the grass *Deschampsia*

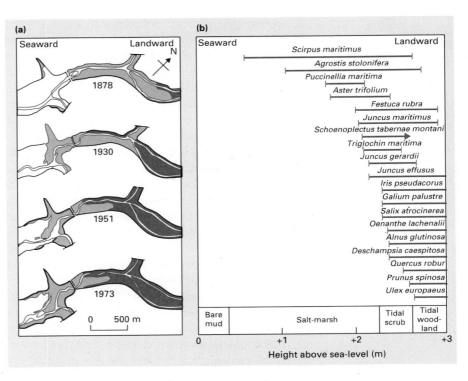

Figure 17.10 (a) Seawards extension of salt-marsh (light shading) and tidal woodland (dark shading) during the last century in the Fal estuary, Cornwall. (b) Vertical zonation of selected species in the salt-marsh to tidal woodland succession. (After Ranwell, 1974.)

695   NATURE OF THE COMMUNITY

and the sedge *Carex* build tussocks, and ants construct mounds. Thus, part of this succession may be autogenic, not simply allogenic; and, of course, the presence of vegetation may speed up silt deposition, further blurring the distinction.

17.4.3 Autogenic succession

primary and secondary successions

Successions that occur on newly exposed landforms, and in the absence of gradually changing abiotic influences, are known as *autogenic* successions. If the exposed landform has not previously been influenced by a community, the sequence of species is referred to as a primary succession. Freshly formed sand-dunes, lava flows and pumice plains caused by volcanic eruptions (like that of Mt St Helens; Del Moral & Bliss, 1993), and substrate exposed by the retreat of a glacier are examples of this. In cases where the vegetation of an area has been partially or completely removed, but where well-developed soil and seeds and spores remain, the subsequent sequence of species is termed a secondary succession. The loss of trees locally as a result of disease, high winds, fire or felling may lead to secondary successions that occur within the mosaic of a forest.

primary succession observed ...

Successions on newly exposed landforms typically take several hundreds of years to run their course. However, a precisely analogous process occurs amongst the seaweeds on recently denuded boulders in the rocky intertidal zone, and this succession takes less than a decade. The research life of an ecologist is sufficient to encompass the seaweed succession but not that following glacial retreat. Fortunately, however, information can sometimes be gained over the longer time scale. Often, successional stages in time are represented by community gradients in space—the transition from salt-marsh to woodland is a case in point. The use of historical maps, carbon dating or other techniques may enable the age of a community since exposure of the landform to be estimated. A series of communities currently in existence, but corresponding to different lengths of time since the onset of succession, can be inferred to reflect succession. However, whether or not different communities that are spread out in space really do represent various stages of succession must be judged with caution. We must remember, for example, that in northern temperate areas the vegetation we see may still be undergoing recolonization and responding to climatic change following the last ice-age (see Chapter 1).

... or inferred

facilitation: early successional species paving the way for later ones—Glacier Bay

An early successional species may so alter conditions or the availability of resources in a habitat that the entry of new species is made possible. This process of *facilitation* (Connell & Slatyer, 1977) may be particularly important in primary successions where conditions are initially severe (e.g. as on a substrate laid bare after the retreat of a glacier). Remarkably rapid glacial retreat has occurred at Glacier Bay in south-eastern Alaska. Since about 1750 the glaciers have retreated almost 100 km. As the glaciers retreat they deposit mineral debris as moraines, the ages of which can be estimated by ring counts of the oldest trees present at the site. The oldest trees on the last morainic ridge are about 200 years old, and the maximum age of trees decreases as the glacier is approached. Information on the last 80 years has been gathered by direct observation.

primary succession and soil development

The succession on the exposed boulder-clay till, with its thin, nutrient-deficient clay soil, proceeds as follows. The first plants to colonize are mosses and a few shallow-rooted herbaceous species, notably *Dryas*. Next, several kinds of willow (*Salix*) appear, prostrate species at first, but later shrubby types. Soon alder (*Alnus crispus*) enters the succession and after about 50 years produces thickets up to 10 m

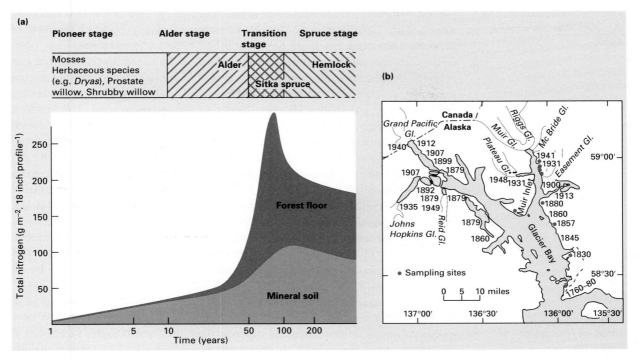

Figure 17.11 (a) Plant succession after glacial retreat in Glacier Bay, Alaska, and changes in total nitrogen content of the soils. (b) History of ice recession at Glacier Bay. (After Crocker & Major, 1955.)

tall with a scattering of cottonwood. The alder is invaded by sitka spruce (*Picea sitchensis*), forming a dense mixed forest which continues to develop as western hemlock (*Tsuga heterophylla*) and mountain hemlock (*T. mertensiana*) become established.

One of the principal forces that drives this succession is the change in soil conditions caused by the early colonists. Both *Dryas* and alder have symbionts that fix atmospheric nitrogen (see Chapter 13) and lead to an accumulation of a large nitrogen capital in the soil (Figure 17.11). Alder also has a strong acidifying effect, lowering the pH of the surface soil from approximately pH 8 to 5 within 50 years. Sitka spruce is then able to invade and displace the alder, using the accumulated nitrogen capital. The gradual accumulation of soil carbon leads to the development of crumb structure in the soil, and an increase in soil aeration and water-retaining capacity.

On less well-drained slopes the succession does not always culminate in a spruce–hemlock forest. Where the ground is flat, or has only a shallow slope, drainage is impeded and the forest is invaded by *Sphagnum* mosses, which accumulate water and greatly acidify the soil. The forest floor becomes waterlogged and deficient in oxygen and most trees die. Occasional, scattered individuals of lodge-pole pine (*Pinus contorta*) are the only trees that can tolerate the poor aeration of the resulting *muskeg* bog. The culminating vegetation of the succession clearly depends on local conditions.

Species of algae replace one another in a rather regular sequence on boulders in the low intertidal zone of the southern Californian coast (Sousa, 1979a). The major

a primary succession of short
duration—algae on a rocky
shore

natural disturbance that clears space in this system is the overturning of boulders by
wave action. Algae may recolonize cleared surfaces through vegetative regrowth of
surviving individuals, or more generally by recruitment from spores. Most commonly, boulders are completely cleared of algae, so the succession that occurs can
properly be described as primary. The typical successional pattern that occurs
naturally on cleared boulders can be mimicked, either by artificially clearing
boulders or by introducing concrete blocks to the intertidal zone. This made it
possible for Sousa to obtain a precise description of succession and also, through
various manipulations, to reveal the mechanisms underlying the successional
pattern.

Bare substrate is colonized within the first month by a mat of the pioneer green
alga *Ulva* (Figure 17.12). During the autumn and winter of the first year, several
species of perennial red algae, including *Gelidium coulteri*, *Gigartina leptorhynchos*,
Rhodoglossum affine and *Gigartina canaliculata* colonized the surface.

Within a period of 2–3 years *G. canaliculata* comes to dominate the community,
covering 60–90% of the rock surface. In the absence of further disturbance this
virtual monoculture persists through vegetative spread, resisting invasion by all
other species.

We have already seen how in the succession that occurs after glacial retreat some
pioneer species actually influence the environment to such an extent that later
species can enter the sequence and suppress them. In contrast to this facilitation of
succession, the pioneer rocky shore species *Ulva* actually inhibits further change.
Inhibition was demonstrated conclusively by comparing the recruitment of *Gigartina*
individuals (both species) to concrete blocks from which *Ulva* had or had not been
removed. *Gigartina* recruitment was dramatically better in the absence of *Ulva*
(Figure 17.13a). Similarly, the middle successional red algal species *Gigartina
leptorhynchos* and *Gelidium coulteri* inhibit, by their presence, the invasion and growth
of the dominant late-successional *G. canaliculata*. Thus, the speed with which open
space is colonized is crucial in early succession. The relatively short-lived and rapidly
growing *Ulva* is an efficient pioneer of open space because it reproduces throughout
the year and can quickly establish itself on exposed substrate. The perennial red

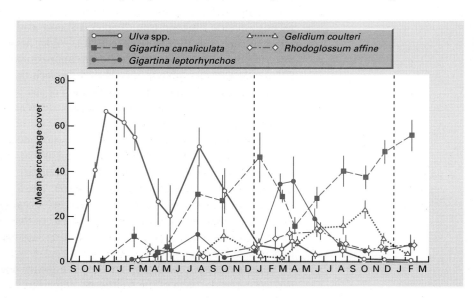

Figure 17.12 Mean percentage
cover (± standard error) of five
algal species that colonized
concrete blocks introduced to
the intertidal zone in
September 1974. (After Sousa,
1979a.)

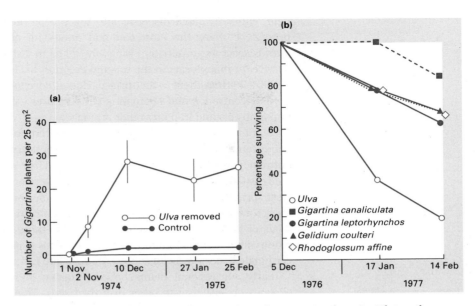

Figure 17.13 (a) Effect of manually removing the early successional species *Ulva* on the recruitment of *Gigartina* spp. over a 4-month period beginning on October 15, 1974. Data are mean numbers (± standard error) of sporelings of *Gigartina leptorhynchos* and *G. canaliculata* which were recruited to four 25 cm² plots on concrete blocks. (After Sousa, 1979a.) (b) Survival curves for five species of algae over a 2-month period, beginning on December 5, 1976, when low tides occurred in the afternoon, creating harsh physical conditions. Thirty plants of each species were tagged initially. (After Sousa, 1979a.)

algae, with their highly seasonal recruitment and slow growth, are inhibited as long as the early colonists remain healthy and undamaged.

This succession is characterized by species, each of which inhibits its replacement by another. The succession occurs mainly because the species that dominate early are more susceptible to the rigours of the physical environment. This was demonstrated by following the mortality of tagged plants during a 2-month period when low tides occurred in the afternoon, creating particularly harsh conditions of intense sunlight and exposure to the air and drying winds (Figure 17.13b). In addition, selective grazing on *Ulva* by the crab *Pachygrapsus crassipes* accelerates succession to the tougher, longer-lived red algae. Late species colonize small openings and grow to maturity when individuals of the early species are killed. They gradually take over and dominate the space until a further large-scale opening of space occurs by wave action.

Successions on old fields have been studied primarily along the eastern part of the USA, where many farms were abandoned by farmers who moved west after the frontier was opened up in the 19th century. Most of the precolonial mixed conifer–hardwood forest had been destroyed, but regeneration was swift. In many places, a series of sites that had been abandoned for different, recorded periods of time are available for study. The typical sequence of dominant vegetation is: annual weeds, herbaceous perennials, shrubs, early successional trees, late successional trees.

The early pioneers of the American West left behind exposed land, which was colonized by pioneers of a very different kind! The pioneer species are those which can establish themselves quickly in the disturbed habitat of a recently cultivated

old field sequences— successions where facilitation is not obvious but inhibition is

field, either by rapid dispersal into the site or from propagules that are already present. Perhaps the most common annual of old field succession is *Ambrosia artemisiifolia*. Its germination is closely linked to disturbance. The seeds of *Ambrosia* survive for many years in the soil and are most likely to germinate when movement of the soil brings them to the surface. Here, they experience unfiltered light, reduced CO_2 concentration and fluctuating temperatures—all conditions that induce germination in this and in many other annuals.

Summer annuals such as *Ambrosia* are often quickly eclipsed in importance by winter annuals (mostly other composites). These have small seeds with little or no dormancy, but disperse over long distances. Their seeds germinate soon after they land, usually in late summer or autumn, and develop into rosettes which overwinter. Next spring they have a head start over summer annuals, and thus they pre-empt the resources of light, water, space and nutrients. In contrast to the pioneer annuals, seeds of later successional plants, especially those found in climax forest, do not require light for germination. Thus, whilst seeds of pioneers could not be expected to germinate beneath a forest canopy, those of late successional species are able to do so.

later species in old field successions tend to grow more slowly ...

Early successional plants have a fugitive lifestyle. Their continued survival depends on dispersal to other disturbed sites. They cannot persist in competition with later species, and thus they must grow and consume the available resources rapidly. A high relative growth rate is a crucial property of the fugitive. The relative growth rate is lower in later successional plants (perhaps because so much of the plant is old support tissue). The rate of photosynthesis also generally declines with position in the succession (Table 17.1).

but tend also to be more shade tolerant

Shade tolerance is one factor in the success of later species. Figure 17.14 illustrates idealized light saturation curves of early, mid and late successional species. At low light intensities, later successional species are able to grow, albeit quite slowly, but still faster than the species they replace.

Table 17.1 Some representative photosynthetic rates (mg CO_2 dm^{-2} h^{-1}) of plants in a successional sequence. Late successional trees are arranged according to their relative successional position. (After Bazzaz, 1979.)

Plant	Rate	Plant	Rate
Summer annuals		Early successional trees	
Abutilon theophrasti	24	*Diospyros virginiana*	17
Amaranthus retroflexus	26	*Juniperus virginiana*	10
Ambrosia artemisiifolia	35	*Populus deltoides*	26
Ambrosia trifida	28	*Sassafras albidum*	11
Chenopodium album	18	*Ulmus alata*	15
Polygonum pensylvanicum	18		
Setaria faberii	38	Late successional trees	
		Liriodendron tulipifera	18
Winter annuals		*Quercus velutina*	12
Capsella bursa-pastoris	22	*Fraxinus americana*	9
Erigeron annuus	22	*Quercus alba*	4
Erigeron canadensis	20	*Quercus rubra*	7
Lactuca scariola	20	*Aesculus glabra*	8
		Fagus grandifolia	7
Herbaceous perennials		*Acer saccharum*	6
Aster pilosus	20		

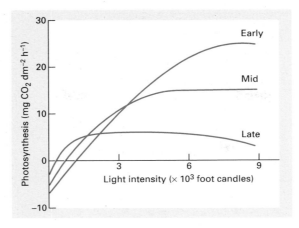

Figure 17.14 Idealized light saturation curves for early, mid and late successional plants. (After Bazzaz, 1979.)

early and late trees—
multilayered and monolayered
species

We can group the trees that take part in the later stages of an old field succession into early successional and late successional classes (Table 17.1). A characteristic of many early successional trees is that their foliage is multilayered. Leaves extend deep into the canopy, where they still receive enough light to be above the light compensation point. Species such as eastern red cedar (*Juniperus virginiana*) are thus able to use the abundant light that is available early in the tree stage of the succession. In contrast, sugar maple (*Acer saccharum*) and American beech (*Fagus grandifolia*) can be described as monolayered species. These possess a single layer of leaves in a shell around the tree and are more efficient in the crowded canopy of late succession (Horn, 1975). If a monolayered and a multilayered tree simultaneously colonize an open site, the multilayered tree will usually grow faster and dominate until it is crowded by its neighbours, when the slower growing monolayered tree emerges to dominance.

The early colonists amongst the trees usually have efficient seed dispersal; this in itself makes them likely to be early on the scene. They are usually precocious reproducers and are soon ready to leave descendants in new sites elsewhere. The late colonists are those with larger seeds, poorer dispersal and long juvenile phases. The contrast is between the lifestyles of the 'quickly come, quickly gone' and 'what I have, I hold'.

experiments unravel the
multiple causes of an old field
succession

A particularly detailed study of old field succession has been performed by Tilman (1987, 1988) and his associates. The Cedar Creek Natural History Area in Minnesota, USA, exists on well-drained and nutrient-poor soil. One consequence of the poor growing conditions is that species replacement is considerably slower than has been reported on rich soils with, for example, annual plants persisting as dominants for up to 40 years after field abandonment and woody plants (mostly vines and shrubs) still only providing 13% of ground cover in 60-year-old fields. In other respects, however, the pattern of species replacement is similar to what has been described elsewhere (Figure 17.15). A significant feature of the old field succession at Cedar Creek is that available nitrogen in the soil increases with time (Figure 17.15e—this provides a parallel with the primary succession at Glacier Bay). It may be that the plant species differ in their competitive abilities with respect to nitrogen availability, and that as nitrogen increases in the soil, first one species then another comes to competitive dominance. Tilman (1987) tested this hypothesis by experimentally increasing nitrogen concentrations and many species responded as

701   NATURE OF THE COMMUNITY

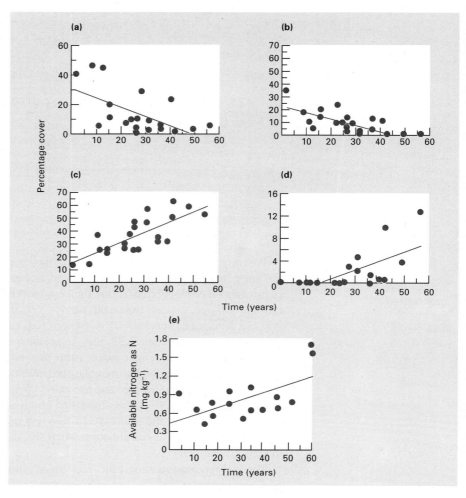

Figure 17.15 Relationships between percentage of land surface covered by (a) bare soil, (b) annuals, (c) herbaceous perennials and (d) woody plants, in relation to time since abandonment of 22 old fields at Cedar Creek surveyed in 1983. (e) Relationship between available nitrogen in the soil and field age. (After Inouye *et al.*, 1987.)

predicted. Thus, the pioneer annual *Ambrosia artemisiifolia*, naturally associated with the nitrogen-poor early successional stage, decreased in relative abundance in a multispecies assemblage following nitrogen fertilization. Near the other end of the time and nitrogen spectrum, the woody blackberry vine, *Rubus* spp., increased in relative abundance in experiments where nitrogen was added to the soil. A second group of species, expected on the basis of the results of the experiment to be dominant in nitrogen-poor soils of newly abandoned fields, exhibited a delay of up to 30 years in the time they took to become dominant. This was probably a result of poor dispersal ability; the dock *Rumex acetosella*, with large seeds lacking dispersal structures, provides an example. A final group of species whose appearance in the succession was also later than expected, possessed particularly low maximal rates of vegetative growth. Thus, for example, the low growth rate of the nitrogen-fixing legume *Lespedeza capitata*, may cause it to be suppressed for a time by species that

Table 17.2 A 50-year tree-by-tree transition matrix from Horn (1981). The table shows the probability of replacement of one individual by another of the same or different species 50 years hence.

Present occupant	Occupant 50 years hence			
	Grey birch	Blackgum	Red maple	Beech
Grey birch	0.05	0.36	0.50	0.09
Blackgum	0.01	0.57	0.25	0.17
Red maple	0.0	0.14	0.55	0.31
Beech	0.0	0.01	0.03	0.96

are poorer competitors for nitrogen but have higher maximum growth rates. Tilman's experimental approach has helped unravel the complexity of processes underlying this secondary succession on a poor soil.

17.4.4 Mechanisms underlying autogenic successions

forest succession can be represented as a tree-by-tree replacement process—Horn's model

A model of succession developed by Horn (1975; 1981) sheds some light on the successional process. Horn recognized that in a hypothetical forest community it would be possible to predict changes in tree species composition given two things. First, one would need to know for each tree species the probability that, within a particular time interval, an individual would be replaced by another of the same species or of a different species. Second, an initial species composition would have to be assumed.

Horn considered that the proportional representation of various species of saplings established beneath an adult tree reflected the probability of the tree's replacement by each of those species. Using this information, he estimated the probability, after 50 years, that a site now occupied by a given species will be taken over by another species or will still be occupied by the same species (Table 17.2). Thus, for example, there is a 5% chance that a location now occupied by grey birch will still support grey birch in 50-years' time, whereas there is a 36% chance that blackgum will take over, a 50% chance for red maple and 9% for beech.

Beginning with an observed distribution of the canopy species in a stand in New Jersey known to be 25 years old, Horn modelled the changes in species composition over several centuries. The process is illustrated in simplified form in Table 17.3 (which deals with only four species out of those present). The progress of this hypothetical succession allows several predictions to be made. Red maple should dominate quickly, whilst grey birch disappears. Beech should slowly increase to

Table 17.3 The predicted percentage composition of a forest consisting initially of 100% grey birch. (After Horn, 1981.)

Age of forest (years):	0	50	100	150	200	∞	Data from old forest
Grey birch	100	5	1	0	0	0	0
Blackgum	0	36	29	23	18	5	3
Red maple	0	50	39	30	24	9	4
Beech	0	9	31	47	58	86	93

predominate later, with blackgum and red maple persisting at low abundance. All these predictions are borne out by what happens in the real succession.

The most interesting feature of Horn's matrix model is that, given enough time, it converges on a stationary, stable composition that is independent of the initial composition of the forest. The outcome is inevitable (it depends only on the matrix of replacement probabilities) and will be achieved whether the starting point is 100% grey birch or 100% beech (as long as adjacent areas provide a source of seeds of species not initially present), 50% blackgum and 50% red maple, or any other combination.

Since the model seems to be capable of generating quite accurate predictions, it may prove to be a useful tool in formulating plans for forest management. For example, Namkoong and Roberds (1974) used this modelling approach to support proposals for the management of Californian coastal redwood forests. However, the model is simplistic and its assumptions that transition probabilities remain constant over time and are not affected by historic factors, such as initial biotic conditions and order of arrival of species, are likely to be wrong in many cases (Facelli & Pickett, 1990). Moreover, an ideal theory of succession should not only predict, it should also explain. To do this, we need to consider the *biological* basis for the replacement values in the model, and here we have to turn to alternative models.

Connell and Slatyer's classification of mechanisms: facilitation, tolerance and inhibition

The examples of succession described in previous sections illustrate various underlying mechanisms, including facilitation and inhibition. An overview of mechanisms has been produced by Connell and Slatyer (1977). They propose three models, of which the first (facilitation) is the classical explanation most often invoked in the past, whilst the other two (tolerance and inhibition) may be equally important and have frequently been overlooked. The fundamental characteristics of these models are illustrated in Figure 17.16.

The essential feature of facilitation succession, in contrast with either the tolerance or inhibition models, is that early occupants change the abiotic environment in a way that makes it comparatively less suitable for themselves and more suitable for the recruitment of others. Thus, the entry and growth of the later species depends on earlier species preparing the ground. The most obvious examples are found in primary successions such as that occurring after glacial retreat. We cannot say how common this state of affairs is. However, it is by no means uncommon for plant species to alter the environment in a way that makes it more, rather than less, suitable for themselves (Wilson & Agnew, 1992).

The tolerance model suggests that a predictable sequence is produced because different species exploit resources in different ways. Later species are able to tolerate lower resource levels and can grow to maturity in the presence of early species, eventually outcompeting them. Old field successions provide our example.

The inhibition model applies when all species resist invasions of competitors, as described earlier for the seaweed–boulder succession. Later species gradually accumulate by replacing early individuals only when they die. An important distinction between the models is in the cause of death of the early colonists. In the case of facilitation and tolerance, they are killed in competition for resources, notably light and nutrients. In the case of the inhibition model, however, the early species are killed by very local disturbances caused by extreme physical conditions or the action of predators.

In his *resource-ratio* hypothesis of succession, Tilman (1988) places strong emphasis on the role of changing relative competitive abilities of plant species as

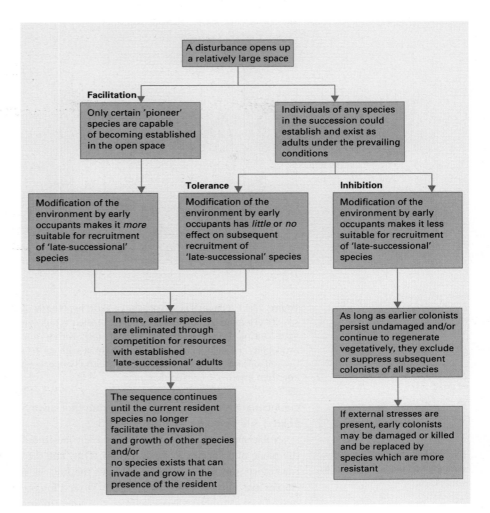

Figure 17.16 Three models of the mechanisms that underlie successions. (After Connell & Slatyer, 1977.)

Tilman's resource-ratio hypothesis emphasizes changing competitive abilities of plant species in a succession

conditions slowly change with time. He argues that species dominance at any point in a terrestrial succession is strongly influenced by the relative availability of two resources: a limiting soil nutrient (often nitrogen) and light. Early in succession, the habitat experienced by seedlings has low nutrient but high light availability. As a result of litter input and the activities of decomposer organisms, nutrient availability increases with time—this can be expected to be particularly marked in primary successions that begin with a very poor soil (or no soil at all). Total plant biomass increases with time and, in consequence, light penetration to the soil surface decreases. Tilman's ideas are illustrated in Figure 17.17 for five hypothetical species. Species A has the lowest requirement for the nutrient and the highest requirement for light at the soil surface. It has a short, prostrate growth form. Species E, which is the superior competitor in high-nutrient, low-light habitats, has the lowest requirement for light and the highest for the nutrient. It is a tall, erect species. Species B, C and D are intermediate in their requirements and each reaches its peak abundance at a different point along the soil nutrient : light gradient. There is plenty of scope for experimental testing of the hypothesis. Of course, many other factors also may influence the pattern of autogenic succession, including differences in species'

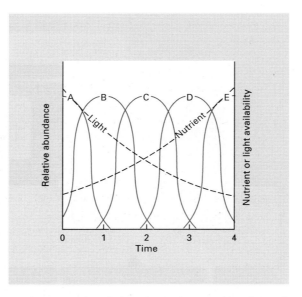

Figure 17.17 Tilman's (1988) resource-ratio hypothesis of succession. Five hypothetical plant species are assumed to be differentiated in their requirements for a limiting soil nutrient and light. During the succession, the habitat starts with a nutrient-poor soil but high light availability, changing gradually into a habitat with a rich soil but low availability of light at the soil surface. Relative competitive abilities change as conditions vary, and first one species and then another comes to dominate.

colonizing ability, maximal growth rate (not always directly related to competitive ability) and response to grazing.

Noble and Slatyer (1979; 1981) have tried to define the qualities that determine the place of a species in a succession. They call these properties *vital attributes*. The two most important relate: (i) to the method of recovery after disturbance (four classes are defined: vegetative spread, V; seedling pulse from a seed bank, S; seedling pulse from abundant dispersal from the surrounding area, D; no special mechanism with just moderate dispersal from only a small seed bank, N); and (ii) the ability of individuals to reproduce in the face of competition (defined in terms of tolerance T at one extreme and intolerance I at the other). Thus, for example, a species may be classed as SI if disturbance releases a seedling pulse from a seedbank, and if the plants are intolerant of competition (being unable to germinate or grow in competition with older or more advanced individuals of either their own or another species). Seedlings of such a species could establish themselves only immediately after a disturbance, when competitors are rare. Of course, a seedling pulse fits well with such a pioneer existence. An example is the annual *Ambrosia artemisiifolia*, already discussed in the context of old field succession.

In contrast, the American beech (*Fagus grandifolia*) could be classed as VT (being able to regenerate vegetatively from root stumps, and tolerant of competition since it is able to establish itself and reproduce in competition with older or more advanced individuals of either its own or another species), or NT (if no stumps remain, it would invade slowly via seed dispersal). In either case, it would eventually displace other species and form part of the 'climax' vegetation. Noble and Slatyer argue that it should be possible to classify all the species in an area according to these two vital attributes (to which relative longevity might be added as a third). Given this

the 'vital attributes' of species strongly influence their role in succession

information, quite precise predictions about successional sequences should be possible.

Consideration of vital attributes from an evolutionary point of view suggests that certain attributes are likely to occur together more often than by chance. We can envisage two alternatives that might increase the fitness of an organism in a succession (Harper, 1977), either: (i) the species reacts to the competitive selection pressures and evolves characteristics that enable it to persist longer in the succession, i.e. it responds to K selection; or (ii) it may develop more efficient mechanisms of escape from the succession, and discover and colonize suitable early stages of succession elsewhere, i.e. it responds to r selection (see Chapter 14). Thus, it may be that good colonizers will generally be poor competitors and vice versa. This is evident in Table 17.4, which lists some physiological characteristics that tend to go together in early and late successional plants.

The structure of communities and the successions within them are usually treated as essentially botanical matters. There are obvious reasons for this. Plants commonly provide most of the biomass and the physical structure of communities; moreover, plants do not hide or run away and this makes it rather easy to assemble species lists, determine abundances and detect change. The massive contribution that plants make to determining the character of a community is not just a measure of their role as the primary producers, it is also a result of their slowness to decompose. The plant population not only contributes biomass to the community,

Table 17.4 Physiological characteristics of early and late successional plants. (After Bazzaz, 1979.)

Attribute	Early successional plants	Late successional plants
Seed dispersal in time	Well dispersed	Poorly dispersed
Seed germination:		
enhanced by		
light	Yes	No
fluctuating temperatures	Yes	No
high NO_3^-	Yes	No
inhibited by		
far-red light	Yes	No
high CO_2 concentration	Yes	No?
Light saturation intensity	High	Low
Light compensation point	High	Low
Efficiency at low light	Low	High
Photosynthetic rates	High	Low
Respiration rates	High	Low
Transpiration rates	High	Low
Stomatal and mesophyll resistances	Low	High
Resistance to water transport	Low	High
Recovery from resource limitation	Fast	Slow
Resource acquisition rates	Fast	Slow?

necromass plays a crucial role
in the late successional
dominance of trees ...

but is also a major contributor of *necromass*. Thus, unless microbial and detritivore activity is fast, dead plant material accumulates as leaf litter or as peat. Moreover, the dominance of trees in so many communities comes about because they accumulate dead material; the greater part of a tree's trunk and branches is dead. The tendency in mesic habitats for shrubs and trees to succeed herbaceous vegetation comes largely from their ability to hold leaf canopies (and root systems) on an extending skeleton of predominantly dead support tissue (the heart wood).

Animal bodies decompose much more quickly, but there are situations in which animal remains, like those of plants, can determine the structure and succession of a community. This happens when the animal skeleton resists decomposition, as is the case in the accumulation of calcified skeletons during the growth of corals. A coral reef, like a forest or a peat bog, gains its structure, and drives its successions, by accumulating its dead past. Reef-forming corals, like forest trees, gain their dominance in their respective communities by holding their assimilating parts progressively higher on predominantly dead support. In both cases, the organisms have an almost overwhelming effect on the abiotic environment, and they 'control' the lives of other organisms within it. The coral reef community (dominated by an animal, albeit one with a plant symbiont) is as structured, diverse and dynamic as a tropical rainforest.

The fact that plants dominate most of the structure and succession of communities does not mean that the animals always follow the communities that plants dictate. This will often be the case, of course, because the plants provide the starting point for all food webs and determine much of the character of the physical environment in which animals live. But, it is also sometimes the animals that determine the nature of the plant community. We have already seen that part of the reason why the green seaweed *Ulva* is gradually replaced on boulders by tougher red species is because crabs feed preferentially on it.

animals are often affected by,
but may also affect, plant
successions

The terrestrial environment of a recently ploughed field in southern England provides another example of a similar phenomenon. The plant succession begins with the appearance of annual forbs (non-grass species) to be replaced, within 1 year or so, by perennial grasses and perennial forbs. Herbivorous insects are associated with the plants, feeding both on leaves (sap-feeding hemipterans) and on roots (chewing larvae of beetles and flies). Reducing leaf herbivory (by application of an insecticide) resulted in large increases in perennial grass growth. Reducing subterranean herbivory prolonged the persistence of annual forbs and greatly increased perennial forb colonization (Brown & Gange, 1992). Leaf-feeding insects thus act to delay succession by slowing grass colonization. In contrast, root-feeding insects accelerate succession by reducing forb persistence and colonization.

leaf-feeding and root-feeding
herbivores affect an old field
succession in different ways

A final, tropical example, on a much larger scale, comes from the savannah at Ndara in Kenya. The vegetation in savannahs is often held in check by grazers (regular burning can also be very important; Deshmukh, 1986). Experimental exclusion of elephants from a plot of savannah led to a more than threefold increase in the density of trees (to approximately 2800 ha^{-1} (0.28 m^{-2})) over a 10-year period (work by Oweyegha-Afundaduula, reported in Deshmukh, 1986).

More often though, animals are passive followers of successions amongst the plants. This is certainly the case for passerine bird species in an old field succession (Figure 17.18). Vesicular–arbuscular mycorrhizal fungi (see Chapter 13, Section 13.7.2), which show a clear sequence of species replacement in the soils associated with an old-field succession (Johnson *et al.*, 1991), may also be passive followers of

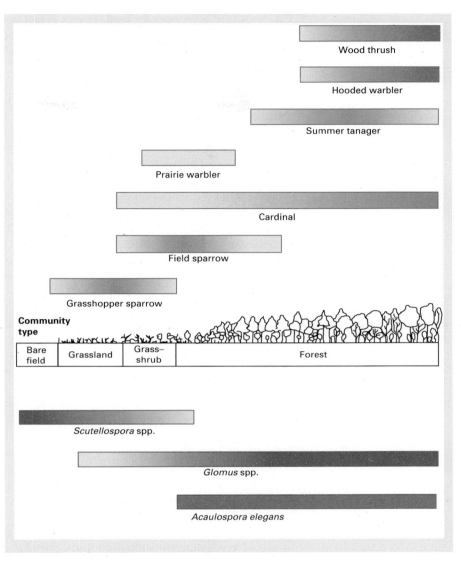

Figure 17.18 Top: Bird species distributions along a plant succession gradient in the Piedmont region of Georgia, USA. Differential shading indicates relative abundance of the birds. (After Johnston & Odum, 1956; from Gathreaux, 1978.) Bottom: Distributions of vesicular–arbuscular mycorrhizae in the soils associated with an old field succession in Minnesota, USA. Differential shading indicates relative abundance of spores of species in the genera *Scutellospora*, *Glomus* and *Acaulospora*. (After Johnson *et al.*, 1991.)

the plants. But, this does not mean that the birds, which eat seeds, or the fungi, which affect plant growth and survival, do not influence the succession in its course. They probably do.

17.4.5 The concept of the climax

Do successions come to an end? It is clear that a stable equilibrium will occur if individuals that die are replaced on a one-to-one basis by young of the same species.

At a slightly more complex level, Horn's model (see Section 17.4.4) tells us that a stationary species composition should, in theory, occur whenever the replacement probabilities (of one species by itself or by any one of several others) remain constant through time.

The concept of the climax has a long history. One of the earliest students of succession, Frederic Clements (1916), is associated with the idea that a single climax will dominate in any given climatic region, being the endpoint of all successions, whether they happened to start from a sand-dune, an abandoned old field or even a pond filling in and progressing towards a terrestrial climax. The *monoclimax* view was challenged by many ecologists, amongst whom Tansley (1939) was prominent. The *polyclimax* school of thought recognized that a local climax may be governed by one factor or a combination of factors: climate, soil conditions, topography, fire, and so on. Thus, a single climatic area could easily contain a number of specific climax types. Later still, Whittaker (1953) proposed his climax-pattern hypothesis. This conceives a continuity of climax types, varying gradually along environmental gradients and not necessarily separable into discrete climaxes. (This is an extension of Whittaker's approach to gradient analysis of vegetation, discussed in Section 17.3.1.)

climaxes may be approached rapidly—or, so slowly that they are rarely if ever reached

In fact, it is very difficult to identify a stable climax community in the field. Usually, we can do no more than point out that the rate of change of succession slows down to the point where any change is imperceptible to us. In this context, the seaweed–boulder succession discussed in Section 17.4.3 is unusual in that convergence to a climax took only a few years. Old field successions might take 100–300 years to reach a 'climax', but in that time the probabilities of fire or hurricanes occurring (e.g. every 70 or so years in New England, USA) are so high that a process of succession may rarely go to completion. If we bear in mind that forest communities in northern temperate regions, and probably also in the tropics, are still recovering from the last glaciation (see Chapter 1), it is questionable whether the idealized climax vegetation is often reached in nature.

Finally, we again return to issues of scale which feature in almost every chapter of this book. A forest, or a rangeland, that appears to have reached a stable community structure when studied on a scale of hectares, will always be a mosaic of miniature successions. Every time a tree falls or a grass tussock dies, an opening is created in which a new succession starts. One of the most seminal papers published in the history of ecology was entitled 'Pattern and process in the plant community' (Watt, 1947). Part of the pattern of a community is caused by the dynamic processes of deaths, replacements and microsuccessions that the broad view may conceal. We return to some of these themes in Chapter 21.

Chapter 18
The Flux of Energy
through Communities

18.1 Introduction

All biological entities require matter for their construction and energy for their
activities. This is true not only for individual organisms, but also for the populations
and communities that they form in nature. It is tempting to imagine that we could
treat the processing of matter and energy by communities in the same way as we
have done for individual organisms in earlier chapters. But, in Chapter 17 we cast
aside the view that 'the community' could be regarded as a sort of superorganism
(the view of Clements, 1916). A major problem is that 'the community' is something
that has quite different significance to different sorts of organism—a worm's-eye-
view of a 'community' would be quite different from a protozoan's-eye-view, or that
of a caterpillar or an elephant. For this reason, 'the community' has no scale—other
than one that we impose upon it. We look at the world through human eyes and
adopt scales that happen to be appropriate for our own species.

the community perspective—in the eye of the beholder

In this chapter, and the next, the intrinsic importance of fluxes of energy (this
chapter) and of matter (Chapter 19) means that community processes are particularly strongly linked with the abiotic environment. When researchers focus on this
level of organization, the term *ecosystem* is used to denote the biological community
together with the abiotic environment in which it is set. Thus, ecosystems normally
include primary producers, decomposers and detritivores, a pool of dead organic
matter, herbivores, carnivores and parasites, plus the physicochemical environment
that provides living conditions and acts both as a source and a sink for energy and
matter.

communities and ecosystems

A strongly developed character of our species is that a wide range of resources
are substitutable for each other. Work may be exchanged for food; one food can be
exchanged for another; food can be exchanged for work; gold can be exchanged for
housing; even, in some communities, cattle can be exchanged for wives. There is
little, if anything, that is comparable in other species. In effect, dollars and pounds
have come to be surrogates for resources. Areas of land come into this category of
resources that can be bought and sold—or swapped for other resources. Even areas
of rainforest worthy of conservation can be exchanged for currency in agreements
with the International Bank (as in Costa Rica) and areas of the oceans are defined,
and even fought for, because of their financial value for fisheries.

the importance of area

For these reasons, it is appropriate to describe the activities of communities and
ecosystems *per unit area*. Of course, it is not the area itself which has value but the
light and water that fall upon it and the nutrients that it supplies. It is these that
farmers, foresters, fishers or conservationists really pay for when they buy an area.

In this chapter, we are concerned with the ways in which areas of land or water receive and process incident radiation. This source of energy places the ultimate limits on the biological activities of any area of land or water.

A classic paper by Lindemann (1942) laid the foundations of a science of ecological energetics. He attempted to quantify the concept of food chains and food webs by considering the efficiency of transfer between trophic levels—from incident radiation received by a community, through its capture by green plants in photosynthesis to its subsequent use by herbivores, carnivores and decomposers. Lindemann's paper was a major (although not the only) catalyst that stimulated the

International Biological Programme (IBP). The subject of the IBP was defined as 'the biological basis of productivity and human welfare'. Recognizing the problem of a rapidly increasing human population, it was considered that scientific knowledge would be required as a basis for rational resource management. Cooperative international research programmes were directed especially to study the productivity and ecological energetics of areas of land, freshwaters and the seas (Worthington, 1975).

The IBP provided the first occasion on which biologists throughout the world were challenged to work together towards a common end. The results produced through the IBP form the basis of some of this chapter. More recently, a further pressing issue has again galvanized the ecological community into action. Deforestation, the burning of fossil fuels and other pervasive human influences are expected to cause dramatic changes to global climate and atmospheric composition, and can be expected in turn to influence patterns of productivity and the composition of vegetation on a global scale. The International Geosphere–Biosphere Programme

(IGBP) has established a core project on Global Change and Terrestrial Ecosystems (GCTE) with the prime objective of predicting the effects of changes in climate, atmospheric composition and land use on terrestrial ecosystems, including agricultural and production forest systems (Steffen *et al.*, 1992). Early results of this effort, and of other independent research, are also presented below.

The bodies of the living organisms within a unit area constitute a *standing crop* of biomass. By *biomass* we mean the mass of organisms per unit area of ground (or water) and this is usually expressed in units of energy (e.g. $J\,m^{-2}$) or dry organic matter (e.g. tonnes ha^{-1}). The great bulk of the biomass in communities is almost always formed by green plants, which are the primary producers of biomass because of their almost unique ability to fix carbon in photosynthesis. (We have to say 'almost unique' because bacterial photosynthesis and chemosynthesis may also contribute to forming new biomass.) Biomass includes the whole bodies of the organisms even though parts of them may be dead. This needs to be borne in mind, particularly when considering woodland and forest communities in which the bulk of the biomass is dead heartwood and bark. We often need to distinguish this mass of dead

material, the *necromass*, from the living, active fraction of the biomass. The latter represents active capital capable of generating interest in the form of new growth, whereas the necromass is incapable of new growth. In practice, we include in biomass all those parts, living or dead, that are attached to the living organism. They cease to be biomass when they fall off and become litter, humus or peat.

The *primary productivity* of a community is the rate at which biomass is produced *per unit area* by plants, the primary producers. It can be expressed either in units of energy (e.g. $J\,m^{-2}\,day^{-1}$) or of dry organic matter (e.g. $kg\,ha^{-1}\,year^{-1}$). The total fixation of energy by photosynthesis is referred to as gross primary productivity

(GPP). A proportion of this is respired away by the plants themselves and is lost from the community as respiratory heat (R). The difference between GPP and R is known as net primary productivity (NPP) and represents the actual rate of production of new biomass that is available for consumption by heterotrophic organisms (bacteria, fungi and animals). The rate of production of biomass by heterotrophs is called *secondary productivity*.

18.2 Patterns in primary productivity

Global terrestrial NPP is estimated to be $110–120 \times 10^9$ tonnes dry weight per year, and in the sea $50–60 \times 10^9$ tonnes per year (Leith, 1975; Whittaker, 1975; Rodin *et al.*, 1975). Thus, although oceans cover about two-thirds of the world's surface, they account for only one-third of its production. The fact that productivity is not evenly spread across the earth is further emphasized in the maps of world NPP in Figure 18.1 and in Table 18.1, which provides estimates of annual NPP and plant standing crop biomass for the principal biomes on land, for freshwater lakes and streams and for marine systems. As we are concerned in this chapter with the flux of energy in communities, it is instructive to compare the maps in Figure 18.1 with Chapter 3, Figure 3.1, which shows the distribution of solar radiation incident on the earth's surface. Although all biological activity is ultimately dependent on received solar radiation, it is obvious that solar radiation alone does not determine primary productivity. In very broad terms, the discrepancies between the maps occur because incident radiation can be captured efficiently only when water and nutrients are available and when temperatures are in the range suitable for plant growth. Many areas of land receive abundant radiation but lack adequate water, and most areas of the oceans are deficient in mineral nutrients.

The comparisons between terrestrial and aquatic systems may be biased against terrestrial communities because of the enormous technical problems in measuring the standing crop of underground parts—measuring productivity below ground is even more of a problem. It is said that one attempt to extract tree roots for measurement from a woodland site in California, by using dynamite, resulted in most of the root material landing in Nevada. Although this story may be a myth, it illustrates the problem. Most measurements of primary productivity in terrestrial systems represent measures of above-ground parts. Yet, primary production below ground may equal or be greater than that above ground.

A large proportion of the globe produces less than $400 \, \mathrm{g \, m^{-2} \, year^{-1}}$. This includes over 30% of the land surface (Figure 18.1, top) and 90% of the ocean (Figure 18.1, bottom). The open ocean is, in effect, a marine desert. At the other extreme, the most productive systems are found amongst swamp and marshland, estuaries, algal beds and reefs and cultivated land.

In the forest biomes of the world, there is a general latitudinal trend of increasing productivity from boreal, through temperate, to tropical conditions. The same trend may exist in the productivity of tundra and grassland communities (Figure 18.2a), and in various cultivated crops (Figure 18.2b), but there is much variation. In aquatic communities this trend is clear in lakes (Figure 18.2c), but not in the oceans (Figure 18.1, bottom). The trend with latitude suggests that radiation (a resource) and temperature (a condition) may be the factors usually limiting the productivity of communities. Other factors can, however, constrain productivity within even narrower limits.

biological activity depends upon, but is not solely determined by, solar radiation

below-ground productivity is almost certainly underestimated

the productivity of forests, grasslands, crops and lakes follows a latitudinal pattern ...

713   FLUX OF ENERGY

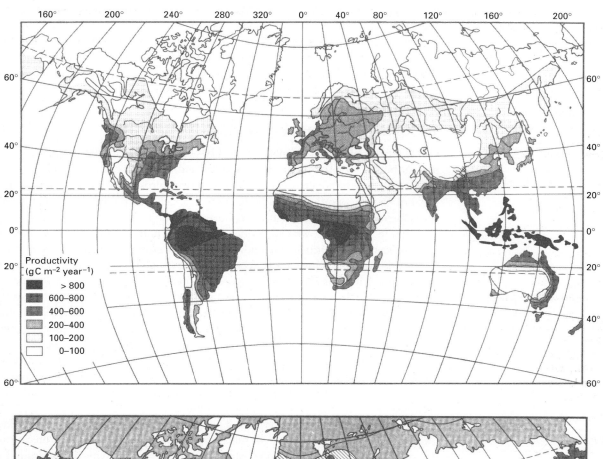

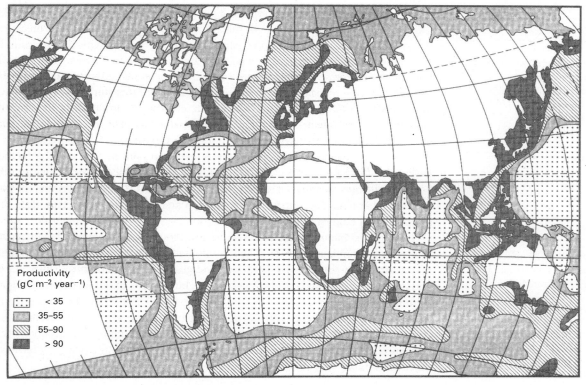

714 CHAPTER 18

Table 18.1 Annual net primary productivity (NPP) and standing crop biomass estimates for contrasting communities of the world. (After Whittaker, 1975.)

Ecosystem type	Area (10^6 km²)	NPP, per unit area (g m⁻² *or* t km⁻²) Normal range	Mean	World NPP (10⁹ t)	Biomass per unit area (kg m⁻²) Normal range	Mean	World biomass (10⁹ t)
Tropical rainforest	17.0	1000–3500	2200	37.4	6–80	45	765
Tropical seasonal forest	7.5	1000–2500	1600	12.0	6–60	35	260
Temperate evergreen forest	5.0	600–2500	1300	6.5	6–200	35	175
Temperate deciduous forest	7.0	600–2500	1200	8.4	6–60	30	210
Boreal forest	12.0	400–2000	800	9.6	6–40	20	240
Woodland and shrubland	8.5	250–1200	700	6.0	2–20	6	50
Savannah	15.0	200–2000	900	13.5	0.2–15	4	60
Temperate grassland	9.0	200–1500	600	5.4	0.2–5	1.6	14
Tundra and alpine	8.0	10–400	140	1.1	0.1–3	0.6	5
Desert and semi-desert shrub	18.0	10–250	90	1.6	0.1–4	0.7	13
Extreme desert, rock, sand and ice	24.0	0–10	3	0.07	0–0.2	0.02	0.5
Cultivated land	14.0	100–3500	650	9.1	0.4–12	1	14
Swamp and marsh	2.0	800–3500	2000	4.0	3–50	15	30
Lake and stream	2.0	100–1500	250	0.5	0–0.1	0.02	0.05
Total continental	149		773	115		12.3	1837
Open ocean	332.0	2–400	125	41.5	0–0.005	0.003	1.0
Upwelling zones	0.4	400–1000	500	0.2	0.005–0.1	0.02	0.008
Continental shelf	26.6	200–600	360	9.6	0.001–0.04	0.01	0.27
Algal beds and reefs	0.6	500–4000	2500	1.6	0.04–4	2	1.2
Estuaries	1.4	200–3500	1500	2.1	0.01–6	1	1.4
Total marine	361		152	55.0		0.01	3.9
Full total	510		333	170		3.6	1841

In the sea, productivity may more often be limited by a shortage of nutrients—very high productivity occurs in marine communities where there are upwellings of nutrient-rich waters, even at high latitudes and low temperatures. In terrestrial systems, the overall trend of increasing productivity with decreasing latitude is upset in areas where water is in short supply, for example the continental interior of Australia.

Small differences in topography can result in large differences in community productivity. For example, in tundra a distance of a few metres, from beach ridge to meadow with impeded drainage, can change primary productivity from less than $10\,\text{g m}^{-2}\,\text{year}^{-1}$ to $100\,\text{g m}^{-2}\,\text{year}^{-1}$ (Devon Island, Canada). In favourable sites in Greenland and in South Georgia (Antarctica), productivities of tundra communities may reach $2000\,\text{g m}^{-2}\,\text{year}^{-1}$, which is higher than that recorded in many temperate communities. Thus, although there is a general latitudinal trend, there is also a broad spectrum of variation at a given latitude resulting from contrasting microclimates.

... which is frequently modified locally

Figure 18.1 (*facing page*) Top: World-wide pattern of net primary productivity on land. (After Reichle, 1970.) Bottom: World-wide pattern of net primary productivity in the oceans. (After Koblentz-Mishke *et al.*, 1970.)

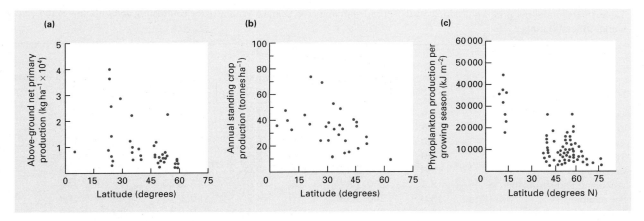

Figure 18.2 Latitudinal patterns in net primary productivity of the following. (a) Grassland and tundra ecosystems. (After Cooper, 1975.) (b) Cultivated crops. (After Cooper, 1975.) (c) Lakes. (After Brylinski & Mann, 1973.)

18.2.1 Aquatic communities: autochthonous and allochthonous material

All biotic communities depend on a supply of energy for their activities. In most terrestrial systems this is contributed *in situ* by the photosynthesis of green plants. Organic matter (and fixed energy) generated within the community is called *autochthonous*. In aquatic communities, the autochthonous input is through the photosynthesis of large plants and attached algae in shallow waters, and by microscopic plankton in the open water. However, a substantial proportion of the organic matter (energetic resource) in such communities often comes into the community as dead material that has been formed outside it. Such *allochthonous* material arrives in rivers or is blown in by the wind. The relative importance of the two autochthonous sources and the allochthonous source of organic material in an aquatic system depends on the dimensions of the body of water and the types of terrestrial community that deposit organic material into it.

A small stream running through a wooded catchment derives most of its energy input from litter shed by surrounding vegetation (Figure 18.3). Shading from the trees prevents any significant growth of planktonic or attached algae, or aquatic higher plants. As the stream widens further downstream, shading by trees is restricted to the margins, and autochthonous primary production increases. Still further downstream, in deeper and more turbid waters, rooted higher plants contribute much less, and the role of the microscopic phytoplankton becomes more important. Where large river channels are characterized by a floodplain, with associated oxbow lakes, swamps and marshes, allochthonous dissolved and particulate organic matter is carried to the river channel during episodes of flooding (Junk *et al.*, 1989; Townsend, 1996a).

The sequence from small, shallow lakes to large, deep ones shares some of the characteristics of the river continuum just discussed (Figure 18.4). A small lake is likely to derive quite a large proportion of its energy from the land because its periphery is large in relation to lake area. Small lakes are also usually shallow, so internal littoral production is more important than that by phytoplankton. In contrast, a large, deep lake will derive only limited organic matter from outside (small periphery relative to lake surface area), and littoral production, limited to the

variations along 'the river continuum' ...

paralleled by lakes

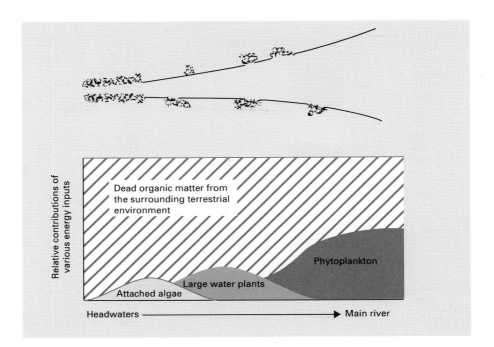

Figure 18.3 Longitudinal variation in the nature of the energy base in stream communities.

the oceans, estuaries and continental shelves

shallow margins, may also be low. The organic inputs to the community may then be due almost entirely to photosynthesis by the phytoplankton.

In a sense, the open ocean can be described as the largest, deepest 'lake' of all. Input of organic material from terrestrial communities is negligible, and the great depth precludes photosynthesis in the darkness of the seabed. The phytoplankton are then all-important as primary producers.

Estuaries are often highly productive systems, but the most important contribution to their energy-base varies. In large estuarine basins, with restricted interchange with the open ocean and with small marsh peripheries relative to basin area, phytoplankton tend to dominate. By contrast, seaweeds dominate in some open basins with extensive connections to the sea.

Finally, continental-shelf communities derive a proportion of their energy from terrestrial sources (particularly via estuaries), and their shallowness often provides for significant production by littoral seaweed communities. Indeed, some of the most productive systems of all are to be found amongst seaweed beds and reefs (Table 18.1).

18.2.2 Variations in the relationship of productivity to biomass

We can relate the productivity of a community to the standing crop that produces it (the interest rate on the capital). Alternatively, we can think of the standing crop as the biomass that is sustained by the productivity (the capital resource that is sustained by earnings). The average values of productivity (P) and standing crop biomass (B) for the community types listed in Table 18.1 are plotted against each other in Figure 18.5. It is evident that a given value of NPP is produced by a smaller biomass when non-forest terrestrial systems are compared with forests, and the biomass involved is smaller still when aquatic systems are considered. Thus, P : B

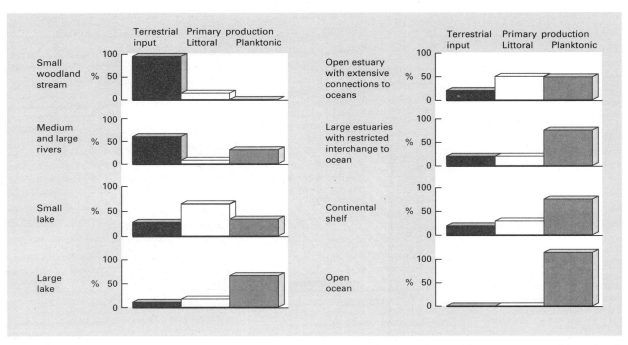

Figure 18.4 Variation in the importance of terrestrial input of organic matter and littoral and planktonic primary production in contrasting aquatic communities.

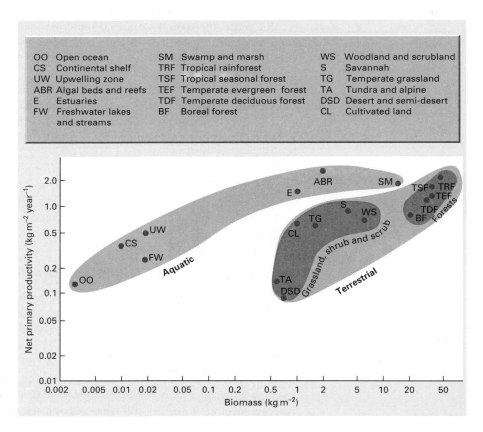

Figure 18.5 The relationship between average net primary productivity and average standing crop biomass for the community types listed in Table 18.1.

P : B ratios are very low in
forests and very high in
aquatic communities

ratios (i.e. kilograms produced per year per kilogram of standing crop) average 0.042 for forests, 0.29 for other terrestrial systems and 17 for aquatic communities. The major reason for this is almost certainly that a large proportion of forest biomass is dead (and has been so for a long time), and also that much of the living support tissue is not photosynthetic. In grassland and scrub, a greater proportion of the biomass is alive and involved in photosynthesis, although 50% or more of the biomass may be roots. In aquatic communities, particularly where productivity is due mainly to phytoplankton, there is no support tissue, no need for roots to absorb water and nutrients, dead cells do not accumulate (they are usually eaten before they die) and photosynthetic output per kilogram of biomass is thus very high indeed. Another factor that helps to account for high P : B ratios in phytoplankton communities is the rapid turnover of biomass. The annual NPP shown in Figure 18.5 is actually produced by a succession of overlapping phytoplankton generations, whilst the standing crop biomass is only the average present at an instant.

A general feature of autogenic successions (see Chapter 17, Section 17.4.3) is that the pioneers are rapidly growing herbaceous species with relatively little support tissue. Thus, early in succession the P : B ratio is high. However, the species that come to dominate later are generally slow growing, but they eventually achieve a large size and come to monopolize the supply of space and light. Their structure involves considerable investment in non-photosynthesizing and dead support tissue, and as a consequence their P : B ratio is low. This principle is illustrated in Figure 18.6.

There is another way of looking at the figures of productivity in relation to biomass. This is to say that biomass as we define it is a quite unrealistic way of measuring the standing crop of a community. For example, is it justifiable to treat the

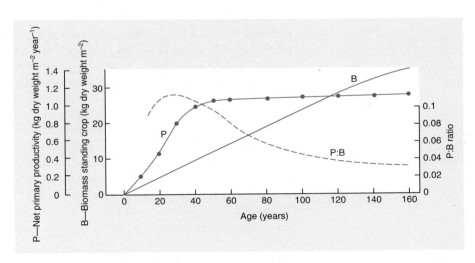

Figure 18.6 Annual above-ground net primary productivity (P), standing crop biomass (B) and P : B ratio in forest succession following fire on Long Island, New York. Production increased rapidly through herb and shrub stages to a forest with a stable production of about 1.05 kg m^{-2} year^{-1} after 40–50 years. Biomass was still increasing in this period and would be expected to achieve a value of about 40 kg m^{-2} in a mature oak forest after about 200 years. The P : B ratio rose to a value of about 0.1 in 20–40 years after the fire, and then declined to a value of only 0.03 after 160 years. (After Whittaker & Woodwell, 1968; 1969.)

dead wood and bark of a tree as if they were in any direct way causal in determining the rate at which that tree makes new dry matter (or fixes energy)? Nearly all the differences in community productivity that we see between community types arise because we choose to define biomass in this way. It might be much more sensible to define biomass in terms of weight of living tissues (if we could find a way to measure it). It is certain that a large part of the difference between communities in their P : B ratio would then disappear. The differences that remain could be much more informative. Unfortunately, such measurements have not been made.

18.3 Factors limiting primary productivity

18.3.1 Terrestrial communities

Sunlight, carbon dioxide (CO_2), water and soil nutrients are the resources required for primary production on land, whilst temperature, a condition, has a strong influence on the rate of photosynthesis. CO_2 is normally present at around 0.03% of atmospheric gases. Turbulent mixing and diffusion prevent CO_2 concentration varying from place to place, except in the immediate neighbourhood of a leaf, and it seems to play no significant role in determining differences between the productivities of different communities (although global increases in CO_2 concentration are expected to have profound effects). On the other hand, the quality and quantity of light, the availability of water and nutrients and temperature all vary dramatically from place to place. They are all candidates for the role of limiting factor. Which of them actually sets the limit to primary productivity?

terrestrial communities use radiation inefficiently

Depending on location, something between 0 and 5 J of solar energy strikes each 1 m^2 of the earth's surface every minute. If all this were converted by photosynthesis to plant biomass (i.e. if photosynthetic efficiency were 100%) there would be a prodigious generation of plant material, one or two orders of magnitude greater than recorded values. Much of this solar energy is unavailable for use by plants. In particular, only about 44% of incident short-wave radiation occurs at wavelengths suitable for photosynthesis. However, even when this is taken into account, productivity still falls well below the maximum possible (see also Chapter 3). Figure 18.7 plots on a log scale, net photosynthetic efficiencies (percentage of incoming photosynthetically active radiation (PAR) incorporated into above-ground NPP). The data were gathered during the IBP, in the USA, from seven coniferous forests, seven deciduous forests and eight desert communities. The conifer communities had the highest efficiencies, but these were only between 1 and 3%. For a similar level of incoming radiation, deciduous forests achieved 0.5–1%, and, despite their greater energy income, deserts were able to convert only 0.01–0.2% of PAR to biomass. These values can be compared with short-term peak efficiencies achieved by crop plants under ideal conditions, when values from 3 to 10% can be achieved (Cooper, 1975).

but productivity may still be limited by a shortage of PAR

However, the fact that radiation is not used efficiently does not in itself imply that it does not limit community productivity. We would need to know whether at increased intensities of radiation the productivity increased or remained unchanged. Some of the evidence given in Chapter 3 shows that the intensity of light during part of the day is below the optimum for canopy photosynthesis. Moreover, at peak light intensities, most canopies still have their lower leaves in relative gloom, and would almost certainly photosynthesize faster if the light intensity were higher. For C_4

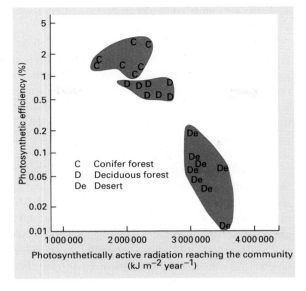

Figure 18.7 Photosynthetic efficiency (percentage of incoming photosynthetically active radiation converted to above-ground net primary productivity) for three sets of terrestrial communities in the USA. (After Webb *et al.*, 1983.)

plants, a saturating intensity of radiation never seems to be reached, and the implication is that productivity may in fact be limited by a shortage of PAR even under the brightest natural radiation.

There is no doubt, however, that what radiation is available would be used more efficiently if other resources were in abundant supply. The much higher values of community productivity from agricultural systems bear witness to this.

shortage of water may be a critical factor

The general relationship between above-ground NPP and precipitation for forests of the world is illustrated in Figure 18.8a. Water is an essential resource both as a constituent of cells and for photosynthesis. Large quantities of water are lost in transpiration—particularly because the stomata need to be open for much of the time for CO_2 to enter. It is not surprising that the rainfall of a region is quite closely correlated with its productivity. In arid regions, there is an approximately linear increase in NPP with increase in precipitation, but in the more humid forest climates there is a plateau beyond which productivity does not continue to rise. Note that a large amount of precipitation is not necessarily equivalent to a large amount of water

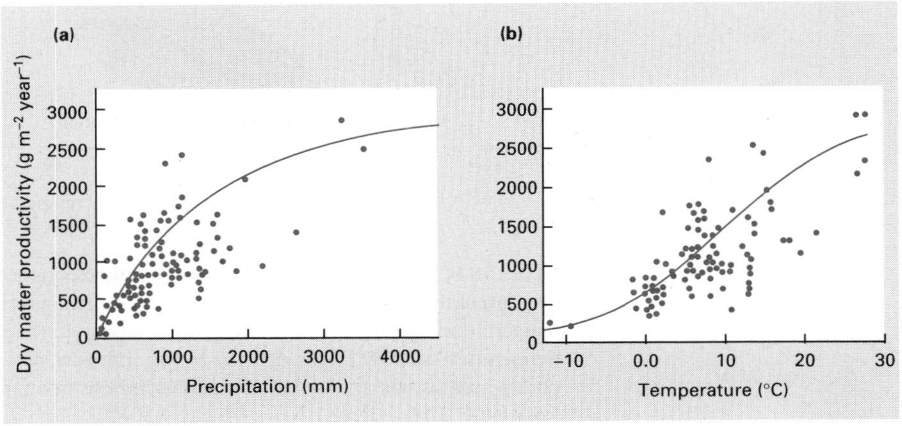

Figure 18.8 Relationships between forest net primary productivity and (a) annual precipitation, and (b) temperature. (After Reichle, 1970.)

available for plants; all water in excess of field capacity will drain away if it can (see Chapter 3, Figure 3.10).

There is a clear relationship between the above-ground NPP and mean annual temperature (Figure 18.8b), but the explanation is complex. Increasing temperature leads to increasing rates of gross photosynthesis, but there is a law of diminishing returns. Respiration, by contrast, tends to increase almost exponentially with temperature. The result is that net photosynthesis is maximal at temperatures well below those for gross photosynthesis (Figure 18.9). Higher temperatures are also associated with rapid transpiration and, thus, they increase the rate at which water shortage may become important.

At several of the study sites illustrated in Figure 18.7, and also for a grassland site, estimates of annual precipitation (mm per year) are available. In addition, *potential evapotranspiration* (PET) was calculated. This is an index of the theoretical maximum rate at which water might evaporate into the atmosphere (mm per year) given the prevailing radiation, average vapour pressure deficit in the air, windspeed and temperature. PET minus precipitation provides, for each site, a crude index of how far the water available for plant growth falls below what might be transpired by actively growing vegetation. Its relationship with above-ground NPP is shown in Figure 18.10. Clearly, drought is one of the characteristic features of some of the communities with low productivity.

Water shortage has direct effects on the rate of plant growth but also leads to the development of less dense vegetation. Vegetation that is sparse intercepts less light (much of which falls on bare ground). This wastage of solar radiation is the main cause of the low productivity in many arid areas, rather than the reduced photosynthetic rate of the droughted plants. This point is made by comparing the productivity per unit weight of leaf biomass instead of per unit area of ground. Coniferous forest produced $1.64 \, \mathrm{g \, g^{-1} \, year^{-1}}$; deciduous forest $2.22 \, \mathrm{g \, g^{-1} \, year^{-1}}$; grassland $1.21 \, \mathrm{g \, g^{-1} \, year^{-1}}$; and desert $2.33 \, \mathrm{g \, g^{-1} \, year^{-1}}$.

Leaf area index (LAI) is defined as the surface area of leaves per unit surface area

temperatures may be too low for rapid dry matter production

the interaction of temperature and precipitation

productivity and the structure of the canopy

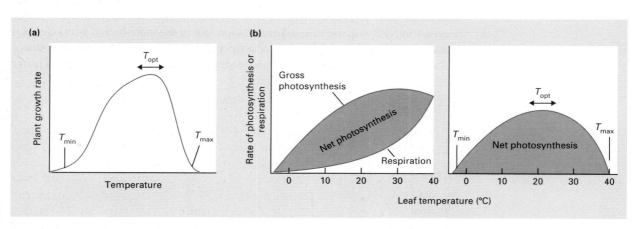

Figure 18.9 Schematic representations of plant responses to temperature. (a) A generalized diagram of the responses of plant growth rate to temperature, illustrating three critical temperatures, i.e. the minimum (T_{min}) and maximum (T_{max}) temperatures, and the optimum temperature range (T_{opt}) for growth. (b) The influence of temperature on gross photosynthesis, respiration and net photosynthesis in a typical plant. (After Pisek *et al.*, 1973; from Fitter & Hay, 1981.)

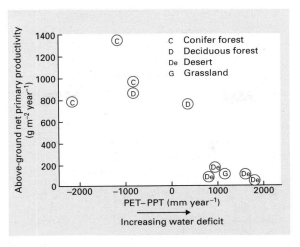

Figure 18.10 Relationship between above-ground net primary productivity and an index of water deficiency (potential evapotranspiration (PET) minus precipitation (PPT)) for several types of community in North America. (After Webb *et al.*, 1983.)

of ground. Desert vegetation possesses a lower LAI than forest, and in the example above this accounts for much of the difference in productivity. In general, as leaves are added to the canopy, the increase in LAI may be expected to lead to greater productivity; but eventually, because of shading, as we have seen in Chapter 3, a point is reached where leaves low in the canopy do not receive enough light to photosynthesize at a rate equal to their rate of respiration. Beyond this level, increasing LAI leads to decreasing productivity.

LAI is not the only structural feature that influences canopy productivity. Two other important attributes are: (i) the angle at which leaves are inclined; and (ii) the pattern of leaf density with depth in the canopy (see Chapter 3).

NPP increases with the length of the growing season

The productivity of a community can be sustained only for that period of the year when the plants bear photosynthetically active foliage. Deciduous trees have a self-imposed limit on the period of the year during which they bear foliage. Evergreen trees hold a canopy throughout the year, but during some seasons it may barely photosynthesize at all or even respire faster than it photosynthesizes. The latitudinal patterns in forest productivity seen earlier (see Figure 18.2) are largely the result of differences in the number of days when there is active photosynthesis. Figure 18.11 shows the results of part of an IBP study, in which the productivity of North American deciduous forest is related to the number of days of photosynthesis along a transect from Wisconsin and New York in the north to Tennessee and Carolina in the south.

A contrast between the productivity of a deciduous beech forest (*Fagus sylvatica*) and the evergreen Norway spruce (*Picea abies*) is shown in Table 18.2. The forests were within 1 km of each other under very similar abiotic conditions. Beech leaves photosynthesize at a greater rate (per gram dry weight) than those of spruce; even the shade leaves of beech, deep in its canopy, have a higher photosynthetic capacity than 1-year-old spruce needles. In addition, beech 'invests' a considerably greater amount of biomass in its leaves each year. However, these factors are easily outweighed by the greater length of the growing season of spruce, which shows positive CO_2 uptake during 260 days of the year compared with 176 days for beech.

NPP may be low because appropriate mineral resources are deficient

No matter how brightly the sun shines, how often the rain falls and how equable is the temperature, productivity must be low if there is no soil in a terrestrial community, or if the soil is deficient in essential mineral nutrients. The geological

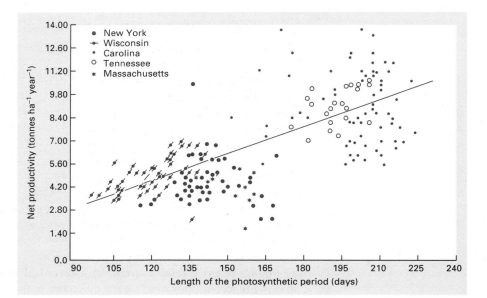

Figure 18.11 Relationship between net primary productivity of deciduous forests in North America and the length of the growing season. (After Leith, 1975.)

conditions that determine slope and aspect also determine whether a soil forms, and they have a large, although not wholly dominant, influence on the mineral content of the soil. For this reason, a mosaic of different levels of community productivity develops within a particular climatic regime. Of all the mineral nutrients, the one that has the most overriding influence on community productivity is fixed nitrogen (and this is invariably partly or mainly biological, not geological, in origin, as a result of nitrogen fixation by microorganisms). There is probably no agricultural system that does not respond to applied nitrogen by increased primary productivity, and this may well be true of natural vegetation as well. Nitrogen fertilizers added to forest soils almost always stimulate forest growth.

The deficiency of other elements can also hold the productivity of a community far below that of which it is theoretically capable. A classic example is deficiency of phosphate and zinc in South Australia, where the growth of commercial forest (Monterey pine, *Pinus radiata*) is made possible only when these nutrients are supplied artificially.

Table 18.2 Characteristics of representative trees of two contrasting species growing within 1 km of each other on the Solling Plateau, Germany. (After Schulze, 1970; Schulze *et al.*, 1977a and b.)

	Beech	Norway spruce
Age (years)	100	89
Height (m)	27	25.6
Leaf shape	Broad	Needle
Annual production of leaves	Higher	Lower
Photosynthetic capacity per unit dry weight of leaf	Higher	Lower
Length of growing season (days)	176	260
Primary productivity (t C ha^{-1} year^{-1})	8.6	14.9

18.3.2 Résumé of factors limiting terrestrial productivity

The ultimate limit on the productivity of a community is determined by the amount of incident radiation that it receives—without this, no photosynthesis can occur.

Incident radiation is used inefficiently by all communities. The causes of this inefficiency can be traced to: (i) shortage of water restricting the rate of photosynthesis; (ii) shortage of essential mineral nutrients which slows down the rate of production of photosynthetic tissue and its effectiveness in photosynthesis; (iii) temperatures that are lethal or too low for growth; (iv) an insufficient depth of soil; (v) incomplete canopy cover, so that much incident radiation lands on the ground instead of on foliage (this may be because of seasonality in leaf production and leaf shedding *or* because of defoliation by grazing animals, pests and diseases); and (vi) the low efficiency with which leaves photosynthesize—under ideal conditions, efficiencies of more than 10% (of PAR) are hard to achieve even in the most productive agricultural systems. However, most of the variation in primary productivity of world vegetation is due to factors (i–v), and relatively little is accounted for by intrinsic differences between the photosynthetic efficiencies of the leaves of the different species.

In the course of 1 year, the productivity of a community may (and probably usually will) be limited by a succession of the factors (i–v). For instance, in a grassland community the primary productivity may be far below the theoretical maximum because the winters are too cold and light intensity is low, the summers are too dry, the rate of nitrogen mobilization is too slow and, for periods, grazing animals may reduce the standing crop to a level at which much incident light falls on bare ground.

An important new approach to unravelling the relative importance of various factors in determining NPP is to build and test simulation models that have an explicit physiological basis. For example, by developing a process-based model of forest NPP in relation to mean annual air temperature, Bonan (1993) was able to investigate the physiological processes responsible for the relationship depicted in Figure 18.8b. His model incorporated the exchange of energy, heat and moisture between forest and atmosphere, moisture and heat flow in the soil, decomposition and nitrogen mineralization, CO_2 assimilation and respiration and whole tree carbon allocation. Using this model, Bonan was able to simulate the observed relationship between forest NPP and mean annual air temperature quite accurately. The results produced new insights; at least for boreal and temperate coniferous forests growing on moist soils. Bonan concluded that NPP was higher in warm than cold climates because of the longer growing seasons, less limitation of growth by available nitrogen and lower maintenance respiration rates for warm-adapted trees in the former.

18.3.3 The primary productivity of aquatic communities

The factors that most frequently limit the primary productivity of aquatic environments are the availability of nutrients, light and the intensity of grazing. The most commonly limiting nutrients are nitrogen (usually as nitrate) and phosphorus (phosphate). Productive aquatic communities occur where, for one reason or another, nutrient concentrations are unusually high.

Lakes receive nutrients by weathering of rocks and soils in their catchment areas, in rainfall and as a result of human activity (fertilizers and sewage input). They vary

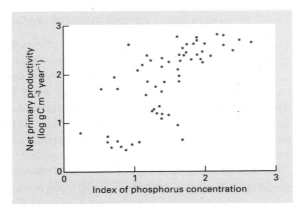

Figure 18.12 Relationship between phytoplankton net primary productivity in lakes throughout the world and estimated steady-state concentration of phosphorus. (After Schindler, 1978.)

considerably in nutrient availability. An analysis of IBP data derived from lakes all over the world points to a significant role in limiting productivity for nutrients, particularly the estimated steady-state concentration of phosphorus (Figure 18.12). We noted earlier a negative relationship between lake primary productivity and latitude (see Figure 18.2c). This must be due in part to the higher light intensities and longer growing seasons at lower latitudes. The supply of nutrients may also be greater in low latitudes, because of a more rapid rate of mineralization, and the concentrations of phosphorus compounds may be higher in tropical rainfall (Schindler, 1978).

In the oceans, locally high levels of primary productivity are associated with high nutrient inputs from two sources. First, nutrients may flow continuously into coastal shelf regions from estuaries. An example is provided in Figure 18.13. Productivity in the inner shelf region is particularly high both because of high nutrient concentrations and because the relatively clear water provides a reasonable depth within which net photosynthesis is positive (the *euphotic zone*). Closer to land, the water is richer in nutrients but highly turbid and its productivity is less. The least productive zones are on the outer shelf (and open ocean) where it might be expected that primary productivity would be high because the water is clear and the euphotic zone is deep. Here, however, productivity is low because of the extremely low concentrations of nutrients.

Ocean upwellings are a second source of high nutrient concentrations. These

estuaries and upwellings provide rich supplies of nutrients in marine environments

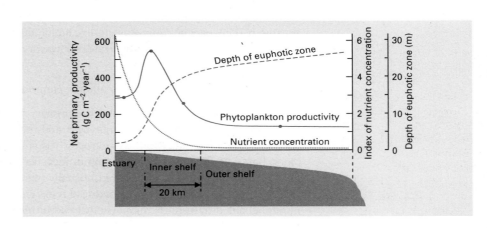

Figure 18.13 Variation in phytoplankton net primary productivity, nutrient concentration and euphotic depth on a transect from the coast of Georgia, USA, to the edge of the continental shelf. (After Haines, 1979.)

726　CHAPTER 18

occur on continental shelves where the wind is consistently parallel to, or at a slight angle to, the coast. As a result, water moves offshore and is replaced by cooler, nutrient-rich water originating from the bottom, where nutrients have been accumulating by sedimentation. Strong upwellings can also occur adjacent to submarine ridges, as well as in areas of very strong currents. Where it reaches the surface, the nutrient-rich water sets off a bloom of phytoplankton production. A chain of heterotrophic organisms takes advantage of the abundant food, and the great fisheries of the world are located in these regions of high productivity. Figure 18.14 illustrates for the Pacific Ocean the general correspondence between phosphate concentration and phytoplankton productivity.

Although the concentration of a limiting nutrient usually determines the productivity of aquatic communities on an areal basis, in any given water body there is also considerable variation with depth as a result of attenuation of light intensity. Figure 18.15a shows how GPP declines with depth. The depth at which GPP is just balanced by phytoplankton respiration (R) is known as the compensation point. Above this, NPP is positive. Light is absorbed by water molecules as well as by dissolved and particulate matter, and it declines exponentially with depth. Near the surface, light is superabundant, but at greater depths its supply is limited and light intensity ultimately determines the extent of the euphotic zone. Very close to the surface, particularly on sunny days, there may even be photoinhibition of photosynthesis. This seems to be due largely to radiation being absorbed by the photosynthetic pigments at such a rate that it cannot be used via the normal photosynthetic channels, and it overflows into destructive photooxidation reactions.

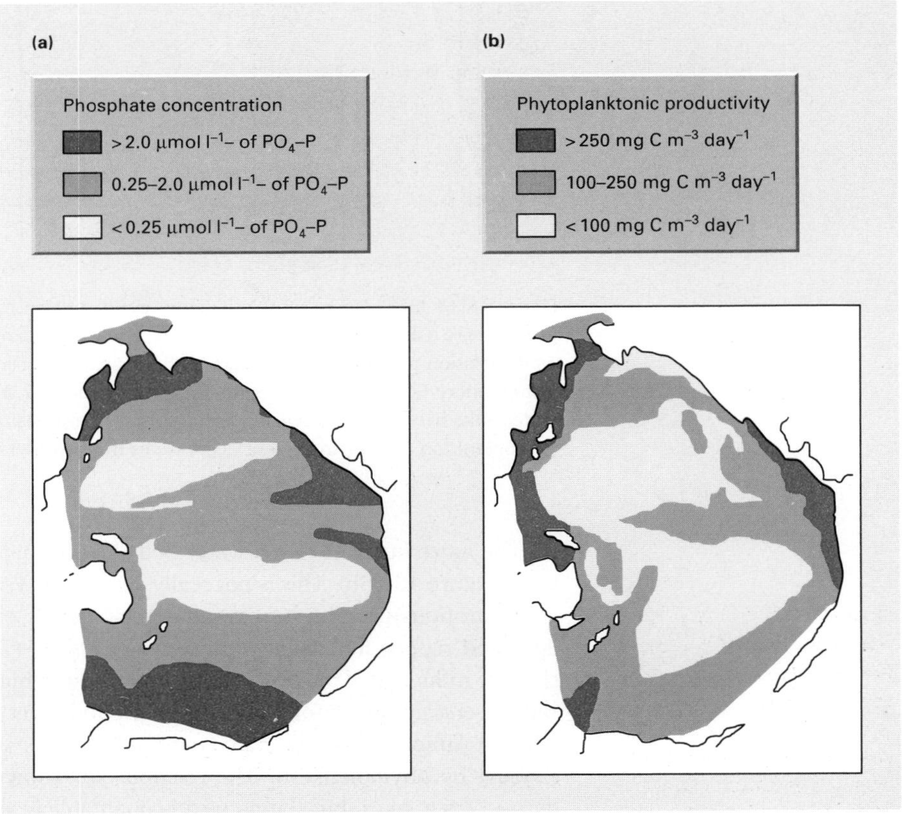

phytoplankton productivity varies with depth

Figure 18.14 The general correlation between (a) distribution of nutrients (phosphates in this case, measured at 100 m depth) and (b) phytoplanktonic productivity in the Pacific Ocean. (After Barnes & Hughes, 1982.)

(a)

Phosphate concentration

▪ > 2.0 μmol l^{-1}– of PO$_4$–P

▪ 0.25–2.0 μmol l^{-1}– of PO$_4$–P

□ < 0.25 μmol l^{-1}– of PO$_4$–P

(b)

Phytoplanktonic productivity

▪ > 250 mg C m^{-3} day^{-1}

▪ 100–250 mg C m^{-3} day^{-1}

□ < 100 mg C m^{-3} day^{-1}

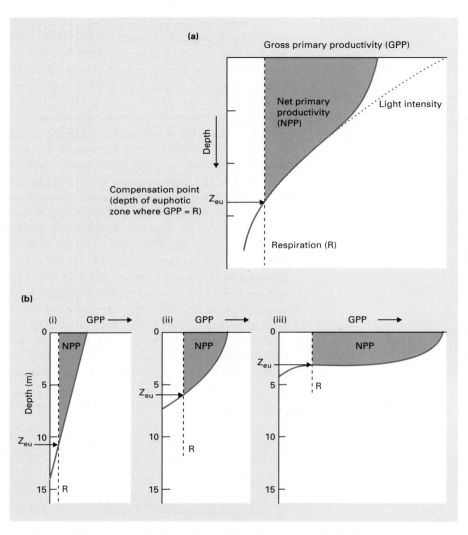

Figure 18.15 (a) The general relationship with depth, in a water body, of gross primary productivity (GPP), respiratory heat loss (R) and net primary productivity (NPP). The compensation point (or depth of the euphotic zone, Eu) occurs at the depth (Z_{eu}) where GPP just balances R and NPP is zero. (b) Total NPP increases with nutrient concentration in the water (lake iii > ii > i). Increasing fertility itself is responsible for greater biomasses of phytoplankton and a consequent decrease in the depth of the euphotic zone.

The more nutrient-rich a water body is, the shallower its euphotic zone is likely to be (Figure 18.15b). This is not really a paradox. Water bodies with higher nutrient concentrations usually possess greater biomasses of phytoplankton, which absorb light and reduce its availability at greater depth. (This is exactly analogous to the shading influence of the tree canopy in a forest, which may remove up to 98% of the radiant energy before it can reach the ground-layer vegetation.) Even quite shallow lakes, if sufficiently fertile, may be devoid of water weeds on the bottom because of shading by phytoplankton. The relationships shown in Figures 18.15a and b are derived from lakes, but the pattern is qualitatively similar in ocean environments.

the productivity of
phytoplankton varies with the
seasons

Typical seasonal variations in the primary productivity of phytoplankton are illustrated in Figure 18.16 for Arctic, north temperate and tropical lakes and seas. For part of the year in Arctic and temperate latitudes, conditions are harsh and water turbulence, particularly in marine environments, constantly forces phytoplankton cells into deep water where GPP does not compensate for respiratory losses: NPP is then zero or negative. As the year progresses, increases in solar radiation, day-length and temperature, coupled with a decrease in wind-generated turbulence, permit phytoplankton production to commence in earnest. This happens progressively later in the year at higher latitudes. Production ceases again, first in Arctic waters, later at temperate latitudes, either because nutrients have become exhausted or as a result of deteriorating conditions later in the year. At lower latitudes, production is maintained throughout the year.

Two peaks of biomass have sometimes been observed within a season in certain temperate lakes and marine environments. There are two possible explanations. Primary productivity may itself pass through two peaks in the year (Figure 18.16, dashed line). Nutrients become limiting in the summer, but are recycled from deep water when the thermally stratified water becomes thoroughly mixed again in the autumn. A second period of phytoplankton production then occurs whilst physical conditions are still favourable. Alternatively, there may be a single peak in the activity of the phytoplankton, but zooplankton grazing may be particularly intense during the summer, causing a dramatic decline in phytoplankton standing crop, followed by an autumnal increase when zooplankton grazing ceases. In freshwaters,

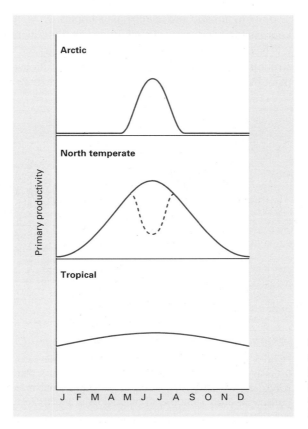

Figure 18.16 Generalized descriptions of seasonal patterns in phytoplankton and primary productivity in aquatic environments at different latitudes.

nutrient depletion is probably the most important factor, whilst zooplankton grazing is generally thought to be more significant in the seas.

18.4 The fate of energy in communities

Secondary productivity is defined as the rate of production of new biomass by heterotrophic organisms. Unlike plants, heterotrophic bacteria, fungi and animals cannot manufacture from simple molecules the complex, energy-rich compounds they need. They derive their matter and energy either directly by consuming plant material, or indirectly from plants by eating other heterotrophs. Plants, the primary producers, comprise the first trophic level in a community; primary consumers occur at the second trophic level; secondary consumers (carnivores) at the third; and so on.

there is a general positive relationship between primary and secondary productivity

Since secondary productivity depends on primary productivity, we should expect a positive relationship between the two variables in communities. Figure 18.17 illustrates this general relationship in aquatic and terrestrial examples. Secondary productivity by zooplankton, which principally consume phytoplankton cells, is positively related to phytoplankton productivity in a range of lakes in different parts of the world (Figure 18.17a). The productivity of heterotrophic bacteria in lakes and oceans also parallels that of phytoplankton (Figure 18.17b); they metabolize dissolved organic matter released from intact phytoplankton cells or produced as a result of 'messy feeding' by grazing animals. Figure 18.17c shows how in a series of African savannah game parks, secondary productivity by large mammalian herbivores is positively related to estimates of the primary productivity of plants.

In both aquatic and terrestrial examples, secondary productivity by herbivores is approximately an order of magnitude less than the primary productivity upon which it is based. This is a consistent feature of all 'grazer' systems (that part of the trophic structure of a community which depends, at its base, on the consumption of *living* plant biomass). It results in a pyramidal structure in which the productivity of plants

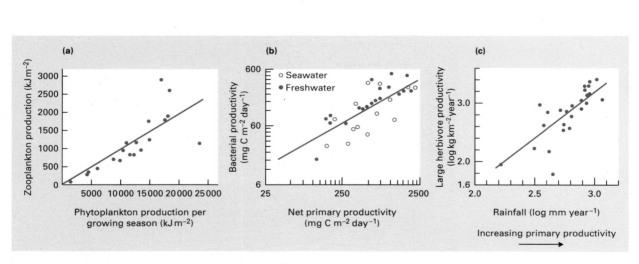

Figure 18.17 The relationship between primary and secondary productivity for the following. (a) Zooplankton in lakes. (After Brylinski & Mann, 1973.) (b) Bacteria in fresh- and seawater. (After Cole *et al.*, 1988.) (c) Large mammalian herbivores in African game parks—primary productivity in these communities is directly related to annual rainfall. (After Coe *et al.*, 1976.)

provides a broad base upon which a smaller productivity of primary consumers depends, with a still smaller productivity of secondary consumers above that. Trophic levels may also have a pyramidal structure when expressed in terms of density or biomass. (Elton, 1927, was the first to recognize this fundamental feature of community architecture. The ideas were later elaborated by Lindemann, 1942.) But, there are many exceptions. Food chains based on trees will certainly have larger numbers (but *not* biomass) of herbivores per unit area than of plants, whilst chains dependent on phytoplankton production may give inverted pyramids of biomass, with a highly productive but small biomass of short-lived algal cells maintaining a larger biomass of longer lived zooplankton.

The productivity of herbivores is invariably less than that of the plants on which they feed. Where has the missing energy gone? First, not all of the plant biomass produced is consumed alive by herbivores. Much dies without being grazed and supports the decomposer community (bacteria, fungi and detritivorous animals). Second, not all the plant biomass that is eaten by herbivores (or herbivore biomass eaten by carnivores) is assimilated and available for incorporation into consumer biomass. Some is lost in faeces, and this also passes to the decomposers. Third, not all the energy which has been assimilated is actually converted to biomass. A proportion is lost as respiratory heat. This occurs because: (i) no energy conversion process is ever 100% efficient (some is lost as unusable random heat, consistent with the second law of thermodynamics); and (ii) because animals do work which requires energy, again released as heat.

These three energy pathways occur at all trophic levels and are illustrated in Figure 18.18.

18.4.1 A comprehensive model of the trophic structure of a community

Figure 18.19 provides a complete description of the trophic structure of a community. It consists of the grazer system pyramid of productivity, but with two additional elements of realism. Most importantly, it adds a decomposer system which is invariably coupled to the grazer system in communities. Second, it recognizes that there are subcomponents of each trophic level in each subsystem which operate in different ways. Thus, a distinction is made between invertebrate and vertebrate categories, between microbes and detritivores which occupy the same trophic level and utilize dead organic matter and between consumers of microbes (microbivores) and of detritivores. Displayed in Figure 18.19 are the possible routes that 1 J of energy, fixed in net primary production, can take as it is dissipated on its path through the community.

A joule of energy may be consumed and assimilated by an invertebrate herbivore which uses part of it to do work and loses it as respiratory heat. Or, it might be consumed by a vertebrate herbivore and later be assimilated by a carnivore that dies and enters the dead organic matter compartment. Here, what remains of the joule may be assimilated by a fungal hypha and consumed by a soil mite, which uses it to do work, dissipating a further part of the joule as heat. At each consumption step, what remains of the joule may fail to be assimilated and pass in the faeces to the dead organic matter, or it may be assimilated and respired or assimilated and incorporated into growth of body tissue (or the production of offspring). The body may die and what remains of the joule may enter the dead organic matter compartment, or it may be captured alive by a consumer in the next trophic level where it meets a further set

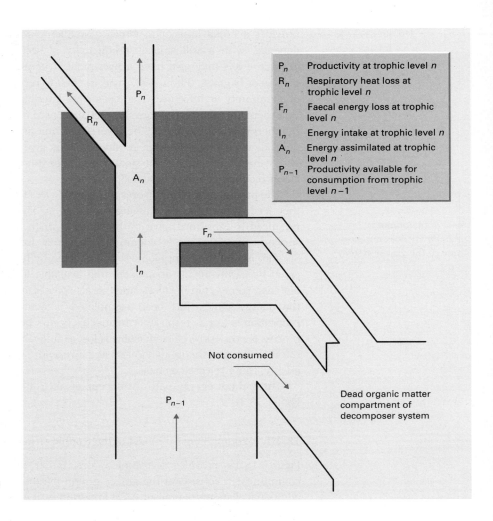

Figure 18.18 The pattern of energy flow through a trophic compartment.

P_n Productivity at trophic level n

R_n Respiratory heat loss at trophic level n

F_n Faecal energy loss at trophic level n

I_n Energy intake at trophic level n

A_n Energy assimilated at trophic level n

P_{n-1} Productivity available for consumption from trophic level $n-1$

Not consumed

Dead organic matter compartment of decomposer system

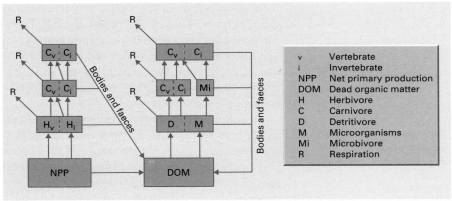

Figure 18.19 A generalized model of trophic structure and energy flow for a terrestrial community. (After Heal & MacLean, 1975.)

v	Vertebrate
i	Invertebrate
NPP	Net primary production
DOM	Dead organic matter
H	Herbivore
C	Carnivore
D	Detritivore
M	Microorganisms
Mi	Microbivore
R	Respiration

of possible branching pathways. Ultimately, each joule will have found its way out of the community, dissipated as respiratory heat at one or more of the transitions in its path along the food chain. Whereas a molecule or ion may cycle endlessly through the food chains of a community, energy passes through just once.

The possible pathways in the grazer and decomposer systems are the same, with one critical exception—faeces and dead bodies are lost to the grazer system (and enter the decomposer system), but faeces and dead bodies from the decomposer system are simply sent back to the dead organic matter compartment at its base. This has a fundamental significance. The energy available as dead organic matter may finally be completely metabolized—and all the energy lost as respiratory heat—even if this requires several circuits through the decomposer system. The exceptions to this are situations: (i) where matter is exported out of the local environment to be metabolized elsewhere (e.g. detritus being washed out of a stream); and (ii) where local abiotic conditions are very unfavourable to decomposition processes, leaving pockets of incompletely metabolized high-energy matter, otherwise known as oil, coal and peat.

consumption, assimilation and production efficiencies determine the relative importance of energy pathways

The proportions of net primary production that flow along each of the possible energy pathways depend on transfer efficiencies in the way energy is used and passed from one step to the next. A knowledge of the values of just three categories of transfer efficiency is all that is required to predict the pattern of energy flow. These are *consumption efficiency* (CE), *assimilation efficiency* (AE) and *production efficiency* (PE).

$$\text{Consumption efficiency, CE} = \frac{I_n}{P_{n-1}} \times 100. \tag{18.1}$$

In words, the consumption efficiency is the percentage of total productivity available at one trophic level (P_{n-1}) that is actually consumed ('ingested') by a trophic compartment 'one level up' (I_n). For primary consumers in the grazer system, the consumption efficiency is the percentage of joules produced per unit time as NPP that find their way into the guts of herbivores. In the case of secondary consumers, it is the percentage of herbivore productivity eaten by carnivores. The remainder dies without being eaten and enters the decomposer chain.

Various reported values for consumption efficiencies of herbivores are given in Table 18.3. All the estimates are remarkably low, perhaps reflecting the unattractiveness of much plant material because of its high proportion of structural support tissue, but also a consequence of generally low herbivore densities (because of the action of their natural enemies). The consumers of microscopic plants achieve greater densities and account for a greater percentage of primary production. When viewing Table 18.3, note that the estimates of consumption efficiency are for particular herbivore groups and do not necessarily represent the total consumed

the variations in consumption efficiency

alive. Reasonable average figures for consumption efficiency are approximately 5% in forests, 25% in grasslands and 50% in phytoplankton-dominated communities.

We know much less about the consumption efficiencies of carnivores feeding on their prey, and any estimates are speculative. Vertebrate predators may consume 50–100% of production from vertebrate prey but perhaps only 5% from invertebrate prey. Invertebrate predators consume perhaps 25% of available invertebrate prey production, but this is a very crude figure of uncertain reliability.

$$\text{Assimilation efficiency, AE} = \frac{A_n}{I_n} \times 100. \tag{18.2}$$

Assimilation efficiency is the percentage of food energy taken into the guts of consumers in a trophic compartment (I_n) which is assimilated across the gut wall (A_n) and becomes available for incorporation into growth or is used to do work. The

Table 18.3 Consumption efficiencies of various herbivore groups for their plant food. (After Brylinski & Mann, 1973; Pimentel *et al.*, 1975.)

Host plant	Taxon of feeding animal	Percentage of productivity consumed
Beech trees	Invertebrates	8.0
Oak trees	Invertebrates	10.6
Maple—beech trees	Invertebrates	5.9–6.6
Tulip poplar trees	Invertebrates	5.6
Grass + forbs	Invertebrates	< 0.5–20
Alfalfa	Invertebrates	2.5
Sericea lespedeza	Invertebrates	1.0
Grass	Invertebrates	9.6
Aquatic plants	Bivalves	11.0
Aquatic plants	Herbivorous animals	18.9
Algae	Zooplankton	25.0
Phytoplankton	Zooplankton	40.0
Phytoplankton	Herbivorous animals	21.2
Marsh grass	Invertebrates	4.6–7.0
Meadow plants	Invertebrates	14.0
Sedge grass	Invertebrates	8.0

remainder is lost as faeces and enters the base of the decomposer system. An 'assimilation efficiency' is much less easily ascribed to microorganisms. Food does not enter an invagination of the outside world passing through the microorganism's body (like the gut of a higher organism), and faeces are not produced. In the sense that bacteria and fungi typically assimilate effectively 100% of the dead organic matter that they digest externally and absorb, they are often said to have an 'assimilation efficiency' of 100%.

the variations in assimilation efficiency

Assimilation efficiencies are typically low for herbivores, detritivores and microbivores (20–50%) and high for carnivores (around 80%). In general, animals are poorly equipped to deal with dead organic matter (mainly plant material) and living vegetation, no doubt partly because of the very widespread occurrence of physical and chemical plant defences, but mainly as a result of the high proportion of complex structural chemicals such as cellulose and lignin in their make-up. As Chapter 11 describes, however, many contain a symbiotic gut microflora which produces cellulase and aids in the assimilation of plant organic matter. In one sense, these animals have harnessed their own personal decomposer system. The way that plants allocate production to roots, wood, leaves, seeds and fruits influences their usefulness to herbivores (see Chapter 3, Figure 3.12b). Seeds and fruits may be assimilated with an efficiency as high as 60–70%, and leaves with about 50% efficiency, whilst the assimilation efficiency for wood may be as low as 15%. The animal food of carnivores (and detritivores such as vultures which consume animal carcasses) poses less problems for digestion and assimilation.

Production efficiency, $\mathrm{PE} = \dfrac{P_n}{A_n} \times 100.$ (18.3)

Production efficiency is the percentage of assimilated energy (A_n) which is incorporated into new biomass (P_n). The remainder is entirely lost to the community as respiratory heat. (Energy-rich secretory and excretory products, which have taken

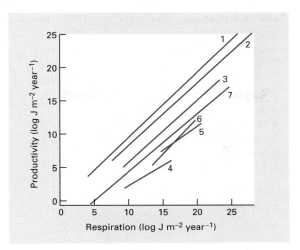

Figure 18.20 Graphs of animal production against animal respiration in seven groups of animal populations. 1, non-social insects; 2, invertebrates other than insects; 3, social insects and fish (not significantly different from each other); 4, insectivores; 5, birds; 6, small mammals; 7, mammals other than insectivores and other small mammals. (After Humphreys, 1979.)

part in metabolic processes, may be viewed as production, P_n, and become available, like dead bodies, to the decomposers.)

Production efficiency varies mainly according to the taxonomic class of the organisms concerned. Invertebrates in general have high efficiencies (30–40%), losing relatively little energy in respiratory heat and converting more assimilate to production. Amongst the vertebrates, ectotherms (whose body temperature varies according to environmental temperature) have intermediate values for production efficiency (around 10%), whilst endotherms, with their high-energy expenditure associated with maintaining a constant temperature, convert only 1–2% of assimilated energy into production. The relationships between productivity and respiration rate for seven classes of consumer are given in Figure 18.20, and average values for production efficiency in each class are shown in Table 18.4. The small-bodied endotherms have the lowest efficiencies, with the tiny insectivores (e.g. shrews) having the lowest production efficiencies of all.

Microorganisms, including protozoa, tend to have very high production efficiencies. They have short lives, small size and rapid population turnover. Unfortunately, available methods are not sensitive enough to detect population changes on scales of time and space relevant to microorganisms, especially in the soil. In general, efficiency of production increases with size in endotherms and decreases very markedly in ectotherms.

Trophic level transfer efficiency, $\text{TLTE} = \dfrac{P_n}{P_{n-1}} \times 100.$ (18.4)

The overall trophic transfer efficiency from one trophic level to the next is simply $\text{CE} \times \text{AE} \times \text{PE}$. In the period after Lindemann's (1942) pioneering work, it was generally assumed that trophic transfer efficiencies were around 10%; indeed, some ecologists referred to a 10% 'law'. However, there is certainly no law of nature that results in precisely one-tenth of the energy that enters a trophic level transferring to

the variations in production efficiency

Table 18.4 Production efficiency (P / A × 100) of various animal groups ranked in order of increasing efficiency. (After Humphreys, 1979.)

Group	P / A (%)
1 Insectivores	0.86
2 Birds	1.29
3 Small mammal communities	1.51
4 Other mammals	3.14
5 Fish and social insects	9.77
6 Non-insect invertebrates	25.0
7 Non-social insects	40.7
Non-insect invertebrates	
8 Herbivores	20.8
9 Carnivores	27.6
10 Detritivores	36.2
Non-social insects	
11 Herbivores	38.8
12 Detritivores	47.0
13 Carnivores	55.6

the next. For example, a recent compilation of trophic studies from a wide range of freshwater and marine environments revealed that trophic level energy transfer efficiencies varied between about 2 and 24%, although the mean was 10.13% (Figure 18.21).

18.4.2 Energy flow through a model community

Given specified values for NPP at a site, and CE, AE and PE for the various trophic groupings shown in the model in Figure 18.19, it is possible to map out the relative importance of different pathways. Heal and MacLean (1975) did this for a hypothetical grassland community. The values they used for assimilation and production efficiencies are shown in diagrammatic form in Figure 18.22, and the resulting fate of each 100 J of NPP is given in Table 18.5. A consumption efficiency of 25% was assumed for the vertebrate herbivores and 4% for invertebrate herbivores. Thus, for every l00 J of NPP, 29 J are ingested by herbivores. A greater

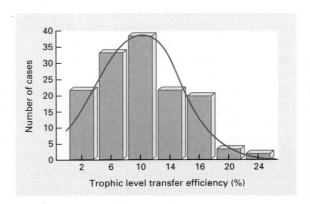

Figure 18.21 Frequency distribution of trophic level transfer efficiencies in 48 trophic studies of aquatic communities. There is considerable variation amongst studies and amongst trophic levels. The mean is 10.13% (standard error = 0.49). (After Pauly & Christensen, 1995.)

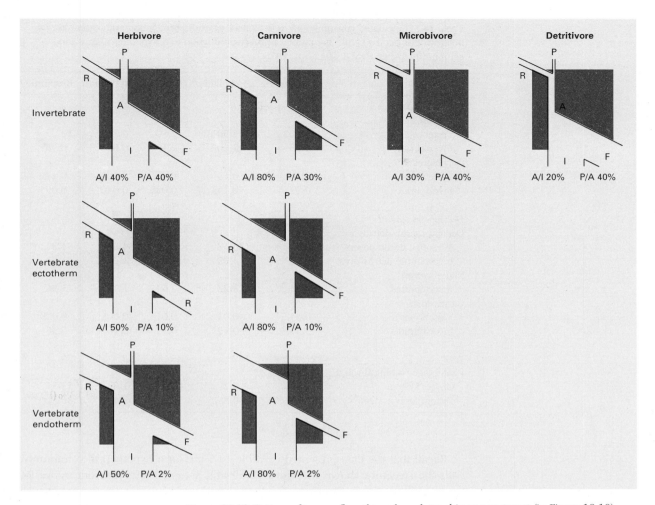

Figure 18.22 Pattern of energy flow through each trophic compartment (in Figure 18.19) and the values of assimilation and production efficiencies used in Heal and MacLean's (1975) grassland model.

proportion of NPP goes to the decomposer system, but, in addition, decomposers consume well in excess of 100 J for every 100 J of NPP! This comes about simply because energy that is consumed is not all assimilated on its first trip through the decomposer chain and so is available for consumption again. The decomposers are responsible for 84.8% of the consumption of matter. However, they carry out 90.8% of the assimilation (mainly because of the importance of microbial activity and their assumed 100% assimilation efficiency). Once again, the apparent discrepancy in total joules assimilated by the decomposers (157 J per 100 J of NPP) results from their ability to 'work over' organic matter on a number of occasions.

the decomposer system is responsible for 98% of secondary productivity in this grassland community

The most significant result of this study is the overwhelming importance of the decomposer system. Even though the grazer system consumes 29% of NPP, it only accounts for 2% of secondary productivity. Of every 100 J of NPP, more than 55 J find their way into decomposer production per year but less than 1 J into grazer production. Overall, in this steady-state community, losses through animal respiration balance NPP so that standing crop biomass (not illustrated) stays the same.

Table 18.5 Calculated consumption, assimilation, egestion, production and respiration by heterotrophs per 100 J m^{-2} net annual primary production in a hypothetical grassland community. (After Heal & MacLean, 1975.)

	Consumption	Assimilation	Egestion	Production	Respiration
Grazer system					
Herbivores					
Vertebrate	25.00	12.50	12.50	0.25	12.25
Invertebrate	4.00	1.60	2.40	0.64	0.96
Carnivores					
Vertebrate	0.16	0.13	0.03	0.003	0.127
Invertebrate	0.17	0.135	0.035	0.040	0.095
Decomposer system					
Decomposers + detritivores					
Microbial decomposers	136.38	136.38	0	54.55	81.83
Invertebrate detritivores	15.15	3.03	12.12	1.21	1.82
Microbivores					
Invertebrates	10.91	3.27	7.64	1.31	1.96
Carnivores					
Vertebrates	0.04	0.03	0.01	0.001	0.029
Invertebrates	0.65	0.52	0.13	0.16	0.36
Total	*192*	*157*	*35*	*58*	*99*
Percentage passing through					
Grazer system	15.2	9.2	42.9	1.6	13.5
Decomposer system	84.8	90.8	57.1	98.4	86.5

Recall that the data presented in Table 18.5 are for a hypothetical community, with the structure shown in Figure 18.19 and assumed (but realistic) values for transfer efficiencies. Is there any way to tell whether the model and assumptions are reasonable—in other words, do real systems operate as predicted? Heal and MacLean (1975) attempted a validation using real data on primary and secondary productivity from 10 tundra, grassland and forest ecosystems. By feeding NPP data into the model (with appropriate herbivore consumption efficiencies: forest, 5%; non-forest, 25%), it was possible to predict secondary productivity of various trophic compartments. These *predicted* values are plotted against the *actual*, measured values in Figure 18.23, and there is good agreement (perfect agreement would have resulted in all the data points falling on the line at 45° to the origin).

In general then, the model receives some support from this exercise. Also in its favour, it incorporates most of the recognized fundamental features of community structure. It is faulty in at least one respect, however. As we saw in Chapter 11, many detritivorous animals consume microbial biomass and dead organic matter together, even if they do so inadvertently. Unfortunately, we have relatively few detailed data on these relationships, so the community structure illustrated in Figure 18.19 must stand for now. We need not be too concerned, however, because the overwhelming importance of microbial consumption and production means that fine adjustment in the model of relationships amongst detritivores and microbivores (often the same animals), microbes and dead organic matter will not affect the overall picture of productivity relationships to any great extent.

the energy flow model receives support from field studies

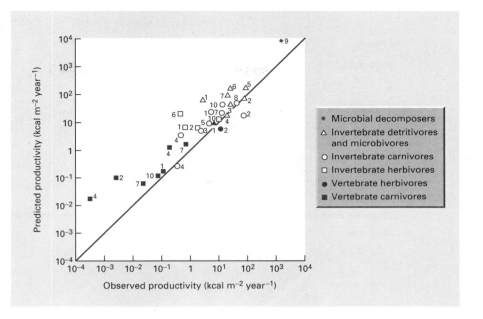

Figure 18.23 Relationship between predicted and observed heterotroph productivity values in a range of sites: tundra (3, 4); cold temperate moorland (2, 7); temperate grassland (1); temperate deciduous forest (5, 6, 8, 9, 10). (After Heal & MacLean, 1975.)

Like all generalizations about trophic structure, our model also suffers the shortcoming that not all consumers can be slotted neatly into a single compartment. Some herbivores eat dead matter on occasion, whilst some carnivores eat both herbivores and detritivores (and the occasional plant). In general, the productivity of omnivores can be split into the different compartments according to the magnitude of their involvement in each. Again, the general picture we have described is not dramatically affected by this technical problem.

18.4.3 Patterns of energy flow in contrasting communities

There have been only a very few studies in which all the community compartments have been studied together, and most of those which appeared in earlier ecological texts are unreliable because they overstressed the role of the grazer system and generally failed to measure, or appreciate, the overwhelming importance of microbial production. However, some generalizations are possible if we compare the gross features of contrasting systems. Figure 18.24 illustrates the patterns of energy flow in a forest, a grassland, a plankton community (of the ocean or a large lake) and the community of a small stream or pond. The decomposer system is probably responsible for the majority of secondary production, and therefore respiratory heat loss, in every community in the world. The grazer system has its greatest role in plankton communities, where a large proportion of NPP is consumed alive and assimilated at quite a high efficiency. Even here, though, it is now believed that very high densities of heterotrophic bacteria in the plankton community subsist on dissolved organic molecules excreted by phytoplankton cells, perhaps consuming more than 50% of primary productivity as 'dead' organic matter in this way (Fenchel, 1987a). The grazer system holds little sway in terrestrial communities because of low herbivore consumption and assimilation efficiencies, and it is almost non-existent in many small streams and ponds simply because primary productivity is so low. The latter depend for their energy base on dead organic matter produced

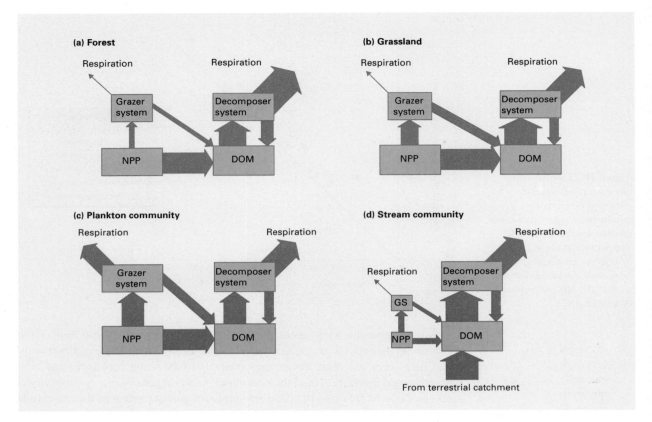

Figure 18.24 General patterns of energy flow for (a) a forest, (b) a grassland, (c) a plankton community in a sea and (d) the community of a stream or small pond. Relative sizes of boxes and arrows are proportional to relative magnitudes of compartments and flows. DOM, dead organic matter; NPP, net primary productivity.

in the terrestrial environment, which falls or is washed or blown into the water. The deep-ocean benthic community has a trophic structure very similar to that of streams and ponds (all can be described as heterotrophic communities). In this case, the community lives in water too deep for photosynthesis to be appreciable or even to take place at all, but it derives its energy base from dead phytoplankton, bacteria, animals and faeces, which sink from the autotrophic community in the euphotic zone above. From a different perspective, the ocean bed is equivalent to a forest floor beneath an impenetrable forest canopy.

A large-scale effort has been made to model energy flow through the world's aquatic systems, to assess the status and sustainability of global fishery resources (Christensen & Pauly, 1993). Researchers working in rivers, lakes, fish culture systems and coastal and ocean-shelf waters have made use of a steady-state multispecies ecosystem model, called ECOPATH 2, developed to provide a basic description of flows of energy (biomass), given inputs such as primary productivity, dietary links in the food web, transfer efficiencies, detritus biomass, P : B ratios and annual fish harvests (Pauly *et al.*, 1993). Recognizing that only incomplete data sets are available for complex aquatic ecosystems, a strength of the ECOPATH model is that it automatically makes ecologically reasonable estimates for unknown

modelling energy flow through fishery ecosystems

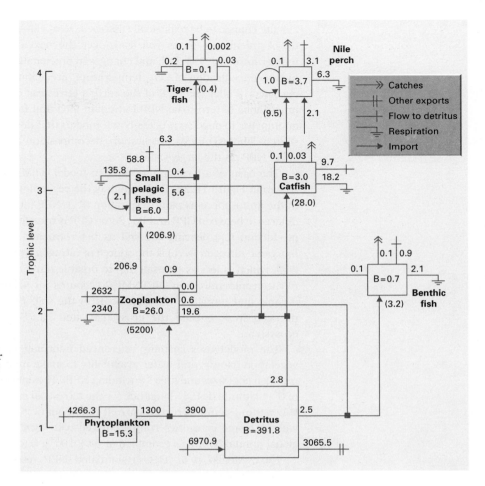

Figure 18.25 Flow diagram for Lake Turkana, Kenya. Each of its state variables (boxes) has inputs (food) and outputs (predation, fisheries catches, respiration and flow to detritus). All flows are in tonnes km^{-2} year^{-1}. (After Kolding, 1993.)

parameters. As an example, Figure 18.25 presents the flow diagram derived for Lake Turkana, Kenya.

One outcome of such modelling efforts, directed at a range of aquatic systems, is that we can estimate the primary production required to sustain the world's annual fisheries harvest (94.3 million tonnes plus 27 million tonnes discarded fish) (Pauly & Christensen, 1995). Prospects differ greatly for different community types. Only 2% of estimated annual primary production is required to sustain the fisheries of open-ocean systems, but estimates ranged from 24 to 35% in freshwater, upwelling and shelf zones. If these latter estimates are accurate then there is legitimate concern about the sustainability of fisheries in a world where human activities may be affecting patterns of primary productivity. We turn to this topic in the next section.

18.5 Modelling terrestrial NPP in relation to global climate change

An enormous amount of effort is expended on developing global vegetation models that can predict NPP and GPP and other characteristics of terrestrial vegetation (as part of the GCTE programme). Models that properly incorporate our understanding of the factors that limit NPP (see Section 18.3.2) will allow the effects of global climate change to be realistically assessed. Most climate models predict that the build up of greenhouse gases will lead, over the next century, to rises in surface air temperature of 1.5–4.5°C, and changes in precipitation and cloud patterns. Changes in the concentration of CO_2, temperature, precipitation and cloud cover can all be expected to alter the NPP of the earth's terrestrial communities. The fundamental importance of terrestrial NPP to human food and forestry production has provided an impetus to develop process-based models that describe how important processes such as photosynthesis, respiration, decomposition and nutrient cycling interact to affect NPP on the large scale.

The basis of a terrestrial ecosystem model (TEM) used by Melillo *et al.* (1993) is shown in Figure 18.26. Carbon enters the vegetation pool (C_V) as GPP and transfers to the atmosphere as plant respiration (R_A). NPP (not indicated in the model) is the difference between GPP and R_A. Some GPP is transferred to the soil pool (C_S) as litter production (L_C), leaving the soil as heterotrophic respiration (R_H). Available soil inorganic nitrogen (N_{AV}) is the source of nitrogen uptake (NUPTAKE) for vegetation (N_V), which is lost as litterfall (L_N) to organic nitrogen in soil and detritus (N_S). Net nitrogen mineralization (NETNMIN) accounts for nitrogen exchanged between the organic and inorganic nitrogen pools in the soil. Finally, NINPUT is the input of nitrogen from outside the ecosystem, and NLOST is nitrogen losses from the ecosystem.

The model uses spatially referenced information on climate, elevation, soils, vegetation (biome) and water availability to make monthly estimates of carbon and nitrogen pool sizes and fluxes (including NPP). The data sets are gridded at a resolution of 0.5° latitude by 0.5° longitude for the terrestrial environments of the whole globe. When the system was run for current conditions (335 ppmv CO_2), NPP predictions for various biome categories compared well with available field measurements. Total global annual NPP was estimated at 51×10^{15} g C (or 418 g C m^{-2} year^{-1}).

Next, Melillo *et al.* (1993) simulated NPP responses to changes in climate, assuming a doubling of CO_2 concentration and concomitant changes, as predicted by four different general circulation models of global climate change (see Melillo *et al.*, 1993), in surface air temperature, precipitation and cloud cover, at the same 0.5° latitude by 0.5° longitude spatial resolution. The result was a predicted increase in global terrestrial NPP of between 20.0 and 26.1% (total = 61.2–64.3 × 10^{15} g C; Table 18.6). The detailed nature of Melillo *et al.*'s process-based model provided

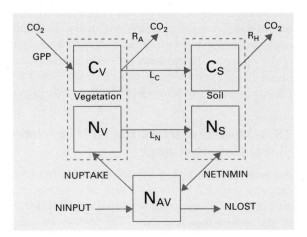

Figure 18.26 The terrestrial ecosystem model used to model patterns in net primary productivity.

Table 18.6 Comparison of current net primary productivity (NPP) estimates (10^{15} g C) for a variety of biomes, with model predictions for current conditions (312.5 ppmv CO_2) and for a future scenario with elevated CO_2 (625.0 ppmv). The range in values was derived from four different general circulation models of global climate change. (After Melillo *et al.*, 1993.)

Biome	Current NPP (field measurements)	Predicted current NPP	Predicted future NPP
Polar desert/alpine tundra	0.4	0.4	0.5
Wet/moist tundra	0.6	0.7	0.7–0.8
Boreal woodland	1.1	1.3–1.4	1.4–1.6
Boreal forest	2.9	3.5–3.8	3.7–4.4
Temperate coniferous forest	1.1	1.1	1.3–1.4
Desert	0.6	0.5–0.6	0.9–1.0
Arid shrubland	1.8	1.8–1.9	2.5–2.7
Short grassland	1.0	1.1–1.2	1.2–1.4
Tall grassland	1.2	1.3–1.5	1.4–1.6
Temperate savannah	2.2	2.3–2.6	2.9–3.1
Temperate deciduous forest	2.2	2.0–2.5	2.4–2.8
Temperate mixed forest	3.3	3.4–3.7	4.0–4.2
Temperate broadleaf evergreen forest	2.2	2.2–2.3	2.4–2.8
Mediterranean shrubland	0.5	0.5	0.6–0.7
Tropical savannah	5.3	5.6–6.0	6.3–6.7
Xeromorphic forest	2.9	2.7–2.9	3.7–4.1
Tropical deciduous forest	3.8	3.3–3.5	4.5–4.6
Tropical evergreen forest	18.0	14.3–16.4	19.3–21.9
Total	51.0	49.8–51.5	61.2–64.3

insights into the reasons for differences between biomes. Thus, for tropical and dry temperate ecosystems, increases in NPP were dominated by the effects of elevated CO_2. On the other hand, for northern and moist temperate ecosystems, NPP increases reflected primarily the effects of elevated temperature in enhancing nitrogen cycling.

The accuracy of predictions from models such as Melillo *et al.*'s depend on how well they have encapsulated the underlying ecological processes as well as the efficacy of the available climate change models. Moreover, an important shortcoming is that their model assumes that the biomes stay where they are, whereas, in reality, vegetation can be expected to be redistributed as a result of climate change. This is the focus of other workers such as Prentice *et al.* (1992). The bringing together of these approaches is an important challenge. A further challenge is to carry out similar studies on aquatic systems, the fisheries resources of which are crucial to human populations around the world (Fields *et al.*, 1993).

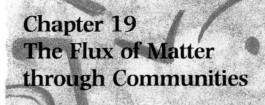

Chapter 19
The Flux of Matter
through Communities

19.1 Introduction

Chemical elements and compounds are vital for the processes of life. Living organisms expend energy to extract chemicals from their environment, hold on to them and use them for a period, then lose them again. Thus, the activities of organisms profoundly influence the patterns of flux of chemical matter in the biosphere. Physiological ecologists focus their attention on the way individual organisms obtain and use the chemicals they need (Townsend & Calow, 1980; Chapin, 1987; see also Chapter 3). However, in this chapter, as in the last, we change the emphasis and consider the ways in which the biota on an area of land, or in a volume of water, accumulates, transforms and moves matter between the various components of the ecosystem (Evans, 1956). The area that we choose may be that of the whole globe, a continent, a farm, a river catchment or simply 1 m².

19.1.1 The fate of matter in communities

The great bulk of living matter in any community is water. The rest is made up mainly of carbon compounds (95% or more) and this is the form in which energy is accumulated and stored. The energy is ultimately dissipated when the carbon compounds are oxidized to carbon dioxide (CO_2) by the metabolism of living tissue or of its decomposers. Although we consider the fluxes of energy and of carbon in different chapters, the two are so intimately bound together in all biological systems that their separation is more a matter of convenience than of logic.

Carbon enters the trophic structure of a community when a simple molecule, CO_2, is taken up in photosynthesis. If it becomes incorporated in net primary productivity, it is available for consumption as part of a sugar, a fat, a protein or, very often, a cellulose molecule. It follows exactly the same route as energy, being successively consumed, defaecated, assimilated and perhaps incorporated into secondary productivity somewhere within one of the trophic compartments. When the high-energy molecule in which the carbon is resident is finally used to provide energy for work, the energy is dissipated as heat (as we have discussed in Chapter 18), whilst the carbon is released again to the atmosphere as CO_2. Here, the tight link between energy and carbon ends.

Once energy is transformed into heat, it can no longer be used by living organisms to do work or to fuel the synthesis of biomass. (Its only possible role is momentary, in helping to maintain a high body temperature.) The heat is eventually lost to the atmosphere and can never be recycled. In contrast, the carbon in CO_2 can be used again in photosynthesis. Carbon, and all other nutrient elements (e.g. nitrogen, phosphorus, etc.) are available to plants as simple organic molecules or ions in the atmosphere (CO_2), or as dissolved ions in water (nitrate, phosphate, potassium, etc.). Each can be incorporated into complex carbon compounds in biomass. Ultimately, however, when the carbon compounds are metabolized to CO_2, the mineral nutrients are released again in simple inorganic form. Another plant may then absorb them, and so an individual atom of a nutrient element may pass repeatedly through one food chain after another. The relationship between energy flow and nutrient cycling is illustrated in Figure 19.1.

By its very nature, energy cannot be recycled. It is made available to living organisms as solar radiation, which can be fixed during photosynthesis. But, once the resulting chemical energy is used, it is dissipated as useless heat. Although energy can pass back and forth between the dead organic matter compartment and the decomposer system this must not be described as cycling. It simply reflects the ability of the decomposer system to 'work over' organic matter several times; but each joule of energy can be *used* only once. Life on earth is possible because a fresh supply of solar energy is made available every day.

On the other hand, chemical nutrients, the building blocks of biomass, simply change the form of molecule of which they are a part of (e.g. nitrate–nitrogen to protein–nitrogen to nitrate–nitrogen). They can be used again, and repeatedly recycled. Unlike the energy of solar radiation, nutrients are not in unalterable supply, and the process of locking some into living biomass reduces the supply remaining to the rest of the community. If plants, and their consumers, were not eventually decomposed, the supply of nutrients would become exhausted and life on earth would cease. The activity of heterotrophic organisms is crucial in bringing about nutrient cycling and maintaining productivity. Figure 19.1 shows the release of nutrients in their simple inorganic form as occurring only from the decomposer system. In fact, a small proportion is also released from the grazer system. However, the decomposer system plays a role of overwhelming importance in nutrient cycling.

energy cannot be cycled and reused—matter can

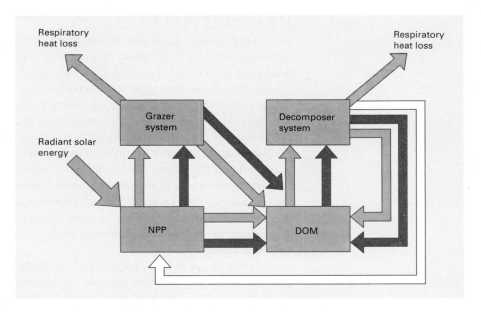

Figure 19.1 Diagram to show the relationship between energy flow (■) and nutrient cycling. Nutrients locked in organic matter (■) are distinguished from the free inorganic state (□). NPP, Net primary production; DOM, dead organic matter.

The picture described in Figure 19.1 is an over-simplification in one important respect. Not all nutrients released during decomposition are necessarily taken up again by plants. Nutrient recycling is never perfect and some nutrients are exported from land by runoff into streams (ultimately to the ocean) and others, such as nitrogen and sulphur which have gaseous phases, can be lost to the atmosphere. Moreover, a community receives additional supplies of nutrients that do not depend directly on inputs from recently decomposed matter, such as minerals dissolved in rain, or derived from weathered rock.

nutrient cycling is never perfect

19.1.2 Biogeochemistry and biogeochemical cycles

We can conceive of pools of chemical elements existing in compartments. Some compartments occur in the *atmosphere* (carbon in CO_2, nitrogen as gaseous nitrogen, etc.), some in the rocks of the *lithosphere* (calcium as a constituent of calcium carbonate, potassium in feldspar) and others in the water of soil, stream, lake or ocean—the *hydrosphere* (nitrogen in dissolved nitrate, phosphorus in phosphate, carbon in carbonic acid (H_2CO_3), etc.). In all these cases, the elements exist in inorganic form. In contrast, living organisms (the biota) and dead and decaying bodies can be viewed as compartments containing elements in organic form (carbon in cellulose or fat, nitrogen in protein, phosphorus in adenosine triphosphate (ATP), etc.). Studies of the chemical processes occurring within these compartments and, more particularly, of the fluxes of elements between them, comprise the science of biogeochemistry.

the 'bio' in biogeochemistry

Many geochemical fluxes would occur in the absence of life, if only because all geological formations above sea-level are eroding and degrading. Volcanoes release sulphur into the atmosphere whether there are organisms present or not. On the other hand, organisms alter the rate of flux and the differential flux of the elements by extracting and recycling some chemicals from the underlying geochemical flow (Waring & Schlesinger, 1985). The term biogeochemistry is apt.

19.1.3 Biogeochemistry of small and large systems

The flux of matter can be investigated at a variety of spatial and temporal scales. Ecologists interested in the gains, uses and losses of nutrients by the community of a small pond or a hectare of grassland can focus on local pools of chemicals. They need not concern themselves with the contribution to the nutrient budget made by volcanoes, or the possible fate of nutrients leached from land to eventually be deposited on the ocean floor. Moving up in scale, we find that the chemistry of streamwater is profoundly influenced by the biota of the area of land it drains (its catchment area) and, in turn, influences the chemistry and biota of the lake into which it flows. The stream plus its catchment area make a natural unit of study and we describe the results of such investigations in Section 19.2. Other investigators are interested in the global scale. With their broad brush, they seek to paint a picture of the contents and fluxes of the largest conceivable compartments—the entire atmosphere, the oceans as a whole, and so on. Global biogeochemical cycles will be discussed in Section 19.4.

the catchment area as a unit of study

global patterns ...

At home, in industry, in agriculture and in medicine, people use a vast number of different chemicals (on one estimate, there are about four million chemical substances produced by human activities; El-Hinnawi & Hashmi, 1982). Some of

... are often upset by human activities

these are released in such quantities that biogeochemical cycles are upset, with consequences for natural and cultivated communities and for world climate. These, too, will be discussed in Sections 19.4 and 19.5.

19.1.4 Nutrient budgets

Nutrients are gained and lost by communities in a variety of ways (Figure 19.2). We can construct a nutrient budget if we can identify and measure all the processes on the credit and debit sides of the equation. For some nutrients, in some communities, the budget may be more or less in balance, so that:

$$\text{inputs} = \text{outputs}. \tag{19.1}$$

In other cases, the inputs exceed the outputs and nutrients accumulate in the compartments of living biomass and dead organic matter. This is especially obvious during community succession (see Chapter 17, Section 17.4), when the equation can be represented as:

$$\text{inputs} - \text{outputs} = \text{storage}. \tag{19.2}$$

Finally, outputs may exceed inputs if the biota is disturbed by an event such as fire, massive defoliation, such as that caused by a plague of locusts, or large-scale deforestation or crop harvesting by humans. Another important source of loss in terrestrial systems occurs where mineral export (e.g. of base cations due to acid

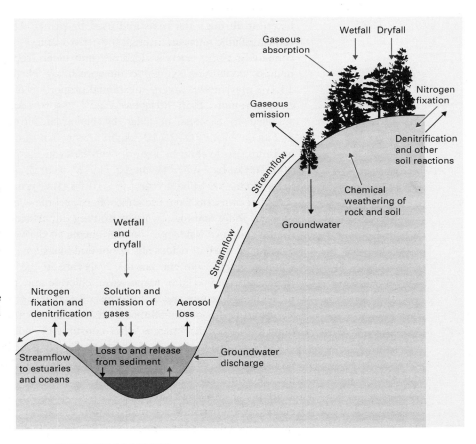

Figure 19.2 Components of the nutrient budgets of a terrestrial and an aquatic system. Note how the two communities are linked by streamflow, which is a major output from the terrestrial system but a major input to the aquatic one. Inputs are shown in orange and outputs in black.

747   FLUX OF MATTER

deposition) exceeds replenishment from weathering. In all these cases, the equation can be written as:

$$\text{output} - \text{input} = \text{loss.} \tag{19.3}$$

The components of nutrient budgets are discussed below.

19.2 Nutrient budgets in terrestrial communities

19.2.1 Nutrient inputs to terrestrial communities

nutrient inputs ...

Weathering of parent bedrock and soil is the dominant source of nutrients such as calcium, iron, magnesium, phosphorus and potassium, which may then be taken up via the roots of plants. Mechanical weathering is caused by processes such as freezing of water and growth of roots in crevices. However, much more important to the release of plant nutrients are chemical weathering processes. Of particular significance is carbonation, in which H_2CO_3 reacts with minerals to release ions, such as calcium and potassium. Simple dissolution of minerals in water also makes

... from the weathering of bedrock and soil

nutrients available from rock and soil and so do hydrolytic reactions involving organic acids released from plant roots, fungi or lichens (Ascaso *et al.*, 1982).

Atmospheric CO_2 is the source of the carbon content of terrestrial communities. Similarly, gaseous nitrogen from the atmosphere provides most of the nitrogen content of communities. Several types of bacteria and blue-green algae possess the enzyme nitrogenase and convert nitrogen to ammonium ions, which can then be taken up through the roots and used by plants. All terrestrial ecosystems receive some available nitrogen through the activity of free-living bacteria, but communities containing plants such as legumes and alder trees (*Alnus* spp.), with their root nodules containing symbiotic nitrogen-fixing bacteria (see Chapter 13, Section 13.10), may receive a very substantial proportion of their nitrogen in this way. For example, more than 80 kg ha^{-1} $year^{-1}$ (8×10^{-3} kg m^{-2} $year^{-1}$) of nitrogen was supplied to a stand of alder by biological nitrogen fixation, compared with 1–2 kg ha^{-1} $year^{-1}$ ($1-2 \times 10^{-4}$ kg m^{-2} $year^{-1}$) in rainfall (Bormann & Gordon, 1984), and nitrogen fixation by legumes can be even more dramatic: values in the range 100–300 kg ha^{-1} $year^{-1}$ ($1-3 \times 10^{-2}$ kg m^{-2} $year^{-1}$) are normal, with tropical clover fixing up to 900 kg ha^{-1} $year^{-1}$ (9×10^{-2} kg m^{-2} $year^{-1}$) (Stolp, 1988).

... from the atmosphere

Other nutrients from the atmosphere become available to communities as *wetfall* (in rain, snow and fog) or *dryfall* (settling of particles during periods without rain). Rain is not pure water but contains chemicals derived from a number of sources: (i) trace gases, such as oxides of sulphur and nitrogen; (ii) aerosols (produced when tiny water droplets from the oceans evaporate in the atmosphere and leave behind particles rich in sodium, magnesium, chloride and sulphate); and (iii) dust particles from fires, volcanoes and windstorms, often rich in calcium, potassium and sulphate. The constituents of rainfall that serve as nuclei for raindrop formation make up the *rainout* component, whereas other constituents, both particulate and gaseous, are cleansed from the atmosphere as the rain falls—these are the *washout* component (Waring & Schlesinger, 1985). The nutrient concentrations in rain are highest early in a rainstorm, but fall subsequently as the atmosphere is progressively cleansed (Schlesinger *et al.*, 1982). Snow scavenges chemicals from the atmosphere less effectively than rain, but tiny fog droplets have particularly high ionic concentrations. Nutrients dissolved in precipitation mostly become available to plants when

Table 19.1 Total annual atmospheric deposition of major ions (mean MEq m^{-2} year^{-1} + standard error for 2 year's data) to an oak forest (*Quercus prinus* and *Q. alba*) in Tennessee. (After Lindberg *et al.*, 1986.)

	Atmospheric deposition					
	SO_4^{2-}	NO_3^-	H^+	NH_4^+	Ca^{2+}	K^+
Wetfall	70 ± 5	20 ± 2	69 ± 5	12 ± 1	12 ± 2	0.9 ± 0.1
Dryfall						
Fine particles	7 ± 2	0.1 ± 0.02	2.0 ± 0.9	3.6 ± 1.3	1.0 ± 0.2	0.1 ± 0.05
Coarse particles	19 ± 2	8.3 ± 0.8	0.5 ± 0.2	0.8 ± 0.3	30 ± 3	1.2 ± 0.2
Vapours	62 ± 7	26 ± 4	85 ± 8	1.3	0	0
Total	160 ± 9	54 ± 4	160 ± 9	18 ± 2	43 ± 4	2.2 ± 0.3

the water reaches the soil and can be taken up by plant roots. However, some are absorbed by leaves directly.

Dryfall is an important process in communities with a long dry season, but dryfall can provide a large proportion of the nutrient budget even in more humid locations. For example, in a mature oak (*Quercus* spp.) forest in Tennessee, USA, dryfall accounted for more than 50% of the atmospheric input to the tree canopy of sulphate, hydrogen ions and nitrate (all mainly as dry vapour uptake), and also of calcium and potassium (mainly as dry particle deposition) (Table 19.1).

... from hydrological inputs

Streamwater plays a major role in the output of nutrients from terrestrial communities (see Section 19.2.2). However, in a few cases, streamflow can provide a significant input to terrestrial communities when material is deposited in floodplains. In forests of swamp cypress (*Taxodium distichum*), for example in south-eastern USA, net primary productivity was significantly correlated with the estimated input of phosphorus from the surrounding upland: 600 g m^{-2} year^{-1} or less in closed bogs without a stream input, rising to 2000 g m^{-2} year^{-1} in forests receiving polluted streamwater (Figure 19.3).

... and from human activities

Of course, human activities contribute significant inputs of nutrients to many communities. For example, the amounts of CO_2 and oxides of nitrogen and sulphur in the atmosphere have been increased by the burning of fossil fuels and by the exhausts of cars, and the concentrations of nitrate and phosphate in streamwater

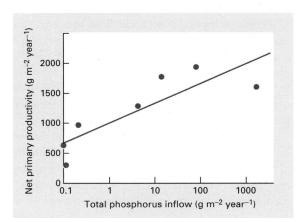

Figure 19.3 Above-ground net primary productivity in cypress (*Taxodium distichum*) swamp forests in relation to the phosphorus received in streamwaters from upland areas. (After Brown, 1981.)

have been raised by agricultural practices and sewage disposal. These changes have far-reaching consequences, which will be discussed later.

19.2.2 Nutrient outputs from terrestrial communities

A particular nutrient atom may be taken up by a plant that is then eaten by a herbivore which then dies and is decomposed, releasing the atom back to the soil from where it is taken up through the roots of another plant. In this manner, nutrients may circulate within the community for many years. Alternatively, the atom may pass through the system in a matter of minutes, perhaps without interacting with the biota at all. Whatever the case, the atom will eventually be lost through one of the variety of processes that remove nutrients from the system (see Figure 19.2). These processes constitute the debit side of the nutrient budget equation.

nutrients are lost ...

Release to the atmosphere is one pathway of nutrient loss. In many communities there is an approximate annual balance in the carbon budget; the carbon fixed by photosynthesizing plants is balanced by the carbon released to the atmosphere as CO_2 from the respiration of plants, microorganisms and animals. Other gases are released through the activities of anaerobic bacteria. Methane is a well-known product of the soils of bogs, swamps and floodplain forests, produced by bacteria in the waterlogged, anoxic zone of wetland soils. However, its net flux to the atmosphere depends on the rate at which it is produced in relation to its rate of consumption by aerobic bacteria in the shallower, unsaturated soil horizons, with as much as 90% consumed before it reaches the atmosphere (Bubier & Moore, 1994). Methane may be of some importance in drier locations too. It is produced by fermentation in the anaerobic stomachs of grazing animals, and even in upland forests, periods of heavy rainfall may produce anaerobic conditions that can persist for some time within microsites in the organic layer of the soil (Sexstone et al., 1985).

... to the atmosphere

In such locations, bacteria such as *Pseudomonas* reduce nitrate to gaseous nitrogen or nitrous oxide in the process of denitrification. Plants themselves may be direct sources of gaseous and particulate release. For example, forest canopies produce volatile hydrocarbons (e.g. terpenes) and tropical forest trees appear to emit aerosols containing phosphorus, potassium and sulphur (Waring & Schlesinger, 1985). Finally, ammonia (NH_3) gas is released during the decomposition of vertebrate excreta and has been shown to be a significant component in the nutrient budget of many systems (Woodmansee, 1978; Sutton et al., 1993).

Other pathways of nutrient loss are important in particular instances. For example, fire can turn a very large proportion of a community's carbon into CO_2 in a very short time. The loss of nitrogen, as volatile gas, can be equally dramatic; during an intense wild fire in a conifer forest in north-west USA, 855 kg ha^{-1} (8.55×10^{-2} kg m^{-2}) (equal to 39% of the pool of organic nitrogen) was lost in this way (Grier, 1975). Substantial losses of nutrients also occur when foresters or farmers harvest and remove their trees and crops.

... and to streams and groundwaters

For many elements, the most substantial pathway of loss is in streamflow. The water that drains from the soil of a terrestrial community into a stream carries a load of nutrients that is partly dissolved and partly particulate. With the exception of iron and phosphorus, which are not mobile in soils, the loss of plant nutrients is predominantly in solution. Particulate matter in streamflow occurs both as dead organic matter (mainly tree leaves) and as inorganic particles. After rainfall or

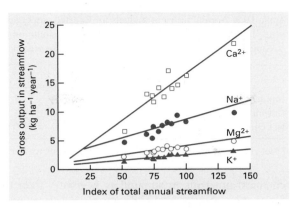

Figure 19.4 Annual loss of major nutrients in streamflow as a function of annual stream discharge during a period of 11 years in Hubbard Brook, New Hampshire, USA. (After Likens *et al.*, 1977.)

snow-melt, the water draining into streams is generally more dilute than during dry periods, when the concentrated waters of soil solution make a greater contribution. However, the effect of high volume more than compensates for lower concentrations in wet periods. Thus, total loss of nutrients is greatest in years when rainfall and stream discharge are high (Figure 19.4). In regions where the bedrock is permeable, losses occur not only in streamflow but also in water that drains deep into the groundwater. This may discharge into a stream or lake after a considerable delay and at some distance from the terrestrial community. Groundwater losses are very difficult to quantify and, as a consequence, the most revealing nutrient budget studies have been those performed in catchment areas with impermeable bedrock where losses to deep groundwater are negligible.

19.2.3 The catchment area as a unit of study

the movement of water links terrestrial and aquatic communities

It is the movement of water, under the force of gravity, that links the nutrient budgets of terrestrial and aquatic communities (see Figure 19.2): terrestrial systems lose dissolved and particulate nutrients into streams and groundwaters; aquatic systems (including the stream communities themselves and, ultimately, the oceans) gain nutrients from streamflow and groundwater discharge.

Because many nutrient losses from terrestrial communities are channelled through streams, a comparison of the chemistry of streamwater with that of incoming precipitation can reveal a lot about the differential uptake and cycling of chemical elements by the terrestrial biota. Just how important is nutrient cycling in relation to the through-put of nutrients? Is the amount of nutrients cycled per year small or large in comparison with external supplies and losses? The most thorough study of this question has been carried out by Likens *et al.* (1971) in the Hubbard

the Hubbard Brook project ...

Brook Experimental Forest, an area of temperate deciduous forest drained by small streams in the White Mountains of New Hampshire, USA. The catchment area—the extent of terrestrial environment drained by a particular stream—was taken as the unit of study because of the role that streams play in nutrient export. Six small catchments were defined and their outflows were monitored. A network of precipitation gauges recorded the incoming amounts of rain, sleet and snow. Chemical analyses of precipitation and streamwater made it possible to calculate the amounts of various nutrients entering and leaving the system, and these are shown in Table 19.2. A similar pattern is found each year. In most cases, the output of

Table 19.2 Annual nutrient budgets for forested catchments at Hubbard Brook (kg ha^{-1} year^{-1}). Inputs are for dissolved materials in precipitation or as dryfall. Outputs are losses in streamwater as dissolved material plus particulate organic matter. (After Likens *et al.*, 1971.)

	NH_4^+	NO_3^-	SO_4^{2-}	K^+	Ca^{2+}	Mg^{2+}	Na^+
Input	2.7	16.3	38.3	1.1	2.6	0.7	1.5
Output	0.4	8.7	48.6	1.7	11.8	2.9	6.9
Net change*	+2.3	+7.6	−10.3	−0.6	−9.2	−2.2	−5.4

* Net change is positive when the catchment gains matter and negative when it loses it.

chemical nutrients in streamflow is greater than their input from rain, sleet and snow. The source of the excess chemicals is parent rock and soil, which are weathered and leached at a rate of about 70 g m^{-2} year^{-1}.

In almost every case, the inputs and outputs of nutrients are small in comparison with the amounts held in biomass and recycled within the system. This is illustrated in Figure 19.5 for one of the most important nutrients—nitrogen. Nitrogen was added to the system not only in precipitation (6.5 kg ha^{-1} year^{-1} (6.5 × 10^{-4} kg m^{-2} year^{-1})), but also through atmospheric nitrogen fixation by microorganisms (14 kg ha^{-1} year^{-1} (1.4 × 10^{-3} kg m^{-2} year^{-1})). (Note that denitrification by other microorganisms, releasing nitrogen to the atmosphere, will also have been occurring but was not measured.) The export in streams of only 4 kg ha^{-1} year^{-1}

inputs and outputs of nutrients are typically low compared with the amounts cycling …

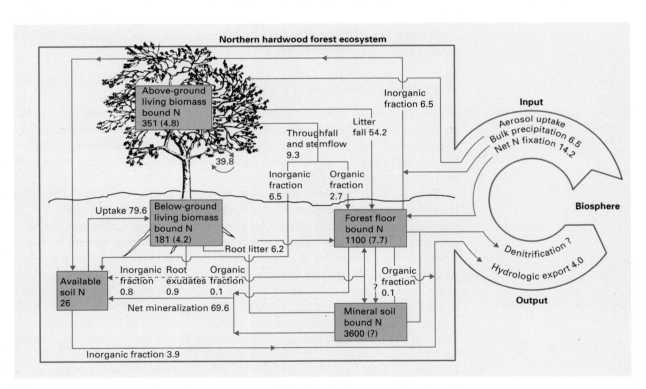

Figure 19.5 Annual nitrogen budget for the undisturbed Hubbard Brook Experimental Forest. Values in boxes are the sizes of the various nitrogen pools in kilograms of nitrogen per hectare. The rate of accretion of each pool (in parentheses) and transfer rates are expressed in kilograms of nitrogen per hectare per year. (After Bormann *et al.*, 1977.)

$(4 \times 10^{-4} \text{ kg m}^{-2} \text{ year}^{-1})$ emphasizes how securely nitrogen is held and cycled within the forest biomass. Stream output represents only 0.1% of the total (organic) nitrogen standing crop held in living and dead forest organic matter.

Nitrogen was unusual in that its net loss in stream runoff was less than its input in precipitation, reflecting the complexity of inputs and outputs and the efficiency of its cycling. However, despite the net loss to the forest of other nutrients, their export was still low in relation to the amounts bound in biomass. In other words, relatively efficient recycling is the norm.

The major exception is sulphur. The amount of sulphur leaving the system annually (about 24 kg ha^{-1} $(2.4 \times 10^{-3} \text{ kg m}^{-2})$) was far in excess of the amount in annual litterfall (5.5 kg ha^{-1} $(5.5 \times 10^{-4} \text{ kg m}^{-2})$). It has been calculated that 50% of the annual input of sulphur can be ascribed to pollution resulting from the burning of fossil fuels. Such pollution has led to *acid rain* (at Hubbard Brook, rain is essentially dilute sulphuric acid with a pH often below 4.0), and this is now recognized as one of the most widespread pollution problems in much of the Northern hemisphere, although nitrogen oxides and ozone may also share the blame. So far, forestry resources (in Germany) and fish stocks (e.g. in Scandinavia and Scotland) have apparently been adversely affected. The large proportions of sulphur leaving the Hubbard Brook system each year, in contrast with all the other macronutrients, must be seen as a consequence of the input of sulphur from pollution.

In a large-scale experiment, all the trees were felled in one of the Hubbard Brook catchments. The overall export of dissolved inorganic nutrients from the disturbed catchment then rose to 13 times the normal rate (Figure 19.6). Two phenomena were responsible. First, the enormous reduction in transpiring surfaces (leaves) led to 40% more precipitation passing through the groundwater to be discharged to the streams, and this increased outflow caused greater rates of leaching of chemicals and weathering of rock and soil. Second, and more significantly, deforestation effectively broke the within-system nutrient cycling by uncoupling the decomposition process from the plant-uptake process. In the absence of nutrient uptake in spring, when the deciduous trees would have started production, the inorganic nutrients released by decomposer activity were available to be leached in the drainage water.

The main effect of deforestation was on nitrate–nitrogen, emphasizing the normally efficient cycling to which inorganic nitrogen is subject. The output of nitrate in streams increased 60-fold after the disturbance. Other biologically important ions were also leached faster as a result of the uncoupling of nutrient cycling mechanisms (potassium—14-fold increase; calcium—sevenfold increase; magnesium—fivefold increase). However, the loss of sodium, an element of lower biological significance, showed a much less dramatic change following deforestation (2.5-fold increase). Presumably it is cycled less efficiently in the forest, and so uncoupling had less effect. Once again, sulphur was exceptional. Its rate of loss actually decreased after deforestation. No satisfactory explanation for this is available.

How far can we generalize from the Hubbard Brook study to other communities? First, it seems likely that, in the majority of terrestrial environments, nutrient cycling within a community will also be shown to be important relative to its inputs and outputs, although nutrient-use efficiency (defined as net primary productivity per unit of nutrient uptake by plants) varies according to species and location. Thus, evergreen plants generally use nutrients more efficiently than deciduous species; this

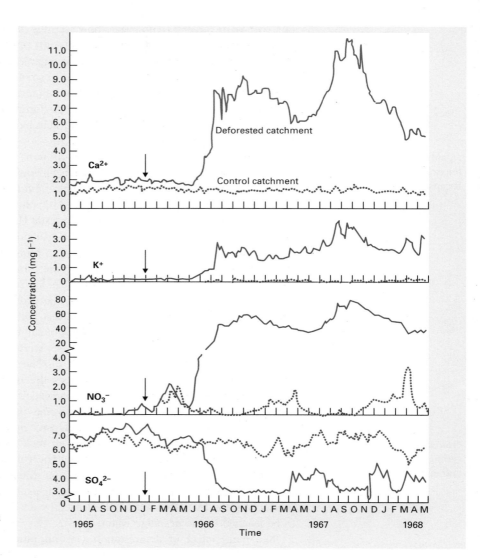

Figure 19.6 Concentrations of ions in streamwater from the experimentally deforested catchment and a control catchment at Hubbard Brook. The timing of deforestation is indicated by arrows. Note that the 'nitrate' axis has a break in it. (After Likens & Bormann, 1975.)

has been shown, for example, in the case of forest trees (Table 19.3), the flora of peatbogs (Reader, 1978), Arctic tundra (Chapin, 1980) and chaparral and coastal scrub (Gray, 1983). Higher nutrient-use efficiency in evergreen species is partly due to the greater longevity of their leaves, with less nutrient being lost per unit time in leaf fall. In addition, nutrient conservation may be enhanced by a combination of a

Table 19.3 Net primary productivity ($kg\ ha^{-1}\ year^{-1}$) per unit of nutrient uptake, as an index of nutrient-use efficiency, to compare deciduous and coniferous forests. (After Cole & Rapp, 1981.)

Forest type	Production per unit nutrient uptake				
	N^{3-}	P^{3+}	K^+	Ca^{2+}	Mg^{2+}
Deciduous	143	1859	216	130	915
Coniferous	194	1519	354	217	1559

... according to vegetation type

low rate of leaching of nutrients from living plants and more efficient nutrient storage and recycling within the plant (some nutrients are translocated from leaves back into the rest of the plant before they are shed; Chapin, 1980). Such properties could help to explain the predominance of evergreen species in many nutrient-poor environments. Another generalization to emerge is that the efficiency with which

... and in relation to overall nutrient availability

key nutrients (particularly nitrogen and phosphorus) are taken up and used by vegetation, whether evergreen or deciduous, is greater in environments where nutrients are less available (Shaver & Melillo, 1984). However, this generalization has received several challenges (Chapin & Kedrowski, 1983; Birk & Vitousek, 1986).

... tree harvest and defoliation by insects have similar effects

An increase in streamflow and nutrient loss after tree harvest, like that recorded in Hubbard Brook, has been reported in other forests (Waring & Schlesinger, 1985; Waring, 1989), and massive defoliation of hardwood forest in North Carolina during an outbreak of caterpillars of *Alsophila pometaria* (the autumn cankerworm) had a similar effect (Swank *et al.*, 1981). Moreover, studies have begun to shed light on the precise influence on stream runoff of the type of vegetation present in a catchment. For example, converting forest cover from hardwoods to conifers (*Pinus strobus*) in a catchment of the Coweeta Forest, in North Carolina, reduced streamflow by 18%. In contrast, a cover of shrubs and tree saplings yielded 35% more runoff than mature hardwoods (Waring *et al.*, 1981).

19.2.4 Lessons for agriculture and forestry

The practices of both agriculture and forestry necessarily break the continuity of recycling that characterizes most natural communities. Because harvesting involves the removal of biomass together with its contained minerals, it depletes the mineral resources locally. The resources are then added somewhere else. The international trade in foods (especially cereals) is one of the largest scale transfers of minerals (e.g. the export of grain from the USA to Russia and in famine aid to third world countries). On a more local scale, the movement of grain from arable crop areas to regions of intensive livestock production depletes the minerals in the former, and

agriculture and forestry break the continuity of recycling

supplements them in the latter.

The act of cultivation and harvesting necessarily reduces the biomass of plants on an area of land and therefore the amounts of mineral resources that are held in living tissues. Minerals that are then released by decomposition (or by the burning of stubble) are free to leach away in drainage waters. Nitrogen is especially likely to be lost in this way, as nitrate ions which move freely in soil water. Perhaps the most instructive experimental study of this phenomenon is the classic long-term experiment made on the Broadbalk field at Rothamsted, England. In 1865, land that had been in cultivation for arable crops was divided into plots receiving different treatments. In one treatment no fertilizers or farmyard manure were applied. In another, farmyard manure was applied each year and, in another, complete artificial

the effects of fertilization of land

fertilizer was applied at rates consistent with conventional farming practice. The treatments were continued every year and a crop of wheat was taken annually. The balance sheet for nitrogen was calculated (Table 19.4).

As part of the experiment, drain gauges had been set up on land that had been kept completely free of vegetation and these showed that over the period of 1870–1915 a third of the nitrogen originally present in the top 22.5 cm of soil had been lost—almost all of this was as nitrate.

The loss of nitrogen from soils causes two sorts of problems. First, it means that

Table 19.4 Nitrogen balance sheet for a cultivated soil at Broadbalk, Rothamsted, England, over the period 1865–1914. (After Russell, 1973.)

	Treatment		
	No manure	Manure	Fertilizer
N in soil in 1865	3070	4860	3630–3550
1914	2920	6700	3600–3640
Total change in 49 years	−150	+1840	−30 to +90
N retained (+) or lost (−) per year	−3.1	+37.5	−0.6 to +1.8
N added in manure, fertilizer, seed and rain per year	+8	+233	+104
N removed in crop per year	−19	−56	−49 to −52
(N unaccounted for per year)	(Gain 8)	(Loss 140)	(Loss 52–53)

farmers wanting to maintain high productivity need to supplement their crops with manure, or more often nowadays with industrially produced fertilizers. The second problem is that the nitrogen leached from land creates problems elsewhere by contributing to the eutrophication of lakes—a matter of vital concern—and finding its way into drinking waters. These two problems cause a convergence in the interests of the farmer and the conservationist—the former does not want to pay for more nitrogen fertilizer than necessary and the latter wants to limit the amounts of nitrogen that drain from farmland into other areas and water supplies.

nitrate pollution of lakes—nitrogen, a resource out of place

The nitrogen economy of farming systems will, of course, vary widely, and the rate of nitrate leaching will depend to some extent on the degree to which precipitation exceeds evaporation (Figure 19.7). The wetter a climate the more expensive it is to apply nitrogen fertilizer because more of it is lost to leaching instead of entering the crop.

For illustration (but not as a generalization) the nitrogen budget of one whole farm is illustrated in Figure 19.8. This is a small Danish dairy farm of 32.3 ha (3.23×10^5 m^2) on a sandy soil in Jutland with a stock of 44 dairy cows and about 60 young animals (Bennekon & Schroll, 1988). The inputs from biological nitrogen fixation (clover in grassland and dry deposition from the atmosphere) are estimated to add 1246 kg of nitrogen to the farm each year, and together with fertilizers the total annual input was 10 019 kg of nitrogen. Only 1515 kg of nitrogen could be accounted for in the meat, milk and cereals exported from the farm, leaving a deficit of 8504 kg (84.9%) unaccounted for. The evaporation of NH_3 and denitrification must account for a small fraction. The rest was presumably lost by leaching.

The nitrogen economy of agricultural land can be controlled in a variety of ways.

nitrate loss may be reduced ...

1 If actively absorbing plant root systems are present throughout the year, at least some of the nitrate ions released by decomposition may be recycled through the vegetation instead of leaking into the drainage waters. Thus, any crop, natural vegetation or weed community that persists as a ground cover, especially during periods when rainfall exceeds precipitation, will make the system less leaky to nitrogen. For example, if annual cereal crops can be sown in autumn and make some growth during the winter, nitrogen cycling is likely to be more efficient than if the land is left bare through the winter months. Moreover, species with deep roots are likely to catch more nitrate as it passes down through the soil.

... by residual vegetation

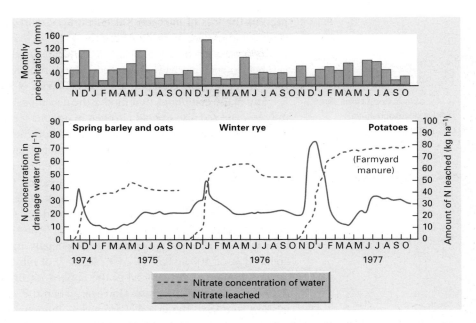

Figure 19.7 The changes in nitrate concentration (mg N l^{-1}) of water draining from an arable field (gley–podsolic soil, medium-fine sand) at Hannover, Germany, and the amounts of nitrate leached (kg N ha^{-1}). The monthly precipitation during the period is shown above the graphs. Note how the period of maximal leaching occurs during the winter months, before the concentration of nitrate has reached its high spring and summer levels. (After Duynisveld *et al.*, 1988.)

2 Any treatment of land that allows organic matter to accumulate will (at least in the medium term) reduce the processes that lead to the liberation of nitrate ions. For example, incorporating the stubble left after a cereal crop has been harvested quite dramatically reduces the rate of nitrification. This is because the incorporated organic matter has a high ratio of carbon to nitrogen (80–120 : 1) and the ratio in the bodies of decomposers is much narrower (8–10 : 1). Hence, as material is consumed by decomposers, freely available mineral nitrogen is quickly absorbed into their tissues instead of being released to leaching.

... by encouraging accumulation of organic matter

3 If land is to be irrigated, it will be important that this is only adequate to make up deficiencies in the soil's water-holding capacity—excess will drain away and carry nitrates with it. However, during the growing season some irrigation may actually

... by careful timing of irrigation

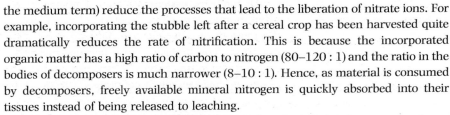

Figure 19.8 The nitrogen budget of a Danish dairy farm in 1984. Values are in kilograms of nitrogen per year for the whole farm. (After Bennekon & Schroll, 1988.)

prevent the loss of nitrogen if it encourages more rapid plant growth and, hence, more rapid uptake of nitrates from the soil.

4 If nitrogen fertilizer is to be applied it will be used most efficiently if its application is timed to coincide with crop needs. If it can be applied as a spray to the foliage, more may enter the crop and less will be at risk of leaching from the soil.

... by abiotic conditions that hinder decomposition

5 Any of the conditions that hinder the decomposition of organic matter in the soil may be expected also to hold nitrogen in organic form where it is much less likely to be leached. Poor drainage and low pH will both have this effect, although they are scarcely likely to be encouraged by the farmer. They may, however, have a place in regimes in which land is set aside from cultivation, when preventing nitrates entering drinking waters becomes more important than raising crops.

6 It might be expected that the management of livestock on permanent grasslands (a managed animal–plant community) would be efficient in cycling nitrogen and other elements. This turns out not to be true—especially when the system is intensive. In intensive grassland management in The Netherlands, an annual application of 400 kg ha^{-1} (4×10^{-2} kg m^{-2}) of nitrogen is recommended. If this is mown, the fate of virtually all of the applied nitrogen can be accounted for in the increased output of the grass. However, when the same fertilizer is applied and the pasture is grazed, the animals are given additional foodstuffs which, added to the fertilizer, means that a total of 500 kg ha^{-1} (5×10^{-2} kg m^{-2}) is put into the system. Of this, 426 kg ha^{-1} (4.26×10^{-2} kg m^{-2}) cannot be accounted for in the nitrogen present in the meat and milk that are produced. The lost nitrogen is primarily accounted for by volatilization (from dung and urine—with downwind consequences in the form of local nitrogen-deposition), denitrification, leaching and surface runoff (Steenvorden, 1988). Of course, under the mowing regime, the harvested herbage is taken away from the field and fed to animals elsewhere and thus lost from the original agricultural system.

problems of waste disposal

All intensive systems of animal management (including the management of urban humans!) produce problems of waste disposal. These problems are amongst the most intractable that can face any ecologist, and solving them can be very expensive. The problems arise because materials harvested over large areas are concentrated in small areas. It has been estimated that a pig-fattening unit of 10 000 places produces as much polluting waste as a town of 18 000 inhabitants (Sprent, 1987).

The excreta of cattle, pigs and humans has a nitrogen content of approximately 2.4%, and the droppings from battery hens and the slurry produced from cattle and pig excreta has a nitrogen content as high as 6%. In areas where fuel is in short supply for cooking, dung is often dried and used for fuel, but in the cooler areas where most conurbations and intensive animal production are found, part of the problems of disposal of excreta is that they contain so much water. Other problems are that, if the sewage or slurry is to be returned to the land, the product is bulky and expensive to transport and may simply, like a fertilizer, create problems as nitrates are released and enter water supplies. For communities living near the sea, the oceans may be used as a waste disposal bin, and human and other effluents may be disposed of by putting them at the mercy of marine food chains. A discussion of the ecology of nitrogen in animal (including human) wastes is given by Sprent (1987).

the management of forests

Many of the features of the nitrogen economy of agricultural systems are also found in managed forests. There are of course also some large differences. In particular, the crop is removed far less frequently, although when it is felled and cleared, up to 700 kg of nitrogen may be removed per hectare (Heal *et al.*, 1982). Moreover, for a large part of its growth cycle, it can provide a complete canopy cover and an extensive root system. The critical period when the system becomes leaky is in the years immediately following felling and whilst a new crop is becoming established. Vitousek *et al.* (1982) studied the response to clear-felling in 17 different forests (including both evergreen and deciduous) from sites ranging across the USA, and recognized four phases in the behaviour of nitrogen. These phases could occur in sequence, or in some situations one or more phases could be bypassed. However, sooner or later nitrate was lost to drainage water.

We have chosen to illustrate nutrient fluxes in agricultural and forest systems by reference to the behaviour of nitrogen. Each mineral element will have its own special cycling behaviour, but of all the elements that may cycle in a terrestrial community, nitrogen (as nitrate) is likely to be the most important because the productivities of most terrestrial grasslands and forests that have been studied to date are nitrogen-limited.

19.3 Nutrient budgets in aquatic communities

When attention is switched from terrestrial to aquatic communities, there are several important distinctions to be made. In particular, aquatic systems receive the bulk of their supply of nutrients from stream inflow (see Figure 19.2). In stream and river communities, and also in lakes with a stream outflow, export in outgoing streamwater is a major factor. By contrast, in lakes without an outflow (or where this is small relative to lake volume), and also in oceans, nutrient accumulation in permanent sediments is often the major export pathway.

19.3.1 Streams

We noted, in the case of Hubbard Brook, that nutrient cycling within the forest was great in comparison with nutrient exchange through import and export. By contrast, only a small fraction of available nutrients takes part in biological interactions in stream and river communities (Winterbourn & Townsend, 1991). The majority flows on, as particles or dissolved in the water, to be discharged into a lake or the sea. Nevertheless, some nutrients do cycle from inorganic form in streamwater, to organic form in biota, to inorganic form in streamwater, and so on. But, because of the inexorable transport downstream, the displacement of nutrients may be best

nutrient 'spiralling' in streams

represented as a spiral (Elwood *et al.*, 1983), where fast phases of inorganic nutrient displacement alternate with periods when the nutrient is locked in biomass at successive locations downstream (Figure 19.9). Bacteria, fungi and microscopic algae, growing on the substratum of the streambed, are mainly responsible for the uptake of inorganic nutrients from streamwater in the biotic phase of spiralling. Nutrients, in organic form, pass on through the food web via invertebrates that graze and scrape microbes from the substratum ('grazer-scrapers'—see Chapter 11, Figure 11.5). Ultimately, decomposition of the biota releases inorganic nutrient molecules and the spiral continues. The concept of nutrient spiralling is equally applicable to

... and riverine wetlands

'wetlands', such as backwaters, marshes and alluvial forests, that occur in the floodplains of rivers. However, in these cases, spiralling can be expected to be much tighter because of reduced water velocity (Ward, 1988).

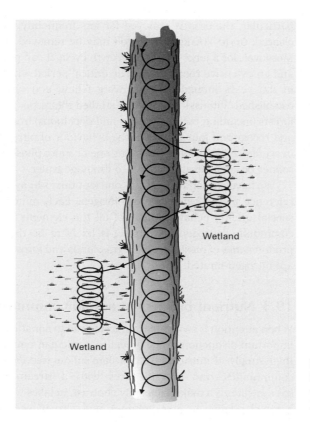

Figure 19.9 Nutrient spiralling in a river channel and adjacent wetland areas. (After Ward, 1988.)

19.3.2 Freshwater lakes

plankton plays a key role in nutrient cycling in lakes

In lakes, it is the plankton that plays the key role in nutrient cycling, as illustrated for phosphorus in Figure 19.10. Unfortunately, we cannot include in the figure any details of the size of pools or flux rates. This is mainly because many of the processes occur so rapidly (e.g. up to 75% of phosphate may be released from phytoplankton cells within a few hours of their death) and show so much seasonal variation that only some general points can be made. For example, during a summer day in a large temperate lake, with relatively little throughflow of streamwater, the most prominent fluxes are: (i) the uptake of dissolved phosphorus by phytoplankton and bacteria; (ii) the loss to grazing by zooplankton; and (iii) recycling to the water column through excretion by plankton and by the decomposition of dead cells of phytoplankton, zooplankton and bacteria. In this case, there is relatively little loss to the sediment and most phosphorus in the lake cycles between the dissolved phosphorus compartment and the biota (Moss, 1989). By contrast, in a small lake with a relatively large throughflow of water, nutrient cycling will be small compared with the throughput of nutrients to the outflowing stream. Finally, the return of nutrients from the sediment can be a major contributor to phosphate in the water column in very shallow lakes in summer, when high rates of decomposition at the sediment surface produce the anaerobic conditions necessary for phosphorus release.

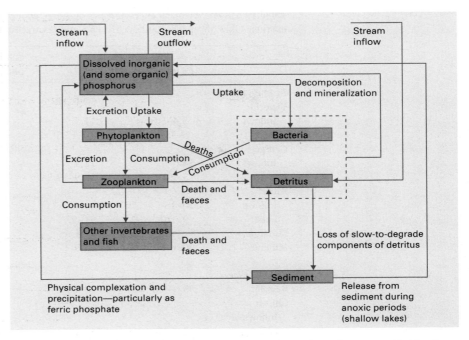

Figure 19.10 Pathways and cycles of phosphorus in a lake. Lake ecologists usually identify three phosphorus components: particulate phosphorus, comprising the phosphorus in plankton, bacteria and detritus; soluble reactive phosphorus (SRP), which includes the important inorganic phosphate component and is readily available for uptake by both phytoplankton and bacteria; and soluble unreactive phosphorus (SUP), consisting mainly of organic compounds that are not readily available to phytoplankton. In this model, the dissolved phosphorus compartment includes both SRP and SUP. Bacteria and detritus are placed together in a decomposition 'blackbox', but the bacteria can take up inorganic phosphorus, and make a contribution to photosynthesis in some circumstances.

19.3.3 Saline lakes and oceans

saline lakes lose water only by evaporation ...

Many lakes in arid regions, lacking a stream outflow, lose water only by evaporation. The waters of these endorheic lakes (internal flow) are thus more concentrated than their freshwater counterparts, being particularly rich in sodium but also in other nutrients such as phosphorus (Table 19.5). Saline lakes should not be considered as oddities; globally, they are just as abundant in terms of numbers and volume as freshwater lakes (Williams, 1988). They are usually very fertile and have dense populations of blue-green algae (e.g. *Spirulina platensis*), and some, such as Lake Nakuru in Kenya, support huge aggregations of plankton-filtering flamingoes (*Phoeniconaias minor*). No doubt, the high level of phosphorus is due in part to the concentrating effect of evaporation. In addition, there may be a tight nutrient cycle in lakes such as Nakuru, in which continuous flamingo feeding and the supply of their excreta to the sediment creates circumstances where phosphorus is continuously regenerated from the sediment to be taken up again by phytoplankton (Moss, 1989).

... and have high nutrient concentrations

The largest of all endorheic 'lakes' is the ocean—a huge basin of water supplied by the world's rivers and losing water only by evaporation. Its great size, in comparison with the input from rain and rivers, leads to a remarkably constant chemical composition.

Table 19.5 Some chemical characteristics of saline, endorheic (no outflow) lakes and exorheic lakes (those with outflows) in East and Central Africa. (After Moss, 1980.)

	Ca (mg l^{-1})	Na (mg l^{-1})	Soluble reactive phosphorus (µg l^{-1})	Total P (µg l^{-1})
Exorheic lakes				
Victoria				
Inflows	5.4–7.0	11.5	—	—
Lake	5–15	10–13.5	13–140	47–140
George				
Inflows	15.6	12.3	79	—
Lake	15–20.2	12.9–13.5	2	412
Albert				
Inflows	10–15	—	—	—
Lake	9–16	91–97	120–180	133–200
Mutanda		15.5	18–93	65–110
Mulehe	20.8–21.7	10.8	220	272
Endorheic lakes				
Rudolf	5.0–5.8	770–810	715	2600
Chilwa				
Inflows			0–24	—
Lake	7.2–18.0	348–2690	0–7520	—
Abiata	3–10	2870–6375	—	890
Natron	Trace	129 000	290 000	—
Nakuru	10	5550–38 000	40–4400	308–12 200
Magadi	10	38 000		11 000
Elmenteita lake	1.1–10	9450	—	2000

ocean mixing links surface waters and deep waters

Geochemists who study the sea find it convenient to divide it into two compartments—the warm surface waters, where most of the ocean's plant life is found, and the cold deep waters (the latter is about 10 times the volume of the former). Again, we turn our attention to phosphorus, because it is a critical plant nutrient often in limited supply. Nutrient in the surface waters comes from two sources—river inputs and water welling up from the deep waters (Figure 19.11). About 30 times more water arrives at the surface via upwelling than from rivers, and the former is about three times as concentrated in phosphorus. A given phosphorus atom, arriving in the surface waters, is taken up by a phytoplankton cell or bacterium and is passed along the food chain. This is very much as described for the lake in Figure 19.10, but marine ecologists place particular emphasis on the importance of the microscopic picoplankton in ocean productivity (perhaps 60% of primary productivity) and nutrient cycling (they are consumed mainly by tiny flagellates and take part in a subcycle that is almost totally divorced from the nutrient cycle involving phytoplankton and their consumers—similar 'microbial loops' are now known from some lake systems too) (Stockner & Antia, 1986). Detrital particles are continuously sinking to the deep waters where the majority are decomposed, releasing soluble phosphorus again that eventually finds its way back to the surface water by upwelling. Only about 1% of detrital phosphorus is lost to the sediment in each oceanic mixing cycle (about every 1000 years the entire amount of any given constituent is sent through the surface-water phase and back to the deep) (Broecker & Peng, 1982). Note that this process of nutrient cycling occurs usually

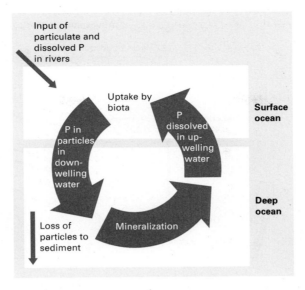

Input of
particulate and
dissolved P
in rivers

Uptake by
biota

P in
particles
in down-
welling
water

P
dissolved
in up-
welling
water

**Surface
ocean**

Loss of
particles to
sediment

Mineralization

**Deep
ocean**

Figure 19.11 A simple two-compartment model showing the major fluxes of phosphorus in the ocean.

over vast distances; a lot of the nutrients arriving in upwelling water in the Antarctic probably derive from oceans in the Northern hemisphere.

19.4 Global biogeochemical cycles

Nutrients are moved over vast distances by winds in the atmosphere and by the moving waters of streams and ocean currents. There are no boundaries, either natural or political. It is appropriate, therefore, to conclude this chapter by moving to an even larger spatial scale to examine global biogeochemical cycles.

Recognizing the need for a thorough understanding of the global picture, the General Council of the International Council of Scientific Unions decided in September 1986 to establish the International Geosphere–Biosphere Programme: A Study of Global Change, whose broad aims were:

the International Geosphere–Biosphere Programme (IGBP)

> to describe and understand the interactive physical processes that regulate the total earth system, the unique environment that it provides for life, the changes that are occurring in this system, and the manner in which they are influenced by human actions (Steffen *et al.*, 1992).

In the remainder of this chapter we explore some of the ways that human activities are upsetting global geochemical cycles.

19.4.1 Perturbation of the hydrological cycle

The hydrological cycle is simple to conceive (although its elements are by no means always easy to measure) (Figure 19.12). The principal source of water is the oceans; radiant energy makes water evaporate into the atmosphere, winds distribute it over the surface of the globe, and precipitation brings it down to earth (with a net movement of atmospheric water from oceans to continents), where it may be stored temporarily in soils, lakes and icefields. Loss occurs from the land through evaporation and transpiration or as liquid flow through stream channels and groundwater aquifers, eventually to return to the sea. The major pools of water occur in the oceans (97.3% of the total for the biosphere—Berner & Berner, 1987),

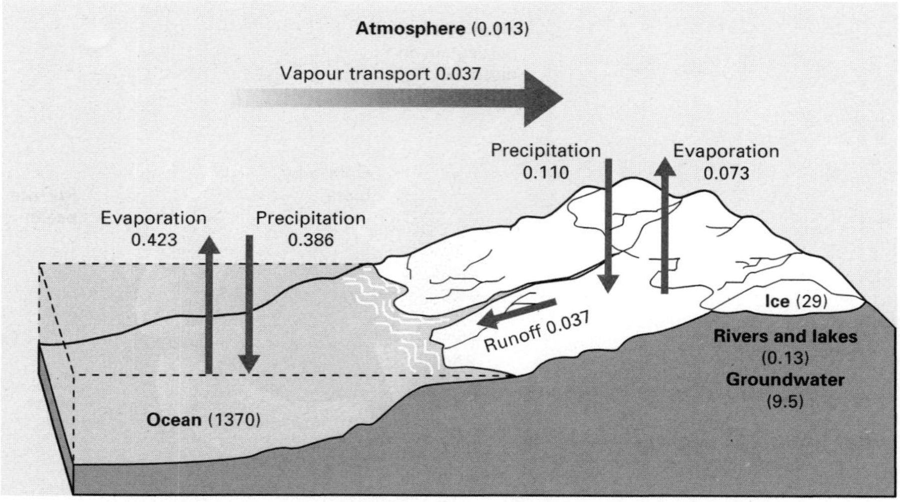

Figure 19.12 The hydrological cycle showing fluxes and sizes of reservoirs ($\times 10^6$ km^3). Values in parentheses represent the size of the various reservoirs. (After Berner & Berner, 1987.)

plants live between two counterflowing movements of water

the ice of polar ice-caps and glaciers (2.06%), deep in the groundwater (0.67%) and in rivers and lakes (0.01%). The proportion that is in transit at any time is very small—water draining through the soil, flowing along rivers and present as clouds and vapour in the atmosphere constitutes only about 0.08% of the total. However, this small percentage plays a crucial role, both by supplying the requirements for survival of living organisms and for community productivity, and because so many chemical nutrients are transported with the water as it moves.

The hydrological cycle would proceed whether or not a biota were present. However, terrestrial vegetation can modify to a significant extent the fluxes that occur. Plants live between two counterflowing movements of water (McCune & Boyce, 1992). One moves within the plant, proceeding from the soil into the roots, up through the stem and out from the leaves as transpiration. The other is deposited on the canopy as precipitation from where it may evaporate or drip from the leaves or flow down the stem to the soil. In the absence of vegetation, some of the incoming water would evaporate from the ground surface, but the rest would enter the streamflow (via surface runoff and groundwater discharge). Vegetation can intercept water at two points on this journey, preventing some from reaching the stream and causing it to move back into the atmosphere: (i) by catching some in foliage from which it may evaporate; and (ii) by preventing some from draining from the soil water by taking it up in the transpiration stream.

We have seen on a small scale how cutting down the forest in a catchment in Hubbard Brook can increase the throughput to streams of water, together with its load of dissolved and particulate matter. It is small wonder that large-scale deforestation around the globe, usually to create new agricultural land, can lead to loss of topsoil, nutrient impoverishment and increased severity of flooding.

Another major perturbation to the hydrological cycle is expected to be global climate change resulting from human activities (see Section 19.4.6). The predicted temperature increase, with its concomitant changes to wind and weather patterns, can be expected to affect the hydrological cycle by causing some melting of polar caps and glaciers, by changing patterns of precipitation and by influencing the details of evaporation, transpiration and streamflow (see, for example, Meyer & Pulliam, 1992).

19.4.2 A general model of global nutrient flux

the major nutrient compartments and fluxes in global biogeochemical cycles

The world's major abiotic reservoirs for nutrients are illustrated in Figure 19.13. The biota of both terrestrial and aquatic habitats obtains some of its nutrient elements predominantly via the weathering of rock. This is the case, for example, for phosphorus. Carbon and nitrogen, on the other hand, derive mainly from the atmosphere—the first from CO_2 and the second from gaseous nitrogen, fixed by microorganisms in soil and water. Sulphur derives from both atmospheric and lithospheric sources. In the following sections, we consider phosphorus, nitrogen, sulphur and carbon in turn, and ask how human activities upset the global biogeochemical cycles of these biologically important elements.

19.4.3 Perturbation of the phosphorus cycle

phosphorus derives mainly from the weathering of rocks

The principal stocks of phosphorus occur in the water of the soil, rivers, lakes and oceans, and in rocks and ocean sediments. The phosphorus cycle may be described as an 'open' (or as a sedimentary) cycle because of the general tendency for mineral phosphorus to be carried from the land inexorably to the oceans, where ultimately it becomes incorporated into the sediments (Figure 19.14a). Sedimentation removes about 13×10^6 tonnes of phosphorus per year from ocean water (Ramade, 1981).

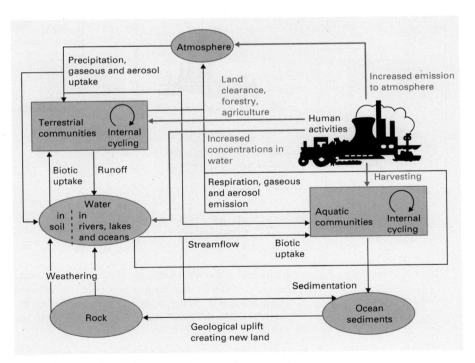

Figure 19.13 The major global pathways of nutrients between the abiotic 'reservoirs' of atmosphere, water (hydrosphere) and rock and sediments (lithosphere), and the biotic 'reservoirs' constituted by terrestrial and aquatic communities. Human activities (in orange) affect nutrient fluxes through terrestrial and aquatic communities both directly and indirectly, via their effects on global biogeochemical cycling through the release of extra nutrients into the atmosphere and water.

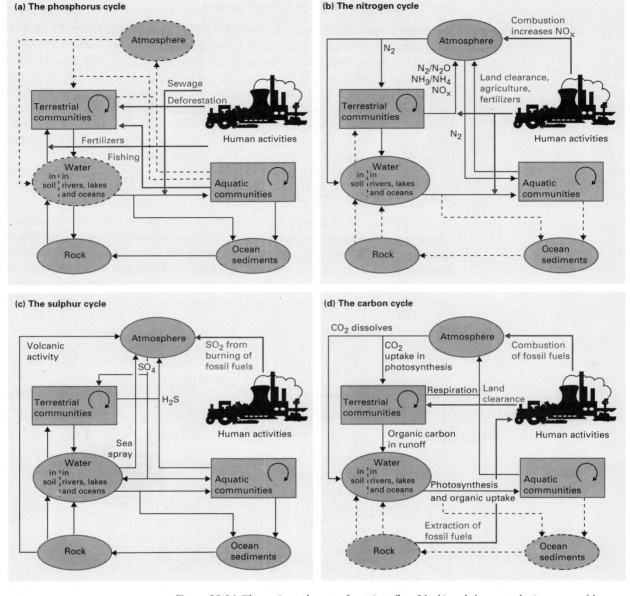

Figure 19.14 The main pathways of nutrient flux (black) and the perturbations caused by human activities (orange) for four important nutrient elements: (a) phosphorus, (b) nitrogen, (c) sulphur and (d) carbon. Insignificant compartments and fluxes are represented by dashed lines. (Based on the model illustrated in Figure 19.10, where further details can be found.)

Referring back to the figure for phosphorus flux in the ocean (see Section 19.3.3; Figure 19.11), we can unravel an intriguing story that starts in a terrestrial catchment area. A typical phosphorus atom, released from the rock by chemical weathering, may enter and cycle within the terrestrial community for years, decades or centuries before it is carried via the groundwater into a stream, where it takes part in the nutrient spiralling described in Section 19.3.1. Within a short time of entering the stream (weeks, months or years), the atom is carried to the ocean. It then makes, on average, about 100 round trips between surface and deep waters, each lasting perhaps 1000 years. During each trip, it is taken up by surface-dwelling organisms, before eventually settling into the deep again. On average, on its 100th descent (after 10 million years in the ocean) it fails to be released as soluble phosphorus, but instead enters the bottom sediment in particulate form. Perhaps 100 million years later, the ocean floor is lifted up by geological activity to become dry land. Thus, our phosphorus atom will eventually find its way back via a river to the sea, and to its existence of cycle (biotic uptake and decomposition) within cycle (ocean mixing) within cycle (continental uplift and erosion).

Human activities affect the phosphorus cycle in a number of ways. Marine fishing transfers about 50×10^6 tonnes of phosphorus from the ocean to the land each year. Since the total oceanic pool of phosphorus is around $120\,000 \times 10^6$ tonnes (Ramade, 1981), this reverse flow has negligible consequences for the ocean compartment. However, phosphorus from the fish catch will eventually move back through rivers to the sea and, thus, fishing contributes indirectly to increased concentrations in inland waters. More than 13×10^6 tonnes of phosphorus are dispersed annually over agricultural land as fertilizer (some derived from the marine fish catch) and a further $2-3 \times 10^6$ tonnes as an additive to domestic detergents. Much of the former reaches the aquatic system as agricultural runoff, whereas the latter arrives in domestic sewage. In addition, deforestation and many forms of land cultivation increase erosion in catchment areas and contribute to artificially high amounts of phosphorus in runoff water. All told, human activities contribute about two-thirds of the annual river outflow of phosphorus to the oceans (Ramade, 1981).

An increase in phosphorus input to the oceans on this scale is insignificant, but as the more concentrated water passes through rivers, estuaries and particularly lakes, its influence can be profound. This is because phosphorus is often the nutrient whose supply limits aquatic plant growth. In many lakes world-wide, the input of large quantities of phosphorus from agricultural runoff and sewage, and also of nitrogen (mainly as runoff from agricultural land), produce ideal conditions for high phytoplankton productivity. In such cases of cultural eutrophication (enrichment), the lake water becomes turbid because of dense populations of phytoplankton (often the blue-green species), and large aquatic plants are outcompeted and disappear along with their associated invertebrate populations. Moreover, decomposition of the large biomass of phytoplankton cells may lead to low oxygen concentrations, which kill fish and invertebrates. The outcome is a productive community, but one with low biotic diversity and low aesthetic appeal. The remedy is to reduce nutrient input, for example by altering agricultural practices and by diverting sewage, or by chemically 'stripping' phosphorus from treated sewage before it is discharged. Where phosphate loading has been reduced in deep lakes, such as Lake Washington in North America, a reversal of the trends described above may occur within a few years (Edmonson, 1970). In shallow lakes, however, phosphorus stored in the sediment may continue to be released and the physical removal of some of the sediment may be called for (Moss *et al.*, 1988).

The effects of agricultural runoff and sewage discharge are localized, in the sense that only those waters that drain the catchment area concerned are affected. But, the problem is pervasive and world-wide.

19.4.4 Perturbation of the nitrogen cycle

the nitrogen cycle has an atmospheric phase of overwhelming importance

The atmospheric phase is predominant in the global nitrogen cycle, in which nitrogen fixation and denitrification by microbial organisms are of particular importance (Figure 19.14b). Atmospheric nitrogen is also fixed by lightning discharges during storms, and reaches the ground as nitric acid dissolved in rainwater, but only about 3–4% of fixed nitrogen derives from this pathway. The magnitude of the flux in streamflow from terrestrial to aquatic communities is relatively small, but it is by no means insignificant for the aquatic systems involved. This is because nitrogen is one of the two elements (along with phosphorus) that most often limits plant growth. Finally, there is a small annual loss of nitrogen to ocean sediments.

human activities impinge on the hydrospheric phase of the nitrogen cycle ...

Human activities have a variety of far-reaching effects on the nitrogen cycle. Deforestation, and land clearance in general, leads to substantial increases in nitrate flux in streamflow and nitrous oxide losses to the atmosphere (see Section 19.2.3; Figure 19.6). In addition, technological processes yield fixed nitrogen as a by-product of internal combustion and in the production of fertilizers. The agricultural practice of planting legume crops, with their root nodules containing nitrogen-fixing bacteria, contributes further to nitrogen fixation. In fact, the amount of fixed nitrogen produced by these human activities is of the same order of magnitude as that produced by natural nitrogen fixation (Rosswall, 1983). The production of nitrogenous fertilizers (more than 50×10^6 tonnes per year) is of particular significance because an appreciable proportion of fertilizer added to land finds its way into streams and lakes. The artificially raised concentrations of nitrogen contribute to the process of cultural eutrophication of lakes discussed in Section 19.4.1.

... and on the atmospheric phase

Human activities impinge on the atmospheric phase of the nitrogen cycle too. For example, fertilization of agricultural soils leads to increased runoff as well as to an increase in denitrification (Bremner & Blackmer, 1978), and the handling and spreading of manure in areas of intensive animal husbandry releases substantial amounts of NH_3 to the atmosphere (Berden et al., 1987). Atmospheric NH_3 is increasingly recognized as a major pollutant when it is deposited downwind of livestock farming areas (Sutton et al., 1993). Since many plant communities are adapted to low nutrient conditions, an increased input of nitrogen can be expected to cause changes to community composition. Lowland heathland is particularly sensitive to nitrogen enrichment (this is a terrestrial counterpart to lake eutrophication—see Section 19.4.3). For example, more than 35% of former Dutch heathland has been replaced by grassland (Bobbink et al., 1992). Further sensitive communities include calcareous grasslands and upland herb and bryophyte floras, where declines in species richness have been recorded (Sutton et al., 1993). The vegetation of other terrestrial communities is less sensitive, because it is normally nitrogen limited. For example, increased nitrogen deposition to nitrogen-limited forests can be expected to result initially in increased forest growth, but at some point the system becomes 'nitrogen-saturated' (Aber, 1992). Further increases in nitrogen deposition can be expected to 'break through' into drainage, with raised concentrations of nitrogen in stream runoff contributing to eutrophication of downstream lakes. The concept of 'critical load' is currently gaining attention in the difficult task of defining community sensitivity to nutrient enrichment and formulating policy about safe limits for pollutant emission (Grennfelt & Thornelof, 1992).

NH_3, NO_x and eutrophication of aquatic and terrestrial communities

differential sensitivity of communities, nitrogen saturation and the concept of critical load

There is clear evidence of increased NH_3 emissions during the past few decades and current estimates indicate that these account for 60–80% of anthropogenic nitrogen input to European ecosystems (Sutton *et al.*, 1993). The other 20–40% derives from oxides of nitrogen (NO$_x$), resulting from combustion of oil and coal in power stations, and from industrial processes and traffic emissions. Atmospheric NO$_x$ is converted, within days, to nitric acid, which contributes, together with NH$_3$, to the acidity of precipitation within and downwind of industrial regions. Sulphuric acid is the other culprit, and we discuss the consequences of acid rain in the next section, after dealing with the global sulphur cycle.

19.4.5 Perturbation of the sulphur cycle

In the global phosphorus cycle we have seen that the lithospheric phase is predominant (Figure 19.14a), whereas the nitrogen cycle has an atmospheric phase of overwhelming importance (Figure 19.14b). Sulphur, by contrast, has atmospheric and lithospheric phases of similar magnitude (Figure 19.14c).

the sulphur cycle has atmospheric and lithospheric phases of the same magnitude

Three natural biogeochemical processes release sulphur into the atmosphere: the formation of sea-spray aerosols (about 44×10^6 tonnes per year), volcanic activity (relatively minor) and anaerobic respiration by sulphate-reducing bacteria (may be substantial, estimates vary from 33 to 230 tonnes per year; Ramade, 1981). Sulphur bacteria release reduced sulphur compounds, particularly hydrogen sulphide, from waterlogged bog and marsh communities and from marine communities associated with tidal flats. A reverse flow from the atmosphere involves oxidation of sulphur compounds to sulphate, which returns to earth as both wetfall and dryfall (see Section 19.2.1). About 21×10^6 tonnes return per year to land and 19×10^6 tonnes to the ocean.

The weathering of rocks provides about 50% of the sulphur draining off land into rivers and lakes, the remainder deriving from atmospheric sources. On its way to the ocean, a proportion of the available sulphur (mainly dissolved sulphate) is taken up by plants, passed along food chains and, via decomposition processes, becomes available again to plants. However, in comparison with phosphorus and nitrogen, a much smaller fraction of the flux of sulphur is involved in internal recycling in terrestrial and aquatic communities.

Finally, there is a continuous loss of sulphur to ocean sediments, mainly through abiotic processes such as the conversion of hydrogen sulphide, by reaction with iron, to ferrous sulphide (which gives marine sediments their black colour).

combustion of fossil fuels releases huge quantities of SO$_2$ into the atmosphere ...

Combustion of fossil fuels is the major human perturbation to the global sulphur cycle (coal contains 1–5% of sulphur and oil 2–3%). The sulphur dioxide (SO$_2$) released to the atmosphere is oxidized and converted to sulphuric acid in aerosol droplets, mostly less than 1 μm in size. Natural and human releases of sulphur to the atmosphere are of similar magnitude. However, whereas natural inputs are spread fairly evenly over the globe, most human inputs are concentrated in and around industrial zones in northern Europe and eastern North America, where they can contribute up to 90% of the total (Fry & Cooke, 1984). Concentrations decline progressively as one moves downwind from sites of production, but they can still be high at distances of several hundred kilometres. Thus, one nation can export its SO$_2$ to other countries; concerted international political action is required to alleviate the problems that arise.

causing acid rain

Water in equilibrium with CO$_2$ in the atmosphere forms dilute H$_2$CO$_3$ with a pH of about 5.6. However, the pH of acid precipitation (rain or snow) can average well below 5.0, and values as low as 2.4 have been recorded in Britain, 2.8 in Scandinavia and 2.1 in the USA. The emission of SO_2 often contributes most to the acid rain problem, although together NO_x and NH_3 account for 30–50% of the problem (Mooney et al., 1987; Sutton et al., 1993).

perturbations can have indirect effects on biogeochemistry

We saw earlier how a low pH can drastically affect the biota of streams and lakes (see Chapter 2, Section 2.6). Acid rain has been responsible for the extinction of fish in thousands of lakes, particularly in Scandinavia. In addition, low pH can have far-reaching consequences for forests and other terrestrial communities. It can affect plants directly, by breaking down lipids in foliage and damaging membranes (Chia et al., 1984), or indirectly, by increasing leaching of some nutrients from the soil and by rendering other nutrients unavailable for uptake by plants (see Chapter 2, Section 2.6). It is important to note that some perturbations to biogeochemical cycles arise through indirect, 'knock-on' effects on other biogeochemical components. For example, alterations in sulphur flux themselves are not always damaging to terrestrial and aquatic communities, but the effect of the ability of sulphate to mobilize metals such as aluminium, to which many organisms are sensitive, may indirectly lead to changes in community composition. (In another context, sulphate in lakes can reduce the ability of iron to bind phosphorus, releasing phosphorus and increasing phytoplankton productivity—Caraco, 1993.)

can acid rain be controlled and are its effects reversible?

Provided that governments show the political will to reduce SO_2 and NO_x emissions (e.g. by making use of technologies already available to remove sulphur from coal and oil), the acid rain problem should be controllable (Wright et al., 1988); indeed, reductions in sulphur emissions have been reported in some parts of the world. However, the conservationist's task is not an easy one. For example, a decision to decrease emissions from British power stations was greeted by an outcry from those who found that the cleaning process required large-scale provision of limestone, most of it from areas of natural beauty or scientific interest. It remains to be seen whether reversibility will be complete, first in chemical terms and, second, in biological terms—will biological recolonization of formerly affected locations occur?

19.4.6 Perturbation of the carbon cycle

the opposing forces of photosynthesis and respiration drive the global carbon cycle

Photosynthesis and respiration are the two opposing processes that drive the global carbon cycle. It is predominantly a gaseous cycle, with CO_2 as the main vehicle of flux between atmosphere, hydrosphere and biota. Historically, the lithosphere played only a minor role; fossil fuels lay as dormant reservoirs of carbon until human intervention in recent centuries (Figure 19.14d).

Terrestrial plants use atmospheric CO_2 as their carbon source for photosynthesis, whereas aquatic plants use dissolved carbonates (i.e. carbon from the hydrosphere). The two subcycles are linked by exchanges of CO_2 between atmosphere and oceans as follows:

$$\text{atmospheric } CO_2 \rightleftharpoons \text{dissolved } CO_2 \tag{19.4}$$

$$CO_2 + H_2O \rightleftharpoons H_2CO_3 \text{ (carbonic acid).} \tag{19.5}$$

In addition, carbon finds its way into inland waters and oceans as bicarbonate resulting from weathering (carbonation) of calcium-rich rocks such as limestone and chalk:

$$CO_2 + H_2O + CaCO_3 \rightleftharpoons CaH_2(CO_3)_2. \tag{19.6}$$

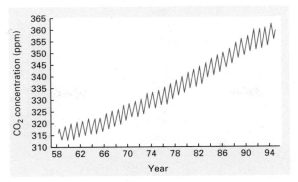

Figure 19.15 Concentration of atmospheric CO_2 at Mauna Loa Observatory, Hawaii, showing the seasonal cycle (resulting from changes in photosynthetic rate) and the long-term increase that is due largely to the burning of fossil fuels. (After Keeling *et al.*, 1995.)

Respiration by plants, animals and microorganisms releases the carbon locked in photosynthetic products back to the atmospheric and hydrospheric carbon compartments.

CO₂ in the atmosphere has increased significantly because of …

The concentration of CO_2 in the atmosphere has increased from about 280 parts per million (ppm) in 1750 to about 350 ppm today, and it is still rising. The pattern of increase recorded at the Mauna Loa Observatory in Hawaii between 1958 and 1994 is shown in Figure 19.15. (Note the cyclical decreases in CO_2 associated with higher rates of photosynthesis during the summer.) The principal cause of the increase in atmospheric CO_2 in recent years has been the combustion of fossil fuels,

… combustion of fossil fuels

which in 1980 released about 5.2 ($\pm$ 0.5) $\times 10^9$ tonnes of carbon in to the atmosphere (Table 19.6). The kilning of limestone, to produce cement, releases a further 0.1×10^9 tonnes. Thus, power generation and industrial activity release to the atmosphere about 5.3×10^9 tonnes of carbon annually.

… and exploitation of tropical forest

The exploitation of tropical forest also causes a significant release of CO_2, but the precise effect depends on whether forest is being cleared for permanent agriculture, shifting agriculture or timber production. The burning that follows most forest clearance quickly converts some of the vegetation to CO_2, whilst decay of the remaining vegetation releases CO_2 over a more extended period. If forests have been cleared to provide for permanent agriculture, the carbon content of the soil is

Table 19.6 Balancing the global carbon budget to account for increase in atmospheric carbon caused by human activities. (In the 'missing' row, the minus sign indicates the need to identify an unknown source of carbon of the size shown, whereas the plus signs indicate the need for a sink.) (After Detwiler & Hall, 1988.)

	Extreme estimate	Median estimate	Extreme estimate
Release to atmosphere			
Fossil fuel combustion	4.7	5.2	5.7
Cement production	0.1	0.1	0.1
Tropical forest clearance	0.4	1.0	1.6
Non-tropical forest clearance	−0.1	0.0	0.1
Total release	5.1	6.3	7.5
Accounted for			
Atmospheric increase	−2.9	−2.9	−2.9
Ocean uptake	−2.5	−2.2	−1.8
Missing?	−0.3	+1.2	+2.8

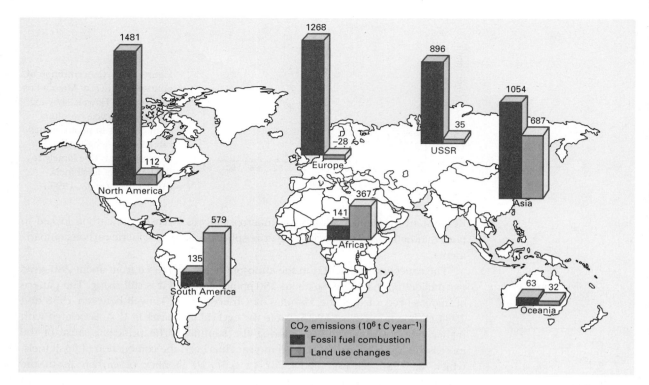

Figure 19.16 Emissions of CO_2 resulting from fossil fuel combustion and land-use changes (mainly forest clearance) in 1980. (After UNEP, 1991.)

reduced by decomposition of organic matter, by erosion and sometimes by mechanical removal of topsoil. Clearance for shifting agriculture has similar effects, but regeneration of ground flora and secondary forest during the fallow period sequesters a proportion of the carbon originally lost. Shifting agriculture and timber extraction involve 'temporary' clearance in which the net release of CO_2 per unit area is significantly less than is the case for 'permanent' clearance for agriculture or pasture. In total, about $1.0\ (\pm 0.6)\times 10^9$ tonnes per year are released through changes in tropical land use (calculated for 1980; Detwiler & Hall, 1988). Forest clearance in Brazil, Indonesia and Columbia is responsible for more than 50% of this total. Changes to land use in non-tropical terrestrial communities seem to have a negligible effect on the net release of CO_2 to the atmosphere (estimated to be $0.0\ (\pm 0.1)\times 10^9$ tonnes per year). The global pattern in CO_2 emissions is illustrated in Figure 19.16.

The total amount of 5.1–7.5×10^9 tonnes of carbon released each year to the atmosphere by human activities can be compared with the 100×10^9 tonnes released naturally by respiration of the world's biota (Mooney *et al.*, 1987). Where does the extra CO_2 go? The observed increase in atmospheric CO_2 accounts for 2.9×10^9 tonnes per year (equal to 39–57% of human inputs; Table 19.6). Much of the rest is dissolved in the oceans. The errors in the various estimates are so great that to account for the fate of CO_2 from human activities, we may need to find either a further small source of extra CO_2 or, and more likely, another quite substantial sink (Table 19.6). It is possible that terrestrial communities act as a net sink for atmospheric CO_2. Thus, part of the increase in CO_2 may serve to 'fertilize' terrestrial

some of the extra CO_2 dissolves in the oceans …

772 CHAPTER 19

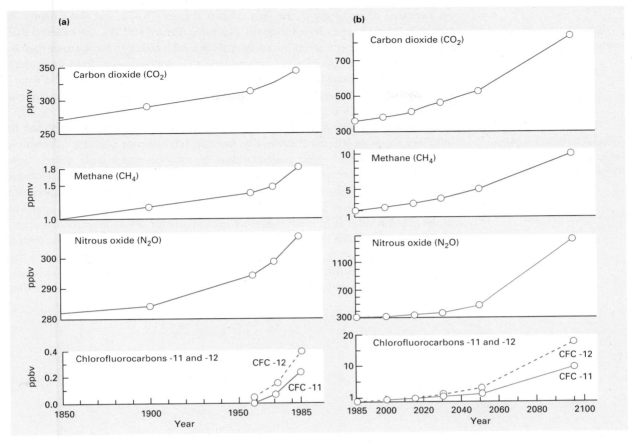

Figure 19.17 The increase in atmospheric concentrations of the principal greenhouse gases (a) recorded from 1850 to 1985 and (b) predicted for 1985–2100. (After World Meteorological Organization, 1985.)

... and some may be stored in terrestrial biomass

communities and be assimilated into extra biomass; we know that increasing the concentration of CO_2 in controlled environments increases rates of photosynthesis and growth of many plant species (Strain, 1987).

19.5 The 'greenhouse' effect

Estimates of future CO_2 emissions, and the concentration to be expected in the atmosphere, vary considerably. However, it seems likely that the concentration will rise to a mean of more than 700 ppm by the year 2100. An increase in the atmospheric concentration of a radiative gas such as CO_2 is expected to cause a warming of the earth through the so-called greenhouse effect (see Chapter 2, Section 2.11). The atmosphere behaves like the glass of a greenhouse; during the day solar radiation warms up the earth's surface, which then reradiates energy outwards, principally as infrared radiation. CO_2 is one of the gases, along with nitrous oxide, methane, ozone and various fluorocarbons, which, like greenhouse glass, absorb infrared radiation and keep the temperature high.

increases in the concentrations of 'greenhouse' gases ...

The average atmospheric concentrations of the principal greenhouse gases have all increased dramatically in the past century (Figure 19.17a; see also Chapter 2, Figure 2.24), and are expected to go on increasing (Figure 19.17b). It is believed that the present air temperature at the surface of the land is $0.5 \pm 0.2°C$ warmer than in pre-industrial times. Doubling of atmospheric CO_2 concentration from its present level is predicted to lead to a further warming of $3.5–4.2°C$. The effects of this would be profound indeed, particularly through a melting of the ice-caps, the consequent raising of sea-level and large-scale alterations to the global climate.

In addition, changes in temperature and global climate can be expected to influence the biogeochemical cycles that humans have already perturbed. Research is underway to improve our understanding of, amongst many other things, how increased CO_2 concentrations will affect primary productivity (with plant biomass acting as a carbon sink—see Chapter 18, Section 18.5) and how a rise in temperature will increase methane production in wetlands (Bubier & Moore, 1994), and soil decomposition rates in general (with soil organic matter acting as carbon sources) (Anderson, 1992). The biosphere is a complex system indeed and we are far from a full understanding of how its components interact and how it will respond to the insults our activities throw at it. The need is pressing for ecologists, environmental physicists, chemists and geologists to respond to the multidisciplinary challenge that is before them.

Chapter 20
The Influence of Competition
on Community Structure

20.1 Introduction

Chapters 20 and 21 should be read together—they discuss two strands of an important debate in ecology

In this chapter and those that follow, we will look at factors that shape and structure communities. In particular, Chapters 20 and 21 should be read together; the ideas they contain are strongly linked and they reflect a debate that has been central to ecology in the last quarter of the 20th century. As we explain below, there are sound theoretical reasons for expecting interspecific competition to be important in shaping communities by determining which, and how many, species can coexist. Indeed, the prevalent view amongst ecologists in the 1970s was that competition was of overriding importance (MacArthur, 1972; Cody, 1975). More recently, however, the conventional wisdom in ecology has moved away from this monolithic view to one giving more prominence to non-equilibrial and stochastic factors, such as physical disturbance and inconstancy in conditions (Diamond & Case, 1986; Gee & Giller, 1987). In this chapter, we consider the role of interspecific competition in theory and practice. In the next chapter, we consider the factors that in some communities and for some organisms make competition much less influential.

The view that interspecific competition plays a central and powerful role in the shaping of communities was first fostered by the competitive exclusion principle (see Chapter 7, Section 7.4.2) which says that if two or more species compete for the same limiting resource, then all but one of them will be driven to extinction. More sophisticated variants of the principle, namely the concepts of limiting similarity, optimum similarity and niche packing (see Chapter 7, Section 7.9), have suggested a limit to the similarity of competing species, and thus, a limit to the number of species that can be fitted into a particular community before niche space is fully saturated. Within this theoretical framework, interspecific competition is obviously important, because it excludes particular species, from some communities, and determines precisely which species coexist in others. The crucial question, however, is: how important are such theoretical effects in the real world?

interspecific competition may determine which, and how many species can coexist

20.2 The prevalence of current competition in natural communities

There is no argument about whether competition *sometimes* affects community structure; nobody doubts that it does. Equally, nobody claims that competition is of overriding importance in each and every case. In a community where the species are competing with one another on a day to day or minute by minute basis, and where the environment is homogeneous, it is indisputable that competition will have a powerful effect on community structure. Suppose instead, though, that other factors (predation, low-quality food, weather) keep densities at a low level where competition itself is negligible. By definition, competition *cannot* then be a potent force. One of the fundamental questions that community ecologists must try to answer, therefore, is: to what extent and in which cases *in practice* is interspecific competition an active force?

Perhaps the most direct way of answering this question is from the results of experimental field manipulations, in which one species is removed from or added to the community, and the responses of the other species are monitored (see Chapter 15, Section 15.7). Schoener (1983) examined the results of all the field experiments he could find on interspecific competition—164 studies in all. He found that approximately equal numbers of studies had dealt with terrestrial plants, terrestrial animals and marine organisms, but that studies of freshwater organisms amounted to only about 50% of the number in the other groups. Amongst the terrestrial studies, however, he found that most were concerned with temperate regions and mainland populations, and that there were relatively few dealing with phytophagous insects (see below). Any conclusions were therefore bound to be subject to the limitations imposed by what ecologists had chosen to look at. Nevertheless, Schoener found that approximately 90% of the studies had demonstrated the existence of interspecific competition, and that the figures were 89, 91 and 94% for terrestrial, freshwater and marine organisms, respectively. Moreover, if he looked at single species or small groups of species (of which there were 390) rather than at whole studies which may have dealt with several groups of species, he found that 76% showed effects of competition at least sometimes, and 57% showed effects in all the conditions under which they were examined. Once again, terrestrial, freshwater and marine organisms gave very similar figures.

Connell (1983) also reviewed field experiments on interspecific competition. His review was less extensive than Schoener's (being limited to six major journals, yielding 72 studies), but it was more intensive (dealing particularly with the relative importance of inter- and intraspecific competition). Connell's 72 studies dealt with a total of 215 species and 527 different experiments; interspecific competition was demonstrated in most of the studies, more than 50% of the species, and approximately 40% of the experiments. He did find, though, that in those cases where the intensities of inter- and intraspecific competition could be compared, interspecific competition was the more intense in only about one-sixth of the studies. In contrast with Schoener, he found that interspecific competition was more prevalent in marine than in terrestrial organisms, and also that it was more prevalent in large than in small organisms.

Taken together, Schoener's and Connell's reviews certainly seem to indicate that active, current interspecific competition is widespread. Its percentage occurrence amongst species is admittedly lower than its percentage occurrence amongst whole studies, but this is to be expected, since, for example, if four species were arranged along a single niche dimension and all adjacent species competed with each other, this would still be only three out of six (or 50%) of all possible pair-wise interactions. Moreover, the fact that intraspecific competition is typically more intense than interspecific competition does not mean that interspecific competition is unimportant; this pattern is to be expected whenever there is niche differentiation (see Chapter 7, Section 7.9).

Connell also found, however, that in studies of just one pair of species, interspecific competition was almost always apparent, whereas with more species the prevalence dropped markedly (from more than 90% to less than 50%). This can be explained to some extent by the argument outlined above, but it may also indicate biases in the particular pairs of species studied, and in the studies that are actually reported (or accepted by journal editors). It is highly likely that many pairs of species are chosen for study because they are 'interesting', that is, because competition between them is suspected, and if none is found this is simply not reported. Judging the prevalence of competition from such studies is rather like judging the prevalence of debauched clergymen from the 'gutter press'. This is a real problem, only partially alleviated in studies on larger groups of species when a number of 'negatives' can be conscientiously reported alongside one or a few 'positives'. Underwood (1986) highlights a further problem: rigorous competition experiments are notoriously difficult to design and execute and some of the studies reviewed by Schoener and Connell can be criticized on the grounds of inadequate experimental design or statistical analysis. Thus, the results of surveys such as those by Schoener and Connell exaggerate, *to an unknown extent*, the frequency and importance of competition.

In addition, there is another kind of bias in such results, namely the relative contributions to the data made by different groups and different types of organisms. This is discussed next.

20.2.1 Phytophagous insects and other possible exceptions

As Schoener himself noted, phytophagous insects were poorly represented in his data. This is particularly regrettable in view of the fact that they account for roughly 25% of all living species (Southwood, 1978b). In order to counter this neglect, Strong *et al.* (1984) have reviewed studies reporting either the presence or absence of interspecific competition amongst phytophagous insects. Only 17 out of 41 studies (or 41%) showed its presence, and in many of these it was demonstrated for only a very small proportion of the possible pair-wise interactions.

In fact, there are other reasons too for thinking that interspecific competition between phytophagous insects may be relatively rare (Strong *et al.*, 1984). The proportion of populations exhibiting intraspecific competition, as indicated by k-value analysis, is only around 20%; and if organisms do not compete with members of their own species, we might not expect them to compete with the less similar members of other species. There are also many examples of 'vacant niches' for phytophagous insects: feeding sites or feeding modes on a widespread plant which are utilized by insects in one part of the world, but not in another part of the world where the native insect fauna is different (Figure 20.1). This failure to saturate the niche space also argues against a powerful role for interspecific competition.

On a more general level, it has been suggested that herbivores *as a whole* are seldom food-limited, and are therefore not likely to compete for common resources (Hairston *et al.*, 1960; Slobodkin *et al.*, 1967; see also Chapter 22 for a detailed discussion). In a similar vein, phytophagous insects have been described as lying 'between the devil and the deep blue sea' (Lawton & McNeill, 1979): between the devil of abundant predators and parasitoids, and the deep blue sea of plant food which is generally low in quality as well as being protected physically and chemically (see Chapter 3). Phytophagous insects may therefore only rarely be able to reach the

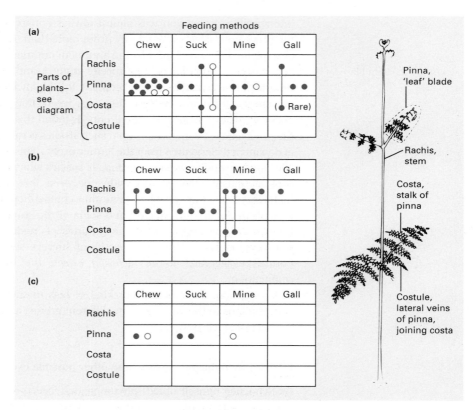

Figure 20.1 Feeding sites and feeding methods of herbivorous insects attacking bracken (*Pteridium aquilinum*) on three continents. (a) Skipwith Common in northern England; the data are derived from both a woodland and a more open site. (b) Hombrom Bluff, a savannah woodland in Papua New Guinea. (c) Sierra Blanca in the Sacramento Mountains of New Mexico and Arizona, USA; as at Skipwith, the data here are derived from both an open and a wooded site. Each bracken insect exploits the frond in a characteristic way. Chewers live externally and bite large pieces out of the plant; suckers puncture individual cells or the vascular system; miners live inside tissues; and gall-formers do likewise but induce galls. Feeding sites are indicated on the diagram of the bracken frond. Feeding sites of species exploiting more than one part of the frond are joined by lines. ●, Open and woodland sites; ○, open sites only. (After Lawton, 1984.)

sorts of densities at which competition becomes important. Finally, Schoener found the proportion of herbivores exhibiting interspecific competition to be significantly lower than the proportions of plants, carnivores or detritivores.

the strength of competition is likely to vary from community to community

Taken overall, therefore, current interspecific competition has been reported in studies on a wide range of organisms, and in some groups its incidence may be particularly obvious (e.g. amongst sessile organisms in crowded situations). However, in other groups of organisms interspecific competition may have little or no influence. It appears to be relatively rare amongst herbivores generally, and particularly rare amongst the large and important group of phytophagous insects. In the next chapter, when the roles of predation and abiotic disturbance are discussed, it will be clear that there are many communities where the densities that induce competition are rarely achieved.

20.2.2 The intensity and the structuring power of competition are not always connected

Atkinson and Shorrocks's simulations

Even when competition is potentially intense, the species concerned may nevertheless coexist. This has been highlighted in theoretical studies of model communities in which species compete for patchy and ephemeral resources, and in which species themselves have aggregated distributions, with each species distributed independently of the others (Atkinson & Shorrocks, 1981; Shorrocks & Rosewell, 1987; see also Chapter 7, Section 7.5). The species exhibited 'current competition' (like that in Schoener's and Connell's surveys), in that the removal of one species led to an increase in the abundance of others (Figure 20.2). But, despite the fact that competition coefficients were high enough to lead to competitive exclusion in a uniform environment, the patchy nature of the environment and the aggregative behaviours of individuals of the two species made coexistence possible without any niche differentiation. Thus, even if interspecific competition is actually affecting the abundance of populations, it need not determine the species composition of the community (Shorrocks et al., 1984).

the ghost of competition past

On the other hand, even when interspecific competition is absent or difficult to detect, this does not necessarily mean that it is unimportant as a structuring force. Species may not compete at present because selection in the past favoured an avoidance of competition, and thus a differentiation of niches (Connell's 'ghost of competition past'—see Chapter 7, Section 7.8.1). Alternatively, unsuccessful competitors may already have been driven to extinction; the present, observed species may then simply be those that are able to exist because they compete very little or not at all with other species. Furthermore, species may compete only rarely (perhaps

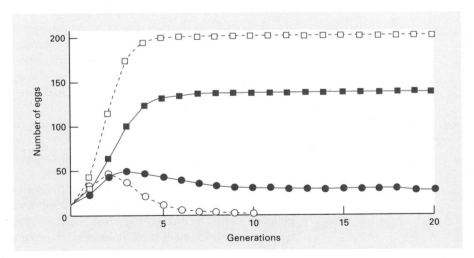

Figure 20.2 Coexistence of competitors without niche differentiation (computer simulations). The niche of species 2 is included entirely within the niche of species 1 and, in competition, species 2 has no realized niche. In a homogeneous environment, species 1 (□) excludes species 2 (○). But, in a patchy environment, over which both species are distributed in an aggregated manner, there is coexistence. In addition, both species 1 (■) and species 2 (●) have densities lower than they would achieve in the absence of interspecific competition (200 in both cases). The aggregated distributions, not niche differentiation, have led to coexistence. (After Atkinson & Shorrocks, 1981.)

779 INFLUENCE OF COMPETITION

during population outbreaks), or only in localized patches of especially high density, but the results of such competition may be crucial to their continued existence at a particular location. In all of these cases, interspecific competition must be seen as a powerful influence on community structure, affecting which species coexist and the precise nature of those species. Yet, this influence will not be reflected in the level of *current* competition. It could be claimed that in those types of organism where competition is typically rare at present (like phytophagous insects), competition is also likely to have been rare in the past and therefore unimportant as a structuring force. Nevertheless, the possibilities of past competition, added to the results of Atkinson and Shorrocks' simulations, stress that the intensity of *current* competition may sometimes be linked only weakly to the structuring power of competition within the community.

enemy-free space

A further complication may arise if species 'appear' to compete for 'enemy-free space' (Holt, 1984; Jeffries & Lawton, 1984). As we saw in Chapter 7, there may be limits to the similarity of coexisting prey that share a predator, in just the same way as there may be limits to the similarity of coexisting consumers that share a resource. Any judgement as to the significance of these effects in the real world must await further study; but in the present context it is important to note that such 'competition' can occur (or could have occurred in the past) without any resource (other than enemy-free space) having been in short supply. Once again, we see that competition may influence the structure of a community without its current power being at all obvious.

20.3 Evidence from community patterns

This weak link between current competition and the structuring power of competition has led a number of community ecologists to carry out studies on competition that do not rely on the existence of current competition. (In all honesty, many more ecologists have been led to such studies because this allows them to avoid the awkward question of whether there is current competition or not.) The approach has been first to predict what a community *should* look like if interspecific competition was shaping it or had shaped it in the past, and then to examine real communities to see whether they conform to these predictions.

the expectations from competition theory

The predictions themselves emerge readily from conventional competition theory (see Chapter 7).

1 Potential competitors that coexist in a community should, at the very least, exhibit niche differentiation.

2 This niche differentiation will often manifest itself as morphological differentiation.

3 Within any one community, potential competitors with little or no niche differentiation should be unlikely to coexist. Their distributions in space should therefore be negatively associated: each should tend to occur only where the other is absent.

There are very real problems in interpreting the data supposed to test these predictions. But, the best way to appreciate the problems is to examine the data themselves.

20.3.1 Niche differentiation

The various types of niche differentiation in animals and plants were outlined in Chapter 7. On the one hand, resources may be utilized differentially. This may express itself directly within a single habitat, or as a difference in microhabitat, geographical distribution or temporal appearance if the resources are themselves separated spatially or temporally. Alternatively, species and their competitive abilities may differ in their responses to environmental conditions. This too can express itself as either microhabitat, geographical or temporal differentiation, depending on the manner in which the conditions themselves vary.

Pyke's bumblebees ...

In one study of niche differentiation and coexistence a number of species of bumblebee were examined in Colorado (Pyke, 1982). The bumblebees fell into four groups in terms of both their proboscis length and the corolla length of the flowers they visited preferentially (Figure 20.3). The long-proboscis bumblebees (*Bombus*

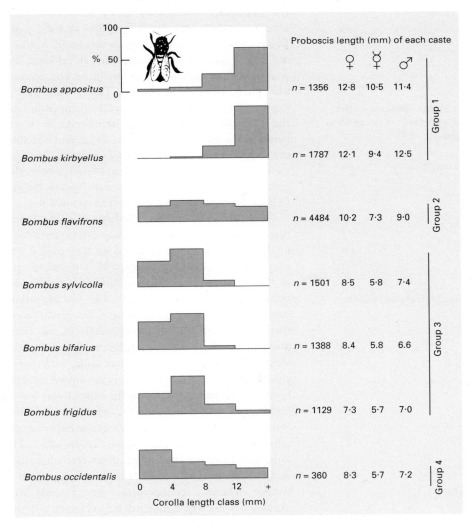

Figure 20.3 Percentage of all observations of each bumblebee species on plants in four classes of flower size. The total number of observations (*n*) is indicated in each case. Also shown for each species are the mean proboscis lengths of each caste (♀, queen; ☿, sterile female worker; ♂, male). The seven species can be divided into three groups according to proboscis length. *Bombus occidentalis* is placed into a fourth group because of its unique mandible structure. (After Pyke, 1982.)

781 INFLUENCE OF COMPETITION

appositus and *B. kirbyellus*) clearly favoured plants with long corollas, particularly *Delphinium barbeyi* (61% and 72% of all observations of visits by the two species, respectively). The short-proboscis species (*B. sylvicolla*, *B. bifarius* and *B. frigidus*) fed most frequently on various species of composite and on *Epilobium angustifolium*, all of which possess quite short corollas. The medium-proboscis species, *B. flavifrons*, fed over the entire range of corolla lengths. Finally, another short-proboscis species, *B. occidentalis*, fed as expected on plants with short corollas, but was also able to obtain nectar from long corollas. It did this by using its large, powerful mandibles to bite through the bases of the corollas and 'rob' them of their nectar. For this reason, it was placed in a group of its own.

Pyke found a clear tendency, in any single locality on an altitudinal gradient, for the bumblebee community to be dominated by one long-proboscis species (*B. appositus* at low altitude, *B. kirbyellus* at high), one medium-proboscis species (*B. flavifrons*) and one short-proboscis species (*B. bifarius* at low, *B. frigidus* at intermediate and *B. sylvicola* at high altitude). In addition, the nectar-robbing *B. occidentalis* was present at sites where its preferred and exclusive nectar source was available (the plant *Ipomopsis aggregata*). This pattern is consistent with what would be expected of communities moulded by competition (predictions (1–3) above). Indeed, the competition hypothesis is reinforced by work on two of the species that commonly coexisted—*B. appositus* and *B. flavifrons* (Inouye, 1978). When one or other of the bee species was temporarily removed, the remaining species quickly increased its own utilization of less-preferred flowers formerly exploited mainly by the other species. The differences in resource utilization, between these species at least, appear to be actively maintained by current competition.

Two further points are highlighted by the bumblebees which will be relevant throughout this chapter. First, they can be considered to be a guild (see Chapter 1), in that they are a group of species that exploit the same class of environmental resource in a similar way (Root, 1967). If interspecific competition is to occur at all, or if it has occurred in the past, then it will be most likely to occur, or to have occurred, within guilds. But, this does *not* mean that guild members do necessarily compete or have necessarily competed: the onus is on ecologists to demonstrate that this is the case. Moreover, although in the bumblebees guild membership is associated with taxonomic affiliation, there are other cases where closely related species do not belong to the same guild, and cases where guild members are not closely related (see Chapter 1). It is not always easy to decide whether species belong to the same guild. For example, competition has been demonstrated between a number of seed-eating desert rodents and seed-eating desert ants (Brown & Davidson, 1977). When either rodents or ants were removed, there was a statistically significant increase in the numbers of the other. The rodents and ants exploit the same resources but they differ, in several ways, in their mode of exploitation. For example, the ants take seeds of various sizes in proportion to their occurrence on the surface, whereas rodents prefer larger seeds and are particularly efficient at exploiting dense aggregations of buried seeds. Whether they belong to the same guild (exploiting the same resources in a similar way) is a moot point (Simberloff & Dayan, 1991).

The second point about the bumblebees is that they demonstrate *niche complementarity*; that is, within the guild as a whole, niche differentiation involves several niche dimensions, and species that occupy a similar position along one dimension tend to differ along another dimension. Thus, *B. occidentalis* differs from

evidence for niche partitioning
in differentiation of feeding
apparatus

the other six species in terms of food-plant species (*I. aggregata*) and feeding method (nectar-robbing). Amongst the others, species differ from each other either in the corolla length (and thus, proboscis length) or in the altitude (i.e. habitat) favoured or in both. Complementary differentiation along several dimensions has also been reported for birds (Cody, 1968), lizards (Schoener, 1974), tree squirrels (Emmons, 1980) and bats (McKenzie & Rolfe, 1986).

Morphological evidence of niche differentiation is perhaps most likely to be seen in the details of feeding apparatus, since organisms that utilize different resources may need different tools, as was the case for the bumblebees. Another compelling example concerns the three weasel species of North America, each of which is sexually dimorphic with females smaller than males. Each species and sex was treated separately in an analysis of the diameter of their canine teeth. Weasels aim their canines at the nape of their prey so that canine size is likely to reflect the size of prey taken. Figure 20.4 shows that in a variety of locations, the spacing in canine size is remarkably even, and the pattern is particularly striking in Michigan where there is a reversal in size sequence (this is the only place where male *Mustela erminea* have larger teeth than female *M. frenata*) (Dayan *et al.*, 1989). Parallel results have been gained for wild-cat species in Israel (Dayan *et al.*, 1990).

Intense competition may, in theory, be avoided by partitioning resources in space, as in the examples above, *or* in time, for example by a staggering of life cycles through the year. A case of apparent seasonal *and* habitat partitioning has been reported for five species of amphipod crustaceans (genus *Gammarus*) that inhabit the brackish waters of Denmark's Limfjord (Fenchel & Kolding, 1979; Kolding & Fenchel, 1979). The geographical distribution of these species is related to salinity. For example, *G. duebeni* was found in the least saline areas (0–5%), whilst at the

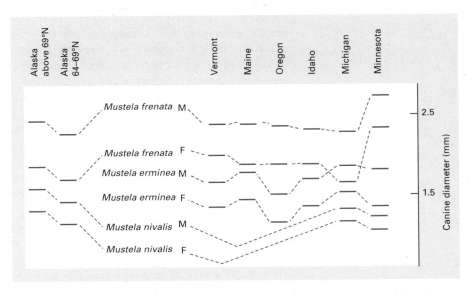

Figure 20.4 The mean diameters of the upper canines at eight locations in North America for each sex of three species of *Mustela* are shown as solid bars. For visual convenience dashed lines connect each sex of each species and the Alaskan populations are shown to the left. Despite considerable variation in number and identity of coexisting species, the spacing of canine sizes is remarkably even. (After Dayan *et al.*, 1989; Pimm & Gittleman, 1990.)

783   INFLUENCE OF COMPETITION

opposite extreme *G. oceanicus* occurred only at the highest salinities (20–33%). At any given location there were usually only two abundant species. Copulation occurs during the moulting of the female, after the male has been linked to her for some time in a precopula pairing. It turns out that there are marked differences in the periods of peak reproductive activity between the pairs of species that normally coexist. These differences in the timing of reproduction mean that the development of juveniles of coexisting species will be staggered to some extent. This, in turn, is likely to relax interspecific competition for refuges in different-sized crevices, and it may allow different-sized food particles to be exploited by coexisting species at any given time. However, the timing of reproduction may well have a further significance. In the laboratory, interspecific precopulation followed by sterile mating can occur. Strong selective forces can be expected to act to prevent such matings, and the displaced breeding periods between neighbouring species along a salinity gradient may be interpreted as the result of such forces (Fenchel, 1987a).

Mantids feature as predators in many parts of the world. Different-sized mantids have different optimal prey sizes owing to mechanical constraints imposed by their grasping forelegs. It is notable that two species that commonly coexist both in Asia and North America, *Tenodera sinensis* and *Mantis religiosa*, have life cycles that are 2–3 weeks out of phase. To test the hypothesis that this asynchrony serves to reduce interspecific competition, the timing of their egg hatch was experimentally synchronized in replicated field enclosures (Hurd & Eisenberg, 1990). *Tenodera sinensis*, which normally hatches earlier, was unaffected by *M. religiosa*. In contrast, survival and body size of *M. religiosa* declined in the presence of *T. sinensis*.

Staggering of life cycles is a common feature in plant communities, too, and conceivably the same function of reducing interspecific competition for resources may be served. As an example, consider bird-pollinated flowers. Competition between plants for the available pollinators might be expected to result in the evolution of staggered flowering times. Figure 20.5 presents data collected for 4 years in the rainforest of Costa Rica, on the flowering times of 10 plant species upon which the hermit humming-bird depends for its food supply (Stiles, 1977). The flowering periods tend to be staggered throughout the year, with only one or two species overlapping, suggesting that minimization of competition for humming-bird pollinators has been an important selective force for these plants.

Many cases of apparent resource partitioning have been reported. It is likely, however, that studies failing to detect such differentiation have tended to go unpublished. It is always possible, of course, that these 'unsuccessful' studies are flawed and incomplete, and that they have failed to deal with the relevant niche dimensions; but, a number, such as those by Rathcke (1976) and Strong (1982) on leaf-eating insects, have been sufficiently beyond reproach to raise the possibility that in certain animal groups resource partitioning is not an important feature.

Strong (1982) studied a group of hispine beetles (Chrysomelidae) that commonly coexist as adults in the rolled leaves of *Heliconia* plants. These long-lived tropical beetles are closely related; they eat the same food and they occupy the same habitat. They would appear to be good candidates for demonstrating resource partitioning. Yet, Strong could find no evidence of segregation, except in the case of just one of the 13 species studied, which was segregated weakly from a number of others. The beetles lack any aggressive behaviour, either within or between species; their host specificity does not change as a function of co-occupancy of leaves with other species that might be competitors; and the levels of food and habitat are commonly not

niche differentiation may be evident from partitioning of resources in time ... in animals

... and plants

phytophagous insects that appear not to partition resources

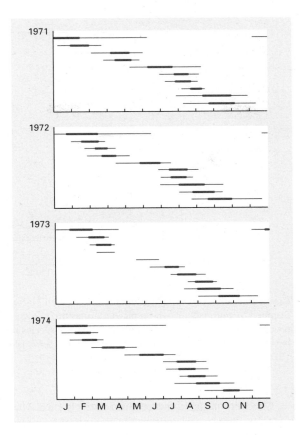

Figure 20.5 Flowering periods (thick line, main flowering; thin line, 'good' flowering) for 10 Costa Rican rainforest plants upon which hermit hummingbirds depend. The flowering periods tend to be staggered through the year, in a pattern that is reasonably consistent from year to year. (After Stiles, 1977; Pimm, 1991.)

limiting for these beetles, which suffer heavily from parasitism and predation. In these species, resource partitioning associated with interspecific competition does not appear to structure the community. As we have seen, this may well be true of many phytophagous insect communities (Lawton & Strong, 1981; Strong *et al.*, 1984). On the other hand, a study of carnivorous tiger beetle assemblages has similarly revealed no evidence of resource partitioning (mandible lengths of coexisting species were not evenly spaced, in contrast to the honeybee probosces and weasel canine teeth discussed earlier) (Pearson & Juliano, 1991). Whilst patterns consistent with a niche differentiation hypothesis are reasonably widespread, they are by no means universal.

evidence for niche partitioning is by no means universal

20.3.2 Negatively associated distributions

A number of studies have used patterns in distribution as evidence for the importance of interspecific competition. Foremost amongst these is Diamond's (1975) survey of the land birds living on the islands of the Bismarck archipelago off the coast of New Guinea. The most striking evidence comes from distributions that Diamond refers to as 'checkerboard'. In these, two or more ecologically similar species (i.e. members of the same guild) have mutually exclusive but interdigitating distributions such that any one island supports only one of the species (or none at all). Figure 20.6 shows this for two small, ecologically similar cuckoo-dove species: *Macropygia mackinlayi* and *M. nigrirostris*.

checkerboard distributions

785   INFLUENCE OF COMPETITION

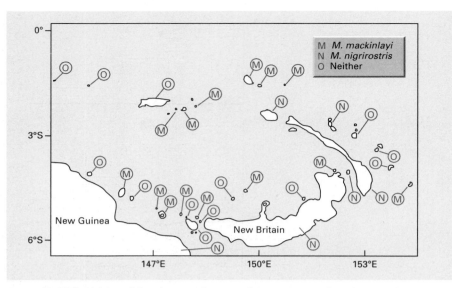

Figure 20.6 Checkerboard distribution of two small *Macropygia* cuckoo-dove species in the Bismarck region. Islands whose pigeon faunas are known are designated as M (*M. mackinlayi* resident), N (*M. nigrirostris* resident) or O (neither species resident). Note that most islands have one of these species, no island has both and some islands have neither. (After Diamond, 1975.)

incidence functions and the
possible role of diffuse
competition in supertramps

niche differences and local
coexistence in a Canadian
pasture

Checkerboards, however, are relatively rare. Much more of Diamond's evidence suggests diffuse competition from a number of species (often unspecified). For a variety of species, Diamond plotted an *incidence function*, in the form of a relationship between island 'size', S (measured as the total number of bird species supported by the island) and the proportion of islands of that size occupied by the species in question (Figure 20.7). Certain species, such as the flycatcher *Monarcha cinerascens* and the honeyeater *Myzomela pammelaena* (Figure 20.7a), appear to be excellent colonizers but poor at persisting in the diverse communities of 'large' islands. Diamond called these 'supertramps' (see p. 179). Other species, such as the pigeon *Chalcophaps stephani* (Figure 20.7b), seem to be competent colonizers but also persist in species-rich communities; whilst others (Figure 20.7c), especially those with specialized diets or restricted habitat requirements, appear to be confined to large islands. The absence of supertramps from large islands providing suitable resources and conditions implies strongly that competition is important.

The apparent niche differences between a number of plant species, and the patterns of co-occurrence of the plants on a very local scale have been studied in pastures in Ontario, Canada (Turkington *et al.*, 1977). The frequencies of physical contact between various species-pairs were measured and compared with expected frequencies derived simply from the abundances of the different species. Particular attention was paid to six species—three grasses and three legumes—and results for these are presented in the first data column of Table 20.1. A ' + ' means that two species occurred in close contact more frequently than expected by chance; a '−' means that contacts were less frequent than expected.

A negative association could mean either that two species exclude one another through competition, or that they have quite different habitat requirements. A

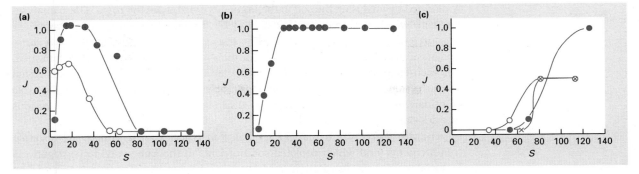

Figure 20.7 Incidence functions for various species in the Bismarcks in which J, the proportion of islands occupied by a given species, is plotted against S, a measure of island 'size' (actually the total number of bird species present). (a) Incidence functions for two 'supertramps': ●, the flycatcher *Monarcha cinerascens*; ○, the honeyeater *Myzomela pammelaena*. (b) Incidence function for the pigeon *Chalcophaps stephani*, a competent colonizer and, apparently, an effective competitor. (c) Incidence functions for three species that are restricted to larger islands: ●, the hawk *Henicopernis longicauda*; ○, the rail *Rallina tricolor*; ×, the heron *Butorides striatus*. (After Diamond, 1975.)

positive association could mean either that two species have similar habitat requirements (possibly bringing them into competition) or that they co-occur because they are indifferent to one another (i.e. they do not compete). In order to distinguish between these alternatives, the local habitats of each of the species were

Table 20.1 Coexisting plant species that live in similar types of soil in a Canadian pasture tend not to occur together (possible competitive exclusion), but species living in different types of soil do occur together (coexistence with possible niche differentiation). BI, *Bromus inermis*; DG, *Dactylis glomerata*; ML, *Medicago lupulina*; PP, *Phleum pratense*; TP, *Trifolium pratense*; TR, *T. repens*. Note that the soil characteristics reflect the consequences of a species growing in the soil and extracting nutrients differentially. (After Turkington *et al.*, 1977.)

Species-pair	Distributions more (+) or less (−) congruent than expected by chance	Number of significant correlations between the soil types of the two species examined for five characteristics		Balance of correlations
		+	−	
TR/TP	−	3	0	+
TR/ML	−	2	1	+
TR/BI	+	1	1	0
TR/DG	−	0	0	0
TR/PP	+	0	1	−
TP/ML	−	2	0	+
TP/BI	−	0	0	0
TP/DG	+	1	0	+
TP/PP	+	1	0	+
ML/BI	+	1	1	0
ML/DG	+	0	0	0
ML/PP	−	1	0	+
BI/DG	−	2	0	+
BI/PP	−	2	1	+
DG/PP	−	2	0	+

examined by analysing the soil in the immediate vicinity of the roots. These analyses indicate which resources are left behind by the growing plant rather than their resource requirements. Samples were tested for their phosphorus, potassium, magnesium and calcium content, and for their pH. The researchers then established, for each factor separately, whether particular species-pairs tended to be associated with the same or different soil conditions (Table 20.1). The final column in the table summarizes the overall similarity or dissimilarity of soil types for the various species-pairs.

Overall, there was a marked tendency for species with similar soil-resource requirements (and which depleted soil nutrients in the same way) to be negatively associated in their distribution, and for species living in different soil types to be positively associated. In particular, all three grass–grass pairs and all three legume–legume pairs lived in similar types of soil but were found only rarely in close contact. This certainly suggests an organizing role for competition in these localized plant communities. However, there are certain problems with the data.

1 Many pairs did not conform to the general tendency, and two pairs (*Trifolium pratense* with *Dactylis glomerata* and with *Phleum pratense*) ran counter to it: they lived in similar soils but were positively associated.

2 It is important to know whether the number of pairs of species where the results suggested competition was greater than would be expected simply by chance.

3 The soils were analysed in only a small number of ways; other factors (other niche dimensions) might have suggested relationships different to those observed.

4 Indeed, it is not known whether the plants actually competed for any of the soil resources considered. Thus, the results can be no more than suggestive of competition.

20.3.3 Conclusions

the results 'suggest' competition, but can do no more

We can now draw a number of conclusions about the evidence for competition discussed in this section.

1 Interspecific competition is a possible and indeed a plausible explanation for many aspects of the organization of many communities—but, it is not often a proven explanation.

2 One of the main reasons for this is that active, current competition has been demonstrated in only a small number of communities. Its actual prevalence overall can be judged only imperfectly from the results and considerations discussed in Section 20.2.

3 As an alternative to current competition, the ghost of competition past can always be invoked to account for present-day patterns. But, it can be invoked so easily because it is impossible to observe directly and therefore difficult to disprove.

4 The communities chosen for study may not be typical. The ecologists observing them have usually been specifically interested in competition, and they may have selected appropriate, 'interesting' systems. Studies that fail to show niche differentiation may often have been considered 'unsuccessful' and are likely to have gone largely unreported.

5 The community patterns uncovered, even where they appear to support the competition hypothesis, often have alternative explanations. Species may differ because of the advantages of avoiding interspecific breeding and the production of low-fitness hybrids (e.g. the *Gammarus* spp.). Alternatively, species that have negatively associated distributions may recently have speciated allopatrically (i.e. in different places), and their distributions may still be expanding into one another's ranges. Whenever there are alternative explanations, however, support for the competition hypothesis is strengthened greatly by demonstrations that there is contemporary competition between the species concerned (e.g. the bumblebees, although not the hispine beetles).

6 The recurring alternative explanation to competition as the cause of community patterns is that these have arisen simply by chance. Niche differentiation may occur because the various species have evolved independently into specialists, and their specialized niches happen to be different. Even niches arranged along a resource dimension at random are bound to differ to some extent. Similarly, species may differ in their distribution because each has been able, independently, to colonize and establish itself in only a small proportion of the habitats that are suitable for it. Ten blue and 10 red balls thrown at random into 100 boxes are almost certain to end up with different distributions. Hence, competition cannot be inferred from mere 'differences' alone. But, what sorts of differences *do* allow the action of competition to be inferred? This problem is dealt with in the following section.

20.4 Neutral models and null hypotheses

A number of workers, notably Simberloff and Strong, have criticized what they see as a tendency to interpret 'mere differences' as confirming the importance of interspecific competition. On the other hand, competition theory actually does more than this. On the basis of particular assumptions (see Chapter 7, Section 7.9), it predicts a limit to the similarity of competing species so that their niches are

the aim of demonstrating that patterns are not generated merely by chance

arranged *regularly rather than randomly* in niche space (i.e. the niches should be overdispersed); and it predicts that species with very similar niches should exclude each other so that their distributions differ *more than would be expected from chance alone.* A more rigorous investigation of the role of interspecific competition, therefore, should address itself to the question: does the observed pattern, even if it appears to implicate competition, differ significantly from the sort of pattern that could arise in the community even in the absence of any interactions between species?

Questions of this type have been the driving force behind a number of analyses that have sought to compare real communities with so-called *neutral models.* These are models of actual communities that retain certain of the characteristics of their real counterparts, but reassemble the components at random (as described below), specifically excluding the consequences of biological interactions. In fact, the neutral model analyses are attempts to follow a much more general approach to scientific investigation, namely the construction and testing of *null hypotheses.* The idea

null hypotheses are intended to ensure statistical rigour

(probably familiar to most readers in a statistical context) is that the data are rearranged into a form (the neutral model) representing what the data *would* look like in the absence of the phenomenon under investigation (in this case species interactions, particularly interspecific competition). Then, if the actual data show a significant statistical difference from the null hypothesis, the null hypothesis is rejected and the action of the phenomenon under investigation is strongly inferred. Rejecting (or falsifying) the absence of an effect is reckoned to be better than confirming its presence, because there are well-established statistical methods for testing whether things are significantly different (allowing falsification) but none for testing whether things are 'significantly similar'.

It will be clear by the end of this section that the application of null hypotheses to community structure is a difficult task. But, once again, the best way to examine these difficulties is by considering some examples of the neutral-model approach.

20.4.1 Neutral models and resource partitioning

a neutral model of resource-use in lizard communities ...

Some of the less controversial applications of the approach have been in the field of differential-resource utilization. For example, Lawlor (1980) has looked at 10 North American lizard communities, consisting of four to nine species, for which he had estimates of the amounts of each of 20 food categories consumed by each species in each community (Pianka, 1973). A number of neutral models of these communities were created (see below), which were then compared with their real counterparts in terms of their patterns of overlap in resource-use. If competition is or has been a significant force in determining community structure, the niches should be spaced out, and overlap in resource-use in the real communities should be less than predicted by the neutral models.

Lawlor's analysis was based on the 'electivities' of the consumer species, where the electivity of species i for resource k was the proportion of the diet of species i which consisted of resource k. Electivities therefore ranged from 0 to 1. These electivities were in turn used to calculate, for each pair of species in a community, an index of resource-use overlap, which itself varied between 0 (no overlap) and 1 (complete overlap). Finally, each community was characterized by a single value: the mean resource overlap for all pairs of species present.

... based on four 'reorganization algorithms'

The neutral models were of four types, generated by four 'reorganization algorithms' (RA1–RA4). Each retained a different aspect of the structure of the original community whilst randomizing the remaining aspects of resource-use.

RA1 retained the minimum amount of original community structure. Only the original number of species and the original number of resource categories were retained. Observed electivities (including zeros) were replaced in every case by random values between 0 and 1. This meant that there were far fewer zeros than in the original community. The niche breadth of each species was therefore increased.

RA2 replaced all electivities, *except zeros*, with random values. Thus, the qualitative degree of specialization of each consumer was retained (i.e. the number of resources consumed to any extent by each species was correct).

RA3 retained not only the original qualitative degree of specialization but also the original consumer niche breadths. No randomly generated electivities were used. Instead, the original sets of values were rearranged. In other words, for each consumer, all electivities, both zeros and non-zeros, were randomly reassigned to the different resource types.

RA4 reassigned only the non-zero electivities. Of all the algorithms, this one retained most of the original community structure.

Each of the four algorithms was applied to each of the 10 communities. In every one of these 40 cases, 100 'neutral-model' communities were generated and the corresponding 100 mean values of resource overlap were calculated. If competition were important in the real community, these mean overlaps should have exceeded the real community value. The real community was therefore considered to have a

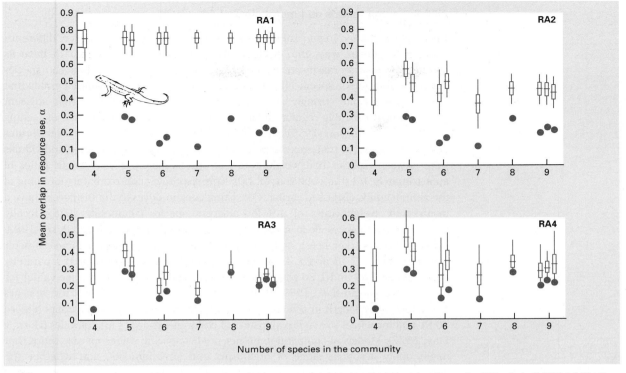

Figure 20.8 The mean indices of resource-use overlap for each of Pianka's (1973) 10 North American lizard communities are shown as solid circles. These can be compared, in each case, with the mean (horizontal line), standard deviation (vertical rectangle) and range (vertical line) of mean overlap values for the corresponding set of 100 randomly constructed communities. The analysis was performed using four different reorganization algorithms (RAs), as described in the text. (After Lawlor, 1980.)

significantly lower mean overlap than the neutral model ($P < 0.05$) if five or fewer of the 100 simulations gave mean overlaps less than the real value.

The results are shown in Figure 20.8. Increasing the niche breadths of all consumers (RA1) resulted in the highest mean overlaps (significantly higher than the real communities). Rearranging the observed non-zero electivities (RA2 and RA4) also always resulted in mean overlaps that were significantly higher than those actually observed. With RA3, on the other hand, where all electivities were reassigned, the differences were not always significant. But, in all communities the algorithm mean was higher than the observed mean. In the case of these lizard communities, therefore, the observed low overlaps in resource-use suggest that niches are segregated, and that interspecific competition plays an important role in community structure.

A similar analysis in a series of five arid grassland grasshopper communities considered overlap in both food resources and microhabitat. In both cases the results were similar: only complete randomization of utilization rates amongst all resource states (RA1) resulted in mean overlap values larger than those observed (Joern & Lawlor, 1980). Whilst this may indicate some resource partitioning in real communities, the evidence is not convincing. Perhaps this is a further example of a community of phytophagous insects in which competition plays no significant role.

the lizards appear to pass the test

a similar model for grasshopper communities— with a different result

20.4.2 Neutral models and morphological differences

Niche differentiation in general is frequently manifested as morphological differentiation. In a similar way, the spacing out of niches can be expected to have its counterpart in a regularity in degree of morphological difference between species belonging to a guild. Specifically, a common feature claimed for animal guilds that appear to segregate strongly along a single-resource dimension is that adjacent species tend to exhibit regular differences in body size or in the size of feeding structures. Hutchinson (1959) catalogued many examples, drawn from vertebrates and invertebrates, of sequences of potential competitors in which average individuals from adjacent species had weight ratios of approximately 2.0 or length ratios of approximately 1.3 (the cube root of 2.0). This also holds approximately for some of the animal guilds discussed earlier. Coexisting cuckoo-doves in the Bismarcks have a mean body weight ratio of 1.9 for adjacent species (Diamond, 1975); locally coexisting bumblebees have a mean proboscis-length ratio of 1.32 (Pyke, 1982, calculated for worker bees); and coexisting weasels have a mean canine-diameter ratio of between 1.13 and 1.23 when species and sexes are considered separately, and between 1.23 and 1.50 when the sexes are combined to give average values for each species (Dayan *et al.*, 1989). It has also been pointed out that this 'rule' appears to hold for body length in the conventional musical ensembles of recorders (Figure 20.9), and for wheel size in larval instars of children's tricycles and bicycles (Horn & May, 1977). Models of competition do not predict specific values for size ratios that might apply across a range of organisms and environments, and whether the apparent regularity in ratios is an empirical quirk remains to be determined. It may be that empirical support for Hutchinson's 'rule' has sometimes involved unwarranted selection of data and that there is a case for more rigorous statistical procedures (Roth, 1981).

The neutral-model approach to this apparent regularity of morphological differences has tended to concentrate on one particular aspect: the fact that coexisting potential competitors may be more different morphologically than might be expected by chance alone. In one example, Bowers and Brown (1982) investigated the relationship between body size and coexistence in granivorous (seed-eating) desert rodents. They did so on two scales—within individual local sites and within

Hutchinson's 'rule' about size ratios: a consequence of competitive interactions?

neutral models of body size and coexistence in desert rodents

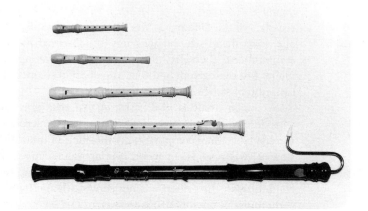

Figure 20.9 The conventional musical ensemble of recorders, which appears to conform to Hutchinson's size-ratio rule. (After Horn & May, 1977.) (Instruments kindly lent by R. Acott, Oxford. Photograph courtesy of B. Roberts.)

much larger geographical areas—using data on species from 95 sites (less than 5 ha $(5 \times 10^4 \, m^2)$ in each case) in three major North American deserts: the Great Basin (33 sites), the Mojave (24 sites) and the Sonoran (38 sites). Analysis was restricted to common species (greater than 5% of total individuals at a site) except for very large rodents (greater than 80 g) which were always included. These criteria were applied in order to exclude from the analysis rare individuals of small species which were transients from other habitats, and to correct for under-representation of large rodents in the samples because of trapping bias. The analysis was straightforward. The proportion of sites occupied by each species was determined in each of the three deserts separately, and also in all three combined. Next, the expected frequency of co-occurrence within a site for each *pair* of species was calculated, assuming that all distributions were independent. These values were compared with the actual frequency of co-occurrence of the species-pairs. Each pair was also scored as differing in body mass by either more or less than a ratio of 1.5, invoked as a conservative estimate of Hutchinson's body mass ratio. The competition hypothesis would predict that similar-sized species would co-occur less often than expected by chance, whereas this would not be so for different-sized species.

The results are shown in Table 20.2. In both the Great Basin and Sonoran Deserts, pairs of species which differed in body size by a ratio of less than 1.5 coexisted less frequently than expected on the basis of chance. The null hypothesis was not rejected for the Mojave Desert, perhaps because of the smaller sample size. But, in the analysis of the largest, most complete set, that for all deserts combined, the null hypothesis was convincingly rejected. Figure 20.10 presents typical size distributions in diverse communities from each desert.

Table 20.2 Some 2×2 contingency tables testing the null hypothesis that local coexistence and geographical overlap of granivorous rodents in North American deserts are independent of body size. Association was scored as either positive ($+$) or negative ($-$) depending on whether observed frequencies of coexistence were greater or less than expected on the basis of chance. In each case the *P*-value indicates the probability that the pattern would arise by chance. All values are statistically significant except for the Mojave Desert. (After Bowers & Brown, 1982.)

Location	Body mass ratio	Association −	Association +
Great Basin	< 1.5	6	0
	> 1.5	15	15
			$P < 0.005$
Mojave	< 1.5	3	1
	> 1.5	11	5
			$P > 0.5$
Sonoran	< 1.5	7	0
	> 1.5	23	15
			$P < 0.05$
Local coexistence (all deserts combined)	< 1.5	27	0
	> 1.5	65	28
			$P < 0.01$
Geographical overlap	< 1.5	44	13
	> 1.5	72	60
			$P = 0.01$

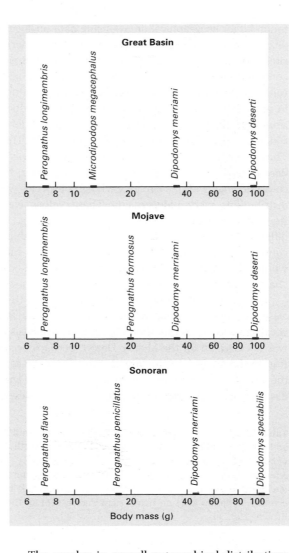

Figure 20.10 Distribution of body sizes of rodents (plotted on a logarithmic scale so that equal spacing represents equal ratios) in arbitrarily chosen diverse communities in three deserts. Note that the size distributions of species are very similar even though the identity of species in many cases is different. (After Bowers & Brown, 1982.)

The overlap in overall geographical distributions also appeared to be related to body size. Pairs of species with size ratios less than 1.5 tended to be negatively associated, resulting in a strong rejection of the null hypothesis. Particularly striking cases of displaced geographical ranges involved rodents of extreme size. Figure 20.11 illustrates this for the tiny pocket mice (body mass less than 11 g) and the larger kangaroo rats (body mass greater than 100 g).

Perhaps of equal significance, was a further analysis that considered a combination of guilds of desert rodents (granivores, folivores, insectivores and omnivores). In this case, the null hypothesis could not be rejected, suggesting that important patterns will be missed if analyses are conducted on assemblages of consumer types that are too diverse, and within which competition is expected to be relatively rare (Bowers & Brown, 1982).

In a study with a similar aim, three data sets were examined: the birds of the Tres Marias islands off the coast of Mexico, the birds of the California Channel Islands and the finches of the Galapagos. In each case, any one island actually contained only a small proportion of the total number of species in the 'source pool'. (The source pools

competition is apparently demonstrated for granivores ...

... but, not for a mixture of guilds

794 CHAPTER 20

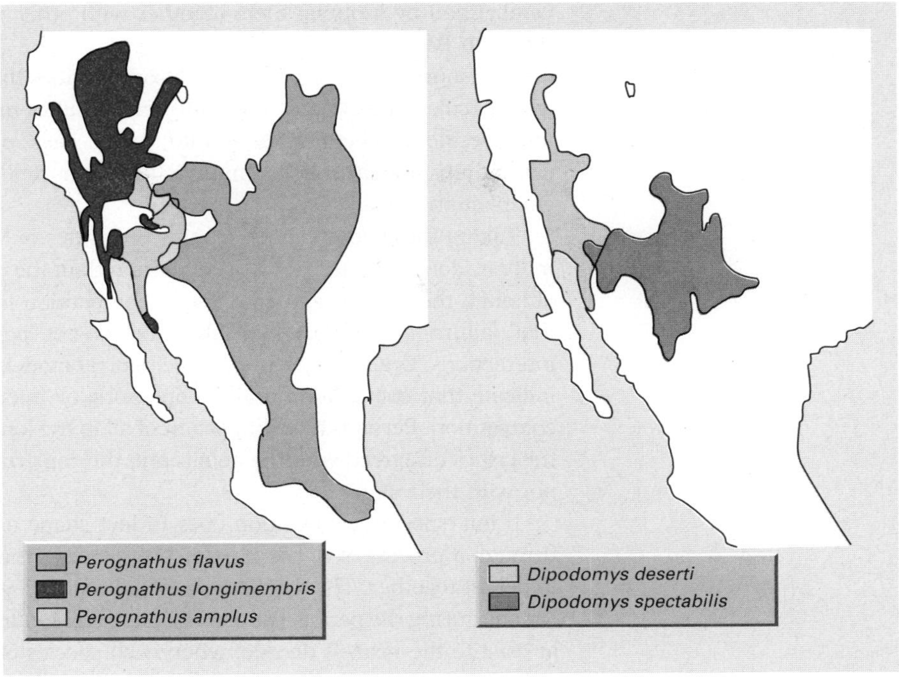

Figure 20.11 Geographical ranges of small pocket mice (*Perognathus* spp., body mass less than 11 g) and of large kangaroo rats (*Dipodomys* spp., body mass greater than 100 g). Note the extremely small overlap in geographical ranges of species of similar size. (After Bowers & Brown, 1982.)

Legend (left map):
- Perognathus flavus
- Perognathus longimembris
- Perognathus amplus

Legend (right map):
- Dipodomys deserti
- Dipodomys spectabilis

neutral models of body size and coexistence of island vertebrates ...

... the generation of a controversy

were all the species living on the nearby mainlands in the cases of the Tres Marias and California Channel Islands, and those living on the archipelago as a whole in the case of the Galapagos.) These actual communities were compared with 100 randomly assembled 'null' communities. The null communities had the same number of species per family on each island as did the actual communities, but the species themselves were drawn at random from the source pool. Using a number of different linear measurements, the ratios in length of pairs of contiguous species within families were compared. ('Contiguous species' were consecutive species in rank: longest compared with second longest, second longest with third longest, and so on.) Interspecific competition was then inferred if the length ratios in the actual communities were, in general, significantly greater than those in the null communities.

The ratios in the real communities were not generally greater and there was no reason to reject the null hypotheses, and no evidence of interspecific competition. These conclusions gave rise to considerable controversy (e.g. Grant & Abbott, 1980; Hendrickson, 1981; Simberloff & Boecklen, 1981; Strong & Simberloff, 1981; Schoener, 1984; Simberloff, 1984), and four aspects of the criticisms levelled at them are particularly worth considering.

1 There were statistical flaws in the methods used: the general point being that the construction and testing of null hypotheses for things as complicated as biological communities is technically very difficult.

2 Comparisons were made of species within families. But, competition is expected to occur within *guilds*, which do not necessarily correspond with taxonomic groupings.

3 Conclusions were drawn on the basis of whole faunas (all bird species). This is likely to dilute the effects of any patterns that *are* generated by interspecific

competition by lumping them together with other patterns that are not. (Compare this with Bowers and Brown's results.)

4 The source pools are themselves communities that may have been influenced by interspecific competition. The 'null' communities may therefore have the effects of interspecific competition built into them. Lack of a difference between a real community and a null community may thus tell us nothing about interspecific competition.

Taking these points into account, the critics of Strong and his colleagues found more evidence of interspecific competition than the original analysis had uncovered, although they could not demonstrate competition in every case. They also stressed that failure to reject a neutral model is not positive proof of the absence of interactions. Even at their best, these neutral models can never do more than either indicate that competition may be operating or infer that there is no good case for competition. Perhaps most important of all in the long term, however, is the fact that the critics disagreed with the details and the construction of the neutral models, but not with their aim.

a study involving records of selective extinctions

If interspecific competition does in fact shape a community, it will often do so through a process of selective extinction. Species that are too similar will simply fail to persist together. The detailed records of ornithologists from the six main Hawaiian islands during the period 1860–1980 allowed Moulton and Pimm (1986) to estimate, at least to the nearest decade, when each species of passerine bird was introduced and if and when it became extinct. (Almost all birds at altitudes below 600 m are introduced—testimony to the devastating effect of habitat modification on the native lowland avifauna.) In the records, overall, there were 18 pairs of congeneric species present at the same time on the same island. Of these, six pairs persisted together; in nine cases one species became extinct; and in three cases both species died out (the last category was ignored in the analysis because the outcome is not compatible with pair-wise competitive exclusion). In cases where one species became extinct the species-pair was morphologically more similar than in cases where both species persisted; average percentage difference in bill length was 9 and 22%, respectively. This statistically significant result is consistent with the competition hypothesis.

a study involving the reconstruction of an evolutionary sequence

This approach was informative because it invoked historical data, providing a glimpse of the elusive workings of 'the ghost of competition past'. An historical perspective is also important in community ecology because *evolutionary* interactions amongst community members can be expected to have played a key role in community patterns. The recent development of 'cladistic analysis' has allowed us to reconstruct phylogenies (evolutionary trees), based on similarities and differences between species in their DNA molecules and/or in morphological (or other biologically meaningful) characteristics.

The results of such an analysis of the *Anolis* lizards of Puerto Rico (Figure 20.12a) are consistent with the hypothesis of divergent evolution in body size (Losos, 1992). The two-species stage in evolution (the first, lowermost node in Figure 20.12a) was composed of species with markedly different snout-vent lengths (SVL—a standard index of size for lizards) of approximately 38 and 64 mm (*A. occultus* and the ancestor of all the remaining types, respectively) whilst sizes at the three-species stage (next node) were 38, 64 and 127 mm. In Jamaica, on the other hand (Figure 20.12b), no such pattern is observed; the two- and three-species stages were composed of species of similar size (61 and 73 mm, then 57, 61 and 73 mm SVL). However, the phylogenies of the two islands show remarkable consistency when viewed from the

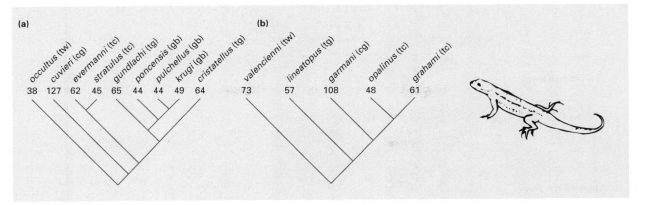

Figure 20.12 Phylogenies of lizards in the genus *Anolis* (a) on Puerto Rico and (b) on Jamaica. For each species, size (snout–vent length, mm) and ecomorph are shown: cg, crown-giant; gb, grass–bush; tc, trunk–crown; tg, trunk–ground; tw, twig. (After Losos, 1992.)

point of view of patterns in 'ecomorphs'—each distinct in morphology, ecology and behaviour. On both islands, the two-morph stage was composed of a short-legged twig (tg) ecomorph, which crawls slowly on narrow supports on the periphery of trees, and a generalist ancestral species. At the three-morph stage, too, both islands possessed the same assembly—a twig ecomorph, one specialized at foraging in the tree crown and a trunk–ground type, the latter being robust and long-legged and using its jumping and running abilities to forage on the ground. At the four-morph stage the patterns were again identical, each having added a trunk–crown type. Only at the five-morph stage was there a difference—the grass–bush morph was the last to evolve on Puerto Rico, but its counterpart has never appeared on Jamaica (Figure 20.13). Note that on each island a morph usually consists of a single species of *Anolis*, but Puerto Rico has several trunk–ground and grass–bush species. This phylogenetic analysis is consistent with the hypothesis that faunal assembly on both Puerto Rico and Jamaica has occurred via sequential microhabitat partitioning, with morphological differences perhaps being related to differences in microhabitat utilization.

Another remarkable set of examples of character displacement is provided by the new forms of Northern hemisphere fishes evolving in lakes formed after the glaciers receded 10 000–15 000 years ago. In a very wide diversity of fish taxa, including sticklebacks (*Gasterosteus* spp.) and smelts (*Osmerus* spp.), speciation and divergence has yielded sympatric pairs of species that partition resources in a remarkably consistent way—in every case one of the species is benthic, eating mainly macroinvertebrates from vegetation and sediment, whilst the other is pelagic and exploits plankton in the open water. The pelagic member of each pair characteristically is smaller, has a narrower mouth, and longer, more numerous gill rakers (which influence the efficiency of prey ingestion) than its benthic counterpart (Schluter & McPhail, 1992, 1993).

20.4.3 Neutral models and distributional differences

The absence of coexistence of morphologically similar species is really only a special case of the failure of presumed competitors to coexist. A more general approach is to

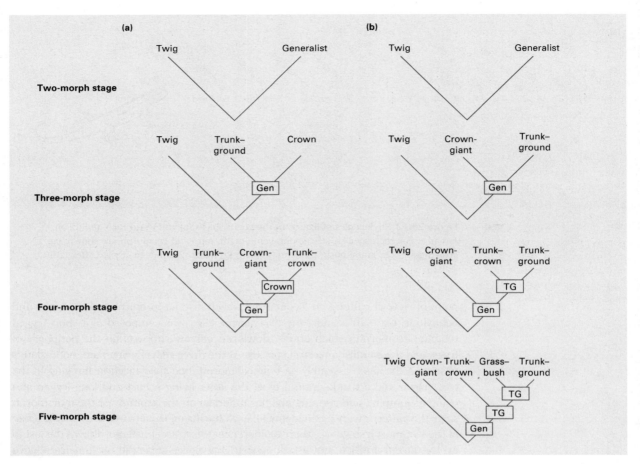

Figure 20.13 The evolution of *Anolis* communities (a) on Jamaica and (b) on Puerto Rico for two-, three-, four- and, in the case of Puerto Rico, five-ecomorph communities. Labels at nodes in the trees are the estimated ecological characteristics of the ancestors. (After Losos, 1992.)

competition as a determinant of plant-community structure on islands?

compare the pattern of species co-occurrences at a suite of locations with what would be expected by chance. An excess of negative associations would be consistent with a role for competition in determining community structure.

Thorough censuses of both native and exotic (introduced) plants occurring on 23 small islands in Lake Manapouri in the South Island of New Zealand (Wilson, 1988b), were the basis for computing a standard index of association for every pair of species:

$$d_{ik} = (O_{ik} - E_{ik}) / SD_{ik} \tag{20.1}$$

where d_{ik} is the difference between the observed (O_{ik}) and the expected (E_{ik}) number of islands shared by species i and k, expressed in terms of the standard deviation of the expected number (SD_{ik}) (Gilpin & Diamond, 1982).

different results for native and exotic plants

The resulting sets of association values for the real communities of native and exotic species are presented as histograms in Figures 20.14a and b. These can be compared with neutral-model communities in which island species richnesses and species frequencies of occurrence were fixed at those observed, but species

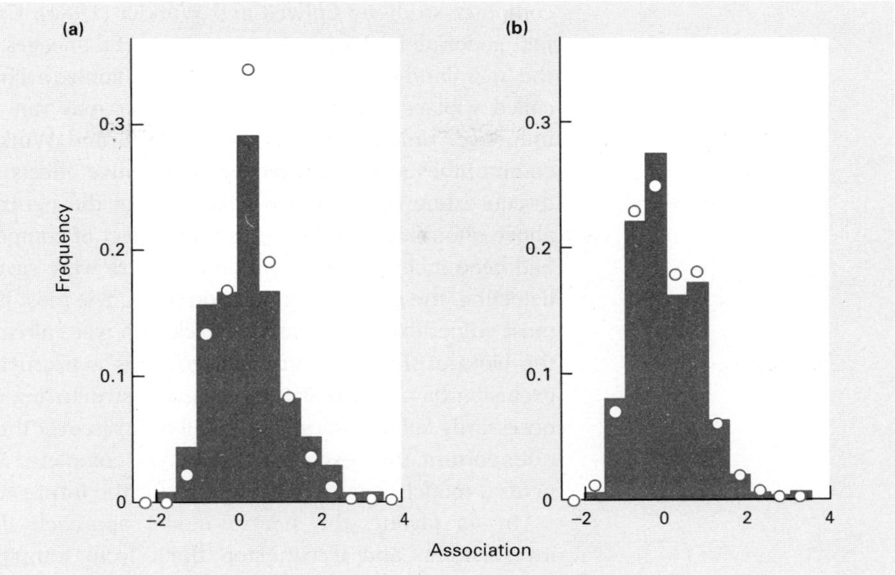

Figure 20.14 A comparison between the observed values of association between pairs of (a) native plant species and (b) exotic plant species on islands in Lake Manapouri (histograms), and the distributions expected on the basis of a neutral model (○). (After Wilson, 1988b.)

occurrences on islands were randomized (Wilson, 1987). One thousand randomizations were performed, yielding a mean frequency in each d_{ik} category (the circles in Figure 20.14). The analysis of native plants showed an excess of negative associations (highly statistically significant for the bottom four categories) and of positive associations (highly significant for the top five categories), with a corresponding deficit of associations near zero. In contrast, the analysis of exotic plants showed no significant departure from the neutral model.

In the case of native species, the excess of negative associations is consistent with the action of competitive exclusion, and this is particularly likely for the woody species. However, we cannot rule out an explanation based on a tendency of particular pairs of species to occur in different habitats, which themselves are not represented on every island (Wilson, 1988b). The most likely explanation for the excess of positive associations amongst native plants is a tendency for certain species to occur in the same habitats. The agreement of exotic species with the neutral model may reflect their generally weedy status and effective colonization abilities, or it may indicate that the exotics have not yet reached an equilibrium distribution (Wilson, 1988b).

20.4.4 Verdict on the neutral-model approach

What then should be our verdict on the neutral-model approach? Perhaps most fundamentally, its aim is undoubtedly worthy. We do need to guard against the temptation to see competition in a community simply because we are looking for it; even the critics of the earlier neutral models have tended to replace them with improved versions of their own. On the other hand, the approach is bound to be of limited use unless it is applied to groups (usually guilds) within which competition may be expected.

Second, the approach can only ever be valid when the effects of competition are truly eliminated from the neutral model itself. This has been highlighted in a

competition as one possible determinant of island distributions ... amongst several

799 INFLUENCE OF COMPETITION

computer study by Colwell and Winkler (1984). Using a program called god, they first generated a biota whose phylogenetic lineages obey specified rules. Subsets of the 'mainland biota' thus created then colonize archipelagos according to a program called wallace, and competition may or may not be introduced as a structuring influence. Unlike field workers, Colwell and Winkler *know* whether or not their communities are structured by competitive effects. The interesting and somewhat disconcerting conclusion is that some of the neutral-model approaches described above often failed to distinguish any effect of competition even though competition had been included in the program. There were several reasons for the difficulty in detecting 'the ghost of competition past'. The most important was that those species most vulnerable to competitive exclusion were already likely to have been lost from the biota of the entire archipelago. This is a crucial point because if competitive exclusion has been really important in structuring communities in nature, we will necessarily fail to observe it; we will only discover those cases in which it has been so unimportant that exclusion is not yet complete. A search for more appropriate neutral models is an important task for the future (Wilson, 1987).

In its favour, the neutral-model approach does concentrate the mind of investigators, and it can stop them from jumping to conclusions too readily. Ultimately, though, it can never take the place of a detailed understanding of the field ecology of the species in question, or of manipulative experiments designed to reveal competition by increasing or reducing species abundances (Law & Watkinson, 1989). It can only be part of the community ecologist's armoury.

20.5 The role of competition: some conclusions

1 The importance of interspecific competition in the organization of communities is easy to imagine, but is usually difficult to establish. There are alternative explanations for many of the patterns that it might appear to generate, and much of its power is presumed to have been exercised in the past rather than at present. These difficulties make caution and scientific rigour particularly necessary.

2 Interspecific competition is certain to vary in importance from community to community: it has no single, general role. For example, it appears frequently to be important in vertebrate communities, particularly those of stable, species-rich environments, and in communities dominated by sessile organisms such as plants and corals; whilst, for example, in phytophagous insect communities it is less often important. A challenge for the future is to understand why some guilds show evidence for a role for competition, such as regularity in size ratios, whilst others do not (Hopf *et al.*, 1993).

3 Even when interspecific competition is important, it may affect only a small proportion of the species interactions within a community. Mostly, it will affect only interactions between members of the same guild; and even within a guild, only the species closest together in niche space are likely to compete to a significant extent. Hence, the effects of competition can easily and misleadingly be lost in sweeping overviews of large and heterogeneous assemblages of species.

The importance of competition will be discussed again in Chapter 24, after the other organizing (and disorganizing) forces within communities have been examined.

Chapter 21
The Influence of Predation and
Disturbance on Community Structure

21.1 Introduction

Biologists are sometimes accused of 'physics envy' because the systems they work with seem not to reveal fundamental rules like the Laws of Thermodynamics, or neat and tidy order like the periodic table of the elements. One reason why simple laws are difficult to perceive in ecology is that the patterns keep changing; organisms are heritably variable and evolve, and thus, at least some of the rules of behaviour and interaction themselves change. The life of the chemist or physicist would be very different if the periodic table kept changing or if gravitational force varied erratically in intensity.

Attempts to see in the living world the simplicity of the physicist's world may deny us a vision of the essence of ecological systems. They constantly undergo change—no environment is uniform, no year is exactly like any other. Moreover, the ultimate units that ecologists work with (individual organisms or their parts) exist in numbers so small compared with the number of molecules in a test-tube that we cannot disregard, as the chemist usually can, the chancy nature of events. One forest fire may make a rare species extinct.

If ecological communities are non-uniform, continually altering and subject to the statistical events of random change, the facts cannot be ignored. A science has to be robust enough to take account of its realities. This is why 'disturbance' is allotted a whole chapter in this book. Amongst the forces of disturbance are included predators, earthquakes, fires, burrowing gophers and even the fall of a raindrop. It is important not to impose a human sense of scale on the communities studied—a raindrop may be lethal to a seedling, and there can be no more serious disturbance to life than sudden death.

A further reason for emphasizing disturbance in a text on ecology is that we ourselves are the source of particularly dramatic disturbances of nature, through agriculture and forestry, fertilizers and pollution, recreation and hunting. Most (perhaps now all) of the communities of the world have been disturbed from their 'pristine' condition by the action of mankind. Ecologists who try to study undisturbed communities (such as virgin forest) are likely to spend their whole life trying to find one!

21.1.1 Disturbance and the diversity of communities

It is possible to conceive of a world with just one species of plant (or herbivore) with supreme performance over an enormous range of tolerance; in this scenario the most

temporal variation

... and disturbance—crucial
aspects of ecological reality

competitive species (the one that is most efficient at converting limited resources into descendants) would be expected to drive all less competitive species to extinction. The diversity we witness in real communities is a clear demonstration of the failure of evolution to produce such supreme species. An extension of this competitive argument holds that diversity can be explained through a partitioning of resources amongst competing species whose requirements do not overlap completely (resource partitioning was discussed in Chapter 20). However, this rests on two assumptions that are not necessarily always valid.

The first assumption is that the organisms are actually competing, which in turn implies that resources are limiting. But, there are many situations where disturbance, such as predation, storms on a rocky shore or frequent fires, may hold down the densities of populations, so that resources are not limiting and individuals do not compete for them.

The second assumption is that when competition is operating and resources are in limited supply, one species will inevitably exclude another. But, in the real world, when no year is exactly like another, and no square centimetre of ground exactly the same as the next, the process of competitive exclusion may not proceed to its monotonous end (see Chapter 7, Section 7.6). Any force that continually changes direction at least delays, and may prevent, an equilibrium or a stable conclusion being reached. Any force that simply interrupts the process of competitive exclusion may prevent extinction and enhance diversity.

equilibrium and non-equilibrium theories

A basic distinction can thus be made between *equilibrium* and *non-equilibrium* theories. An equilibrium theory, like the one concerned with niche differentiation, helps us to focus attention on the properties of a system at an equilibrium point—time and variation are not the central concern. A non-equilibrium theory, on the other hand, is concerned with the transient behaviour of a system away from an equilibrium point, and specifically focuses our attention on time and variation. Of course, it would be naive to think that any real community has a precisely definable equilibrium point, and it is wrong to ascribe this view to those who are associated with equilibrium theories. The truth is that investigators who focus attention on equilibrium points have in mind that these are merely states towards which systems tend to be attracted, but about which there may be greater or lesser fluctuation. In one sense, therefore, the contrast between equilibrium and non-equilibrium theories is a matter of degree. However, this difference of focus is instructive in unravelling the important role of temporal heterogeneity in communities.

21.1.2 What is disturbance?

disturbance defined in a community context

Disturbance implies interruption of the course of a process or interference with a settled state. In community ecology, the underlying process is usually taken to be interspecific competition and the settled state is the structure the community would assume if conditions remained constant. A disturbance is then any relatively discrete event in time that removes organisms (Townsend & Hildrew, 1994) or otherwise disrupts the community by influencing the availability of space or food resources, or by changing the physical environment (Pickett & White, 1985). A general consequence is likely to be the opening up of space, or freeing up of resources, that can be taken over by new individuals.

We have grouped together predation and disturbance in this chapter because the activity of predators is often a disturbance in the 'normal' course of a competitive interaction, and because population explosions of pests and pathogens (which are included here with predators) often have dramatic effects in truly interfering with an otherwise settled condition. Moreover, the result of a predator opening up a gap for colonization is sometimes indistinguishable from that of battering by waves on a rocky shore or a hurricane in a forest.

We discuss first the ways in which disturbances caused by predators, parasites and disease can affect community structure. We then consider the effects of temporal heterogeneity and physical disturbance. A non-equilibrial theory of community structure is developed in which disturbance plays a key role. This contrasts with theories involving competitive equilibrium discussed in Chapters 7 and 20.

21.2 The effects of predation on community structure

21.2.1 The effect of grazers

grazing can enhance plant species richness

Lawn-mowers are relatively unselective predators capable of maintaining a close-cropped sward of vegetation. Darwin (1859) was the first to notice that the mowing of a lawn could maintain a higher diversity of species than occurred in its absence. He wrote that:

> If turf which has long been mown, and the case would be the same with turf closely browsed by quadrupeds, be let to grow, the most vigorous plants gradually kill the less vigorous, though fully grown plants; thus out of 20 species growing on a little plot of mown turf (3 feet by 4 feet) nine species perished from the other species being allowed to grow up freely.

Rabbits are more choosy than lawn-mowers, and this is clearly demonstrated by the occurrence in the neighbourhood of rabbit burrows of plants that are unacceptable as food (including *Atropa belladonna*, *Urtica dioica*, *Solanum dulcamara* and *Sambuca nigra*). Nevertheless, they seem in many areas to have a similar general effect to lawn-mowers. The rabbit is not native to Great Britain, and its introduction (probably in the 12th century) and subsequent spread must have been a major disturbance to the vegetation. Its presence and its effects subsequently became part of the normal state of affairs. Deliberate exclusion of rabbits from areas of vegetation was then a disturbance of the new status quo. Tansley and Adamson (1925) excluded rabbits from areas of species-rich chalk grassland on the English South Downs, and these areas quickly became dominated by just a few species of grass. In 1954, the viral disease myxomatosis was introduced to Great Britain and drastically reduced rabbit populations. The immediate response of the vegetation was an increase in the number of flowering perennial plants observed (e.g. orchids). These had been present but not recognized because they had been repeatedly nibbled. Later, a few species of grass became dominant (as they had in Tansley and Adamson's experiment), and later still the number of grass species also fell as the community became dominated by the taller, more severely self-shading types.

rabbit grazing: 'natural' and artificial experiments

Grazing by rabbits had apparently kept the aggressive, dominant grasses in check and allowed a greater diversity of plants to persist. However, at very high intensities of grazing, diversity may actually be reduced as the rabbit is forced to turn from heavily grazed, preferred plant species to less preferred species. Plant species may then be driven to extinction. Figure 21.1 shows the relationship between the intensity of rabbit grazing (on a scale from 0 to 5) and the richness of the flora in a series of plots on sand-dunes in the Frisian Islands of The Netherlands.

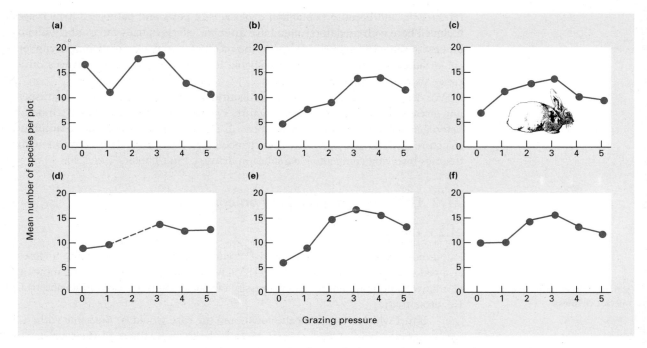

Figure 21.1 Relationship between plant species richness and intensity of rabbit grazing (on a scale of 0–5) in 1 m² plots on five sand-dunes (a–e) and all dunes combined (f). (After Zeevalking & Fresco, 1977.)

The figure y-axis label is "Mean number of species per plot" and x-axis "Grazing pressure".

exploiter-mediated coexistence

When predation promotes the coexistence of species amongst which there would otherwise be competitive exclusion (because the densities of some or all of the species are reduced to levels at which competition is relatively unimportant), this is known generally as 'exploiter-mediated coexistence'.

herbivores: periwinkles and algae

Along the rocky shores of New England in the USA, the most abundant and important herbivore in mid- and low-intertidal zones is the periwinkle snail *Littorina littorea*. The snail will feed on a wide range of algal species but shows a strong preference for small, tender species and in particular for the green alga *Enteromorpha*. The least-preferred foods are much tougher (e.g. the perennial red alga *Chondrus crispus* and brown algae such as *Fucus* spp.). They are either never eaten by *L. littorea* or are eaten only if no other food has been available for some time.

Lubchenco (1978) noticed that the algal composition of tide pools in the rocky intertidal region varied from almost pure stands of *Enteromorpha intestinalis* to the opposite extreme, preponderance of *C. crispus*. To test whether grazing by *L. littorea* was responsible for these differences, she removed all periwinkles from a *Chondrus* pool and added them to an *Enteromorpha* pool, monitoring subsequent changes for 17 months. During that time, periwinkles in a control *Chondrus* pool were observed to feed on microscopic plants and the young stages of many ephemeral algae that settled on the *Chondrus* (including *Enteromorpha*). No changes in algal composition occurred in the control pool. However, in the *Chondrus* pool from which periwinkles had been removed, *Enteromorpha* and several other seasonal, ephemeral algae immediately settled or grew from microscopic sporelings or germlings, and became abundant. *Enteromorpha* achieved dominance and the plants of *Chondrus* became bleached and then disappeared. It is clear that the presence of *L. littorea* was responsible for the dominance of *Chondrus* in *Chondrus* pools. Addition of *L. littorea* to *Enteromorpha* pools led, in 1 year, to a decline in percentage cover of the green alga from almost 100% to less than 5%. *Chondrus* colonizes slowly, but eventually comes to dominate pools where *L. littorea* has eaten out its competitor.

Why then do some pools contain *L. littorea* whilst others do not? The periwinkle colonizes pools whilst it is in an immature, planktonic stage. Although planktonic periwinkles are just as likely to settle in *Enteromorpha* pools as in *Chondrus* pools, the crab *Carcinus maenas*, which can shelter in the *Enteromorpha* canopy, feeds on the young periwinkles and prevents them from establishing a new population. The final thread in this tangled web of interactions is the effect of gulls which prey on crabs where the dense, green algal canopy is absent. Thus, there is no bar to continuing periwinkle recruitment in *Chondrus* pools.

If we were to predict the effect that periwinkle grazing would have on algal species richness or diversity in tide pools, we might postulate that when *L. littorea* is absent or rare, *Enteromorpha* would competitively exclude several other species and algal diversity would be low. A survey of a number of pools with different densities of *L. littorea* demonstrated just this (Figure 21.2a). When *L. littorea* was present in intermediate densities, the abundance of *Enteromorpha* and other ephemeral algal species was reduced, competitive exclusion was prevented and many species, both ephemeral and perennial, coexisted. But, at very high densities of *L. littorea*, all palatable algal species were consumed to extinction and prevented from reappearing, leaving an almost pure stand of the tough *Chondrus*. Note the similarity between the humped curves in Figure 21.1 and Figure 21.2a.

As a generalization, selective predation may be expected to favour a higher community diversity if the preferred prey are competitively dominant. In a study of grazing on grassland, principally by sheep, Jones (1933) came to the same conclusion. Species-rich communities were maintained by continuous grazing because it was the competitive dominants, *Lolium perenne* and *Trifolium repens*, that were most severely grazed.

The picture is quite different when the preferred prey species is competitively inferior to other prey, and this is again well illustrated by Lubchenco's work on periwinkles. In the rocky intertidal zone Lubchenco studied in New England, the competitive dominance of the most abundant tide-pool plants is actually reversed when the plant species interact on emergent substrata rather than in the tide pools. Here, it is perennial brown and red algae that predominate, whilst a number of ephemeral species manage to maintain a precarious foothold, at least in sites where *L. littorea* is rare or absent. In this situation, an increase in grazing pressure, however, tended to cause a decrease in algal diversity, as the preferred species were consumed totally and prevented from re-establishing (Figure 21.2b).

A similar situation is suggested by work on sheep grazing in upland habitats in Wales (Milton, 1940, 1947). The dominant plant species in these sites, *Festuca rubra*, *Agrostis tenuis* and *Molinia caerulea*, were relatively unpalatable and overgrazing consolidated the dominance of these few.

Note that the explanation offered in the examples above begs an important question. If the unpalatable species are truly competitively dominant, why do they not exclude the other, more palatable species in the absence of predation? The answer is that other mechanisms for coexistence must be operating. Resource partitioning amongst the plants is one possibility. Others will be discussed in Sections 21.3–21.6.

... and crabs ... and gulls

predation and humped curves of diversity

diversity is decreased when competitive inferiors are preferred by predators

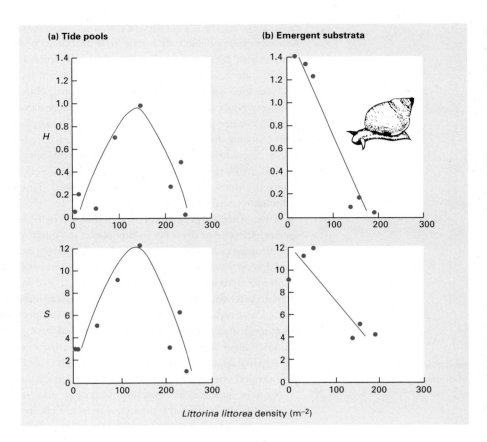

Littorina littorea density (m⁻²)

Figure 21.2 Effect of *Littorina littorea* density on species richness (*S*) and species diversity (Shannon index *H* calculated on the basis of percentage cover, see p. 683): (a) in tide pools and (b) on emergent substrata. (After Lubchenco, 1978.)

21.2.2 The effect of carnivores

Paine's starfish on a rocky shore ...

The rocky intertidal zone also provided the location for pioneering work by Paine (1966) on the influence of a top carnivore on community structure. The starfish *Pisaster ochraceus* preys on sessile filter-feeding barnacles and mussels, and also on browsing limpets and chitons and a small carnivorous whelk. These species, together with a sponge and four macroscopic algae, form predictable associations on rocky shores of the Pacific coast of North America. Paine removed all starfish from a typical piece of shoreline about 8 m long and 2 m deep, and continued to exclude them for several years. At irregular intervals, the density of invertebrates and cover of benthic algae were assessed in the experimental area and in an adjacent control site. The latter remained unchanged during the study. Removal of *Pisaster*, however, had dramatic consequences. Within a few months, the barnacle *Balanus glandula* settled successfully. Later, it was crowded out by mussels (*Mytilus californianus*), and eventually the site became dominated by these. All but one of the species of algae disappeared, apparently through lack of space, and the browsers tended to move away, partly because space was limited and partly due to lack of suitable food. Overall, the removal of starfish led to a reduction in number of species from 15 to eight.

... another example of exploiter-mediated coexistence

The main influence of the starfish *Pisaster* appears to be to make space available for competitively subordinate species. It cuts a swathe free of barnacles and, most importantly, free of the dominant mussels which would otherwise outcompete other invertebrates and algae for space. Once again, there is exploiter-mediated coexistence. Physical disturbance by storm waves can have a directly analogous effect by generating gaps which are invaded by rapidly colonizing but competitively inferior species (Paine & Levin, 1981; see Section 21.5.1). Note that this argument applies specifically to the primary space occupiers, such as mussels, barnacles and macroalgae. In contrast, the number of less conspicuous species associated with living and dead mussel shells would be expected to increase in the bed that develops after *Pisaster* removal (more than 300 species of animals and plants occur in mussel beds—Suchanek, 1992).

21.2.3 Diet switching and frequency dependent selection

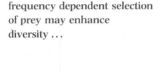
frequency dependent selection of prey may enhance diversity...

Predators seldom simply take potential prey species from a community in turn, bringing each to extinction before turning to the next. Selection is moderated by the time or energy spent in search for the preferred prey, and many species take a mixed diet. However, others switch sharply from one type of prey to another, taking disproportionately more of the most common acceptable types of prey. In theory, such behaviour could lead to the coexistence of a large number of relatively rare species (a frequency dependent form of exploiter-mediated coexistence). In practice, diet switching has been demonstrated only rarely. However, there is evidence that predation on seeds of tropical trees is often more intense where seeds are more dense (beneath and near the adult that produced them) (Connell, 1979); a herbivore species, the butterfly *Battus philenor*, forms search images for leaf shape when foraging for its two larval host plants, and concentrates on whichever happens to be the more common (Rauscher, 1978); and the zooplanktivorous fish *Rutilus rutilus* switches from large planktonic waterfleas, its preferred prey, to small sediment-dwelling waterfleas when the density of the former falls below about 40 per litre (Figure 21.3).

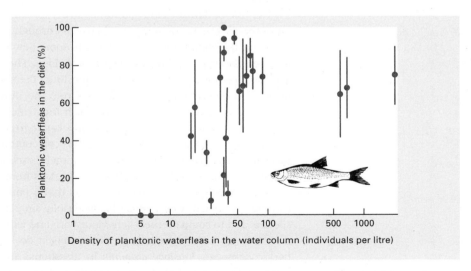

Figure 21.3 Young-of-the-year roach, *Rutilus rutilus*, living in a small English lake prefer planktonic waterfleas as prey, and these comprise the majority of their diet on most occasions (shown as mean percentage ± standard error). However, when these become relatively scarce in the environment the roach switch to sediment-dwelling waterfleas. (After Townsend *et al.*, 1986.)

807   INFLUENCE OF PREDATION AND DISTURBANCE

Despite the occurrence of such frequency dependent selection in some consumers, it is not a general rule and may not be common. For one thing, some species are so highly specialized that switching is not an option—giant pandas are specialists on bamboo shoots, the flea *Cediopsylla tepolita* feeds only on the blood of the rare and endangered Mexican volcano rabbit (*Romerolagus diazi*) and specialization in diet is equally extreme amongst many phytophagous insects (see Chapter 9, Section 9.2). Moreover, in other cases a predator may be sustained by one prey type whilst exterminating others. This has been claimed for the introduced snake *Boiga irregularis*, on the small island of Guam, about half-way between Japan and New Guinea. Coincident with its arrival in the early 1950s, and its subsequent spread through Guam, most of the 18 native bird species have declined dramatically and seven are now extinct. Savidge (1987) argues that by including abundant small lizards in its diet, *B. irregularis* has maintained high densities whilst exterminating the more vulnerable bird species.

21.2.4 Outbreaks of parasites and disease

The incidence of a parasite, like that of other types of exploiter, may determine whether or not a host species occurs in an area. Thus, the extinction of nearly 50% of the endemic bird fauna of the Hawaiian Islands has been attributed in part to the introduction of bird pathogens such as malaria and bird pox (Warner, 1968; van Riper *et al.*, 1986); and recent changes in the distribution of the North American moose (*Alces alces*) have been associated with the parasitic nematode *Pneumostrongylus tenuis* (Anderson, 1981). Probably the largest single change wrought in the structure of communities by a parasite has been the destruction of the chestnut (*Castanea dentata*) in North American forests, where it had been a dominant tree over large areas until the introduction of the fungal pathogen *Endothia parasitica*, probably from China.

The role of parasites in community structure can be seen in its full drama when an outbreak of disease occurs and sweeps through a community. Sooner or later, however, the community settles down to some new state in which the interactions between parasite and host become less obvious. The reduction in host density will itself reduce the rate of spread and intensity of damage. It is then very much more difficult to determine just how large a role the disease plays in the community structure. We need experiments in which fungicides or insecticides are applied to natural vegetation before we can really begin to see how important pests and diseases are in determining the structure of communities. Just as we can only measure the effect of predators on vegetation by excluding them, so we might gain a proper measure of the role of pests and diseases from exclusion experiments.

Another way to determine the effects that a parasite or disease might have on community structure is to assemble artificially simple communities, of one or a few species, and to compare their behaviour with and without a parasite or disease. This can show us what a parasite or disease *might* do in nature. The presence of the barley pathogen *Erysiphe graminis* in glasshouse mixtures of barley and wheat, significantly reduced the competitiveness of barley in relation to that of wheat (Figure 21.4; Burdon & Chilvers, 1977; Burdon, 1987). Similarly, the hemiparasite *Rhinanthus minor*, which obtains inorganic resources by attaching to the roots of other plants, caused the yield of a preferred host (*Lolium perenne*) to decline to the advantage of a non-preferred host (*Holcus lanatus*) in a mixed grass sward (Gibson &

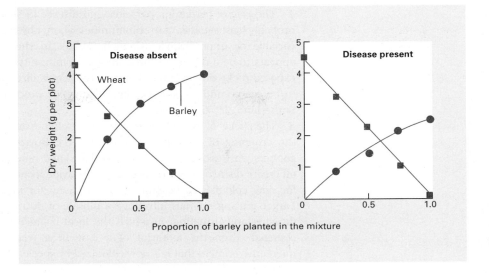

Figure 21.4 Replacement series for competition between barley and wheat showing the effect, after 67 days of growth, of powdery mildew. In the presence of disease, the performance of barley is adversely affected. (After Burdon & Chilvers, 1977.)

Watkinson, 1991). Finally, in his elegant laboratory study of the effect of a parasite on competitive ability of animals, Park (1948) showed that the protozoan parasite *Adelina triboli* was able to reverse the outcome of competition between two species of flour beetle, *Tribolium castaneum* and *T. confusum.* It is worth noting that the rate of increase of parasites and disease is usually greater at high than at low host density. Thus, their impact on a host species is likely to be *proportionately* greater when the host is abundant in a community than when it is rare. This is the phenomenon of frequency dependence referred to above, and increases the chance that host species may coexist.

21.2.5 Effects of predators, parasites and disease: finale

1 Selective predators are likely to act to enhance diversity in a community if their preferred prey are competitively dominant. It seems likely that there is some general correlation between palatability to predators and high growth rates. If the production of chemical and physical defences by prey requires a sacrifice of resources used in growth and reproduction, we might expect species that are competitive dominants in the absence of predators (and hence, which devote resources to competition rather than defence) to suffer excessively from their presence. Thus, selective predators may frequently enhance diversity. If the predators act in a frequency dependent manner their action should be even stronger.

2 Even very generalist predators may be expected to have the effect of increasing community diversity through exploiter-mediated coexistence, because even if the prey are attacked simply in proportion to their abundance, it will be those species that are assimilating resources and producing biomass and offspring most rapidly (the competitive dominants) that will be most abundant, and will therefore be most severely set-back by predation.

3 An intermediate intensity of predation is most likely to be associated with high prey diversity, since too low an intensity may not prevent competitive exclusion of inferior prey species, whilst too high an intensity may itself drive preferred prey to extinction. (Note, however, that 'intermediate' is difficult to define *a priori.*)

4 The role of predators, parasites and disease in shaping community structure is probably least significant in communities where physical conditions are more severe, variable or unpredictable (Connell, 1975). In sheltered coastal sites, predation appears to be a dominant force shaping community structure (Paine, 1966), but in exposed rocky tidal communities where there is direct wave action, predators seem to be scarce and to have negligible influence on community structure (Menge & Sutherland, 1976; Menge *et al.*, 1986).

5 The effects of animals on a community often extend far beyond just those due to the cropping of their prey (Huntly, 1991). Burrowing animals (such as moles, gophers, earthworms, rabbits and badgers) and mound-builders (ants and termites) all create disturbances. Their activities provide local heterogeneities, including sites for new colonists to become established and for microsuccessions to take place. Larger grazing animals introduce a mosaic of nutrient-rich patches, as a result of dunging and urinating, in which the local balance of other species is profoundly changed. Even the footprint of a cow in a wet pasture may so change the microenvironment that it is now colonized by species that would not be present were it not for the disturbance (Harper, 1977). The predator is just one of the many agents disturbing community equilibrium. Indeed, disturbance of one sort or another is so much the norm in natural communities that it is an open question whether truly equilibrial situations ever occur in nature.

21.3 Temporal variation in conditions

If physical conditions are continually changing and relative competitive status is not constant then competitive exclusion of one species by another is not inevitable.

The idea was first invoked by Hutchinson (1941, 1961) to account for the high species richness of phytoplankton communities in lakes and oceans (see Chapter 7, Section 7.5.3). Such communities are often extremely diverse—more so than can reasonably be attributed to their limited opportunities for resource partitioning, or to the influence of predators. Physical conditions such as temperature and light intensity, and chemical factors such as nutrient concentrations, are known to vary from hour to hour and from day to day. It seems reasonable to suppose that the high species richness of phytoplankton is partly due to repeated interruptions of the process of competitive exclusion. A similar phenomenon may occur amongst higher

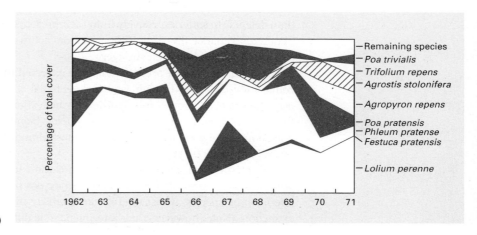

Figure 21.5 Changes in the species composition of a sown pasture in West Germany subject to varying intensities of flooding and silt deposition. (After Müller & Foerster, 1974.)

plants in communities whose conditions vary from year to year. For example, grasslands in the Rhine valley are subject to various intensities of flooding and silt deposition. This is reflected in changes in the relative importance of herbaceous species (Figure 21.5).

the role of competitive
exclusion must be compared
with the death rate ...

The important variable in this non-equilibrium explanation is the rate at which competitive exclusion occurs. Figure 21.6a shows the results of a Lotka–Volterra simulation in which competitive equilibrium is quickly reached and one species goes extinct. Figure 21.6b is the result of a simulation in which competitive exclusion is prevented by a periodic density independent reduction (by 50%) in both populations (which could be caused by either predators or physical disturbance). Now the outcome of competition is quite different. The species coexist for much longer, although eventually species 2 goes extinct because its low rate of increase (r) does not permit its population to recover sufficiently between disturbances. In this model, species 2 is the one to triumph under equilibrium conditions because of its higher carrying capacity (K) (Huston, 1979; see also Shorrocks & Begon, 1975; Caswell, 1978).

Huston (1979) has also modelled communities, consisting of six species, in which disturbances either never occur, occur with intermediate frequency or are very common (Figure 21.7). Competitive exclusions take place relatively rapidly in the absence of disturbances (one species has already gone extinct by the end of the run). At intermediate frequencies of disturbance, diversity is higher for much longer, because the rate of competitive exclusion is slowed dramatically. However, at a high frequency of disturbance, diversity is reduced by extinctions of species that are unable to recover sufficiently between disturbances (Figure 21.7c).

... and depends on the rate of
population growth

Not unexpectedly, the rate of competitive exclusion is significantly higher when rates of population increase are themselves high (Figure 21.8b). In other words, coexistence may in theory be extended indefinitely at appropriate rates of disturbance, but only when population growth rates are not too high. Does this prediction

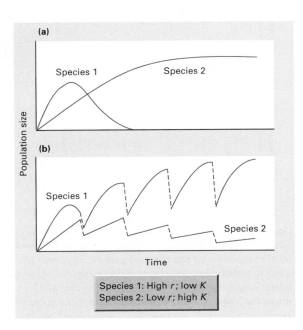

Figure 21.6 Effect of non-equilibrium conditions on the outcome of competition: (a) simulation in which competitive exclusion is reached; (b) competitive exclusion is prevented by periodic density independent reductions of the population. (After Huston, 1979.)

811   INFLUENCE OF PREDATION AND DISTURBANCE

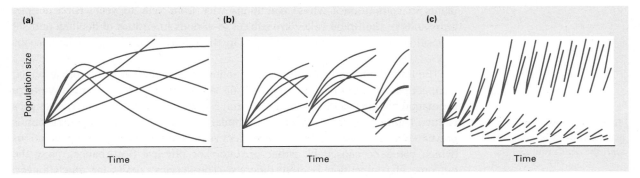

Figure 21.7 Effect of frequency of population reduction on the maintenance of diversity in a hypothetical community. (a) No population reductions; diversity is reduced as the system approaches competitive equilibrium. (b) Periodic reductions; high diversity is maintained for longer than in (a). (c) High frequency of reductions; diversity is reduced as populations with low *r* are unable to recover between reductions. (After Huston, 1979.)

gain any support from real communities? In fact, it offers an explanation for the reduction in species richness noted in certain productive environments, including fertilized grasslands and eutrophic lakes (see Chapter 24). On this view, highly productive environments contain populations which approach and reach the stage of competitive exclusion much more reliably than in unproductive environments. Thus, unproductive environments may be expected to be richer in species, as exemplified by the remarkable plant diversity associated with the very nutrient-poor soils of the Fynbos of South Africa and the heath scrublands of Australia (Chapter 24).

The relationships shown in Figures 21.8a and b are combined in Figure 21.8c.

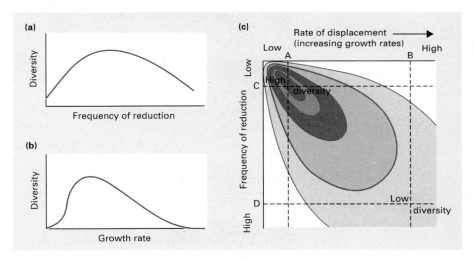

Figure 21.8 Relationships derived from Huston's model communities. (a) Predicted relationship between diversity and the frequency of population reduction. (b) Predicted relationship between diversity and population growth rates in non-equilibrium systems with a low to intermediate frequency of population reduction. (c) Generalized contour map of the relationship between the rate of competitive displacement (equivalent to population growth rates), the frequency of population reduction and species diversity. Diversity is shown on the axis perpendicular to the page and is represented by contour lines. (After Huston, 1979.)

The dashed line transects on Figure 21.8c demonstrate the predicted changes in diversity when one parameter is held constant and the other is varied.

21.4 Disturbances and the patch-dynamics concept

Disturbances that open up gaps are common in all kinds of community. In forests, they may be caused by high winds, lightning, earthquakes, elephants, lumberjacks or simply by the death of a tree through disease or old age. Agents of disturbance in grassland include frost, burrowing animals and the teeth, feet or dung of grazers. On rocky shores or coral reefs, gaps in algal or sessile animal communities may be formed as a result of severe wave action during hurricanes, tidal waves, battering by logs or moored boats, by the fins of careless scuba-divers or by the action of predators (see Section 21.2). The formation of gaps is of considerable significance to sessile or sedentary species that have a requirement for open space. It matters much less in the lives of mobile animal species for whom space is not the limiting factor.

With these facts in mind, an important class of model has been developed that views communities as consisting of a number of patches (all identical in conditions and available resources) that are colonized at random by individuals of a number of species. Species interactions within patches proceed according to Lotka–Volterra principles with local cases of extinction. However, the stochastic elements of the model, the patchy nature of the community and the opportunities in each generation for dispersal between patches, allows species-rich communities to persist without the need to invoke niche differentiation (Yodzis, 1986). Implicit in the 'patch-dynamics' view is a critical role for disturbance as a reset mechanism.

Patch-dynamics models of community organization differ in a fundamental respect from models described earlier. Both the equilibrium model concerned with resource partitioning and niche differentiation (see Chapter 20) and the non-equilibrium model that incorporates temporal variation in conditions (see Section 21.3) dealt with *closed* systems. If a closed community contains, for example, two species competing for the same resource, the outcome will be as predicted by the Lotka–Volterra equations. Any extinction occurs once and for all in this 'Lotka–Volterra in a bottle' view of community organization. Very often, however, real communities are better described as *open* systems, consisting of a mosaic of patches within which interactions proceed but with a vital extra feature—migration occurs from patch to patch or, more specifically, the number of recruits to a patch is decoupled from the fecundity of the population on the patch. This is the essence of the 'patch-dynamics' view (Pickett & White, 1985). A single patch without migration is, by definition, a closed system, and any extinction would be final. However, extinction within a patch in an open system is not necessarily the end of the story because of the possibility of reinvasion from other patches. Note the strong link between the patch-dynamics view of community organization and metapopulation theory, which deals with the effects on the dynamics of populations of dividing them into fragments (Hanski, 1994a, b; see also discussions in Chapters 15 and 25).

Exploiter-mediated coexistence in open systems can be readily demonstrated in simple models that explore the influence of a species of predator on competition between prey species (e.g. Caswell, 1978). In such open, non-equilibrium models, the rate of competitive exclusion can be slowed down to such an extent that indefinite coexistence of the species results. The structure of the models is

open and closed systems

exploiter-mediated coexistence
in patches

biologically quite realistic; for example, they correspond closely to Paine's starfish-dominated community (see Section 21.2). In the models, and in reality, the predator opens up otherwise closed cells for colonization by the inferior competitor. In real communities it is not only predators that can have this effect. As we have seen, gaps are generated in many kinds of community by physical disturbances. Indeed, when a physical disturbance acts as if it were a 'predator', the coexistence effect should be even more dramatic because there is no chance of the 'predator' becoming extinct (Caswell, 1978).

the fundamental role of dispersal and recruitment

Fundamental to the patch-dynamics perspective is recognition of the importance of migration between habitat patches. This may involve adult individuals, but very often the process of most significance is the dispersal of immature propagules (seeds, spores, larvae) and their recruitment to populations within habitat patches. The order of arrival and relative recruitment levels of individual species may determine or modify the nature and outcome of population interactions in the community (Booth & Brosnan, 1995).

patch-dynamics models of competition

Within the patch-dynamics framework, two fundamentally different kinds of community organization can be recognized according to the type of competitive relationships exhibited by the component species (Yodzis, 1986). Situations in which some species are strongly competitively superior can be described as *dominance controlled* (see Section 21.5), whereas those in which all species have similar competitive abilities are *founder controlled* (see Section 21.6). The dynamics of these two situations are quite different, and we deal with them in turn.

21.5 Dominance-controlled communities

In patch-dynamics models where some species are competitively superior to others, an initial colonizer of a patch cannot necessarily maintain its presence there. Dispersal between patches, or growth of an individual within a patch, will bring about a reshuffle and species may be competitively excluded locally (although they persist in the patchy environment, as a whole, in what Shorrocks & Rosewell, 1987, call 'probability refuges'). Such a community can be called dominance controlled (Yodzis, 1986). In these cases, disturbances that open up gaps lead to reasonably predictable species sequences, strongly reminiscent of successions in which a sequence is produced because different species have different strategies for exploiting resources—early species are good colonizers and fast growers, whereas later species can tolerate lower resource levels and grow to maturity in the presence of early species, eventually outcompeting them (see Chapter 17, Section 17.4.4). The effect of the disturbance is to knock the community back to an earlier stage of succession (Figure 21.9). The open space is colonized by one or more of a group of opportunistic, early successional species (p_1, p_2, etc., in Figure 21.9). As time passes, more species invade, often those with poorer powers of dispersal. These eventually reach maturity, dominating mid-succession (m_1, m_2, etc.) and many or all of the pioneer species are driven to extinction. Later still, the community regains the climax stage when the most efficient competitors (c_1, c_2, etc.) oust their neighbours. In this sequence, diversity starts at a low level, increases at the mid-successional stage and usually declines again at the climax. The gap essentially undergoes a mini-succession. Table 21.1 lists some species characteristic of pioneer, mid-successional and climax stages after disturbances in four communities in which the scale of the disturbance and the timing of recovery are dramatically different.

dominance-controlled communities and ecological succession

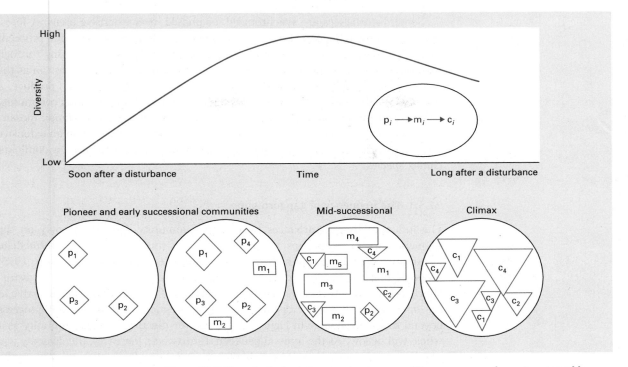

Figure 21.9 Hypothetical mini-succession in a gap. The occupancy of gaps is reasonably predictable. Diversity begins at a low level as a few pioneer (p_i) species arrive; reaches a maximum in mid-succession when a mixture of pioneer, mid-successional (m_i) and climax (c_i) species occur together; and drops again as competitive exclusion by climax species takes place.

Table 21.1 Some species characteristic of pioneer, mid-successional and climax stages after disturbances in four contrasting communities. The sequence may take hundreds of years after glacial retreat or abandonment of an old field (Crocker & Major, 1955; Tilman, 1988, respectively—see Chapter 17, Section 17.4.3), but only a few years in the case of boulders overturned on a stormswept beach (Sousa, 1979a—see Chapter 17, Section 17.4.3), or a few days in stream pools scoured out by a flash flood (Power & Stewart, 1987).

Disturbance	Pioneer species, p_1	Mid-successional species, m_1	Climax species, c_1
Glacial advance and retreat	Mosses, *Dryas*, prostrate and shrubby willows	Alder, cottonwood, sitka spruce	Western and mountain hemlock
Old field abandonment	Summer and winter annuals	Herbaceous perennials, red cedar	Sugar maple, American beech
Overturned boulder	*Ulva*	*Gelidium*, *Rhodoglossum*, *Gigartina leptorhynchos*	*Gigartina canaliculata*
Stream flood	Diatoms and blue–greens	*Rhizoclonium*	*Spirogyra*

Some disturbances are synchronized, or phased, over extensive areas. A forest fire may destroy a huge tract of a climax community. The whole area then proceeds through a more or less synchronous succession, with diversity increasing through the early colonization phase and falling again through competitive exclusion as the climax is approached. Other disturbances are much smaller and produce a patchwork of habitats. If these disturbances are unphased, the resulting community comprises a mosaic of patches at different stages of succession. A climax mosaic, produced by unphased disturbances, is much more rich in species than an extensive area undisturbed for a very long period and occupied by just one or a few dominant climax species.

21.5.1 The frequency of gap formation

The influence that disturbances have on a community depends strongly on the frequency with which gaps are opened up. In this context, the intermediate

disturbance hypothesis (Connell, 1978; see also the earlier account by Horn, 1975) proposes that the highest diversity is maintained at intermediate levels of disturbance. Soon after a severe disturbance, propagules of a few pioneer species arrive in the open space. If further disturbances occur frequently, gaps will not progress beyond the pioneer stage in Figure 21.10, and the diversity of the community as a whole will be low. As the interval between disturbances increases, the diversity will also increase because time is available for the invasion of more species. This is the situation at an intermediate frequency of disturbance. At very low frequencies of disturbance, most of the community for most of the time will reach and remain at the climax, with competitive exclusion having reduced diversity. This is shown diagrammatically in Figure 21.10, which plots the pattern of species richness to be expected as a result of unphased high, intermediate and low frequencies of gap formation, in separate patches and for the community as a whole.

The influence of the frequency of gap formation has been studied in southern

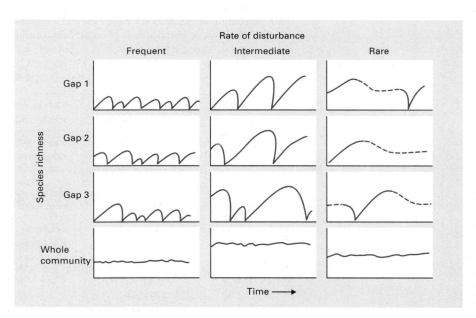

Figure 21.10 Diagrammatic representation of the time course of species richness in three gaps, and in the community as a whole, at three frequencies of disturbance. The disturbance is unphased. Dashed lines indicate the phase of competitive exclusion as the climax is approached.

California by Sousa (1979a, b), in an intertidal algal community associated with boulders of various sizes. Wave action disturbs small boulders more often than large. Using a sequence of photographs, Sousa estimated the probability that a given boulder would be moved during the course of 1 month. A class of mainly small boulders (which required a force less than 49 N to move them) had a monthly probability of movement of 42%. An intermediate class (which required a force of 50–294 N) had a much smaller monthly probability of movement, 9%. Finally, the class of mainly large boulders (greater than 294 N) moved with a probability of only 0.1% per month. The 'disturbability' of the boulders had to be assessed in terms of the force required to move them, rather than simply in terms of top surface area, because some rocks which appeared to be small were actually stable portions of larger, buried boulders, and a few large boulders with irregular shapes moved when a relatively small force was applied. The three classes of boulder (less than 49 N, 50–294 N and greater than 294 N) can be viewed as patches exposed to a decreasing frequency of disturbance when waves caused by winter storms overturn them.

The successional sequence that occurs in the absence of disturbance was established by studying both large boulders which had been experimentally cleared and implanted concrete blocks. The basic succession has already been described in Chapter 17, Section 17.4.3. To recap, within the first month the surfaces were colonized by a mat of the ephemeral green alga, *Ulva* spp. In the autumn and winter of the first year, several species of perennial red alga became established, including *Gelidium coulteri*, *Gigartina leptorhynchos*, *Rhodoglossum affine* and *Gigartina canaliculata*. The last-named species gradually came to dominate the community, holding 60–90% of the surface area after 2–3 years. What is essentially an algal monoculture persists through clonal growth, and resists invasion by all other species. Species richness increased during early stages of succession through a process of colonization, but declined again at the climax because of competitive exclusion by *G. canaliculata*. It is important to note that the same succession occurred on small boulders that had been artificially made stable. Thus, variations in the communities associated with the surfaces of boulders of different size were not simply an effect of size, but rather of differences in the frequency with which they were disturbed.

Populations on unmanipulated boulders in each of the three size/disturbability classes were assessed on four occasions. Table 21.2 shows that the percentage of bare space decreased from small to large boulders, indicating the effects of the greater frequency of disturbance of small boulders. Mean species richness was lowest on the regularly disturbed small boulders. These were dominated most commonly by *Ulva* spp. (and barnacles, *Chthamalus fissus*). The highest levels of species richness were consistently recorded on the intermediate boulder class. Most held mixtures of three to five abundant species from all successional stages. The largest boulders had a lower mean species richness than the intermediate class, although a monoculture was achieved on only a few boulders. *G. canaliculata* covered most of the rock surfaces.

These results offer strong support for the intermediate disturbance hypothesis as far as frequency of appearance of gaps is concerned. However, we must be careful not to lose sight of the fact that this is a highly stochastic process. By chance, some small boulders were not overturned during the period of study. These few were dominated by the climax species *G. canaliculata*. Conversely, two large boulders in the May census had been overturned, and these became dominated by the pioneer

boulders on a rocky shore varying in disturbability ...

... offering strong support for the hypothesis

Table 21.2 Seasonal patterns in bare space and species richness on boulders in each of three classes, categorized according to the force (in newtons) required to move them. (After Sousa, 1979b.)

Census date	Boulder class (N)	Percentage bare space	Species richness		
			Mean	Standard error	Range
November 1975	< 49	78.0	1.7	0.18	1–4
	50–294	26.5	3.7	0.28	2–7
	> 294	11.4	2.5	0.25	1–6
May 1976	< 49	66.5	1.9	0.19	1–5
	50–294	35.9	4.3	0.34	2–6
	> 294	4.7	3.5	0.26	1–6
October 1976	< 49	67.7	1.9	0.14	1–4
	50–294	32.2	3.4	0.40	2–7
	> 294	14.5	2.3	0.18	1–6
May 1977	< 49	49.9	1.4	0.16	1–4
	50–294	34.2	3.6	0.20	2–5
	> 294	6.1	3.2	0.21	1–5

Ulva. However, on average, species richness and species composition followed the predicted pattern.

This study deals with a single community conveniently composed of identifiable patches (boulders) which become gaps (when overturned by waves) at short, intermediate or long intervals. Recolonization occurs mainly from propagules derived from other patches in the community. Because of the pattern of disturbance, this mixed boulder community is more diverse than would be one with only large boulders.

further support from a study of small streams

Disturbances in small streams often take the form of bed movements during periods of high discharge. Because of differences in flow regimes and in the substrates of stream beds, some stream communities are disturbed more frequently and to a larger extent than others. This variation was assessed in 54 stream sites in the Taieri River in New Zealand (Townsend *et al.*, in press) by recording the frequency at which at least 40% (chosen arbitrarily) of the bed moved and the average percentage that moved (assessed on five occasions during 1 year, using painted particles of sizes characteristic of the stream in question). The pattern of richness of invertebrate species conformed with the intermediate disturbance hypothesis (Figure 21.11). It is likely that low richness at high frequencies and intensities of disturbance reflects the inability of many species to persist in such situations. Whether low richness at low frequencies and intensities of disturbance is due to competitive exclusion, as proposed in the intermediate disturbance hypothesis, remains to be tested.

21.5.2 The formation and filling of gaps

size of gaps

Gaps of different sizes may influence community structure in different ways because of contrasting mechanisms of recolonization. The centres of very large gaps are most likely to be colonized by species producing propagules which travel relatively great distances. Such mobility is less important in small gaps, since most recolonizing propagules will be produced by adjacent established individuals. The smallest gaps of

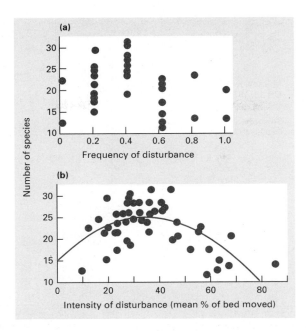

Figure 21.11 Relationship between invertebrate species richness and (a) frequency of disturbance—assessed as the number of occasions in 1 year when more than 40% of the bed moved (analysis of variance significant at $P < 0.0001$), and (b) intensity of disturbance—average percentage of the bed that moved (polynomial regression fitted, relationship significant at $P < 0.001$) assessed at 54 stream sites in the Taieri River. The patterns are essentially the same; intensity and frequency of disturbance are strongly correlated.

gap formation and filling-in in a mussel bed

all may be filled simply by lateral movements of individuals around the periphery.

The processes of formation and filling-in of gaps in intertidal beds of mussels, *Mytilus californianus*, have been monitored on waveswept rocky shores on the north-west coast of the USA (Paine & Levin, 1981). In the absence of disturbance, mussel beds may persist as extensive monocultures. More often, they are an ever-changing mosaic of many species which inhabit gaps formed by the action of waves. Gaps can appear virtually anywhere, and may exist for years as islands in a sea of mussels. The size of these gaps at the time of formation range from the dimensions of a single mussel to 38 m² (Figure 21.12a). In general, a mussel or group of mussels becomes infirm or damaged through disease, predation, old age or, most often, the effects of storm waves or battering by logs. Sometimes, mussels are dislodged immediately. In other cases, the shells persist during the summer to be dislodged as wave action becomes more severe in winter. Once disruption has started, erosion of the bed is swift and the gap is formed. Gaps generated in summer tend to be smaller and less abundant than those formed in winter.

Gaps begin to fill as soon as they are formed. The recovery process is initiated by the mussels on the perimeter leaning towards the centre of the gap. Although it lasts only briefly, this process initially reduces the size of the gap very rapidly (0.2 cm day⁻¹), and it can completely fill very small gaps within days. The lifetime of small gaps (less than 100 cm²) is too short for any appreciable invasion by other species.

Intermediate-sized gaps in a mussel bed (less than 3500 cm²) may persist for several years before they are recolonized by mussels. In the meantime, the

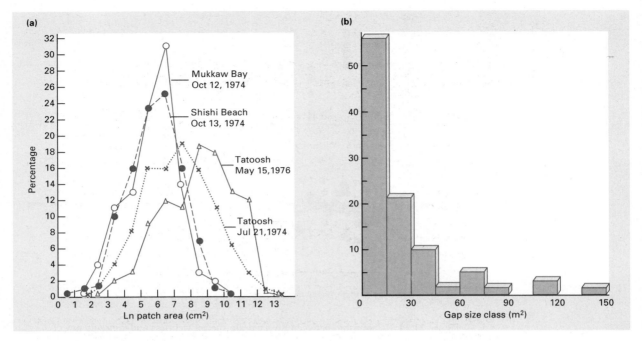

Figure 21.12 Frequency distributions of gap size (a) at the time of their formation in mussel beds at four locations on the north-west coast of North America. (After Paine & Levin, 1981.) (b) In a wind-exposed tropical cloud forest. (After Lawton & Putz, 1988.)

opportunity to invade the bare rock is taken by many other species. In the first season after formation, a major portion of the gaps was held by the alga *Porphyra pseudolanceolata*. By the second year, the herbivore-resistant alga *Coralina vancouveriensis*, had become established, together with barnacles (*Pollicipes polymerus*, *Balanus glandula* and *B. cariosus*) and mussels (*M. edulis* and *M. californianus*). Within 4 years, bare space had been reduced to only 1%, and *M. californianus* was again the dominant species.

In the largest gaps of all (greater than 3500 cm^2), leaning and lateral movement by peripheral mussels are of relatively little importance. Mussels recruit mainly as larvae from the plankton, although some dislodged adult mussels may be washed on to the patch and become entrapped. The filling process is mainly due to growth of individual mussels and takes many years. Associated with the opening of space, there is an initial burst of diversity, which persists for about 4 years, followed by a winnowing of species as they are eliminated through competition, eventually yielding a bed of mussels.

The species that take advantage of the open space are essentially the same for both intermediate-sized and large gaps. However, large gaps provide islands of diversity which persist for longer, and this is their special contribution. Overall, it is the young stages of large gaps that contribute most to the richness of the community as a whole.

The pattern of colonization of gaps in mussel beds is repeated in almost every detail in the colonization of gaps in grassland caused by burrowing animals or patches killed by urine. Initially, leaves lean into the gap from plants outside it. Then colonization begins by clonal spread from the edges, and a very small gap may close

a similar picture from grasslands ...

up quickly. In larger gaps, new colonists may enter as dispersed seed, or germinate from the seed bank in the soil. Over 2–3 years the vegetation begins to acquire the character that it had before the gap was formed.

... and forests

The gaps produced in forests vary greatly in size, but usually there is a preponderance of small gaps produced by the fall of one or a few trees (Figure 21.12b). Regrowth occurs from three sources: seeds, plants established prior to gap formation and lateral growth of branches from trees on the gap periphery. An early response is the more rapid growth of the surrounding trees as they start to extend their branches into the sunlit space that has been created. Nearer ground level, colonization often takes place by the fast growth of saplings that were already present but suppressed. These are released and grow up rapidly into the gap; for example, the growth rate of saplings of sugar maple (*Acer saccharum*) and American beech (*Fagus grandifolia*) can be an order of magnitude greater in gaps than beneath an intact canopy (Canham, 1988). There is something of a race to fill the gap, and if the gap is small the race will usually be won by lateral growth of the trees at its edge. In larger gaps, buried seeds may make a big contribution to regeneration, and juveniles of many neotropical canopy trees are significantly more abundant in gaps and at their edges (Figure 21.13). Organisms other than plants can also be overrepresented in gaps. In a study of tropical rainforest in Costa Rica, Levey (1988) found that nectarivorous and frugivorous birds were much more abundant in treefall gaps, reflecting the fact that understorey plants in gaps tend to produce more fruit over a longer period than conspecifics fruiting under a closed canopy.

21.6 Founder-controlled communities

In the dominance-controlled communities discussed in Section 21.5 there was the familiar *r*- and *K*-selection dichotomy in which colonizing ability and competitive status are inversely related. In founder-controlled communities, on the other hand, all species are both good colonists and essentially equal competitors; thus, within a patch opened by disturbance, succession is not expected. If a large number of species are approximately equivalent in their ability to invade gaps, are equally tolerant of the abiotic conditions and can hold the gaps against all-comers during their lifetime, then the probability of competitive exclusion may be much reduced in an environment where gaps are appearing continually and randomly. This can be referred to as a competitive lottery. A further condition for coexistence is that the number of young which invade and occupy gaps should not be consistently greater for parent populations that produce more offspring, otherwise the most productive species would come to monopolize space even in a continuously disturbed environment.

in founder-controlled communities all species are more or less equivalent

If these conditions are met, it is possible to envisage how the occupancy of a series of gaps will change through time (Figure 21.14). On each occasion that an organism dies (or is killed) the gap is reopened for invasion. All conceivable replacements are possible and species richness will be maintained at a high level. Some tropical reef communities of fish may conform to this model (Sale, 1977, 1979; Sale & Douglas, 1984). They are extremely diverse. For example, the number of species of fish on the Great Barrier Reef ranges from 900 in the south to 1500 in the north, and more than 50 resident species may be recorded on a single patch of reef 3 m in diameter. Only a proportion of this diversity is likely to be attributable to resource partitioning of food and space—indeed, the diets of many of the coexisting species are very similar.

fish coexisting on coral reefs

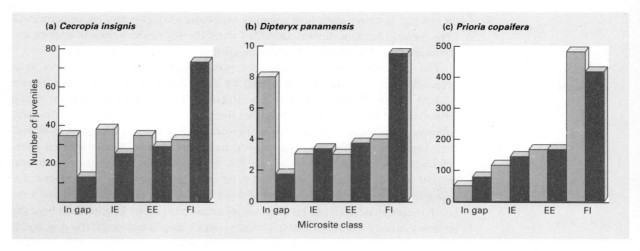

Figure 21.13 Distribution of juveniles of three canopy trees in a neotropical forest in four microsite classes: in gaps (individuals per 5 × 5 m plot), at the interior edge of gaps (IE), exterior edge of gaps (EE) and in the forest interior (FI). Light bars give the distributions in each microsite and dark bars give the expected distribution based on the proportion of total area made up by each class. (a) *Cecropia insignis*, a species that is overrepresented in gaps and on the interior edges. (b) *Dipteryx panamensis*, a species overrepresented only in the middle of gaps. (c) *Prioria copaifera*, a species overrepresented in the forest interior. (After Hubbell & Foster, 1986.)

In this community, vacant living space seems to be a crucial limiting factor, and it is generated unpredictably in space and time when a resident dies or is killed. The life styles of the species match this state of affairs. They breed often, sometimes year-round, and produce numerous clutches of dispersive eggs or larvae. It can be argued that the species compete in a lottery for living space in which larvae are the tickets, and the first arrival at the vacant space wins the site, matures quickly and holds the space for its lifetime.

Three species of herbivorous pomacentrid fish co-occur on the upper slope of Heron Reef, part of the Great Barrier Reef off eastern Australia. Within rubble patches, the available space is occupied by a series of contiguous and usually non-overlapping territories, each up to 2 m² in area, held by individuals of

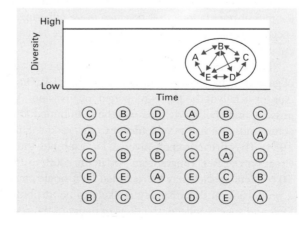

Figure 21.14 Hypothetical competitive lottery. Occupancy of gaps which periodically become available. Each of species A–E is equally likely to fill a gap, regardless of the identity of its previous occupant (this is illustrated in the inset). Species richness remains high and relatively constant. (Compare with Figure 21.12.)

Eupomacentrus apicalis, Plectroglyphidodon lacrymatus and *Pomacentrus wardi*. Individuals hold territories throughout their juvenile and adult life and defend them against a broad range of chiefly herbivorous species, including conspecifics. There seems to be no particular tendency for space initially held by one species to be taken up, following mortality, by the same species. Nor is any successional sequence of ownership evident (Table 21.3). *Pomacentrus wardi* both recruited and lost individuals at a higher rate than the other two species, but all three species appear to have recruited at a sufficient level to balance their rates of loss and maintain a resident population of breeding individuals. Thus, the maintenance of high diversity depends, at least in part, on the unpredictability of the supply of living space; and as long as all species win some of the time and in some places, they will continue to put larvae into the plankton, and hence, into the lottery for new sites.

An analogous situation has been postulated for the highly diverse chalk grasslands of Great Britain (Grubb, 1977). Any small gap that appears is rapidly exploited, very often by a seedling. In this case, the tickets in the lottery are seeds, either in the act of dispersal or as components of a persistent seed bank in the soil. Which seeds develop to established plants, and therefore which species comes to occupy the gap, depends on a strong random element since the seeds of many species overlap in their requirements for germination. The successful seedling rapidly establishes itself amidst the short turf and retains the patch for its lifetime, in a similar way to the reef fish described above. However, the analogy with tropical reefs should not be taken too far. Three extra factors are important in the grassland.

1 Gaps are not necessarily identical. They may vary in the nature of the soil surface, the amount of litter that is present and whether the gap is also occupied by other organisms, including bacteria, fungi, viruses and animals. The orientation of the gap and even its size (Figure 21.15) can influence physical conditions such as temperature. Subtle differences in the germination requirements of particular species mean that not all those seeds that are present at the time the gap appears will stand an equal chance of establishing themselves.

2 Gaps that appear only a short-time apart may be subjected to quite different microclimatic regimes in terms of temperature, light intensity and water potential of the atmosphere. Daily, even hourly, fluctuations are likely to result in a constant shifting of the identity of species for which the prevailing conditions are most nearly optimal for breaking dormancy, for photosynthesis and for growth. Identical gaps with identical arrays of seeds may nevertheless be taken over by different species as a result of very small fluctuations in weather (e.g. Figure 21.16).

Table 21.3 Numbers of individuals of each species observed occupying sites, or parts of sites, that had been vacated during the immediately prior interperiod between censuses through the loss of residents of each species. Sites vacated through loss of 120 residents have been reoccupied by 131 fish. Species of new occupant is not dependent on species of previous resident.

	Reoccupied by:		
Resident lost	E. apicalis	P. lacrymatus	P. wardi
E. apicalis	9	3	19
P. lacrymatus	12	5	9
P. wardi	27	18	29

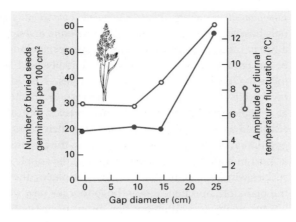

Figure 21.15 Comparison of soil temperature fluctuations (○) and germination (●) of the grass *Holcus lanatus*, in gaps of different diameters in the canopy, in a sown pasture of low productivity in northern England. (After Grime, 1979.)

3   It is not surprising that in seasonal environments the species which will colonize a gap varies with the season. For example, the colonists of bare ground that is created during the summer include small winter annuals and many grasses, such as *Festuca ovina*, *Helictotrichon pratense* and *Koeleria cristata*, which produce many seedlings in the early autumn; whilst colonization of gaps created during winter is often delayed until the appearance the following spring of quite different species such as *Linum catharticum*, *Pimpinella saxifraga* and *Viola riviniana* (Grime, 1979).

   The formation of gaps is an inherent feature of all communities where space is monopolized by individual organisms—each is certain to die at some time. If interactions between species take the form of a race for unchallenged dominance in recently created openings, rather than direct competitive interactions between

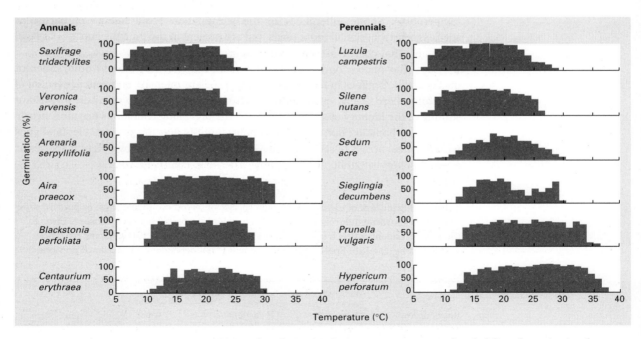

Figure 21.16 The relationship between temperature and probability of germination for several annual and perennial herbaceous plants of common occurrence in calcareous grassland in northern England. (After Grime, 1979.)

established adults, then species richness will be higher because the likelihood of competitive exclusion is reduced. An additional factor is that the identity of neighbouring individuals is unpredictable in space and time. This will also act to reduce the likelihood of competitive exclusion of one species by another.

21.7 Appraisal of non-equilibrium models

No community is truly the homogeneous, temporally invariant system described by simple Lotka–Volterra mathematics and exemplified by laboratory microcosms, although some are less variable than others. In most real communities, population dynamics will be spatially distributed (patch-dynamics models) and temporal variation will be present, providing a variety of ways in which the probability of coexistence is enhanced and diversity increased. In this chapter, we have seen many examples of communities that conform reasonably closely with the assumptions incorporated into non-equilibrium models, which therefore provide a clear insight into the reasons why species-rich communities persist in nature. Coexistence under a stochastic, non-equilibrium mechanism can be just as strong and stable as that occurring under a deterministic, niche-differentiation model (Chesson, 1986).

21.7.1 The importance of scale

In a closed system, composed of a single patch, species extinctions can occur for two very different reasons: (i) as a result of biotic instability caused by competitive exclusion, overexploitation and other strongly destabilizing species interactions; or (ii) as a result of environmental instability caused by unpredictable disturbances and changes in conditions. By integrating unstable patches of either of these types into the open system of a larger landscape (consisting of many patches out of phase with each other), persistent species-rich communities can result (DeAngelis & Water-house, 1987). This is the principal message to emerge from the patch-dynamics perspective (and its larger scale counterpart, 'landscape ecology'—Wiens et al., 1993), stressing the importance of the spatial scale at which we view communities and the open nature of most of them.

21.7.2 Pluralism in community ecology

It would be wrong to replace one monolithic view of community organization (the overriding importance of competition and niche differentiation) with another (the overriding importance of temporal variation and disturbance). For one thing, taxonomic shortcomings may lead us to the wrong conclusion. For example, recent taxonomic research on corals, polychaetes and other coral reef invertebrates has repeatedly shown that what were previously considered to be single generalist species are, in fact, several sibling species with differentiated niches (Knowlton & Jackson, 1994). Thus, enthusiasm amongst marine ecologists for models dominated by chance and disturbance may sometimes be mistaken.

Certainly, communities structured by competition are not a general rule, but neither necessarily are communities structured by any single agency. Even in the same community, some patches or taxonomic groups may be structured by dominance control and others by founder control (Townsend, 1989). Most communities are probably organized by a mixture of forces—competition, predation,

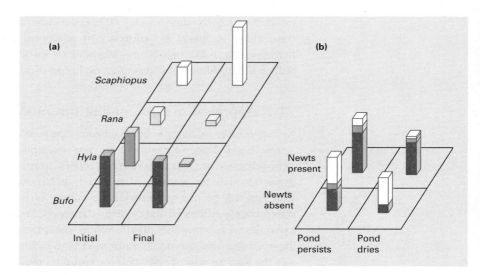

Figure 21.17 (a) Relative abundance of tadpoles of each of four species introduced at high density into ponds (initial), and the relative abundances of metamorphs (final) at the end of the experiment. (b) Number of metamorphs of four species in the absence and presence of predatory salamanders, and in ponds which persist or which dry up 100 days into the experiment. (After Wilbur, 1987; Townsend, 1991a.)

disturbance and recruitment—although their relative importance may vary systematically, with competition and predation figuring more prominently in communities where recruitment levels are high (Menge & Sutherland, 1987) and in less disturbed environments (Menge & Sutherland, 1976; Buss, 1986; Townsend, 1991a).

In an elegant series of experiments, Wilbur (1987) investigated the interactions between competition, predation and disturbance as they influenced four species of frog and toad that occur in North American ponds. In the absence of predators, tadpoles of *Scaphiopus holbrooki* were competitively dominant whilst, at the opposite extreme, *Hyla chrysoscelis* had a very low competitive status (Figure 21.17a). The presence of predatory salamanders, *Notophthalmus viridescens*, did not alter the total number of tadpoles reaching metamorphosis, but relative abundances were shifted because *Scaphiopus*, the competitive dominant, was selectively eaten (Figure 21.17b). Finally, Wilbur subjected his tadpole communities, in the presence and absence of predators, to water loss, to simulate a natural drying regime (disturbance). The influence of competition was to slow growth and retard the timing of metamorphosis, thus increasing the risk of desiccation in drying ponds. *Scaphiopus* had the shortest larval period and made up a greater proportion of metamorphs in the drying experiment. The presence of predators ameliorated the impact of competition, allowing surviving tadpoles of several species to grow rapidly enough to metamorphosize before the ponds dried up.

21.7.3 Relevance of non-equilibrium theory to ecological management

Good theory not only helps to explain, it also predicts and can thus be used to control events. Disturbance (non-equilibrium) theory suggests ways in which communities might be manipulated to desired ends—such as nature conservation, agriculture, forestry and wildlife management. In particular, it suggests that if we are keen to preserve natural diversity, we should not prevent disturbances. Indeed, disturbance may be the most powerful way in which we can generate diversity. Just as the recurrent disturbances of the ice-ages, and the break-up of continents and formation of islands, appear to have been powerful forces in the origin of species diversity, so the creation of gaps, new successions and patchwork mosaics within communities may be the most powerful way in which we might generate and maintain ecological diversity.

The disturbances that are most effective in allowing a diversity of species to invade an area are those that hurt community dominants. This suggests that patchwork deforestation should increase species diversity—although there will be an optimal size for the patches.

Agriculture involves repeated disturbance. It is not surprising that it brings with it such species diversity that it keeps chemical manufacturers in lucrative business, developing the variety of herbicides needed to control the diversity of weeds. Practices that reduce the frequency of disturbance, such as cropping without cultivation, result in much reduced problems in the long term. Disturbances are less prominent in forestry practice, and forestry usually involves establishing a potential community dominant. Commercial forestry short-circuits the normal progress of succession, taking the community directly to the dominant. The communities that are supported beneath such forests tend to be poor in species, especially when the monotony of a single-species canopy is exaggerated by having a uniform age structure. A mixed-age population generates some of its own diversity through the formation of gaps and through regeneration cycles.

We can begin to see dimly that we might find ways of measuring the force of different types of disturbance—is a forest fire a stronger perturbation than the introduction of deer? We might then begin to compare the response of different types of community, and measure their resistance to change or compare their rates of recovery. In the meantime, the theoretical models suggest persuasively that we may be on the edge of an understanding of the role of the accidents, uncertainties and hazards that dominate the life of most communities.

Chapter 22
Food Webs

22.1 Introduction

In the previous two chapters we began to consider the ways in which population interactions can shape the structure of communities. Our focus was on interactions between species that occupy the same trophic level (interspecific competition—see Chapter 20), or between members of adjacent trophic levels (carnivore–herbivore, herbivore–plant and parasite–host interactions—see Chapter 21). It has already become clear that the structure of communities cannot be understood solely in terms of direct interactions between species. In the case of exploitation competition for living resources, the interaction between competitors necessarily involves one or more further species—those whose individuals are being consumed. Moreover, a recurrent effect of predators has been shown to be consumer-mediated coexistence, where predation altered the relative competitive status of species on the next trophic level down, leading to coexistence of species that would otherwise be competitively excluded.

The influence of a species can be expected to ramify even further. Thus, the effects of a carnivore on its herbivorous prey may also be felt by any plant population upon which the herbivore feeds, by other predators and parasites of the herbivore, by other consumers of the plant, by competitors of the herbivore and the plant and by the myriad of species linked even more remotely in the food web.

This chapter is about food webs. In essence, we are shifting the focus to systems usually with at least three trophic levels, and emphasis will be placed on both direct and, especially, indirect effects that a species may have on others on the same trophic level or on levels below or above it. We will pay special attention to the properties and effects of 'keystone' species—those with particularly profound and far-reaching consequences in the web of interactions (see Section 22.2).

An important area of research in recent years has concerned the question of whether the control of food webs is 'top-down' or 'bottom-up'. Top-down control refers to situations where the structure (abundance, biomass, diversity) of lower trophic levels depends on the effects of consumers from higher trophic levels. Bottom-up control refers to dependence of community structure on factors, such as nutrient concentration and prey availability, that influence a trophic level from below. In Section 22.3, we enquire whether there are patterns in top-down and bottom-up control; does one or the other predominate in particular places, at particular times or in certain types of community?

Finally, we will consider the important topic of community stability (see Section 22.4). Ecologists have shown an interest in community stability for two reasons. The

key questions about food webs

the importance of indirect effects?

the role of keystone species?

top-down or bottom-up control of food webs?

food webs and community stability?

first is practical, and indeed pressing. Since people are perturbing natural and agricultural communities at an ever-increasing rate, it is essential to know how communities react to such perturbations and how they are likely to respond in the future. The stability of a community measures its sensitivity to disturbance. The second reason for being interested in stability is less practical but more fundamental. Stable communities are, by definition, those that persist. The communities we actually see today are thus likely to possess properties conferring stability. The most fundamental question in community ecology is: why are communities the way they are? At least part of the answer is therefore likely to be: because they possess certain stabilizing properties.

22.2 Indirect effects in food webs

22.2.1 Unexpected effects

The experimental removal of a species can be a powerful tool in unravelling the workings of a food web. If a predator species is removed from a community we expect an increase in the density of its prey. If a competitor species is removed, an increase is expected in the success of those with which it competes. Not surprisingly, there are plenty of examples of such expected results (Connell, 1983; Schoener, 1983; Sih *et al.*, 1985).

indirect, 'unexpected' effects ... Sometimes, however, when a species is removed, a competitor may actually decrease in abundance, whilst removal of a predator can lead to a decrease in a prey species population. Such unexpected effects arise when direct effects (on prey or competitors) are less important than effects that occur through indirect pathways. Thus, removal of a species might increase the density of one competitor, which in turn causes another competitor to decline. Or removal of a predator might increase the abundance of a prey species that is competitively superior to another, leading to a decrease in the density of the latter.

In a survey of more than 100 experimental studies of predation, more than 90% demonstrated statistically significant results, and of these about one in three showed unexpected effects (Sih *et al.*, 1985). It seems that unexpected effects, and therefore ... are common in nature the indirect interactions that are responsible for them, are common in nature.

22.2.2 Strong interactors and keystone species

Some species are more intimately and tightly woven into the fabric of the food web than others. A species whose removal would produce a significant effect (extinction or a large change in density) in at least one other species may be thought of as a strong interactor. Some strong interactors would lead, through their removal, to significant changes spreading throughout the food web—we refer to these as keystone species.

keystones in food-web architecture A keystone is the wedge-shaped block at the highest point of an arch, that locks the other pieces together. Its early use in food-web architecture referred to a top predator (such as the starfish *Pisaster* on a rocky shore; see p. 806) that has an indirect beneficial effect on a suite of inferior competitors by depressing the abundance of a superior competitor (Paine, 1966). Removal of the keystone predator, just like removal of the keystone in an arch, leads to collapse of the structure. More precisely, it leads to extinction or large changes in abundance of several species, producing a community with a very different species composition and, to our eyes, an obviously different physical appearance.

plants as keystone species?

Should we restrict the use of 'keystone' to predators, as Paine originally intended? If so, what are we to call a plant species that is intertwined, like a top predator, in a web of direct and indirect interactions such that its effects spread out to other plant species, up to herbivores and beyond them to carnivores, and down to detritivores and decomposers? Removal of a particular plant species (e.g. all the oak trees in an oak forest) would certainly lead to dramatic changes in densities, and extinctions amongst a variety of species; it seems equally deserving of the epithet 'keystone'. (We are moving onto architecturally shaky ground here. A more appropriate phrase would now be 'foundationstone species', since like a foundationstone the plant occurs at the base of the food-web structure. However, we will keep to the term keystone.)

keystone herbivores ...

It is now widely accepted that keystone species can occur at any trophic level (Hunter & Price, 1992). For example, lesser snow geese (*Chen caerulescens caerulescens*) are herbivores that breed in large colonies in coastal brackish and freshwater marshes along the west coast of Hudson Bay in Canada. At their nesting sites in spring, before the onset of above-ground growth of vegetation, adult geese grub for roots and rhizomes of graminoid plants in dry areas and eat the swollen bases of shoots of sedges in wet areas. Their activity creates bare areas (1–5 m^2) of peat and sediment. Since there are few pioneer plant species able to recolonize these patches, recovery is very slow. Furthermore, in areas of intense summer grazing of ungrubbed brackish marshes, grazing 'lawns' of *Carex* and *Puccinellia* have become established. Here, the presence of high densities of grazing geese is essential to maintain above-ground production and species composition of the vegetation (Kerbes *et al.*, 1990). Certainly, it seems reasonable to consider the lesser snow goose as a keystone species.

... and detritivores

In Chapter 11, we discussed the significance to community functioning of decomposers (bacteria and fungi) and detritivorous animals. Can species in these groups play a keystone role? No doubt, if all the bacteria or fungi were removed from a community (were that possible to achieve), its workings would grind to a halt. However, it is probably unhelpful to think in terms of a keystone decomposer because microbial ecologists are generally not in a position to identify and study the ecology of single species, but it is not hard to envisage a detritivorous animal species exerting indirect and widespread effects. This is illustrated by an experimental study in a Venezuelan stream of the effects on the insect community of fish that feed in contrasting ways (Flecker, 1993). Paradoxically, a detritivorous species, *Prochilodus mariae*, proved to be exerting a stronger indirect effect on insects than the direct effects of insectivorous fish. By grazing on detritus on the surface of stones, *Prochilodus* dramatically reduced resources used by stream insects (both detritus and attached algae), leading to much lower insect densities. On the other hand, the effects of insectivorous fish were compensated by rapid insect recolonization of stone surfaces possessing abundant resources.

22.2.3 Multiple trophic-level interactions

The multitude of direct and indirect effects that one species has on others needs to be viewed in the context of the food web with its several trophic levels. With this in mind, Wootton (1992) experimentally manipulated bird predation pressure over a 2-year period in an intertidal community on the north-west coast of the USA to determine the consequences for three limpet species and their algal food. Thus,

experimental unravelling of an intertidal food web ...

glaucous-winged gulls (*Larus glaucescens*) and oystercatchers (*Haematopus bachmani*) were excluded by means of wire cages from large areas (each $10\,m^2$) in which limpets were common.

It became evident that birds reduced the overall abundance of one of the limpet species, *Lottia digitalis*, as might have been expected, but a second limpet species increased in abundance when birds had access (*L. strigatella*) and the third, *L. pelta*, which was the one most frequently consumed, did not vary in abundance. The reasons are complex and go well beyond the direct effects of consumption of limpets.

Lottia digitalis, a light-coloured limpet, tends to occur on light-coloured goose barnacles (*Pollicipes polymerus*), whilst *L. pelta*, a dark-coloured species, occurs primarily on dark Californian mussels (*Mytilus californianus*). Both limpets show strong habitat selection for these cryptic locations. Predation by gulls reduced the area covered by goose barnacles (to the detriment of *L. digitalis*), leading through competitive release to an increase in the area covered by mussels (benefitting *L. pelta*). The third limpet species, *L. strigatella*, is competitively inferior to the others and increased in density because of competitive release. The effects of bird predation cascaded down to the plant trophic level, because by consuming limpets (limpet biomass was much lower in the presence of birds, which select the largest individuals) grazing pressure on fleshy algae, such as *Halosaccion glandiforme* and *Porphyra* spp., was reduced. In addition, space was freed up for algal colonization through the removal of barnacles (Figure 22.1). Short-term studies of pair-wise interactions would have failed to reveal the rich array of direct and indirect interactions in this food web spanning the terrestrial and marine environments.

... and a food web of microorganisms in marine sediment

In another marine environment, Walters and Moriarty (1993) focused on a community of microscopic organisms where the detection of multiple trophic level effects is fraught with difficulty because of the tiny spatial scale and rapid patterns of change. By manipulating animals in the top 1 cm of sediment associated with an intertidal seagrass bed, they were able to uncover something of the nature of the interactions between meiofauna (size range 63–500 µm; mainly nematode worms and copepods), protists (unicellular ciliates and flagellates) and bacteria. The meiofauna are generalist feeders on protists, bacteria and algae. The protists, in turn, consume a variety of microbes. Gentle raking was used to reduce meiofaunal density before pushing experimental chambers into the sediment. In other chambers, the meiofauna was supplemented so that densities in replicates with supplemented meiofauna were at least three times greater than those with reduced densities. A multiple trophic level interaction was detected in early summer, when bacterial densities increased in treatments containing enhanced numbers of meiofauna, most likely because of a significant reduction in grazing by protists, whose numbers had been reduced through predation by copepods. The lack of a clear pattern at other times of the year may be due in part to the ability of bacteria to divide rapidly, compensating for consumption losses through increased population growth.

22.3 Top-down or bottom-up control of food webs?

Our examples of multiple trophic level interactions have emphasized the potential for consumption by individuals at one trophic level to influence the numbers and biomass of individuals at the next level down. Such top-down control is a basic feature of our understanding of predator–prey interactions. On the other hand, when we view communities from the point of view of fluxes of energy and nutrients,

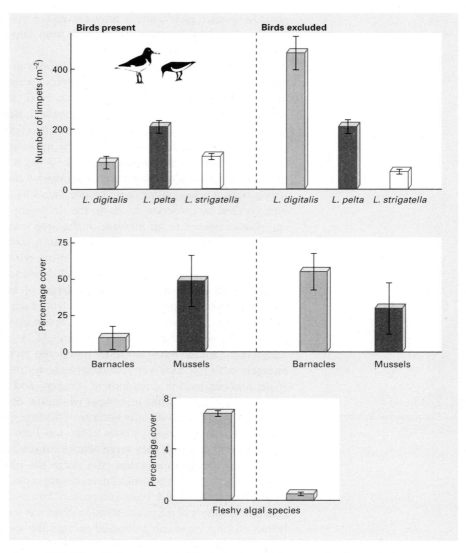

Figure 22.1 When birds are excluded from the intertidal community, barnacles increase in abundance at the expense of mussels, and three limpet species show marked changes in density, reflecting changes in the availability of cryptic habitat and competitive interactions as well as the easing of direct predation. Algal cover is much reduced in the absence of effects of birds on intertidal animals (means ± standard errors are shown). (After Wootton, 1992.)

fundamental properties of a given trophic level are expected to be influenced by the trophic level below not the one above. Bottom-up control occurs when resource availability is the key process; in this case, populations within the trophic level are affected predominantly by competition not predation.

22.3.1 The influence of number of trophic levels

It is helpful to dissect the question of whether predators or resources will dominate

number of trophic levels and
the top-down, bottom-up
debate

the dynamics of populations by considering first a hypothetical community with just
one (plant) trophic level and then successively adding extra trophic levels, one at a
time.

One trophic level

Unless physical conditions are particularly harsh, we expect the rate of production of
biomass and the standing stock biomass of plants to be higher where nutrients and
light are available in greater amounts, at least up to some threshold level when
available space or the ability to incorporate the resources becomes limiting. In a
one-trophic level community, by definition, predation will be absent, control will be
bottom-up and competition will be the predominant population interaction (Figure
22.2).

Two trophic levels

Whilst it is not easy in practice to envisage a community consisting of only one
trophic level, two-trophic level systems can be invoked if we are permitted to focus
on a limited but significant component of a real system. Grazing by the giant tortoise,
on the most remote island on earth, Aldabra, crops the turf down to less than 5 mm
over extensive areas (Strong, 1992). Experimental fences that keep out tortoises
allow many species of trees, shrubs and grasses, normally excluded by grazing, to
grow and dominate the plant community. No predators of the giant tortoises, or their
eggs or young, have arrived on the island. In this two-trophic level system, we see
clear evidence of top-down control of the bottom trophic level, where the effects of
predation predominate. Were there several species in the herbivore trophic level,
competition and not predation would be the process dominating their dynamics
(Figure 22.2). The destructive foraging by lesser snow geese in coastal areas of
Hudson Bay in Canada provides a parallel example (see Section 22.2.2).

Three trophic levels

The saline Great Salt Lake of Utah in the USA, provides an intriguing case when
what is essentially a two-trophic level (zooplankton–phytoplankton) system is

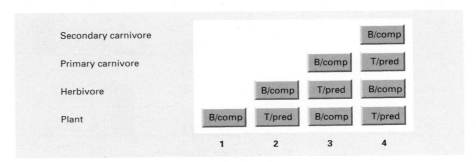

Figure 22.2 Diagrammatic representation of communities with one, two, three or four
trophic levels, illustrating for each trophic level whether control is predicted to be bottom-up
(B) or top-down (T), and whether population dynamics are determined primarily by
competition (comp) or predation (pred).

833   FOOD WEBS

augmented by a third trophic level (a predacious insect, *Trichocorixa verticalis*) in unusually wet years when salinity is lowered (Wurtsbaugh, 1992). Normally, the zooplankton, dominated by brine shrimp (*Artemia franciscana*), is capable of keeping phytoplankton biomass at a low level, producing high water clarity. But, when salinity declined from above $100\,g\,l^{-1}$ to $50\,g\,l^{-1}$ in 1985, *Trichochorixa* invaded and *Artemia* biomass was reduced from 720 to $2\,mg\,m^{-3}$, leading to a 20-fold increase in chlorophyll *a* concentration and a fourfold decrease in water clarity (Figure 22.3). In such three-trophic level communities, the plants are subject to bottom-up control, having been released from heavy grazing pressure by the effects of predators on the grazers. The grazers are thus subject to top-down and the predators to bottom-up control (see Figure 22.2).

Four trophic levels

It can be seen from Figure 22.2, that as each trophic level is added, the controlling process for the plant trophic level shifts alternately from bottom-up (odd number of trophic levels) to top-down (even number of trophic levels). Moreover, when we move vertically up from the bottom trophic level to those above, there is again an alternation between top-down and bottom-up control. Thus, we expect for a four-trophic level system that the plants and the primary carnivores will be limited

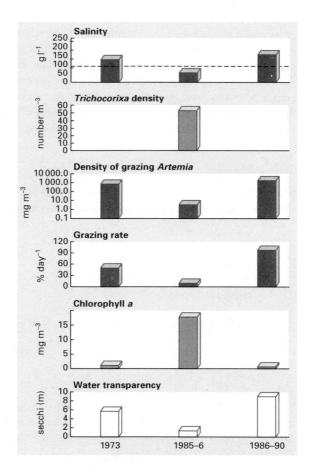

Figure 22.3 Variation in the pelagic ecosystem of the Great Salt Lake during three periods that differed in salinity. (After Wurtsbaugh, 1992.)

top-down, whilst the herbivores and secondary carnivores will be limited bottom-up. This is precisely what Mary Power (1990) found in her experimental study of the food web in Eel River, northern California. Large fish (roach, *Hesperoleucas symmetricus*, and steelhead trout, *Oncorhynchus mykiss*) reduced the abundance of fish fry and invertebrate predators, allowing their prey, tuft-weaving midge larvae (*Pseudochironomus richardsoni*) to attain high density and to exert intense grazing pressure on filamentous algae (*Cladophora*), whose biomass was thus kept low.

four trophic levels sometimes function as three

On the other hand, in a study of a four-trophic level stream community in New Zealand (brown trout (*Salmo trutta*), predatory invertebrates, grazing invertebrates and algae), the presence of the top predator did not lead to reduced algal biomass because the fish influenced not only the predatory invertebrates but also directly affected the activity of the herbivorous species (Flecker & Townsend, 1994). They did this both by consuming grazers and by constraining the foraging behaviour of the survivors; in the presence of trout, grazing nymphs of mayflies such as *Nesameletus ornatus* hide under stones during the day (McIntosh & Townsend, 1994). A similar situation has been reported for a four-trophic level terrestrial community in the Bahamas, consisting of lizards (mainly *Anolis sagrei*), web spiders (mainly *Metepeira datona*), herbivorous arthropods and sea grape shrubs (*Coccoloba uvifera*) (Spiller & Schoener, 1994). The results of experimental manipulations indicated a strong interaction between top predators (lizards) and herbivores, but a weak effect of intermediate predators (spiders). Consequently, the net effect of top predators on plants was positive and there was less leaf damage in the presence of lizards. In essence, these four-trophic level communities function as if they had only three levels (compare Figures 22.4b, c with Figure 22.4a).

22.3.2 How green is our world?

the world is green

The top-down view was first introduced in a famous paper by Hairston *et al.* (1960), with its frequently cited 'the world is green' proposition. They argued that green plant biomass accumulates in the world because predators keep herbivores in check. This is essentially the three-trophic level case in Section 22.3.1. The argument was later extended to systems with fewer or more than three trophic levels (Fretwell, 1977; see Figure 22.2).

the world is prickly and tastes bad

Murdoch (1966) challenged the idea that mature, terrestrial communities are green because of herbivore control by predators. Murdoch's view (described by Pimm, 1991, as 'the world is prickly and tastes bad') emphasized that many plants have evolved physical and chemical defences that make life difficult for herbivores (see Chapter 3); thus, the world may be green because plants are protected and not because herbivores are limited by their predators. Moreover, even though herbivores are rare, they may, nevertheless, compete for resources of limited quality and their predators may, in turn, compete for scarce herbivores. Most of the examples that have a bearing on the ideas of Hairston *et al.* and Murdoch come from aquatic environments; these tend to lend weight to the former. Whether terrestrial examples, as they come to light, will generally conform to the hypothesis remains to be seen.

sometimes, the world is white ...

Oksanen (1988) noted that the world is not always green—particularly if the observer is standing in the middle of a desert or on the northern coast of Greenland. Oksanen's contention is that in extremely unproductive or 'white' ecosystems, grazing will be light because there is not enough forage to support effective

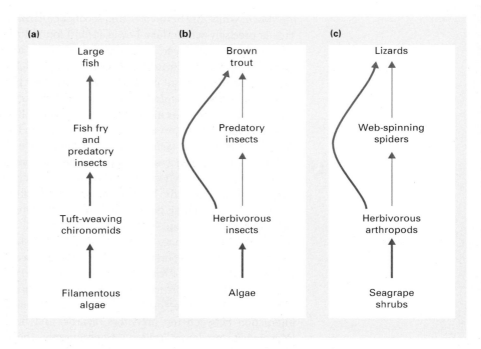

Figure 22.4 Three examples of food webs, each with four trophic levels. (a) The absence of omnivory (feeding at more than one trophic level) in this North American stream community means it functions, as predicted in Figure 22.3, as a four-trophic level system. On the other hand, web (b) from a New Zealand stream community and web (c) from a terrestrial Bahamanian community both function as three-trophic level webs. This is because of the strong direct effects of omnivorous top predators on herbivores and their less influential effects on intermediate predators. (After Power, 1990; Flecker & Townsend, 1994; Spiller & Schoener, 1994, respectively.)

... or yellow

populations of grazers. At the highest levels of plant productivity, in 'green' ecosystems, there will also be light grazing because of limitation by predators (as argued by Hairston *et al.*, 1960). Finally, between these extremes, ecosystems may be 'yellow' where plants are limited by grazing because there are insufficient herbivores to support effective populations of predators. This hypothesis remains to be critically tested.

There are further pointers that the level of primary productivity is influential in determining whether top-down or bottom-up control is predominant. Thus, zooplankton may have weaker top-down effects on phytoplankton in lakes with high productivity (McQueen *et al.*, 1990). A similar observation was made in a study involving trout, invertebrates and algae in a stream (Figure 22.4b); when nutrient concentrations were augmented, the presence of trout no longer resulted in an increase in algal biomass. Perhaps, in these two freshwater situations, high plant productivity obscures any cascading effects from above because the predators cannot increase their rate of consumption of herbivores to keep up with increased herbivore production. Alternatively, plant species composition may shift towards less palatable species in more nutrient-rich conditions. Thus, in eutrophic lakes less edible species of blue–green algae often assume significance (McQueen *et al.*, 1990) and in the trout study, higher nutrient concentrations led to dominance of filamentous algae that were more difficult for the herbivores to handle. To some extent, such results support the proposition associated with Murdoch (1966) that 'the world tastes bad'.

Finally, energetic considerations might lead us to expect that in more productive environments there will be more trophic levels (Fretwell, 1977) (in other words, there would be a shift to the right with increasing productivity in Figure 22.2). In fact, as we shall see later (see Section 22.5.1), the evidence gathered so far does not support this idea.

22.3.3 Incorporating heterogeneity into the top-down, bottom-up debate

injecting realism into the top-down, bottom-up debate

The theory describing top-down and bottom-up effects may be criticized for being oversimplistic, treating primary producers, herbivores and carnivores as undifferentiated units and making predictions based on characteristics of food chains (number of levels odd or even) rather than food webs. Some realism can be injected into the debate by incorporating heterogeneity amongst species *within* a trophic level (some may be resource-limited but others predator-limited, depending on their relative vulnerability to predators), heterogeneity in space (a species' relationships with resources and predators will vary from place to place) and heterogeneity in time (species interactions will vary in response to changing conditions and circumstances) (Hunter & Price, 1992). Thus, the question that should be asked is to what extent variation within and between trophic levels, or in abiotic factors, can influence the relative strengths of bottom-up and top-down forces?

Bartell *et al.* (1988) addressed the question of temporal heterogeneity by constructing a mathematical model of a lake community consisting of several species of phytoplankton, zooplankton, zooplanktivorous fish and piscivorous fish, subject to realistic seasonality in available radiant energy, nutrients and feeding rates. In a simplistic, unvarying world, the pattern would be constant with, in this four-trophic level situation, the phytoplankton (trophic level 1 in the four-trophic level case of Figure 22.2) controlled by the top-down forces of predation by zooplankton. Focusing on one attribute of the phytoplankton, their species richness, the model community demonstrated periods of decreasing species richness when competitive interactions (bottom-up forces) predominated. On the other hand, the highest values of phytoplankton species richness occurred at times of year when top-down regulation by the consumers was most evident. In the model simulation, and no doubt in reality, the relative importance of various factors shows considerable variation.

Given that in real food webs a great variety of biotic and abiotic factors influence the relative importance of predator control and resource limitation, under what circumstances might we expect a top-down cascade effect to be most clearly evident?

A clear top-down effect is more likely to be observed where primary producers are relatively short lived and fast growing so that they can respond rapidly to the effects of their consumers; in contrast, we are unlikely to detect top-down effects in an oak forest readily. Furthermore, consumers should be capable of dramatically and quickly reducing their resources; in such situations, interspecific competition can be expected to be observably intense. More generally, one species or guild per trophic level should have sufficiently strong potential effects on resources in the next level down to produce chain-like, rather than indeterminate web-like responses as a result of effects of higher trophic levels (Power, 1992).

These characteristics often apply closely to pelagic communities of lakes and to benthic communities of streams and rocky shores. Indeed, trophic cascades are generally aquatic and restricted to rather low-diversity places where great influence can emanate from one or a few species (Strong, 1992). This is not to say that top-down forces are absent in more species-rich systems, but rather that patterns of consumption are so differentiated that their overall effects are buffered; such effects may be represented as trophic trickles rather than cascades. In more complex communities, trophic structure can only properly be described as a web in which trophic levels are much less distinct than the aquatic examples above.

22.4 Community stability and food-web structure

For the remainder of this chapter, we turn our attention to the question of what determines the stability or instability of a community. To accomplish this, we will explore the theory that links food-web structure to stability and assess the extent to which real food webs are structured as theory predicts. Before proceeding further, however, it is necessary to define 'stability', or rather to identify the various different types of stability.

The first distinction that can be made is between the resilience of a community (or any other system) and its resistance. *Resilience* describes the speed with which a community returns to its former state after it has been perturbed and displaced from that state. *Resistance* describes the ability of the community to avoid displacement in the first place. (Figure 22.5 provides a figurative illustration of these and other aspects of stability.)

The second distinction is between local stability and global stability. *Local stability* describes the tendency of a community to return to its original state (or something close to it) when subjected to a small perturbation. *Global stability* describes this tendency when the community is subjected to a large perturbation.

A third aspect of stability is related to the local/global distinction, but concentrates more on the environment of the community. The stability of any community depends on the environment in which it exists, as well as on the densities and characteristics of the component species. A community that is stable only within a

narrow range of environmental conditions, or for only a very limited range of species' characteristics, is said to be *dynamically fragile*. Conversely, one that is stable within a wide range of conditions and characteristics is said to be *dynamically robust*.

Lastly, it remains for us to specify the aspect of the community which is having its stability assessed. Usually, ecologists have taken a demographic approach. They have concentrated on the composition of a community: the number and identity of the component species and their densities. However, as we shall see below (see Section 22.6), it is also possible to deal with other properties of a community (e.g. the rate of biomass production, or the amount of calcium it contains). We begin though, in Sections 22.4 and 22.5, with the demographic approach.

22.4.1 The 'conventional wisdom'

The aspect of community structure that has received most attention from the point of view of stability has been community complexity. During the 1950s and 1960s, the 'conventional wisdom' in ecology was that increased complexity within a community leads to increased stability. Increased complexity, then as now, was variously taken to mean more species, more interactions between species, greater average strength

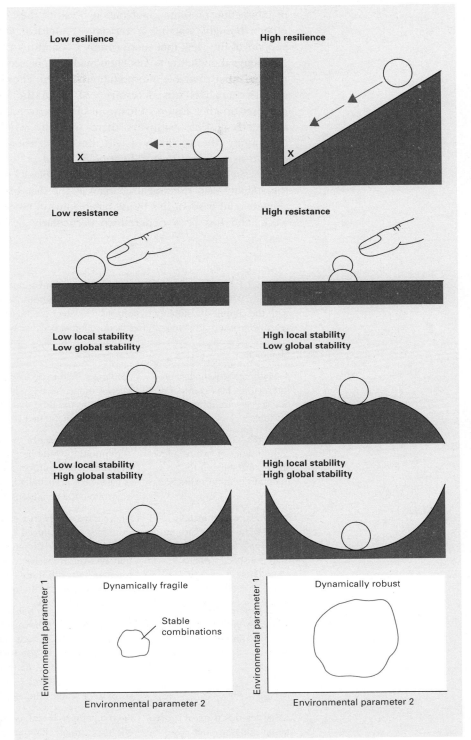

Figure 22.5 Various aspects of stability, used in this chapter to describe communities, illustrated here in a figurative way. In the resilience diagrams, X marks the spot from which the community has been displaced.

of interaction or some combination of all of these things. Elton (1958), amongst others, brought together a variety of empirical and theoretical observations in support of the view that more complex communities are more stable (Table 22.1). Nowadays though, the points Elton made can be seen as either untrue or else liable to some other plausible interpretation. (Indeed, Elton himself pointed out that more extensive analysis was necessary.) At about the same time, MacArthur (1955) proposed another argument in favour of the conventional wisdom. He suggested that the more possible pathways there were by which energy passed through a community, the less likely it was that the densities of constituent species would change in response to an abnormally raised or lowered density of one of those species. In other words, the greater the complexity (more pathways), the greater is the stability (less numerical change) in the face of a perturbation. However, the conventional wisdom has by no means always received support from more recent work, and has been undermined particularly by the analysis of mathematical models.

Table 22.1 Summary of Elton's arguments (1958) in support of the 'conventional wisdom' prior to 1970 that complexity begets stability in communities. Each observation is consistent with the complexity/stability hypothesis. However, each can be explained in terms of reasonable alternative hypotheses or questioned because of lack of a control.

Statement	Assessment
1 Mathematical models of interactions between two or a few species are inherently unstable	No longer held to be true after recent development of two-species models (see Chapter 10). In any case, there was no evidence at the time that multispecies models would be any more stable (none had been developed)
2 Simple laboratory communities of two or a few species are difficult to maintain without extinctions	True, but no evidence that multispecies laboratory communities would be more stable. The probable reason that laboratory cultures are difficult to set up and maintain is because it is virtually impossible to reproduce the natural environmental conditions
3 Islands, which usually possess few species, are more vulnerable to invading species than are species-rich continents	There have also been well-documented and remarkable examples of introduced species assuming pest proportions on continents (see Chapter 17). Nevertheless, statement 3 may be true. However, vulnerability to invasion has only a tenuous link with more conventional definitions of stability and resilience
4 Crop monocultures are peculiarly vulnerable to invasions and destruction by pests	Differences between natural and agricultural communities could derive from the long periods of coevolution to which natural associations may have been subject (Maynard Smith, 1974). Note also that arable crops are typically early successional species, naturally subject to rapid change. Natural monocultures such as salt-marsh and bracken seem to be stable (May, 1972)
5 Species-rich tropical communities are not noted for insect outbreaks when compared to their temperate and boreal counterparts	Such a pattern, if real, could be due to the destabilizing effects of climatic fluctuations in temperate and boreal communities (Maynard Smith, 1974). Wolda (1978) has produced evidence suggesting that insect abundances fluctuate just as markedly in the tropics

22.4.2 Complexity and stability in model communities

There have been a number of attempts to explore mathematically the relationship between community complexity and community stability (reviewed by May, 1981c), but most of these have come to similar conclusions. The approach taken by May (1972) can be used to illustrate the general nature of this mathematical work, the conclusions that can be drawn from such work and its shortcomings. May constructed model food webs comprising a number of species, and concerned himself with the way in which the population size of each species changed in the neighbourhood of its equilibrium abundance (i.e. he was concerned with *local* stability of a type similar to the 'stability' of one- and two-species models examined in Chapters 6, 7 and 10). Each species was influenced by its interaction with all other species, and the term βij was used to measure the effect of species j's density on species i's rate of increase. Thus, βij would be zero when there was no effect, both βij and βji would be negative for two competing species, and βij would be positive and βji negative for a predator (i) and its prey (j).

in randomly assembled food webs, local stability declines with complexity

May 'randomly assembled' food webs. He set all self-regulatory terms (βii, βjj, etc.) at -1, but distributed all other β values at random, including a certain number of zeros. The model webs so constructed could then be described by three parameters: S, the number of species; C, the 'connectance' of the web (the fraction of all possible pairs of species that interacted directly, i.e. with βij non-zero); and β, the average 'interaction strength' (i.e. the average of the non-zero β values, disregarding sign). What May found was that these food webs were only likely to be stable (i.e. the populations would return to equilibrium after a small disturbance) if:

$$\beta(SC)^{1/2} < 1. \tag{22.1}$$

Otherwise, they tended to be unstable (i.e. disturbed populations failed to return to equilibrium).

In other words, increase in number of species, increase in connectance and increase in interaction strength all tend to increase instability (because they increase the left-hand side of the inequality above). Yet, each of these represents an increase in complexity. Thus, this model (along with others) suggests that complexity leads to *instability*. This clearly runs counter to the conventional wisdom of Elton and MacArthur, and it certainly indicates that there is no necessary, unavoidable connection linking stability to complexity. It is possible, though, that the connection between complexity and instability is an artefact arising out of the particular characteristics of the model communities or the way they have been analysed.

In the first place, randomly assembled food webs often contain biologically unreasonable elements (e.g. predators without prey or loops of the type: A eats B eats C eats A). Analyses of food webs that are constrained to be reasonable (Lawlor, 1978; Pimm, 1979a) show: (i) that they are more stable than their unreasonable counterparts; and (ii) that whilst stability still declines with complexity, there is no sharp transition from stability to instability (compared with the inequality in Equation 22.1).

... but if the models are altered, the conclusions become less clear-cut

Second, the results of model analyses are altered if it is assumed that consumer populations are affected by their food supply, but that the food supply is not affected by the consumers ($\beta ij > 0$, $\beta ji = 0$: so-called 'donor-controlled' systems in which bottom-up forces predominate—see Section 22.3). In this type of food web, stability is unaffected by or actually increases with complexity (DeAngelis, 1975). The most obvious group of organisms to which this condition applies is the detritivores (see Chapter 11), but consumers of nectar and seeds are also unlikely, in practice, to have any influence on the supply of their food resource (Odum & Biever, 1984). Furthermore, few of the multitude of rare phytophagous insects are likely to have any impact upon the abundance of their host plants (Lawton, 1989) and conditions for donor control are also probably widespread in parasitoid–host systems (Hawkins, 1992). These are clearly not unimportant groups of 'exceptions'.

The picture also alters if, instead of considering local stability, larger perturbations are examined. Pimm (1979b) has attempted to do this by using 'species-deletion stability'. His model communities were subjected to a single, large and persistent perturbation: the deletion of one species. The system was then said to be species-deletion stable if all of the remaining species were retained at locally stable equilibria. Figure 22.6 shows the results of simulated deletions of species from a simple six-species community containing two top predators, two intermediate predators and two basal species (plants or categories of dead organic matter). Interaction strengths were ascribed at random, whilst connectance was varied systematically; complexity could therefore be equated with connectance. In general, stability decreased with increasing complexity; but, this trend was reversed when basal species were removed. These simulations therefore agree with most other models when perturbing 'from above' (removal of a top predator), but they conform with the conventional wisdom when perturbing 'from below' (removal of a species from the base of the food web). Interestingly, MacArthur's original argument, since it dealt with energy flowing through a community, also envisaged perturbations coming from below.

Finally, the relationship between complexity and stability in models becomes more complicated if attention is focused on the resilience of those communities

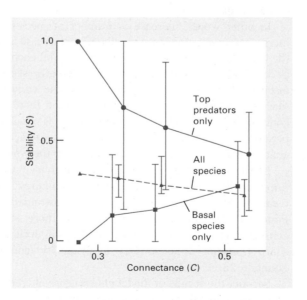

Figure 22.6 Species-deletion stability (S) plotted against connectance (C) for six-species models. Results are for systems where: (i) only top predators are deleted; (ii) only basal species are deleted; and (iii) all species are deleted in turn. The bars show the range in species-deletion stability for all models of a given connectance. (After Pimm, 1979b.)

which *are* stable. Pimm (1979a), like other modellers, found that as complexity (actually connectance) increased, the proportion of stable communities *decreased*. However, within the small subset of stable communities, resilience (a crucial aspect of stability) actually increased with complexity (Figure 22.7).

Overall, most models indicate that stability tends to decrease as complexity increases. This is sufficient to undermine the conventional wisdom prior to 1970. However, the conflicting results amongst the models at least suggest that no single relationship will be appropriate in all communities. It would be wrong to replace one sweeping generalization with another.

Even if complexity and instability are connected in models, this does not mean that we should expect to see an association between complexity and instability in real communities. For one thing, the models often refer to randomly constructed communities, but real communities are far from randomly constructed. Their complexity may be of a particular type and *could* enhance stability. One task facing theoretical ecologists, therefore, is to suggest ways in which special structures in food webs may enhance stability.

Furthermore, unstable communities will fail to persist when they experience environmental conditions that reveal their instability. But, the range and predictability of environmental conditions will vary from place to place. In a stable and predictable environment, a community will experience only a limited range of conditions, and a community that is dynamically fragile may still persist. However, in a variable and unpredictable environment, only a community that is dynamically robust will be able to persist. Such an argument requires predictability to be defined in a sensible way, and measured. This is far from being a straightforward task. However, what we might expect to see are: (i) complex and fragile communities in stable and predictable environments, with simple and robust communities in variable and unpredictable environments; and (ii) approximately the same recorded stability (in terms of population fluctuations, etc.) in all communities, since this will depend on the inherent stability of the community combined with the variability of the environment.

on balance, though, the conventional wisdom is undermined

stability may vary between environments

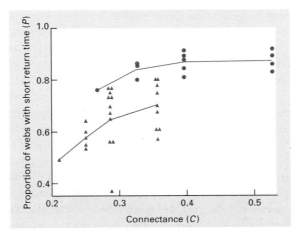

Figure 22.7 Complexity and resilience in model food webs. As complexity (actually connectance) increases, the proportion of stable food-web models that have short return times (*P*) after a perturbation also increases (a short return time indicates a high resilience). ●, Models with three trophic levels; ▲, models with four trophic levels. (After Pimm, 1982.)

843   FOOD WEBS

This line of argument, however, carries a further, very important implication for the likely effects of unnatural, man-made perturbations on communities. We might expect these to have their most profound effects on the dynamically fragile, complex communities of stable environments, which are relatively unused to perturbations. Conversely, they should have least effect on the simple, robust communities of variable environments, which have previously been subjected to repeated (albeit natural) perturbations. In this sense too, complexity and instability may be associated.

What is the evidence from real communities?

22.4.3 Complexity and stability in practice

A number of studies have sought to investigate the general importance of dynamic/stability constraints on community structure by examining the relationship between S, C and β in real communities. The argument they use runs as follows. If communities are only stable for $\beta(SC)^{1/2} < 1$, then increases in S will lead to decreased stability unless there are compensatory decreases in C and/or β. Data on interaction strengths for whole communities are unavailable. It is therefore usually assumed, for simplicity, that β is constant. In such a case, communities with more species will only retain stability if there is an associated decline in average connectance (so that SC is approximately constant).

A group of 40 food webs was gleaned from the literature by Briand (1983), including terrestrial, freshwater and marine examples. For each community, a single value for connectance was calculated as the total number of identified interspecies links as a proportion of the total possible number of pair-wise interactions. Both predator–prey and presumed competitive interactions (where consumers shared the same prey) were included. Connectance is plotted against S in Figure 22.8a. As predicted, connectance decreases with species number.

Briand's pioneering compilation suffers from the drawback that the data on which it is based were not collected for the purpose of quantitative study of food-web properties and, moreover, the level of taxonomic resolution varied substantially from web to web. More recent studies, in which food webs were much more rigorously documented, indicate that C may decrease with S (as predicted) (Figure 22.8b), that C may be independent of S (Figure 22.8c) or may even increase with S (Figure 22.8c). Thus, the stability argument does not receive consistent support from food-web analyses. The significance of connectance is considered further in Section 22.5.1.

The relationship between complexity and stability can also be investigated experimentally. Thus, the prediction that complex communities are less stable when given an experimental perturbation was tested on two sets of plant communities. In the first experiment, the perturbation consisted of an addition of plant nutrients to the soil, whilst in the second it involved the action of grazing animals. In each case, the effect of the perturbation was monitored in both species-rich and species-poor plant communities. Each perturbation significantly reduced the diversity of the species-rich but not the species-poor community (Table 22.2), consistent with the hypothesis that more complex communities are less likely to return to their prior state after a perturbation.

In contrast, in a 'natural experiment' involving a set of eight varied grassland communities, those that were more diverse in species were more resistant (more

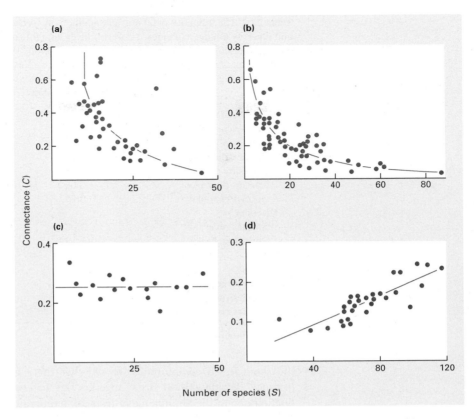

Figure 22.8 The relationships between connectance (C) and species richness (S). (a) For a compilation from the literature of 40 food webs from terrestrial, freshwater and marine environments. (After Briand, 1983.) (b) For a compilation of 95 insect-dominated webs from various habitats. (After Schoenly *et al.*, 1991.) (c) For seasonal versions of a food web for a large pond in northern England, varying in species richness from 12 to 32. (After Warren, 1989.) (d) For food webs from swamps and streams in Costa Rica and Venezuela. (After Winemiller, 1990.) (After Hall & Raffaelli, 1993.)

and another that is *more* stable

stable in community composition) in the face of a severe summer drought in Yellowstone National Park (Figure 22.9). Enhanced stability may have resulted from the greater heterogeneity in soil conditions in the diverse communities. Once again, these 'experimental' studies provide conflicting results in relation to the predicted complexity/stability relationship.

the importance of trophic level

The idea that the effect of complexity on stability depends on which trophic level is perturbed, gains some support from a study of simple, laboratory protozoan communities set up by Hairston *et al.* (1968). The first trophic level comprised one, two or three species of bacterium. These were consumed by one, two or three *Paramecium* spp. The percentage of cultures showing no extinctions of *Paramecium* increased with diversity of bacteria. However, the extinction of the rarest *Paramecium* spp. was most likely when three species of *Paramecium* were cultured together as opposed to only two.

perceived stability is apparently constant

Finally, there have been a number of studies directed at the question of whether the level of 'perceived stability' (i.e. observed fluctuations in abundance) is roughly

Table 22.2 The influence of (a) nutrient addition on species richness, equitability ($H / \ln S$) and diversity (Shannon's index, H) in two fields, and (b) grazing by African buffalo on species diversity in two areas of vegetation. (After McNaughton, 1977.)

	Control plots	Experimental plots	Statistical significance
(a) *Nutrient addition*			
Species richness per 0.5 m² plot			
Species-poor plot	20.8	22.5	n.s.
Species-rich plot	31.0	30.8	n.s.
Equitability			
Species-poor plot	0.660	0.615	n.s.
Species-rich plot	0.793	0.740	$P < 0.05$
Diversity			
Species-poor plot	2.001	1.915	n.s.
Species-rich plot	2.722	2.532	$P < 0.05$
(b) *Grazing*			
Species diversity			
Species-poor plot	1.069	1.357	n.s.
Species-rich plot	1.783	1.302	$P < 0.005$

n.s., not significant.

the same in all communities, or whether there are any noticeable trends. For instance, Wolda (1978) drew together an extensive array of data on annual fluctuations in the abundance of tropical, temperate and sub-Arctic insect populations. The old conventional wisdom assumed that the more diverse insect communities of the tropics were more stable (i.e. fluctuated less) than their depauperate temperate and sub-Arctic counterparts. However, Wolda concluded that, on

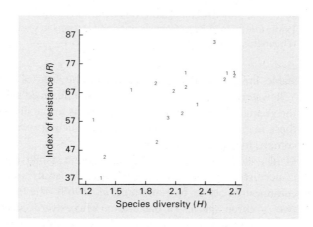

Figure 22.9 Relationship between resistance in grassland community composition and species diversity (Shannon's index, H) for a number of grassland areas in Yellowstone National Park, USA. The resistance measure (R) is designed to be inversely related to the cumulative differences in species abundances at sites between 1988 (a year of severe drought) and 1989 (a year of normal rainfall). Thus, a high value for R indicates that relative abundances changed little in the face of the drought, whilst a low value means they changed considerably. (After Frank & McNaughton, 1991.)

average, populations of tropical insects have the same annual variability as those of insects from temperate zones. Even the few sub-Arctic studies show a similar spread of values. Similar conclusions have been reached by Leigh (1975) for herbivorous vertebrates and Bigger (1976) for crop pests.

22.4.4 Appraisal

The relationship between the complexity of a community and its inherent stability is not clear-cut. It appears to vary with the precise nature of the community, with the way in which the community is perturbed and with the way in which stability is assessed.

However, it seems likely (and we have presented some evidence to support this) that the complex, fragile communities of relatively constant environments (e.g. equatorial regions) are more susceptible to outside, unnatural disturbance (and are more in need of protection) than the simpler, more robust communities that are more 'used' to disturbance (e.g. in more temperate regions).

Lastly, it is worth noting that there is likely to be an important parallel between the properties of a community and the properties of its component populations. In stable environments, populations will be subject to a relatively high degree of K selection (see Chapter 14); in variable environments they will be subject to a relatively high degree of r selection. The K-selected populations (high competitive ability, high inherent survivorship but low reproductive output) will be *resistant* to perturbations, but once perturbed will have little capacity to recover (low resilience). The r-selected populations, by contrast, will have little resistance but a higher resilience. The forces acting on the component populations will therefore reinforce the properties of their communities, namely fragility (low resilience) in stable environments and robustness in variable ones.

22.5 Empirical properties of food webs

Interest in the properties of food webs and how they might explain community dynamics has prompted two research approaches. The first, which we have just discussed, involves searching for certain food-web properties predicted by models to confer stability (see Section 22.4). The second involves studying and describing real food webs in a quest for generalizations about which of all conceivable food-web structures are actually represented in nature. The repeated occurrence of certain food-web patterns can then be considered in terms of underlying causes and implications for community persistence (Pimm, 1991).

It should be emphasized at the outset that progress in our understanding of food webs depends critically on the quality of data that are gathered from natural communities. Recently, several authors have called this into doubt, particularly for the earlier studies, pointing out that taxa have often been resolved grossly and extremely unevenly—even in the same web, different taxa may have been grouped at the level of kingdom (plants), family (Diptera) and species (polar bear). Some of the most thoroughly described food webs have been examined for the effects of such uneven resolution by progressively lumping web elements into coarser and coarser taxa (see Hall & Raffaelli, 1993, for a review). The uncomfortable conclusion is that most food-web properties seem to be sensitive to the level of taxonomic resolution that is achieved. For example, Hall and Raffaelli (1991) found for their well-documented estuarine food web that connectance increased with decreasing taxonomic resolution. These limitations should be borne in mind as we explore the evidence for food-web patterns in the following sections.

22.5.1 Connectance and biological constraints

We have already noted that for model food webs to remain stable, an increase in number of species (S) requires a concomitant decrease in connectance (C) (see Section 22.4.3). However, when these things are measured in real food webs, the hypothesized pattern is found in some cases, whilst in others C is independent of, or increases with S (see Figure 22.8). We ask here whether another hypothesis might do better in accounting for the recorded patterns in connectance.

the degree of diet specialization influences connectance

Morphological, physiological and behavioural features restrict the number of conceivable prey that a consumer can actually exploit; it is difficult to be a 'jack of all trades'. The consequences for connectance depend on how these restrictions are manifested (Warren, 1994). If each species is adapted to feed on a fixed number of other species, then SC turns out to be constant and, just as in the stability argument, C should show a hyperbolic decrease with increasing S (and, thus, empirical food-web data would not allow us to judge between the two situations). On the other hand, if each species feeds upon anything whose characteristics, such as size, toughness or chemistry, falls within the range to which it is adapted, then as the total number of species increases, a greater number are likely to be within the acceptable range. In this case, which is biologically more realistic, connectance neither decreases nor increases with S, but is roughly constant. Moreover, if webs are made up of species adapted to a small range of prey (specialists), overall connectance will be low (and constant), whereas webs composed of generalists will have high (and constant) connectance. A real problem in interpreting compilations of food webs, such as those in Figure 22.8, is that some may be made up, on average, of more specialist species than others. Thus, the observed unpredictable 'pattern' may actually be a mixture of patterns that on their own would be predictable. A challenge for the future will be to look for particular sets of food webs where greater specialization or generalization might be the rule.

22.5.2 The number of trophic levels

A fundamental feature of food webs is the number of trophic links in pathways from basal species to top predators. A *maximal food chain* is defined as a sequence of species running from a basal species to another species that feeds on it, to yet another species that feeds on the second, and so on up to a top predator (fed on by no other species). For instance, starting with basal species 1 in Figure 22.10, we can trace four possible trophic pathways via species 4 to a top predator. These are 1–4–11–12, 1–4–11–13, 1–4–12 and 1–4–13. This provides four values for the length of a maximal food chain: 4, 4, 3 and 3. Figure 22.10 lists a total of 21 further estimates for chains, starting from basal species 1, 2 and 3. The average of all the possible maximal food-chain lengths is 3.32. This parameter defines the number of trophic levels that can be assigned to the food web in question. Most communities described so far typically consist of between two and five trophic levels, and they have been reported most often to have three or four. Ecologists interested in community structure have asked why food chains are as short as this.

many communities have only three or four trophic levels

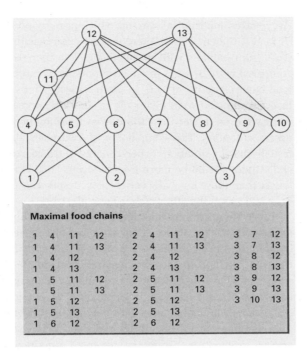

Maximal food chains

1	4	11	12		2	4	11	12		3	7	12
1	4	11	13		2	4	11	13		3	7	13
1	4	12			2	4	12			3	8	12
1	4	13			2	4	13			3	8	13
1	5	11	12		2	5	11	12		3	9	12
1	5	11	13		2	5	11	13		3	9	13
1	5	12			2	5	12			3	10	13
1	5	13			2	5	13					
1	6	12			2	6	12					

Figure 22.10 Community matrix for an exposed intertidal rocky shore in Washington State, USA. The pathways of all possible maximal food chains are listed. 1, detritus; 2, plankton; 3, benthic algae; 4, acorn barnacles; 5, *Mytilus edulis*; 6, *Pollicipes*; 7, chitons; 8, limpets; 9, *Tegula*; 10, *Littorina*; 11, *Thais*; 12, *Pisaster*; 13, *Leptasterias*. (After Briand, 1983.)

Once again, the stability properties of various model communities may tell us something about the pattern to be expected in real food webs. However, the 'stability' hypothesis is not the only one to consider. Other plausible hypotheses based on fundamental ecological principles must also be examined.

The energy flow hypothesis

It has long been argued that energetic considerations set a limit to the number of trophic levels that an environment can support. Of the radiant energy which reaches the earth, only a small fraction is fixed by photosynthesis and is available either as live food for herbivores or as dead food for detritivores.

However, the amount of energy available for consumption is considerably less than that fixed by the plants, because of work done by the plants (in growth and maintenance), and because of losses due to inefficiencies in all energy-conversion processes (see Chapter 18). Each feeding link amongst the heterotrophs is characterized by the same phenomenon: at most, 30%, and sometimes as little as 1%, of energy consumed at one trophic level is available as food to the next.

a testable prediction based on energy flow receives little support

In theory, therefore, the observed pattern of just three or four trophic levels could be due to energetic limitations—a further trophic level just could not be supported by the available energy. This hypothesis provides a testable prediction. Systems that possess a greater primary productivity should be able to support a larger number of trophic levels. A comprehensive test of this hypothesis involved an analysis of 32 published accounts of food webs in habitats ranging from desert and woodland to

849 FOOD WEBS

Arctic lakes and tropical seas, for which primary productivity has also been reported (Briand & Cohen, 1987). There was no difference in length of food chains when 22 webs from habitats with productivities less than 100 g of carbon $m^{-2}\,year^{-1}$ were compared with 10 webs in which productivity exceeded 1000 g $m^{-2}\,year^{-1}$ (median food-chain lengths of 3.0 in both cases). In a similar vein, a survey of 95 insect-dominated webs (Schoenly *et al.*, 1991) revealed that food chains in tropical webs were no longer than those from (presumably) less productive temperate and desert situations. This compilation of food webs allows a second prediction to be tested—that in situations where energy is transferred more efficiently up the food chain there should be more trophic levels. Food chains composed of insects are no longer than those involving vertebrates, despite the lower energy transfer efficiencies associated with the latter (Schoenly *et al.*, 1991).

Given these various findings, the classical energy explanation for the lengths of food chains has been rejected by a number of ecologists. However, it should be borne in mind that species richness is usually significantly higher in productive regions (see Chapter 24), and that each consumer probably feeds on only a limited range of species at a lower trophic level. The amount of energy flowing up a single food chain in a productive region (a large amount of energy, but divided amongst many subsystems) may not be very different from that flowing up a single food chain in an unproductive region (having been divided amongst fewer subsystems). Thus, the energy explanation cannot be accepted uncritically—but, neither should it be rejected altogether. Indeed, in an experiment using water-filled containers as analogues of natural tree-holes, a 10-fold or 100-fold reduction from a 'natural' level of energy input (leaf litter) reduced maximal food-chain length by one link, because in this simple community of larvae of mosquitos, midges, beetles and mites, the principal predator, a chironomid midge *Anatopynia pennipes*, was usually absent from the less productive habitats (Jenkins *et al.*, 1992).

... except from one experimental study

The dynamic fragility of model food webs

stability theory predicts the short food chains found in nature ...

In a study of the stability properties of variously structured Lotka–Volterra models, webs with long food chains (more trophic levels) typically underwent population fluctuations so severe that the extinction of top predators was more likely than with shorter food chains. Some of these results are shown in Figure 22.11. Return times after a perturbation were very much shorter in four-species models with only two trophic levels, than in those arranged into three or four levels. Because less resilient systems are unlikely to persist in an inconstant environment, it may be argued that only systems with few trophic levels will commonly be found in nature (Pimm & Lawton, 1980). Note that the three model food webs differ not only in the number of trophic levels but also in the number of links between levels 1 and 2. Short return times may conceivably result from the larger number of links rather than the smaller number of trophic levels.

Constraints on predator design and behaviour

To feed on prey at a given trophic level, a predator has to be large enough, manoeuvrable enough and fierce enough to effect a capture. In general, predators are larger than their prey (not true, though, of grazing insects and parasites), and body size increases at successive trophic levels. There may well be a limit above

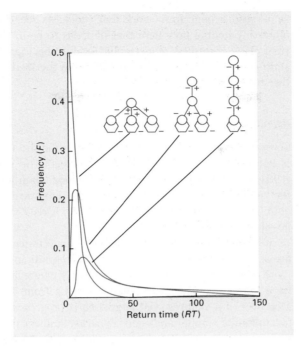

Figure 22.11 Frequency (*F*) of return times (*RT*, arbitrary scale) for the two-, three- and four-trophic level models illustrated. A small number of trophic levels is associated with shorter return times, i.e. the systems are more resilient. (After Pimm & Lawton, 1980.)

but, short food chains may result from predator design constraints

which design constraints rule out another link in the food chain. In other words, it may be impossible to design a predator that is both fast enough to catch an eagle and big and fierce enough to kill it. Of course, it is not possible to carry out a critical test of this hypothesis since we can only judge what is possible in evolution from what has actually evolved.

A further question relates to the density of top predators. Since they are generally large and much less than 100% efficient in finding and converting prey into predator biomass, they are necessarily much rarer than their prey and need to forage over a larger home range. There must be some limit to the necessary home-range size and density that would still permit members of the opposite sex to meet occasionally. If the lengths of food chains were limited by home-range considerations, we might expect communities that are restricted to small areas, such as islands, to have fewer trophic levels. In fact, there have been a number of reports of an absence of large predatory birds on small islands—and other birds often seem remarkably tame as a result.

We need also to consider the question of the optimal choice of diet. Consider the arrival in a community of a new carnivore species. Would it do best to feed on the herbivores or the carnivores already there? The herbivore trophic level offers the richer energy environment because there have been only a minimum number of energy-losing steps. Thus, an advantage to feeding low down in the food chain can readily be seen. On the other hand, if all species did this, competition would intensify, and feeding higher in the food chain could reduce competition.

In more general terms, it is difficult to imagine a top predator sticking religiously to a rule that it should prey only on the trophic level immediately below it, especially as the prey there are likely to be larger, fiercer and rarer than species at lower levels. In fact, it is unlikely that animals conceive of their world as the same rigid trophic framework that ecologists see. More probably, carnivores will attack a potential prey item that happens to be in the right size range and in the right place, regardless of its trophic level. If this is so, food chains will frequently be shorter than modellers conceive (compare the food webs in Figures 22.4b, c with that in Figure 22.4a).

Appraisal

(another) pattern with no agreed explanation ...

Several plausible hypotheses to account for the shortness of food chains are on offer. Is it possible to determine which is correct? The simple answer is 'no'. As May (1981c) has remarked, the situation for food-chain length is typical of the situation for community structure generally: 'the empirical patterns are widespread and abundantly documented, but instead of an agreed explanation there is only a list of possibilities to be explored'.

... and a need for more rigorous descriptions of food webs

It is important to note that, as with connectance, estimates of food-chain length are sensitive to the degree of taxonomic resolution. This may be why many of the more recently documented webs, especially where larger webs have been described, have longer average chain lengths ranging from 5 to 7 (Hall & Raffaelli, 1993). Moreover, if a well-resolved large web is progressively simplified by lumping taxa together (in a manner analogous to earlier studies), the estimate of food-chain length declines (Martinez, 1993). There is clearly a need for rigorous studies of many more food webs before acceptable generalizations can be reached.

22.5.3 The prevalence of omnivory in food webs

Our understanding of the prevalence in food webs of omnivory, defined as feeding on more than one trophic level, is also in an uncertain state. Compilations of early descriptions of food webs indicated that omnivores are usually uncommon (Pimm, 1982). This was taken to support expectations from simple Lotka–Volterra model communities where omnivory is found to be destabilizing (Pimm, 1982). In cases of omnivory, it is argued that intermediate species both compete with and are preyed upon by top species and, in consequence, are unlikely to persist long. A more complex and realistic model incorporates population age structure and 'life-history omnivory', in which different life-history stages of a species feed on different trophic levels (Pimm & Rice, 1987). Life-history omnivory also reduces stability, but much less than single life-stage omnivory does. Intriguingly, omnivory is not destabilizing in donor-control models and some preliminary evidence suggests that omnivores may be particularly common in decomposer food webs (Hildrew *et al.*, 1985; Walter, 1987), to which donor-control dynamics can be applied.

omnivory tends to destabilize model food webs ...

... but, the patterns in nature are inconsistent

In contrast to expectation, omnivory has also been found to be common in plankton food webs in North American glacial lakes (Sprules & Bowerman, 1988) (Figure 22.12). It may be that the assumptions in the models are inappropriate for plankton communities, for some as yet unknown reason. On the other hand, the apparent conformity of earlier food webs with the expected low level of omnivory could conceivably have been an artefact of poorly described food webs. Sprules and Bowerman (1988) identified all their zooplankton to species, and their measures are much more reliable as a result. Polis (1991) also argues that real food webs are much more complex than the early literature suggested, finding abundant evidence of omnivory in his detailed study of a desert sand community.

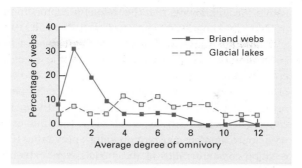

Figure 22.12 The prevalence of omnivory in glacial lakes in north-east North America (Sprules & Bowerman, 1988) is much greater than that observed in Briand's set of food webs (see Figure 22.8a). The degree of omnivory in a web is quantified as the number of closed omnivorous links divided by the number of top predators. A closed omnivorous link exists when a feeding path can be traced to a prey more than one trophic level away, and from that prey back to the predator through at least one other prey occupying an intermediate trophic level.

22.5.4 The ratio of predators to prey in food webs

in communities there is an approximate constancy in the ratio ...

An approximately constant ratio of species of predator to prey has been observed in several communities, although the value of the ratio may vary with type of habitat and the kinds of organisms in the web. An example, drawn from freshwater invertebrate communities, is shown in Figure 22.13a. For once we can take some

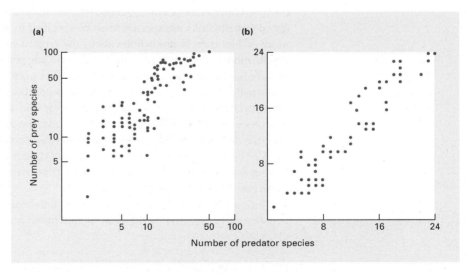

Figure 22.13 (a) The relationship between number of species of predators and prey in a variety of communities of freshwater invertebrates is described well by a straight line passing through the origin, corresponding to a constant predator : prey ratio of 0.36 predators per prey species. (b) Plot of final numbers of prey and predator species in model food webs generated at random and subject to Lotka–Volterra dynamics. A straight line through the origin (indicating a constant predator : prey ratio) also describes these data. (After Mithen & Lawton, 1986.)

comfort from knowing that the ratio of predators to prey is quite robust to variations in level of taxonomic resolution (Hall & Raffaelli, 1993).

Could such an apparently constant feature of food webs result from dynamic constraints in the interactions amongst the species? Mithen and Lawton (1986) created some simple model communities with two trophic levels by drawing species of predators and prey at random from a species pool, simulating their interactions using Lotka–Volterra equations. Despite varying initial conditions, all the webs converged over successive periods of invasion and extinction to locally stable systems with approximately constant predator : prey ratios (Figure 22.13b). The mechanisms creating constant ratios in the models were exploitation competition amongst the predators for victims, and competition amongst the prey for 'enemy-free space' (see Chapter 7, Section 7.6). These mechanisms may also be responsible for constant ratios in real food webs. However, the majority of reported predator : prey ratios have values that are close to one, a feature that may be an arithmetical artefact due to the way the ratio is defined (Closs *et al.*, 1993). Thus, when predator : prey ratios are calculated, taxa may be recorded as both predator and prey (i.e. be double-counted). In many webs the proportion of species that are double-counted is large so that the ratio will inevitably be roughly equal to one.

A related generalization that can be made about real food webs is that the ratio of basal species (those that only serve as prey) to intermediate species (that are both predators and preyed upon) to top species (that only serve as predators) is approximately constant at 0.19 : 0.52 : 0.29 (Hall & Raffaelli, 1993). A model that lacks equations describing species interactions, and is therefore very different from those described earlier in this chapter, does indeed generate the observed proportions (Cohen & Newman, 1985). In it, feeding links are assigned at random, but are subject to two constraints. First, species are arranged in a cascade such that each can only feed on species below it. Second, the number of species fed on by each predator (the density of links per species) must be specified in advance. The model also makes more or less correct predictions about the lengths of food chains. Why should this simple model work? It seems, that given one key property of food webs (the density of links per species, i.e connectance in another guise), many other properties follow necessarily (i.e. what seems to be several properties is actually just one). However, we are still left without an explanation for the key property. An approximate constancy in density of links per species (d) may be due to biological constraints that limit the diet-breadths of species (as discussed in Section 22.5.1). On the other hand, the key parameter d may be the product of dynamic interactions between species (Cohen & Newman, 1988), something that is taken as given and not explored in the static cascade model.

Once again, we have an empirical pattern to be explained and a range of possible explanations suggested by different modelling approaches.

22.5.5 Compartments in communities

A food web is compartmentalized if it is organized into subunits within which interactions are strong, but between which interactions are weak. (The most perfectly compartmentalized community possesses only linear food chains.)

In studies where habitat divisions are major and unequivocal, not surprisingly there is a clear tendency for compartments to parallel habitats. For instance, Figure 22.14 shows the major interactions within and between three interconnected

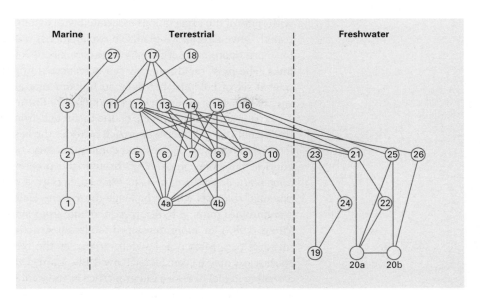

Figure 22.14 The major interactions within and between three interconnected habitats on Bear Island in the Arctic Ocean. 1, plankton; 2, marine animals; 3, seals; 4a, plants; 4b, dead plants; 5, worms; 6, geese; 7, Collembola; 8, Diptera; 9, mites; 10, Hymenoptera; 11, seabirds; 12, snow bunting; 13, purple sandpiper; 14, ptarmigan; 15, spiders; 16, ducks and divers; 17, Arctic fox; 18, skua and glaucous gull; 19, planktonic algae; 20a, benthic algae; 20b, decaying matter; 21, protozoa a; 22, protozoa b; 23, invertebrates a; 24, Diptera; 25, invertebrates b; 26, microcrustacean; 27, polar bear. (After Pimm & Lawton, 1980.)

habitats on Bear Island in the Arctic Ocean (Summerhayes & Elton, 1923); there is a significantly smaller number of interactions between habitats than would be expected by chance (Pimm & Lawton, 1980). On the other hand, when habitat divisions are more subtle, the evidence for compartments is typically poor. There are even greater difficulties in providing a clear demonstration of compartments (or the lack of them) *within* habitats. Such analyses as have been carried out, however, suggest that food webs within habitats are approximately as compartmentalized as would be expected by chance alone (Pimm & Lawton, 1980; Pimm, 1982).

... predicted by *some* models to promote stability

A number of theoretical studies have indicated that communities will have increased stability if they are 'compartmentalized' (e.g. May, 1972; Goh, 1979). On the other hand, when Pimm (1979a) excluded biologically unreasonable elements from his model food webs, he found no clear connection between stability and compartmentalization. There is certainly no weight of evidence to support the idea that compartmentalization has promoted the persistence of particular food webs because of the stability it confers.

22.5.6 Appraisal

If we want to understand how communities function, so that they can be managed, protected or restored, we surely need to understand the food webs that are at their heart. This quest, however difficult, is not one that ecologists can afford to give up on.

an important quest
... but, the study of food webs has a long way to go

The theory of food webs has a long way to go before it reaches maturity. We have repeating patterns in food webs (which may or may not be real) that are consistent with some of the expectations associated with dynamical constraints (but, which also usually have alternative plausible explanations). The most pressing requirement is for more rigorous descriptions of the components of food webs and of their feeding links. The most rapid progress can be expected from analyses of food webs from a related set of habitats that differ in fundamental aspects (e.g. degree of consumer specialization, or productivity or frequency of disturbance).

Progress may also be made on the theoretical front. It is questionable whether the traditional Lotka–Volterra approach provides the best way forward. Another class of food-web theory recognizes that communities are not created instantly as complete entities, but come into being through the progressive invasion of species. Such *community assembly models*, in essence, specify a model food web and a pool of potential colonists which challenge the system (Hall & Raffaelli, 1993). In this way, communities may be built up sequentially from simple starting points (e.g. Post & Pimm, 1983), or more developed food-web structures may be challenged by new invaders (e.g. Mithen & Lawton, 1986), or the reconstruction of food webs after species loss may be considered (Law & Blackford, 1992). Drake (1988; 1990) used an assembly model to derive the properties of endpoint communities that were resistant to further invasion, and found striking similarities with data from real communities. Taking his work a step further, the role of historical factors and assembly mechanics in producing pattern was examined in a freshwater laboratory ecosystem of bacteria, protozoans, algae and invertebrates (Drake, 1991). By varying the sequence of species invasions a variety of alternative community states could be produced. It is clear that to fully understand a community, we cannot rely only on knowledge of the existing food web; we need also to understand the historical processes by which it was created. Community assembly models are likely to become of increasing practical importance because they deal explicitly with invasions, something we need to understand if we are to cope with and combat the ever-increasing incursions of exotic species from one part of the globe to another (Townsend, 1991b).

22.6 Non-demographic stability

As has been suggested already, community stability can be viewed from perspectives other than a purely demographic one.

First, recall McNaughton's (1978) studies from p. 844. Table 22.2 showed how species-rich grassland communities responded to perturbations with a significant drop in species diversity, not recorded in the case of species-poor communities. The effect of the perturbations was quite different, however, when viewed in terms of an

the stability of productivity and biomass

aspect of functioning (primary productivity) and a composite aspect of structure (standing-crop biomass). The addition of fertilizer significantly increased primary productivity in the species-poor field in New York State (+53%), but only slightly and insignificantly changed productivity in the species-rich field (+16%). Similarly, grazing in the Serengeti significantly reduced the standing-crop biomass in the species-poor grassland (–69%), but only slightly reduced that of the species-rich field (–11%). In other words, a more complex structure seems to have enhanced the stability of these communities when criteria other than demographic ones are used.

This argument gains further weight from a study in Minnesota, USA, which took advantage of the most severe drought in 50 years to test the hypothesis that grasslands richer in species should exhibit more stability in biomass in the face of a major disturbance. Species-poor plots were both more greatly harmed by the

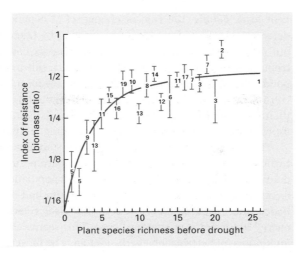

Figure 22.15 Relationship between an index of community resistance and plant species richness in a series of grasslands in Minnesota. Resistance is measured as the ratio of biomass in 1988, at the height of a severe drought, to that recorded in 1986, before the drought hit. Species richness was recorded before the drought. Mean, standard error and number of plots with a given species richness are shown. (After Tilman & Downing, 1994.)

drought (less resistant—Figure 22.15) and took longer to recover their predrought biomass (less resilient) than species-rich communities.

resilience in the face of energy and nutrient perturbations

Several studies have revealed that the structure of the food web can influence its resilience (speed of return to equilibrium) in response to perturbations of energy and nutrient supplies. O'Neill (1976) considered the community as a three-compartment system consisting of active plant tissue (P), heterotrophic organisms (H) and inactive dead organic matter (D) (illustrated in Figure 22.16). The rate of change in standing crop in these compartments depends on transfers of energy between them. Thus, the rate of change in P depends on one input (net primary productivity) and two outputs (fraction consumed by heterotrophs and fraction lost to D as litter). The rate of change of H depends on two inputs (consumption of living plant biomass and consumption of dead organic matter) and two outputs (defaecation and respiratory

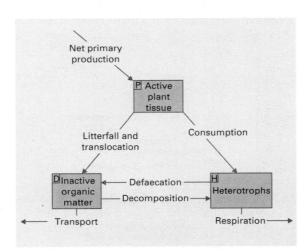

Figure 22.16 A simple model of a community. The three boxes represent components of the system and arrows represent transfers of energy between the system components. (After O'Neill, 1976.)

857   FOOD WEBS

heat loss). Finally, the rate of change of D depends on two inputs (litterfall and defaecation) and two outputs (consumption by heterotrophs and physical transport out of the system). Inserting real data from six communities representing tundra, tropical forest, temperate deciduous forest, a salt-marsh, a freshwater spring and a pond, O'Neill subjected the *models* of these communities to a standard perturbation, consisting of a 10% decrease in the initial standing crop of active plant tissue. He then monitored the rates of recovery towards equilibrium, and plotted these as a function of the energy input per unit standing crop of living tissue (Figure 22.17).

The pond system, with a relatively low-standing crop and a high rate of biomass turnover, was the most resilient to the standard perturbation. Most of its populations doubtless have short lives and rapid rates of population increase. The salt-marsh and forests had intermediate values, whilst tundra had the slowest rate of recovery and, hence, the lowest resilience. There is a clear relationship between resilience and energy input per unit standard crop. This seems to depend in part on the relative importance of heterotrophs in the system. The most resilient system, the pond, has a biomass of heterotrophs 5.4 times that of autotrophs (reflecting the short life and rapid turnover of phytoplankton, the dominant plants in this system), whilst the least resilient tundra has a heterotroph : autotroph ratio of only 0.004.

resilience seems to depend on the rate of flux through the community …

Thus, the flux of energy through the system has an important influence on resilience. The higher this flux, the more quickly will the effects of a perturbation be 'flushed' from the system. An exactly analogous conclusion has been reached by DeAngelis (1980), but for nutrient cycling rather than energy flow.

… and on the nutrient concerned

A striking pattern has emerged from studies based on residence times of various nutrients in different components of woodland communities. Table 22.3 shows the residence times of nitrogen and calcium in four components: soil, forest biomass, litter and detritivorous heterotrophic organisms. For example, a unit of nitrogen will typically exist in the soil for an average of 109 years, be taken up into forest biomass for 88 years, drop to the litter and remain there for up to 5 years, pass through a detritivore in a matter of days, only to be taken up into the forest biomass for a second time. Nitrogen, because of its tight recycling, tends to reside in the system for a very long time (1815 years) before being leached out and lost. Calcium is also a long-term resident (445 years). Both of these contrast with less tightly cycled elements, such as caesium.

Models involving both calcium and caesium, and based on the structure and measured flux rates in a Puerto Rican tropical rainforest, were subjected to a

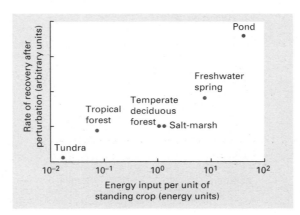

Figure 22.17 The rate of recovery (index of resilience) after perturbation (as a function of energy input per unit standing crop) for models of six contrasting communities. The pond community was most resilient to perturbation, tundra least so. (After O'Neill, 1976.)

Table 22.3 Estimated residence times, in years, of a unit of nitrogen and of calcium in four compartments, and in total, in temperate deciduous forest. (After Reichle, 1970.)

Component	Nitrogen	Calcium
Soil	109	32
Forest biomass	88	8
Litter	< 5	< 5
Decomposers	0.02	0.02
Total	1815	445

perturbation consisting of a doubling of the steady-state nutrient input for 10 years. Their return to equilibrium was then simulated for a further 30 years (Jordan *et al.*, 1972). The extra caesium in the system was quickly flushed out (Figure 22.18). This conclusion has received validation from a study of the half-life of radioactive caesium-137 (derived from fallout associated with the testing of nuclear bombs), which was essentially lost from the recycling system of a tropical rainforest as soon as the leaves fell (Kline, 1970). By contrast with caesium, the calcium model showed much less resilience (see Figure 22.17).

Self-contained, well-studied ecosystems such as lakes are ideal for the investigation of stability, resilience and resistance. Thus, Carpenter *et al.* (1992a) assessed the fluxes of phosphorus through the community of Tuesday Lake in Wisconsin, USA, in 1984 when planktivorous minnows were at the top of the food web (together with predatory invertebrates, zooplankton, phytoplankton and other suspended organic matter), and again in 1986 when piscivorous largemouth bass had been added to the system. They modelled the cycling of phosphorus under both circumstances and found that the short food-chain system was more resilient (faster rate of return to equilibrium phosphorus level) than its long food-chain counterpart. Intriguingly, this non-demographic outcome is just what Pimm and Lawton (1980) predicted from their demographic model (see Figure 22.11).

short food chains more resilient to nutrient perturbation?

Detritus is expected to play an important role in nutrient dynamics and has implications for system stability. In particular, a large detrital compartment can

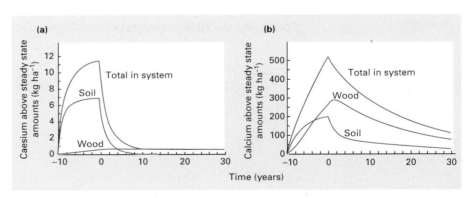

Figure 22.18 (a) Response of the caesium content components of a model of the Puerto Rican tropical rainforest during 30 years following a perturbation (for 10 years) involving a doubling of the steady-state input rate of this element. (b) Response of the calcium content when subjected to the equivalent perturbation. (After Jordan *et al.*, 1972.)

859   FOOD WEBS

buffer the living components of the system against large fluctuations in nutrient availability (DeAngelis, 1992). Such resistance has often frustrated efforts to restore shallow nutrient-enriched lakes by cutting their external phosphorus loads; high concentrations are maintained in the water column because of resuspension from a large detrital pool (Carpenter *et al.*, 1992b). This effect is similar to that of large soil nutrient pools in terrestrial systems.

Somewhat surprisingly, fish may have a bottom-up stabilizing effect with respect to phosphorus similar to that of the detritus pool, because fish constitute a large nutrient pool that turns over much more slowly than the other living compartments (through excretion, defaecation and decomposition after death). Paradoxically, as we saw in Section 22.3.2, fish can also exert a profound demographic effect through a top-down influence on the rest of the food web.

22.7 Finale

In essence, the top-down view is based primarily on population dynamics, whereas the bottom-up view rests primarily on an ecosystem perspective. Undeniably, the productivity of individuals, populations or trophic levels will generally be largely determined by the food resources available. Where nutrients and light are more abundant, other things being equal, the plant trophic level will produce biomass at a faster rate. Similarly, where primary productivity is higher, the rate of production of biomass of herbivores can be expected to reflect this, and so on up the chain. However, whether or not the rate of production of biomass is reflected in the density of organisms (the domain of population dynamics) depends not only on their pattern of production but also the rate of their consumption by predators. Bottom-up control will commonly be seen as predominant if our view of the food web is defined in terms of productivity at each trophic level. But, when we focus on numbers or biomass present at an instant, or averaged over a period of time, top-down forces will often appear to predominate. We find common ground in the top-down/bottom-up debate when the arguments are not limited to demographic or energetic considerations but when both are given due weight.

It is clear, in short, that food-web dynamics and community stability are complex topics. The challenge for the future is to develop a full understanding of both demographic and non-demographic aspects and of the way these interact to determine the fundamental properties of communities.

Side notes (left margin):

compartments with slow turnover may increase resistance to nutrient perturbation ... detritus

... and fish in lakes

top-down and bottom-up control revisited

bringing together demographic and non-demographic arguments

Chapter 23
Islands, Areas and Colonization

23.1 Introduction: species–area relationships

It has long been recognized that islands contain fewer species than apparently comparable pieces of mainland. It is also well established that the number of species on islands decreases as island area decreases. Such *species–area* relationships can be seen in Figure 23.1 for a variety of groups of organisms on a variety of islands. They are usually displayed with both species number and area plotted on log scales, although plotting species number itself against log area sometimes gives the best straight line (fully discussed by Williamson, 1981). In all cases, however, an arithmetic plot of species number against area is curved, with the number of species increasing more slowly at larger areas (e.g. Figure 23.1c).

'Islands' need not mean islands of land in a sea of water. Lakes are islands in a 'sea' of land; mountain tops are high-altitude islands in a low-altitude ocean; gaps in a forest canopy where a tree has fallen (see Chapter 21, Section 21.5.2) are islands in a sea of trees; and there can be islands of particular geological types, soil types or vegetation types surrounded by dissimilar types of rock, soil or vegetation. Species–area relationships can be equally apparent for these types of islands (Figure 23.2).

In fact, species–area relationships are not restricted to islands at all. They can also be seen when the numbers of species occupying different-sized arbitrary areas of the same geographical region are compared (Figure 23.3). The question therefore arises: is the impoverishment of species on islands more than would be expected in comparably small areas of mainland? In other words, does the characteristic isolation of islands contribute to their impoverishment of species? If the answer to these questions is 'yes', another question follows: can we understand how the degree of impoverishment is related to the degree of isolation or any other quality of the island?

These are important questions for an understanding of community structure. There are many oceanic islands, many lakes, many mountain tops, many woodlands surrounded by fields and many isolated trees. On a smaller scale, an individual animal or a leaf is an island from the point of view of its parasites (see Chapter 12). Equally, small patches in a heterogeneous habitat are islands from the point of view of the organisms colonizing them (see Chapter 21); the patch-dynamics concept of community structure (see Chapter 21, Section 21.4), metapopulation theory (see Chapter 15, Section 15.6) and island theory have much in common. In short, there can be few natural communities lacking at least some element of 'islandness'. Hence, we cannot hope to understand community structure without an understanding of

species–area relationships on oceanic islands ...

habitat islands and areas of mainland

'island effects' and community structure

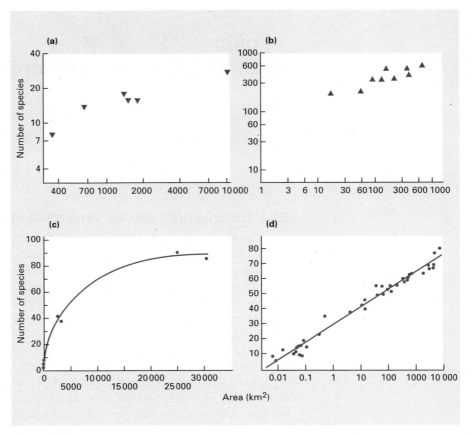

Figure 23.1 Species–area relationships for organisms on land islands. (a) Native land and freshwater birds on Hawaiian islands. (b) Vascular plants on the Azores. (After Eriksson *et al.*, 1974.) (c) Amphibians and reptiles on West Indian islands. (After MacArthur & Wilson, 1967.) (d) Birds on the Solomon Islands. (After Diamond & May, 1976.)

island biogeography. In addition, as we shall see in Chapter 25, island biogeography is of considerable potential importance for an enlightened approach to nature conservation.

In this chapter we examine three approaches to the understanding of island biogeography. These have sometimes been treated as alternative explanations for the species richness of island communities, but they are really complementary. The first concentrates on the suitability of islands as habitats for various species ('habitat diversity'—see Sections 23.2.1 and 23.2.2). The second concentrates on the balance between the rate at which islands are colonized by species new to the island and the rate at which resident species go extinct on the island ('the equilibrium theory'—see Sections 23.2.3 and 23.2.4). The third takes a more evolutionary approach, and considers the balance between colonization from outside the island and evolution within it (see Section 23.4).

These three approaches will be followed for islands of various types and, where appropriate, for arbitrarily defined areas of mainland that are not islands. The approaches will also be applied, in a slightly different way, to one particular group of

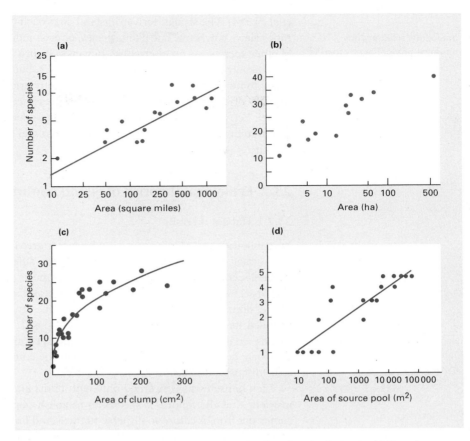

Figure 23.2 Species–area relationships for islands other than land islands. (a) Boreal mammals living on the montane islands of the Great Basin of North America. (After Brown, 1971.) (b) Breeding birds on isolated woodlots in Illinois. (After Blake & Karr, 1987.) (c) Invertebrates inhabiting clumps of intertidal mussels (*Brachidontes rostratus*) off the south coast of Victoria, Australia. (After Peake & Quinn, 1993.) (d) Fish living in Australian desert springs having source pools of different sizes. (After Kodric-Brown & Brown, 1993.)

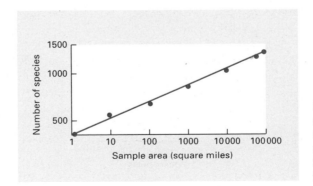

Figure 23.3 The species–area curve for the number of flowering plants found in different-sized sample areas of England. (After Williams, 1964; from Gorman, 1979.)

organisms, namely communities of insects feeding on the living tissues of higher plants. These comprise an estimated 35% of all animal species, and they have received careful and considerable attention from ecologists (fully reviewed by Strong

863   ISLANDS, AREAS AND COLONIZATION

et al., 1984). The straightforward 'island' approach to phytophagous insect communities views patches of one plant species, or even individual plants, as islands in a sea of other species. As an alternative, however, Figure 23.4 provides examples that are superficially like those in Figures 23.1 and 23.2, but in this case the 'area' of a plant refers to how widespread it is (i.e. in how many grid squares on a map it is found). 'Islands of differing size' are therefore represented by 'plant species of differing range'. Yet, as we shall see, the searches for an understanding of the structures of the communities of animals on such plants, and on real islands, run remarkably parallel courses.

23.2 Ecological theories of island communities

23.2.1 Habitat diversity

Probably the most obvious reason why larger areas should contain more species is that larger areas typically encompass more different types of habitat. In the particular context of island biogeography, the major proponent of this view was Lack (1969b, 1976; discussed by Williamson, 1981). Lack argued that the number of species found on an island simply reflects the 'type' of island, within which he included its climate and the habitats it provides. He believed essentially that large islands contain more species because they contain more habitats. This must obviously be true to some degree, although the conjecture has received few critical tests (discussed in Section 23.3.1).

Lack himself was concerned only with birds, and this is important since he was insistent that the failure of species to establish populations on a particular island comes not from a failure to disperse to the island but from a failure to find the right habitat. The major problem with Lack's bird-dominated view, therefore, is that it takes no account of the rather low probability that many other kinds of organisms can disperse to all suitable islands. Even amongst birds, it takes no account of the great differences in dispersal ability between species. In addition, this theory is a purely ecological rather than an evolutionary one. It does not concern itself with the

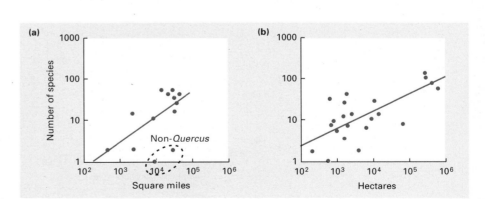

Figure 23.4 Species–area relationships for phytophagous insects on their host plants—the areas are actually the ranges of different plant species. (a) Cynipid gall wasps on Californian oaks—all oaks are species in the genus *Quercus* except those indicated. (After Cornell & Washburn, 1979.) (b) Insect pests on cacao—each point is for a different country. (After Strong, 1974.)

extent to which the community on an island may reflect evolution that has occurred *within* the island itself (cf. Section 23.4).

23.2.2 Habitat diversity and phytophagous insects

two ways in which plants can vary in their habitat diversity

There are two ways in which the habitat diversity argument can be applied to species–area relationships for insects on plants. The first, related directly to data sets like those in Figure 23.4, proposes simply that widely distributed plants themselves live in a wide variety of habitats. Thus, they in turn offer a wide variety of habitats to insects, since the habitat of the insect includes not only the presence of the plant itself, but also climatic conditions, the presence of other plants, and so on.

The second way of applying the arguments concentrates on the size, the structure, the variety of parts or the 'architecture' of different plant species: 'complex' plant species might be expected to offer a wider range of conditions and resources and so support more insect species than 'simple' ones.

23.2.3 MacArthur and Wilson's 'equilibrium' theory

The essence of MacArthur and Wilson's (1967) 'equilibrium theory of island biogeography' is very simple. It is that the number of species on an island is determined by a balance between immigration and extinction, and that this balance is dynamic, with species continually becoming extinct and being replaced (through immigration) by the same or by different species.

The theory is depicted in Figure 23.5. Taking immigration first, imagine an island that as yet contains no species at all. The rate of immigration of *species* will be high, because any colonizing individual represents a species new to that island. However, as the number of resident species rises, the rate of immigration of new, unrepre-

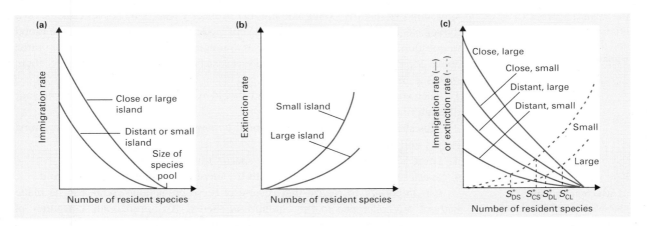

Figure 23.5 MacArthur and Wilson's (1976) equilibrium theory of island biogeography. (a) The rate of species immigration on to an island, plotted against the number of resident species on the island, for large and small islands and for close and distant islands. (b) The rate of species extinction on an island, plotted against the number of resident species on the island, for large and small islands. (c) The balance between immigration and extinction on small and large, and on close and distant islands. In each case, S^* is the equilibrium species richness. C, close; D, distant; L, large; S, small.

sented species diminishes. The immigration rate reaches zero when all species from the 'source pool' (i.e. from the mainland or from other nearby islands) are present on the island in question (Figure 23.5a).

The immigration graph is drawn as a curve, because immigration rate is likely to be particularly high when there are low numbers of residents and many of the species with the greatest powers of dispersal are yet to arrive. In fact, the curve should really be a blur rather than a single line, since the precise curve will depend on the exact sequence in which species arrive, and this will vary by chance. In this sense, the immigration curve can be thought of as the 'most probable' curve.

The exact immigration curve will depend on the degree of remoteness of the island from its pool of potential colonizers (Figure 23.5a). The curve will always reach zero at the same point (when all members of the pool are resident); but, it will generally have higher values on islands close to the source of immigration than on more remote islands, since colonizers have a greater chance of reaching an island the closer it is to the source. It is also likely (although not specified in the original formulation of MacArthur and Wilson's theory) that immigration rates will generally be higher on a large island than on a small island, since the larger island represents a larger 'target' for the colonizers (Figure 23.5a).

The rate of species extinction on an island (Figure 23.5b) is bound to be zero when there are no species there, and it will generally be low when there are few species. However, as the number of resident species rises, the extinction rate is assumed by the theory to increase, probably at a more than proportionate rate. This is thought to occur because with more species, competitive exclusion becomes more likely, and the population size of each species is on average smaller, making it more vulnerable to chance extinction. Similar reasoning suggests that extinction rates should be higher on small than on large islands—population sizes will typically be smaller on small islands (Figure 23.5b). As with immigration, the extinction curves are best seen as 'most probable' curves.

the balance between
immigration and extinction,
and the predictions of the
equilibrium theory ...

In order to see the net effect of immigration and extinction, the two curves can be superimposed (Figure 23.5c). The number of species where the curves cross (S^*) is a dynamic equilibrium, and should be the characteristic species richness for the island in question. Below S^*, richness increases (immigration rate exceeds extinction rate); above S^*, richness decreases (extinction exceeds immigration). The theory, then, makes a number of predictions.

1 The number of species on an island should eventually become roughly constant through time.

2 This should be a result of a continual *turnover* of species, with some becoming extinct and others immigrating.

3 Large islands should support more species than small islands.

4 Species number should decline with the increasing remoteness of an island.

... which are not all
characteristic of this theory
alone

It is important to realize that several of these predictions could also be made without any reference to the equilibrium theory. An approximate constancy of species number would be expected if richness were determined simply by island type (see Section 23.2.1). Similarly, a higher richness on larger islands would be expected as a consequence of larger islands having more habitat types. One test of the equilibrium theory, therefore, would be whether richness increases with area at a rate greater than could be accounted for by increases in habitat diversity alone.

The effect of island remoteness can be considered quite separately from the equilibrium theory. Merely recognizing that many species are limited in their dispersal ability, and have not yet colonized all islands, leads to the prediction that more remote islands are less likely to be saturated with potential colonizers. However, the final prediction arising from the equilibrium theory—constancy as a result of turnover—is truly characteristic of the equilibrium theory.

Comparing the equilibrium theory and the habitat diversity theory, the equilibrium theory deals mostly with *numbers* of species and lays relatively little stress on the nature of species' requirements. It does, though, recognize that dispersal abilities are generally limited and vary from species to species. Like the habitat diversity theory, it pays no attention to evolution.

23.2.4 The equilibrium theory and phytophagous insects

Applying the equilibrium theory to insects on plants was suggested first by Janzen (1968). Relatively widespread plants can be seen as comparatively 'large' islands in a sea of other vegetation. Indeed, Southwood (1961) had previously proposed that more widespread plants represented a larger target for potential colonizing insects. In addition, a plant species may be seen as 'remote' from other species if it is morphologically, biochemically or otherwise biologically unusual. Thus, the equilibrium theory predicts that insect species richness will be higher for plants with large ranges, lower for plants that are geographically isolated or rare and lower for plants that are morphologically or biochemically 'isolated'. However, the problems of disentangling the predictions of the equilibrium theory from those of other theories are the same here as they are for islands generally. In particular, widespread species are not only 'large islands', they also occupy a wide diversity of habitats.

23.3 Evidence for the ecological theories

23.3.1 Habitat diversity alone—or a separate effect of area?

The most fundamental question in island biogeography is whether there is an 'island effect' as such, or whether islands simply support few species because they are small areas containing few habitats.

an example where habitat diversity is paramount

At one extreme are examples where the diversity of available habitats is paramount. In the case of fish occupying desert springs in the Dalhousie Basin in South Australia (see Figure 23.2d), species composition was highly predictable with each of five species only occurring in springs above a characteristic threshold size. Springs with the smallest source pools predictably contained only the tiny Dalhousie goby (*Chlamydogobius* sp.), which inhabits extremely shallow water. At the other end of the spectrum, the large, predatory spangled perch (*Leiopotherapon unicolor*), which requires a deep, open-water situation, was found only in the largest springs, where it co-occurred with the goby and each of three intermediate species (Kodric-Brown & Brown, 1993). In this case, the relationship between species richness and spring area was due entirely to a strong correlation between spring area and the number of specific fish habitat types present.

partitioning variation between habitat diversity and area itself

Some studies have attempted to partition species–area variation on islands into that which can be entirely accounted for in terms of habitat heterogeneity, and that which remains and must be accounted for by island area in its own right. In the case of species richness of beetles on the Canary Islands, the relationship with plant species richness (an important component of habitat diversity for beetles) is much

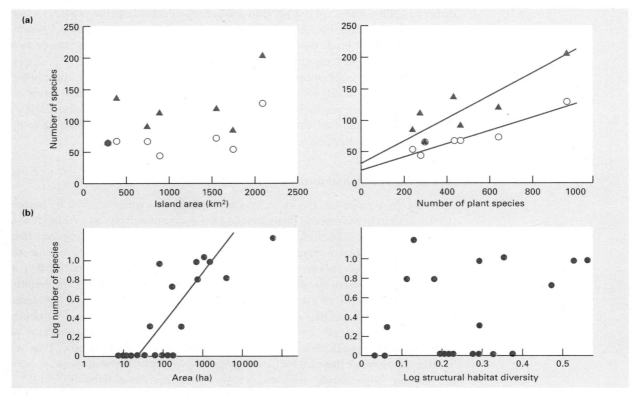

Figure 23.6 (a) The relationships between species richness (S) of herbivorous (○) and carnivorous (▲) beetles of the Canary Islands and both island area and plant species richness. (After Becker, 1992.) (b) The relationship between species richness (S) of birds on islands off the coast of Western Australia and both island area and structural habitat diversity (on log scales). (After Abbott, 1978.)

stronger than that with island area, and this is particularly marked for the herbivorous beetles, presumably because of their particular food-plant requirements (Figure 23.6a). At the opposite extreme, Abbott (1978) found a positive correlation between bird species richness and the area of islands off the coast of Western Australia, but he found no relationship between species richness and habitat diversity (Figure 23.6b). In fact, there was no relationship between island area and habitat diversity. In this case, area *per se* appears to play a very major role. In contrast, the relationship between species number, island area and habitat diversity for birds of the Aegean islands indicates that the diversity of habitats on an island was more important than the area of the island (Watson, 1964; described in Williamson, 1981).

An experiment was carried out to try to separate the effects of habitat diversity and area on some small mangrove islands in the Bay of Florida (Simberloff, 1976). These consist of pure stands of the mangrove species *Rhizophora mangle*, which support communities of insects, spiders, scorpions and isopods. After a preliminary faunal survey, some islands were reduced in size by means of a power saw and brute force! Habitat diversity was not affected, but arthropod species richness on three islands nonetheless diminished over a period of 2 years (Figure 23.7). A control

experimental reductions in the size of mangrove islands

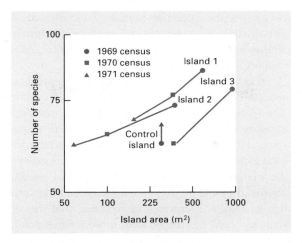

Figure 23.7 The effect on the number of arthropod species of artificially reducing the size of mangrove islands. Islands 1 and 2 were reduced in size after both the 1969 and 1970 censuses. Island 3 was reduced only after the 1969 census. The control island was not reduced, and the change in its species richness was attributable to random fluctuations. (After Simberloff, 1976.)

island, the size of which was unchanged, showed a slight *increase* in richness over the same period, presumably as a result of random events.

Some islands that were once part of a land-bridge to the mainland (sections of which have since become submerged in the ocean) provide data consistent with the equilibrium theory. If an equilibrium number of species is determined partly by a relationship between extinction rate and island area, these new islands would be expected to lose species until they reached a new equilibrium appropriate to their size. The process is usually referred to as 'relaxation' and several instances have been reported. For example, Figure 23.8 shows the relationship between the length of time for which islands in the Gulf of California have been separated from the mainland and the number of lizard species they support. Having made corrections for island area and latitude, analysis showed that the number of species present had fallen from 50–75 species to 25 or so during a period of 4000 years. Of course, such data must be interpreted with caution. Since such long time periods are involved, we need to be confident that major climatic changes have not played a role. Another

the 'relaxation' of richness on land-bridge islands

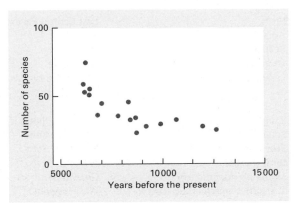

Figure 23.8 Species richness of lizards on former land-bridge islands in the Gulf of California, plotted against the length of time for which they have been isolated. (After Wilcox, 1978.)

concern is the influence humans have exerted, by destroying habitat or by the introduction of competitors and predators (Richman *et al.*, 1988). Nevertheless, these studies lend weight to the 'equilibrium' interpretation of island diversity (i.e. they support a separate effect of area).

Another way of trying to distinguish a separate effect of island area is to compare species–area graphs for islands with those for arbitrarily defined areas of mainland. The species–area relationships in the latter should be due almost entirely to habitat diversity alone. All species will be well able to 'disperse' between such areas, and the continual flow of individuals across the arbitrary boundaries will therefore mask local extinctions (i.e. what would be an extinction on an island is soon reversed by the exchange of individuals between local areas). An arbitrarily defined area of mainland should thus contain more species than an otherwise equivalent island, and this is usually interpreted as meaning that the slopes of species–area graphs for islands should be steeper than those for mainland areas (since the effect of island isolation should be most marked on small islands, where extinctions are most likely). The difference between the two types of graph would then be attributable to the island effect in its own right. Table 23.1 shows that despite considerable variation, the island graphs do typically have steeper slopes.

species–area relationships for islands are, on average, steeper than those for mainland areas

Table 23.1 Values of the slope z, of species–area curves ($\log S = \log C + z \log A$, where S is species richness, A is area and C is a constant giving the number of species when A has a value of 1), for (a) arbitrary areas of mainland, (b) oceanic islands and (c) habitat islands. (After Preston, 1962; May, 1975b; Gorman, 1979; Browne, 1981.)

Taxonomic group	Location	z
(a) *Arbitrary areas of mainland*		
Flowering plants	England	0.10
Land plants	Britain	0.16
Birds	Mediterranean	0.13
Birds	Neotropics	0.16
Birds	Neoarctic	0.12
Savannah vegetation	Brazil	0.14
(b) *Oceanic islands*		
Birds	New Britain Islands	0.18
Birds	New Guinea Islands	0.22
Birds	West Indies	0.24
Birds	East Indies	0.28
Birds	East Central Pacific	0.30
Ants	Melanesia	0.30
Beetles	West Indies	0.34
Land plants	Galapagos	0.31
Land plants	Californian islands	0.37
(c) *Habitat islands*		
Zooplankton (lakes)	New York State	0.17
Snails (lakes)	New York State	0.23
Fish (lakes)	New York State	0.24
Birds (Paramo vegetation)	Andes	0.29
Mammals (mountains)	Great Basin, USA	0.43
Terrestrial invertebrates (caves)	West Virginia	0.72

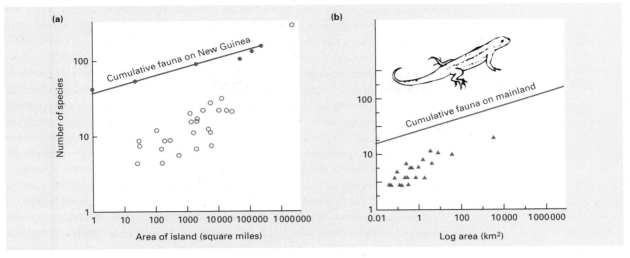

Figure 23.9 (a) The species–area graph for ponerine ants on various Moluccan and Melanesian islands compared with a graph for different-sized sample areas on the very large island of New Guinea. (After Wilson, 1961.) (b) The species–area graph for reptiles on islands off the coast of South Australia compared with the mainland species–area relationship. In this case, the islands were formed within the last 10 000 years as a result of rising sea-level. (After Richman *et al.*, 1988.)

Note that a reduced number of species per unit area on islands should also lead to a lower value for the intercept on the *S*-axis of the species–area graph. Figure 23.9a illustrates both an increased slope and a reduced value for the intercept for the species–area graph for ant species on isolated Pacific islands, compared with the graph for progressively smaller areas of the very large island of New Guinea. Figure 23.9b gives a similar relationship for reptiles on islands off the coast of South Australia.

Overall, it is clear that smaller areas typically contain a lower diversity of habitats and thus support fewer species. But, it is also apparent that there is often a recognizable island effect, reducing still further the species richness of communities that are isolated from other, similar communities.

23.3.2 Remoteness

It follows from the above argument that the island effect and the species impoverishment of an island should be greater for more remote islands. (Indeed, the comparison of islands with mainland areas is only an extreme example of a comparison of islands varying in remoteness, since local mainland areas can be thought of as having minimal remoteness.) Remoteness, however, can mean two things. First, it can simply refer to the degree of physical isolation. Alternatively, a single island can also itself vary in remoteness, depending on the type of organism being considered: the same island may be remote from the point of view of land mammals but not from the point of view of birds.

The effects of remoteness can be demonstrated either by plotting species richness against remoteness itself, or by comparing the species–area graphs of groups of islands (or for groups of organisms) that differ in their remoteness (or powers of

871 ISLANDS, AREAS AND COLONIZATION

bird species richness on Pacific
islands decreases with
remoteness

colonization). In either case, there can be considerable difficulty in extricating the effects of remoteness from all the other characteristics by which two islands may differ. Nevertheless, the direct effect of remoteness can be seen in Figure 23.10 for non-marine, lowland birds on tropical islands in the South-West Pacific. With increasing distance from the large source island of New Guinea, there is a decline in the number of species, expressed as a percentage of the number present on an island of similar area but close to New Guinea. Species richness decreases exponentially with distance, approximately halving every 2600 km. The species–area graphs in Figure 23.11a also show that remote islands of a given size possess fewer species than their counterparts close to a land mass. In addition, Figure 23.11b contrasts the species–area graphs of two classes of organisms in two regions: the relatively remote Azores (in the Atlantic, far to the west of Portugal) and the Channel Islands (close to the north coast of France). Whereas the Azores are indeed far more remote than the Channel Islands from the point of view of the birds, the two island groups are apparently equally remote for ferns, which are particularly good dispersers because of their light, wind-blown spores. Finally, it is noteworthy that the two steepest slopes in Table 23.1 relate to: (i) mammals from mountains surrounded by desert in the Great Basin of North America; and (ii) terrestrial invertebrates inhabiting limestone caves in West Virginia. In both cases, rates of migration are dramatically low, and the habitat islands are therefore extremely isolated from one another, irrespective of their distance apart. Thus, on the basis of all these examples, the species' impoverishment caused by the island effect does indeed appear to increase as the degree of isolation of the island increases.

A more direct effect on the species impoverishment of islands, especially remote islands, arises from the fact that many islands lack species that they could potentially support, simply because there has been insufficient time for the species to colonize the island. An example is provided by the island of Surtsey, which emerged in 1963 as a result of a volcanic eruption (Fridriksson, 1975). The new island, 40 km south-west of Iceland, was reached by bacteria and fungi, some seabirds, a fly and seeds of several beach plants within 6 months of the start of the eruption. Its first established vascular plant was recorded in 1965, and the first moss colony in 1967.

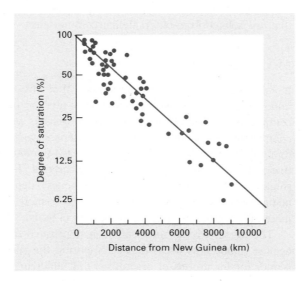

Figure 23.10 The number of resident, non-marine, lowland bird species on islands more than 500 km from the larger source island of New Guinea expressed as a proportion of the number of species on an island of equivalent area but close to New Guinea, and plotted as a function of island distance from New Guinea. (After Diamond, 1972.)

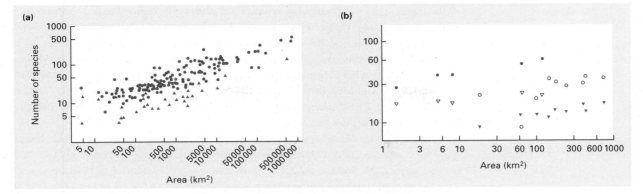

Figure 23.11 Remoteness increases the species impoverishment of islands. (a) A species–area plot for the land birds of individual islands in tropical and subtropical seas. ▲, islands more than 300 km from the next largest land mass or the very remote Hawaiian and Galapagos archipelagos; ●, islands less than 300 km from source. (After Williamson, 1981.) (b) Species–area plots in the Azores and the Channel Islands for land and freshwater breeding birds (▼, Azores; ●, Channel Islands) and for native ferns (○, Azores; ▽, Channel Islands). The Azores are more remote for birds but not for ferns. (After Williamson, 1981.)

By 1973, 13 species of vascular plant and more than 66 mosses had become established (Figure 23.12). Colonization is continuing still.

islands may lack species simply because there has been insufficient time for colonization

The general importance of this example is that the communities of many islands can be understood *neither* in terms of simple habitat suitability *nor* as a characteristic equilibrium richness. Rather, they stress that many island communities have not reached equilibrium and are certainly not fully 'saturated' with species.

23.3.3 Diversity, area and remoteness for phytophagous insects

widespread plants support many insects because they occupy a variety of habitats …

We can now retrace our steps, and look at species–area graphs, habitat diversity and remoteness in phytophagous insect communities. In the first place, widespread plant species undoubtedly owe much of their rich insect fauna to the high diversity of habitats within which they grow. In fact, there is an important parallel between the habitat diversity of plant species and the habitat diversity of islands generally. Many

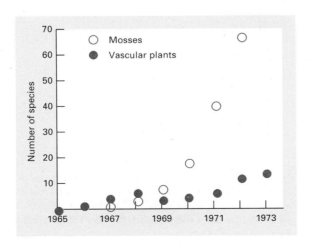

Figure 23.12 The number of species of mosses and vascular plants recorded on the new island of Surtsey from 1965 to 1973. (After Fridriksson, 1975.)

873   ISLANDS, AREAS AND COLONIZATION

of the insect species supported by widespread plants are confined to only a small part of their host-plant's range. Thus, a number of studies have shown that widespread and common plant species do not necessarily support more insect species at a particular locality even though they support more species overall (Claridge & Wilson, 1976, 1978; Futuyma & Gould, 1979; Karban & Ricklefs, 1983). This argues against a separate effect of area in addition to the simple effect of habitat diversity. But, it also means that a *particular* community of phytophagous insects will not necessarily be richer just because it exists on a widespread plant.

... but there is also evidence for a separate 'area effect'

Better evidence for an area effect comes from the study of phytophagous insect communities associated with the bracken fern (*Pteridium aquilinum*), already touched on in Chapter 20, Section 20.2.1. Communities were compared at both open and wooded sites in northern England and New Mexico, at a wooded site in Papua New Guinea and an open site in Hawaii. Bracken occurs naturally at all these locations. A total of 27 species feed on bracken in Great Britain, with another eight possibly or occasionally doing so. Approximately 30 species feed on it in Papua New Guinea, only five species in New Mexico and only one or two in Hawaii. The niches occupied by each species were described both in terms of the method of feeding (chewers, suckers, miners and gall-formers) and the part of the plant upon which the insects feed (main stem, pinnae (leaves), main stalks arising from the stem, main leaf-veins of the pinnae). The results, presented in Figure 20.1, indicate a large number of apparently unexploited niches in New Mexico. (Equivalent data for the two Hawaiian species have not been presented, but this community must have the greatest number of unexploited niches of all.)

Figure 23.13 gives the species–area graph for herbivorous insects on bracken from the four continents mentioned above, and also the relationship for New Zealand. 'Area' refers to an estimate of the area occupied by bracken in each case. Although based on only five points, the species–area relationship is statistically

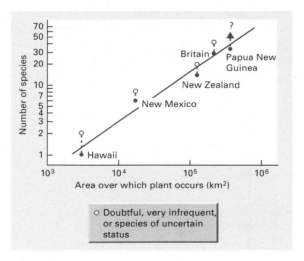

Figure 23.13 Species–area relationships for herbivorous insects on bracken. Those parts of the world where bracken grows more widely appear to support more species. The regression line is fitted to the solid dots (open circles include species which are infrequent or of uncertain host-plant status). (After Lawton, 1984; except for New Zealand which is after Winterbourn, 1987.)

where bracken is
comparatively rare, the
community of insects on it is
comparatively unsaturated

highly significant. In short, there seem to be few phytophagous insects on bracken in New Mexico and Hawaii because the plant itself is comparatively rare. Since the effect is apparent in comparisons of individual sites, it seems to be rarity itself that is important and not just habitat diversity.

The bracken study also indicates that the New Mexico and Hawaiian communities, at least, are limited in richness not by competitive exclusion but by exhaustion of the species pool (itself small because the plant is not widespread). This then is another example in which the island nature of a community plays an important part in ensuring that it is far from fully saturated with species.

The effect of plant *structural* diversity on phytophagous insect communities can be seen in Figures 23.14a and b. From Figure 23.14a it is clear that, comparing plants with similar geographical ranges, trees tend to support more insect species than shrubs, which support more species than herbs. This is a sequence of plant types showing a decrease in size, a decrease in the variety of microclimates offered, a decrease in the number of modules (e.g. leaves) per individual, a decrease in the variety of plant parts and a decrease in the range of resources available to insects. Figure 23.14b plots the number of phytophagous insect species known to attack different species of American *Opuntia* cacti against the 'architectural rating' of each cactus species. Cacti with a high rating (larger, more modules and a greater variety of plant parts) clearly support more insect species.

Plant architecture obviously has a major influence on the richness of the phytophagous insect community, but, as with species on islands, it is difficult to

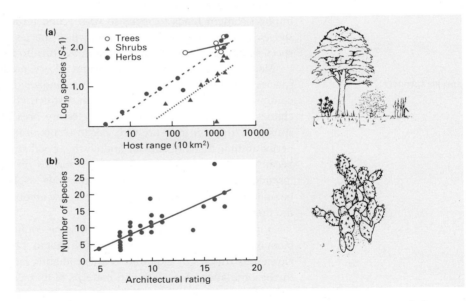

Figure 23.14 The species richness of phytophagous insects on plants increases with the complexity of 'plant architecture'. (a) The 'species–area' relationships for genera of British trees, shrubs and herbs in the family Rosaceae; host range is expressed in terms of the number of 10-km map grid squares in which the plant is found. (After Leather, 1986.) (b) The number of species associated with *Opuntia* cactus species in North and South America as a function of their 'architectural rating'. Architectural rating is the sum of five variables each measured on a scale from 1 to 4: height of mature plant, mean number of cladodes on a large plant, cladode size, degree of development of woody stems and the cladode complexity from smooth to strongly tuberculate. (After Moran, 1980.)

the diversity of plant
architecture influences the
richness of phytophagous
insects

decide how much is due simply to size (including the *number* of modules) and how much is due to environmental and resource heterogeneity. In the case of the cacti, size alone accounted for 35% of the variance in richness, whereas all components of plant architecture together accounted for 69% (Moran, 1980). However, the cacti cover a size range much smaller than the tree–herb contrast in Figure 23.14a, and a partitioning between size and heterogeneity in these cases awaits further work.

The influence of taxonomic or biochemical 'remoteness' on the species impoverishment of plants has not always been apparent in comparisons of different species, but some studies suggest such an effect. For instance, two 'non-*Quercus*' oaks in California supported fewer gall-forming cynipid wasps than might be expected from the data on the more numerous *Quercus*-oak species (see Figure 23.4b); whilst British monocotyledonous plant species with fewer other species in the same genus supported fewer species of insect than otherwise expected (Lawton & Schröder, 1977).

In summary then, a greater number of phytophagous insect species are to be found on plant species that are larger and/or with a more complex architecture, on common and widespread plants, and perhaps on plants that live in the same area as related species.

23.3.4 Which species? Turnover

So far, we have seen that smaller islands support fewer species, partly but not entirely because of their lower habitat diversity; and we have seen that this species impoverishment tends to increase with island remoteness. But, this emphasis on *species number* deflects attention from *which* species live on different islands, *which* species immigrate and *which* go extinct. We turn to these important questions next.

MacArthur and Wilson's equilibrium theory predicts not only a characteristic species richness for an island, but also a *turnover* of species in which new species continually colonize whilst others become extinct. This implies a significant degree of chance regarding precisely which species are present at any one time. However, studies of turnover itself are rare, because communities have to be followed over a period of time (usually difficult and costly). Good studies of turnover are rarer still, because it is necessary to count every species on every occasion so as to avoid 'pseudo-immigrations' and 'pseudo-extinctions'. Note that any results are bound to be underestimates of actual turnover, because an observer cannot be everywhere all the time.

the turnover of species on
islands lends indeterminacy to
community structure

A study that can be examined with some confidence concerns the build-up of biota on Rakata Island, the remnant of Krakatau which erupted massively in 1883. Numerous surveys by explorers of this romantic landscape have documented both the accumulation of species, and patterns of immigration and extinction (the latter defined as the absence of a previously recorded species from two successive surveys of suitable habitat). Figure 23.15 presents data for vascular plants, birds and butterflies; in no case are the idealized smooth patterns of change observed, but immigration rate has a tendency to decline and extinction rate to increase at high species richness. The rate of immigration of vascular plants fell initially, as expected by theory, but then rose to fall again. The rise coincided with forest formation, and was paralleled by a rise for birds and butterflies. The plant immigration curve may in fact be two curves superimposed: (i) an early successional phase of colonization, in which the pool of potential pioneer species quickly becomes depleted; and (ii) a

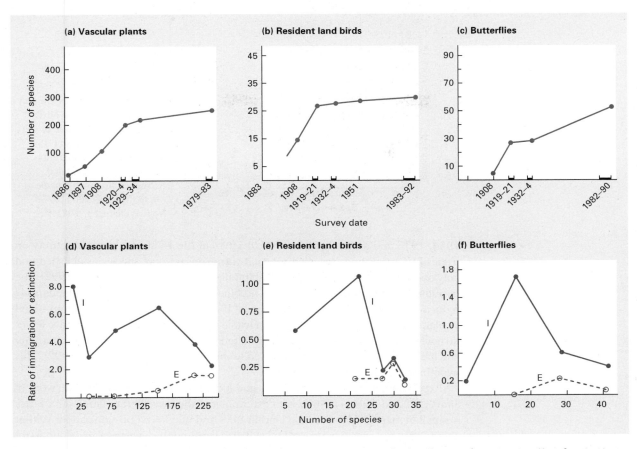

Figure 23.15 Species accumulation curves (a–c) and rates of immigration (I) and extinction (E) as species per year for intersurvey periods (d–f) for vascular plants, resident land birds and butterflies on Rakata Island between 1883 and 1992. (After Thornton *et al.*, 1993; plant data from Whittaker *et al.*, 1989; butterfly data from Bush & Whittaker, 1991.)

subsequent phase of late successional colonizers, whose potential pool becomes depleted more slowly (Thornton *et al.*, 1993). It seems that a full understanding of the Krakatau story requires the application both of island biogeography and successional theory.

Another revealing study involved censuses from 1949 to 1975 of the breeding birds in a small oak wood (Eastern Wood) in southern England. In all, 44 species bred in the wood over this period, and 16 of them bred every year. The number breeding in any one year varied between 27 and 36, with an average of 32 species. The immigration and extinction 'curves' are shown in Figure 23.16. Their most obvious feature is the scattering of points compared to the assumed simplicity of the MacArthur–Wilson model (see Figure 23.5). Nevertheless, whilst the positive correlation in the extinction graph is statistically insignificant, the negative correlation in the immigration graph is highly significant; and the two lines do seem to cross at roughly 32 species, with three new immigrants and three extinctions each year. There is clearly a considerable turnover of species, and consequently considerable year to year variation in the bird community of Eastern Wood despite its approximately constant species richness. In contrast, a long-term study (surveys in

877 ISLANDS, AREAS AND COLONIZATION

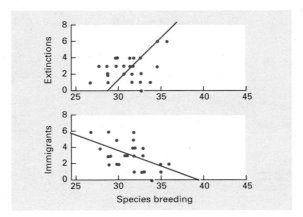

Figure 23.16 Immigration and extinction of breeding birds at Eastern Wood. The line in the extinction diagram is at 45°. The line in the immigration diagram is the calculated regression line with a slope of −0.38. (After Beven, 1976; from Williamson, 1981.)

1954, 1976 and annually from 1984 to 1990) of the 15-strong bird community on tropical Guana Island, revealed no such turnover—no new species established and only one went extinct, as a result of habitat destruction (Mayer & Chipley, 1992). The position of Guana Island within an archipelago of numerous small islands may reduce the likelihood of local extinctions if there is continuous dispersal from island to island. On the other hand, it is conceivable that tropical birds really do have lower turnover rates—because they are more often sedentary, have lower adult mortality and are more often resident, as opposed to migratory (Mayer & Chipley, 1992). This hypothesis deserves further attention.

Experimental evidence of turnover and indeterminacy is provided by the work of Simberloff and Wilson (1969), who exterminated the invertebrate fauna on a series of small mangrove islands in the Florida Keys and monitored recolonization. Within about 200 days, species richness had stabilized around the level prior to defaunation, but with many differences in species composition. Since then, the rate of turnover of species on the islands has been estimated as 1.5 extinctions and colonizations per year (Simberloff, 1976).

Thus, the idea that there is a turnover of species leading to a characteristic equilibrium richness on islands—but, an indeterminacy regarding particular species—appears to be correct, at least approximately.

23.3.5 Which species? Disharmony

It has long been recognized (J. Hooker, 1866; discussed by Turrill, 1964) that one of the main characteristics of island biotas is 'disharmony', by which it is meant that the relative proportions of different taxa are not the same on islands as they are on the mainland. We have already seen from the species–area relationships in Figure 23.11 that groups of organisms with good powers of dispersal (like ferns and, to a lesser extent, birds) are more likely to colonize remote islands than are groups with relatively poor powers of dispersal (most mammals); even within groups, powers of dispersal vary. For example, Figure 23.17 shows the eastern, seaward limits in the Pacific of various groups of land and freshwater birds found in New Guinea. On a more anecdotal level, most of the species of land snail on Pacific islands are very small, and so easily transported (Vagvolgyi, 1975), and most of the beetles on the island of St Helena are wood-borers or bark-clingers that are most likely to have been carried across the sea on floating trees (MacArthur & Wilson, 1967).

some groups, and some species, are better suited than others to reaching islands and persisting on them

878 CHAPTER 23

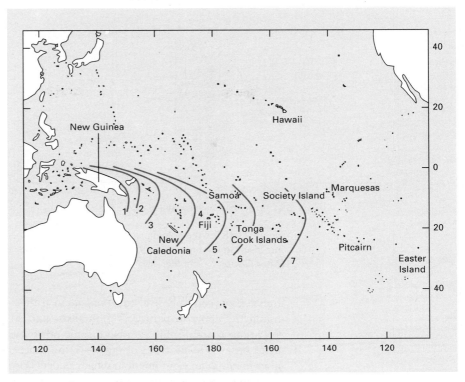

Figure 23.17 Eastern limits of families and subfamilies of land and freshwater breeding birds found in New Guinea. The decline in taxa is fairly smooth, and shows both differences in dispersal ability and a general decline in island size to the east. (1) Not beyond New Guinea: pelicans, storks, larks, pipits, birds of paradise and nine others. (2) Not beyond New Britain and the Bismark Islands: cassowaries, quails and pheasants. (3) Not beyond the Solomon Islands: owls, rollers, hornbills, drongos and six others. (4) Not beyond Vanuatu and New Caledonia: grebes, cormorants, ospreys, crows and three others. (5) Not beyond Fiji and Niufo'ou: hawks, falcons, turkeys and wood swallows. (6) Not beyond Tonga and Samoa: ducks, thrushes, waxbills and four others. (7) Not beyond the Cook and Society Islands: barn owls, swallows and starlings. Beyond (7), the Marquesas and Pitcairn group: herons, rails, pigeons, parrots, cuckoos, swifts, kingfishers, warblers and flycatchers. (After Firth & Davidson, 1945; from Williamson, 1981.)

species may differ in their risk of extinction …

However, variation in dispersal ability is not the only factor leading to disharmony. Species may vary in their risk of extinction. Thus, herbivorous beetles, by virtue of their greater dietary specialization, may be more prone to extinction than generalist carnivorous beetles—certainly, carnivorous beetles are relatively richer in species than their herbivorous counterparts on islands, when compared with their proportions on nearby mainlands (Becker, 1992; see also Figure 23.6a). Moreover, species that naturally have low densities per unit area are bound to have only small populations on islands, and a chance fluctuation in a small population is quite likely to eliminate it altogether. Vertebrate predators, which generally have relatively small populations, are notable for their absence on many islands. For example, the birds on the Atlantic island of Tristan da Cunha have no bird, mammal or reptile predators apart from those released by humans.

Specialist predators are also liable to be absent from islands because their

879   ISLANDS, AREAS AND COLONIZATION

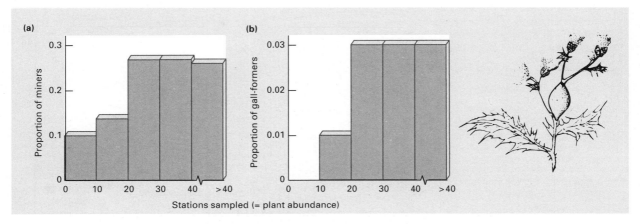

Figure 23.18 As the abundance of host plants within the European Cynareae (thistles) increases, so the proportion of (a) leaf- and stem-miners, and (b) gall-formers also increases. (After Lawton & Schröder, 1977.)

... or likelihood of establishment

incidence functions and assembly rules emphasize that island communities are characterized by more than just species richness

immigration can only lead to colonization if their prey have arrived first. Similar arguments apply to parasites, mutualists, and so on. In other words, for many species an island is only suitable if some other species is present, and disharmony arises because some types of organism are more 'dependent' than others. On the other hand, in an experiment in which spiders (*Metepeira datona*) were introduced to small Bahamanian islands, the likelihood of establishment of a population was significantly reduced if predatory lizards were already present (Schoener, 1986).

The development by Diamond (1975) of *incidence functions* and *assembly rules* for the birds of the Bismark archipelago (see Chapter 20, Section 20.3.2) is probably the fullest attempt to understand island communities by combining ideas on dispersal and extinction differentials with those on sequences of arrival and habitat suitability. Constructing such incidence functions (see Chapter 20, Figure 20.7) allowed Diamond to contrast 'supertramp' species (high rates of dispersal but a poorly developed ability to persist in communities with many other species), with 'high-*S*' species (only able to persist on large islands with many other species), and to contrast these in turn with intermediate categories. Faaborg (1976) made similar detailed studies of the birds on West Indian islands. Such work illustrates particularly clearly that it takes far more than a count of the number of species present to characterize the community of an island.

In phytophagous insect communities, an unusually large proportion of the species that colonize a newly introduced plant are polyphagous, and chewing and sucking insects, feeding externally on the plant, are more likely to colonize new hosts than are leaf-miners and gall-formers (Strong *et al.*, 1984). It is particularly pertinent that at least within the European Cynareae (the thistles and their relatives), the proportion of miners and gall-formers is higher on more widespread species (Figure 23.18; Lawton & Schröder, 1977). Here too, therefore, 'small islands' (plants with small ranges) support a community that is disharmonic compared to those of larger 'islands'.

Island communities then are not merely impoverished—the impoverishment affects particular types of organism disproportionately. Those with poor dispersal ability, a high probability of extinction and/or a dependence on the prior arrival of

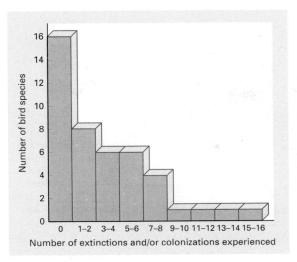

Figure 23.19 In Eastern Wood in southern England many species never become extinct, but others become extinct and recolonize repeatedly. The frequency histogram shows the number of extinctions plus recolonizations over 26 years. The average number per species of bird is 3.4. (After Beven, 1976.)

some other species all tend to be relatively rare. Large predators are therefore especially affected, and this is likely to have repercussions throughout the whole community (see Chapter 22). Moreover, although turnover introduces an element of chance into the community structure of an island, this tends to affect only a proportion of the species, whilst a core of other species remain unaffected. This can be seen in Figure 23.19 for the birds of Eastern Wood.

23.4 Evolution and island communities

rates of evolution on islands can be faster than rates of colonization

No aspect of ecology can be fully understood without reference to evolutionary processes taking place over evolutionary time scales, and this is particularly true for an understanding of island communities. On isolated islands, the rate at which new species evolve may be comparable with or even faster than the rate at which they arrive as new colonists. Clearly, the communities of many islands will be incompletely understood by reference only to ecological processes.

To begin with an extreme example, the remarkable numbers of species of *Drosophila* found on the remote Hawaiian islands and described in Chapter 1 (see Figure 1.8) have evolved, almost entirely, on the islands themselves. The communities of which they are a part are clearly much more strongly affected by evolution and speciation *in situ* than by the processes of invasion and extinction described in previous sections of this chapter.

endemism is more likely on remoter islands and within groups with poorer dispersal

A more modest but more widespread aspect of the influence of evolution is the very common occurrence, especially on 'oceanic' islands, of *endemic* species (i.e. species that are found nowhere else). The Hawaiian *Drosophila* themselves are endemics, as are all the species of land birds on the island of Tristan da Cunha. A more complete illustration of the balance between colonization and the evolution of endemics is shown in Figure 23.20. Norfolk Island is a small island (about 70 km²) approximately 700 km from New Caledonia and New Zealand, but about 1200 km from Australia. The ratio of Australian species : New Zealand and New Caledonian species within a group can therefore be used as a measure of the group's dispersal ability, and as Figure 23.20 shows, the proportion of endemics on Norfolk Island is

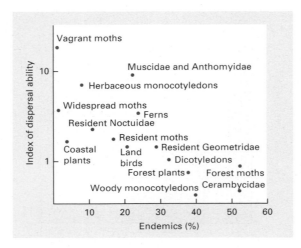

Figure 23.20 Poorly dispersing groups on Norfolk Island have a higher proportion of endemic species, and are more likely to contain species that have reached Norfolk Island from either New Caledonia or New Zealand than species from Australia, which is further away. The converse holds for good dispersers. (After Holloway, 1977.)

highest in groups with poor dispersal ability and lowest in groups with good dispersal ability.

In a similar vein, Lake Tanganyika, one of the ancient and deep Great Rift lakes of Africa, contains 214 species of cichlid fish, many of which show exquisite specializations in the manner and location of their feeding. Of these 214 species, 80% are endemic. With an estimated age of the lake of 9–12 million years, together with evidence that the various endemic groups diverged some 3.5–5 million years ago, it is likely that this uniquely diverse, endemic fish fauna evolved within the lake from a single ancestral lineage (Meyer, 1993). By contrast, Lake Rudolph, which has only been an isolated water body for 5000 years, since its connection to the River Nile system was broken, contains only 37 species of cichlid of which only 16% are endemic (Fryer & Iles, 1972).

communities may be unsaturated because of insufficient time for evolution

Communities of phytophagous insects show particularly clearly that low rates of colonization combined with insufficient time for evolution can lead to especially marked impoverishment on 'isolated' plant species. One study examined three kinds

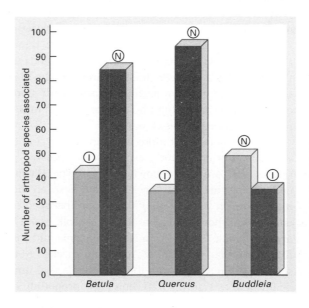

Figure 23.21 Relative species richness of phytophagous arthropods associated with three species of tree in Great Britain (■) and South Africa (▣). In each case, fewer species are associated with the tree where it has been introduced (I) compared to where it is native (N). (After Southwood *et al.*, 1982.)

of tree that are found in both Great Britain and South Africa: *Betula pendula*, *Quercus robur* and *Buddleia* spp. *Betula* and *Quercus* are native to Great Britain but were introduced to South Africa, whilst for *Buddleia* the converse is true. It is apparent that the species richness of phytophagous arthropods in each case is greater in the location where the tree is native (Figure 23.21).

The conclusion is clear: a community may be less than fully saturated not only because there has been insufficient time for colonization, but also because there has been insufficient time for evolution.

Chapter 24
Patterns in Species Richness

24.1 Introduction

In this penultimate chapter, we turn to some central questions which present themselves not only to community ecologists but to anybody who observes and ponders the natural world. Why do some communities contain more species than others? Are there patterns or gradients of species richness? If so, what are the reasons for these patterns? There are plausible and sensible answers to these questions, but conclusive answers are virtually lacking. Yet, this is not so much a disappointment as a challenge to ecologists of the future. Much of the fascination of ecology lies in the fact that many of the problems are blatant and obvious, whilst the solutions are not.

An understanding of patterns in biological diversity, and their underlying causes, has more than just academic interest. In fact, the conservation of biodiversity is one of the major challenges facing ecologists as we enter the 21st century. *Biodiversity* is a term intended to encompass all of nature's variety. Its most basic component is the number of species present in an area—species richness. But, as we shall see in Chapter 25, conservation strategies often invoke more sophisticated measures that take into account the diversity of functional roles represented by species in a community, or that weight species according to their taxonomic uniqueness or endemism. Moreover, elements of biodiversity occur at other ecological levels including, at one end of the scale, genetic variation within species, and at the other, variation in community types within a region. We restrict our discussion in the present chapter to species richness, partly because of its fundamental nature but mainly because so much more data are available for this than for any other aspect of biodiversity.

There are a number of factors to which the species richness of a community can be related, and these are of several different types. First, there are factors that can be referred to broadly as 'geographical', notably latitude, altitude and, in aquatic environments, depth. These have often been correlated with species richness, as we shall discuss below, but they cannot presumably be causal agents in their own right. If species richness changes with latitude, then there must be some other factor changing with latitude, exerting a direct effect on the communities.

A second group of factors does indeed show a tendency to be correlated with latitude, and so on—but, they are not perfectly correlated. To the extent that they are correlated at all, they may play a part in explaining latitudinal and other gradients. But, because they are not perfectly correlated, they serve also to blur the relationships along these gradients. Such factors include climatic variability, the input of energy, the productivity of the environment, possibly the 'age' of the environment and the 'harshness' of the environment.

A further group of factors vary geographically but not in a consistent manner (i.e. they vary quite independently of latitude, etc.). They therefore tend to blur or counteract relationships between species richness and other factors. This is true of the amount of physical disturbance a habitat experiences (see Chapter 21), the isolation or 'islandness' of a habitat (see Chapter 23) and the extent to which it is physically and chemically heterogeneous.

secondary factors

Finally, there is a group of factors that are biological attributes of a community, but are also important influences on the structure of the community of which they are part of. Notable amongst these are the amount of predation in a community (see Chapter 21), the amount of competition (see Chapter 20), the spatial or architectural heterogeneity generated by the organisms themselves and the successional status of a community (see Chapter 17). These should be thought of as 'secondary' factors in that they are themselves the consequences of influences outside the community. Nevertheless, they can all play powerful roles in the final shaping of a community.

A number of these factors of various types have been examined in detail in previous chapters: competition, predation, disturbance and 'islandness'. In this chapter we continue (see Section 24.3) by examining the relationships between species richness and a number of other factors which can be thought of as exerting an influence in their own right, namely productivity, spatial heterogeneity, climatic variation, environmental harshness and the age of the environment (i.e. the time over which evolution/colonization has been possible). We will then be in a position (see Section 24.4) to consider trends with latitude, altitude, depth, succession and position in the fossil record. We begin, though, by constructing a simple theoretical framework (following MacArthur, 1972) to help us think about variations in species richness.

Before proceeding, it is worth repeating some definitions. Species richness is the number of species present at a site. In contrast, species diversity takes into account both species richness and the evenness with which individuals in the community are distributed amongst the species (see Chapter 17, Section 17.2.1). Our treatment in this chapter is largely restricted to patterns in species richness.

24.2 A simple model

We will assume, for simplicity, that the resource available to a community can be depicted as a one-dimensional continuum, R units long (Figure 24.1). Each species in the community utilizes only a portion of this resource continuum, and these portions define the niche breadths (n) of the various species: the average niche breadth within the community is $\bar{n}$. Some, at least, of these niches are likely to overlap, and each overlap between adjacent species can be measured by a value o. The average niche overlap within the community is then $\bar{o}$.

With this simple background, it is possible to consider why some communities should contain more species than others. First, for given values of n and o, a community will contain more species the larger the value of R; that is, the greater the range of resources (Figure 24.1a). This is true when the community is dominated by competition and the species 'partition' the resources (see Chapter 20). But, it will also presumably be true when competition is relatively unimportant. Wider resource spectra provide the means for existence of a wider range of species, whether or not those species interact with one another.

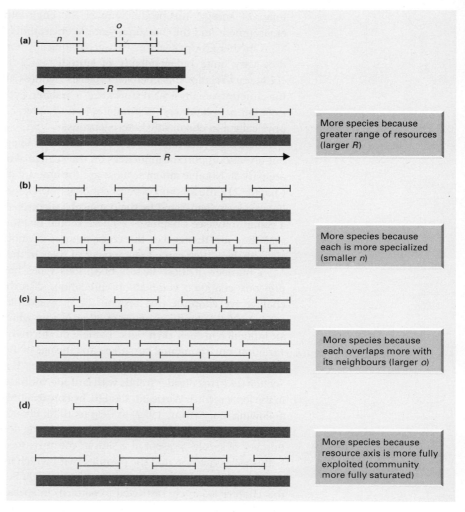

Figure 24.1 A simple model of species richness. Each species utilizes a portion n of the available resource dimension R, overlapping with adjacent species by an amount o. More species may occur in one community than in another: (a) because a greater range of resources is present (larger R); (b) because each species is more specialized (smaller n); (c) because each species overlaps more with its neighbours (larger o); and (d) because the resource dimension is more fully exploited.

Second, for a given range of resources, more species will be accommodated if n is smaller; that is, if the species are more specialized in their use of resources (Figure 24.1b). Alternatively, if species overlap to a greater extent in their use of resources (greater o), then more may coexist along the same resource continuum (Figure 24.1c). Finally, a community will contain more species the more fully saturated it is; conversely, it will contain fewer species when more of the resource continuum is unexploited (Figure 24.1d). Note that the utility of this model does not depend on any assumption about why species have particular niche breadths or overlaps. More species may occur in species-rich locations because those locations permit narrower niches (or greater overlaps) directly, or the locations may support a greater abundance of individuals, allowing species with narrower niches (or greater overlaps) to persist, and hence, more species to be packed in (Turner *et al.*, 1996). Either way, the model is useful in providing a framework for describing the basis of patterns in species richness.

the role of competition

We can now reconsider the various factors and processes described in previous chapters. If a community is dominated by interspecific competition (see Chapter 20), the resources are likely to be fully exploited. Species richness will then depend on the range of available resources, the extent to which species are specialists and the permitted extent of niche overlap (Figures 24.1a–c). We will examine presently certain influences on these three things.

the role of predation

Predation is capable of exerting a variety of effects in communities.

1 Many field studies from a wide range of habitats have demonstrated that predators can exclude certain prey species (Holt, 1984; Jeffries & Lawton, 1984, 1985). A community in the absence of these species may then be less than fully saturated, in the sense that some available resources may go unexploited (Figure 24.1d).

2 Predation may tend to keep species below their carrying capacities for much of the time, reducing the intensity and importance of direct interspecific competition for resources, and permitting much more niche overlap and a greater richness of species than in a community dominated by competition (Figure 24.1c). (Note that physical disturbances can exert a similar effect—see Chapter 21.)

3 Predation may generate patterns in the structure of communities similar to those produced by competition. In theory, prey species may compete for 'enemy free space' (Holt, 1977, 1984; Jeffries & Lawton, 1984, 1985). Such 'apparent competition' means that invasion and stable coexistence of prey in a habitat are favoured by prey being sufficiently different from other prey species already present. In other words, there may be a limit to the similarity of prey that can coexist (equivalent to the presumed limits to similarity of coexisting competitors; Chapter 7, Section 7.6).

the role of islandness

4 The depauperate communities of islands (see Chapter 23) can be seen in the present context as partly reflecting the reduced range of resources offered by smaller areas (Figure 24.1a), and partly reflecting a reduced level of saturation (Figure 24.1d) as a result of a higher tendency for species to go extinct, combined with a higher probability that not all supportable species will have colonized the island.

24.3 Richness relationships

In the following sections, we examine the relationships between species richness and a number of factors that may, in theory, influence the composition of communities. It will become clear that it is often extremely difficult to come up with unambiguous predictions and 'clean' tests of hypotheses when dealing with something as complex as a community.

24.3.1 Richness of resources and productivity as determinants of species richness

variations in productivity

For plants, the productivity of the environment can depend on whichever resource or condition is most limiting to growth. There is a general increase in primary productivity from the poles to the tropics (see Chapter 18) as light levels, average temperatures and the length of the growing season all increase. With increasing altitude in terrestrial environments, the declines in both temperature and the length of the growing season lead to a general drop in productivity; whilst in aquatic environments, productivity typically declines with depth as temperature and light levels fall. In addition, there is often a striking decrease in productivity with decreasing availability of water, especially in relatively dry environments where water supply may limit growth; and there is always likely to be an increase in productivity associated with an increase in the supply rates of essential nutrients like nitrogen, phosphorus and potassium. Broadly speaking, the productivity of the environment for animals follows these same trends, both as a result of the changes in resource levels at the base of the food chain, and as a result of the changes in temperature and other conditions.

increased productivity might be expected to lead to increased richness ...

If higher productivity is correlated with a wider range of available resources, then this is likely to lead to an increase in species richness (Figure 24.1a). However, a more productive environment may contain larger amounts or supply rates of resources without this affecting the variety of resources. This might lead to more individuals per species rather than more species. On the other hand, it is possible, even if the overall variety of resources is unaffected, that rare resources or low-productivity segments of the resource spectrum, that are insufficient to support species in an unproductive environment, may become abundant enough in a productive environment for extra species to be added. A similar line of argument would suggest that if the community were dominated by competition, then an increase in the quantity of resources would allow greater specialization (i.e. a decrease in n) without the individual specialist species being driven to very low densities (Figures 24.1b and 24.2).

In general then, we might expect species richness to increase with the richness of available resources and productivity. Strong indirect support for this contention comes from a detailed analysis of the species richness of trees in North America in relation to a variety of climatic variables (Currie & Paquin, 1987). Two measures of

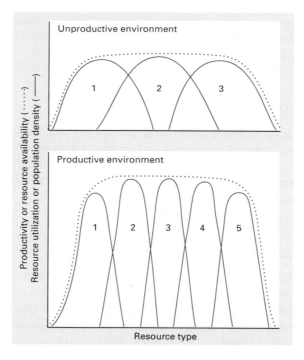

Figure 24.2 A more productive environment may support a greater number of more specialized species (smaller n at equilibrium).

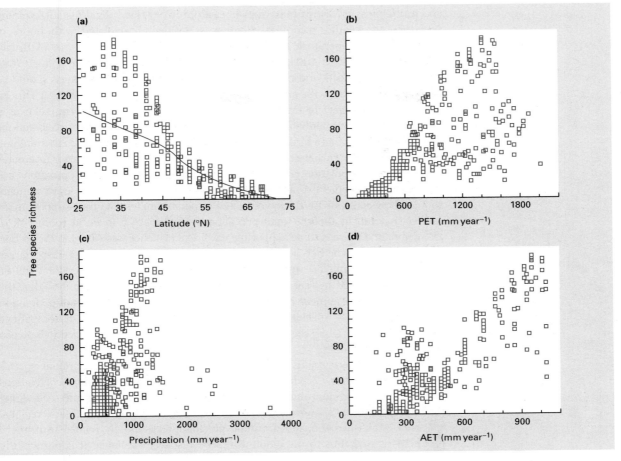

Figure 24.3 Species richness of trees in North America north of the Mexican border (in which the continent has been divided into 336 quadrats following lines of latitude and longitude) in relation to: (a) latitude; (b) potential evapotranspiration (PET); (c) precipitation; and (d) actual evapotranspiration (AET). The relationship with AET is strongest. (After Currie & Paquin, 1987; Currie, 1991.)

available environmental energy figure prominently in the analysis: *potential evapo-transpiration* (PET) is the amount of water that evaporates or is transpired from a saturated surface, independent of water availability, and can be interpreted as a crude integrated measure of available energy; *actual evapotranspiration* (AET) is the amount of water that actually evaporates from an area, and represents the joint availability of energy and water, such that PET and precipitation both set upper limits to AET. Tree species richness was most strongly correlated with AET (Figure 24.3), which can be regarded as a good correlate of the energy that can be captured by plants and therefore of primary productivity.

When this work was extended to four vertebrate groups, species richness was found to be correlated to some extent with tree species richness, at least in low-richness environments (Currie, 1991). However, the best correlations were consistently with PET (Figure 24.4), whilst other energy variables—temperature and solar radiation—were nearly as good at predicting species richness. Furthermore, the species richness of British butterflies (measured in 900 km^2 quadrats surrounding each of 144 weather-recording stations) is positively correlated with hours of sunshine and temperature in summer (Turner *et al.*, 1987), and that of British insectivorous birds (in 400 km^2 quadrats) with temperature (Turner *et al.*, 1988). These intriguing results raise the question of why animal species richness should be positively correlated with crude atmospheric energy? In simple terms, it can be argued that the effect of extra atmospheric warmth for an ectotherm, such as a butterfly, would be to enhance the intake and utilization of food resources, and for an endotherm, such as a bird, to require less expenditure of resources in maintaining body temperature. Turner *et al.* (1996) suggest that this will lead to faster or greater individual and population growth; thus, they argue, because large populations are less likely to become extinct through random processes, warmer environments should allow species with narrower niches to persist and should thus support more species in total. In theory, small populations are expected to be at more risk of extinction, and there is some supporting empirical evidence (see Chapter 25, Section 25.3). However, whether both premises upon which this argument rests (larger populations and smaller extinction probabilities in warmer environments) will be generally supported in real communities remains to be seen. Note also that the antithesis of this argument has also been presented—that high rates of population growth may reduce species richness by increasing the likelihood of competitive exclusion (Huston, 1979; see below). We lack the data to distinguish clearly between these contrasting ideas.

Sometimes, there seems to be a direct relationship between animal species richness and primary productivity. Thus, there are strong positive correlations between species richness and precipitation for both seed-eating ants and seed-eating rodents in the south-western deserts of the USA (Figure 24.5a). In such arid regions, it is well established that mean annual precipitation is closely related to primary productivity, and thus to the amount of seed resource available. It is particularly noteworthy that in the species-rich sites, the communities contain more species of very large ants (which consume large seeds) and more species of very small ants (which take small seeds) (Davidson, 1977). There are also more very small rodents at these sites. It seems that either the range of sizes of seeds is greater in the more productive environments, or that the abundance of seeds becomes sufficient to support extra consumer species. There is also a positive relationship between the species richness of lizards in the deserts of the south-western USA and the length of the growing season—an important aspect of productivity in desert environments (Figure 24.5b); whilst the diversity of chydorid cladocerans (a kind of zooplankton) in 11 unpolluted Indiana lakes is positively related to the total productivity of those lakes (Figure 24.5c).

On the other hand, an increase in diversity with productivity is by no means universal. This is particularly well illustrated by the unique 'Parkgrass' experiment which has been running from 1856 to the present day at Rothamsted in England. An 8-acre (3.2×10^4 m^2) pasture was divided into 20 plots, two serving as controls and the others receiving a fertilizer treatment once a year. Whilst the unfertilized areas

---

Figure 24.4 (*facing page*) Species richness of birds, mammals, amphibians and reptiles in North America in relation to tree species diversity (a–d) and potential evapotranspiration (PET) (e–h). In every case, PET is the best predictor of animal species richness. (After Currie, 1991.)

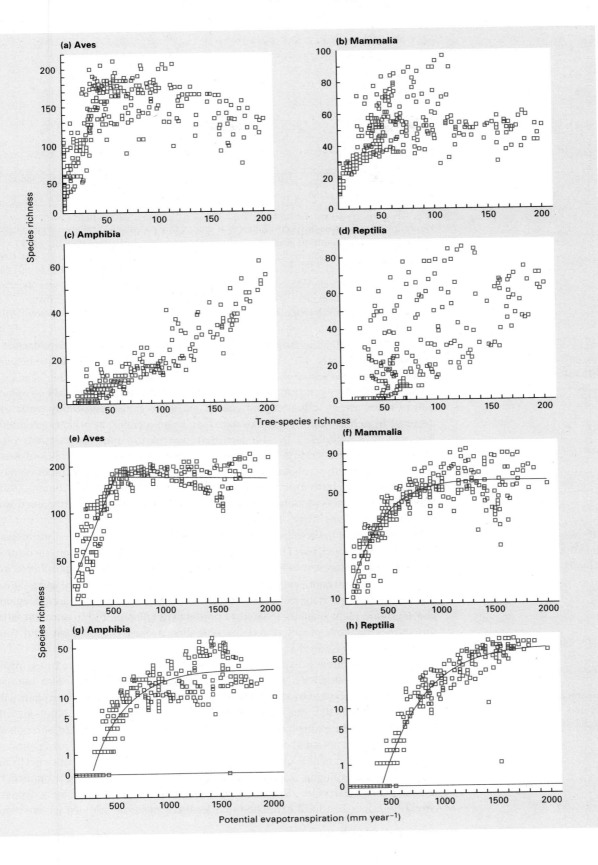

(a) Aves

(b) Mammalia

(c) Amphibia

(d) Reptilia

Tree-species richness

(e) Aves

(f) Mammalia

(g) Amphibia

(h) Reptilia

Species richness

Potential evapotranspiration (mm year^{-1})

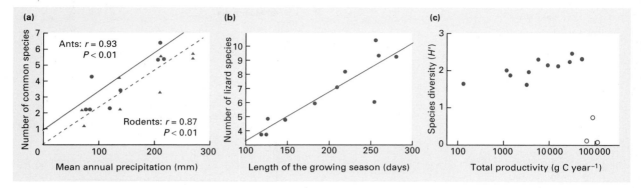

Figure 24.5 (a) Patterns of species richness of seed-eating rodents (▲) and ants (●) inhabiting sandy soils in a geographical gradient of precipitation and productivity. (After Brown & Davidson, 1977.) (b) Species richness of lizards at 11 sites in the south-western USA plotted against length of the growing season. (After Pianka, 1967.) (c) Relationship between species diversity (*H*) of chydorid cladocerans and primary productivity (expressed as grams of carbon per year per entire lake) in 14 lakes in Indiana, USA. Open circles show three heavily polluted lakes. (After Whiteside & Harmsworth, 1967; reanalysed in Brown & Gibson, 1983.)

but other evidence shows richness declining with productivity ...

remained essentially unchanged, the fertilized areas showed a progressive decline in diversity (see Chapter 17, Figure 17.3). This decline, referred to by Rosenzweig (1971) as 'the paradox of enrichment', has been found in several other studies of species richness in plant communities. It can be seen when the cultural eutrophication of lakes, rivers, estuaries and coastal marine regions consistently leads to a decrease in richness of phytoplankton (as well as an increase in primary productivity); and it is paralleled by the fact that two of the most species-rich plant communities in the world occur on very nutrient-poor soils (the Fynbos of South Africa and the heath scrublands of Australia), whereas nearby communities on more nutrient-rich soils have much lower plant richness (all reviewed by Tilman, 1986). High species richness of animals in the open ocean also tends to be associated with regions of low productivity (Angel, 1993).

and further evidence suggests a 'humped' relationship

Several studies have suggested that species richness may be highest at intermediate levels of productivity. For instance, there are humped curves when the number of Malaysian rainforest woody species is plotted against an index of phosphorus and potassium concentration, expressing the resource richness of the soil (Figure 24.6a), and when species richness of South African Fynbos plants is plotted against above-ground plant biomass (related to productivity) (Figure 24.6b) and also when the species richness of desert rodents is plotted against precipitation (and thus, productivity) along a gradient in Israel (Figure 24.6c).

explanations for declines in richness with increasing productivity

As we have already explained, it is easy to see why species richness might increase with productivity (either generally or in the left-hand limb of a humped curve). But, a decrease in richness with productivity is less easy to explain. One possible solution is that high productivity leads to high rates of population growth, bringing about the extinction of some of the species present because of a speedy conclusion to any potential competitive exclusion (Huston, 1979). At lower productivities, the environment is more likely to have changed before competitive exclusion is achieved (see Chapter 21, Section 21.3). This argument can be extended by recalling that the species richness of a habitat may be maintained by a dynamic equilibrium between local extinction and local colonization—an idea we have met

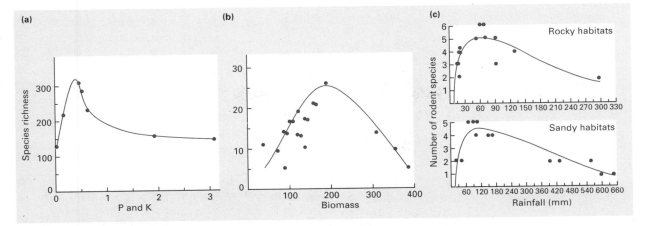

Figure 24.6 (a) Species richness of woody species in several Malaysian rainforests plotted against an index of phosphorus and potassium concentration. (After Tilman, 1982.) (b) Species richness of Fynbos plants plotted against above-ground plant biomass. (After Bond, 1983.) (c) Species richness of desert rodents in Israel plotted against rainfall for both rocky and sandy habitats. (After Abramsky & Rosenzweig, 1983.)

before in MacArthur and Wilson's (1967) theory of island biogeography (see Chapter 23, Section 23.2.3). Thus, as well as a faster extinction rate, in more productive environments there may be a lower colonization rate. In an 11-year experiment in which plots in grassland fields were supplied with extra nitrogen, decreases in species richness (Figure 24.7a) were caused as much by lower rates of species gain as by greater rates of loss of existing species (Figure 24.7b). In more productive grasslands, accumulated litter, and possibly lower light penetration, seemed to inhibit germination and/or survival of seedlings, and thus decreased rates of establishment of new species.

Overall, it seems that when increased productivity means an increased range of resources, an increase in species richness is to be expected (and has, at least sometimes, been found). In particular, a more productive and diverse community of plants is likely to support a greater richness of herbivores, and so on up the food web. On the other hand, if increased productivity means an increased supply but not a greater range of resources, then there are theoretical grounds for expecting both an increase and a decrease in species richness. The evidence, especially from plants, suggests that a decrease in species richness with resource enrichment is most common, or at least that a humped curve of species richness will be found if the whole productivity range is examined.

In this context, it is worthwhile reconsidering the nature of light as a resource for plants. In the productive environment of a tropical forest the abundant radiant energy is reflected and diffused down a long column of vegetation. Hence, there is not only a high rate of supply, but also a long and gradual gradient of light intensities (extending from very high to very low intensities) and probably also a wide range of frequency spectra. A high rate of supply, therefore, seems to lead necessarily to a wide range of light regimes, to an increased opportunity for specialization, and thus to an increased species richness. A further consequence is that the tallest species must be able to operate over the whole range of light intensities, as they grow up from ground level to the upper canopy. On the other hand, increasing productivity in

more light may lead to more light regimes

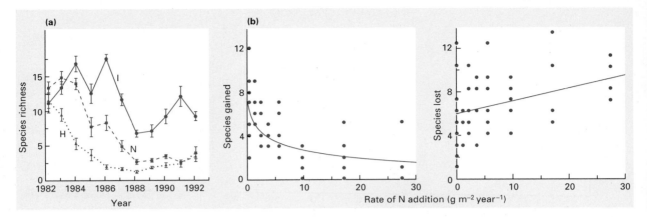

Figure 24.7 (a) Mean number of plant species per plot of grassland ($\pm$ standard error) in an 11-year experiment in which nitrogen was supplied at an intermediate rate (I, 5.4 g m^{-2} year^{-1}), a high rate (H, 27 g m^{-2} year^{-1}) or not at all (N). (b) Dependence of species gains and losses on a range of rates of nitrogen addition, including those illustrated in (a), based on comparisons of 1982 and 1983 data with 1984 and 1985 data. (After Tilman, 1993.)

grassland seems generally to be associated with increasing dominance by fewer and fewer species that monopolize, rather than take a share of, the incident radiation.

24.3.2 Spatial heterogeneity

We have already seen how the patchy nature of an environment, coupled with aggregative behaviour, can lead to coexistence of competing species (Atkinson & Shorrocks, 1981). In addition, environments that are more spatially heterogeneous can be expected to accommodate extra species, precisely because they provide a greater variety of microhabitats, a greater range of microclimates, more types of places to hide from predators, and so on. In effect, the extent of the resource spectrum is increased (see Figure 24.1a). (Environments that experience patchy disturbance are likely to be spatially heterogeneous as a consequence—but, it is convenient to think of this as 'disturbance' (see Chapter 21). Here, we deal separately with spatial heterogeneity in its own right.)

richness and the heterogeneity of the abiotic environment

In some cases, it has been possible to relate species richness to the spatial heterogeneity of the abiotic environment. For instance, a study of freshwater molluscs in a large number of sites, including road-side ditches, swamps, rivers and lakes, revealed a positive correlation between species richness and an estimate of the number of types of mineral and organic substrates present (Figure 24.8). Similarly, a plant community covering a range of soil types and a variety of topographies is almost certain (other things being equal) to contain more species than one covering a flat area of homogeneous soil.

animal richness and plant spatial heterogeneity

Most studies of spatial heterogeneity, however, have related the species richness of animals to the structural diversity of the plants in their environment. For example, Figure 24.9a shows a positive correlation between the species richness of freshwater fish and an index of plant spatial heterogeneity in lakes in northern Wisconsin; whilst Figure 24.9b shows a similar correlation for birds in Mediterranean-type habitats in California, central Chile and south-west Africa. The difficulty with such studies is that the species richness of the animals may actually have been caused by

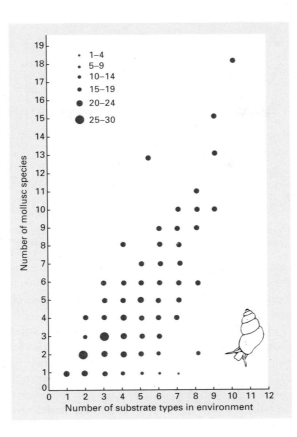

Figure 24.8 Relationship between species richness of freshwater molluscs and spatial heterogeneity, assessed in terms of the number of substrate types present. The size of the plotted points reflects the number of study sites which fall into that category. (After Harman, 1972.)

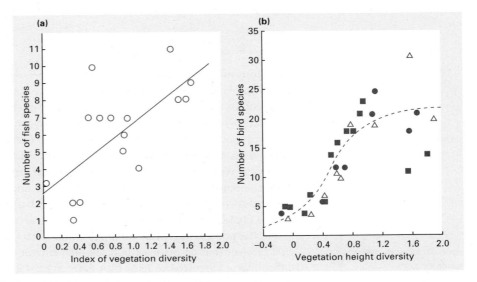

Figure 24.9 Relationships between animal species richness and an index of structural diversity of the vegetation for the following. (a) Freshwater fish in 18 Wisconsin lakes. (After Tonn & Magnuson, 1982.) (b) Birds in Mediterranean-type habitats in California (△), central Chile (●) and South-West Africa (■). (After Cody, 1975.)

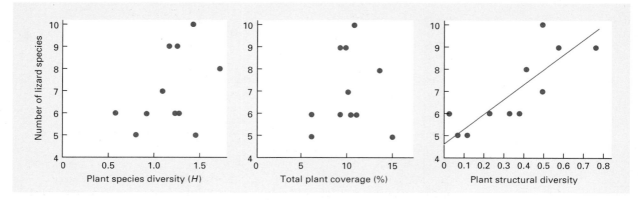

Figure 24.10 Relationship between lizard species richness and three measures of vegetation in desert sites in the south-west USA. (After Pianka, 1967.)

the same factor that led to the richness of the vegetation. For instance, for the data in Figure 24.9a, vegetation diversity was itself positively correlated with both lake size and plant nutrient concentration.

Thus, evidence that spatial heterogeneity is important in its own right in animal communities is much more convincing when the correlation of animal richness with plant *structural* heterogeneity is far stronger than the correlation with plant *species* richness. A number of studies of birds have shown this to be the case, and the same pattern has been described for lizards in the deserts of the south-west USA (Figure 24.10). Lizards of different species forage at different heights in the vegetation, and use perches at characteristic locations to watch for food, competitors, mates and predators. Whether spatial heterogeneity arises intrinsically from the abiotic environment, or is provided by other biological components of the community, it is capable of promoting an increase in species richness.

24.3.3 Climatic variation

The effects of climatic variation on species richness depend on whether the variation is predictable or unpredictable (measured on time scales that matter to the organisms involved). In a predictable, seasonally changing environment, different species may be suited to conditions at different times of the year.

temporal niche differentiation in seasonal environments

More species might therefore be expected to coexist in a seasonal environment, rather than in a completely constant one. For instance, different annual plants in temperate regions germinate, grow, flower and produce seeds at different times during a seasonal cycle; whilst phytoplankton and zooplankton pass through a seasonal succession in large, temperate lakes, with a variety of species dominating in turn as changing conditions and resources become suitable for each.

specialization in non-seasonal environments

On the other hand, there are opportunities for specialization in a non-seasonal environment that do not exist in a seasonal environment. For example, it would be difficult for a long-lived obligate fruit-eater to exist in a seasonal environment when fruit is available for only a very limited portion of the year. But, such specialization is found repeatedly in non-seasonal, tropical environments where fruit of one type or another is available continuously.

Unpredictable climatic variation (climatic instability) could have a number of

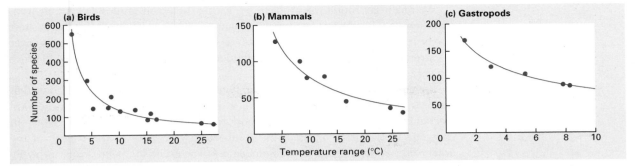

Figure 24.11 Relationships between species richness and the range of monthly mean temperature at sites along the west coast of North America for (a) birds, (b) mammals and (c) gastropods. (After MacArthur, 1975.)

climatic instability may
increase or decrease
richness …

effects on species richness. On the one hand: (i) stable environments may be able to support specialized species that would be unlikely to persist where conditions or resources fluctuated dramatically (see Figure 24.1b); (ii) stable environments are more likely to be saturated with species (see Figure 24.1d); and (iii) theoretical considerations (see p. 902) suggest that a higher degree of niche overlap will be found in more stable environments (see Figure 24.1c). All of these processes could increase species richness. On the other hand: (i) populations in a stable environment are more likely to reach their carrying capacities; (ii) the community is more likely to be dominated by competition; and (iii) species are therefore more likely to be excluded through competition (*o* smaller—see Figure 24.1c). It is therefore reasonable to argue that unpredictable climatic variation is a form of disturbance, and species richness may be highest at 'intermediate' levels; that is, species richness may increase or decrease with climatic instability.

… but there is no good
evidence either way

Some studies have seemed to support the notion that species richness increases as climatic variation decreases. For example, there is a significant negative relationship between species richness and the range of monthly mean temperatures for birds, mammals and gastropods that inhabit the west coast of North America (from Panama in the south to Alaska in the north) (Figure 24.11). However, there are many other things that change between Panama and Alaska, and this correlation does not prove causation. Other studies of climatic variation have been similarly unable to provide unequivocal conclusions.

24.3.4 Environmental harshness

what is harsh?

Environments dominated by an extreme abiotic factor—often called harsh environments—are more difficult to recognize than might be immediately apparent. An anthropocentric view might describe as 'extreme' both very cold and very hot habitats, unusually alkaline lakes and grossly polluted rivers. However, species have evolved that live in all such environments; and what is very cold and extreme for us must seem benign and unremarkable to a penguin.

A less arbitrary definition might state that for factors that can be classified along a continuum, values close to the minimum and maximum are extreme. But, is a relative humidity close to 100% (saturated air) as extreme as one of 0%? Is the minimum concentration of a pollutant 'extreme'? Certainly not.

We might get round the problem by 'letting the organism decide'. An environment may be classified as 'extreme' if organisms, by their failure to live there, show it to be so. But, if the claim is to be made—as it often is—that species richness is lower in extreme environments, then this definition is circular, and it is designed to prove the very claim we wish to test.

Perhaps the most reasonable definition of an extreme condition is one that requires, of any organism tolerating it, a morphological structure or biochemical mechanism which is not found in most related species, and is costly, either in energetic terms, or in terms of the compensatory changes in the biology of the organism that are needed to accommodate it. For example, plants living in highly acidic soils (low pH) may be affected through direct injury by hydrogen ions, or indirectly via deficiencies in the availability and uptake of important resources such as phosphorus, magnesium and calcium. In addition, aluminium, manganese and heavy metals may have their solubility increased to toxic levels, and mycorrhizal activity and nitrogen fixation may be impaired. Plants can only tolerate low pH if they have specific structures or mechanisms allowing them to avoid or counteract these effects.

In unmanaged grasslands in northern England, the mean number of plant species recorded per 1 m² quadrat was lowest in soils of low pH (Figure 24.12a). Similarly, the species richness of benthic stream invertebrates in the Ashdown Forest (southern England) was markedly lower in the more acidic streams (Figure 24.12b). Further examples of extreme environments that are associated with low species richness include hot springs, caves and highly saline water bodies such as the Dead Sea. The problem with these examples, however, is that they are also characterized by other features associated with low species richness. Many are unproductive and (perhaps as a consequence) most have low spatial heterogeneity. In addition, many occupy small areas (caves, hot springs) or are at least rare compared to other types of habitat (only a small proportion of the streams in southern England are acidic). Hence,

it seems that harsh environments are species-poor, but the concept of 'harsh' is difficult to define and the evidence is difficult to interpret

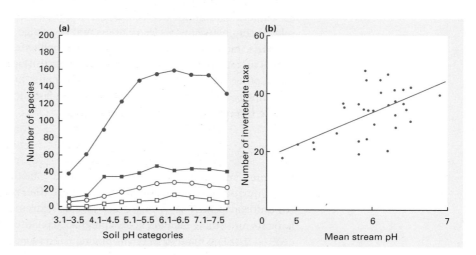

Figure 24.12 (a) Mean (○), maximum (■) and minimum (□) plant species richness per square metre, and total number (●) of species in categories of surface soil pH in unmanaged grasslands in northern England. (After Grime, 1973.) (b) Number of taxa of invertebrates in streams in Ashdown Forest, southern England, plotted against pH of the streamwater. (After Townsend *et al.*, 1983.)

'extreme' environments can often be seen as small and isolated islands. Although it appears reasonable that intrinsically extreme environments should as a consequence support few species, this has proved an extremely difficult proposition to establish.

24.3.5 Environmental age: evolutionary time

The idea that a community may have a relatively low species richness because there has been insufficient time for colonization or evolution has been discussed already in the context of island communities (see Chapter 23). In addition, we have seen that the non-equilibrium nature of many communities in disturbed habitats is the result of incomplete recolonization following a disturbance (see Chapter 21). It has often been suggested, however, that communities occupying large areas, and which are 'disturbed' only on very extended time scales, may also lack species because they have yet to reach an ecological or an evolutionary equilibrium (e.g. Stanley, 1979). Thus, communities may differ in species richness because some are closer to equilibrium and are therefore more saturated than others (see Figure 24.1d).

variable recovery following the glaciations

The idea has been proposed most frequently with regard to the recovery of communities following the Pleistocene glaciations. For example, the low richness of temperate forest trees in Europe compared to North America has been attributed to the fact that the major mountain ranges in Europe run east to west (the Alps and the Pyrenees), whereas in North America they run north to south (the Appalachians, the Rockies and the Sierra Nevada). The suggestion is that in Europe the trees were trapped by the glaciers against the mountains (and, indeed, against the Mediterranean Sea) and became extinct, whereas in North America they simply retreated southwards and survived (an hypothesis attributed to Reid, 1935—although even then it was not a new idea). Recent studies of vegetation history offer little support for the mountain/Mediterranean barrier idea (Huntley, 1993). Accumulated evidence of historic climate and vegetation patterns indicates that the greater differential extinction in Europe was a simple consequence of the much reduced area of forests that persisted there during the glaciations. There has subsequently been insufficient evolutionary time for the European trees to recover their equilibrium richness. In fact, even in North America it is unlikely that equilibrium is regained during interglacial periods, because of the slow rate of post-glacial spread of the displaced species (see Chapter 1).

the unchanging tropics and the recovering temperate zones?

More generally, it has often been proposed that the tropics are richer in species than are more temperate regions (see below), at least in part because the tropics have existed over long and uninterrupted periods of evolutionary time, whereas the temperate regions are still recovering from the Pleistocene glaciations (or even more ancient events). It seems, however, that the long-term stability of the tropics has in the past been greatly exaggerated by ecologists. Whilst the climatic and biotic zones of the temperate region moved towards the Equator during the glaciations, the tropical forest appears to have contracted to a limited number of small refuges surrounded by grasslands (see Chapter 1, Figure 1.7; see Section 1.2.2 for a more complete discussion). A simplistic contrast between the unchanging tropics and the disturbed and recovering temperate regions is therefore untenable. If the lower species richness of communities nearer the poles is to be attributed partly to their being well below an evolutionary equilibrium, then some complex (and unproven) argument must be constructed. (Perhaps movements of temperate zones to a different latitude caused many more extinctions than the contractions of the tropics into smaller areas at the same latitude.) The matter could be settled if there were a detailed fossil record showing that the tropics have always had approximately the same species richness, and that the temperate regions either had a very much higher species richness in the past, or that their species richness is currently increasing markedly. Unfortunately, no such fossil record has been uncovered. Thus, whilst it seems likely that some communities are further below equilibrium than others, it is impossible at present to pinpoint those communities with certainty or even confidence.

24.4 Gradients of richness

As Section 24.3 makes plain, *explanations* for variations in species richness are difficult to formulate and test. However, it is easy to find *patterns* in species richness and these are discussed below.

24.4.1 Latitude

Perhaps the most widely recognized pattern in species richness is the increase that occurs from the poles to the tropics. This can be seen in a wide variety of groups, including trees (see Figure 24.3), marine fish and invertebrates, ants, lizards and birds (Figure 24.13). The pattern can be seen, moreover, in terrestrial, marine and freshwater habitats.

richness decreases with latitude

Such increases in richness are apparent not only over extensive geographical regions, but also in small communities. A 1 ha $(1 \times 10^4\,\text{m}^2)$ area of equatorial rainforest may contain 40–100 different tree species, whilst comparable areas of deciduous forest in eastern North America and coniferous forest in northern Canada usually have 10–30 species and 1–5 species, respectively (Brown & Gibson, 1983). There are, of course, exceptions. Particular groups like penguins and seals are most diverse in polar regions; whilst coniferous trees and ichneumonid parasitoids both reach peaks of richness at temperate latitudes. For each of these exceptions, however, there are many groups found only in the tropics; for example, the New World fruit bats and the Indo-Pacific giant clams.

A number of explanations have been put forward for the general latitudinal trend in species richness, but not one of these is without problems. In the first place, the richness of tropical communities has been attributed to a greater intensity of predation; thus, natural enemies could be a key factor in maintaining a high richness of tree species in tropical forests (Janzen, 1970; Connell, 1971). The prediction, that disproportionately high mortality of young trees will occur close to adults because these harbour host-specific consumers, has received support from a number of studies (Clark & Clark, 1984; see, for example, Figure 24.14). If there is a low probability of recruitment close to conspecific adults, the likelihood of establishment of non-conspecifics is enhanced, and a higher richness of tree species can be expected to result. Note, however, that whilst host-specific predation may contribute to tropical richness, it cannot be the root cause, since the richness of predators is itself an attribute of the community. Other reports also indicate that predation may be more intense in the tropics. Thus, for example, predation (mainly by birds) on web-spinning spiders is twice as great in tropical Peruvian and Gabonese forests as in their temperate counterparts in North America (Rypstra, 1984), and nest predation is so high on bulbuls (*Andropadus latirostris*) in Gabon that they may need to lay up to

predation as a 'secondary' explanation

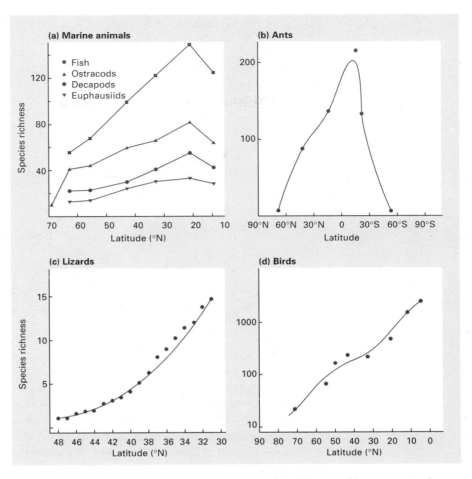

Figure 24.13 Latitudinal patterns in species richness of the following. (a) Marine animals in the north-east Atlantic—at the time of sampling a major front between south Atlantic and north Atlantic waters occurred at 18°N, so that the peak in species richness there resulted from a mixing of different faunal elements. (After Angel, 1993.) (b) Ants. (After Kusnezov, 1957.) (c) Lizards in the USA. (After Pianka, 1983.) (d) Breeding birds in North and Central America. (After Dobzhansky, 1950.)

five clutches just to rear one young per year (Brosset, 1981). In theory, more intense predation could reduce the importance of competition, permitting greater niche overlap and promoting higher richness (see Figure 24.1c).

Second, increasing species richness may be related to an increase in the income of energy as one moves from the poles to the Equator. For heterotrophic organisms, one possibility is that increased richness may arise because higher temperatures lead to larger populations that are less likely to go extinct (Turner *et al.*, 1996), or because of a wider range of resources (i.e. more different types of resource existing in exploitable amounts). The latter, resource-range hypothesis depends on an underlying increase in the richness of plants at lower latitudes. Can this be explained by increasing productivity?

If increasing productivity nearer the tropics means 'more of the same', as it does for light, then this would be expected to lead to a lower not a higher species richness

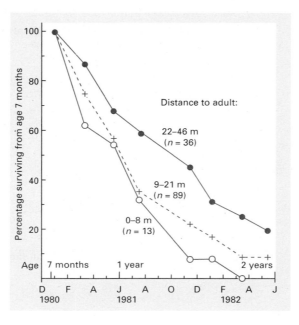

Figure 24.14 Survivorship of seedlings of the tropical rainforest tree *Dipteryx panamensis*, from 7 months to 2 years after germination, when situated at different distances from the nearest adult. (After Clark & Clark, 1984.)

Labels within figure:
Percentage surviving from age 7 months

Distance to adult:

22–46 m (*n* = 36)

9–21 m (*n* = 89)

0–8 m (*n* = 13)

Age | 7 months | 1 year | 2 years

D F A J A O D F A J
1980   1981   1982

productivity as a complex explanation

(see Section 24.3.1). More light may, however, mean a wider range of light regimes, and thus greater richness—but, this is only conjecture (see p. 893). On the other hand, light is not the only determinant of plant productivity. Tropical soils tend, on average, to have lower concentrations of plant nutrients than temperate soils. The species-rich tropics might therefore be seen as reflecting their *low* productivity. The tropical soils are poor in nutrients because most of the nutrients are locked up in the large tropical biomass, and because decomposition and release of nutrients are relatively rapid in the tropics (see Chapter 19). A 'productivity' argument would therefore have to run as follows. The light, temperature and water regimes of the tropics lead to high plant biomass (not necessarily diverse). This leads to nutrient-poor soils and perhaps a wide range of light regimes. These in turn lead to high plant species richness. At higher trophic levels, too, there is evidence that the environment is not perceived by individual species as highly productive. For example, a comparison of foraging rates in tropical and temperate guilds of small, foliage-gleaning, insectivorous birds revealed that attack rates were four to six times lower in the tropics (Thiollay, 1988). There is certainly no simple 'productivity explanation' for the latitudinal trend in richness.

Some ecologists have invoked the climate of low latitudes as a reason for their high species richness. Certainly equatorial regions are less seasonal than temperate regions (although rainfall may follow a marked seasonal cycle in the tropics in general), and for many organisms they are probably also more predictable (although this conjecture is extremely difficult to test because of the effects of body size and generation time on what constitutes 'predictability' to a particular species). The argument that a less seasonally variable climate allows species to be more specialized (i.e. have narrower niches; see Figure 24.1b) has now received a number of tests. Thus, a comparison of bird communities in temperate Illinois and tropical Panama showed that both tropical shrub and tropical forest habitats contained very many more breeding species than their temperate counterparts, and between 25 and 50%

climatic variation as a complex explanation

of the increase in richness consisted of specialist fruit-eaters in the tropical habitats (Karr, 1971). Further species exploited large insects, which were available throughout the year only in the tropics. Thus, the consistent availability of certain food resources creates extra opportunities for specialization in the tropical avian communities. In contrast, a group of insects, the bark and ambrosia beetles (Coleoptera: Scotytidae and Platypodidae), are less host-specific in the tropics than in temperate regions, even though there are considerably more species present in the tropics (Figure 24.15). Finally, contrasting patterns can be discerned amongst the parasites of marine fish. One group, the digenean trematodes, are more specialized closer to the Equator, as indicated by the proportion of species that are restricted to a single host (Figure 24.16). However, monogenean parasites are just as specific at all latitudes, despite the greater species richness nearer the Equator. There is, then, a plausible place for climatic variation in any explanation of the latitudinal trend in richness, but its exact role is uncertain.

The greater evolutionary 'age' of the tropics has been proposed as a reason for their greater species richness, and another line of argument suggests that the repeated fragmentation and coalescence of tropical forest refugia promoted genetic differentiation and speciation, accounting for much of the high richness in tropical regions (Connor, 1986). These ideas are plausible but very far from proven (Flenley, 1993).

Overall, the latitudinal gradient lacks a clear and unequivocal explanation. This is hardly surprising. The components of a possible explanation—trends with productivity, climatic stability, and so on—are themselves understood only in an incomplete

the full explanation must be complex ...

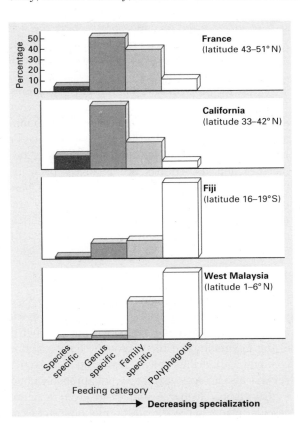

Figure 24.15 Number of species of ambrosia beetles and bark beetles (combined) in various feeding categories. These species are generally more specialized in their diets at high latitudes. (After Beaver, 1979.)

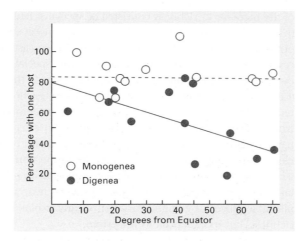

Figure 24.16 The percentage of monogenean and digenean parasite species that are restricted to a single species of fish host, at various locations on a latitudinal gradient. (After Rohde, 1978.)

... or perhaps simple

and rudimentary way, and the latitudinal gradient intertwines these components with one another, and with other, often opposing forces. Nevertheless, the explanation may ultimately prove to be simple, for the following reason. For example, if there was a single extrinsic factor promoting a latitudinal gradient in richness amongst the plants, then the increases in resource richness and diversity, and in heterogeneity, would promote increased richness amongst herbivores. This would increase the predation pressure on the plants (promoting further increased richness), whilst providing a rich and diverse resource for carnivores—which would in turn increase the predation pressure on the herbivores, and so on. In short, a subtle extrinsic force could promote a cascade effect, leading eventually to a marked richness gradient. As yet, however, we lack a totally convincing explanation that could 'set the ball rolling'.

Part of the problem lies with the many exceptions to the general latitudinal trend. It is, of course, as important to explain the exceptions as it is to explain the generalities. Islands are one large class of communities coming into this category. In addition, deserts, even near the tropics, are species-poor, probably because they are highly unproductive (being arid) and because the climate is very extreme. Salt-marshes and hot springs are relatively poor in species, despite being highly productive, seemingly because they represent harsh or extreme abiotic environments (and in the case of springs, because they are small 'islands'). As Chapter 21 made clear, neighbouring communities may differ in richness simply because of differences in the amount of physical disturbance they experience.

24.4.2 Altitude

In terrestrial environments, a decrease in species richness with altitude is a phenomenon almost as widespread as a decrease with latitude. Someone climbing a mountain in an equatorial region will pass through an equatorial habitat at the base of the mountain, and then through climatic and biotic zones that have much in common with Mediterranean, temperate and Arctic environments. If the mountaineer was also an ecologist, they would probably notice that there were fewer species as they climbed higher. Examples are given in Figure 24.17 for birds, mammals and vascular plants in the Himalayan mountains of Nepal.

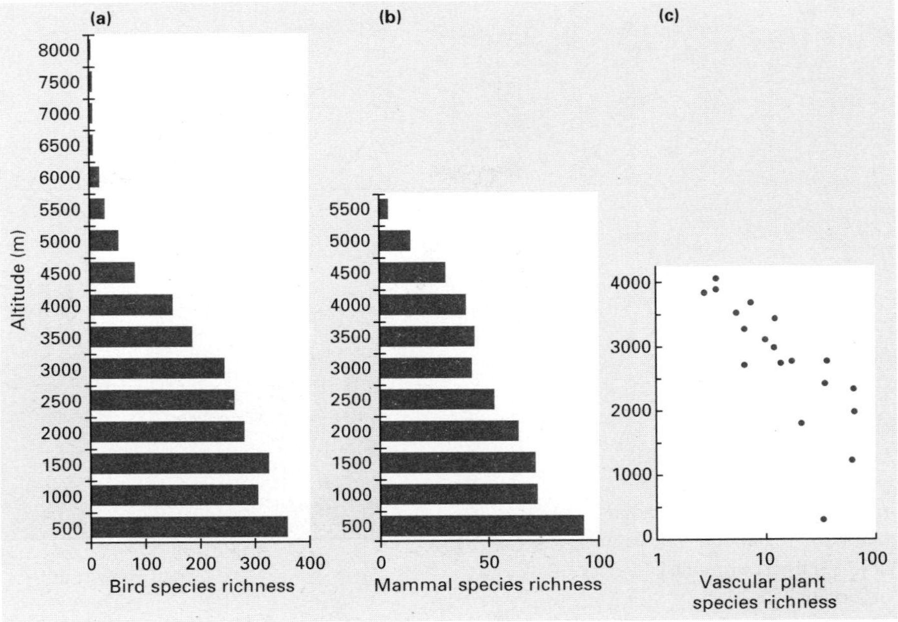

Figure 24.17 Relationship between species richness and altitude in the Nepalese Himalayas for the following. (a) Breeding bird species. (b) Mammals. (After Hunter & Yonzon, 1992.) (c) Vascular plants. (After Whittaker, 1977; from data of K. Yoda, unpublished observations.)

This suggests that at least some of the factors instrumental in the latitudinal trend in richness are also important as explanations for this altitudinal trend (although this is unlikely for evolutionary age, and less likely for climatic stability). Of course, the problems that exist in explaining the latitudinal trend are also equally pertinent to altitude. In addition, however, high-altitude communities almost invariably occupy smaller areas than lowlands at equivalent latitudes; and they will usually be more isolated from similar communities than lowland sites, which will often form part of a continuum. These effects of area and isolation are certain to contribute to the decrease in species richness with altitude.

On an almost absurdly small altitudinal scale, Figure 24.18 shows how species richness can vary dramatically between the troughs and peaks of ridge-and-furrow grassland in Great Britain. It is worth emphasizing that great variations in community composition and richness can occur over very small distances within what may be classified as a single community.

24.4.3 Depth

In aquatic environments, the change in species richness with depth shows strong similarities to the terrestrial gradient with altitude. In larger lakes, the cold, dark, oxygen-poor abyssal depths contain fewer species than the shallow surface waters. Likewise, in marine habitats, plants are confined to the photic zone (where they can photosynthesize) which rarely extends below 30 m. In the open ocean, therefore, there is a rapid decrease in richness with depth, reversed only by the variety of often bizarre animals living on the ocean floor. Interestingly, however, the effect of depth on the species richness of benthic invertebrates is to produce not a single gradient, but a peak of richness at about 2000 m—the approximate limit of the continental slope (Figure 24.19). This has been said to reflect an increase in the predictability of the environment between 0 and 2000 m (Sanders, 1968). At greater depths, beyond

905 PATTERNS IN SPECIES RICHNESS

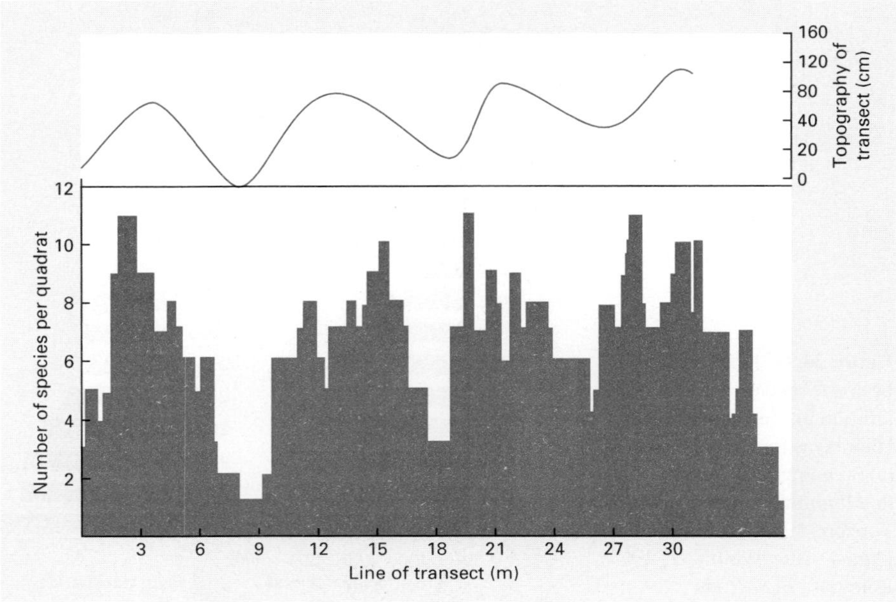

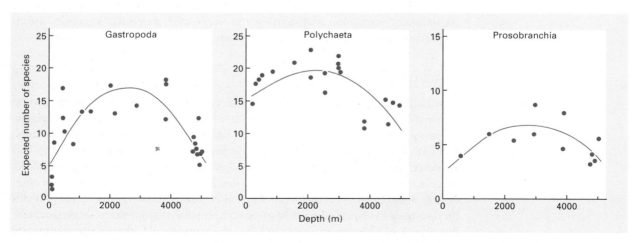

Figure 24.18 Histogram showing the relationship between number of species present and topography of ridge-and-furrow grassland at Marston, England. (From data of J.L. Harper, unpublished observations.)

the continental slope, species richness declines again, probably because of the extreme paucity of food resources in abyssal regions.

24.4.4 Succession

Several studies of plant communities have indicated a gradual increase in species richness during succession. During an old-field succession, the increase in richness is accompanied by a shift from a geometrical type of rank-abundance curve, with a high level of dominance, to a more even distribution of ground cover amongst the species, and lowered dominance (Figure 24.20). Richness either continues to increase through to the climax or reverses to some extent as some late successional

Figure 24.19 Variation in species richness (for samples of 50 individuals) of bottom-dwelling animals along the depth gradient of the ocean. (After Rex, 1981.)

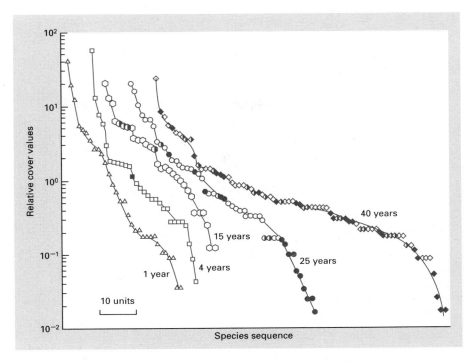

Figure 24.20 Rank-abundance curves for plants in old fields of five different ages in succession in old fields in southern Illinois, USA. Symbols are open for herbs, half-open for shrubs and closed for trees. Richness increases from 31 species at 1 year to more than 80 species at 40 years. (After Bazzaz, 1975.)

species disappear. The few studies which have been carried out on animals in successions indicate a parallel increase in species richness. Figure 24.21 illustrates this for birds and insects associated with different old-field successions.

To a certain extent, the successional gradient is a necessary consequence of the gradual colonization of an area by species from surrounding communities that are at later successional stages (i.e. later stages are more fully saturated with species—see Figure 24.1d). However, this is a small part of the story, since succession involves a process of replacement of species and not just the mere addition of new ones.

a cascade effect

As with the other gradients, there is bound to be a cascade effect with succession. In fact, it is possible to think of succession as being this cascade effect in action. The earliest species will be those that are the best colonizers and the best competitors for open space. They immediately provide resources that were not previously present, and they introduce heterogeneity that was not previously present. For example, the earliest plants generate resource-depletion zones in the soil which inevitably increase the spatial heterogeneity of plant nutrients. The plants themselves provide a new variety of microhabitats, and for the animals that might feed on them they provide a much greater range of food resources. The increase in herbivory and predation may then feed back to promote further increases in species richness, which provides further resources and more heterogeneity, and so on. In addition, temperature, humidity and wind speed are much less variable within a forest than in an exposed early successional stage, and the enhanced constancy of the environment may provide a stability of conditions and resources that permits specialist species to

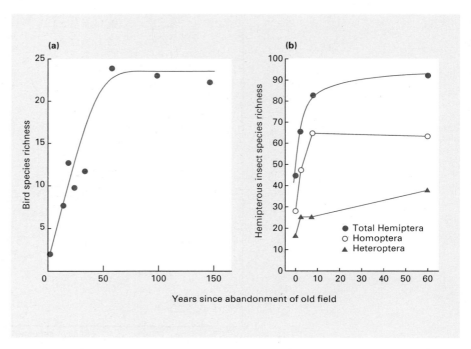

Figure 24.21 The increase in species richness during old-field successions. (a) Birds. (After Johnston & Odum, 1956). (b) Hemipterous insects. (After Brown & Southwood, 1983.)

build up populations and persist (see, for example, Parrish & Bazzaz, 1979, 1982; Brown & Southwood, 1983). As with the other gradients, the interaction of many factors makes it difficult to disentangle cause from effect. But, with the successional gradient of richness, the tangled web of cause and effect appears to be of the essence.

24.4.5 Appraisal of contemporary patterns in species richness

There are many generalizations that can be made about the species richness of communities. We saw in Chapter 21 how richness may peak at an intermediate level of disturbance frequency, and in Chapter 23 how richness declines with a reduction in island area or an increase in island remoteness. In the present chapter, we find that species richness decreases with increasing latitude, altitude and (after an initial rise) depth in the ocean. In more productive environments, a wider range of resources may promote species richness whereas just more of the same can lead to a reduction in richness. A positive correlation between species richness and temperature has also been documented. Species richness increases with an increase in spatial heterogeneity but may decrease with an increase in temporal heterogeneity (increased climatic variation). It increases during the course of succession and, as we shall see in the next section, with the passage of evolutionary time. However, for many of these generalizations important exceptions can be found, and for most of them the current explanations are not entirely adequate.

The search for patterns in species richness, and for explanations of those patterns, has emphasized some of the general difficulties we have in testing ecological theories. To answer these problems, we will need to use evolutionary arguments (for example, see Chapters 1–3, and 14), and to invoke our knowledge of population dynamics (see Chapters 4–7, 10, 15 and 16), of population interactions (see Chapters 8, 9, 11–13, 17 and 20–23) and of energy and nutrient fluxes (see Chapters 18 and 19). The study of community ecology is one of the most difficult and challenging areas of modern ecology. Clear, unambiguous predictions and tests of ideas are often very difficult to devise and will require great ingenuity on the part of future generations of ecologists.

24.4.6 Patterns in faunal and floral richness in the fossil record

The imperfection of the fossil record has always been the greatest impediment to the palaeontological study of evolution. Incomplete preservation, inadequate sampling and taxonomic problems combine to make the accurate assessment of long-term evolutionary patterns in richness a formidable task (Benton, 1994). Nevertheless, some general patterns have emerged, and our knowledge of six important groups of organisms is summarized in Figure 24.22.

the Cambrian explosion—exploiter-mediated coexistence?

About 600 million years ago, almost all the phyla of marine invertebrates entered the fossil record within the space of only a few million years (Figure 24.22a). Before that, for the previous 2500 million years, the world was populated virtually only by bacteria and algae. Of course, we can never be sure why this Cambrian 'explosion' took place. Stanley (1976), recognizing that the introduction of a higher trophic level can increase richness at a lower level, sees the first single-celled herbivorous protist as the 'hero' of the story. The opening up of space by cropping of the algal monoculture, coupled with the availability of recently evolved eukaryotic cells, may have caused the biggest burst of evolutionary activity the world has known (Gould, 1981). Speculating on the reason for the equally dramatic decline in the number of families of shallow-water invertebrates at the end of the Permian (Figure 24.22a), Schopf (1974) noted that the Late Permian was the time when the earth's continents coalesced to produce the single supercontinent of Pangaea. This joining of continents produced a marked reduction in the area occupied by shallow seas (which occur around the periphery of continents), and thus a marked decline in the area of habitat available to shallow-water invertebrates. Moreover, at this time the world was subject to a prolonged period of global cooling in which huge quantities of water were locked up in enlarged polar caps and glaciers, causing widespread marine regression and reduction of warm, shallow-sea environments (Maxwell, 1989). Thus, the well-known species–area relationship may be invoked to account for the reduction in richness of this fauna. The increase in faunal richness after the Permian may reflect the process in reverse—global warming coupled with an increasing area of shallow seas as the continents spread apart again.

the Permian decline—a species–area relationship?

competitive displacement amongst the major plant groups?

The analysis of fossil remains of vascular land plants (Figure 24.22b) reveals four distinct evolutionary phases: (i) a Silurian–Mid-Devonian proliferation of early vascular plants; (ii) a subsequent Late Devonian–Carboniferous radiation of fern-like lineages; (iii) the appearance of seed plants in the Late Devonian and the adaptive radiation to a gymnosperm-dominated flora; and (iv) the appearance and rise of flowering plants in the Cretaceous and Tertiary. It seems that following initial invasion of the land (made possible by the appearance of roots—Harper *et al.*, 1991), the diversification of each plant group coincided with a decline in species numbers of the previously dominant group. In two of the transitions (early plants to gymno-

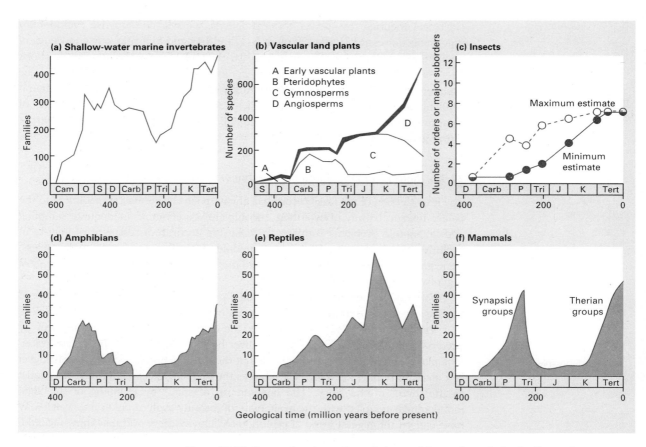

Figure 24.22 Curves showing patterns in taxon richness through the fossil record.
(a) Families of shallow-water invertebrates. (After Valentine, 1970.) (b) Species of vascular
land plants in four groups—early vascular plants, pteridophytes, gymnosperms and
angiosperms. (After Niklas *et al.*, 1983.) (c) Major orders and suborders of insects. The
minimum values are derived from definite fossil records, whilst the maximum values include
'possible' records. (After Strong *et al.*, 1984.) (d–f) Vertebrate families of amphibians, reptiles
and mammals, respectively. (After Webb, 1987.) Cam, Cambrian; O, Ordovician; S, Silurian;
D, Devonian; Carb, Carboniferous; P, Permian; Tri, Triassic; J, Jurassic; K, Cretaceous; Tert,
Tertiary.

sperms, and gymnosperms to angiosperms), this pattern may reflect the competitive
displacement of older, less specialized taxa by newer and presumably more
specialized taxa (Niklas *et al.*, 1983). On the other hand, the transition from
pteridophytes to gymnosperms coincides with major environmental changes, and it
may be that the partial extinction of the pteridophyte group allowed the radiation of
the gymnosperms into the vacated ecological space.

The first undoubtedly phytophagous insects are known from the Carboniferous.
Thereafter, modern orders appeared steadily (Figure 24.22c) with the Lepidoptera
arriving last on the scene, at the same time as the rise of the angiosperms. This
continuous rise in the number of orders, implying a concomitant increase in the
number of species, provides no reason to believe that insect species richness has
reached a present-day plateau or equilibrium level (Strong *et al.*, 1984). Reciprocal
evolution and counterevolution between plants and phytophagous insects has almost

910 CHAPTER 24

certainly been, and still is, an important mechanism driving the increase in richness observed in both land plants and insects through their evolution.

the vertebrate dynasties

In its broadest outline, the history of the tetrapod vertebrates may be viewed as a progression of four dynasties (Webb, 1987). First came the amphibian dynasty (Late Devonian and Carboniferous), dominated by large predators at the top of freshwater food webs (Figure 24.22d). The rise of the synapsid mammal dynasty in the Permo-Triassic (Figure 24.22f), many species of which were terrestrial herbivores, may be linked to the success of major new seed plants and the appearance of a productive ground layer of plants in more open terrestrial habitat. By the Late Triassic, diapsid reptiles (Figure 24.22e) had replaced synapsid mammals as the dominant herbivores and carnivores on all continents. One explanation for this shift is that climates were becoming increasingly dry, favouring reptiles that were physiologically better equipped to tolerate aridity. The last transition, to therian mammals, occurred at the end of the Cretaceous and was extremely rapid. If the mass extinction of the dinosaurs is linked to comet impact and a consequent dust cloud that dramatically reduced light and lowered temperatures, their demise may have taken less than 1 year. An alternative, and less likely 'gradualist' hypothesis, that extinction of dinosaurs and radiation of therian mammals was the result of gradual climatic changes and competitive interactions, would indicate a period of transition of perhaps a thousand to a few million years. The heralds of the new dynasty were small mammalian insectivores and carnivores, already living inconspicuously amongst the dinosaurs. In general, it seems that the triggers for the dynastic changes were climate—or vegetation—related.

progressive enrichment or stasis over evolutionary time?

What does the fossil record tell us about the extent to which present-day communities are 'saturated' with species? We can compare the observed patterns against two extreme perspectives. First, does evolution tend steadily to enrich communities with more and more species (arguing against any limits to species richness or an important role for species interactions)? Alternatively, does the number of niches stay roughly constant as different species fill them in turn (a pattern that is consistent with ecological interactions constraining species richness)? Taken overall, the answer appears to lie between these alternatives. Evolutionary 'breakthroughs' may have opened up major new niche dimensions, permitting dramatic jumps in species richness, followed by remarkably long periods of approximately constant richness.

extinctions of large animals in the Pleistocene—prehistoric overkill?

Towards the end of the last ice-age the continents were much richer in large animals than they are today. For example, Australia was home to many genera of giant marsupials, North America had its mammoths, giant ground sloths and more than 70 other genera of large mammals, whilst New Zealand and Madagascar were home to giant flightless birds, the moas (Dinorthidae) and elephant bird (Aepyornis), respectively (Martin, 1984). Over the past 11 000 years or so, a major loss of this biotic diversity has occurred over much of the globe. The extinctions particularly affected large terrestrial animals (Figure 24.23a), they were more pronounced in some parts of the world than others and they occurred at different times in different places (Figure 24.23b). The extinctions mirror patterns of human migration. Thus, the arrival in Australia of ancestral aborigines occurred between 40 000 and 30 000 years ago, stone spear points became abundant throughout the USA about 11 500 years ago and humans have been in both Madagascar and New Zealand for 1000 years. It can be convincingly argued that the arrival of efficient human hunters led to the rapid overexploitation of vulnerable and profitable large prey. Africa, where

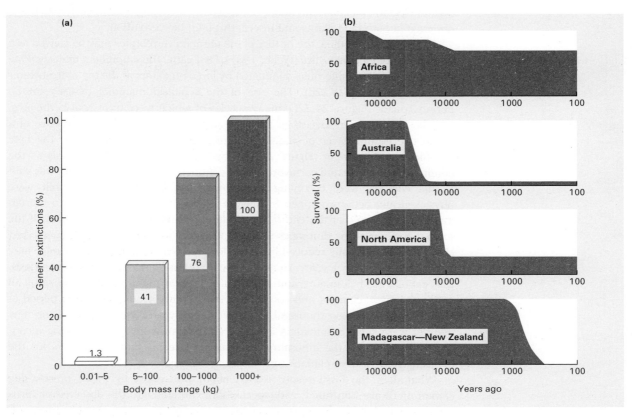

Figure 24.23 (a) The percentage of genera of large mammalian herbivores that have gone extinct in the last 130 000 years is strongly size dependent (data from North and South America, Europe and Australia combined). (After Owen-Smith, 1987.) (b) Percentage survival of large animals on three continents and two large islands (New Zealand and Madagascar). (After Martin, 1984.)

humans originated, shows much less evidence of loss, perhaps because coevolution of large animals alongside early humans provided ample time for them to develop effective defences (Owen-Smith, 1987).

The Pleistocene extinctions herald the modern age in which the influence upon natural communities of human activities has been increasing dramatically. We turn to an assessment of these adverse effects, and how they may be remedied, in our final chapter.

Chapter 25
Conservation and Biodiversity

25.1 A rationale for conservation

Extinction has always been a fact of life, but the arrival on the scene of humans has injected some novelty into the list of its causes. Overexploitation by hunting was probably the first of these, but more recently a large array of other impacts have been brought to bear, including habitat destruction, introduction of exotic pest species and pollution. Not surprisingly, conservation of the world's remaining species has come to assume great importance. This will be the topic of our final chapter.

25.1.1 What should we conserve?

the meanings of biodiversity

The term 'biodiversity' makes frequent appearances in both the popular media and scientific conservation journals, often without an unambiguous definition. This is unacceptable; it is important to be clear just what biodiversity is. At its simplest it is species richness, the number of species present in an arbitrarily defined geographical unit. As we have seen in Chapter 17, more complex indices of species diversity can be devised to take into account the relative abundances, biomasses or productivities of the coexisting species. Biodiversity can also be viewed at scales smaller and larger than that of the species. For example, we may wish to retain as much genetic diversity as possible in each species, perhaps seeking to conserve genetically distinct subpopulations and subspecies. At a higher level, we may focus attention on phylogenetic distinctiveness to ensure that species without close relatives are afforded special protection so that the overall evolutionary variety of the world's biota is maintained as large as possible. Another measure of biodiversity at the large scale is the variety of community types present in a region—swamps, deserts, early and late stages in a woodland succession, and so on.

There are many nuances in the debate about what constitutes biodiversity: is a field with two grass species as biodiverse as one with a grass and a rabbit; does a plant with polymorphic leaves add more biodiversity than one that is monomorphic; does a tree species add more biodiversity than an animal, because the former contributes so many more resources for other species? We will not pursue this argument any further here (see Hawksworth, 1994, and references therein, for a fuller discussion). Suffice it to say that when people talk generally about biodiversity they are concerned with everything that contributes to variety in the living world. On the other hand, it will be necessary to be specific if the term is to be of any practical use. Thus, ecologists must define precisely what it is they mean to conserve in their particular circumstances, and how to measure whether this has been achieved.

ecologists need to be specific

25.1.2 The scale of the problem

Most often the focus of concern is the rate of extinction of species in the face of human influence. To judge the scale of this problem we need to know the total number of species that occur in the world, the rate at which these are going extinct and how this rate compares with that of pre-human times. Unfortunately, there are considerable uncertainties in our estimates of all these things.

how many species on earth?

About 1.8 million species have so far been named, but the real number must be much larger. Estimates have been derived in a variety of ways (see May, 1990a, for details and related references). One approach is based on a general observation that for every temperate or boreal mammal or bird there are approximately two tropical counterparts. If this is assumed also to hold for insects, the grand total would be around 3–5 million. Another approach uses information on the rate of discovery of new species to project forward, group by group, to a total estimate of up to 6–7 million species in the world. A third approach is based on a species size–species richness relationship, taking as its starting point that as one goes down from terrestrial animals whose characteristic linear dimensions are a few metres to those of around 1 cm, there is an approximate empirical rule that for each 10-fold reduction in length there are 100 times the number of species. If the observed pattern is arbitrarily extrapolated down to animals of a characteristic length of 0.2 mm, we arrive at a global total of around 10 million species of terrestrial animal. A fourth approach is based on estimates of beetle species richness (more that 1000 species recorded in one tree) in the canopies of tropical trees (about 50 000 species), and assumptions about the proportion of non-beetle arthropods that will also be present in the canopy plus other species that do not occupy the canopy; this yields an estimate of about 30 million tropical arthropods. The uncertainties in estimating global species richness are profound and our best guesses range from 3 to 30 million or more.

modern extinction rates ...

An analysis of recorded extinctions during the modern period of human history shows that the majority have occurred on islands and that birds and mammals have been particularly badly affected (Figure 25.1). The percentage of extant species involved appears at first glance to be quite small, and moreover, the extinction rate appears to have dropped in the latter half of the 20th century, but how good are these data?

Once again, these estimates are bedevilled by uncertainty. First, the data are much better for some taxa and in some places than others, so the patterns in Figure 25.1 must be viewed with a good deal of scepticism. For example, there may be serious underestimates even for the comparatively well-studied birds and mammals because many tropical species have not received the careful attention needed for fully certified extinction. Second, a very large number of species have gone unrecorded and we will never know how many of these have become extinct. Finally, the drop in recorded extinctions in the latter half of the 20th century may signal some success for the conservation movement. But, it might equally well be due to the convention that a species is only denoted extinct if it has not been recorded for 50 years. Or, it may indicate that many of the most vulnerable species are already extinct.

... compared with historic extinction rates

An important lesson from the fossil record (see Chapter 24) is that the vast majority, probably all, of extant species will become extinct eventually—more than 99% of species that ever existed are now extinct (Simpson, 1952). However, given that individual species are believed, on average, to have lasted about 1–10 million years (Raup, 1978), and if we take the total number of species on earth to be 10

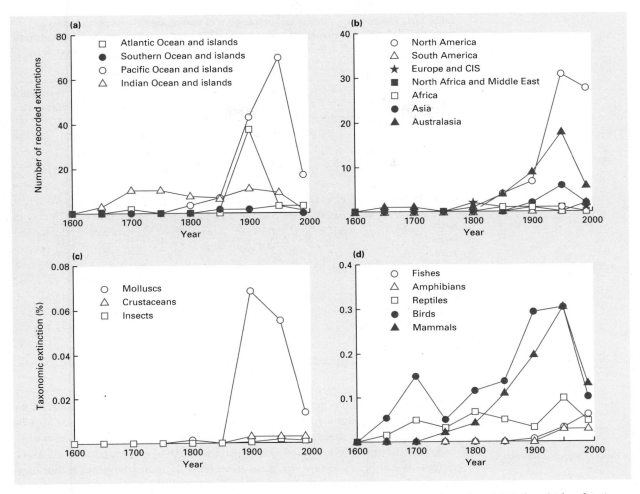

Figure 25.1 Trends in recorded animal species extinctions since 1600, for which a date is known, in (a) the major oceans and their islands, (b) major continental areas, for (c) invertebrates and (d) vertebrates. (After Smith *et al.*, 1993.)

a human-induced mass extinction?

million, we would predict that only an average of between 100 and 1000 species (0.001–0.01%) would go extinct each century. The current observed rate of extinction of birds and mammals of about 1% per century is 100–1000 times this 'natural' background rate. Furthermore, the scale of the most powerful human influence, that of habitat destruction, continues to increase and the list of endangered species in many taxa is alarmingly long (Table 25.1). We cannot be complacent. The evidence, whilst inconclusive because of the unavoidable difficulty of making accurate estimates, suggests that our children and grandchildren may live through a period of species extinction comparable to the five 'natural' mass extinctions evident in the geological record (see Chapter 24).

25.1.3 Why conserve?

To most people, biological diversity is undeniably of value, but it does not always lend itself to the economic valuation upon which policy decisions are usually based.

Table 25.1 The current numbers and percentages of named animal and plant species in major taxa judged to be threatened with extinction. The higher values associated with plants, birds and mammals may reflect our greater knowledge of these taxa. (After Smith *et al.*, 1993.)

Taxons	Number of threatened species	Approximate total species	Percentage threatened
Animals			
Molluscs	354	10^5	0.4
Crustaceans	126	4.0×10^3	3
Insects	873	1.2×10^6	0.07
Fishes	452	2.4×10^4	2
Amphibians	59	3.0×10^3	2
Reptiles	167	6.0×10^3	3
Birds	1029	9.5×10^3	11
Mammals	505	4.5×10^3	11
Total	3565	1.35×10^6	0.3
Plants			
Gymnosperms	242	758	32
Monocotyledons	4421	5.2×10^4	9
Monocotyledons: palms	925	2820	33
Dicotyledons	17 474	1.9×10^5	9
Total	22 137	2.4×10^5	9

a new ecological economics?

Standard economics has generally failed to assign value to ecological resources, so that the costs of environmental damage or depletion of living resources have usually been disregarded. A major challenge for the future is the development of a new *ecological economics* (Costanza, 1991) in which the worth of species, communities and ecosystems can be set against the gains to be made in industrial projects that may damage them. Some feel that this kind of economic valuation is inappropriate because features of the natural world are often unique and thus priceless (Ehrenfeld, 1988). However, as ecologists we need to remain part of the mainstream of the current economically driven world if we wish to be able to influence policy decisions.

Value can be divided into three main components: (i) direct value of products that are harvested; (ii) indirect value where aspects of biodiversity bring economic benefit without the need to consume the resource (McNeeley, 1988); and (iii) ethical value.

species may have direct economic value ...

Many species are recognized, now, as having actual *direct value* as living resources; many more species are likely to have a potential value which as yet remains untapped (Miller, 1988).

Wild meat and plants remain a vital resource in many parts of the world, whilst most of the world's food is derived from plants that were originally domesticated from wild plants in tropical and semi-arid regions. In future, wild strains of these species may be needed for their genetic diversity in attempts to breed for improved yield, pest resistance, drought resistance, and so on, and quite different species of plants and animals may be found that are appropriate for domestication.

In another context, we have seen in Chapter 16 the potential benefits that could come from natural enemies if they can be used as biological control agents for pest species. Most natural enemies of most pests remain unstudied and often unrecognized.

About 40% of the prescription and non-prescription drugs used throughout the world have active ingredients extracted from plants and animals. Aspirin, probably the world's most widely used drug, was derived originally from the leaves of the tropical willow, *Salix alba*. The nine-banded armadillo (*Dasypus novemcinctus*) has been used to study leprosy and prepare a vaccine for the disease; the Florida manatee (*Trichechus manatus*), an endangered mammal, is being used to help understand haemophilia; whilst the rose periwinkle (*Catharanthus roseus*), a marine snail from Madagasgar, has yielded two potent drugs effective in treating blood cancer. These are by no means isolated cases and a large-scale world-wide search is under way to discover organisms with new medicinal applications, including the treatment of acquired immune deficiency syndrome (AIDS). The vast majority of the world's animals and plants have yet to be screened—the potential value of any that go extinct can never be realized. By conserving species, we maintain their *option value*—the potential to provide benefit in the future.

... indirect economic value

Non-consumptive, *indirect economic value* is sometimes relatively easy to calculate. For example, a multitude of wild insect species are responsible for pollinating crop plants. The value of these pollinators could be assigned either by calculating the extent to which the insects increase the value of the crop or by the expenditure necessary to hire hives of honeybees to do the pollinating (Primack, 1993). In a related context, the monetary value of recreation and ecotourism, often called *amenity value*, is becoming ever more considerable. For example, each year nearly 200 million adults and children in the USA take part in nature recreation and spend about $4 billion on fees, travel, lodging, food and equipment. Moreover, ecotourists, who visit a country wholly or partly to experience its biological diversity, spend approximately $12 billion per year world-wide on their enjoyment of the natural world (Primack, 1993). On a smaller scale, a multitude of natural history films, books and educational programmes are 'consumed' annually without harming the wildlife upon which they are based. More ingenuity is required to assign value to other indirect economic benefits that accrue as a result of natural ecosystems; for example, biological communities can be of vital importance in maintaining the chemical quality of natural waters, in buffering ecosystems against floods and droughts, in protecting and maintaining soils, in regulating local and even global climate and in breaking down or immobilizing organic and inorganic wastes.

.... ethical value

The final category is *ethical value*. Many people believe that there are ethical grounds for conservation, arguing that every species is of value in its own right, and would be of equal value even if people were not here to appreciate or exploit it. From this perspective even species with no conceivable economic value require protection. So-called 'deep ecologists' (Sessions, 1987) urge that everyone should act as if nature really mattered, and argue that existing political, economic, technological and idealogical structures need to change.

Of these three main reasons for conserving biodiversity, the first two, direct and indirect economic value, have a truly objective basis. The third, ethics, on the other hand, is subjective and is faced with the problem that a subjective reason will inevitably carry less weight with those not committed to the conservationist cause. The arguments in favour of conservation cannot be seen in isolation—not that there are really arguments against conservation as such, but there are arguments in favour of the human activities that make conservation a necessity: agriculture, the felling of

framing conservation arguments in cost–benefit terms

trees, the harvesting of wild animal populations, the exploitation of minerals, the burning of fossil fuels, irrigation, the discharge of wastes, and so on. To be effective,

917   CONSERVATION AND BIODIVERSITY

it may be that the arguments of conservationists must ultimately be framed in cost–benefit terms because governments will always determine their policies against a background of the money they have to spend and (sometimes) the priorities accepted by their electorates.

25.1.4 Where should we focus conservation effort?

From the point of view of species richness, we have already set out in Chapter 24 where the greatest concentrations of species can be found. On a global scale, this is often in tropical regions, whilst on a more local scale species richness varies according to factors such as productivity, spatial heterogeneity, habitat (or island) area, stage of succession and frequency of disturbance. If our task were simply to conserve the most species possible, we could begin by looking to these patterns to illuminate our choice of priority areas.

Of course, the 'where' of conservation biology depends on the 'why' and the 'what'. If a primary concern is to maintain the 'option value' of species that may turn out to be economically important as foods, biological control agents or medicines then our strategy should be to maintain the maximum species richness—but, we would only have to bother to maintain a single viable population of each species somewhere in the world. From another point of view, many would accept that some species are more important than others because of their phylogenetic uniqueness. Certain biogeographical regions have high concentrations of endemic species, so we may identify centres of endemism for special treatment (Pressey *et al.*, 1993). However, these approaches would be unlikely to satisfy the ethical requirements of those who hold that every species is of equal value.

the need to set priorities

Given the reality of a world in which limited money is devoted to conservation, the strategy must be to set objectives (the why and the what of conservation), and to identify for conservation action those priority species and areas that will allow the most effective use of available resources (the where of conservation). This point will be taken up again in Section 25.6.3.

25.2 Threats to species

25.2.1 Natural and human-induced rarity

The aim of biological conservation is to prevent individual species, or sometimes whole communities, from becoming extinct either regionally or globally. Several categories of risk of species extinction have been defined (Mace & Lande, 1991). A species can be described as *vulnerable* if there is considered to be a 10% probability of extinction within 100 years, *endangered* if the probability is 20% within 20 years or 10 generations, whichever is longer and *critical* if within 5 years or two generations the risk of extinction is at least 50% (Figure 25.2). Based on these criteria, 43% of vertebrate species have been classified as threatened (i.e. they fell into one of the above categories) (Mace, 1994). Species that are at high risk of extinction are almost always rare, although, as we shall see, not all rare species are at risk. We need to ask, at least in a qualitative sense, 'what precisely do we mean by rare?'

the classification of threat to species

Abundance is not just a question of density within an inhabited area—an aspect we may refer to as *intensity*. The concept must also take into account the number and size of inhabited areas within the region as a whole (i.e. the *prevalence*—see

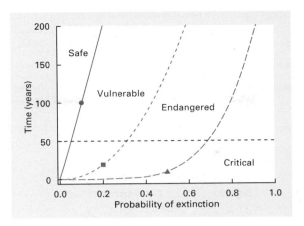

Figure 25.2 Levels of threat as a function of time and probability of extinction. (After Akçakaya, 1992.)

Chapter 4). Of the world's millions of species of animals and perhaps 300 000 species of plant, most are absent from most places for most of the time; the majority are strongly clumped in their distributions. When we ask what determines the abundance of a species in a particular place, we are really concerned with the tip of the tail of the problem. It is the zeros—the absences from most localities—that are the overwhelming importance in any species list that we prepare.

By distinguishing between the prevalence of a species and the intensity with which its populations are concentrated, it can be recognized that the terms 'common' and 'rare' are very unsatisfactory without qualification. A species may be rare: (i) in the sense that its geographical range is narrow; (ii) in the sense that its habitat range is narrow (both of these are aspects of prevalence); or (iii) in the sense that local populations, even where they do occur, are small and non-dominant (low intensity). Thus, as set out in Table 25.2 (following Rabinowitz, 1981), there are eight types of commonness or rarity, only one of which it would be impossible to consider 'rare'—namely species with a large geographical range, wide habitat specificity and large local populations.

Species that are rare on all three counts, such as the giant panda (*Ailuropoda melanoleuca*) and the giant cave weta (*Gymnoplectron giganteum*, a primitive orthopteran insect from New Zealand) are intrinsically vulnerable to extinction. However, species need only be rare in one sense in order to become endangered. For example, the peregrine falcon (*Falco peregrinus*) is broadly distributed across habitats and geographical regions, yet, because it exists always at low densities, local populations in the USA have become extinct and have had to be re-established with individuals bred in captivity. The grand Otago skink (*Leiolopisma grande*), on the other hand, is found in only one part of New Zealand and is restricted to schist gullies containing native tussock grasses—but, where it occurs, it does so at relatively high densities. It is at risk because changes in land use threaten its special habitats.

Nevertheless, rare species, just by virtue of their rarity, are not necessarily at risk of extinction. It is clear that many, probably most, species are naturally rare. The population dynamics of such species may follow a characteristic pattern. For example, out of a group of four species of *Calochortus* lilies in California, one is abundant and three are rare (Fiedler, 1987). The rare species have larger bulbs but produce fewer fruits per plant and have a smaller probability of survival to reproductive age than the common species. The rare species can all be categorized as

seven types of rarity

... any of which might require conservation measures

Table 25.2 A classification of types of commonness and rarity. (After Rabinowitz, 1981.)

Geographical range	Large		Small	
Habitat specificity	Wide	Narrow	Wide	Narrow
Local populations, large — Plant examples	Fat hen (*Chenopodium album*)	Red mangrove (*Rhizophora mangle*) found only in mangrove swamps	Pygmy cypress (*Cupressus pygmaea*)	Double coconut (*Lodoicea seychellarum*)
Local populations, large — Animal examples	Brown rat (*Rattus norvegicus*) Starling (*Sturnus vulgaris*)	Mudskipper (Periophthalmidae) found only in mangrove swamps		
Local populations, small — Plant examples	Knotroot bristle grass (*Setaria geniculata*)—always inconspicuous	Canadian yew (*Taxus canadensis*)		Stinking cedar (*Torreya taxifolia*)
Local populations, small — Animal examples	Peregrine falcon (*Falco peregrinus*)—widespread but, like many top predators, always at low density	Osprey (*Pandion haliaetus*)—another top predator bird, but feeding only on fish		Giant panda (*Ailuropoda melanoleuca*)

climax species that are restricted to unusual soil types, whilst the common one is a colonizer of disturbed habitats. Rare taxa may generally have a tendency towards asexual reproduction, lower overall reproductive effort and poorer dispersal abilities (Kunin & Gaston, 1993). In the absence of human interference there is no reason to expect that the rarer types would be substantially more at risk of extinction.

species in double jeopardy

However, other things being equal, it will be easier to make a rare species extinct, simply because a localized effect may be sufficient to push it to the brink. The problem is compounded because there appears to be a general, positive correlation between local abundance and size of geographical range (see Figure 25.3 for an example involving birds), within a variety of taxa, including plants, birds, mammals, fish, molluscs, mites and insects, species restricted to a narrow geographical area tend to have lower local abundances (Lawton *et al.*, 1994—although, this is not invariably true, as exemplified by the peregrine falcon above). Thus, many species are subject to double jeopardy if human activities increase their rates of mortality or decrease their natality.

some species have rarity thrust upon them

Some species are born rare, others have rarity thrust upon them. The actions of humans have undoubtedly affected both the prevalence and intensity aspects of abundance. The main influences are discussed in turn below.

25.2.2 Overexploitation

large species prone to overexploitation ...

We have already reviewed the evidence that in prehistoric times humans were responsible for the extinction of many large animals by overexploiting them (see Chapter 24). The so-called megaherbivores were vulnerable and profitable and an annual harvest level of greater than about 10% was likely to be unsustainable

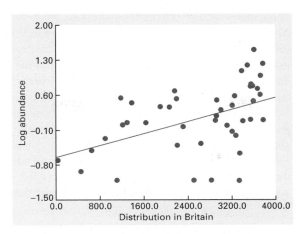

Figure 25.3 An example of a positive correlation between local abundance and size of geographical range for British breeding birds. Abundances are average numbers of singing territorial males in a 16-ha (1.6×10^5 m^2) wood in eastern England. Distribution is the total number of 10-km squares with possible, probable or confirmed breeding in Great Britain and Ireland. (After Lawton, 1993.)

(Owen-Smith, 1987). In more recent times, the history of the great whales (see Chapter 16, Section 16.13) has followed a similar pattern, whilst today we are still taking our toll of other vulnerable giants. Sharks provide an interesting example. Amongst the most feared of species (although attacks are much rarer than held in the popular imagination), large numbers are taken for sport, many others to make shark-fin soup, whilst a large proportion of the estimated annual 200 million shark kills are accidental by-catches of commercial fishing. Evidence is mounting that many species of shark have been declining in abundance, a trend that should come as no surprise given their late ages of maturity, slow reproductive cycles and low fecundities (Manire & Gruber, 1990). Sharks are amongst the most important predators in the marine environment, and their enforced rarity may have widespread repercussions in ocean communities.

... and so are some small ones

Overexploitation is not restricted to large vertebrates. The freshwater mussel *Margaritifera auricularia*, was once widespread in the large rivers of western Europe and Morocco. Now, there is only one known population, in the lower Ebro River in Spain. The species has long been collected for its attractive shell and pearls (even being used for ornamentation in the Stone Age), and disappeared from central Europe in the 15th and 16th centuries (Altaba, 1990). Like most endangered species, it has more than one problem to contend with. The larval stage is an ectoparasite of certain fish species, with the sturgeon (*Acipenser sturio*) believed to be principally involved. This fish is itself seriously threatened because of overfishing (for its eggs, also known as caviar), and both fish and mussel have been adversely affected by water pollution and habitat destruction. The outlook is bleak indeed for the mollusc whose only remaining population is made up entirely of old individuals with no recent signs of successful reproduction.

the threat posed by collectors

A feature of animals that are collected for ornamentation, whether for their body parts or as exotic pets, is that their value to collectors goes up as they become rarer; thus, instead of the normal safeguard of a density dependent reduction in consumption rate at low density (see Chapter 10) the very opposite occurs. The

921   CONSERVATION AND BIODIVERSITY

phenomenon is not restricted to animals. Thus, New Zealand's endemic mistletoe (*Trilepidia adamsii*), parasitic on a few forest understorey shrubs and small trees, was undoubtedly overcollected to provide herbarium specimens. Always a rare species, its extinction (recorded from 1867 to 1954, but not seen since—Norton, 1991) was due to overcollecting combined with forest clearance and perhaps an adverse effect on fruit dispersal because of reductions in bird populations.

The essence of overexploitation is that populations are harvested at a rate that, given their natural rates of mortality and capacities for reproduction, is unsustainable (see Chapter 16). Overexploitation is avoidable given the will to gather the relevant data, build the appropriate models and adequately police the exploiters so that recommended harvest levels are not exceeded. However, for any species this is an expensive undertaking and much easier said than done.

25.2.3 Habitat disruption

Habitats may be adversely affected by human influence in three main ways.

1 A proportion of the habitat available to a particular species may simply be destroyed, for urban and industrial development or for the production of food and other natural resources such as timber.

2 Habitat may be degraded by pollution to the extent that conditions become untenable for certain species.

3 Habitat may be disturbed by human activities to the detriment of some of its occupants.

habitat may be destroyed ...

Forest clearance has been, and is still, the most pervasive of the forces of habitat destruction, with annual deforestation rates often at the level of 1% or more per annum in tropical rainforest areas, whilst much of the native temperate forests in the developed world were destroyed long ago. More generally, in excess of 50% of wildlife habitat has been destroyed in 49 out of 61 Old World tropical countries, with particularly high losses of 80% or more in certain tropical Asian and sub-Saharan countries (Primack, 1993). Again, although quantitative data are not available, the losses of wildlife habitat in developed temperate areas are likely to be as large or larger. The process of habitat destruction often results in the habitat available to a particular species being more fragmented than was historically the case. This can have several repercussions for the populations concerned, a point that we take up again in Section 25.3.3.

... or degraded

Degradation by pollution can take many forms, from the application of pesticides that harm non-target organisms, to acid rain with its adverse effects on organisms as diverse as forest trees, amphibians in ponds and fish in lakes, to global climate change that may turn out to have the most pervasive influence of all (see Chapters 2 and 19). Aquatic environments are particularly vulnerable to pollution. Water, inorganic chemicals and organic matter enter from drainage basins, with which streams, rivers, lakes and continental shelves are intimately connected. Land-use changes, waste disposal and water impoundment and abstraction can profoundly affect their patterns of water flow and the quality of their water (Allan & Flecker, 1993).

... or disturbed

Habitat disturbance is not such a pervasive influence as destruction or degradation, but certain species are particularly vulnerable. For example, censuses of grey bats (*Myotis grisescens*) in caves in Alabama/Tennessee in 1968–1970 and again in 1976 indicated population declines that ranged from 0 to 100%. Disturbance by

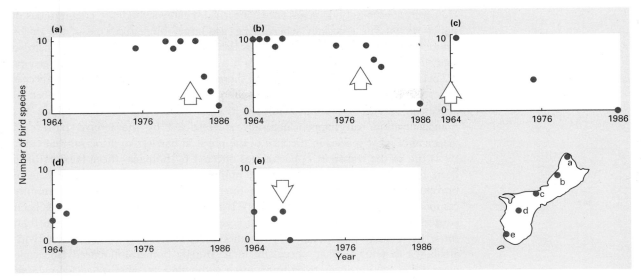

Figure 25.4 Decline in the number of forest bird species at five locations on the island of Guam. Large arrows indicate the first sightings of Boiga at each location (in location d, Boiga was first sighted in the early 1950s). (After Savidge, 1987.)

visitors to the caves seems to have been responsible; where visits were fewer than one per month only 0–20% of bats were lost, but caves with more than four visits per month suffered population declines of between 86 and 95% (Tuttle, 1979). Nature recreation, ecotourism and even ecological research are not without risk of disturbance and decline of the populations concerned.

25.2.4 Introduced species

a flood of introduced species …

Invasions of exotic species into new geographical areas sometimes occur naturally and without human agency. However, human actions have increased this trickle to a flood. Human-caused introductions may occur accidentally as a consequence of human transport, intentionally but illegally to serve some private purpose, and legitimately to procure some hoped-for public benefit by bringing a pest under biotic control, producing new agricultural products or providing novel recreational opportunities. Many introduced species are assimilated into communities without much obvious effect. However, some have been responsible for dramatic changes to native species and natural communities.

… that reduce biodiversity

For example, the accidental introduction of the brown tree-snake *Boiga irregularis*, onto Guam, an island in the Pacific, has through nest predation reduced 10 endemic forest bird species to the point of extinction (Savidge, 1987). The gradual spread of the snake from its bridgehead population in the centre of the island has been paralleled by the timing of the loss of bird species to the north and south (Figure 25.4). Similarly, the introduction as a source of human food of the predaceous Nile perch (*Lates nilotica*) to the enormously species-rich Lake Victoria in East Africa has driven most of its 350 endemic species of fish to extinction or near extinction (Kaufman, 1992).

The effects of introduced species seem to have been particularly harshly felt on islands and in freshwater communities, which are also insular in their own way.

However, conservation biologists are concerned wherever there are communities of native organisms, especially where these are largely endemic. One of the major reasons for the world's huge diversity is the occurrence of centres of endemism so that similar habitats in different parts of the world are occupied by different groups of species that happen to have evolved there. If every species naturally had access to everywhere on the globe, we might expect a relatively small number of successful species to become dominant in each biome. The extent to which this homogenization can happen naturally is restricted by the limited powers of dispersal of most species in the face of the physical barriers to dispersal that exist. By virtue of the transport opportunities offerred by humans, these barriers have been breached by an ever-increasing number of exotic species. The effects of introductions have been to convert a hugely diverse range of local community compositions into something much more homogeneous. In this context it has been suggested that Old World species have a generally higher invasive potential than those from other parts of the world (di Castri, 1989). Old World species with attributes that made them successful in a strongly human-influenced landscape were the ones most likely to be transported with colonists, whether accidentally or deliberately, and to be successful in the receiving communities, which quickly became human-influenced in a way that paralleled the Old World situation (Townsend, 1996b).

It would be wrong to conclude that introducing species to a region will inevitably cause a decline in species richness there. For example, there are numerous species of plants, invertebrates and vertebrates found in continental Europe but absent from the British Isles (many because they have so far failed to recolonize after the last glaciation). Their introduction would be likely to augment British biodiversity. The significant detrimental effect noted above arises where aggressive species, often of Old World origin, provide a novel challenge to endemic biotas ill-equipped to deal with them.

25.2.5 Genetic considerations

Theory tells conservation biologists to beware genetic problems in small populations that may arise through loss of genetic variation. Genetic variation is determined primarily by the joint action of natural selection and genetic drift (where the frequency of genes in a population is determined by chance rather than evolutionary advantage). The relative importance of genetic drift is higher in small isolated populations, which as a consequence are expected to lose genetic variation. The rate at which this happens depends on the *effective population size* (N_e). This is the size of the 'genetically idealized' population to which the actual population (N) is equivalent in genetic terms. As a first approximation, N_e is equal to or less than the number of breeding individuals. N_e is usually less, often much less, than N, for a number of reasons (detailed formulae can be found in Lande & Barrowclough, 1987):

1   if the sex ratio is not $1:1$, for example with 100 breeding males and 400 breeding females $N = 500$, but $N_e = 320$;

2   if the distribution of progeny from individual to individual is not random, for instance if 500 individuals each produce one individual for the next generation on average $N = 500$, but the variance in progeny production is 5 (with random variation this would be 1), then $N_e = 100$;

3   if population size varies from generation to generation, then N_e is disproportion-

ately influenced by the smaller sizes, for example for the sequence 500, 100, 200, 900 and 800, mean $N = 500$, but $N_e = 258$.

loss of evolutionary potential

The preservation of genetic diversity is important because of the long-term evolutionary potential it provides. Rare forms of a gene (alleles), or combinations of alleles, may confer no immediate advantage but could turn out to be well suited to changed environmental conditions in the future. Small populations that have lost rare alleles, through genetic drift, have less potential to adapt.

the risk of inbreeding depression

A more immediate potential problem is *inbreeding depression*. When populations are small there is a tendency for individuals breeding with one another to be related. Inbreeding reduces the heterozygosity of the offspring far below that of the population as a whole. More important, though, is that all populations carry recessive alleles which are deleterious, or even lethal, when homozygous; individuals that are forced to breed with close relatives are more likely to produce offspring where the harmful alleles are derived from both parents so the deleterious effect is expressed. There are many examples of inbreeding depression—breeders have long been aware of reductions in fertility, survivorship, growth rates and resistance to disease—although high levels of inbreeding may be normal and non-deleterious in some animal species (Wallis, 1994) and many plants.

50, 500 or 5000—magic genetic numbers?

How many individuals are needed to maintain genetic variability? Franklin (1980) suggested that an effective population size of about 50 would be unlikely to suffer from inbreeding depression, whilst 500 might be needed to maintain longer term evolutionary potential. Such rules of thumb should be applied cautiously and, bearing in mind the relationship between N_e and N, the minimum population size N should probably be set at five to 10 times N_e, i.e. 2500–5000 individuals (Nunney & Campbell, 1993).

25.2.6 A review of risks

Some species are at risk for a single reason but often, as in the cases of the freshwater mussel and the New Zealand mistletoe discussed earlier, a combination of factors is at work. A review of the factors responsible for recorded vertebrate extinctions showed that habitat loss, overexploitation and species introductions were all of considerable significance, although habitat loss was less prominent in the case of reptiles, and overexploitation hardly figured in the case of fishes (Table 25.3). As far as threatened extinctions are concerned, habitat loss is most commonly the major hazard, whilst risk of overexploitation remains very high, especially for mammals and reptiles.

relative risks—single factors or in combination

An extinction category that makes no appearance in Table 25.3 is 'secondary extinction'. The web of interactions that ties species together (see Chapter 23) means that the loss of one species, such as a specialized pollinator or prey or host, may lead to a chain of extinctions of other species that are highly dependent on the first. Thus, the well-known extinction in 1914 of the passenger pigeon (*Ectopistes migratorius*) was certainly accompanied by the extinction of at least two species of passenger pigeon louse (*Columbicola extinctus* and *Campanulotes defectus*) (Stork & Lyal, 1993).

It is interesting that no example of extinction due to genetic problems has come to light. Perhaps inbreeding depression has occurred, although undetected, as part of the 'death rattle' of some dying populations (Caughley, 1994). Thus, a population may have been reduced to a very small size by one or more of the processes described above and this may have led to an increased frequency of matings amongst

extinction vortices

Table 25.3 Review of the factors responsible for recorded extinctions of vertebrates and an assessment of risks currently facing species categorized globally as endangered, vulnerable or rare by the International Union for the Conservation of Nature (IUCN). (After Reid & Miller, 1989.)

	Percentage due to each cause*					
Group	Habitat loss	Overexploitation†	Species introduction	Predators	Other	Unknown
Extinctions						
Mammals	19	23	20	1	1	36
Birds	20	11	22	0	2	37
Reptiles	5	32	42	0	0	21
Fishes	35	4	30	0	4	48
Threatened extinctions						
Mammals	68	54	6	8	12	—
Birds	58	30	28	1	1	—
Reptiles	53	63	17	3	6	—
Amphibians	77	29	14	—	3	—
Fishes	78	12	28	—	2	—

* The values indicated represent the percentage of species that are influenced by the given factor. Some species may be influenced by more than one factor, thus, some rows may exceed 100%.
† Overexploitation includes commercial, sport, and subsistence hunting and live animal capture for any purpose.

relatives and the expression of deleterious recessive alleles in offspring, leading to reduced survivorship and fecundity, and causing the population to become smaller still—the so-called extinction vortex (Figure 25.5).

25.3 Uncertainty and the risk of extinction

25.3.1 The dynamics of small populations

Much of conservation biology is a crisis discipline. Inevitably we are confronted with too many problems and too few resources. Should we focus attention on the various forces that bring species to extinction and attempt to persuade governments to act to reduce their prevalence; or should we restrict our activities to identifying areas of high species richness where we can set up and protect reserves; or should we identify species at most risk of extinction and work out ways of keeping them in existence? Ideally, we should do all these things. However, the greatest pressure is often in the area of species preservation. The remaining population of pandas in China (or yellow-eyed penguins in New Zealand, or spotted owls in North America) has become so small that if nothing is done the species may become extinct within a few years or decades. Responding to the crisis requires that we devote scarce resources to identifying some special solutions; more general approaches may have to be put on the back-burner.

uncertainty and small populations

The dynamics of small populations are governed by a high level of uncertainty, whereas large populations can be described as being governed by the law of averages

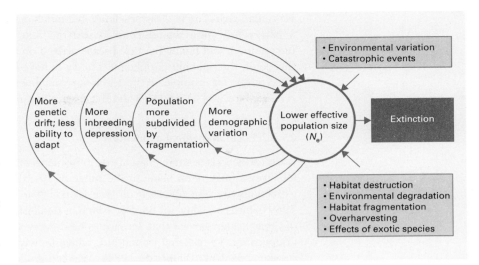

Figure 25.5 Extinction vortices may progressively lower population sizes leading inexorably to extinction. (After Primack, 1993.)

(Caughley, 1994). Three kinds of uncertainty or variation can be identified that are of particular importance to the fate of small populations.

1 Demographic uncertainty: random variations in the number of individuals that are born male or female, or in the number that happen to die or reproduce in a given year or in the quality (genotypic/phenotypic) of the individuals in terms of survival/reproductive capacities can matter very much to the fate of small populations. Suppose a breeding pair produces a clutch consisting entirely of females—such an event would go unnoticed in a large population but would be the last straw for a species down to its last pair.

2 Environmental uncertainty: unpredictable changes in environmental factors, whether 'disasters' (such as floods, storms or droughts of a magnitude that occurs very rarely—see Chapter 2) or more minor (year to year variation in average temperature or rainfall) can also seal the fate of a small population. Even where the average rainfall of an area is known accurately, because of records going back centuries, we cannot predict whether next year will be average or extreme, nor whether we are in for a number of years of particularly dry conditions. A small population is more likely than a large one to be reduced by adverse conditions to zero (extinction), or to numbers so low that recovery is impossible (quasi-extinction).

3 Spatial uncertainty: many species consist of an assemblage of subpopulations that occur in more or less discrete patches of habitat (habitat fragments). The overall population is a *metapopulation* (Levins, 1970). Since the subpopulations are likely to differ in terms of demographic uncertainty, and the patches they occupy in terms of environmental uncertainty, the patch dynamics of extinction and local recolonization can be expected to have a large influence on the chance of extinction of the metapopulation. We return to these ideas in Section 25.4.6.

To illustrate some of these ideas, take the demise in North America of the heath hen (*Tympanychus cupido cupido*) (see Simberloff, 1988). This bird was once extremely common from Maine to Virginia. Being highly edible and easy to shoot (and also susceptible to introduced cats and affected by conversion of its grassland habitat to farmland), by 1830 it had disappeared from the mainland and was only found on the island of Martha's Vineyard. In 1908 a reserve was established for the remaining 50 birds and by 1915 the population had increased to several thousand. However, 1916

demographic

environmental

spatial

the case of the heath hen

927 CONSERVATION AND BIODIVERSITY

was a bad year. Fire (a disaster) eliminated much of the breeding ground, there was a particularly hard winter coupled with an influx of goshawks (environmental uncertainty), and finally poultry disease arrived on the scene (another disaster). At this point, the remnant population is likely to have become subject to demographic uncertainty; for example, of the 13 birds remaining in 1928 only two were females. A single bird was left in 1930 and the species became extinct in 1932.

25.3.2 Local and global extinctions

Local extinctions of small populations in insular habitat patches are common events for diverse taxa, often being in the range of 10–20% per year (Figure 25.6). Such extinctions are also observed on true islands. The detailed records from 1954 to 1969 of birds breeding on Bardsey Island, a small island (1.8 km^2) off the west coast of Great Britain, reveal that 16 species bred every year, two of the original species disappeared, 15 flickered in and out, whilst four were initially absent but became regular breeders (Diamond, 1984). We can build a picture of frequent local extinctions, which in some cases are countered by recolonization from the mainland or other islands. Examples such as these provide a rich source of information about the factors affecting the fate of small populations in general, and the understanding gained is entirely applicable to species in danger of global extinction, since a global extinction is nothing more nor less than the final local extinction. A local extinction of an endemic species on a remote island is precisely equivalent to a global extinction, since recolonization is impossible. This is a principal reason for the high recorded rates of global extinction on islands (see Figure 25.1).

global extinction is simply the final local extinction

Of the high-risk factors associated with local extinctions, habitat or island area is probably the most pervasive. Figure 25.7 shows the negative relationships between annual extinction rate and area, for examples ranging from zooplankton in lakes, through vascular plants in mainland Sweden, to birds on a variety of island groups.

the importance of habitat or island area

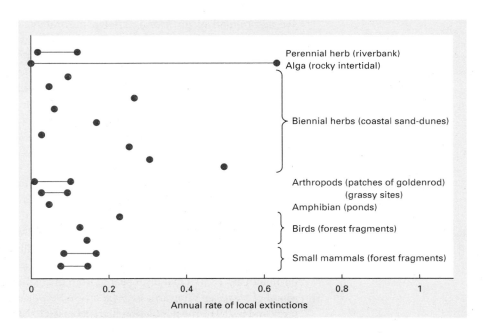

Figure 25.6 Fractions of local populations in habitat patches becoming extinct each year. (After Fahrig & Merriam, 1994.)

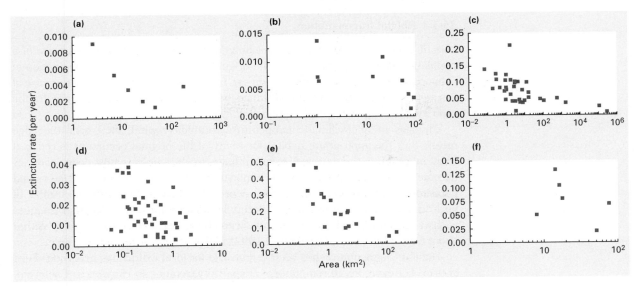

Figure 25.7 Percentage extinction rates as a function of habitat area for (a) zooplankton in lakes in the north-eastern USA lakes, (b) birds of the Californian Channel Islands, (c) birds on northern European islands, (d) vascular plants in southern Sweden, (e) birds on Finnish islands and (f) birds on the islands in Gatun Lake, Panama. (Data assembled by Pimm, 1991)

No doubt the main reason for the vulnerability of populations in small areas is the fact that the populations themselves are small. This is illustrated in Figure 25.8 for bird species on islands and for bighorn sheep in various desert areas in south-west USA.

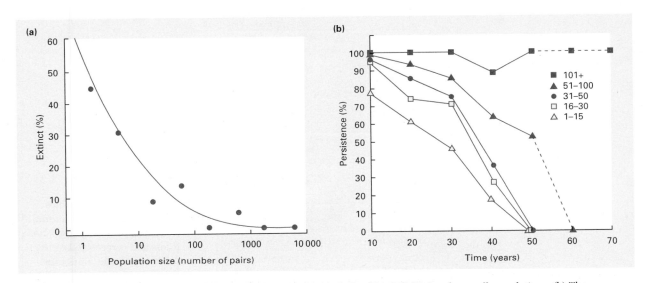

Figure 25.8 (a) Extinction rate of island birds is higher for small populations. (b) The percentage of populations of bighorn sheep in North America that persists over a 70-year period reduces with initial population size. (After Berger, 1990.)

929   CONSERVATION AND BIODIVERSITY

25.3.3 Habitat fragmentation

fragmentation and the metapopulation

edge effects

Loss of habitat frequently results not only in a reduction in the absolute size of a population but also the division of the original population into a metapopulation of subpopulations. Further fragmentation can result in a decrease in the average size of fragments, an increase in the distance between them and an increase in the proportion of edge habitat (Burgman *et al.*, 1993).

Changes in microclimate (temperature, humidity, wind, etc.) near the edge means that the appropriate habitat for many of the original occupants is reduced even more than the physical fragmentation might imply, whilst 'edge' species become more successful. Moreover, opportunities are opened up for predators and parasites which are more successful near the edge. The majority of studies of bird-nesting success in fragmented forest landscapes have revealed that enemies such as hawks and brood parasites take a greater toll near the edges, usually within about 50 m of the boundary (Paton, 1992). Figure 25.9 provides an example.

Habitat fragmentation has been responsible for local extinctions in a wide range of taxa. For example, in remnants of chaparral vegetation in canyons in California, species richnesses of plants, birds and rodents were reduced in comparison with that expected from similar-sized plots embedded in unfragmented habitat. Figure 25.10a presents results for chaparral birds. In all cases, species richness was significantly related to remaining habitat area (and thus, no doubt, to population size) and time

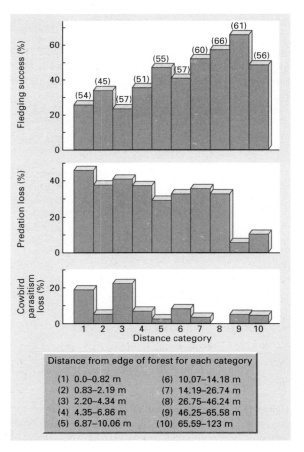

Figure 25.9 Relationships between percentage fledging success, percentage predation loss and percentage loss to brood parasites in nests at various distances from a forest edge in Michigan, USA. Sample sizes are in parentheses. (After Gates & Gysel, 1978.)

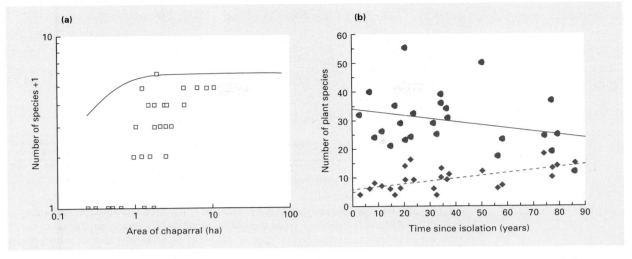

Figure 25.10 (a) A comparison of observed species richness of chaparral birds in habitat fragments of various sizes (□) with what would be expected in similar-sized plots of unfragmented chaparral vegetation (——). (After Bolger *et al.*, 1991.) (b) Native plant species richness decreases (and exotic species increase) as chaparral fragments get older. (After Soulé *et al.*, 1992.)

since the fragment became isolated (e.g. Figure 25.10b). The latter effects may be related to negative interactions with exotic species and, for animals, the reduction through time of true chaparral habitat within the fragments. Degree of isolation of the fragment (distance to nearest fragment or to undisturbed habitat) had no significant effect on species richness; apparently, the intervening barriers to dispersal, such as highways, were too great to be overcome by most of the species involved.

a fragmented population may be more at risk ...

A question of fundamental importance is whether a species is more at risk simply because its population is subdivided. In other words, would a single population of a given size be less or more at risk than one divided into a number of subpopulations in habitat fragments? The simplest models of metapopulation dynamics make predictions about the effects of fragment number, fragment size and isolation of habitat patches on metapopulation persistence.

In models that only include demographic uncertainty (the chance events of birth and death that *are not* correlated between individuals), subdivision into small fragments produces a metapopulation that is predicted to go extinct more quickly than an undivided population (Quinn & Hastings, 1987). When a proportion of individuals are allowed each generation to migrate between fragments, although the probability of extinction is reduced for a metapopulation of any particular degree of subdivision, undivided populations still persist longer (Figure 25.11a).

... or less at risk

A different outcome is obtained if environmental uncertainty (the chance events of birth and death that *are* correlated between individuals) is incorporated in the model. In this case, the probability of extinction decreases more slowly with population size so that a metapopulation of independent subpopulations may well persist longer than the equivalent undivided population because of a degree of spreading of risk (Figure 25.11b).

In their model, Quinn and Hastings (1987) ignore dispersal and assume that extinctions in different population fragments are uncorrelated. More realistically, we

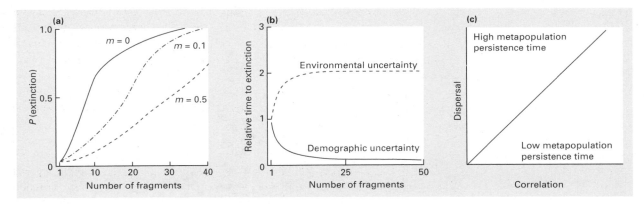

Figure 25.11 (a) Fate of metapopulations of identical initial size subdivided into various numbers of identical fragments, when there is no migration (*m* = 0) or when migration is permitted in each generation (*m* = 0.1 and *m* = 0.5; *m* is the probability that an individual will migrate to another fragment). The probability of extinction (*P*) is the proportion of 100 replicate simulations that go extinct in 50 generations. (After Burkey, 1989.) (b) The relative time to extinction increases with subdivision when environmental uncertainty is at play, in contrast to the effect of demographic uncertainty. (After Quinn & Hastings, 1987.) (c) Metapopulation persistence time can be expected to increase with between-patch dispersal rate but to decrease with increasing correlation in the dynamics of subpopulations. (After Hanski, 1989.)

may expect: (i) that where the probability of dispersal between fragments is high (because the organisms are intrinsically good dispersers, or the fragments are close together, or barriers to dispersal are absent or dispersal corridors of appropriate habitat are present), metapopulations will persist for longer; and (ii) that if extinction events are correlated (because environmental variation acts identically in all fragments), metapopulations will be more at risk (Figure 25.11c).

the case of the checkerspot butterfly

A long-term study of the checkerspot butterfly (*Euphydryas editha*) at Jasper Ridge (western USA) reveals that these phenomena are important in real

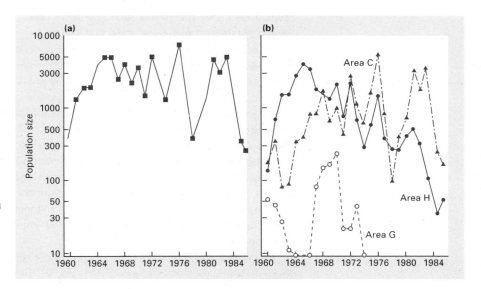

Figure 25.12 Abundance of checkerspot butterflies (a) when viewed as a single population and (b) when subdivided into three spatially distinct units. (After Ehrlich & Murphy, 1987.)

populations. Figure 25.12a shows that if we view the whole population (metapopulation) as a unit, a fair amount of year to year variation in abundance is evident, but with a suggestion of density dependence and an equilibrium density. However, when the population is divided into spatially distinct subpopulations we see evidence that environmental variation is not well correlated between subpopulations (patterns of increase and decrease are not very similar), and in area G the subpopulation went extinct twice, recolonizing the first time but not the second (Figure 25.12b).

fragmentation and population genetic structure

Population fragmentation clearly has marked effects on the dynamics of populations. It may also influence their genetic structure if fragmentation produces very small subpopulations. Subdivision may result in an increase in the overall effective population size (N_e) and in the expectation that more total genetic variation will be maintained. However, this may not confer the expected benefit if high variation is due to the differentiation of a number of relatively homozygous subpopulations; the advantages of genetic variation discussed in Section 25.3.5 involve either diverse genotypes being present in the same population or individual heterozygosity (Nunney & Campbell, 1993). As we have seen, inbreeding depression may also become a greater problem if a large population becomes subdivided into several smaller ones.

25.4 Population viability analysis

Given the environmental circumstances and species characteristics of a particular rare species, what is the chance it will go extinct in a specified period? Alternatively, how big must its population be to reduce the chance of extinction to an acceptable level? These are frequently the crunch questions in conservation biology. The ideal classical approach of experimentation, which might involve setting up and monitoring for several years a number of populations of various sizes, is unavailable to those concerned with species at risk, because the situation is usually too urgent and there are inevitably too few individuals to work with. How then are we to decide what constitutes the minimum viable population (MVP)? Three approaches will be discussed in turn: (i) a search for patterns in evidence already gathered in long-term studies; (ii) subjective assessment based on expert knowledge; and (iii) development of population models, both general and specific. All the approaches have their limitations, which we will explore by looking at particular examples.

trying to determine the minimum viable population (MVP) ...

25.4.1 Clues from long-term studies of biogeographical patterns

Data sets such as the one displayed in Figure 25.8b are unusual because they depend on a long-term commitment to monitoring a number of populations—in this case, bighorn sheep in desert areas. If we set an arbitrary definition of the necessary MVP as one that will give at least a 95% probability of persistence for 100 years, we can explore the data on the fate of bighorn populations to provide an approximate answer. Populations of fewer than 50 individuals all went extinct within 50 years whilst only 50% of populations of 51–100 sheep lasted for 50 years. Evidently, for our MVP we require more than 100 individuals; in the study, such populations demonstrated close to 100% success over the maximum period studied of 70 years.

... from available biogeographical data ...

A similar analysis of long-term records of birds on the Californian Channel Islands indicates an MVP of between 100 and 1000 pairs of birds (needed to provide a

Table 25.4 The relationship for a variety of bird species on the Californian Channel Islands between initial population size and probability of populations persisting. (After Thomas, 1990.)

Population size (pairs)	Time period (years)	Percentage persisting
1–10	80	61
10–100	80	90
100–1000	80	99
1000 +	80	100

probability of persistence of between 90 and 99% for the 80 years of the study—Table 25.4).

Studies such as these are rare and valuable. The long-term data are available because of the extraordinary interest people have in hunting (bighorn sheep) and ornithology (Californian birds). Their value for conservation, however, is limited because they deal with species that are generally not at risk. It is at our peril that we use them to produce recommendations for management of endangered species. There will be a temptation to report to a manager 'if you have a population of more than 100 pairs of your bird species you are above the minimum viable threshold.' Indeed, such a statement would not be without value. But, it will only be a safe recommendation if the species of concern and the ones in the study are sufficiently similar in their vital statistics, and if the environmental regimes are similar, something that it would rarely be safe to assume.

... a risky approach

25.4.2 Subjective expert assessment

in the minds of experts—decision analysis

Information that may be relevant to a conservation crisis exists not only in the scientific literature but also in the minds of experts. By bringing experts together in a conservation workshop, well-informed decisions can be reached. The recommended actions may not be entirely correct, but when a decision about an endangered species has to be made without the possibility of further research, this approach may provide an acceptable way forward.

the case of the Sumatran rhinocerus

To illustrate the strengths and weaknesses of the approach, we take as our example the results of a workshop concerning the Sumatran rhinocerus (*Dicerorhinus sumatrensis*). The species persists only in small, isolated subpopulations in an increasingly fragmented habitat in Sabah (East Malaysia), Indonesia and West Malaysia, and perhaps also in Thailand and Burma. Unprotected habitat is threatened by timber harvest, human resettlement and hydroelectric development. There are only a few designated reserves, which are themselves subject to exploitation, and only two individuals were held in captivity at the time of the workshop.

The vulnerability of the Sumatran rhinocerus, the way this vulnerability varies with different management options and the most appropriate management option given various criteria were assessed by a technique known as *decision analysis*. A decision tree is shown in Figure 25.13, based on the estimated probabilities of the species becoming extinct within a 30-year period (equivalent to approximately two rhinocerus generations). The tree was constructed in the following way. The two small squares are decision points: the first distinguishes between intervention on the rhinocerus' behalf and non-intervention (status quo); the second distinguishes the

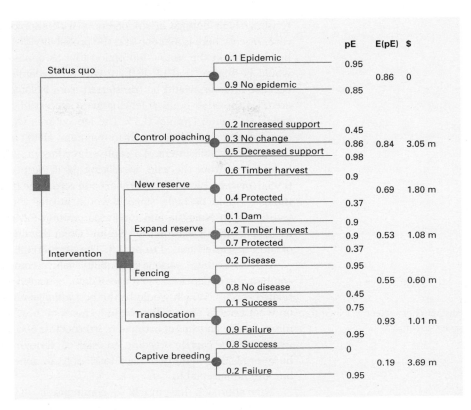

Figure 25.13 Decision tree for management of the Sumatran rhinoceros. ■, Decision points; ●, random events. Probabilities of random events are estimated for a 30-year period; pE, probability of species extinction within 30 years; E(pE), expected value of pE for each alternative. Costs are present values of 30-year costs discounted at 4% per year; m, million. (After Maguire *et al.*, 1987.)

various management options. For each option, the line branches at a small circle. The branches represent alternative scenarios that might occur, and the numbers on each branch indicate the probabilities estimated for the alternative scenarios. Thus, for the status quo option, there was estimated to be a probability of 0.1 that a disease epidemic will occur in the next 30 years, and hence, a probability of 0.9 that no epidemic will occur.

If there is an epidemic, the probability of extinction (pE) is estimated to be 0.95 (i.e. 95% probability of extinction in 30 years), whereas with no epidemic pE is 0.85. The overall estimate of species extinction for an option, E(pE), is then given by:

E(pE) = probability of first option × pE for first option
    + probability of second option × pE for second option   (25.1)

which, in the case of the status quo option, is:

E(pE) = (0.1 × 0.95) + (0.9 × 0.85) = 0.86.   (25.2)

The values of pE and E(pE) for the other options are calculated in a similar way. The final column in Figure 25.13 then lists the estimated costs of the various options.

We will consider. here two of the interventionist management options in a little more detail The first is to fence an area in an existing or new reserve, managing the resulting high density of rhinoceruses with supplemental feeding and veterinary care. Disease here is a major risk: the probability of an epidemic was estimated to be higher than in the status quo option (0.2 as opposed to 0.1), because the density would be higher, and pE if there was an epidemic (0.95) was considered higher because animals would be transferred from isolated subpopulations to the fenced area. On the other hand, if fencing were successful, pE was expected to fall to 0.45, giving an overall E(pE) of 0.55. The fenced area would cost around US$60 000 to establish and $18 000 per year to maintain, giving a 30-year total of $0.60 million.

For the establishment of a captive breeding programme, animals would have to be captured from the wild, increasing pE if the programme failed to an expected 0.95. However, pE would clearly drop to zero if the programme succeeded. The cost, though, would be high, since it would involve the development of facilities and techniques in Malaysia and Indonesia (around $2.06 million) and the extension of those that already exist in the USA and Great Britain ($1.63 million). The probability of success was estimated to be 0.8. The overall E(pE) is therefore 0.19.

strengths of subjective expert assessment ...

Where do these various probability values come from? The answer is from a combination of data, educated use of data, educated guesswork and experience with related species. Which would be the best management option? The answer depends on what criteria are used in the judgement of 'best'. Suppose we wanted simply to minimize the chances of extinction, irrespective of cost. The proposal for best option would then be captive breeding. In practice, though, costs are most unlikely ever to be ignored. We would then need to identify an option with an acceptably low E(pE) but also an acceptable cost.

... and weaknesses

This approach has much to commend it. It makes use of available data, knowledge and experience in a situation when a decision is needed and time for further research is unavailable. Moreover, it explores the various options in a systematic manner and does not duck the regrettable but inevitable truth that unlimited resources will not be available. However, it also runs a risk. In the absence of all necessary data, the recommended best option may simply be wrong. With the benefit of hindsight (and in all probability some rhinocerus experts who were not part of the workshop would have suggested this alternative outcome), we can now report that about $2.5 million have been spent catching Sumatran rhinoceruses; three died during capture, six died post-capture, and of the 21 rhinoceruses now in captivity only one has given birth and she was pregnant when captured (data of N. Leader-Williams reported in Caughley, 1994). Leader-Williams suggests that $2.5 million could effectively have been used to protect 700 km^2 of prime rhinocerus habitat for nearly two decades. This could in theory hold a population of 70 Sumatran rhinoceruses which, with a rate of increase of 0.06 per individual per year (shown by other rhinocerus species given adequate protection), might give birth to 90 calves during that period.

25.4.3 A general mathematical model of population persistence time

a general modelling approach ...

At its simplest, the likely persistence time of a population, T, can be expected to be influenced by its size, N, its intrinsic rate of increase, r, and the variance in r resulting from variation in environmental conditions through time, V. Demographic uncertainty is only expected to be influential in very small populations; persistence time increases from a low level with population size when numbers are tiny, but approaches infinity at a relatively small population (dashed curve in Figure 25.14).

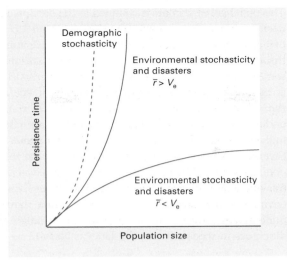

Figure 25.14 Relationships between population persistence time and population size, both on arbitrary scales, when the population is subject to demographic uncertainty or to environmental uncertainty/disasters. (After Lande, 1993.)

Various researchers have manipulated the mathematics of population growth, allowing for uncertainty in the expression of the intrinsic rate of increase, to provide an explicit estimate, T, of mean time to extinction as a function of carrying capacity, K (briefly reviewed by Caughley, 1994). Making a number of approximations (e.g. that demographic uncertainty is inconsequential, and that r is constant except if the population is at the carrying capacity when r is zero), Lande (1993) has produced one of the more accessible equations:

$$T = \frac{2}{Vc}\left(\frac{K^c - 1}{c} - \ln K\right) \tag{25.3}$$

where:

$$c = 2r/V - 1 \tag{25.4}$$

r is the intrinsic rate of increase, V is the variance in r resulting from variation in environmental conditions through time.

This equation is the basis for the solid curves in Figure 25.14, which indicate that mean time to extinction is higher for larger maximum population sizes (K), for greater intrinsic rates of population growth and when environmental influences on the expression of r are smaller. In contrast to earlier claims that random disasters pose a greater threat than smaller environmental variation, it turns out that what really matters is the relationship between the mean and variance of r (Lande, 1993; Caughley, 1994). The relationship between persistence time and population size curves sharply upwards (i.e. is influential only for small or intermediate population sizes) if the mean rate of increase is greater than the variance, whereas if the variance is greater than the mean the relationship is convex—so that, even at large population sizes, environmental uncertainty still has an influence on likely persistence time. This all makes intuitive good sense but can it be put to practical use?

... put to the test

In their study of the Tana River crested mangabey (*Cercocebus galeritus galeritus*) in Kenya, Kinnaird and O'Brien (1991) used a similar equation to estimate the population size (K) needed to provide a 95% probability of persistence for 100 years.

This endangered primate is confined to the floodplain forest of a single river where it declined in numbers from 1200 to 700 between 1973 and 1988 despite the creation of a reserve. Its naturally patchy habitat has become progressively more fragmented through agricultural expansion. The model parameters, estimated on the basis of some real population data, were taken to be $r = 0.11$ and $V = 0.20$. The latter was particularly uncertain, because only a few year's data were available. Substituting in the model yielded a MVP of 8000. Using the standard rule of thumb described earlier, that to avoid genetical problems an effective population size of 500 individuals is needed, an actual population of about 5000 individuals was indicated. Given the available habitat it was concluded that the mangabeys could not attain a population size of 5000–8000. Moreover, Kinnaird and O'Brien think it unlikely that this naturally rare and restricted species ever has. Either the data were deficient (e.g. environmental variation in r may be smaller than estimated if they are able to undergo dietary shifts in response to habitat change) or the model is too general to be much use in specific cases. The latter is likely to be true.

25.4.4 Simulation models

the species-specific approach— simulation modelling

Simulation models provide an alternative, more specific way of gauging viability. Usually, these encapsulate survivorships and reproductive rates in age-structured populations. Random variations in these elements or in K can be employed to represent the impact of environmental variation, including that of disasters of specified frequency and intensity. Density dependence can be introduced where required, as can population harvesting or supplementation. The engine of the simulation is a Monte Carlo algorithm and every individual is treated separately in terms of the probability, with its imposed uncertainty, that it will survive or produce a certain number of offspring in the current time period. The programme is run many times, each giving a different population trajectory because of the random elements involved. The outputs, for each set of model parameters used, are estimates of the probability of extinction (the proportion of simulated populations that go extinct) over a specified time interval and the mean time to extinction of populations that went extinct.

the case of the Florida panther

This approach, using a model called Vortex, has been applied to the Florida panther (*Felis concolor coryi*), whose population is believed to be 30–50 animals and declining. The basic population data are estimated, on the basis of limited real data, to be: age at first reproduction for both sexes of 2–3 years; percentages breeding each year of 10% at 2 years old and 50% for older individuals; litter sizes of one to six with a mean of 2.7; lifetime reproduction of six to 15 per female; mortality rates of 50% per year for infants (up to 1 year old), 30% for juveniles and 27% for adults (mostly car kills and deliberate illegal shootings); and maximum longevity of 15–18 years. The results of running a series of models (1000 replicates of each), in which uncertain biological parameters were systematically varied, are shown in Table 25.5. The only model that produced a population with a 95% probability of persisting for 100 years had the lower age of first reproduction (2 years), the lower adult mortality (20%) and the larger litter size (three)—probability of extinction after

importance of vital statistics

100 years was 0.04 in this case. Clearly, the values of these vital statistics are truly vital and the better our information the more likely we are to arrive at an acceptable outcome.

Further simulations were used to explore the effect of continuing loss of habitat

Table 25.5 Probabilities of extinction and mean times to extinction (of those populations going extinct within 200 years) for 1000 simulated populations of Florida panthers using some plausible demographic parameters. The initial population was always 45, assumed equal to the carrying capacity (*K*) which was constant. No environmental variation or disasters are modelled and there is no harvest or supplementation of the population. Inbreeding is assumed not to affect survivorship or fecundity. (After IUCN, 1989.)

Input demographic parameters					Simulation results				
			Initiation		Probability of extinction (years)				Mean time to extinction
Age of reproduction	Adult mortality	Litter size	N	K	25	50	100	200	(years)
3	25	3.0	45	45	0.24	0.77	0.99	1.00	39
3	25	2.5	45	45	0.38	0.94	1.00	1.00	30
3	20	3.0	45	45	0.04	0.28	0.71	0.95	77
3	20	2.5	45	45	0.11	0.57	0.96	1.00	51
2	25	3.0	45	45	0.04	0.17	0.45	0.77	96
2	25	2.5	45	45	0.11	0.51	0.91	0.99	56
2	20	3.0	45	45	0.00	0.01	0.04	0.11	114
2	20	2.5	45	45	0.01	0.09	0.27	0.56	107

(by incorporating a diminishing value for carrying capacity), but this had little impact on expected time to extinction, a surprising result of the fact that the rate of habitat loss (1–2% per year) was less than the current decline in the panther population (6% per year). The management objectives drawn from this analysis included the need to set up a captive breeding programme so that subpopulations at a number of reintroduction sites would total 500 animals by the year 2010.

the case of the African elephant

Overall numbers of African elephants (*Loxodonta africana*) are in decline and few are expected to survive over the next few decades outside high-security areas, mainly because of habitat loss and poaching for ivory. For their simulation models Armbruster and Lande (1992) chose to represent the elephant population in 12 5-year age classes through discrete 5-year time steps. Values for age-specific survivorship and density dependent reproductive rates were derived from a thorough data set from Tsavo National Park in Kenya, because its semi-arid nature has the general characteristics of land planned for game reserves now, and in the future. Environmental stochasticity, perhaps most appropriately viewed as disasters, was modelled as drought events affecting sex- and age-specific survivorship—again realistic data from Tsavo were used, based on a mild drought cycle of approximately 10 years superimposed on a more severe 50-year drought and an even more severe 250-year drought. Table 25.6 gives the survivorship of females under 'normal' conditions and the three drought conditions. The relationship between habitat area and the probability of extinction was examined in 1000-year simulations with and without a culling regime; at least 1000 replicates were performed for each model with many more (up to 30 000) to attain acceptable statistical confidence in the smaller extinction probabilities associated with larger habitat areas. Extinctions were taken to have occurred when no individuals remained or when only a single sex was represented.

The results imply that an area of 500 mile^2 (1300 km^2) is required to yield a 99% probability of persistence for 1000 years (Figure 25.15); this conservative outcome

939   CONSERVATION AND BIODIVERSITY

Table 25.6 Survivorship for 12 elephant age classes in normal years (occur in 47% of 5-year periods), and in years with 10-year droughts (41% of 5-year periods), 50-year and 250-year droughts (10% and 2% of 5-year periods, respectively). (After Armbruster & Lande, 1992.)

Age class (years)	Female survivorship			
	Normal years	10-year droughts	50-year droughts	250-year droughts
0–5	0.500	0.477	0.250	0.01
5–10	0.887	0.877	0.639	0.15
10–15	0.884	0.884	0.789	0.20
15–20	0.898	0.898	0.819	0.20
20–25	0.905	0.905	0.728	0.20
25–30	0.883	0.883	0.464	0.10
30–35	0.881	0.881	0.475	0.10
35–40	0.875	0.875	0.138	0.05
40–45	0.857	0.857	0.405	0.10
45–50	0.625	0.625	0.086	0.01
50–55	0.400	0.400	0.016	0.01
55–60	0.000	0.000	0.000	0.00

was chosen because of the difficulty of re-establishing viable populations in isolated areas where extinctions have occurred and because of the elephant's long generation time (about 31 years). In fact, the authors recommend to managers an even more conservative minimum area of 1000 mile^2 (2600 km^2) for reserves; the data are least reliable for survivorship in the youngest age class and for the long-term drought regime, and a 'sensitivity analysis' shows extinction probability to be particularly sensitive to slight variations in these parameters. Of the parks and game reserves in Central and Southern Africa, only 35% are larger than 1000 mile^2 (2600 km^2).

Simulations revealed that up to 50% of carrying capacity could be culled every 5 years without much affecting extinction probabilities. (It is interesting to note that the maximum sustainable yield—see Chapter 16, Section 16.12.1—occurs when

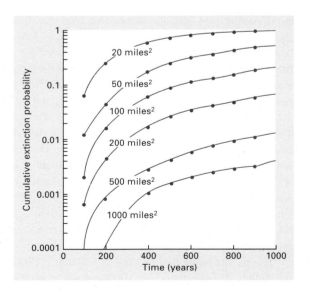

Figure 25.15 Cumulative probability of elephant population extinction over 1000 years for six habitat areas (without culling). (After Armbruster & Lande, 1992.)

elephants are culled to 20% of carrying capacity, which in the long term would increase the probability of extinction.) It seems that limited exploitation of elephant populations for their ivory, although fraught with ethical problems and potential for corruption, may be a biologically feasible way of generating income. The African elephant was listed in the Convention on International Trade in Endangered Species (CITES) in 1989, because of the disastrous effect of poaching. However, many African countries are now in favour of lifting this ban, arguing that it hinders their conservation efforts by preventing them from exploiting ivory as a sustainable resource.

the economics of ivory—to trade or not to trade?

A further example of the application of simulation modelling concerns Furbish's lousewort (*Pedicularis furbishiae*), an endangered plant that is endemic to northern Maine (USA) and adjacent New Brunswick (Canada). Assumed extinct until rediscovery in 1976, it is now known to exist in 28 subpopulations along a 140-mile (225-km) stretch of the St John River. This metapopulation is characterized by frequent local extinctions due to ice scour and bank slumping. The same disturbances seem also to be essential to the species by preventing establishment of competitively superior trees and shrubs. Individual subpopulations are short lived, and the viability of the metapopulation depends on dispersal and establishment of new populations (Menges, 1990). The kinds of simulation model discussed above lend themselves readily to extention to a metapopulation. Such a model was developed for the lousewort; each of 100 replicate runs started with 500 individuals, was followed for 100 years and incorporated environmental uncertainty (population parameters being chosen randomly from a known range). The results emphasize that disasters affecting many populations simultaneously, such as periodic ice damage, are of particular concern. A range of edaphic conditions (saturated wet, mesic, dry) in potential recolonization sites is expected to offer the greatest buffering from variations in disturbance intensity and frequency.

adding metapopulation structure—the case of an endangered plant

Akçakaya and Ginzburg (1991) have also used a simulation model to explore the dynamics of a metapopulation—involving three more or less isolated subpopulations of mountain gorillas in the Virunga mountains of south-west Uganda. They posed two questions: how does the fate of the metapopulation change when the degree of correlation of environmental variation in the three habitat fragments varies from 0 (no correlation) to 1 (complete correlation), and when the probability of recolonization/dispersal ranges from 0 to 1% per year. The probability of metapopulation extinction in these 500-year simulations of realistically age-structured populations is much lower when environmental variations are out of phase. At these lower levels of correlation, increasing the chance of recolonization greatly decreases the probability of extinction (Figure 25.16). When the environments of the three habitat fragments are strongly correlated, extinction probability is similar for all levels of recolonization.

a gorilla metapopulation analysis

25.5 Communities and ecosystems

So far in this chapter we have focused attention on individual species, treating them as though they were largely independent entities and applying what we know about population dynamics. However, it hardly needs to be pointed out that conservation of biodiversity also requires a broader perspective in which we apply our knowledge of communities and ecosystems. There are many reasons for this and these are discussed below.

applying our knowledge of communities and ecosystems

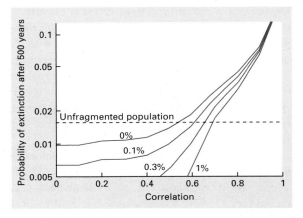

Figure 25.16 Probability of gorilla metapopulation extinction after 500 years at different levels of recolonization probability (0–1% per year) and environmental correlation (0–1). (After Akçakaya & Ginzburg, 1991.)

25.5.1 Chains of extinctions

the case of the flying fox

If we ignore community interactions, a chain of extinctions may follow inexorably from the extinction of particular native species, which therefore deserves special attention. Flying foxes in the genus *Pteropus* occur on South Pacific islands, where they are the major, and sometimes the only pollinators and seed dispersers for hundreds of native plants (many of which are of considerable economic importance, providing medicines, fibre, dyes, prized timber and foods). Features such as night flowering testify to the coevolution of the bats and certain of the plants. Flying foxes are highly vulnerable to human hunters and there is widespread concern about declining numbers. For example, on Guam the two indigenous species are either extinct, or virtually so, and there are already indications of reductions in fruiting and dispersal (Cox *et al.*, 1991).

25.5.2 New species in complex food webs

new species for old

A similarly unwanted outcome may occur if a particularly influential exotic species is introduced into a community without regard to food-web consequences. Our example is provided by the introduction into Flathead Lake and its tributaries in Montana, USA, of opossum shrimp (*Mysis relicta*) to provide food for kokanee salmon (*Oncorhynchus nerka*) (itself a previous introduction). Instead of benefitting the fishery as intended, the shrimps heavily exploited the zooplankton food of the salmon whilst avoiding predation themselves because of their behavioural adaptation of migrating to deep water during the day (Spencer *et al.*, 1991). The collapse of the kokanee population has apparently dramatically reduced numbers of bald eagles and grizzly bears that gather each autumn to feed on spawning salmon and their carcasses (Figure 25.17). Ecotourism has also suffered a serious decline because the charismatic megafauna are now less in evidence.

25.5.3 Habitat fragmentation and community structure

the notion of functional integrity in a reserve

When habitats become fragmented we need to pay attention not only to the fate of particular species but also to whether food webs maintain their functional integrity. We have already noted that species richness declines with habitat area when previously continuous habitat becomes fragmented. Mikkelson (1993) posed the

942 CHAPTER 25

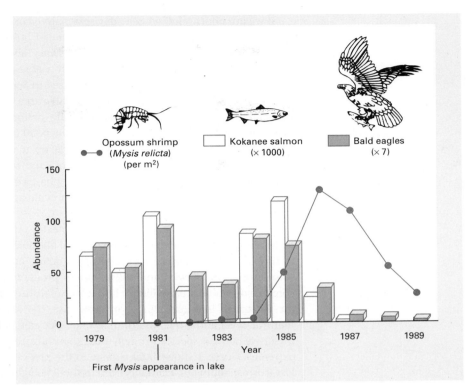

Figure 25.17 As the opposum shrimp population increased and exploited zooplankton in Flathead Lake, kokanee salmon declined, as did the eagle population that relied on the salmon. (After Spencer *et al.*, 1991.)

question: how does trophic organization change in the process of fragmentation? He gathered together information from five contrasting situations where fragments of different sizes were available for study—one set of mountain tops and two sets of land-bridge islands, each of which comprised continuous habitat 6000–14 000 years ago, and two groups of nature reserves in areas of Australia where habitat fragmentation began at the turn of the century. Three hypotheses were tested: that proportional representation of producers, herbivores and carnivores does not change with fragmentation any more than expected through random extinctions and colonizations (hypothesis 1), or that it changes more than (hypothesis 2) or less than (hypothesis 3) this random model. In fact, trophic organization varied very little between different-sized fragments and much of the evidence pointed to hypothesis 3, indicating a process in which extinctions and/or colonizations in different trophic categories depend on one another in such a way that the same trophic structure persists on a given fragment whilst overall species richness declines.

species composition in relation to fragment size

Several studies of species composition in habitat fragments have shown that species typically occur in patches exceeding an approximate threshold size; this threshold varies amongst species, producing the more or less regular decrease in species with fragment area, and resulting in a pattern in which species typical of small fragments are a nested subset of those present in large fragments (Hanski, 1994b). This picture is rather depressing, indicating as it does that large fragments are indispensible if we wish to conserve high biodiversity. On the other hand, Mikkelson's conclusion provides some cheer since it implies that, up to a point at least, fragmentation need not lead to a dislocation of one fundamental ecological process.

943   CONSERVATION AND BIODIVERSITY

If we are not careful, we may devise a plan to conserve some endangered species but forget to allow for the ecological functional links required to keep the system going. Conservation managers need to be wary about the scale of their plans in relation to what is required for functionally viable units. Where the focus is on a particular species (e.g. the elephants discussed in Section 25.4.4) the recommended MVP may, by chance, define a minimum viable area that is big enough to contain all the necessary functional links. Ecologically defined boundaries may, however, extend out further than required from just the point of view of any particular species. For example, there is little point in setting up a wetland, riverine or lacustrine reserve unless it extends sufficiently far into the surrounding catchment area to provide the necessary habitat for terrestrial phases of aquatic insects, and to ensure that terrestrial energy and nutrient fluxes provide appropriate inputs to the aquatic ecosystem.

In the community and ecosystem contexts, our focus on the importance of maintaining biodiversity begs a fundamental question. How important is biodiversity in terms of the underlying processes that define our system? Can we lose a few, several or many species and still maintain productivity, nutrient cycling, intrinsic community stability or resilience in the face of disturbance? Three possible scenarios are illustrated in Figure 25.18. In each case, the ecosystem process operates at a 'natural' level at the high (natural) level of species richness and the process reduces to zero when no species are left. But, there are three curves between these two extremes. Curve 1 shows little change to the process until many species have been lost, corresponding to a situation where there is high ecological redundancy. Loss of biodiversity is of least significance in this scenario. Curve 2 has every species contributing to ecosystem function; if this corresponded to reality we should beware any loss of biodiversity but the system may be able to cope with a certain level of loss. Curve 3 has the system collapsing with even the loss of one or a few (perhaps keystone) species. Which of these models is closest to reality? The simple answer is that we have almost no idea. This is an area that needs research effort, both in the well-controlled conditions of laboratory experiments and in comparative studies of natural communities that differ in species richness or some other aspect of biodiversity.

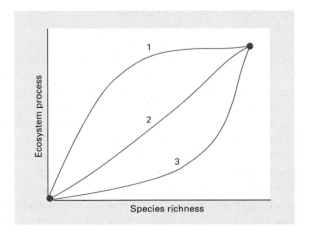

Figure 25.18 Hypothetical relationships between a basic ecosystem process (such as productivity, nutrient cycling, stability in composition) and species richness.

25.6 Conservation in practice

25.6.1 Species management plans

In an ideal world, a population viability analysis (PVA) (see Section 25.4) would enable us to produce a specific and reliable recommendation for an endangered species of the population size, or reserve area, that would permit persistence for a given period with a given level of probability. But, this is rarely achievable because the biological data are hardly ever good enough. The modellers know this and it is important that conservation managers also appreciate it. Within the inevitable constraints of lack of knowledge and lack of time and opportunity to gather data, the model-building exercise is no more nor less than a rationalization of the problem and quantification of ideas. Moreover, even though such models produce quantitative outputs, common sense tells us to trust the results only in a qualitative fashion. Nevertheless, the examples in Section 25.4.4 show how models allow us to make the very best use of available data and may well give us the confidence to make a choice between various possible management options and to identify the relative importance of factors that put a population at risk. The sorts of management interventions that may then be recommended include translocation of individuals to augment target populations, raising the carrying capacity by artificial feeding, restricting dispersal by fencing, fostering of young (or cross-fostering of young by related species), reducing mortality by controlling predators or poachers, or through vaccination and, of course, habitat preservation.

Previously, conservation biologists worked with gross rules of thumb. As we have seen, the recognition of potential genetic problems indicates safe population sizes of between 2500 and 5000 (Nunney & Campbell, 1993). General demographic models, on the other hand, have produced results suggesting that some hundreds to a few thousands would constitute a rule of thumb for an MVP. It is surely better to make use of all the data available for each species, building simulation models that consider the full gamut of questions of genetical and demographic consequences in defined scenarios.

In most cases, it will be appropriate to incorporate matapopulation structure into the models because existence in population fragments appears to be more the rule than the exception. Even when there is only one local population left, as there may be in the case of the most critically endangered species, a species survival plan will be likely to incorporate a metapopulation structure to enhance the probability of persistence.

A point of particular significance is that local extinctions are common events. It follows that recolonization of habitat fragments is critical for the survival of fragmented populations. Thus, we need to pay particular attention to the relationships amongst the landscape elements, including dispersal corridors, in relation to the dispersal characteristics of the species in question (Fahrig & Merriam, 1994). There has been considerable debate (in the absence of much data) about the importance of protecting or creating corridors to provide recolonization rates that can balance the extinction side of the metapopulation persistence equation. Although there are cons—for example, corridors could increase the correlation amongst fragments of catastrophic effects such as the spread of fire or disease (Hess, 1994)—the pros of the argument are persuasive. Indeed, high recolonization potential (even if this means the conservation managers themselves moving organisms around) may be indispensible

acknowledging the limitations of population viability analyses (PVAs)

beware rules of thumb

incorporating metapopulation structure

local extinction, recolonization and habitat corridors

to the success of conservation of endangered metapopulations. Fahrig and Merriam (1994) suggest that populations that naturally experience high rates of local extinction can be expected to have high dispersal rates. If so, then human fragmentation of the landscape, producing subpopulations that are more and more isolated, is likely to have had the strongest effect on populations with naturally low rates of dispersal. For example, the widespread declines of the world's amphibians may be due in significant measure to their poor potential for dispersal (Blaustein *et al.*, 1994).

25.6.2 *Ex situ* conservation

the place of captive propagation in the conservationists' armoury …

Captive propagation is attractive in its directness, its immediacy and the opportunity it provides to preserve options. However, by itself it clearly offers no long-term solutions. Specifically, captive propagation can fulfill four functions in conservation: (i) providing material for basic research in population biology, reproductive biology, etc.; (ii) providing material for basic research in care and management techniques; (iii) providing demographic or genetic reservoirs for enhancing existing natural populations or establishing new ones; and (iv) providing a final refuge for species with no immediate hope of survival in the wild. Thus, the Père David deer (*Elaphurus davidianus*) and the Mongolian or Przewalski's wild horse (*Equus przewalski*), well established in captivity, are believed extinct in nature; whilst, the endangered but still surviving wild populations of blackbuck (*Antilope cervicapra*), the peregrine falcon (*Falco peregrinus*), the giant tortoise (*Geochelone elephantopus*) and the European otter (*Lutra lutra*), amongst others, have all been augmented from captive-bred populations.

the case of the Arabian oryx

The Arabian oryx (*Oryx leucoryx*) provides a notable success story. It became extinct in the wild in 1972 because of overhunting (Stanley Price, 1989). By good fortune, a breeding colony with nine founders had been established in 1963 at Phoenix Zoo in the USA and a further one was set up in Los Angeles Zoo. The global captive population gradually increased as surplus individuals were dispersed to other zoos. After surveying Oman to identify a site where reintroduction was likely to be successful, two social groups totalling 21 individuals were released in 1982 and 1984. Monitoring of the population since that time shows it has been increasing at $r = 0.20$ per year.

zoo or ark?

Zoos have become the centre of a spirited debate. At one extreme, some see zoological and botanical gardens (and germ banks and seed banks) as small arks that provide refuge for endangered species from a flood of species extinction; others see them as living museums—once a species enters a zoo it is essentially dead (Ginsberg, 1993). The cases above demonstrate the merit of the ark perspective. However, captive breeding for conservation should involve three stages: (i) planning a conservation strategy; (ii) captive breeding; and (iii) reintroduction, and these need to be integrated with habitat protection and *in situ* conservation. Regrettably, many species are simply inappropriate for captive breeding and eventual reintroduction (e.g. breeding technology unavailable, appropriate habitat unavailable, causes for endangerment irreversible) and available resources are insufficient to build an 'ark' big enough for a significant proportion of the world's biota (Ginsberg, 1993).

25.6.3 Protected areas

Producing individual species survival plans may be the best way to deal with species recognized to be in deep trouble and identified to be of special importance (e.g.

getting the most from the conservation dollar

keystone species, evolutionarily unique species, charismatic megafauna that are easy to 'sell' to the public). However, there is no possibility that all endangered species could be dealt with one at a time. For instance, the Fish and Wildlife Service calculated it would need to spend about $4.6 billion over 10 years to fully recover all gazetted species in the USA (US Department of the Interior, 1990), whereas the annual budget for 1993 was $60 million (Losos, 1993). In contrast, we can expect to conserve the greatest biodiversity if we protect whole communities by setting aside protected areas (Grumbine, 1993). Primack (1993) provides an extensive review of the establishment, design and management of protected areas.

protected areas—limits to growth

Protected areas of various kinds (national parks, nature reserves, multiple-use management areas, etc.) have grown in number and area throughout the 20th century, with the greatest expansion occurring since 1970 (Table 25.7). However, the 4500 protected areas in existence in 1989 still only represented 3.2% of the world's land area. At best, and given the political will, perhaps 6% of land area could eventually be provided protection—the rest would be considered necessary to provide the natural resources needed by the human population (Primack, 1993).

marine protected areas— conservation and sustainable use can go hand in hand

Priorities for marine conservation, which have lagged behind terrestrial efforts, are now being urgently addressed. Most of the world's biota is found in the sea (32 of the 33 known phyla are marine, 15 exclusively so) and marine communities are subject to a number of potentially adverse influences, including overfishing, habitat disruption and particularly pollution from land-based activities. Many of the new generation of marine protected areas are designed as multiple-use reserves, accommodating many different users (environmentalists, cultural harvesters, recreational and commercial fishers, etc.) (Agardy, 1994); it is clear that conservation and sustainable use can often proceed hand in hand as long as planning has a scientific basis and the negotiated objectives are clear.

centres of diversity, endemism and extinction . . .

It is important to devise priorities so that the restricted number of new protected areas, in terrestrial and marine settings, can be evaluated systematically and chosen with care. We know that the biotas of different locations vary in species richness (centres of diversity), the extent to which the biota is unique (centres of endemism) and the extent to which the biota is endangered (hotspots of extinction, for instance because of imminent habitat destruction). One or more of these criteria could be used

Table 25.7 The pattern of establishment of protected areas prior to and during the 20th century. Only protected areas greater that 1000 ha (10 km^2) are included. The dates of establishment are unknown for 711 areas covering 194 395 km^2. (After Reid & Miller, 1989.)

Date established	Number of areas	Total area protected (km^2)
Unknown	711	194 395
Pre-1900	37	51 455
1900–1909	52	131 385
1910–1919	68	76 983
1920–1929	92	172 474
1930–1939	251	275 381
1940–1949	119	97 107
1950–1959	319	229 025
1960–1969	573	537 924
1970–1979	1317	2 029 302
1980–1989	781	1 068 572

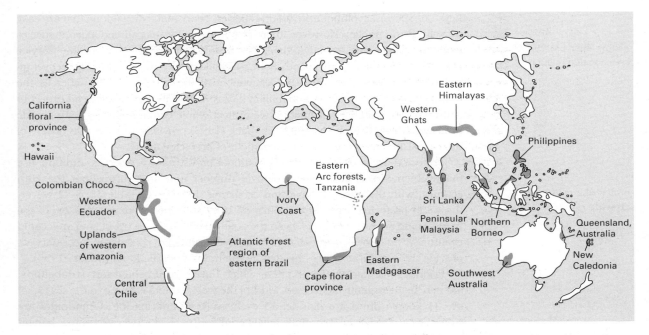

Figure 25.19 Hotspots of high endemism and significant threat of imminent extinction for tropical rainforest and other biomes. (After Myers, 1988b, 1991.)

to prioritize potential areas for protection (Figure 25.19). Moreover, if we were to give less weight to the 'existence' value of species (every species equal) and more weight to the 'option' value of species that may provide future benefit (for food, domestication, medical products, etc.), we could give priority to locations that contained more species likely to be useful (centres of utility).

... and of utility

But, biodiversity encompasses more than just species richness. The selection of new areas should also try to ensure that we protect representatives of as many types of community and ecosystem as possible. Two key principles here are 'complementarity' and 'irreplaceability' (Pressey *et al.*, 1993).

the key concepts of complementarity ...

With limited resources, the ideal strategy is to assess the content of candidate areas and to proceed in a stepwise fashion, selecting at each step the site that is most *complementary* to the others in the features it contains. A number of algorithms are now available to carry out this procedure efficiently; for example, one gives more weight to uniqueness of community or land system, another to the average rareness of the land systems present in different locations (Figure 25.20a).

... and irreplaceability

A related but subtly different approach identifies *irreplaceability* as a fundamental measure of the conservation value of a site. Irreplaceability is an index of the potential contribution that a site will make to a defined conservation goal and the extent to which the options for conservation are lost if the site is lost (Figure 25.20b). The notions of complementarity and irreplaceability can equally well be applied to strategies designed to maximize species richness.

the design of nature reserves

A rather surprising application of island biogeography theory (see Chapter 23) is in nature conservation. This is because many conserved areas and nature reserves are surrounded by an 'ocean' of habitat made unsuitable, and therefore hostile, by humans. Can the study of islands in general provide us with 'design principles' that can be used in the planning of nature reserves? The answer is a qualified 'yes' (see

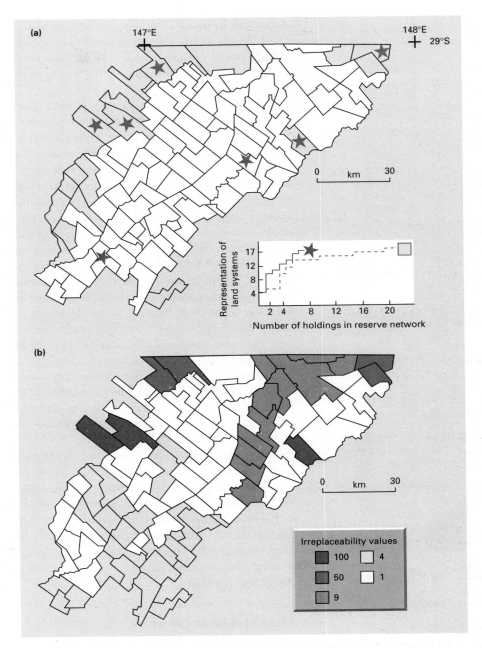

Figure 25.20 (a) A map of 95 pastoral holdings in New South Wales, Australia, showing two sets of holdings needed to represent all 17 land systems at least once. Stars indicate a minimum set identified by a complementarity algorithm that selects sites with unique land systems, and then proceeds stepwise to select the site with the rarest unrepresented land system. Shading indicates the set required if all holdings are scored according to the average rarity of the land systems they contain. (b) A landscape of conservation value for each holding derived by predicting irreplaceability levels. (After Pressey *et al.*, 1993.)

949   CONSERVATION AND BIODIVERSITY

discussions in Diamond & May, 1976; Simberloff & Abele, 1976; Soulé & Wilcox, 1980; Soulé, 1986). Certain general recommendations can be made, but these generalizations must be tempered by the realization that there is no substitute for a painstaking study of the ecology of the species or community whose conservation is intended.

The most obvious point to emerge from island biogeography in the present context is that an area that is part of a continuum will, if it is turned into an isolated reserve, support fewer species than it had previously. A less straightforward question is whether to construct one large reserve or several small ones adding up to the same total area. If each of the small reserves supported the same species, then it would clearly be preferable to construct the larger reserve, and hence, conserve more species (although this conclusion ignores epidemiological aspects of management; many scattered reserves would be less susceptible to the ravages of epidemic diseases—Dobson & May, 1986). On the other hand, if the region as a whole is heterogeneous, then each of the small reserves may support a different group of species and the total conserved might exceed that in a large reserve. In their analysis of this question, Quinn and Harrison (1988) consistently found that collections of small islands contained more species than a comparable area composed of one or a few large islands. For example, in the Galapagos archipelago, Isabella is the single largest island and supports 344 land plant species. A collection of smaller islands, with the same combined area, was found to contain 609 species. The pattern is similar for habitat islands and, most significantly, for national parks. Thus, several small parks contained more species than larger ones of the same area for mammals and birds of East African parks, for mammals and lizards in Australian reserves and for large mammals in national parks in the USA. It seems likely that habitat heterogeneity is a general feature of considerable importance in determining species richness.

A few further, fairly uncontroversial recommendations can also be made. First, since a certain number of extinctions in island reserves are inevitable, the conservationist can help by increasing the amount of counterbalancing immigration. This can be done by the careful juxtaposition of scattered reserves, or by providing corridors or stepping-stones of natural habitat between them. In fact, long, thin reserves may be optimal for catching immigrants, but other reasoning suggests that reserves should be approximately circular, since this minimizes the chances of creating 'islands within islands' on semi-isolated peninsulars, and reduces 'edge effects' to a minimum.

25.6.4 Restoration ecology

when things have gone too far

Conservation is an appropriate aim when there is something to conserve. In many cases, however, matters have gone too far. Species, or, quite likely, whole communities have disappeared, been destroyed or have been altered beyond recognition. Restoration, rather than conservation, may then seem appropriate.

restoring a 'natural' community

But, what precisely is to be restored? First, there may be an attempt to restore the plants and animals that are considered the natural inhabitants of that area of land—'natural' in the sense that they were the species living there prior to the disturbance that ultimately led to the need for restoration. This approach may be appropriate, for example, in attempts to replace either unwanted or overexploited agricultural land with an authentic natural community. The implication here is that the abiotic conditions are fundamentally the same as they have ever been. Hence, it should in principle be possible to re-establish each of the original species in their original combination, although it will surprise nobody to be told that the gap between principle and practice is often vast. There are three main problems:

1 we may have no clear idea what the authentic, original species are;

2 their natural establishment (e.g. through a natural succession—see Chapter 17) may take years;

3 short-cutting these natural processes, even if possible in principle, would require detailed knowledge of the ecologies of many species, which we rarely if ever have.

A community may have been changed dramatically because of the effects of one or more introduced species. In such cases, restoration involves removing and not just adding species. New Zealand's offshore islands provide some significant success stories. Over the centuries domestic stock were purposefully introduced to many islands. Successful removals of cattle, goats, pigs and cats occurred between 1916 and 1936. The much more daunting task of tackling the smaller, faster reproducing and less conspicuous species, such as Norway rats, ship rats and mice, has occupied teams of New Zealand ecologists for the past two decades, and with considerable success; over 120 pest-eradication campaigns have been conducted to restore island habitats in a way that improves them for endangered species already present and that opens up opportunities to translocate individuals from more vulnerable settings on the mainland (Towns & Ballantine, 1993). Where eradication attempts involve employing non-specialists to carry out the trapping operation, such as the successful removal from England of South American coypus (*Myocastor coypus*), it may be necessary as part of the management plan to devise an incentive bonus scheme to overcome the problem that the trappers might be reluctant to work themselves out of a job (Gosling & Baker, 1987). Conservation managers require diverse skills.

An alternative approach is necessary where abiotic conditions have been profoundly, perhaps irreversibly, altered. For instance, this occurs when mine wastes of various sorts are deposited above ground. Here, restoration as such may not be realistic. Bradshaw (1984) has suggested the terms 'rehabilitation' for the establishment of a community that is similar to but not the same as the original, and 'replacement' for the establishment of a quite different community. The pragmatic approach of replacement will often be necessary, establishing whichever species can be established, in order to provide aesthetic or recreational facilities, or perhaps even a productive environment in terms, say, of forestry or grassland.

Pollution by human activities may occur on both the local (mine wastes, eutrophication of rivers and lakes) and the global scale (acid rain, global climate change). Restoration of habitats degraded by such physicochemical changes will often need us to clean up our act. The River Thames in England provides a graphic example (Allan & Flecker, 1993). As early as 1620, comments by the Bishop of London brought attention to the foulness of the Thames water. Pollution by human and animal wastes continued unabated and the river became so foul that 1858 was recorded as the 'Year of the Great Stink'. Improved treatment in the early decades of the 20th century failed to cope with the increasing volume of sewage, and 1955 recorded an all-time low—odours released from anaerobic sections were again a public nuisance and a 70 km section around London was devoid of fish other than eels. This spurred a concerted effort in the treatment of organic wastes and within a decade oxygen levels were high enough to permit fish passage, and there has been a startling recovery in species richness of fish and waterfowl.

removing unwanted species

community rehabilitation or replacement

pollution—cleaning up our act

951   CONSERVATION AND BIODIVERSITY

restoration ecology is an acid
test of ecological
understanding

Whether the aim is pragmatic or romantic (pristine restoration), the exercise is one in community rather than population ecology. There is a need, in other words, for an understanding of the principles and processes described in earlier chapters. For example, reversing the process of nutrient enrichment (eutrophication) of lakes may, depending on circumstances, be approached either by a 'bottom-up' approach of diverting nutrient inputs or increasing flushing rates, or by a 'top-down' approach of manipulating upper trophic levels to increase grazing on nuisance phytoplankton (see Chapter 22, Section 22.3). Two general, complementary points are worth emphasizing. The first is that restoration is an 'acid test' for ecology (Bradshaw, 1983, 1987). The second point is that ecology must increasingly go beyond observation, description and correlation in the development of a manipulative, experimental approach to ecological research—such as that represented by the practice of restoration ecology (Harper, 1987).

An extensive review of *The Restoration of Land* (especially derelict land) is provided by Bradshaw and Chadwick (1980) and of *Restoration Ecology* by Jordan *et al.* (1987).

25.6.5 Finale: a healthy approach to conservation

In desperate times, painful decisions have to be made about priorities. Thus, wounded soldiers arriving at field hospitals in the First World War were subjected to a *triage* evaluation: priority 1—those who were likely to survive but only with rapid intervention; priority 2—those who were likely to survive without rapid intervention; priority 3—those who were likely to die with or without intervention. Conservation managers are often faced with the same kind of choices and need to demonstrate some courage in giving up on hopeless cases, and prioritizing those species and habitats where something can be done.

Some argue that the species approach is very much an emergency room strategy when what is needed is better preventative and primary care—something that can most appropriately be accomplished by focusing on communities and ecosystems (Losos, 1993). We have also seen that intensive care (captive breeding in zoos or botanical gardens) and reconstructive surgery (restoration ecology of communities) need to figure in the biodiversity health-care compendium. All these approaches have their place in protecting and enhancing the health of the living world.

The spectrum of opinions on conservation is complete. It ranges from the environmental terrorist, who is prepared to destroy property and put human life at risk for what is seen as unacceptable exploitation of animals, to the other extreme of the exploitational terrorist, who is prepared to destroy rare habitat just as it is about to achieve protected status. There are zealots on both sides of the spectrum too. On the one hand, there are the industrialists, fishers, farmers and foresters who accept none of the conservationist case and are not prepared to look objectively at the scientific evidence; whilst, on the other hand, are the environmental zealots— preservationists who seem unwilling to accept any exploitation of the natural world, some even pronouncing that fishing, hunting or logging are intrinsically wrong. The middle ground is occupied by both exploiters and conservationists whose basic philosophy holds that natural resources can be used, but this should be in a sustainable and balanced manner. A thorough understanding of the principles and applications of ecological science should enable all to pay healthy regard to the scientific aspects of what, in its broader context, is very much an ethical and sociopolitical problem.

Glossary

This glossary is an attempt to help the reader by describing what we mean by a number of the more specialist words. We do not intend it to be a set of formal definitions. We have used our personal interpretations much as Humpty Dumpty 'When I use a word it means just what I choose it to mean—neither more nor less'. 'The question is' said Alice, 'whether you *can* make words mean different things.' 'The question is,' said Humpty Dumpty, 'which is to be master—that's all.'

There are many words in everyday speech that carry very special meanings in ecology and many that are commonly used in science but are never properly defined in that context. Words such as *adverse, benign, harsh, equable, stressful, harm, favourable, beneficial* are difficult to avoid, but the reader who encounters them in our text, or any other, would do well to stop and think what they are really supposed to mean.

Abaptation The process by which the present match between organisms and their environment, and the constraints on this match, have been determined by evolutionary forces acting on ancestors. The prefix 'ab-' emphasizes that the heritable characteristics of an organism are consequences of the past and *not* anticipation of the present or future.

Abiotic Non-living; usually applied to the physical and chemical aspects of an organism's environment.

Abundance The number of organisms in a population, combining 'intensity' (density within inhabited areas) and 'prevalence' (number and size of inhabited areas).

Acclimation The habituation of an organism's physiological response to environmental conditions (usually applied to laboratory environments).

Acclimatization The habituation of an organism's physiological response to environmental conditions (usually applied to natural environments).

'Acid rain' Rain with a very low pH (often below 4.0) resulting from emissions to the atmosphere of oxides of sulphur and nitrogen.

Adaptation A confusing word used to mean quite different things. (i) Characteristics of organisms evolved as a consequence of natural selection in its evolutionary past and which result in a close match with features of the environment and/or constrain the organism to life in a narrow range of environments. The prefix 'ad-' is unfortunate as it implies that the process anticipates the present or the future (*see* Abaptation). (ii) Changes in the form or behaviour of an organism *during its life* as a response to environmental stimuli, e.g. the formation of sun and shade leaves on the same tree and the acquisition of cold tolerance as a result of prior experience of low temperatures. (iii) Changes in the excitability of a sense organ as a result of continuous stimulation.

Adenosine triphosphate (ATP) Molecule composed of adenine, ribose and three phosphate groups bound by high-energy linkages and associated with energy transfer in living cells.

Adiabatic expansion Expansion taking place without heat entering or leaving the system.

Adventitious roots Roots that arise in 'abnormal' positions, e.g. from a stem or leaf. The contrast is with the primary roots that develop from the axis of a seedling and roots that arise from other roots.

Aerobic decomposition The process of breakdown of organic molecules to simple inorganic constituents when oxygen is in free supply.

Aesthetic injury level The level of pest abundance above which aesthetic or sociological considerations suggest control measures should be taken against the pest.

Aestivation A state of dormancy during the summer or dry season. (The word is also used by botanists to describe the arrangement of the parts in a flower bud.)

Aggregated distribution The distribution of organisms in which individuals are closer together than they would be if they were randomly or evenly distributed.

Aggregation of risk A pattern in which prey vary, from prey patch to prey patch, more than would be expected by chance alone in their risk of being attacked by a predator (especially applied to hosts varying in their risk of being attacked by parasitoids).

Aggregative response The response of a predator through which it spends more time either in habitat patches with higher densities of prey, leading to higher densities of predators in patches with higher densities of prey, or in habitat patches with lower densities of prey, leading to higher densities of predators in patches with lower densities of prey. (*See also* Aggregation of risk.)

Alcohol dehydrogenase (ADH) An enzyme catalysing the conversion of alcohols into aldehydes and ketones, and the reverse.

Allelochemical A substance produced by one organism that is toxic or inhibitory to the growth of another.

Allochthonous material Organic matter entering a stream, lake or ocean but derived from an adjacent terrestrial system.

Allogenic succession A temporal succession of species at a location that is driven by external influences which alter conditions.

Allometry The study of the changing proportions of the parts of an organism as size changes, either in individual growth (ontogenetic) or comparing related organisms of different sizes (phylogenetic).

Allopatry Occurring in different places; usually refers to geographical separation of species.

Amensalism An interaction in which one organism (or species) adversely affects a second organism (or species), but the second has no effect (good or bad) on the first.

Anaerobic decomposition A process of breakdown of organic molecules to simpler inorganic constituents, that occurs in the absence of oxygen.

Analogous structures Organs of different evolutionary origin which perform the same role in different organisms.

Angiosperms Flowering plants. Strictly, those seed-bearing plants that develop their seeds from ovules *within* a closed cavity, the ovary (*cf.* gymnosperms, in which the ovules are naked).

Annual A species with a life cycle which takes approximately 12 months or rather less to complete, whose life cycle is therefore directly related to the annual cycle of weather, and whose generations are therefore discrete.

Anoxic Without oxygen.

Antagonistic resources A pair of resources for which increased consumption by an organism of one resource leads to an increased requirement by that organism for the other.

Anthropocentric Regarding humans as the central fact of the universe and interpreting everything in relation to them.

Aposematism Conspicuous appearance of organism that is noxious or distasteful.

Apparent competition An interaction in which the organisms (or species) have adverse effects on one another by virtue of the beneficial effects that each has on a predatory organism (or species) which they share.

Arthropod A member of the animal phylum Arthropoda, which includes the insects, crustaceans (e.g. crabs, shrimps, barnacles), spiders, scorpions, mites, millipedes and centipedes.

Assimilation efficiency The percentage of energy ingested by an animal that is absorbed across the gut wall.

Asymmetric competition Competition between two organisms (or species) in which one is much more severely affected than the other.

Autochthonous material Organic matter produced within a community (to contrast with allochthonous material that is produced outside it).

Autocidal control A type of pest control in which the pest is manipulated so that it contributes significantly to its own control, or literally 'kills itself'.

Autogenic succession A temporal succession of species at a location that is driven by processes operating within the community (in contrast with allogenic succession).

Autotroph An organism that is independent of outside sources for *organic* food materials and manufactures its own organic material from inorganic sources.

Autotrophic succession A temporal succession of species at a location principally involving plants.

Awn A stiff bristle, especially on the grains of cereals and grasses.

Axillary bud A bud produced in the angle between a leaf and the stem that bears it, i.e. the normal position of a lateral bud.

Bacteroids In legume nodules the symbiotic *Rhizobium* bacteria that have entered an active nitrogen-fixing state and have usually ceased to divide and commonly become banded and branched.

Balanced preference A preference by a consumer for food items based on the need of the consumers to obtain a balanced diet of complementary food items.

Basic reproductive rate (R_0) The average number of offspring produced by individuals in a population over the course of their life.

Benthic communities The plants, microorganisms and animals that inhabit the bed of aquatic environments.

Bicentric distribution The presence of a species (or other taxonomic class) in two widely separated geographic areas.

Biodiversity In its most general sense, biodiversity refers to all aspects of variety in the living world. Specifically, the term may be used to describe the number of species, the amount of genetic variation or the number of community types present in an area.

Biogeochemical cycling The movement of chemical elements between organisms and non-living compartments of atmosphere, lithosphere and hydrosphere.

Biogeography The study of the geographical distribution of organisms.

Biological control The use of a pest's natural enemies in order to control that pest.

Biological oxygen demand The rate at which oxygen disappears from a sample of water—a measure of deoxygenating ability commonly used as an index of the quality of sewage effluent.

Biological pesticides A preparation used to provide immediate control of a pest, and which consists of biological as opposed to chemical material.

Biomagnification The increasing concentration of a compound in the tissues of organisms as the compound passes along a food chain, resulting from the accumulation of the compound at each trophic level prior to its consumption by organisms at the next trophic level.

Biomass The weight of living material. Most commonly used as a measure per unit area of land or per volume of water. Commonly includes the dead parts of living organisms, e.g. the bark and heart wood of trees and the hair, claws etc. of animals which are strictly 'necromass'.

Biome One of the major categories of the world's distinctive plant assemblages, e.g. the tundra biome, the tropical rainforest biome.

Biorational insecticides Insecticides which have no, or relatively limited, adverse effects on other, non-pest organisms in the pest's environment.

Biota The fauna and flora together; all the living organisms at a location.

Biotic Living; usually applied to the biological aspects of an organism's environment, i.e. the influences of other organisms.

Biotrophic A word used of parasites that can complete their development on only a living host (*cf.* Necroparasite).

Biotype A physiologic race or a group of individuals having distinctive genetic characters in common.

Boreal Northern.

Botanical insecticides Chemicals extracted from plants for the control of insect pests.

Boundary layer The relatively still layer of water just above the bed of a river.

Brackish Saline water with a concentration between freshwater and seawater.

Brood parasitism The act of leaving eggs or progeny to be reared by an individual that is not the parent—usually a member of another species.

Browsers Vertebrate herbivores that feed from trees or shrubs.

Caecum A blindly ending sac at the junction between the small and large intestines.

Calcareous Composed of, or containing lime or limestone.

Capture–recapture A method for estimating the size of populations of mobile organisms (usually animals), in which one or more samples are captured, marked and released, and one or more samples examined for the proportion of recaptured marks (broadly speaking, high in small populations and low in large).

Carbohydrase An enzyme that acts on carbohydrate.

Carbon (C) 3 plants Those in which the assimilation of atmospheric carbon dioxide is directly via the enzyme ribulose-1,5-bisphosphate carboxylase in the cells of the leaf mesophyll.

Carbon (C) 4 plants Species of higher plants in which the assimilation of atmospheric carbon dioxide in the photosynthetic process is indirect, via the enzyme phosphoenol pyruvate carboxylase in the sheaths surrounding the veins of the leaves.

Carboniferous period Geological period from *c.* 270 to 220 million years ago.

Carnivory The consumption by an organism of living animals or parts of living animals.

Carrying capacity The maximum population size that can be supported indefinitely by a given environment, at which intraspecific competition has reduced the *per capita* net rate of increase to zero. An idealized concept not to be taken literally in practice.

Catabolism The decomposition by living organisms of complex organic molecules to simpler forms, with the release of energy.

Catastrophe A major change in the environment that causes extensive damage and usually widespread death, and occurs so infrequently that the effects of natural selection by similar events in the past (if they have ever occurred) do not remain in the 'genetic memory' of the species. The Mt St Helens volcanic eruption was in this sense a catastrophe (cf. the recurrent hurricanes in eastern USA which may be defined as 'disasters').

Catch per unit effort Usually applied to the harvesting of a natural resource (e.g. marine fish), the total catch (in terms of numbers or biomass) divided by the total harvesting effort (e.g. a product of the total number and size of ships and the number of days that they fished).

Cellulolytic enzymes Enzymes that act on cellulose.

Cellulose A complex polymer of glucose molecules. The fundamental cell wall constituent in all green plants.

Census In ecology, an attempt to count every member of a population.

Chamaephyte *see* Table 1.2 (p. 33).

Chaos Applied to a time-series of population abundance or density, a pattern in which elements of the pattern are never repeated exactly, and where two very similar abundances in the time-series follow trajectories subsequently which diverge from one another exponentially. Nonetheless, the series is not random and fluctuates within definable limits.

Chaparral A thicket of low evergreen oaks or dense tangled brushwood.

Character displacement A measurable physical difference between two species which has arisen by natural selection as a result of the selection pressures on one or both from competition with the other.

Chemosynthesis The synthesis of organic molecules by certain bacteria that use the energy released by specific inorganic molecules.

Chlorophyll Green pigment(s) found in almost all plants and playing a crucial role in the capture of radiant energy in the process of photosynthesis.

Chloroplast The inclusions (plastids) within plant cells which contain chlorophyll.

Classification A mathematical procedure for categorizing communities in which communities with similar species compositions are grouped together in subsets.

Climax The presumed endpoint of a successional sequence; a community that has reached a steady state.

Clonal dispersal The movement or growth away from each other of the parts of a modular organism (commonly, but not necessarily, becoming detached from each other).

Clone The whole product of growth from a single zygote, in organisms that grow by the repeated iteration of units of structure (modular organisms) and in which the units have, at least potentially, the capacity for physiological independence. The parts of a clone are genetically identical except for what are probably rare somatic mutations.

Clumped distribution The distribution of organisms in which individuals are closer together than if they were distributed at random or equidistant from each other. (= Aggregated distribution.)

Coefficient of interference A measure of the extent to which interference amongst consumers increasingly depresses individual consumption rates as consumer density increases.

Coefficient of variation A statistical term, referring to the standard deviation of a distribution divided by the distribution's mean, and hence providing a standardized measure of the variation in a distribution, which does not increase simply because the mean itself increases or because the units of measurement change.

Coevolution The process by which members of two (or more) species contribute reciprocally to the forces of natural selection that they exert on each other, e.g. parasites and their hosts.

Coexistence The living together of two species (or organisms) in the same habitat, such that neither tends to be eliminated by the other. This begs lots of questions about the meaning of 'living together' and the 'same habitat'.

Cohort generation time (T_c) An approximation to the true generation length, which takes no account of the fact that some offspring may themselves develop and give birth during the reproductive life of the parent; hence, simply the average time between the birth of a parent and the birth of its offspring.

Cohort life table A life table constructed by monitoring a group of individuals all born during the same short period, from this time of birth through to the death of the last surviving individual.

Collector–filterers Aquatic animals that filter small particles of organic matter from the water flowing over them.

Collector–gatherers Aquatic animals that gather small particles of organic matter from the sediment.

Colonization The entry and spread of a species (or genes) into an area, habitat or population from which it was absent.

Commensalism An interaction in which one organism (or species) beneficially affects a second organism (or species), but the second has no effect (good or bad) on the first.

Community The species that occur together in space and time.

Community stability The tendency of a community to return to its original state after a disturbance.

Community structure The list of species and their relative abundances in a community.

Compartmentalization in communities A tendency in communities to be organized into subunits within which interactions are strong but between which interactions are weak.

Compensation point The intensity of radiation at which photosynthesis equals and balances respiration.

Competition An interaction between two (or more) organisms (or species), in which, for each, the birth and/or growth rates are depressed and/or the death rate increased by the other organisms (or species).

Competition coefficient In interspecific competition, a measure of the competitive effect of one species on another relative to the competitive effect of the second species on itself.

Competitive exclusion The elimination from an area or habitat of one species by another through interspecific competition.

Competitive exclusion principle *see* Gause's principle.

Competitive release The expansion of the niche of a species associated with the lack of competition with other species; for instance, because it occurs on an island where competitors are absent.

Complementarity A condition of two resources of which less is required when they are taken together than when consumed separately.

Complementary resources A pair of resources for which consumption by the consumer of one resource reduces the consumer's requirements for the other.

Conservation The principles and practice of the science of preventing species extinctions.

Conspecific Belonging to the same species.

Consumption efficiency The percentage of energy available that is actually consumed at a trophic level. In the case of herbivores it is the percentage of net primary productivity that is ingested.

Contest competition Intraspecific competition in which mortality compensates exactly for increases in density, so that there are a constant (or approximately constant) number of survivors irrespective of initial density.

Continental drift The separation and movement of land masses in geological time.

Continentality Climatic conditions associated with locations in the middle of large continents.

Control action threshold (CAT) The combination of pest density and the densities of the pest's natural enemies beyond which it is necessary to intervene and take control measures against the pest, to prevent its population rising to a level at which it will cause economic damage.

Convergent evolution The process by which organisms of different evolutionary lineages come to have similar form or behaviour.

Coprophagy The consumption of faeces.

Correlogram A statistical procedure for the analysis of time-series data and to test for the presence of cyclic phenomena.

Coupled oscillations Linked fluctuation in the abundance of two species, broadly speaking a 'predator' and a 'prey', in which low prey abundance leads to low predator abundance which leads to high prey abundance, which leads to high predator abundance which leads to low prey abundance, and so on.

Crassulacean acid metabolism (CAM) A pathway for the assimilation of carbon dioxide by plants, in which carbon dioxide is fixed into organic acids during the night and released during the day for photosynthesis. Characteristic of succulent desert plants.

Cretaceous period A geological era; the final part of the Mesozoic, from approximately 140 to 170 million years ago.

Crypsis Form or behaviour of an organism that makes it difficult to detect, e.g. camouflage from a predator.

Cryptophyte *see* Table 1.2 (p. 33).

Cuticle In plants, a layer of waxy substance on the outer surface of epidermal cell walls. In animals, the outermost layer of many invertebrates.

Cyanogenic Releasing hydrogen cyanide.

Cytoplasm The living matter within a cell, excluding the genetic material.

Decomposition The breakdown of complex, energy-rich organic molecules to simple inorganic constituents.

Degradative succession A temporal succession of species that occurs on a degradable resource.

Demographic process A process capable of changing the size of a population, viz. birth, death or migration.

Density dependence The tendency for the death rate in a population to increase, or the birth or growth rate to decrease, as the density of the population increases.

Density independence The tendency for the death, birth or growth rate in a population neither to rise nor fall as the density increases.

Deoxyribonucleic acid (DNA) The carrier of genetic information in cells; capable of self-replication as well as coding for RNA synthesis.

Desert A desolate and barren region, usually deficient in available water, and with scant vegetation.

Deterministic forces Forces which can be characterized exactly with no element of chance or probability (*cf.* Stochastic forces).

Detritivory Consumption of dead organic matter (detritus) usually together with associated microorganisms.

Developmental threshold The body temperature of an organism below which no development occurs.

Devonian period A geological era lasting from approximately 400 to 360 million years ago.

Diapause A state of arrested development or growth, accompanied by greatly decreased metabolism, often correlated with the seasons, usually applied only to insects.

Dicotyledon A member of one of the two classes of flowering plants, distinguished by having seedlings with usually a pair of seedling leaves (cotyledons) and commonly with floral parts in fours or fives, leaves with net venation and ability to form wood by secondary (cambial) cell division within the tissues.

Differential resource utilization Normally used only in the context of interspecific competition, the use of different resources by two different species, or the use of the same resource at a different time, in a different place, or generally in a different manner.

Diffusion coefficient A measure of the rate of movement of solutes or gases in response to a concentration gradient in the medium in which they are dissolved.

Dimorphism The existence of two distinct forms of an organism or organ, e.g. winged and wingless generations in the life of aphids, winged and wingless seeds produced from the same flower.

Disaster Major disturbances in the life of a community or population which occur sufficiently often to leave their record in the 'genetic memory' of the population (*cf.* Catastrophe).

Discrete generations A series of generations in which, strictly, each one finishes before the next begins. Commonly, however, the early life cycle stages of the succeeding generation overlap with the end of the final stages of the preceding generation.

Disease The disturbed or altered condition of an organism (malfunctioning) caused by the presence of an antagonist (toxin or pathogen) or the absence of some essential (e.g. micronutrient or vitamin).

Disjunct distribution The geographical distribution of a species or other taxonomic class of which parts are widely separated.

Dispersal The spreading of individuals away from each other, e.g. of offspring from their parents and from regions of high density to regions of lower density.

Dispersal polymorphism Two or more types of dispersal structures found within a species or among the progeny of an individual.

Distribution The spatial range of a species, usually on a geographic but sometimes on a smaller scale, or the arrangement or spatial pattern of a species over its habitat.

Disturbance In community ecology, an event that removes organisms and opens up space which can be colonized by individuals of the same or different species.

Diversity *see* Species diversity.

Diversity index A mathematical index of species diversity in a community.

Dominant species Species which make up a large proportion of community biomass or numbers.

Donor-controlled models Mathematical models of predator–prey interactions in which the donor (prey) controls the density of the recipient (predator) but not the reverse.

Dormancy An extended period of suspended or greatly reduced activity, e.g. aestivation and hibernation.

Dynamic equilibrium The state of a system when it remains unchanged because two opposing forces are proceeding at the same rate.

Dynamically fragile Describes a community which is stable only within a narrow range of environmental conditions.

Dynamically robust Describes a community which is stable within a wide range of environmental conditions.

Ecological energetics The branch of ecology in which communities are studied from the point of view of the energy flowing through them.

Ecological niche A term with alternative definitions, not all of them synonymous. To state two: (i) the 'occupation' or 'profession' of an organism or species; or (ii) the range of conditions, resource levels and densities of other species allowing the survival, growth and reproduction of an organism or species. Hence, if each condition, resource or other species is seen as a dimension, the niche is an *n*-dimensional hypervolume.

Economic injury level (EIL) The level of pest abundance above which it costs less to control the pest than is saved by pest control, but below which it costs more than is saved.

Ecophysiology The study of physiology and tolerance limits of species that enhances understanding of their distribution in relation to abiotic conditions.

Ecosystem A holistic concept of the plants, the animals habitually associated with them and all the physical and chemical components of the immediate environment or habitat which together form a recognizable self-contained entity. The concept is due to Tansley (1935).

Ecotype A subset of individuals within a species with a characteristic ecology.

Ectomycorrhiza An association of a fungus and the root of a plant (usually a tree) in which the fungus forms a sheath around the root and penetrates *between* the cells of the host.

Ectoparasite A parasite that lives on the surface of its host.

Ectotherm An organism in which the body temperature relies on sources of heat outside itself.

Effective population size (N_e) The size of a genetically idealized population with which an actual population can be equated genetically.

Electivity A measure of the preference, or lack of it, shown by a consumer species for its range of prey.

Emergent properties Properties not possessed by individuals or populations that become apparent only when the community is the focus of attention.

Emigration The movement of individuals out of a population or from one area to another.

Endemic Having their habitat in a specified district or area, or the presence of a disease at relatively low levels, all the time.

Endobiont (endosymbiont) An organism which lives *within the cells* of a host organism in a mutualistic relationship or doing no apparent harm.

Endogenous rhythm A metabolic or behavioural rhythm that originates within the organism and persists even when external conditions are kept constant.

Endotherm An organism which is able to generate heat within itself to raise its body temperature significantly.

Environmental 'noise' Extraneous background signals that tend to mask biotic processes.

Enzyme denaturation Structural change in an enzyme, involving an unfolding of peptide chains and rendering the enzyme less soluble, produced by mild heat or various chemicals.

Ephemerals Organisms with a short life cycle, especially plants whose seeds germinate, grow to produce new seeds and then die all in a short period within a year.

Epidemic The outbreak of a disease which affects a large number and/or proportion of individuals in a population at the same time.

Epidemiology The study of the occurrence of infectious diseases, their origins and pattern of spread through a population.

Equilibrium theory A theory of community organization that focuses attention on the properties of the system at an equilibrium point, to which the community tends to return after a disturbance.

Equitability The evenness with which individuals are distributed among species in a community.

Eukaryote Organism with cells possessing a membrane-bound nucleus in which the DNA is complexed with histones and organized into chromosome, i.e. protozoans, algae, fungi, plants and animals.

Euphotic zone The surface zone of a lake or ocean within which net primary productivity occurs.

Eutrophication Enrichment of a water body with plant nutrients; usually resulting in a community dominated by phytoplankton.

Evapotranspiration The water loss to the atmosphere from soil and vegetation. The *potential* evapotranspiration may be calculated from physical features of the environmental such as incident radiation, wind speed and temperature. The *actual* evapotranspiration will commonly fall below the potential depending on the availability of water from precipitation and soil storage.

Even distribution The distribution of organisms in which the average distance between them is greater than would occur if they were distributed at random.

Evolutionarily stable strategy (ESS) A strategy which, if adopted by most of a population, cannot be bettered by any other strategy, and will therefore tend to become established by natural selection.

Evolutionary trees Lineages designed to show the evolutionary history of relationships between groups of organisms.

Exact compensation Density dependence in which increases in initial density are exactly counterbalanced by increases in death rate and/or decreases in birth rate and/or growth rate, such that the outcome is the same irrespective of initial density.

Exogenous Originating outside an organism.

Exploitation competition Competition in which any adverse effects on an organism are brought about by reductions in resource levels caused by other, competing organisms.

Exploiter-mediated coexistence A situation where predation promotes the coexistence of species amongst which there would otherwise be competitive exclusion.

Exponential growth Growth in the size of a population (or other entity) in which the rate of growth increases as the size of the population increases.

Extinction The condition that arises from the death of the last surviving individual of a species, group, or gene, globally or locally.

Extrafloral nectaries Nectar secreting glands found on the leaves and other vegetative parts of plants.

Extrinsic factors Literally, factors acting from outside. In ecology, physical and chemical features of the environment, and other organisms, are all extrinsic factors acting on an organism.

Facilitation (successional) In which the influence of early species in a community succession is to facilitate establishment of later ones by changing the conditions encountered.

Facultative annual An organism (the phrase is usually used of plants) which may in some circumstances complete its life cycle within 12 months.

Facultative mutualism The condition in which one or both species in a mutualistic association may survive and maintain populations in the absence of the other partner.

Fecundity The number of eggs, or seeds, or generally offspring in the first stage of the life cycle, produced by an individual.

Fecundity schedule A table of data displaying the lifetime pattern of birth amongst individuals of different ages within a population.

Field capacity The condition of a soil when it contains all the water retained after draining freely under gravity.

Fitness The contribution made to a population of descendants by an individual *relative* to the contribution made by others in its present population. The relative contribution that an individual makes to the gene pool of the next generation.

Floristic equilibrium The species composition of a flora when there is no further internally or externally generated change.

Folivores Animals that eat leaves.

Food chain An abstract representation of the links between consumers and consumed populations, e.g. plant–herbivore–carnivore.

Food web Representation of feeding relationships in a community that includes all the links revealed by dietary analysis.

Frequency dependence Of a predator, describes the tendency to take disproportionately more of the commoner prey species.

Functional response The relationship between a predator's consumption rate of prey and the density of those prey.

Fundamental net reproductive rate (R) The average number of individuals that each existing individual in a population gives rise to one time interval later; hence, the multiplication rate relating size of a population to its size one time interval earlier.

Fundamental niche The largest ecological niche that an organism or species can occupy in the absence of interspecific competition and predation.

Gamete A reproductive cell (haploid) which unites with another in fertilization to produce a zygote from which a new individual (genet) arises.

Gause's principle The idea that if two competing species coexist in a stable environment, then they do so as a result of differentiation of their realized niches; but if there is no such differentiation, or if it is precluded by the habitat, then one competing species will eliminate or exclude the other.

Gene A unit of inherited material—a hereditary factor.

Gene flow The consequence of cross-fertilization between members of a species across boundaries between populations, or within populations, which results in the spread of genes across and between populations.

Generation length (*T*) Not quite the average length of time between the birth of a parent and the birth of its offspring (the 'cohort generation time'): the production by offspring of their own offspring during the life of the original parent makes the average, computed over many generations, less than the cohort generation time.

Genet The organism developed from a zygote. The term is used especially for modular organisms and members of a clone to define the *genetic individual* and to contrast with 'ramet' the potentially physiologically independent part that may arise from the iterative process by which modular organisms grow.

Genetic drift Random changes in gene frequency within a population resulting from sampling effects rather than natural selection, and hence of greatest importance in small populations.

Genetic engineering Any change in the genetic constitution of an organism brought about by artificial means other than simple artificial selection and which would not usually occur in nature, such as the introduction of a gene from one species to another.

Genotype All the genetic characteristics that determine the structure and functioning of an organism.

Geometric series A series of numbers in which each is obtained by multiplying the preceding term by a constant factor, e.g. 1, 3, 9, 27, etc.

'Ghost of competition past' A term coined by J.H. Connell to stress that interspecific competition, acting as an evolutionary force in the past, has often left its mark on the behaviour, distribution or morphology of species, even when there is no present-day competition between them.

Global stability The tendency of a community to return to its original state even when subjected to a large perturbation.

Global warming The predicted warming of the planet resulting from increasing atmospheric concentrations of radiative gases such as carbon dioxide, methane, nitrous oxide and chlorofluorocarbons.

Glycoside A derivative of glucose (or another sugar) in which one hydrogen atom is replaced by an organic radical. (= Glucosinolate.)

Gonad Organs of animals which produce gametes.

Gradient analysis The analysis of species composition along a gradient of environmental conditions.

Granivores Animals that eat seeds.

Grazer A consumer which attacks large numbers of large prey during its lifetime, but removes only a part of each prey individual, so that the effect, although often harmful, is rarely lethal in the short term, and never predictably lethal.

Grazer–scrapers Aquatic animals that graze the organic layer of algae, microorganisms and dead organic matter on stones and other substrates.

Greenhouse effect Warming of the earth's atmosphere as a result of increases in CO_2 and other gases.

Gross primary production (GPP) The total fixation of energy by photosynthesis in a region.

Group selection The evolutionary process which is supposed to act through the different numbers of descendants left by groups rather than by individuals.

Guild A group of species that exploit the same class of environmental resources in a similar way.

Habitat Place where a microorganism, plant or animal lives.

Habitat diversity The range of habitats present in a region.

Halophyte A plant that tolerates very salty soil.

Handling time The length of time a predator spends in pursuing, subduing and consuming a prey item and then preparing itself for further prey searching.

Hardening A process by which the tolerance of extreme conditions, e.g. cold or drought, is increased by prior exposure to the same but less extreme conditions.

Haustoria Branches of parasitic plants or fungi which enter the tissues or cells of the host.

Hemicryptophyte *see* Table 1.2 (p. 33)

Hemiparasite Plants which are photosynthetic but form connections with the roots or stems of other plant species, drawing most or all of their water and mineral nutrient resources from their host.

Herbicide A chemical or biological preparation which kills plants.

Herbivory The consumption of living plant material.

Heritable variation The proportion of variation in a trait due to the effects of genetic factors.

Heterotroph An organism with a requirement for energy-rich organic molecules (animals, fungi and most bacteria).

Heterotrophic succession A temporal succession of species at a location, principally involving animals.

Heterozygote An organism carrying different alleles at the corresponding sites on homologous chromosomes.

Hibernate To remain dormant during the winter period.

Holoparasite Parasitic plants which lack chlorophyll and are therefore wholly dependent on their host plant for the supply of water, nutrients and fixed carbon.

Homeostasis Maintenance of relatively constant internal conditions in the face of a varying external environment.

Homeotherm An organism which maintains an approximately constant body temperature, usually above that of the surrounding medium.

Homing To return accurately to the place of origin, e.g. the return of salmon, after migration to the sea, to the same river in which they originally hatched.

Homologous structures Similarity in structure assumed to result from a common ancestry, e.g. the wing of a bird and the foreleg of a mammal.

Homozygote An organism carrying identical alleles at the corresponding sites on homologous chromosomes.

Host An organism which is parasitized by a parasite.

Host density dependence Applied to the aggregation of risk to hosts from parasitoid attacks: hosts in either higher or lower host density patches tend to be at a greater risk of attack (direct and inverse host density dependence, respectively). Could also be applied to other predator–prey interactions. (*See also* Aggregation of risk *and cf.* Host density independence.)

Host density independence Applied to the aggregation of risk to hosts from parasitoid attacks: risk varies from patch to patch but is not related to host density in a patch. Could also be applied to other predator–prey interactions. (*See also* Aggregation of risk *and cf.* Host density dependence.)

Hydrological cycle The movement of water from ocean, by evaporation, to atmosphere, to land and back, via river flow, to ocean.

Hydrophilia An overwhelming desire for water.

Hydrospheric Pertaining to the water in soil, river, lake and ocean.

Immigration Entry of organisms to a population from elsewhere.

Immune response A response mounted by an animal in which specific antibodies (vertebrates only) and/or cytotoxic cells are produced against invading microorganisms, parasites, transplanted tissue, and many other substances which are recognized as foreign by the body.

Inbreeding depression A loss of vigour amongst offspring occurring when closely related individuals mate, resulting from the expression of numbers of deleterious genes in a homozygous state and from a generally low level of heterozygosity.

Incidence function The relationship between island size and the proportion of islands of that size occupied by the species in question.

Incubation period The period during which the embryo in an egg develops before hatching. Also, in epidemiology, the period between infection and the appearance of symptoms of disease.

Individualistic concept Concept of the community as an association of species that occur together simply because of similarities in requirements and not as a result of a long coevolutionary history.

Inflorescence The organ bearing an aggregation of flowers in a flowering plant.

Inhibition (successional) The tendency of early successional species to resist the invasion of later species.

Inoculation The introduction of a disease organism or vaccine, usually through a deliberately made wound.

Intensity of abundance The number of individuals per habitable site in a community (*cf.* Prevalence (of abundance)).

Intensity of infection The number of parasites per host in a population (*cf.* Prevalence of infection).

Interference coefficient *see* Coefficient of interference *and also* Mutual interference.

Interference competition Competition between two organisms in which one physically excludes the other from a portion of habitat and hence from the resources that could be exploited there.

Interglacial period Period between glaciations, during which species begin to recolonize locations occupied previously.

Interspecific competition Competition between individuals of different species.

Intraspecific competition Competition between individuals of the same species.

Intrinsic rate of natural increase (*r*) The per capita rate of increase of a population which has reached a stable age structure without competitive or other restraints.

Inverse cube law A law which states that the intensity of an effect at a point B due to a source at A varies inversely as the cube of their distance apart.

Inverse square law A law which states that the intensity of an effect at a point B due to a source at A varies inversely as the square of their distance apart.

Island biogeography The study of distribution of species and community composition on islands.

Isocline A line linking points, each of which gives rise to the same rate of population increase for the species being considered. The points may represent combinations of species densities or combinations of resource levels.

Isotherm A line on a map that joins places having the same mean temperature.

Isotonic Having the same osmotic pressure.

Isotopes Atoms of the same element having the same chemical properties but differing in mass and in the physical properties that depend on their mass.

Iteroparity Where organisms produce offspring in a series of separate events during and after each of which the organisms maintain themselves in a condition that favours survival to reproduce again subsequently (though the 'separate events' may merge into continuous reproduction).

***K* selection** Selection of life-history traits which promote an ability to make a large proportionate contribution to a population which stays close to its carrying capacity—the traits being, broadly, large size, delayed reproduction, iteroparity, a small reproductive allocation, much parental care, and the production of few but large offspring.

***k*-value** The loss of individuals from a given stage of a life cycle when the numbers at the beginning and end of the stage are both expressed as common logarithms, i.e. $\log_{10}$ (before) – $\log_{10}$ (after).

Key-factor analysis A statistical treatment of population data designed to identify factors most responsible for change in population size.

Killing power Synonymous with *k*-value (see above).

Leaf area index The area of leaves exposed over a unit area of land surface.

Leptokurtic A symmetrical frequency distribution which differs from a normal distribution by deficiency in the shoulders compared to the tails and the top.

Life cycle The sequence of stages through which an organism passes in development from a zygote to the production of progeny zygotes.

Life form Characteristic structure of a plant or animal.

Life history An organism's lifetime pattern of growth, differentiation, storage and reproduction.

Life table A summary of the age- or stage-related survivorship of individuals in a population.

Light compensation point *see* Compensation point.

Lignin Complex organic material deposited within the cell walls of plants associated with cellulose, especially in wood and fibres.

Limiting similarity The level of similarity between two competing species which cannot be exceeded if the species are to coexist.

Lipase Enzyme(s) that split true fats into alcohols and acids.

Littoral zone The zone at the edge of a lake or ocean which is periodically exposed to the air and periodically immersed.

Local stability The tendency of a community to return to its original state when subjected to a small perturbation.

Logistic equation A much-used equation for the growth of a single-species population without discrete generations under the influence of competition, which is simple and captures the essence of actual examples but very few of the details of actual examples.

Longevity Length of life.

Macrofauna Large animals in a community within an arbitrary size range, e.g. between 2 mm and 20 mm body width, in soil invertebrates.

Macroinvertebrate An invertebrate with a body length greater than 2 mm.

Macroparasite A parasite which grows in its host but multiplies by producing infective stages that are released from the host to infect new hosts. They are often intercellular or live in body cavities, rather than within the host cells.

Marginal value theorem A proposed decision-rule (derived from theoretical exploration) for a predator foraging from patches of prey which it depletes, which states that the predator should leave all patches at the same rate of prey extraction, namely the maximum average overall rate for that environment as a whole.

Marsupial mammal A mammal in which the young are born in a very immature state and migrate to a pouch where they are suckled until relatively mature.

Masting The production, in some years, of especially large crops of seeds by trees and shrubs.

Maturation The process of becoming fully differentiated, fully functional and hence fully reproductive.

Maximal food chain A sequence of species running from a basal species (plant) to another species that feeds on it and so on to a top predator (fed on by no other).

Maximum sustainable yield (MSY) The maximum crop or yield that can be removed repeatedly from a population without driving it towards extinction.

Mean intensity of infection The mean number of parasites per host in a population (including those hosts that are not infected).

Megafauna The largest arbitrary size categorization of animals in a community, e.g. > 20 mm body width in soil invertebrates.

Megaherbivore A term referring to very large terrestrial grazing animals (> 1000 kg), such as elephants, also including the many species that went extinct in the last 30 000 years or so.

Megaphyte Plants with normally unbranched stems or trunks and bearing a crown of very large leaves. The inflorescence is commonly also massive.

Meristem A part of a plant in which cell division is concentrated (especially the shoot apices, lateral buds and a zone near the tips of roots and rootlets).

Mesofauna Animals in the size-range 100 μm to 2 mm body length.

Mesophyll The internal non-vascular tissue of a leaf. In green plants it is in these cells that most chloroplasts are found and in which most photosynthesis occurs.

Metabolism The sum of all chemical reactions occurring within a cell or an organism

Metamorphosis Abrupt transition between life stages; for example from larval to adult form.

Metapopulation A population perceived to exist as a series of subpopulations, linked by migration between them. However, the rate of migration is limited, such that the dynamics of the metapopulation should be seen as the sum of the dynamics of the individual subpopulations.

Microbes Microorganisms: any microscopic organism, including bacteria, viruses, unicellular algae and protozoans, and microscopic fungi such as yeasts.

Microbivores Animals that feed on microorganisms.

Microclimate The climate within a very small area or in a particular, often tightly defined, habitat.

Microfauna The smallest arbitrary size categorization of animals in a community.

Microflora Bacteria, fungi and microscopic algae.

Microparasite A parasite which multiplies directly within its host, usually within host cells.

Microsite The small subset of environments within a habitat that provide the specialized resources and conditions required for a phase in the life of an organism, e.g. the cracks or crevices which provide conditions suitable for germination of the seed of a particular species.

Microtopography Very small scale (roughly, 'organism-sized') variations in the height and roughness of the ground surface.

Migration The movement of individuals, and commonly whole populations from one region to another.

Mimicry The resemblance of an organism (the mimic) either to another organism or to a non-living object (the model), presumably conferring a benefit on the mimic in natural selection.

Miocene A geological era lasting from approximately 25 to 5 million years ago.

Modular organism One that grows by the repeated iteration of parts, e.g. the leaves, shoots and branches of a plant, the polyps of a coral or bryozoan. Modular organisms are almost always branched, though the connections between branches may separate or decay and the separated parts may in many cases then become physiologically independent, e.g. *Hydra* spp. and duckweeds (*Lemna* spp.). (*See also* Ramet *and cf.* Unitary organisms.)

Moisture gradient A spatial gradient in the availability of water in soil.

Monoclimax theory The concept that all successional sequences lead to a single characteristic climax in a given region.

Monocotyledons One of the two main groups of flowering plants (*see* Angiosperms). Characterized normally by the presence of just a single seedling leaf (cotyledon) and commonly by floral parts arranged in threes, parallel leaf veins and the inability to form secondary tissues, e.g. wood, from cell division within the tissues.

Monoculture A large area covered by a single species (or, for crops, a single variety) of plant; or, in experiments, plants of the same species grown alone without any other species.

Monogenean An ectoparasitic trematode flatworm, parasitic on fish or amphibia, and having only one host in its life cycle.

Monomorphic Occurring in only one form.

Monophagy Where an organism consumes only a single type of food item.

Monotreme mammal A primitive mammal belonging to one of only three genera, laying eggs but having hair and secreting milk.

Morphogenesis The development of size, form and other structural features of organisms.

Morphology The form and structure of an organism.

Motile organism An organism capable of spontaneous movement.

Multiple resistance (to pesticides) Resistance of an organism to a number of pesticides requiring different mechanisms to counteract their effects.

Mutual antagonism Describes two species with reciprocal negative effects on each other (either interspecific competition or mutual predation).

Mutual interference Interference amongst predators leading to a reduction in the consumption rate of individual predators which increases with predator density.

Mutualism An interaction between the individuals of two (or more) species in which the growth, growth rate and/or population size of both are increased in a reciprocal association. (*See also* Facultative mutualism *and* Obligate mutualism.)

Mycorrhiza A commonly mutualistic and intimate association between the roots of a plant and a fungus. (*See also* Ectomycorrhiza *and* Vesicular arbuscular mycorrhiza.)

***n*-dimensional hypervolume** *see* Ecological niche.

'n'-shaped curve More correctly, 'unimodal': a curve with a single maximum.

Natural selection The force that causes some individuals in a population to contribute more descendants (and genes) than others to subsequent generations and so leads to changes in the genetic composition of populations over time (evolution). (*See also* Fitness.)

Necromass The weight of dead organisms, usually expressed per unit of land or volume of water. The term is sometimes used to include the dead parts of living organisms, e.g. the bark and heartwood of trees, the hair and claws of animals.

Necroparasite A parasite that kills its host (or a part of it) and continues growth on the dead resource.

Net primary production (NPP) The total energy accumulated by plants during photosynthesis (gross primary production minus respiration).

Neutral models Models of communities that retain certain features of their real counterparts but exclude the consequences of biotic interactions; they are used to evaluate whether real communities are structured by biotic forces.

Neutralism The lack of an interaction between two organisms (or species): neither has any effect on the other.

Niche The limits, for all important environmental features, within which individuals of a species can survive, grow and reproduce.

Niche complementarity The tendency for coexisting species which occupy a similar position along one niche dimension, e.g. altitude, to differ along another, e.g. diet.

Niche differentiation The tendency for coexisting species to differ in their niche requirements.

Niche packing The tendency for coexisting species between them to fill the available 'space' along important niche dimensions.

Nitrification The conversion of nitrites to nitrates, usually by microorganisms. The term is commonly used to describe the process of conversion of ammonium ions via nitrites to nitrates.

Nitrogen fixation The conversion of gaseous nitrogen (N_2) into more complex molecules. The process is used industrially to produce nitrogen fertilizers. Biological nitrogen fixation is accomplished by both free-living and symbiotic microorganisms (prokaryotes). The process is more properly called 'dinitrogen fixation'.

Node The place on a stem where one or more leaves arise.

Non-equilibrium theory In community ecology, concerned with the transient behaviour of a system away from any equilibrium point, it specifically focuses attention on time and variation.

Null hypothesis The hypothesis that an observed pattern of data and an expected pattern are effectively the same, differing only by chance, not because they are truly different. A statistical significance test is then generally applied to the data to test whether the hypothesis can be rejected. If so, the observed and expected patterns are said to be significantly different. Tests do not establish that the null hypothesis is true. 'Expected' patterns may be derived from theory or from other, related data sets.

Nunatak Islands standing out in a 'sea of ice' during periods of glaciation and in which species may have persisted.

Nutrient cycling The transformation of chemical elements from inorganic form in the environment to organic form in organisms and, via decomposition, back to inorganic form.

Obligate mutualism A condition in which a mutualistic relationship with other species is essential for a species to survive.

Oligophagous Consuming a small range of types of food items.

Omnivory Feeding on prey from more than one trophic level.

Ontogenetic Occurring during the course of an organism's development.

Opportunistic species One that is capable of exploiting spasmodically occurring environments.

Optimal diet model Mathematical formulation of prey selection that would, in theory, provide the best energy returns to the consumer: used as a yardstick to assess actual performance.

Optimum similarity The level of similarity between competing species, which, if either exceeded or reduced, would lead to a lowering of fitness for individuals in at least one of the species.

Ordination Mathematical system for categorizing communities on a graph so that those that are most similar in species composition appear closest together.

Osmoregulation Regulation of the salt concentration in cells and body fluids.

Osmosis Diffusion of water across a semi-permeable membrane.

Osmotic pressure The tendency of water to move across a semi-permeable membrane into a solution.

Outbreeding In which genetically dissimilar organisms mate with each other.

Overcompensating density dependence Density dependence which is so intense that increases in initial density lead to reductions in final density.

Overexploitation Exploitation of (removal of individuals or biomass from) a natural population at a rate greater than the population is able to match with its own recruitment, thus tending to drive the population towards extinction.

Overgrowth competition Competition amongst sessile organisms where the mechanism is the growth of one individual over another, preventing the efficient capture of light, suspended food or some other resource.

Palaeoarctic The biogeographic region comprising the land-mass of Europe and Asia from its northern border to the Sahara and Himalayas.

Pampa(s) The treeless plains of South America, south of the Amazon.

Parallel evolution The evolution along similar lines of systematic groups that had been separated geographically at an earlier stage in their history.

Parasite An organism that obtains its nutrients from one or a very few host individuals causing harm but not causing death immediately.

Parasitoid Insects (mostly wasps and flies) in which the adults are free-living, but eggs are laid in, on or near an insect host (or rarely, a spider or isopod), after which the parasitoid larva develops in the host (itself usually a pre-adult), initially doing little apparent harm, but eventually consuming and killing the host before or during the pupal stage.

Partial refuges Areas of prey habitat in which their consumption rate by predators is less than the average for the habitat as a whole, as a result of the predators' behavioural responses to the prey's spatial distribution.

Passive dispersal Movement of seeds, spores or dispersive stages of animals caused by external agents such as wind current.

Patch dynamics The concept of communities as consisting of a mosaic of patches within which abiotic disturbances and biotic interactions proceed.

Patchy habitat A habitat within which there are significant spatial variations in suitability for the species under consideration.

Pathogen A microorganism or virus that causes disease.

Permafrost Layer of permanently frozen soil.

Permanent wilting point The condition of a soil in which water is sufficiently unavailable to cause plants growing in it to wilt irrecoverably.

Perturbation approach In community ecology, an experimental approach in which artificial disturbances are used to unravel species interactions.

Pest species Simply: any species which we, as humans, consider undesirable. More explicitly: a species which competes with humans for food, fibre or shelter, transmits pathogens, feeds on people, or otherwise threatens human health, comfort or welfare.

Petiole The stalk of a leaf.

pH A scale of acidity (1–7) or alkalinity (7–14) derived from the logarithm of the concentration of hydrogen ions (10^{-1}–10^{-14}).

Phagocyte White blood-corpuscle capable of destroying harmful bacteria.

Phanerophyte *see* Table 1.2 (p.33).

Phenology Strictly the study of periodic biological events; in practice often applied to periodic phenomena themselves, such as the lifetime pattern in an organism of growth, development and reproduction in relation to the seasons.

Phenotype A visible, or otherwise measurable, physical or biochemical characteristic of an organism, resulting from the interaction between the genotype and the environment.

Pheromones Chemicals released, usually in minute amounts, by one animal, that are detected by, and act as a signal to other members of the same species.

Phloem A plant tissue in the veins (vascular bundles) of plants that is responsible for most of the transport of organic solutes.

Photoperiod Length of the period of daylight each day.

Photosynthate The energy-rich organic molecules produced during photosynthesis.

Photosynthesis Utilization of the energy of sunlight to combine CO_2 and water into sugars.

Photosynthetically active radiation (PAR) Those wavelengths in the spectrum of radiation that are effective in photosynthesis.

Phyletic lines Links drawn between present and past groups of organisms which imply their evolutionary relationships and derivation.

Phyllosphere The microenvironment on or in the immediate neighbourhood of a leaf.

Phylogeny Evolutionary history of a taxonomic group.

Physiological time A measure combining time and temperature and applied to ectothermic and poikilothermic organisms, reflecting the fact that growth and development in particular are dependent on environmental temperature and therefore require a period of time–temperature rather than simply time for their completion.

Physiology Study of the internal processes and activities of organisms.

Phytoalexins Complex organic compounds produced by plants in response to infection and that are inhibitors of further growth by the pathogen.

Phytophagous Feeding on plant material.

Placental mammal Mammals which develop a persistent placenta, i.e. all mammals other than marsupials and monotremes.

Pleistocene A geological era lasting from approximately 2 million to 10 000 years ago.

Pliocene A geological era lasting from approximately 5 to 2 million years ago.

Pogonophoran A marine invertebrate of the phylum Pogonophora.

Poikilotherm An organism whose body temperature is strongly correlated with that of its external environment.

Polycentric distribution The presence of a population, species or other taxonomic group in several widely separated places.

Polyclimax theory The idea that succession leads to one of a variety of climaxes, depending on local environmental conditions.

Polymorphism The existence within a species or population of different forms of individuals, beyond those that are the result simply of recurrent mutation.

Polyphagous Consuming a wide range of types of food items.

Polysaccharide A carbohydrate polymer made up of a chain of monosaccharides, e.g. starch, cellulose.

Population A group of individuals of one species in an area, though the size and nature of the area is defined, often arbitrarily, for the purposes of the study being undertaken.

Population cycle Changes in the numbers of individuals in a population which repeatedly oscillate between periods of high and low density.

Population density The numbers in a population per unit area, or sometimes 'per unit volume', 'per leaf' or whatever seems appropriate.

Population dynamics The variations in time and space in the sizes and densities of populations.

Population ecology The study of the variations in time and space in the sizes and densities of populations, and of the factors causing those variations.

Population fluctuations Variations over time in the size of a population.

Population pyramid A means of illustrating the age structure of a population diagrammatically, by placing the youngest age class at the base and stacking successive age classes above it.

Population regulation A tendency in a population for some factor to cause density to increase when it is low and to decrease when it is high.

Population vulnerability analysis (PVA) An analysis, generally applied to populations or species in danger of extinction, of the population's chances of extinction.

Potential evapotranspiration *see* Evapotranspiration.

Prairie grassland A local, North American, name for the temperate grassland biome.

Precocity Reproduction occurring early in the life and growth of an organism relative to other organisms of the same or related species that, relatively, delay reproduction.

Predation The consumption of one organism, in whole or in part, by another, where the consumed organism is alive when the consumer first attacks it.

Predator An organism that consumes other organisms, divisible into true predators, grazers, parasites and parasitoids.

Prevalence (of abundance) The proportion or percentage of habitable sites or areas in which a particular species is present.

Prevalence of infection The proportion, or percentage, of a population that is infected with a specific parasite.

Prey An individual liable to be, or actually, consumed (and hence killed) by a predator.

Primary productivity The rate at which biomass is produced per unit area by plants.

Production efficiency The percentage of energy assimilated by an organism that becomes incorporated into new biomass.

Productivity The rate at which biomass is produced per unit area by any class of organisms.

Prokaryote A cell lacking a membrane-bound nucleus; a bacterium or cyanobacterium.

Propagule A term used for a structure in a plant (occasionally used for invertebrates) from which a new individual may arise, e.g. seed, corm, bulb, cyst and which may often also be a unit of dispersal.

Protoplasm Living matter.

Protozoan Single-celled animal.

Pseudo-interference A pattern of declining predator consumption rate with increasing predator density, reminiscent of the effects of mutual interference, but resulting from the aggregative response of the predator.

Pteridophytes A division of the plant kingdom comprising ferns, horsetails, clubmosses and their allies. Plants with true stems, leaves and roots (diploid) reproducing by spores and alternating with a free-living inconspicuous sexual (haploid) generation (prothalli).

Q_{10} The quotient of two reaction rates at temperatures differing by 10°C.

Quadrat A sampling unit used to assess density of organisms.

r **selection** Selection of life-history traits which promote an ability to multiply rapidly in numbers—the traits being, broadly, small size, precocious reproduction, semelparity, a large reproductive allocation and the production of many but small offspring.

Radiative evolution An evolutionary process that involves the branching of an evolutionary lineage and so leads to greater systematic diversity.

Ramet An offshoot or module formed by vegetative growth in some plants and modular invertebrates that is actually or potentially independent physiologically, e.g. the runners of the strawberry, the tubers of the potato, the polyps on a colonial hydroid.

Random distribution Lacking pattern or order. The result of (or indistinguishable from the consequence of) chance events.

Rank–abundance diagram A graphical plot of differential abundances of species in a community.

Ranked preference A preference exhibited by a consumer between food items all of which can be classified on the same simple scale (usually energy value).

Rarity *see* Prevalence (of abundance) *and* Intensity of abundance.

Realized niche That portion of its potential (fundamental) niche occupied by a species when competitors or predators are present.

Reciprocal predation An interaction between two species (or individuals) in which each preys upon the other, so that the interaction is, in essence, competitive.

Recombination The formation in offspring of combinations of genes not present in either parent. This results from the assortment of chromosomes and their genes during the production of gametes and the subsequent union of different sorts of gametes at fertilization.

Recruitment Additions to a population, either through birth or immigration, or, in the case of net recruitment, the differences between such additions and the losses resulting from death or emigration.

Regression analysis Analysis of the mathematical relationship between two variables.

Regular distribution The arrangement of individuals with respect to each other which has some pattern or order that ensures that they are more widely separated from each other than would be expected by chance.

Regulation *see* Population regulation.

Relative humidity Very roughly, the dampness of the air; more correctly, the percentage saturation of the air with water vapour; better still, the mass of water vapour per unit volume of air as a percentage of the same measure for saturated air at the same temperature.

Relict population An often very locally distributed residue from a large population that has declined.

Remoteness In island biogeography, the distance an island is from a mainland source of colonizing organisms.

Reproduction The production of new individuals, usually by sexual means, through the production of a zygote from which the new individual grows; though organisms that fall apart as they grow are often misleadingly said to undergo asexual reproduction.

Reproductive allocation Strictly, the proportion of an organism's available resource input that is allocated to reproduction over a defined period of time; often in practice, the proportion of an organism's mass or volume that is reproductive tissue.

Reproductive cost The decrease in survivorship and/or rate of growth, and hence the decrease in the potential for future reproduction, suffered by an individual as a result of increasing its current allocation to reproduction.

Reproductive isolation The isolation from each other, in space or time, of two parts of a population of which the individuals would be capable of interbreeding were this not prevented by their isolation. Such isolation is believed to be normal precondition for the evolution of new species to occur, especially in animals.

Reproductive output The production of offspring by an individual or population.

Reproductive rate The number of offspring produced by an organism per unit time or over a defined period of time.

Reproductive value The expected relative contribution of an individual to the population of its descendants, by reproduction, now and in the future.

Residual reproductive value (RRV) The expected relative contribution of an individual to its population, by reproduction, for all stages of its life cycle subsequent to the present.

Resilience The speed with which a community returns to its former state after it has been disturbed.

Resistance The ability of a community to avoid displacement from its present state by a disturbance.

Resource That which may be consumed by an organism and, as a result, becomes unavailable to another, e.g. food, water, nesting sites, etc.

Resource depletion zone The region around a consumer in which the availability of a resource is reduced, e.g. the zone around the absorbing surface of a root from which nutrients and water are absorbed.

Resource partitioning The differential use by organisms of resources such as food and space.

Respiration Any or all of the processes used by organisms to generate metabolically usable energy.

Restoration ecology The science concerned with the deliberate colonization and revegetation of derelict land, especially after major damage from activities such as mining and waste disposal and after land has been released from agricultural use.

Resurgence in pests A rapid increase in pest number, after the immediate impact of a control measure has passed, resulting from adverse effects on the natural enemies of the pest.

Rhizosphere The surface and immediate neighbourhood of a root which provides a specialized environment for microorganisms. The term is also used to define the microflora that lives in this region.

Ring-barking The removal of bark and the immediately underlying living tissues from a ring around the stem or trunk of a tree or shrub. This severs the route of transport of leaf products down to the roots.

Ruderal A plant of waste places, usually associated with human disturbance. The word distinguishes this group of plants from 'weeds' which are plants that are a nuisance to human activities—the ruderal is not necessarily a nuisance.

Ruminant Herbivorous mammals such as cows that chew the cud and have complex stomachs containing microorganisms that break down the cellulose in plant material.

Salt pan A basin or pool containing an accumulation of salty water or salt.

Saprophyte An organism that carries out external digestion of non-living organic matter and absorbs the products across the plasma membrane of its cells (e.g. fungi).

Savannah The tropical grassland biome.

Scramble competition The most extreme form of overcompensating density dependence in the effect of intraspecific competition on survivorship where all competing individuals are so adversely affected that none of them survive.

Searching efficiency The instantaneous probability that a given predator will consume a given prey (also called 'the attack rate').

Secondary productivity The rate at which biomass is produced per unit area by heterotrophic organisms.

Seed bank The population of viable dormant seeds that accumulates in and on soil and in sediments under water.

Selective pressure A force acting on populations that determines that some individuals leave more descendants (or genes) than others to subsequent generations (and so gives direction to the process of evolution).

Self-limitation A process where intraspecific competition leads to a reduction in reproduction and/or survival at higher densities.

Self-thinning The progressive decline in density which accompanies and interacts with the increasing size of individuals in a population of growing individuals.

Semelparity Where organisms produce all of their offspring in a single reproductive event over one relatively short period.

Serotiny The retention by a plant (usually by trees) of seeds in hard enclosing structures, e.g. ovaries or cones so that they are not dispersed and free to germinate until after some disaster, especially forest fire.

Serpentine soil Soil formed by the weathering of serpentine rock which contains high concentrations of various heavy metals. Serpentine soils are commonly localized and bear a specialized flora of species tolerant of these metals and of locally specialized 'ecotypes' (see above) of species that are also found elsewhere.

Sessile organism Literally a 'seated' organism. One whose position is fixed in space except during a dispersal phase, e.g. a rooted plant, barnacles, mussels (*Mytilus*), corals.

Sexual recombination The process by which DNA is exchanged between homologous chromosomes by chromosome pairing and crossing-over at meiosis during gamete formation.

Shredders Aquatic animals that feed on coarse particles of organic matter.

Sigmoid curve An 'S-shaped' curve in which there is an initial acceleration phase followed by a subsequent deceleration phase leading to a plateau.

Silurian A geological era lasting from approximately 438 to 408 million years ago.

Social facilitation An increase in consumption rate with increasing consumer density, resulting, for instance, from an increased time available for feeding when less time is required for vigilance against predators.

Somatic polymorphism The presence on the same genetic individual (genet) of organs of two or more different forms, e.g. several different leaf shapes on the same plant, two distinct types of seed produced on the same flower or inflorescence, modules of very different form in the body of a salp.

Species–area relationship A common pattern in which the number of species on islands decreases as island area decreases.

Species-deletion stability Tendency in a model community for the remaining species to remain at locally stable equilibria after a species is made extinct.

Species diversity An index of community diversity that takes into account both species richness and the relative abundance of species.

Species richness The number of species present in a community.

Spiracle A hole in the sides of insects through which the tracheal respiratory system connects with the exterior, and which can be opened and closed.

Stable equilibrium In an ecological context, a level of a population, or populations, or of resources, which is returned to after slight displacements from that level.

Stable limit cycles A regular fluctuation in abundance, the path of which is returned to after slight displacements from the path.

Standing crop The biomass of living organisms within a unit area.

Static life table A life table constructed from the age structure of a population at a single moment in time.

Steppe Treeless plains of southeastern Europe and Siberia.

Stochastic forces Random processes that affect community structure.

Stoloniferous Bearing stolons.

Stolons Horizontally growing, short-lived, stems that root at the nodes. Usually at or on the surface of the soil.

Stoma (*pl.* stomata) Pore in the epidermis of plants through which gas exchange takes places; especially abundant on leaves.

Stress Physics has a strict definition of 'a force per unit area' and producing 'strain' in the body to which the force is applied. Biology has a wide variety of meanings, e.g. any condition that results in reduced growth, any condition that prevents an organism from realizing its 'genetic potential'. The word is often redundant, e.g. the effects of drought stress/the effects of drought. The word is also often used confusingly in two senses, both to describe force and the condition induced in the organism by the force—a confusion of stimulus and response. We have tried not to use the word.

Structural diversity Range of types of physical structure in a community that may provide habitats for species.

Stylet Slender, elongated mouth part, usually of an insect, used for stinging or piercing prey or sucking sap.

Succession The non-seasonal, directional and continuous pattern of colonization and extinction on a site by populations.

Succulents Plants with fleshy or juicy tissues with high water content characteristic of desert and saline environments.

Superorganism concept The view of communities as consisting of member species that are tightly bound together both now and in their common evolutionary history.

'Supertramp species' Species with good colonizing ability that arrive earliest in a new habitat.

Surplus yield model A simple model of the impact of harvesting on a population, in which the population is represented by its size or biomass, undifferentiated into any internal structure.

Surrogate resource A resource, not in itself in limited supply, which is competed for because of the access it provides to some other resource, which is or may become limited in supply.

Survivorship The probability of a representative newly born individual surviving to various ages.

Survivorship curve A plot of the declining size of a cohort, or presumed cohort, as the individuals die, usually with time on the horizontal axis and $\log_{10} l_x$ on the vertical axis (where l_x is the proportion of the orginal cohort still alive).

Switching Of a predator, the tendency to switch between prey categories according to their relative abundance in the environment.

Symbiosis A close association between the individuals of pairs of species. The term 'mutualism' is reserved for symbioses for which there is evidence that the association brings mutual gains.

Sympatry The presence of two or more species living in such proximity that breeding between them should be possible, though their continued existence as separate species indicates that it does not normally happen. This contrasts with allopatry in which regional or geographical isolation normally denies the possibility of interbreeding.

Synergism The situation in which the combined effect of two forces, e.g. treatment with two drugs, is greater than the sum of their separate effects.

Taiga The coniferous forest that extends across much of North America and Eurasia bounded by tundra to the North and by steppe to the south.

Tannins Complex astringent substances containing phenolic compounds found in plants and usually making the plant material less readily digested. Especially common in bark, unripe fruits and galls on woody plants.

Taxonomy The study of the rules, principles and practice of classifying living organisms.

Tectonic plate An area of the earth's crust which moves during geological time resulting in continental drift and other major changes in the topography of the surface of the globe.

Temperature coefficient *see* Q_{10}.

Temporal variation Variability in conditions, such as temperature, on an hourly, daily or seasonal basis.

Terpenoids Aromatic hydrocarbons formed by many plants and responsible for many of the plant scents.

Territoriality The establishment by an animal or animals of an area from which other individuals are partially or totally excluded.

Tertiary A geological era lasting from approximately 65 to 2 million years ago.

Thermoneutral zone The range of environmental temperatures for an endotherm over which it has to exert the minimum metabolic effort in order to maintain a constant body temperature.

Therophyte *see* Table 1.2 (p. 33).

Thinning line The line on a plot of log mean individual weight against log density along which self-thinning populations of a sufficiently high biomass tend to progress, and beyond which they cannot pass.

Third generation insecticides Synthesized, usually organic insecticides, which are nevertheless produced with the aim of minimizing the impact on non-pest, non-target species.

Tiller A branch formed at or near ground level by grasses and sedges.

Time-delay population model A model of population growth in which the net reproductive rate of individuals currently is determined by the population size some time previously.

Tolerance (successional) Where modification of the environment by early occupants has little or no effect on subsequent performance of late-successional species.

Topography Representation of the physical structure of an environment.

Transfer efficiency Efficiency with which energy is passed through various steps in the trophic structure of a community.

Transgenic An organism containing a gene which has been artificially transferred from a member of another species.

Transient polymorphism The occurrence of two (or more) forms of a species or of genes (alleles) within a population while one form is being replaced by another.

Transmission threshold The basic reproductive rate of a parasite that is a necessary condition for a disease to spread.

Transpiration The evaporation of water from a plant surface.

Triassic A geological era lasting from approximately 250 to 213 million years ago.

Trichome Single- or many-celled outgrowth from the epidermis of a plant. A plant hair.

Trophic level Position in the food chain assessed by the number of energy-transfer steps to reach that level.

Trophic structure The organization of a community described in terms of energy flow through its various trophic levels.

True predator A predator which kills other organisms (their prey) more or less immediately after attacking them, killing several or many over the course of its lifetime.

Tundra The biome that occurs around the Arctic circle, characterized by lichens, mosses, sedges and dwarf trees.

Turbulence Fluid flow in which the motion at any point varies rapidly in direction and magnitude.

Turgor The distension of living tissue due to internal pressures.

Undercompensating density dependence Density dependence in which death rate increases, or birth or growth rate decreases, less than initial density increases, so that increases in initial density still lead to (smaller) increases in final density.

Unit leaf rate The dry weight increment made by a plant over a period of time expressed as a function of the plant's leaf area.

Unitary organisms Those that proceed by a determinate pathway of development of a tightly canalized adult form, e.g. all arthropods and vertebrates. The contrast is with 'Modular organisms' (see above) in which growth occurs by the indeterminate iteration of repeated units of structure (modules).

Unstable equilibrium In an ecological context, a level of a population, or populations, or of resources, from which slight displacements lead to larger displacements.

Vacuole Membrane-bound, fluid-filled sac within the cytoplasm of a cell.

Vector (i) Any agent (living or otherwise) that acts as a carrier for a pathogenic organism and transmits it to a susceptible host. (ii) A physical quantity with a direction as well as a strength.

Veld Open pasture land in South Africa.

Vesicular arbuscular mycorrhiza (VAM) An intimate and perhaps usually mutualistic association between a fungus and a plant root in which the fungus enters the host cells and also usually extends widely into the surrounding soil (*cf.* Ectomycorrhiza).

Vital attributes (species) Properties of species that determine their place in a succession.

Weed species Plants that threaten human welfare by competing with other plants that have food, timber or amenity value.

Zero isocline or **Zero net growth isocline (ZNGI)** An isocline along which the rate of population growth is zero.

Zonation The characteristic distributions of species along environmental gradients.

Zygote Diploid cell resulting from the fusion of female and male gametes.

References

Abbott, I. (1978) Factors determining the number of land bird species on islands around south-western Australia. *Oecologia*, **33**, 221–223.
23.3.1

Abbott, R.J., Chapman, H.M., Crawford, R.M.M. & Forbes, D.G. (1995) Molecular diversity and derivations of populations of *Silene acaulis* and *Saxifraga oppositifolia* from the high Arctic and more southerly latitudes. *Molecular Ecology*, **4**, 199–207.
1.2.2

Aber, J.D. (1992) Nitrogen cycling and nitrogen saturation in temperate forest ecosystems. *Trends in Ecology and Evolution*, **7**, 220–224.
19.4.4

Aber, J.D. & Federer, C.A. (1992) A generalised, lumped-parameter model of photosynthesis, evapotranspiration and net primary production in temperate and boreal forest ecosystems. *Oecologia*, **92**, 463–474.
3.4

Abrahamson, W.G. (1975) Reproductive strategies in dewberries. *Ecology*, **56**, 721–726.
8.2.2

Abrams, P. (1976) Limiting similarity and the form of the competition coefficient. *Theoretical Population Biology*, **8**, 356–375.
7.9

Abrams, P. (1983) The theory of limiting similarity. *Annual Review of Ecology and Systematics*, **14**, 359–376.
7.5.5, 7.9

Abrams, P.A. (1990) Ecological vs evolutionary consequences of competition. *Oikos*, **57**, 147–151.
7.9

Abramsky, Z. & Rosenzweig, M.L. (1983) Tilman's predicted productivity–diversity relationship shown by desert rodents. *Nature*, **309**, 150–151.
24.3.1

Abramsky, Z. & Sellah, C. (1982) Competition and the role of habitat selection in *Gerbillus allenbyi* and *Meriones tistrami*: a removal experiment. *Ecology*, **63**, 1242–1247.
7.8.1

Addicott, J.F. (1979) A multispecies aphid–ant association: density–dependence and species-specific effects. *Canadian Journal of Zoology*, **57**, 558–569.
13.12

Addicott, J.F. & Freedman, H.I. (1984) On the structure and stability of mutualistic systems: analysis of predator–prey and competition models as modified by the action of a slow growing mutualist. *Theoretical Population Biology*, **26**, 320–339.
13.12

Agardy, M.T. (1994) Advances in marine conservation: the role of marine protected areas. *Trends in Ecology and Evolution*, **9**, 267–270.
25.6.3

Agricultural Research Council (1965) *The Nutritional Requirements of Farm Livestock. 2. Ruminants*. Agricultural Research Council, London.
8.4

Akçakaya, H.R. (1992) Population viability analysis and risk assessment. In: *Proceedings of Wildlife 2001: Populations* (D.R. McCullough, ed.). Elsevier, Amsterdam.
25.2.1

Akçakaya, H.R. & Ginzburg, L.R. (1991) Ecological risk analysis for single and multiple populations. In: *Species Conservation: A Population–Biological Approach* (A. Seitz & V. Loeschcke, eds), pp. 78–87. Birkhauser Verlag, Basel, Germany.
25.4.4

Albertson, F.W. (1937) Ecology of mixed prairie in west central Kansas. *Ecological Monographs*, **7**, 481–547.
3.5

Alexander, R. McN. (1991) Optimization of gut structure and diet for higher vertebrate herbivores. In: *The Evolutionary Interaction of Animals and Plants* (J.L. Harper & J.H. Lawton, eds), pp. 73–79. The Royal Society, London; also in *Philosophical Transactions of the Royal Society of London, Series B*, **333**, 249–255.
3.7.2

Alicata, J.E. & Jindrak, K. (1970) *Angiostrongylosis in the Pacific and Southeast Asia*. C.C. Thomas, Springfield, IL.
12.3.3

Allan, J.D. & Flecker, A.S. (1993) Biodiversity conservation in running waters. *Bioscience*, **43**, 32–43.
23.2.3, 25.6.4

Allee, W.C. (1931) *Animal Aggregations. A Study in General Sociology*. University of Chicago Press, Chicago.
10.4

Allen, K.R. (1972) Further notes on the assessment of Antarctic fin whale stocks. *Report of the International Whaling Commission*, **22**, 43–53.
6.4

Alliende, M.C. & Harper, J.L. (1989) Demographic studies of a dioecious tree. I. Colonization, sex and age-structure of a population of *Sarex cinerea*. *Journal of Ecology*, **77**, 1029–1047.
6.4

Alphey, T.W. (1970) Studies on the distribution and site location of *Nippostrongylus brasiliensis* within the small intestine of laboratory rats. *Parasitology*, **61**, 449–460.
12.3.3

Altaba, C.R. (1990) The last known population of the freshwater mussel *Margaritifera auricularia* (Bivalvia, Unionida): a conservation priority. *Biological Conservation*, **52**, 271–286.
25.2.2

Anderson, J.M. (1975) Succession, diversity and trophic relationships of some soil animals in decomposing leaf litter. *Journal of Animal Ecology*, **44**, 475–495.
11.3.1

Anderson, J.M. (1978) Inter- and intrahabitat relationships between woodland Cryptostigmata species diversity and diversity of soil and litter micro-habitats. *Oecologia*, **32**, 341–348.
11.2.2

Anderson, J.M. (1992) Responses of soils to climate change. *Advances in Ecological Research*, **22**, 163–210.
19.5

Anderson, P.K. (1989) Dispersal in rodents: a resident fitness hypothesis. *American Society of Mammalologists, Special Publication*, 9.
5.4.3, 15.4.5

Anderson, R.M. (1979) The influence of parasitic infection on the dynamics of host population growth. In: *Population Dynamics* (R.M. May, ed.), pp. 318–355. Blackwell Scientific Publications, Oxford.
12.4

Anderson, R.M. (1981) Population ecology of infectious disease agents. In: *Theoretical Ecology: Principles and Applications*, 2nd edn. (R.M. May, ed.), pp. 318–355. Blackwell Scientific Publications, Oxford.
12.5, 21.2.4

Anderson, R.M. (1982) Epidemiology. In: *Modern Parasitology* (F.E.G. Cox, ed.), pp. 205–251. Blackwell Scientific Publications, Oxford.
12.3.1, 12.3.7, 12.5.2, 12.5.4

Anderson, R.M. (1991) Populations and infectious diseases: ecology or epidemiology? *Journal of Animal Ecology*, **60**, 1–50.
12.5

Anderson, R.M. & May, R.M. (1978) Regulation and stability of host–parasite population interactions. I. Regulatory processes. *Journal of Animal Ecology*, **47**, 219–249.
12.5.2, 12.6.3

Anderson, R.M. & May, R.M. (1980) Infectious diseases and population cycles of forest insects. *Science*, **210**, 658–661.
12.6

Anderson, R.M. & May, R.M. (1981) The population dynamics of microparasites and their invertebrate hosts. *Philosophical Transactions of the Royal Society of London*, **291**, 451–524.
12.6.1

Anderson, R.M. & May, R.M. (1991) *Infectious Diseases of Humans: Dynamics and Control*. Oxford University Press, Oxford.
12.5.1, 12.5.2, 12.6.2

Anderson, R.M., Jackson, H.C., May, R.M. & Smith, A.D.M. (1981) Population dynamics of fox rabies in Europe. *Nature*, **289**, 765–771.
12.6.2

Andrewartha, H.G. & Birch, L.C. (1954) *The Distribution and Abundance of Animals*. University of Chicago Press, Chicago.
15.2.2, 15.6.1, 15.6.3

Andrewartha, H.G. & Birch, L.C. (1960) Some recent contributions to the study of the distribution and abundance of insects. *Annual Review of Entomology*, **5**, 219–242.
15.2.2, 15.6.3

Andrewartha, H.G. & Birch, L.C. (1984) *The Ecological Web*. The University of Chicago Press, Chicago.
15.2.2, 15.6.3

Andrews, J.H. (1991) *Comparative Ecology of Microorganisms and Macroorganisms*. Springer-Verlag, New York.
4.2.1

Andrews, P., Lorde, J.M. & Nesbit Evans, E.M. (1979) Patterns of ecological diversity in fossil and mammalian faunas. *Biological Journal of the Linnean Society*, **11**, 177–205.
1.4.2

Angel, M.V (1993) Biodiversity of the pelagic ocean. *Conservation Biology*, **7**, 760–772.
24.3.1, 24.4.1

Antonovics, J. & Alexander, H.M. (1992) Epidemiology of anther-smut infection of *Silene alba* (= *Silene latifolia*) caused by *Ustilago violacea*: patterns of spore deposition in experimental populations. *Proceedings of the Royal Society, Series B*, **250**, 157–163.
12.5.3

Antonovics, J. & Bradshaw, A.D. (1970) Evolution in closely adjacent plant populations. VIII. Clinical patterns at a mine boundary. *Heredity*, **25**, 349–362.
1.5.1

Arcese, P. & Smith, J.N.M. (1988) Effects of population density and supplemental food on reproduction in song sparrows. *Journal of Animal Ecology*, **57**, 119–136.
6.2

Armbruster, P. & Lande, R. (1992) A population viability analysis for African elephant (*Loxodonta africana*): how big should reserves be? *Conservation Biology*, **7**, 602–610.
25.4.4

Arnold, G.W. (1964) Factors within plant associations affecting the behaviour and performance of grazing animals. In: *Grazing in Terrestrial and Marine Environments* (D.J. Crisp, ed.), pp. 133–154. Blackwell Scientific Publications, Oxford.
9.2.2

Arnold, S.J. (ed.) (1994) Sexual selection in plants and animals. A symposium organised by Stevan J. Arnold. *American Naturalist*, **144**, S1–S149.
13.4.2

Arthur, W. (1982) The evolutionary consequences of interspecific competition. *Advances in Ecological Research*, **12**, 127–187.
7.8.1, 7.8.3

Ascaso, C., Galvan, J. & Rodriguez-Oascual, C. (1982) The weathering of calcareous rocks by lichens. *Pedobiologica*, **24**, 219–229.
19.2.1

Ashmole, N.P. (1971) Sea bird ecology and the marine environment. In: *Avian Biology, Volume 1* (D.S. Farner & J.R. King, eds), pp. 224–286. Academic Press, New York.
14.12

Askew, R.R. (1961) On the biology of inhabitants of oak galls of the Cynipidae (Hymenoptera) in Britain. *Transactions of the*

British Society for Entomology, **14**, 237–286.
12.3.3, 12.3.5

Askew, R.R. (1971) *Parasitic Insects*. American Elsevier, New York.
12.3.5

Atkinson, W.D. & Shorrocks, B. (1981) Competition on a divided and ephemeral resource: a simulation model. *Journal of Animal Ecology*, **50**, 461–471.
7.5.5, 11.4, 20.2.2, 24.3.2

Auer, C. (1971) Some analyses of the quantitative structure in populations and dynamics of larch bud moth 1949–1968. In: *Statistical Ecology*, Vol. 2 (G.P. Patil, E.C. Pielou & W.E. Walters, eds), pp. 151–173. Pennsylvania State University Press, University Park, PA.
12.6

Augspurger, C.K. (1983) Seed dispersal of the tropical tree *Platypodium elegans*, and the escape of its seedlings from fungal pathogens. *Journal of Ecology*, **71**, 759–771.
12.3.9

Augspurger, C.K. (1984) Seedling survival of tropical tree species: interactions of dispersal distance, light gaps and pathogens. *Ecology*, **65**, 1705–1712.
12.3.9

Augspurger, C.K. (1990) Spatial patterns of damping-off disease during seedling recruitment in tropical forests. In: *Pests, Pathogens and Plant Communities* (J.J. Burdon & S.R. Leather, eds), pp. 131–144. Blackwell Scientific Publications, Oxford.
12.3.9

Augspurger, C.K. & Hogan, K.P. (1983) Wind dispersal of fruits with variable seed number in a tropical tree (*Lonchocarpus pentaphyllus*: Leguminosae). *American Journal of Botany*, **70**, 1031–1037.
5.3.4

Ausmus, B.S., Edwards, N.T. & Witkamp, M. (1976) Microbial immobilisation of carbon, nitrogen, phosphorus and potassium: implications for forest ecosystem processes. In: *The Role of Terrestrial and Aquatic Organisms in Decomposition Processes* (J.M. Anderson & A. MacFadyen, eds), pp. 397–416. Blackwell Scientific Publications, Oxford.
11.2.3

Ayal, Y. (1994) Time-lags in insect response to plant productivity: significance for plant–insect interactions in deserts. *Ecological Entomology*, **19**, 207–214.
8.4

Ayala, F.J. (1969) Evolution of fitness. IV. Genetic evolution of interspecific competitive ability in *Drosophila*. *Genetics*, **61**, 737–747.
7.8.3

Bach, C.E. (1994) Effects of herbivory and genotype on growth and survivorship of sand-dune willow (*Salix cordata*). *Ecological Entomology*, **19**, 303–309.
8.2.4

Bailey, J.A. & Mansfield, J.W. (1982) *Phytoalexins*. Blackie, Glasgow.
12.3.5

Bailey, N.T.J. (1975) *The Mathematical Theory of Infectious Diseases and its Applications*, 2nd edn. Macmillan, New York.
12.5.2

Bailey, W.S. (1972) *Spirocerca lupi:* a continuing enquiry. *Journal of Parasitology*, **58**, 3–22.
12.3.3

Baker, A.J. (1992) Molecular genetics of *Calidris*, with special reference to knots. *Wader Study Group Bulletin, 64, Supplement, 28–35*. Joint Nature Conservation Committee, Publications Branch, Peterborough.
5.7.3

Baker, R.R. (1978) *The Evolutionary Ecology of Animal Migration*. Hodder & Stoughton, London.
5.3.1

Baker, R.R. (1982). *Migration: Paths Through Time and Space*. Hodder & Stoughton, London.
5.3.1, 5.7.3

Bakker, K. (1961) An analysis of factors which determine success in competition for food among larvae of *Drosophila melanogaster*. *Archieves Néerlandaises de Zoologie*, **14**, 200–281.
6.6

Baltensweiler, W., Benz, G., Bovey, P. & Delucchi, V. (1977) Dynamics of larch budmoth populations. *Annual Review of Ecology and Systematics*, **22**, 79–100.
8.2.2

Barkalow, F.S., Hamilton, R.B. & Soots, R.F. (1970) The vital statistics of an unexploited gray squirrel population. *Journal of Wildlife Management*, **34**, 489–500.
14.3

Barkman, J.J. (1988) New systems of plant growth forms and phenological plant types. In: *Plant Form and Vegetation Structure. Adaptation, Plasticity and Relation to Herbivory* (M.J.A. Werger, P.J.M. van der Aart, H.J. During & J.T.A. Verhoeven, eds), pp. 9–44. SPB Academic Publishing, The Hague.
1.4.2

Bärlocher, F. & Kendrick, B. (1975) Assimilation efficiency of *Gammarus pseudolimnaeus* (Amphipoda) feeding on fungal mycelium or autumn-shed leaves. *Oikos*, **26**, 55–59.
11.3.1

Barnard, C. & Thompson, D.B.A. (1985) *Gulls and Plovers: The Ecology and Behaviour of Mixed Species Feeding Groups*. Croon Helm, London.
9.6

Barnes, H. (1957) The northern limits of *Balanus balanoides* (L.). *Oikos*, **8**, 1–15.
2.2.2

Barnes, R.S.K. & Hughes, R.N. (1982) *An Introduction to Marine Ecology*. Blackwell Scientific Publications, Oxford.
18.3.3

Baross, J.A. & Deming, J.W. (1995) Growth at high temperatures: isolation and taxonomy, physiology, ecology. In: *Microbiology of Deep-Sea Hydrothermal Vent Habitats* (D.M. Karl, ed.). CRC Press, New York.
2.4

Barrett, J.A. (1983) Plant–fungus symbioses. In: *Coevolution* (D.J. Futuyma & M. Slatkin, eds), pp. 137–160. Sinauer Associates, Sunderland, MA.
16.10

Bartell, S.M., Brenkert, A.L., O'Neill, R.V. & Gardner, R.H. (1988) Temporal variation in regulation of production in a pelagic food web model. In: *Complex Interactions in Lake Communities* (S.R. Carpenter, ed.), pp. 101–118. Springer-Verlag, New York.
22.3.3

Bartholomew, G.A. (1982) Body temperature and energy metabolism. In: *Animal Physiology* (M.S. Gordon, ed.), pp.

46–93. Macmillan, New York.
2.5, 2.5.1

Barton, K.A., Whiyely, H.R. & Yang, N. (1987) *Bacillus thuringiensis* delta-endotoxin expressed in transgenic *Nicotiana tabacum* provides resistance to lepidopteran insects. *Plant Physiology*, **85**, 1103–1109.
16.10

Baskin, J.M. & Baskin, C.C. (1979) Studies on the autecology and population biology of the monocarpic perennial *Grindelia lanceolata*. *American Midland Naturalist*, **41**, 290–299.
4.8

Bass, M. & Cherrett, J.M. (1994) The role of leaf-cutting ant workers (Hymenoptera: Formicidae) in fungus garden maintenance. *Ecological Entomology*, **19**, 215–220.
13.3.3

Bateson, P.P.G. (1978) Sexual imprinting and optimal outbreeding. *Nature*, **273**, 659–660.
5.4.4

Bateson, P.P.G. (1980) Optimal outbreeding and the development of sexual preferences in the Japanese quail. *Zeitschrift für Tierpsychologie*, **53**, 231–244.
5.4.4

Battarbee, P.W. (1984) Diatom analysis and the acidification of lakes. *Philosophical Transactions of the Royal Society of London, Series B*, **305**, 451–477.
2.10

Batzli, G.O. (1983) Responses of arctic rodent populations to nutritional factors. *Oikos*, **40**, 396–406.
10.3.2, 15.4.3

Batzli, G.O., White, R.G., MacLean, S.F., Pitelka, F.A. & Collier, B.D. (1980) The herbivore based trophic system. In: *An Arctic Ecosystem: The Coastal Tundra at Barrow, Alaska* (J. Brown, P.C. Miller, L.L. Tieszen & F.L. Bunnell, eds), pp. 335–410. Dowden, Hutchinson & Ross, Stroudsburg, PA.
15.4.2, 15.4.3

Bazzaz, F.A. (1975) Plant species diversity in old-field successional ecosystems in southern Illinois. *Ecology*, **56**, 485–488.
24.4.4

Bazzaz, F.A. (1979) The physiological ecology of plant succession. *Annual Review of Ecology and Systematics*, **10**, 351–371.
17.4.3, 17.4.4

Bazzaz, F.A. (1990) The response of natural ecosystems to the rising global CO_2 levels. *Annual Review of Ecology and Systematics*, **21**, 167–196.
3.3.4

Bazzaz, F.A. & Reekie, E.G. (1985) The meaning and measurement of reproductive effort in plants. In: *Studies on Plant Demography: A Festschrift for John L. Harper* (J. White, ed.), pp. 373–387. Academic Press, London.
14.2

Bazzaz, F.A. & Williams, W.E. (1991) Atmospheric CO_2 concentrations within a mixed forest: implications for seedling growth. *Ecology*, **72**, 12–16.
3.3

Bazzaz, F.A., Miao, S.L. & Wayne, P.M. (1993) CO_2-induced growth enhancements of co-occurring tree species decline at different rates. *Oecologia*, **96**, 478–482.
3.3.4

Beattie, A. (1989) Myrmecotrophy: plants fed by ants. *Trends in Ecology and Evolution*, **4**, 172–176.
13.2.1

Beaver, R.A. (1979) Host specificity of temperate and tropical animals. *Nature*, **281**, 139–141.
24.4.1

Becker, P. (1992) Colonization of islands by carnivorous and herbivorous Heteroptera and Coleoptera: effects of island area, plant species richness, and 'extinction' rates. *Journal of Biogeography*, **19**, 163–171.
23.3.1, 23.3.5

Beddington, J.R. & May, R.M. (1977) Harvesting natural populations in a randomly fluctuating environment. *Science*, **197**, 463–465.
16.15.2

Beddington, J.R., Free, C.A. & Lawton, J.H. (1978) Modelling biological control: on the characteristics of successful natural enemies. *Nature*, **273**, 513–519.
10.5.1

Beebee, T.J.C. (1991) Purification of an agent causing growth inhibition in anuran larvae and its identification as a unicellular unpigmented alga. *Canadian Journal of Zoology*, **69**, 2146–2153.
7.3

Beerling, D.J. (1994) Modelling palaeophotosynthesis: late Cretaceous to present. *Philosophical Transactions of the Royal Society of London, Series B*, **346**, 421–432.
3.3.4

Begon, M. (1976) Temporal variations in the reproductive condition of *Drosophila obscura* Fallén and *D. subobscura* Collin. *Oecologia*, **23**, 31–47.
4.3, 5.5.1

Begon, M. (1985) A general theory of life-history variation. In: *Behavioural Ecology* (R.M. Sibly & R.H. Smith, eds), pp. 91–97. Blackwell Scientific Publications, Oxford.
14.6

Begon, M. (1990) *Ecological Food Production*. IPPR, London.
16.11

Begon, M. & Bowers, R.G. (1995) Beyond host–pathogen dynamics. In: *Ecology of Infectious Diseases in Natural Populations* (B.T. Grenfell & A.P. Dobson, eds), pp. 478–509. Cambridge University Press, Cambridge.
12.6.3

Begon, M. & Mortimer, M. (1986) *Population Ecology: A Unified Study of Animals and Plants*, 2nd edn. Blackwell Scientific Publications, Oxford.
4.6.1, 7.7.2

Begon, M. & Wall, R. (1987) Individual variation and competitor coexistence: a model. *Functional Ecology*, **1**, 237–241.
7.5.5

Begon, M., Firbank, L. & Wall, R. (1986) Is there a self-thinning rule for animal populations? *Oikos*, **46**, 122–124.
6.13

Begon, M., Harper, J.L. & Townsend, C.R. (1990) *Ecology: Individuals, Populations and Communities*, 2nd edn. Blackwell Scientific Publications, Oxford.
9.7.2, 10.3.1

Begon, M., Sait, S.M. & Thompson, D.J. (1995a) Persistence of a predator–prey system: refuges and generation cycles? *Proceedings of the Royal Society of London, Series B*, **260**, 131–137.
10.2.4

Begon, M., Bowers, R.G., Sait, S.M. & Thompson, D.J. (1995b) Population dynamics beyond two species: hosts, parasitoids

46–93. Macmillan, New York.
2.5, 2.5.1

and pathogens. In: *Frontiers of Population Ecology* (R.B. Floyd & A.W. Sheppard, eds). CSIRO Press, Melbourne. (In press).
10.7

Bekoff, M. (1977) Mammalian dispersal and the ontogeny of individual behavioural phenotypes. *American Naturalist*, **111**, 715–732.
5.4.3, 15.4.4

Bell, G. (1980) The costs of reproduction and their consequences. *American Naturalist*, **116**, 45–76.
14.7.2

Bell, G. & Burt, A. (1991) The comparative biology of parasite species diversity: internal helminths of freshwater fish. *Journal of Animal Ecology*, **60**, 1047–1063.
12.3.8

Bell, G. & Koufopanou, V. (1986) The cost of reproduction. *Oxford Surveys in Evolutionary Biology*, **3**, 83–131.
14.4

Bellows, T.S. Jr. (1981) The descriptive properties of some models for density dependence. *Journal of Animal Ecology*, **50**, 139–156.
6.2, 6.8.2

Belt, T. (1874) *The Naturalist in Nicaragua*. J.M. Dent, London.
13.2.1

Benjamin, R.K. & Shorer, L. (1952) Sex of host specificity and position specificity of certain species of *Laboulbenia* on *Bembidion picipes*. *American Journal of Botany*, **39**, 125–131.
12.3.3

Bennekon, G. & Schroll, H. (1988) Nitrogen budget for a dairy farm. *Ecological Bulletins*, **39**, 134–135.
19.2.4

Bennet, K.D. (1986) The rate of spread and population increase of forest trees during the postglacial. *Philosophical Transactions of the Royal Society, Series B*, **314**, 523–531; and in *Quantitative Aspects of the Ecology of Biological Invasions* (H. Kornberg & M.H. Williamson, eds), pp. 21–27. The Royal Society, London.
1.2.2

Benson, J.F. (1973a) The biology of Lepidoptera infesting stored products, with special reference to population dynamics. *Biological Reviews*, **48**, 1–26.
6.6

Benson, J.F. (1973b) Population dynamics of cabbage root fly in Canada and England. *Journal of Applied Ecology*, **10**, 437–446.
6.6

Bentley, B.L. (1977) Extrafloral nectaries and protection by pugnacious bodyguards. *Annual Review of Ecology and Systematics*, **8**, 407–427.
13.2.1

Bentley, S. & Whittaker, J.B. (1979) Effects of grazing by a chrysomelid beetle, *Gastrophysa viridula*, on competition between *Rumex obtusifolius* and *Rumex crispus*. *Journal of Ecology*, **69**, 79–90.
8.2.6

Benton, M.J. (1994) Palaeontological data and identifying mass extinctions. *Trends in Ecology and Evolution*, **9**, 181–185.
24.4.6

Berden, M., Nilsson, S.I., Rosen, K. & Tyler, G. (1987) *Soil Acidification—Extent, Causes and Consequences*. National Swedish Environmental Protection Board, Report 3292.
19.4.4

Bergelson, J. (1994) The effects of genotype and the environment on costs of resistance in lettuce. *American Naturalist*, **143**, 349–359.
12.4

Bergelson, J.M. (1985) A mechanistic interpretation of prey selection by *Anax junius* larvae (Odonata: Aeschnidae). *Ecology*, **66**, 1699–1705.
9.2.3

Berger, J. (1990) Persistence of different-sized populations: an empirical assessment of rapid extinctions in bighorn sheep. *Conservation Biology*, **4**, 91–98.
25.3.2

Berner, E.K. & Berner, R.A. (1987) *The Global Water Cycle: Geochemistry and Environment*. Prentice-Hall, New Jersey, USA.
19.4.1

Berner, R.A. (1991) A model for atmospheric CO_2 over Phanerozoic time. *American Journal of Science*, **291**, 339–376.
3.3.4

Bernstein, C., Kacelnik, A. & Krebs, J.R. (1988) Individual decisions and the distribution of predators in a patchy environment. *Journal of Animal Ecology*, **57**, 1007–1026.
9.10

Bernstein, C., Kacelnik, A. & Krebs, J.R. (1991) Individual decisions and the distribution of predators in a patchy environment. II. The influence of travel costs and structure of the environment. *Journal of Animal Ecology*, **60**, 205–225.
9.10

Berry, J.A. & Björkman, O. (1980) Photosynthetic response and adaptation to temperature in higher plants. *Annual Review of Plant Physiology*, **31**, 491–543.
2.4

Berthold, P. (1994) *Bird Migration: A General Survey*. Oxford University Press, Oxford.
5.7.3

Beven, G. (1976) Changes in breeding bird populations of an oak-wood on Bookham Common, Surrey, over twenty-seven years. *London Naturalist*, **55**, 23–42.
23.3.3, 23.3.5

Beverly, S.M. & Wilson, A.C. (1985) Ancient origin for Hawaiian Drosophilinae inferred from protein comparisons. *Proceedings of the National Academy of Sciences of the USA*, **82**, 4753–4757.
1.2.3

Beverton, R.J.H. (1993) The Rio Convention and rational harvesting of natural fish resources: the Barents Sea experience in context. In: *Norway/UNEP Expert Conference on Biodiversity* (O.T. Sandlund & P.J. Schei, eds), pp. 44–63. DN/NINA, Trondheim.
16.13.4

Beverton, R.J.H. & Holt, S.J. (1957) On the dynamics of exploited fish populations. *Fishery Investigations*, (Series II), **19**, 1–533.
16.14

Beverton, R.J.H. & Holt, S.J. (1959) A review of the lifespans and mortality rates of fish in nature and in relation to growth and other physiological characteristics. In: *CIBA Foundation. Colloquia on Ageing. The Lifespan of Animals*, pp. 142–177. Churchill, London.
14.7.2

Bigger, M. (1976) Oscillations of tropical insect populations. *Nature*, **259**, 207–209.
22.4.3

Biggers, J.D. (1981) *In vitro* fertilisation and embryo transfer in

human beings. *New England Journal of Medicine*, **304**, 336–342.
4.5.1

Birk, E.M. & Vitousek, P.M. (1986). Nitrogen availability and nitrogen use efficiency in loblolly pine stands. *Ecology*, **67**, 69–79
19.2.3

Birks, H.J.B. (1993) Is the hypothesis of survival on glacial nunataks necessary to explain the present-day distributions of Norwegian mountain plants? *Phytocoenologia*, **23**, 399–426.
1.2.2

Björkman, O. (1981) Responses to different flux densities. In: *Encyclopedia of Plant Physiology* (O.L. Lange, P.S. Nobel, C.B. Osmond & H. Ziegler, eds), New Series, 12A, pp. 57–107. Springer-Verlag, Berlin.
3.2.1

Black, J.N. (1963) The interrelationship of solar radiation and leaf area index in determining the rate of dry matter production of swards of subterranean clover (*Trifolium subterraneum*). *Australian Journal of Agricultural Research*, **14**, 20–38.
8.3

Blake, J.G. & Karr, J.R. (1987) Breeding birds of isolated woodlots: area and habitat relationships. *Ecology*, **68**, 1724–1734.
23.1

Blaustein, A.R., Wake, D.B. & Sousa, W.P. (1994) Amphibian declines: judging stability, persistence, and susceptibility of populations to local and global extinctions. *Conservation Biology*, **8**, 60–71.
25.6.1

Block, W. (1990) Cold tolerance of insects and other arthropods. *Proceedings of the Royal Society*, **B 326**, 613–633; also in *Life at Low Temperatures* (R.M. Laws & F. Franks, eds), pp. 97–117. The Royal Society, London.
2.1

Blueweiss, L., Fox, H., Kudzma, V., Nakashima, D., Peters, R. & Sams, S. (1978) Relationships between body size and some life history parameters. *Oecologia*, **37**, 257–272.
14.12.1

Bobbink, R., Boxman, D., Fremstad, E., Heil, G., Houdijk, A. & Roelofs, J. (1992) Critical loads for nitrogen eutrophication of terrestrial and wetland ecosystems based upon changes in vegetation and fauna. In: *Critical Loads for Nitrogen* (P. Grennfelt & E. Thornelof, eds), pp. 111–160. Nordic Council of Ministers, Copenhagen.
19.4.4

Boden, T.A., Kanciruk, P. & Fartell, M.P. (1990) *Trends '90. A Compendium of Data on Global Change.* Carbon Dioxide Analysis Center, Oak Ridge National Laboratory, Tennessee.
1.2.2

Bolger, D.T., Alberts, A.C. & Soulé, M.E. (1991) Occurrence patterns of bird species in habitat fragments: sampling, extinction, and nested species subsets. *American Naturalist*, **137**, 155–166.
25.3.3

Bonan, G.B. (1993) Physiological derivation of the observed relationship between net primary production and mean annual air temperature. *Tellus*, **45B**, 397–408.
18.3.2

Bond, W. (1983) Alpha diversity of southern cape fynbos. In:

Mediterranean-Type Ecosystems (F.J. Kruger, D.T. Mitchell & J.U.M. Jarvis, eds), pp. 337–356. Springer-Verlag, Berlin.
24.3.1

Boonstra, R. & Boag, P.T. (1992) Spring declines in *Microtus pennsylvanicus* and the role of steroid hormones. *Journal of Animal Ecology*, **61**, 339–352.
15.4.4

Booth, D.J. & Brosnan, D.M. (1995) The role of recruitment dynamics in rocky shore and coral reef fish communities. *Advances in Ecological Research*, **26**, 309–385.
21.4

Booth, D.T., Clayton, D.H. & Block, B.A. (1993) Experimental demonstration of the energetic cost of parasitism in free-ranging hosts. *Proceedings of the Royal Society, Series B*, **253**, 125–129.
12.4

Boots, M. & Begon, M. (1993) Trade-offs with resistance to a granulosis virus in the Indian meal moth, examined by a laboratory evolution experiment. *Functional Ecology*, **7**, 528–534.
14.4

Boray, J.C. (1969) Experimental fascioliasis in Australia. *Advances in Parasitology*, **7**, 85–210.
12.4

Bormann, B.T. & Gordon, J.C. (1984) Stand density effects in young red alder plantations: productivity, photosynthate partitioning and nitrogen fixation. *Ecology*, **65**, 394–402.
19.2.1

Bormann, F.H., Likens, G.E. & Melillo, J.M. (1977) Nitrogen budget for an aggrading northern hardwood forest ecosystem. *Science*, **196**, 981–983.
19.2.3

Bosch, R. van den, Leigh, T.F., Falcon, L.A., Stern, V.M., Gonzales, D. & Hagen, K.S. (1971) The developing program of integrated control of cotton pests in California. In: *Biological Control* (C.B. Huffaker, ed.), pp. 377–394. Plenum Press, New York.
16.11

Bossema, I. (1979) Jays and oaks: an eco–ethological study of a symbiosis. *Behaviour*, **70**, 1–117.
5.3.6

Botkin, D.B., Jordan, P.A., Dominski, A.S., Lowendorf, H.S. & Hutchinson, G.E. (1973) Sodium dynamics in a northern ecosystem (moose, wolves, plants). *Proceedings of the National Academy of Science of the USA*, **70**, 2745–2748.
14.2

Boucher, D.H., James, S. & Kresler, K. (1984) The ecology of mutualism. *Annual Review of Ecology and Systematics*, **13**, 315–347.
13.1

Bowers, M.A. & Brown, J.H. (1982) Body size and coexistence in desert rodents: chance or community structure? *Ecology*, **63**, 391–400.
20.4.2

Bowers, R.G., Begon, M. & Hodgkinson, D.E. (1993) Host–pathogen population cycles in forest insects? Lessons from simple models reconsidered. *Oikos*, **67**, 529–538.
12.6.1

Box, E.O. (1981) *Macroclimate and Plant Forms: An Introduction to Predictive Modelling in Phytogeography.* Junk, The Hague.
1.4.2

Boyce, C.C.K. & Boyce, J.L. (1988a) Population biology of *Microtus arvalis*. I. Lifetime reproductive success of solitary and grouped breeding females. *Journal of Animal Ecology*, **571**, 711–722.
15.4.2, 15.4.3

Boyce, C.C.K. & Boyce, J.L. (1988b) Population biology of *Microtus arvalis*. II. Natal and breeding dispersal of females. *Journal of Animal Ecology*, **57**, 723–736.
15.4.3

Boyce, C.C.K. & Boyce, J.L. (1988c) Population biology of *Microtus arvalis*. III. Regulation of numbers and breeding display of females. *Journal of Animal Ecology*, **57**, 737–754.
15.4.3

Boyce, M.S. (1984) Restitution of *r*- and *K*-selection as a model of density-dependent natural selection. *Annual Review of Ecology and Systematics*, **15**, 427–447.
14.9

Bradley, D.J. (1977) Human pest and disease problems: contrasts between developing and developed countries. In: *Origins of Pest, Parasite, Disease and Weed Problems* (J.M. Cherrett & G.R. Sagar, eds), pp. 329–346. Blackwell Scientific Publications, Oxford.
12.4

Bradshaw, A.D. (1959) Population differentiation in *Agrostis tenuis* Sibth. II. The incidence and significance of infection by *Epichloe typhina*. *New Phytologist*, **58**, 310–315.
12.3.5

Bradshaw, A.D. (1965) Evolutionary significance of phenotypic plasticity in plants. *Advances in Genetics*, **13**, 115–155.
1.6

Bradshaw, A.D. (1983) The reconstruction of ecosystems. *Journal of Applied Ecology*, **20**, 1–17.
25.6.4

Bradshaw, A.D. (1984) Ecological principles and land reclamation practice. *Landscape Planning*, **11**, 35–48.
25.6.4

Bradshaw, A.D. (1987) The reclamation of derelict land and the ecology of ecosystems. In: *Restoration Ecology* (W.R. Jordan III, M.E. Gilpin & J.D. Aber, eds), pp. 53–74. Cambridge University Press, Cambridge.
2.10, 25.6.4

Bradshaw, A.D. & Chadwick, M.J. (1980) *The Restoration of Land*. Blackwell Scientific Publications, Oxford.
25.6.4

Bradshaw, A.D. & McNeilly, T. (1981) *Evolution and Pollution*. Edward Arnold, London.
2.10

Branch, G.M. (1975) Intraspecific competition in *Patella cochlear* Born. *Journal of Animal Ecology*, **44**, 263–281.
6.5, 6.6

Bremner, J.M. & Blackmer, A.M. (1978) Nitrous oxide: emmissions from soils during nitrification of fertilizer nitrogen. *Science*, **199**, 295–296.
19.4.4

Breton, L.M. & Addicott, J.F. (1992) Density-dependent mutualism in an aphid–ant interaction. *Ecology*, **73**, 2175–2180.
13.12

Breznak, J.A. (1975) Symbiotic relationships between termites and their intestinal biota. In: *Symbiosis* (D.H. Jennings & D.L. Lee, eds), pp. 559–580. Symposium 29, Society for Experi-

mental Biology, Cambridge University Press, Cambridge.
13.5.2

Briand, F. (1983) Environmental control of food web structure. *Ecology*, **64**, 253–263.
22.4.3, 22.5.2

Briand, F. & Cohen, J.E. (1987) Environmental correlates of food chain length. *Science*, **238**, 956–960.
22.5.2

Brittain, J.E. & Eikeland, T.I. (1988) Invertebrate drift—a review. *Hydrobiologia*, **166**, 77–93.
5.3.7

Broecker, W.S. & Peng, T. (1982) *Tracers in the Sea*. Lamont-Doherty Geological Observatory, Columbia University, New York.
19.3.3

Broekhuizen, N., Evans, H.F. & Hassell, M.P. (1993) Site characteristics and the population dynamics of the pine looper moth. *Journal of Animal Ecology*, **62**, 511–518.
4.3

Bronstein, J.L. (1988) Mutualism, antagonism and the fig–pollinator interaction. *Ecology*, **69**, 1298–1302.
13.4.2

Brooke, M. de L. & Davies, N.B. (1988) Egg mimicry by cuckoos *Culculus canorus* in relation to discrimination by hosts. *Nature*, **335**, 630–632.
12.8

Brosset, A. (1981) La periodicitie de la reproduction chez un bulbul de foret equatoriale Africaine: *Andropadus latirostris*, ses incidences demographique. *Review Ecologie (Terre et Vie)*, **35**, 109–129.
24.4.1

Brower, L.P. & Corvinó, J.M. (1967) Plant poisons in a terrestrial food chain. *Proceedings of the National Academy of Science of the USA*, **57**, 893–898.
3.7.3

Brown, H.T. & Escombe, F. (1900) Static diffusion of gases and liquids in relation to the assimilation of carbon and translocation in plants. *Philosophical Transactions of the Royal Society, Series B*, **193**, 223–291.
3.4

Brown, J., Everett, K.R., Webber, P.J., MacLean, S.F. & Murray, D.F. (1980) The coastal tundra at Barrow. In: *An Arctic Ecosystem: The Coastal Tundra at Barrow* (J. Brown, P.C. Miller, L.L. Tieszen & F.L. Bunnell, eds). Dowden, Hutchinson & Ross, Stroudsburg, PA.
15.4.2

Brown, J.H. (1971) Species richness of boreal mammals living on the montane islands of the Great Basin. *American Naturalist*, **105**, 467–478.
23.1

Brown, J.H. & Davidson, D.W. (1977) Competition between seed-eating rodents and ants in desert ecosystems. *Science*, **196**, 880–882.
7.8.1, 20.3.1, 24.3.1

Brown, J.H. & Gibson, A.C. (1983) *Biogeography*. C.V. Mosby, St Louis.
24.3.1, 24.4.1

Brown, J.S., Kotler, B.P., Smith, R.J. & Wirtz, W.O. III (1988) The effects of owl predation on the foraging behaviour of heteromyid rodents. *Oecologia*, **76**, 408–415.
9.4

Brown, K.M. (1982) Resource overlap and competition in pond snails: an experimental analysis. *Ecology*, **63**, 412–422.
7.5.4

Brown, S. (1981) A comparison of structure, primary productivity, and transpiration of cypress ecosystems in Florida. *Ecological Monographs*, **51**, 403–427.
19.2.1

Brown, V.K. & Gange, A.C. (1992) Secondary plant succession: how is it modified by insect herbivory? *Vegetatio*, **101**, 3–13.
17.4.4

Brown, V.K. & Southwood, T.R.E. (1983) Trophic diversity, niche breadth, and generation times of exopterygote insects in a secondary succession. *Oecologia*, **56**, 220–225.
24.4.4

Browne, R.A. (1981) Lakes as islands: biogeographic distribution, turnover rates, and species composition in the lakes of central New York. *Journal of Biogeography*, **8**, 75–83.
23.3.1

Bryant, J.P. & Kuropat, P.J. (1980) Selection of winter forage by subarctic browsing vertebrates: the role of plant chemistry. *Annual Review of Ecology and Systematics*, **11**, 261–285.
8.2.2, 9.2.2

Brylinski, M. & Mann, K.H. (1973) An analysis of factors governing productivity in lakes and reservoirs. *Limnology and Oceanography*, **18**, 1–14.
18.2, 18.4, 18.4.1

Bubier, J.L. & Moore, T.R. (1994) An ecological perspective on methane emissions from northern wetlands. *Trends in Ecology and Evolution*, **9**, 460–464.
19.2.2, 19.5

Buchanan, G.A., Crowley, R.H., Street, J.E. & McGuire, J.A. (1980) Competition of siclepod (*Cassia obtusifolia*) and redroot pigweed (*Amaranthus retroflexus*) with cotton (*Gossypium hirsutum*). *Weed Science*, **28**, 258–262.
7.7.2

Bull, C.M. & Burzacott, D. (1993) The impact of tick load on the fitness of their lizard hosts. *Oecologia*, **96**, 415–419.
12.4

Bullock, J.M., Mortimer, A.M. & Begon, M. (1994a) Physiological integration among tillers of *Holcus lanatus*: age-dependence and responses to clipping and competition. *New Phytologist*, **128**, 737–747.
4.2.1

Bullock, J.M., Mortimer, A.M. & Begon, M. (1994b) The effect of clipping on interclonal competition in the grass *Holcus lanatus*—a response surface analysis. *Journal of Ecology*, **82**, 259–270.
7.7.2

Buntin, G.D., Bruckner, P.L. & Johnson, J.W. (1990) Management of the Hessian fly (Diptera: Cecidomyiidae) in Georgia by delayed planting of winter wheat. *Journal of Economic Entomology*, **83**, 1025–1033.
16.9

Burdon, J.J. (1980) Intraspecific diversity in a natural population of *Trifolium repens*. *Journal of Ecology*, **68**, 717–735.
1.5.2, 12.7

Burdon, J.J. (1987) *Diseases and Plant Population Biology.* Cambridge University Press, Cambridge.
2.2.2, 12.5.3, 12.7, 16.10, 21.2.4

Burdon, J.J. & Chilvers, G.A. (1975) Epidemiology of damping off disease (*Pythium irregulare*) in relation to density of

Lepidium sativum seedlings. *Annals of Applied Biology*, **81**, 135–143.
12.3.1

Burdon, J.J. & Chilvers, G.A. (1977) The effect of barley mildew on barley and wheat competition in mixtures. *Australian Journal of Botany*, **25**, 59–65.
21.2.4

Burgman, M.A., Ferson, S. & AkÇakaya, H.R. (1993) *Risk Assessment in Conservation Biology.* Chapman & Hall, London.
25.3.3

Burkey, T.V. (1989) Extinction in nature reserves: the effect of fragmentation and the importance of migration between reserve fragments. *Oikos*, **55**, 75–81.
25.3.3

Burnett, T. (1958) Dispersal of an insect parasite over a small plot. *Canadian Entomologist*, **90**, 279–283.
9.6

Burrows, F.J. & Milthorpe, F.L. (1976) Stomatal conductance in the control of gas exchange. In: *Water Deficits and Plant Growth* (T.T. Kozlowski, ed.), Vol. 4, pp. 103–152. Academic Press, London.
3.4

Burrows, F.M. (1986) The aerial motion of seeds, fruits, spores and pollen. In: *Seed Dispersal* (D.R. Murray, ed.). Academic Press, Sydney.
5.3.4

Bush, M.B. & Whittaker, R.J. (1991) Krakatau: colonization patterns and hierarchies. *Journal of Biogeography*, **18**, 341–356.
23.3.3

Buss, L.W. (1979) Byrozoan overgrowth interactions—the interdependence of competition for food and space. *Nature*, **281**, 475–477.
7.3

Buss, L.W. (1986) Competition and community organization on hard surfaces in the sea. In: *Community Ecology* (J. Diamond & T.J. Case, eds), pp. 517–536. Harper & Row, New York.
21.7.2

Cain, M.L., Pacala, S.W., Silander, J.A. & Fortin, M-J. (1995) Neighbourhood models of clonal growth in the white clover *Trifolium repens*. *American Naturalist*, **145**, 888–917.
3.10, 13.10.1

Caldwell, M.M. & Richards, J.H. (1986) Competing root systems: morphology and models of absorption. In: *On the Economy of Plant Form and Function* (T.J. Givnish, ed.), pp. 251–273. Cambridge University Press, Cambridge.
3.4

Callaghan, T.V. (1976) Strategies of growth and population dynamics of plants: 3. Growth and population dynamics of *Carex bigelowii* in an alpine environment. *Oikos*, **27**, 402–413.
4.6.4

Calow, P. (1981) Resource utilization and reproduction. In: *Physiological Ecology: An Evolutionary Approach to Resource Use* (C.R. Townsend & P. Calow, eds), pp. 245–270. Blackwell Scientific Publications, Oxford.
14.7.1

Canham, C.D. (1988) Growth and canopy architecture of shade-tolerant trees: response to canopy gaps. *Ecology*, **69**, 786–795.
21.5.2

Caraco, N.F. (1993) Disturbance of the phosphorus cycle: a case of indirect effects of human activity. *Trends in Ecology and Evolution*, **8**, 51–54.
19.4.5

Caraco, T. & Kelly, C.K. (1991) On the adaptive value of physiological integration of clonal plants. *Ecology*, **72**, 81–93.
4.2.1

Carne, P.B. (1969) On the population dynamics of the eucalypt-defoliating chrysomelid *Paropsis atomaria* OI. *Australian Journal of Zoology*, **14**, 647–672.
8.3

Carpenter, S.R., Kraft, C.E., Wright, R., He, X., Soranno, P.A. & Hodgson, J.R. (1992a) Resilience and resistance of a lake phosphorus cycle before and after food web manipulation. *American Naturalist*, **140**, 781–798.
22.6

Carpenter, S.R., Cottingham, K.L. & Schindler, D.E. (1992b) Biotic feedbacks in lake phosphorus cycles. *Trends in Ecology and Evolution*, **7**, 332–336.
22.6

Carroll, G.C. (1988) Fungal endophytes in stems and leaves: from latent pathogen to mutualistic symbiont. *Ecology*, **69**, 2–9.
13.7

Carson, H.L. & Kaneshiro, K.Y. (1976) *Drosophila* of Hawaii: systematics and ecological genetics. *Annual Review of Ecology and Systematics*, **7**, 311–345.
1.2.3

Carson, R. (1962) *Silent Spring*. Houghton Mifflin, Boston.
16.4

Caswell, H. (1978) Predator-mediated coexistence: a non-equilibrium model. *American Naturalist*, **112**, 127–154.
21.3, 21.4

Caswell, H. (1989) Life-history strategies. In: *Ecological Concepts* (J.M. Cherrett, ed.), pp. 285–307. Blackwell Scientific Publications, Oxford.
14.1, 14.3

Caughley, G. (1994) Directions in conservation biology. *Journal of Animal Ecology*, **63**, 215–244.
25.2.6, 25.3.1, 25.4.2, 25.4.3

Caughley, G. & Lawton, J.H. (1981) Plant–herbivore systems. In: *Theoretical Ecology: Principles and Applications*, 2nd edn. (R.M. May, ed.), pp. 132–166. Blackwell Scientific Publications, Oxford.
10.3.2, 10.5.6, 15.7.6

Cerling, T.E. (1991) Carbon dioxide in the atmosphere: evidence from Cenozoic and Mesozoic paleosols. *American Journal of Science*, **291**, 377–400.
3.3.4

Chapin, F.S. (1980) The mineral nutrition of wild plants. *Annual Review of Ecology and Systematics*, **11**, 233–260.
19.2.3

Chapin, F.S. (1987) Adaptations and physiological responses of wild plants to nutrient stress. In: *Genetic Aspects of Plant Mineral Nutrition* (H.W. Gabelman & B.C. Loughman, eds), pp. 69–76. Martinus Nijoff Publishers, Dordrecht.
19.1

Chapin, F.S. & Kedrowski, R.A. (1983) Seasonal changes in nitrogen and phosphorus fractions and autumn retranslocations in evergreen and deciduous taiga trees. *Ecology*, **64**, 376–391.
19.2.3

Charlesworth, B. (1980) *Evolution in Age-Structured Populations*. Cambridge University Press, London.
14.3

Charnov, E.L. (1976a) Optimal foraging: attack strategy of a mantid. *American Naturalist*, **110**, 141–151.
9.3.1

Charnov, E.L. (1976b) Optimal foraging: the marginal value theorem. *Theoretical Population Biology*, **9**, 129–136.
9.9.1

Charnov, E.L. & Berrigan, D. (1991) Dimensionless numbers and the assembly rules of life histories. *Philosophical Transactions of the Royal Society of London, Series B*, **332**, 41–48.
14.7.2

Charnov, E.L. & Finnerty, J. (1980) Vole population cycles; a case for kin-selection? *Oecologia*, **45**, 1–2.
15.4.4

Charnov, E.L. & Krebs, J.R. (1974) On clutch size and fitness. *Ibis*, **116**, 217–219.
14.8.1

Cherrett, J.M., Powell, R.J. & Stradling, D.J. (1989) The mutualism beween leaf-cutting ants and their fungus. In: *Insect/Fungus Interactions* (N. Wilding, N.M. Collins, P.M. Hammond & J.F. Webber, eds), pp. 93–120. Royal Entomological Society Symposium No 14. Academic Press, London.
13.3.3

Cherrill, A.J. (1988) The shell size–prey size relationship in mudsnails. *Oikos*, **51**, 110–112.
7.8.1

Cherrill, A.J. & James, R. (1987) Character displacement in *Hydrobia*. *Oecologia*, **71**, 618–623.
7.8.1

Chesson, P. & Murdoch, W.W. (1986) Aggregation of risk: relationships among host–parasitoid models. *American Naturalist*, **127**, 696–715.
10.5.3

Chesson, P.L. (1986) Environmental variation and the coexistence of species. In: *Community Ecology* (J. Diamond & T.J. Case, eds), pp. 240–256. Harper & Row, New York.
21.7

Chesson, P.L. & Case, T.J. (1986) Overview: non-equilibrium community theories, chance, variability, history and coexistence. In: *Community Ecology* (J. Diamond & T.J. Case, eds), pp. 229–239. Harper & Row, New York.
7.5.5

Chia, L.S., Mayfield, C.I. & Thompson, J.E. (1984) Simulated acid rain induces lipid peroxidation and membrane damage in foliage. *Plant, Cell and Environment*, **7**, 333–338.
19.4.5

Choudhary, M., Strassman, J.E., Queller, D.C., Turilazzi, S. & Cervo, R. (1994) Social parasites in polistine wasps are monophyletic: implications for sympatric speciation. *Proceedings of the Royal Society, Series B*, **257**, 31–35.
12.8

Christensen, B.M. (1978) *Dirofilaria immitis*: effects on the longevity of *Aedes trivittatus*. *Experimental Parasitology*, **44**, 116–119.
12.4

Christensen, V. & Pauly, D. (eds) (1993) *Trophic Models of Aquatic Ecosystems*. International Center for Living Aquatic Resources Management, Manila, Phillipines.
18.4.3

Christian, J.J. (1950) The adreno-pituitary system and population cycles in mammals. *Journal of Mammalogy*, **31**, 247–259.
15.4.4

Christian, J.J. (1970) Social subordination, population density, and mammalian evolution. *Science*, **168**, 84–90.
5.4.3, 15.4.4

Christian, J.J. (1980) Endocrine factors in population regulation. In: *Biosocial Mechanisms of Population Regulation* (M.N. Cohen, R.S. Malpass, & H.G. Klein, eds), pp. 55–115. Yale University Press, New Haven, Connecticut.
15.4.4

Christie, G.C. & Regier, H.A. (1988) Measures of optimal thermal habitat for four commercial fish species. *Canadian Journal of Fisheries and Aquatic Sciences*, **45**, 301–314.
2.5

Clapham, W.B. (1973) *Natural Ecosystems*. Collier–Macmillan, New York.
7.2.4

Claridge, M.F. & Wilson, M.R. (1976) Diversity and distribution patterns of some mesophyll-feeding leaf-hoppers of temperate woodland canopy. *Ecological Entomology*, **1**, 231–250.
23.3.3

Claridge, M.F. & Wilson, M.R. (1978) British insects and trees: a study in island biogeography or insect/plant coevolution? *American Naturalist*, **112**, 451–456.
23.3.3

Clark, C.W. (1976) *Mathematical Bioeconomics: The Optimal Management of Renewable Resources*. Wiley-Interscience, New York.
16.12

Clark, C.W. (1981) Bioeconomics. In: *Theoretical Ecology: Principles and Applications*, 2nd edn. (R.M. May, ed.), pp. 387–418. Blackwell Scientific Publications, Oxford.
16.12, 16.12.1, 16.13.1

Clark, C.W. (1985) *Bioeconomic Modelling and Fisheries Management*. Wiley, New York.
16.15

Clark, D.A. & Clark, D.B. (1984) Spacing dynamics of a tropical rain forest tree: evaluation of the Janzen–Connell model. *American Naturalist*, **124**, 769–788.
24.4.1

Clark, L.R. (1962) The general biology of *Cardiaspina albitextura* (Psyllidae) and its abundance in relation to weather and parasitism. *Australian Journal of Zoology*, **10**, 537–586.
10.6.1

Clark, L.R. (1964) The population dynamics of *Cardiaspina albitextura* (Psyllidae). *Australian Journal of Zoology*, **12**, 362–380.
10.6.1

Clarke, B.C. & Partridge, L. (eds) (1988) Frequency dependent selection. *Philosophical Transactions of the Royal Society of London, Series B*, **319**, 457–645.
1.5.2

Clatworthy, J.N. & Harper, J.L. (1962) The comparative biology of closely related species living in the same area. V. Inter- and intraspecific interference within cultures of *Lemna* spp. and *Salvinia natans*. *Journal of Experimental Botany*, **38**, 307–324.
7.7.1

Clausen, J., Keck, D.D. & Hiesey, W.M. (1941) Experimental studies on the nature of species. IV. Genetic structure of ecological races. *Carnegie Institute of Washington Publication, No. 242*. Washington, DC.
1.5.1

Clay, K. (1990) Fungal endophytes of grasses. *Annual Review of Ecology and Systematics*, **21**, 275–297.
13.7

Clayton, D.H. (1990) Mate choice in experimentally parasitized rock doves: Lousy males lose. *American Zoologist*, **30**, 251–262.
12.4

Clements, F.E. (1916) *Plant Succession: Analysis of the Development of Vegetation*. Carnegie Institute of Washington Publication, No. 242. Washington, DC.
17.3.3, 17.4.5, 18.1

Closs, G., Watterson, G.A. & Donnelly, P.J. (1993) Constant predator–prey ratios: an arithmetical artifact? *Ecology*, **74**, 238–243.
22.5.4

Clutton-Brock, T.H. & Harvey, P.H. (1979) Comparison and adaptation. *Proceedings of the Royal Society of London, Series B*, **205**, 547–565.
14.12.1

Clutton-Brock, T.H., Major, M., Albon, S.D. & Guinness, F.E. (1987) Early development and population dynamics in red deer. I. Density-dependent effects on juvenile survival. *Journal of Animal Ecology*, **56**, 53–67.
6.1

Cody, M.L. (1968) On the methods of resource division in grassland bird communitites. *American Naturalist*, **102**, 107–148.
20.3.1

Cody, M.L. (1975) Towards a theory of continental species diversities. In: *Ecology and Evolution of Communities* (M.L. Cody & J.M. Diamond, eds), pp. 214–257. Belknap, Cambridge, MA.
20.1, 24.3.2

Cody, M.L. & Mooney, H.A. (1978) Convergence versus non-convergence in Mediterranean climate ecosystems. *Annual Review of Ecology and Systematics*, **9**, 265–321.
1.4.2

Coe, M.J. (1978) The decomposition of elephant carcasses in the Tsavo (East) National Park, Kenya. *Journal of Arid Environments*, **1**, 71–86.
11.3.3

Coe, M.J., Cumming, D.H. & Phillipson, J. (1976) Biomass and production of large African herbivores in relation to rainfall and primary production. *Oecologia*, **22**, 341–354.
18.4

Cofrancesco, A.F. Jr., Stewart, R.M. & Sanders, D.R. Sr. (1985) The impact of *Neochetina eichhorniae* (Coleoptera: Curculionidae) on water hyacinth in Louisiana. *Proceedings of the VIth International Symposium on Biological Control of Weeds*, pp. 525–535.
16.8

Cohen, D. & Levin, S.A. (1987) The interaction between dispersal and dormancy strategies in varying and heterogeneous environments. In: *Mathematical Topics in Population Biology, Morphogenesis and Neurosciences, Proceedings, Kyoto 1985* (E. Teramoto & M. Yamaguti, eds), pp. 110–122. Princeton University Press, Princeton, New Jersey.
5.3.2, 5.5

Cohen, D. & Mitro, U. (1989) More on optimal rates of dispersal: taking into account the cost of the dispersal mechanism. *American Naturalist*, **134**, 659–663.
5.3.2

Cohen, J.E. & Newman, C.M. (1985) A stochastic theory of community food webs I. Models and aggregated data. *Proceedings of the Royal Society of London, Series B*, **224**, 421–448.
22.5.4

Cohen, J.E. & Newman, C.M. (1988) Dynamic basis of food web organization. *Ecology*, **69**, 1655–1664.
22.5.4

Cole, D.W. & Rapp, M. (1981) Elemental cycling in forest ecosystems. In: *Dynamic Principles of Forest Ecosystems* (D.E. Reichle, ed.), pp. 341–409. Cambridge University Press, London.
19.2.3

Cole, J.J., Findlay, S. & Pace, M.L. (1988) Bacterial production in fresh and salt water ecosystems: a cross-system overview. *Marine Ecology Progress Series*, **4**, 1–10.
18.4

Collins, M.D., Ward, S.A. & Dixon, A.F.G. (1981) Handling time and the functional response of *Aphelinus thomsoni*, a predator and parasite of the aphid, *Drepanosiphum platanoidis*. *Journal of Animal Ecology*, **50**, 479–487.
9.5.3

Colwell, R.K. & Winkler, D.W. (1984) A null model for null models in biogeography. In: *Ecological Communities: Conceptual Issues and the Evidence* (D.R. Strong, D. Simberloff, L.G. Abele & A.B. Thistle, eds), pp. 344–359. Princeton University Press, Princeton, NJ.
20.4.4

Comins, H.N., Hassell, M.P. & May, R.M. (1992) The spatial dynamics of host–parasitoid systems. *Journal of Animal Ecology*, **61**, 735–748.
10.5.5

Connell, J.H. (1961) The influence of interspecific competition and other factors on the distribution of the barnacle *Chthamalus stellatus*. *Ecology*, **42**, 710–723.
7.2.3

Connell, J.H. (1970) A predator–prey system in the marine intertidal region. I. *Balanus glandula* and several predatory species of *Thais*. *Ecological Monographs*, **40**, 49–78.
4.6.1

Connell, J.H. (1971) On the role of natural enemies in preventing competitive exclusion in some marine animals and in rain forest trees. In: *Dynamics of Populations* (P.J. den Boer & G.R. Gradwell, eds), pp. 298–310. *Proceedings of the Advanced Study Institute in Dynamics of Numbers in Populations, Oosterbeck*. Centre for Agricultural Publishing and Documentation. Wageningen.
24.4.1

Connell, J.H. (1975) Some mechanisms producing structure in natural communities: a model and evidence from field experiments. In: *Ecology and Evolution of Communities* (M.L. Cody & J.M. Diamond, eds), pp. 460–490. Belknap, Cambridge, MA.
21.2.5

Connell, J.H. (1978) Diversity in tropical rainforests and coral reefs. *Science*, **199**, 1302–1310.
21.5.1

Connell, J.H. (1979) Tropical rain forests and coral reefs as open nonequilibrium systems. In: *Population Dynamics* (R.M. Anderson, B.D. Turner & L.R Taylor, eds), pp. 141–163. Blackwell Scientific Publications, Oxford.
21.2.3

Connell, J.H. (1980) Diversity and the coevolution of competitors, or the ghost of competition past. *Oikos*, **35**, 131–138.
7.3, 7.8.1, 7.8.2

Connell, J.H. (1983) On the prevalence and relative importance of interspecific competition: evidence from field experiments. *American Naturalist*, **122**, 661–696.
7.3, 7.6, 20.2, 22.2.1

Connell, J.H. (1990) Apparent versus 'real' competition in plants. In: *Perspectives on Plant Competition* (J.B. Grace & D. Tilman, eds), pp. 9–26. Academic Press, New York.
7.6

Connell, J.H. & Slatyer, R.O. (1977) Mechanisms of succession in natural communities and their role in community stability and organisation. *American Naturalist*, **111**, 1119–1144.
17.4.3, 17.4.4

Connor, E.F. (1986) The role of Pleistocene forest refugia in the evolution and biogeography of tropical biotas. *Trends in Ecology and Evolution*, **1**, 165–169.
24.4.1

Connor, E.F. & McCoy, E.D. (1979) The statistics and biology of the species–area relationship. *American Naturalist*, **113**, 791–833.
12.3.8

Conway, G. (1981) Man versus pests. In: *Theoretical Ecology*, 2nd edn. (R.M. May, ed.), pp. 356–386. Blackwell Scientific Publications, Oxford.
16.2

Cooper, J.P. (ed.) (1975) *Photosynthesis and Productivity in Different Environments*. Cambridge University Press, Cambridge.
18.2, 18.3.1

Copplestone, J.F. (1977) A global view of pesticide safety. In: *Pesticide Management and Insecticide Resistance* (D.L. Watson & A.W.A. Brown, eds), pp. 147–155. Academic Press, New York.
16.7

Corey, S., Dall, D. & Milne, W. (1993) *Pest Control and Sustainable Agriculture*. CSIRO, East Melbourne.
16.8, 16.11

Cornell, H.V. & Washburn, J.O. (1979) Evolution of the richness-area correlation for cynipid gall wasps on oak trees: a comparison of two geographic areas. *Evolution*, **33**, 257–274.
23.1

Costanza, R. (ed.) (1991) *Ecological Economics: The Science and Management of Sustainability*. Columbia University Press, New York.
25.1.3

Cottam, D.A., Whittaker, J.B. & Malloch, A.J.C. (1986) The effects of chrysomelid beetle grazing and plant competition on the growth of *Rumex obtusifolius*. *Oecologia*, **70**, 452–456.
8.2.6

Cousens, R. (1985) A simple model relating yield loss to density. *Annals of Applied Biology*, **107**, 239–252.
7.7.2

Cowan, I.R. & Farquhar, G.D. (1977) Stomatal function in

relation to leaf metabolism and environment. *Symposium of the Society of Experimental Biology*, **31**, 471–505.
3.2.4

Cowie, R.J. (1977) Optimal foraging in great tits *Parus major*. *Nature*, **268**, 137–139.
9.9.2

Cox, C.B., Healey, I.N. & Moore, P.D. (1976) *Biogeography*, 2nd edn. Blackwell Scientific Publications, Oxford.
1.4.1, 2.2, 2.7.1

Cox, F.E.G. (ed.) (1982) Immunology. In: *Modern Parasitology*, pp. 173–203. Blackwell Scientific Publications, Oxford.
12.3.5

Cox, P.A., Elmquist, T., Pierson, E.D. & Rainey, W.E. (1991) Flying foxes as strong interactors in South Pacific island ecosystems: a conservation hypothesis. *Conservation Biology*, **5**, 448–454.
25.5.1

Crawley, M.J. (1983) *Herbivory: The Dynamics of Animal–Plant Interactions*. Blackwell Scientific Publications, Oxford.
8.2.1, 8.2.5, 9.2.2, 9.5.2, 10.1

Crawley, M.J. (1989) Insect herbivores and plant population dynamics. *Annual Review of Entomology*, **34**, 531–564.
8.2.4, 8.3

Crawley, M.J. & May, R.M. (1987) Population dynamics and plant community structure: competition between annuals and perennials. *Journal of Theoretical Biology*, **125**, 475–489.
3.10, 7.5.1

Crawley, M.J., Hails, R.S., Rees, M., Kohn, D. & Buxton, J. (1993) Ecology of transgenic oilseed rape in natural habitats. *Nature*, **363**, 620–623.
16.10

Crisp, D.J. (1976) The role of the pelagic larva. In: *Perspectives in Experimental Biology* (P. Spencer Davies, ed.), Vol. I (Zoology), pp. 145–155. Pergamon Press, Oxford.
5.3.7

Crisp, M.D. & Lange, R.T. (1976) Age structure distribution and survival under grazing of the arid zone shrub *Acacia burkitti*. *Oikos*, **27**, 86–92.
4.6.1

Crocker, R.L. & Major, J. (1955) Soil development in relation to vegetation and surface age at Glacier Bay, Alaska. *Journal of Ecology*, **43**, 427–448.
17.4.3, 21.5

Croll, N.A., Anderson, R.M., Gyorkos, T.W. & Ghadirian, E. (1982) The population biology and control of *Ascaris lumbricoides* in a rural community in Iran. *Transactions of the Royal Society of Tropical Medicine and Hygiene*, **76**, 187–197.
12.3.5

Cromartie, W.J. (1975) The effect of stand size and vegetational background on the colonization of cruciferous plants by herbivorous insects. *Journal of Applied Ecology*, **12**, 517–533.
9.7.1

Crombie, A.C. (1945) On competition between different species of graminivorous insects. *Proceedings of the Royal Society of London, Series B*, **132**, 362–395.
6.4

Cullen, J.M. & Hasan, S. (1988) Pathogens for the control of weeds. *Philosophical Transactions of the Royal Society of London, Series B*, **318**, 213–224.
16.8

Cummins, K.W. (1974) Structure and function of stream ecosystems. *Bioscience*, **24**, 631–641.
11.2.2, 11.3.1

Currie, D.J. (1991) Energy and large-scale patterns of animal and plant species richness. *American Naturalist*, **137**, 27–49.
24.3.1

Currie, D.J. & Paquin, V. (1987) Large-scale biogeographical patterns of species richness in trees. *Nature*, **29**, 326–327.
24.3.1

Daan, S., Dijkstra, C. & Tinbergen, J.M. (1990) Family planning in the kestrel (*Falco tinnunculus*): the ultimate control of covariation of laying date and clutch size. *Behavior*, **114**, 83–116.
14.11

DaCosta, C.P. & Jones, C.M. (1971) Cucumber beetle resistance and mite susceptibility controlled by the bitter gene in *Cucumis sativus*. *Science*, **172**, 1145–1146.
16.10

Dahl, E. (1979) Deep-sea carrion feeding amphipods: evolutionary patterns in niche adaptation. *Oikos*, **33**, 167–175.
11.3.3

Daly, H.V., Doyer, J.T. & Ehrlich, P.R. (1978) *Introduction to Insect Biology and Diversity*. McGraw Hill, New York.
3.7.1

Darlington, C.D. (1960) Cousin marriage and the evolution of the breeding system in man. *Heredity*, **14**, 297–332.
5.4.4

Darwin, C. (1859) *The Origin of Species by Means of Natural Selection*, 1st edn. John Murray, London.
1.1.4, 8.2.4, 13.4.2, 21.2.1

Darwin, C. (1888) *The Formation of Vegetable Mould Through the Action of Worms*. John Murray, London.
11.2.2

Dash, M.C. & Hota, A.K. (1980) Density effects on the survival, growth rate, and metamorphosis of *Rama tigra* tadpole. *Ecology*, **61**, 1025–1028.
6.5

Davidson, D.W. (1977) Species diversity and community organization in desert seed-eating ants. *Ecology*, **58**, 711–724.
7.8.1, 24.3.1

Davidson, D.W. (1978) Size variability in the worker caste of a social insect (*Veromessor pergandei* Mayr) as a function of the competitive environment. *American Naturalist*, **112**, 523–532.
7.8.1

Davidson, D.W. & McKey, D. (1993) Ant–plant symbioses: stalking the Chuyachaqui. *Trends in Ecology and Evolution*, **8**, 326–332.
13.2.1

Davidson, D.W., Samson, D.A. & Inouye, R.S. (1985) Granivory in the Chihuahuan Desert: interactions within and between trophic levels. *Ecology*, **66**, 486–502.
8.2.4

Davidson, J. (1944) On the relationship between temperature and rate of development of insects at constant temperatures. *Journal of Animal Ecology*, **13**, 26–38.
2.2

Davidson, J. & Andrewartha, H.G. (1948a) Annual trends in a natural population of *Thrips imaginis* (Thysanapotera). *Journal of Animal Ecology*, **17**, 193–199.
15.2

Davidson, J. & Andrewartha, H.G. (1948b) The influence of rainfall, evaporation and atmospheric temperature on fluctuations in the size of a natural population of *Thrips imaginis* (Thysanoptera). *Journal of Animal Ecology*, **17**, 200–222.
15.2, 15.2.2

Davies, N.B. (1977) Prey selection and social behaviour in wagtails (Aves: Motacillidae). *Journal of Animal Ecology*, **46**, 37–57.
9.2.2

Davies, N.B. & Houston, A.I. (1984) Territory economics. In: *Behavioural Ecology: An Evolutionary Approach*, 2nd edn. (J.R. Krebs & N.B. Davies, eds), pp. 148–169. Blackwell Scientific Publications, Oxford.
6.12

Davis, M.B. (1976) Pleistocene biogeography of temperate deciduous forests. *Geoscience and Man*, **13**, 13–26.
1.2.2

Davis, M.B., Brubaker, L.B. & Webb, T. III (1973) Calibration of absolute pollen inflex. In: *Quaternary Plant Ecology* (H.J.B. Birks & R.G. West, eds), pp. 9–25. Blackwell Scientific Publications, Oxford.
1.2.2

Dawkins, R. (1976) *The Selfish Gene*. Oxford University Press, Oxford.
5.2.1

Dawkins, R. (1982) *The Extended Phenotype*. Oxford University Press, Oxford.
5.2.1

Dawson, P.S. (1979) Evolutionary changes in egg-eating behaviour of flour beetles in mixed-species populations. *Evolution*, **33**, 585–594.
7.8.3

Dayan, T., Simberloff, E., Tchernov, E. & Yom-Tov, Y. (1989) Inter- and intraspecific character displacement in mustelids. *Ecology*, **70**, 1526–1539.
20.3.1, 20.4.2

Dayan, T., Simberloff, E., Tchernov, E. & Yom-Tov, Y. (1990) Feline canines: community-wide character displacement among the small cats of Israel. *American Naturalist*, **136**, 39–60.
20.3.1

de Jong, G. (1982) The influence of dispersal pattern on the evolution of fecundity. *Netherlands Journal of Zoology*, **32**, 1–30.

de Jong, G. (1990) Quantitative genetics of reaction norms. *Journal of Evolutionary Biology*, **3**, 447–468.
14.11

Dean, A.M. (1983) A simple model of mutualism. *American Naturalist*, **121**, 409–417.
13.12

DeAngelis, D.L. (1975) Stability and connectance in food web models. *Ecology*, **56**, 238–243.
22.4.2

DeAngelis, D.L. (1980) Energy flow, nutrient cycling and ecosystem resilience. *Ecology*, **61**, 764–771.
22.6

DeAngelis, D.L. (1992) *Dynamics of Nutrient Cycling and Food Webs*. Chapman & Hall, London.
22.6

DeAngelis, D.L. & Waterhouse, J.C. (1987) Equilibrium and nonequilibrium concepts in ecological models. *Ecological Monographs*, **57**, 1–21.
21.7.1

DeBach, P. (ed.) (1964) *Biological Control of Insect Pests and Weeds*. Chapman & Hall, London.
10.1

Deecher, D.C. & Soderlund, D.M. (1991) RH3421, an insecticidal dihydropyrazole, inhibits sodium-channel dependent sodium uptake in mouse brain preparations. *Pesticide Biochemical Physiology*, **39**, 130–137.
16.6.4

Deevey, E.S. (1947) Life tables for natural populations of animals. *Quarterly Review of Biology*, **22**, 283–314.
4.5.1

Del Moral, R. & Bliss, L.C. (1993) Mechanisms of primary succession: insights resulting from the eruption of Mount St Helens. *Advances in Ecological Research*, **24**, 1–66.
17.4.3

Deming, J.W. & Baross, J.A. (1993) Deep-sea smokers: windows to a subsurface biosphere? *Geochimica et Cosmochimica Acta*, **57**, 3219–3230.
2.4

Dempster, J.P. (1983) The natural control of populations of butterflies and moths. *Biological Reviews*, **58**, 461–481.
10.2.5, 15.2.3

Dempster, J.P. & Lakhani, K.H. (1979) A population model for cinnabar moth and its food plant, ragwort. *Journal of Animal Ecology*, **48**, 143–164.
10.1

Dempster, J.P. & Pollard, E. (1986) Spatial heterogeneity, stochasticity and the detection of density dependence in animal populations. *Oikos*, **46**, 413–416.
10.2.5

Denholm-Young, P.A. (1978) *Studies on Decomposing Cattle Dung and its Associated Fauna*. D. Phil thesis, Oxford University.
11.3.2

Denno, R.F. & Roderick, G.K. (1992) Density related dispersal in planthoppers: effects of interspecific crowding. *Ecology*, **73**, 1324–1334.
5.4.2

Deriso, R.B. (1987) Optimal $F_{0.1}$ criteria and their relationship to maximum sustainable yield. *Canadian Journal of Fisheries and Aquatic Sciences*, **44** (Suppl. 2), 339–348.
16.15.2

Deshmukh, I. (1986) *Ecology and Tropical Biology*. Blackwell Scientific Publications, Oxford.
6.4, 17.4.4

Detwiler, R.P. & Hall, C.A.S. (1988) Tropical forests and the global carbon cycle. *Science*, **239**, 42–47.
19.4.6

di Castri, F. (1989) History of biological invasions with special emphasis on the old world. In: *Biological Invasions: A Global Perspective* (J.A. Drake, H.A. Mooney, F. di Castri, R.H. Groves, F.J. Kruger, M. Rejmanek & M. Williamson, eds). John Wiley & Sons, New York.
25.2.4

Diamond, J. & Case, T.J. (eds) (1986) *Community Ecology*. Harper & Row, New York.
20.1

Diamond, J.M. (1972) Biogeographic kinetics: estimation of

relaxation times for avifaunas of South-West Pacific islands. *Proceedings of the National Academy of Science of the USA*, **69**, 3199–3203.
23.3.1

Diamond, J.M. (1973) Distributional ecology of New Guinea birds. *Science*, **179**, 759–769.
5.3.1

Diamond, J.M. (1975) Assembly of species communities. In: *Ecology and Evolution of Communities* (M.L. Cody & J.M. Diamond, eds), pp. 342–444. Belknap, Cambridge, MA.
20.3.2, 20.4.2, 23.3.5

Diamond, J.M. (1983) Taxonomy by nucleotides. *Nature*, **305**, 17–18.
1.2.1

Diamond, J.M. (1984) 'Normal' extinctions of isolated populations. In: *Extinctions* (M.H. Nitecki, ed.), pp. 191–245. University of Chicago Press, Chicago.
25.3.2

Diamond, J.M. & May, R.M. (1976) Island biogeography and the design of natural reserves. In: *Theoretical Ecology: Principles and Applications* (R.M. May, ed.), pp. 228–252. Blackwell Scientific Publications, Oxford.
5.3.1, 23.1, 25.6.3

Dickson, R.E. & Cicerone, R.J. (1986) Future global warming from atmospheric trace gases. *Nature*, **319**, 109–114.
2.11.1

Dijkstra, C., Bult, A., Bijlsma, S., Daan, S., Meijer, T. & Zijlstra, M. (1990) Brood size manipulations in the kestrel (*Falco tinnunculus*): effects on offspring and parent survival. *Journal of Animal Ecology*, **59**, 269–286.
14.8.1

Dirzo, R. & Harper, J.L. (1982) Experimental studies of slug–plant interactions. IV. The performance of cyanogenic and acyanogenic morphs of *Trifolium repens* in the field. *Journal of Ecology*, **70**, 119–138.
1.5.2, 3.7.3

Dixon, A.F.G. (1971) The role of aphids in wood formation. II. The effect of the lime aphid, *Eucallipterus tiliae* L. (Aphididae), on the growth of lime *Tilia × vulgaris* Hayne. *Journal of Applied Ecology*, **8**, 383–399.
8.2.3

Dobben, W.H. van (1952) The food of the cormorants in The Netherlands. *Ardea*, **40**, 1–63.
12.4

Dobson, A.P. & Hudson, P.J. (1992) Regulation and stability of a free-living host–parasite system: *Trichostrongylus tenuis* in red grouse. II. Population models. *Journal of Animal Ecology*, **61**, 487–498.
12.6, 12.6.3, 15.4.1

Dobson, A.P. & Keymer, A.E. (1990) Population dynamics and community structure of parasitic helminths. In: *Living in a Patchy Environment* (B. Shorrocks & I.R. Swingland, eds), pp. 107–126. Oxford University Press, Oxford.
12.3

Dobson, A.P. & May, R.M. (1986) Disease and conservation. In: *Conservation Biology: The Science of Scarcity and Diversity* (M.E. Soulé, ed.), pp. 345–365. Sinauer Associates, Sunderland, MA.
25.6.3

Dobson, A.P. & Pacala, S.W. (1992) The parasites of *Anolis* lizards in the northern Lesser Antilles. II. The structure of the parasite community. *Oecologia*, **91**, 118–125.
12.3, 12.3.8

Dobson, A.P., Pacala, S.W., Roughgarden, J.D., Carper, E.R. & Harris, E.A. (1992) The parasites of *Anolis* lizards in the northern Lesser Antilles. I. Patterns of distribution and abundance. *Oecologia*, **91**, 110–117.
12.3, 12.3.8

Dobson, F.S. (1982) Competition for mates and predominant juvenile male dispersal in mammals. *Animal Behaviour*, **30**, 1183–1192.
5.4.3

Dobzhansky, T. (1950) Evolution in the tropics. *American Scientist*, **38**, 209–221.
24.4.1

Dodd, A.P. (1940) *The Biological Campaign Against Prickly Pear*. Commonwealth Prickly Pear Board. Government Printer, Brisbane.
8.2.6, 15.7.6

Dohmen, G.P., McNeil, S. & Bell, J.N.B. (1984) Air pollution increases *Aphis fabae* pest potential. *Nature*, **307**, 52–53.
8.7

Donald, C.M. (1951) Competition among pasture plants. I. Intra-specific competition among annual pasture plants. *Australian Journal of Agricultural Research*, **2**, 335–341.
6.5

Donaldson, A.I. (1983) Quantitative data on air borne foot-and-mouth disease virus: its production, carriage and deposition. *Philosophical Transactions of the Royal Society, Series B*, **302**, 529–534.
12.3.1

Dong, M. (1993) Morphological plasticity of the clonal herb *Lamiastrum galeobdolon* (L.) Ehrend. & Polatschek in response to partial shading. *New Phytologist*, **124**, 291–300.
4.2.1

Doube, B.M. (1987) Spatial and temporal organization in communities associated with dung pads and carcasses. In: *Organization of Communities: Past and Present* (J.H.R. Gee & P.S. Giller, eds), pp. 255–280. Blackwell Scientific Publications, Oxford.
11.3.3, 11.4, 17.4.1

Doube, B.M., Macqueen, A., Ridsdill-Smith, T.J. & Weir, T.A. (1991) Native and introduced dung beetles in Australia. In: *Dung Beetle Ecology* (I. Hanski & Y. Cambefort, eds), pp. 255–278. Princeton University Press, Princeton, USA.
11.3.2

Douglas, A. & Smith, D.C. (1984) The green hydra symbiosis. VIII. Mechanisms in symbiont regulation. *Proceedings of the Royal Society of London, Series B*, **221**, 291–319.
13.8

Douglas, A.E. (1989) Mycetocyte symbiosis in insects. *Biological Reviews*, **64**, 409–434.
13.6

Douglas, A.E. (1992) Microbial brokers of insect–plant interactions. In: *Proceedings of the 8th Symposium on Insect–Plant Relationships* (S.B.J. Menken, J.H. Visser & P. Harrewijn, eds), pp. 329–336. Kluwer Academic Publishing, Dordrecht.
13.5.2

Douglas, A.E. (1993) The nutritional quality of phloem sap utilized by natural aphid populations. *Ecological Entomology*, **18**, 31–38.
3.7.2, 13.6

Douglas, A.E. (1994) *Symbiotic Interactions.* Oxford University Press, Oxford.
2.4, 13.1, 13.6, 13.11

Downs, C. & McQuilkin, W.E. (1944) Seed production of Southern Appalachian oaks. *Journal of Forestry*, **42**, 913–920.
4.6.3

Downton, W.J.S., Berry, J.A. & Seeman, J.R. (1984) Tolerance of photosynthesis to high temperatures in desert plants. *Plant Physiology*, **74**, 786–790.
2.4

Drake, J.A. (1988) Models of community assembly and the structure of ecological landscapes. In: *Mathematical Ecology* (T. Hallam, L. Gross & S. Levin, eds), pp. 584–604. World Press, Singapore.
22.5.6

Drake, J.A. (1990) The mechanics of community assembly and succession. *Journal of Theoretical Biology*, **147**, 213–234.
22.5.6

Drake, J.A. (1991) Community-assembly mechanics and the structure of an experimental species ensemble. *American Naturalist*, **137**, 1–26.
22.5.6

Drift, J. van der & Witkamp, M. (1959) The significance of the breakdown of oak litter by *Enoicyla pusilla* Burm. *Archives Néerlandaises de Zoolofie*, **13**, 485–492.
11.2.3

Duarte, C.M. (1992) Nutrient concentration of aquatic plants: patterns across species. *Limnology and Oceanography*, **37**, 882–889.
11.2.4

Ducrey, M. & Labbé, P. (1985) Étude de la régénération naturale contrôlée en forêt tropicale humide de Guadeloupe. I. Revue bibliographique, milieu naturel et élaboration d'un protocole expérimental. *Annales Scientifiques Forestière*, **42**, 297–322.
2.9.1

Ducrey, M. & Labbé, P. (1986) Étude de la régénération naturale contrôlée en forêt tropicale humide de Guadeloupe. II. Installation et croissance des semis après les coupes d'ensemencement. *Annales Scientifiques Forestière*, **43**, 299–326.
2.9.1

Dunn, E. (1977) Predation by weasels (*Mustela nivalis*) on breeding tits (*Parrus* spp.) in relation to the density of tits and rodents. *Journal of Animal Ecology*, **46**, 634–652.
6.12

Duynisveld, W.H.M., Strebel, O. & Bottcher, J. (1988) Are nitrate leaching from arable and nitrate pollution of groundwater avoidable? *Ecological Bulletins*, **39**, 116–125.
19.2.4

Dwyer, G., Levin, S.A. & Buttel, L. (1990) A simulation model of the population dynamics and evolution of myxomatosis. *Ecological Monographs*, **60**, 423–447.
12.7

Dytham, C. (1994) Habitat destruction and competitive coexistence: a cellular model. *Journal of Animal Ecology*, **63**, 490–491.
7.5.1

Earth System Sciences Committee, NASA (1988) *Earth System Science: a Program for Global Change.* NASA Advisory Council, Washington, DC, USA.
19.4.6

Edmonson, W.T. (1970) Phosphorus, nitrogen and algae in Lake Washington after diversion of sewage. *Science*, **169**, 690–691.
19.4.3

Edwards, P.B. & Aschenborn, H.H. (1987) Patterns of nesting and dung burial in *Onitis* dung beetles: implications for pasture productivity and fly control. *Journal of Applied Ecology*, **24**, 837–851.
11.3.2

Ehleringer, J.R. & Monson, R.K. (1993) Evolutionary and ecological aspects of photosynthetic pathway variation. *Annual Review of Ecology and Systematics*, **24**, 411–439.
3.3.4

Ehleringer, J.R. & Mooney, H.A. (1984) Photosynthesis and productivity of desert and Mediterranean climate plants. In: *Encyclopedia of Plant Physiology* (O.J. Lange, P.S. Nobel, C.B. Osmond & H. Ziegler, eds), New Series, 12D, pp. 205–231. Springer-Verlag.
3.2.5

Ehleringer, J.R., Sage, R.F., Flanagan, L.B. & Pearcy, R.W. (1991) Climate change and the evolution of C_4 photosynthesis. *Trends in Ecology and Evolution*, **6**, 95–99.
3.3.4

Ehrenfeld, D. (1988) Why put a value on biodiversity? In: *Biodiversity* (E.O. Wilson & F.M. Peter, eds), pp. 212–216. National Academy Press, Washington, DC.
25.1.3

Ehrlich, P. & Raven, P.H. (1964) Butterflies and plants: a study in coevolution. *Evolution*, **18**, 586–608.
3.7.3, 9.3.1

Ehrlich, P.R. & Murphy, D.D. (1987) Conservation lessons from long-term studies of checkerspot butterflies. *Conservation Biology*, **1**, 122–131.
25.3.3

Einarsen, A.S. (1945) Some factors affecting ring-necked pheasant population density. *Murrelet*, **26**, 39–44.
6.4

Eis, S., Garman, E.H. & Ebel, L.F. (1965) Relation between cone production and diameter increment of douglas fir (*Pseudotsuga menziesii* (Mirb). Franco), grand fir (*Abies grandis* Dougl.) and western white pine (*Pinus monticola* Dougl.). *Canadian Journal of Botany*, **43**, 1553–1559.
14.4

El-Hinnawi, E. & Hashmi, M. (1982) *Global Environmental Issues.* Published for the United Nations Environment Programme by Tycooly International Publishing, Dublin.
19.1.3

El Titi, A. (1989) Farming systems research at Lautenbach Oedhein, FRGermany. In: *Current Status of Integrated Farming Systems Research in Western Europe* (P. Vereijken & D.J. Royle, eds), pp. 21–35. IOBC WPRS Bulletin, 12 (5).
16.11

Elliott, J.M. (1976) Body composition of brown trout (*Salmo trutta* L.) in relation to temperature and ration size. *Journal of Animal Ecology*, **45**, 273–289.
4.2

Elliott, J.M. (1984) Numerical changes and population regulation in young migratory trout *Salmo trutta* in a Lake District stream 1966–83. *Journal of Animal Ecology*, **53**, 327–350.
6.4

Elliott, J.M. (1993) The self-thinning rule applied to juvenile

sea-trout, *Salmo trutta. Journal of Animal Ecology*, **62**, 371–379.
6.13

Elner, R.W. & Hughes, R.N. (1978) Energy maximisation in the diet of the shore crab *Carcinus maenas* (L.). *Journal of Animal Ecology*, **47**, 103–116.
9.2.2

Elton, C. (1927) *Animal Ecology*. Sidgwick & Jackson, London.
18.4

Elton, C.S. (1958) *The Ecology of Invasion by Animals and Plants.* Methuen, London.
22.4.1

Elton, C.S. & Nicholson, A.J. (1942) The ten year cycle in numbers of the lynx in Canada. *Journal of Animal Ecology*, **11**, 215–244.
15.4.1

Elwood, J.W., Newbold, J.D., O'Neill, R.V. & van Winkle, W. (1983) Resource spiralling: an operational paradigm for analyzing lotic ecosystems. In: *Dynamics of Lotic Ecosystems* (T.D. Fontaine & S.M. Bartell, eds), pp. 3–28. Ann Arbor Science Publishers, Ann Arbor, MI.
19.3.1

Emiliani, C. (1966) Isotopic palaeotemperatures. *Science*, **154**, 851–857.
1.2.2

Emmons, L.H. (1980) Ecology and resource partitioning among nine species of African rain forest squirrels. *Ecological Monographs*, **50**, 31–54.
20.3.1

Enriquez, S., Duarte, C.M. & Sand-Jensen, K. (1993) Patterns in decomposition rates among photosynthetic organisms: the importance of detritus C : N : P content. *Oecologia*, **94**, 457–471.
11.2.4

Ens, B.J., Kersten, M., Brenninkmeijer, A. & Hulscher, J.B. (1992) Territory quality, parental effort and reproductive success of oystercatchers (*Haematopus ostralegus*). *Journal of Animal Ecology*, **61**, 703–715.
6.12

Eriksson, O., Hansen, A. & Sunding, P. (1974) *Flora of Macronesia: Checklist of Vascular Plants*. University of Umeå, Sweden.
23.1

Evans, F.C. (1956) Ecosystem as the basic unit in ecology. *Science*, **123**, 1127–1128.
19.1

Evenari, M., Koller, D. & Gutterman, Y. (1966) Effects of the environment of the mother plant on germination by control of seed-coat permeability to water in *Ononis sicula* Guss. *Australian Journal of Biological Sciences*, **19**, 1007–1016.
14.11

Faaborg, J. (1976) Patterns in the structure of West Indian bird communities. *American Naturalist*, **111**, 903–916.
23.3.5

Facelli, J.M. & Pickett, S.T.A. (1990) Markovian chains and the role of history in succession. *Trends in Ecology and Evolution*, **5**, 27–30.
17.4.4

Fahrig, L. & Merriam, G. (1994) Conservation of fragmented populations. *Conservation Biology*, **8**, 50–59.
25.3.2, 25.6.1

Fajer, E.D. (1989) The effects of enriched CO_2 atmospheres on plant–insect–herbivore interactions: growth responses of larvae of the specialist butterfly, *Junonia coenia* (Lepidoptera: Nymphalidae). *Oecologia*, **81**, 514–520.
3.3.4

Fajer, E.D., Bowers, M.D. & Bazzaz, F.A. (1989) The effects of enriched carbon dioxide atmospheres on plant–insect–herbivore interactions. *Science*, **243**, 1198–1200.
3.3.4

Fajer, E.D., Bowers, M.D. & Bazzaz, F.A. (1991) The effects of enriched CO_2 atmospheres on the buckeye butterfly *Junonia coenia. Ecology*, **72**, 751–754.
3.3.4

Farquhar, G.D. & Sharkey, T.D. (1982) Stomatal conductance and photosynthesis. *Annual Review of Plant Physiology*, **33**, 317–345.
3.2.4

Feeny, P. (1976) Plant apparency and chemical defence. *Recent Advances in Phytochemistry*, **10**, 1–40.
3.7.3

Fenchel, T. (1975) Character displacement and coexistence in mud snails (Hydrobiidae). *Oecologia*, **20**, 19–32.
7.8.1

Fenchel, T. (1987a) *Ecology—Potentials and Limitations*. Ecology Institute, Federal Republic of Germany.
18.4.3, 20.3.1

Fenchel, T. (1987b) Patterns in microbial aquatic communities. In: *Organization of Communities: Past and Present* (J.H.R. Gee & P.S. Giller, eds), pp. 281–294. Blackwell Scientific Publications, Oxford.
11.2.1, 18.4.3

Fenchel, T. (1988) Hydrobia and Ockham's razor: a reply to A.H. Cherrill. *Oikos*, **B51**, 113–115.
7.8.1

Fenchel, T. & Kofoed, L. (1976) Evidence for exploitative interspecific competition in mud snails (Hydrobiidae). *Oikos*, **27**, 367–376.
7.8.1

Fenchel, T. & Kolding, S. (1979) Habitat selection and distribution patterns of five species of the amphipod genus *Gammarus. Oikos*, **33**, 316–322.
20.3.1

Fenner, F. (1983) Biological control, as exemplified by smallpox eradication and myxomatosis. *Proceedings of the Royal Society, Series B*, **218**, 259–285.
12.7

Fenner, F. & Ratcliffe, R.N. (1965) *Myxomatosis*. Cambridge University Press, London.
12.7

Fernando, M.H.J.P. (1977) *Predation of the Glasshouse Red Spider Mite by* Phytoseiulus persimilis *A.-H*. Ph.D. thesis, University of London.
9.6

Fiedler, P.L. (1987) Life history and population dynamics of rare and common Mariposa lilies (*Calochortus* Pursh: Liliaceae). *Journal of Ecology*, **75**, 977–995.
25.2.1

Fields, P.A., Graham, J.B., Rosenblatt, R.H. & Somero, G.N. (1993) Effects of expected global climate change on marine faunas. *Trends in Ecology and Evolution*, **8**, 361–367.
18.5

Firbank, L.G. & Watkinson, A.R. (1985) On the analysis of

competition within two-species mixtures of plants. *Journal of Applied Ecology*, **22**, 503–517.
7.7.2

Firbank, L.G. & Watkinson, A.R. (1990) On the effects of competition: from monocultures to mixtures. In: *Perspectives on Plant Competition* (J.B. Grace & D. Tilman, eds), pp. 165–192. Academic Press, New York.
7.7.2

Firth, R. & Davidson, J.W. (1945) *Pacific Islands. Vol. 1. General Survey*. Naval Intelligence Division, London.
23.3.5

Fischhoff, D.A., Bowdish, K.S., Perlack, F.J. *et al.* (1987) Insect tolerant transgenic tomato plants. *Biotechnology*, **5**, 807–813.
16.10

Fisher, R.A. (1930) *The Genetical Theory of Natural Selection*. Clarendon Press, Oxford.
14.3

Fitter, A.H. & Hay, R.K.M. (1981) *Environmental Physiology of Plants*, 2nd edn. Academic Press, London.
18.3.1

FitzGibbon, C.D. (1990) Anti-predator strategies of immature Thomson's gazelles: hiding and the prone response. *Animal Behaviour*, **40**, 846–855.
8.3

FitzGibbon, C.D. & Fanshawe, J. (1989) The condition and age of Thomson's gazelles killed by cheetahs and wild dogs. *Journal of Zoology*, **218**, 99–107.
8.3

Flecker, A.S. (1993) Fish trophic guilds and the organisation of neotropical streams: weak direct versus strong indirect effects. *Ecology*, **73**, 927–940.
22.2.2

Flecker, A.S. & Townsend, C.R. (1994) Community-wide consequences of trout introduction in New Zealand streams. *Ecological Applications*, **4**, 798–807.
22.3.1, 22.3.2

Flenley, J. (1993) The origins of diversity in tropical rain forests. *Trends in Ecology and Evolution*, **8**, 119–120.
24.4.1

Fletcher, J.P., Hughes, J.P. & Harvey, I.F. (1994) Life expectancy and eggload affect oviposition decisions of a solitary parasitoid. *Proceedings of the Royal Society, Series B*, **258**, 163–167.
12.3.2

Flint, M.L. & van den Bosch, R. (1981) *Introduction to Integrated Pest Management*. Plenum Press, New York.
16.3, 16.4, 16.6.1, 16.6.3, 16.6.4, 16.8, 16.11

Flower, R.J. & Battarbee, R.W. (1983) Diatom evidence for recent acidification of two Scottish lochs. *Nature*, **305**, 130–133.
2.10

Folstad, I. & Karter, A.J. (1992) Parasites, bright males and the immunocompetence handicap. *American Naturalist*, **139**, 603–622.
12.4

Fonseca, C.R. (1994) Herbivory and the long-lived leaves of an Amazonian ant-tree. *Journal of Ecology*, **82**, 833–842.
13.2.1

Ford, E.B. (1940) Polymorphism and taxonomy. In: *The New Systematics* (J. Huxley, ed.), pp. 493–513. Clarendon Press, Oxford.
1.5.2

Ford, R.G. & Pitelka, F.A. (1984) Resource limitation in the California vole. *Ecology*, **65**, 122–136.
15.4.3

Forrester, N.W. (1993) Well known and some not so well known insecticides: their biochemical targets and role in IPM and IRM programmes. In: *Pest Control and Sustainable Agriculture* (S. Corey, D. Dall & W. Milne, eds), pp. 28–34. CSIRO, East Melbourne.
16.5.1, 16.6.4

Fowler, M.C. (1985) The present status of grass carp (*Ctenopharyngodon idella*) for the control of aquatic weeds in England and Wales. In: *Proceedings of the VIth International Symposium on Biological Control of Weeds*, pp. 537–542.
16.8

Fowler, S.V. & Lawton, J.H. (1985) Rapidly induced defenses and talking trees: the devil's advocate position. *American Naturalist*, **126**, 181–195.
8.2.2

Franco, M., Harper, J.L. & Silvertown, J. (1996) Plant life histories: ecological correlates and phylogenetic constraints. *Philosophical Transactions of the Royal Society, Series B*. (In press).
1.4.2

Frank, D.A. & McNaughton, S.J. (1991) Stability increases with diversity in plant communities: empirical evidence from the 1988 Yellowstone drought. *Oikos*, **62**, 360–362.
22.4.3

Frank, F. (1953) Zur entstehung öbernormaler Populationsdichten im Massenwechselder Fjelmaus, *Micrtus arvalis*, Pallas. *Zoologisches Jahrbuecher Abteilung Systematik*, **81**, 610–624.
15.4.3

Frank, P.W. (1957) Coactions in laboratory populations of two species of *Daphnia*. *Ecology*, **38**, 510–519.
7.7.1

Franklin, I.A. (1980) Evolutionary change in small populations. In: *Conservation Biology, An Evolutionary-Ecological Perspective* (M.E. Soulé & B.A. Wilcox, eds), pp. 135–149. Sinauer Associates, Sunderland, MA.
25.2.5

Franklin, R.T. (1970) Insect influences in the forest canopy. In: *Analysis of Temperate Forest Ecosystems* (D.E. Reichle, ed.), pp. 86–99. Springer-Verlag, New York.
8.2.3

Franks, F., Mathias, S.F. & Hatley, R.H.M. (1990) Water, temperature and life. *Philosophical Transactions of the Royal Society, Series B*, **326**, 517–533; also in *Life at Low Temperatures* (R.M. Laws & F. Franks, eds), pp. 97–117. The Royal Society, London.
2.3

Free, C.A., Beddington, J.R. & Lawton, J.H. (1977) On the inadequacy of simple models of mutual interference for parasitism and predation. *Journal of Animal Ecology*, **46**, 543–554.
9.10, 10.5.3

Fretwell, S.D. (1977) The regulation of plant communities by food chains exploiting them. *Perspectives in Biology and Medicine*, **20**, 169–185.
22.3.2

Fretwell, S.D. & Lucas, H.L. (1970) On territorial behaviour and other factors influencing habitat distribution in birds. *Acta Biotheoretica*, **19**, 16–36.
9.10

Fricke, H.W. (1975) The role of behaviour in marine symbiotic animals. In: *Symbiosis* (D.H. Jennings & D.L. Lee, eds), pp. 581–594. Symposium 29, Society for Experimental Biology. Cambridge University Press, Cambridge.
13.2.4, 13.2.5

Fridriksson, S. (1975) *Surtsey: Evolution of Life on a Volcanic Island*. Butterworths, London.
23.3.2

Fry, G.L.A. & Cooke, A.S. (1984) Acid deposition and its implications for nature conservation in Britain. *Focus on Nature Conservation*, No. 7. Nature Conservancy Council, Attingham Park, Shrewsbury, England.
19.4.5

Fryer, G. & Iles, T.D. (1972) *The Cichlid Fishes of the Great Lakes of Africa*. Oliver & Boyd, Edinburgh.
23.4

Futuyma, D.J. (1983) Evolutionary interactions among herbivorous insects and plants. In: *Coevolution* (D.J. Futuyma & M. Slatkin, eds), pp. 207–231. Sinauer Associates, Sunderland, MA.
3.7.3

Futuyma, D.J. & Gould, F. (1979) Associations of plants and insects in a deciduous forest. *Ecological Monographs*, **49**, 33–50.
23.3.3

Futuyma, D.J. & May, R.M. (1992) The coevolution of plant–insect and host–parasite relationships. In: *Genes in Ecology* (R.J. Berry, T.J. Crawford & G.M. Hewitt, eds), pp. 139–166. Blackwell Scientific Publications, Oxford.
9.3.1

Futuyma, D.J. & Slatkin, M. (eds) (1983) *Coevolution*. Sinauer, Massachusetts.
3.7.3, 9.3.1

Gadgil, M. (1971) Dispersal: population consequences and evolution. *Ecology*, **52**, 253–261.
5.3.2, 15.6.1

Gaines, M.S. & McClenaghan, L.R. (1980) Dispersal in small mammals. *Annual Review of Ecology and Systematics*, **11**, 163–196.
5.3.3, 15.4.4

Gaines, M.S., Vivas, A.M. & Baker, C.L. (1979) An experimental analysis of dispersal in fluctuating vole populations: demographic parameters. *Ecology*, **60**, 814–828.
15.4.4

Gaines, S.D. & Denny, M.W. (1993) The largest, smallest, highest, lowest, longest, and shortest: extremes in ecology. *Ecology*, **74**, 1677–1692.
2.2.1, 2.9.1

Gallun, R.L. (1977) Genetic basis for Hessian fly epidemics. *Annals of the New York Academy of Sciences*, **287**, 223–229.
16.10

Gallun, R.L. & Khush, G.S. (1980) Genetic factors affecting expression and stability of resistance. In: *Breeding Plants Resistant to Insects* (F.G. Maxwell & P.R. Jennings, eds), pp. 63–85. Wiley, New York.
16.10

Garrod, D.J. & Jones, B.W. (1974) Stock and recruitment relationships in the N.E. Atlantic cod stock and the implications for management of the stock. *Journal Conseil International pour l'Exploration de la Mer*, **173**, 128–144.
16.14

Gartside, D.W. & McNeilly, T. (1974) The potential for evolution of heavy metal tolerance in plants. II. Copper tolerance in normal populations of different plant species. *Heredity*, **32**, 335–348.
2.10

Gates, J.E. & Gysel, L.W. (1978) Avian nest dispersion and fledging success in field–forest ecotones. *Ecology*, **59**, 871–883.
25.3.3

Gause, G.F. (1934) *The Struggle for Existence*. Williams & Wilkins, Baltimore (reprinted 1964 by Hafner, New York).
7.2.4

Gause, G.F. (1935) Experimental demonstration of Volterra's periodic oscillation in the numbers of animals. *Journal of Experimental Biology*, **12**, 44–48.
7.2.4

Gauthreaux, S.A. (1978) The structure and organization of avian communities in forests. In: *Proceedings of the Workshop on Management of Southern Forests for Nongame Birds* (R.M. DeGraaf, ed.), pp. 17–37. Southern Forest Station, Asheville, NC.
17.4.4

Gebhardt, M.D. & Stearns, S.C. (1988) Reaction norms for development time and weight at eclosion in *Drosophila mercatorum*. *Journal of Evolutionary Biology*, **1**, 335–354.
14.11

Gee, J.H.R. & Giller, P.S. (eds) (1987) *Organization of Communities: Past and Present*. Blackwell Scientific Publication, Oxford.
20.1

Geiger, R. (1955) *The Climate Near the Ground*. Harvard University Press, Cambridge, MA.
2.2

GER (1993) *Global Environmental Change: The U.K. Framework 1993*. U.K. Global Environmental Research (GER) Office, Swindon.
2.11.2

Gibson, A.H. & Jordan, D.C. (1983) Ecophysiology of nitrogen-fixing systems. In: *Physiological Plant Ecology III. Encyclopedia of Plant Physiology*, New Series, 12c (O.L. Lange, P.S. Nobel, C.B. Osmond & H. Ziegler, eds), pp. 300–390. Springer-Verlag, Berlin.
13.10

Gibson, C.C. & Watkinson, A.R. (1991) Host selectivity and the mediation of competition by the root hemiparasite *Rhinanthus minor*. *Oecologia*, **86**, 81–87.
21.2.4

Gilbert, J.M. (1959) Forest succession in the Florentine Valley, Tasmania. *Papers of the Proceedings of the Royal Society of Tasmania*, **93**, 129–151.
2.4

Gilbert, N. (1984) Control of fecundity in *Pieris rapae*. II. Differential effects of temperature. *Journal of Animal Ecology*, **53**, 589–597.
2.2

Gill, F.B. & Wolf, L.L. (1975) Economics of feeding territoriality in the golden-winged sunbird. *Ecology*, **56**, 333–345.
6.12

Gill, L.S. (1957) Dwarf mistletoe on lodgepole pine. *United States Department of Agricultural and Forest Service Forest Pest Leaflet*, 18.
12.2.2

Gilligan, C.A. (1990) Comparison of disease progress curves. *New Phytologist*, **115**, 223–242.
12.5.3

Gilman, M.P. & Crawley, M.J. (1990) The cost of sexual reproduction in ragwort (*Senecio jacobaea* L.). *Functional Ecology*, **4**, 585–589.
14.4.1

Gilpin, M. (1991) The genetic effective size of a metapopulation. In: *Metapopulation Dynamics* (M.E. Gilpin & I. Hanski, eds), pp. 165–175. Academic Press, London.
15.6.4

Gilpin, M.E & Diamond, J.M. (1982) Factors contributing to non-randomness in species co-occurrences on islands. *Oecologia*, **52**, 75–84.
20.4.3

Ginsberg, J.R. (1993) Can we build an Ark? *Trends in Ecology and Evolution*, **8**, 4–6.
25.6.2

Ginzburg, L.R. & Taneyhill, D.E. (1994) Population cycles of forest Lepidoptera: a maternal effect hypothesis. *Journal of Animal Ecology*, **63**, 79–92.
12.6.1

Gleason, H.A. (1926) The individualistic concept of the plant association. *Torrey Botanical Club Bulletin*, **53**, 7–26.
17.3.3

Gleick, J. (1987) *Chaos: Making a New Science*. Viking, New York.
6.9

Glick, P.A. (1939) The distribution of insects, spiders and mites in the air. *Technical Bulletin, United States Department of Agriculture*, **673**.
5.3.1

Gliddon, C. & Trathan, P. (1985) Interaction between white clover and perennial rye grass in an old permanent pasture. In: *The Structure and Functioning of Plant Populations II* (J. Haeck & J.W. Woldendorp, eds), pp. 161–169. North Holland Publishing Company, Amsterdam.
1.5.2

Godfray, H.C.J. (1987) The evolution of clutch size in invertebrates. *Oxford Surveys in Evolutionary Biology*, **4**, 117–154.
14.8.1

Godfray, H.C.J. (1994) *Parasitoids: Behavioral and Evolutionary Ecology*. Princeton University Press, Princeton.
8.1

Godfray, H.C.J. & Grenfell, B.T. (1993) The continuing quest for chaos. *Trends in Ecology and Evolution*, **8**, 43–44.
6.9

Godfray, H.C.J. & Hassell, M.P. (1989) Discrete and continuous insect populations in tropical environments. *Journal of Animal Ecology*, **58**, 153–174.
10.2.3

Godfray, H.C.J. & Pacala, S.W. (1992) Aggregation and the population dynamics of parasitoids and predators. *American Naturalist*, **140**, 30–40.
10.5.5, 15.6.2

Goh, B.S. (1979) Robust stability concepts for ecosystem models. In: *Theoretical Systems in Ecology* (E. Halfon, ed.), pp. 467–487. Academic Press, New York.
22.5.5

Gold, C.S., Altieri, M.A. & Bellotti, A.C. (1989) The effects of intercropping and mixed varieties of predators and parasitoids of cassava whiteflies (Hemiptera: Aleyrodidae) in Colombia. *Bulletin of Entomological Research*, **79**, 115–121.
16.9

Goldman, J.C., Caron, D.A. & Dennett, M.R. (1987) Regulation of gross growth efficiency and ammonium regeneration in bacteria by substrate C : N ratio. *Limnology and Oceanography*, **32**, 1239–1252.
11.2.4

Goodall, D.W. (1967) Computer simulation of changes in vegetation subject to grazing. *Journal of the Indian Botanical Society*, **46**, 356–362.
3.4

Gordon, D.M., Nisbet, R.M., De Roos, A., Gurney, W.S.C. & Stewart, R.K. (1991) Discrete generations in host–parasitoid models with contrasting life-cycles. *Journal of Animal Ecology*, **60**, 295–308.
10.2.3

Gorman, M.L. (1979) *Island Ecology*. Chapman & Hall, London.
23.1, 23.3.1

Gosling, L.M. & Baker, S.J. (1987) Planning and monitoring an attempt to eradicate coypus from Britain. *Symposium of the Zoological Society of London*, **58**, 99–113.
25.6.4

Goss-Custard, J.D. (1970) Feeding dispersion in some overwintering wading birds. In: *Social Behaviour in Birds and Mammals* (J.H. Crook, ed.), pp. 3–34. Academic Press, New York.
9.7.1

Gould, S.J. (1966) Allometry and size in ontogeny and phylogeny. *Biological Reviews*, **41**, 587–640.
14.12.1

Gould, S.J. (1981) Palaeontology plus ecology as palaeobiology. In: *Theoretical Ecology: Principles and Applications*, 2nd edn. (R.M. May, ed.), pp. 295–317. Blackwell Scientific Publications, Oxford.
24.4.6

Gould, S.J. & Lewontin, R.C. (1979) Spandrels of San Marco and the Panglossian paradigm—a critique of the adaptationist program. *Proceedings of the Royal Society, Series B*, **205**, 581–598.
1.1.5

Grace, J.B. & Wetzel. R.G. (1981) Habitat partitioning and competitive displacement in cattails (*Typha*): experimental field studies. *American Naturalist*, **118**, 463–474.
7.3

Grafen, A. & Godfray, H.C.J. (1991) Vicarious selection explains some paradoxes in dioecious fig-pollinator systems. *Proceedings of the Royal Society of London, Series B*, **245**, 73–76.
13.4.2

Grant, P.R. & Abbott, I. (1980) Interspecific competition, island biogeography and null hypotheses. *Evolution*, **34**, 332–341.
20.4.2

Gray, J.S. (1981) *The Ecology of Marine Sediments: An Introduction to the Structure and Function of Benthic Communities*. Cambridge University Press, Cambridge.
17.3.2

Gray, J.T. (1983) Nutrient use by evergreen and deciduous shrubs in southern California. I. Community nutrient cycling and nutrient use efficiency. *Journal of Ecology*, **71**, 21–41.
19.2.3

Green, T.R. & Ryan, C.A. (1972) Wound-induced proteinase

inhibitor in plant leaves: a possible defense mechanism against insects. *Science*, **175**, 776–777.
8.2.2

Greenbank, D.O. (1957) The role of climate and dispersal in the initiation of outbreaks of the spruce budworm in New Brunswick. II. The role of dispersal. *Canadian Journal of Zoology*, **35**, 385–403.
5.3.3

Greenwood, P.J. (1980) Mating systems, philopatry and dispersal in birds and mammals. *Animal Behaviour*, **28**, 1140–1162.
5.4.1

Greenwood, P.J. (1983). Mating systems and the evolutionary consequences of dispersal. In: *The Ecology of Animal Movement* (I.R. Swingland & P.J. Greenwood, eds), pp. 116–131. Clarendon Press, Oxford, England.
5.4.4

Greenwood, P.J., Harvey, P.H. & Perrins, C.M. (1978) Inbreeding and dispersal in the great tit. *Nature*, **271**, 52–54.
5.3.3, 5.4.4

Gregor, J.W. (1930) Experiments on the genetics of wild populations. I. *Plantago maritima. Journal of Genetics*, **22**, 15–25.
1.5.1

Greig-Smith, P. (1983) *Quantitative Plant Ecology*, 3rd edn. Blackwell Scientific Publications, Oxford.
5.2

Grenfell, B.T. & Dobson, A.P. (1995) *Ecology of Infectious Diseases in Natural Populations*. Cambridge University Press, Cambridge.
12.5, 12.6.3

Grennfelt, P. & Thornelof, E. (eds) (1992) *Critical Loads for Nitrogen*. Nordic Council of Ministers, Copenhagen.
19.4.4

Gribbin, S.D. & Thompson, D.J. (1990) Asymmetric intraspecific competition among larvae of the damselfly *Ischnura elegans* (Zygoptera: Coenagrionidae). *Ecological Entomology*, **15**, 37–42.
6.11

Grier, C.C. (1975) Wildfire effects on nutrient distribution and leaching in a coniferous forest ecosystem. *Canadian Journal of Forest Research*, **5**, 599–607.
19.2.2

Griffith, D.M. & Poulson, T.M. (1993) Mechanisms and consequences of intraspecific competition in a carabid cave beetle. *Ecology*, **74**, 1373–1383.
6.1

Griffiths, K.J. (1969) Development and diapause in *Pleolophus basizonus* (Hymenoptera: Ichneumonidae). *Canadian Entomologist*, **101**, 907–914.
9.5.1

Griffiths, M. & Barker, R. (1966) The plants eaten by sheep and kangaroos grazing together in a paddock in south-western Queensland. *CSIRO Wildlife Research*, **11**, 145–167.
1.3

Griffiths, R.A., Denton, J. & Wong, A.L.-C. (1993) The effect of food level on competition in tadpoles: interference mediated by prototphecan algae? *Journal of Animal Ecology*, **62**, 274–279.
7.3

Grime, J.P. (1973) Control of species density in herbaceous

vegetation. *Journal of Environmental Management*, **1**, 151–167.
24.3.4

Grime, J.P. (1974) Vegetation classification by reference to strategies. *Nature*, **250**, 26–31.
14.10.1

Grime, J.P. (1979) *Plant Strategies and Vegetation Processes*. Wiley, Chichester.
14.10.1, 21.6

Grime, J.P. & Lloyd, P.S. (1973) *An Ecological Atlas of Grassland Plants*. Edward Arnold, London.
17.3.1

Grime, J.P., Hodgson, J.G. & Hunt, R. (1988) *Comparative Plant Ecology*. Allen & Unwin, London.
14.10.1

Groves, R.H. & Williams, J.D. (1975) Growth of skeleton weed (*Chondrilla juncea* L.) as affected by growth of subterranean clover (*Trifolium subterraneum* L.) and infection by *Puccinia chondrilla* Bubak and Syd. *Australian Journal of Agricultural Research*, **26**, 975–983.
7.3

Grubb, P. (1977) The maintenance of species richness in plant communities: the importance of the regeneration niche. *Biological Reviews*, **52**, 107–145.
21.6

Grubb, P.J. (1986) The ecology of establishment. In: *Ecology and Design in Landscape* (A.D. Bradshaw, D.A. Goode & E. Thorpe, eds). *Symposia of the British Ecological Society*, **24**, 83–97. Blackwell Scientific Publications, Oxford.
13.10.3

Grumbine, R.E. (1993) What is ecosystem management? *Conservation Biology*, **8**, 27–38.
25.6.3

Guégan, J-F., Lambert, A., Lévêque, C., Combes, C. & Euzet, L. (1992) Can host body size explain the parasite species richness of tropical freshwater fishes? *Oecologia*, **90**, 197–204.
12.3.8

Gulland, F.M.D. (1995) The impact of infectious diseases on wild animal populations—a review. In: *Ecology of Infectious Diseases in Natural Populations* (B.T. Grenfell & A.P. Dobson, eds), pp. 20–51. Cambridge University Press, Cambridge.
12.4

Gulland, J.A. (1971) The effect of exploitation on the numbers of marine animals. In: *Dynamics of Populations* (P.J. den Boer & G.R. Gradwell, eds), pp. 450–468. Centre for Agricultural Publishing and Documentation, Wageningen.
16.13.1

Haefner, P.A. (1970) The effect of low dissolved oxygen concentrations on temperature–salinity tolerance of the sand shrimp *Crangon septemspinosa. Physiological Zoology*, **43**, 30–37.
2.12

Hagner, S. (1965) Cone crop fluctuations in Scots pine and Norway spruce. *Studia Forestalia Suecica Skogshogski*, **33**, 1–21.
8.4

Haines, E. (1979) Interaction between Georgia salt marshes and coastal waters: a changing paradigm. In: *Ecological Processes in Coastal and Marine Systems* (R.J. Livingston, ed.). Plenum Press, New York.
18.3.3

Hainsworth, F.R. (1981) *Animal Physiology*. Addison-Wesley, Reading, MA.
2.5, 2.5.1

Hairston, N.G. (1980) The exponential test of an analysis of field distributions: competition in terrestrial salamanders. *Ecology*, **61**, 817–826.
7.2.1

Hairston, N.G., Smith, F.E. & Slobodkin, L.B. (1960) Community structure, population control, and competition. *American Naturalist*, **44**, 421–425.
20.2.1, 22.3.2

Hairston, N.G., Allan, J.D., Colwell, R.K. *et al.* (1968) The relationship between species diversity and stability: an experimental approach with protozoa and bacteria. *Ecology*, **49**, 1091–1101.
22.4.3

Haldane, J.B.S. (1949) Disease and evolution. *La Ricerca Scienza*, **19** (Suppl.), 3–11.
4.5.1

Haldane, J.B.S. (1953) Animal populations and their regulation. *New Biology*, **15**, 9–24.
15.2.3, 15.6.1

Hall, D.C., Norgard, R.B. & True, P.K. (1975) The performance of independent pest management consultants. *California Agriculture*, **29**, 12–14.
16.11

Hall, S.J. & Raffaelli, D.G. (1991) Food-web patterns: lessons from a species-rich web. *Journal of Animal Ecology*, **60**, 823–842.
22.5

Hall, S.J. & Raffaelli, D.G. (1993) Food webs: theory and reality. *Advances in Ecological Research*, **24**, 187–239.
22.4.3, 22.5, 22.5.2, 22.5.4, 22.5.6

Hamilton, W.D. (1971) Geometry for the selfish herd. *Journal of Theoretical Biology*, **31**, 295–311.
5.2.1

Hamilton, W.D. & Howard, J.C. (eds) (1994) Infection, polymorphism and evolution. *Philosophical Transactions of the Royal Society of London, Series B*, **346**, 267–385.
12.7

Hamilton, W.D. & May, R.M. (1977) Dispersal in stable habitats. *Nature*, **269**, 578–581.
5.3.2

Hamilton, W.D. & Zuk, M. (1982) Heritable true fitness and bright birds: a role for parasites? *Science* (Washington, D.C.), **218**, 384–387.
12.4

Hanek, G. (1972) *Microecology and Spatial Distribution of the Gill Parasites Infesting* Lepomis gibbosus *(L.) and* Ambloplitis rupestris *(Raf.) in the Bay of Quinto, Ontario*. Ph.D. thesis, University of Waterloo.
12.3.3

Hanlon, R.D.G. & Anderson, J.M. (1980) The influence of macroarthropod feeding activities on fungi and bacteria in decomposing oak leaves. *Soil Biology and Biochemistry*, **12**, 255–261.
11.2.3

Hanski, I. (1981) Coexistence of competitors in patchy environments with and without predation. *Oikos*, **37**, 306–312.
7.5.5

Hanski, I. (1989) Metapopulation dynamics: does it help to

have more of the same? *Trends in Ecology and Evolution*, **4**, 113–114.
25.3.3

Hanski, I. (1991) Single-species metapopulation dynamics: concepts, models and observations. In: *Metapopulation Dynamics* (M.E. Gilpin & I. Hanski, eds), pp. 17–38. Academic Press, London.
15.6.5

Hanski, I. (1994a) Metapopulation ecology. In: *Spatial and Temporal Aspects of Population Processes* (O.E. Rhodes Jr., ed.). University of Georgia, Georgia.
15.6.5, 21.4

Hanski, I. (1994b) Patch occupancy dynamics in fragmented landscapes. *Trends in Ecology and Evolution*, **9**, 131–135.
15.6.5, 21.4, 25.5.3

Hanski, I. & Gyllenberg, M. (1993) Two general metapopulation models and the core-satellite hypothesis. *American Naturalist*, **142**, 17–41.
15.6.5

Hanski, I. & Kuusela, S. (1977) An experiment on competition and diversity in the carrion fly community. *Annales Zoologi Fennici*, **43**, 108–115.
7.5.5

Hanski, I., Hansson, L. & Henttonen, H. (1991) Specialist predators, generalist predators, and the microtine rodent cycle. *Journal of Animal Ecology*, **60**, 353–367.
10.4, 15.4.3

Hanski, I., Turchin, P., Korpimaki, E. & Henttonen, H. (1993) Population oscillations of boreal rodents: regulation by mustelid predators leads to chaos. *Nature*, **364**, 232–235.
6.9, 15.4.2, 15.4.3

Hanski, I., Pakkala, T., Kuussaari. M. & Lei, G. (1995) Metapopulation persistence of an endangered butterfly in a fragmented landscape. *Oikos*, **72**, 21–28.
15.6.5

Hansson, L. & Stenseth, N.C. (eds) (1988) Modelling small rodent population dynamics. *Oikos*, **52**, 137–229.
15.4.3

Harcourt, D.G. (1971) Population dynamics of *Leptinotarsa decemlineata* (Say) in eastern Ontario. III. Major population processes. *Canadian Entomologist*, **103**, 1049–1061.
5.3.3, 15.3.1

Harley, J.L. & Smith, S.E. (1983) *Mycorrhizal Symbiosis*. Academic Press, London.
13.7.1

Harman, W.N. (1972) Benthic substrates: their effect on freshwater molluscs. *Ecology*, **53**, 271–272.
24.3.2

Harper, D.G.C. (1982) Competitive foraging in mallards: 'ideal free ducks'. *Animal Behaviour*, **30**, 575–584.
9.10

Harper, J.L. (1961) Approaches to the study of plant competition. In: *Mechanisms in Biological Competition* (F.L. Milthorpe, ed.), pp. 1–39. Symposium No. 15, Society for Experimental Biology. Cambridge University Press, Cambridge.
6.5, 7.5.2

Harper, J.L. (1977) *The Population Biology of Plants*. Academic Press, London.
5.3.4, 5.5, 7.3, 7.7.2, 7.10, 8.2.4, 12.3.3, 13.4.2, 14.2, 14.9.1, 17.4.4, 21.2.5, 25.6.4

Harper, J.L. (1981) The meanings of rarity. In: *The Biological*

Aspects of Rare Plant Conservation (H. Synge, ed.), pp. 189–203. Wiley, Chichester.
15.2.3

Harper, J.L. (1985) Modules, branches and the capture of resources. In: *Population Biology and Evolution of Clonal Organisms* (J.B.C. Jackson, L.W. Buss & R.E. Cook, eds), pp. 1–33. Yale University Press, New Haven.
3.10

Harper, J.L. (1987) The heuristic value of ecological restoration. In: *Restoration Ecology* (W.R. Jordon III, M.E. Gilpin & J.D. Aber, eds), pp. 35–45. Cambridge University Press, Cambridge.
25.6.4

Harper, J.L. & Ogden, J. (1970) The reproductive strategy of higher plants: I. The concept of strategy with special reference to *Senecio vulgaris* L. *Journal of Ecology*, **58**, 681–698.
14.2, 14.7.2

Harper, J.L. & White, J. (1974) The demography of plants. *Annual Review of Ecology and Systematics*, **5**, 419–463.
14.9.1

Harper, J.L., Lovell, P.H. & Moore, K.G. (1970) The shapes and sizes of seeds. *Annual Review of Ecology and Systematics*, **1**, 327–356.
5.3.5

Harper, J.L., Rosen, R.B. & White, J. (eds) (1986a) The growth and form of modular organisms. *Philosophical Transactions of the Royal Society of London, Series B*, **313**, 1–250.
4.2.1

Harper, J.L., Rosen, R.B. & White, J. (1986b) Preface. *Philosophical Transactions of the Royal Society of London, Series B*, **313**, 3–5.
4.2.1

Harper, J.L., Jones, M. & Sackville Hamilton, N.R. (1991) The evolution of roots and the problems of analysing their behaviour. In: *Plant Root Growth: An Ecological Perspective* (D. Atkinson, ed.), pp. 3–22. Special Publication of the British Ecological Society, No. 10. Blackwell Scientific Publications, Oxford.
3.4, 4.2.1, 24.4.6

Harris, M.P. (1973) The Galapagos avifauna. *Condor*, **75**, 265–278.
1.2.3

Harris, P., Thompson, L.S., Wilkinson, A.T.S. & Neary, M.E. (1978) Reproductive biology of tansy ragwort, climate and biological control by the cinnabar moth in Canada. In: *Proceedings of the 4th International Symposium on the Biological Control of Weeds* (T.E. Freeman, ed.), pp. 163–173. CEPIFAS, Gainesville.
8.2.4

Harrison, R.G. (1980) Dispersal polymorphisms in insects. *Annual Review of Ecology and Systematics*, **11**, 95–118.
5.4.2

Hart, A. & Begon, M. (1982) The status of general reproductive-strategy theories, illustrated in winkles. *Oecologia*, **52**, 37–42.
14.2

Hart, J.S. (1956) Seasonal changes in insulation of the fur. *Canadian Journal of Zoology*, **34**, 53–57.
1.6

Harvey, P.H. (1995) Why ecologists need to be phylogenetically challenged. *Journal of Ecology*, **83**, 535–536.
1.4.2

Harvey, P.H. & Pagel, M.D. (1991) *The Comparative Method in Evolutionary Biology*. Oxford University Press, Oxford.
1.4.3, 12.3.8, 14.12.1, 14.12.2

Harvey, P.H. & Partridge, L. (1988) Of cuckoo clocks and cowbirds. *Nature*, **355**, 566–567.
12.8

Harvey, P.H. & Zammuto, R.M. (1985) Patterns of mortality and age at first reproduction in natural populations of mammals. *Nature*, **315**, 319–320.
14.12.2

Harvey, P.H., Leigh Brown, J. & Maynard Smith, J. (eds) (1995) New uses for new phylogenies. *Philosophical Transactions of the Royal Society of London, Series B*, **349**, 1–118.
1.4.2

Hassall, M. & Jennings, J.B. (1975) Adaptive features of gut structure and digestive physiology in the terrestrial isopod *Philoscia muscorum* (Scop.). *Biological Bulletin*, **149**, 348–364.
11.3.1

Hassell, M.P. (1971) Mutual interference between searching insect parasites. *Journal of Animal Ecology*, **40**, 473–486.
9.7.1

Hassell, M.P. (1976) *The Dynamics of Competition and Predation*. Edward Arnold, London.
6.2

Hassell, M.P. (1978) *The Dynamics of Arthropod Predator–Prey Systems*. Princeton University Press, Princeton.
8.4, 10.3.1, 9.5.1, 9.6, 10.5.2

Hassell, M.P. (1981) Arthropod predator–prey systems. In: *Theoretical Ecology: Principles and Applications*, 2nd edn (R.M. May, ed.), pp. 105–131. Blackwell Scientific Publications, Oxford.
10.4

Hassell, M.P. (1982) Patterns of parasitism by insect parasitoids in patchy environments. *Ecological Entomology*, **7**, 365–377.
9.10

Hassell, M.P. (1985) Insect natural enemies as regulating factors. *Journal of Animal Ecology*, **54**, 323–334.
10.2.5, 10.5.6

Hassell, M.P. (1987) Detecting regulation in patchily distributed animal populations. *Journal of Animal Ecology*, **56**, 705–713.
10.2.5, 15.2.3

Hassell, M.P. & Anderson, R.M. (1989) Predator–prey and host–pathogen interactions. In: *Ecological Concepts* (J.M. Cherrett, ed.), pp. 147–196. Blackwell Scientific Publications, Oxford.
10.7

Hassell, M.P. & May, R.M. (1973) Stability in insect host–parasite models. *Journal of Animal Ecology*, **43**, 567–594.
10.5.3

Hassell, M.P. & May, R.M. (1974) Aggregation of predators and insect parasites and its effect on stability. *Journal of Animal Ecology*, **43**, 567–594.
9.7.1

Hassell, M.P. & Pacala, S.W. (1990) Heterogeneity and the dynamics of host–parasitoid interactions. *Philosophical Transactions of the Royal Society of London, Series B*, **330**, 203–220.
10.5.3

Hassell, M.P. & Varley, G.C. (1969) New inductive population model for insect parasites and its bearing on biological control. *Nature*, **223**, 1133–1136.
9.6

Hassell, M.P., Lawton, J.H. & May, R.M. (1976) Patterns of dynamical behaviour in single species populations. *Journal of Animal Ecology*, **45**, 471–486.
6.9

Hassell, M.P., Latto, J. & May, R.M. (1989) Seeing the wood for the trees: detecting density dependence from existing life-table studies. *Journal of Animal Ecology*, **58**, 883–892.
15.2.3

Hassell, M.P., May, R.M., Pacala, S.W. & Chesson, P.L. (1991) The persistence of host–parasitoid associations in patchy environments. I. A general criterion. *American Naturalist*, **138**, 568–583.
10.5.3

Hastings, A., Hom, C.L., Ellner, S., Turchin, P. & Godfray, H.C.J. (1993) Chaos in ecology: is mother nature a strange attractor? *Annual Review of Ecology and Systematics*, **24**, 1–33.
15.2.4

Haukioja, E. & Neuvonen, S. (1985) Induced long-term resistance of birch foliage against defoliators: defensive or accidental? *Ecology*, **66**, 1303–1308.
8.2.2

Hawkes, C. (1974) Dispersal of adult cabbage root fly (*Erioischia brassicae* (Bouché)) in relation to brassica crop. *Journal of Applied Ecology*, **11**, 83–94.
3.7.3

Hawkins, B.A. (1992) Parasite–host food webs and donor control. *Oikos*, **65**, 159–162.
22.4.2

Hawksworth, D.L. (1994) Biodiversity: measurement and estimation. *Philosophical Transactions of the Royal Society of London, Series B*, **345**, 5–136.
25.1.1

Headley, T.J. (1975) The economics of pest management. In: *Introduction to Insect Pest Management* (R.L. Metcalf & W.L. Luckmann, eds), pp. 75–89. Wiley, New York.
16.3

Heads, P.A. & Lawton, J.H. (1983) Studies on the natural enemy complex of the holly leaf-miner: the effects of scale on the detection of aggregative responses and the implications for biological control. *Oikos*, **40**, 267–276.
10.5.6

Heal, O.W. & MacLean, S.F. (1975) Comparative productivity in ecosystems—secondary productivity. In: *Unifying Concepts in Ecology* (W.H. van Dobben & R.H. Lowe–McConnell, eds), pp. 89–108. Junk, The Hague.
18.4.1, 18.4.2

Heal, O.W., Swift, M.J. & Anderson, J.M. (1982) Nitrogen cycling in United Kingdom forests: the relevance of basic ecological research. *Philosophical Transactions of the Royal Society of London, Series B*, **296**, 427–444.
19.2.4

Heal, O.W., Menaut, J.C. & Steffen, W.L. (eds) (1993) *Towards a Global Terrestrial Observing System (GTOS): Detecting and Monitoring Change in Terrestrial Ecosystems*. MAB Digest **14** and IGBP Global Change Report **26**, UNESCO, Paris and IGBP, Stockholm.
1.4.1

Heed, W.B. (1968) Ecology of Hawaiian Drosophiladae. *University of Texas Publications*, **6861**, 387–419.
1.2.3

Heie, O.E. (1987) Palaeontology and phylogeny. In: *Aphids:*

Their Biology, Natural Enemies and Control. World Crop Pests, Vol. 2A (A.K. Minks & P. Harrewijn, eds), pp. 367–391. Elsevier, Amsterdam.
13.6

Heinrich, B. (1977) Why have some animals evolved to regulate a high body temperature? *American Naturalist*, **111**, 623–640.
2.4

Heliövaara, K. & Väisänen, R. (1986) Industrial pollution and the pine bark bug, *Aradus cinnamomeus* (Het. Aradidae). *Journal of Applied Entomology*, **101**, 469–478.
8.2.7

Hendrickson, J.A. Jr. (1981) Community-wide character displacement reexamined. *Evolution*, **35**, 794–810.
20.4.2

Hendrix, S.D. (1979) Compensatory reduction in a biennial herb following insect defloration. *Oecologia*, **42**, 107–118.
8.2.1

Hengeveld, R. (1990) *Dynamic Biogeography*. Cambridge University Press, Cambridge.
2.2

Henkelem, W.F. van (1973) Growth and lifespan of *Octopus cyanea* (Mollusca: Cephalopoda). *Journal of Zoology*, **169**, 299–315.
4.8

Hess, G.R. (1994) Conservation corridors and contagious disease: a cautionary note. *Conservation Biology*, **8**, 256–262.
25.6.1

Hilborn, R. & Walters, C.J. (1992) *Quantitative Fisheries Stock Assessment*. Chapman & Hall, New York.
4.3, 16.12, 16.13.1, 16.15, 16.15.1, 16.15.2, 16.16

Hildebrand, M. (1974) *Analysis of Vertebrate Structure*. Wiley, New York.
1.3

Hilder, V.A., Gatehouse, A.M.R., Sheerman, S.E., Barker, R.F. & Boulter, D. (1987) A novel mechanism of insect resistance engineered into tobacco. *Nature*, **330**, 160–163.
16.10

Hildrew, A.G. (1985) A quantitative study of the life history of a fairy shrimp (Branchiopoda: Anostraca) in relation to the temporary nature of its habitat, a Kenyan rainpool. *Journal of Animal Ecology*, **54**, 99–110.
4.5.2

Hildrew, A.G. & Townsend, C.R. (1980) Aggregation, interference and the foraging by larvae of *Plectrocnemia conspersa* (Trichoptera: Polycentropodidae). *Animal Behaviour*, **28**, 553–560.
9.8

Hildrew, A.G., Townsend, C.R., Francis, J. & Finch, K. (1984) Cellulolytic decomposition in streams of contrasting pH and its relationship with invertebrate community structure. *Freshwater Biology*, **14**, 323–328.
2.6

Hildrew, A.G., Townsend, C.R. & Hasham, A. (1985) The predatory Chironomidae of an iron-rich stream: feeding ecology and food web structure. *Ecological Entomology*, **10**, 403–413.
22.5.3

Hill, D.S. (1987) *Agricultural Insect Pests of Temperate Regions and their Control*. Cambridge University Press, Cambridge.
16.2

Hill, R.B. (1926) The estimation of the number of hookworms

harboured by the use of the dilution egg count method. *American Journal of Hygiene*, **6** (Suppl.), 19–41.
12.3.4

Hintikka, V. (1963) Ueber das Grossklima einiger Pflanzenareale, durch zwei Klimakoordinatensystemen dargestellt. *Annales Botanici Societatis Zoologicae Botanicae Fennicae 'Vanamo'*, **34**, 1–65.
2.2

HMSO (1991) *The Potential Effects of Climate Change in the United Kingdom*. First Report of the United Kingdom Climate Change Impacts Review Group. Her Majesty's Stationery Office, London.
2.11.2

Hochberg, M.E. & Lawton, J.H. (1990) Spatial heterogeneities in parasitism and population dynamics. *Oikos*, **59**, 9–14.
10.5.3

Hochberg, M.E., Thomas, J.A. & Elmes, G.W. (1992) A modelling study of the population dynamics of a large blue butterfly, *Maculinea rebeli*, a parasite of red ant nests. *Journal of Animal Ecology*, **61**, 397–409.
13.3.2

Hockland, S.H., Dawson, G.W., Griffiths, D.C., Maples, B., Pickett, J.A. & Woodcock, C.M. (1986) The use of aphid alarm pheromone (*E*-β-farnesene) to increase effectiveness of the entomophilic fungus *Verticillium lecanii* in controlling aphids on chrysanthemums under glass. In: *Fundamental and Applied Aspects of Invertebrate Pathology* (R.A. Sampson, J.M. Vlak & D. Peters, eds), p. 252. Foundation of the Fourth International Colloquium of Invertebrate Pathology, Wageningen.
16.5.1

Hodgkinson, K.C. (1991) Shrub recruitment response to intensity and season of fire in a semi-arid woodland. *Journal of Applied Ecology*, **28**, 60–70.
2.4

Hodgkinson, K.C. (1992) Water relations and growth of shrubs before and after fire in a semi-arid woodland. *Oecologia*, **90**, 467–473.
2.4

Hodgkinson, K.C. & Griffin, G.F. (1982) Adaptation of shrub species to fire in the arid zone. In: *Evolution of the Flora and Fauna of Arid Australia* (W.R. Barker & P.J.H. Greenslade, eds), pp. 145–152. Peacock Publications, Frewville, South Australia.
2.4

Hogan, M.E., Veivers, P.C., Slaytor, M. & Czolij, R.T. (1988) The site of cellulose breakdown in higher termites (*Nasutitermes walkeri* and *Nasutitermes exitosus*). *Journal of Insect Physiology*, **34**, 891–899.
13.5.2

Holling, C.S. (1959) Some characteristics of simple types of predation and parasitism. *Canadian Entomologist*, **91**, 385–398.
9.5, 9.5.1, 9.5.3

Holloway, J.D. (1977) *The Lepidoptera of Norfolk Island, their Biogeography and Ecology*. Junk, The Hague.
23.4

Holmes, E.E., Lewis, M.A., Banks J.E. & Veit, R.R. (1994) Partial differential equations in ecology: spatial interactions and population dynamics. *Ecology*, **75**, 17–29.
3.10

Holt, R.D. (1977) Predation, apparent competition and the structure of prey communities. *Theoretical Population Biology*, **12**, 197–229.
7.6, 24.2

Holt, R.D. (1984) Spatial heterogeneity, indirect interactions, and the coexistence of prey species. *American Naturalist*, **124**, 377–406.
7.6, 20.2.2, 24.2

Holt, R.D. & Hassell, M.P. (1993) Environmental heterogeneity and the stability of host–parasitoid interactions. *Journal of Animal Ecology*, **62**, 89–100.
10.5.5

Hopf, F.A., Valone, T.J & Brown, J.H. (1993) Competition theory and the structure of ecological communities. *Evolutionary Ecology*, **7**, 142–154.
20.5

Horn, D.S. (1988) *Ecological Approach to Pest Management*. Elsevier, London.
16.5, 16.6.1, 16.10

Horn, H.S. (1975) Markovian processes of forest succession. In: *Ecology and Evolution of Communities* (M.L. Cody & J.M. Diamond, eds), pp. 196–213. Belknap, Cambridge, MA.
17.4.3, 17.4.4, 21.5.1

Horn, H.S. (1981) Succession. In: *Theoretical Ecology: Principles and Applications* (R.M. May, ed.), pp. 253–271. Blackwell Scientific Publications, Oxford.
17.4.4

Horn, H.S. & May, R.M. (1977) Limits to similarity among coexisting competitors. *Nature*, **270**, 660–661.
20.4.2

Horton, K. (1964) Deer prefer jack pine. *Journal of Forestry*, **62**, 497–499.
9.2.1

Howarth, S.E. & Williams, J.T. (1972) *Chyrsanthemum segetum*. *Journal of Ecology*, **60**, 473–584.
14.2

Howe, H.F. & Westley, L.C. (1986) Ecology of pollination and seed dispersal. In: *Plant Ecology* (M.J. Crawley, ed.), pp. 185–215. Blackwell Scientific Publications, Oxford.
13.4.2

Howe, W.L. & Smith, O.F. (1957) Resistance to the spotted alfalfa aphid in Lahontan alfalfa. *Journal of Economic Entomology*, **50**, 320–324.
16.10

Hubbard, S.F. & Cook, R.M. (1978) Optimal foraging by parasitoid wasps. *Journal of Animal Ecology*, **47**, 593–604.
9.9.2

Hubbell, S.P. & Foster, R.B. (1986) Canopy gaps and the dynamics of a neotropical forest. In: *Plant Ecology* (M.J. Crawley, ed.), pp. 77–96. Blackwell Scientific Publications, Oxford.
21.5.2

Hudson, P.J., Dobson, A.P. & Newborn, D. (1992a) Do parasites make prey vulnerable to predation? Red grouse and parasites. *Journal of Animal Ecology*, **61**, 681–692.
12.4

Hudson, P.J., Newborn, D. & Dobson, A.P. (1992b) Regulation and stability of a free-living host–parasite system: *Trichostrongylus tenuis* in red grouse. I. Monitoring and parasite reduction experiments. *Journal of Animal Ecology*, **61**, 477–486.
12.6.3, 15.4.1

Huffaker, C.B. (1958) Experimental studies on predation:

dispersion factors and predator–prey oscillations. *Hilgardia*, **27**, 343–383.
9.7.3

Huffaker, C.B. (ed.) (1971) *Biological Control*. Plenum Press, New York.
16.8

Huffaker, C.B. & Messenger, P.S. (1976) *Theory and Practice of Biological Control*. Academic Press, New York.
16.8

Huffaker, C.B., Shea, K.P. & Herman, S.G. (1963) Experimental studies on predation. *Hilgardia*, **34**, 305–330.
9.7.3

Hughes, P.R. (1983) Increased success of the Mexican bean beetle on field-grown soybeans exposed to sulphur dioxide. *Journal of Environmental Quality*, **12**, 565–568.
8.2.7

Hughes, R.G. & Thomas, M.L. (1971) The classification and ordination of shallow-water benthic samples from Prince Edward Island, Canada. *Journal of Experimental Marine Biology and Ecology*, **7**, 1–39.
17.3.1

Hughes, R.N. (1989) *A Functional Biology of Clonal Animals*. Chapman and Hall, London.
4.2.1

Hughes, R.N. & Croy, M.I. (1993) An experimental analysis of frequency-dependent predation (switching) in the 15-spined stickleback, *Spinachia spinachia*. *Journal of Animal Ecology*, **62**, 341–352.
9.2.3

Hughes, R.N. & Gliddon, C.J. (1991) Marine plants and their herbivores: coevolutionary myth and precarious mutualisms. In: *The Evolutionary Interaction of Animals and Plants* (W.G. Chaloner, J.L. Harper & J.H. Lawton, eds), pp. 55–63. The Royal Society, London. Also in *Philosophical Transactions of the Royal Society of London, Series B*, **333**, 231–239.
3.7.3

Hughes, R.N. & Griffiths, C.L. (1988) Self-thinning in barnacles and mussels: the geometry of packing. *American Naturalist*, **132**, 484–491.
6.13

Hughes, T.P. & Connell, J.H. (1987) Population dynamics based on size or age? A reef-coral analysis. *American Naturalist*, **129**, 818–829.
4.2.1

Hughes, T.P., Ayre, D. & Connell, J.H. (1992) The evolutionary ecology of corals. *Trends in Ecology and Evolution*, **7**, 292–295.
4.2.1

Humphreys, W.F. (1979) Production and respiration in animal populations. *Journal of Animal Ecology*, **48**, 427–454.
18.4.1

Hungate, R.E. (1975) The rumen microbial ecosystem. *Annual Review of Ecology and Systematics*, **6**, 39–66.
13.5.1

Hungate, R.E., Reichl, J. & Prins, R.A. (1971) Parameters of rumen fermentation in a continuously fed sheep: evidence of a microbial rumination pool. *Applied Microbiology*, **22**, 1104–1113.
13.5.1

Hunter, J.R., Argue, A.W., Bayliff, W.H. *et al.* (1986) The *Dynamics of Tuna Movement: An Evaluation of Past and Future Research*. FAO Fisheries Technical Paper No. 277.
16.16

Hunter, M.D. & Price, P.W. (1992) Playing chutes and ladders: heterogeneity and the relative roles of bottom-up and top-down forces in natural communities. *Ecology*, **73**, 724–732.
22.2.2, 22.3.3

Hunter, M.L. & Yonzon, P. (1992) Altitudinal distributions of birds, mammals, people, forests, and parks in Nepal. *Conservation Biology*, **7**, 420–423.
24.4.2

Huntley, B. (1993) Species-richness in north-temperate zone forests. *Journal of Biogeography*, **20**, 163–180.
24.3.5

Huntly, N. (1991) Herbivores and the dynamics of communities and ecosystems. *Annual Review of Ecology and Systematics*, **22**, 477–503.
21.2.5

Hurd, L.E. & Eisenberg, R.M. (1990) Experimentally synchronized phenology and interspecific competition in mantids. *American Midland Naturalist*, **124**, 390–394.
20.3.1

Huston, M. (1979) A general hypothesis of species diversity. *American Naturalist*, **113**, 81–101.
21.3, 24.3.1

Hutchings, M.J. & de Kroon, H. (1994) Foraging in plants: the role of morphological plasticity in resource acquisition. *Advances in Ecological Research*, **25**, 159–238.
4.2.1, 9.9.3

Hutchings, M.J. (1983) Ecology's law in search of a theory. *New Scientist*, **98**, 765–767.
6.13

Hutchinson, G.E. (1941) Ecological aspects of succession in natural populations. *American Naturalist*, **75**, 406–418.
21.3

Hutchinson, G.E. (1957) Concluding remarks. *Cold Spring Harbour Symposium on Quantitative Biology*, **22**, 415–427.
2.12

Hutchinson, G.E. (1959) Homage to Santa Rosalia, or why are there so many kinds of animals? *American Naturalist*, **93**, 145–159.
1.1.3, 20.4.2

Hutchinson, G.E. (1961) The paradox of the plankton. *American Naturalist*, **95**, 137–145.
7.5.3, 21.3

Hutchinson, G.E. (1965) *The Ecological Theatre and the Evolutionary Play*. Yale University Press, New Haven.
1.1.3

Huxley, C.R. & Cutler, D.F. (1991) *Ant–Plant Interactions*. Oxford University Press, Oxford.
13.2.1

IGBP (International Geosphere–Biosphere Programme) (1990) *Global Change. Report No 12. The Initial Core Projects. The International Geosphere–Biosphere Programme: A Study of Global Change of the International Council of Scientific Unions (ICSU)*. I.G.B.P., Stockholm, Sweden.
1.2.2, 2.11.1

Iles, T.D. (1973) Interaction of environment and parent stock size in determining recruitment in the Pacific sardine as revealed by analysis of density-dependent 0-group growth.

Rapports et Procés-verbaux, Conseil International pour l'Exploration de la Mer, **164**, 228–240.
16.12.1

Iles, T.D. (1981) A comparison of two stock/recruitment hypotheses as applied to North Sea herring and Gulf of St Lawrence cod. *International Council for the Exploitation of the Sea*, CM.1981/H:48.
16.13.3

Ims, R.A. (1990) On the adaptive value of reproductive synchrony as a predator swamping strategy. *American Naturalist*, **136**, 485–498.
5.2.1

Inghe, O. (1989) Genet and ramet survivorship under different mortality regimes: a cellular automata model. *Journal of Theoretical Biology*, **138**, 257–270.
7.5.1

Inglesfield, C. & Begon, M. (1981) Open ground individuals and population structure in *Drosophila subobscura* Collin. *Biological Journal of the Linnean Society*, **15**, 259–278.
2.2.2

Inglesfield, C. & Begon, M. (1983) The ontogeny and cost of migration of *Drosophila subobscura* Collin. *Biological Journal of the Linnean Society*, **19**, 9–15.
14.4

Inouye, D.W. (1978) Resource partitioning in bumblebees: experimental studies of foraging behaviour. *Ecology*, **59**, 672–678.
20.3.1

Inouye, D.W. & Taylor, O.R. (1979) A temperate region plant–ant–seed predator system: consequences of extra-floral nectary secretion by *Helianthella quinquinervis*. *Ecology*, **60**, 1–7.
13.2.1

Inouye, D.W., Huntly, N.J., Tilman, D., Tester, J.R., Stillwell, M.A. & Zinnel, K.C. (1987) Old field succession on a Minnesota sand plain. *Ecology*, **68**, 12–26.
17.4.3

Inouye, D.W., Gill, D.E., Dudash, M.R. & Fenster, C.B. (1994) A model and lexicon for pollen fate. *American Journal of Botany*, **81**, 1517–1530.
13.4.2

International Organisation for Biological Control (1989) *Current Status of Integrated Farming Systems Research in Western Europe* (P. Vereijken & D.J. Royle, eds). IOBC WPRS Bulletin 12(5).
16.11

IUCN (1989) *Florida Panther* (Felis concolor coryi): *Viability Analysis and Species Survival Plan*. Captive Breeding Specialist Group, Species Survival Commission, International Union for the Conservation of Nature.
25.4.4

IUCN/UNEP/WWF (1991) *Caring for the Earth. A Strategy for Sustainable Living*. Gland, Switzerland.
16.17

Ives, A.R. (1992a) Density-dependent and density-independent parasitoid aggregation in model host–parasitoid systems. *American Naturalist*, **140**, 912–937.
10.5.3

Ives, A.R. (1992b) Continuous-time models of host–parasitoid interactions. *American Naturalist*, **140**, 1–29.
10.5.4, 10.5.5

Ives, A.R. & May, R.M. (1985) Competition within and between species in a patchy environment: relations between microscopic and macroscopic models. *Journal of Theoretical Biology*, **115**, 65–92.
7.5.5

Ives, J.D. (1974) Biological refugia and the nunatak hypothesis. In: *Arctic and Alpine Environments* (J.D. Ives & R.D. Berry, eds). Methuen, London.
1.2.2

Iwasa, Y., Kazunori, S. & Nakashima, S. (1991) Dynamic modelling of wave regeneration (Shimagare) in subalpine *Abies* forests. *Journal of Theoretical Biology*, **152**, 143–158.
3.10

Jackson, J.B.C. (1979) Overgrowth competition between encrusting cheilostome ectoprocts in a Jamaican cryptic reef environment. *Journal of Animal Ecology*, **48**, 805–823.
7.3

Jackson, J.B.C., Buss, L.W. & Cooke, R.E. (eds) (1985) *Population Biology and Evolution of Clonal Organisms*. Yale University Press, New Haven.
4.2.1

Jacob, F. (1977) Evolution and tinkering. *Science*, **196**, 1161–1166.
1.1.4, 1.1.5

Janis, C.M. (1993). Tertiary mammal evolution in the context of changing climates, vegetation and tectonic events. *Annual Review of Ecology and Systematics*, **24**, 467–500.
1.2.1

Jannasch, H.W. & Mottl, M.J. (1985) Geomicrobiology of deep-sea hydrothermal vents. *Science*, **229**, 717–725.
2.4

Janzen, D.H. (1967) Interaction of the bull's-horn acacia (*Acacia cornigera* L.) with an ant inhabitant (*Pseudomyrmex ferruginea* F. Smith) in eastern Mexico. *University of Kansas Science Bulletin*, **47**, 315–558.
13.2.1

Janzen, D.H. (1968) Host plants in evolutionary and contemporary time. *American Naturalist*, **102**, 592–595.
12.3, 23.2.4

Janzen, D.H. (1970) Herbivores and the number of tree species in tropical forests. *American Naturalist*, **104**, 501–528.
24.4.1

Janzen, D.H. (1973) Host plants as islands: II. Competition in evolutionary and contemporary time. *American Naturalist*, **107**, 786–790.
12.3

Janzen, D.H. (1976) Why bamboos take so long to flower. *Annual Review of Ecology and Systematics*, **7**, 347–391.
5.2.1

Janzen, D.H. (1979) How to be a fig. *Annual Review of Ecology and Systematics*, **10**, 13–51.
13.4.2

Janzen, D.H. (1980) Specificity of seed-eating beetles in a Costa Rican deciduous forest. *Journal of Ecology*, **68**, 929–952.
9.2

Janzen, D.H. (1981) Evolutionary physiology of personal defence. In: *Physiological Ecology: An Evolutionary Approach to Resource Use* (C.R. Townsend & P. Calow, eds), pp. 145–164. Blackwell Scientific Publications, Oxford.
3.7.3, 11.3.1

Janzen, D.H., Juster, H.B. & Bell, E.A. (1977) Toxicity of secondary compounds to the seed-eating larvae of the bruchid beetle

Callosobruchus maculatus. Phytochemistry, **16**, 223–227.
3.8.2

Jarvis, P.G. & McNaughton, K.G. (1986) Stomatal control of transpiration: scaling up from leaf to region. *Advances in Ecological Research*, **15**, 1–49.
3.4

Jeffries, M.J. & Lawton, J.H. (1984) Enemy-free space and the structure of ecological communities. *Biological Journal of the Linnean Society*, **23**, 269–286.
7.6, 20.2.2, 24.2

Jeffries, M.J. & Lawton, J.H. (1985) Predator–prey ratios in communities of freshwater invertebrates: the role of enemy free space. *Freshwater Biology*, **15**, 105–112.
7.6, 24.2

Jenkins, B., Kitching, R.L. & Pimm, S.L. (1992) Productivity, disturbance and food web structure at a local spatial scale in experimental container habitats. *Oikos*, **65**, 249–255.
22.5.2

Jenkins, D., Watson, A. & Miller, C.R. (1967) Population fluctuations in the red grouse *Lagopus lagopus scoticus. Journal of Applied Ecology*, **36**, 97–122.
10.3.2

Joern, A. & Lawlor, L.R. (1980) Food and microhabitat utilization by the grasshoppers from arid grasslands: comparisons with neutral models. *Ecology*, **61**, 591–599.
20.4.1

Johnson, C.G. (1967) International dispersal of insects and insect-borne viruses. *Netherlands Journal of Plant Pathology*, **73** (Suppl. 1), 21–43.
5.3.3

Johnson, C.G. (1969) *Migration and Dispersal of Insects by Flight.* Methuen, London.
5.3.3

Johnson, M.L. & Gaines, M.S. (1990) Evolution of dispersal; theoretical models and empirical tests using birds and mammals. *Annual Review of Ecology and Systematics*, **21**, 449–480.
5.4.3, 15.4.4, 15.4.5

Johnson, N.C., Zak, D.R., Tilman, D. & Pfleger, F.L. (1991) Dynamics of vesicular–arbuscular mycorrhizae during old field succession. *Oecologia*, **86**, 349–358.
17.4.4

Johnson, N.C., Tilman, D. & Wedin, D. (1992) Plant and soil controls on mycorrhizal fungal communities. *Ecology*, **73**, 2034–2042.
13.7.2

Johnston, D.W. & Odum, E.P. (1956) Breeding bird populations in relation to plant succession on the piedmont of Georgia. *Ecology*, **37**, 50–62.
17.4.4, 24.4.4

Johnston, I.A. (1990) Cold adaptation in marine organisms. *Philosophical Transactions of the Royal Society, Series B*, **326**, 665–667; also in *Life at Low Temperatures* (R.M. Laws & F. Franks, eds), pp. 139–151. The Royal Society, London.
2.3.2

Johnston, R.J., Gregory, D. & Smith, D.M. (1994) *The Dictionary of Human Geography.* Basil Blackwell Ltd., Oxford.
4.9

Jones, C.G. & Firn, R.D. (1991) On the evolution of plant secondary chemical diversity. In: *The Evolutionary Interaction of Animals and Plants* (W.G. Chaloner, J.L. Harper & J.H.

Lawton, eds), pp. 97–104. The Royal Society, London. Also in *Philosophical Transactions of the Royal Society, Series B*, **333**, 273–280.
3.7.3

Jones, M. & Harper, J.L (1987) The influence of neighbours on the growth of trees. I. The demography of buds in *Betula pendula. Proceedings of the Royal Society of London, Series B*, **232**, 1–18.
6.3

Jones, M.B. (1978) Aspects of the biology of the big-handed crab, *Heterozius rohendifrons* (Decapoda: Brachyura), from Kaikoura, New Zealand. *New Zealand Journal of Zoology*, **5**, 783–794.
14.2

Jones, M.G. (1933) Grassland management and its influence on the sward. *Empire Journal of Experimental Agriculture*, **1**, 43–367.
21.2.1

Jordan, C.F., Kline, J.R. & Sasscer, D.S. (1972) Effective stability of mineral cycles in forest ecosystems. *American Naturalist*, **106**, 237–253.
22.6

Jordan, W.R. III, Gilpin, M.E. & Aber, J.D. (1987) *Restoration Ecology.* Cambridge University Press, Cambridge.
25.6.4

Jordano, P. (1995) Angiosperm fleshy fruits and seed dispersers: a comparative analysis of adaptation and constraints in plant–animal interactions. *American Naturalist*, **145**, 163–191.
13.4.1

Juhren, M., Went, F.W. & Phillips, E. (1956) Ecology of desert plants. IV. Combined field and laboratory work on germination of annuals in the Joshua Tree National Monument, California. *Ecology*, **37**, 318–330.
4.5.3

Julien, M.H. (ed.) (1982) *Biological Control of Weeds. A World Catalogue of Agents and their Target Weeds.* Commonwealth Agricultural Bureaux, Farnham Royal.
16.8

Juniper, S.K., Tunnicliffe, V. & Southward, E.C. (1992) Hydrothermal vents in turbidite sediments on a Northeast Pacific spreading centre: organisms and substratum at an ocean drilling site. *Canadian Journal of Zoology*, **70**, 1792–1809.
2.4

Junk, W.J., Bayley, P.B. & Sparks, R.E. (1989) The flood–pulse concept in river-floodplain systems. *Canadian Special Publications in Fisheries and Aquatic Sciences*, **106**, 110–127.
18.2.1

Kalela, O.(1962) On the fluctuations in the numbers of arctic and boreal small rodents as a problem of production biology. *Annales Academiae Scientiarum Fennicae (A IV)*, **66**, 1–38.
15.4.3

Kaplan, R.H. & Salthe, S.N. (1979) The allometry of reproduction: an empirical view in salamanders. *American Naturalist*, **113**, 671–689.
14.12.1

Karban, R. & Ricklefs, R.E. (1983) Host characteristics, sampling intensity, and species richness of Lepidopteran larvae on broad-leaved trees in southern Ontario. *Ecology*, **64**, 636–641.
23.3.3

Kareiva, P. (1990) Population dynamics in spatially complex environments: theory and data. *Philosophical Transactions of the Royal Society of London, Series B*, **330**, 175–190; and In: *Population, Regulation and Dynamics* (M. Hassell & R.M. May, eds), pp. 53–68. The Royal Society, London.
15.6.2

Kareiva, P. (1994) Space: the final frontier for ecological theory. *Ecology*, **75**, 1.
3.10

Kareiva, P. & Odell, G. (1987) Swarms of predators exhibit preytaxis if individual predators use area restricted search. *American Naturalist*, **130**, 233–270.
15.6.2

Karplus, I., Tsurnamal, M. & Szlep, M. (1972) Associative behaviour of the fish *Cryptocentrus cryptocentrus* (Gogiidae) and the pistol shrimp *Alpheus djiboutensis* (Alpheidae) in artificial burrows. *Marine Biology*, **15**, 95–104.
13.2.4

Karr, J.R. (1971) Structure of avian communities in selected Panama and Illinois habitats. *Ecological Monographs*, **41**, 207–229.
24.4.1

Kaufman, L. (1992) Catastrophic change in a species-rich freshwater ecosystem: Lessons from Lake Victoria. *Bioscience*, **42**, 846–858.
25.2.4

Kawata, M. (1987) The effect of kinship on spacing among female red-backed voles, *Clethrionomys rufocanus bedfordiae*. *Oecologia*, **72**, 115–122.
15.4.4

Kays, S. & Harper, J.L. (1974) The regulation of plant and tiller density in a grass sward. *Journal of Ecology*, **62**, 97–105.
4.2.1, 6.5

Kearns, C.A. & Inouye, D.W. (1993) *Techniques for Pollination Biologists*. University Press of Colorado, Niwot, Colorado.
13.4.2

Keddy, P.A. (1982) Experimental demography of the sand-dune annual, *Cakile edentula*, growing along an environmental gradient in Nova Scotia. *Journal of Ecology*, **69**, 615–630.
15.5

Keddy, P.A. & Shipley, B. (1989) Competitive heirarchies in herbaceous plant communities. *Oikos*, **54**, 234–241.
7.3

Keeling, C.D., Whorf, T.P., Wahlen, M. & van der Plicht, J. (1995) Interannual extremes in the rate of rise of atmospheric carbon dioxide since 1980. *Nature*, **375**, 666–670.
19.4.6

Keith, L.B. (1983) Role of food in hare population cycles. *Oikos*, **40**, 385–395.
10.2.4, 10.5.6

Keith, L.B., Cary, J.R., Ronstad, O.J. & Brittingham, M.C. (1984) Demography and ecology of a declining snowshoe hare population. *Wildlife Monographs*, **90**, 1–43.
10.2.4

Kendrick, W.B. & Burges, A. (1962) Biological aspects of the decay of *Pinus sylvestris* leaf litter. *Nova Hedwiga*, **4**, 313–342.
17.4.1

Kerbes, R.H., Kotanen, P.M. & Jefferies, R.L. (1990) Destruction of wetland habitats by lesser snow geese: a keystone species on the west coast of Hudson Bay. *Journal of Applied Ecology*, **27**, 242–258.
22.2.2

Kershaw, K.A. (1973) *Quantitative and Dynamic Plant Ecology*, 2nd edn. Edward Arnold, London.
4.3

Kigel, J. (1980) Analysis of regrowth patterns and carbohydrate levels in *Lolium multiforum* Lam. *Annals of Botany*, **45**, 91–101.
8.2.1

Kingsland, S.E. (1985) *Modeling Nature*. University of Chicago Press, Chicago.
7.9

Kingston, T.J. (1977) *Natural Manuring by Elephants in the Tsavo National Park, Kenya*. PhD thesis, University of Oxford.
11.3.2

Kingston, T.J. & Coe, M.J. (1977) The biology of a giant dung-beetle (*Heliocorpis dilloni*) (Coleoptera: Scarabaeidae). *Journal of Zoology*, **181**, 243–263.
11.3.2

Kinnaird, M.F. & O'Brien, T.G. (1991) Viable populations for an endangered forest primate, the Tana River crested mangabey (*Cerocebus galeritus galeritus*). *Conservation Biology*, **5**, 203–213.
25.4.3

Kira, T., Ogawa, H. & Shinozaki, K. (1953) Intraspecific competition among higher plants. I. Competition–density–yield inter-relationships in regularly dispersed populations. *Journal of the Polytechnic Institute, Osaka City University*, **4**(4), 1–16.
6.5

Kitting, C.L. (1980) Herbivore–plant interactions of individual limpets maintaining a mixed diet of intertidal marine algae. *Ecological Monographs*, **50**, 527–550.
9.2.2

Klemow, K.M. & Raynal, D.J. (1981) Population ecology of *Melilotusalba* in a limestone quarry. *Journal of Ecology*, **69**, 33–44.
4.8

Klepac, D. (1955) Effect of *Viscum album* on the increment of silver fir stands. *Sumarski List*, **79**, 231–244.
12.2.2

Kline, J.R. (1970) Retention of fallout radionuclides by tropical forest vegetation. In: *A Tropical Rain Forest* (H.T. Odum, ed.). Division of Technical Information, US Atomic Energy Commission, Washington, DC.
22.6

Kneidel, K.A. (1984) Competition and disturbance in communities of carrion breeding Diptera. *Journal of Animal Ecology*, **53**, 849–865.
11.3.3

Knoch, T.R., Faeth, S.H. & Arnott, D.L. (1993) Endophytic fungi alter foraging and dispersal by desert seed-harvesting ants. *Oecologia*, **95**, 470–473.
13.7

Knowlton, N. & Jackson, J.B.C. (1994) New taxonomy and niche partitioning on coral reefs: jack of all trades or master of some? *Trends in Ecology and Evolution*, **9**, 7–9.
21.7.2

Koblentz-Mishke, I.J., Volovinsky, V.V. & Kabanova, J.B. (1970) Plankton primary production of the world ocean. In: *Scientific Exploration of the South Pacific* (W.S. Wooster, ed.).

National Academy of Sciences, Washington, D.C.
18.2

Kodric-Brown, A. & Brown, J.H. (1993) Highly structured fish communities in Australian desert springs. *Ecology*, **74**, 1847–1855.
23.1, 23.3.1

Kok, L.T. & Surles, W.W. (1975) Successful biocontrol of musk thistle by an introduced weevil, *Rhinocyllus conicus*. *Environmental Entomology*, **4**, 1025–1027.
8.3

Kolding, J. (1993) Trophic interrelationships and community structure at two different periods of Lake Turkana, Kenya: a comparison using the ECOPATH II box model. In: *Trophic Models of Aquatic Ecosystems* (V. Christensen & D. Pauly, eds), pp. 116–123. International Center for Living Aquatic Resources Management, Manila, Phillipines.
18.4.3

Kolding, S. & Fenchel, T.M. (1979) Coexistence and life-cycle characteristics of five species of the amphipod genus *Gammarus*. *Oikos*, **33**, 323–327.
20.3.1

Koller, D. & Roth, N. (1964) Studies on the ecological and physiological significance of amphicarpy in *Gymnarhena micrantha* (Compositae). *American Journal of Botany*, **51**, 26–35.
5.4.2

Koricheva, J. & Haukioja, E. (1995) Variations in chemical composition of birch foliage under air pollution stress and their consequences for *Eriocrania* miners. *Environmental Pollution*, **88**, 41–50.
8.2.7

Kornberg, H. & Williamson, M.H. (eds) (1986) Quantitative aspects of the ecology of biological invasions. *Philosophical Transactions of the Royal Society of London, Series B*, **314**, 501–742.
16.2

Kozlowski, J. (1993) Measuring fitness in life-history studies. *Trends in Ecology and Evolution*, **7**, 155–174.
14.3

Krebs, C.J. (1972) *Ecology*. Harper & Row, New York.
Introduction

Krebs, C.J. (1985) Do changes in spacing behaviour drive population cycles in small mammals? In: *Behavioural Ecology* (R.M. Sibly & R.H. Smith, eds), pp. 295–312. Blackwell Scientific Publications, Oxford.
15.4.2, 15.4.4

Krebs, C.J. (1989) *Ecological Methodology*. Harper & Row, New York.
4.3

Krebs, C.J. & Myers, J.H. (1974) Population cycles in small mammals. *Advances in Ecological Research*, **8**, 267–399.
15.4.2

Krebs, C.J., Keller, B.L. & Tamarin, R.H. (1969) *Microtus* population biology: demographic changes in fluctuating populations of *M. ochrogaster* and *M. pennsylvanicus* in southern Indiana. *Ecology*, **50**, 587–607.
5.3.3, 15.4.4

Krebs, C.J., Boonstra, R., Boutin, S. *et al.* (1992) What drives the snowshoe hare cycle in Canada's Yukon. In: *Wildlife 2001: Populations* (D.R. McCullough & R.H. Barrett, eds), pp. 886–896. Elsevier, New York.
10.2.4

Krebs, J.R. (1971) Territory and breeding density in the great tit, *Parus major* L. *Ecology*, **52**, 2–22.
6.12

Krebs, J.R. (1978) Optimal foraging: decision rules for predators. In: *Behavioural Ecology: An Evolutionary Approach* (J.R. Krebs & N.B. Davies, eds), pp. 23–63. Blackwell Scientific Publications, Oxford.
5.4.3, 9.2.2, 9.3, 9.3.1, 9.9.2, 15.4.4

Krebs, J.R. & Davies, N.B. (1993) *An Introduction to Behavioural Ecology*, 3rd edn. Blackwell Scientific Publications, Oxford.
9.9.3

Krebs, J.R. & Kacelnik, A. (1991) Decision-making. In: *Behavioural Ecology. An Evolutionary Approach*, 3rd edn. (J.R. Krebs & N.B. Davies, eds), pp. 105–136. Blackwell Scientific Publications, Oxford.
9.3, 9.9.3

Krebs, J.R. & McCleery, R.H. (1984) Optimization in behavioural ecology. In: *Behavioural Ecology: An Evolutionary Approach*, 2nd edn. (J.R. Krebs & N.B. Davies, eds), pp. 91–121. Blackwell Scientific Publications, Oxford.
9.3.1, 9.9.3

Krebs, J.R., Erichsen, J.T., Webber, M.I. & Charnov, E.L. (1977) Optimal prey selection in the great tit (*Parus major*). *Animal Behaviour*, **25**, 30–38.
9.3.1

Krebs, J.R., Stephens, D.W. & Sutherland, W.J. (1983) Perspectives in optimal foraging. In: *Perspectives in Ornithology* (A.H. Brush & G.A. Clarke, Jr., eds), pp. 165–216. Cambridge University Press, New York.
9.3, 9.5.1

Kreitman, M., Shorrocks, B. & Dytham, C. (1992) Genes and ecology: two alternative perspectives using *Drosophila*. In: *Genes in Ecology* (R.J. Berry, T.J. Crawford & G.M. Hewitt, eds), pp. 281–312. Blackwell Scientific Publications, Oxford.
7.5.5

Krueger, D.A. & Dodson, S.I. (1981) Embryological induction and predation ecology in *Daphnia pulex*. *Limnology and Oceanography*, **26**, 219–223.
3.7.3

Kullenberg, B. (1946) Über verbreitung und Wanderungen von vier *Sterna*–Arten. *Arkiv für Zoologie: utgifvet af K. Svenska vetenskapsakademien*, bd 1, **38**, 1–80.
5.7.3

Kunin, W.E. & Gaston, K.J. (1993) The biology of rarity: patterns, causes and consequences. *Trends in Ecology and Evolution*, **8**, 298–301.
25.2.1

Kuno, E. (1991) Some strange properties of the logistic equation defined with r and K: inherent defects or artifacts? *Researches on Population Ecology*, **33**, 33–39.
6.8.1

Kusnezov, M. (1957) Numbers of species of ants in faunas of different latitudes. *Evolution*, **11**, 298–299.
24.4.1

Labbé, P. (1994) Régénération après passage du cyclone Hugo en forêt dense humide de Guadeloupe. *Acta Ecologica*, **15**, 301–315.
2.9.1

Lack, D. (1947a) *Darwin's Finches*. Cambridge University Press, Cambridge.
1.2.3

Lack, D. (1947b) The significance of clutch size. *Ibis*, **89**, 302–352.
14.8.1

Lack, D. (1954) *The Natural Regulation of Animal Numbers*. Clarendon Press, Oxford.
5.7.3

Lack, D. (1963) Cuckoo hosts in England. (With an appendix on the cuckoo hosts in Japan, by T. Royama.) *Bird Study*, **10**, 185–203.
12.8

Lack, D. (1969a) Subspecies and sympatry in Darwin's finches. *Evolution*, **23**, 252–263.
1.2.3

Lack, D. (1969b) The numbers of bird species on islands. *Bird Study*, **16**, 193–209.
23.2.1

Lack, D. (1971) *Ecological Isolation in Birds*. Blackwell Scientific Publications, Oxford.
7.2.5

Lack, D. (1976) *Island Birds*. Blackwell Scientific Publications, Oxford.
23.2.1

Lambers, H. (1985) Respiration in intact plants and tissues: its regulation and dependence on environmental factors, metabolism and invaded organisms. In: *Encyclopedia of Plant Physiology* (O.L. Lange, P.S. Nobel, C.B. Osmond & H. Zieglerm, eds), New Series, Vol. 18. Springer-Verlag, Berlin.
3.2.5

Lambin, X. & Krebs, C.J. (1991a) Can changes in female relatedness influence microtine population dynamics? *Oikos*, **61**, 126–132.
15.4.4

Lambin, X. & Krebs, C.J. (1991b) Spatial organisation and mating system of the Townsend's vole, *Microtus townsendii*. *Behavioural Ecology and Sociobiology*, **28**, 353–363.
15.4.4

Lambin, X. & Krebs, C.J. (1993) Influence of female relatedness on the demography of Townsend's vole populations in spring. *Journal of Animal Ecology*, **62**, 536–550.
15.4.4

Lambin, X., Krebs, C.J. & Scott, B. (1992) Spacing system of the tundra vole (*Microtus oeconomus*) during the breeding season in Canada's western Arctic. *Canadian Journal of Zoology*, **70**, 2068–2072.
15.4.2

Lanciani, C.A. (1975) Parasite-induced alterations in host reproduction and survival. *Ecology*, **56**, 689–695.
12.4

Lande, R. (1993) Risks of population extinction from demographic and environmental stochasticity, and random catastrophes. *American Naturalist*, **142**, 911–927.
25.4.3

Lande, R. & Barrowclough, G.F. (1987) Effective population size, genetic variation, and their use in population management. In: *Viable Populations for Conservation* (M.E. Soulé, ed.), pp. 87–123. Cambridge University Press, Cambridge.
25.2.5

Lappe, F. (1971) *Diet for a Small Planet*. Ballantine, New York.
3.8.2

Larcher, W. (1980) *Physiological Plant Ecology*, 2nd edn. Springer-Verlag, Berlin.
2.6, 3.1

Large, E.C. (1958) *The Advance of the Fungi*. Jonathan Cape, London.
16.4

Laties, G. (1982) The cyanide resistant alternative path in higher plant respiration. *Annual Review of Plant Physiology*, **33**, 519–555.
3.2.5

Law, R. & Blackford, J.C. (1992) Self-assembling food webs: a global viewpoint of coexistence of species in Lotka–Volterra communities. *Ecology*, **73**, 567–578.
22.5.6

Law, R. & Lewis, D.H. (1983) Biotic environments and the maintenance of sex—some evidence from mutualistic symbiosis. *Biological Journal of the Linnean Society*, **20**, 249–276.
13.13

Law, R. & Watkinson, A.R. (1987) Response-surface analysis of two-species competition: an experiment on *Phleum arenarium* and *Vulpia fasciculata*. *Journal of Ecology*, **75**, 871–886.
7.7.2

Law, R. & Watkinson, A.R. (1989) Competition. In: *Ecological Concepts* (J.M. Cherrett, ed.), pp. 243–284. Blackwell Scientific Publications, Oxford.
20.4.4

Law, R., Bradshaw, A.D. & Putwain, P.D. (1977) Life history variation on *Poa annua*. *Evolution*, **31**, 233–246.
14.7.2

Lawler, S.P. & Morin, P.J. (1993) Temporal overlap, competition, and priority effects in larval anurans. *Ecology*, **74**, 174–182.
7.10

Lawlor, L.R. (1978) A comment on randomly constructed ecosystem models. *American Naturalist*, **112**, 445–447.
22.4.2

Lawlor, L.R. (1980) Structure and stability in natural and randomly constructed competitive communities. *American Naturalist*, **116**, 394–408.
20.4.1

Laws, R.M. (1984) Seals. In: *Antarctic Ecology* (R.M. Laws, ed.). Academic Press, London.
1.4.3

Lawton, J.H. (1984) Non-competitive populations, non-convergent communitites, and vacant niches: the herbivores of bracken. In: *Ecological Communities Conceptual Issues and the Evidence* (D.R. Strong, D. Simberloff, L.G. Abele & A.B. Thistle, eds), pp. 67–100. Princeton University Press, Princeton, NJ.
20.2.1, 23.3.3

Lawton, J.H. (1989) Food webs. In: *Ecological Concepts* (J.M. Cherrett, ed.), pp. 43–78. Blackwell Scientific Publications, Oxford.
22.4.2

Lawton, J.H. (1990) *Red Grouse Populations and Moorland Management*. Field Studies Council, Shrewsbury.
12.6.3

Lawton, J.H. (1993) Range, population abundance and conservation. *Trends in Ecology and Evolution*, **8**, 409–413.
25.2.1

Lawton, J.H. & Hassell, M.P. (1981) Asymmetrical competition in insects. *Nature*, **289**, 793–795.
7.3

Lawton, J.H. & May, R.M. (1984) The birds of Selborne. *Nature*, **306**, 732–733.
15.2

Lawton, J.H. & McNeill, S. (1979) Between the devil and the deep blue sea: on the problems of being a herbivore. In: *Population Dynamics* (R.M. Anderson, B.D. Turner & L.R. Taylor, eds), pp. 223–244. Blackwell Scientific Publications, Oxford.
20.2.1

Lawton, J.H. & Schröder, D. (1977) Effects of plant type, size of geographical range and taxonomic isolation on a number of insect species associated with British plants. *Nature*, **265**, 137–140.
23.3.3, 23.3.5

Lawton, J.H. & Strong, D.R. Jr. (1981) Community patterns and competition in folivorous insects. *American Naturalist*, **118**, 317–338.
20.3.1

Lawton, J.H. & Woodroffe, G.L. (1991) Habitat and the distribution of water voles: why are there gaps in a species' range? *Journal of Animal Ecology*, **60**, 79–91.
15.6.3

Lawton, J.H., Nee, S., Letcher, A.J. & Harvey, P.H. (1994) Animal distributions: patterns and processes. In: *Large-Scale Ecology and Conservation Biology* (P.J. Edwards, R.M. May & N.R. Webb, eds), pp. 41–58. Blackwell Scientific Publications, Oxford.
25.2.1

Lawton, R.O. & Putz, F.E. (1988) Natural disturbance and gap-phase regeneration in a wind-exposed tropical cloud forest. *Ecology*, **69**, 764–777.
21.5.2

Le Cren, E.D. (1973) Some examples of the mechanisms that control the population dynamics of salmonid fish. In: *The Mathematical Theory of the Dynamics of Biological Populations* (M.S. Bartlett & R.W. Hiorns, eds), pp. 125–135. Academic Press, London.
6.2

Leather, S.R. (1986) Insect species richness of the British Rosaceae: the importance of host range, plant architecture, age of establishment, taxonomic isolation and species–area relationships. *Journal of Animal Ecology*, **55**, 841–860.
23.3.3

Lee, A.K. & Cockburn, A. (1985a) *Evolutionary Ecology of Marsupials*. Cambridge University Press, Cambridge, UK.
15.4.4

Lee, A.K. & Cockburn, A. (1985b) Spring declines in small mammal populations. *Acta Zoologica Fennica*, **173**, 75–76.
15.4.4

Lee, H., León, J. & Raskin, I. (1995) Biosynthesis and metabolism of salicylic acid. *Proceedings of the National Academy of Science of the USA*, **92**, 4076–4079.
12.3.5

Lee, K.E. & Wood, T.G. (1971) *Termites and Soils*. Academic Press, London.
11.3.1

Leggett, W.C. (1977) The ecology of fish migrations. *Annual Review of Ecology and Systematics*, **8**, 285–308.
5.7.4

Leigh, E. (1975) Population fluctuations and community structure. In: *Unifying Concepts in Ecology* (W.H. van Dobben & R.H. Lowe–McConnell, eds), pp. 67–88. Junk, The Hague.
22.4.3

Leishman, M.R. & Westoby, M. (1992). Classifying plants into groups on the basis of associations of individual traits—evidence from Australian semi-arid woodlands. *Journal of Ecology*, **80**, 417–424.
1.4.2

Leith, H. (1975) Primary productivity in ecosystems: comparative analysis of global patterns. In: *Unifying Concepts in Ecology* (W.H. van Dobben & R.H. Lowe–McConnell, eds), pp. 67–88. Junk, The Hague.
18.2, 18.3.1

Leneteren, J.C. van & Woets, J. (1988) Biological and integrated pest control in greenhouses. *Annual Review of Entomology*, **33**, 239–269.
16.8

Lessells, C.M. (1985) Parasitoid foraging: should parasitism be density dependent? *Journal of Animal Ecology*, **54**, 27–41.
9.7.1, 10.5.6

Lessells, C.M. (1991) The evolution of life histories. In: *Behavioural Ecology*, 3rd edn. (J.R. Krebs & N.B. Davies, eds), pp. 32–68. Blackwell Scientific Publications, Oxford.
14.1, 14.3, 14.4, 14.4.2, 14.8.1, 14.11

Leverich, W.J. & Levin, D.A. (1979) Age-specific survivorship and reproduction in *Phlox drummondii*. *American Naturalist*, **113**, 881–903.
4.5.1, 14.3

Levey, D.J. (1988) Tropical wet forest treefall gaps and distributions of understorey birds and plants. *Ecology*, **69**, 1076–1089.
21.5.2

Levin, D.A. & Kerster, H.W. (1974) Gene-flow in seed plants. *Evolutionary Biology*, **7**, 179–220.
5.4.4

Levin, S.A. (1983) Coevolution. Some approaches to modelling of coevolutionary interactions. In: *Coevolution* (M. Ntecki, ed.), pp. 21–65. University of Chicago Press, Chicago.
12.7

Levin, S.A. (1992) The problem of pattern and scale in ecology. *Ecology*, **73**, 1943–1967.
5.2

Levins, R. (1968) *Evolution in Changing Environments*. Princeton University Press, Princeton.
14.6

Levins, R. (1969) Some demographic and genetic consequences of environmental heterogeneity for biological control. *Bulletin of the Entomological Society of America*, **15**, 237–240.
15.6.4

Levins, R. (1970) Extinction. In: *Lectures on Mathematical Analysis of Biological Phenomena*, pp. 123–138. Annals of the New York Academy of Sciences, Vol. 231.
15.6.4, 25.3.1

Lewis, D.H. (1974) Microorganisms and plants. The evolution of parasitism and mutualism. *Symposium of the Society for General Microbiology*, **24**, 367–392.
13.7.3

Lewis, J.R. (1976) *The Ecology of Rocky Shores*. Hodder & Stoughton, London.
2.2.2, 2.8

Lewontin, R.C. & Levins, R. (1989) On the characterization of density and resource availability. *American Naturalist*, **134**, 513–524.
5.2, 6.3, 15.1

Lidicker, W.Z. Jr. (1975) The role of dispersal in the demography of small mammal populations. In: *Small Mammals: Their Productivity and Population Dynamics* (K. Petruscwicz, F.B. Golley & L. Ryszkowski, eds), pp. 103–128. Cambridge University Press, New York.
15.4.4

Lidicker, W.Z. Jr. (1988) Solving the enigma of microtine 'cycles'. *Journal of Mammalogy*, **69**, 225–235.
15.4.2

Lidicker, W.Z. Jr. & Patton, J.L. (1987) Patterns of dispersal and genetic structure in populations of small mammals. In: *Mammalian Dispersal Patterns* (B.D. Chepko-Sade & Z.T. Halpin, eds), pp. 141–161. University of Chicago Press, Chicago.
5.4.3, 15.4.5

Likens, G.E. (1987) *Long-Term Studies in Ecology*. Springer-Verlag, New York.
15.1

Likens, G.E. & Bormann, F.G. (1975) An experimental approach to New England landscapes. In: *Coupling of Land and Water Systems* (A.D. Hasler, ed.), pp. 7–30. Springer-Verlag, New York.
19.2.3

Likens, G.E., Bormann, F.H., Pierce, R.S. & Fisher, D.W. (1971) Nutrient–hydrologic cycle interaction in small forested watershed ecosystems. In: *Productivity of Forest Ecosystems* (P. Duvogneaud, ed.). UNESCO, Paris.
19.2.3

Likens, G.E., Bormann, F.H., Pierce, R.S., Eaton, J.S. & Johnson, N.M. (1977) *Biogeochemistry of a Forested Ecosystem*. Springer-Verlag, Berlin.
19.2.2

Limbaugh, C. (1961) Cleaning symbiosis. *Scientific American*, **205**, 42–49.
13.2.2

Lindberg, S.E., Lovett, G.M., Richter, D.D. & Johnson, D.W. (1986) Atmospheric deposition and canopy interactions of major ions in a forest. *Science*, **231**, 141–145.
19.2.1

Lindemann, R.L. (1942) The trophic–dynamic aspect of ecology. *Ecology*, **23**, 399–418.
18.1, 18.4, 18.4.1

Little, C. (1990) *The Terrestrial Invasion: An Ecophysiological Approach to the Origins of Land Animals*. Cambridge University Press, Cambridge.
2.8

Lively, C.M. & Apanius, V. (1995) Genetic diversity in host–parasite interactions. In: *Ecology of Infectious Diseases in Natural Populations* (B.T. Grenfell & A.P. Dobson, eds), pp. 421–449. Cambridge University Press, Cambridge.
12.7

Lockhart, J.A.R., Holmes, J.C. & MacKay, D.B. (1982) The evolution of weed control in British agriculture. In: *Weed Control Handbook: Principles*, 7th edn. (H.A. Roberts, ed.), pp. 37–63. Blackwell Scientific Publications, Oxford.
16.4, 16.6.3

Logan, J.A. & Allen, J.C. (1992) Nonlinear dynamics and chaos in insect populations. *Annual Review of Entomology*, **37**, 455–477.
6.9

Loik, M.E. & Nobel, Park S. (1993). Freezing tolerance and water relations of *Opuntia fragilis* from Canada and the United States. *Ecology*, **74**, 1722–1732.
1.5.1, 2.3.4

Long, J.N. & Smith, F.W. (1984) Relation between size and density in developing stands: a description and possible mechanisms. *Forest Ecology and Management*, **7**, 191–206.
6.13

Lonsdale, W.M. (1990) The self-thinning rule: dead or alive? *Ecology*, **71**, 1373–1388.
6.13

Lonsdale, W.M. & Watkinson, A.R. (1983) Light and self-thinning. *New Phytologist*, **90**, 431–435.
6.13

Losos, E. (1993) The future of the US Endangered Species Act. *Trends in Ecology and Evolution*, **8**, 332–336.
25.6.3, 25.6.5

Losos, J.B. (1992) The evolution of convergent structure in Caribbean *Anolis* communities. *Systematic Biology*, **41**, 403–420.
20.4.2

Lotka, A.J. (1907) Studies on the mode of growth of material aggregates. *American Journal of Science*, **24**, 199–216.
4.7.2

Lotka, A.J. (1932) The growth of mixed populations: two species competing for a common food supply. *Journal of the Washington Academy of Sciences*, **22**, 461–469.
7.4.1, 10.2.1

Louda, S.M. (1982) Distributional ecology: variation in plant recruitment over a gradient in relation to insect seed predation. *Ecological Monographs*, **52**, 25–41.
8.3

Louda, S.M. (1983) Seed predation and seedling mortality in the recruitment of a shrub, *Haplopappus venetus* (Asteraceae), along a climatic gradient. *Ecology*, **64**, 511–521.
8.3

Lovett Doust, J. & Lovett Doust, L. (eds) (1988) *Plant Reproductive Ecology: Patterns and Strategies*. Oxford University Press, New York.
13.4.2

Lovett Doust, J., Schmidt, M. & Lovett Doust, L. (1994) Biological assessment of aquatic pollution: a review, with emphasis on plants as biomonitors. *Biological Reviews*, **69**, 147–186.
2.10

Lovett Doust, L. (1981) Population dynamics and local specialization in a clonal perennial (*Ranunculus repens*). I. The dynamics of ramets in contrasting habitats. *Journal of Ecology*, **69**, 743–755.
4.2.1

Lovett Doust, L. & Lovett Doust, J. (1982) The battle strategies of plants. *New Scientist*, **95**, 81–84.
5.6

Lowe, V.P.W. (1969) Population dynamics of the red deer (*Cervus elaphus* L.) on Rhum. *Journal of Animal Ecology*, **38**, 425–457.
4.6.1

Lubchenco, J. (1978) Plant species diversity in a marine intertidal community: importance of herbivore food preference and algal competitive abilities. *American Naturalist*, **112**, 23–39.
21.2.1

Lubchenko, J., Olson, A.M. *et al.* (1991) The sustainable biosphere initiative: an ecological research agenda. *Ecology*, **72**, 371–412.
16.17

Luckmann, W.H. & Decker, G.C. (1960) A 5-year report on observations in the Japanese beetle control area of Sheldon, Illinois. *Journal of Economic Entomology*, **53**, 821–827.
16.6.1

Ludwig, J.W., Bunting, E.S. & Harper, J.L. (1957) The influence of environment on seed and seedling mortality. III. The influence of aspect on maize germination. *Journal of Ecology*, **45**, 205–224.
2.2

Lundberg, P. (1988) Functional response of a small mammalian herbivore: the disc equation revisited. *Journal of Animal Ecology*, **57**, 999–1006.
9.5.1

Lussenhop, J. (1992) Mechanisms of microarthropod–microbial interactions in soil. *Advances in Ecological Research*, **23**, 1–33.
11.2.3

MacArthur, J.W. (1975) Environmental fluctuations and species diversity. In: *Ecology and Evolution of Communities* (M.L. Cody & J.M. Diamond, eds), pp. 74–80. Belknap, Cambridge, MA.
24.3.3

MacArthur, R.H. (1955) Fluctuations of animal populations and a measure of community stability. *Ecology*, **36**, 533–536.
22.4.1

MacArthur, R.H. (1962) Some generalized theorems of natural selection. *Proceedings of the National Academy of Science of the USA*, **48**, 1893–1897.
14.9

MacArthur, R.H. (1972) *Geographical Ecology*. Harper & Row, New York.
2.2, 20.1, 24.1

MacArthur, R.H. & Levins, R. (1964) Competition, habitat selection and character displacement in a patchy environment. *Proceedings of the National Academy of Sciences*, **51**, 1207–1210.
5.2

MacArthur, R.H. & Levins, R. (1967) The limiting simularity, convergence and divergence of coexisting species. *American Naturalist*, **101**, 377–385.
7.9

MacArthur, R.H. & Pianka, E.R. (1966) On optimal use of a patchy environment. *American Naturalist*, **100**, 603–609.
9.3.1

MacArthur, R.H. & Wilson, E.O. (1967) *The Theory of Island Biogeography*. Princeton University Press, Princeton, NJ.
12.3, 14.9, 15.6.4, 23.1, 23.2.3, 23.3.5, 24.3.1

McBrayer, J.F. (1973) Exploitation of deciduous litter by *Apheloria montana* (Diplopoda: Eurydesmidae). *Pedobiologia*, **13**, 90–98.
11.3.2

McClure, H.E. (1966) Flowering, fruiting and animals in the canopy of a tropical rain forest. *Malaysian Forester*, **29**, 192–203.
13.4.1

McCune, D.C. & Boyce, R.L. (1992) Precipitation and the transfer of water, nutrients and pollutants in tree canopies. *Trends in Ecology and Evolution*, **7**, 4–7.
19.4.1

Macdonald, D.W. (1980) *Rabies and Wildlife*. Oxford University Press, Oxford.
12.6.2

McDonald, I.R., Lee, A.K., Than, K.A., & Martin, R.W. (1988) Concentration of free glucocorticoids in plasma and mortality in the Australian bush rat (*Rattus fuscipes* Waterhouse). *Journal of Mammalology*, **69**, 740–748.
15.4.4

Mace, G.M. (1994) An investigation into methods for categorizing the conservation status of species. In: *Large-Scale Ecology and Conservation Biology* (P.J. Edwards, R.M. May & N.R. Webb, eds), pp. 293–312. Blackwell Scientific Publications, Oxford.
25.2.1

Mace, G.M. & Lande, R. (1991) Assessing extinction threats: toward a reevaluation of IUCN threatened species categories. *Conservation Biology*, **5**, 148–157.
25.2.1

McGaughey, W.H. & Whalon, M.E. (1992) Managing insect resistance to *Bacillus thuringiensis* toxins. *Science*, **258**, 1451–1455.
16.10

McGrorty, S. & Goss-Custard, J.D. (1993) Population dynamics of the mussel *Mytilus edulis* along environmental gradients: spatial variations in density dependent mortalities. *Journal of Animal Ecology*. **62**, 415–427.
15.3.2

McIntosh, A.R. & Townsend, C.R. (1994) Interpopulation variation in mayfly anti-predator tactics: differential effects of contrasting fish predators. *Ecology*, **75**, 2078–2090.
22.3.1

McKenzie, N.L. & Rolfe, J.K. (1986) Structure of bat guilds in the Kimberley mangroves, Australia. *Journal of Animal Ecology*, **55**, 401–420.
20.3.1

Mackerras, M.I. & Saunders, D.F. (1955) The life history of the rat lung-worm, *Angiostrongylus cantonensis* (Chen) (Nematode: Metastrongylidae). *Australian Journal of Zoology*, **3**, 1–21.
12.3.3

Mackie, G.L., Qadri, S.U. & Reed, R.M. (1978) Significance of litter size in *Musculium securis* (Bivalvia: Sphaeridae). *Ecology*, **59**, 1069–1074.
6.2

MacLachlan, A.J., Pearce, L.J. & Smith, J.A. (1979) Feeding interactions and cycling of peat in a bog lake. *Journal of Animal Ecology*, **48**, 851–861.
11.3.2

MacLulick, D.A. (1937) Fluctuations in numbers of the varying hare (*Lepus americanus*). *University of Toronto Studies, Biology Series*, **43**, 1–136.
10.1

McMahon, T. (1973) Size and shape in biology. *Science*, **179**, 1201–1204.
14.12.1

McMillan, C. (1957) Nature of the plant community. III. Flowering behaviour within two grassland communities under reciprocal transplanting. *American Journal of Botany*, **44**, 143–153.
1.5.1

McNaughton, K.G. & Jarvis, P.G. (1991) Effects of spatial scale on stomatal control of transpiration. *Agricultural and Forest Meteorology*, **54**, 279–301.
3.4

McNaughton, S.J. (1975) r- and k-selection in *Typha*. *American Naturalist*, **109**, 251–261.
14.9.1

McNaughton, S.J. (1977) Diversity and stability of ecological communities: a comment on the role of empiricism in ecology. *American Naturalist*, **111**, 515–525.
22.4.3

McNaughton, S.J. (1978) Stability and diversity of ecological communities. *Nature*, **274**, 251–253.
22.6

McNeeley, J.A. (1988) *Economics and Biological Diversity: Developing and Using Economic Incentives to Conserve Biological Resources*. IUCN, Gland, Switzerland.
25.1.3

McNeill, S. (1973) The dynamics of a population of *Leptoterna dolabrata* (Heteroptera: Miridae) in relation to its food resources. *Journal of Animal Ecology*, **42**, 495–507.
6.6

McQueen, D.J., Johannes, M.R.S., LaFontaine, N.R., Young, A.S., Longbotham, E. & Lean, D.R.S. (1990) Effects of planktivore abundance on chlorophyll *a* and secchi depth. *Hydrobiologia*, **200/201**, 337–343.
22.3.2

Madge, D.S. (1969) Field and laboratory studies on the activities of two species of tropical earthworms. *Pedobiologia*, **9**, 188–214.
11.2.2

Maguire, L.A., Seal, U.S. & Brussard, P.F. (1987) Managing critically endangered species: the Sumatran rhino as a case study. In: *Viable Populations for Conservation* (M.E. Soulé, ed.), pp. 141–158. Cambridge University Press, Cambridge.
25.4.2

Maillette, L. (1982) Structural dynamics of silver birch. I. The fates of buds. *Journal of Applied Ecology*, **19**, 203–218.
4.2.1

Manire, C.A. & Gruber, S.H. (1990) Many sharks may be headed toward extinction. *Conservation Biology*, **4**, 10–11.
25.2.2

Margulis, L. (1975) Symbiotic theory of the origin of eukaryotic organelles. In: *Symbiosis* (D.H. Jennings & D.L. Lee, eds), pp. 21–38. Symposium 29, Society for Experimental Biology. Cambridge University Press, Cambridge.
13.11

Marshall, C. & Sagar, G.R. (1968) The interdependence of tillers in *Lolium multiflorum* Lam.: a quantitative assessment. *Journal of Experimental Botany*, **19**, 785–794.
8.2.1

Marshall, I. D. & Douglas, G.W. (1961) Studies in the epidemiology of infectious myxomatosis of rabbits. VIII. Further observations on changes in the innate resistance of Australian wild rabbits exposed to myxomatosis. *Journal of Hygiene*, **59**, 117–122.
12.7

Martin, M.M. (1991) The evolution of cellulose digestion in insects. In: *The Evolutionary Interaction of Animals and Plant* (W.G. Chaloner, J.L. Harper & J.H. Lawton, eds), pp. 105–112. The Royal Society, London. Also in *Philosophical Transactions of the Royal Society, Series B*, **333**, 281–288.
3.7.2, 11.3.1

Martin, P.S. (1984) Prehistoric overkill: the global model. In: *Quaternary Extinctions: A Prehistoric Revolution* (P.S. Martin & R.G. Klein, eds). University of Arizona Press, Tuscon, Arizona.
24.4.6

Martinez, N.D. (1993) Effects of resolution on food web structure. *Oikos*, **66**, 403–412.
22.5.2

Marzusch, K. (1952) Untersuchungen über di Temperaturabhängigkeit von Lebensprozessen bei Insekten unter besonderer Berücksichtigung winter-schlantender Kartoffelkäfer. *Zeitschrift für vergleicherde Physiologie*, **34**, 75–92.
2.2

Maxwell, W.D. (1989) The end Permian mass extinction. In: *Mass Extinctions: Processes and Evidence* (S.K. Donovan, ed.), pp. 152–173. Columbia University Press, New York.
24.4.6

May, R.M. (1972) Will a large complex system be stable? *Nature*, **238**, 413–414.
22.4.1, 22.4.2, 22.5.5

May, R.M. (1973) On relationships among various types of population models. *American Naturalist*, **107**, 46–57.
7.9, 10.5.5

May, R.M. (1975a) Biological populations obeying difference equations: stable points, stable cycles and chaos. *Journal of Theoretical Biology*, **49**, 511–524.
6.8.2

May, R.M. (1975b) Patterns of species abundance and diversity. In: *Ecology and Evolution of Communities* (M.L. Cody & J.M. Diamond, eds), pp. 81–120. Belknap, Cambridge, MA.
23.3.1

May, R.M. (1976) Estimating *r*: a pedagogical note. *American Naturalist*, **110**, 496–499.
4.7.2, 4.9

May, R.M. (1977) Thresholds and breakpoints in ecosystems with a multiplicity of stable states. *Nature*, **269**, 471–477.
10.6

May, R.M. (1978) Host–parasitoid systems in patchy environments: a phenomenological model. *Journal of Animal Ecology*, **47**, 833–843.
10.5.2, 10.5.3

May, R.M. (1980) Nations and numbers, 1980. *Nature*, **287**, 482–483.
4.9

May, R.M. (ed.) (1981a) Models for single populations. In: *Theoretical Ecology: Principles and Applications*, 2nd edn., pp. 5–29. Blackwell Scientific Publications, Oxford.
10.2.1, 10.2.4

May, R.M. (1981b) Models for two interacting populations. In: *Theoretical Ecology: Principles and Applications*, 2nd edn. (R.M. May, ed.), pp. 78–104. Blackwell Scientific Publications, Oxford.
13.12

May, R.M. (1981c) Patterns in multi-species communitites. In: *Theoretical Ecology: Principles and Applications*, 2nd edn. (R.M.

May, ed.), pp. 197–227. Blackwell Scientific Publications, Oxford.
22.4.2, 22.5.2

May, R.M. (1987) Chaos and the dynamics of biological populations. *Proceedings of the Royal Society of London, Series A*, **413**, 27–44.
6.9

May. R.M. (1989) Detecting density dependence in imaginary worlds. *Nature*, **338**, 16–17.
15.2.3

May, R.M. (1990a) How many species? *Philosophical Transactions of the Royal Society of London, Series B*, **330**, 293–304.
25.1.2

May, R.M. (1990b) Population biology and population genetics of plant–pathogen associations. In: *Pests, Pathogens and Plant Communities* (J.J. Burdon & S.R. Leather, eds), pp. 309–325. Blackwell Scientific Publications, Oxford.
12.5.3

May, R.M. (1993) Resisting resistance. *Nature*, **361**, 593–594.
16.10

May, R.M. & Anderson, R.M. (1979) Population biology of infectious diseases. *Nature*, **280**, 455–461.
12.2

May, R.M. & Anderson, R.M. (1983) Epidemiology and genetics in the coevolution of parasites and hosts. *Proceedings of the Royal Society of London, Series B*, **219**, 281–313.
12.7

Mayer, G.C. & Chipley, R.M. (1992) Turnover in the avifauna of Guana Island, British Virgin Islands. *Journal of Animal Ecology*, **61**, 561–566.
23.3.4

Maynard Smith, J. (1972) *On Evolution*. Edinburgh University Press, Edinburgh.
5.3.2

Maynard Smith, J. (1974) *Models in Ecology*. Cambridge University Press, Cambridge.
22.4.1

Maynard Smith, J. & Slatkin, M. (1973) The stability of predator–prey systems. *Ecology*, **54**, 384–391.
6.8.2

Meeusen, R.L. & Warren, G. (1989) Insect control with genetically engineered crops. *Annual Review of Entomology*, **34**, 373–381.
16.10

Meinhardt, H. (1982) *Models of Biological Pattern Formation*. Academic Press, New York.
15.6.2

Melillo, J.M., McGuire, A.D., Kicklighter, D.W., Moore, B., Vorosmarty, C.J. & Schloss, A.L. (1993) Global climate change and terrestrial net primary production. *Nature*, **363**, 234–240.
18.5

Menge, B.A. & Sutherland, J.P. (1976) Species diversity gradients: synthesis of the roles of predation, competition, and temporal heterogeneity. *American Naturalist*, **110**, 351–369.
21.2.5, 21.7.2

Menge, B.A. & Sutherland, J.P. (1987) Community regulation: variation in disturbance, competition, and predation in relation to environmental stress and recruitment. *American Naturalist*, **130**, 730–757.
21.2.5. 21.7.2

Menge, B.A., Lubchenco, J., Gaines, S.D. & Ashkenas, L.R. (1986) A test of the Menge–Sutherland model of community organization in a tropical rocky intertidal food web. *Oecologia*, **71**, 75–89.
21.2.5

Menges, E.S. (1990) Population viability analysis for an endangered plant. *Conservation Biology*, **4**, 52–62.
25.4.4

Metcalf, R.L. (1980) Changing role of insecticides in crop protection. *Annual Review of Entomology*, **25**, 219–256.
16.4, 16.6.4, 16.7, 16.8

Metcalf, R.L. (1982) Insecticides in pest management. In: *Introduction to Insect Pest Management*, 2nd edn. (R.L. Metcalf & W.L. Luckmann, eds), pp. 217–277. Wiley, New York.
16.6.1

Meyer, A. (1993) Phylogenetic relationships and evolutionary processes in East African cichlid fishes. *Trends in Ecology and Evolution*, **8**, 279–284.
23.4

Meyer, J.L. & Pulliam, W.M. (1992) Modification of terrestrial–aquatic interactions by a changing climate. In: *Global Climate Change and Freshwater Ecosystems* (P. Firth & S.G. Fisher, eds), pp. 177–191. Springer-Verlag, New York.
19.4.1

Mikkelson, G.M. (1993) How do food webs fall apart? A study of changes in trophic structure during relaxation on habitat fragments. *Oikos*, **67**, 539–547.
25.5.3

Milinski, M. & Parker, G.A. (1991) Competition for resources. In: *Behavioural Ecology. An Evolutionary Approach*, 3rd edn. (J.R. Krebs & N.B. Davies, eds), pp. 137–168. Blackwell Scientific Publications, Oxford.
9.10

Miller, D. (1970) *Biological Control of Weeds in New Zealand 1927–48*. New Zealand Department of Scientific and Industrial Research Information Series, Vol. 74, pp. 1–104.
10.1

Miller, G.T. Jr. (1988) *Environmental Science*, 2nd edn. Wadsworth, Belmont.
16.6.1, 16.6.4, 16.7, 25.1.3

Miller, G.T. Jr. (1993) *Environmental Science*, 4th edn. Wadsworth, Belmont CA.
4.9

Miller, P.J. (1993) Grading of gobies and disturbing of sleepers. *Natural Environment Research Council News*, **27**, 16–19.
1.1.4

Mills, J.N. (1983) Herbivory and seedling establishment in post-fire southern California chaparral. *Oecologia*, **60**, 267–270.
8.2.4

Milne, L.J. & Milne, M. (1976) The social behaviour of burying beetles. *Scientific American*, **August**, 84–89.
11.3.3

Milton, W.E.J. (1940) The effect of manuring, grazing and liming on the yield, botanical and chemical composition of natural hill pastures. *Journal of Ecology*, **28**, 326–356.
21.2.1

Milton, W.E.J. (1947) The composition of natural hill pasture, under controlled and free grazing, cutting and manuring. *Welsh Journal of Agriculture*, **14**, 182–195.
21.2.1

Minorsky, P.V. (1985) An heuristic hypothesis of chilling injury in plants: a role for calcium as the primary physiological transducer of injury. *Plant Cell and Environment*, **8**, 75–94.
2.3.1

Minot, E.O. & Perrins, C.M. (1986) Interspecific interference competition—nest sites for blue and great tits. *Journal of Animal Ecology*, **50**, 375-385.
15.7.2

Mithen, R. & Lawton, J.H. (1986) Food-web models that generate constant predator–prey ratios. *Oecologia*, **69**, 542–550.
22.5.4, 22.5.6

Mithen, R., Harper, J.L. & Weiner, J. (1984) Growth and mortality of individual plants as a function of 'available area'. *Oecologia*, **62**, 57–60.
6.11

Monro, J. (1967) The exploitation and conservation of resources by populations of insects. *Journal of Animal Ecology*, **36**, 531–547.
9.7.2, 10.5.6, 15.7.6

Montague, J.R., Mangan, R.L. & Starmer, W.T. (1981) Reproductive allocation in the Hawaiian Drosophilidae: egg size and number. *American Naturalist*, **118**, 865–871.
14.4.2

Mooney, H.A. & Gulmon, S.L. (1979) Environmental and evolutionary constraints on the photosynthetic pathways of higher plants. In: *Topics in Plant Population Biology* (O.T. Solbrig, S. Jain, G.B. Johnson & P.H. Raven, eds), pp. 316–337. Columbia University Press, New York.
3.2.5

Mooney, H.A., Harrison, A.T. & Morrow, P.A. (1975) Environmental limitations of photosynthesis on a California evergreen shrub. *Oecologia*, **19**, 293–301.
3.2.5

Mooney, H.A., Vitousek, P.M. & Matson, P.A. (1987) Exchange of materials between terrestrial ecosystems and the atmosphere. *Science*, **238**, 926–932.
19.4.5, 19.4.6

Moran, N.A., Munson, M.A., Baumann, P. & Ishikawa, H. (1993) A molecular clock in endosymbiotic bacteria is calibrated using the insect hosts. *Proceedings of the Royal Society of London, Series B*, **253**, 167–171.
13.6

Moran, P.A.P. (1952) The statistical analysis of gamebird records. *Journal of Animal Ecology*, **21**, 154–158.
15.4.1

Moran, P.A.P. (1953) The statistical analysis of the Canadian lynx cycle. *Australian Journal of Zoology*, **1**, 163–173.
15.4.1

Moran, V.C. (1980) Interactions between phytophagous insects and their *Opuntia* hosts. *Ecological Monographs*, **5**, 153–164.
23.3.3

Moreau, R.E. (1952) The place of Africa in the palaearctic migration system. *Journal of Animal Ecology*, **21**, 250–271.
5.7.3

Morris, R.F. (1959) Single-factor analysis in population dynamics. *Ecology*, **40**, 580–588.
15.3.1

Morris, R.F. (ed.) (1963) The dynamics of epidemic spruce budworm populations. *Memoirs of the Entomological Society of Canada*, **31**, 1–332.
10.6.1

Morrison, G. & Strong, D.R. Jr. (1981) Spatial variations in egg density and the intensity of parasitism in a neotropical chrysomelid (*Cephaloleia consanguinea*). *Ecological Entomology*, **6**, 55–61.
9.10

Mortimer, A.M. (1989) The biology of weeds. In: *Weed Control Handbook: Principles*, 8th edn. Blackwell Scientific Publications, Oxford.
16.2

Moss, B. (1980) *Ecology of Fresh Waters*. Blackwell Scientific Publications, Oxford.
19.3.3

Moss, B. (1989) *Ecology of Fresh Waters: Man and Medium*, 2nd edn. Blackwell Scientific Publications, Oxford.
19.3.2, 19.3.3

Moss, B., Balls, H., Booker, I., Manson, K. & Timms, M. (1988) Problems in the construction of a nutrient budget for the River Bure and its Broads (Norfolk) prior to its restoration from eutrophication. In: *Algae and the Aquatic Environment* (F.E. Round, ed.), pp. 326–352. Biopress, Bristol.
19.4.3

Moss, G.D. (1971) The nature of the immune response of the mouse to the bile duct cestode, *Hymenolepis microstoma*. *Parasitology*, **62**, 285–294.
12.3.4

Moss, R., Welch, D. & Rothery, P. (1981) Effects of grazing by mountain hares and red deer on the production and chemical composition of heather. *Journal of Applied Ecology*, **18**, 487–496.
9.7

Moss, R., Parr, R. & Lambin, X. (1994) Effects of testosterone on breeding density, breeding success and survival of red grouse. *Proceedings of the Royal Society of London, Series B*, **258**, 175–180.
3.10, 15.4.1

Moulton, M.P. & Pimm, S.L. (1986) The extent of competition in shaping an introduced avifauna. In: *Community Ecology* (J. Diamond & T.J. Case, eds), pp. 80–97. Harper & Row, New York.
20.4.2

Mountford, M.D. (1988) Population regulation, density dependence, and heterogeneity. *Journal of Animal Ecology*, **57**, 845–858.
15.2.3

Muller, C.H. (1969) Allelopathy as a factor in ecological process. *Vegetatio*, **18**, 348–357.
7.3

Müller, G. & Foerster, E. (1974) Entwicklung von Weideansaaten in Überflutungsbereich des Rheines bei Kleve. *Acker und Pflanzenban*, **140**, 161–174.
21.3

Müller, H.J. (1970) Food distribution, searching success and predator–prey models. In: *The Mathematical Theory of the Dynamics of Biological Populations* (R.W. Hiorns, ed.), pp. 87–101. Academic Press, London.
5.5

Murdie, G. & Hassell, M.P. (1973) Food distribution, searching success and predator–prey models. In: *The Mathematical Theory of the Dynamics of Biological Populations* (R.W. Hiorns, ed.), pp. 87–101. Academic Press, London.
9.5.3

Murdoch, W.W. (1966) Community structure, population control and competition—a critique. *American Naturalist*, **100**, 219–226.
22.3.2

Murdoch, W.W. & Stewart-Oaten, A. (1975) Predation and population stability. *Advances in Ecological Research*, **9**, 1–131.
9.2.3, 9.7.3

Murdoch, W.W. & Stewart-Oaten, A. (1989) Aggregation by parasitoids and predators: effects on equilibrium and stability. *American Naturalist*, **134**, 288–310.
10.5.4

Murdoch, W.W., Avery, S. & Smith, M.E.B. (1975) Switching in predatory fish. *Ecology*, **56**, 1094–1105.
9.2.3

Murdoch, W.W., Reeve, J.D., Huffaker, C.B. & Kennett, C.E. (1984) Biological control of olive scale and its relevance to ecological theory. *American Naturalist*, **123**, 371–392.
10.5.6

Murdoch, W.W., Chesson, J. & Chesson, P.L. (1985) Biological control in theory and practice. *American Naturalist*, **125**, 343–366.
10.5.6

Murdoch, W.W., Briggs, C.J., Nisbet, R.M., Gurney, W.S.C. & Stewart-Oaten, A. (1992) Aggregation and stability in meta-population models. *American Naturalist*, **140**, 41–58.
10.5.5

Murphy, G.J. (1977) Clupeoids. In: *Fish Population Dynamics* (J.A. Gulland, ed.), pp. 283–308. Wiley-Interscience, New York.
16.13.3

Murray, J.D. (1989) *Mathematical Biology*. Springer, Berlin.
5.2

Murton, R.K. (1971) The significance of a specific search image in the feeding behaviour of the wood pigeon. *Behaviour*, **40**, 10–42.
9.2.3

Murton, R.K., Isaacson, A.J. & Westwood, N.J. (1966) The relationships between wood pigeons and their clover food supply and the mechanism of population control. *Journal of Applied Ecology*, **3**, 55–93.
5.2.1, 9.7.1

Murton, R.K., Westwood, N.J. & Isaacson, A.J. (1974) A study of wood-pigeon shooting: the exploitation of a natural animal population. *Journal of Applied Ecology*, **11**, 61–81.
8.3

Muscatine, L. & Pool, R.R. (1979) Regulation of numbers of intracellular algae. *Proceedings of the Royal Society of London, Series B*, **204**, 115–139.
13.8

Myers, J.H. (1988a) Can a general hypothesis explain population cycles of forest Lepidoptera. *Advances in Ecological Research*, **18**, 179–242.
12.6.1, 15.4

Myers, J.M. & Krebs, C.J. (1971) Genetic, behavioural and reproductive attributes of dispersing field voles *Microtus pennsylvanicus* and *Microtus ochrogaster*. *Ecological Monographs*, **41**, 53–78.
5.4.3, 15.4.4

Myers, N. (1988b) Threatened biotas: 'hotspots' in tropical forests. *Environmentalist*, **8**, 1–20.
25.6.3

Myers, N. (1991) The biodiversity challenge: expanded 'hotspots' analysis. *Environmentalist*, **10**, 243–256.
25.6.3

Namkoong, G. & Roberds, J.H. (1974) Extinction probabilities and the changing age structure of redwood forests. *American Naturalist*, **108**, 355–368.
17.4.4

National Research Council (1990) *Alternative Agriculture*. National Academy of Sciences, Academy Press, Washington, D.C.
16.11

Naveh, Z. (1975) The evolutionary significance of fire in the Mediterrranean region. *Vegetatio*, **29**, 199–208.
2.4, 2.9.1

Nedergaard, J. & Cannon, B. (1990) Mammalian hibernation. *Philosophical Transactions of the Royal Society, Series B*, **326**, 669–686; also in *Life at Low Temperatures* (R.M. Laws & F. Franks, eds), pp. 153–170. The Royal Society, London.
2.5.1

Nee, S. & May, R.M. (1992) Dynamics of metapopulations: habitat destruction and competitive coexistence. *Journal of Animal Ecology*, **61**, 37–40.
12.7

Nei, M. (1978) Estimation of average heterozygosity and genetic distance from a small number of individuals. *Genetics*, **89**, 583–590.
1.2.2

Nelson, J.S. (1994) *Fishes of the World*, 3rd edn. John Wiley & Sons, New York.
1.1.4

NERC (1990) *Our Changing Environment*. Natural Environment Research Council, London. (NERC acknowledges the significant contribution of Fred Pearce to the document.)
2.10

Neumann, R.L. (1967) Metabolism in the Eastern chipmunk (*Tamias striatus*) and the Southern flying squirrel (*Glaucomys volans*) during the winter and summer. In: *Mammalian Hibernation III* (K.C. Fisher, A.R. Dawe, C.P. Lyman, E. Schönbaum & F.E. South, eds), pp. 64–74. Oliver & Boyd, Edinburgh & London.
2.5.1

Newsham, K.K., Fitter, A.H. & Watkinson, A.H. (1994) Root pathogenic and arbuscular mycorrhizal mycorrhizal fungi determine fecundity of asymptomatic plants in the field. *Journal of Ecology*, **82**, 805–814.
13.7.2

Nicholson, A.J. (1933) The balance of animal populations. *Journal of Animal Ecology*, **2**, 131–178.
15.2.3

Nicholson, A.J. (1954a) Compensatory reactions of populations to stress, and their evolutionary significance. *Australian Journal of Zoology*, **2**, 1–8.
15.2.3

Nicholson, A.J. (1954b) An outline of the dynamics of animal populations. *Australian Journal of Zoology*, **2**, 9–65.
6.6, 8.3, 15.2.3, 16.12

Nicholson, A.J. (1957) The self adjustment of populations to change. *Cold Spring Harbor Symposium of Quantitative Biology*, **22**, 153–172.
15.2.3

Nicholson, A.J. (1958) Dynamics of insect populations. *Annual Review of Entomology*, **3**, 107–136.
15.2.3

Nicholson, A.J. & Bailey, V.A. (1935) The balance of animal populations. *Proceedings of the Zoological Society of London*, **3**, 551–598.
10.2.2

Nielsen, B.O. & Ejlerson, A. (1977) The distribution of herbivory in a beech canopy. *Ecological Entomology*, **2**, 293–299.
8.2.1

Niklas, K.J., Tiffney, B.H. & Knoll, A.H. (1983) Patterns in vascular land plant diversification. *Nature*, **303**, 614–616.
24.4.6

Nilsson, L.A. (1988) The evolution of flowers with deep corolla tubes. *Nature*, **334**, 147–149.
13.4.2

Nilsson, L.A. (1992) Orchid pollination biology. *Trends in Ecology and Evolution*, **7**, 255–259.
13.4.2

Nitecki, M.H. (1983) *Coevolution*. University of Chicago Press, Chicago.
3.7.3

Nixon, M. (1969) The lifespan of *Octopus vulgaris* Lamarck. *Proceedings of the Malacological Society of London*, **38**, 529–540.
4.8

Nobel, P.S. (1983) Spine influences on PAR interception, stem temperature and nocturnal acid accumulation by cacti. *Plant, Cell and Environment*, **6**, 153–159.
2.4

Noble, J.C. & Slatyer, R.O. (1979) The effect of disturbance on plant succession. *Proceedings of the Ecological Society of Australia*, **10**, 135–145.
17.4.4

Noble, J.C. & Slatyer, R.O. (1981) Concepts and models of succession in vascular plant communities subject to recurrent fire. In: *Fire and the Australian Biota* (A.M. Gill, R.H. Graves & I.R. Noble, eds). Australian Academy of Science, Canberra.
17.4.4

Noble, J.C., Bell, A.D. & Harper, J.L. (1979) The population biology of plants with clonal growth. I. The morphology and structural demography of *Carex arenaria*. *Journal of Ecology*, **67**, 983–1008.
4.2.1, 5.5.3

Norton, D.A. (1991) *Trilepidea adamsii*: an obituary for a species. *Conservation Biology*, **5**, 52–57.
25.2.2

Norton, I.O. & Sclater, J.G. (1979) A model for the evolution of the Indian Ocean and the breakup of Gondwanaland. *Journal of Geophysical Research*, **84**, 6803–6830.
1.2.1

Nowak, C.L., Nowak, R.S., Tausch, R.J. & Wigand, P.E. (1994) Tree and shrub dynamics in northwestern Great Basin woodland and shrub steppe during the Late Pleistocene and Holocene. *American Journal of Botany*, **8**, 265–277.
1.2.2

Nowak, M.A. & May, R.M. (1994) Superinfection and the evolution of parasite virulence. *Proceedings of the Royal Society, Series B*, **255**, 81–89.
12.7

Noy-Meir, I. (1975) Stability of grazing systems: an application of predator–prey graphs. *Journal of Ecology*, **63**, 459–483.
10.6

Nunney, L. & Campbell, K.A. (1993) Assessing minimum viable population sizes: demography meets population genetics. *Trends in Ecology and Evolution*, **8**, 234–239.
25.2.5, 25.3.3, 25.6.1

Nutman, P.S. (1963) In: *Symbiotic Associations* (Society for General Microbiology, eds). Cambridge University Press, Cambridge.
13.12

Nye, P.H. & Tinker, P.B. (1977) *Solute Movement in the Soil–Root System*. Blackwell Scientific Publications, Oxford.
3.5

O'Dor, R.K. & Wells, M.J. (1978) Reproduction versus somatic growth: hormonal control in *Octopus vulgaris*. *Journal of Experimental Biology*, **77**, 15–31.
4.8

O'Neill, R.V. (1976) Ecosystem persistence and heterotrophic regulation. *Ecology*, **57**, 1244–1253.
22.6

Oakeshott, J.G., May, T.W., Gidson, J.B. & Willcocks, D.A. (1982) Resource partitioning in five domestic *Drosophila* species and its relationship to ethanol metabolism. *Australian Journal of Zoology*, **30**, 547–556.
11.3.3

Obeid, M., Machin, D. & Harper, J.L. (1967) Influence of density on plant to plant variations in fiber flax, *Linum usitatissimum*. *Crop Science*, **7**, 471–473.
6.11

Oborny, B. (1994) Growth rules in clonal plants and environmental predictability—a simulation study. *Journal of Ecology*, **82**, 341–351.
3.10

Odum, E.P. & Biever, L.J. (1984) Resource quality, mutualism, and energy partitioning in food chains. *American Naturalist*, **124**, 360–376.
22.4.2

Ødum, S. (1965) Germination of ancient seeds; floristical observations and experiments with archaeologically dated soil samples. *Dansk Botanisk Arkiv*, **24**, 1–70.
5.5.2

Ogden, J. (1968) *Studies on Reproductive Strategy with Particular Reference to Selected Composites*. Ph.D. thesis, University of Wales.
14.9.1

Ogden, J. (1993) Population increase and nesting patterns in the black noddy *Anous minutus* in *Pisonia* forest on Heron Island: observations in 1978, 1979 and 1992. *Australian Journal of Ecology*, **18**, 395–403.
4.3

Oinonen, E. (1967) The correlation between the size of Finnish bracken (*Pteridium aquilinum* (L.) Kuhn) clones and certain periods of site history. *Acta Forestalia Fennica*, **83**, 3–96.
5.6

Oka, U.M.I. & Pimentel, D. (1976) Herbicide (2,4-D) increases insect and pathogen pests on corn. *Science*, **193**, 239–240.
16.6.1

Oksanen, L. (1988) Ecosystem organisation: mutualism and cybernetics of plain Darwinian struggle for existence. *American Naturalist*, **131**, 424–444.
22.3.2

Okubo, A. (1980) *Diffusion and Ecological Problems: Mathematical Models.* Springer-Verlag, Berlin.
5.2

Ollason, J.G. (1980) Learning to forage—optimally? *Theoretical Population Biology*, **18**, 44–56.
9.9.3

Ong, C.K., Marshall, C. & Sagar, G.R. (1978) The physiology of tiller death in grasses. 2. Causes of tiller death in a grass sward. *Journal of the British Grassland Society*, **33**, 205–211.
8.2.1

Orshan, G. (1963) Seasonal dimorphism of desert and Mediterranean chamaephytes and its significance as a factor in their water economy. In: *The Water Relations of Plants* (A.J. Rutter & F.W. Whitehead, eds), pp. 207–222. Blackwell Scientific Publications, Oxford.
1.6, 3.2.3

Orzack, S.H. & Tuljapurkar, S. (1989) Population dynamics in variable environments. VII. The demography and evolution of iteroparity. *American Naturalist*, **133**, 901–923.
14.10.2

Osawa, A. & Allen, R.B. (1993) Allometric theory explains self-thinning relationships of mountain beech and red pine. *Ecology*, **74**, 1020–1032.
6.13

Osborn, R.G. (1975) *Models of Lemming Demography and Avian Predators Near Barrow, Alaska.* M.S. thesis, San Diego State University.
15.4.3

Osmond, C.B., Austin, M.P., Berry, J.A. *et al.* (1987) Stress physiology and the distribution of plants. *Bioscience*, **37**, 38–48.
2.7

Owen-Smith, N. (1987) Pleistocene extinctions: the pivotal role of megaherbivores. *Paleobiology*, **13**, 351–362.
24.4.6, 25.2.2

Pacala, S.W. & Crawley, M.J. (1992) Herbivores and plant diversity. *American Naturalist*, **140**, 243–260.
8.2.6

Pacala, S.W. & Hassell, M.P. (1991) The persistence of host–parasitoid associations in patchy environments. II. Evaluation of field data. *American Naturalist*, **138**, 584–605.
9.7.1, 10.5.3, 10.5.6

Pacala, S.W., Hassell, M.P. & May, R.M. (1990) Host–parasitoid associations in patchy environments. *Nature*, **344**, 150–153.
10.5.3

Pagel, M.D. & Harvey, P.H. (1988) Recent developments in the analysis of comparative data. *Quarterly Review of Biology*, **63**, 413–440.
14.12.1

Paine, R.T. (1966) Food web complexity and species diversity. *American Naturalist*, **100**, 65–75.
21.2.2, 21.2.5, 22.2.2

Paine, R.T. (1979) Disaster, catastrophe and local persistence of the sea palm *Postelsia palmaeformis. Science*, **205**, 685–687.
7.5.1

Paine, R.T. (1994) *Marine Rocky Shores and Community Ecology: An Experimentalist's Perspective.* Ecology Institute, Oldendorf/Luhe, Germany.
2.8

Paine, R.T. & Levin, S.A. (1981) Intertidal landscapes: distur-bance and the dynamics of pattern. *Ecological Monographs*, **51**, 145–178.
21.2.2, 21.5.2

Painter, E.L. & Detling, J.K. (1981) Effects of defoliation on net photosynthesis and regrowth of western wheatgrass. *Journal of Range Management*, **34**, 68–71.
8.2.1

Palmblad, I.G. (1968) Competition studies on experimental populations of weeds with emphasis on the regulation of population size. *Ecology*, **49**, 26–34.
6.6

Park, T. (1948) Experimental studies of interspecific competition. I. Competition between populations of the flour beetle *Tribolium confusum* Duval and *Tribolium castaneum* Herbst. *Ecological Monographs*, **18**, 267–307.
12.4, 12.6, 21.2.4

Park, T. (1954) Experimental studies of interspecific competition. II. Temperature, humidity and competition in two species of *Tribolium. Physiological Zoology*, **27**, 177–238.
7.4.3

Park, T. (1962) Beetles, competition and populations. *Science*, **138**, 1369–1375.
7.4.3

Park, T., Mertz, D.B., Grodzinski, W. & Prus, T. (1965) Cannibalistic predation in populations of flour beetles. *Physiological Zoology*, **38**, 289–321.
7.4.3

Parker, G.A. (1970) The reproductive behaviour and the nature of sexual selection in *Scatophaga stercoraria* L. (Diptera: Scatophagidae) II. The fertilization rate and the spatial and temporal relationships of each sex around the site of mating and oviposition. *Journal of Animal Ecology*, **39**, 205–228.
9.10

Parker, G.A. (1984) Evolutionarily stable strategies. In: *Behavioral Ecology: An Evolutionary Approach*, 2nd edn. (J.R.Krebs & N.B. Davies, eds), pp. 30–61. Blackwell Scientific Publications, Oxford.
5.3.2

Parker, G.A. & Stuart, R.A. (1976) Animal behaviour as a strategy optimizer: evolution of resource assessment strategies and optimal emigration thresholds. *American Naturalist*, **110**, 1055–1076.
9.9.1

Parlange, J.-Y. & Waggoner, P.E. (1970) Stomatal dimensions and resistance to diffusion. *Plant Physiology*, **46**, 337–342.
3.4

Parrish, J.A.D. & Bazzaz, F.A. (1979) Differences in pollination niche relationships in early and late successional plant communities. *Ecology*, **60**, 597–610.
24.4.4

Parrish, J.A.D. & Bazzaz, F.A. (1982) Competitive interactions in plant communities of different successional ages. *Ecology*, **63**, 314–320.
24.4.4

Parsons, P.A. & Spence, G.E. (1981) Ethanol utilization: threshold differences among three *Drosophila* species. *American Naturalist*, **117**, 568–571.
11.3.1

Partridge, L. & Farquhar, M. (1981) Sexual activity reduces lifespan of male fruitflies. *Nature*, **294**, 580–581.
14.4

Paton, P.W. (1992) The effect of edge on avian nest success: how strong is the evidence? *Conservation Biology*, **8**, 17–26.
25.3.3

Pauly, D. & Christensen, V. (1995) Primary production required to sustain global fisheries. *Nature*, **374**, 255–257.
18.4.1, 18.4.3

Pauly, D., Soriano-Bartz, M.L. & Palomares, M.L.D. (1993) Improved construction, parametrization and interpretation of steady-state ecosystem models. In: *Trophic Models of Aquatic Ecosystems* (V. Christensen & D. Pauly, eds), pp. 1–13. International Center for Living Aquatic Resources Management, Conference Proceedings 26, Manila, Phillipines.
18.4.3

Payne, C.C. (1988) Pathogens for the control of insects: where next? *Philosophical Transactions of the Royal Society of London, Series B*, **318**, 225–248.
16.8

Payne, R.B. (1977) The ecology of brood parasitism in birds. *Annual Review of Ecology and Systematics*, **8**, 1–28.
12.8

Peake, A.J. & Quinn, G.P. (1993) Temporal variation in species–area curves for invertebrates in clumps of an intertidal mussel. *Ecography*, **16**, 269–277.
23.1

Pearce, D.W. & Turner, R.K. (1990) *Economics of Natural Resources and the Environment*. Harvester Wheatsheaf, New York.
16.15.1

Pearcy, R.W., Björkman, O., Caldwell, M.M., Keeley, J.E., Monson, R.K. & Strain, B.R. (1987) Carbon gain by plants in natural environments. *Bioscience*, **37**, 21–29.
3.2.5

Pearl, R. (1927) The growth of populations. *Quarterly Review of Biology*, **2**, 532–548.
6.4

Pearl, R. (1928) *The Rate of Living*. Knopf, New York.
4.5.1

Pearson, D.L. & Juliano, S.A. (1991) Mandible length ratios as a mechanism for co-occurrence: evidence from a world-wide comparison of tiger beetle assemblages (Cicindelidae). *Oikos*, **61**, 223–233.
20.3.1

Pease, J.L., Vowles, R.H. & Keith, L.B. (1979) Interaction of snowshoe hares and woody vegetation. *Journal of Wildlife Management*, **43**, 43–60.
10.2.4

Pemadasa, M.A., Grieg-Smith, P. & Lovell, P.H. (1974) A quantitative description of the distribution of annuals in the dune system at Aberffraw, Anglesey. *Journal of Ecology*, **62**, 379–402.
17.3.2

Penman, H.L. (1948) Natural evaporation from open water, bare soil and grass. *Proceedings of the Royal Society, Series A*, **193**, 120–145.
3.4

Perrins, C.M. (1965) Population fluctuations and clutch size in the great tit, *Parus major* L. *Journal of Animal Ecology*, **34**, 601–647.
4.6.3

Perry, J.N., Woiwod, I.P. & Hanski, I. (1993) Using response-surface methodology to detect chaos in ecological time series. *Oikos*, **68**, 329–339.
6.9

Peterman, R.M., Clark, W.C. & Hollings, C.S. (1979) The dynamics of resilience: shifting stability domains in fish and insect systems. In: *Population Dynamics* (R.M. Anderson, B.D. Turner & L.R. Taylor, eds), pp. 321–341. Blackwell Scientific Publicatons, Oxford.
10.6.1, 16.13.3

Peters, B. (1980) *The Demography of Leaves in a Permanent Pasture*. Ph.D. thesis, University of Wales.
15.3.2

Peters, R.H. (1983) *The Ecological Implications of Body Size*. Cambridge University Press, Cambridge.
14.12.1, 15.4.3

Petranka, J.W. (1989) Chemical interference competition in tadpoles: does it occur outside laboratory aquaria? *Copeia*, **1989**, 921–930.
7.3

Pévet, P., Masson-Pévet, M., Hermes, M.L.H.J., Buijs, R.M. & Canguilhem, B. (1989) Photoperiod, pineal gland, vasopressinergic innervation and timing of hibernation. In: *Living in the Cold*. II (A. Malan & B. Canguilhem, eds), pp. 43–51. John Libbey, London.
2.5.1

Phillipi, T. & Seger, J. (1989) Hedging one's evolutionary bets, revisited. *Trends in Ecology and Evolution*, **4**, 41–44.
14.10.2

Pianka, E.R. (1967) On lizard species diversity: North American flatland deserts. *Ecology*, **48**, 333–351.
24.3.1, 24.3.2

Pianka, E.R. (1970) On r- and k-selection. *American Naturalist*, **104**, 592–597.
14.9, 14.12.2

Pianka, E.R. (1973) The structure of lizard communities. *Annual Review of Ecology and Systematics*, **4**, 53–74.
20.4.1

Pianka, E.R. (1983) *Evolutionary Ecology*, 3rd edn. Harper & Row, New York.
24.4.1

Pickett, J.A. (1988) Integrating use of beneficial organisms with chemical crop protection. *Philosophical Transactions of the Royal Society of London, Series B*, **318**, 203–211.
16.5.1

Pickett, S.T.A. & White, P.S. (eds) (1985) *The Ecology of Natural Disturbance as Patch Dynamics*. Academic Press, New York.
21.1.2, 21.4

Pierce, N.E. & Young, W.R. (1986) Lycaenid butterflies and ants: two species in stable equilibria in mutualistic, commensal, and parasitic interactions. *American Naturalist*, **128**, 216–227.
13.3.2

Piersma, T. & Davidson, N.C. (1992a) *The Migration of Knots, Wader Study Group Bulletin 64, Supplement, April 1992*, pp. 1–209. Joint Nature Conservation Committee, Publications Branch, Peterborough.
5.7.3

Piersma, T. and Davidson, N.C. (1992b) The migrations and annual cycles of five subspecies of Knots in perspective. In: *The Migration of Knots, Wader Study Group Bulletin 64, Supplement, April 1992*, pp. 187–197. Joint Nature Conservation Committee, Publications Branch, Peterborough.
5.7.3

Pijl, L. van der (1969) *Principles of Dispersal in Higher Plants*. Springer-Verlag, Berlin.
5.3.6

Pimentel, D. (1979) *Pesticide and Toxic Chemical News*, **Jan 17**, 11–12.
16.7

Pimentel, D. (1993) Cultural controls for insect pest management. In: *Pest Control and Sustainable Agriculture* (S. Corey, D. Dall & W. Milne, eds), pp. 35–38. CSIRO, East Melbourne.
16.2, 16.7, 16.9, 16.11

Pimentel, D., Levins, S.A. & Soans, A.B. (1975) On the evolution of energy balance in some exploiter–victim systems. *Ecology*, **56**, 381–190.
18.4.1

Pimentel, D., Krummel, J., Gallahan, D. *et al.* (1978) Benefits and cost of pesticide use in U.S. food production. *Bioscience*, **28**, 777–784.
16.7

Pimentel, D., McLaughlin, L., Zepp, A. *et al.* (1991) Environmental and economic impacts of reducing U.S. agricultural pesticide use. In: *Handbook of Pest Management in Agriculture*, Vol. 1 (D. Pimentel, ed.), pp. 679–719. CRC Press, Boca Raton.
16.9

Pimm, S.L. (1979a) Complexitiy and stability: another look at MacArthur's original hypothesis. *Oikos*, **33**, 351–357.
22.4.2, 22.5.5

Pimm, S.L. (1979b) The structure of food webs. *Theoretical Population Biology*, **16**, 144–158.
22.4.2

Pimm, S.L. (1982) *Food Webs*. Chapman & Hall, London.
11.1, 22.4.2, 22.5.3, 22.5.5

Pimm, S.L. (1991) *The Balance of Nature: Ecological Issues in the Conservation of Species and Communities*. University of Chicago Press, Chicago and London.
20.3.1, 22.3.2, 22.5, 25.3.2

Pimm, S.L. & Gittleman, J.L. (1990) Carnivores and ecologists on the road to Damascus. *Trends in Ecology and Evolution*, **5**, 70–73.
20.3.1

Pimm, S.L. & Lawton, J.H. (1980) Are food webs divided into compartments? *Journal of Animal Ecology*, **49**, 879–898.
22.5.2, 22.5.5, 22.6

Pimm, S.L. & Rice, J.C. (1987) The dynamics of multi-species, multi-life-stage models of aquatic food webs. *Theoretical Population Biology*, **32**, 303–325.
22.5.3

Pisek, A., Larcher, W., Vegis, A. & Napp-Zin, K. (1973) The normal temperature range. In: *Temperature and Life* (H. Precht, J. Chridtopherson, H. Hense & W. Larcher, eds), pp. 102–194. Springer-Verlag, Berlin.
2.12, 18.3.1

Pitcher, T.J. & Hart, P.J.B. (1982) *Fisheries Ecology*. Croom Helm, London.
16.13.2, 16.14

Plante, C.J., Jumars, P.A. & Baross, J.A. (1990) Digestive associations between marine detritivores and bacteria. *Annual Review of Ecology and Systematics*, **21**, 93–127.
11.2.3

Podoler, H. & Rogers, D.J. (1975) A new method for the identification of key factors from life-table data. *Journal of Animal Ecology*, **44**, 85–114.
15.3.1, 15.3.2

Pojar, T.M. (1981) A management perspective of population modelling. In: *Dynamics of Large Mammal Populations* (D.W. Fowler & T.D. Smith, eds), pp. 241–261. Wiley-Interscience, New York.
16.13.2

Polis, G.A. (1991) Complex trophic interactions in deserts: an empirical critique of food-web theory. *American Naturalist*, **138**, 123–155.
22.5.3

Poole, R.W. (1978) *An Introduction to Quantitative Ecology*. McGraw Hill, New York.
15.2.2, 15.4.1

Possingham, H.J.P. & Roughgarden, J. (1990) Spatial population dynamics of a marine organism with a complex life cycle. *Ecology*, **71**, 973–985.
15.6.2

Post, W.M. & Pimm, S.L. (1983) Community assembly and food web stability. *Mathematical Biosciences*, **64**, 162–192.
22.5.6

Potts, G.R., Tapper, S.C. & Hudson, P.J. (1984) Population fluctuations in red grouse: analysis of bag records and a simulation model. *Journal of Animal Ecology*, **53**, 21–36.
12.4, 12.6, 15.4.1

Poulin, R. (1991) Group living and the richness of the parasite fauna in Canadian freshwater fishes. *Oecologia*, **86**, 390–394.
12.3.8

Poulin, R. (1995) Phylogeny, ecology, and the richness of parasite communities in vertebrates. *Ecological Monographs*, **65**, 283–302.
12.3.8

Poulson, M.E. & DeLucia, E.H. (1993) Photosynthetic and structural acclimation to light direction in vertical leaves of *Silphium terebinthaceum*. *Oecologia*, **95**, 393–400.
3.2.1

Power, M.E. (1990) Effects of fish in river food webs. *Science*, **250**, 411–415.
22.3.1

Power, M.E. (1992) Top-down and bottom-up forces in food webs: do plants have primacy? *Ecology*, **73**, 733–746.
22.3.3

Power, M.E. & Stewart, A.J. (1987) Disturbance and recovery of an algal assemblage following flooding in an Oklahoma stream. *The American Midland Naturalist*, **117**, 333–345.
21.5

Prance, G.T. (1981) Discussion. In: *Vicariance Biogeography: A Critique* (D. Nelson & D.E. Rosen, eds), pp. 395–405. Columbia University Press, New York.
1.2.2

Prance, G.T. (1985) The changing forests. In: *Amazonia* (G.T. Prance & T.E. Lovejoy, eds). Pergamon Press, Oxford.
1.2.2

Prance, G.T. (1987). Biogeography of neotropical plants. In: *Biogeography and Quaternary History of Tropical America* (T.C. Whitmore & G.T. Prance, eds), pp. 46–65. Oxford Monographs on Biogeography, 3.
1.2.2

Prentice, I.C., Cramer, W., Harrison, S.P., Leemans, R., Monserud, R.A. & Solomon, A.M. (1992) A global model based on

plant physiology and dominance, soil properties and climate. *Journal of Biogeography*, **19**, 117–134.
18.5

Pressey, R.L., Humphries, C.J., Margules, C.R., Vane-Wright, R.I. & Williams, P.H. (1993) Beyond opportunism: key principles for systematic reserve selection. *Trends in Ecology and Evolution*, **8**, 124–128.
25.1.4, 25.6.3

Preston, F.W. (1962) The canonical distribution of commoness and rarity. *Ecology*, **43**, 185–215, 410–432.
23.3.1

Preszler, R.W. & Price, P.W. (1993) The influence of *Salix* leaf abscission on leaf-miner survival and life-history. *Ecological Entomology*, **18**, 150–154.
8.2.2

Price, M.V. & Waser, N.M. (1979) Pollen dispersal and optimal outcrossing in *Delphinium nelsoni*. *Nature*, **277**, 294–297.
5.4.4

Price, P.W. & Clancy, K.M. (1983) Patterns in number of helminth parasite species in freshwater fishes. *Journal of Parasitology*, **69**, 449–454.
12.3.8

Price, P.W., Westoby, M., Rice, B. *et al.* (1986) Parasite mediation in ecological interactions. *Annual Review of Ecology and Systematics*, **17**, 487–505.
12.6.3

Primack, R.B. (1993) *Essentials of Conservation Biology*. Sinauer Associates, Sunderland, MA.
25.1.3, 25.2.3, 25.2.6, 25.6.3

Privalov, P.L. (1983) Stability of proteins: small globular proteins. *Advances in Protein Chemistry*, **33**, 167–241.
2.4

Prokopy, R.J., Johnson, S.A. & O'Brien, M.T. (1990) Second stage integrated management of apple arthropod pests. *Entomologia Experimentalis et Applicata*, **54**, 9–19.
16.9

Pugh, C.J.F. (1980) Strategies in fungal ecology. *Transactions of the British Mycological Society*, **75**, 1–14.
11.2.1

Pulliam, H.R. (1988) Sources, sinks and population regulation. *American Naturalist*, **132**, 652–661.
15.6.5

Pulliam, H.R. & Curaco, T. (1984) Living in groups: is there an optimal group size? In: *Behavioural Ecology: An Evolutionary Approach*, 2nd edn. (J.R. Krebs & N.B. Davies, eds), pp. 122–147. Blackwell Scientific Publications, Oxford.
5.2.1

Putman, R.J. (1978a) Patterns of carbon dioxide evolution from decaying carrion. Decomposition of small carrion in temperate systems. *Oikos*, **31**, 47–57.
11.2.3

Putman, R.J. (1978b) Flow of energy and organic matter from a carcass during decomposition. Decomposition of small mammal carrion in temperate systems 2. *Oikos*, **31**, 58–68.
11.3.3

Putman, R.J. (1983) *Carrion and Dung: The Decomposition of Animal Wastes*. Edward Arnold, London.
11.3.2, 11.3.3

Putwain, P.D. (1989) Resistance of plants to herbicides. In: *Weed Control Handbook: Principles*, 8th edn (R.J. Hance & K.

Holly, eds), pp. 217–242. Blackwell Scientific Publications, Oxford.
16.6.4

Pyke, G.H. (1982) Local geographic distributions of bumblebees near Crested Butte, Colorada: competition and community structure. *Ecology*, **63**, 555–573.
20.3.1, 20.4.2

Quinn, J.F. & Harrison, S.P. (1988) Local geographic distributions of bumblebees near Crested Butte, Colorado: competition and community structure. *Ecology*, **63**, 555–573.
25.6.3

Quinn, J.F. & Hastings, A. (1987) Extinction in subdivided habitat. *Conservation Biology*, **1**, 198–208.
25.3.3

Rabinowitz, D. (1981) Seven forms of rarity. In: *The Biological Aspects of Rare Plant Conservation* (H. Synge, ed.), pp. 205–217. Wiley, Chichester.
25.2.1

Rabotnov, T.A. (1964) On the biology of monocarpic perennial plants. *Bulletin of the Moscow Society of Naturalists*, **71**, 47–55. (In Russian.)
5.5.3

Rafes, P.M. (1970) Estimation of the effects of phytophagous insects on forest production. In: *Analysis of Temperate Forest Systems* (D.E. Reichle, ed.), pp. 100–106. Springer-Verlag, New York.
8.2.3

Raffa, K.F. & Berryman, A.A. (1983) The role of host plant resistance in the colonization behavior and ecology of bark beetles (Coleoptera: Scolytidae). *Ecological Monographs*, **53**, 27–49.
8.2.2

Rai, B., Freedman, H.I. & Addicott, J.F. (1983) Analysis of three species models of mutualism in predator–prey and competitive systems. *Mathematical Biosciences*, **65**, 13–50.
13.12

Rainey, R.C., Browning, K.A., Cheke, R.A. & Haggis. M.J. (eds) (1990) *Migrant Pests: Progress, Problems and Potentialities*. The Royal Society, London.
5.7.5

Ramade, F. (1981) *Ecology of Natural Resources*. Wiley, Chichester.
19.4.3, 19.4.5

Rand, D.A. & Wilson, H.B. (1995) Using spatio-temporal chaos and intermediate scale determinism to quantify spatially extended ecosystems. *Proceedings of the Royal Society, Series B*, **259**, 111–117.
15.2.4

Randall, M.G.M. (1982) The dynamics of an insect population throughout its altitudinal distribution: *Coleophora alticolella* (Lepidoptera) in northern England. *Journal of Animal Ecology*, **51**, 993–1016.
2.2.2

Ranta, E. (1992) Gregariousness versus solitude: another look at parasite faunal richness in Canadian freshwater fishes. *Oecologia*, **89**, 150–152.
12.3.8

Ranwell, D.S. (1972) *Ecology of Salt Marshes and Sand Dunes*. Chapman & Hall, London.
2.8, 17.4.2

Ranwell, D.S. (1974) The salt marsh to tidal woodland transi-

tion. *Hydrobiological Bulletin (Amsterdam)*, **8**, 139–151.
17.4.2

Rathcke, B.J. (1976) Competiton and coexistence within a guild of herbivorous insects. *Ecology*, **57**, 76–87.
20.3.1

Rathcke, B.J. & Poole, D.O. (1975) Coevolutionary race continues: butterfly larval adaptation to plant trichomes. *Science*, **187**, 175.
3.7.3

Rätti, O., Dufva, R. & Alatalo, R.V. (1993) Blood parasites and male fitness in the pied flycatcher. *Oecologia*, **96**, 410–414.
12.4

Raunkiaer, C. (1934) *The Life Forms of Plants.* Oxford University Press, Oxford. (Translated from the original published in Danish, 1907.)
1.4.2

Raup, D.M. (1978) Cohort analysis of generic survivorship. *Paleobiology*, **4**, 1–15.
25.1.2

Rausher, M.D. (1978) Search image for leaf shape in a butterfly. *Science*, **200**, 1071–1073.
21.2.3

Raushke, E., Haar, T.H. von der, Bardeer, W.R. & Paternak, M. (1973) The annual radiation of the earth–atmosphere system during 1969–70 from Nimbus measurements. *Journal of the Atmospheric Science*, **30**, 341–346.
3.1

Read, A.F. & Harvey, P.H. (1989) Life history differences among the eutherian radiations. *Journal of Zoology*, **219**, 329–353.
14.12.2

Reader, R.J. (1978) Contribution of overwintering leaves to the growth of three broad-leaved, evergreen shrubs belonging to the Ericaceae family. *Canadian Journal of Botany*, **56**, 1248–1261.
19.2.3

Redfern, M., Jones, T.H. & Hassell, M.P. (1992) Heterogeneity and density dependence in a field study of a tephritid–parasitoid interaction. *Ecological Entomology*, **17**, 255–262.
10.5.6

Reece, C.H. (1985) The role of the chemical industry in improving the effectiveness of agriculture. *Philosophical Transactions of the Royal Society of London, Series B*, **310**, 201–213.
16.5

Reekie, E.G. & Bazzaz, F.A. (1989) Competition and patterns of resource use among seedlings of five tropical trees grown at ambient and elevated CO_2. *Oecologia*, **79**, 212–222.
3.3.4

Reichle, D.E. (1970) *Analysis of Temperate Forest Ecosystems.* Springer-Verlag, New York.
18.2, 18.3.1

Reid, E.M. (1935) British floras antecedent to the Great Ice Age. Discussion on the origin and relationship of the British flora. *Proceedings of the Royal Society of London, Series B*, **118**, 197.
24.3.5

Reid, W.V. & Miller, K.R. (1989) *Keeping Options Alive: The Scientific Basis for Conserving Biodiversity.* World Resources Insititute, Washington, DC.
25.2.6, 25.6.3

Retief, J.D., Krajewski, C., Westerman, M., Winkfein, R.J. &

Dixon, G.D. (1995) Molecular phylogeny and evolution of marsupial protamine P1 genes. *Proceedings of the Royal Society, Series B*, **259**, 7–14.
1.3

Rex, M.A. (1981) Community structure in the deep sea benthos. *Annual Review of Ecology and Systematics*, **12**, 331–353.
24.4.3

Reynoldson, T.B. & Bellamy, L.S. (1971) The establishment of interspecific competition in field populations with an example of competition in action between *Polycelis nigra* and *P. tenuis* (Turbellaria, Tricladidae). In: *Dynamics of Populations* (P.J. den Boer & G.R. Gradwell, eds), pp. 282–297. Proceedings of the Advanced Study Institute in Dynamics of Numbers in Populations, Oosterbeck. Centre for Agricultural Publishing and Documentation, Wageningen.
15.7.2

Reznick, D.N. (1982) The impact of predation on life history evolution in Trinidadian guppies: genetic basis of observed life history patterns. *Evolution*, **36**, 1236–1250.
14.7.2

Reznick, D.N. (1985) Cost of reproduction: an evaluation of the empirical evidence. *Oikos*, **44**, 257–267.
14.4

Reznick, D.N., Bryga, H. & Endler, J.A. (1990) Experimentally induced life history evolution in a natural population. *Nature*, **346**, 357–359.
14.7.2

Rhoades, D.F. & Cates, R.G. (1976) Towards a general theory of plant antiherbivore chemistry. *Advances in Phytochemistry*, **110**, 168–213.
3.7.3

Ribble, D.O. (1992) Dispersal in a monogamous rodent, *Peromyscus californicus. Ecology*, **73**, 859–866.
5.4.3, 15.4.4

Rice, E.L. (1984) *Allelopathy*, 2nd edn. Academic Press, New York.
7.3

Richards, B.N. (1974) *Introduction to the Soil Ecosystem.* Longman, Harlow.
17.4.1

Richards, O.W. & Davies, R.G. (1977) *Imm's General Textbook of Entomology*, Vol. 1. Structure, Physiology and Development. Vol. 2. Classification. Biology. Wiley, New York.
3.7.1

Richards, O.W. & Waloff, N. (1954) Studies on the biology and population dynamics of British grasshoppers. *Anti-Locust Bulletin*, **17**, 1–182.
4.5.1, 5.5.1, 6.4

Richman, A.D., Case, T.J. & Schwaner, T.D. (1988) Natural and unnatural extinction rates of reptiles on islands. *American Naturalist*, **131**, 611–630.
23.3.1

Richman, S. (1958) The trasnformation of energy by *Daphnia pulex. Ecological Monographs*, **28**, 273–291.
8.4

Rickards, J., Kelleher, M.J. & Storey, K.B. (1987) Strategies of freeze avoidance in larvae of the goldenrod gall moth *Epiblema scudderiana*: winter profiles of a natural population. *Journal of Insect Physiology*, **33**, 581–586.
2.3.3

Ridley, H.N. (1930) *The Dispersal of Plants Throughout the World*. L. Reeve and Company, Ashford.
5.3.6

Ridsdill-Smith, T.J. (1991) Competition in dung-breeding insects. In: *Reproductive Behaviour of Insects* (W.J. Bailey & T.J. Ridsdill-Smith, eds), pp. 264–294. Chapman & Hall, London.
11.3.2

Rieck, A.F., Belli, J.A. & Blaskovics, M.E. (1960) Oxygen consumption of whole animal tissues in temperature acclimated amphibians. *Proceedings of the Society of Experimental Biology and Medicine*, **103**, 436–439.
2.3.3

Riemer, J. & Whittaker, J.B. (1989) Air pollution and insect herbivores: observed interactions and possible mechanisms. In: *Insect–Plant Interactions* (E.A. Bernays, ed.), pp. 73–105. CRC Press, Boca Raton.
8.2.7

Rigler, F.H. (1961) The relation between concentration of food and feeding rate of *Daphnia magna* Straus. *Canadian Journal of Zoology*, **39**, 857–868.
9.5.2

Riper, C. van, Riper, S.G. van, Goff, M.L. & Laird, M. (1986) The epizootiology and ecological significance of malaria in Hawaiian land birds. *Ecological Monographs*, **56**, 327–344.
21.2.4

Ripper, W.E. (1956) Effects of pesticide on balance of arthropod populations. *Annual Review of Entomology*, **1**, 296–307.
16.6.2

Roberts, H.A. (1964) Emergence and longevity in cultivated soil of seeds of some annual weeds. *Weed Research*, **4**, 296–307.
4.5.2

Robertson, J.H. (1947) Responses of range grasses to different intensities of competition with sagebrush (*Artemisia tridentata* Nutt.). *Ecology*, **28**, 1–16.
7.6

Robinson, S.P., Downton, W.J.S. & Millhouse, J.A. (1983) Photosynthesis and ion content of leaves and isolated chloroplasts of salt-stressed spinach. *Plant Physiology*, **73**, 238–242.
2.7

Rodin, L.E. *et al.* (1975) Primary productivity of the main world ecosystems. In: *Proceedings of the First International Congress of Ecology*, pp. 176–181. Centre for Agricultural Publications, Wageningen.
18.2

Roff, D. (1981) On being the right size. *American Naturalist*, **118**, 405–422.
14.2

Rohani, P., Godfray, H.C.J. & Hassell, M.P. (1994a) Aggregation and the dynamics of host–parasitoid systems: a discrete-generation model with within-generation redistribution. *American Naturalist*, **144**, 491–509.
10.5.4

Rohani, P., Miramontes, O. & Hassell, M.P. (1994b) Quasi-periodicity and chaos in population models. *Proceedings of the Royal Society, Series B*, **258**, 17–22.
15.2.4

Rohde, K. (1978) Latitudinal differences in host-specificity of marine Monogenea and Digenea. *Marine Biology*, **47**, 125–134.
24.4.1

Room, P.M., Harley, K.L.S., Forno, I.W. & Sands, D.P.A. (1981) Successful biological control of the floating weed *Salvinia*. *Nature*, **294**, 78–80.
15.7.6

Room, P.M., Sands, D.P.A., Forno, I.W., Taylor, M.F.J. & Julien, M.H. (1985) A summary of research into biological control of salvinia in Australia. *Proceedings of the VIth International Symposium on Biological Control of Weeds*, pp. 543–549.
16.8

Room, P.M., Maillette, L. & Hanan, J.S. (1994) Module and metamer dynamics and virtual plants. *Advances in Ecological Research*, **25**, 105–157.
4.2.1

Root, R. (1967) The niche exploitation pattern of the blue-grey gnatcatcher. *Ecological Monographs*, **37**, 317–350.
1.3, 20.3.1

Rose, M.R., Service, P.M. & Hutchinson, E.W. (1987) Three approaches to tradeoffs in life-history evolution. In: *Genetic Constraints on Evolution* (V. Loeschke, ed.). Springer, Berlin.
14.4

Rosenzweig, M.L. (1971) Paradox of enrichment: destabilization of exploitation ecosystems in ecological time. *Science*, **171**, 385–387.
24.3.1

Rosenzweig, M.L. & MacArthur, R.H. (1963) Graphical representation and stability conditions of predator–prey interactions. *American Naturalist*, **97**, 209–223.
10.2

Rosewell, J., Shorrocks, B. & Edwards, K. (1990) Competition on a divided and ephemeral resource: testing the assumptions. I. Aggregation. *Journal of Animal Ecology*, **59**, 977–1001.
7.5.5

Ross, M.A. & Harper, J.L. (1972) Occupation of biological space during seedling establishment. *Journal of Ecology*, **60**, 77–88.
6.11

Ross, R. & Thomson, D. (1910) A case of sleeping sickness studied by precise enumerative methods: regular periodical increase of the parasite described. *Proceedings of the Royal Society of London, Series B*, **82**, 411–415.
12.7

Rosswall, T.H. (1983) *The Major Biogeochemical Cycles and their Interactions*, pp. 46–50. Wiley, Chichester.
19.4.4

Roth, G.D. (1981) *Collins Guide to the Weather*. Collins, London.
20.4.2

Rotheray, G.E. (1979) The biology and host searching behaviour of a cynipid parasite of aphidophagous syrphid larvae. *Ecological Entomology*, **4**, 75–82.
9.8

Rothstein, S.I. (1990) A model system for coevolution: avian brood parasitism. *Annual Review of Ecology and Systematics*, **21**, 481–508.
12.8

Roughgarden, J. (1979) *Theory of Population Genetics and Evolutionary Ecology: An Introduction*. Macmillan, New York.
15.6.4

Roughgarden, J. (1992) *Anolis Lizards of the Caribbean: Ecology, Evolution and Plate Tectonics*. Oxford University Press, Oxford.
12.3.8

Roughgarden, J., Gaines, S.D. & Isawa, Y. (1984) Dynamics and

evolution of marine populations with pelagic larval dispersal. In: *Exploitation of Marine Communities* (R.M. May, ed.), pp. 111–128. Dahlem Konferenzen, Springer Verlag, Berlin.
2.3.2

Roush, R.T. & McKenzie, J.A. (1987) Ecological genetics of insecticide and acaricide resistance. *Annual Review of Entomology*, **32**, 361–380.
16.6.4

Rowell, J.B. (1984) Controlled infection by *Puccinia graminis* f. sp. *tritici* under artificial conditions. In: *The Cereal Rusts*, Vol. 1 (W.R. Bushell & A.P. Roelfs, eds), pp. 291–332. Academic Press, New York.
2.2.2

Rubenstein, D.I. (1981) Individual variation and competition in the Everglades pygmy sunfish. *Journal of Animal Ecology*, **50**, 337–350.
6.11

Russell, E.W. (1973) *Soil Conditions and Plant Growth*, 10th edn. Longman, London.
19.2.4

Russell, N.J. (1990) Cold adaptation of microorganisms. *Philosophical Transactions of the Royal Society Series B*, **326**, 595–661; also in *Life at Low Temperatures* (R.M. Laws & F. Franks, eds), pp. 79–95. The Royal Society, London.
2.3

Ryan, C.A. & Jagendorf, A. (1995) Self defense by plants. *Proceedings of the National Academy of Science of the USA*, **92**, 4075.
12.3.5

Ryan, C.A., Lamb, C.J., Jagendorf, A.T. & Kolattukudy, P.E. (eds) (1995) Self-defense by plants: induction and signalling pathways. *Proceedings of the National Academy of Science of the USA*, **92**, 4075–4205.
12.3.5

Ryle, G.J.A. (1970) Partition of assimilates in an annual and perennial grass. *Journal of Applied Ecology*, **7**, 217–227.
8.2.1

Rypstra, A.L. (1984) A relative measure of predation on web-spiders in temperate and tropical forests. *Oikos*, **43**, 129–132.
24.4.1

Sackville Hamilton, N.R., Schmid, B. & Harper, J.L. (1987) Life history concepts and the population biology of clonal organisms. *Proceedings of the Royal Society of London, Series B*, **232**, 35–57.
5.6

Sagar, G.R. & Harper, J.L. (1960) Factors affecting the germination and early establishment of plantains (*Plantago lanceolata, P. media* and *P. major*). In: *The Biology of Weeds* (J.L. Harper, ed.), pp. 236–244. Blackwell Scientific Publications, Oxford.
15.7.1

Sait, S.M., Begon, M. & Thompson, D.J. (1994) Long-term population dynamics of the Indian meal moth *Plodia interpunctella* and its granulosis virus. *Journal of Animal Ecology*, **63**, 861–870.
12.6

Sakai, A. & Otsuka, K. (1970) Freezing resistance of alpine plants. *Ecology*, **51**, 665–671.
2.3.3

Sakai, K.I. (1958) Studies on competition in plants and

animals. IX. Experimental studies on migration in *Drosophila melanogaster*. *Evolution*, **12**, 93–101.
5.4.3

Sale, P.F. (1977) Maintenance of high diversity in coral reef fish communities. *American Naturalist*, **111**, 337–359.
21.6

Sale, P.F. (1979) Recruitment, loss and coexistence in a guild of territorial coral reef fishes. *Oecologia*, **42**, 159–177.
21.6

Sale, P.F. & Douglas, W.A. (1984) Temporal variability in the community structure of fish on coral patch reefs and the relation of community structure to reef structure. *Ecology*, **65**, 409–422.
21.6

Salisbury, E.J. (1942) *The Reproductive Capacity of Plants*. Bell, London.
14.9.1

Salonen, K., Jones, R.I. & Arvola, L. (1984) Hypolimnetic retrieval by diel vertical migrations of lake phytoplankton. *Freshwater Biology*, **14**, 431–438.
5.7.1

Saloniemi, I. (1993) An environmental explanation for the character displacement pattern in *Hydrobia* snails. *Oikos*, **67**, 75–80.
7.8.1

Salthe, S.N. (1969) Reproductive modes and the number and sizes of ova in the urodeles. *American Midland Naturalist*, **81**, 467–490.
14.8

Sances, F.V., Toscana, N.C., Johnson, M.W. & LaPre, L.F. (1981) Pesticides may reduce lettuce yields. *Californian Agriculture*, **11–12**, 4–5.
16.6.1

Sanders, H.L. (1968) Marine benthic diversity: a comparative study. *American Naturalist*, **102**, 243–282.
24.4.3

Sarukhán, J. (1974) Studies on plant demography: *Ranunculus repens* L., *R. bulbosus* L. and *R. acris* L. II. Reproductive strategies and seed population dynamics. *Journal of Ecology*, **62**, 151–177.
8.2.5, 15.2

Sarukhán, J. & Harper, J.L. (1973) Studies on plant demography: *Ranunculus repens* L., *R. bulbosus* L. and *R. acris* L. I. Population flux and survivorship. *Journal of Ecology*, **61**, 675–716.
15.2

Saurola, P. (1985) Finnish birds of prey: status and population change. *Ornis Fennica*, **62**, 64–72.
15.4.3

Savidge, J.A. (1987) Extinction of an island forest avifauna by an introduced snake. *Ecology*, **68**, 660–668.
21.2.3, 25.2.4

Schaefer, M.B. (1954) Some aspects of the dynamics of populations important to the management of marine fisheries. *Bulletin of the Inter-American Tropical Tuna Commission*, **1**, 27–56.
16.16

Schaffer, W.M. (1974) Optimal reproductive effort in fluctuating environments. *American Naturalist*, **108**, 783–790.
14.10.2

Schaffer, W.M. & Kot, M. (1986) Chaos in ecological systems:

the coals that Newcastle forgot. *Trends in Ecology and Evolution*, **1**, 58–63.
6.9

Schaffer, W.M. & Rosenzweig, M. (1977) Selection for optimal life histories. II. Multiple equilibria and the evolution of alternative reproductive strategies. *Ecology*, **58**, 60–72.
14.7.3

Schall, J.J. (1992) Parasite-mediated competition in *Anolis* lizards. *Oecologia*, **92**, 58–64.
12.4

Scheiner, S.M. & Lyman, R.F. (1991) The genetics of phenotypic plasticity. II. The response to selection. *Journal of Evolutionary Biology*, **4**, 23–50.
14.11

Schenck, N.C. & Pérez, Y. (1990) *Manual for the Identification of VA Mycorrhizal Fungi*, 3rd edn. Synergistic Publications, Gainesville, Florida.
13.7.2

Schindler, D.W. (1978) Factors regulating phytoplankton production and standing crop in the world's freshwaters. *Limnology and Oceanography*, **23**, 478–486.
18.3.3

Schluter, D. & McPhail, J.D. (1992) Ecological character displacement and speciation in sticklebacks. *American Naturalist*, **140**, 85–108.
7.8.1, 20.4.2

Schluter, D. & McPhail, J.D. (1993) Character displacement and replicate adaptive radiation. *Trends in Ecology and Evolution*, **8**, 197–200.
7.8.1, 20.4.2

Schlyter, F. & Anderbrant, O. (1993) Competition and niche separation between two bark beetles: existence and mechanisms. *Oikos*, **68**, 437–447.
7.10

Schmidt-Nielsen, K. (1983) *Animal Physiology*. Cambridge University Press, Cambridge.
2.2

Schmidt-Nielsen, K. (1984) *Scaling: Why is Animal Size so Important?* Cambridge University Press, Cambridge.
14.12.1

Schmitt, R.J. (1987) Indirect interactions between prey: apparent competition, predator aggregation, and habitat segregation. *Ecology*, **68**, 1887–1897.
7.6

Schnell, R.C. & Vali, G. (1976) Biogenic ice nuclei. I. Terrestrial and marine sources. *Journal of Atmospheric Sciences*, **33**, 1554–1564.
2.3.2

Schlesinger, W.H., Gray, J.T. & Gilliam, F.S. (1982) Atmospheric deposition processes and their importance as sources of nutrients in a chapparal ecosystem of southern California. *Water Resource Research*, **18**, 623–629.
19.2.1

Schoener, T.W. (1974) Resource partitioning in ecological communities. *Science*, **185**, 27–39.
20.3.1

Schoener, T.W. (1983) Field experiments on interspecific competition. *American Naturalist*, **122**, 240–285.
7.6, 7.7, 20.2, 22.2.1

Schoener, T.W. (1984) Size differences among sympatric, bird-eating hawks: a worldwide study. In: *Ecological Communities: Conceptual Issues and the Evidence* (D.R. Strong, D. Simberloff, L.G. Abele & A.B. Thistle, eds), pp. 254–281. Princeton University Press, Princeton, NJ.
20.4.2

Schoener, T.W. (1986) Patterns in terrestrial vertebrate versus arthropod communities: do systematic differences in regularity exist? In: *Community Ecology* (J. Diamond & T.J. Case, eds), pp. 556–586. Harper & Row, New York.
23.3.5

Schoenly, K., Beaver, R.A. & Heumier, T.A. (1991) On the trophic relations of insects: a food-web approach. *American Naturalist*, **137**, 597–638.
22.4.3, 22.5.2

Schopf, Y.J.M. (1974) Permo–Triassic extinctions: a relation to sea-floor spreading. *Journal of Geology*, **82**, 129–143.
24.4.6

Schultz, J.C. (1988) Plant responses induced by herbivory. *Trends in Ecology and Evolution*, **3**, 45–49.
8.2.2

Schulze, E.D. (1970) Dre CO_2-Gaswechsel de Buche (*Fagus sylvatica* L.) in Abhäbgigkeit von den Klimafaktoren in Freiland. *Flora, Jena*, **159**, 177–232.
18.3.1

Schulze, E.D., Fuchs, M.I. & Fuchs, M. (1977a) Spatial distribution of photosynthetic capacity and performance in a mountain spruce forest in northern Germany. I. Biomass distribution and daily CO_2 uptake in different crown layers. *Oecologia*, **29**, 43–61.
18.3.1

Schulze, E.D., Fuchs, M.I. & Fuchs, M. (1977b) Spatial distribution of photosynthetic capacity and performance in a mountain spruce forest in northern Germany. III.The significance of the evergreen habit. *Oecologia*, **30**, 239–249.
18.3.1

Schulze, E.D., Robichaux, G.H., Grace, J., Rundel, P.W. & Ehrelinger, J.R. (1987) Plant water balance. *Bioscience*, **37**, 30–37.
3.4

Schwerdtfeger, F. (1941) Uber die Ursachen des Massenweschels der Insekten. *Zeischrift für angewandte Entomologie*, **28**, 254–303.
10.6.1

Seaton, A.P.C. & Antonovics, J. (1967) Population interrelationships. I. Evolution in mixtures of *Drosophila* mutants. *Heredity*, **22**, 19–33.
7.8.3

Sessions, G. (1987) The deep ecology movement: a review. *Environmental Review*, **11**, 105–125.
25.1.3

Sexstone, A.Y., Parkin, T.B. & Tiedje, J.M. (1985) Temporal response of soil denitrification rates to rainfall and irrigation. *Soil Science Society of America Journal*, **49**, 99–103.
19.2.2

Sharitz, R.R. & McCormick, J.F. (1973) Population dynamics of two competing annual plant species. *Ecology*, **54**, 723–740.
7.10

Shaver, G.R. & Melillo, J.M. (1984) Nutrient budgets of marsh plants: efficiency concepts and relation to availability. *Ecology*, **65**, 1491–1510.
19.2.3

Sherman, P.W. (1981) Reproductive competition and infanticide in Belding's ground squirrels and other animals. In: *Natural Selection and Social Behaviour: Recent Research and New Theory* (R.D. Alexander & D.W. Tinkle, eds), pp. 311–331. Chiron Press, New York.
6.12

Shorrocks, B. & Begon, M. (1975) A model of competiton. *Oecologia*, **20**, 363–367.
21.3

Shorrocks, B. & Rosewell, J. (1987) Spatial patchiness and community structure: coexistence and guild size of drosophilids on ephemeral resources. In: *Organization of Communities: Past and Present* (J.H.R. Gee & P.S. Giller, eds), pp. 29–52. Blackwell Scientific Publications, Oxford.
7.5.5, 20.2.2, 21.5

Shorrocks, B., Atkinson, W. & Charlesworth, P. (1979) Competition on a divided and ephemeral resource. *Journal of Animal Ecology*, **48**, 899–908.
7.5.5

Shorrocks, B., Rosewell, J., Edwards, K. & Atkinson, W. (1984) Competition may not be a major organizing force in many communities of insects. *Nature*, **310**, 310–312.
20.2.2

Shorrocks, B., Rosewell, J. & Edwards, K. (1990) Competition on a divided and ephemeral resource: testing the assumptions. II. Association. *Journal of Animal Ecology*, **59**, 1003–1017.
7.5.5

Shykoff, J.A. & Bucheli, E. (1995) Pollinator visitation patterns, floral rewards and the probability of transmission of *Microbotryum violaceum*, a venereal disease of plants. *Journal of Ecology*, **83**, 189–198.
13.4.2

Sibly, R.M. & Calow, P. (1983) An integrated approach to life-cycle evolution using selective landscapes. *Journal of Theoretical Biology*, **102**, 527–547.
14.5, 14.6

Sibly, R.M. & Calow, P. (1985) Classification of habitats by selection pressures: a synthesis of life-cycle and r/K theory. In: *Behavioural Ecology* (R.M. Sibly & R.H. Smith, eds), pp. 75–90. Blackwell Scientific Publications, Oxford.
14.6

Sibly, R.M. & Calow, P. (1986) *Physiological Ecology of Animals*. Blackwell Scientific Publications, Oxford.
14.2, 14.3

Sih, A. (1982) Foraging strategies and the avoidance of predation by an aquatic insect, *Notonecta hoffmanni*. *Ecology*, **63**, 786–796.
9.4

Sih, A., Crowley, P., McPeek, M., Petranka, J. & Strohmeier, K. (1985) Predation, competition and prey communities: a review of field experiments. *Annual Review of Ecology and Systematics*, **16**, 269–311.
22.2.1

Silander, J.A. & Antonovics, J. (1982) A perturbation approach to the analysis of interspecific interactions in a coastal plant community. *Nature*, **298**, 557–560.
15.7.3

Silvertown, J.W. (1980) The evolutionary ecology of mast seeding in trees. *Biological Journal of the Linnean Society*, **14**, 235–250.
8.4

Silvertown, J.W. (1982) *Introduction to Plant Population Ecology*. Longman, London.
15.2, 15.3.2

Silvertown, J.W. (1984) Phenotypic variety in seed germination behavior: the ontogeny and evolution of somatic polymorphism in seeds. *American Naturalist*, **124**, 1–16.
14.11

Silvertown, J.W., Holtier, S., Johnson, J. & Dale, P. (1992a) Cellular automaton models of interspecific competition for space—the effect of pattern on process. *Journal of Ecology*, **80**, 527–533.
3.10, 7.5.5

Silvertown, J.W., Franco, M. & McConway, K. (1992b) A demographic interpretation of Grime's triangle. *Functional Ecology*, **6**, 130–136.
14.10.1

Silvertown, J.W., Franco, M., Pisanty, I. & Mendoza, A. (1993) Comparative plant demography—relative importance of life-cycle components to the finite rate of increase in woody and herbaceous perennials. *Journal of Ecology*, **81**, 465–476.
14.10.1

Simberloff, D. (1984) Properties of coexisting bird species in two archipelagoes. In: *Ecological Communities: Conceptual Issues and the Evidence* (D.R. Strong, D. Simberloff, L.G. Abele & A.B. Thistle, eds), pp. 234–253. Princeton University Press, Princeton, NJ.
20.4.2

Simberloff, D. (1988) The contribution of population and community biology to conservation science. *Annual Review of Ecology and Systematics*, **19**, 473–511.
25.3.1

Simberloff, D. & Boecklen, W. (1981) Santa Rosalia reconsidered: size ratios and competition. *Evolution*, **35**, 1206–1228.
20.4.2

Simberloff, D. & Dayan, T. (1991) The guild concept and the structure of ecological communities. *Annual Review of Ecology and Systematics*, **22**, 115–143.
20.3.1

Simberloff, D.S. (1976) Experimental zoogeography of islands: effects of island size. *Ecology*, **57**, 629–648.
23.3.1, 23.3.4

Simberloff, D.S. & Abele, L.G. (1976) Island biogeography theory and conservation practice. *Science*, **191**, 285–286.
25.6.3

Simberloff, D.S. & Wilson, E.O. (1969) Experimental zoogeography of islands: the colonization of empty islands. *Ecology*, **50**, 278–296.
23.3.4

Simpson, G.G. (1952) How many species? *Evolution*, **6**, 342.
25.1.2

Sinclair, A.R.E. (1973) Regulation, and population models for a tropical ruminant. *East African Wildlife Journal*, **11**, 307–316.
15.3.2

Sinclair, A.R.E. (1975) The resource limitation of trophic levels in tropical grassland ecosystems. *Journal of Animal Ecology*, **44**, 497–520.
8.4

Sinclair, A.R.E. (1989) The regulation of animal populations.

In: *Ecological Concepts* (J. Cherrett, ed.), pp. 197–241. Blackwell Scientific Publications, Oxford.
15.2.3

Sinclair, A.R.E. & Norton-Griffiths, M. (1982) Does competition or facilitation regulate migrant ungulate populations in the Serengeti? A test of hypothesis. *Oecologia*, **53**, 354–369.
6.4

Sinclair, A.R.E., Krebs, C.J., Smith, J.N.M. & Boutin, S. (1988) Population biology of snowshoe hares. III. Nutrition, plant secondary compounds, and food limitation. *Journal of Animal Ecology*, **57**, 787–806.
10.2.4

Sinervo, B. (1990) The evolution of maternal investment in lizards: an experimental and comparative analysis of egg size and its effects on offspring performance. *Evolution*, **44**, 279–294.
14.4.2

Skaf, R., Popov, G.B. & Roffey, J. (1990) The desert locust: an international challenge. *Philosophical Transactions of the Royal Society of London*, **328**, 525–538; and In: *Migrant Pests: Progress, Problems and Potentialities* (R.C. Rainey, K.A. Browning, R.A. Cheke & M.J. Haggis, eds), pp. 7–20. The Royal Society, London.
5.7.5

Skidmore, D.I. (1983) *Population Dynamics of* Phytophthora infestans *(Mont.) de Bary*. Ph.D. thesis, University of Wales.
12.3.9

Skogland, T. (1983) The effects of dependent resource limitation on size of wild reindeer. *Oecologia*, **60**, 156–168.
6.5

Slade, A.J. & Hutchings, M.J. (1987) Clonal integration and plasticity in foraging behaviour in *Glechoma hedereacea*. *Journal of Ecology*, **75**, 1023–1036.
4.2.1

Slobodkin, L.B., Smith, F.E. & Hairston, N.G. (1967) Regulation in terrestrial ecosystems, and the implied balance of nature. *American Naturalist*, **101**, 109–124.
20.2.1

Smith, D.C. (1979) From extracellular to intracellular: the establishment of a symbiosis. *Proceedings of the Royal Society of London, Series B*, **204**, 131–139.
13.5

Smith, F.D.M, May, R.M., Pellew, R., Johnson, T.H. & Walter, K.R. (1993). How much do we know about the current extinction rate? *Trends in Ecology and Evolution*, **8**, 375–378.
25.1.2

Smith, F.E. (1961) Density dependence in the Australian thrips. *Ecology*, **42**, 403–407.
15.2.2

Smith, J.N.M. (1974) The food searching behaviour of two European thrushes. II. The adaptiveness of the search patterns. *Behaviour*, **49**, 1–61.
9.8

Smith, J.N.M., Krebs, C.J., Sinclair, A.R.E. & Boonstra, R. (1988) Population biology of snowshoe hares. II. Interactions with winter food plants. *Journal of Animal Ecology*, **57**, 269–286.
10.2.4

Smith, R.H. (1991) Genetic and phenotypic aspects of life-history evolution in animals. *Advances in Ecological Research*, **21**, 63–120.
14.1, 14.3

Smith, R.H. & Bass, M.H. (1972) Relation of artificial removal to soybean yields. *Journal of Economic Entomology*, **65**, 606–608.
8.2.1

Smith, S.D., Dinnen-Zopfy, B. & Nobel, P.S. (1984) High temperature responses of North American cacti. *Ecology*, **6**, 643–651.
2.4

Snaydon, R.W. & Bradshaw, A.D. (1969) Differences between natural populations of *Trifolium repens* L. in response to mineral nutrients. II. Calcium, magnesium and potassium. *Journal of Applied Ecology*, **6**, 185–202.
1.5.1

Snodgrass, R.E. (1944) The feeding apparatus of biting and sucking insects affecting man. *Smithsonian Miscellaneous Collections*, **194**(7), 113.
3.7.1

Snyman, A. (1949) The influence of population densities on the development and oviposition of *Plodia interpunctella* Hubn. (Lepidoptera). *Journal of the Entomological Society of South Africa*, **12**, 137–171.
6.6

Solangaarachchi, S.M. & Harper, J.L. (1989) The growth and asymmetry of neighbouring plants of white clover (*Trifolium repens* L.). *Oecologia*, **78**, 208–213.
6.3

Solbrig, O.T. & Simpson, B.B. (1974) Components of regulation of a population of dandelions in Michigan. *Journal of Ecology*, **62**, 473–486.
14.7.1

Solomon, M.E. (1949) The natural control of animal populations. *Journal of Animal Ecology*, **18**, 1–35.
9.5

Sommer, U. (1990) Phytoplankton nutrient competition—from laboratory to lake. In: *Perspectives on Plant Competition* (J.B. Grace & D. Tilman, eds), pp. 193–213. Academic Press, New York.
7.10.2

Sorensen, A.E. (1978) Somatic polymorphism and seed dispersal. *Nature*, **276**, 174–176.
5.3.6

Soulé, M.E. (1986) *Conservation Biology: The Science of Scaricity and Diversity*. Sinauer Associates, Sunderland, MA.
25.6.3

Soulé, M.E. & Wilcox, B.A. (eds) (1980) *Conservation Biology: An Evolutionary–Ecological Perspective*. Sinauer Associates, Sunderland, MA.
25.6.3

Soulé, M.E., Alberts, A.C. & Bolger, D.T. (1992) The effects of habitat fragmentation on chaparral plants and vertebrates. *Oikos*, **63**, 39–47.
25.3.3

Sousa, M.E. (1979a) Experimental investigation of disturbance and ecological succession in a rocky intertidal algal community. *Ecological Monographs*, **49**, 227–254.
17.4.3, 21.5, 21.5.1

Sousa, M.E. (1979b) Disturbance in marine intertidal boulder fields: the nonequilibrium maintenance of species diversity. *Ecology*, **60**, 1225–1239.
21.5.1

Southern, H.N. (1970) The natural control of a population of

tawny owls (*Strix aluco*). *Journal of Zoology*, **162**, 197–285.
10.1, 15.3.2

Southwood, T.R.E. (1961) The numbers of species of insect associated with various trees. *Journal of Animal Ecology*, **30**, 1–8.
23.2.4

Southwood, T.R.E. (1977) Habitat, the templet for ecological strategies? *Journal of Animal Ecology*, **46**, 337–365.
14.1, 14.6

Southwood, T.R.E (1978a) *Ecological Methods*, 2nd edn. Chapman & Hall, London.
4.3, 5.2

Southwood, T.R.E. (1978b) The components of diversity. *Symposium of the Royal Entomological Society of London*, **9**, 19–40.
20.2.1

Southwood, T.R.E. (1988) Tactics, strategies and templets. *Oikos*, **52**, 3–18.
14.1

Southwood, T.R.E., Moran, V.C. & Kennedy, C.E.J. (1982) The richness, abundance and biomass of the arthropod communities on trees. *Journal of Animal Ecology*, **51**, 635–649.
23.4

Southwood, T.R.E., Hassell, M.P., Reader, P.M. & Rogers, D.J. (1989) The population dynamics of the viburnum whitefly (*Aleurotrachelus jelinekii*). *Journal of Animal Ecology*, **58**, 921–942.
10.6.1

Spencer, C.N., McClelland, B.R. & Stanford, J.A. (1991) Shrimp stocking, salmon collapse, and eagle displacement. *Bioscience*, **41**, 14–21.
25.5.2

Spiller, D.A. & Schoener, T.W. (1994) Effects of a top and intermediate predators in a terrestrial food web. *Ecology*, **75**, 182–196.
22.3.1

Spooner, G.M. (1947) The distribution of *Gammarus* species in estuaries. *Journal of the Marine Biological Association*, **27**, 1–52.
2.7.1

Spradbery, J.P. (1970) Host findings of *Rhyssa persuasoria* (L.), an ichneumonid parasite of siricid woodwasps. *Animal Behaviour*, **18**, 103–114.
9.6

Sprent, J.I. (1979) *The Biology of Nitrogen Fixing Organisms*. McGraw Hill, London.
13.10.1

Sprent, J.I. (1987) *The Ecology of the Nitrogen Cycle*. Cambridge University Press, Cambridge.
13.10, 19.2.4

Sprent, J.I. & Sprent, P. (1990) *Nitrogen Fixing Organisms: Pure and Applied Aspects*. Chapman & Hall, London.
13.10, 13.10.1, 13.10.3

Sprules, W.G. & Bowerman, J.E. (1988) Omnivory and food chain length in zooplankton food webs. *Ecology*, **69**, 418–426.
22.5.3

Srinivasan, K. & Moorthy, P.N. (1991) Indian mustard as a trap crop for management of major lepidopterous pests on cabbage. *Tropical Pest Management*, **37**, 26–32.
16.9

Staaland, H., White, R.G., Luick, J.R. & Holleman, D.F. (1980) Dietary influences on sodium and potassium metabolism of reindeer. *Canadian Journal of Zoology*, **58**, 1728–1734.
9.2.2

Stafford, J. (1971) Heron populations of England and Wales 1928–70. *Bird Study*, **18**, 218–221.
10.6.1

Stahler, A.N. (1960) *Physical Geography*, 2nd edn. Wiley, Chichester.
1.6

Stanley Price, M.R. (1989) *Animal Re-introductions: The Arabian Oryx in Oman*. Cambridge University Press, Cambridge.
25.6.2

Stanley, S.M. (1976) Ideas on the timing of metazoan diversification. *Paleobiology*, **2**, 209–219.
24.4.6

Stanley, S.M. (1979) *Macroevolution*. W.H. Freeman, San Francisco.
24.3.5

Stanton, M.L. (1994) Male–male competition during pollination in plant populations. *American Naturalist*, **144**, S40–S68.
13.4.2

Stapledon, R.G. (1928) Cocksfoot grass (*Dactylis glomerata* L.): ecotypes in relation to the biotic factor. *Journal of Ecology*, **16**, 72–104.
1.5.1

Stauffer, B. (1988) Ziele und Ergebnisse klimatologisch-glaziologischer Forschungsprojekte in der Arctis und Antarktis. In: *Swiss Commission for Polar Research (SKP), Die Polarregionen und die Schweizerische Forschung*, pp. 95–110. Schweizerische Naturforschenden Gesellschaft, Switzerland.
2.11.1

Stearns, S.C. (1977) The evolution of life history traits. *Annual Review of Ecology and Systematics*, **8**, 145–171.
14.9.1

Stearns, S.C. (1980) A new view of life-history evolution. *Oikos*, **35**, 266–281.
14.7.1

Stearns, S.C. (1983) The impact of size and phylogeny on patterns of covariation in the life history traits of mammals. *Oikos*, **41**, 173–187.
14.12.2

Stearns, S.C. (1992) *The Evolution of Life Histories*. Oxford University Press, Oxford.
14.1, 14.4.2, 14.7.2, 14.7.3, 14.8.1, 14.10.2, 14.11, 14.12.1, 14.12.2

Steenvorden, J.H.A.M. (1988) How to reduce nitrogen losses in intensive grassland management. In: *Ecological Implications of Contemporary Agriculture* (H. Eijsackers & A. Quispel, eds). *Ecological Bulletins*, **39**, 126–130.
19.2.4

Steffen, W.L., Walker, B.H., Ingram, J.S.I. & Koch, G.W. (eds) (1992) *Global Change and Terrestrial Ecosystems: The Operational Plan*. IGBP, ICSU, Stockholm, Sweden.
18.1, 19.4

Stemberger, R.S. & Gilbert, J.J. (1984) Spine development in the rotifer *Keratella cochlearis*: induction by cyclopoid copepods and *Asplachna*. *Freshwater Biology*, **14**, 639–648.
3.7.3

Steneck, R.S. (1991) Plant herbivore coevolution: a reappraisal from the marine realm and its fossil record. In: *Plant–Animal*

Interactions in the Marine Benthos (D.M. John, S.J. Hawkins & J.H. Price, eds). Systematics Association.
3.7.3

Stenseth, N.T. (1993) *The Biology of Lemmings*. Academic Press, New York.
15.4.5

Stephens, D.W. & Krebs, J.R. (1986) *Foraging Theory*. Princeton University Press, Princeton.
9.3, 9.9.3

Stephens, G.R. (1971) The relation of insect defoliation to mortality in Connecticut forests. *Connecticut Agricultural Experimental Station Bulletin*, **723**, 1–16.
8.2.4

Steponkus P.L., Lynch, D.V. & Uemura, M. (1990) The influence of cold acclimation on the lipid composition and cryobehaviour of the plasma membrane of isolated rye protoplasts. *Philosophical Transactions of the Royal Society, Series B*, **326**, 571–583; also in *Life at Low Temperatures* (R.M. Laws & F. Franks, eds), pp. 55–77. The Royal Society, London.
2.3.3

Stern, D.L. (1995) Phylogenetic evidence that aphids, rather than plants, determine gall morphology. *Proceedings of the Royal Society, Series B*, **260**, 85–89.
12.3.5

Stern, V.M., Smith, R.F., van den Bosch, R. & Hagen, K.S. (1959) The integration of chemical and biological control of the spotted alfalfa aphid. *Hilgardia*, **29**, 81–154.
16.11

Steven, D. De (1983) Reproductive consequences of insect seed predation in *Hamamelis virginiana*. *Ecology*, **64**, 89–98.
8.4

Stiles, G.F. (1977) Coadapted competitors: the flowering seasons of hummingbird-pollinated plants in a tropical forest. *Science*, **198**, 1177–1178.
20.3.1

Stiling, P. (1988) Density-dependent processes and key-factors in insect populations. *Journal of Animal Ecology*, **57**, 581–594.
15.2.3

Stiling, P.D. (1987) The frequency of density-dependence in insect host–parasitoid systems. *Ecology*, **68**, 844–856.
9.7.1, 10.5.6

Stockner, J.G. & Antia, N.J. (1986) Algal picoplankton from marine and freshwater ecosystems. *Canadian Journal of Fish and Aquatic Science*, **43**, 2472–2503.
19.3.3

Stolp. H. (1988) *Microbial Ecology*. Cambridge University Press, Cambridge.
2.4, 2.6, 19.2.1

Stoner, D.S. (1992) Vertical distribution of a colonial ascidian on a coral reef: the roles of larval distribution and life-history variation. *American Naturalist*, **139**, 802–824.
15.6.2

Storey, K.B. (1990) Biochemical adaptation for cold hardiness in insects. *Philosophical Transactions of the Royal Society, Series B*, **326**, 635–654; also in *Life at Low Temperatures* (R.M. Laws & F. Franks, eds), pp. 119–138. The Royal Society, London.
2.3.3

Stork, N.E. & Lyal, C.H.C. (1993) Extinction or 'co-extinction' rates? *Nature*, **366**, 307.
25.2.6

Stout, J. & Vandermeer, J. (1975) Comparison of species richness for stream-inhabiting insects in tropical and mid-latitude streams. *American Naturalist*, **109**, 263–280.
17.2

Stowe, L.G. & Teeri, J.A. (1978) The geographic distribution of C_4 species of the Dicotyledonae in relation to climate. *American Naturalist*, **112**, 609–623.
3.3.2

Strain, B.R. (1987) Direct effects of increasing atmospheric CO_2 on plants and ecosystems. *Trends in Ecology and Evolution*, **2**, 18–21.
19.4.6

Straube, W.L., Deming, J.W., Somerville, C.C., Colwell, R.R. & Baross, J.A. (1990) Particulate DNA in smoker fluids: evidence for existence of microbial populations in hot hydrothermal systems. *Applied and Environmental Microbiology*, **56**, 1440–1447.
2.4

Stribley, D.P., Tinker, P.B. & Snellgrove, R.C. (1980) Effect of vesicular–arbuscular mycorrhizal fungi on the relations of plant growth, internal phosphorus concentration and soil phosphate analysis. *Journal of Soil Science*, **31**, 655–672.
13.7.2

Strickland, E.H. (1945) Could the widespread use of DDT be a disaster? *Entomological News*, **26**(4), 85–88.
16.4

Strobeck, C. (1973) N species competition. *Ecology*, **54**, 650–654.
7.4.1

Strobel, G.A. & Lanier, G.N. (1981) Dutch elm disease. *Scientific American*, **245**, 40–50.
8.2.6

Strong, D.R. (1992) Are trophic cascades all wet? Differentiation and donor-control in speciose ecosystems. *Ecology*, **73**, 747–754.
22.3.1, 22.3.3

Strong, D.R. Jr. (1974) Rapid asymptotic species accumulation in phytophagous insect communities: the pests of Cacao. *Science*, **185**, 1064–1066.
23.1

Strong, D.R. Jr. (1982) Harmonious coexistence of hispine bettles on *Heliconia* in experimental and natural communities. *Ecology*, **63**, 1039–1049.
20.3.1

Strong, D.R. Jr., & Simberloff, D.S. (1981) Straining at gnats and swallowing ratios: character displacement. *Ecology*, **35**, 810–812.
20.4.2

Strong, D.R. Jr., Lawton, J.H. & Southwood, T.R.E. (1984) *Insects on Plants: Community Patterns and Mechanisms*. Blackwell Scientific Publications, Oxford.
13.3.2, 20.2.1, 20.3.1, 23.1, 23.3.5, 24.4.6

Stubbs, M. (1977) Density dependence in the life-cycles of animals and its importance in *k*- and *r*-strategies. *Journal of Animal Ecology*, **46**, 677–688.
15.3.2

Suberkropp, K., Godshalk, G.L. & Klug, M.J. (1976) Changes in the chemical composition of leaves during processing in a woodland stream. *Ecology*, **57**, 720–727.
11.2.1

Suchanek, T.H. (1992) Extreme biodiversity in the marine environment: mussel bed communities of *Mytilus califor-*

nianus. The Northwest Environmental Journal, **8**, 150.
21.2.2

Suhonen, J., Alatalo, R.V. & Gustafsson, L. (1994) Evolution of foraging ecology in Fennoscandian tits. *Proceedings of the Royal Society, Series B*, **258**, 127–131.
1.4.3

Sullivan, T.P. & Sullivan, D.S. (1982) Population dynamics and regulation of the Douglas squirrel (*Tamiasciurus douglassi*) with supplemental food. *Oecologia*, **53**, 264–270.
15.7.2

Summerhayes, V.S. & Elton, C.S. (1923) Contributions to the ecology of Spitsbergen and Bear Island. *Journal of Ecology*, **11**, 214–286.
22.5.5

Sunderland, K.D., Hassall, M. & Sutton, S.L. (1976) The population dynamics of *Philoscia muscorum* (Crustacea, Oniscoidea) in a dune grassland ecosystem. *Journal of Animal Ecology*, **45**, 487–506.
4.5.3

Sunderland, N. (1960) Germination of the seeds of angiospermous root parasites. In: *The Biology of Weeds* (J.L, Harper, ed.). Blackwell Scientific Publications, Oxford.
5.5.4

Sutcliffe, J. (1977) *Plants and Temperature*. Edward Arnold, London.
2.3.3, 2.4

Sutherland, W.J. (1982) Do oystercatchers select the most profitable cockles? *Animal Behaviour*, **30**, 857–861.
9.5.1

Sutherland, W.J. (1983) Aggregation and the 'ideal free' distribution. *Journal of Animal Ecology*, **52**, 821–828.
9.10

Sutton, M.A., Pitcairn, C.E.R. & Fowler, D. (1993) The exchange of ammonia between the atmosphere and plant communities. *Advances in Ecological Research*, **24**, 302–393.
19.2.2, 19.4.4, 19.4.5

Swank, W.T., Waide, J.B., Crossley, D.A. & Todd, R.L. (1981) Insect defoliation enhances nitrate export from forest ecosystems. *Oecologia*, **51**, 297–299.
19.2.3

Swift, M.J. (1987) Organization of assemblages of decomposer fungi in space and time. In: *Organization of Communities: Past and Present* (J.H.R. Gee & P.S. Giller, eds), pp. 229–254. Blackwell Scientific Publications, Oxford.
17.4.1

Swift, M.J., Heal, O.W. & Anderson, J.M. (1979) *Decomposition in Terrestrial Ecosystems*. Blackwell Scientific Publications, Oxford.
11.2.2, 11.2.4, 11.3.1

Symonides, E. (1977) Mortality of seedlings in the natural psammophyte populations. *Ekologia Polska*, **25**, 635–651.
15.3.2

Symonides, E. (1979) The structure and population dynamics of psammophytes on inland dunes. II. Loose-sod populations. *Ekologia Polska*, **27**, 191–234.
6.6, 15.2, 15.3.2

Szarek, S.R., Johnson, H.B. & Ting, I.P. (1973) Drought adaption in *Opuntia basilaris*. Significance of recycling carbon through Crassulacean acid metabolism. *Plant Physiology*, **52**, 539–541.
3.3.3

Tamm, C.O. (1956) Further observations on the survival and flowering of some perennial herbs. *Oikos*, **7**, 274–292.
6.11

Tansley, A.G. (1917) On competiton between *Galium sylvestre* Poll. (*G. asperum* Schreb.) on different types of soil. *Journal of Ecology*, **5**, 173–179.
7.2.2

Tansley, A.G. (1935) The use and abuse of vegetational concepts and terms. *Ecology*, **16**, 284–307.
17.1

Tansley, A.G. (1939) *The British Islands and their Vegetation*. Cambridge University Press, Cambridge.
17.4.5

Tansley, A.G. & Adamson, R.S. (1925) Studies of the vegetation of the English chalk. III. The chalk grasslands of the Hampshire–Sussex border. *Journal of Ecology*, **5**, 173–179.
15.7.4, 21.2.1

Taper, M.L. & Case, T.J. (1994) Coevolution among competitors. In: *Oxford Surveys in Evolutionary Biology*, Vol. 8 (D.J. Futuyma & J. Antonovics, eds), pp. 63–109. Oxford University Press, Oxford.
7.8.1

Taylor, A.D. (1990) Metapopulations, dispersal and predator–prey dynamics: an overview. *Ecology*, **71**, 429–433.
15.6.2

Taylor, A.D. (1993) Heterogeneity in host–parasitoid interactions: the CV^2 rule. *Trends in Ecology and Evolution*, **8**, 400–405.
10.5.3

Taylor, F.J.R. (1982) Symbioses in marine microplankton. *Annales de l'Institut Oceanographique, Paris*, **58**(S), 61–90.
13.8

Taylor, L.R. (1987) Objective and experiment in long-term research. In: *Long-Term Studies in Ecology* (G.E. Likens, ed.), pp. 20–70. Springer-Verlag, New York.
15.1

Teeri, J.A. & Stowe, L.G. (1976) Climatic patterns and the distribution of C_4 grasses in North America. *Oecologia*, **23**, 112.
3.3.2

Ter Braak, C.J.F. & Prentice, I.C. (1988) A theory of gradient analysis. *Advances in Ecological Research*, **18**, 272–317.
17.3.2

Thiegles, B.A. (1968) Altered polyphenol metabolism in the foliage of *Pinus sylvestris* associated with European pine sawfly attack. *Canadian Journal of Botany*, **46**, 724–725.
8.2.2

Thiollay, J.-M. (1988) Comparative foraging success of insectivorous birds in tropical and temperate forests: ecological complications. *Oikos*, **53**, 17–30.
24.4.1

Thomas, C.D. (1990) What do real population dynamics tell us about minimum viable population sizes? *Conservation Biology*, **4**, 324–327.
25.4.1

Thomas, C.D. & Harrison, S. (1992) Spatial dynamics of a patchily distributed butterfly species. *Journal of Applied Ecology*, **61**, 437–446.
15.6.3, 15.6.5

Thomas, C.D. & Jones, T.M. (1993) Partial recovery of a skipper butterfly (*Hesperia comma*) from population refuges: lessons

for conservation in a fragmented landscape. *Journal of Animal Ecology*, **62**, 472–481.
15.6.5

Thomas, C.D., Thomas, J.A. & Warren, M.S. (1992) Distributions of occupied and vacant butterfly habitats in fragmented landscapes. *Oecologia*, **92**, 563–567.
15.6.3

Thomas, P.A. & Room, P.M. (1985) Towards biological control of salvinia in Papua New Guinea. *Proceedings of the VIth International Symposium on Biological Control of Weeds*, pp. 567–574.
16.8

Thomas, S.C. & Weiner, J. (1989) Growth, death and size distribution change in an *Impatiens pallida* population. *Journal of Ecology*, **77**, 524–536.
6.11

Thompson, D.J. (1975) Towards a predator–prey model incorporating age-structure: the effects of predator and prey size on the predation of *Daphnia magna* by *Ischnura elegans*. *Journal of Animal Ecology*, **44**, 907–916.
9.5.1

Thompson, D.J. (1989) Sexual size dimorphism in the damselfly *Coenagrion puella* (L.). *Advances in Ódonatology*, **4**, 123–131.
14.2

Thompson, J.N. (1982) *Interaction and Coevolution*. Wiley-Interscience, New York.
8.1

Thompson, J.N. (1988) Evolutionary ecology of the relationship between oviposition preference and performance of offspring in phytophagous insects. *Entomologia Experimentia et Applicata*, **47**, 3–14.
9.2.2

Thompson, J.N. (1995) *The Coevolutionary Process*. University of Chicago Press, Chicago.
13.4.2

Thompson, K. & Grime, J.P. (1979) Seasonal variation in seed banks of herbaceous species in ten contrasting habitats. *Journal of Ecology*, **67**, 893–921.
4.5.2

Thórhallsdóttir, T.E. (1990) The dynamics of five grasses and white clover in a simulated mosaic sward. *Journal of Ecology*, **78**, 909–923.
3.10, 7.5.5

Thornthwaite, C.W. (1948) An approach towards a rational classification of climate. *Geographical Reviews*, **38**, 55–94.
3.4

Thornton, I.W.B., Zann, R.A. & van Balen, S. (1993) Colonization of Rakata (Krakatau Is.) by non-migrant land birds from 1983 to 1992 and implications for the value of island equilbrium theory. *Journal of Biogeography*, **20**, 441–452.
23.3.4

Tilman, D. (1977) Resource competition between planktonic algae: an experimental and theoretical approach. *Ecology*, **58**, 338–348.
7.10.2

Tilman, D. (1982) *Resource Competition and Community Structure*. Princeton University Press, Princeton, NJ.
3.1, 3.8, 7.10.2, 17.2.1, 24.3.1

Tilman, D. (1986) Resources, competition and the dynamics of

plant communities. In: *Plant Ecology* (M.J. Crawley, ed.), pp. 51–74. Blackwell Scientific Publications, Oxford.
7.10.2, 24.3.1

Tilman, D. (1987) Secondary succession and the pattern of plant dominance along experimental nitrogen gradients. *Ecological Monographs*, **57**, 189–214.
17.4.3

Tilman, D. (1988) *Plant Strategies and the Dynamics and Structure of Plant Communities*. Princeton University Press, Princeton, NJ.
17.4.3, 17.4.4, 21.5

Tilman, D. (1990) Mechanisms of plant competition for nutrients: the elements of a predictive theory of competition. In: *Perspectives on Plant Competition* (J. B. Grace & D. Tilman, eds), pp. 117–141. Academic Press, New York.
7.10, 7.10.1, 7.10.2

Tilman, D. (1993) Species richness of experimental productivity gradients: how important is colonization limitation? *Ecology*, **74**, 2179–2191.
24.3.1

Tilman, D. (1994) Competition and biodiversity in spatially structured habitats. *Ecology*, **75**, 2–16.
3.10

Tilman, D. & Downing, J.A. (1994) Biodiversity and stability in grasslands. *Nature*, **367**, 363–365.
22.6

Tilman, D. & Wedin, D. (1991a) Plant traits and resource reduction for five grasses growing on a nitrogen gradient. *Ecology*, **72**, 685–700.
7.10.1

Tilman, D. & Wedin, D. (1991b) Dynamics of nitrogen competition between successional grasses. *Ecology*, **72**, 1038–1049.
7.10.1

Tilman, D., Mattson, M. & Langer, S. (1981) Competition and nutrient kinetics along a temperature gradient: an experimental test of a mechanistic approach to niche theory. *Limnology and Oceanography*, **26**, 1020–1033.
7.2.6

Tinbergen, L. (1960) The natural control of insects in pinewoods. 1: factors influencing the intensity of predation by songbirds. *Archives néerlandaises de Zoologie*, **13**, 266–336.
9.2.3

Tinker, P.H.B. (1975) Effects of vesicular–arbuscular mycorrhizas on higher plants. In: *Symbiosis* (D.H. Jennings & D.L. Lee, eds). Symposium 29, Society for Experimental Biology. Cambridge University Press, Cambridge.
13.7.2

Tisdale, W.H. (1919) Physoderma disease of corn. *Journal of Agricultural Research*, **16**, 137–154.
12.2.1

Tjallingii, W.F. & Hogen Esch, Th. (1993) Fine structure of aphid stylet routes in plant tissues in correlation with EPG signals. *Physiological Entomology*, **18**, 317–328.
3.7.1

Toft, C.A. & Karter, A.J. (1990) Parasite–host coevolution. *Trends in Ecology and Evolution*, **5**, 326–329.
12.4

Tokeshi, M. (1993) Species abundance patterns and community structure. *Advances in Ecological Research*, **24**, 112–186.
17.2.2

Tongeren, O. van & Prentice, I.C. (1986) A spatial simulation model for vegetation dynamics. *Vegetatio*, **65**, 163–173.
3.10

Tonn, W.M. & Magnuson, J.J. (1982) Patterns in the species composition and richness of fish assemblages in northern Wisconsin lakes. *Ecology*, **63**, 137–154.
24.3.2

Toth, J.A., Papp, L.B. & Lenkey, B. (1975) Litter decomposition in an oak forest ecosystem (*Quercetum petreae* Cerris) in northern Hungary studied in the framework of 'Sikfökut Project'. In: *Biodegradation et Humification* (G. Kilbertus, O. Reisinger, A. Mourey & J.A. Cancela da Foneseca, eds), pp. 41–58. Pierrance Editeur, Sarregue Mines.
11.2.1

Towns, D.R. & Ballantine, W.J. (1993) Conservation and restoration of New Zealand island ecosystems. *Trends in Ecology and Evolution*, **8**, 452–457.
25.6.4

Townsend, C.R. (1980) *The Ecology of Streams and Rivers.* Edward Arnold, London.
2.9, 5.3.7

Townsend, C.R. (1989) The patch dynamics concept of stream community ecology. *Journal of the North American Benthological Society*, **8**, 36–50.
21.7.2

Townsend, C.R. (1991a) Community organisation in marine and freshwater environments. In: *Fundamentals of Aquatic Ecology* (R.S.K. Barnes & K.H. Mann, eds), pp. 125–144. Blackwell Scientific Publications, Oxford.
21.7.2

Townsend, C.R. (1991b) Exotic species management and the need for a theory of invasion ecology. *New Zealand Journal of Ecology*, **15**, 1–3.
22.5.6

Townsend, C.R. (1996a) Concepts in river ecology: pattern and process in the catchment hierarchy. *Archiv fur Hydrobiologie* (In press).
18.2.1

Townsend, C.R. (1996b) Invasion biology and ecological impacts of brown trout (*Salmo trutta*) in New Zealand. *Biological Conservation* (In press).
25.2.4

Townsend, C.R. & Calow, P. (eds) (1980) *Physiological Ecology: An Evolutionary Approach to Resource Use.* Blackwell Scientific Publications, Oxford.
19.1

Townsend, C.R. & Hildrew, A.G. (1980) Foraging in a patchy environment by a predatory net-spinning caddis larva: a test of optimal foraging theory. *Oecologia*, **47**, 219–221.
9.8

Townsend, C.R. & Hildrew, A.G. (1994) Species traits in relation to a habitat templet for river systems. *Freshwater Biology*, **31**, 265–275.
14.6, 21.1.2

Townsend, C.R. & Hughes, R.N. (1981) Maximizing net energy returns from foraging. In: *Physiological Ecology: An Evolutionary Approach to Resource Use* (C.R. Townsend & P. Calow, eds), pp. 86–108. Blackwell Scientific Publications, Oxford.
9.3, 9.9.3

Townsend, C.R. & Winfield, I.J. (1985) The application of optimal foraging theory to feeding behaviour in fish. In: *Fish*

Energetics—A New Look (P. Calow & P. Tyler, eds), pp. 67–98. Croom Helm, Beckenham.
9.4

Townsend, C.R., Hildrew, A.G. & Francis, J.E. (1983) Community structure in some southern English streams: the influence of physicochemical factors. *Freshwater Biology*, **13**, 521–544.
17.3.2, 24.3.4

Townsend, C.R., Winfield, I.J., Peirson, G. & Cryer, M. (1986) The response of young roach *Rutilus rutilus* to seasonal changes in abundance of microcrustacean prey: a field demonstration of switching. *Oikos*, **46**, 372–378.
21.2.3

Townsend, C.R., Scarsbrook, M.R. & Dolédec, S. (in press) The intermediate disturbance hypothesis, refugia and biodiversity in streams. *Limnology and Oceanography*.
21.5.1

Tracy, C.R. (1976) A model of the dynamic exchanges of water and energy betweeen a terrestrial amphibian and its environment. *Ecological Monographs*, **46**, 293–326.
2.5

Tregenza, T. (1995) Building on the ideal free distribution. *Advances in Ecological Research*, **26**, 253–307.
9.10

Trench, R.K. (1975) Of 'leaves that crawl': functional chloroplasts in animal cells. In: *Symbiosis* (D.H. Jennings & D.L. Lee, eds), pp. 229–266. Symposium 29, Society for Experimental Biology. Cambridge University Press, Cambridge.
13.8

Trexler, J.C., McCulloch, C.E. & Travis, J. (1988) How can the functional response best be determined? *Oecologia*, **76**, 206–214.
9.5.1

Tripp, M.R. (1974) A final comment on invertebrate immunity. In: *Contemporary Topics in Immunobiology* (E.L. Cooper, ed.), Vol. 4, pp. 289–290. Plenum Press, New York.
12.3.5

Trostel, K., Sinclair, A.R.E., Walters, C.J. & Krebs, C. (1987) Can predation cause the 10-year hare cycle? *Oecologia*, **74**, 185–192.
10.2.4

Trumble, J.T., Kolodny-Hirsch, D.M. & Ting, I.P. (1993) Plant compensation for arthropod herbivory. *Annual Review of Entomology*, **38**, 93–119.
8.2.1

Turchin, P. (1993) Chaos and stability in rodent population dynamics: evidence from non-linear time-series analysis. *Oikos*, **68**, 167–172.
6.9

Turchin, P. & Taylor, A.D. (1992) Complex dynamics in ecological time series. *Ecology*, **73**, 289–305.
6.9

Turchin, P.D. (1990) Rarity of density dependence or population regulation with lags? *Nature*, **344**, 660–663.
15.2.3, 15.3.2

Turesson, G. (1922a) The species and variety as ecological units. *Hereditas*, **3**, 100–113.
1.5.1

Turesson, G. (1922b) The genotypical response of the plant species to the habitat. *Hereditas*, **3**, 211–350.
1.5.1

Turkington, R. & Harper, J.L. (1979) The growth, distribution and neighbour relationships of *Trifolium repens* in a permanent pasture. IV. Fine scale biotic differentiation. *Journal of Ecology*, **67**, 245–254.
1.5.2

Turkington, R. & Mehrhoff, L.A. (1990) The role of competition in structuring pasture communities. In: *Perspectives on Plant Competition* (J.B. Grace & D. Tilman, eds), pp. 307–340. Academic Press, New York.
7.8.2

Turkington, R.A., Cavers, P.B. & Aarsen, L.W. (1977) Neighbour relationships in grass–legume communities: I. Interspecific contacts in four grassland communities near London, Ontario. *Canadian Journal of Botany*, **55**, 2701–2711.
20.3.2

Turnbull, A.L. (1962) Quantitative studies of the food of *Linyphia triangularis* Clerck (Aranaea: Linyphiidae). *Canadian Entomologist*, **96**, 568–579.
8.4

Turner, J.R.G., Gatehouse, C.M. & Corey, C.A. (1987) Does solar energy control organic diversity? Butterflies, moths and the British climate. *Oikos*, **48**, 195–205.
24.3.1

Turner, J.R.G., Lennon, J.J. & Lawrenson, J.A. (1988) British bird species distributions and the energy theory. *Nature*, **335**, 539–541.
24.3.1

Turner, J.R.G., Lennon, J.J. & Greenwood, J.J.D. (1996) Does climate cause the global biodiversity gradient? In: *Aspects of the Genesis and Maintenance of Biological Diversity* (M. Hochberg, J. Claubert & R. Barbault, eds). Oxford University Press, London, New York.
24.2, 24.3.1, 24.4.1

Turrill, W.B. (1964) *Joseph Dalton Hooker*. Nelson, London.
23.3.5

Tuttle, M.D. (1979) Status, causes of decline, and management of endangered gray bats. *Journal of Wildlife Management*, **43**, 1–17.
25.2.3

Uchmanski, J. (1985) Differentiation and frequency distributions of body weights in plants and animals. *Philosophical Transactions of the Royal Society of London, Series B*, **310**, 1–75.
6.11

Underwood, T. (1986) The analysis of competition by field experiments. In: *Community Ecology: Past and Present* (J. Kikkawa & D.J. Anderson, eds), pp. 240–268. Blackwell Scientific Publications, Oxford.
20.2

UNEP (1991) *Environmental Data Report*, 3rd edn. Basil Blackwell, Oxford.
19.4.6

Urquhart, F.A. (1960) *The Monarch Butterfly*. University of Toronto Press, Toronto.
5.7.4

US Department of the Interior (1990) *Audit Report: The Endangered Species Program*. US Fish and Wildlife Service Report 90–98.
25.6.3

Utida, S. (1957) Cyclic fluctuations of population density intrinsic to the host–parasite system. *Ecology*, **38**, 442–449.
10.2.4

Vaeck, M., Reynaerts, A., Hofte, H. *et al.* (1987) Transegenic plants protected from insect attack. *Nature*, **328**, 33–37.
16.10

Vagvolgyi, J. (1975) Body size, aerial dispersal, and origin of the Pacific land snail fauna. *Systematic Zoology*, **24**, 465–488.
23.3.5

Valentine, J.W. (1970) How many marine invertebrate fossil species? A new approximatio. *Journal of Paleontology*, **44**, 410–415.
24.4.6

Vandermeer, J.H. (1972) Niche theory. *Annual Review of Ecology and Systematics*, **3**, 107–132.
2.12

Vandermeer, J.H. & Boucher, D.H. (1978) Varieties of mutualistic interaction in population models. *Journal of Theoretical Biology*, **74**, 549–558.
13.12

Vanderplank, J.E. (1963) *Plant Diseases: Epidemics and Control*. Academic Press, New York.
12.5.3

Vannier, G. (1987) Mesure de la thermotorpeur chez les insectes. *Bulletin de la Societé Ecophysiologique*, **12**, 165–186.
2.1

Varley, G.C. (1947) The natural control of population balance in the knapweed gall-fly (*Urophora jaceana*). *Journal of Animal Ecology*, **16**, 139–187.
10.2.5

Varley, G.C. & Gradwell, G.R. (1968) Population models for the winter moth. *Symposium of the Royal Entomological Society of London*, **9**, 132–142.
6.8.2, 15.3.1, 15.3.2

Varley, G.C. & Gradwell, G.R. (1970) Recent advances in insect population dynamics. *Annual Review of Entomology*, **15**, 1–24.
4.5.1

Varley, G.C., Gradwell, G.R. & Hassell, M.P. (1973) *Insect Population Ecology*. Blackwell Scientific Publications, Oxford.
10.2.5, 15.2, 15.2.2

Varley, M.E. (1967) *British Freshwater Fishes*. Fishing News Books, London.
2.2.2

Veldkamp, H., Gemerden, H. van, Harder, W. & Laanbroek, H. J. (1984) Competition among bacteria: an overview. In: *Current Perspectives in Microbial Ecology* (M.J. Klug and C.A. Reddy, eds), pp. 279–299. American Society for Microbiology, Washington.
7.7.1

Venable. D.L., & Brown, J.S. (1988) The selective interactions of dispersal, dormancy and seed size as adaptations for reducing risk in variable environments. *American Naturalist*, **131**, 360–384.
5.4.2, 5.5.2

Venable, D.L., Burquez, A., Corral, G., Morales, E. & Espinosa, F. (1987) The ecology of seed heteromorphism in *Hetrosperma pinnatum* in Central Mexico. *Ecology*, **68**, 65–76.
5.4.2

Verhulst, P.F. (1838) Notice sur la loi que la population suit dans son aaccroissement. *Correspondances Mathématiques et Physiques*, **10**, 113–121.
6.10

Via, S. & Lande, R. (1985) Genotype–environment interaction

and the evolution of phenotypic plasticity. *Evolution*, **39**, 505–522.
14.11

Vickerman, K. & Cox, F.E.G. (1967) *The Protozoa*. John Murray, London.
12.2.1

Vitousek, P.M. (1994) Beyond global warming: ecology and global change. *Ecology*, **75**, 1861–1876.
2.11.2

Vitousek, P.M., Gosz, J.R., Greir, C.L., Mellilo, J.M. & Reiners, W.A. (1982) A comparative analysis of potential nitrification and nitrate mobility in forest ecosystems. *Ecological Monographs*, **52**, 155–177.
19.2.4

Vitt, L.J. & Congdon, J.D. (1978) Body shape, reproductive effort and relative clutch mass in lizards: resolution of a paradox. *American Naturalist*, **112**, 595–608.
14.7.1

Volterra, V. (1926) Variations and fluctuations of the numbers of individuals in animal species living together. (Reprinted in 1931. In: R.N. Chapman, *Animal Ecology*. McGraw Hill, New York.)
7.4.1, 10.2.1

Waage, J.K. (1979) Foraging for patchily-distributed hosts by the parasitoid *Nemeritis canescens*. *Journal of Animal Ecology*, **48**, 353–371.
9.9.3

Waage, J.K. & Godfray, H.C.J. (1985) Reproductive strategies and population ecology of insect parasitoids. In: *Behavioural Ecology* (R.M. Sibly & R.H. Smith, eds), pp. 449–470. Blackwell Scientific Publications, Oxford.
14.8.1

Waage, J.K. & Greathead, D.J. (1988) Biological control: challenges and opportunities. *Philosophical Transactions of the Royal Society of London, Series B*, **318**, 111–128.
10.5.6, 16.8

Walde, S. & Murdoch, W.W. (1988) Spatial density-dependence in parasitoids. *Annual Review of Entomology*, **33**, 441–466.
9.7.1, 10.5.6

Wall, R. & Begon, M. (1985) Competition and fitness. *Oikos*, **44**, 356–360.
6.1

Wall, R. & Begon, M. (1987) Individual variation and the effects of population density in the grasshopper *Chorthippus brunneus*. *Oikos*, **49**, 15–27.
14.4

Wallace, J.B. & O'Hop, J. (1985) Life on a fast pad: waterlily leaf beetle impact on water lilies. *Ecology*, **66**, 1534–1544.
8.2.3

Walley, K., Khan, M.S.I. & Bradshaw, A.D. (1974) The potential for evolution of heavy metal tolerance in plants. I. Copper and zinc tolerancein *Agrostis tenuis*. *Heredity*, **32**, 309–319.
2.10

Wallis, G.P. (1994) Population genetics and conservation in New Zealand: a hierarchical synthesis and recommendations for the 1990s. *Journal of the Royal Society of New Zealand*, **24**, 143–160.
25.2.5

Walsby, A.E. (1980) A square bacterium. *Nature*, **283**, 69–71.
1.4.2

Walsh, J.A. (1983) Selective primary health care: strategies for control of disease in the developing world. IV. Measles. *Reviews of Infectious Diseases*, **5**, 330–340.
12.5.2

Walter, D.E. (1987) Trophic behaviour of 'mycophagous' microarthropods. *Ecology*, **68**, 226–228.
22.5.3

Walters, K. & Moriarty, D.J.W. (1993) The effects of complex trophic interactions on a marine microbenthic community. *Ecology*, **74**, 1475–1489.
22.2.3

Ward, J.V. (1988) Riverine–wetland interactions. In: *Freshwater Wetlands and Wildlife* (R.R. Sharitz & J.W. Gibbon, eds), pp. 385–400. Office of Science and Technology Information, US Department of Energy, Oak Ridge, TN.
19.3.1

Ward, S.A. (1992) Assesssing functional explanations of host-specificity. *American Naturalist*, **139**, 883–891.
12.3.2

Ware, G.W. (1983) *Pesticides Theory and Application*. W.H. Freeman, New York.
16.5

Waring, R.H. (1989) Ecosystems: fluxes of matter and energy. In: *Ecological Concepts* (J.M. Cherrett, ed.), pp. 17–42. Blackwell Scientific Publications, Oxford.
19.2.3

Waring, R.H. & Schlesinger, W.H. (1985) *Forest Ecosystems: Concepts and Management*. Academic Press, Orlando, FL.
19.1.2, 19.2.1, 19.2.2, 19.2.3

Waring, R.H., Roger, J.J. & Swank, W.T. (1981) Water relations and hydrologic cycles. In: *Dynamic Properties of Forest Ecosystems* (D.E. Reichle, ed.), pp. 205–264. Cambridge University Press, London.
19.2.3

Warkowska-Dratnal, H. & Stenseth, N.C. (1985) Dispersal and the microtine cycle: comparison of two hypotheses. *Oecologia*, **65**, 468–477.
15.4.2, 15.4.4

Warner, R.E. (1968) The role of introduced diseases in the extinction of endemic Hawaiian avifauna. *Condor*, **70**, 101–120.
21.2.4

Warren, P.H. (1989) Spatial and temporal variation in the structure of a freshwater food web. *Oikos*, **55**, 299–311.
22.4.3

Warren, P.H. (1994) Making connections in food webs. *Trends in Ecology and Evolution*, **9**, 136–141.
22.5.1

Warrington, S., Cottam, D.A. & Whittaker, J.B. (1989) Effects of insect damage on photosynthesis, transpiration and SO_2 uptake by sycamore. *Oecologia*, **80**, 136–139.
8.2.7

Warrington, S., Mansfield, T.A. & Whittaker, J.B. (1978) Effect of SO_2 on the reproduction of pea aphids, *Acyrthosiphon pisum*, and the impact of SO_2 and aphids on the growth and yield of peas. *Environmental Pollution*, **48**, 285–294.
8.2.7

Waser, P.M. & Jones, W.T. (1989) Heritability of dispersal in banner-tailed kangaroo rats, *Didomys spectabilis*. *Animal Behaviour*, **37**, 987–991.
5.4.3, 15.4.4

Waterhouse, D.F. (1974) The biological control of dung.

Scientific American, **230**, 100–108.
11.3.2

Watkins, C.V. & Harvey, L.A. (1942) On the parasites of silver foxes on some farms in the South West. *Parasitology*, **34**, 155–179.
12.3.7

Watkinson, A.R. (1984) Yield–density relationships: the influence of resource availability on growth and self-thinning in populations of *Vulpia fasciculata*. *Annals of Botany*, **53**, 469–482.
6.5, 15.5

Watkinson, A.R. & Davy, A.J. (1985) Population biology of salt marsh and sand dune animals. *Vegetatio*, **62**, 487–497.
6.2

Watkinson, A.R. & Harper, J.L. (1978) The demography of a sand dune annual: *Vulpia fasciculata*. I. The natural regulation of populations. *Journal of Ecology*, **66**, 15–33.
6.2, 15.2.3

Watkinson, A.R. & Sutherland, W.J. (1995) Sources, sinks and pseudo-sinks. *Journal of Animal Ecology*, **64**, 126–130.
15.6.5

Watson, A. & Moss, R. (1972) A current model of population dynamics in red grouse. In: *Proceedings of the XVth International Ornithological Congress* (K.H. Voous, ed.), pp. 139–149.
10.3.2

Watson, A. & Moss, R. (1980) Advances in our understanding of the population dynamics of red grouse from a recent fluctuation in numbers. *Areda*, **68**, 103–111.
15.4.1

Watson, G.E. (1964) *Ecology and Evolution of Passerine Birds on the Islands of the Aegean Sea*. Ph.D. thesis, Yale University (Dissertation microfilm 65–1956).
23.3.1

Watt, A.S. (1947) Pattern and process in the plant community. *Journal of Ecology*, **35**, 1–22.
17.4.5

Watt, A.S. (1970) Factors controlling the floristic composition of some plant communities in Breckland. In: *The Scientific Management of Animal and Plant Communities for Conservation* (E. Duffey & A.S. Watt, eds), pp. 137–152. Blackwell Scientific Publications, Oxford.
5.2

Way, M.J. & Cammell, M. (1970) Aggregation behaviour in relation to food utilization by aphids. In: *Animal Populations in Relation to their Food Resource* (A. Watson, ed.), pp. 229–247. Blackwell Scientific Publications, Oxford.
9.7.2

Weaver, J.E. & Albertson, F.W. (1943) Resurvey of grasses, forbs and underground plant parts at the end of the great drought. *Ecological Monographs*, **13**, 63–117.
3.5

Webb, S.D. (1987) Community patterns in extinct terrestrial invertebrates. In: *Organization of Communities: Past and Present* (J.H.R. Gee & P.S. Giller, eds), pp. 439–468. Blackwell Scientific Publications, Oxford.
24.4.6

Webb, W.L., Lauenroth, W.K., Szarek, S.R. & Kinerson, R.S. (1983) Primary production and abiotic controls in forests, grasslands and desert ecosystems in the United States. *Ecology*, **64**, 134–151.
18.3.1

Webster, J. (1970) *Introduction to Fungi*. Cambridge University Press, Cambridge.
11.2.1, 12.2.2

Wegener, A. (1915) *Entstehung der Kontinenter und Ozeaner*. Samml. Viewig, Braunschweig. English translation (1924) *The Origins of Continents and Oceans*. Translated by J.G.A. Skerl. Metheun, London.
1.2.1

Weiner, J. (1986) How competition for light and nutrients affects size variability in *Ipomoea tricolor* populations. *Ecology*, **67**, 1425–1427.
6.11

Weiner, J.(1990) Asymmetric competition in plant populations. *Trends in Ecology and Evolution*, **5**, 360–364.
6.11

Weiner, J. & Thomas, S.C. (1986) Size variability and competition in plant monocultures. *Oikos*, **47**, 211–222.
6.11

Weiser, C.J. (1970) Cold resistance and injury in woody plants. *Science*, **169**, 736–750.
2.3.4

Weisner, S.E.B. (1993) Long-term competitive displacement of *Typha latifolia* by *Typha angustifolia* in a eutrophic lake. *Oecologia*, **94**, 451–456.
7.3

Weller, D.E. (1987) A reevaluation of the $-3/2$ power rule of plant self-thinning. *Ecological Monographs*, **57**, 23–43.
6.13

Weller, D.E. (1990) Will the real self-thinning rule please stand up?—A reply to Osawa and Sugita. *Ecology*, **71**, 1204–1207.
6.13

Weller, D.E. (1991) The self-thinning rule: dead or unsupported?—A reply to Lonsdale. *Ecology*, **72**, 747–750.
6.13

Werner, E.E., Gilliam, J.F., Hall, D.J. & Mittlebach, G.G. (1983b) An experimental test of the effects of predation risk on habitat use in fish. *Ecology*, **64**, 1540–1550.
9.4

Werner, E.E., Mittlebach, G.G., Hall, D.J. & Gilliam, J.F. (1983a) Experimental tests of optimal habitat use in fish: the role of relative habitat profitability. *Ecology*, **64**, 1525–1539.
9.4

Werner, H.H. & Hall, D.J. (1974) Optimal foraging and the size selection of prey by the bluegill sunfish *Lepomis macrochirus*. *Ecology*, **55**, 1042–1052.
9.3.1

Werner, P.A. (1975) Predictions of fate from rosette size in teazel (*Dipsacus fullonum* L.). *Oecologia*, **20**, 197–201.
4.8

Werner, P.A. & Platt, W.J. (1976) Ecological relationships of co-occurring golden rods (*Solidago*: Compositae). *American Naturalist*, **110**, 959–971.
14.4.2

Wesson, G. & Wareing, P.F. (1969) The induction of light sensitivity in weed seeds by burial. *Journal of Experimental Biology*, **20**, 413–425.
5.5.2

West, C. (1985) Factors underlying the late seasonal appearance of the lepidopterous leaf-mining guild on oak. *Ecological Entomology*, **10**, 111–120.
8.2.2

West, H.M., Fitter, A.H. & Watkinson, A.R. (1993) Response of *Vulpia ciliata* ssp. *ambigua* to removal of mycorrhizal infection and to phosphate application under natural conditions. *Journal of Ecology*, **81**, 351–358.
13.7.2

Westoby, M., Leishman, M.R. & Lord, J.M. (1995) On misinterpreting the 'phylogenetic correction'. *Journal of Ecology*, **83**, 531–534.
1.4.2

White, J. (1980) Demographic factors in populations of plants. In: *Demography and Evolution in Plant Populations* (O.T. Solbrig, ed.), pp. 21–48. Blackwell Scientific Publications, Oxford.
6.13

White, T.C.R. (1978) The importance of relative shortage of food in animal ecology. *Oecologia*, **33**, 71–86.
8.4

White, T.C.R. (1984) The abundance of invertebrate herbivores in relation to the availability of nitrogen in stressed food plants. *Oecologia*, **63**, 90–105.
8.2.7, 8.4

White, T.C.R. (1993) *The Inadequate Environment: Nitrogen and the Abundance of Animals*. Springer, Berlin.
8.2.7, 8.4

Whiteside, M.C. & Harmsworth, R.V. (1967) Species diversity in chydroid (Cladocera) communities. *Ecology*, **48**, 664–667.
24.3.1

Whitfield, P.J. (1982) *The Biology of Parasitism: An Introduction to the Study of Associating Organisms*. Edward Arnold, London.
12.2.2

Whittaker, R.H. (1953) A consideration of climax theory: the climax as a population and pattern. *Ecological Monographs*, **23**, 41–78.
17.4.5

Whittaker, R.H. (1956) Vegetation of the Great Smoky Mountains. *Ecological Monographs*, **23**, 41–78.
17.3.1

Whittaker, R.H. (1975) *Communities and Ecosystems*, 2nd edn. Macmillan, London.
1.4.1, 18.2

Whittaker, R.H. (1977) Evolution of species diversity in land communities. *Evolutionary Biology*, **10**, 1–67.
24.4.2

Whittaker, R.H. & Feeney, P.P. (1971) Allelochemics: chemical interactions between species. *Science*, **171**, 757–770.
7.3

Whittaker, R.H. & Woodwell, G.M. (1968) Dimension and production relations of trees and shrubs in the Brookhaven Forest, New York. *Journal of Ecology*, **56**, 1–25.
18.2.2

Whittaker, R.H. & Woodwell, G.M. (1969) Structure production and diversity of the oak–pine forest at Brookhaven, New York. *Journal of Ecology*, **57**, 157–176.
18.2.2

Whittaker, R.J., Bush, M.B. & Richards, K. (1989) Plant recolonization and vegetation succession on the Krakatau Islands, Indonesia. *Ecological Monographs*, **59**, 59–123.
23.3.4

Wiebes, J.T. (1979) Coevolution of figs and their insect pollinators. *Annual Review of Ecology and Systematics*, **10**, 1–12.
13.4.2

Wiebes, J.T. (1982) Fig wasps (Hymenoptera). *Monographiae Biologicae*, **42**, 735–755.
13.4.2

Wiens, D. (1984) Ovule survivorship, life history, breeding systems and reproductive success in plants. *Oecologia*, **64**, 47–53.
4.5.1

Wiens, J.A. (1989) *The Ecology of Bird Communities. Volume 2. Processes and Variations*. Cambridge University Press, Cambridge.
15.6.1, 15.7.5

Wiens, J.A., Rotenberry, J.T. & Van Horne, B. (1987) Habitat occupancy patterns of North American shrubsteppe birds: the effects of spatial scale. *Oikos*, **48**, 132–147.
15.6.1

Wiens, J.A., Stenseth, N.C., Van Horne, B. & Ims, R.A. (1993) Ecological mechanisms and landscape ecology. *Oikos*, **66**, 369–380.
21.7.1

Wigglesworth, V.B. (1945) DDT and the balance of nature. *Atlantic Monthly*, **176**, 107.
16.4

Wilbur, H.M. (1987) Regulation of structure in complex systems: experimental temporary pond communities. *Ecology*, **68**, 1437–1452.
21.7.2

Wilcox, B.A. (1978) Supersaturated island faunas: a species–age relationship for lizards on post-pleistocene land-bridge islands. *Science*, **199**, 996–998.
23.3.1

Williams, C.B. (1944) Some applications of the logarithmic series and the index of diversity to ecological problems. *Journal of Ecology*, **32**, 1–44.
12.3.7

Williams, C.B. (1964) *Patterns in the Balance of Nature and Related Problems in Quantitative Ecology*. Academic Press, New York.
23.1

Williams, G.C. (1966) *Adaptation and Natural Selection*. Princeton University Press, Princeton, NJ.
14.3

Williams, K.S., Smith, K.G. & Stephen, F.M. (1993) Emergence of 13-year periodical cicadas (Cicadidae: *Magicicada*): phenology, mortality and predator satiation. *Ecology*, **74**, 1143–1152.
5.2.1

Williams, W.D. (1988) Limnological imbalances: an antipodean viewpoint. *Freshwater Biology*, **20**, 407–420.
19.3.3

Williamson, G.B. (1990) Allelopathy, Koch's postulates, and the neck riddle. In: *Perspectives on Plant Competition* (J. B. Grace & D. Tilman, eds), pp. 143–162. Academic Press, New York.
7.3

Williamson, M.H. (1972) *The Analysis of Biological Populations*. Edward Arnold, London.
7.2.4

Williamson, M.H. (1981) *Island Populations*. Oxford University Press, Oxford.
1.2.3, 23.1, 23.2.1, 23.3.1, 23.3.2, 23.3.4, 23.3.5

Willson, M.F. (1994) Sexual selection in plants: perspective and overview. *American Naturalist*, **144**, S13–S39.
13.4.2

Wilson, D.S. (1986) Adaptive indirect effects. In: *Community Ecology* (J. Diamond & T.J. Case, eds), pp. 437–444. Harper & Row, New York.
11.3.3

Wilson, E.O. (1961) The nature of the taxon cycle in the Melanesian ant fauna. *American Naturalist*, **95**, 169–193.
23.3.1

Wilson, E.O. (1975) *Sociobiology: The New Synthesis*. Harvard University Press, Cambridge, Massachusetts.
2.5

Wilson, J.B. & Agnew, A.D.Q. (1992) Positive-feedback switches in plant communities. *Advances in Ecological Research*, **23**, 263–336.
17.4.4

Wilson, J.B. (1987) Methods for detecting non-randomness in species co-occurrences: a contribution. *Oecologia*, **73**, 579–582.
20.4.3, 20.4.4

Wilson, J.B. (1988a) Shoot competition and root competition. *Journal of Applied Ecology*, **25**, 279–296.
7.3

Wilson, J.B. (1988b) Community structure in the flora of islands in Lake Manapouri, New Zealand. *Journal of Ecology*, **76**, 1030–1042.
20.4.3

Wilson, L.J. (1993) Dynamic thresholds for a dynamic pest: spider mites in cotton. In: *Pest Control and Sustainable Agriculture* (S. Corey, D. Dall & W. Milne, eds), pp. 407–410. CSIRO, East Melbourne.
16.3

Winemiller, K.O. (1990) Spatial and temporal variation in tropical fish trophic networks. *Ecological Monographs*, **60**, 331–367.
22.4.3

Winfield, I.J., Peirson, G., Cryer, M. & Townsend, C.R. (1983) The behavioural basis of prey selection by underyearling bream (*Abramis brama* (L.)) and roach (*Rutilus rutilus* (L.)). *Freshwater Biology*, **13**, 139–149.
3.7.1

Winterbourn, M.J. (1987) The arthropod fauna of bracken (*Pteridium aquilinum*) on the Port Hills, South Island, New Zealand. *New Zealand Entomologist*, **10**, 99–104.
23.3.3

Winterbourn, M.J. & Townsend, C.R. (1991) Streams and rivers: one-way flow systems. In: *Fundamentals of Aquatic Ecology* (R.S.K. Barnes & K.H. Mann, eds), pp. 230–244. Blackwell Scientific Publications, Oxford.
19.3.1

Wit, A.K.H. (1985) The relation between partial defoliation during the pre-heading stages of spring cabbage and yield, as a method to assess the quantitative damage induced by leaf mining insects. *Zeitschrift Angewandte Entomologie*, **100**, 96–100.
8.2.3

Wit, C.T. de (1960) On competition. *Verslagen van landbouwkundige onderzoekingen*, **660**, 1–82.
7.7.2

Wit, C.T. de (1965) Photosynthesis of leaf canopies. *Verslagen van Landbouwkundige Onderzoekingen*, **663**, 1–57.
3.2.1

Wit, C.T. de, Tow, P.G. & Ennik, G.C. (1966) Competition between legumes and grasses. *Verslagen van landbouwkundige onderzoekingen*, **112**, 1017–1045.
7.7.2, 13.10.1

Woiwod, I.P. & Hanski, I. (1992) Patterns of density dependence in moths and aphids. *Journal of Animal Ecology*, **61**, 619–629.
15.2.3

Wolda, H. (1978) Fluctuations in abundance of tropical insects. *American Naturalist*, **112**, 1017–1045.
22.4.1, 22.4.3

Wolfram, S. (1983) Statistical mechanics of cellular automata. *Reviews of Modern Physics*, **55**, 601–643.
3.10

Wolfram, S. (1984) Computer software in science and mathematics. *Scientific American*, **251**, 188–203.
3.10

Wood, R.K.S. & Way, M.J. (eds) (1988) Biological control of pests, pathogens and weeds: development and prospects. *Philosophical Transactions of the Royal Society of London, Series B*, **318**, 109–376.
16.8

Woodmansee, R.G. (1978) Additions and losses of nitrogen in grassland ecosystems. *Bioscience*, **28**, 448–453.
19.2.2

Woodward, F.I. (1987) *Climate and Plant Distribution*. Cambridge University Press, Cambridge.
1.4.1, 2.2

Woodward, F.I. (1990) The impact of low temperatures in controlling the geographical distribution of plants. *Philosophical Transactions of the Royal Society of London, Series B*, **326**, 585–593, and also in *Life at Low Temperatures* (R.M. Laws & F. Franks, eds), pp. 69–77. The Royal Society, London.
2.3.4

Woodward, F.I. (1994) Predictions and measurements of the maximum photosynthetic rate, Amax, at the global scale. In: *Ecophysiology: Photosynthesis* (E.D. Schulze & M.M. Caldwell, eds), pp. 491–509. Springer Verlag, Berlin.
3.2.6

Woodwell, G.M., Whittaker, R.H. & Houghton, R.A. (1975) Nutrient concentrations in plants in the Brookhaven oak pine forest. *Ecology*, **56**, 318–322.
3.5

Woollhead, A.S. (1983) Energy partitioning in semelparous and iteroparous triclads. *Journal of Animal Ecology*, **52**, 603–620.
14.11

Wootton, J.T. (1992) Indirect effects, prey susceptibility, and habitat selection: impacts of birds on limpets and algae. *Ecology*, **73**, 981–991.
22.2.3

World Meteorological Organization (1985) Trace gas effects on climate. *Atmospheric Ozone 1985*. Global Ozone Research and Monitoring Project, Report No. 16, Volume III, Chapter 15.
19.5

Worthen, W.B. & McGuire, T.R. (1988) A criticism of the aggregation model of coexistence: non-independent distribution of dipteran species on ephemeral resources. *American Naturalist*, **131**, 453–458.
7.5.5

Worthington, E.B. (ed.) (1975) *Evolution of I.B.P.* Cambridge University Press, Cambridge.
18.1

Wright, D.H. (1989) A simple, stable model of mutualism incorporating handling time. *American Naturalist*, **134**, 664–667.
13.12

Wright, J.L. & Lemon, E.R. (1966) Photosynthesis under field conditions. IX. Vertical distribution of photosynthesis within a corn crop. *Agronomy Journal*, **58**, 265–268.
3.3

Wright, R.F., Lotse, E. & Semb, A. (1988) Reversability of acidification shown by whole-catchment experiments. *Nature*, **334**, 670–675.
19.4.5

Wurtsbaugh, W.A. (1992) Food-web modification by an invertebrate predator in the Great Salt Lake (USA). *Oecologia*, **89**, 168–175.
22.3.1

Wynne-Edwards, V.C. (1962) *Animal Dispersion in Relation to Social Behaviour*. Oliver and Boyd, Edinburgh.
6.12

Wynne-Edwards, V.C. (1977) Intrinsic population control and introduction. In: *Population Control by Social Behaviour* (F.J. Ebling & D.M. Stoddart, eds), pp. 1–22. Institute of Biology, London.
6.12

Wynne-Edwards, V.C. (1986) *Evolution Through Group Selection*. Blackwell Scientific Publications, Oxford.
6.12

Wynne-Edwards, V.C. (1993) A rationale for group selection. *Journal of Theoretical Biology*, **162**, 1–22.
5.2.1

Yoda, K., Kira, T., Ogawa, H. & Hozumi, K. (1963) Self thinning in overcrowded pure stands under cultivated and natural conditions. *Journal of Biology, Osaka City University*, **14**, 107–129.
6.2, 6.13

Yodzis, P. (1986) Competiton, mortality and community struc-

ture. In: *Community Ecology* (J. Diamond & T.J. Case, eds), pp. 480–491. Harper & Row, New York.
21.4

Young, R.G., Huryn, A.D. & Townsend, C.R. (1994) Effects of agricultural development on processing of tussock leaf litter in high country New Zealand streams. *Freshwater Biology*, **32**, 413–428.
11.2.4

Young, T.P. (1985) *Lobelia telekii* herbivory, mortality, and size at reproduction: variation with growth rate. *Ecology*, **66**, 1879–1883.
4.8

Young, T.P. (1990) Evolution of semelparity in Mount Kenya lobelias. *Evolutionary Ecology*, **4**, 157–172.
14.7.3

Young, T.P. & Augspurger, C.K. (1991) Ecology and evolution of long-lived semelparous plants. *Trends in Ecology and Evolution*, **6**, 285–289.
14.7.3

Yu, L.M. (1995) Elicitins from *Phytophthora* and basic resistance in tobacco. *Proceedings of the National Academy of Science of the USA*, **92**, 4088–4094.
12.3.5

Zadoks, J.S. & Schein, R.D. (1979) *Epidemiology and Disease Management*. Oxford University Press, Oxford.
12.5, 12.5.3

Zahn, R. (1994) Fast flickers in the tropics. *Nature (London)*, **372**, 621–622.
1.2.2

Zeevalking, H.J. & Fresco, L.F.M. (1977) Rabbit grazing and diversity in a dune area. *Vegetatio*, **35**, 193–196.
21.2.1

Zeide, B. (1987) Analysis of the 3/2 power law of self-thinning. *Forest Science*, **33**, 517–537.
6.13

Ziv, Y., Abramsky, Z., Kotler, B.P. & Subach, A. (1993) Interference competition and temporal and habitat partitioning in two gerbil species. *Oikos*, **66**, 237–246.
7.10

Organism Index

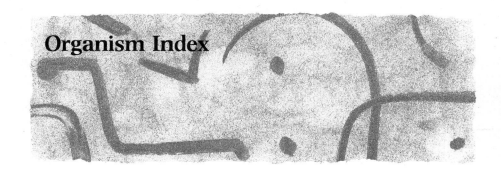

Numbers in *italic* refer to figures and those in **bold** refer to tables

Abies alba, 88
Abramis brama, 127
Abutilon theophrasti, **700**
Acacia, 96, 199, 483
  burkittii, 158, *159, 160, 160*
  cornigera, 483, *484*
acacia, horn (*Acacia cornigera*), 483, *484*
Acanthaster, 587
acanthocephalans, 432
Acari, *407, 408,* 409
Acaulospora elegans, 709
Acer
  campestre, 98
  pseudoplatanus, 325
  rubrum, 13
  saccharum, 13, **700**, *701, 821*
Acetobacterium, 291
Achillea, **40**
Achnanthes microcephala, 83
Acipenser sturio, 921
Acmaea scutum, 336
Aconitum, 493
Acremonium, 503
Acropora, 141
Actinomyces scabies, 448
actinomycetes, 65, 271, 406
Acythosiphon pisum, 325, 502
Adelina triboli, 457, 469, 809
Adenostoma fasciculatum, 322
Adonis annua, 633
Aedes trivittatus, 455
Aeschylanthus sp., *492*
Aepyornis, 911
Aesculus
  glabra, **700**
  octandra, 687
African buffalo (*Syncerus caffer*), 585–7
African dung beetle (*Heliocopris dilloni*), *421, 422*
Agalinis, 617
Agaricus campestris, 119
Agave, **27**
Agrilus, 441
Agriolimax reticulatus, **124**
Agrobacterium tumefaciens, 447, 647
Agropyron

  repens, 307, 308, 810
  smithii, 317
Agrostemma githago, 633
Agrostis, 156, 695
  capillaris, 134
  scabra, 306–7, *307, 308*
  stolonifera, 286, 695, *810*
  tenuis, 81, 613, 805
Ailuropoda melanoleuca, 919, **920**
Alabama argillacea, 634
albatross, 560, 561
Alces alces, 808
alder (*Alnus*), 512, 515, 697
alderfly (*Sialis*), *411*
  fuliginosa, 691
Aleurotrachelus jelinekii, 399
alfalfa (*Medicago sativa*), **97**, 624, **734**
algae, 39, 45, 75, 76, 77, 109, 200, 520, 607–8, 716, **734**, *759, 849*
  blue–green, 748
    Spirulina platensis, 74, 761
  Cladophora, 835
  Coralina vancouveriensis, 820
  green, 76
    Enteromorpha, 804
  Porphyra pseudolanceolata, 820
  red, 76, 698, 804
    Chondrus crispus, 804
Allionia linearis, 112
Allium spp., 621
almond moth (*Ephestia cautella*), *233*
Alnus, 512, 516, 695, 748
  glutinosa, 695
Aloe, **27**
Alopax lagopus, 595
Alpheus, 485
  djiboutensis, 486
Alsophila pometaria, 755
Altica sublicata, 321
Amaranthaceae, **97**
Amaranthus retroflexus, 293, 293, **700**
Amaryllidaceae, **27**
Amblyomma limbatum, 458
Ambrosia
  artemisiifolia, 700
  psilostachya, 112
Ambystoma
  opacum, 563, 563, 564
  tigrinum, 563, 563, 564
American eel (*Anguilla rostrata*), 202
amoeboids, 517

amphibian, 58, 69, 200, *862,* 890, *891,* 911, **916**, 922, **926**, 946
  urodele, 550
amphipod, *408*
  crustacean, 75, *75*
Amphiprion, 486
Anabaenopsis arnoldii, 74
Anatopynia pennipes, 850
Anax junius, 339
anchoveta, Peruvian (*Engraulis ringens*), 656
anchovies, 658
Andropadus latirostris, 900
Andropogon, 617
  gerardi, 306, 307, *307, 308*
Androsace septentrionalis, 233, 571, *572, 586,* 587–9
Anemone hepatica, 252, 252
angiosperm, **27**, 503, 910
Angiostrongylus cantonensis, 442
Angraecum sesquipedale, 493
Anguilla
  anguilla, 202
  rostrata, 202
Anisoptera, **179**
Anolis, 452, 457, 796, 797, *797, 798*
  gingivinus, 457
  occultus, 796
  sagrei, 835
  wattsi, 457
Anomoeoneis vitrea, 81, 83
Anopheles
  albimanus, 636, **637**
  annulipes, 478
Anous minutus, 145
ant, 295–6, 418, 483–5, 487, 488–9, 520, 782, **870**, 890, *892,* 900, *901*
  Atta sexdens, 489
  black (*Monomorium minimum*), *189*
  fire (*Solenopsis geminata*), 638
  Formica cinerea, 519
  harvester (*Veromessor pergandei*), 295, 296, *296*
  leaf-cutting (*Atta*)
    cephalotes, 488, 489
    vollenweideri, 488, 489
  Myrmica schenkii, 487
  ponerine, 871, *871*
  Pseudomyrmex ferruginea, 483
  stinging (*Pseudomyrmex concolor*), 484

anteater (*Mymecophaga*), *26*
Antechinus, 598
antelope, 424
  pronghorn, 658
Anthoceros, 515
Anthonomus grandis, 634
Anthoxanthum odoratum, 39, *688*
Antilope cervicapra, 946
Aphelinus thomsoni, 350
Apheloria montana, 420
aphid, 115, *116*, 120, 173, 174, 178,
    189, 314, 320, 324–5, 432,
    501, 629, 642
  Acythosiphon pisum, 325
  Aphis
    fabae, 116, 324
    varians, 519
  apple (*Aphis pomi*), *189*
  cabbage (*Brevicoryne brassicae*), 355,
    634
  cotton (*Aphis gossypii*), 634
  leaf root (*Pemphigus bursarius*), 458
  lime (*Eucallipterus tiliae*), 144, 320
  Myzus, 441
    persicae, *502*
  raspberry (*Aphis idaei*), *118*
  rubus (*Aphis rubi*), *118*
  spotted alfalfa (*Therioaphis trifolii*),
    625, 646
  sycamore (*Drepanosiphum platanoidis*),
    350
Aphis
  fabae, 116, 324
  gossypii, 634
  idaei, *118*
  pomi, *189*
  rubi, *118*
  varians, 519
aphyines, 9
Apion ulicis, 369
Apis mellifera, *177*, 493, **631**, 785
Aponomma hydrosauri, 458
apple tree, 645
Aquilegia, 493
Arachnothera, 492
Aradus cinnamomeus, 324
Araneida, *408*
Archaea, 5
archaebacteria, 5, 66, 74, 511
Arctic fox (*Alopax lagopus*), 595
Arctic tern (*Sterna paradisaea*), 202
Arctocephalus gazella, *37*
Arctostaphylos, 515, 516
Ardea cinerea, *400*
Arecaceae, *24*
Aristida purpurea, *112*
Arizona pocket mouse (*Perognathus
    amplus*), 346
armadillo, 127
  nine-banded (*Dasypus novemcinctus*),
    917
Armillaria mellea, 439
army-worm (*Spodoptera exempta*), 205
Artemia, 338, 339
  franciscana, *834*, *834*
Artemisia, 290
arthropod, 58, *412*, 882
Artocarpus lanceifolius, 492

Arvicola terrestris, 606
Asarina, 25
Ascaris, *444*, 467
ascidians, colonial, 137
Asclepias spp., *125*
Ascomycetes, 406, 502, 503
Ascophyllum nodosum, 76
Asellus aquaticus, 154
aspen (*Populus tremuloides*), 198
aspen sunflower (*Helianthella
    quinquenervis*), 484
Aspergillus fumigatus, 65
Asphodelus ramosus, 331, *331*
Asphondylia, 441
Aspidiotiphagus citrinus, 354
Asplachna priodonta, 122
Aster, 617
  pilosus, **700**
  pripolium, *695*
Asteraceae, 25
Asterionella, 278
  formosa, 268, 269, 311, *311*
Astraea undosa, 288, **288**, *289*
Atriplex
  hastata, **97**
  rosea, **97**
Atropa belladonna, 803
Atta
  cephalotes, 488, 489
  sexdens, 489
  vollenweideri, 488, 489
Attamyces bromatificus, 489
Australian bush rat (*Rattus fascipes*),
    598
Austroicetes cruciata, 53, *54*
Azolla, 515
Azorhizobium, 512
Azotobacter, 511
Azotococcus, 511
azuki bean weevil (*Callosobruchus
    chinensis*), 376, *376*

Bacillus
  acidocaldarius, 74
  anthracis, 448, 461
  thuringiensis, **631**, 642, 647
backswimmer (*Notonecta hoffmanni*),
    344
bacteria, 32, 47, 66, 74, 109, 116,
    117, 122, 271, 417, 482, 500,
    501, 504, 509, 511–12, 516,
    520, 730, *730*, 739, 748, 831,
    872
  common crown gall (*Agrobacterium
    tumefaciens*), 647
  Erwinia herbicola, 60
  Pseudomonas syringae, 60
Bacteroides succinogenes, 498
baculoviruses, 642
badger, 351, 385, 424
  honey (*Melliovora capensis*), 485
Baetis, 78
Bailey's pocket mouse (*Perognathus
    baileyi*), 346
Balaenicops rex, 117
Balaenoptera musculus, 671

Balansieae, 503
Balanus, 271, 279
  balanoides, 56, 266, *267*
  cariosus, 820
  glandula, 158, 166, **166**, 806, 820
balsam fir, 398
bamboo, 47, 177, 548
banana, 59
bandicoot, 25, *27*
bank vole, *370*
barberry (*Berberis vulgaris*), 436, 437,
    467–8
barbet (*Megalaima* spp.), *492*
barley, 129, 486, 808
barn owl (*Tyto alba*), 346
barnacle, 39, 78, 131, 132, 215, 263,
    279, 290, 806, *832*
  acorn, *849*
  Balanus
    balanoides, 56, 266, *267*
    cariosus, 820
    glandula, 158, 166, **166**, 806, 820
  Chthamalus
    fissus, 817
    stellatus, 266, *267*
  Pollicipes polymerus, 820, 831
  Semibalanus balanoides, 263
basidiomycetes, 406, 503, 694
bass, largemouth, 344, *345*
bat, 33, 69, 71, 200, 494
  grey (*Myotis grisescens*), 922
Battus philenor, 807
Bdellocephala punctata, 559, *559*
bean
  broad (*Vicia faba*), 116
  field (*Vicia faba*), **97**
  French (*Phaseolus multiflorus*), **97**
  mung, *119*
  Phaseolus
    multiflorus, **97**
    vulgaris, 122
  tic, 339
  Vicia faba, **97**, *116*, 324, 448
bear, 71, **72**
  black, 45
  grizzly, 942
Beauveria bassiana, 642
bedstraw, 270
bee, 68, 69, 90, 494, 628, 633
  see also honeybee
beech, **703**, **734**
  American (*Fagus grandifolia*), 701,
    706, 821
  Fagus, 13, **97**, 613
    sylvatica, 88, 723
  mountain (*Nothofagus solandri*), 262
beet army-worm (*Spodoptera exigua*),
    649, *650*, *651*
beet leafhopper (*Circulifer tenellus*), 189
beetle, 115, 127, 407, 488–9, 501,
    870, 878, 879
  African dung (*Heliocopris dilloni*), 421,
    422
  ambrosia, 489, *903*
  bark, *903*
  bombardier (*Brachinus crepitans*),
    125
  bruchid, 121, 130, 535–6, *535*

Callosobruchus maculatus, 535–6, 535
burying (*Necrophorus* spp.), 425, 426, 426
carrion, 187
chrysomelid dock (*Gastrophysa viridula*), 323, 324
Colorado potato (*Leptinotarsa decemlineata*), 581–4
dermestid, 425
dung, 187
  African, 422
flea (*Altica sublicata*), 321
flour (*Tribolium*)
  castaneum, 280, **280**, **281**, 469, 809
  confusum, 217, 241, 280, **280**, **281**, 809
  hispine, 784, 789
  Japanese (*Popillia japonica*), 631
  ladybird (*Rodolia cardinalis*), 369, 625, 626, 640
  Lasioderma serricorne, 241
  Mexican bean (*Epilachna varivestis*), 325
  Phyllotreta cruciferae, 354
  raspberry (*Byturus tormentosus*), 116, 118
  Rhizopertha dominica, 228
  scolytid, 520
  Scolytidae, 489
  Scolytus, 441
  Stegobium panaceum, 241
  tiger, 785
  vedalia (*Rodolia cardinalis*), 626
  waterlily leaf (*Pyrrhalta nymphaeae*), 320, *320*
Belding's ground squirrel (*Spermophilus beldingi*), 255
Bellis perennis, 622
Bembidion picipes, 442
Berberis vulgaris, 436
Betula, 13, **97**, 199
  alleghaniensis, 687
  lenta, 687
  pendula, 88, 143, 222, *222*, 325, 883
  pubescens, 58, 318, 325
Bignonia, 25
Bignoniaceae, 25
Bioga irregularis, 808, 923
birch, 13, 92, **97**
  grey, **703**
  silver (*Betula pendula*), 88, 143, 222, *222*, 325, 883
  white (*Betula pubescens*), 58, 318, 325
birds, 10, 12, 38, 43, 69, 115, *117*, 157, 173, 200, 205, 492, 546, 562, 568, **631**, 735, **736**, 794, 862, 863, **870**, 877–8, *877*, *878*, 890, 891, 895, 896, 897, *897*, 900, 901, 902, *905*, 914, 916, 920, 923, **926**, 929, **934**
bird of paradise, *879*
  black noddy (*Anous minutus*), 144, *145*
  elephant (*Aepyornis*), 911
black stem rust, 467, 468
blackbird (*Turdus merula*), 358

blackbuck (*Antilope cervicapra*), 946
blackfly, 642
  Simulium, 411
blackgum, **703**, 704
Blarina, 350
Blasia, 515
blowfly, 425
  Australian sheep (*Lucilia cuprina*), 327, 402, 448, 652, **652**
blue tit (*Parus caeruleus*), 268
bluebottle fly (*Calliphora vomitoria*), 350
bluegill sunfish (*Lepomis macrochirus*), 342, *342*, 344, *345*
Boiga irregularis, 808, 923
boll weevil (*Anthonomus grandis*), 634
bollworm, false pink (*Sacadodes pyralis*), 634
bombardier beetle (*Brachinus crepitans*), 125
Bombus
  appositus, 781–2
  bifarius, 782
  flavifrons, 782
  frigidus, 782
  kirkbyellus, 782
  occidentalis, 781, 782
  sylvicolla, 782
Boraginaceae, 27
Borrichia, 617
Botrytis, 441
  anthophila, 492
  cinerea, 402
  fabi, 448
Bougainvillea, 87
Bouteloua gracilis, 112
Brachidontes rostratus, 863
Brachinus crepitans, 125
bracken fern (*Pteridium aquilinum*), 125, *125*, 198, *778*, 874, *874*
Bradirhizobium, 512
bramble (*Rubus fruticosus*), 493
Brassica, 25, 354
Brazil nut, 119
bream (*Abramis brama*), 127
Bremia lactucae (downy mildew), 458
Brevicoryne brassicae, 353, 355, 634
brine shrimp, 187
brittle star (*Acanthaster*), 587
Briza media, 688
broadbean (*Vicia faba*), 116
broccoli, 25
Bromus
  inermis, **787**
  madritensis, 283
  rigidus, 283, *283*
broomrape (*Orobanche*), 432, 435
Bruchus, 441
Brugia, 435
brush rabbit (*Sylvilagus bachmani*), 322
bryozoans, 39, 137, 155, 271, 272
Buceros rhinoceros, 492
Buchloe dactyloides, 112, 138
Buchnera aphidicola, 501
buckeye butterfly (*Junonia coenia*), 106
buckwheat, *119*
Buddleia, 883
budworm, spruce (*Choristoneura fumiferana*), 205

buffalo, African (*Syncerus caffer*), 585–7
buffalo fly (*Haematobia irritans exigua*), 423
buffalo grass (*Buchloe dactyloides*), 112, 138
Bufo
  calamita, 271
  woodhousii, 304
bulbul
  Andropadus latirostris, 900
  Chloropsis spp., 492
bullfinch, 123
bumblebee (*Bombus*), 68, 781–2, 789
  appositus, 781–2
  flavifrons, 782
  frigidus, 782
  kirkbyellus, 782
  occidentalis, 781, 782
  sylvicolla, 782
Bupalus piniaria, 144, 145
bushfly (*Musca vetustissima*), 423
Butorides striatus, 787
buttercup, 560
  creeping (*Ranunculus repens*), 139, 140, 570, 571, **571**, 588
  Ranunculus spp., 322
butterfly, 204, 494, 568, 877, 890
  Battus philenor, 807
  bordered white, 577, 579
  buckeye (*Junonia coenia*), 106
  cabbage white (*Pieris rapae*), 54, 354
  checkerspot (*Euphydryas editha*), 932, *932*
  clouded yellow (*Colias croceus*), 204
  Glanville fritillary (*Melitaea cinxia*), 608, *608*, 609, 610
  large blue (*Maculinea* spp.), 487, 516
  Maculinea arion, 487
  monarch (*Danaus plexippus*), 125, 127, 130, 204
  painted lady (*Vanessa cardui*), 204
  Phlebejus argus, 606, 609, *610*
  Pieris rapae, 354
  red admiral (*Vanessa atalanta*), 204
  silver-spotted skipper (*Hesperia comma*), 608
buzzard, 385
Byturus tormentosus, 116, *118*

cabbage, *119*
cabbage aphid (*Brevicoryne brassicae*), 355
cabbage looper (*Trichoplusia ni*), 649
cabbage moth (*Mamestra brassicae*), *314*
cabbage root fly
  Delia brassicae, 126
  Eriosichia brassicae, 234
cabbage white butterfly (*Pieris rapae*), 54, 354
cacao, 864
Cactoblastis, 323, **355**
  cactorum, 618, 620, 643
Cactospiza
  heliobates, 20, 21
  pallidus, 20, 21
cactus, 65, 876
  Opuntia fragilis, 63

cactus (*cont.*):
  prickly pear (*Opuntia inermis*), 395
  prickly pear (*Opuntia stricta*), 395
  saguaro, 55
caddis (*Glossosoma*), 411
caddis-fly, 204
  Enoicyla pusilla, 413, *414*
  Plectrocnemia conspersa, 357, 691
Caenolestes fuliginosus, 27
Cakile edentula, 601, 602, *602*, 609
Calamus, 24
Caldariella acidophila, 66
Calidris canutus, 202, *203*
Callaspidia defonscolumbei, 357
Calliphora vomitoria, 350
Callosciurus nigrovittatus, 492
Callosobruchus
  chinensis, 376, *376*
  maculatus, 535–6, *535*
Calluna, 613
  vulgaris, 383
Calochortus, 919
Calothrix, 511
Camarhynchus
  parculus, 21
  pauper, 20, 21
  psittacula, 20, 21
Campanotus, 433
Campanula uniflora, 15, *15*
Campanulaceae, **27**
Campanularia spp., 138
Campanulotes defectus, 925
campion
  Silene acaulis, 16, *16*
  white (*Silene alba*), 464–5, *466*,
    492
Canadian lynx (*Lynx canadensis*), 369,
    370, 371, 377
Canadian pond weed, 199
canary grass (*Phalaris paradoxa*), 636
Canis, 26
cankerworm, autumn (*Alsophila
    pometaria*), 755
Cape hake (*Merluccius capensis* and
    M. paradoxus), 144, *145*
Cape sugarbird (*Promerops cafer*), *493*
capercaillie, 337
Capitalle, 691
Capsella bursa-pastoris, 155, 235, **700**
Capsodes infuscatus, 331, *331*
Carcinus maenas, 336, 805
Cardiaspina albitextura, 398, *399*
Carduus nutans, 326
Carex, 830
  arenaria, 141, *141*, 197
  bigelowii, 161, *163*
caribou (*Rangifer tarandus*), 202, 337
carp, **57**
carrion beetle, 187
Carya, 13
  alba, 687
Cassia obtusifolia, 293, *293*
cassowaries, 10, *12*, 879
Castanea, 13, 14, 420
  dentata, 808
cat
  domestic, 385
  native (*Dasyurus*), 26, 27

caterpillar, 55, 121, 123
catfish (*Synodontis multipunctatus*), *119*,
    480
Catharanthus roseus, 917
cattail, 272, 290, 305, 555, **555**
  Typha
    angustifolia, 271, 272, 555, **555**
    domingensis, 555, **555**
    latifolia, 271, 272
cattle, 200
cauliflower, 25
cauliflower mosaic virus, 432
cave beetle (*Neapheanops tellkampfi*),
    215, *216*
Ceanothus greggii, 322
Cecidiophyopsis, 441
Cecropia
  insignis, 822
  obtusifolia, 105
cedar, eastern red (*Juniperus virginiana*),
    700, 701
Cediopsylla tepolita, 808
Centaurea cyanus, 633
Centropus violaceus, 179, *180*
Ceonothus, 515
Cercocebus galeritus galeritus, 937
Cercosporella, 441
cereals, 755
Certhidia olivacea, 20, 21
Cervus elaphus, 157, **158**, **160**, 542
Chaerophyllum prescottii, 197
Chaetoceros tetratichon, 508
Chaitophorus viminalis, 502
Chalcophora stephani, 786, *787*
Chama, 289
chamise (*Adenostoma fasciculatum*), 322
Chaoborus, 122
  astictopus, 631, *632*
cheetah, 327, *328*
Chelonethi, 408
Chen caerulescens caerulescens, 830
Chenopodiaceae, 32, 97, 189
Chenopodium album, 140, 196, **700**,
    920
chervil, Prescott (*Chaerophyllum
    prescottii*), 197
chestnut (*Castanea*), 13, 14
  dentata, 808
Chiasognathus grantii, 177
Chilopoda, *408*
chipmunk, eastern (*Tamias striatus*), 71
Chironomus lugubris, 421
chitons, 806
Chlamydogobius, 867
Chlorella, 507
Chloroflexus aurantiacus, 65
Chloropsis spp., 492
Chondrilla juncea, 273, *273*
Chondrus, 805
  crispus, 804
Choristoneura fumiferana, 181, 205, 398,
    399
Chorthippus brunneus, 147–9, **150**, 154,
    195, 225, 264, 535
Chrysanthemum segetum, 530
Chrysocharis
  gemma, 394
  pubicornis, 394

Chthamalus, 271, 279
  fissus, 817
  stellatus, 266, *267*
Chydorus sphaericus, 421
Chytrids, 441
cicadas, *178*, 548
ciliates, 498, 499, 508, 517, 831
  holotrich, 498
cinnabar moth (*Tyria jacobaeae*), 321,
    369, *370*, *371*
Circulifer tenellus, 189
citricola scale (*Coccus
    pseudomagnoliarum*), 641
Citrobacter freundii, 500
Citrus limonum, 88
cladoceran, 187
  chydorid, *892*
  Daphnia
    magna, 291
    pulicaria, 291
Cladochytrium replicatum, 405
Cladophora, 835
clam
  fingernail (*Musculium securis*), 220
  giant, 66
  giant Indo-Pacific, 900
Claviceps, 441, 502
Clavicipitaceae, 502
Clavicularia, 515
Clematis, 24, 25
Clostridium, 512
  locheadii, 498
clouded yellow butterfly (*Colias croceus*),
    204
clover
  hare's foot (*Trifolium arvense*), 138
  subterranean (*Trifolium subterraneum*),
    228, 230, 273, *273*, 327, *327*
  white (*Trifolium repens*), 41, 42, 43,
    97, 124, **124**, 222–3, 298,
    299, 515, **787**, 805, *810*
  white sweet (*Melilotus alba*), 168, *168*
clown fish (*Amphiprion*), 486
Clupea harengus, 659
Clytostoma, 25
coal tit (*Parus ater*), 268
Cobaea, 24
Cobaeaceae, 24
Coccinella septempunctata, 353
Coccoloba uvifera, 835
Coccophagoides utilis, 354
Coccus pseudomagnoliarum, 641
Cochliomyia hominivorax, 646
cockle, 347
cockroach, 417, 501
  wood-eating (*Cryptocercus*), 499
cockspur (*Echinochloa crus-galli*), **97**
coconut, double (*Lodoicea seychellarum*),
    920
cod, 121
  Arctic, 663, *663*
  Arcto-Norwegian, 663
  Bering Sea Pacific (*Gaddus
    macrocephalus*), 667
Codium fragile, 509
coelenterates, 507, 516
Coenagrion puella, 529
Coenorhinus germanicus, 118

Coffea
  arabica, 55, 56
  robusta, 55, 56
coffee (*Coffea*)
  arabica, 55, 56
  robusta, 55, 56
Coleophora alticolella, 56
Coleoptera, *117*, 407, *408*, 903
Colias croceus, 204
Collema, 510
Collembola, 407, *408*, 409
Colorado potato beetle (*Leptinotarsa decemlineata*), 54, *182*, 581–4, 601, 642
Columba
  livia, 457
  palumbus, 177, 326, 339, *353*
Columbicola extinctus, 925
Combretaceae, 25
Combretum, 25
common crane fly (*Tipula pallidosa*), *118*
Compositae, **27**
Coniosporium, 693, *693*
Connochaetes taurinus, *228*, 332
Conyza, 617
copepods, 831
coral, 38, 67, 137, 139, 157, 482, 825
  gorgonian, 140
  stony, 508
Coralina vancouveriensis, 820
Cordulegaster, 411
Cordyline, **27**
Coregonus clupeaformis, 68
cormorant, 455, *879*
corn (*Zea mays*), 53, 62, 92, **97**, 100, 216, 231, *431*, 486, 638
corn cockle (*Agrostemma githago*), 633
corn leaf blight, southern (*Helminthosporium maydis*), 459
corn rootworm (*Diabrotica*), 638
cornflower (*Centaurea cyanus*), 633
Cornus
  florida, 687
  stolonifera, 63
Corophium volutator, *353*
cotton, 293, 626
cotton aphid (*Aphis gossypii*), 634
cotton bollworm (*Heliothis zea*), 634, 636, 649, *650*, 651
cottony cushion-scale (*Icerya purchasi*), 369, 626, 640, 641
cow, 119
cowpea (*Vigna unguiculata*), 626, 647
coypus, South American (*Myocastor coypus*), 951
crab, 42, 66, 200, 509, 542, 708
  Carcinus maenas, 336
crane fly, common (*Tipula pallidosa*), *118*, 411
Crangon septemspinosa, 88, 89
Crassulaceae, 104
creeping buttercup (*Ranunculus repens*), 139, 140, 570, 571, **571**, 588
Crenicichla, **545**
  alta, 546
cress (*Lepidium sativum*), *439*
cricket, 215
Crocidolomia binotalis, 645

crocus, 131
crow, 385, 424, *879*
Cruciferae, 189
crustacea, **916**
  aquatic, 114
Cryptocentrus, 9, 485–6
  cryptocentrus, 486
Cryptocercus, 499
Cryptochaetum spp., 640
Cryptosula spp., 138
Ctenopharyngodon idella, 644
cuckoo, *879*
  Centropus violaceus, 179, *180*
  Cuculus canorum, 480, 481
Culex, 117
Cupressus, 199
  pygmaea, **920**
Cyanidium caldarius, 74
cyanobacteria, 74, 496, 508, *508*, 510, 511, 512, 516, 517, 520
cycad, **27**
Cyclops vicinus, 127, *127*
Cyclotella
  arentii, 83
  kutzingiana, 83
  meneghiniana, 311, *311*
Cydia pomonella, 622
Cymadothea, 42
Cymbella gracilis, 83
Cynometra inaequifolia, 492
Cynosurus cristatus, *134*, 286
Cyperus
  papyrus, **97**
  rotundus, **97**
cypress, pygmy (*Cupressus pygmaea*), **920**
cyprinid, 450, *451*
Cyrtobagous, 618, 619
  salviniae, 643
Cyzenis, 585

Dactylis glomerata, **97**, 249, *249*, **787**, 788
daisy (*Bellis perennis*), 622
damselfly
  Coenagrion puella, 529
  Ischmura elegans, 251
Danaus plexipus, 125, 127, 130, 204
dandelion (*Taraxacum officinale*), 543, *543*, 622
Daphnia, 342, 346
  magna, 291, 349, *349*
  pulex, 122
    var. *pulicaria*, 329
  pulicaria, 291
Dasypus novemcinctus, 917
Dasyurus, 26
  hallucatus, 27
deer, 70, **335**
  mule, 200, 658
  Père David (*Elaphurus davidianus*), 946
  red (*Cervus elaphus*), 157, **158**, *160*, 542
deer mouse (*Peromyscus*), 350
Delia
  brassicae, 126

radicum, 354
Delphinium
  ajacis, 633
  barbeyi, 782
  nelsoni, 194, *194*
delphinium, wild (*Delphinium ajacis*), 633
Dendroceras, 43
Depressaria pastinacella, 317, *317*
Deschampsia, 695
  caespitosa, 695
  flexuosa, 156, 688
desert honeysweet (*Tidestromia oblongifolia*), 64
Desmazierella, 693, 694
Desmoncus, 25
Desulfotomaculum spp., 512
dewberries, 318
Diabrotica, 638
Diacaeum livundinaceum, 85, 490
diatom, 81, 83, 109, 278, 508
  Asterionella formosa, 268, 269
  Chaetoceros tetratichon, 508
  Synedra ulna, 268, 269
Dicerorhinus sumatrensis, 934, 935, 936
Dichothrix, 511
dicotyledons, 23, **27**, *27*, 102, **916**
Dicrostonyx torquatus, 592
Didelphis marsupialis, 27
Digitalis purpurea, 156, 536
Dimorphotheca phuvialis, 189
dinophyceae, 508
dinosaurs, 73, 911
Dioscorea, 24
Dioscoreaceae, 24
Diospyros virginiana, **700**
Dipetalonema, 435
Diphyllobothrium latum, 434, 435
Diplopoda, 407
Diplura, *408*
Dipodomys, 795
  deserti, 794
  merriami, 346, 794
  spectabilis, 794
Dipsacus
  fullonum, 169
  sylvestris, 183
Diptera, *117*, **182**, 314, 407, 642
Dipteryx panamensis, 822, 902
Dirofilaria immitis, 455
Ditylenchus, 407
Diuraphis noxia, 502
Diurnea, 441
dock (*Rumex*)
  crispus, 323, 324, *324*
  obtusifolius, 323, 324, *324*
dodder, 432
dog, 424
  wild, 327, *328*
dogwood, red-osier (*Cornus stolonifera*), 63
dolphin, 22, 114, 546
dormice, 69, 71
Doryphorophaga doryphorea, 584
Douglas fir (*Pseudotsuga menziesii*), 533, 534
Douglas squirrel (*Tamiasciurus douglasii*), 613, *614*

downy mildew (*Bremia lactucae*), 432, 458
Dracaena, **27**
dragonfly, 69
   Anax junius, 339
   Cordulegaster, 411
Drepanosiphum platanoidis, *350*
Dromiciops australis, 27
drongo, *879*
Drosophila, 18, 146, 191, *192*, 300, 301, 337, 338, *338*, 339, 419, 533, 535, *538*, 547, 550, 881
   adiastola, 19, 20
   attigua, *19*
   busckii, 419
   glabriapex, *19*
   grimshawi, *19*
   hydei, 419
   immigrans, 419
   melanogaster, *233*, 235, 534
   mercatorum, 558, *558*
   nubulosa, 299, *300*
   obscura, 195
   picture-winged, 18, *19*
   planitidia, *19*
   primaeva, *19*
   punalua, *19*
   serrata, 299, *300*
   setosimentum, 20
   simulans, 419
   subobscura, 58, *534*
Dryas, 515, 516, 697
   octopetala, *52*
duck, 365, *879*
   eider, 542
duck-billed platypus (*Ornithorhynchus anatinus*), 23
duckweed (*Lemna* sp.), 78, 137, *138*
Dugesia lugubris, 559, *559*
dung beetle, 187
dungfly (*Scatophaga stercoraria*), 423
dunnant, 27
dunnock (*Prunella modularis*), 480

eagle, bald, 942, *943*
earthworm, 58, 121, 407, *408*–9, 410, *412*, 693
Echinocactus fendleri, **97**
Echinochloa crus-galli, **97**
Echium, **27**
Ectopistes migratorius, 925
eel, 78, 173, 203–4, *205*
   American (*Anguilla rostrata*), 202
   European (*Anguilla anguilla*), 202
eelworm *see* nematode
Eichhornia, 137
   crassipes, 618, 643, *644*
Elacatinus, 9
Elaphurus davidianus, 946
Elassomia evergladei, 251
elephant, 7, 45, 73, 121, 421, 564, 708, *940*, **940**
   African (*Loxodonta africana*), 939, 941
elephant bird (*Aepyornis*), 911
elephant seal, *546*
elk, 200, 658

elm, 323
Elodea canadensis, 618
Elysia atroviridis, *509*
Embellisia chlamydospora, 506
Empdera somemsos, 784
Empoasca fabae, 122
emu, 10, *12*
Encarsia formosa, 352, 641
Encephalartos, 515
Enchytraeidae, 407, *408*
Endothia parasitica, 808
Endymion non-scriptus, 156
English oak (*Quercus robur*), 318, 447, 883
Engraulis ringens, 656
Enoicyla pusilla, 413, *414*
Entamoeba histolytica, 430
Enterobacter, 500
Enteromorpha, 804
   intestinalis, 804
entodiniomorph, 498, 499
Entrophosphora, 504
Eotetranychus sexmaculatus, 356, *356*
Ephemera, 411
Ephestia cautella, *233*, 353, 362
Epiblema scudderiana, 62
Epichloe, 502
   typhina, 447
Epilachna varivestis, 325
Epirrita autumnata, 318
Equisetum, 507
Equus przewalski, 946
Eragrostis, 617
Ergasilus caeruleus, 442
Erigeron
   annuus, **700**
   canadensis, **700**
Eriocrania, 325
Eriophyllum wallacei, 156
Eriosichia brassicae, *234*
Erithacus rubecula, 202
ermine (*Mustela erminea*), 595
Erwinia herbicola, 60
Erysiphe graminis, 808
Escherichia coli, *502*
Esox lucius, 68
Espeletia spp., 23, **27**
eubacteria, 517
Eucallipterus tiliae, 320
Eucalyptus, 8, 47, 197, 328
   regnans, 65, *183*
eucalyptus psyllid (*Cardiaspina albitextura*), 398, *399*
Eunotia
   alpina, *83*
   pectinalis, *83*
   tenella, *83*
   veneris, 81, *83*
Euphorbia, 617
   maculata, **97**
Euphrasia spp., 433
Euphydryas editha, 932, *932*
Eupomacentrus apicalis, 823, **823**
European eel (*Anguilla anguilla*), 202
European hamster, *72*
European pine sawfly (*Neodiprion sertifer*), *347*, 347, *350*

Eutermes, 417
Eutintinnus pinguis, *508*
Exerastes cintipes, *314*

Fagus, 13, **97**, 613
   grandifolia, **700**, 701, 821
   sylvatica, 88, 723
fairy bluebird (*Irena puella*), 492
fairy shrimp (*Streptocephalus vitreus*), 155
Falco
   peregrinus, 919, **920**, 946
   tinnunculus, 560
falcon, *879*
   peregrine (*Falco peregrinus*), 919, **920**, 946
Fasciola hepatica, 443
fat hen (*Chenopodium album*), 140, 196, **700**, **920**
Felis, 26
   concolor coryi, 938, *939*
feral rock dove (*Columba livia*), 457
fern, 109, 503, 507, 872, 873, 878
   aquatic (*Salvinia molesta*), 618, *619*
   bracken (*Pteridium aquilinum*), 125, *125*, 198, 778, *874*, *874*
   tree, *27*
Ferocactus acanthoides, **97**
fescue grass *see* Festuca
Festuca
   octoflora, 138
   ovina, 175, 688, 824
   pratensis, 810
   rubra, 613, 695, 805
Ficedula hypoleuca, 457, *457*
Ficus, 24, 490, 494
   glabella, 492
   rubinervis, 492
   sumatrana, 492
field bean (*Vicia faba*), **97**
field grasshopper, common (*Chorthippus brunneus*), 147–9, **150**, 154, 195
fig (*Ficus*), 490, 494
fig wasp, 494, *495*
Fimbristylis, 615, *617*
finch
   Darwin's, 20, *21*
   Galapagos, 794
Fiorinia externa, 354
fire-gobies (oxymetopontines), 9
fish, 6, 38, **57**, 68, *68*, 69, 144, 146, 344, *450*, 451, *451*, **631**, 735, **736**, *863*, 895, 900, **916**, 920, 922, 925, **926**
   cichlid, 115, 882
   cleaner, 485
   clown (*Amphiprion*), 486
   cyprinid, 450, *451*
   cyprinodont, 542
   gobiid, 516
   gobioid, 9
   goby, 8, 9, 485–6, 867
   mosquito (*Gambusia affinis*), 544
   Pagothenia borchgrevinki, 60
   percid, 450, *451*
   pomacentrid, 822
   salmonid, 450, *451*

tunny (*Thunnus thynnus*), 202
flagellate, 499, 500, 517, 831
flamingo
    greater (*Phoenicapterus ruber*), 117
    Phoeniconaias minor, 761
flatworm, 507
    Bdellocephala punctata, 559, *559*
    Dugesia lugubris, 559, *559*
flax (*Linum usitatissimum*), 248, *248*
flea, 432, 442
    Cediopsylla tepolita, 808
flea beetle (*Altica sublicata*), 321
flour beetle (*Tribolium*)
    castaneum, 280, **280**, **281**, 469, 809
    confusum, 217, 218, 241, 280, **280**, **281**, 809
flour moth (*Ephestia cautella*), 362
fluke, 432
fly, 314, 407, 441, 872
    agromyzid, 446
    bluebottle (*Calliphora vomitoria*), *350*
    buffalo (*Haematobia irritans exigua*), 423
    cabbage root
        Delia brassicae, 126
        Eriosichia brassicae, *234*
    calliphorid, 425
    carrion, 285
    Hessian (*Mayetiola destructor*), 645, 646
    house (*Musca domestica*), 634, 636
    Oscinella, 441
    Phytomyza, 441
    scatophagid, *336*
    screw-worm (*Cochliomyia hominivorax*), 646
    tephritid, 326
flycatcher, *879*
    Monarcha cinerascens, 786, *787*
    pied (*Ficedula hypoleuca*), 457, *457*
flying fox (*Pteropus*), 942
Formica cinerea, 519
fox, 70, 122
    Arctic (*Alopax lagopus*), 424, 595
    red (*Vulpes vulpes*), 472, *473*
foxglove (*Digitalis purpurea*), 156, 536
Fragilaria
    construens, *83*
    virescens, *83*
Frankia, 512, 515, 516, 520
Fratercula corniculata, 117
French bean (*Phaseolus multiflorus*), **97**
frog, 68, 200, 568, 826
    common (*Rana temporaria*), 271
    Rana
        pipiens, 61
        tigrina, *229*
fruit bat, 900
fruit-fly *see Drosophila*
fucoid, 78
Fucus, 804
    serratus, 76
    spiralis, 76
fulmar, 560
Fungi Imperfecti, 406
fungus, 65, 67, 69, 113–14, 116, 117, 122, 137, 184, 271, 405, 432, 435, 437, 439, 442, 482, 489,

496, 503, *505*, 515, 520, 730, 872
Agaricus campestris, *199*
Ascomycete, 502, 503
Aspergillus fumigatus, 65
Attamyces bromatificus, 489
Basidiomycete, 503
black stem rust, 57, 436, *437*, 467, 468
Botrytis, *441*
    anthophila, 492
    cinerea, 402
    fabi, 448
Claviceps, *441*, 502
Endothia parasitica, 808
Epichloe, 502
    typhina, 447
ergot, 502
Fusarium oxysporum, 465, 506
Fusarium wilt, 626
Helminthosporium maydis, 459
Mucor, 65, 404
    pusillus, 65
Phytophthora infestans, 454, 646
Pilobolus, 184
powdery mildew, 626, 809
Puccinia
    chondrillina, 643
    graminis, 436, *437*, 467, 468
Pythium, 114, 439, 448
    irregulare, *439*
rust, 47, 436, 467–8
Taphrina, 441
Ustilago
    longissima, 441
    tritici, 443
    violacaea, 464, 466
yeast (*Saccharomyces cerevisiae*), *349*
fur seal, 666, 667
Fusarium oxysporum, 465, 506
Fusarium wilt, 626
Fusicoccum, 693, *693*

Gaddus macrocephalus, 667
Galinsoga parviflora, *189*
Galium
    hercynicum, 266, 278
    palustre, 695
    pumilum, 266, 278
    saxatile, 266
    sylvestre, 266
Gambusia affinis, 544
Gammarus, 338, 339, 410, 783, 789
    duebeni, 783
    hercynicum, 278
    locusta, 75, *75*
    oceanicus, 784
    pseudolimnaeus, 419
    pulex, 75, *75*
    pumilum, 278
    zaddachi, 75, *75*
Garrulus glandarius, 185–6, *186*
Gasterosteus, 797
    aculeatus, 296, *297*
Gastrophysa viridula, 323, *324*
gastropod, **288**, *897*, *897*, *906*
Gaumanniomyces, 441

Gaura pulchella, 105
gazelle, 129
    Thomson's, 327, *328*
Gelidium coulteri, 698, *698*, 817
gentian, crossleaved (*Gentiana cruciata*), 487
Gentiana cruciata, 487
Genychromis meuto, 115
Geochelone elephantopus, 946
Geospiza
    conirostris, 20, *21*
    difficilis, 20, *21*
    fortis, 20, *21*
    fuliginosa, 20, *21*
    magnirostris, 20, *21*
    scandens, 20, *21*
Geotrichum candidum, 65
Geotrupes spiniger, 423
Geranium, 182
gerbil
    Gerbillus
        allenbyi, 295, 305
        pyramidus, 305
    Meriones tristrami, 295
Gerbillus
    allenbyi, 295, 305
    pyramidus, 305
Gerris, 191
Geum reptans, 88
ghost swift moth (*Hepiallis humuli*), *118*
gibbon (*Hylobates lar*), 492
Gigartina, 699
    canaliculata, 698, *698*, 699, 817
    leptorhynchos, 698, *698*, 699, 817
Gigaspora, 504
    calospora, 505
Ginsburgellus, 9
Glaucomys, 26, 33
Glechoma hederacea, 143, *144*
Glomus, 504, *709*
    mosseae, 505, 506
Glossina, 432
Glossiphonia, 411
Glossosoma, 41
glutton, 424
Glycine
    javanica, 292, *292*
    max, 317
    soja, 219, 325, 514, *514*
gnat (*Chaoborus astictopus*), 631, *632*
goat, 200
gobiodontines, 9
goby, 8
    Cryptocentrus, 485–6
    Dalhousie (*Chlamydogobius*), 867
goby sandeels (microdesmines), 9
golden-winged sunbird (*Nectarinia reichenowi*), 255
goldenrod *see Solidago*
Gomphocerripus rufus, 52
goose, *119*
gopher, 197
Gordius, 447
Gorgonia spp., *138*
gorilla, mountain, 941, *942*
gorse (*Ulex europaeus*), 369
goshawk, 928

grape phylloxera (*Viteus vitifoliae*), 190, 646

grass, 99, 137, 157, **734**
  Agropyron
    repens, 307, *308*, 810
    smithii, 317
  Agrostis, 156, 695
    capillaris, 134
    scabra, 306–7, *307*, *308*
    stolonifera, 286, 695, 810
  Andropogon, 617
    gerardi, 306, 307, *307*, *308*
  annual meadow (*Poa annua*), 155, 156, 323, 544, *545*
  bent (*Agrostis tenuis*), 81, *613*, 805
  Briza media, 688
  Bromus
    inermis, **787**
    madritensis, 283
    rigidus, 283, *283*
  buffalo (*Buchloe dactyloides*), 112, 138
  canary (*Phalaris paradoxa*), 636
  Cynosurus cristatus, 134
  Deschampsia, 695
    caespitosa, 695
    flexuosa, 156, 688
  fescue (*Festuca*)
    octoflora, 138
    ovina, 175, *688*, 824
    pratensis, 810
    rubra, 613, 695, 805
  Helictotrichon pratense, 824
  Holcus mollis, 156
  knotroot bristle (*Setaria geniculata*), **920**
  Koeleria cristata, 824
  marsh, **734**
  Minuartia uniflora, 305, *305*
  Molinia caerulea, 805
  orchard (*Dactylis glomerata*), *97*, 249, *249*, **787**, 788
  Panicum
    maximum, 292, *292*
    miliaceum, *97*
  Paspalum, 514, *514*, 515, 617
  pasture (*Holcus lanatus*), *134*, 142, *142*, 286, *688*, 808, 824
  Phleum
    arenarium, 294
    pratense, **787**, 788, 810
  Poa
    alpina, **40**
    pratensis, 308, 810
    trivialis, 134, 156, 286, 810
  rye (*Secale cereale*), *97*
  rye grass (*Lolium*)
    perenne, *97*, 231, *231*, 257, *257*, 258, 498
    multiflorum, 316, *316*
  sea, 75, *76*
  sedge, **734**
  Setaria, 615
    faberii, **700**
    italica, **97**
  sorghum (*Sorghum vulgare*), **97**
  Spartina, 76, 103
    patens, 615, *617*

sweet vernal (*Anthoxanthum odoratum*), 39, *688*
tussock, 199
Vulpia fasciculata, 230, *230*, *231*
Zea mays, 53, 62, 92, **97**, 100, *431*, 486, 638
Zerna erecta, 688

grass carp (*Ctenopharyngodon idella*), 644
grass mirid (*Leptoterna dolabrata*), 234
grasshopper, 127, 214, 533, 791
  Austroicetes cruciata, 53, 54
  common field (*Chorthippus brunneus*), 147–9, **150**, 154, 195, 225, 264, *535*
  rufous (*Gomphocerripus rufus*), 52
  Senegalese (*Oedaleus senegalensis*), 205

great tit (*Parus major*), 161, **161**, *193*, 253, 255, 268, 342, *342*
grebe, *879*
  western, 631, *632*
Grindelia lanceolata, 168, 169
ground hog (*Marmota*), *26*
ground squirrel, 383
  African, 127
groundsel (*Senecio vulgaris*), 155, 636
grouse, 337, 378
  red, 589–91
    Lagopus lagopus lagopus, 131, *132*, 383, 455, 469, 474–6, 589, *591*
gull, 805
  glaucous-winged (*Larus glaucescens*), 595, 831
Gunnera, 515
guppy, 337, 338, *338*, 339
  Poecilia reticulata, **545**, 550
Gurania, 129
Gymnarrhena, 189, 191
Gymnoplectron giganteum, 919
gymnosperm, 503, 909–10, **916**
gypsy moth (*Lymantria dispar*), 321

Haematobia irritans exigua, 423
Haematopus
  bachmani, 831
  ostralegus, 255, *256*, 347
hake, Cape, 144, *145*
Halesia monticola, 687
halibut, 666
  Pacific (*Hippoglossus stenolepis*), 658, 667
Halosaccion glandiforme, 831
Hamamelis virginiana, 330, *330*, 687
hamster, 71
Haplopappus
  squarrosus, 326
  venetus, 326
hare, 337, 497
  snowshoe (*Lepus americanus*), 318, 332, 369, *370*, *371*, 377, 394
hare's foot clover (*Trifolium arvense*), 138
hawk, *879*, 930
  Henicopernis longicauda, 787
hawkmoth, 493
heath, 506

heath hen (*Tympanuchus cupido cupido*), 927
heath rush (*Juncus squarrosus*), 56
heather (*Calluna vulgaris*), 383
hedgehog, 69, 122, 127
hedgehog tick (*Ixodes hexagonus*), *433*
Helianthella quinquenervis, 484
Helicoma, 693, 694
Heliconia, 784
Heliconius, 129, 130, 343
Helictotrichon pratense, 824
Heliocopris, 422
  dilloni, 421, *422*
Heliothis zea, 634, 636, 649, 650, 651
helminth worms, 432
Helminthosporium maydis, 459
Hemizonia, **40**
hemlock (*Tsuga*), 13, 14
  heterophylla, 697
  mountain (*Tsuga mertensiana*), 697
Henicopernis longicauda, 787
Hepiallis humuli, 118
Heptagenia, 411
heron, 570
  Ardea cinerea, 400
  Butorides striatus, 787
herring, 658
  North Sea (*Clupea harengus*), 659, *659*
Hesperia comma, 608
Hesperoleucas symmetricus, 835
Hessian fly (*Mayetiola destructor*), 645, 646, **647**
Heteromastus spp., 691
Heteromeles arbutifolia, 99
Heterospilus prosopidis, 376, *376*
Heterozius rohendifrons, *528*
hickory (*Carya*), 13
Himanthalia, 75
Hippoglossus stenolepis, 658, 667
Hippophaë, 515, 516
Hirondella, 427
Hirundo rustica, 201
Holcus
  lanatus, 134, 142, *142*, 286, *688*, 808, 824
  mollis, 156
holly (*Ilex aquifolium*), 52
holly leaf-miner (*Phytomyza ilicis*), 394
Homo sapiens, 27, 45, **72**, 170, 456, 486
Homoptera, 487
honey badger (*Melliovora capensis*), 485
honey guide (*Indicator indicator*), 485
honeybee (*Apis mellifera*), 68–9, *117*, 493, **631**, 785, 917
honeyeater (*Myzomela pammelaena*), 786, 787
hookworm, human (*Necator*), 467
horn acacia (*Acacia cornigera*), 483, 484
hornbill, *879*
  Buceros rhinoceros, 492
horse, 121, 546
  Przewalski's, 946
horsetail (*Equisetum*), 507
human *see Homo sapiens*
Humicola lanuginosa, 65
humming bird, 72, 351
  hermit, 784

Hydra, 138, 187
  viridis, 507
Hydraphantes tenuabilis, 455, 456
Hydrocotyle, 617
hydroids, 39, 137
Hydrometra myrae, 455, 456
Hydropsyche, 411
Hydrurga leptonyx, 37
hyena, 424
Hyla
  chrysoscelis, 826
  crucifer, 304
Hylobates
  lar, 492
  syndactylus, 492
Hymenolepis microstoma, 443, 444, 444
Hymenoptera, 314, 642
Hyphomycetes, 405

Icerya purchasi, 369, 626, 640, 641
ichneumonid, 900
ichthyosaur, *22*
Ilex aquifolium, 52
Impatiens pallida, 250, 250, 253
Indian meal moth (*Plodia interpunctella*), *376, 377, 469, 534*
Indicator indicator, 485
insect, 28, 58, 68, 69, *117*, 120, **179**, **182**, 204, 221, 245, 369, 493, 626, 640, 641, 834, **916**, 920
Iomopsis aggregata, 782, 783
Ipomoea tricolor, 251, 251
Irena puella, 492
Iris pseudacorus, 695
Ischmura elegans, 251
Isoetes howellii, 104
Isoodon macrouris, 27
isopod, 58, 407, *408*, 413
  Asellus aquaticus, 154
  Philoscia muscorum, 157
Isoptera, *408*
Italian millet (*Setaria italica*), **97**
Italian rye grass (*Lolium multiflorum*), 316, *316*
Ixodes hexagonus, 433

Jack pine, **335**
jackal, 424
jaegar, 595
jay
  blue, 125, 127
  Garrulus glandarius, 185–6, 186
jellyfish, 139, 508
Juncus
  bufonius, 156
  effusus, 156, 695
  maritimus, 695
  squarrosus, 56
Juniperus
  osteospermum, 14
  *virginiana, **700**, 701*
Junonia coenia, 106

kangaroo, 25, 27, 423

*Kelletia kelletii, 288, **288**, 289*
kelp, 67, 78
Keratella cochlearis, 122
kestrel, 552, 559
  Falco tinnunculus, 560
killifish (*Rivulus hartii*), 546
kingfisher, *879*
kite, 424
  Everglades (*Rostrahamus sociabilis*), 335
kiwi
  brown, *12*
  greater spotted, *12*
  little spotted, *12*
knots, *203*
  Calidris canutus, 202, 203
koala bear, 8, 47
  Phascolarctos cineueus, 27
Koeleria cristata, 824
Kraemerlines, *9*
Kuhnia glutinosa, 112

Laboulbenia, 442
lactobacilli, 404
Lactuca
  sativa, 458
  *scariola, **700***
ladybird (*Rodolia cardinalis*), 369, 625
lagomorph, 121
Lagopus lagopus scoticus, 131, 132, 383, 455, 468, 469, 474–6, 589, 591
Lambrosia artemisiifolia, 706
Lamiastrum galeobdolon, 143, 144
Laminaria, 75
Lampronia rubiella, 116, 118
Lantana, 626
Lapsana communis, 252, 254
larch budmoth (*Zeiraphera diniana*), 318, 332, 468, 469, 470–2
Larix decidua, 88
lark, *879*
Larus glaucescens, 831
Lasioderma serricorne, 241
Lates nilotica, 923
Lathyrus, 25
leaf root aphid (*Pemphigus bursarius*), 458
leafhopper
  Empoasca fabae, 122
  Ossiannilssonola callosa, 325
leafroller (*Crocidolomia binotalis*), 645
leafworm, Alabama (*Alabama argillacea*), 634
leech (*Glossiphonia*), *411*
legume, 109, 182, 748
Leguminosae, 105
Leiolopisma grande, 919
Leiopotherapon unicolor, 867
lemming, 245, 383, 592, 593, 595, 596
  brown (*Lemmus sibericus*), 592, **593**
*Lemmus sibericus, 592, **593***
Lemna, 78, 137, 138
  polyrhiza, 291
Leonotis, 255
Lepidium sativum, 439
Lepidoptera, 642

Lepomis
  globosus, 442
  macrochirus, 342, 342, 344, 345
Leptasterias, 849
Leptinotarsa decemlineata (Colorado potato beetle), 54, *182*, 581–4, 601, 642
Leptonychotes weddellii, 37, 37
Leptoterna dolabrata, 234
Lepus americanus, 318, 369, 370, 371, 377, 394
Lespedeza capitata, 702
lettuce (*Lactuca sativa*), 119, 458, *458*
lettuce necrotic yellow virus, 432
Leucojum vernum, 88
Leuctra nigra, 691
lice, 432, 442, 457
  Pediculus humanis capitis, 449
  sucking, 501
lichens, 28, 76, 337, 482, 510–11, 516, 520
  Prochloron, 496
Ligula intestinalis, 455
Liliaceae, 25, **27**
lily (*Calochortus*), 919
lime (*Tilia vulgaris*), 320
lime aphid (*Eucallipterus tiliae*), 144, 320
Limonium, 617
limpet, 78, 249, 806, *832*
  Lottia
    digitalis, 831
    pelta, 831
    strigatella, 831
  Patella cochlear, 228, 229, 234
  plate (*Acmaea scutum*), 336
Linum
  catharticum, 824
  usitatissimum, 248, 248
Linyphia triangularis, 329
lion, 129, 341
  mountain, 115
*Liriodendron tulipifera, 687, **700***
Littonia, 25
Littorina, 849
  littorea, 804, 805, 806
liver fluke, 115
  Fasciola hepatica, 443
liverworts, 77
lizard, 68, 72, 131, 452, 790, 791, 869, 880, 890, 892, 896, 896, 900, 901, 950
  Anolis, 457, 796
    gingivinus, 457
    sagrei, 835
    wattsi, 457
  Sceloporus occidentalis, 538, 538
  sleepy (*Tiliqua rugosa*), 458
Loa, 435
loaches, 78
*Lobelia, 23, **27**, 548, 549*
  keniensis, 548, 549
  telekii, 169, 170, 548, 549
Lobodon carcinophagus, 37, 37, 38
lobster (*Panulirus interruptus*), 288, **288**, 289
locust, 173, 204, 205, 206, 622, 747
  desert (*Schistocerca gregaria*), 204, 206

lodge-pole pine (*Pinus contorta*), 697
Lodoicea seychellarum, **920**
Lolium
　multiflorum, 316, *316*
　perenne, **97**, *134*, 156, 231, *231*,
　　257, *257*, 258, 286, 298, *498*,
　　808, *810*
Lonchocarpus pentaphylla, *183*
Lophodemium, 693, *693*
Lottia
　digitalis, 831
　pelta, 831
　strigatella, 831
louse, pigeon
　Campanulotes defectus, 925
　Columbicola extinctus, 925
lousewort, Furbish's (*Pedicularis*
　furbishiae), 941
Loxodonta africana, 939, 941
Lucilia, 425
　cuprina, 327, 402, 448, 652, **653**
Luciogolbius, 9
Lupinus texensis, 105
Lutra lutra, 946
lycopods, 503, 507
Lygodesmia juncea, 112
lygus bug (*Lygus hesperus*), 649, *650*,
　651
Lygus hesperus, 649, *650*, 651
Lymantria dispar, 321, 354
Lymnaea elodes, 284
lynx, Canadian (*Lynx canadensis*), 369,
　377, 589, 591

Macrolepidoptera, **179**
Macropus
　agilis, 27
　eugenii, 27
　giganteus, 27
　rufogriseus, 27
　rufus, 27
Macropygia
　mackinlayi, *180*, 785, 786
　nigrirostris, 785, 786
Macrotermitineae, 499
Maculinea, 487, 516
　arion, 487
　rebeli, 487
madder (*Rubia peregrina*), 51, *51*
Malvastrum coccineum, *112*
Mamestra brassicae, 314
mammals, 33, 35, 69, 546, 735, **736**,
　863, **870**, 872, 890, 891, 897,
　897, 905, 911, 914, **916**, 920,
　926, 950
　marsupial, 23, *26*
　placental, 23, *26*
Manduca quinquemaculata, 117
mangabey (*Cercocebus galeritus galeritus*),
　937
mangrove (*Rhizophora mangle*), 868,
　920
Mansonella, 435
mantids, 784
Mantis religiosa, 784
maple, 181, **734**
　Acer campestre, 98

red (*Acer rubrum*), *13*, **703**, 704
　sugar (*Acer saccharum*), *13*, 701
Margaritifera auricularia, 921
marmot, 71, **82**
　Marmota, 26
marsh tit (*Parus palustris*), 268
marsupial, 493
Mastoterma paradoxa, 500
Matricaria matricarioides, 156
Mayetiola destructor, 645, 646
mayfly
　Baetis, 78
　Ephemera, *411*
　Heptagenia, *411*
　Nesameletus ornatus, 835
meadow grass, annual (*Poa annua*),
　155, *156*, 323, 544, *545*
mealworm, *342*
mealy bug, 439
Mechanitis isthmia, 123
Medicago
　lupulina, **787**
　sativa, **97**, 624, **734**
Megadrili, 407, *408*
Megalaima spp., 492
Melanerpes formicovorus, 187
Melanorrhoea inappendiculata, 492
Melaphis rhois, 502
Melilotus alba, 168, *168*
Melitaea cinxia, 608, *608*, 609,
　610
Melliovora capensis, 485
Melosira distans, 83
Melospiza melodia, 220
Mercurialis, 613
Meriones tristrami, 295
Merluccius
　capensis, 144, *145*
　paradoxus, 144, *145*
Merriam's kangaroo rat (*Dipodomys*
　merriami), 346
Mesodinum rubrum, 508
Metarhizium anisopliae, 642
Metepeira datona, 835, 880
Methanobacterium, 66
　ruminanthium, 498
Methanococcus jannaschii, 66
Mexican bean beetle (*Epilachna*
　varivestis), 325
Microbotryum violaceum, 492
microdesmines, 9
Microdipodops megacephalus, 794
Micropus apus, 570
Microtus, 592
　ochrogaster, 181
　pennsylvanicus, 181, 192, 598, 599,
　　600
　townsendii, 600
midge, 114
　Anatopynia pennipes, 850
　Asphondylia, 441
　phantom (*Chaoborus*), 122
　Pseudochironomus richardsoni, 835
mildew, powdery, 626, *809*
Milium effusum, 156
milkweed (*Asclepias* spp.), 125
millet
　Italian (*Setaria italica*), **97**

Panicum miliaceum, **97**
millipede, 127
　Apheloria montana, 420
　pill, 127
Milyeringa, 9
Mindarus victoriae, 502
Minuartia uniflora, 305, *305*
mirid bug (*Capsodes infuscatus*), 331,
　331
Mirounga leonina, 37
mistletoe, 127, 435, 455
　Trilepidia adamsii, 922, 925
mistletoe bird, 185
　Australian (*Dicaeum livundinaceum*),
　　185, 490
mite, 395, 432, 920
　Cecidiophyopsis, 441
　Eotetranychus sexmaculatus, 356,
　　356
　household, 403
　Hydraphantes tenuabilis, 455, 456
　litter, 407
　oribatid, 419, *420*
　Poecilochirus necrophori, 426, *426*
　raspberry leaf-bud (*Phyroptus placilis*),
　　118
　red spider (*Tetranchyus urticae*), 647
　soil, 409
　two-spotted spider (*Tetranchyus*
　　urticae), 352, 624
　Typhlodromus, 441
　　occidentalis, 356, *356*
moa, 911
mole, 127, 197
　common (*Talpa*), 26
　marsupial (*Notoryctes typhlops*), 26,
　　27
Molinia caerulea, 805
mollusc, 41, 58, 407, *408*, 507, 894,
　895, **916**, 920
　bivalve, 66
　gastropod, 508
monarch butterfly (*Danaus plexippus*),
　125, 127, 130, 204
Monarcha cinerascens, 786, *787*
monito del monte, 27
monkey, leaf (*Prescytis obscura*),
　492
monocotyledons, 23, **27**, 102, 876,
　916
Monodelphis domestica, 27
monogeneans, 442, 450, 451
Monomorium minimum, 189
monotremes, 23
Monotropa, 507
Monterey pine (*Pinus radiata*), 723
moose, 337, 529, 808
　Alces alces, 808
Mora, 184, 526
Moraceae, 24
morning glory (*Ipomoea tricolor*), 251,
　251
mosquito, 115, 155, 438, 642
　Aedes trivittatus, 455
　Anopheles annulipes, 478
　anopheline, 432
　Culex, 117

malaria vector (*Anopheles albimanus*), 636, **637**
mosquito fish (*Gambusia affinis*), 544
moss, 28, 38, 77, 482, 503, 872, 873
Motacilla alba yarrellii, 336
moth, 69, 204, 326, 494, 568, 629
  almond (*Ephestia cautella*), 233, 353
  bordered white, 577
  cabbage (*Mamestra brassicae*), 314
  Cactoblastis cactorum, 618, 620
  cinnabar (*Tyria jacobaeae*), 321, 369, 370, 371
  codling (*Cydia pomonella*), 622
  Coleophora alticolella, 56
  diamondback (*Trichoplusia ni*), 645, 649
  flour (*Ephestia cautella*), 362
  geometrid (*Epirrita autumnata*), 318
  ghost swift (*Hepiallis humuli*), 118
  goldenrod gall (*Epiblema scudderiana*), 62
  gypsy (*Lymantria dispar*), 321
  Indian meal (*Plodia interpunctella*), 233, 367, 376, 377, 396, 469, 534
  oak eggar, 122
  pine beauty (*Panolis flammaea*), 399
  pine looper (*Bupalus piniaria*), 144, 145
  Plodia interpunctella, 233
  raspberry (*Lampronia rubiella*), 116, 118
  tineid, 425
  winter (*Operophtera brumata*), 125, 189, 241, 380, 585, 601
mountain beech (*Nothofagus solandri*), 262
mountain hemlock (*Tsuga mertensiana*), 697
mouse, 27, 71, **72**, 323, 564
  Arizona pocket (*Perognathus amplus*), 346
  Bailey's pocket (*Perognathus baileyi*), 346
  deer (*Peromyscus*), 350
  pocket (*Perognathus* spp.), 794, 795
  wood, 370
Mucor, 65, 404
  pusillus, 65
mudskippers, 9, **920**
Muhlenbergia, 617
mule deer, 200
mung bean, 119
Mus musculus, 27
Musca
  domestica, 634, 636
  vetustissima, 423
Musculium securis, 220
mussel, 39, 123, 131, 132, 263, 542, 806, 807, 832, 925
  Brachidontes rostratus, 863
  Californian (*Mytilus californianus*), 282, 338, 806, 819, 820, 831
  Margaritifera auricularia, 921
  Mytilus edulis, 336, 337, 338, 588, 820, 849

mustard, 113
Mustela
  erminea, 595, 783, *783*
  frenata, 783, *783*
  nivalis, 592, 595, *783*
mustelids, 385
Mutisia, 25
Mya arenaria, 688
mycetobacteria, 521
Myocastor coypus, 951
Myotis grisescens, 922
Myrica, 515, 516
Myristica gigantea, 492
Myrmecophaga, 26
Myrmica, 487
  laevonoides, 487
  scabrinoides, 487
  schenkii, 487
Mysis relicta, 942, 943
Mysticeti, 202
Mytilus
  californianus, 282, *338*, 806, 819, 820, 831
  edulis, 336, 337, 338, 588, 820, 849
myxoma virus, 477, 478, 479, 616
Myzomela pammelaena, 786, 787
Myzus, 441
  persicae, 502

Nassarius obsoletus, 688
natterjack toad (*Bufo calamita*), 271
Navicula
  cocconeiformis, 83
  hofleri, 83
Neapheanops tallkampfi, 215, 217
Necator, 467
Necrophorus
  germanicus, 425
  tomentosus, 426
  vespilloides, 425
Nectarinia reichenowi, 255
nematode, 58, 65, 115, 155, 198, 407, 409, 412, 498, 642, 643, 831
  Angiostrongylus cantonensis, 442
  filarial, 435
  Nippostrongylus brasiliensis, 443
  Pneumostrongylus tenuis, 808
  Spirocerca lupi, 443
  Trichostrongylus tenuis, 455, 468, 469, 474–6
Nemurella, 411
Neochetina eichhorniae, 643, 644
Neodiprion sertifer, 347, *347*, 350
neon-gobies (*Elacatinus*), 9
Neopanope texana, 688
Neottia, 507
Nereis virens, 688
Nesameletus ornatus, 835
Neuroterus, 441
newts, 114, 200, 550
Nippostrongylus brasiliensis, 443
Nitzchia parminuta, 83
nodding thistle (*Carduus nutans*), 326
Norway spruce (*Picea abies*), 330, 723
Nostoc, 511, 515, 516
Nothofagus solandri, 262
Notonecta hoffmanni, 344

Notophthalmus viridescens, 826
Notoryctes, 26
  typhlops, 27

oak (*Quercus*), 13, 14, 89, **97**, 138, 161, *162*, **734**, 830
  English (*Quercus robur*), 318, 447, 695
  white (*Quercus alba*), 111, *162*, 318, 406, 687, **700**, **749**, 883
oak eggar moth, 122
oak gall wasp, 121
oats, 456, 486
Obelia, 139
ocelot (*Felis*), 26
octopus
  common (*Octopus vulgaris*), 169
  Octopus bimaculatus, 288, **288**, 289
Octopus
  bimaculatus, 288, **288**, 289
  vulgaris, 169
Odontites verna, 433, 435
Oedaleus senegalensis, 205
Oenanthe lachenalii, 695
Oenothera, 617
Olea ilex, 88
oligochaete worm (*Tubifex*), 411
olive, 87
olive scale (*Parlatoria oleae*), 354, 394–5
Ommatophoca rossii, 37, *37*
Onagraceae, 105
Onchocerca, 435
  volvulus, 435
Oncorhynchus
  mykiss, 835
  nerka, 203, 942, 943
onions, 504
Onitis
  uncinatus, 423
  viridualus, 423
Ononis sicula, 558
Ooencytus kuvanai, 354
Operophtera brumata, 125, 189, 241, 380, 585, 601
Opiliones, *408*
opossum, 127
  American, 27
Opuntia, 323, **355**, 875
  basilaris, 104
  fragilis, **40**, 63
  inermis, 395, 618, 620, 643
  polyacantha, **97**
  stricta, 395, 618, 643
orchard grass (*Dactylis glomerata*), **97**, 249, *249*, **787**, 788
orchid, 7, 184, 494, 506, 526, 803
  long-tubed (*Platanthera* spp.), 494
  Madagascar star (*Angraecum sesquipedale*), 493
Ornithorhynchus anatinus, 23, 26, 27
Orobanche, 435
Oryctolagus cuniculus, 478
oryx, Arabian (*Oryx leucoryx*), 946
Oryx leucoryx, 946
Oscinella, 441
Osmerus spp., 797
osprey, 879

osprey (cont.):
  Pandion haliaetus, **920**
Ossiannilssonola callosa, 325
ostrich, *12*
Otiorrhynchus, *118*
  clavipes, *118*
  suicatus, *118*
otter, 564
  European (*Lutra lutra*), 946
owl, *879*
  barn (*Tyto alba*), 346, *879*
  short-eared, 595
  snowy, 595
  spotted, 926
  tawny (*Strix aluco*), *369, 370, 371, 385, 585, 586, 587,* 601
Oxydendrum arboreum, 687
Oxyria digyna, 88
oyster, 173
oystercatcher (*Haematopus*)
  bachmani, 831
  ostralegus, 255, 256, 347

Pachygrapsus crassipes, 699
Pacific salmon (*Oncorhynchus nerka*), 203
Pagothenia borchgravinki, 60
painted lady butterfly (*Vanessa cardui*), 204
palms, **27, 916**
pampas, 28
panda, giant (*Ailuropoda melanoleuca*), 47, 808, 919, **920**, 926
Pandion haliaetus, **920**
Panicum
  maximum, *292, 292*
  miliaceum, **97**
Panolis flammaea, 399
panther, Florida (*Felis concolor coryi*), *938, 939*
Panulirus interruptus, 288, **288**, *289*
Papaver
  argemone, *123, 124*
  dubium, 123, *183*
  hybridum, 123
  rhoeas, 123
  somniferum, 124
Paralicella, 427
Paramecium, 267–8, 270, 291, 845
  aurelia, *267, 268, 278*
  bursaria, *267, 268, 278*
  caudatum, *267, 268, 278*
Parasponia, 512
Parlatoria oleae, 354, 394–5
parrots, *879*
parsnip, wild (*Pastinaca sativa*), 317, *317*
parsnip webworm (*Depressaria pastinacella*), 317, *317*
Parthenocissus, *24*
Parus, 38
  ater, 268
  caeruleus, 268
  major, 161, **161**, *193*, 268, 342, *342, 361, 362*
  montanus, 268
  palustris, 268
Paspalum, 514, *514, 515, 617*
Passiflora, 123, 129, 130

caea, 343
Pastinaca sativa, 317, *317*
Patella cochlear, *228, 229, 234*
pea
  maple, 339
  Pisum sativum, 325
peacock, 7
Pebbarua spp., *138*
Pedicularis furbishiae, 941
Pediculus humanis capitis, 449
pelicans, *879*
Pelodera, 407
Pemphigus
  betae, 502
  bursarius, 458
penguin, 22, 72, 900
  yellow-eyed, 926
Penicillium, 404, 693
Perameles gunnit, 27
perch, 122
  Nile (*Lates nilotica*), 923
  spangled (*Leiopotherapon unicolor*), 867
Percidae, 450, *451*
Perga affinis affinis, 328
Pergonathus, 795
  amplus, 346
  baileyi, 346
  flavus, 794
  formosus, 794
  longimembris, 794
  penicillatus, 794
periwinkle, rose (*Catharanthus roseus*), 917
Peromyscus, 350
Peronospora, 441
Peruvian anchoveta (*Engraulis ringens*), 656
Petaurus, 26
petrels, 560
phalanger, flying (*Petaurus*), 26
Phalaris paradoxa, 636
Phascolarctos cineueus, 27
Phaseolus
  multiflorus, **97**
  vulgaris, 122
pheasant, *879*
  ring-necked, 226, 227
pheasant's eye (*Adonis annua*), 633
Philodendron, 69
Philoscia muscorum, 157, 418
Phlebejus argus, 606, 609, 610
Phleum
  arenarium, 294
  pratense, **787**, *788, 810*
Phlox drummondii, 151, 152, **152**, 153, *153, 154, 531*
Phoenicapterus ruber, 117
Phoeniconaias minor, 761
Phycomycetes, 406
Phyllobius argentatus, 316
Phyllonorycter spp., 318, 319
Phyllotreta
  cruciferae, 354
  striolata, 354
phylloxera, grape (*Vitreus vitifoliae*), 190, 626
Phyroptus placilis, 118

Physa gyrina, 284
Physobrachia, 486
Physoderma zeae-maydis, 430, *431*
Phytomyza, 441
  ilicis, 394
Phytophthora, 441
  infestans, 454, 646
Phytoseiulus persimilis, *352*, 641, *641*
Picea, 13, 14, 92
  abies, 88, 330, 723
  sitchensis, 697
Pieris rapae, 54, 354
pig, 200
pigeon, *879*
  Chalcophaps stephani, *786, 787*
  Macropygia mackinlayi, 180
  passenger (*Ectopistes migratorius*), 925
pigweed, redroot (*Amaranthus retroflexus*), 293, *293*
pike, **57**
  northern (*Esox lucius*), *68*
pill millipede, 127
Pilobolus, 184
Pimpinella saxifraga, 824
Pinaroloxias inornata, 20
pine, 3, 13, 92, **97**, 181, 197
  jack, 14, **335**
  lodge-pole (*Pinus contorta*), 697
  Monterey (*Pinus radiata*), 723
  red, 14, **335**
    Pinus densiflora, *262, 263*
  Scots (*Pinus sylvestris*), 330, 504, *693, 693*
  white, 14, **335**
pine bark bug (*Aradus cinnamomeus*), 324
pine beauty moth (*Panolis flammaea*), 399
pine looper moth (*Bupalus piniaria*), 144, *145*
Pinus
  cembra, 88
  contorta, 697
  densiflora, *262, 263*, **335**
  monophylla, 14
  radiata, 723
  rigida, 111, 687
  strobus, 687, 755
  sylvestris, 330, 504, *693, 693*
  virginiana, 687
Piper auritum, 105
pipits, *879*
Pisaster, 807, 829, 849
  ochraceus, 806
Pistia, 137
Pisum sativum, 99, 325
pitcher-plant, 115
Plactonema nostocorum, 74
plant-hoppers, 501
Plantago
  lanceolata, 106, 611, *612, 613*
  major, *234*, 611, *612, 613, 613*
  maritima, **40**
  media, 611, *612, 613*
plantain *see Plantago*
Plasmodiophora, 441, 461
  brassicae, 430, 447
Plasmodium, 430, *431, 432*

azurophilum, 457
Platanthera spp., 494
platyhelminths, 432
Platypodidae, 903
Platypodium elegans, 454
platypus (*Ornithorhynchus anatinus*), 23, 26, 27
Platyspiza crassirostris, 20, 21
Plectrocnemia conspersa, 357, 358, 691
*Plectroglyphidodon lacrymatus, 823, **823***
Pleolophus basizonus, 347, 347
Plethodon
  glutinosus, 265
  jordani, 265
Plodia interpunctella, 233, 367, 376, 377, 396, 469, 534
plum, *119*
Plumbago capensis, 490
Pneumostrongylus tenuis, 808
Poa
  *alpina, **40***
  annua, 155, 156, 323, 544, 545
  pratensis, 308, 810
  trivialis, 134, 156, 286, 810
Pocillopora, 141
*Poecilia reticulata, **545***
Poecilochirus necrophori, 426, 426
pogonophorans, 66, 509
polar bear, 72, 72, 847
Pollicipes, 849
  polymerus, 820, 831
polychaetes, 825, *906*
Polydora, 691
Polygonum
  aviculare, 156
  *pensylvanicum, **700***
Pomacea, 335
*Pomacentrus wardi, 823, **823***
pomarine skua, 595
pondweed, Canadian (*Elodea canadensis*), 618
Popillia japonica, 631
poppy (*Papaver*)
  argemone, 123, 124
  dubium, 123, 183
  hybridum, 123
  oil-seed, *123*
  rhoeas, 123
  somniferum, 124
Populus
  *deltoides, **700***
  tremuloides, 198
porcupine, 176, 322
  African, 127
Porites, 141
Porphyra, 831
  pseudolanceolata, 820
Portulacaceae, **97**
Posidonia, 75, 76
Postelsia palmaeformis, 282, 283
pot worm, *407*
Potamogeton natans, 46
potato, *119*
potato leaf roll virus, 432
powdery mildew, 432, 626, 809, *809*
prairie dogs, 127
Prescott chervil (*Chaerophyllum prescottii*), 197

Prescytis obscura, 492
prickly pear cactus (*Opuntia*), 618, 620
  inermis, 395
  stricta, 395
primates, 33
Prioria copaifera, 822
Prochilodus mariae, 830
Prochloron, 496
Prokelisia spp., 191
Promerops cafer, 493
Prosobranchia, *906*
Protea eximia, 493
Proteus vulgaris, 502
protozoa, 65, 198, 407, 408, 409, 417, 430, 498, 499, 500, 508
Protura, 408
Prunella modularis, 480
Prunus
  laurocerasus, 88
  spinosa, 695
Przewalski's wild horse (*Equus przewalski*), 946
Pseudanthonomus hamamelidis, 330
Pseudochironomus richardsoni, 835
Pseudochirops cupreus, 27
Pseudomonas, 750
  syringae, 60
Pseudomyrmex
  concolor, 484, 485
  ferruginea, 483
Pseudopeziza medicaginis, 448
Pseudotsuga menziesii, 533, 534
Psoralia tenuiflora, 112
ptarmigan, 337
Pteridium aquilinum, 125, 125, 198, 778, 874, 874
pteridophytes, 910
Pteropus, 942
Puccinellia, 695, 830
  maritima, 695
Puccinia, 441
  chondrillina, 643
  graminis, 436, 437, 467, 468
  recondita, 465
  striiformis, 466
puffin, tufted (*Fratercula corniculata*), *117*
Pullularia, 693, 694
Pyrococcus furiosus, 66
Pyrodictium occultum, 66
Pyrrhalta nymphaeae, 320, 320
Pythium, 115, 439, 448
  irregulare, 439

Quadraspidiotus perniciosus, 634
quails, *879*
*Quercus, 13, 14, **97**, 138, 613, 695, 749, 864, 876*
  *alba, 111, 162, 318, 406, 687, **700**, 749, 883*
  cerris, 406
  coccinea, 687
  ilex, 88
  ilicifolia, 111
  montana, 687
  nigra, 162
  *prinus, 162, **749***

pubescens, 88
robur, 318, 447, 695, 883
*rubra, **700***
*velutina, **700***

rabbit, 115, 127, 139, 161, 477, 497, 568, 616, 803–4
  brush (*Sylvilagus bachmani*), 322
  European (*Oryctolagus cuniculus*), 478
  jungle (*Sylvilagus brasiliensis*), 477
  Mexican volcano (*Romerolagus diazi*), 808
rabbit flea (*Spilopsyllus cuniculi*), 478
Radianthus, 486
radiolarians, 508
Rafflesia arnoldii, 433
ragwort (*Senecio jacobaea*), 321, 369, 370, 371, 536, 537
rail (*Rallina tricolor*), 787
Rallina tricolor, 787
Rana
  pipiens, 61
  temporaria, 271
  tigrina, 229
Rangifer tarandus, 177, 202
Ranunculaceae, 24, 25
Ranunculus, 322
  aquatilis, 46, 46
  bulbosus, 493
  circinatus, 46
  ficaria, 156, 493
  flabellaris, 46
  fluitans, 46, 77
  glacialis, 88
  hederaceus, 46
  omiophyllus, 46
  *repens, 139, 140, 570, 571, **571**, 588*
  subgenus *Batrachium, 46*
  trichophyllus, 46
Rapistrum rugosum, 189
raspberry, *118, 493*
  Rubus idaeus, 118
  wild, 115
raspberry aphid (*Aphis idaei*), *118*
raspberry beetle (*Byturus tormentosus*), 116, *118*
raspberry leaf-bud mite (*Phyroptus placilis*), 118
raspberry moth (*Lampronia rubiella*), 116, *118*
rat, 72, 622, **631**
  Australian bush (*Rattus fascipes*), 598
  brown (*Rattus norvegicus*), 622, **920**
  kangaroo (*Dipodomys* spp.), 794, 795
  Merriam's kangaroo (*Dipodomys merriami*), 346
Rattus
  fascipes, 598
  *norvegicus, 622, **920***
Ratufa
  affinis, 492
  bicolor, 492
red admiral butterfly (*Vanessa atalanta*), 204

red deer (*Cervus elaphus*), 157, **158**, **160**, 216, *217*, 542
red grouse (*Lagopus lagopus scoticus*), 383, 455, 468, 469, 474–6, 589, 591
red pine (*Pinus densiflora*), 262, *263*, **335**
red spider mite (*Tetranchyus urticae*), 647
redshank (*Tringa totanus*), 353
redwood (*Sequoia sempervirens*), 260
reindeer (*Rangifer tarandus*), 177, 229
reptiles, 69, *862*, 871, *871*, 890, *891*, **916**, 925, **926**
   diapsid, 911
Reticulitermes, 417
   flavipes, 500
rhea, 10, *12*
Rhinanthus minor, 433, 808
rhinocerus, Sumatran (*Dicerorhinus sumatrensis*), *934*, 935, *936*
Rhinocyllus conicus, 326
Rhizobium, 292, *292*, 447, 512, *513*, 514, *514*, 519
Rhizopertha dominica, 228
Rhizophora mangle, 868, **920**
Rhizopus, 65, 404
Rhodalosiphon padi, 502
Rhodoglossum affine, 698, *698*, 817
Rhopalomyia californica, 354
Rhopalosiphum maidis, 502
Rhyacichthys, 9
Rhynchops nigra, 117
Rhyssa persuasoria, 351
rice, 486
Richelia, 515
Risor, 9
Rivulus, **545**
   hartii, 546
roach, 455, 456
   Hesperoleucas symmetricus, 835
   Rutilus rutilus, 127, 807, *807*
robin (*Erithacus rubecula*), 202
rodent, 72, 191, 493, 494, 497, 782, 792, *793*, **793**, *794*, 890, *892*, 893, 894
Rodolia cardinalis, 369, 625, 626, 640
rollers, *879*
Romerolagus diazi, 808
Rostrahamus sociabilis, 335
rotifer, 122, 187, 407, *408*, 409
   Keratella cochlearis, 122
roundworm (*Ascaris*), 432, 444, 467
Rubia peregrina, 51, *51*
Rubus, 702
   fruticosus, 493
   idaeus, 118
rubus aphid (*Aphis rubi*), 118
Rumex
   acetosella, 613, 702
   crispus, 323, 324, *324*
   obtusifolius, 323, 324, *324*
Ruminobacter amylophilus, 502
Ruminococcus flavefaciens, 498, *498*
rust, 432
   black stem (*Puccinia graminis*), 57, 436, *437*, 467, 468
   Puccinia chondrillina, 643

yellow stripe (*Puccinia striiformis*), 466
Rutilus rutilus, 127, 807, *807*
rye (*Secale cereale*), **97**
rye grass
   Italian (*Lolium multiflorum*), 316, *316*
   Lolium perenne, **97**, 231, *231*, 257, *257*, 258, 498

Sabatia, 615, 617
Sacadodes pyralis, 634
Saccharomyces cerevisiae, 349
Saccharum officinale, **97**
Sagittaria sagittifolia, 46
salamander, 279, 290, 550, 561, 563, 563
   Notophthalmus viridescens, 826
   Plethodon
      glutinosus, 265
      jordani, 265
Salicornia, 76, 615
Salix, 695
   alba, 917
   atrocinerea, 695
   cordata, 321, *321*
   lasiolepis, 319
   sachalinensis, 61
Salmo
   salar, 203
   trutta (brown trout), **97**, 225, *226*, 835
   trutta (sea trout), *263*, 264
salmon, 169, 203, 205, 549, 658, 659
   Atlantic (*Salmo salar*), 203
   kokanee (*Oncorhynchus nerka*), 942, *943*
   Pacific (*Oncorhynchus nerka*), 203
salmonid, 450, *451*
Salvelinus namaycush, 68
Salvia glutinosa, 490
Salvinia, 199
   auriculata, 643
   molesta, 291, 618, *619*, 643
Sambuca nigra, 803
San Jose scale insect (*Quadraspidiotus perniciosus*), 634
sand-dune willow (*Salix cordata*), 321, *321*
sand sedge (*Carex arenaria*), 197
Santiria laevigata, 492
sardine, 658
   Pacific, 659
Sassafras albidum, **700**
sawfly, 318
   Perga affinis affinis, 328
Saxifraga acaulis, 16
saxifrage (*Saxifraga acaulis*), 16
scale insect, 395
Scaphiopus holbrooki, 826
Scarabeidae, 422
scarlet tanager, 175
Scatophaga stercoraria, 423
Sceloporus occidentalis, 538, *538*
Schistocephalus solidus, 433
Schistocerca gregaria, 204, *205*
schistosomes, human, 435, *436*, 467

Schizachyrium scoparium, 306, 307, *307*, 308
Schizophis graminum, 502
Schlectendalia chinensis, 502
Schoenoplectus tabemae montani, 695
Scirpus, 695
   maritimus, 695
Scolytidae, 489
Scolytus, 441
Scots pine (*Pinus sylvestris*), 330, 504, 693, *693*
Scotytidae, 903
screw-worm fly (*Cochliomyia hominivorax*), 646
Scrophulariaceae, 25
Scutellispora spp., 504, *709*
Scytonema, 511
sea anemone, 508
   Physobrachia, 486
   Radianthus, 486
sea buckthorn (*Hippophaë*), 515
sea fan, *138*
sea grass, 75, 76
sea palm (*Postelsia palmaeformis*), 282, 283
sea trout (*Salmo trutta*), *263*, 264
seal, 900
   Antarctic, 36, *37*
   crabeater (*Lobodon carcinophagus*), 37, 37, 38
   elephant (*Mirounga leonina*), *37*
   fur (*Arctocephalus gazella*), 37, **667**
   leopard (*Hydrurga leptonyx*), 37
   Ross (*Ommatyophoca rossii*), 37, *37*
   Weddell (*Leptonychotes weddellii*), 37, 37
Secale cereale, **97**
sedge, 28
   Carex
      arenaria, 141, *141*, 197
      bigelowii, 161, *163*
Sedum smallii, 305, *305*
Semibalanus balanoides, 263
Senecio, **27**
   jacobaea, 183, 321, 369, *370*, *371*, 536, *537*
   vulgaris, 155, *155*, 530, 547, *547*
Senegalese grasshopper (*Oedaleus senegalensis*), 205
Senna multijuga, 105
Sequoia sempervirens, 260
Sericea lespediza, **734**
sesame, *119*
Setaria, 615
   faberii, **700**
   geniculata, **920**
   italica, **97**
shark, 22, 73, 921
she-oak (*Casuarina*), 515
sheep, 25, *82*, 121, 161, 200, 424, 805
   bighorn, 929, *929*, 934
sheep blowfly (*Lucilia cuprina*), 402, 448
shepherd's purse (*Capsella bursa-pastoris*), 235
shoebill (*Balaenicos rex*), 117
shrimp, 7, *119*, 121, 516

Alpheus, 485
  djiboutensis, 486
  brine, 187
    Artemia franciscana, 834, *834*
  fairy (*Streptocephalus vitreus*), 155
  opossum (*Mysis relicta*), 942, *943*
  sand (*Crangon septemspinosa*), 88, 89
Sialis fuliginosa, 411, 691
sicklepod (*Cassia obtusifolia*), 293, *293*
Sicydiines, 9
Sideranthus spinulosus, 112
Silene
  acaulis, 16, *16*
  alba, 464–5, *466*, 492
Simulium, 411
sitka spruce (*Picea sitchensis*), 697
skeleton weed (*Chondrilla juncea*), 273,
  273
skimmer, black (*Rhynchops nigra*), 117
skink, grand Otago (*Leiolopisma grande*),
  919
Skrjabinodon spp., 452, *453*
skua, 424
skunk, 127
sloth, giant ground, 911
slug, 121, 200, 322, 407, 425
  Agriolimax reticulatus, **124**
  Dendroceras, 43
smelts (*Osmerus* spp.), 797
Smilacaceae, 25
Smilax, 25
Sminthopsis crassicaudata, 27
smut fungus, 432
  loose (*Ustilago tritici*), 443
  Ustilago violacaea, 464, *466*
snail, 121, 127, 200, 213, 337, 407,
  425, 467, **870**, 878, 917
  brown-shelled, 8
  land, 878
  marine, 917
  mud (*Hydrobia*)
    ulvae, 297, *297*, 298
    ventrosa, 297, *297*, 298
  periwinkle (*Littorina littorea*), 804,
    805, *806*
  predatory, 337
  pulmonate, 284
  shore, 337
  yellow-shelled, 8
snake, brown tree- (*Bioga irregularis*),
  808, 923
snow geese (*Chen caerulescens
  caerulescens*), 830
snowshoe hare (*Lepus americanus*), 318,
  332, 369, *370*, 371, 377, 394
Solanum dulcamara, 803
Solenopsis geminata, 638
Solidago, 538, 550, 617
  caesia, 538
  canadensis, 538
  gigantea, 538
  graminifolia, 538
  missouriensis, 538
  mollis, 112
  nemoralis, 538
  rigida, 538
  rugosa, 538
  speciosa, 538

song thrush (*Turdus philomelos*), 358
Sorex, 350
sorghum (*Sorghum vulgare*), **97**
southern corn leaf blight
  (*Helminthosporium maydis*), 459
soybean (*Glycine*), 218–19
  max, 317
  soja, 219, 325, 514, *514*
sparrow
  field, *709*
  grasshopper, *709*
  song (*Melospiza melodia*), 220
Spartina, 76, 102
  patens, 615, *617*
speedwell (*Veronica*), 155
Spermophilus beldingi, 255
Sphagnum, 611, *613*, 697
Sphegigaster pallicornis, 394
spider, 89, 181
  Linyphia triangularis, *329*
  Metepeira datona, 835, 880
  web-spinning, 900
spider mite (*Tetranychus urticae*), 641,
  641
Spilopsyllus cuniculi, 478
Spinachia spinachia, 339
spiny anteaters, 23
Spirillum lipiferum, 512
Spirocerca lupi, 443
spirochaete, 499, 500, *500*, 517
Spirulina
  platensis, 74, 761
spittle bugs, 642
Spodoptera
  exempta, 205
  exigua, 649, 650, 651
sponge, 39, 67, 137, 155, 187, 507
springtail, 407, 409
  Tomocerus, 418
spruce, 13, 14, 92
  Norway (*Picea abies*), 330, 723
  sitka (*Picea sitchensis*), 697
  white, **335**
spruce budworm (*Choristoneura
  fumiferana*), 181, 205, 398, *399*
squid, 37, 38
  pelagic, 37
squirrel, 33, 90, 115, 322, 323
  Belding's ground (*Spermophilus
    beldingi*), 255
  Douglas (*Tamiasciurus douglasii*), 613,
    614
  flying, 26, 33
  forest (*Callosciurus nigrovittatus*), 492
  grey, 531
  ground, 71, 383
  Ratufa bicolor, 492
  red, 72
Staphylothermus marinus, 66
starfish, 173
starling, 879
  Sturnus vulgaris, **920**
Stegobium panaceum, 241
Stellaria media, 155
Stenophylax spp., 410
Sterculia parviflora, 492
Sterna paradisaea, 202
stickleback, 122, 144, *338*

Gasterosteus spp., 797
  three-spined (*Gasterosteus aculeatus*),
    296, *297*, 433
Stigonema, 511
Stizostenodon vitreum vitreum, 68, 667
stoat, 385
stonefly, 204
  Leuctra nigra, 691
  Nemurella, 411
stork, *879*
strawberry, 139
strawberry thynchites (*Coenorhinus
  germanicus*), 118
Streptocephalus vitreus, 155
Strix aluco, 369, *370*, 371, 385, 585,
  586, 587, 601
sturgeon (*Acipenser sturio*), 921
Sturnus vulgaris, **920**
subterranean clover (*Trifolium
  subterraneum*), 228, 230, 273,
  273, 327
sugar cane (*Saccharum officinale*), **97**
sugar maple (*Acer saccharum*), 701, 821
Sulfolobus acidocaldarius, 66
sunbird
  Arachnothera spp., *492*
  golden-winged (*Nectarinia reichenowi*),
    255
sunfish
  bluegill (*Lepomis macrochirus*), 342,
    342, 344, 345
  pygmy (*Elassomia evergladei*), 251
sunflower, 24, 92, *92*
  aspen (*Helianthella quinquenervis*), 484
swallow
  Hirundo rustica, 201
  wood, *879*
sweet gale (*Myrica*), 512, 515
sweet vernal grass (*Anthoxanthum
  odoratum*), 39
swift, 526, *879*
  Micropus apus, 570
sycamore (*Acer pseudoplatanus*), 325
sycamore aphid (*Drepanosiphum
  platanoidis*), 350
Sylvilagus
  bachmani, 322
  brasiliensis, 477
Symbiodinium, 507
Symphyla, 408
Sympodiella, 693, 694
Syncerus caffer, 585–7
Synedra, 278
  ulna, 268, 269
Synodontis multipunctatus, 119, 480

Tabanus atratus, 117
Tabellaria
  binalis, 83
  flocculosa, 81, *83*
  quadriseptata, 81, *83*
Tachigali myrmecophila, 484, 485
Taenia saginata, 434, *435*
Taenionines, 9
Talpa, 26
Tamias striatus, 71
Tamiasciurus, 613

Tamiasciurus (cont.):
  douglasii, 613, 614
tanager, summer, 709
tapeworm, 314, 432, 435
  Hymenolepis mecrostoma, 443, 444
  Ligula intestinalis, 455
Taphrina, 441
Taraxacum officinale, 543, 543, 622
tardigrades, 187
tawny owl (Strix aluco), 369, 370, 371,
  385, 585, 586, 587, 601
Taxodium distichum, 749, 749
Taxus
  baccata, 88
  canadensis, **920**
teazel (Dipsacus fullonum), 169
Tegula, 849
  aureotincta, 288, **288**, 289
  eiseni, 288, **288**, 289
teleost, 8
Terellia serratulae, 395, 396
termite, 68, 410, 417, 418, 499, 501
  Reticulitermes flavipes, 500
tern, Arctic (Sterna paradisaea), 202
Tetranychus urticae, 352, 352, 624, 641,
  641, 647
Tetrasticus spp., 354
Teucrium polium, 47, 96
Thais, 849
Theileria, 442
Thelandros cubensis, 452, 453
Therioaphis trifolii, 625, 646
Thermus aquaticus, 65
Thiobacillus ferroxidans, 74
thistle, 396, 880, 880
  nodding (Carduus nutans), 326
Thlaspi arvense, 155
Thomson's gazelle, 327, 328
thrip, apple-blossom (Thrips imaginis),
  572, 573, 574, 574, 575, 575,
  580
Thrips imaginis, 572, 573, 574, 574,
  575, 575, 580
thrush, 358, 879
Thunnus
  albacares, 669, 669, 670, **671**
  thynnus, 202
Thylacinus, 26
thyme, 487
tick, 442
  Amblyomma limbatum, 458
  Aponomma hydrosauri, 458
Tidestromia oblongifolia, 64
Tilia
  cordata, 51
  heterophylla, 687
  vulgaris, 320
Tiliqua rugosa, 458
Tillandsia recurvata, **97**
tinamous, 10, 12
Tipula pallidosa, 118, 411
tit, 38, 279
  blue (Parus caeruleus), 268
  coal (Parus ater), 268
  great (Parus major), 161, **161**, 193,
    253, 255, 268, 342, 342, 361,
    362
  marsh (Parus palustris), 268

willow (Parus montanus), 268
toad, 200, 826
  natterjack (Bufo calamita), 271
Tomocerus, 418
Torreya taxifolia, **920**
tortoise, giant, 833
  Geochelone elephantopus, 946
Townsend's vole (Microtus townsendii),
  600
Toxicara canis, 449
Trebouxia, 511
tree, 13–14, 28, 32, 42, 55, 78, 137,
  157, 888, 889, 900, 907, 922
  Abies alba, 88
  Acacia
    burkittii, 158, 159, 160, 160
    cornigera, 483 484
  Acer
    campestre, 98
    pseudoplatanus, 325
    rubrum, 13
    saccharum, 13, **700**, 701, 821
  Aesculus
    glabra, **700**
    octandra, 687
  Alnus glutinosa, 695
  apple, 645
  balsam fir, 398
  Carya, 13
    alba, 687
  Cupressus pygmaea, **920**
  Douglas fir (Pseudotsuga menziesii),
    533, 534
  grass (Xanthorrhoeaceae), 27
  Jack pine, **335**
  Juniperus virginiana, **700**, 701
  Monterey pine (Pinus radiata), 723
  oak see Quercus
  redwood, 173
    Sequoia sempervirens, 260
trematodes, digenean, 903
Trialeurodes vaporariorum, 352, 641,
  641
Tribolium, 456
  castaneum, 280, **280**, **281**, 469, 809
  confusum, 217, 218, 241, 280, **280**,
    281, 809
Trichocorixa, 834, 834
  verticalis, 834
Trichoderma, 693
Trichogramma, 314
  evanescens, 551
  pretiosum, 366, 367
Trichomonas termopsidis, 499
Trichoplusia ni, 645, 649
Trichospermum mexicanum, 105
Trichostrongylus tenuis, 455, 468, 469,
  474–6
Trichosurus vulpecula, 27
Trifolium, 492
  arvense, 138
  pratense, **787**, 788
  repens, 41, 42, 43, **97**, 124, **124**,
    222–3, 223, 298, 299, **787**,
    805, 810
  subterraneum, 228, 230, 273, 273,
    327
Triglochin, 615, 695

maritima, 695
Trilepidia adamsii, 922, 925
Tringa totanus, 353
Triplasis, 617
Tripleurospermum maritimum, 156
Triticum vulgare, 97
trout, 136, 218, 219, 836
  brown (Salmo trutta), **57**, 225, 226,
    835
  lake (Salvelinus namaycush), 68
  sea (Salmo trutta), 263, 264
  steelhead (Oncorhynchus mykiss), 835
Trybliographa rapae, 354
Trypanosoma, 457, 457
trypanosomes, 430, **474**, **477**
tsetse fly (Glossina), 432
Tsuga, 13, 14
  canadensis, 687
  heterophylla, 697
  mertensiana, 697
Tubifex, 411
tuna, yellowfin (Thunnus albacares),
  669, 669, 670, **671**
tunny fish (Thunnus thynnus), 202
Turdus
  merula, 358
  philomelos, 358
turkey, 879
turtle, 114
Tussilago farfara, 183
Tympanuchus cupido cupido, 927
Typha, 555, **555**
  angustifolia, 271, 273, 555, **555**
  domingensis, 555, **555**
  latifolia, 271, 273
Typhleotris, 9
Typhlodromus, 441
  occidentalis, 356, 356
Tyria jacobaeae, 321, 369, 370, 371
Tyto alba, 346, 879

Ulex europaeus, 369, 695
Ulmus alata, **700**
Ulva spp., 698, 698, 699, 699, 708,
  817, 818
Umbilicus rupestris, 63, 64
Uniola, 615, 617
Uroleuconsonchi, 502
Urophora stylata, 395, 396
Urtica dioica, 43, 156, 803
Ustilago
  longissima, 441
  tritici, 443
  violacaea, 464, 466

Vaccinium vacillans, 111
Vanderhorstia, 9
Vanessa
  atalanta, 204
  cardui, 204
Veillonella alcalescens, 291
Venturia, 364
  canescens, 353, 362, 363, 376, 377,
    396, 440
Veromessor pergandei, 295, 296, 296
Veronica, 155

vertebrates, 245
    tetrapod, 911
Verticillium lecanii, 642
viburnum whitefly (*Aleurotrachelus jelinekii*), 399
Vicia faba, **97**, 116, 324, 448
Vigna unguiculata, 647
vine weevil (*Otiorrhynchus sulcatus*), 118
Viola riviniana, 824
virus, 534, 561, 642
    cauliflower mosaic, 432
    granulosis, 642
    larch bud-moth granulosis, 468, 469, 470, **472**
    lettuce necrotic yellow, 432
    measles, 314
    myxoma, 477, 478, 479, 616
    nuclear polyhedrosis, 642
    potato leaf roll, 432
Vitaceae, 24
Viteus vitifoliae, 189, 190, 626
vole, 173, 206, 245, 323, 383, 384, 592, 597, 597
    bank, 370
    meadow (*Microtus pennsylvanicus*), 181, 192, 598, 599, 600
    Microtus ochrogaster, 181
    Townsend's (*Microtus townsendii*), 600
    water (*Arvicola terrestris*), 606
Vulpes vulpes, 472, 473
Vulpia
    ciliata ssp. *ambigua*, 506
    fasciculata, 220, 230, 230, 231, 294, 577, 579
vulture, 424

wagtail, pied (*Motacilla alba yarrellii*), 336
Wallabia bicolor, 27
wallabies, 27
walleye, Lake Erie (*Stizostenodon vitreum vitreum*), 68, 667
walnut, 123
warbler, 89, 879
    hooded, 709
    prairie, 709
warthog, 564
wasp, 314, 440, 628
    Aphelinus thomsoni, 350
    cecidomyid, 446
    chalcicoid, 326
    cynipid, 446, 864
    fig, 494, 495
    gall, 45
    Neuroterus, 441
    oak gall, 121
    parasitic, 628
    parasitoid, 440
    wood, 318

water beetles, **179**
water bug (*Hydrometra myrae*), 455, 456
water crowfoot
    Ranunculus fluitans, 46, 77
    subgenus *Batrachium*, 46
water fern (*Salvinia*), 199
water flea, 115, 807, *807*
    cladoceran, 122
    Daphnia pulex, 122
        var. *pulicaria*, 329
water hyacinth, 199
    Eichhornia crassipes, 137, 618, 643, 644
water lettuce (*Pistia*), 137
water strider (*Gerris*), 191
water vole (*Arvicola terrestris*), 606
waterlily (*Nuphar luteum*), 320, *320*
waterlily leaf beetle (*Pyrrhalta nymphaeae*), 320, *320*
watermelon, 626
waxbill, *879*
weasel, 127, 385, 597, 783, *783*, 785, 792
    least (*Mustela nivalis*), 592, 595
weevil, 121
    Apion ulicis, 369
    azuki bean (*Callosobruchus chinensis*), 376, *376*
    black long-snouted (*Cyrtobagous* spp.), 618, *619*
    boll (*Anthonomus grandis*), 634
    clay-coloured (*Otiorrhynchus* sp.), 118
    Cyrtobagous salviniae, 643
    Phyllobius argentatus, 316
    Pseudanthonomus hamamelidis, 330
    red-legged (*Otiorrhynchus clavipes*), 118
    Rhinocyllus conicus, 326
    vine (*Otiorrhynchus sulcatus*), 118
weta, giant cave (*Gymnoplectron giganteum*), 919
whale, 15, 45, 72, **97**, 114, 561, 658
    Antarctic baleen, 656
    Antarctic fin, 227
    baleen, 38, 202
    blue (*Balaenoptera musculus*), 671
    great, 921
wheat, 57, 65, **97**, 129, 486, **647**, *647*
    rusts, 622
    winter, 502
whelk, 806
    Kelletia kelletii, 288, **288**, 289
white campion (*Silene alba*), 464–5, 466, 492
white clover (*Trifolium repens*), 41, 42, 43, **97**, 124, **124**, 222–3, 223, 298, 299, 515, **787**, 805, 810
white oak (*Quercus alba*), 111, 162, 406, 687
white pine, **335**

white spruce, **335**, 398
white sweet clover (*Melilotus alba*), 168, *168*
whitefish, lake (*Coregonus clupeaformis*), 68
whitefly, 352, 641, *641*, 642, 645
    Trialeurodes vaporariorum, 352
    viburnum (*Aleurotrachelus jelinekii*), 399
wildebeest (*Connochaetes taurinus*), 228, 332, 332
willow
    Salix
        alba, 917
        lasiolepis, 319
        sachalinensis, 61
    sand-dune (*Salix cordata*), 321, *321*
willow tit (*Parus montanus*), 268
winter moth (*Operophtera brumata*), 125, 189, 241, 380, 585, 586
witch-hazel (*Hamamelis virginiana*), 330, *330*
wolf, 72
    Canis, 26
    Tasmanian (*Thylacinus*), 26
wolverine, 72
wood mouse, 370
wood thrush, 709
wood wasp, 318
woodlice, 407, 413, 418
woodpecker (*Melanerpes formicovorus*), 187
woodpigeon (*Columba palumbus*), 177, 326, 339, 353
worm, 67
    gordian (*Gordius*), 447
    polychaete, 66, 509
    tubificid, 337, 338, *338*, 339
Wucheria bancrofti, 435, *435*

Xanthorrhoeaceae, **27**
Xylopia stenopetala, 492

yeast, 349, 407, 419, 550, 558
    Saccharomyces cerevisiae, 349
Yucca, 27
    flaccida, 119

Zea mays, 53, 62, 92, **97**, 100, 431, 486, 638
zebra, 129
Zeiraphera diniana, 318, 468, 469, 470–2
Zerna erecta, 688
Zoothamnium pelagicum, 508
zorapteran (*Zorotypu hubbardi*), 189
Zorotypu hubbardi, 189
Zostera, 76
    marina, 688

Subject Index

Numbers in *italic* refer to figures and those in **bold** refer to tables

abaptation (exaptation), 6–8
abiotic
  conditions, 951
  environment, 711, 894, 896
  environmental factor, 48
abundance, 244, 567–20
  and crowding, 576–80
  and dispersal, 601–3
  intensity, 918
  manipulating, 621–73
  prevalence, 918–19
  see also conservation; culling;
     harvesting; population
acclimation, 60
  to low temperatures, *61*, 63
acclimatization, 58, 60, 61
acetic acid, 498
acetylcholinesterase, 635
achenes, 189
acid rain, 81, 324, 753, 769, 769–70,
  922
acidity, 73
  see also pH
acorns, 90, 115
acquired immune deficiency syndrome
  (AIDS), 917
actinorhiza, 515
action thresholds, 623–5
actual evapotranspiration (AET), 889
actual net reproductive rate, 237, 238
acyanogenic plants, **124**
adaptation, 6–8
adenosine triphosphate (ATP), 98, 513,
  746
advective energy, 4, 108
aeciospore, 437
aerosols, 748, 750, 769
aestivation, 45
age at maturity, 548
age classes, 151
Age of Pesticides, 627, 634
age-specific survival rates, 159, 167
Agent Orange, 633
aggregated
  dispersion, 174
  distributions, 284–7
aggregation, 176–8, 448, 449, 450

for coexistence, 286
heterogeneity, and spatial variation,
  394–7
and interference, 364–8
parameter, 474
of risk, 388–91
aggression, 191, 598
agricultural revolution, 625
agriculture, 917
  continuity of recycling, 755–9
  disturbance, 827
  see also crop
air
  adiabatic expansion, 50
  pollution, and insect herbivores,
    324–5
alcohol dehydrogenase (ADH), 419
aldrin, 628
algae
  benthic, *849*
  Cambrian explosion, 909
  chloroplasts, *509*
  fossil, 909
  marine, 75
  microscopic, 759
  monoculture, 817
  mutualism, 507
  planktonic, 109, 200
  on rocky shore, 607–8
  symbiotic, 507, 508
  unicellular, 45
alkaloids, 124, 626
  toxic, 503
Allee effect, 384–6
allelochemicals, 629
allelopathy, 271
Allen's rule, 70
allochthonous material, 716–17
allogenic succession, 694–6
allometric (/allometry)
  constraints, 560–6
  definition, 561
  relationships, 561, *562*, *563*
  slopes, *562*
allopatry, 294, 295, 298, 789
alternative stable states *see* multiple
    equilibria
altitude, 50
  and species richness, 904–5
aluminium, 73, 109, 770, 898
amensalism, 212, 213

amino acids, 60
aminobutyric acid-gated chlorine
  channel, 635
ammonia, 498, 499, 750
ammonium sulphamate, 629
amoeboids, 517
amphibian dynasty, 911
analogous structures, 22, 23
analysis of variance, nested, **565**
animal–resource interaction, 264
animals, 69, 810
  aquatic and oxygen, 114
  bodies, composition, 121
  bottom-dwelling, and species
    richness, *906*
  consequential dormancy, 198
  detritivorous, 114, 122
  dispersal, 178
  as food resources, *119*
  fungivorous, 407
  marine, *901*
  mimics, 127
  nutritional content, 116–22
  and plant successions, 708
  populations, wild, harvesting, 917
  and predators, 126–8
  predictive dormancy, 195–6
  production and respiration, *735*
  self-thinning, 262, 263
  and sodium chloride, 49–50
  species extinctions, *915*
  species richness, 892, 896
    and index of structural diversity,
      895
    and plant spatial heterogeneity,
      894–5, 896
    and primary productivity, 890
  and temperature, 50–9
  terrestrial, and water balance, 58
  water losses, 57
  vectors, 492
annual species, 32–3, 99, 147–57
  crop plants, 96
  desert, 96, 156
  facultative, 156–7, 157
  individual of, 282
  iteroparous, *148*
  model fugitive, *282*
  pioneer, **700**, 702
  rhythms, 93
  semelparous, *148*

annual species (*cont.*):
  simple: cohort life tables, 147–54
  summer/winter, **700**
ant–plant mutualisms, 483–5
anthrax, 448
antibiotics, 427, 501
antifreezes, 60
ants
  desert, 782
  farming caterpillars, 487
  farming fungi, 488–9
  queen, 520
  seed-eating, 890
  worker, 295–6
aphids
  fumigation, 324–5
  pheromone, 629
  stylets, *116*, 120, 320
aposematism, 126, 127
aposymbiont, 507
apparent competition, 287–90, *287*,
  887
apparent species, 125
aquatic communities, 716–17
  nutrient budgets, 759–63
  primary productivity, 725–9
aquatic depth, and species richness,
  905–6
aquatic plants, **33**
aquifers, 763
arable field, nitrate concentration, *757*
arbuscules, *505*
archaic mammalian lineages, 10
area, 711, 861–3, 874
area-restricted search, 357
arsenicals, 629
arthropods, *412*
  phytophagous, *882*
  temperature, *49*
  terrestrial, 58
  thermobiology, *49*
artificial feeding, 945
asparagine, 512
assembly rules, 880
assimilation efficiency (AE), 733, 734
asulam, 629
asynchrony, 392
atmosphere, relative humidity, 57
attack rate, 371–2
augmentation, 639
autochthonous material, 716–17
autocidal control, 645
autocombustible products, 65
autocorrelation, 589
  coefficients, 591
autogenic succession, 694, 696–703,
  719
autotrophic community, 740
autotrophic succession, 694
autotrophs, 114, 122, 858

B lymphocytes, *445*
β-glucosidase enzyme, 124
β-phenylethyl alcohol, 204
babesiosis, 442
backswimmer, foraging, 344
bacteria, 117, 122, **415**, 759, 760, 909

aerobic, 750
anaerobic, 750
C : N ratio, 116
Cambrian explosion, 909
chemolithotropic sulphur-oxidizing,
  74
chemophilic, 66
chemosynthetic, 509
as decomposers, 114, 404–7
endospore-forming, *500*
heterotrophic, 730, 739
insect-pathogenic, 643
mycetocyte, 501, 504, 520
nitrogen-fixing, 47, 109, 500,
  511–12, 748
rhizobial, 482
in 'smokers', 66
square, 32
symbiotic, 417, 509
and temperature, 65
thermophilic, 66
tuberculosis, 314
bacteroids, 512, 513
Bardsey Island, birds, 928
barnacle
  cohort life table, **166**
  competition between, 266–7
Bartell's model of lake community, 837
basic reproductive rate (R_o), 151, 152,
  163–8, 531
basidiospore, 437
beech–maple woodland, 681
bees, colonies and toxins, 633
beetles
  carnivorous, 879
  farming of fungi, 488–9
  herbivorous, 879
behaviour
  and aggregated distributions, 357–9
  aggregative, 894
  cyclic, 44
  dispersal, 191–3
behavioural defences, 127
Beltian bodies, 483, *484*
benefit : cost ratio, 638
benthic
  community, 77, 740
  invertebrates, 898, 905
  species, 296, 797
Bergmann's rule, 70
bet-hedging, 557
bicarbonate, 770
biennial life cycles, 168–70
bilharzia, 435, *436*, 467
biodiversity, 884, 913–52, 944, 948
  conservation, 884
  health-care compendium, 952
  meanings, 913
biogeochemical cycles, global, 763–73
biogeochemistry, 680, 746–7, 774
  see also carbon; nitrogen;
    phosphorus; sulphur
biogeographical patterns, 933–4
biogeography, 5, 28, 52, 53
biological control, 622, 639–45
  in the 20th century, 626
  agent, 394, *641*, **690**
  and Chinese, 625

heterogeneity in, 395
by importation, 640
of pests, 478
value, 644
biological criteria, 667–8
biological oxygen demand, 114
biological status, 151
biomass, 31, 90, 231, 712, 744, 745
  of decomposers, 416
  decomposition of litter, 411
  definition, 719
  forest, 858
  imperfect measure, 413
  productivity, 717–19
  terrestrial, 773
  yield, 661
biomathematics, 459
biomes, **743**, 924
  forest, 713
  marine and freshwater, 31
  temperate forest, 680, *681*
  terrestrial, 28–31
biosphere, 744, 774
biota, 744, 746, 747, 750, 764, 800
  decomposition, 759
  mainland, 800
biotype, 543, 646
bipyridyliums, 630
bird-pollinated flowers, 784
bird pox, 808
birds
  abundance and range, *921*
  aquatic, mouth parts, *117*
  breeding, *863*, 877, *878*, *901*, *905*
  chaparral, 930
    species richness, *931*
  communities, 902
  endotherm, 69
  flightless, 10, 12
  forest, decline, *923*
  fruit eating, *492*
  habitat, 200
  insectivorous, 890, 902
  krill as food, 38
  land, 173, *877*
  migratory, 205, 878
  monogamous, 189
  plumage, 43
  populations, *934*
  predation, 830, 831
  on remote islands, *872*
  seed-eating, 115
  specialist dispersing, 185
  species richness, 872
  supertramp, 179, *180*
  temperate-region, 157
birth, 147
  see also fecundity
birth rate, 151, 171
Bismark archipelago, 880
boom and bust cycle, 454
borates, 629
Bordeaux mixture, 626
boron, 109
botanical insecticides, 627
bottom-up control, 828, 831–8, 860,
  952
boulders

removal of, 817
species richness on, *818*
Boyle's Law, 235
bracken, cyanide and tannin, *125*
breeding
  experiments, 533–4
  periods, displaced, 784
  seasons, repeated, 157–63
browsers, 806
buds
  axillary, 137
  and predators, 123
bulbs, **33**

C_3
  pathway, 101–2
  physiology, 96, **97**, 101–2
  plants, 32, 100
C_4
  pathway, 102
  physiology, 96, **97**, 102
  plants, 32, 101, *103*, 721
cactus, desert, 65
caecum, 121, 497
caesium, 858, 859
  radioactive, *82*
  in tropical rainforest, *859*
calcium, 746, 748, 788, 858, **859**, 898
  in fungal hyphae, 503
  and injury, 59
  in plants, 109
Calvin–Benson cycle *see* C_3, pathway
cancer, blood, 917
cannibalism, 280
canopy productivity, 723
capitulum, *190*
capsules, explosive, 184
captive breeding, 936, 952
  for conservation, 946
capture–recapture, 146
carbamates, 626, 628, 629, 635
carbaryl, 628, **631**, 649
carbohydrates, 406, 498
  soluble, 503
  stored, and herbivory, 316
carbon, 744, 745, 746, **771**
  carbon-14, 86, 504
  carbon : nitrogen ratios, 116, 417
  cycle, 770–3
  dating, 696
  sink, 774
carbon dioxide, 84, *85*, 95, 499, 742, 744
  atmospheric, 14, 86, 748, *771*
  emissions, *772*
  plants and changing, 104–5
  as resource, 100–5
carbon tetrachloride, 84
carbonation, 748
carbonic acid, 746
cardiac glycosides, 125
carnivores, 33, 410, **736**, **738**, 806–7, 904
  C : N ratios, 116
  consumer, *411*
  and digestion, 121–2
  energy requirements, 117–20

food resources, 90
heterotroph, **738**
interspecific competition, 778
marine, 22, *22*
predators, 313
carrion, 410
  consumption, 423–7
carrion-feeders, in sea, 427
carrying capacity (K_N), 224, 227, 381, 470, 653, **671**
  and budworm recruitment, 399
  definition, 224
  K-selected population, 552
  limits, 517
  peripheral species, 302
  and population density, *382*, *385*
  population size, 237, 238
  stable, 225, 247, 269
  stable equilibrium points, 277
cascade effect, 904, 907
catch per unit effort (CPUE), *145*, 669, *670*, **671**
catchment area, 751–5
Cedar Creek Natural History Area, 701–2
cellular automata, 133, 286
cellulases, 418
cellulolysis, 417, *418*
cellulose, 694, 734, 746
  concentration, 121
  digestion, 120, 404, 406, 417, 496
  ruminant, 498
  termites, 499
census, **221**
  data, interpretation, 567–70
cephalodium, *510*
cercariae, 467
cerci, long tail, 78
chains of extinctions, 942
chamaephytes, **33**
chaos, 243–6, 374, 580–1
chaparral vegetation, 29, 32, 930, 931
character displacement, 295–6, 297–8, *297*
checkerboard distributions, 785–6, *786*
chemical
  control, of pests, 626
  defences, by organisms, 124, 126
  nutrients, 745
  pesticides, 627–30
    problems, 631–7
    virtues, 637–9
chemosynthesis, 509
chemosynthetic microbial activity, 67
Chernobyl nuclear accident, *82*
chilling injury, 59
chitin, 404
chlordane, 628
chloride, 748
chlorinated hydrocarbons, 628, 631, 638
chlorofluorocarbons (CFCs), 84
chloroplasts, 100
  kidnapping, 508–9
ciliates
  blooms, 508
  cellulolytic, 499
cladistic analysis, 796

classification
  of communities, 690–1
  of organisms, 32
cleaning stations, fish, 485
climatic
  changes, 10
  instability, 896–7
  variation
    complex, 902
    species richness, 884–5, 896–7
climax
  communities, 179
  concept of, 709–10
  mosaic, 816
  vegetation, 906
clones, 144
  age structure of shoots, *141*
  dispersal, 198–9
club-root, 461
clutch size, **161**, 536, 559
coarse particulate organic matter (CPOM), *412*
cobalt, 109
coefficient
  of interference, 352, 366, *366*
  of variation (CV), *296*, 390
coevolution, 47, 122, 125, 185, 186, 494
  of mutualism, 520
  predator–prey, 343
coexistence, 270, 283, 310, **787**, 792, 811, 825
  of competitors, 279, 296, 303, 310
  consumer-mediated, 828
  exploiter-mediated, 804, 806–7, 809, 813
  niche differentiation, 288, 300–4
  Paramecia, 267
  positive association, 286
  of species, 37, 304, **787**
  stable, 278, 302
cohort, 168, 217, 257
  of acacia, *159*
  generation time (T_c), 165
  life, 149
  life tables, 147–54, *150*, **152**, 157–8, *159*, 165, *166*, *167*
  survivorship curves, 168, *169*
  true, 155
cold tolerance, 17, 60
  and corn, 62–3
  and genetic variation, 61–4
collagen, 425
collagenase, 425
collector–filterers/gatherers, 410, *411*
colon, 121
colonization, 281–7, 861–83
  local, 892
Colorado potato beetle, 581–4, 642
  life-table data, *572*
  populations
    k-values, *583*
    life-table analysis, **582**
  spread, *182*
  spring adults, 581
  summer adults, 581, *584*
commensalism, 212, 520
commonness and rarity, **920**

community (/communities), 941–4
  actual, 795
  age, 696
  assembly models, 856
  compartments in, 854–5
  complex, 838, 844, 847
  complexity and stability, 844–7
  composition, 681–5
  definition, 679
  dispersion, *175*
  diversity of matches, 35–8
  ecology, 679–80, 692, 825–6
    problems of boundaries, 691–2
  emergent properties, 679
  energy through, 711–43
  flux of matter through, 744–71
  fragile, complex, 844
  hypothetical, 214
  level of organization, 692
  life-form spectra, 32–5
  managed animal–plant, 758
  matrix, *849*
  mixed, of endotherms and
      ectotherms, 72
  model, *857*
    complexity and stability, 841–4
  natural, and food webs, 847
  nature, 679–710
  null, 795, 796
  ordination and classification, 686–91
  patterns, 788
    and competition, 780–9
    in space, 686–92
    in time, 692–710
  primary productivity, 712
  real, 795
  resistance, *857*
  stability, 828
    and food-web structure, 838–47
  structure, 942–4
    and competition, 775–800
  terrestrial, nutrient outputs, 750–1
  trophic levels, *833*
  unsaturated, 882–3
  unstable, 843
  see also aquatic communities;
      competition; energy, flow;
      island(s); nutrient, budgets
compartmentalization, 854–5
compensation, 232, *317*, 327
  perfect, 240
  point, 98, 727
competing species
  coexistence, 894
  separation, 304
competition, 885
  apparent, 287–90
  asymmetrical, 248–53
    between cattail species, *272*
  coefficient, *273*, *294*, *302*, 779
  and community structure, 775–800
  current, 775–80
  definition, 214
  density dependent, 217
  diffuse, 783
  and environment, 281
  environment, free, 234
  evidence, 788–9

  experiments, *307*
  and guilds, 782
  hypothesis, 793, 796
  increased, and size inequality, 249
  intensity and structuring power,
      779–780
  interspecific, 265–312
    and module numbers, 231
    and niche differentiation, 825
  one-sided, 216
  overgrowth, 271–2
  between *Paramecium* spp., 267–8
  pre-emptive, 281–7
  range, 240–2
  for resources, 272
  response surface of, *294*
  role of, conclusions, 800
  root and shoot, *273*
  scramble, 235, 237
  and species richness, 887
  strength, 778
  theory, 780, 782
  timing, *283*
  between *Trilobium* spp., **281**
  and two species, *291*
competitive
  ability, 299, *300*
  asymmetries, 251
  displacement, 295
  dominants, 809
  effect, of species, 293
  equilibrium, 811
  exclusion, 273–81, 286, 295, *310*,
      811, 892
    among parasites, 452
    compared with death rate, 811
    in non-equilibrium models, 813
  exclusion principle, 278–80, 301,
      775
  lottery, 821, *822*
  release, 295
  strategy, 557
  suppression or release, 298
competitor
  coexistence without niche
      differentiation, *779*
  inferior/superior, 283, 285, 309
  potential, coexisting, 792
  removal, 615
complementarity, 948
complementary differentiation, 783
complexity/stability hypothesis, **840**
composer system, *738*
condition(s), 48–89
  definition, 48
  extreme, 898
    and distributions, 55–6
  intertidal areas, 75–7
  limiting, 56
  maps of average, 56
  spatial mosaics, 52
  as stimuli, 58–9
conidia, *405*
connectance, 841, 842, 844, 854
  and biological constraints, 848
  and species richness, *845*
conservation, 913–52
  cost–benefit terms, 918

  crisis, expert assessment, 934–6
  dollar, 947
  ethical value, 917
  focus of effort, 918
  genetic considerations, 924–5
  in practice, 945–52
  rationale, 913–18
  see also captive breeding
conspecifics, 823, 900
constant final yield, *230*
consumer(s), 402
  density: mutual interference, 351–2
  and food patches, 352–7
  generalists or specialists, 115
  niche breadths, 790
consumption
  conditions, 211–12
  on consumers, 328–33
  efficiency (CE), 733, **734**, 736
  rate, 329, 346–51
contagious dispersion, 174
contemporary reproductive output, *532*
continental shelf communities, 717
continuous breeding, 246–7
  see also overlapping generations
continuum models, 605
control action threshold (CAT), 624,
      625, 638, 649
Convention on International Trade in
      Endangered Species (CITES), 941
convergence of form, 22
convergent systems, 243
convergents and parallels, 22–8
coordinates of slope and aspect, 53
copper, *49*, 80, 109, 626
coprophagy, 421
correlogram, 589, *591*
corticosterone, 598
cost of reproduction (CR), 536–7, 539,
      541
costs
  defence/energy, 128
  dispersal, 184
  ectothermy, 71
  endothermy, 71
  of hibernation, **72**
  insulation, 72
  migration, 205
  nitrogen fixation, 513
cotton, and IPM, 648–9
counting individuals, 144–7
coupled oscillations, 374, 376, *376*
covariation, 566
crassulacean acid metabolism (CAM)
  pathway, 103–4
  physiology, 96, **97**
Cretaceous, 22, 104, 911
crop
  growth, and LAI, *327*
  harvesting, 747
  pathogens, 463–5
  rotation, 645, 649
cross-resistance, 635
crowding, 191, 238, 318, 381–4, 400
  in practice, 383–4
  on seed production, 603
  susceptibility, 247, 294
crytophytes, **33**

crystalline lattices, 393
culling, 652–5, 940, 942
    see also harvesting
cultivars, resistant, 453
cultural control, 626, 645
cumulative energy intake, *360*
current competition, 788
$CV^2 > 1$ rule, 390, 394
cyanide, 124
cyanobacteria
    epiphytic, *508*
    mutualistic, 520
cyanogenic
    glycoside, *43*
    glycosinolates, 125
    plants, **124**
cycles, 371–81
    plant–herbivore, 371–81
    predator–prey, 371–81
cysteines, *26*

damped oscillations, 240, 244, 245
Darwin, Charles, 322
    finches, *20*
    theory of evolution, 7
DDD, 631, *632*
DDT, 626, 628, **631**, 636, **637**, 641
de Wit replacement series, 291, 292
dead organic matter (DOM), *745*
death, quantification, 147
death before birth, 153
deciduous plants, **415**
decision
    analysis, 934
    tree, 934, *935*
decomposer(s), 114, 120, 122, 402–28,
        745
    chemical composition, 415–17
    specialist, 403
    system, 737
decomposition, 403
    aerobic, 405
    anaerobic, 114
    of animals, 708
    defined, 403
    see also decomposer(s); detritivores;
            saprophytes
deer
    feeding preference, **335**
    red
        cohort life table, 157, **158**
        static life table, 159, **160**
        survivorship curves, *158*
defaecation, 858
    see also dung
defoliation, *316*, 747, 755
    clover, 327
    herbivory and plant growth, 319–21
    repeated, 321
    timing, 323
deforestation, 84, 712, 747, 764, 768,
        922
    and loss of nutrients, 753
    patchwork, 827
deforested catchment, *754*
degree of aggregation, 286
deltamethrin, 627

demographic
    processes, 135
    significance of dispersal, 181
    uncertainty, 927
dendrogram, *16*
denitrification, 750, 752, 758, 768
density
    definition, 569
    dilutions, 453
    and growth rate and size, *229*
    meaning, 221
    see also abundance; density
            dependent (/dependence);
            density independent
            (/independence)
density or crowding?, 220–3
density dependent (/dependence), 354,
        662–3, 938
    birth rates, *224*
    birth/death, *576*
    chaos, 589
    delayed, 379–81, 397, 578
    direct, 354, *354*, *367*, *368*, 389,
        578
    effect, 389
    and equilibrium density, 933
    exactly compensating, 218, 219
    factor, 575
    growth, 228–31
    inverse, 330, *330*, 334, 351, 354,
        354, *367*, 368
    inverse spatial, 389
    mortality, *224*, *241*
        and *k*-values, *233*
        undercompensating, *218*, 219
    mortality and fecundity, 217–20
    overcompensating, 218, 219, *220*,
        243, 601
    responses, in parasites, *444*
    spatial, 388–9, *389*
    type, 294
    undercompensating, 239, 584
density governance, 577–8
density independent (/independence),
        389
    aggregation, *354*
    birth, *576*
    death, *576*
    mortality, 217, *218*
    noise, 579
deoxyribosenucleic acid *see* DNA
depletion zones, 110, 112
desert succulents, 65, 104
deserts, 29, 50, *896*, 904
detoxification, 635
detritivores, 90, 117, 120, 402–28,
        651, **738**, 842
    aquatic, 410
    assimilation efficiency, *734*
    C : N ratio, 116–17
    chemical composition, 415–17
    faeces, 415
    interspecific competition, 778
    invertebrate, **415**
    and microbivores, 407–11, 731
    and microflora, 411–15
detritivorous heterotrophic organisms,
        858

detritivory, 212, 312
developed (/developing) countries, 171,
        172
diapause, 59, 115, 528
    in animals, 195–6
    facultative, 195
    in insects, 42
    obligatory, 195
    predictive, 196
    resistant, 195
diatoms, competition between, 268–9
diazinon, 628
dichlorodiphenyldichloroethane *see*
    DDD
dichlorodiphenyltrichloroethane *see*
    DDT
dichlorophenoxyacetic acid (2,4-D),
        626, 629, 633
diclofopmethyl, 636
dieldrin, 205, 628, 631, 635
diet
    optimal choice, 851
    predators, 334–9
    specialization, 848
    switching, 807–8
diet-width model, 340–3
difference equations, 246
differential resource utilization, 304,
        311, 790
diffusion coefficients, 113
diflubenzuron, 628, **631**
dimorphism, *190*
    in aphids, 190–1
    seed, 190
dinitro-*o*-cresol (DNOC), 626
dinitrogen, 511, 515
dinosaurs, 911
dioxin (TCCD), 633
diphenamid, 629
diphtheria, 462
diquat, 630, 633
discrete generations, 147, 163
disease, 429–81, 809–10, 936, 945
    arthropods as vectors, 626
    in crop plants, 57
    damping-off, 454
    eradication, 463
    exponential phase, 464
    immunity, 455
    and parasites, 808–9
    second phase, 464
    sexually-transmitted, 463
    terminal phase, 464
    transmission rate, 460
    vectors, 492
    wind-borne, 439
    see also bacteria; fungi; parasite(s);
            virus
dispersal, 6, 176, 178–88, 392, 814,
        924
    active and passive, 181
    clonal, 198–9
    corridors, 945
    costs and constraints, 184
    definition, 173
    demographic significance, 181
    distance, 605
    as escape and discovery, 178–80

dispersal (*cont*.):
  limited, 400
  in metapopulations, 393
  and microtines, 598–600
  and outbreeding, 193–4
  passive, 39, 181–4
    by active agent, 185–7
    in water, 187–9
  phases, 205–7
  polymorphism, 189–91
  in space and time, 173–207
  units, 184
  variation, 189–94
dispersers, 20, 599
dispersion
  clumped, 174
  of organisms or communities, *175*
  in space and time, 173–207
dispersive strategy, 180
dissolved organic matter (DOM), *412*
distribution
  bicentric, 15
  and extreme conditions, 55–6
  and temperature, 56–8
disturbance, **815**, 894
  on community structure, 801–27
  and diversity of communities, 801–2
  and patch-dynamics concept,
    813–14
  species characteristic, **815**
  what is it?, 802–3
diurnal rhythms, 93, 199–200
divergent evolution, 796
divergent systems, 243
diversity, 805, *812*
  indices, 681–3
  of matches, in communities, 35–8
  see also biodiversity
djenkolic acid, 130
DNA, 66
  hybridization techniques, *12*
  sequences, *26*
dominance
  control, 825
  controlled communities, 814, 821
dominance–decay model, 685
dominance–pre-emption model, 685
donor-controlled models, 402, 841
donor population, 63
dormancy, 61, 115, 195–8, 528
  breaking, 58
    mechanisms, 108
  consequential, 195, 196, 198
  induced, 196–7
  innate, 195
  non-seed, 197–8
  predictive, 195
drought, 321
  resistance, 916
drugs, 917
dryfall, 748, 749, **749**
DSMA, 629, *630*
dung, 65, 120, 649, 758, 768
  herbivore, 421
Dutch elm disease, 323, 622
dynamic pool models, 663

ecological
  diversity, of mammalian faunas, 33
  economics, 916
  fact of life, 135–6, 147
    equations for, 139
  food production (EFP), 651
  methodology, 146
  niche, 23, 87–9
    resource dimensions, 130
  radiation, of gobioid fish, 9
  redundancy, 944
  theories, evidence, 867–81
Ecological Society of America, 672
ecomorphs, 797
economic injury level (EIL), 623, *623*,
    624, 638, 646, 648
economically optimum yield (EOY), 665
ECOPATH, 2, 740–1
ecophysiology, 133
ecosystem, 71, 679, 941–4
  buffering, 917
  ecology, 679–80
  green, 835–6
  model, terrestrial, *742*
  pelagic, variation, *834*
  process, relationships, *944*
  white, 835
  yellow, 836
ecotourism, 917, 942
ecotypes, 39–40
  differences in higher plants, **40**
  tolerant to heavy metals, 39
  variation, for frost tolerance, 17
ectoparasites, 442, 942
ectotherms, 67–9, 69, 735, 890
  avenues of heat exchange, *68*
edaphic conditions, 941
effective competitor, *787*
effective density of organisms, 174
effective population size (N_e), 924, 933
El Niño, 659, 660
elastin, 425
electrophoretic analysis, 599
elements, essential, 109
Elton's conventional wisdom, **840**
emigration, density dependent, *584*
endemism, 881, 884, 924, 947
  high, hotspots, *948*
  species, 918
endophytes, 503, *505*
endosymbionts, 507, 520
  aphids and primary, *502*
endotherms, 67, 69–73, 735, 890
  thermostatic heat production, *70*
endothermy, 561
  costs, 71
endrin, 636
enemy-free space, 287–90, 780, 854,
    887
energy, 745, *745*
  allocation, 559
  assimilated, 734
  flow, 849
    in aquatic systems, 740
    hypothesis, 849–50
    model, 738
    patterns, *740*
    reradiated, 84

  through model community, 736–9
    in trophic compartment, *737*
  and heat, 744
  input, and species richness, 885
  maximization premise, 340
  pathways, 731–3
  radiant, 91, 849
  resources, 90
environment (/environmental)
  age, 885, 899–900
  conditions, for life, 5
  continuous homogeneous, 285
  cyclic changes, 42, 43
  directional changes, 42
  erratic change, 42
  fluctuating, 283
  gradients of conditions or resources,
    36
  grain, 174
  harshness, 897–9
    and species richness, 885
  inconstant (unpredictable), 281
  instability, 825
  nature, 4–5
  patchy, 36
  productive, *888*
  productive/unproductive, 547
  productivity, 885
  resource-poor, 99
  stochasticity, 939
  uncertainty, 927
  varying, 42–7
  see also pollution; temperature
enzyme(s)
  activity, and acidity, 73
  and high temperatures, 64
  polymorphism, 16
  and temperature, 48–9
  see also β-glucosidase enzyme;
    cellulases
ephemeral patches, 283
ephemerals, 156–7
  semelparous, 156
epidemics, 466
epidemiological approach, 460
epilimnion, 200
epiphytes, 503, 521
equation
  difference, 371
  differential, 371
  exponential difference, 246
  general, for population increase, 164
  implicit, 165
  logistic, 246–7, 247, 273
  Lotka–Volterra predator/prey, 372
equilibrium
  abundance, *623*
  populations, 577, 602
  theory, 802, 862, 869, 876, 877
    of island biogeography, 865–7
    and phytophagous insects, 867
    predictions, 866
equitability, 682–3
estuary communities, 717
ethanol, 419
ethylene, 114
eukaryote, 7, 65, 74, 495, 511, 517,
    521

euphotic zone, 726, 727, 728, 740
eutrophication, 756, 767, 768, 892,
    952
evaporation, 4, 58
evergreens, 754
evolution
    convergent, *21, 22*
    and island communities, 881–3
    by natural selection, 7
    perfection, 8
    radiative, 37
    tinkering, 8
    see also coevolution
evolutionarily
    stable strategy (ESS), 180
    unique species, 947
evolutionary
    convergence, 32
    divergence, 32, 202
    effect of competition, 279
    tree, *19, 19*
exact compensation, 233, *239*
experiments
    additive, 293, *293*, 294
    and interspecific competition, 290–4
    natural, 294–8
    selection, 299–300
    single-generation, 291–4
    substitutive, 291, 293, 294
exploitation, 215, *216*, 271
    competition, 131, *287*, 351, 828
    and niche differentiation, 304–12
    pressure, 221
    terrorist, 952
exploitationists, 483
exploitative competition, 290, 301–2,
    303, 306
    theory, 307
exploited population, 657
exploiter-mediated coexistence, 804,
    806–7, 809, 813
exploratory dispersal, 179
exposure, 76, 907
    on seashores, 78
extinction, 81, 655, 813, 865, *877*,
    931
    animal and plant species, **916**
    curves, 877
    global, 928–9
    hotspots, 947
    human-induced, 915
    imminent, *948*
    levels of threat, *919*
    local, 892, 928–9, 941, 945
    probabilities and mean times, **939**
    rate
        and habitat area, *929*
        of island birds, *929*
        modern, 914
    and recolonization, *881*
    risk, 926–33
    vertebrates, **926**
    vortices, *927*
    see also conservation

facilitation, 352, 696
facilitation model, 704

faeces
    animals which eat own, 121
    decomposition, 403
    feeding, 420–3
    ruminant, 512
    and seed dispersal, 185
    see also coprophagy; defaecation;
        dung; refaecation
farming
    of caterpillars by ants, 487–9
    of fungi by beetles and ants, 488–9
faunas, ecologically matched, 34
fecundity, 147, 156, 215, 234, 246–7,
    293
    age-specific, 152
    density independence, *220*
    exactly compensating density
        dependence, *220*
    host, 474, 475
    and mortality, patterns of, 152
    schedules, 147, 151, 161–2, **167**
        estimating variables, 165–8
    and survivorship, 151–2
feeding preference, **335**
fertilization, effect on land, 755
fertilizer, 856
    nitrogen, 758
field capacity, 106, 108, 111
fine particulate organic matter (FPOM),
    412
fire, 747, 816, 945
    carbon, 750
    ecological hazard, 65, 78, 801, 802
fish
    Canadian freshwater, *450, 451, 451*
    cleaning stations, 485
    coexisting on coral, 821
    extinction, 770
    freshwater, 895
    gills, 144
    habitat divergence, 8
    insectivorous, 830
    in lakes, 860
    living in water, 6
    marine, 900
    migratory, 203
    mosquito, high-CR habitats, 544
    parasites, 442, 450–1
    solitary, *451*
    temperature and oxygen, *57*
    temperature regulation, *68*
    zooplanktivorous, 344
fishing, 652–5, 947
    effort, 667
    mortality rate, 667
fitness, 8–9, 48, 270, 531
    contours, 538–9, *540, 543*, 549
    and outbreeding, 194
fixed quota, 653, 655–6
flagellate, anaerobic, 499
flavonoids, 124
flight, defence mechanism, 127
floral specialization, benefits, 493
flowering
    delay, 323
    periods, *785*
flowers, insect-pollinated, 492
fluorocarbons, 773

foetus, 136, 198
folivores, 794
food
    chains, 114, 850, 851, 852
    density, 346–51
    high-quality, shortage, 332
    international trade, 755
    patches, and consumers, 352–7
    preferences, 335
        ranked and balanced, 335–7
    resources, 90
        and consumers, 122–8
    selection, 121
    webs, 759, 828–60
        complex, 942
        empirical properties, 847–56
        four trophic levels, *836*
        indirect effects, 829–31
        intertidal, 830–1
        model, complexity and resilience,
            843
        model, dynamic fragility, 850
        plankton, 852
        real, 847
        theory, 855–6
    see also predation
foot-and-mouth disease, 439
foraging, 334, *360*
    broader, 344–6
    and diet width, 339–44
    theory, 339–40
forb
    annual, 32, 708
    perennial, 32, 708
forest
    birch and spruce, *92*
    clearance, 922
    composition, **703**
    deciduous, 858
    exploitation, 771
    gallery, *17*
    gaps, 821
    litter decomposition, *412*
    management, 758–9
    net primary productivity, *721*
    northern coniferous (taiga), 28
    oak, 749
    oak hickory, 174
    pine, *92*
    self-thinning, *261*
    temperate, 28, 100
    tropical, *17*, 100, 858
forestry, 827
    continuity of recycling, 755–9
fossil fuels, 81, 84, 712, 753
    combustion, 749, 769, 771, *771*,
        772, 917
fossil record, *910, 914*
    faunal and floral richness, 909–12
founder-controlled communities, 821–7
fox rabies, 472–4
freeze-avoiding strategy, 61
freezing
    injury, 59–60
    tolerance, 60, *61*
frequency dependent
    selection, 41
        and diet switching, 807–8

frequency dependent (*cont.*):
  transmission, 463
frequency distribution, *143*
frequency of disturbance, *819*
frost
  injury, 55, 321
  tolerance, 17
frost-free period, 55
fructivores, 33
fruits, fleshy, 490, *491*
fucoids, midshore, 76
fumigation, of aphids, 324–5
functional response, 346–9, 349–51,
    385, 389, 518, 619
  and Allee effect, 384–6
fundamental niche, 89, 278, 279
fungal macroparasites, 432
fungal population, changes, *693*
fungi, 114, 116, 117, 122, 404–7,
    415, 759
  aeciospore, 437
  biotrophic, 446
  conidia, 405
  ectotrophic mycorrhizal, *496*
  eukaryotic, 66
  farming, 488–9
  garden, *488, 489*
  haustoria, 435, 442, *496*
  host-specific rust, 47
  hyphae, *496*
  mycelium, 67, 503
  mycorrhiza, 113–14, 184
    root-infecting, 520
    sheathing, 503
    vesicular arbuscular, *505*
  resting spores, 65
  rhizomorph, 439
  rhizomycelium, *405*
  rhizosphere, *496*
  sporangia, 184, 515
  staphylae, 489
fungicides, 468, 506, 626, 808

gall-formers, 880
gametes, 194, 207
gap, 827
  in forest, 821
  formation, 816–18, 818–21
    frequency, 816
  in grassland, 820
  hypothetical minisuccession, *815*
  intermediate-sized, 719–20
  not equal, 823
  size, *820*
  unpredictable, 282–3, 283–4
gaseous exchange, and acidity, 73
gases, industrial, and greenhouse
    effect, 84–5
gene, δ-endotoxin, 647
gene flow, 38
gene-for-gene effects, 646
generalists, 334, 341, 385, 492
generation
  increase, inverse, *237*
  lengths, 163–8
  time, 166
genes

dispersal, 206
  neutral marker, 207
  recessive, 646
  tolerant, 80
genet, 139, 141, 144, 198, *231*, 403,
    480
  and modular behaviour, 143
genetic
  comparisons, 533–4
  control, 645–8
  differences, in dispersal behaviour,
    191–3
  diversity, 913
  drift, 924, 925
  engineering, 207
  exchange, 18
  fingerprinting, 41, 192, 599, 600
    techniques, 192
  manipulation, 646
  material, exchange, 38
  polymorphism, 40–2, 192
  variation, 924
    and cold tolerance, 61–4
genetic–behavioural polymorphism
    hypothesis, 599
genotype, 38
  diversity, *42*
  intolerant, 39
  mutant, 180
  non-dispersive, 180
  recombinant, 190
  of white clover, *43*
geocryptophytes, **33**
geomagnetic measurements, 10
geometrical series model, 685
geophytes, **33**
gerbils, competitive release, 295
germination, 196, *824*
  delayed, 197
  speed, and aspect, *53*
ghost of competition past, 270, 295,
    779, 788, 800
giving-up times, 359
glaciation, cycles, 13
glaciers, 86, 797
  plant succession, 696, *697*
global
  climate, 901, 917, 922
  environment, temperature and
    rainfall, *30*
  environmental change, 81–7
  stability, 838
  temperatures, computer models, 86,
    87
  warming, 14, 86–7, 87
Global Change and Terrestrial
    Ecosystems (GCTE), 712
glucosinolates, 124, 126
glycerol, 60, 61, *62*
glycogen, 61, *62*
glycopeptides, 60
gongylidia, 489
gonophores, 139
gradient analysis, 686
granivores, **793**, 794
grass
  cohorts, 158
  and grazing, 320

perennials, 708
  seedling, 214
  and space, 133, 134
  species, distribution gradients, *688*
grassland, 28
  chalk, diverse, 823
  community, 737, *737, 738*
  gaps, 820
  management, 758
  plant species numbers, *894*
  ridge-and-furrow, 905, *906*
  savannah, 28
  unmanaged, *898*
grazer–scrapers, 410, *411*, 759
grazer systems, 730, 731, 733, **738**,
    739
grazers, 212, 708, 745, 803–6
  changes, in species, 810
  prey for, 115, 313–14
  resources, 114
grazing, *324*, 413, 810
  and grasses, 320
  in Serengeti, 856
greenhouse effect, 14, 773–4
  enhanced, 86
  and industrial gases, 84–5
  principal gases, *773*
greenness of our world, 835–7
gross primary productivity (GPP),
    712–13, *728*, 729, 741, 742
group selection, 177
growth
  and development, 528
  exponential, 236
  increments, 539
  and light, 143
  rate, 129
    and herbivory, *321*
  regulators, 637
guild, 25–7, 800
  animal, 792
  and competition, 782
  competition within, 795
  contemporary, 38
  mixture, 794
  tropical and temperate, 902
gut, as culture chamber, 495
gut limitation, 348

habitable sites, 176, 606
habitat(s), 89
  abnormal, 611
  anoxic, 405
  classification, 539–42, 555–7
  and diet, 345
  dispersal stages, *188*
  disruption, 922–3, 947
  disturbance, 922–3
  diversity, 862, 864–5, 867–71
    and phytophagous insects, 865
    theory, 867
  favourabilities, *303*
  fragmentation, 930–3, 942–4
  high-CR, 541, 542, 548
  interconnected, *855*
  locally suitable, 15
  loss, 925

low-CR, 541, 542, 548, 549
marine, 74
normal, 611
occupancy patterns, of birds, *604*
preservation, 945
seasonal movement, 200
tropical, 902
within hosts, 440–3
haemoglobin, 512
haemophilia, 917
halophytes, 74, 76
handling time, 341, 347, 348, 349
harvestable resources, managing, 664–8
harvested populations, 660–4
instability, 658–9, 659–60
harvester, as predator, 666
harvesting, 652–5, 755
efficiency, 212, 670, **671**
fixed-quota, 653, *654*, 655–6, *657*
harvesting effort, 656–8, 670
Hatch–Slack cycle *see* C$_4$, pathway
haustoria *see* fungi, haustoria
heathland species, 99
helophytes, **33**
hemicellulose, 406, 499
hemicryptophytes, **33**
hemiparasites, 433
herbicide(s), 615, 629–30, 827
chemical structure, *630*
economic benefits, 638
glyphosphate, 630
resistance, 636
herbivores (/herbivory), 117, 730, **734**, **736**, **738**, 750, 804, 904
aggregations, 355
alimentary canal, 120
C$_4$ plants, 102
C : N ratio, 116
consumption efficiencies, *734*
defoliation and plant growth, 319–21
diets, 334, 335
food resources, 90
changing, 141
grazing, 25
on individual plants, 315–25
insect, 105
interactive effects, 323–4
and interspecific competition, 778
large, 121
leaf, 485
mammalian, *912*
and mutualism, 322
as parasites, 314
and plant competition, 323–4
and plant fecundity, 322–3
and plant survival, 321–2
productivity, 731
ranked preference, 336
seldom food-limited, 777
small, as specialists, 121
as specialist, 115
as vectors, 323
and water, 106
herbs, *907*
and germination, 58
temperate, Arctic and alpine, 58
herd immunity, 462

herpes, 461
heterogeneity, 36, 110, 281–7, 386–8, 837–8
in continuous-time models, 391
spatial or architectural, 885
heterotroph(s) (/heterotrophic), 730, **738**, 857
biomass, 858
net annual primary production, *738*
organisms, 745
productivity values, *739*
resources for, 114
succession, 693
heterozygotes, superior fitness, 41
hibernation, 45, 70, 198
cost of rewarming, **72**
hierarchy, competitive, 533
holdfast, 76, 78
Holling's disc equation, 348, 352
holoparasites, 433
homeotherms, 69
homeothermy, 566
homing, in migratory animals, 204
homologous structures, 22, 23, *24*
homozygote, 192
honeybee, and hive temperatures, 68–9
horizons of infection, 439
host(s)
annual nutrient budgets, **752**
density dependence (HDD), 389, 390, 391, 396
density independence (HDI), 389, 390, 391, 395, 396
as habitats, 437–55
specificity, 440
patches, dispersion and colonization, 438–40
populations, effects of diversity, 453–5
as reactive environments, 444–7
host–parasite dynamics, 474, 475
host–parasitoid
encounters, 375
interactions, 354, 371
host–pathogen cycles, 470
host-specific races, 481
Hubbard Brook Experimental Forest project, 751–2, 753, 759, 764
nitrogen budget, **752**
human(s)
activities
and atmospheric carbon, **771**, *771*
and chemicals, 746–7
global climate change, 764
and nitrogen cycles, 768
and nutrients, 749
phosphorus cycles, 767
demography, 170–1
disturbance, and communities, 844
fragmentation of landscape, 946
influence, on habitats, 922
migration, patterns, 911
see also agriculture; parasite(s); pollution
hunter–gatherers, 486
Hutchinson's rule, 792, *792*, 793
hybrids, low-fitness, 789

hydration, 106
hydrogen
cyanide, 41, *42*, *43*
ions, 749
peroxide, 125
sulphide, 114
hydrological cycles, perturbation, 763–4
hydroquinone, 125
hydrosphere, 746, *765*
hydrothermal
activity, *67*
vents, 509
hypolimnion, 200

ice
extracellular, 59, 60
gas trapped, *85*
intracellular, 60
ice-age, 911
ice-caps, melting, 774
ice-cores, gas content, 86
ichneumonid parasitoids, 900
ideal free distribution, 365
ideal free theory, *365*
immigration, 193, 865
counterbalancing, 950
curve, 866, *877*
and extinction, balance, 866
graph, 366
rates, *877*
immobilization, 403
importation, 639
improved fit, 294
inbreeding, 495
depression, 194, 925, 933
in great tits, 193
incidence functions, 786, 880
for species, *787*
index of abundance, 146
index of density, 154
Index Kewensis, 32, 33
index of resilience, *858*
index of water deficiency, *723*
indirect
economic value, 917
interactions, *287*
negative impacts, 288
indiscriminate mortality, 544
individual(s)
competing, 15
concept of community, 692
condition, 136
and density, 569
fitness, 216
quality, 136
what is it?, 136–44
infecteds, 438, *471*
infection
endemic phase, 461
epidemic phase, 461
mean intensity, 449
microparasitic, *471*
population dynamics of, 458–68
prevalence, 448
transient, 445
infectious disease, 464

infectious disease (*cont.*):
    theory, 459
inhibition, 698
    model, 704
inoculation, 639
    against glasshouse pests, 641, *641*
    by fish, 644
inoculum, 205
inorganic insecticides, 627, 629, 922
insect(s)
    abundance, **182**
    as biological control agents, 640, **640**
    and body temperatures, 61, 68
    defoliation, 755
    diapause, 59
    dispersive phase, 179
    growth regulators, 628
    herbivorous, 146, 324–5, *778*
    life cycle, 136
    migration/flight, **179**
    mining, 432
    mouth parts, *117*
    mycetocyte, 501, 520
    non-social, *735*, **736**
    parasitoids, 212
    pathogens, 642
    pest control, 626
    phytophagous, 237, 484, 776–8,
        791, 842, *875*, 910
        and resource partitioning, 784
    poikilothermic, 566
    r-selected, 566
    resistant to insecticides, *635*
    semelparous, 170
    social, *735*
    univoltine, 578
    see also phytophagous insects
insecticides, 467, 627, 808
    chemical structure, *628*
    plant yield depression, 632
    quantity produced, *637*
    resistance, 635, *637*
    third-generation, 629
insectivores, 33, 735, 794, 911
instability, biotic, 825
integrated
    farming systems (IFS), 651
    pest management (IPM), 648–52,
        651
intensity
    of disturbance, *819*
    of infection, 449
intercept constant, 260
interference, 215, *216*, 271
    coefficients, 365, 367
    competition, 131, *287*
intermediate-density plots, 354
intermediate disturbance hypothesis,
    817, 818
International Biological Programme
    (IBP), 712, 720, 723, 726
International Commission for the
    Conservation of Atlantic Tunas
    (ICCAT), 669
International Council of Scientific
    Unions, 84
International Geosphere–Biosphere
    Programme (IGBP), 712, 763

International Union for the
    Conservation of Nature (IUCN),
    926
International Whaling Commission,
    656, 668
interspecific
    aggression, 281
    breeding, 788
    brood parasitism, 480
    competition, 265–312, 775–6, 788,
        789, 795, 828
        asymmetrical, 271–2
        community, 796, 887
        ecological effects, 290–4
        evolutionary effects, 294–300
        examples of, 265–9
        general features, 269–73
        and intraspecific competition, 276
        logistic model, 273–8
        for refuges, 784
        see also competition; intraspecific,
            competition
    competitors
        strong, 276
        weak, 276
    precopulation, 784
intertidal
    areas, 75
    community, *832*
interventionist management options,
    935
intraspecific
    aggregation, 133
    brood parasitism, 480
    competition, 142, 214–64, 470, 653,
        776, 777
        and density dependent growth,
            228–31
        general aspects, *225*
        quantifying, 231–5
        reduced, 326
        and regulation of population size,
            223–8
intrinsic
    community stability, 944
    dispersal ability, 18
    photosynthetic capacity, 93, 97
    rate of natural increase, 165, 246,
        275
inundation, 639
invading fronts, 439
inverse square law, 184
invertebrate(s), 66, 792
    consumer, *411*
    detritivores, **415**
    drift, 187
    faeces, 420–1
    freshwater, communities, *853*
    Permian decline, 909
    predatory, 859
    shallow-water, 909
    and symbionts, 507–9
ions, 128
    atmospheric deposition, *749*
    in streamwater, *754*
IPM *see* integrated, pest management
iron, 73, 109, 748
irreplaceability, 948

irrigation, 757, 917
island(s), 603–10, 861–83
    biogeography, 603, 861–2, 950
        theory, 948
    biotas, 18
    communities
        ecological theories, 864–7
        and evolution, 881–3
    disharmony, 878–81
    effects, and community structure,
        861
    habitat, 861, 868
    land-bridge, 869, *869*
    models, 605
    oceanic, 861
    patterns, 18–21
    and predation, 887
    remoteness, 871–3
    reserves, extinctions, 950
    species–area relationships, *863*, 870
    species impoverishment, *873*
    species turnover, 876–8
    within islands, 950
isobutyl acid, 498
isoclines, 309, 518
    and consumption vectors, 311
    growth, *128*
    model for mutualistic interactions,
        519
    zero, 130, 275, 276, 308, *309*, 372,
        399
isofemale line, *558*
isopod, terrestrial, 58
isotherms, 52, 53, 87
isotope radioactive decay technique,
    83
isovaleric acid, 498
iteroparity, 154, 528, 548, 552, 554
    continuous, 170–1
    modular, 162
    overlapping, *148*, 157
ivory, 942

jay, memory, 186
juvenile migration rates, 236
juvenile survivorship, and reproductive
    value, 547

K-selected population, 552–5, 821,
    847
K species
    extinction, 935
    richness, of arthropods, *882*
    and succession, 707
k-values, 149, 238, *380*, 578, 583
    analysis, 777
    defined, 232
    and density dependent mortality,
        233
    for density dependent reductions in
        fecundity/growth, *234*, *235*
    of predator-induced mortality, 379
    in quantification, 231
keratin, 404, 425
key-factor analysis, 581–9
keystone species, 828, 947

and strong interactors, 829–30
kidney, *119*
kipukas, 18
Krakatau, 876, 877
Krebs cycles, 517
krill, 72, 115

Lack clutch size, *551*
Lake Turkana, flow diagram, *741*
lakes, 716, 905
    endorheic, 761, **762**
    entrophic, 812, 836
    exorheic, **762**
    freshwater, 760–1
    glacial, omnivory, *853*
    and phosphorus, 859
    saline, 761–3
land clearance, 768
land-use changes, *722*, 922
larvae, *62*
latitude, and species richness, 900–4
Lautenbach project, 651
law of constant final yield, 229
leaching, 753, 758
lead, 80, *83*
leaf
    area index (LAI), 227, 260, *261*,
        327, 722–3
    area per unit area of ground (*L*), 260
    diurnal movements, 42
    herbivory, *485*
    shade, 95, 97
    sun, 95, 97
    surface, 96, 97
leapfrog territories, 255
learning
    and migration, 367
    in switching, 339
leghaemoglobin, 512
legumes, 109, 748
    mutualism, 512–15
    nodulated, 213
lemmings, 592–3
    activity, **593**
    avian predators, 596
    densities, *592*
leptokurtic transmission, 439
lethal temperature experience, 55
lettuce, isogenic lines, 458
lichens, 76, 510–11
    mutualisms, 511
    structure, *510*
life cycles
    asynchronous, 784
    and death and birth, 147
    range, *148*
    stages, 136
    staggering, 783
life-form spectra, 33, 35
life history
    components, 527–30
    omnivory, 852
    patterns, 526
    variation, 526–66
life-history traits, of cattails, **555**
life tables, 147
    data, **582**

estimating variables, 165–8
and fecundity schedule, 149
static, 158–61
light
    compensation point, 102
    as resource, 893–4
    saturation curves, *701*
    stimulus, 58
lignin
    access to, 404
    and basidiomycetes, 694
    concentration, 121
    as resource, 120, 406, 734
    termites feeding, 417, 499
lignotubers, 65
lime–sulphur sprays, 634
limiting condition, 56
limiting resource, 295
limiting similarity, 302, 775
linamarin, 124
lindane, 628, **631**, 635
lipids, 406, 515
lithosphere, 746, 765, 770
litter, *416*, 858, 859
    oak-leaf, decomposition, *406*
liver, *119*
living resources, 828
lizards, 452
    desert, 72
    high-CR habitats, 544
    North American, *791*
    predatory, 880
    snout–vent lengths, 796–7
local coexistence, 788
local stability, 838, 841
locusts,
    migration-plagues, 204
    solitaria phase, 204, 205
    transiens phase, 204
logistic equation, 246–7
Lotka–Volterra
    approach, 825, 856
    dynamics, *853*
    equations, *277*, *372*, *392*, 517, 854
    isoclines, 386
    model, 281, 371–4, 376, 387, 391
        coexistence of competitors, 301
        crowding, 381–3
        of interspecific communities, 274,
            277, 280
        variants, 302
        zero isoclines, *275*
    model communities, 852
    principles, 813
    simulation, 811
low input sustainable agriculture
    (LISA), 651
lower input farming and environment
    (LIFE), 651
Lyapunov exponent, 243, 245
lymphokines, *445*

MacArthur and Wilson's equilibrium
    theory, 865–7
MacArthur and Wilson's theory of
    island biogeography, 893
macroalgae, 807

macrofauna, 407, 409, *409*
macroinvertebrates, 797, *819*
macronutrients, 109
macroparasites, 432–7, 460
    of animals, *433*
    definition, 430
    directly transmitted, 467
    fungal, 432
macroparasitic infection, dynamics,
    476
magnesium, 109, 748, 788, 898
maintenance, 329, *329*, 332
malaria, 467, 638, 642, 808
malathion, 635
malic acid, 104
mammals
    boreal, *863*
    breeding cycles, 43
    dynasty, 911
    first reproduction and life
        expectancy, *565*
    forest, *35*
    fruit eating, *492*
    homeothermic, 566
    K-selected, 566
    locomotion of forest, 33
    plant eaters, 33
    small, and endothermy, 70–1
manganese, 73, 109, 898
maps, of population density, *393*
marginal value
    behaviour, 362–4
    theorem, 359–61, *360*
        experimental tests, 361–2
marine
    conservation, 947
    habitats, 74, 905
    sedimentology, 86
marsh plants, **33**
marsupials
    and dormancy, 198
    giant, 911
    phylogeny, 34
mass migration, 191
masting, 161, *330*, 331, 536
    satiation of seed predators, 330
    synchronous, 177, 329
mating
    behaviour, 131
    success, *194*
matter
    in communities, 744–74
    fate, 744–6
maturity, age at, 544–8
maximal food chain, 848
maximum likelihood techniques, 79
maximum sustainable yield (MSY), 653,
    654, 655–60, 940
measles, 461, 462, *462*, 607
megafauna, 407, 947
megaherbivores, 920
megaphytes, 23, 25, **27**
    dicotyledonous, 23
meiofauna, 831
mellibiose, 120
membrane permeability, breakdown,
    59
membrane protectants, 74

mercury, 135
meristems, 24, 32, 320
  and predators, 123, 322
mesofauna, 407, 409, *409*
metabolic rate, *71*
metamorphosis, 136, 149, 826, *826*
metapopulation(s), 603–10, 927, 931, 941, *942*, 945
  dynamics, 479, 607–10, 941
    Levin's model, 607–8, 609
  effect, 395
  of identical initial size, *932*
  persistence time, *932*
  perspective, 391–4
  of subpopulations, 930
  theory, development, 606–7, 861
methabenzthiazuron, 636
methane, 84, *85*, 114, 499, 750, 773
methanogenesis, 405
methoprene, 628, **631**, 636
methoxychlor, 628
methribuzin, 630
methyl-4-chlorophenoxyacetic acid (MCPA), 626
methyl parathion, 636, 649
methylbutyric acid, 2, 498
microalgae, 336–7
microbes, 120, *496*
  rumen, *496*
microbial
  insecticides, 642
  loops, 762
  respiration, 413, *414*
microbivores, 731, 734, **738**
  specialist, 407–11
microclimates, 52, 894, 930
  variation, 50
microenvironment, inhabited, 52
microfauna, 407, 409, *409*
microflora, 90
  and detritivores, 411–15
  exogenous, 418
microhabitat, 346, 781, 907
  differentiation, 304
  species-specific, 615
  switching, 343
  temperature, 52
  variety, 894
micrometeorologists, 108
micronutrients, *49*
  essential, 50, 74
microorganisms, 69, 735
  community, 831
microparasites, 430–2, 448
  definition, 430
  directly transmitted, 438, 460–3, 463
  dynamics of infection, *471*
microplankton, symbiotic associations, *508*
microsites, 135, *822*
microtines, 385, 591–4, 598
  abundance
    extrinsic theories, 594–8
    intrinsic theories, 598–601
  cycles, 592
    and food quality, 594
    predators and parasites, 595

dynamics, 593
  hormonal changes, 598
  hypoglycaemic shock, 594–5
  sex, mating and territories, 600
migration, 45, 135, 205–7
  abundance, 601
  definition, 173
  genetic links, 605
  global change, 82
  and learning, 367
  locust, *206*
  long-distance, 200–2
  one-return journey, 202–3, 204–5
  patch dynamics, 813, 814
  patterns, 199–205, *201*
  random, 191
  in space, 195
  in space and time, 173–207
mineralization, 403
minerals, 747
  exploitation, 917
  in four plant species, *111*
  nutrients, 31
    as resource, 109–14
minimum viable population (MVP), 933
minimum viable threshold, 934, 938
Miocene, *11*, 36, 104
miracidia, 467
mismatches, 40
mitochondria, 517
mixed feeders, 33
model(s)
  Bartell's, 837
  Charnov's optimal diet, 341, 342
  Comin's dispersal, 392, *393*, 394
  Connell and Slatyer's, 704
  of continuous breeding, *246*
  diet-width, 340–3
  with discrete breeding seasons, 236–42
  dominance–decay, 685
  dominance–preemption, 689
  donor controlled, 402, 841
  dynamic pool, 663
  energy flow, 738
  equations, *471*
  equilibrium, 813
  facilitation, 704
  geometrical series, 685
  Holling's, 518
  Horn's matrix, 703, 704, 710
  of intraspecific competition, 238
  Levin's, 607–8, 609
  logistic, of interspecific competition, 273–8
  Lotka–Volterra *see* Lotka–Volterra, model
  mathematical, 235–6
  May's, 387, *388*, 390
  multiple regression, 574, 575
  neutral, 789–800
  Nicholson–Bailey, 374–5, 376, 383, 387–8
  parasitoid–host, *380*
  of perfectly compensating density dependence, *247*
  population, 167
    increase, 238

of population increase with time, *237*
Quinn and Hastings, 931–2
random fraction, 685
rank–abundance, 683–5
regression, *451*
Schaefer, 670
simulation, 938–41
  of species richness, 885–7
stepping-stone, 605
surplus yield, 660, 661
Tokeshi's niche-orientated, 685
tolerance, 704
trophic level, 844, *851*
Waage's mechanistic, 364
modular growth schedule, 162
modularity
  components, 137
  importance, 161–2
modules, 137
  and intraspecific competition, *231*
  reconstructed static life table, *163*
molluscicides, 467
molluscs, freshwater, 894, *895*
molybdenum, 109
momentum of population growth, 171
monoclimax, 710
monoculture, *261*, 454, 698, 817
monophagous feeders, 115, 334, 440
Monte Carlo algorithm, 938
monuron, 629
morphological differentiation, 780
mortality, 293
  asymmetrical, 252
  factor, 583
  intensity of, 149
  and intraspecific competition, 232, 234
  nesting, 193
  rate
    age-specific daily, *153*
    and colony size, *141*
  and seed production, *602*
moth, pest, 629
multilines, 454
multiple equilibria, 397–400
  in harvesting, *658*
  in nature?, 398–400
multiple resistance, 636
multiple trophic-level interactions, 830–31
mussels
  bed, gap formation and filling-in, 819–20
  freshwater, 925
mutants, 454
mutual
  antagonism, 280–1
  dependence, 47
  interference, 351–2
mutualism, 47, 212, 213, *287*, 407, 482–525
  algae–animal, 507
  in animal tissues and cells, 500–2
  ant–acacia, 519
  apparent competition, 288
  beetle/mite, 426
  behaviour, 483–6

culture of crops or livestock, 486–9
cyanobacteria, 520
dispersal of seeds and pollen, 490–5
endosymbiotic, 520
facultative, 417
floral, 494
gut inhabitants, 495–500
herbivory, 322
in higher plants and fungi, 502–7
invertebrate/bacteria, 66
legumes/bacteria, 109
lichen, 511
models, 517–20
mycorrhiza, 114, 184
nitrogen fixation, 511–16
pollination, 490–5
seed dispersal, 490, *491*
mutualists, 402
facultative, *519*
features of lives, 520–1
nitrogen-fixing, 516
mycetocytes, 501
mycoherbicides, 643
mycorrhiza, 184
ectotrophic, *496*
fungi, 113–14
sheathing, 503–4, 520
symbioses, 506–7
vesicular arbuscular, *496*, 504–6,
505, 708
myxomatosis, 477, 608, 616, 803

natural
community, restoring, 950–1
diversity, 827
enemies, 639
experiments, 298–9
populations, dynamics, 242
selection, 634–5, 924
adaptation or abaptation?, 6–8
and cold tolerance, 63
perfection, 8
nature reserves, design, 948–50
necromass, 708, 712
nectar, 90, 490, 493, 783, 842
see also pollination mutualisms
nectaries, 483
extrafloral, 484
negative association, 786, 799
negative binomial distribution, 285,
285
nematocides, 591
nematodes
cysts, 65
and mutualistic bacteria, 643
nerve poisons, 635
nested analysis of variance, 564, *565*
net nitrogen mineralization
(NETNMIN), 742
net primary productivity (NPP), *714*,
715, *728*, 733, 745, **754**
above-ground, 719, 720, 721, 722
average, *718*
in biomes, **743**
current, **743**
decomposer, 737
deficient mineral resources, 723

forest, 713, **715**, *721*, *724*, *749*
grazer, 739
gross primary productivity, *728*
on land, *714–15*
latitudinal patterns, *716*
negative, *729*
in oceans, *714–15*
phytoplankton, *726*, *727*
simulation models, 725
specified values, 736
terrestrial, modelling, 741–3
net recruitment, 225, 327
curves, *227*
net reproductive rate, 236, 242
moderate fundamental, 243
neutral models, 789–800
and distributional differences,
797–9
and morphological differences,
792–8
and resource partitioning, 790–1
neutral stability, 374
niche
breadths, 791
complementarity, 782
differences, 786, 788
differentiation, 279–80, 291, 780–5,
789, 792, 813
and coexistence, 288, 779
and coexisting competitors,
300–4
competition, 776
and mechanisms of exploitation,
304–12
patterns of, 290
in *Rhizobium*, 292
subtle, 310
temporal, 896
divergence, 298
expansion, 295
packing, 775
partitioning, 783
realized, 89, 278, 279, 280, 302,
346
similarity, 304
theory, 278
vacant, 777
Nicholson–Bailey
equations, 391, 392
model, 374–5, 376, 383
heterogeneity, 387–8
nicotine, 626, 627
nitrate, 111, 112, 113, 745, 749–50
loss, 756
pollution, 756
nitric acid, 769
nitrification, 513, 757
nitroanilines, 629
nitrogen, 109, **415**, 745, 752, 858,
859, 888
in aquatic communities, 725
balance sheet, at Rothamsted, **756**
budget, 752
control, 756–8
cycles, perturbation, 768–9
and decomposition, 416
fixation, 697, 724, *747*, *752*, 768
costs, 513

and mutualisms, 511–16
in non-legume mutualisms,
515–16
in fungal hyphae, 503, 506
gaseous, 748, 750
input (NINPUT), 742
loss, 755–6
losses from ecosystem (NLOST), 742
in nutrient cycling, 746
oxides of, 324, 753, 769
plant highest concentrations, 121
saturated system, 768
nitrogen dioxide, 324
nitrogen-fixing rhizobia, 520
nitrogenase, 512, 515, 748
nitrogenous fertilizers, 768
nitrophenols (DNOC), 630
nitrous oxide, 84, 750, 773
nod genes, 512
nodes, 143
non-equilibrium
condition, *811*
model, 813, 825–7
theory, 802, 826–7
null hypotheses, 789–800, 793, **793**,
794, 795
numerical change, 144
nunataks, 16
nutrient
availability, 755
budgets, 747–8, *747*
in aquatic communities, 759–63
in terrestrial communities, 748–59
cycling, 745, *745*, 746, 751, 944
mechanisms, 753
distribution, *727*
element composition, **415**
flux
global, 765
pathways, *766*
global pathways, *765*
inputs, 752
loss, 750, *751*
outputs, 750–1, 752
recycling, 403
spiralling, 759, *760*
nutrient-use efficiency, 306, 753, 754,
754

oak hickory forest, 174
oak-leaf litter, decomposition, *406*
obligate
anaerobes, 498
mutualism, 417, 493, 518
symbionts, 501
ocean, 761–3
buffering atmospheric CO_2, 84
communities, 717
floor, animals, 905
primary productivity, 726
submarine ridges, 727
upwellings, 726–7
see also aquatic communities; marine,
habitats
offspring, 7
number and fitness, 537–8
number/clutch size, 551–2

offspring (*cont.*):
  size and number, 550–2
  size and parental care, 528–9
  size-insensitive/sensitive, 542
old field
  rank–abundance curves, *907*
  succession, 699–700, 701, 708, 906
oligophagous feeders, 334
omasum, 497
omnivores, 336, 794
omnivory, in food webs, 852–3
onchocerciasis, 467, 642
one-generation cycles, 374, 375–6
opportunism, 110, 544
opportunistic plague migrations,
    locusts, 205
optimal foraging and patch use,
    359–64
optimum similarity, 775
option value, 917
options sets, 538–9, *540, 543*
  concave, 549
ordination, 688, 690, *690*, 691
organic matter, 922
  decomposition, 758
organism-weighted density, 221, **221**
organisms, 404–17
  acclimation to low temperatures,
    60–1
  aquatic, and salt, 74–5
  average density, 174
  best performance conditions, 48
  with cellulases, 120
  classifying, 115
  cold-blooded, 69
  dispersion, *175*
  dispersive properties, 180–1
  diversity and patchiness, 5–6
  effective density, 174
  and environments, 4–47
  as food resources, 114–28
  heat tolerant, 66
  and high temperatures, 64–7
  historical factors, 9–21
  hypotonic/isotonic, 74
  and intensity of conditions, *49*
  iteroparous, 158
  match environment, 39, 42–7
  mobility, 38, 157
  modular, 136–44
    range, *138*
  specializations and food resources,
    117
  terrestrial, 67, 69
  thermophilic, 65, 66
  unitary, 136–44
    and growth, 228
  warm-blooded, 69
  zonation, 76
organochlorines, 631, 634, 635
organophosphates, 626, 628, 634
osmoregulation, 544
  and acidity, 73
  and salinity, 74
outbreeding, 492, 494
  and dispersal, 193–4
overexploitation, 671–2, 920–2, 922,
    925

overlapping generations, 163, 165,
    166, 168, 246–7
oviposition, 337, 347, 364, 581, 645
ovipositor, 315, 494
oxalic acids, 124
oxygen
  consumption, 54
  diffusion rate, 689
  isotope ratios, *13*
  isotopes, 10–11
  as resource, 114
ozone, 84, *85*, 324, 753, 773

p-oxon, 634
pair-wise interactions, 272
parallel evolution, 23, *26*
  divergence, 34
  of mutualisms, 494
paraquat, 630, 633
  resistance, 636
parasite(s), 90, 212, 314, 809–10, 930
  acquired, 452
  benign, 457
  biotrophic, 444, 455, 461
  brood, 930
  brood and social, 479–81
  cestode, *435–6*
  classes, 465–8
  communities, diversity, 449–53
  competition between, 443
  digenean, *904*
  and disease, 808–9
  distribution within host, 448–9
  diversity, 429–37
  epidemic foliage, 464
  on humans, 456
  inherited, 452
  monogenean, 903, *904*
  necrotrophic, 404, 444, 448, 461,
    492
  population dynamics of hosts,
    468–76
  survivorship growth and host
    fecundity, 455–8
  and temperature, 56
  toleration of, 457
  see also bacteria; fungi; virus
parasite–host interaction, 212
parasitism, 114, 213, 429–81, 494
  by aphids, 446
  definition, 429
parasitoid–host model, *380*
parasitoids, 314–15, 334, 335, 374,
    640
  and host density, *354*
  searching efficiency, 375
parathion, 628, **631**, 635
parental
  care, 537
  effort, 560
parthenogenesis, 191
pastoral holdings, *949*
patch, 603–10
  definition, 352
  ephemeral, 283
  habitable, 605, 606
  location, 357

pseudo-sink, 609
patch-dynamics concept, 814, 825,
    861
  and disturbances, 813–14
patch-occupancy distributions, 609,
    611
patch-size variation, 609
patchiness, 174, 175–6, 603
  and time, 355–7
patchwork mosaics, 827
patchy populations, 569, 571
pathogenicity, 470
  intermediate host depression, 470
pathogens, 57, 462, 643
pathotypes, 642
patterns
  of abundance, 369–71
  of distribution: dispersion, 174–8
pentosans, 417
peptides, 60
perennial species, 32
  competitively superior, *282*
  desert, 99
  and G–L–F triangle, *556*
  ramet of, *282*
periodic table of elements, *110*
permafrost, 28
permanent wilting point, 107, *107*
permethrin, 627, **631**, 632
persistent infections, 445
pertussis, 462, *462*
pest
  actual, 623
  control, 621–3, 672
    aims, 623–5
    history, 625–7
    mechanical and physical, 626
  eradication, 951
  hypothetical, *623*
  outbreaks, secondary, 634
  potential, 623
  resistance, 916
  what is it?, 621, 622
pesticide, 672, 922
  application, 647
  chemical, 627–30
    virtues, 637–9
  natural, 625
  problems, in cotton pests, *651*
  treadmill, 637
pH, 48, *49*, 689, 770, 788, 898
  soil, 898
  of soil and water, 73–4
  tolerance, of diatoms, 81
  tolerance limits, 73–4
phagocytes, 445, 446
phanerophytes, **33**
phenol derivatives, 630
phenol metabolism, 318
phenotypic correlations, 535
phenotypic plasticity, 557–60
pheromones, 629
phosphate (/phosphorus), 109, 112,
    113, **415**, 506, *506*, 724,
    745, 859, 888, 892, *893*,
    898
  from aerosols, 750
  aquatic communities, 725

availability to plants, 745
concentrations, 749–50
cycles, 765–7
detrital, 762
fluxed, model, *763*
in lakes, 750
in lithosphere, 746
in pasture, 788
pathways and cycles, *761*
radioactive, *113*, 503, 504
stripping, 767
from weathering, 748
phosphoenolpyruvate (PEP)
        carboxylase, 102, 103
photoperiod, 46, 58, 60, 195
photorespiration, 102, 103, 104
photosynthate distribution, 316
photosynthesis, 5, 91, 744, 745, 770,
      849
  and controlled water loss, 96–7
  gross, 98
  net, 98–9
  and PAR, 100
photosynthetic
  activity, 96–8
  biochemical pathways, 101
  capacity, 99
  efficiency, *721*
  rate, in successional sequence, **700**
photosynthetically active radiation
      (PAR), 91–2, 96, 720, 725
  component, 93
  intensity, 98, *98*
  and photosynthesis, 100
phyllosphere, 403
phylogenetic constraints, 560–6
phylogenetic sequences, *9*
phylogenetic-subtraction method, 566
phylogeny, 7, 28, 796
  effects, 564–6
  and size, 561
physical forces, 77–9
physiological time, 54, *54*
phytoalexins, 446
phytophagous insects, 776, 777–8,
      842, 864, 910
  and ants, 484
  diet, 489
  diversity, area and remoteness,
      873–6
  and equilibrium theory, 867
  species–area relationships, *864*
phytoplankton, 760, 761, 767, 810,
      836, 859, 896
  primary productivity, 717, 720, *726*,
      727, 729, *729*
  respiration, 727
picoplankton, 762
pine forest, *92*
piroplasms, 442
plankton, 716, 760, *849*
  see also phytoplankton
planktotrophic strategy, 188
plant(s), 69, 920
  aquatic, 45, *46*
  calcifuge, 74
  carnivorous, 313
  and changing carbon dioxide, 104–5

chemicals, noxious, 124
climbing organs, 23
compensation, 315–17
cushion, **33**
defences, 124, 125
defensive mechanisms, 123
defensive responses, 317–19
defoliated, 317
desert
  somatic polymorphism, 47
  variations, in productivity, 331
detritus, consumption, 417–20
ecotypes in higher, **40**
endangered, 941
ephemeral, 125
fecundity, and herbivory, 322–3
fern-like, 909, 911
flowering, **870**
  and absolute minimum
    temperature, *51*
  Cretaceous/Tertiary, 909
  as food resources, *119*
  Fynbos, 892, *893*
  genetically engineered, 647
green
  body composition, 90
  resources, 122
  growth, 319–21, 325
higher, specialized niches, *441*
life cycle, 555–6
litter, 694
mature, relative abundances of, *156*
morphological features, *24*
non-seed dormancy, 197–8
nutritional content, 116–22
optimal foraging, 364
parasitic flowering, 433
pathogens, 314
pathologists, 464
physical and chemical defences, 835
recruitment, 326
remote, 876
rhizomatous, 137, *138*
rosette, **33**
seed, Devonian, 909
seed dormancy, 196–7
seeds, 65
semelparous, 169
shade, 94, 99
species, 87
species richness, *857*
  and rabbit grazing, *804*
stoloniferous, 137, *138*
structural diversity, 875
succession gradient, *709*
successional, **707**
survival, and herbivory, 321–2
and temperature, 50–9
triangular classification, 555–7
variation in architecture, 143
varieties, resistant to pests, 646
vascular, *905*, 909, *929*
water loss, 108
waterlogged, 108
weight, mean, 841
Pleistocene, 10, 13, 899, 911, 912
Pliocene, 36, 104
poikilothermy, 69, 566

Poisson distribution, **355**, 375
pollen
  analysis, 86
  dispersal, 194
  as food resource, 493
  records, 13
  samples, 10
pollination mutualisms, 490–5
pollinators, *493*, 917, 942
  insect, 47
  long-lived, 494
pollutants, 48, *49*
pollution, 753, 922, 947
  degradation, 922
  environmental, 79–80, *82*
  by humans, 951
  and lichens, 511
pollution-tolerant forms, 80
polyclimax view, 710
polyculture, 645
polygamous mammals, 189
polygynous mating systems, 546
polyhydroxy compounds (PHCs), 60
polymorphism
  dispersal, 189–91, *191*
  and genetic change in parasites and
    hosts, 477–9
  seed, 190
  sequential, 96
  somatic, 45, 46, 47, 189
  specific, 41
  transient, 40
polyols, 74
polyphagy, 343, 489, 880
polyphenism, 557–8
polyps, 139
polysaccharides, 415, 417
population
  actual, 924
  behaviour, 580, 594
  change, 575
  cycles, 589–601
  definition, 569
  density, 144, 146, *192*
  dispersal behaviour, 191–3
  dynamics
    of creeping buttercup, 570–1,
      571
    idealized diagrams, *580*
    of infection, 458–68
    of interactions, 355
    long-term, *377*
    of predation, 369–401
  experimental perturbation, 610–20
  field, 39
  fluctuation or stability, 570–81
  fluctuations, 239, 601
    range, *242*
  flux, **571**
  fragmentation, 931, 933
  genetically highly diverse, 14
  genetically idealized, 924
  growth, 937
    momentum, 171
  heterogeneities in, 38
  high-density, 228, *229*
  increase
    exponential, 246

population (*cont.*):
  sigmoidal pattern, *225*
  speed, 246
  and time, rate of, 164
limitation on size, 573
local
  colonization and extinction, 603–5
  dispersal, 605
low-density, 228, *229*
model, cohort life table and fecundity
    schedule, **167**
mortality, 585
patchy, 569, 571
persistence, 937, *937*
persistence time, mathematical
    model, 936–8
pyramid, 170, *171*
reduction, frequency, *812*
regulation, 576, 579
  in practice, *226*
  and territoriality, 253
resource-limited, 216
to sample, 146
size, 237, 573–6, *575, 577*
  changes, 572
  increase and time, 164
  and persistence time, 937, *937*
  and population change, 575
  regulation, 223–8
  regulation and determination, 579
skewed weight distributions, 248
small
  dynamics, 926–8
  extinction, 890
theory, 568
variation in dispersal, 189–94
viability analysis, 933–41, 945
see also carrying capacity;
    equilibrium, populations;
    harvested populations; maps, of
    population density
potassium, 109, 745, 788, 888, 892,
    893
  aerosols containing, 750
  in hydrosphere, 746
  weathering and, 748
potential evapotranspiration (PET), 722,
    723, 889, *889*, 890, *891*
power law, −3/2, 258, 262
prairie, 28
  root systems of plants, 28, *112*
  short-grass, *112*
pre-reproductive death, 169
precipitation, *721*, 722, 742; 764, 890,
    892
predation, 115, 255, 312, 313–33,
    518, 885
  and community structure, 801–27,
    887
  definition, 313
  population dynamics, 369–401
  on prey populations, 325–8
  reciprocal, 280, **280**
  and reduced competition, 326–7
  see also carnivores; diet; herbivores
predator(s), 114, 123, 212, 287,
    809–10, 894, 930
  behaviour, 334–68

classifications, 313
competitive inferiors, 805
controlling, 945
crowding, 381
density, **288**, *289*
design and behaviour, 850–2
eating profitable prey, *336*
efficient, *384*
inefficient, 383, *384*
introduction, 618–20
and lemmings, 595
removal, 615–18
satiation, 331, 548
specialist, 596, 879
true, 313
types, 313–15
zero isocline, *373, 374, 381, 382,
    384*, 397
predator-free space, *289*
predator-induced mortality, 289, 379
predator–prey
  abundance, 369
  covariance, 391
  cycles, 376–81
  in food webs, 853–4
  interactions, 212, 355, 369, 380,
    381
    mathematical modelling, 213
  metapopulation models, 392
  model, mathematical, 597
  patches, 357
  and plant–herbivore systems,
    371–81
  ratios, 854
  spiral, anticlockwise, 666
  zero isocline model, *398*
preservationists, 952
prey
  density, 355
    aggregative responses, 353–4
  image and nonimage, 338
  isocline, *386*
  patches, high-density, 366
  species population, decrease, 829
  zero isocline, *373, 381, 382, 384*,
    385, 397
prickly pear cactus, control, 618, *620*
primary
  productivity, *730*, 850, 856, 887,
    889, 892
    biomass of herbivores, 860
    and carbon dioxide, 774
    factors limiting, 720–9
    patterns, 713–19
    and trophic levels, 849
  succession, 696, 698
probability
  distributions, 56
  structure of extreme values, 79
  values, 936
production efficiency (PE), 733, 734,
    735, **736**
  variations, 734
productivity, 717, 887–94, 944
  below-ground, 713
  higher, species richness increase,
    888
  richness decrease, 892

progeny, dispersing/non-dispersing,
    189
proglottids, 435
prokaryotes, 74, 114, 495, 500
  cellulolytic, 120
  heat tolerant, 66
promiscuous mating, 546
promitochondria, 517
propanil, 629
propionic acid, 499
propoxur, **637**
protease inhibitors, 318
protected areas, 946–50, **947**
protectionists, 483
protein, 746
  ice-nucleating, 60
  thermal hysteresis, 61
prothalli, 517
protists, 69, 137, 831
  herbivorous, 909
protohemicryptophytes, **33**
pseudo-extinctions, 876
pseudo-immigrations, 876
pseudo-interference, 367, 388, 389,
    391, 395
pseudopods, 508
pupa, 55
pure scramble competition, 233, 235
pyrazolines, 636
pyrethroids, 635
  synthetic, 627, 636, 637
pyrethrum, 627

quinone, 125

r-selected populations, 552–3, 847
r species, 404, 552–5, 821
  and succession, 707
r/K concept, 553–5, 556
rabbit
  grazing, 803–4
  removal, 616
rabies, 472–5
radiation, 91, 713, 849
  diurnally and seasonally variable
    resource, 93–5
  incident, 95, 99, 713, 725
  inefficient use, 720
  intensity, and photosynthesis, 94–5
  intercepted, 95
  photosynthetically active, 31
  as resource, 91–100
  résumé, 99–100
  and seed dormancy, 197
  and water, 95–6
  see also temperature
rainfall, *44*, 748
rainforest, 105
  tropical, 29, 59, *902, 948*
ramet, 137, 140, 142
random
  dispersion, 174
  encounters model, 375
  migration, 191
random-fraction model, 685
rank–abundance models, 683–5

rapidly inducible defences, 318
raspberry plant, life cycle and
    phenology, *118*
rates of increase, 164–8
Raunkiaer, 32, **33**, **34**
    life-form classification, 32, **33**
reaction norm, 558
reactive habitat, vertebrate host, *445*
real community, 790–1
reciprocal predation, 280, **280**
recolonization, 79
    of habitat fragments, 945
recreation, 917
recruitment, 662, 814
    rate, 653
refaecation, 497
refuge, 386
    partial, 387, 474
refuges, probability, 814
refugia, 16
    tropical forest, 903
regression
    coefficient, 583
    equation, 575
    model, *451*
regular dispersion, 174
regulation, 243, 380, 576
rehabilitation community, 951
relative humidity, 48, 58
reorganization algorithms, 790, *791*
replacement
    community, 951
    diagrams, *292*
    series, 514
        experiment, 300, *301*
reproduction
    cost, 536–7
    precocious, 79
reproductive
    allocation (RA), 307, 529, 543–4,
        545, 552
        and success rate, 549
        and timing, 543–50
    effort, 529
    isolation, 18
    potential (m_x), 152
    precociousness, and delay, 528
    rate, basic (R_0), 163–8, 243, 460
    value, 530–32, *531*, *532*
reservoirs, biotic, *765*
residual reproductive value (RRV), 531,
    532, 539, 552
    and C-R habitats, 541
    and competitive ability, 542
    in kestrels, 559–60
    and reproductive costs, 536
resilience
    of community, 838
    energy and nutrient perturbations,
        857
    flux through community, 858
    to perturbations, 847
resistance, 626, 645–8
    of community, 838
    evolved, 634–7
    management, 636, 647–8
    to perturbations, 847
resource, 90–134, 402, 529, 711

antagonistic, 129–30
augmenting, 613–14
capture and protection, 530
categories, 129–30
classification, 128–30
complementary, 129, 336
continuum, 885, 886
crucial, 529
decomposing, 406
depletion zones, 215
dimensions, 130
    of ecological niche, 130
essential, 128, 129
inhibition, 130
input, differences, *535*
intake, decreased rates, 215
limited, 529
management, 212, 668-71
partitioning, 80, 296, 298, 304, 784,
    805, 813, 821, 885
patches, **221**
perfectly substitutable, 129, 336
primary, 415
range, 886
renewal, 309
richness, 887–94
seasonal availability, 115
single, exploitation, 306–8
space as, 131–4
substitutable, 128
systematic variation, 93
two, exploitation, 308–12
see also energy; nutrient; oxygen;
    soil; water
resource-dependent growth isoclines,
    128
resource-depletion zone(s) (RDZ), 93,
    107, 110, 111, 112, 113
    for nitrates, 111
    in soil, 907
    soil bound ions, 111–13
resource-range hypothesis, 901
resource-use overlap, *791*
resource-utilization curve, 302, *302*,
    303
resource-utilization efficiency, 306, 307
resource-weighted density, 221, **221**,
    223
respiration, 91, 98–9, 770, 771
    anaerobic, 405, 769
respiratory
    heat (R), 713
    heat loss, *728*, *858*
    metabolism, pathway, 98
response
    curve, 48, 49
    surface, 244
        analysis, 293, 294
reticulum, 497
rhizobia, 512, 521
    mutualism, 512–15
    nitrogen-fixing, 520
    nodule forming, 520
rhizome, **33**, 55, 162, 197, 198
rhizomorph, 439
rhizomycelium, *405*
rhizosphere, 403
richness relationships, 887–900

River Thames, pollution, 951
rivers, temperature and oxygen, of fish,
    57
rodents
    body sizes, *794*
    desert, 72, 782, 792, *793*, 893, 894
    dispersal, 191
    granivorous, **793**
    seed-eating, 890, *892*
Rogers Lake, *13*
root(s)
    access to mineral resources, 112
    allocations, 307
    capture of water, 106–7
    competing, 251
    evolution, 106
    exploration precedes exploitation,
        110
    growth, cessation, 320
    hairs, 113
    modular, 137
    nodule, development, *513*
    as opportunists, 110
    suckers, 198
    tubers, **33**
    and water, 106
rosette plants, **33**
rosette treelets *see* megaphytes
rosettes, 169, *170*, 570, 700
rotation of crops, 626
Rothamsted
    Broadbalk field, 755, **756**
    insect survey data, 578
    Parkgrass experiment, *683*, 685,
        890–1
Rubisco, 102, 104
ruderal strategy, 557
rule, −3/2, 262
rule, CV^2, 390, 394
rumen, 497–9
    external, 418
    microbial digestion, *497*
ruminant, 84, 120, 121, 497
    faeces, 512

S-shaped (sigmoidal) curve, 227, *228*,
    238, 247, 349, 464, 465
salamanders
    and allometry, 562–3
    competition between, 265–6
salinity, 48, 74–5
salmon
    fisheries, 659
    migration, 203
salt
    pans, 74, 107
    tolerance, 74
salt-marsh, 78, 695, 858, 904
sand-dune plants, *576*, 587
sand-dune study, 688–9
sanitation, 645
saponins, 124
saprophytes, 448
saturation/presaturation dispersal, 191
    hypothesis, 599
schistosomiasis, 435, *436*, 467
sea-level, *85*, *86*

seabirds, 872
search image, for food, 338
searchers, as generalists, 341
searching efficiency, 348, 349, *350*, 351, 371–2
seaweed, 38, 137
secondary
    extinction, 925
    productivity, 713, 730, *730*, 744
    succession, 696
seed(s), 121, 842
    banks, 155–6
        buried, 331
    composition, 121
    dimorphism, 190
    dispersal, 701
    dispersal curves, 183
    dispersal mutualisms, 490, *491*
    dispersers, 942
    dormancy, 196–7
    dormant, 155
    germination
        and fungi, 506–7
        and temperatures, 64
    polymorphism, 190
    population, 151
    predation, 322
        pre-dispersal, 330
    and predators, 123
        satiation, 330
    rain, 181–4
    stay-at-home, 190
    temperature response, 63
    wind-dispersed, 182
seed-eaters, benefit to plants, 323
selection
    experiments, 534
    frequency dependent, 41
selective
    advantage, 177
    extinction, 796
    forces, 41
        gradients, 41
    predation, 805, 809
selenium, 109
self-shading, and herbivory, 315
self-thinning, 255–64, *263*
    cohorts, 258
selfish herd, 178
semelparity, *148*, 154, 528, 536, 548–50, 553
    overlapping and continuous, 169–70
    suicidal, 549
semelparous
    plants, indeterminate, 169
    species, 147, 168, 323
semiochemicals, 628–9, 637
Serengeti Park, *332*
serial endosymbiosis theory of the evolution of the eukaryote cell, 516
sewage, 95, 767
sexual
    dimension, in dispersal, 189
    diseases, transmission, 492
    reproduction, 38
shading, 327, 700, 723
Shannon diversity index, 683, **846**

shifting agriculture, 772
shooting, 652–5
shoots, 615
    and roots competing, 251
shredders, 410, *411*, *412*, 414
shrubs, 32, *907*
silicon, 109
Simpson's diversity index, 682–3
sink patches, 609
size, of individual, 527–8
size and allometry, 561–4
skewed distributions, 251
smallpox, 463
smokers, 66
social
    cohesion hypothesis, 599–600
    subordination hypothesis, 191, 192, 599
sodium, 337, 529, 689, 748
    channel, voltage-dependent, 635, 636
sodium chlorate, 629
sodium chloride, 49, 60
soil, 858, 917
    colloids, 113
    field capacity, 106
    formation, 694
    moisture, 689
    pH, 73–4
    and productivity, 723
    radioautograph, *113*
    resource-depletion zones, 907
    temperature fluctuation, *824*
    water-holding capacity, 757
soil-resource requirements, 788
solar energy, 5, 720
solar radiation, 84, 91, 108, 722, 745, 890
    daily totals, *94*
soluble
    reactive phosphorus (SRP), *761*
    unreactive phosphorus (SUP), *761*
sorbitol, 60, 61
source patches, 609
source pool, 794, 795, 796, 866
space
    pre-empted, 249, *249*, *252*, 283–4
    as resource, 131–4
spatial
    chaos, 393
    heterogeneity, 894–6, 898
    separation, 304–5
    uncertainty, 927
    variation, 578
specialists, 334, 341, 385, 492
specialization, 896
species
    abundance, theories, 572–3
    accumulation curves, *877*
    annual–biennial, 157
    arboreal, 33
    boundary line, 258, *263*
    co-occurrences, 798
    coexisting, 37
    cogeneric, 796
    competing, 304, 894
    composition, and fragment size, 943
    contiguous, 795

on continuously distributed resource, *285*
    critical, 918
    deciduous, 754
    distribution and abundance, 48
    diversity, *806*, 827, 885
        and equitability, *683*
    edge, 930
    endangered, 918
    endemic, 881, *882*, 918
    extinction, 14
        scale of problem, 914–15
        within families, 795
    fugitive, 282, 615
    herbaceous, *616*
    introduced, 923–4
    introduction of new, 611–12
    island, impoverishment, *873*
    iteroparous, 147, 161
    keystone, 828, 829–3, 947
    limnetic, 296
    low-profile, 77
    luxury, 20
    management plans, 945–6
    monocarpic, 147
    monomorphic, 46
    new, in complex food webs, 942
    numbers occupying sites, *823*
    numbers on earth, 914
    pairs, 47
    pelagic, 797
    pest, 579
    plant, co-existing, **787**
    polycarpic, 147
    polygynous, 546
    populations, 16
    with predominant dispersal by one sex, **189**
    preservation, 926
    priority, 918
    rarity, 918–20
    removal treatments, *615–16*
    reproductive rate, 294
    restricted patterns, 5–6
    rhizomatous, 142
    richness, **818**
    risk, 925–6
    scansorial, 33
    scatter-hoarding, 323
    seed-caching, 323
    semelparous, 147, 168, 323
    shade, 93, 94, 98
    specialization within, 38–42
    stoloniferous, 142
    sun, 93, 94, 98
    sympatric, 191
    threats, 918–26
    toleration, 80
    unapparent, 125
    unwanted, removing, 951
    value, 916
    vulnerable, 918
species–area
    curve, *863*, **870**
    graph, 870, *871*, 872, 874
    relationships, 861–4, 872, 878
        for herbivorous insects, *874*
species-deletion stability, 842, *842*

species-pairs, 786, 788
species-rich locations, 886
species richness, 681, 682, *806*
  of beetles, *868*
  butterfly, 890
  contemporary patterns, 908–9
  in gaps, *816*
  geographical factors, 884
  gradients, 900–12
  intermediate levels of productivity, 892
  patterns, 884–912
  and plant architecture, *875*
  and precipitation, 890
  and spatial heterogeneity, *895*
  and success, 906–8
  and temperature, *897*
sperm, 136
spiral waves, 393
sponge, gemmules, 155
spore, 194, 432
  asexual, 504
spring
  freshwater, 858
  hot, 904
sprint speed, 538
stability, 380, *839*
  decrease, 392, 843
  decreased/enhanced, 845
  dynamically fragile/robust, 838
  hypothesis, 849
  non-demographic, 856–60
  perceived, 845
  theory, 844
stable
  age structure, 165
  coexistence, 301
  equilibrium, 240, 277
  limit cycles, 242, 374
stable-point equilibria, 374
stage-specific mortality rate, 149
standard index of association, 798
standing crop biomass, 712, *713*, **715**, 717, *719*, 737
  average, *718*, 856
static life table, 158, 160, 162
stem diameter (*D*), 260
stem tubers, **33**
steppe, 28
stock assessment, 668
stolon, 198, 364, 570
stomata, 95, 96
  and high temperatures, 64
  movement, 96
  restriction, 96
  sunken, 96
strange attractor, 243, 245, *245*
stream communities, energy base, *717*
streamflow, 750
  loss of nutrients, *751*
streams, 759–60
  disturbances, 818
streamwater, 749
stridulation, 427
stylets, 120
subpopulations, 927, 946
succession, 692–710, 827
  definition, 692–3

degradative, 693–4
successional sequence, 817
sugars, 60
sulphate, 748, 749
sulphur, 109, 626, 746, 748, 750, 753
  cycles, perturbation, 769–70
sulphur dioxide, 79–80, 81, 324, 769
sulphuric acid, 125, 769
superorganism, 692
supertramps, 786, *787*, 880
surplus yield models, 660, 661
surrogate resource, 215
survival, probability of, 149
survivorship, 215
  for elephants, *940*, *941*
  high, 156
  schedule, 216
survivorship curves, 147, 153, 160, 588
  classification, 153–4, *154*
  logarithms, 153
susceptibles, 438, *471*
sustainability, 671–3
  definition, 672
switching, 337–9, 349, 667
  behaviour, predator, *385*
  and optimal diets, 343–4
symbionts, 697
  intracellular, 500–1
  invertebrates, 507–9
  mycetocyte, 501
  and temperature, 56
symbiosis, 482–525
  in animal tissues and cells, 500–2
  ant–plant, 493
  aphid–bacterial, 501
  evolution of subcellular structures, 516–17
  facultative, 418
  fig–wasp, 494
  morphological integration, *496*
  see also mutualism
symbiotic relationships, 120
sympatry, 191, 294, 295, 298
synchronous breeding, 548
synchrony, 177
synergism, 325, 428

T lymphocytes, *445*
tadpoles, 200, 826, *826*
taiga, 28, 592
tannins, 124, 125
taproot, 108
target pest resurgence, 633–4
taxa
  competitive displacement, 910
  rare, 920
teliospores, 437
temperature, 4, 48, *49*, 713, *721*, *722*, 742, 773, 774, 890, 907
  absolute minimum, and plants, *51*
  accumulated, 574
  arithmetic mean, 53
  carbon dioxide and, 773
  changes
    in hamster, *72*
    North Sea, *11*

coefficient (Q_{10}), 53
competitors, 56
day-degrees, 54
and distribution of plants and animals, 50–9
effective, 575
germination, *824*
high, and organisms, 64–7
higher critical, *70*
increasing, 57
land and sea, *86*
low, 59–64
lower critical, *70*
and NPP, 722, 742
and parasites, 56
and physical conditions, 57
plant responses, *722*
productivity, 713
symbionts, 56
threshold, *54*
variations, in glacial cycles, *13*
see also ectotherms; endotherms
temporal
  separation, 304–5
  variation, 825
    in conditions, 810–13
    and disturbance, 825
terminal encounter rates, *363*
termites, 410
  and colony temperatures, 68
  gut, 499–500
terpene defences, insect-stimulated, 319
terpenes, 750
terpenoids, 124
terrestrial
  communities, 720–4
    nutrient inputs, 748–50
  ecosystem model (TEM), 742
  productivity, factors limiting, 725
territorial behaviour, 131
territoriality, 253–6
territories
  experimental modification, 132, 618
  resident, 255
territory defence, 255
testosterone, 132
thallus, 78
thamniscophysalodophagy, 507
The Theory of Island Biogeography, MacArthur and Wilson, 606
thermal hysteresis proteins, 61
thermoneutral zone, 69, *70*
thermophiles, 66
therophytes, **33**
Thiessen polygons, 253, *254*
thinning
  line, 258, 262
    dynamic, 258, 259, 260
  slopes, 258, 260
thiocarbamates (EPTC), 629
threshold density, 461, 467, 470
tillage, 645
tillers, 162, *231*, 615
  defoliated, 316
  static life table, **163**
Tilman's resource–ratio hypothesis, 704–5, *706*

timber extraction, 772
time delays, 374
time-series analysis, 670
tits, coexistence between, 268, 279
tolerance model, 704
top-down control, 828, 860, 952
    of food webs, 831–8
toxaphene, 628, 635, 636
toxicity, 631–3
    heavy metal, 39, 73, 898
    to non-target organisms, **631**
toxins, *49*, 73, 79–80, 337
    in developing countries, 639
    and food preference, 337
    heavy metals, 39, 73, 898
trade-off curve, 539, *540*
trade-offs, 533–38
transmission threshold, 460, 461, 467
transovarial transmission, 501
transpiration, 91, 96, 106, 109, 764
trap crops, 645
trapping bias, 793
tree(s), 137–9
    canopy, juveniles, *822*
    coniferous, 900
    of contrasting species, **724**
    deciduous, 42
    distribution, 13–14
    dwarf, 28
    felling, 917
    forest, 137, 922
    growth of isolated, *261*
    harvest, 755
    multilayered and monolayered, 701
    rings, 86
    species, in Great Smoky Mountains,
        687
    species diversity, 890, *891*
    species richness, *889*
    temperate forest, 899
    and temperature, 50
tree-by-tree transition matrix, **703**
triage evaluation, 952
triazines, 630, 636
trichlorophenoxyethanoic acid, 629,
    633
trichomes, 122
trifluralin, 629
trophic
    cascade, 837
    level transfer efficiency, 735, *736*
    levels, 844, *851*
        and food webs, 848–52
        number, 832–5
    structure
        of community, 731–6
        and energy flow, *732*

trickle, 838
tropics
    and high plant biomass, 902
    increasing productivity, 901–2
    richer in species, 899
tuber, 121
tundra, 28, 858
two-dimensional sampling
        programmes, 397

undercompensation, 240
underdispersion, 174
unit leaf rate (ULR), 317
unstable equilibrium points, 277
unsystematic variation, of resource, 93
upwelling, 762
urine, 758

vaccination, 945
    critical level, *463*
    and rabies, 474
values of association, *799*
variance, nested analysis, **565**
variants, 41
variation, 7, 40
vector, 465, 466, 478
vector addition, *276*
vector-transmitted protozoan parasites,
    430, 438, 466
vegetation, and temperature, 50
vents
    chimneys, 66, *67*
    hydrothermal, 509
vertebrates, 792
    extinction, **926**
    faeces, 421–3
    herbivorous, 121
    predators, 879
vines, 87
virus, 642
    resistance, 534
volatilization, 758
volcanoes, 48, 746, 769
vole–weasel populations, 597

waste, discharge, 917
waste disposal, 758, 922
water, 721, 744, 769, 888, 889, 922
    cold, low-salinity, *661*
    conductance, 97
    currents, 78
    and dispersal, 187–9
    impoundment and abstraction, 922
    ionization, 59

loss, controlled, 96–8
movement, 751
natural, chemical quality, 917
pH, 73–4
phases, 4
products, 117
and radiation, 95–6
as resource, 106–9
shortage, 722
supercooled, 59
surface runoff, 758
vapour, atmospheric, 84
weasels
    coexisting, 792
    teeth, 783, *783*
weather, 579
    and population size, 575
weed(s)
    annual, 96
    biological control agent, **640**
    control, 621–3, 626
    killers, 629
    what is it?, 622
wetfall, 748, **749**
wetlands, 759
wind speed, 907
    return times for deviations, *80*
winnowing, 626
woodlands, semi-acid, 32
woody species, species richness, *893*
worm, parasitic, 67

yield, 229, 294, 916
yield-against-effort curve, 665
yield–density relationship, 257
    in cohort, 262
yield–effort relationship, 669
yield per recruit, 667, 668, *668*

zero isocline, 275, 276, 308, *309*, 372,
    399
zinc, 49, 80, 109, 724
zonation
    as result of exposure, 76
    on sand-dunes, 688–9
    seashore, of animals and plants, 77
zonation of individuals, *305*
zooids, 139, *508*
zooplankton, 730, *730*, 760, 834, 836,
    859, **870**, 896, *929*, 942
zoos, 946
zoosporangia, *405*
zoospores, *431*
zooxanthellae, 516
zygote, 136, 137, 153